T0180633

Advances in Intelligent Systems and Computing

Volume 1080

The series "Advances in Intelligent Systems and Computing" contains publications on theory, applications, and design methods of Intelligent Systems and Intelligent Computing. Virtually all disciplines such as engineering, natural sciences, computer and information science, ICT, economics, business, e-commerce, environment, healthcare, life science are covered. The list of topics spans all the areas of modern intelligent systems and computing such as: computational intelligence, soft computing including neural networks, fuzzy systems, evolutionary computing and the fusion of these paradigms, social intelligence, ambient intelligence, computational neuroscience, artificial life, virtual worlds and society, cognitive science and systems, Perception and Vision, DNA and immune based systems, self-organizing and adaptive systems, e-Learning and teaching, human-centered and human-centric computing, recommender systems, intelligent control, robotics and mechatronics including human-machine teaming, knowledge-based paradigms, learning paradigms, machine ethics, intelligent data analysis, knowledge management, intelligent agents, intelligent decision making and support, intelligent network security, trust management, interactive entertainment, Web intelligence and multimedia.

The publications within "Advances in Intelligent Systems and Computing" are primarily proceedings of important conferences, symposia and congresses. They cover significant recent developments in the field, both of a foundational and applicable character. An important characteristic feature of the series is the short publication time and world-wide distribution. This permits a rapid and broad dissemination of research results.

**** Indexing: The books of this series are submitted to ISI Proceedings, EI-Compendex, DBLP, SCOPUS, Google Scholar and Springerlink ****

More information about this series at http://www.springer.com/series/11156

Natalya Shakhovska · Mykola O. Medykovskyy
Editors

Advances in Intelligent Systems and Computing IV

Selected Papers from the International Conference on Computer Science and Information Technologies, CSIT 2019, September 17–20, 2019, Lviv, Ukraine

Editors
Natalya Shakhovska
Department of Artificial Intelligence
Lviv Polytechnic National University
Lviv, Ukraine

Mykola O. Medykovskyy
Institute of Computer Science
and Information Technologies
Lviv Polytechnic National University
Lviv, Ukraine

ISSN 2194-5357 ISSN 2194-5365 (electronic)
Advances in Intelligent Systems and Computing
ISBN 978-3-030-33694-3 ISBN 978-3-030-33695-0 (eBook)
https://doi.org/10.1007/978-3-030-33695-0

This Springer imprint is published by the registered company Springer Nature Switzerland AG
The registered company address is: Gewerbestrasse 11, 6330 Cham, Switzerland

Organization

Proceedings Chairs

General Chairs

Natalia Shakhovska
Mykola Medykovskyy

Proceedings Committee

Natalia Shakhovska	Lviv Polytechnic National University, Ukraine
Mikhail Alexandrov	Autonomous University of Barcelona, Spain
Sergiy Bogomolov	Australian National University, Australia
Jaime Campos	Linnaeus University, Sweden
John Cardiff	Technological University Dublin, Ireland
Paweł Czarnul	Gdansk University of Technology, Poland
Roman Danel	Institute of Technology and Businesses in České Budějovice, Czech Republic
Yannick Estève	University of Avignon, France
Mehmet Koç	Bilecik Seyh Edebali University, Turkey
Tatjana Lange	FH Merseburg, Germany
Rimvydas Laužikas	Vilnius University, Lithuania
Piotr Lipinski	Lodz University of Technology, Poland
Serhii Lupenko	Ternopil Ivan Puluj National Technical University, Ukraine
Vasyl Lytvyn	Lviv Polytechnic National University, Ukraine
Mykola Medykovskyy	Lviv Polytechnic National University, Ukraine
Sergio Montenegro	Julius-Maximilians-Universität Würzburg, Germany
Jan Rogowski	Lodz University of Technology, Poland
Tetiana Shestakevych	Lviv Polytechnic National University, Ukraine

Volodymyr Stepashko	IRTC ITS NASU, Ukraine
Victoria Vysotska	Lviv Polytechnic National University, Ukraine
Vitaliy Yakovyna	Lviv Polytechnic National University, Ukraine

Contents

ICT in High Education and Social Networks

Software Engineering

Cyber-Physical Systems

Mathematical Modeling

Mathematical Models and Program of Resource Feedbacks in the Systems «Production, Retail»

Taisa Borovska(✉), Dmitry Grishin, Irina Kolesnik, and Victor Severilov

Vinnytsia National Technical University, Vinnytsia, Ukraine
taisaborovska@gmail.com, dmitriygrishin2@gmail.com, iraskolesnyk@gmail.com, severilovvictor0@gmail.com

Abstract. The research is devoted to the development of new mathematical models and programs for the analysis and synthesis of systems «production, retail» taking into account information and resource feedbacks. A specific feature of current scientific and practical problems in this area is the substantial nonlinearity, nonstationarity, stochasticity, and high dimensionality of the feedbacks. The classification of the feedbacks in the systems «production, retail» has been performed. To solve the control problems, the methods of optimal aggregation are used. Mathematical models and programs for binary operators of aggregation of structures with feedback are developed. Solutions of optimization problems for parallel and sequential structures with the feedbacks. The study of the dynamics and steady states of the structures with the feedbacks «returning the cost of production», «recycling of waste» has been carried out. To study the dynamics of the retail the simulation model of «producers, consumers» is used. Examples of modeling are given.

Keywords: Optimal aggregation · Feedback · Production function · Information technology

1 Introduction

Linear and nonlinear, positive and negative feedbacks are widely used in various technical systems. For billions of years the feedbacks in biological systems determined the processes of ecosystems development. In particular, the rapid growth of a certain species or ecosystem is usually stopped catastrophically: without the renewal or with the renewal after a long time interval in the same place or in another region. A simple example: saury livestock in the Far East reaches a maximum once every twenty years for about two years. A popular mathematical model of the ecosystem is «foxes, hares». The authors created a mathematical model of «grass, hares, wolves» [3, 19, 20]. By adjusting the parameters, we obtained a model where the livestock of the «wolves» reproduced the «grass» yields, and the «hares» were the technological transformer of grass into feed. Scientists have long been working on the exclusion of «cows» from the technological chain of milk production.

N. Shakhovska and M. O. Medykovskyy (Eds.): CCSIT 2019, AISC 1080, pp. 3–16, 2020.
https://doi.org/10.1007/978-3-030-33695-0_1

Modern systems for the production of automobiles, airliners, clothing and food products are constantly faced with the problems of ensuring the «smart sustainable development» of socio-technical-ecological systems (STES). One of the problems is resource feedback (RFB). Examples of RFB: – agricultural enterprise (processing of plant and animal waste into biohumus, biogas, electricity); – all enterprises (return of expenses in the form of payment for goods and services). The last example of RFB is related to the following areas: mathematical modeling and financial mathematics. According to the theorem of the economist Coase, the field of economics is a transaction, an arrangement agreements on buying and selling, as well as hiring and firing [25]. Modern production systems require efficient mathematical models for optimal control of closed systems «production, consumption, utilization». In this article, in order to solve this complex, interdisciplinary problem, new information technologies are proposed to develop rational mathematical models and programs for managing production systems, taking into account resource feedbacks. The possibility of obtaining non-trivial results is stipulated by a certain amount of previous research and work in the field of optimal aggregation methodology [1, 2, 17].

The aim of the research is to develop models and programs for solving problems of optimal aggregation of binary structures with resource feedback (RFB), development of the methods and program for optimal aggregation of arbitrary structures with resource feedback (RFB).

To achieve this goal, it is necessary to solve the following tasks: classification of typical binary resource structures with RFB in production systems; development of program modules for the implementation of binary operators of optimal aggregation of typical RFB structures; conducting research on production system models with regard to RFB.

Practical use of the research results: embedding software modules in automated production management systems (CAM); development of automated decision support systems (ASR) for analytics and training.

2 Mathematical Models of Resource Structures Based on the Methodology of Optimal Aggregation

Today, the concept of modeling for large systems is changing: instead of «model – mapping of object properties essential for the researcher», reality leads to a new postulate: «model is a means for searching and experimentally verifying design solutions, object is a model implementation». Changes occur in mathematics. The achievements of classical mathematics are integral, differential, calculus of variations created in the pre-computer era, but their intellectual potential is far from being fully discovered. The growth of computational efficiency of computer systems has not yet allowed to defeat the «curse of dimension». Bellman defined the direction of research in the computer age as «the replacement of tasks for finding a point in the high-dimensional phase space with a system of tasks for finding a point in phase spaces of a lower dimension, preferably one-dimensional» [8, 13]. The dynamic programming method is an example of the concretization of this approach.

Next, we will successively consider the solution of the problems, stated in this article, on the basis of the optimal aggregation methodology, the main point of which is the replacement of the multidimensional nonlinear programming problem by the system of one-dimensional optimization problems. The features of optimal aggregation methods are the novelty and the three-dimensional nature of elementary objects; therefore, we present theoretical results in parallel with analogs and visual representations.

Figure 1 shows the «feedback» structures in automation (speed, temperature, pressure regulators) and in production. Interpretation of RFB in this case is the following: waste from primary production is processed into a resource for primary production.

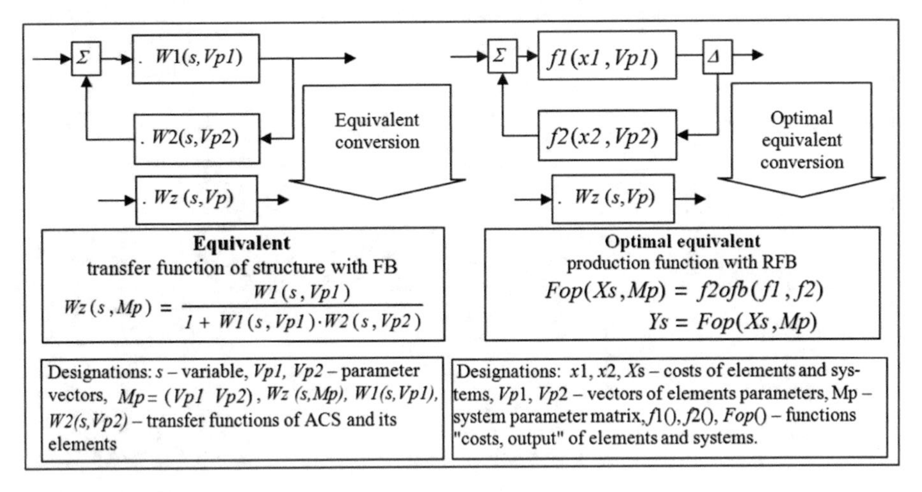

Fig. 1. Elementary feedback circuits (OS) in automation and resource feedback (RFB) in production systems

We presented the structures with the feedback from the automatic control theory (ACT) and optimal aggregation methodology for two reasons: – the «cybernetics» direction emerged after the works of Norbert Wiener, who demonstrated the unity and importance of the concept of feedback in biology and engineering and continued in the works of J. Forrester [4]; – it is necessary to allocate the differences of the goals and methods in ACT and the theory of optimal aggregation. In the theory of automatic control (ACT), objects are mostly linear, operands are functions of time and frequency, feedbacks are mainly negative, and feedback (FB) objectives are to provide the required state in the presence of disturbances. In the methodology of optimal aggregation, objects are the production systems, operands are the functions of the «cost, output» class, the methodology for which (interindustry balance methods, etc.) was created by Leontiev [7], essentially nonlinear, resource feedback (ROS) objectives—optimality production systems. In general, in large production systems, optimal aggregation provides strategic management [1, 11], and automatic control theory (ACT) provides the implementation of optimal strategies [16, 17, 26]. The method of optimal aggregation has its advantages relatively to the classical methods. The main difficulty in the application of this method is the formulation of the problem in accordance with the methodology of optimal aggregation. This is really an intellectual component, after which the algebraic routine remains.

Figure 2 presents two alternative resource structures for optimal aggregation of production systems: based on production functions; and on the basis of integrated systems «production, development».

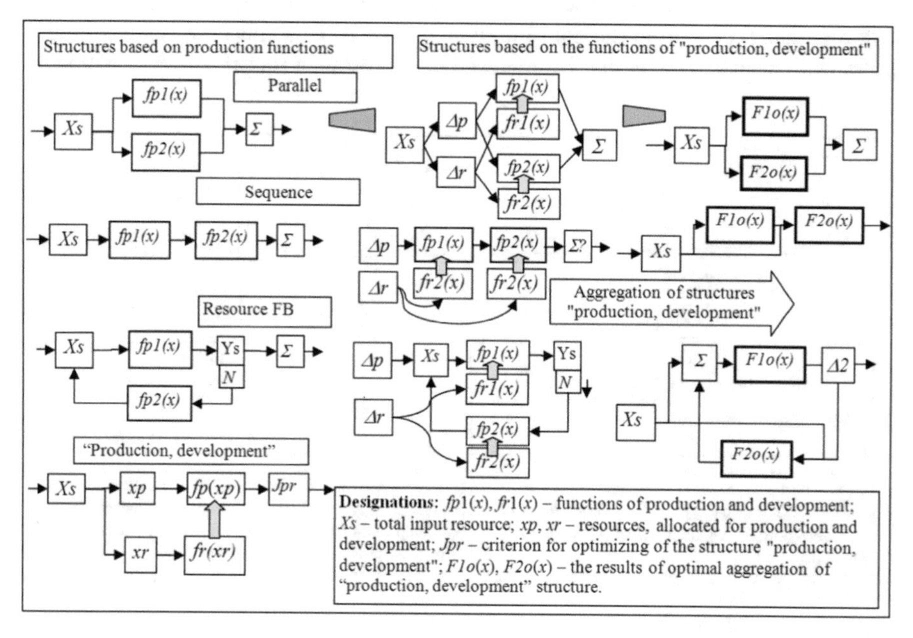

Fig. 2. Typical resource structures, optimal aggregation «production, development»

In the left part of Fig. 2 the following structures are presented: parallel and sequential structures, resource structure with feedback and «production, development» structure. Let us briefly name the optimization tasks that can be put on the basis of these structures: parallel – optimal distribution of the production resource between the subsystems; sequential – distribution of the production resource is set by conveyor production technologies; resource with the feedback (RFB) – here the production in the subsystems is determined by the technologies; the structure «production, development» consists of two elements – «production» and «development» subsystems. The need to integrate these subsystems is imposed by the practice of modern production.

For example, modern machining center replaces a metalworking machine shop. This is one or more modern machines with digital control. The main point of the development of such machines is the software. This is really intellectual component of production. The programmer must be a «metalworker», be close to the machining center, conduct research on a computer and on the machining center. This is one of many examples of the integration of high-tech production with developments. In the above structure, the subsystem «development» creates information and material innovations that increase the efficiency of the subsystem production. In this case, the subsystems are integrated. It is natural to apply optimal aggregation to such a structure. The theoretical foundations of optimal aggregation are discussed in articles and monographs [1–3, 15]. The connection between the subsystems «production» and «development» is parametric: the subsystem «development» changes the parameters of the production function.

In the center of Fig. 2 the expansion of the structures of the «production» class into «production, development» structures is shown; in the right part the results of optimal aggregation of expanded structures are shown. The peculiarity of resource feedbacks (RFB) in production systems is that they are not a tribute to trends, but a necessary condition for survival. Sometimes they «spontaneously» act in the desired direction, and do not attract the attention of researchers and managers.

Figure 3 presents two classes of resource feedbacks (ROS), which are the objects of research in this article. Briefly, this is recycling and returning production costs of the manufacturer to consumers of the products of production. We will give the examples of these classes of resource feedbacks (ROS). Example 1. Replacement of spontaneous resource feedbacks (ROS) with controlled ones. In the United States, one firm has developed an installation for domestic processing of newspapers in log fireplaces. Example 2. Production of styrene – large-capacity. Styrene is formed on platinum catalysts, is separated from unreacted components, which are returned to the reactor. Example 3. Regulation of the global system «production, consumption» by means of «quantitative easing»: distribution among the consumers the funds for the purchase of products and promotion of the machine «production». Another typical means of managing the «production, consumption» cycle is the price mechanism. These are ever-relevant topics for scientific and applied research of an object, the complexity of which is constantly ahead of the mathematical apparatus of researchers. The methodology of optimal aggregation allows to solve problems without distracting from a real object to purely mathematical problems of search, convergence, convex and integer, and other kinds of programming.

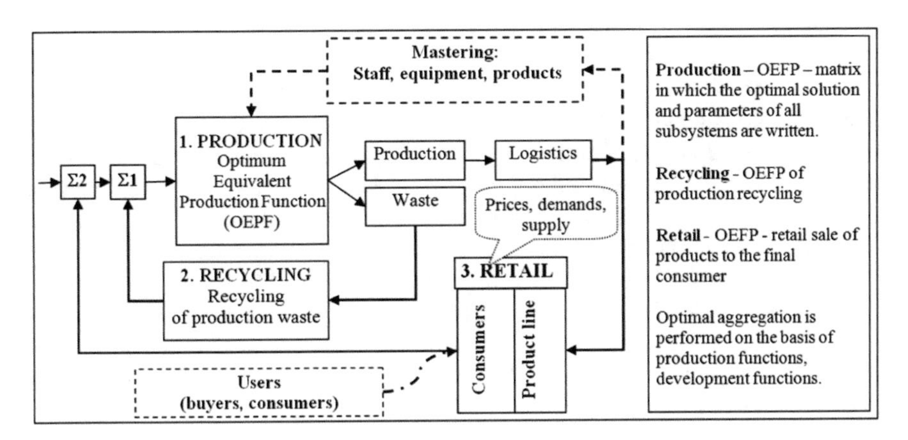

Fig. 3. Aggregated resource structure of production with RFB

The diagram in Fig. 3 is the test object for binary operators of optimal aggregation. Such a choice is stipulated by the relevance and a large number of descriptive information, scattered and mostly descriptive information about the object. The main elements of the object are «production», «recycling», «retail», these are the systems of high dimensionality. They are represented by optimal equivalent production functions. Formally, these are one-dimensional dependencies «total output on total costs», implemented in a program and software environment as matrices. Besides the Table of values, they contain the optimal resource allocations for each value of the total input, the values of the parameters of the production functions of elements, the vectors of current prices of resources and products of production.

In Fig. 3 dotted line indicates the blocks «development», «users», «buyers» and «prices, demand, supply». These are the essential parts of the system, the informational and intellectual components. Prices, demand, supply in modern conditions are the tasks of applied system analysis. These blocks will be discussed in the following articles. This article is devoted to the analysis and optimization of resource links «recycling» and «retail». In Fig. 3, we have a system with two positive feedbacks: the higher the efficiency of recycling and retail is, the higher is the efficiency of the production system. We solve the problem applying the mathematical methods, taking into account the dimensions of the subsystems of orders of 10–10.000 elements and without simplifications, due to the chosen mathematical method. For one «recycling» circuit, the problem of optimal aggregation was solved for a particular case – a bioreactor for an agricultural enterprise. The problem of optimal aggregation of dual-circuit resource feedback (RFB) is solved for the first time [20–24]. For the «retail» object, the problem of optimal aggregation of the product line has been solved, taking into account the random selection with training for the consumer. In accordance with the technology of solving optimization problems on the basis of optimal aggregation methods, after the analysis of binary structures, a binary optimal aggregation tree is constructed [1].

Figure 4 shows the binary tree of optimal aggregation, corresponding to the system diagram in Fig. 3. This is the next step in setting and solving the problem of optimizing of essentially nonlinear system of large dimensionality applying the method of optimal aggregation. The basis of the structure in Fig. 4 is executable working formula for optimal aggregation with the addition of equal signs in the nodes (for ease of the analysis). At the lower level of the binary tree, there are three operations of optimal aggregation «production, development» for the blocks «production», «recycling», «retail». The equal sign in a binary operation (Fig. 4): $Fop1 = f2opr\ (fp1, fp2)$ means calculation buy two functions «costs, output» of the result - the functions «expenses, output». Analogues of such an operation in algebra are the addition and multiplication of numbers. The principal difference in algebra of optimal aggregation is the optimization of resource allocation between operands embedded in binary operators.

In Fig. 4, two structures are presented: a binary tree of optimal aggregation and structural schemes that correspond to two levels of aggregation. Note the non-obvious informational aspect of the methodology of optimal aggregation: visually, three structural diagrams look like simplifications of the initial structure. With optimal aggregation, the original system information is stored in the user functions *Fprod* (*Xs*, *Mp*1), *Fmrk* (*Xs*, *Mp*2). Below the aggregation operations at the number level are considered.

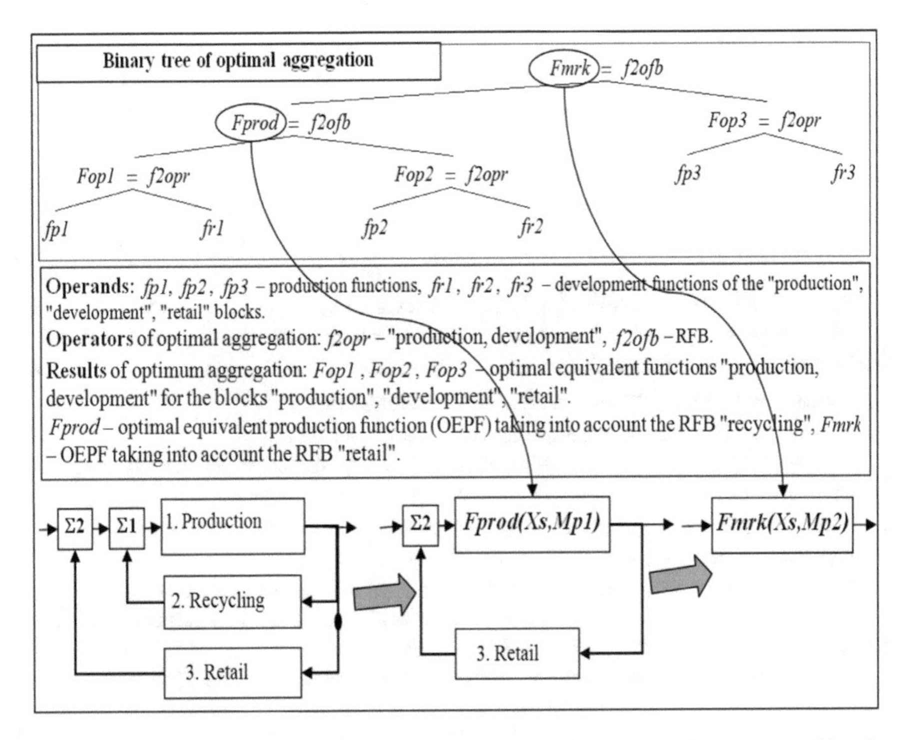

Fig. 4. Solution of the problem of optimal aggregation of the production system taking into account the RFB

3 Development and Program Implementation of Optimal Aggregation Operators

In applied sciences, there exists two information technologies for developing new mathematical models for new production management tasks: models based on statistical data, models based on «generating mechanisms». The natural task of textbooks and dissertations is the selection of the optimal technology of development and justification of the choice. The obvious answer is the effective combination of technology capabilities, taking into account the specific tasks and problems of the life cycle of production and products. In papers [1–3, 6, 11, 16–19], the approach was developed and proposed on the basis of simulation models and obtaining «virtual reality statistics». One of the results of this approach is that the statistics of modern production systems are essentially non-Gaussian, classical statistical methods are of little use and insufficient for modern systems, mostly non-stationary and with many non-linear connections.

The next step in the problem solution is the development or modification of binary operators of optimal aggregation. It also requires the expansion of the alternative information technologies for constructing new models for new tasks [1, 12]. This is the base for software versions. In Figs. 5 and 6, the versions of working mathematical models of binary structures «production, development» and «resource feedback» are presented. They are typed in the environment of the mathematical package and can be copied and imbedded into the program, calculated in numbers and in symbolic form.

Functions of production and development: $fp(xp) = fc(xp, vPp(xr, vPp0));\ fr(xr) = fc(xr, vPr)$. (1)

Efficiency Functions – "output / cost" relationship $efp(xp) = \frac{fp(xp)}{xp};\ efr(xr) = \frac{fr(xr)}{xr}$. (2)

Version of the parametric connection of production and development $vPp = VP2(\alpha, \Delta xs, vPp0, vPr)$ (3)

Initial state of the system: $xp0$ - production rate; $vPp0$ - parameter vector of the PF; vPr - parameter vector of the DF in the model of the first approximation – constant. Optimization variable: $0 \le \alpha \le 1$ (4)

The task of optimizing the distribution of a resource quantum Δxs between:
production costs $dxp = \alpha \cdot \Delta xs$ and development costs $xr = (1 - \alpha) \cdot \Delta xs$. (5)

On the basis of (1) – (5), we write the equations of the system state "production, development" after using a quantum of resource: resource allocation to production $xp = xp0 + \alpha \cdot \Delta xs$ and development $xr = (1 - \alpha) \cdot \Delta xs$. (6)

Mapping of the development costs to change the parameters of the production function: $yr = fr(xr, vPr)$ taking into account (6), we obtain the equation "costs, output": $yr = fr[(1 - \alpha) \cdot \Delta xs, vPr]$.

We use the standard set of parameters of the "costs, output" functions for PF and DF, and **collect in a user function** the specific dependencies of the parameters of the PF model for specific production segments and technologies common to different versions $vPp = VP2(\alpha, \Delta xs, vPp0, vPr)$. (7)

The rate of production after using the resource quantum will be with the account of (6) and (7):

$$yp = fp(xp, vPp) = fp[(xp0 + \alpha \cdot \Delta xs), (vPp0 + \delta vPp)] \quad (8)$$

Taking into account (4), (9) we form the function of the user - "new rate of the output"

$$yp(\Delta xs, \alpha) = fp[(xp0 + \alpha \cdot \Delta xs), VP2(\alpha, \Delta xs, vPp0, vPr)] \quad (9)$$

and user function "increment of the output" (this is the version of the optimization criterion in the basic task of sub-optimization of the integrated system "production, development"):

$$\delta yp(\Delta xs, \alpha) = yp(\Delta xs, \alpha) - yp0 = yp(\Delta xs, \alpha) - fp(xp0, vP0) \quad (10)$$

Fig. 5. Working mathematical model of the resource structure «production, development»

The peculiarity of the binary structure «production, development» is the parametric connection between the subsystems: the resources spent in the subsystem «development» change the parameters of the function «costs, output» of the subsystem «production». The resource allocation scenario chosen in this version is: a «resource quantum» and its usage period is set. It is required to find the proportion of resource allocation between production and development that maximizes the criterion of total output for the period. Obviously, the production of ships (3–7 years) and dairy products require different models. For the production of one segment, it is possible to create a single parameterized aggregation model for products of one class, for example: household refrigerators, diesel engines, «kitchens for apartments», «installations for 3D printing», etc. The objects of these studies are ecologized agro systems, metallurgy enterprises, the organization of the development of production management programs and logistics systems for mass service. The presence of binary operators versions is due to two factors, the search and implementation of effective adequacy of mathematical models of resource structures and development of computationally efficient mapping of mathematical models into program modules.

Similarly, the formulation of the problem of developing mathematical models of resource feedbacks is performed. Previous developments based on the modification of the feedback models in automation [11] did not lead to satisfactory solutions. The reasons are significant differences in the objects. In classical models, the main objects are information, measurements, assessments, sustainability. For RFB, the content of links is the balance, distribution, summation, and processing of the resource flows. In classical science, the main functions of mathematics are simplifications, reduction, and development of searching methods in multidimensional spaces. In optimal aggregation, mathematics is focused on solving problems of applied system analysis, obtaining models of resource links based on «generating mechanisms» [4]. The resource approach stipulates the necessity of optimal aggregation of integrated systems «production, development». The main factor in the integration is that the subsystems of production, transport, maintenance have the built-in computer systems for diagnostics and control, operation management and communication with other subsystems. Critical component of such subsystems is software that is completely protected from failures and break-ins. At the same time, the software must be open to constant changes in the direction of increasing efficiency. Resource feedback is specific. In this paper, we consider feedbacks of two classes, the RFB «recycling of production wastes» and RFB «return of costs of the manufacturer and retailer». We start the development of the models of a new class of interrelations between some subsystems with linguistic models of these connections. We will consider two basic scenarios for the RFB.

Scenario 1. A certain investment project is considered, the total cost of the creation of production capacities of the main production is given. Also, we have costs for recycling of the main production waste into a resource for the main production cycle. So, the production system consists of two subsystems, i.e. of the main production and the production of recycling. The production and development functions for these industries are specified. It is required to construct the function of optimal distribution of the investment resource, which gives the maximum of a certain efficiency criterion.

Scenario 2. Ecological modernization (greening). The production of a certain target product/products with a given range of target product output rates is considered. It is required to construct the distribution function of the investment resource, which gives the maximum at some efficiency criterion. Construct the function of the optimal complement of industrial waste processing of the target production.

Figure 6 presents the version of the mathematical model for the binary structure «resource feedback».

Production functions of resource feedback contour subsystems (RFB)

$$fp1 = fc(x1p, vP1(x1r));\ fp2 = fc(x2p, vP2(x2r))$$

Developmental functions $fr1 = fc(x1r, vrP1)$; $fr2 = fc(x1p, vrP2)$.

Efficiency functions $ef1(x) = \dfrac{fp1(x)}{x}$; $ef2(x) = \dfrac{fp2(x)}{x}$.

Relationship of production functions (PF) and development (DF) $A1p = fr1(x1r)$; $A2p = fr2(x2r)$.

Definition of parameter vectors $vP1 = \psi 1(A1p)$; $vP2 = \psi 2(A2p)$

$$\begin{pmatrix} vP1_1 \\ vP1_2 \\ vP1_3 \end{pmatrix} = \begin{pmatrix} A1p \\ \psi(A1p)_2 \\ \psi(A1p)_3 \end{pmatrix};\quad \begin{pmatrix} vP2_1 \\ vP2_2 \\ vP2_3 \end{pmatrix} = \begin{pmatrix} A1p \\ \psi(A1p)_2 \\ \psi(A1p)_3 \end{pmatrix}.$$

Restrictions on the investment resources $x1r + x2r = Xr$; resource change interval $0 \le Xr \le Xm$

Feedback. We write the expression for the system output with positive FB:

$$y(x) = fp1(x + f2(y(x)))\ \text{from where}\ y(x) = fp1(x + fp2(y(x))).$$

Optimality criterion $J(y(Xr))$ - should reflect the requirements for a real system. Several alternatives are considered. Optimization variables $0 \le \alpha \le 1$; $x1r = \alpha \cdot Xr$; $x2r = (1 - \alpha) \cdot Xr$.

To solve the problem of obtaining the equivalent function "costs, output for RFB", we create a user function $Nors(\cdot)$. Two versions of the user function, differing by the set of parameters are developed

$$NorS(x, \alpha, Xr)\ \text{and}\ NorSp(x, VP1, VP2)$$

Fig. 6. Working mathematical model of the structure «resource feedback»

Figure 7 shows the sequence of optimal aggregation operations for the example of: the system of four integrated subsystems «production, development».

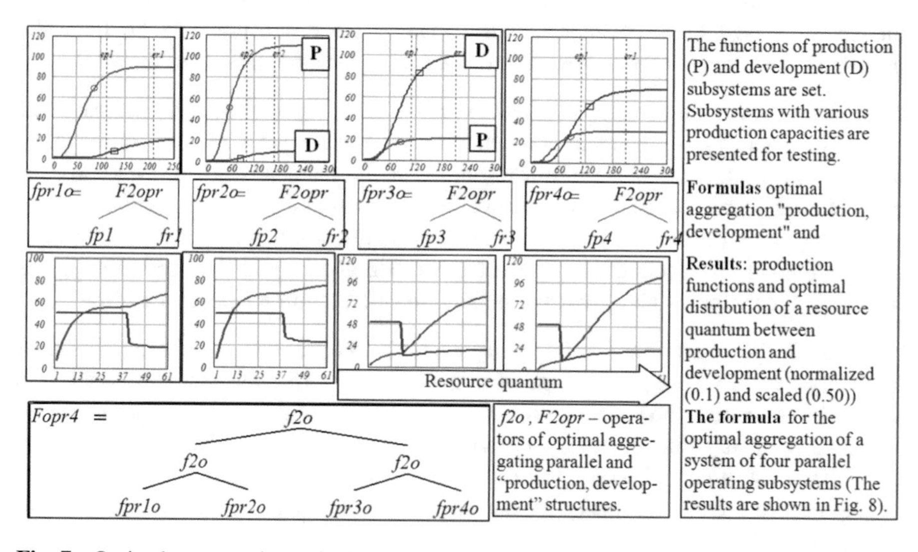

Fig. 7. Optimal aggregation of the system with «production, development» subsystems. Example

Figure 8 shows the results of the optimal aggregation of the system as a whole (the lower formula in Fig. 7). The result of the optimal aggregation *Fopr4* is a matrix, in columns of which there are data for the graphs, presented in Fig. 8. For a new method

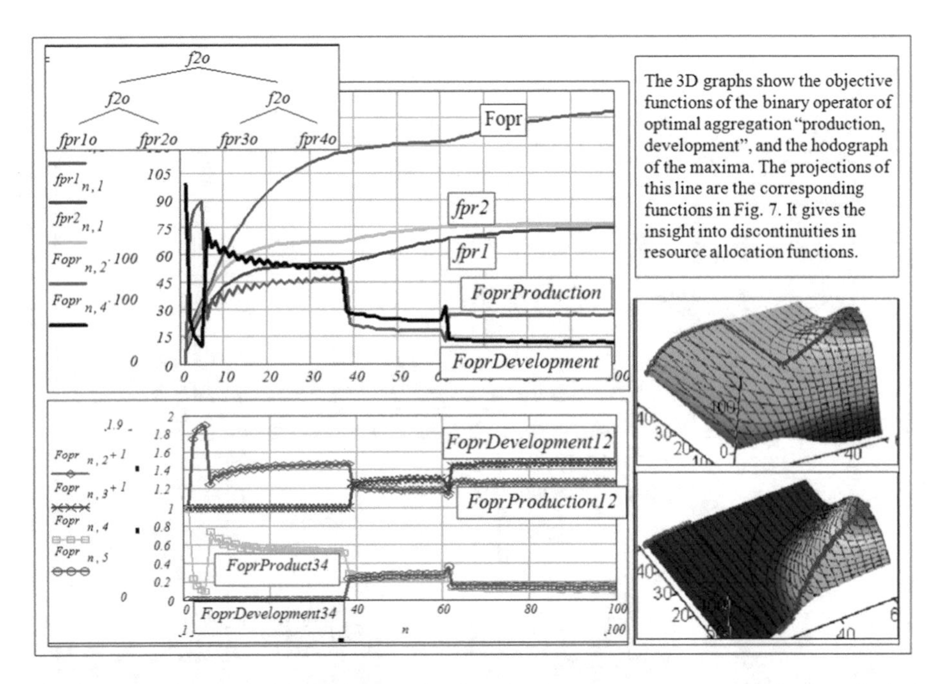

Fig. 8. Optimal aggregation of the «production, development» structure. Optimal equivalent production function. Example

and a new task – the development of a binary operator of optimal aggregation with parametric coupling, the natural question is about the presence of programming errors. We see that the functions of the optimal resource distribution have gaps (lower graphs), and the corresponding production functions (*fr1, fpr2, Fopr*) have no gaps. In the right part of Fig. 8 two 3D graphics for explaining the gaps in resource distribution graphs (Figs. 7 and 8) are shown. It should be noted that the well-known generalizations of the parametric resource connections [5, 9, 10, 12, 13, 15] present a verbal analysis without corresponding working models.

Figure 9 shows the example of the testing results of a single loop of the distributed circuit at the variation of the parameters of the production functions of the subsystems *Vp1*, *Vp2* (vary: production capacity, efficiency, starting costs). No analogs of mathematical models or empirical data were found for this development. Therefore, all the elements of the model have been tested. Figure 9 shows three sets of system parameters and three sets of the function graphs were calculated for these sets: 1 – «Production», 2 – «Recycling», 3 – «Result RFB».

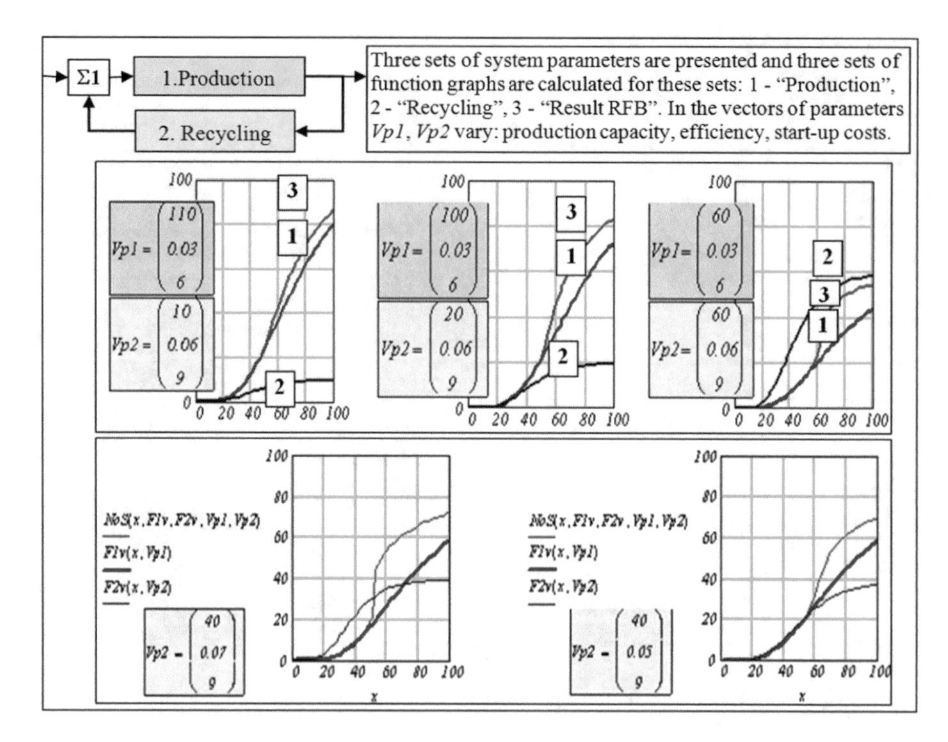

Fig. 9. Optimal aggregation of the system with resource feedback. Testing

4 Conclusion and Future Developments

The research is devoted to the development of new mathematical models and program for the analysis and synthesis of the systems «production, retail», taking into account information and resource feedbacks. The peculiarity of current scientific and practical problems in this area is the substantial nonlinearity, non-stationarity, stochasticity, and high dimensionality of the objects, which are the production systems and systems for selling production results to the end consumers. In such systems, in the closed circuits «production, consumption, utilization», failures and unstable modes often occur. In particular, the problem of recycling of food packaging does not have the system solution. No comprehensive mathematical models for the systems «production, consumption, utilization» were found in the available sources.

The goal of the research is to create mathematical models and software modules to solve problems of optimal aggregation of the binary structures with resource feedbacks, as well as development of the methods and program for optimal aggregation of random structures with RFB.

To achieve this goal, authors classified the typical binary resource structures with RFB in production systems. Next, we developed a software for the implementation of the binary operators of optimal aggregation of structures with RFB. The models of production systems with the account RFB were studied.

To solve the problems, the methods of optimal aggregation were used. Practical implementation of the results of the research is an embedding of the software modules in automated control systems, as well as the development of ASP for analytics and for training.

The prospect for further research is to conduct research of the systems «production, recycling, retail» on the simulation models of the «producers, products, consumers» class.

References

1. Borovska, T.M., Kolesnik, I.S., Severilov, V.A.: Method of optimal aggregation in optimization problems: monograph. UNIVERSUM–Vinnytsia, Vinnytsia (2009). 229 p.
2. Borovska, T.M., Badiora, S.P., Severilov, V.A., Severilov, P.V.: Modeling and optimization of the processes of the production systems development taking into consideration the usage of the external resources and development effects: monograph; under the general editorship of Borovska, T.M., VNTU, Vinnytsia (2009). 255 p.
3. Borovska, T.M., Vernigora, I.V., Grishin, D.I., Severilov, V.A., Gromaszek, K., Aizhanova, A.: Adaptive production control system based on optimal aggregation methods. In: Proceedings of SPIE, Photonics Applications in Astronomy, Communications, Industry, and High-Energy Physics Experiments 2018, vol. 10808, 108086O, 1 October 2018. https://doi.org/10.1117/12.2501520
4. Forrester, J.: Fundamentals of cybernetics of the enterprise (Industrial dynamics): Translated from English. Probgress (1971). 340 p.
5. Opoitsev, V.I.: Equilibrium and stability in models of collective behavior. USSR: World, Moscow (1977). 346 p.
6. Borovska, T.M., Vernigora, I.V., Wójcik, W., Gromaszek, K., Smailova, S., Orazbekov, Z.: Mathematical models of production systems development based on optimal aggregation methodology. In: Proceedings of the SPIE, Photonics Applications in Astronomy, Communications, Industry, and High Energy Physics Experiments, vol. 10445, 104452P, 7 August 2017. https://doi.org/10.1117/12.2281222
7. Leontiev, V.: Theoretical assumptions and nonobservable facts. Economy, ideology, politics, USA, no. 9, p. 15 (1972)
8. Bellman, R., Gliksberg, I., Gross, O.: Certain problems of mathematical control theory. Publishing House of Foreign Literature (1962). 233 p.
9. Rüttimann, B.: Introduction to Modern Manufacturing Theory. Springer International Publishing AG 2018 (2016). 149 p. https://doi.org/10.1007/978-3-319-58601-4
10. Nersessian, N.J., Chandrasekharan, S.: Hybrid analogies in conceptual innovation in science. Cogn. Syst. Res. **10**, 178–188 (2009). https://doi.org/10.1016/j.cogsys.2008.09.009
11. Borovska, T.M., Severilov, V.A., Badiora, S.P., Kolesnik, I.S.: Modeling of the investments management problems: manual for graduate students. VNTU, Vinnytsia (2009). 178 p.
12. Vasylska, M.V., Kolesnik, I.S., Severilov, V.A.: Models-predictors: problems of development and adequacy. Bull. Vinnytsia Polytech. Inst., no. 4, pp. 114–121 (2011)
13. Peschel, M.: Modeling of Signals and Systems. Translated from German. Mir (1981). 302 p.
14. Mesarovych, M., Macko, D., Takahara, I.: Theory of hierarchic multilevel systems, Mir (1973). 344 p.

15. Fagin, R., Kumar, R., Sivakumar, D.: Efficient similarity search and classification via rank aggregation. In: Proceedings of the 2003 ACM SIGMOD international Conference on Management of Data (San Diego, California), SIGMOD 2003, pp. 301–312. ACM Press, New York (2003). . https://doi.org/10.1145/872794.872795
16. Severilov, P.V., Borovska, T.N., Dmytryk, Y.N., Khomyn, E.P.: Modeling and optimization of agrarian systems with waste recycling in bioreactors. Nauka i studia (Poland), no. 16 (126), pp. 42–50 (2014). ISSN 1561-6894
17. Borovska, T.N., Kolesnik, I.S., Severilov, V.A., Shulgan, I.V.: Optimal aggregation of the integrated systems «production-development». Information technologies and computer engineering, no. 2(30), pp. 18–28 (2014). ISSN 1999-9941
18. Borovska, T.M.: Optimal aggregation of the production systems with parametric couplings. Eastern Euro. J. Adv. Technol. **4**(11(70)), 9–19 (2014). https://doi.org/10.15587/1729-4061.2014.26306
19. Borovska, T.M., Kolesnik, I.S., Severilov, V.A., Severilov, P.V.: Models of optimal innovation development of production systems. East-Euro. J. Adv. Technol. Math. Inf. Support Comput. Integr. Control Syst. **5**(71), 42–50 (2014). https://doi.org/10.15587/1729-4061.2014.28030
20. Boubaker, O., Babary, J.P., Ksouri, M.: Variable structure estimation and control of nonlinear distributed parameter bioreactors systems, man, and cybernetics. In: IEEE International Conference on Digital Object Identifier, vol. 4, pp. 3770–3774 (1998). https://doi.org/10.1109/icsmc.1998.726674
21. Baader, W., Dohne, E., Brenndörfer, M.: Biogas in Theorie und Praxis (translated from German, preface of Serebrian, M.I.). Kolos (1982). 540 p.
22. Pollock, J., Coffman, J., Ho, S.V., Farid, S.S.: Integrated continuous bioprocessing: economic, operational, and environmental feasibility for clinical and commercial antibody manufacture. Biotechnol. Prog. **33**(4), 854–866 (2017)
23. Brethauer, S., Studer, M.H.: Consolidated bioprocessing of lignocellulose by a microbial consortium. Energy Environ. Sci. **7**(4), 1446–1453 (2014). https://doi.org/10.1039/c3ee41753k
24. El-Moslamy, S.H., El-Morsy, E.S.M., Mohaisen, M.T., Rezk, A.H., Abdel-Fattah, Y.R.: Industrial bioprocessing strategies for cultivation of local Streptomyces violaceoruber strain SYA3 to fabricate Nano-ZnO as anti-phytopathogens agent. J. Pure Appl. Microbiol. **12**(3), 1133–1145 (2018)
25. Coase, R.H.: The problem of social cost. J. Law Econ. **3**, 1–44 (1960)
26. Borovska, T., Vernigora, I., Severilov, V., Kolesnik, I., Shestakevych, T.: Model of innovative development of production systems based on the methodology of optimal aggregation. In: Advances in Intelligent Systems and Computing III, pp. 171–181. Springer, Cham (2019)

Modeling the Dynamics of Knowledge Potential of Agents in the Educational Social and Communication Environment

Andrii Bomba[1], Mariia Nazaruk[1(✉)], Nataliia Kunanets[2], and Volodymyr Pasichnyk[2]

[1] Informatics and Applied Mathematics Department, State Humanitarian University, Rivne, Ukraine
abomba@ukr.net, marinazaruk@gmail.com
[2] Information Systems and Networks Department, Lviv Polytechnic National University, Lviv, Ukraine
nek.lviv@gmail.com, vpasichnyk@gmail.com

Abstract. The processes of information processing in the form of knowledge are at the forefront when considering the educational social and communication environment as a holistic system. In this paper, the authors examine the issue of modeling the personal educational (curriculum) program of a person who is studying during all life. The models of information processes for the redistribution of a knowledge potential of agents are created taking into account the units of its components. In particular, a multicomponent two-dimensional array of discrete values has been introduced to characterize procedures for the formation of agents' professional competencies that are appropriate to their abilities, interests, motivations, psychodynamic and emotional characteristics, age and level of knowledge potential.

Keywords: Lifelong learning · Knowledge potential · Competence · Agent

1 Introduction

According to expert estimates, professionals who can study throughout life, think critically, set goals and achieve them, work in a team, communicate in a multicultural environment and possess other modern knowledge and skills will be the most successful in the labor market in the near future [1].

New content of education is based on the formation of competencies that are necessary for successful self-realization in society (Fig. 1).

The concept of lifelong learning in various environments has been researched in works [2–4], especially it is proposed the hybrid agent-oriented methods for knowledge assessment in the context of lifelong learning based on the use of binary classification and artificial neural networks (Table 1) [5–9]. In the works of scientists, the probabilistic and statistical approach to the analytical modeling of the educational process has been developed. However, mastering or forgetting a knowledge unit is considered as a random event, and learning is characterized by parameters that are functionally

N. Shakhovska and M. O. Medykovskyy (Eds.): CCSIT 2019, AISC 1080, pp. 17–24, 2020.
https://doi.org/10.1007/978-3-030-33695-0_2

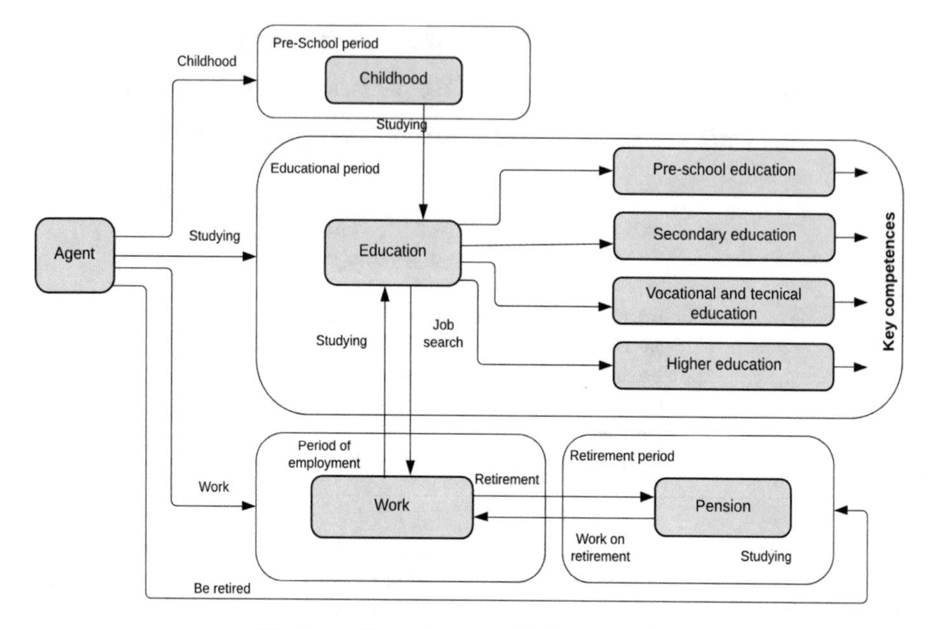

Fig. 1. A block diagram of lifelong learning

related to time. In a more general form, the stochastic learning process throughout life is seen as semi-Markov, when the probability of a transition from one state to another depends on both the initial state and the state where the transition takes place [10–12]. Considering the learning nature, it is thought that only transitions to neighboring states are practically possible, which may depend on time. As a result, the systems of differential equations with time-dependent coefficients are obtained that do not always have analytical solutions.

Table 1. Characteristics of methods

Method	Purpose	Disadvantages
Neural Networks	Evaluating e-learning participants	Long time learning and sensitivity to noise
Branches and boundaries	Building a model for knowledge dissemination	Not taken into account the degree of assimilation of new knowledge by individuals
Multi-level algorithmic knowledge quanta	Construction of an individual trajectory of the learning course	Considering only general characteristics
Markov chains	Dynamics description of the quality indicators of the educational process	A large amount of accumulated data is required to determine the probability of transitions

The research analysis makes it possible to conclude that the proposed models do not allow to consider a number of unknown variables that influence the result. Firstly, they are the characteristics of the agent groups including the features that determine the speed of intra-group knowledge dissemination. Secondly, the mentioned methods and models do not describe the process of transferring knowledge from one group to another one. Furthermore, it is necessary to study the modeling of the dynamics of agents' knowledge potential and the formation of their personal educational (training) curriculum, which would ensure the acquisition of professional competences in accordance with the psychodynamic characteristics and age.

2 Main Part

2.1 Some Features of the Base Model

The numerical characteristic is suggested in our previous works to determine as a knowledge potential (φ) and with its help a certain level of knowledge of a person accumulated during the education and the life experience is recorded. The diffusion-like models of information processes for the redistribution of knowledge potential in the city's educational social and communication environment have been constructed. In particular, an emphasis is put on the process description (modeling) of the knowledge potential redistribution in the formation of the system of professional competences [13, 14]. Competence is a dynamic combination of knowledge, ways of thinking, views, values, skills, abilities and other personal qualities that determines an ability of a person to conduct professional and/or further learning activities successfully [15, 16].

In the references of the European Parliament and the Council of Europe on the development of life-long learning competences, there are defined eight groups of key competences:

- linguistic competence $\left(key_{compet_1}\right)$;
- mathematical competence $\left(key_{compet_2}\right)$;
- competences in natural sciences and technologies $\left(key_{compet_3}\right)$;
- ecological competence $\left(key_{compet_4}\right)$;
- information and digital competence $\left(key_{compet_5}\right)$;
- civil and social competencies $\left(key_{compet_6}\right)$;
- cultural competence $\left(key_{compet_7}\right)$;
- entrepreneurship and financial literacy $\left(key_{compet_8}\right)$.

Thus, the formal model of educational results of an applicant (an agent) will be presented in the form:

$$Competence = key_{competence_1} \cup key_{competence_2} \cup \ldots \cup key_{competence_8}$$

It is assumed that one or another competence is characterized by a certain set of components and their units, and we will understand the combination of components' units of the knowledge potential as a competence.

Thus, $\varphi_{q,l,k,m,\varepsilon}$ is a knowledge potential of q-unit of l-component of k-agent at the m- moment of time ($l = \overline{1, l_*}$, $q = \overline{1, q_*}$, $k = \overline{1, k_*}$). Then, the set of components l of units q for a given situational state can be characterized by some matrix (two-dimensional array):

$$\begin{pmatrix} \varphi_{1,1,k,m,\varepsilon}, & \varphi_{2,1,k,m,\varepsilon}, & \ldots, & \varphi_{q_*,1,k,m,\varepsilon} \\ \varphi_{1,2,k,m,\varepsilon}, & \varphi_{2,2,k,m,\varepsilon}, & \ldots, & \varphi_{q_*,2,k,m,\varepsilon} \\ \ldots\ldots\ldots, & \ldots\ldots\ldots, & \ldots\ldots\ldots, & \ldots\ldots\ldots \\ \varphi_{1,l_*,k,m,\varepsilon}, & \varphi_{2,l_*,k,m,\varepsilon}, & \ldots, & \varphi_{q_*,l_*,k,m,\varepsilon} \end{pmatrix}$$

A set of possible subsets of such an array will be associated with a set of all competences acquired by agents during studying.

2.2 Modeling the Dynamics of Agents' Knowledge Potential

The model describing the redistribution of knowledge potentials, considering the components of their units, can be represented as:

$$\begin{aligned} \varphi_{q,l,k,m+1,\varepsilon} = {} & \varphi_{q,l,k,m,\varepsilon} + \sum_{\tilde{m}}^{m} \alpha_{\tilde{m}} f_{q,l,k,\tilde{m}} + \sum_{\tilde{m}}^{m} \beta_{\tilde{m}} g_{q,l,k,\tilde{m}} \\ & + \sum_{\tilde{m}}^{m} \sum_{\tilde{k}}^{k_*} \xi_{q,l,\tilde{k},k,\tilde{m}} (\varphi_{q,l,k,\tilde{m},\varepsilon} - \varphi_{q,l,\tilde{k},\tilde{m},\varepsilon}) \\ & + \varepsilon(\alpha \sum_{\tilde{k}}^{k_*} \xi_{q,l,\tilde{k},k,m+1} (\varphi_{q,l,k,m+1,\varepsilon} - \varphi_{q,l,\tilde{k},m+1,\varepsilon}) \\ & + \beta f_{q,l,k,m+1,\varepsilon} + g_{q,l,k,m+1,\varepsilon}) \end{aligned} \tag{1}$$

Where $f_{q,l,k,m+1,\varepsilon}$ is an intensity of the source of the knowledge transmission (of a teacher), $g_{q,l,k,m+1,\varepsilon}$ is the functions characterizing the interdependence (mutual influence) between the units and components of the knowledge potential, $\xi_{q,l,\tilde{k},k,\tilde{m}}$ is a coefficient of perception ("bandwidth") of $\tilde{k}$-agent of k-agent, ε is a small parameter characterizing the knowledge transfer between agents at some point of time, α, β are weight coefficients.

The function $\varphi_{q,l,k,m+1,\varepsilon}$ is represented in the form of series by schemes of small parameter ε [17]:

$$\varphi_{q,l,k,m+1,\varepsilon} = \varphi_{q,l,k,m+1,0} + \varphi_{q,l,k,m+1,1} + \varphi_{q,l,k,m+1,2} + \ldots \tag{2}$$

Where $\varphi_{q,l,k,m+1,0}$ is calculated by the formula:

$$\varphi_{q,l,k,m+1,0} = \varphi_{q,l,k,m,0} + f_{q,l,k,m} + g_{q,l,k,m} \tag{3}$$

$\varphi_{q,l,k,m+1,1}$ is represented as follows:

$$\varphi_{q,l,k,m+1,1} = \varphi_{q,l,k,m,1} + f_{q,l,k,m} + g_{q,l,k,m} + p_{q,l,k,m,1} \tag{4}$$

Where, $p_{q,l,k,m,1} = \varepsilon(\alpha \sum_{\tilde{k}}^{k_*} \xi_{q,l,\tilde{k},k,m+1}(\varphi_{q,l,k,m+1,0} - \varphi_{q,l,\tilde{k},m+1,0}))$ etc.

In this case, it is meant that some initial distributions of agents' knowledge potential are set, namely: $\varphi_{q,l,k,0,\varepsilon} = \varphi*_{q,l,k}$, as well as some limit values or other information about the nature of the knowledge perception by individual agents.

The different variants of the representation of these functions of mutual influence are possible as in [18], in particular:

$$g_{q,l,k,m,\varepsilon} = g_{*q,l,k,m,\varepsilon} + \sum_{\tilde{m}}^{m} \left(\sum_{\tilde{l}=1,\tilde{q}=1}^{l_*,q_*} \begin{pmatrix} \alpha_{q,l,\tilde{q},\tilde{l},k,m,\varepsilon} \\ + \beta_{q,l,\tilde{q},\tilde{l},k,m,\varepsilon} \varphi_{q,l,k,\tilde{m},\varepsilon} \end{pmatrix} \varphi_{\tilde{q},\tilde{l},k,\tilde{m},\varepsilon} \right) \tag{5}$$

Where $g_{*q,l,k,m,\varepsilon}$, $\alpha_{q,l,\tilde{q},\tilde{l},k,m,\varepsilon}$, $\beta_{q,l,\tilde{q},\tilde{l},k,m,\varepsilon}$ are coefficients characterizing the units of components of the knowledge potential established on the basis of previous experiments(experience) using the ideas of constructing neural networks. An important advantage of using neural networks to process data arrays is a significant increase in the speed of the process compared with traditional mathematical methods, the ability to study the neural network according to reference samples, as well as to change the network topology (selection of input parameters that guarantee obtaining the model of the highest accuracy), based on the requirements of solvable problem [19].

In the case of the component dependence of the knowledge potential and their units at a certain point of time, we will have:

$$g_{q,l,k,m,\varepsilon} = g_{*q,l,k,m,\varepsilon} + \sum_{\tilde{l}=1,\tilde{q}=1}^{l_*,q_*} \begin{pmatrix} \alpha_{q,l,\tilde{q},\tilde{l},k,m,\varepsilon} \\ + \beta_{q,l,\tilde{q},\tilde{l},k,m,\varepsilon} \varphi_{q,l,k,m,\varepsilon} \end{pmatrix} \varphi_{\tilde{q},\tilde{l},k,m,\varepsilon}$$

It is possible to get a case of the function dependence $g_{*q,l,k,m,\varepsilon}$ on the interaction of agents among themselves:

$$g_{q,l,k,m,\varepsilon} = g_{*q,l,k,m,\varepsilon} + \sum_{\tilde{l}=1,\tilde{q}=1,\tilde{k}=1}^{l_*,q_*,k_*} \begin{pmatrix} \alpha_{q,l,\tilde{q},\tilde{l},\tilde{k},m,\varepsilon} \\ + \beta_{q,l,\tilde{q},\tilde{l},\tilde{k},m,\varepsilon} \varphi_{q,l,\tilde{k},m,\varepsilon} \end{pmatrix} \varphi_{\tilde{q},\tilde{l},\tilde{k},m,\varepsilon}.$$

The matrix (two-dimensional array) for the formation of agents using the four components of the four elements' knowledge potential is represented as:

$$\begin{pmatrix} \varphi_{1,1,k,m,\varepsilon}, & \varphi_{2,1,k,m,\varepsilon}, & \varphi_{3,1,k,m,\varepsilon}, & \varphi_{4,1,k,m,\varepsilon} \\ \varphi_{1,2,k,m,\varepsilon}, & \varphi_{2,2,k,m,\varepsilon}, & \varphi_{3,2,k,m,\varepsilon}, & \varphi_{4,2,k,m,\varepsilon} \\ \varphi_{1,3,k,m,\varepsilon}, & \varphi_{2,3,k,m,\varepsilon}, & \varphi_{3,3,k,m,\varepsilon}, & \varphi_{4,3,k,m,\varepsilon} \\ \varphi_{1,4,k,m,\varepsilon}, & \varphi_{2,4,k,m,\varepsilon}, & \varphi_{3,4,k,m,\varepsilon} & \varphi_{4,4,k,m,\varepsilon} \end{pmatrix}$$

Let's assume that some of the components and units of the knowledge potential are important for the formation of specialists of two types (correspondingly A, B, C) that is (1.1), (1.2), (2.1), (2.1) for A and (2,2), (2,3), (3,2), (3,3) for B and (3,1), (3,2), (4,1), (4,2) for C. The corresponding situational condition is natural portrayed as:

$$\begin{pmatrix} \varphi_{1,1,k,m,\varepsilon}, & \varphi_{2,1,k,m,\varepsilon}, & \varphi_{3,1,k,m,\varepsilon}, & \varphi_{4,1,k,m,\varepsilon} \\ \varphi_{1,2,k,m,\varepsilon}, & \varphi_{2,2,k,m,\varepsilon}, & \varphi_{3,2,k,m,\varepsilon}, & \varphi_{4,2,k,m,\varepsilon} \\ \varphi_{1,3,k,m,\varepsilon}, & \varphi_{2,3,k,m,\varepsilon}, & \varphi_{3,3,k,m,\varepsilon}, & \varphi_{4,3,k,m,\varepsilon} \\ \varphi_{1,4,k,m,\varepsilon}, & \varphi_{2,4,k,m,\varepsilon}, & \varphi_{3,4,k,m,\varepsilon} & \varphi_{4,4,k,m,\varepsilon} \end{pmatrix}$$

A B C

The function g (for fixed k, m) can be represented as: $g_{q,l} = g_{*q,l} + \alpha_{1,1}\varphi_{1,1} + \alpha_{1,2}\varphi_{1,2} + \alpha_{1,3}\varphi_{1,3} + \ldots + \alpha_{3,3}\varphi_{3,3}) + (\beta_{1,1}\varphi_{1,1} + \ldots + \beta_{3,3}\varphi_{3,3})\varphi_{q,l}$ and under certain conditions (in particular, the target function) it is solved the problem of picking up $\alpha_{i,j}$, $\beta_{i,j}$ so that it, in accordance with the values $\varphi_{2,2,k,m,\varepsilon}$ (for A and B), takes optimal values.

Thus, the general logical sequence of the formation of the agent training trajectory is reduced to the following basic steps (Fig. 2):

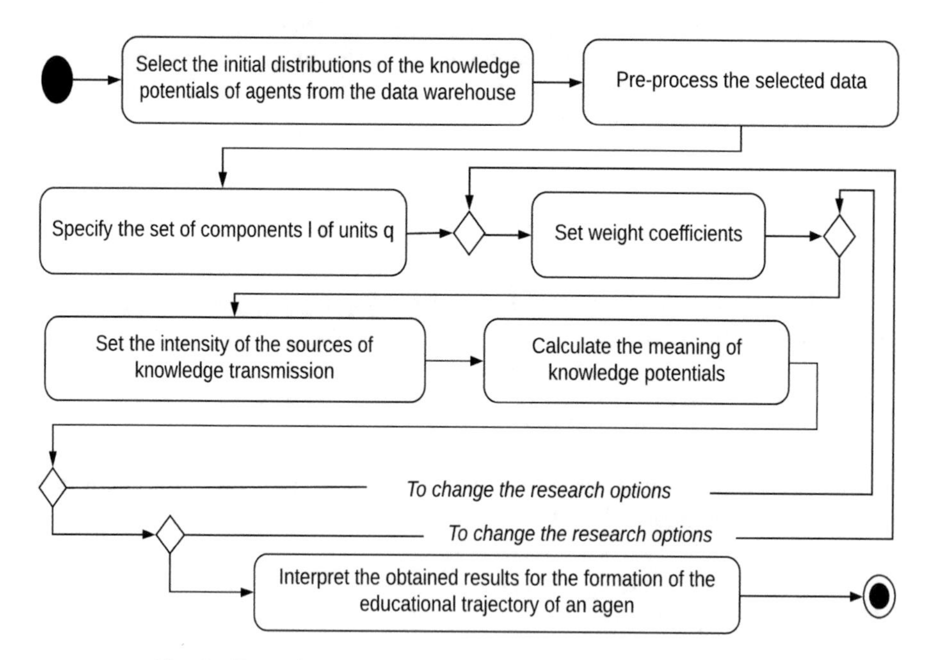

Fig. 2. Formation of an individual learning trajectory of an agent

Step 1. Select the initial distribution of the agent knowledge potential from the data warehouse.
Step 2. Pre-process the selected data.
Step 3. Specify the set of components l of units q for $(m = 0)$ for different values k, as well as the capacity of the main sources of information (teachers).
Step 4. Set the value of the parameters $g_{*q,l,k,m,\varepsilon}$, $\alpha_{q,l,\tilde{q},\tilde{l},k,m,\varepsilon}$, $\beta_{q,l,\tilde{q},\tilde{l},k,m,\varepsilon}$.
Step 5. Create procedures for calculation g, q, k, m according to formula (5).
Step 6. According to (1–3), calculate the meaning of knowledge potentials recurrently at the next moments of time $m = 1, 2, \ldots$.
Step 7. Interpret the obtained results for the formation of the educational trajectory of an agent.

Note that the correction of model parameters according to formulas (1–5) can be obtained, for example, by the method of least squares [20].

3 Conclusions

The models of information processes for the redistribution of knowledge potential with the units of its components are constructed. A multicomponent two-dimensional array of discrete values, as well as its subsets for the characteristics of procedures for the formation of agent's professional competencies have been introduced.

Thus, the characterization of the agents' knowledge potential, taking into account the units of its components and the interactions in the learning process, provided an opportunity to specify these models (consideration of dependencies between the parameters characterizing interpersonal relationships), and, consequently, modeling of more real situational states that has considerable practical value.

In the future, further research is the introduction of target functions as functions of subsets of this two-dimensional array and solving various problems of optimization and control.

References

1. New Ukrainian School: Conceptual principles of secondary school reform (2016)
2. Chen, Z., Liu, B.: Lifelong Machine Learning, p. 127. Morgan & Claypool Publishers, San Rafael (2016)
3. Fei, G., Wang, S., Liu, B.: Learning cumulatively to become more knowledgeable. In: Proceedings of the 22nd International Conference on Knowledge Discovery and Data Mining, pp. 1565–1574 (2016)
4. Meijers, F.: A dialogue worth having: vocational competence, career identity and a learning environment for twenty-first century success at work. In: de Bruijn, E., Billett, S., Onstenk, J. (eds.) Enhancing Teaching and Learning in the Dutch Vocational Education System, vol. 18, pp. 139–155. Springer, Cham (2017)
5. Khetarpal, K., Sodhani, S., Chandar, S., Precup, D.: Environments for lifelong reinforcement learning. In: 2nd Continual Learning Workshop, Neural Information Processing Systems (2018)

6. Machado, M., Bellemare, M., Talvitie, E., Veness, J., Hausknecht, M., Bowling, M.: Revisiting the arcade learning environment: evaluation protocols and open problems for general agents. J. Artif. Intell. Res. **61**, 523–562 (2018)
7. Nonaka, I., Krogh, G., Voelpel, S.: Organizational knowledge creation theory: evolutionary paths and future advances. Organ. Stud. **27**(8), 1179–1208 (2006)
8. Dobrynina, N.: Mathematical models of knowledge dissemination and management of the learning process of students. Sci.Theoret. J. Fundam. Res. **7**, 7–9 (2009)
9. Artemenko, V.: Hybrid of an agent-based knowledge assessment model by distance learning participants. Educ. Technol. Soc. **2**, 423–434 (2011)
10. Petrash, A.: Methodology of an informative mathematical model for the process. In: Innovative Computer Technology at the School, pp. 128–132 (2011)
11. Saviour, A., Mahama, F., Kuadey, N., Ankorah, C.: Mathematical model of knowledge management system in an organization. Global J. Manag. Bus. Res. Adm. Manag. **16** (2016)
12. Ibatullin, R., Anisimova, E.: Construction of individual educational trajectory of students based on e-learning. In: IEEE 10th International Conference on Application of Information and Communication Technologies (2016)
13. Bomba, A., Nazaruk, M., Kunanets, N., Pasichnyk, V.: Constructing the diffusion-like model of biocomponent knowledge potential distribution. Int. J. Comput. **16**(2), 74–81 (2017)
14. Kunanets, N., Nazaruk, M., Pasichnyk, V., Nebesnyi, R.: Information technologies of personalized choice of professionals in smart cities. Inf. Technol. Learn. Tools **65**(3), 277–290 (2018)
15. About Higher Education: The Law of Ukraine 01.07.2014, 1556-VII (2015)
16. Rashkevych, Yu.: Bologna Process and New Paradigm of Higher Education: Monograph, p. 168 (2014)
17. Bomba, A., Safonyk, A.: Mathematical simulation of the process of aerobic treatment of wastewater under conditions of diffusion and mass transfer perturbations. J. Eng. Phys. Thermophys. **91**(2), 318–323 (2018)
18. Pasichnyk, V., Bomba, A., Nazaruk, M., Kunanets, N., Bilak, Y.: Modeling the redistribution processes of knowledge potential in the formation of the professional competency system. In: IEEE 14th International Scientific and Technical Conference on Computer Sciences and Information Technologies (2019)
19. Galushkyn, A.: Theory of neural networks: a textbook for universities. Publishing Enterprise of the Radiotekhnika Magazine (2000). 215 p.
20. Flach, P.: Machine Learning: The Art and Science of Algorithms that Make Sense of Data, p. 396. Cambridge University Press, New York (2012)

Parallel Solving of Fredholm Integral Equations of the First Kind by Tikhonov Regularization Method Using OpenMP Technology

Lesia Mochurad[1(✉)], Khrystyna Shakhovska[1], and Sergio Montenegro[2]

[1] Lviv Polytechnic National University, Lviv 79013, Ukraine
Lesia.I.Mochurad@lpnu.ua,
kristin.shakhovska@gmail.com
[2] Julius-Maximilians-University Würzburg, Am Hubland, 97074 Würzburg, Germany
sergio.montenegro@uni-wuerzburg.de

Abstract. The paper considers the software approach to solving the Fredholm integral equation of the first kind by the Tikhonov regularization method, which uses the properties of modern architectures of computing systems as multi-core and the standard of parallel programming of OpenMP. A numerical algorithm for solving a Fredholm integral equation of the first kind is proposed such as the Tikhonov regularization method. It is based on the method of collocation under conditions of piecewise constant approximation of the desired function. When choosing even this most economical method of approximation, we obtain systems of linear algebraic equations of large dimensions with densely filled matrices to achieve the necessary accuracy. The variation of the error is analyzed, depending on the choice of the regularization parameter, which is selected experimentally. In order to optimize the computational process, the procedure for solving the integral equation is compared with the use of the OpenMP parallel programming technique. Improved acceleration and performance. The results, which indicate the possibility of further optimization of the computational process due to the variation in the number of parallel streams and computer processor cores, are obtained. A series of numerical experiments that confirm the effectiveness and feasibility of the proposed approach to numerical solution of Fredholm integral equations of the first kind.

Keywords: Incorrectness problem · Poorly conditioned matrix · Regularizing operator · Multicore · OpenMP software standard

1 Introduction

When solving various applications, as well as in studies in the field of mathematical theory, so-called correctly and incorrectly set tasks [1]. The latter include:

- problems of numerical differentiation of the experimentally obtained function;

N. Shakhovska and M. O. Medykovskyy (Eds.): CCSIT 2019, AISC 1080, pp. 25–35, 2020.
https://doi.org/10.1007/978-3-030-33695-0_3

- the problem of smoothing the function;
- calculation of Fourier series with inaccurately given coefficients;
- solving systems of linear algebraic equations with poorly induced matrices;
- the establishment of a continuous spectrum in the obstructed spectroscopy problem;
- inverse geological exploration tasks;
- reproduction of the transmitted signal received in the presence of obstacles in the atmosphere;
- pattern recognition tasks, etc.

Incorrect tasks are characterized by the fact that arbitrarily small changes in the original data lead to arbitrarily large changes in the solutions of these problems. Often the source data is the result of some experimental activity, that is known about. Therefore, there is a need for the study and application of methods for solving incorrect tasks. The basis for research in this area is the works of the scientific school A.M. Tikhonov, who created a mathematical theory of incorrectly set tasks. This includes the method of regularizing A.M. Tikhonov, method of replacing M. M. Lavrentyev, method of selection and quasi-solution V.K. Ivanov and other methods. Foreign designs are presented by the methods of optimal filtration of Kalman-Bucy and Wiener, controlled linear filtration methods (Beykus-Hilbert), and others. While these methods are more precise, the methods proposed by Soviet scientists (in particular, the Tikhonov method) require much less up-to-date information about the solution, and therefore they are more widely used in solving incorrect tasks. However, it is known [2] that the methods of regularizing some tasks require a large amount of computing on a computer. This, in turn, leads to instability of the calculations. Therefore, there is a need for the development of an effective numerical algorithm, followed by its parallelization on modern architectures of computing systems.

2 Review of the Literature

Nowadays, the integral equations have begun to be used extensively for solving many tasks of simulation of dynamic objects and systems [2–6]. The most important advantages of the method of integral equations include the reduction of the dimension of the problem per unit and the application for non-bounded domains. Issues of the restoration of signals in spectroscopy, increasing the accuracy of monitoring systems, reduction of long signals, the decay of cells and radioactive elements, medical diagnostics, in particular, the inverse problem of electrocardiography and other virtually essential tasks can be expressed through the operator equation of the first kind [7–9]

$$A[x, u(\xi)] \equiv \int_a^b K(x, \xi) u(\xi) d\xi = f(x), \tag{1}$$

where $x \in [c, d]; K(x, \xi) \in C\{[c, d] \times [a, b]\}$ – from a mathematical point of view, the core of the integral equation, $f(x)$ – known function built on the results of the experiment, $u(\xi)$ – desired function. This equation has properties of incorrectness (at least one of Hadamard's correctness conditions is violated), therefore it can not be solved by classical methods in their traditional form. So there was a need to develop effective methods for solving them. As a result, the concept of a regularizing operator was introduced $A_\alpha[x, u_\alpha(\xi)]$ and formulated one of the most effective methods for solving incorrect tasks - a method α - Tikhonov's regularization. A series of regularizing decomposition schemes was proposed, stable methods were developed for solving problems in various mana areas of mathematics: optimal control, summation of Fourier series, integral equation of the first kind of convolution, systems of linear algebraic equations, operator equations, etc.

The choice of the parameter of the result is well analyzed in [10] when considering the problem of determining the fractional composition of powdered substances of the method using the effects of light scattering. This compares the proposed new method with known methods of selecting these parameters, such as the method of the *L*-curve and the method of generalized cross-checking.

Sustained methods for solving incorrect tasks are described in many monographs and publications, but their presence does not exclude the need to improve existing methods with bringing them to practical algorithms, especially to the development of software.

3 Problem Statement

There is a large number of effective methods for solving Fredholm integral equations of the first kind. The basis of most construction is (1) after sampling to systems of linear algebraic equations (SLAE)

$$Au = f. \tag{2}$$

When SLAE (2) is degenerate or poorly determined, then the inverse operator does not exist or unlimited, and the task is incorrect. This leads to the fact that the solution is not uniform or unstable. Therefore, the actual task is to improve the regularization procedures using parallelization. It is about this task, which previously was an unsolved part of the general problem, is mentioned in the article.

In this case, if it is necessary to improve the accuracy of the output integral equation (IE) (1), increasing the size of the SLAE, this, in turn, will lead to a decrease in the speed of solving SLAE, and in some cases, to a shortage of computer resources (memory). In such cases, modern effective means of parallelism should be used.

When using regularization for Tikhonov there is a need to find a function $u(\xi)$, which minimizes the smoothing functionality of the look:

$$M[\alpha, f, u] \equiv \int_c^d \{A[x, u(\xi)] - f(x)\}^2 dx + \alpha\,\Omega_n[u] \to \min,$$

where $\alpha > 0$ – numeric parameter, $\Omega_n[u]$ – stabilizer n-stage $\Omega_n[u] = \int\limits_a^b \sum\limits_{k=0}^{n} p_k(\xi)$ $\left(\frac{d^k u}{d\xi^k}\right)^2 d\xi$, $p_k(\xi)$ – continuous functions (weights): $p_k(\xi) \geq 0, \xi \in [a, b]$; $p_n(\xi) \geq p_0 > 0$. Often, the second-stage stabilizer is used $(n = 1)$: $\Omega_1[u] = \int\limits_a^b p_0(\xi)u^2(\xi) + p_1(\xi)\left(\frac{du}{d\xi}\right)^2 d\xi$, here $p_0(\xi) \equiv p_1(\xi) \equiv 1$, if there is no information about the character of the behavior of the function.

4 Materials and Methods

For the numerical solution of IE (1) we use the most economical method of collocation under the conditions of piecewise constant approximation of the desired function $u(\xi)$.

- Divide the interval $[a,\ b]$ on n elements of equal length;
- In the middle of each element, we select some points $\{\xi_i\}_{i=1}^n$: $\xi_i = a + \frac{h}{2}(2i - 1)$, where $h = \frac{b-a}{n}$;
- Then our equation is replaced by the following:

$$\sum_{i=1}^{n} \int\limits_{\xi_i - \frac{h}{2}}^{\xi_i + \frac{h}{2}} u(\xi)K(x, \xi)dx = f(x), x \in [a, b];$$

- Choose x from $\{x_j\}_{j=1}^n$: $x_j = a + \frac{h}{2}(2j - 1)$, $j = \overline{1,\ n}$. Then:

$$\sum_{i=1}^{n} u_i \int\limits_{\xi_i - \frac{h}{2}}^{\xi_i + \frac{h}{2}} K(\xi, x_j)dx = f(x_j). \tag{3}$$

In result we receive (3) – SLE with bad-conditioned matrix A:

$$v(A) \equiv \|A\| \left\|A^{-1}\right\| >> 1 \text{ or } v(A) \approx 1.$$

- Therefore, we use the regularization method. It consists in the fact that solving SLAE is reduced to minimizing the functional $\|Au - f\|^2 + \alpha\|\xi - \xi_0\|^2 \rightarrow \min$, where $\alpha \in \{10^{-15};\ 10^{-14};\ \ldots;\ 0,01;\ 0,1;\ 1\}$, ξ_0 – arbitrary vector. Minimizing the specified functionality, we obtain the so-called normal solution of the system $Au = f$.

- Since we do not have any a priori information about the desired solution, we can take it $x_0 = 0$. Parameter α chosen by experiment. So, the unknown value u_α looking solving SLAE by Gaussian form $(\alpha I + A^T A)u_\alpha = A^T f$, where

$$A = \begin{pmatrix} \int\limits_{\xi_1-\frac{h}{2}}^{\xi_1+\frac{h}{2}} K(\xi,x_1)dx & \int\limits_{\xi_2-\frac{h}{2}}^{\xi_2+\frac{h}{2}} K(\xi,x_1)dx & \cdots & \int\limits_{\xi_n-\frac{h}{2}}^{\xi_n+\frac{h}{2}} K(\xi,x_1)dx \\ \int\limits_{\xi_1-\frac{h}{2}}^{\xi_1+\frac{h}{2}} K(\xi,x_2)dx & \int\limits_{\xi_2-\frac{h}{2}}^{\xi_2+\frac{h}{2}} K(\xi,x_2)dx & \cdots & \int\limits_{\xi_n-\frac{h}{2}}^{\xi_n+\frac{h}{2}} K(\xi,x_2)dx \\ \vdots & \vdots & \vdots & \vdots \\ \int\limits_{\xi_1-\frac{h}{2}}^{\xi_1+\frac{h}{2}} K(\xi,x_n)dx & \int\limits_{\xi_2-\frac{h}{2}}^{\xi_2+\frac{h}{2}} K(\xi,x_n)dx & \cdots & \int\limits_{\xi_n-\frac{h}{2}}^{\xi_n+\frac{h}{2}} K(\xi,x_n)dx \end{pmatrix} \quad (4)$$

Given large values n $(n \geq 1000)$, which allows to significantly improve the accuracy of the calculations, to implement this algorithm takes a lot of time, or for the program is not enough RAM, which leads to instability of calculations [12]. Therefore, the actual task is to parallelize the above numerical algorithm.

For small tasks, it often turns out that a parallel version runs slower than single-processor. The noticeable effect of parallelization has to be observed when solving systems with more and more unknowns. During the launch, it is recommended to use SLAE: $P = M * N/10^6$, where $M \times N$ − dimension of the matrix. In other words, the number of processors should be such that the processor has a block of matrix size approximately 1000×1000. The increase in the efficiency of parallelization in the case of increasing the size of the solvable system of equations is explained very simply: in the case of increasing the dimension of the system of equations, the amount of computational work increases proportionally n^3, and the volume of exchanges between processors is proportional n^2. This reduces the relative share of communication costs in case of an increase in the size of the system of equations.

5 Materials and Techniques

In recent years, the OpenMP programming system has become increasingly popular [13–15]. The interface of this technology is conceived as the standard for programming in the general memory model. OpenMP includes a specification for the compiler directives, procedures, and environment variables. He implements the idea of "incremental paralleling." The developer does not create a new parallel program but adds to the text of the sequential program OpenMP-directive. At the same time, the OpenMP programming system provides the developer with significant control over the behavior of parallel applications. The whole program is divided into successive and parallel areas. All following domains execute the mainstream that generates when the program is started, and when entering the parallel region, the mainstream generates additional flows. It is assumed that OpenMP - the program without any modification should work on both multi-processor systems and a single processor. In the latter case, OpenMP directives are ignored.

In the paper, such a property of modern computers as multi-core is used. The program implementation was carried out in the Microsoft Visual Studio 2017 environment in C++, with the organization of the OpenMP parallel programming standard.

The algorithm for the program realization of the Fredholm integral equation of the kind by the Tikhonov regularization method is presented in the form of a sequence of the following actions:

1. Announcement of variables, constants (integral limits, step, etc.);
2. Calculation of the elements of the SLAE matrix (4). At the same step, the exact solution is calculated, if any;
3. Create an expanded matrix:
   a. Transposition of the matrix;
   b. Multiplication of matrices;
   c. Adding matrices;
4. Search SLAE solution by Gauss.

To improve efficiency, the following parts of the algorithm were parallelized:

- Gauss method – direct and reverse by using the pragma parallel for.
  #pragma omp parallel – creates a group of threads, while #pragma omp for divides the iteration of loops between generated flows. Using #pragma omp parallel for allows to do these two actions simultaneously.

```
void gaus_method(double **a, double *x, int n)
{
	for (int k = 0; k < n - 1; k++)
	{
		a = max_element(a, k, n);
		for (int i = k + 1; i < n; i++)
		{
			double m = -a[i][k] / a[k][k];
#pragma omp for
			for (int j = k; j < n + 1; j++)
				a[i][j] = a[i][j] + m * a[k][j];
		}

	}
	x[n - 1] = a[n - 1][n] / a[n - 1][n - 1];
	for (int k = n - 2; k >= 0; k--)
	{
		double sum = 0;
#pragma omp for
		for (int j = n - 1; j > k; j--)
			sum += a[k][j] * x[j];
		x[k] = (a[k][n] - sum) / a[k][k];
	}
}
```

- Transposition of the matrix. Use #pragma omp parallel for in the outer loop.

```
#pragma omp parallel for
    for (int i = 0; i < n; i++)
        for (int j = 0; j < n; j++)
            transp[j][i] = matr[i][j];
```

- Multiplication of matrices. Use #pragma omp parallel for in the outer loop.

```
#pragma omp parallel for
    for (int i = 0; i < n; i++) {
        for (int l = 0; l < n; l++) {
            double s = 0;
            for (int j = 0; j < n; j++) {
                s += transp[i][j] * matr[j][l];
            }
            left[i][l] = s;
        }
    }
```

- Multiplying the matrix by vector. Use #pragma omp parallel for in the outer loop.

```
#pragma omp parallel for
    for (int i = 0; i < n; i++)
        for (int j = 0; j < n; j++)
            rozsh[i][n] += transp[i][j] * f[j];
```

- Adding matrices. Use #pragma omp parallel for in the outer loop.

```
#pragma omp parallel for
            for (int i = 0; i < n; i++)
        for (int j = 0; j < n; j++)
            left[i][j] += one[i][j];
```

Also, #pragma omp parallel for collapse (2) can be used. It collapses two nested loops in one.

6 Experiment

The program was executed on a dual-core processor. System configuration: Intel Core i7-7500U with 16.00 GB of RAM with 2.7–3.5 GHz processor. The results of the program were tested on two examples. The change of error is analyzed, depending on the choice of the parameter of regularization α. For the first example, there is a known exact solution that made it possible to check the reliability of the results obtained.

Example 1

Let the equation be given $\int\limits_0^1 \ln\frac{1}{|t-s|}\varphi(s)ds = 1$. Exact solution is known: $\varphi(t) = \frac{1}{2\pi\ln(2)\sqrt{t(1-t)}}$.

First of all, it is necessary to choose the parameter of regularization in an experimental way. Take a small number for this. So, Table 1 shows the change in the error of the solution, depending on the choice of the parameter α.

Table 1. Change the error of the solution at $n = 10$.

Parameter value α	Execution time, s	Error
0.1	0.01	0.1714619175
0.01	0.01	0.1479883005
0.001	0.01	0.1454157417
0.0000001	0.01	0.1451272652
0.0000000001	0.01	0.1451272363
0.00000000000001	0.01	0.1451272363

Consequently, at small sizes a small error and time of execution can be observed. However, the increase of the SLAE dimension leads to significant change error on the best and worst regularization parameter (see Table 2).

Table 2. Change the error for large n.

Value of n	Execution time, s	Error at $\alpha = 10^{-1}$	Error at $\alpha = 10^{-14}$
1000	13.24	0.6058095796	0.0839762662
2000	196.34	0.6624032341	0.0869721057
3000	476.83	0.6918825840	0.1155640507
4000	1193.88	0.7125021101	0.0552941082
5000	2279.91	0.7263497234	0.0526534119

By selecting the regularization parameter $\alpha = 10^{-14}$ Table 3 shows a comparison of the execution time of a sequential and parallel algorithm.

Table 3. Comparison of the runtime of the sequential and parallel algorithm for Example 1.

Value of n	Sequential execution time, s	Parallel execution time, s
1000	13.24	7.07
2000	196.34	100.58
3000	476.83	265.26
4000	1193.88	631.03
5000	2279.91	1312.25

Example 2
In order to check the validity of the results, a numerical solution of the Fredholm integral equation of the first kind of form:

$$\int_{0}^{\pi} \varphi(x)\sin\left(\frac{x+y}{n}\right)dx = 2n\cdot\sin\left(\frac{\pi}{n}\right)\cdot\sin\left(\frac{y}{n}+\frac{\pi}{2n}\right).$$

is provided. Timelines for performing sequential and parallel algorithms are presented in Table 4.

Table 4. Comparison of the runtime of the sequential and parallel algorithm for Example 2.

Value of n	Sequential execution time, s	Parallel execution time, s
1000	13.28 s	6.97 s
2000	210.35 s	98.51 s
3000	490.08 s	261.75 s
4000	1108.89 s	629.34 s
5000	2425.45 s	1354.42 s

7 Results and Discussion

It is seen from Tables 3 and 4 that the execution time of a parallel algorithm decreases almost in 2 times. The following theoretical assessments are made:

- The complexity of the sequential algorithm: $T_1 = 4n^3 + 7n^2$;
- The complexity of the parallel algorithm: $T_p = n^3 + 7n^2/p + 3n^3/p$;
- Acceleration $S_p(n) = T_1(n)/T_p(n) = \frac{p(4n+7)}{pn+3n+7}$;
- Efficiency $E_p(n) = T_1(n)/(pT_p(n)) = S_p(n)/p = \frac{(4n+7)}{pn+3n+7}$.

To increase the speed and efficiency, there is a need to increase the number of processor cores. Check of the possibility of accelerating the computational process can be resulted using the 4-and 8-core processor. Figure 1 shows a significant influence of the time of execution depending on the number of cores. At small dimensions, the difference in the time of execution is low, while in large-scale dimensions the amount of the core has a significant effect on the time.

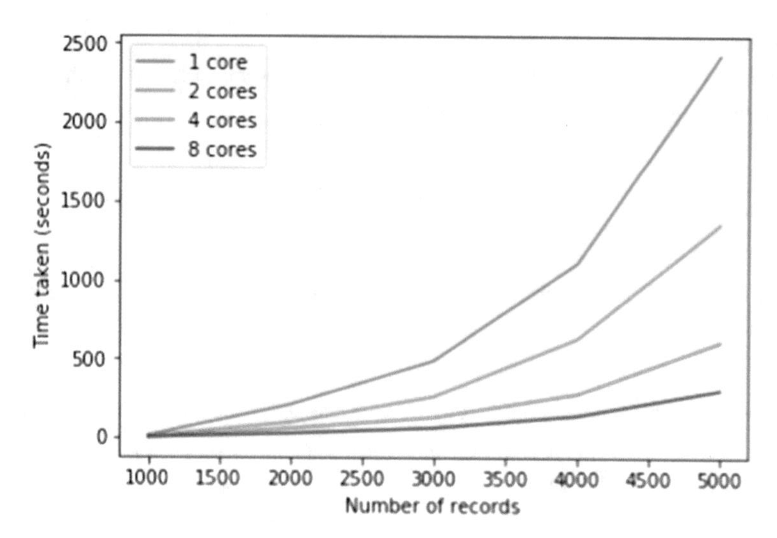

Fig. 1. Dependence of the execution time on the dimension with different number of cores.

8 Conclusions

In order to reduce the execution is proposed an application of the well-known regularization Tikhonov method with modern trends in the development of the computer system. To implement a parallel algorithm, the multi-core architecture of the processors and such property as multithreading are used. The optimization of the computational process in solving the Fredholm integral equation of the first kind, which is due to new requirements for the speed of computation, is realized using the technology of parallel programming of OpenMP. Several numerical experiments were conducted. The advantages of calculating the speed of calculations based on acceleration and efficiency indicators are analyzed, which can be significantly improved due to variations in the number of parental flows and processor cores of the computer.

References

1. Okhrimenko, M.G., Fartushny, I.D., Kulik, A.B.: Incorrectly set tasks and methods for solving them, vol. 225. Politekhnika Publishing House (2016). (In Ukrainian)
2. Miao, B., Zhou, F., Jiang, Ch., Chen, X., Yang, Sh.: A comparative study of regularization method in structure load identification. Shock Vib. 8 p. (2018). Article ID 9204865
3. Glushkov, V.M., Ivanov, V.V., Yanenko, V.M.: On the new class of dynamic models and its application in biology. Cybernetics **4**, 133–141 (1979). (In Russian)
4. Samara, A.A., Vabishcheva, P.N.: Numerical Methods for Solving Inverse Problems of Mathematical Physics, vol. 480. Publisher LKI, Moscow (2009). (In Russian)
5. Mochurad, L.I.: Method of reduction model for calculation of electrostatic fields of electronic optics systems. Science Journal Radioelektronika, Informatics, Management **1** (48), 29–40 (2019). (In Ukrainian)

6. Mochurad, L.I., Ostudin, B.A.: Flat variant of substantially spatial problem of electrostatics and some aspects of its solution, related to specifics of input information. J. Numer. Appl. Math. **2**(105), 98–110 (2011)
7. Chapko, R., Tomas Johansson, B., Savka, Yu.: On the use of an integral approach for the numerical solution of a Cauchy problem for Laplace equation in a doubly connected planar domain. Inverse Prob. Sci. Eng. **22**(1), 130–149 (2014)
8. Baranov, V.L., Zhukov, I.A., Zasyadko, A.A.: Using the method of the main criterion for solving the problem of recovery of signals. Visnyk NAU **1**, 9–13 (2003). (In Ukrainian)
9. Vasin, V.V., Serezhnikova, T.I.: Regular algorithm for approximation of nonsmooth solutions for Fredholm integral equations of the first kind. Comput. Technol. **15**(2), 15–23 (2010). (In Russian)
10. Sizikov, V.S.: Sustainable methods of processing measurement results. Study Guide, St. Petersburg: SpecLit, 240 (1999). (In Russian)
11. Goldak, A., Sivers, V.: Selection of the parameter of regularization on the spectrum of the derivative in Tikhonov's method. Meas. Equip. Metrol. **67**, 42–47 (2007). (In Ukrainian)
12. Balandin, M., Shurina, E.: The Methods for Solving High-dimensional SLAE. NSTU, Novosibirsk (2000). (In Russian)
13. Nemnyugin, S.A., Stesik, O.L.: Parallel Programming for Multiprocessor Computing Systems, vol. 400. BHV-Petersburg, Saint Petersburg (2002). (In Russian)
14. Voss, M.J. OpenMP share memory parallel programming, vol. 270, Toronto, Kanada (2003)
15. Antonov, A.: Parallel Programming Using the OpenMP Technology, tutorial. Moscow State University. Moscow (2009). (In Russian)

Method of Detection the Consistent Subgroups of Expert Assessments in a Group Based on Measures of Dissimilarity in Evidence Theory

Alyona Shved, Igor Kovalenko, and Yevhen Davydenko(✉)

Petro Mohyla Black Sea National University, Mykolaiv, Ukraine
{avshved, ihor.kovalenko, davydenko}@chmnu.edu.ua

Abstract. The procedure for dividing the initial set of group expert assessments into subgroups containing homogeneous, consistent estimates has been proposed in this paper. The proposed approach allows to distinguish groups of experts, with a "close" opinion, to analyze them in order to develop a final (group) assessment that takes into account the opinions (arguments) of each expert. For dividing the expert group into subgroups with consistent opinions, the mathematical apparatus of the Dempster-Shafer theory was used. In contrast to existing approaches, this theory allows to take into account specific forms of un-factors, such as combination of uncertainty and vagueness arising in the process of interaction between expert judgments (evidence). Metrics of evidence theory characterizing the degree of proximity of expert assessments are taken as a measure of consistency. The Jousselme's, Euclidean, Wang, Bhattacharyya's and Tessem's distances have been analyzed in this paper. Expert evidence is considered to belong to one group if the value of the specified metric (distance) for all evidence of this group does not exceed a specified threshold value. The farthest neighbor and average linkage methods have been used in paper to determine the order of grouping expert evidence in accordance with the degree of their similarity (dissimilarity). Numerical calculations of the proposed procedure of formation of consistent groups of expert evidences are provided. The results obtained allow to carry out a more profound analysis of the obtained expert information aimed at a synthesis an effective and substantiated group decisions.

Keywords: Evidence theory · Dissimilarity measures · Expert evidence · Uncertainty

1 Introduction

The analysis of group expert assessments is primarily aimed at determining consistency of the set of individual expert judgments which is the basis for forming a collective decision.

However, quite often, the expert group includes experts whose assessments in magnitude may differ from the estimates of the main group. The presence of such

N. Shakhovska and M. O. Medykovskyy (Eds.): CCSIT 2019, AISC 1080, pp. 36–53, 2020.
https://doi.org/10.1007/978-3-030-33695-0_4

assessments in the total set of group expert estimates violates its homogeneity (consistency).

[1] states that the results of the expert survey can lead to next situations. Situation 1 is characterized by homogeneity (consistency) of expert assessments. Situation 2 implies the existence of a certain number of subgroups in a group of experts, within which expert assessments can be considered as homogeneous. That indicates the experts' presence in the committee, who have different (but similar within the same subgroup) points of view on solving the problem under consideration. Situation 3 leads to the formation of a significant number of small subgroups of experts. That allows us to speak about the heterogeneity of expert assessments. As a result, the task of dividing the expert group into several subgroups of experts with similar (agreed) assessments arises for their subsequent analysis and the calculation of the aggregated expert judgment.

There are several important circumstances that we should take into account for solving this problem and choosing the appropriate methods for its solution: type of measurement scale of expert judgments (absolute, ordinal, nominal, interval, etc.); limited number of experts n ($n \leq 30$) in groups; various types of un-factors (incompleteness, uncertainty, ambiguity, roughness, vagueness, etc.), under the influence of which the expert evidence is formed and processed.

Analysis of the methods that could be applied to solve the problems of structuring (dividing) group expert assessments into subgroups that contain homogeneous, in a certain sense, expert assessments, showed that their effective implementation is not always possible. For example, for the analysis of expert assessments formed within the framework of numerical scales (absolute, ball), widespread use obtained: cluster analysis methods based on calculations of distance functions, for example, Euclidean distance, Manhattan distance, Chebyshev distance, etc.; clustering based on mathematical programming models (dynamic and integer); clustering based on estimation of probability density functions. For the analysis of expert assessments formed in scales of relations or order, clustering methods of non-numeric data can be used, for example, the Kemeny median [2–4]. Calculation the Kemeny distance is a very complex task that is characterized by a rather high computational complexity.

The justified choice and use of considered methods for solving the problem of splitting group expert assessments in order to search for homogeneous subgroups can be carried out with due consideration of various types of un-factors (such as incompleteness, uncertainty, vagueness, inconsistency, heterogeneity, etc.), that arise in the process of obtaining and processing expert information. It is also necessary to take into account the possible structure of expert judgments (agreed, compatible, arbitrary, etc.), and the possible ways of their interaction.

To solve this problem, effective results can be obtained using the evidence distance measures in Dempster-Shafer theory (DST, evidence theory) [5–7]. The mathematical apparatus of the evidence theory allows operating expert evidence correctly in the format of rankings (orderings) formed under uncertainty, inconsistency (conflicting, contradictory expert judgments) and taking into account various ways of their interaction (union, intersection, absorption) [8–11].

2 Measures of Dissimilarity in Evidence Theory

Let $\mathrm{A} = \{a_i | i = \overline{1, n}\}$ be a finite set of n exclusive and exhaustive elements [5–7]. The power set 2^{A} of A allows to construct a set of focal elements $\mathrm{B} = \{B_j | j = \overline{1, s}\}$, $s = 2^{|\mathrm{A}|}$, each of which represents a focal element, on the basis of which a degree of confidence is assigned that the best choice is in a selected subset.

Any subset $B_j \subseteq \mathrm{A}$ can be constructed on the basis of next rules:

$$\begin{array}{l} 1.\ B_j = \{\varnothing\}; \\ 2.\ B_j = \{a_i\}; \\ 3.\ B_j = \{a_i | i = \overline{1, p}\},\ p < n; \\ 4.\ B_j = \mathrm{A} = \{a_i | i = \overline{1, n}\}. \end{array} \tag{1}$$

Case 1 – none of the alternatives can be satisfied with expert choice; case 2 – the expert singled out one alternative ($a_i \in \mathrm{A}$); case 3 – expert selected p alternatives; case 4 – the expert finds it difficult to choose any of the proposed alternatives, i.e. all alternatives are equivalent.

One of the important functions in the evidence theory is the basic probability assignment (*bpa*), $\forall\ B \subseteq \mathrm{A}$, m: $2^{\mathrm{A}} \rightarrow [0, 1]$:

$$0 \leq m(B_j) \leq 1,\ \forall (B_j \in 2^{\mathrm{A}}),\ m(\varnothing) = 0,\ \sum_{B_j \in 2^{\mathrm{A}}} m(B_j) = 1. \tag{2}$$

Over the last years, the measures of dissimilarity between two groups of evidence B_1 and B_2 has been regularly studied in the literature. This measure of conflict and distances evaluate the dissimilarity between two *bpa's* and can characterize the relationship among their focal elements and conflict between several mass functions $m(\cdot)$.

Let us consider some of them [12–19]:

1. Jousselme's distance [12–14]:

$$d_j(\mathrm{m}_1, \mathrm{m}_2) = \sqrt{\frac{1}{2}(\mathrm{m}_1 - \mathrm{m}_2)^{\mathrm{T}} \underline{\underline{D}}(\mathrm{m}_1 - \mathrm{m}_2)}, \tag{3}$$

where m_i is a 2^{A}-dimensional column vector, with elements being $m_j(B)$, $B \subseteq \mathrm{A}$; $(\mathrm{m}_i)^{\mathrm{T}}$ is the transpose of vector m_i; $(m_1 - m_2)$ is the subtraction of two vectors; $\underline{\underline{D}}$ is a $2^{\mathrm{A}} \times 2^{\mathrm{A}}$ matrix with

$$D(B_i, B_j) = \begin{cases} 1, & \text{if } B_i = B_j; \\ S(B_i, B_j), & \forall B_i, B_j \in \mathrm{A}. \end{cases} \tag{4}$$

The Jousselme's measure uses the Jaccard's coefficient:

$$S(B_i, B_j) = |B_i \cap B_j| / |B_i \cup B_j| \tag{5}$$

where $|\cdot|$ is the cardinality of the corresponding subsets.

[15] reviewed different similarity functions between focal elements instead of Jaccard's coefficient, Table 1.

Table 1. Alternative similarity functions.

Name	Dice	Sokal & Sneath	Kulczynski	Ochiai	Fixsen & Mahler
Similarity function	$\frac{2\lvert A\cap B\rvert}{\lvert A\rvert+\lvert B\rvert}$	$\frac{\lvert A\cap B\rvert}{2\lvert A\cup B\rvert-\lvert A\cap B\rvert}$	$\frac{\lvert A\cap B\rvert}{2\lvert A\rvert}+\frac{\lvert A\cap B\rvert}{2\lvert B\rvert}$	$\frac{\lvert A\cap B\rvert}{\sqrt{\lvert A\rvert\lvert B\rvert}}$	$\frac{\lvert A\cap B\rvert}{\lvert A\rvert\lvert B\rvert}$

2. Euclidean distance [14, 16]:

$$d_E(\mathrm{m}_1, \mathrm{m}_2) = \sqrt{\sum_{B\subseteq\Omega} [m_1(B) - m_2(B)]^2}. \tag{6}$$

3. Wang dissimilarity measure [14, 17]:

$$d_W(\mathrm{m}_1, \mathrm{m}_2) = \sum_{B\subseteq\Omega} \frac{|m_1(B) - m_2(B)|}{2}. \tag{7}$$

4. Bhattacharyya's distance [14, 18]:

$$d_B(\mathrm{m}_1, \mathrm{m}_2) = \sqrt{1 - \sum_{B\subseteq\Omega} \sqrt{m_1(B) \cdot m_2(B)}}. \tag{8}$$

5. Tessem's distance [14, 19]:

$$d_T(\mathrm{m}_1, \mathrm{m}_2) = \max_{a\in \mathrm{A}} |BetP_1(a) - BetP_2(a)|, \tag{9}$$

where $BetP_k(a)$ is a pignistic transformation:

$$BetP_k(a) = \sum_{\substack{B\subseteq \mathrm{A}, \\ a\in B}} \frac{1}{|B|} \frac{m(B)}{1 - m(\varnothing)}, \quad m(\varnothing) \neq 1. \tag{10}$$

Measures (3)–(9) can be used to assign the conflict measure between 2 experts in a group of experts $\mathrm{E} = \{E_j | j = \overline{1,t}\}$ [20]:

$$Conf(1,\ 2) = d(\mathrm{m}_1, \mathrm{m}_2). \tag{11}$$

To determine the degree of conflict between expert E_j and the other group of experts $\mathrm{E} = \{E_j | j = \overline{1,t}\}$ the $Conf(E_j, \mathrm{E})$ measure can be used [20]:

$$Conf(i, \mathrm{E}) = \frac{1}{t-1} \sum_{j=1, i \neq j}^{t} Conf(i,j), \tag{12}$$

where $Conf(i, j) = d(\mathrm{m}_i, \mathrm{m}_j)$; $d(\mathrm{m}_i, \mathrm{m}_j)$ is one of the listed measures (3)–(9).

Let us consider an original set of alternatives $\mathrm{A} = \{a_i | i = \overline{1,n}\}$ and a group (set) of experts $\mathrm{E} = \{E_j | j = \overline{1,t}\}$. Then a system of subsets $\mathrm{B} = \{B_j | j = \overline{1,t}\}$ will be formed, where B_j is a 2^{A}-dimensional vector reflecting the preferences (choice) of the expert E_j, each element of which is based on the rules (1). For each subset B_j, $j = \overline{1,t}$ a vector $\mathrm{m}_j = \{m_i | i = \overline{1,\ s}\}$ will be constructed, $s = 2^{|\mathrm{A}|}$, with elements which satisfy the necessary conditions (2).

It is necessary to distinguish a number of expert subgroups that have a similar judgments from the initial set of expert assessments $E \Rightarrow \{G_1\}, \{G_2\}, \ldots, \{G_j\}, \ldots, \{G_{t-1}\} (G_p \subseteq E, \{G_p\} = \{E_1, \ldots, E_r\}, t \geq r \geq 1)$, or highlight those E_j, whose assessments stand out sharply in relation to the estimates of the other experts, i.e. $E_j, \subseteq G_k$ provided that $|G_k| = 1$ (if there are such). We will assume that the opinions of experts who fall into one group $E_j \subseteq G_p$, $t \geq j \geq 2$ are recognized as agreed (consistent); opinions of the experts $E_j \subseteq G_k$, $|G_k| = 1$ are recognized as atypical, i.e. significantly different (conflicting) from the opinions of the rest of the group of experts. If there is a situation, when t subgroups has been formed ($\forall G_j$: $|G_j| = 1$), then conducting further analysis does not make sense.

To solve this problem, we will use the approach consisting of the following. The division of the expert group into subgroups, within which opinions can be considered agreed (consistent), takes place in two stages.

First the degree of similarity of expert evidence among themselves is assessed, since expert evidence cannot be expressed by a numerical indicator and it is possible to establish that the source objects (experts) belong to any groups (classes) only on the basis of distances between them. The paper proposes to evaluate the degree of similarity (dissimilarity) of expert evidence among themselves, based on the evidence theory metrics. The results are saved as a pairwise distance matrix as follows:

$$\begin{pmatrix} - & d(\mathrm{m}_1, \mathrm{m}_2) & \ldots & d(\mathrm{m}_1, \mathrm{m}_t) \\ d(\mathrm{m}_2, \mathrm{m}_1) & - & \ldots & d(\mathrm{m}_2, \mathrm{m}_t) \\ \ldots\ldots\ldots & \ldots\ldots\ldots & - & \ldots\ldots\ldots \\ d(\mathrm{m}_t, \mathrm{m}_1) & d(\mathrm{m}_t, \mathrm{m}_2) & \ldots & - \end{pmatrix} \tag{13}$$

where $d(\mathrm{m}_i, \mathrm{m}_j) = d(\mathrm{m}_j, \mathrm{m}_i)$, $\forall\ i,j = \overline{1,t}$, $i \neq j$; t is the number of compared objects; $d(\mathrm{m}_i, \mathrm{m}_j)$ are values of one of the measures (3)–(9).

The choice of a metric is one of the main factors influencing the results of splitting the initial body of expert evidence and the formation of expert subgroups with fairly close judgments. As a rule, the choice of a metric is sufficiently subjective and is determined by an expert (analyst). The Jousselme's, Euclidean, Wang, Bhattacharyya's and Tessem's measures has been analyzed in this paper.

In the second stage, expert subgroups are formed $G_p \subseteq \text{E}$, $p = \overline{1, t-1}$, where $[\cdot]$ denotes a whole part. To determine the order of their grouping, the farthest neighbor and average linkage methods were used in the paper.

The resulting expert groups were formed in such a way that $\forall\ E_j \in G_p$, $j = \overline{1, r}$, $t \geq r \geq 1$ should be under the next condition $\forall (i,j) = \overline{1, r}$, $i \neq j$, $l_{p-1} < d(\text{m}_i, \text{m}_j) \leq l_p$. Here l_{p-1}, l_p are some predetermined values that are responsible for belonging of expert E_j to the group G_p.

3 Example

As an example, that shows how proposed approach can be used in practical applications, let us consider the problem of selecting the optimal welding process technology in shipbuilding.

One of the most important stages of welding process technology in shipbuilding is the determination and justification of the optimal (most acceptable for this shipyard) version of gas cutting and welding. This is due, first of all, to the fact that the proportion of these works in the total labor input of building ships is quite large (reaches 35–37%), and is a main influencing factor for price and duration of construction.

Making decisions on choosing an effective technological process from the available multivariate set, for example, cutting and welding of ship hull structures, is connected with the need to take into account various factors that influence decision making. As a result of this, such a task can be solved within the framework of the organizational subsystem, in the form of a permanent expert advisory group under the general director or his deputies. The structure of such group includes representatives of headquarters divisions (the chief designer department, the chief technologist department, material and technical supply department, labor and costs department, etc.), as well as representatives of line divisions (production departments, sections, etc.).

If expert group have a high level of agreement, the aggregation of judgments is carried out for the entire group of experts. In the case when the degree of consistency is deemed unacceptable, then it is advisable to divide the initial group of expert assessments into subgroups with close to each other expert assessments. Aggregation of expert judgments in this case is carried out separately for each of the highly agreed subgroups of experts. In this case, the result of the examination will be not one, but several judgments corresponding to the selected subgroups.

3.1 Example 1: Equivalent Experts' Evidence

Let $\text{A} = \{a_i | i = \overline{1, n}\}$, $n = 4$ be a finite non-empty set of original data (alternatives) that represents welding technology options. A group of experts consisting of 10 experts, $\text{E} = \{E_j | j = \overline{1, t}\}$, was asked to evaluate 4 welding technologies listed in Table 2 according to a number of criteria (technological cost of welding, the complexity of welding, etc.).

Then for each expert there will be formed a set of focal elements $X_j = \{B_i | i = \overline{1, s}\}$, $s = 2^{|A|}$. Let us consider a situation where all experts have formed equivalent evidence: $B_1 = \{a_1\}$, $B_2 = \{a_2\}$, $B_3 = \{a_3\}$, $B_4 = \{a_4\}$. The obtained results of the expert survey are given in Table 3.

Table 2. Types of the technology for welding the mounting butt joint of the side skin or set.

Process variant	Process operation list
a_1	1. Mechanized welding with Sv-08G2S in an environment of CO_2 (MAG). 2. Gas gouging of a seam root. 3. Cutting after gouging (MAG)
a_2	1. Mechanized welding with Sv-08G2S in Ar + CO_2 mixture (MIG). Composition of gas mixtures can include 80% Ar and 20% CO_2. 2. Gas gouging of a seam root. 3. Cutting after gouging (MIG)
a_3	1. Mechanized CO_2 welding using flux-cored wire MEGAFIL–713R (MAG). 2. Gas gouging of a seam root. 3. Cutting after gouging (MAG)
a_4	1. Mechanized welding using flux-cored wire MEGAFIL–713R in gas mixture (MIG). 2. Gas gouging of a seam root. 3. Cutting after gouging (MIG)

Let us form a set of metrics $Q = \{q_l | l = \overline{1, k}\}$, $k = 5$ on the basis of which we investigate the degree of difference in the expert's judgments, and the sensitivity of the indicator of the degree of similarity (dissimilarity) between the m_1 and m_2 to the selected metric. Calculate the values of metrics (3)–(9), that characterize the measure of the differences between the selected groups of expert evidence.

Table 3. Basic probability assignments of focal elements $m_j(a_i)$.

Expert E_j	$m_j(a_1)$	$m_j(a_2)$	$m_j(a_3)$	$m_j(a_4)$
E_1	0.1	0.5	0.3	0.1
E_2	0.2	0.3	0.4	0.1
E_3	0.3	0.2	0.2	0.3
E_4	0.5	0.1	0.1	0.3
E_5	0.1	0.1	0.6	0.2
E_6	0.1	0.3	0.2	0.4
E_7	0.3	0.1	0.4	0.2
E_8	0.3	0.3	0.3	0.1
E_9	0.1	0.2	0.1	0.6
E_{10}	0.1	0.6	0.2	0.1

Let us compute the *Conf*(E_j, E), with the Eq. (12). This measure must quantify how much expert E_j is in conflict with the rest of the set E\E_j (Table 4).

Table 4. Values of measure *Conf*(*j*, *t*).

Conf(E_j, E)	Jousselme distance	Euclidean distance	Wang distance	Bhattacharyya distance	Tessem distance
Conf(E_1, E)	0.290	0.411	0.344	0.282	0.311
Conf(E_2, E)	0.245	0.347	0.300	0.244	**0.256**
Conf(E_3, E)	**0.240**	**0.339**	**0.289**	**0.238**	0.267
Conf(E_4, E)	0.350	0.496	**0.433**	0.348	0.356
Conf(E_5, E)	0.349	0.493	0.378	0.325	0.378
Conf(E_6, E)	0.267	0.377	0.311	0.265	0.289
Conf(E_7, E)	0.266	0.377	0.322	0.270	0.289
Conf(E_8, E)	0.241	0.341	**0.289**	0.246	**0.256**
Conf(E_9, E)	**0.375**	**0.530**	**0.433**	**0.350**	**0.422**
Conf(E_{10}, E)	0.343	0.485	0.367	0.315	0.378

Results obtained (Table 4) by the different approaches (distances) allow us to conclude that, regardless of chosen metrics, the expert E_9 evidence is the most conflicting with respect to the evidence of the rest of the expert group (the level of conflict between expert E_9 and the rest of the expert group ranges from 0.35 to 0.53 depending on the selected metric, and its value is significant). The Wang metric also highlights an expert E_4 with the same conflict value as expert E_9. The evidence of expert E_3 were found to be the least conflicting with respect to the rest of the expert group, and the level of conflict varies in the range of 0.238 to 0.339 (except the Tessem metric). The Wang metric identifies experts E_3 and E_8 with the lowest level of conflict; Tessem metric – experts E_2 and E_8.

The resulting expert subgroups were formed under the condition 0 < d(m$_i$, m$_j$) 0.3, which implies a low level of dissimilarity between the groups of evidences m$_i$ and m$_j$, and allows us to conclude that there is a consistency inside the selected groups of evidence.

According to the results of the splitting of the original expert group $\mathrm{E} = \{E_j | j = \overline{1,t}\}$, the following expert subgroups were identified:

3.1.1 The Farthest Neighbor Method

Jousselme's Distance. $G_1 = \{E_2, E_5, E_7, E_8\}$ – when the group are formed, the maximum degree of dissimilarity is achieved by merging subgroups $\{E_2, E_7, E_8\}$ and $\{E_5\}$, and reaches d_J (m$_i$, m$_j$) = 0.3; the minimum degree of dissimilarity is achieved by merging subgroups $\{E_2\}$ and $\{E_8\}$ with d_J (m$_i$, m$_j$) = 0.1.

$G_2 = \{E_1, E_{10}\}$ – the degree of dissimilarity reaches d_J (m$_i$, m$_j$) = 0.1.

$G_3 = \{E_3, E_4\}$ and $G_4 = \{E_6, E_9\}$ – the degree of dissimilarity reaches d_J (m$_i$, m$_j$) = 0.173.

Euclidean Distance. $G_1 = \{E_2, E_7, E_8\}$ – when the group is formed, the maximum degree of dissimilarity is achieved by merging subgroups $\{E_2, E_8\}$ and $\{E_7\}$, and reaches d_E (m$_i$, m$_j$) = 0.245.

$G_2 = \{E_1, E_{10}\}$ – the degree of dissimilarity reaches d_E (m$_i$, m$_j$) = 0.141.

$G_3 = \{E_3, E_4\}$ and $G_4 = \{E_6, E_9\}$ – the degree of dissimilarity reaches d_E (m$_i$, m$_j$) = 0.245.

$G_5 = \{E_5\}$ – the closest subgroup is $\{E_2, E_7, E_8\}$, d_E (m$_{G1}$, m$_{G5}$) = 0.424.

Wang Distance. $G_1 = \{E_2, E_7, E_8\}$ – when the group is formed, the maximum degree of dissimilarity is achieved by merging subgroups $\{E_7\}$ and $\{E_2, E_8\}$, and reaches d_W (m$_i$, m$_j$) = 0.2.

$G_2 = \{E_1, E_{10}\}$ – the degree of dissimilarity reaches d_W (m$_i$, m$_j$) = 0.1.

$G_3 = \{E_3, E_4\}$ and $G_4 = \{E_6, E_9\}$ – the degree of dissimilarity reaches d_W (m$_i$, m$_j$) = 0.2; the union of these subgroups occurs when d_W (m$_i$, m$_j$) = 0.4.

$G_5 = \{E_5\}$ – the closest subgroup is $\{E_1, E_{10}\}$, d_W (m$_{G2}$, m$_{G5}$) = 0.4.

Bhattacharyya's Distance. $G_1 = \{E_1, E_2, E_8, E_{10}\}$ – when the group is formed, the maximum degree of dissimilarity is achieved by merging subgroups $\{E_1, E_{10}\}$ and $\{E_2, E_8\}$, and reaches d_B (m$_i$, m$_j$) = 0.24; the minimum degree of dissimilarity is achieved by merging subgroups $\{E_2\}$ and $\{E_8\}$ with d_B (m$_i$, m$_j$) = 0.086.

$G_2 = \{E_5, E_7\}$ – the degree of dissimilarity reaches d_B (m$_i$, m$_j$) = 0.192.

$G_3 = \{E_3, E_4\}$ – the degree of dissimilarity reaches d_B (m$_i$, m$_j$) = 0.173.

$G_4 = \{E_6, E_9\}$ – the degree of dissimilarity reaches d_B (m$_i$, m$_j$) = 0.154.

Tessem's Distance. $G_1 = \{E_2, E_3, E_4, E_7, E_8\}$ – when the group is formed, the maximum degree of dissimilarity is achieved by merging subgroups $\{E_2, E_3, E_7, E_8\}$ and $\{E_4\}$, and reaches d_T (m$_i$, m$_j$) = 0.3; the minimum degree of dissimilarity is achieved by merging subgroups $\{E_2\}$ and $\{E_8\}$ with d_T (m$_i$, m$_j$) = 0.1.

$G_2 = \{E_1, E_{10}\}$ – the degree of dissimilarity reaches d_T (m$_i$, m$_j$) = 0.1.

$G_3 = \{E_6, E_9\}$ – the degree of dissimilarity reaches d_T (m$_i$, m$_j$) = 0.2.

$G_4 = \{E_5\}$ – the degree of dissimilarity between $\{E_5\}$ and $\{E_1, E_2, E_3, E_4, E_6, E_7, E_8\}$ is d_T (m$_i$, m$_j$) = 0.5.

$G_4 = \{E_5\}$ – the degree of dissimilarity between $\{E_5\}$ and $\{E_1, E_2, E_3, E_4, E_6, E_7, E_8\}$ is d_T (m$_i$, m$_j$) = 0.5.

From the obtained results, it can be observed that regardless of which metric is used, a subgroup $\{E_6, E_9\}$ is allocated with a degree of dissimilarity varying from 0.154 to 0.245 (depending on the metric used). The results obtained for the Euclidean and Wang metrics coincide. The only difference is in the corresponding d(m$_i$, m$_j$) values of the metrics considered.

The Jousselme, Euclidean, Wang and Bhattacharyya metrics distinguish the subgroup $\{E_3, E_4\}$ with the degree of dissimilarity ranging from 0.173 to 0.245 (depending on the metric used). The Tessem metric assigns this subgroup to the subgroup $\{E_2, E_3, E_4, E_7, E_8\}$: $\{E_4\}$ merges with the subgroup $\{E_2, E_3, E_7, E_8\}$ with d_T (m$_i$, m$_j$) = 0.3; $\{E_3\}$ is combined with the subgroup $\{E_2, E_7, E_8\}$ with d_T (m$_i$, m$_j$) = 0.2.

The Tessem, Jousselme, Euclidean and Wang metrics distinguish the subgroup $\{E_1, E_{10}\}$ with the degree of dissimilarity ranging from 0.1 to 0.141 (depending on the metric used). The Bhattacharyya metric assigns the subgroup $\{E_1, E_{10}\}$ to the subgroup $\{E_1, E_2, E_8, E_{10}\}$: $\{E_1, E_{10}\}$ and $\{E_2, E_8\}$ could be combined with d_B (m_i, m_j) = 0.24.

The Tessem, Euclidean and Wang metrics distinguish the subgroup $\{E_5\}$ with a significant value of the degree of dissimilarity ranging from 0.4 to 0.5 (depending on the metric used).

3.1.2 The Average Linkage Method

Jousselme's Distance. $G_1 = \{E_2, E_5, E_7, E_8\}$ – when the group is formed, the maximum degree of dissimilarity is achieved by merging subgroups $\{E_2, E_7, E_8\}$ and $\{E_5\}$, and reaches d_J (m_i, m_j) = 0.241; the minimum degree of dissimilarity is achieved by merging subgroups $\{E_2\}$ and $\{E_8\}$ with d_J (m_i, m_j) = 0.1.

$G_2 = \{E_1, E_{10}\}$ – the degree of dissimilarity reaches d_J (m_i, m_j) = 0.1.

$G_3 = \{E_3, E_4, E_6, E_9\}$ – the degree of dissimilarity between $\{E_3, E_4\}$ and $\{E_6, E_9\}$ reaches d_J (m_i, m_j) = 0.283.

Euclidean Distance. $G_1 = \{E_2, E_7, E_8\}$ – when the group is formed, the maximum degree of dissimilarity is achieved by merging subgroups $\{E_2, E_8\}$ and $\{E_7\}$, and reaches d_E (m_i, m_j) = 0.245.

$G_2 = \{E_1, E_{10}\}$ – the degree of dissimilarity reaches d_E (m_i, m_j) = 0.141.

$G_3 = \{E_3, E_4\}$ and $G_4 = \{E_6, E_9\}$ – the degree of dissimilarity reaches d_E (m_i, m_j) = 0.245; the union of subgroups $\{E_3, E_4\}$ and $\{E_6, E_9\}$ occurs when d_E (m_i, m_j) = 0.436.

$G_5 = \{E_5\}$ – the closest subgroup is $\{E_2, E_7, E_8\}$, d_E (m_{G1}, m_{G5}) = 0.341.

Wang Distance. $G_1 = \{E_2, E_5, E_7, E_8\}$ – when the group is formed, the maximum degree of dissimilarity is achieved by merging subgroups $\{E_5\}$ and $\{E_2, E_7, E_8\}$, and reaches d_W (m_i, m_j) = 0.3; the minimum degree of dissimilarity is achieved by merging subgroups $\{E_2\}$ and $\{E_8\}$ with d_W (m_i, m_j) = 0.1.

$G_2 = \{E_1, E_{10}\}$ – the degree of dissimilarity reaches d_W (m_i, m_j) = 0.1.

$G_3 = \{E_3, E_4\}$ and $G_4 = \{E_6, E_9\}$ – the degree of dissimilarity reaches d_W (m_i, m_j) = 0.2; the union of these subgroups occurs when d_W (m_i, m_j) = 0.32.

Bhattacharyya's Distance. $G_1 = \{E_1, E_2, E_8, E_{10}\}$ – when the group is formed, the maximum degree of dissimilarity is achieved by merging subgroups $\{E_1, E_{10}\}$ and $\{E_2, E_8\}$, and reaches d_B (m_i, m_j) = 0.24; the minimum degree of dissimilarity is achieved by merging subgroups $\{E_2\}$ and $\{E_8\}$ with d_B (m_i, m_j) = 0.086.

$G_2 = \{E_5, E_7\}$ – the degree of dissimilarity reaches d_B (m_i, m_j) = 0.192.

$G_3 = \{E_3, E_4\}$ – the degree of dissimilarity reaches d_B (m_i, m_j) = 0.173.

$G_4 = \{E_6, E_9\}$ – the degree of dissimilarity reaches d_B (m_i, m_j) = 0.154.

Tessem's Distance. $G_1 = \{E_2, E_3, E_4, E_7, E_8\}$ – when the group is formed, the maximum degree of dissimilarity is achieved by merging subgroups $\{E_2, E_3, E_7, E_8\}$ and $\{E_4\}$, and reaches d_T (m_i, m_j) = 0.3; the minimum degree of dissimilarity is achieved by merging subgroups $\{E_2\}$ and $\{E_8\}$ with d_T (m_i, m_j) = 0.1.

$G_2 = \{E_1, E_{10}\}$ – the degree of dissimilarity reaches d_T (m_i, m_j) = 0.1.

$G_3 = \{E_6, E_9\}$ – the degree of dissimilarity reaches d_T (m$_i$, m$_j$) = 0.2.

$G_4 = \{E_5\}$ – the degree of dissimilarity between $\{E_5\}$ and $\{E_1, E_2, E_3, E_4, E_6, E_7, E_8\}$ is d_T (m$_i$, m$_j$) = 0.5.

From the obtained results, it can be observed that regardless of which metric is used, a subgroup $\{E_6, E_9\}$ is allocated within the same range of variation of the degree of dissimilarity as in the farthest neighbor method.

The results obtained for Euclidean, Tessem and Bhattacharyya metrics coincide with the results obtained using the farthest neighbor method. The only difference is in the corresponding d(m$_i$, m$_j$) values of the metrics considered.

The association of subgroups $\{E_3, E_4\}$ and $\{E_6, E_9\}$ was obtained based on the average linkage method by using the Jousselme metric with d_J (m$_i$, m$_j$) = 0.283.

The association of subgroups $\{E_2, E_7, E_8\}$ and $\{E_5\}$ was obtained based on the average linkage method by using the Wang metric with d_W(m$_i$, m$_j$) = 0.3.

After analyzing the results obtained on the basis of the farthest neighbor and the average linkage methods, we can draw the following conclusions. By using the Jousselme metric, the merging of the subgroups $\{E_2, E_7, E_8\}$ and $\{E_5\}$ occurs when d_J (m$_i$, m$_j$) = 0.3; while $Conf(E_5, \{E_2, E_7, E_8\}) = 0.2$. By using the Euclidean metric, the merging of the subgroups $\{E_2, E_7, E_8\}$ and $\{E_5\}$ could be occurred with d_E (m$_{G1}$, m$_{G5}$) = 0.424 based on the farthest neighbor method (d_E (m$_{G1}$, m$_{G5}$) = 0.341 based on the average linkage method), which exceeds the established acceptable boundary level differences on 0.124 (0.041); while $Conf(E_5,\{E_2, E_7, E_8\}) = 0.341$. Such value of the degree of conflict (dissimilarity) between the expert E_5 and the rest of expert group $\{E_2, E_7, E_8\}$ is insignificant.

By using the Wang metric, the merging of the subgroups $\{E_2, E_7, E_8\}$ and $\{E_5\}$ based on the farthest neighbor method could be occurred with d_W (m$_{G1}$, m$_{G5}$) = 0.5, wherein the subgroup $\{E_1, E_2, E_5, E_7, E_8, E_{10}\}$ will be formed (the merging of the subgroups $\{E_2, E_7, E_8\}$ and $\{E_5\}$ based on the average linkage method happens at $d_W = 0.3$), which exceeds the established acceptable boundary difference level by 0.124 (if the average linkage method will be applied, such a union is valid for established conflict level. With $Conf(E_5,\{E_2, E_7, E_8\})$ value equal to 0.341. Such value of the degree of conflict between the expert E_5 and the rest of expert group $\{E_2, E_7, E_8\}$ is insignificant. The results obtained based on Tessem and Bhattacharyya metrics slightly differ from the results obtained based on Jousselme, Euclidean, Wang metrics.

However, the results obtained by using Tessem metric are somewhat similar to the results obtained by Euclidean metric: the subgroups $\{E_1, E_{10}\}$, $\{E_6, E_9\}$ and $\{E_5\}$ are formed.

By Tessem metric, the subgroups $\{E_2, E_7, E_8\}$ and $\{E_3, E_4\}$ are merged; such a union is possible based on the Jousselme metric when merging the whole expert group with $d_J = 0.34$ by the average linkage method and $d_J = 0.48$ by the farthest neighbor method; based on the Euclidean metric when merging the whole expert group with $d_E = 0.48$ by the average linkage method and $d_E = 0.68$ by the farthest neighbor method; based on the Wang metric when merging the whole expert group with $d_W = 0.41$ by the average linkage method and $d_W = 0.6$ by the farthest neighbor method; based on the Bhattacharyya metric when merging the whole expert group with $d_B = 0.33$ by the average linkage method and $d_B = 0.46$ by the farthest neighbor method.

In this example, the boundary value, which is responsible for the optimal level of conflict within the subgroup of experts, was assumed to be 0.3.

Thus, in accordance with the established boundary value of $d(m_i, m_j)$ the expert committee was divided into next subgroups: $E \Rightarrow \{G_1, G_2, G_3, G_4, G_5\}$, containing similar judgments in a certain sense, where $G_1 = \{E_2, E_7, E_8\}$; $G_2 = \{E_1, E_{10}\}$; $G_3 = \{E_3, E_4\}$; $G_4 = \{E_6, E_9\}$, $G_5 = \{E_5\}$.

Such a situation arises, for example, due to the presence among experts of representatives of several scientific schools. In this case, additional procedures are sometimes proposed to bring together the opinions of different subgroups. Or, under the assumption that judgments are stable and final, the aggregation of judgments is performed separately for each of the obtained subgroups of experts.

3.2 Example 2: Disjoint Experts' Evidence

Using the input data of example 1, let us consider the situation with disjoint expert evidence. Let $A = \{a_i | i = \overline{1,n}\}$, $n = 4$ be the frame of discernment (with Shafer's model) to be analyzed by a set of experts $E = \{E_j | j = \overline{1,t}\}$, $t = 10$.

Table 5 gives the basic probability assignments (*bpa*) of generated focal elements, defined by Eq. (2).

Calculate the values of metrics (3)–(9), that characterize the measure of the differences between the selected groups of expert evidence.

Let us compute the $Conf(E_j, E)$, with the Eq. (12). This measure must quantify how much expert E_j is in conflict with the rest of the set $E \backslash E_j$ (Table 6).

Table 5. Basic probability assignments of focal elements *mj(ai)*

Expert E_j	$mj(a_1)$	$mj(a_2)$	$mj(a_3)$	$mj(a_4)$
E_1	–	0.5	0.3	0.2
E_2	0.2	0.3	0.4	0.1
E_3	–	0.6	–	0.4
E_4	0.5	0.1	0.1	0.3
E_5	–	0.3	0.4	0.3
E_6	0.1	0.3	0.2	0.4
E_7	0.3	–	0.5	0.2
E_8	0.4	0.3	0.3	–
E_9	–	0.4	–	0.6
E_{10}	–	0.6	0.2	0.2

Results obtained (Table 6) by the different approaches (distances) allows us to conclude that, regardless of chosen metrics, the expert E_6 evidences are the most closest with the respect to the evidence of the rest of the expert group (the level of conflict between expert E_6 and the rest of the expert group ranges from 0.274 to 0.388 depending on the selected metric, and its value is not significant).

Table 6. Values of measure *Conf*(*j*, *t*).

$Conf(E_j, \mathrm{E})$	Jousselme distance	Euclidean distance	Wang distance	Bhattacharyya distance	Tessem distance
$Conf(E_1, \mathrm{E})$	0.290	0.410	0.344	0.394	0.311
$Conf(E_2, \mathrm{E})$	0.298	0.421	0.378	0.408	0.300
$Conf(E_3, \mathrm{E})$	0.372	0.525	0.456	0.517	0.367
$Conf(E_4, \mathrm{E})$	**0.414**	**0.586**	**0.511**	0.508	**0.433**
$Conf(E_5, \mathrm{E})$	0.294	0.416	0.356	0.407	0.356
$Conf(E_6, \mathrm{E})$	**0.274**	**0.388**	**0.333**	**0.365**	**0.289**
$Conf(E_7, \mathrm{E})$	0.410	0.580	0.522	**0.592**	0.422
$Conf(E_8, \mathrm{E})$	0.365	0.516	0.456	0.560	0.378
$Conf(E_9, \mathrm{E})$	0.397	0.561	0.489	0.528	0.411
$Conf(E_{10}, \mathrm{E})$	0.321	0.454	0.400	0.403	0.378

The expert E_4 evidence is recognized as the most conflicting with the respect to the evidences of the rest of the expert group (except for the Bhattacharyya metric, which identifies expert E_7); the level of conflict varies in the range from 0.414 to 0.586.

The resulting expert subgroups were formed under the condition $0 < d(\mathrm{m}_i, \mathrm{m}_j)$ 0.3, which implies a low level of dissimilarity between the groups of evidences m_i and m_j, and allows us to conclude that there is a consistency inside the selected groups of evidence.

According to the results of the splitting of the original expert group $\mathrm{E} = \{E_j | j = \overline{1, t}\}$, the following expert subgroups were identified:

3.2.1 The Farthest Neighbor Method

Jousselme's Distance. $G_1 = \{E_1, E_5, E_6, E_{10}\}$ – when the group is formed, the maximum degree of dissimilarity is achieved by merging subgroups $\{E_1, E_{10}\}$ and $\{E_5, E_6\}$, and reaches d_J (m_i, m_j) = 0.265; the minimum degree of dissimilarity is achieved by merging subgroups $\{E_1\}$ and $\{E_{10}\}$ with d_J (m_i, m_j) = 0.1.

$G_2 = \{E_3, E_9\}$ – the degree of dissimilarity reaches d_J (m_i, m_j) = 0.2.

$G_3 = \{E_2, E_7, E_8\}$ – when the group is formed, the maximum degree of dissimilarity is achieved by merging subgroups $\{E_2, E_8\}$ and $\{E_7\}$, and reaches d_J (m_i, m_j) = 0.3; the minimum degree of dissimilarity is achieved by merging subgroups $\{E_2\}$ and $\{E_8\}$ with d_J (m_i, m_j) = 0.173.

$G_4 = \{E_4\}$ – the closest subgroup is $\{E_2, E_7, E_8\}$, d_J (m_{G3}, m_{G5}) = 0.361.

Euclidean Distance. $G_1 = \{E_1, E_{10}\}$ – the degree of dissimilarity reaches d_E (m_i, m_j) = 0.141.

$G_2 = \{E_5, E_6\}$ – the degree of dissimilarity reaches d_E (m_i, m_j) = 0.245.

$G_3 = \{E_3, E_9\}$ – the degree of dissimilarity reaches d_E (m_i, m_j) = 0.283.

$G_4 = \{E_2, E_8\}$ – the degree of dissimilarity reaches d_E (m_i, m_j) = 0.245.

$G_5 = \{E_7\}$ – the closest subgroup is $\{E_2, E_8\}$, d_E (m_{G4}, m_{G5}) = 0.424.

$G_6 = \{E_4\}$ – the closest subgroup is $\{E_2, E_7, E_8\}$, d_E = 0.51.

Wang Distance. $G_1 = \{E_1, E_{10}\}$ – the degree of dissimilarity reaches d_W (m_i, m_j) = 0.1.

$G_2 = \{E_5, E_6\}$ – the degree of dissimilarity reaches d_W (m_i, m_j) = 0.2.

$G_3 = \{E_3, E_9\}$ – the degree of dissimilarity reaches d_W (m_i, m_j) = 0.2.

$G_4 = \{E_2, E_8\}$ – the degree of dissimilarity reaches d_W (m_i, m_j) = 0.2.

$G_5 = \{E_7\}$ – the closest subgroup is $\{E_2, E_8\}$, d_W (m_{G4}, m_{G5}) = 0.4.

$G_6 = \{E_4\}$ – the closest subgroup is $\{E_2, E_7, E_8\}$, $d_W = 0.5$.

Bhattacharyya's Distance. $G_1 = \{E_1, E_5, E_6, E_{10}\}$ – when the group is formed, the maximum degree of dissimilarity is achieved by merging subgroups $\{E_1, E_5, E_{10}\}$ and $\{E_6\}$, and reaches d_B (m_i, m_j) = 0.3; the minimum degree of dissimilarity is achieved by merging subgroups $\{E_1\}$ and $\{E_{10}\}$ with d_B (m_i, m_j) = 0.086.

$G_2 = \{E_3, E_9\}$ – the degree of dissimilarity reaches d_B (m_i, m_j) = 0.142.

$G_3 = \{E_2, E_8\}$ – the degree of dissimilarity reaches d_B (m_i, m_j) = 0.266.

$G_4 = \{E_4\}$ – the closest subgroup is $\{E_7\}$, $d_B = 0.38$.

$G_5 = \{E_7\}$.

Tessem's Distance. $G_1 = \{E_1, E_3, E_{10}\}$ – when the group is formed, the maximum degree of dissimilarity is achieved by merging subgroups $\{E_1, E_{10}\}$ and $\{E_3\}$, and reaches d_T (m_i, m_j) = 0.3; the minimum degree of dissimilarity is achieved by merging subgroups $\{E_1\}$ and $\{E_{10}\}$ with d_T (m_i, m_j) = 0.1.

$G_2 = \{E_6, E_9\}$ – the degree of dissimilarity reaches d_T (m_i, m_j) = 0.2.

$G_3 = \{E_2, E_7, E_8\}$ – when the group is formed, the maximum degree of dissimilarity is achieved by merging subgroups $\{E_2, E_8\}$ and $\{E_7\}$, and reaches d_T (m_i, m_j) = 0.3; the minimum degree of dissimilarity is achieved by merging subgroups $\{E_2\}$ and $\{E_8\}$ with d_T (m_i, m_j) = 0.2.

$G_4 = \{E_4\}$ – the closest subgroup is $\{E_2, E_7, E_8\}$, $d_T = 0.4$.

$G_5 = \{E_5\}$ – the closest subgroup is $\{E_2, E_4, E_7, E_8\}$, $d_T = 0.5$.

From the obtained results, it can be observed that regardless of which metric is used, a subgroup $\{E_4\}$ is allocated with a degree of dissimilarity varying from 0.2 to 0.4 (depending on the metric used). The results obtained for the metrics of Euclidean and Wang coincide. The only difference is in the corresponding d(m_i, m_j) values of the metrics considered.

The Jousselme, Euclidean, Wang and Bhattacharyya metrics distinguish the subgroup $\{E_3, E_9\}$ with the degree of dissimilarity ranging from 0.142 to 0.283 (depending on the metric used).

The Euclidean, Wang and Bhattacharyya metrics distinguish the subgroup $\{E_2, E_8\}$ with the degree of dissimilarity ranging from 0.2 to 0.266 (depending on the metric used). The Jousselme and Tessem metrics combine it with a subgroup $\{E_7\}$ with the marginal level of conflict equal to 0.3; while the $Conf_E(E_7,\{E_2, E_8\})$ measure equal to 0.385; the $Conf_W(E_7,\{E_2, E_8\})$ measure equal to 0.35; the $Conf_B(E_7,\{E_2, E_8\})$ measure equal to 0,462.

The Jousselme and Bhattacharyya metrics distinguish the subgroup $\{E_1, E_5, E_6, E_{10}\}$. The Euclidean and Wang metrics distinguish subgroups $G_1 = \{E_1, E_{10}\}$ and $G_2 = \{E_5, E_6\}$, their union can occur at a conflict level equal to 0.37 for Euclidean metric; Wang metric allows merging of subgroups $\{E_1, E_{10}\}$ and $\{E_3, E_9\}$ with $d_W = 0.4$, and next $\{E_1, E_3, E_9, E_{10}\}$ and $\{E_5, E_6\}$ with $d_W = 0.5$.

So we have:

$Conf_E(E_1,\{E_5, E_6, E_{10}\}) = 0.234$; $Conf_W(E_1,\{E_5, E_6, E_{10}\}) = 0.2$;
$Conf_T(E_1,\{E_5, E_6, E_{10}\}) = 0.167$.
$Conf_E(E_5,\{E_1, E_6, E_{10}\}) = 0.288$; $Conf_W(E_5,\{E_1, E_6, E_{10}\}) = 0.3$;
$Conf_T(E_5,\{E_1, E_6, E_{10}\}) = 0.333$.
$Conf_E(E_6,\{E_1, E_5, E_{10}\}) = 0.312$; $Conf_W(E_6,\{E_1, E_5, E_{10}\}) = 0.267$;
$Conf_T(E_6,\{E_1, E_5, E_{10}\}) = 0.233$.
$Conf_E(E_{10},\{E_1, E_5, E_6\}) = 0.297$; $Conf_W(E_{10},\{E_1, E_5, E_6\}) = 0.3$;
$Conf_T(E_{10},\{E_1, E_5, E_6\}) = 0.333$.

The Euclidean, Wang and Bhattacharyya metrics distinguish the subgroup $\{E_7\}$ with the significant value of degree of dissimilarity ranging from 0.38 to 0.424 (depending on the metric used). By Euclidean and Wang metrics subgroup $\{E_7\}$ is as close as possible to $\{E_2, E_8\}$ with $d_E = 0.424$ and $d_W = 0.4$.

3.2.2 The Average Linkage Method

Jousselme's Distance. $G_1 = \{E_1, E_3, E_5, E_6, E_9, E_{10}\}$ – when the group is formed, the maximum degree of dissimilarity is achieved by merging subgroups $\{E_1, E_5, E_6, E_{10}\}$ and $\{E_3, E_9\}$, and reaches d_J (m_i, m_j) = 0.298; the minimum degree of dissimilarity is achieved by merging subgroups $\{E_1\}$ and $\{E_{10}\}$ with d_J (m_i, m_j) = 0.1.

$G_2 = \{E_2, E_7, E_8\}$ – when the group is formed, the maximum degree of dissimilarity is achieved by merging subgroups $\{E_2, E_8\}$ and $\{E_7\}$, and reaches d_J (m_i, m_j) = 0.273; the minimum degree of dissimilarity is achieved by merging subgroups $\{E_2\}$ and $\{E_8\}$ with d_J (m_i, m_j) = 0.173.

$G_3 = \{E_4\}$ – the closest subgroup is $\{E_2, E_7, E_8\}$, d_J (m_{G3}, m_{G5}) = 0.331.

Euclidean Distance. $G_1 = \{E_1, E_{10}\}$ – the degree of dissimilarity reaches d_E (m_i, m_j) = 0.141.

$G_2 = \{E_5, E_6\}$ – the degree of dissimilarity reaches d_E (m_i, m_j) = 0.245.
$G_3 = \{E_3, E_9\}$ – the degree of dissimilarity reaches d_E (m_i, m_j) = 0.283.
$G_4 = \{E_2, E_8\}$ – the degree of dissimilarity reaches d_E (m_i, m_j) = 0.245.
$G_5 = \{E_7\}$ – the closest subgroup is $\{E_2, E_8\}$, d_E (m_{G4}, m_{G5}) = 0.385.
$G_6 = \{E_4\}$ – the closest subgroup is $\{E_2, E_7, E_8\}$, $d_E = 0.4685$.

Wang Distance. $G_1 = \{E_1, E_{10}\}$ – the degree of dissimilarity reaches d_W (m_i, m_j) = 0.1.

$G_2 = \{E_5, E_6\}$ – the degree of dissimilarity reaches d_W (m_i, m_j) = 0.2.
$G_3 = \{E_3, E_9\}$ – the degree of dissimilarity reaches d_W (m_i, m_j) = 0.2.
$G_4 = \{E_2, E_8\}$ – the degree of dissimilarity reaches d_W (m_i, m_j) = 0.2.
$G_5 = \{E_7\}$ – the closest subgroup is $\{E_2, E_8\}$, d_W (m_{G4}, m_{G5}) = 0.35.
$G_6 = \{E_4\}$ – the closest subgroup is $\{E_2, E_7, E_8\}$, $d_W = 0.43$.

Bhattacharyya's Distance. $G_1 = \{E_1, E_5, E_6, E_{10}\}$ – when the group is formed, the maximum degree of dissimilarity is achieved by merging subgroups $\{E_1, E_5, E_{10}\}$ and $\{E_6\}$, and reaches d_B (m_i, m_j) = 0.287; the minimum degree of dissimilarity is achieved by merging subgroups $\{E_1\}$ and $\{E_{10}\}$ with d_B (m_i, m_j) = 0.086.

$G_2 = \{E_3, E_9\}$ – the degree of dissimilarity reaches d_B (m_i, m_j) = 0.142.

$G_3 = \{E_2, E_8\}$ – the degree of dissimilarity reaches d_B (m_i, m_j) = 0.266.
$G_4 = \{E_4\}$ – the closest subgroup is $\{E_7\}$, $d_B = 0.38$.
$G_5 = \{E_7\}$.

Tessem's Distance. $G_1 = \{E_1, E_3, E_6, E_9, E_{10}\}$ – when the group is formed, the maximum degree of dissimilarity is achieved by merging subgroups $\{E_1, E_3, E_{10}\}$ and $\{E_6, E_9\}$, and reaches d_T (m_i, m_j) = 0.3; the minimum degree of dissimilarity is achieved by merging subgroups $\{E_1\}$ and $\{E_{10}\}$ with d_T (m_i, m_j) = 0.1.

$G_2 = \{E_2, E_5, E_7, E_8\}$ – when the group is formed, the maximum degree of dissimilarity is achieved by merging subgroups $\{E_2, E_8\}$ and $\{E_7\}$ and $\{E_5\}$, and reaches d_T (m_i, m_j) = 0.3; the minimum degree of dissimilarity is achieved by merging subgroups $\{E_2\}$ and $\{E_8\}$ with d_T (m_i, m_j) = 0.2.

$G_3 = \{E_4\}$ – the closest subgroup is $\{E_2, E_5, E_7, E_8\}$, $d_T = 0.375$.

From the obtained results, it can be observed that the results obtained for Euclidean, Wang and Bhattacharyya metrics based on the farthest neighbor and average linkage methods coincide. The only difference is in the corresponding d(m_i, m_j) values of the metrics considered.

The results obtained by Jousselme metric make it possible to unite the subgroups $\{E_1, E_5, E_6, E_{10}\}$ and $\{E_3, E_9\}$ with d_J (m_i, m_j) = 0.298; based on the farthest neighbor method, such an merging is possible with the value of d_J (m_i, m_j) equal to 0.361.

The results obtained by Tessem metric make it possible to unite the subgroups $\{E_1, E_3, E_{10}\}$ and $\{E_6, E_9\}$ with a maximum permissible conflict level equal to 0.3; based on the farthest neighbor method, such a merging is possible with the value of d_T (m_i, m_j) equal to 0.4.

The subgroup $\{E_2, E_5, E_7, E_8\}$ is also formed by merging the subgroups $\{E_2, E_8\}$, $\{E_7\}$ and $\{E_5\}$ with d_T (m_i, m_j) = 0.3; on the farthest neighbor method, such a merging is possible only when merging subgroups $\{E_2, E_7, E_8\}$, $\{E_4\}$ and $\{E_5\}$ with d_T (m_i, m_j) = 0.5.

After analyzing the results obtained on the basis of the farthest neighbor and the average linkage methods, we can draw the following conclusions. According to the established boundary value of d(m_i, m_j) the expert committee was divided into next subgroups: E $\Rightarrow$ $\{G_1, G_2, G_3, G_4, G_5, G_6\}$, containing similar judgments in a certain sense, where $G_1 = \{E_1, E_{10}\}$; $G_2 = \{E_5, E_6\}$; $G_3 = \{E_3, E_9\}$; $G_4 = \{E_2, E_8\}$, $G_5 = \{E_4\}$, $G_6 = \{E_7\}$. In discussed example, the boundary value, which is responsible for the optimal level of conflict within the subgroup of experts, was assumed to be 0.3.

However, considering the values of measure (12) that obtained on the basis of Euclidean, Wang and Bhattacharyya metrics it is possible to merge the subgroups $G_1 = \{E_1, E_{10}\}$ and $G_2 = \{E_5, E_6\}$. Thus, we have next subgroups: E $\Rightarrow$ $\{G_1, G_2, G_3, G_4, G_5\}$, where $G_1 = \{E_1, E_{10}, E_5, E_6\}$; $G_2 = \{E_3, E_9\}$; $G_3 = \{E_2, E_8\}$, $G_4 = \{E_4\}$, $G_5 = \{E_7\}$.

The next stage is the aggregation of the individual expert judgments to produce the collective judgments (decision).

Given the high degree of importance of choosing the optimal technology, the multialternative of the cutting and welding technological processes currently used in shipbuilding, as well as the representativeness of the expert commission (10 or more experts), we can conclude that a detailed analysis of expert opinions is necessary, which ultimately should lead to an effective solution.

4 Conclusion

The approach for dividing the initial set of group expert assessments, presented in the format of rankings (orderings) into subgroups, within which the expert assessments are close to each other has been proposed in this paper. Unlike existing methods of clustering expert judgments, this approach allows to process expert judgments with different structure: consistent, equivalent, arbitrary, etc.; take into account possible union and intersection of expert evidences.

The proposed approach is based on the mathematical apparatus of evidence theory metrics, which makes it possible to evaluate the degree of difference (conflict) between the selected groups of expert evidence, given their structure. Various metrics of evidence theory have been analyzed, and their impact on the results of the partition was studied.

Expert evidence is considered consistent (homogeneous), if the value of the specified metric for all evidence of this group does not exceed a specified threshold value. To determine the order of grouping expert evidence in accordance with the degree of their similarity (dissimilarity) the farthest neighbor and average linkage methods have been used this in paper.

Examples of practical use of the proposed approach for two cases have been analyzed: case 1 – each expert evaluated all the available alternatives; case 2 – there are experts that evaluated only some alternatives according to their preferences. In the second case, the situation was considered the presence of incomplete expert information – conflicting (contradictory, mismatched) expert judgments.

Acknowledgment. This research was partially supported by the state research project "Development of information and communication decision support technologies for strategic decision-making with multiple criteria and uncertainty for military-civilian use" (research project no. 0117U007144, financed by the Government of Ukraine).

References

1. Kiseleeva, N.E., Pankova, L.A., Shneiderman, M.V.: Strukturnyi podkhod k analizu I obrabotke dannykh ekspertnogo oprosa. Avtomatika i Telemekhanika **4**, 64–70 (1975). (in Russian)
2. Orlov, A.I.: Nechislovaia statistika. MZ-Press, Moscow (2004). (in Russian)
3. Kemeni, D.zh., Snell, D.zh.: Kiberneticheskoe modelirovanie: Nekotorye prilozheniia. Sovetskoe radio, Moscow (1972)
4. Kovalenko, I.I., Shved, A.V.: Clustering of group expert estimates based on measures in the theory of evidence. Naukovyi Visnyk Natsionalnoho Hirnychoho Universytetu **4**(154), 71–78 (2016)
5. Shafer, G.A.: A Mathematical Theory of Evidence. Princeton University Press, Princeton (1976)
6. Dempster, A.P.: A generalization of Bayesian inference. J. R. Stat. Soc. B **30**, 205–247 (1968)
7. Beynon, M.J.: The Dempster-Shafer theory of evidence: an alternative approach to multicriteria decision modeling. Omega **28**(1), 37–50 (2000)

8. Kovalenko, I.I., Shved, A.V.: Development the technology of structuring of group expert estimates under different types of uncertainty. Eastern-Eur. J. Enterp. Technol. **3/4**(93), 60–68 (2018). https://doi.org/10.15587/1729-4061.2018.133299
9. Shved, A., Davydenko, Y.e.: The analysis of uncertainty measures with various types of evidence. In: 1th International Conference on Data Stream Mining & Processing, pp. 61–64. IEEE Press, Lviv, Ukraine (2016). https://doi.org/10.1109/DSMP.2016.7583508
10. Sentz, K., Ferson, S.: Combination of evidence in Dempster-Shafer theory. Sandia National Laboratories, Albuquerque (2002)
11. Kovalenko, I., Davydenko, Y.e., Shved, A.: Development of the procedure for integrated application of scenario prediction methods. Eastern-Eur. J. Enterp. Technol. **2/4**(98), 31–37 (2019). https://doi.org/10.15587/1729-4061.2019.163871
12. Jousselme, A.L., Grenier, D., Bossé, E.: A new distance between two bodies of evidence. Inf. Fus. **2**, 91–101 (2001)
13. Jousselme, A.L., Grenier, D., Bossé, E.: Analyzing approximation algorithms in the theory of evidence. Sens. Fus. Architect. Algorithms Appl. **VI**(4731), 65–74 (2002)
14. Florea, M.C., Bosse, E.: Crisis management using dempster-shafer theory: using dissimilarity measures to characterize sources' reliability. In: C3I in Crisis, Emergency and Consequence Management, RTO-MP-IST-086, Bucharest, Romania (2009)
15. Bouchon-Meunie, B., Diaz, J., Rifqui, M.: A similarity measure between basic belief assignments. In: 9th International conference on Information Fusion, pp. 1–6. IEEE Press, Florence, Italy (2006)
16. Cuzzolin, F.: A geometric approach to the theory of evidence. Trans. Syst. Man Cybern. Part C: Appl. Rev. **38**(4), 1–13 (2007)
17. Wang, Z.: Theory and technology research of target identification fusion of C4ISR system. Ph.D. Thesis. National Defense Science and Technology University, China (2001)
18. Bhattacharyya, A.: On a measure of divergence between two statistical populations defined by their probability distribution. Bull. Calcutta Math. Soc. **35**, 99–110 (1943)
19. Tessem, B.: Approximations for efficient computation in the theory of evidence. Artif. Intell. **61**, 315–329 (1993)
20. Martin, A., Jousselme, A.L., Osswald, C.: Conflict measure for the discounting operation on belief functions. In: 11th International Conference on Information Fusion, pp. 1–8. IEEE Press, Cologne, Germany (2008)

Flood Monitoring Using Multi-temporal Synthetic Aperture Radar Images

Olena Kavats[1], Volodymyr Hnatushenko[1,2](✉), Yuliya Kibukevych[1], and Yurii Kavats[1]

[1] Department of Information Technology and Systems, National Metallurgical Academy of Ukraine, Av. Gagarin, 4, Dnipro, Ukraine
alena.kavats@gmail.com, vvgnatush@gmail.com

[2] Department of Information Systems and Technologies, Dnipro University of Technology, Av. Dmytra Yavornytskoho, 19, Dnipro 49005, Ukraine
Iuliia.kibukevich@gmail.com, yukavats@gmail.com

Abstract. The rise of water level in the basins of rivers is the most common cause of floods in the world. This can be occurs due to heavy rains, spring snowmelt, wind surge, destruction of dams, rubbish, straightening of rivers or deforestation. Satellite observing systems are the main tool for solving problems such as monitoring of flooded areas. Since it is often not possible to use optical images due to high cloud cover, data from radar satellite imagery is used to ensure all-weather satellite monitoring of the dynamics and effects of flooding. The paper proposes a computer information technology for satellite monitoring of floods based on multi-temporal synthetic aperture radar (SAR) data. The technology allows you to determine flood areas and quickly get floodplain area. In this paper we consider the possibility of remote detection of flooding in the Transcarpathian region in December 2017. Four multi-temporal radar images from the satellite Sentinel-1 were used for flood detection. Dates of shooting are 14, 17, 18 and 20 December 2017. The first image was taken before the flood, the next two – in the midst of a natural disaster, the last – after the decline of the water level. All presented radar images have the dual polarization VV+VH. The studies have shown that satellite radar images provide an opportunity to qualitatively determine the flooded area in the Transcarpathian region of Ukraine. The proposed methodology made it possible to qualitatively determine more than 2000 ha and identify damage areas after the flood. Using archival data in the methodology allows you to build cartograms for the previous period and identify regularity of flood areas.

Keywords: Flood monitoring · Radar data processing · Satellite multi-temporal images · Polarization · Sentinel-1

1 Introduction

The modern level of development of Earth remote sensing systems allows solving a wide range of scientific and applied problems. The list of problems includes such tasks like monitoring the environment, forests, oceans, floods, fires, agricultural processes, the tasks of hydrometeorology and weather forecast [1–6].

N. Shakhovska and M. O. Medykovskyy (Eds.): CCSIT 2019, AISC 1080, pp. 54–63, 2020.
https://doi.org/10.1007/978-3-030-33695-0_5

In recent decades the number of water-related disasters in the world has increased significantly [7]. The greatest damage from all natural disasters causes floods (40%), in the second place – tropical cyclones (20%), on the third and fourth places (by 15%) – earthquakes and droughts. As a result of this natural disaster, buildings, bridges, roads and railways are being collapsed, crops are being destroyed, landslides are being occurred and there are often casualties among the population and animals.

The rise of water level in the basins of such rivers as the Dnieper, Dniester, Western Bug and Tisza is the most common cause of floods in Ukraine. This can be occurs due to heavy rains, spring snowmelt, wind surge, destruction of dams, rubbish, straightening of rivers or deforestation. The formation of floods is non-seasonal and can appear at any time of the year. The rate of flooding ranges from tens of minutes to several hours. Water flows entail boulders, pulled out trees with roots, activate landslide and mudflow processes.

The use of Earth remote sensing data is a perspective direction of river monitoring in Ukraine, because there is a rare network of points of hydrometeorological observations and lack of relevant spatial data on the territory. Due to the dense hydrographic network Transcarpathian and Ivano-Frankivsk region most often suffer from floods in Ukraine. Identification of damaged areas after floods is crucial for understanding and overview the process, for prediction of possible events and also for the timely introduction of preventive measures and prevention of people who are at risk.

Over the past decades, Ukraine has experienced several large-scale floods, which have caused the destruction of roads and bridges, limited the flow of water, gas and electricity to settlements, flooded and partially destroyed thousands of hectares of agricultural land. In December 2017, in Western Ukraine, as a result of the increasing air temperature and precipitation in the form of rain and snow, a large-scale flood has been registered over the past 20 years. Only for the first day the water level in the rivers rose by 1.5–2.5 m, 49 basements and 2 thousand hectares of farmland were flooded, 2 automobile bridges of city importance were destroyed.

2 State of the Art

Natural phenomena in the form of floods are a natural danger to the population and entail significant losses for people who work in the agricultural infrastructure. It is important to foresee such dangerous processes, identify the main causes and develop preventive measures. The main auxiliary tool is high-resolution images and building-up of flood risk maps. It is possible to track changes and state of the earth's surface after floods with the help of remote sensing, mapping or aerial photography [8–12]. The last two methods are quite hard and time-consuming. Optical data can be obtained quickly and analyzed in time. But due to the fact that floods are characterized by high clouds or fogs, the informativeness of such data is significantly reduced and studies may not be reliable. In this case, the alternative is radar data. Many studies have demonstrated that SAR systems are suitable tools for floodwater [13–18]. High spatial and temporal resolution satellite data such as Sentinel-1 are particularly useful for analysis of the flooded areas [19, 20]. The contribution of Sentinel-1 to the application of flood mapping arises from the sensitivity of the backscatter signal to open water [21].

Previously, radar data with one polarization was mainly used (VV or HH). Dual or full polarization polarimetric radar data are now increasingly being used for this purpose. It was found in [22] that among the polarizations offered by the Sentinel-1 sensor, VV-polarized SAR images result in more accurate flood maps than cross-polarized products. In the same context, [21] suggested that HH polarization realizes the highest accuracy in terms of flood mapping.

One of the main tasks of the article is the development of information technology for satellite monitoring of floods using multi-temporal radar data. Besides, the task of the article is to determine the total area of the flood using Sentinel-1 satellite radar images.

Solving the task of monitoring floods on multi-temporal satellite radar data will allow to get information on hard-to-reach areas timely and to minimize the effects of weather. The building of cartograms over the past years will make it possible to identify areas of constant flood, which in future will allow the implementation of appropriate measures to prevent them.

3 Methodology

3.1 The Input Data

In this study satellite flood monitoring was carried out with the help of radar data for the following reasons. First of all, radar imaging is weather-resistant, which makes it possible to conduct observations in the dark, because radar pulses are able to penetrate through the rain clouds, which in 90% of cases close the study area. Secondly, the Sentinel-1 satellite provides a regular shooting of the Earth's surface, which helps to identify any changes in time. Another important aspect of radar images is that it is easy to decipher and distinguish the contrasts of soil and water. The signal received from the sensor is mirrored from the flat surface of the earth, which leads to decreasing the amount of reflected radiation. This causes relatively dark pixels on the radar image of water areas that contrast with other surfaces of the earth (such as forests, fields, or urban areas) [12].

In this paper a satellite flood monitoring using radar images of the Sentinel-1 satellite in Ukraine, Transcarpathian region, VelykiKom'yaty, was proposed. The observation period of the territory is December 2017. It was used a series of four multi-temporal images – 14, 17, 18 and 20 December 2017. The first image was taken before the flood, the next two - in the midst of a natural disaster, the last – after the decline of the water level. All presented radar images have the dual polarization VV+VH. Different polarizations observed by Sentinel-1A/B also allow us to deduce more information about the land cover, as the cross-polarization channel has a stronger return over areas with volume scattering, such as vegetated areas. Furthermore co-polarization tends to have stronger returns over urban areas, due to man-made structures prominence in double bounce return [16].

3.2 The Proposed Computer Technology

The proposed computer technology uses satellite radar data, which has a number of advantages, in particular:

- authenticity;
- visibility and informativeness;
- relevance and efficiency;
- high frequency;
- no clouds;
- Multipolarization shooting using synthesized aperture radars.

The input images were pre-processed for the further research. The main steps of image pre-processing (Fig. 1):

- Data Download – downloading Sentinel-1 data in Interferometric Wide mode using Copernicus resource [23];
- Subset area of interest – cropping the area of interest to reduce image processing time;
- Radiometric Calibration – bringing the original data into physical units for comparison with the other shooting data;
- Ellipsoid Correction – resampling of image with changing the system design on UTM/WGS84;
- Speckle Filtering – removing speckle noise in the image;
- Composite RGB image – formation of the RGB image for the visual assessment;
- Creation flooding mask – create an image of the same size as the original, in which the flooded areas have a value of 1, the rest – 0.

In this work the radar images have been radiometric calibrated for quantitative analysis. This step is necessary in order that the pixel values of the image directly reflect the value of the radar backscattering of the reflection surface.

The radar images can be distorted due to the topographic variation of the scene and the inclination of the satellite sensor. Data will have some distortion if it is not directly at the sensor location in Nadir.

It should be noted that at the stage of geometric correction it is necessary to use a digital elevation model (DEM) on the shooting area. DEM represent the exact data of the height of the earth's surface directly above the sea level, including buildings, vegetation and other high-rise objects. In the work the data of Shuttle Radar Topography Mission SRTM-1 with a spatial resolution of 30 m, mainly based on stereoscopic optical and interferometric radar space surveys was used.

In this work the Ellipsoid Correction function based on bilinear interpolation was used to ensure that the geometric representation of the image was as close as possible to the real world. Also, using the Range Doppler orthorectification method, the image was geocoded from a single design system to the UTM/WGS84 system. In this method available vector information of the orbit state, radar time records and ground-range conversion parameters were used to obtain accurate geolocation information.

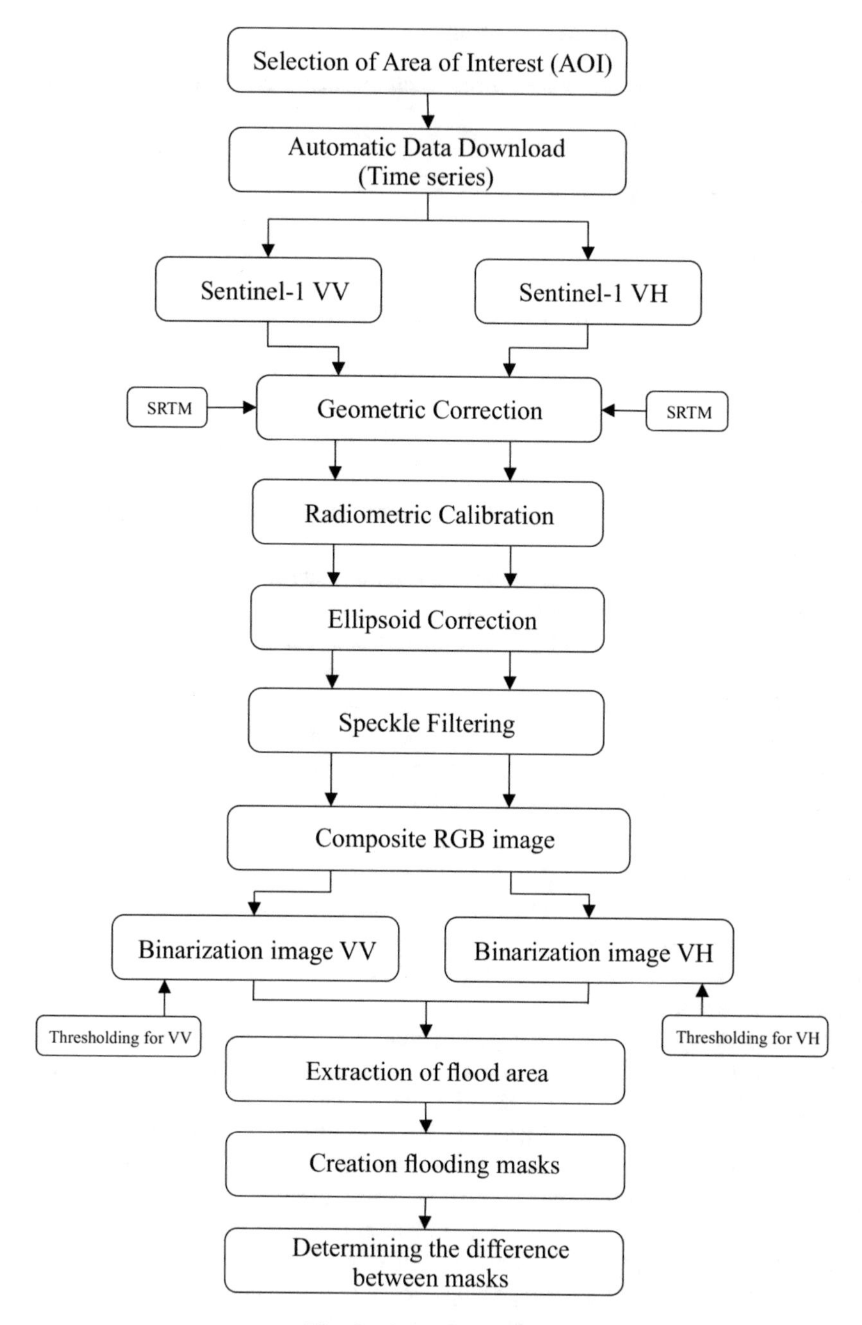

Fig. 1. General workflow

Since synthetic aperture radar is the active system, speckle noise is often present in the images. Speckle is a systematic phenomenon and the result of the interaction of the impulse radar with different scattering, which greatly reduces the informativeness of radar data [13–15]. Therefore, it is proposed to use the function Speckle Filtering to remove speckle noise. The study applied a Lee filter with a 5 by 5 window.

For a visual assessment the pseudocolor RGB-composites was formed. The image in VV polarization corresponds to the red channel, the green channel corresponds to the image in VH polarization, the blue channel corresponds to the image division in VV and VH polarization.

The threshold method [11] was applied to create masks that correspond to the flooding zones, that is, to create an image of the same size as the original, in which the flooding areas have a value of 1, the rest – 0. According to this method, the damaged area as a result of the flood will have a value of 1 if

$$DN \leq TW. \tag{1}$$

where *DN*— is the digital value of the radar image channel VH and *TW*—is the water threshold.

Figures 2 and 3 show a series of multi-temporal fragments of the original composite RGB images after radiometric calibration, ellipsoid correction and removal of speckle noise.

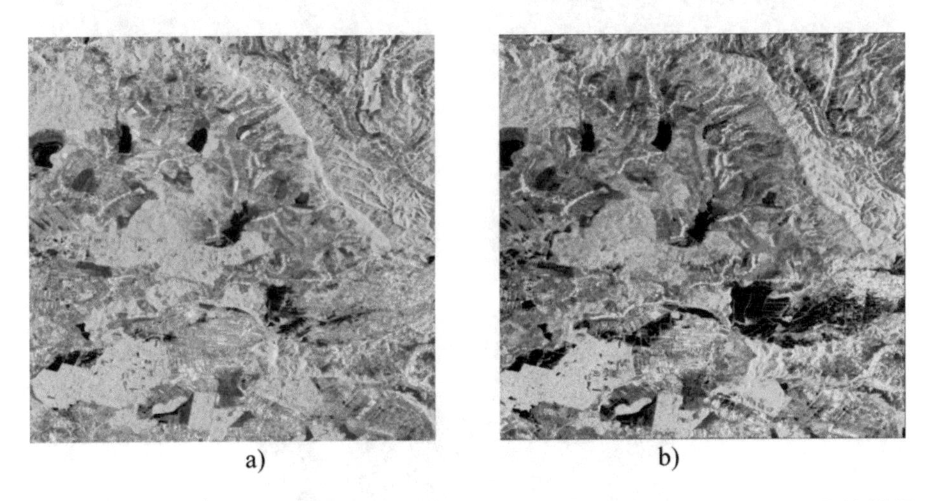

Fig. 2. RGB composite images: (a) shooting date – 14.12.2017; (b) shooting date – 17.12.2017

One of the last stages of flood monitoring is the construction of a two-class mask. The use of threshold values for binarization allows for the separation of water and the determination of flooded areas. The result of the received masks for the corresponding dates is shown in Fig. 4.

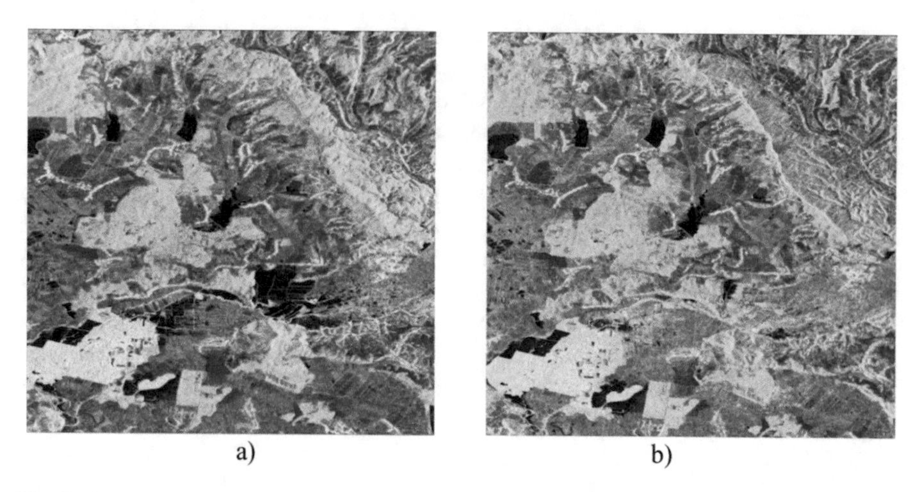

Fig. 3. RGB composite images: (a) shooting date – 18.12.2017; (b) shooting date – 20.12.2017

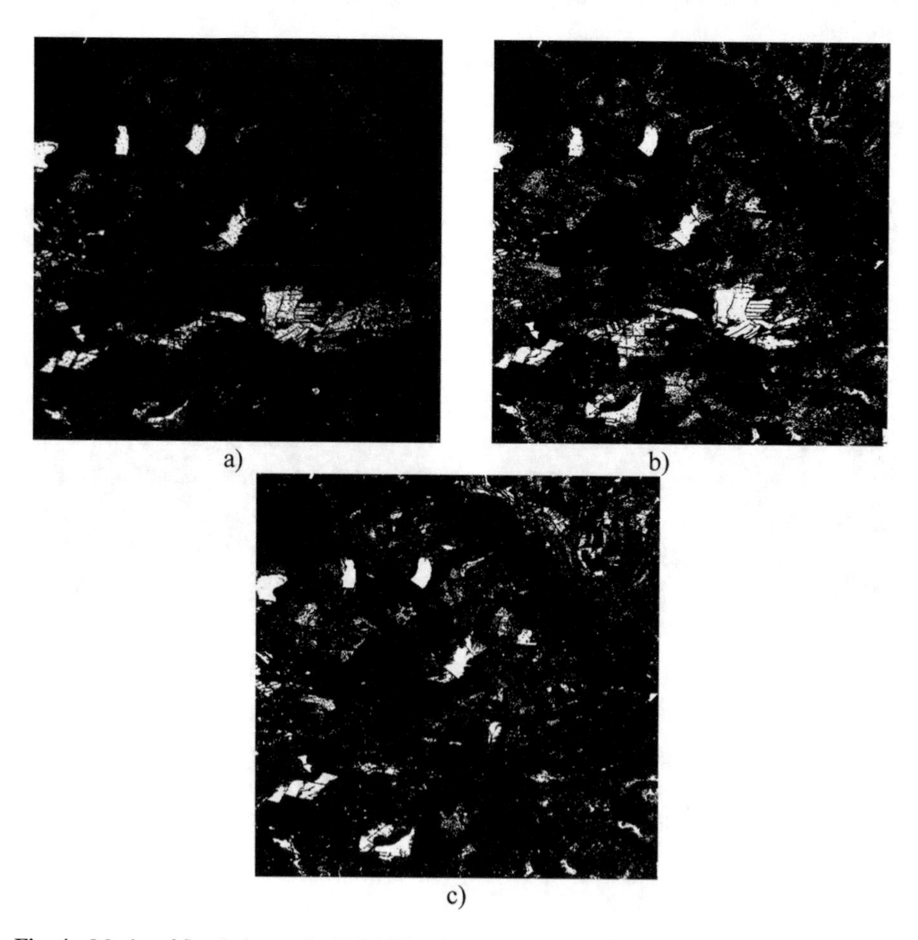

Fig. 4. Masks of flooded areas in VelykiKom'yaty: (a) the 2nd day of the flood; (b) the 3rd day of the flood; (c) after the flood

4 Results

The derived flood extent was validated against cloud free optical data and modelled flood maps. Validation of the received flood map took place on the cloudless data of the high spatial resolution of WorldView-4 and on the official data received from the Ministry of Emergency Situations of Ukraine (MES). In the corresponding report of the MES it is stated that "on December 14–16, 2017 in the territory of the Transcarpathian region as a result of… raising the water levels in the rivers Latoritsa, Borzhava, their inflows, small rivers, streams… there occurred local flooding with a surface runoff of households, dwelling houses and agricultural lands, landslides, damage and destruction of bridges of local significance, disconnection of power supply of some settlements, damage to dams and banks, car bridge destruction, roads of state and communal power ness, poured in some areas flood waters across national importance."

Matching and linking the geographic vector map with certain flood areas from the Ministry of Emergencies and map received by purposed information technology with masks, allowed us to calculate the accuracy of the class water, which was 87%.

The flooded areas were calculated for the obtained multi-temporal radar data of each of the dates (Fig. 5). Only on the presented stage in VelykiKom'yaty, 780 ha were flooded within a day from the beginning of the flood. In the next two days, the flood damaged another 831 ha. In total, more than 2000 ha were flooded during the disaster. Almost 800 ha of damaged territories belong to commercial agricultural production. This flood led to significant losses for entrepreneurs and workers of the agricultural industry.

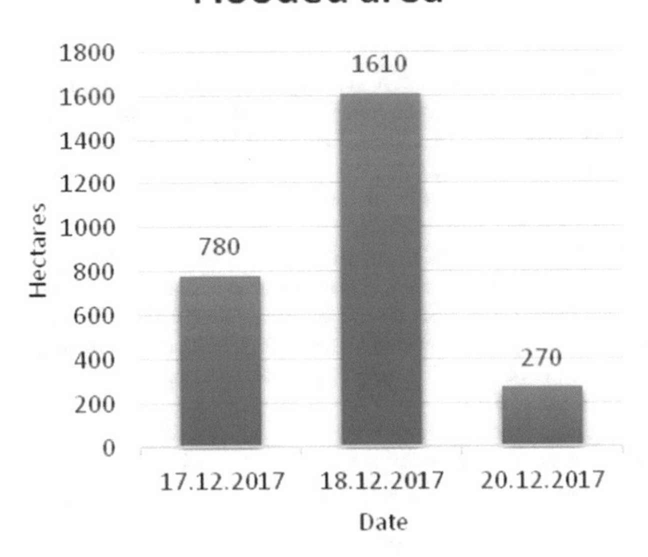

Fig. 5. Flooded areas

5 Conclusion

Satellite observing systems are the main tool for solving problems such as monitoring of flooded areas. Images acquired before and after flooding support assessments of property and environmental damage. Since it is often not possible to use optical images due to high cloud cover, data from radar satellite imagery is used to ensure all-weather satellite monitoring of the dynamics and effects of flooding. Sentinel-1 radar sensor ability to see through clouds and in darkness, makes it particularly useful for monitoring floods.

In this article a computer analysis of radar images for flood monitoring in the Transcarpathian region of Ukraine was proposed. For this case study, Sentinel-1 SAR data provided by European Space Agency were used. All presented radar images have the dual polarization VV+VH. The flooded areas show up particularly strongly in Sentinel-1 data when we combine images from different dates into composite images. The aim of this study is to determine the total area of the flood in the Transcarpathian region in December 2017. The proposed methodology made it possible to qualitatively determine more than 2000 ha and identify damage areas after the flood. Using archival data in the methodology allows you to build cartograms for the previous period and identify regularity of flood areas. In the future, the proposed information technology could benefit from recent SAR sensors of a higher spatial resolution. It is expected that synthetic aperture radar images could improve the accuracy of the classification in flooded areas.

Disclosure Statement. No potential conflict of interest was reported by the authors.

References

1. Shedlovska, Y., Hnatushenko, V.: A very high resolution satellite imagery classification algorithm. In: 2018 IEEE 38th International Conference on Electronics and Nanotechnology (ELNANO), pp. 654–657 (2018). https://doi.org/10.1109/ELNANO.2018.8477447
2. Hnatushenko, V.V., Mozgovoy, D.K., Vasyliev, V.V., Kavats, O.O.: Satellite monitoring of consequences of illegal extraction of amber in Ukraine. Sci. Bull. Nat. Min. Univ. Dnipropetrovsk. **2**(158), 99–105 (2017)
3. Garkusha, I.N., Hnatushenko, V.V., Vasyliev, V.V.: Using sentinel-1 data for monitoring of soil moisture. In: 2017 IEEE International Geoscience and Remote Sensing Symposium (IGARSS), Fort Worth, TX, USA (2017). https://doi.org/10.1109/IGARSS.2017.8127291
4. Hnatushenko, V.V., Mozgovoy, D.K., Vasyliev, V.V.: Satellite monitoring of deforestation as a result of mining. In: Scientific bulletin of National Mining University, Dnipropetrovsk, no. 5(161), pp. 94–99 (2017)
5. Garkusha, I.N., Hnatushenko, V.V., Vasyliev, V.V.: Research of influence of atmosphere and humidity on the data of radar imaging by Sentinel-1. In: IEEE 37th International Conference on Electronics and Nanotechnology (ELNANO) (2017). https://doi.org/10.1109/ELNANO.2017.7939787
6. Hnatushenko, V.V., Mozgovoy, D.K., Serikov, I.Ju., Vasyliev, V.V.: Automatic vegetation classification using multispectral aerial images and neural network. In: System Technologies Dnipro, no. 6(107), pp. 66–72 (2016)

7. Dadhich, G., Miyazaki, H., Babel, M.: Applications of Sentinel-1 synthetic aperture radar imagery for floods damage assessment: a case study of Nakhon SI Thammarat, Thailand. In: The International Archives of the Photogrammetry, Remote Sensing and Spatial Information Sciences, vol. XLII-2/W13, pp. 1927–1931 (2019). https://doi.org/10.5194/isprs-archives-XLII-2-W13-1927-2019
8. Zoka, M., Psomiadis, E., Dercas, N.: The complementary use of optical and SAR data in monitoring flood events and their effects. In: Proceedings, vol. 2, p. 644 (2018)
9. Brown, K.M., Hambridge, C.H., Brownett, J.M.: Progress in operational flood mapping using satellite synthetic aperture radar (SAR) and airborne light detection and ranging (LiDAR) data. Progress Phys. Geogr. **40**(2), 196–214 (2016). https://doi.org/10.1177/0309133316633570
10. Ban, H.-J., Kwon, Y.-J., Shin, H., Ryu, H.-S., Hong, S.: Flood monitoring using satellite-based RGB composite imagery and refractive index retrieval in visible and near-infrared bands. Remote Sens. **9**, 313 (2017)
11. Mozgovoy, D.K., Hnatushenko, V.V., Vasyliev, V.V.: Automated recognition of vegetation and water bodies on the territory of megacities in satellite images of visible and IR bands. In: SPRS Annals of Photogrammetry, Remote Sensing and Spatial Information Sciences, vol. IV-3, pp. 167–172 (2018). https://doi.org/10.5194/isprs-annals-IV-3-167-2018
12. Gan, T.Y., Zunic, F., Kuo, C.-C., Strobl, T.: Flood mapping of Danube river at Romania using single and multi-date ERS2SAR images. Int. J. Appl. Earth Obs. Geoinf. **18**, 69–81 (2012)
13. Tsyganskaya, V., Martinis, S., Marzahn, P., Ludwig, R.: SAR-based detection of flooded vegetation–a review of characteristics and approaches. Int. J. Remote Sens. **39**, 2255–2293 (2018)
14. Rahman Md, R.: Flood inundation mapping and damage assessment using multi-temporal RADARSAT and IRS 1C LISS III image. Asian J. Geoinf. **6**(2), , 11–21 (2006)
15. Landuyt, L., Wesemael, A., Van Coillie, F.M.B. Van, Verhoest, N.E.C.: Pixel-based flood mapping from SAR imagery: a comparison of approaches. In: Geophys. Res. Abstr. vol. 19, EGU2017-14060 (2017)
16. Jo, M.J., Osmanoglu, B., Zhang, B., Wdowinski, S.: Flood extent mapping using dual-polarimetric sentinel-1 synthetic aperture Radar imagery. ISPRS Archives **42**(3), 711–713 (2018)
17. Psomiadis, E.: Flash flood area mapping utilising SENTINEL-1 radar data. In: Proceedings of SPIE, vol. 10005, p. 100051G, October 2016
18. Amitrano, D., Guida, R., Ruello, G.: Multitemporal SAR RGB processing for Sentinel-1 GRD products: methodology and applications. IEEE J. Sel. Top. Appl. Earth Obs. Remote Sens. **12**(5), 1497–1507 (2019)
19. Faour, G.: Potential of Sentinel-1 and 2 to assess flooded areas. Hydrol. Current Res. **9**, 308 (2018). https://doi.org/10.4172/2157-7587.1000308
20. Son, N.-T., Chen, C.-F., Chen, C.-R.: Flood assessment using multi-temporal remotely sensed data in Cambodia. In: Geocarto International (2019). https://doi.org/10.1080/10106049.2019.1633420
21. Henry, J., Chastanet, P., Fellah, K., Desnos, Y.: Envisat multi-polarized ASAR data for flood mapping. Int. J. Remote Sens. 1921–1929 (2006). https://doi.org/10.1080/01431160500486724
22. Twele, A., Cao, W., Plank, S., Martinis, S.: Sentinel-1-based flood mapping: a fully automated processing chain. Int. J. Remote Sens. **37**, 2990–3004 (2016)
23. Copernicus Open Access Hub. https://scihub.copernicus.eu/dhus/#/home

Mathematical Modelling of Non-stationary Processes in the Piecewise-Homogeneous Domains by Near-Boundary Element Method

Liubov Zhuravchak(✉)

Software Department of Institute of Computer Science and Information Technologies, Lviv Polytechnic National University, Bandery street 12, Lviv, Ukraine
lzhuravchak@ukr.net

Abstract. A numerical-analytical approach to finding the physical scalar values (temperature, potential, pressure) or vector values (components of an electromagnetic field) in the piecewise-homogeneous domain of arbitrary shape with mixed boundary conditions and ideal contact conditions at the interface of the media has been considered. Using the indirect near-boundary element technique and time sequence scheme of the initial conditions, the developed software, computational experiments have been carried out to estimate the errors of the discretization of the near-boundary domains and the approximation of the mathematical model. The influence of piezoelectricity coefficients on pressure distribution in composite reservoirs has been studied.

Keywords: Non-stationary processes · Mathematical modelling · Indirect near-boundary element method · Piecewise-homogeneous porous media

1 Introduction

In modern conditions, a significant number of practically important tasks is a key to study non-stationary phenomena. The simplest mathematical models of time-dependent processes of heat conduction, diffusion, hydraulics, electro-magnetism are described by researchers as solutions of linear parabolic equations of the second order for scalar or vector values at the given boundary and initial conditions.

Solving the problem of detecting, allocating and exploration of deposits of minerals in search geophysics at the present stage of development of geophysical research requires theoretical and methodological substantiation of different types of field observations data. Physical modelling and experimental studies of these processes need large material costs. Incomplete certainty of the parameters of the initial state and a small number of experimental installations often complicate the physical experiment to obtain the necessary results. Therefore, the solution of the above problem causes the construction of effective mathematical models of the corresponding physical processes and development of new numerical methods. As a consequence, the scientific studies can be conducted on this basis.

N. Shakhovska and M. O. Medykovskyy (Eds.): CCSIT 2019, AISC 1080, pp. 64–77, 2020.
https://doi.org/10.1007/978-3-030-33695-0_6

By the classical methods one can find analytical solutions of non-stationary problems, which simulate the specified physical processes, only for homogeneous media and with the inclusion of a canonical form or close to this one. For mathematical modelling in heterogeneous objects in recent years researchers are increasingly using numerical methods, oriented at modern high-speed computers.

The most commonly used finite difference methods and finite element methods are appropriate in the simulation of physical processes in continuously heterogeneous objects of finite size and give high accuracy of the results, but require a grid covering of the entire domain, occupied by the object, and significant volumes of random access memory (RAM), long time of calculation and programs of solving systems of linear algebraic equations of large dimension. The use of the method of boundary integral equations and direct or indirect methods of boundary elements, created on its basis, has a number of indisputable advantages in the simulation of processes in piecewise-homogeneous domains, since they allow precisely to satisfy the initial equations of the model.

By boundary integral equations method and all its numerical modifications we discretize only the boundary of the object and the interface of the media, which saves the amount of RAM during the operation of the algorithm and gives a relatively high accuracy of calculations at internal points [1–6]. However, when calculating unknown values near the boundaries or on the interface of the media, the accuracy of calculations is sharply reduced, and finding the derivatives on coordinates and normal from the unknown values requires a preliminary analytical allocation of the singularity (the principal value). In this regard, in many cases it is advisable to use the near-boundary element method (NBEM), which can be considered as one of the variants of the source method and attributed to indirect methods of research, since the unknown components of fictitious sources introduced for the finding of problem solution are not physical variables [7–10].

2 Mathematical Model

The piecewise-homogeneous object, modelled by the domain of arbitrary shape $\Omega \in \mathbf{R}^n, n \in \{2,3\}$, $\Omega = \Omega_1 \cup \Omega_2 \cup \Gamma_{12}$, where $\Gamma_{12} = \Gamma_1 \cap \Gamma_2$ is the interface of the media, Γ_1, Γ_2 are the boundaries of the multi-binding domains Ω_1, Ω_2, has been considered. Each domain Ω_m (m = 1, 2) has constant but different physical characteristics: a_m, λ_m or μ_m, σ_m [11].

It is assumed that unknown physical scalar value $u_0^{(m)}(x,\tau)$ (temperature or potential or pressure) or vector value $\mathbf{u}^{(m)}(x,\tau) = (u_1^{(m)}(x,\tau),\ldots,u_n^{(m)}(x,\tau))$ (components of an electromagnetic field) (m = 1, 2) satisfies the system of differential equations:

$$\mathbf{P}_{0\tau}^{(m)}(u_j^{(m)}(x,\tau)) = \Delta u_j^{(m)}(x,\tau) - \frac{1}{a_m}\frac{\partial u_j^{(m)}(x,\tau)}{\partial \tau} = 0,\ (x,\tau) \in \Omega_m \times \mathbf{T}, \qquad (1)$$
$$\Omega_m \subset \mathbf{R}^n, j = 0,\ldots,n, n = 2,3, \mathbf{T} = \{\tau : 0<\tau<\infty\},$$

where Δ is the Laplace operator, $x = (x_1, \ldots, x_n)$ are the Cartesian coordinates, τ is time.

The system is supplemented by initial conditions

$$u_j^{(m)} = u_{j0}^{(m)}(x), \quad x \in \Omega_m, \ \tau = 0, \tag{2}$$

ideal contact conditions between components

$$\mu_1 u_j^{(1)}(x,\tau) = \mu_2 u_j^{(2)}(x,\tau), \sigma_1^{-1} rot_c \mathbf{u}^{(1)}(x,\tau) = \sigma_2^{-1} rot_c \mathbf{u}^{(2)}(x,\tau), (x,\tau) \in \Gamma_{12} \times \mathbf{T}, \tag{3}$$

or

$$u_0^{(1)}(x,\tau) = u_0^{(2)}(x,\tau), \quad \lambda_1 \frac{\partial u_0^{(1)}(x,\tau)}{\partial \mathbf{n}^{(1)}(x)} = -\lambda_2 \frac{\partial u_0^{(2)}(x,\tau)}{\partial \mathbf{n}^{(2)}(x)}, \quad (x,\tau) \in \Gamma_{12} \times \mathbf{T}, \tag{4}$$

and different types of boundary conditions:

$$u_j^{(m)}(x,\tau) = f_{j\Gamma}^{(1)}(x,\tau), \ (x,\tau) \in \partial\Omega^{(1)} \times \mathbf{T}, \tag{5}$$

$$rot_c \mathbf{u}^{(m)}(x,\tau) = f_{c\Gamma}^{(2)}(x,\tau), \ (x,\tau) \in \partial\Omega^{(2)} \times \mathbf{T}, \tag{6}$$

or

$$\frac{\partial u_0^{(m)}(x,\tau)}{\partial \mathbf{n}^{(m)}(x)} = f_{0\Gamma}^{(2)}(x,\tau), \ (x,\tau) \in \partial\Omega^{(2)} \times \mathbf{T}, \tag{7}$$

where $rot_c \mathbf{u}^{(m)}(x,\tau) = \frac{\partial u_j^{(m)}(x,\tau)}{\partial x_l} - \frac{\partial u_l^{(m)}(x,\tau)}{\partial x_j}$, $c = 1, \ldots, n$; there is a certain relationship between the indexes j, l and c: for c = 1 we choose l = 3, j = 2, for c = 2 we choose l = 1, j = 3, for c = 3 we choose l = 2, j = 1; $\mathbf{n}^{(m)}(\mathbf{x}) = (n_1^{(m)}(\mathbf{x}), \ldots, n_n^{(m)}(\mathbf{x}))$ is uniquely defined unit of outer normal vector to the boundary $\partial\Omega_m$ of the domain Ω_m, $\partial\Omega^{(1)} \cup \partial\Omega^{(2)} = \partial\Omega, \partial\Omega^{(1)} \cap \partial\Omega^{(2)} = \varnothing, \partial\Omega$ is the boundary of the domain Ω, $a_m = 1/(\mu_m \sigma_m)$.

It is clear that for the description of the scalar value, for example, the non-stationary thermal field, the index j is equal 0 in the Eq. (1), in the initial conditions (2) and in the boundary condition (5). In this case, we also take into account the conditions (4) and (7). We describe the vector value by the Eq. (1), the initial conditions (2), the conditions (3) and (5), (6).

3 Integral Representation of Solutions. Boundary Integral Equations

To construct the algorithm for solving the problem (1)–(7), we consider a set $\mathbf{R}^n(2)$, composed of two planes $\mathbf{R}_m^2$ or two spaces $\mathbf{R}_m^3$, which possesses the following properties [7, 8]:

$$\mathbf{R}_m^n \cap \mathbf{R}^n = \Omega_m \cup \partial\Omega_m, \mathbf{R}_1^n \cap \mathbf{R}_2^n = \Gamma_{12}, \mathbf{R}^n(2) \cap \mathbf{R}^n = \Omega \cup \partial\Omega.$$

We introduce external near-boundary single- or dual-binding domains $G^m = B_m \backslash \Omega_m \quad (\Omega_m \subset B_m \subset \mathbf{R}_m^n, \quad \partial\Omega_m \cap \partial B_m = \varnothing)$ with unknown functions $\varphi_j^{(m)}(x, \tau)$ which describe the distribution of fictitious sources. In order to provide a monotonous change of functions $u_{j0}^{(m)}(x)$ during the transition through the boundaries $\partial\Omega_m$, we also introduce single- or dual-binding domains of extended initial conditions $\Omega_{0m} \quad (\Omega_m \subset \Omega_{0m} \subset \mathbf{R}_m^n, \partial\Omega_m \cap \partial\Omega_{0m} = \varnothing)$ [3] with known continuous functions $f_{j0}^{(m)}(x)$. These functions coincide with $u_{j0}^{(m)}(x)$ in $\Omega_m \cup \partial\Omega_m$ and are equal to zero outside Ω_{0m}, in the regions $\Omega_{0m} \backslash (\Omega_m \cup \partial\Omega_m)$ they are chosen in a form that is convenient for integration. Note that Ω_{0m} and G_m should not coincide, although they can.

Expanding the definition domain of function $u_j^{(m)}(x, \tau)$ to the entire plane or space $\mathbf{R}_m^n$, instead of (1), taking into account the extended initial condition, we obtain

$$\mathbf{P}_{0\tau}^{(m)}(u_j^{(m)}(x, \tau)) = -\varphi_j^{(m)}(x, \tau)\chi_{Gm} - f_{j0}^{(m)}(x)\delta(\tau), \ (x, \tau) \in \mathbf{R}_m^n \times \mathbf{T}, \tag{8}$$

where χ_{Gm} is a characteristic function of domain G_m that is equal to unit within this domain and to zero on its boundary and beyond, $\delta(\tau)$ is the Dirac delta function.

Since the operator $\mathbf{P}_{0\tau}^{(m)}(u_j^{(m)}(x, \tau))$ has a well-known fundamental solution (FS) $U^{(m)}(x, \tau, \xi, \xi_4)$, the integral representations of the solutions of Eq. (8) and derivatives of them on coordinates and normal, taking into account the initial distribution (2), have the form:

$$u_j^{(m)}(x, \tau) = \mathbf{F}_{j\tau}^{(m)}(x, \tau, U^{(m)}) + b_{j\tau}^{(m)}(x, \tau, U^{(m)}), \tag{9}$$

$$\partial u_j^{(m)}(x, \tau)/\partial x_l = \mathbf{F}_{j\tau}^{(m)}(x, \tau, Q_l^{(m)}) + b_{j\tau}^{(m)}(x, \tau, Q_l^{(m)}), \tag{10}$$

$$\partial u_j^{(m)}(x, \tau)/\partial \mathbf{n}^{(m)}(x) = \mathbf{F}_{j\tau}^{(m)}(x, \tau, Q^{(m)}) + b_{j\tau}^{(m)}(x, \tau, Q^{(m)}), \tag{11}$$

where $\mathbf{F}_{j\tau}^{(m)}(x, \tau, \Phi^{(m)}) = \int\limits_0^\tau \int\limits_{G_m} \Phi^{(m)}(x, \tau, \xi, \xi_4)\varphi_j^{(m)}(\xi, \xi_4)dG_m(\xi)d\xi_4$,

$b_{j\tau}^{(m)}(x, \tau, \Phi^{(m)}) = \int\limits_{\Omega_{0m}} \Phi^{(m)}(x, \tau, \xi, 0)f_{j0}^{(m)}(\xi)d\Omega_{0m}(\xi), \ \Phi^{(m)} \in \{U^{(m)}, Q_l^{(m)}, Q^{(m)}\}$,

$U^{(m)}(x, \xi, \tau, \xi_4) = \frac{1}{[4\pi a_m(\tau-\xi_4)]^{n/2}}\exp(-\frac{r^2}{4a_m(\tau-\xi_4)}), \xi = (\xi_1, \ldots, \xi_n) \in \mathbf{R}_m^n$,

$$\xi_4 \in \mathbf{T}, r^2 = \sum_{i=1}^{n} y_i^2, y_i = x_i - \xi_i,$$
$$Q_l^{(m)}(x,\tau,\xi,\xi_4) = \frac{\partial U^{(m)}(x,\tau,\xi,\xi_4)}{\partial x_l} = -\frac{U^{(m)}(x,\tau,\xi,\xi_4)y_l}{2a_m(\tau-\xi_4)},$$
$$Q^{(m)}(x,\tau,\xi,\xi_4) = \sum_{l=1}^{n} Q_l^{(m)}(x,\ \tau,\ \xi,\ \xi_4)\mathbf{n}_l^{(m)}(x).$$

Directing x in (9)–(11) from the middle of the domain Ω_m to its boundary, we obtain boundary integral equations that connect unknown functions $\varphi_j^{(m)}(\xi,\xi_4)$ with the known ones $f_{j0}^{(m)}(\xi)$ and those $f_{j\Gamma}^{(1)}(x,\tau)$, $f_{c\Gamma}^{(2)}(x,\tau)$, $f_{0\Gamma}^{(2)}(x,\tau)$, given on the outer boundary, to satisfy the ideal contact conditions (3), (4) and boundary conditions (5)–(7):

$$\begin{aligned}
&\mu_1\mathbf{F}_{j\tau}^{(1)}(x,\tau,U^{(1)}) - \mu_2\mathbf{F}_{j\tau}^{(2)}(x,\tau,U^{(2)}) = -\mu_1 b_{j\tau}^{(1)}(x,\tau,U^{(1)}) + \mu_2 b_{j\tau}^{(2)}(x,\tau,U^{(2)}),\\
&\frac{1}{\sigma_1}[\mathbf{F}_{j\tau}^{(1)}(x,\tau,Q_l^{(1)}) - \mathbf{F}_{l\tau}^{(1)}(x,\tau,Q_j^{(1)})] - \frac{1}{\sigma_2}[\mathbf{F}_{j\tau}^{(2)}(x,\tau,Q_l^{(2)}) - \mathbf{F}_{l\tau}^{(2)}(x,\tau,Q_j^{(2)})]\\
&= -\frac{1}{\sigma_1}[b_{j\tau}^{(1)}(x,\tau,Q_l^{(1)}) - b_{l\tau}^{(1)}(x,\tau,Q_j^{(1)})] + \frac{1}{\sigma_2}[b_{j\tau}^{(2)}(x,\tau,Q_l^{(2)}) - b_{l\tau}^{(2)}(x,\tau,Q_j^{(2)})],\\
&\qquad (x,\tau) \in \Gamma_{12}\times\mathbf{T},
\end{aligned} \tag{12}$$

$$\begin{aligned}
&\mathbf{F}_{0\tau}^{(1)}(x,\tau,U^{(1)}) - \mathbf{F}_{0\tau}^{(2)}(x,\tau,U^{(2)}) = -b_{0\tau}^{(1)}(x,\tau,U^{(1)}) + b_{0\tau}^{(2)}(x,\tau,U^{(2)}),\\
&\lambda_1\mathbf{F}_{0\tau}^{(1)}(x,\tau,Q^{(1)}) + \lambda_2\mathbf{F}_{0\tau}^{(2)}(x,\tau,Q^{(2)}) = -\lambda_1 b_{0\tau}^{(1)}(x,\tau,Q^{(1)}) - \lambda_2 b_{0\tau}^{(2)}(x,\tau,Q^{(2)}),\\
&\qquad (x,\tau) \in \Gamma_{12}\times\mathbf{T},
\end{aligned} \tag{13}$$

$$\mathbf{F}_{j\tau}^{(m)}(x,\tau,U^{(m)}) = f_{j\Gamma}^{(1)}(x,\tau) - b_{j\tau}^{(m)}(x,\tau,U^{(m)}), \quad (x,\tau) \in \partial\Omega^{(1)}\times\mathbf{T} \tag{14}$$

$$\mathbf{F}_{j\tau}^{(m)}(x,\tau,Q_l^{(m)}) - \mathbf{F}_{l\tau}^{(m)}(x,\tau,Q_j^{(m)}) = f_{c\Gamma}^{(2)}(x,\tau) - b_{j\tau}^{(m)}(x,\tau,Q_l^{(m)}) + b_{l\tau}^{(m)}(x,\tau,Q_j^{(m)}), \tag{15}$$

$$\mathbf{F}_{0\tau}^{(m)}(x,\tau,Q^{(m)}) = f_{0\Gamma}^{(2)}(x,\tau) - b_{0\tau}^{(m)}(x,\tau,Q^{(m)}), \quad (x,\tau) \in \partial\Omega^{(2)}\times\mathbf{T}. \tag{16}$$

4 Spatial-Time Discretization of the Mathematical Model

Since it is practically impossible to carry out analytical integration in Eqs. (12)–(16) for applied problems due to the arbitrariness of the domain Ω and functions $\varphi_j^{(m)}(\xi,\xi_4)$ and $f_{j0}^{(m)}(\xi)$, we perform spatial-time discretization by means of such steps. We discretize the boundaries $\partial\Omega_m$ and domains G_m into V_m boundary elements Γ_{mv} and near-boundary elements G_{mv}, respectively. The elements do not intersect each other and $\cup_{v=1}^{V_m}\Gamma_{mv} = \partial\Omega_m$, $\cup_{v=1}^{V_m}G_{mv} = G_m$. Unknown function that describe the distribution of fictitious source onto the near-boundary element G_{mv}, we denote by $\varphi_{jv}^{(m)}(x,\tau)$.

The domains Ω_{0m} are also discretized into the second order elements [3] $\Omega_{0mq}(q = 1, \ldots, Q_{0m})$.

For convenience of description of dependence of functions $\varphi_{jv}^{(m)}(x, \tau)$ on time, we divide the interval **T** into the same intervals $T_k = [t_{k-1}, t_k]$ $(k = 1, 2, \ldots, t_0 = 0)$, and within each T_k these functions are approximated by constants d_{jv}^{mk}. We will construct a discrete-continuum model for a step-by-step time scheme: sequence scheme of initial conditions [2, 3, 7, 8]. Note that each time step T_k is considered as a new problem, i.e. the local time $\tilde{\tau} = \tau - (k-1)\Delta\tau$ ($\Delta\tau$ is the value of the time increment at each step) is introduced and the values $u_j^{(m)}(x, (k-1)\Delta\tau)$ at the end of the $(k-1)$-th time interval in the internal points are used as initial values for the next k-th step. The time integration of the FS and its derivatives is carried out analytically.

Operators $\mathbf{F}_{j\tau}^{(m)}(x, \tau, \Phi^{(m)})$ and $b_{j\tau}^{(m)}(x, \tau, \Phi^{(m)})$ after spatial-time discretization for the K-th step $((K-1)\Delta\tau < \tau \leq K\Delta\tau,\ 0 < \tilde{\tau} \leq \Delta\tau)$ are obtained as follows:

$$\mathbf{F}_j^{mK}(x, \tilde{\tau}, \Phi_\tau^{(m)}) = \sum_{v=1}^{V_m} A_v^{(m)}(x, \tilde{\tau}, \Phi_\tau^{(m)}) d_{jv}^{mK},$$

$$b_j^{mK}(x, \tilde{\tau}, \Phi^{(m)}) = \sum_{q=1}^{Q_{0m}} \int_{\Omega_{0mq}} \Phi^{(m)}(x, \tilde{\tau}, \xi, 0) f_{j0}^{(mK)}(\xi) d\Omega_{0mq}(\xi),$$

where $f_{j0}^{(m1)}(x) = f_{j0}^{(m)}(x),\ f_{j0}^{(mk)}(x) = u_j^{(m)}(x, (k-1)\Delta\tau), k > 1,\ x \in \Omega_m \cup \partial\Omega_m$;

$$A_v^{(m)}(x, \tilde{\tau}, \Phi_\tau^{(m)}) = \int_{G_{mv}} \Phi_\tau^{(m)}(x, \tilde{\tau}, \xi) dG_{mv}(\xi);\ \Phi_\tau^{(m)}(x, \tilde{\tau}, \xi) = \int_0^{\tilde{\tau}} \Phi^{(m)}(x, \tilde{\tau}, \xi, \xi_4) d\xi_4,$$

$0 < \xi_4 \leq \Delta\tau.$

To find unknown constants d_{jv}^{mK}, a system of linear algebraic equations (SLAE) has been constructed, requiring satisfaction of the ideal contact conditions (12), (13) and boundary conditions (14)–(16) in the collocation sense:

$$
\begin{aligned}
&\mu_1 \sum_{v=1}^{V_1} A_v^{(1)}(x^{1w}, \Delta\tau, U_\tau^{(1)}) d_{jv}^{1K} - \mu_2 \sum_{v=1}^{V_2} A_v^{(2)}(x^{1w}, \Delta\tau, U_\tau^{(2)}) d_{jv}^{2K} \\
&\quad = -\mu_1 b_j^{1K}(x^{1w}, \Delta\tau, U^{(1)}) + \mu_2 b_j^{2K}(x^{1w}, \Delta\tau, U^{(2)}), \\
&\frac{1}{\sigma_1}[\sum_{v=1}^{V_1} A_v^{(1)}(x^{1w}, \Delta\tau, Q_{\tau l}^{(1)}) d_{jv}^{1K} - \sum_{v=1}^{V_1} A_v^{(1)}(x^{1w}, \Delta\tau, Q_{\tau j}^{(1)}) d_{lv}^{1K}] \\
&-\frac{1}{\sigma_2}[\sum_{v=1}^{V_2} A_v^{(2)}(x^{1w}, \Delta\tau, Q_{\tau l}^{(2)}) d_{jv}^{2K} - \sum_{v=1}^{V_2} A_v^{(2)}(x^{1w}, \Delta\tau, Q_{\tau j}^{(2)}) d_{lv}^{2K}] \\
&\quad = -\frac{1}{\sigma_1}[b_j^{1K}(x^{1w}, \Delta\tau, Q_l^{(1)}) - b_{l\tau}^{1K}(x^{1w}, \Delta\tau, Q_j^{(1)})] \\
&+\frac{1}{\sigma_2}[b_j^{2K}(x^{1w}, \Delta\tau, Q_l^{(2)}) - b_l^{2K}(x^{1w}, \Delta\tau, Q_j^{(2)})], x^{1w} \in \Gamma_{1w} \subset \Gamma_{12},
\end{aligned}
\tag{17}
$$

$$
\begin{aligned}
&\sum_{v=1}^{V_1} A_v^{(1)}(x^{1w}, \Delta\tau, U_\tau^{(1)}) d_{0v}^{1K} - \sum_{v=1}^{V_2} A_v^{(2)}(x^{1w}, \Delta\tau, U_\tau^{(2)}) d_{0v}^{2K} \\
&\quad = -b_0^{1K}(x^{1w}, \Delta\tau, U^{(1)}) + b_0^{2K}(x^{1w}, \Delta\tau, U^{(2)}), \\
&\lambda_1 \sum_{v=1}^{V_1} A_v^{(1)}(x^{1w}, \Delta\tau, Q_\tau^{(1)}) d_{0v}^{1K} + \lambda_2 \sum_{v=1}^{V_2} A_v^{(2)}(x^{1w}, \Delta\tau, Q_\tau^{(2)}) d_{0v}^{2K} \\
&= -\lambda_1 b_0^{1K}(x^{1w}, \Delta\tau, Q^{(1)}) - \lambda_2 b_0^{2K}(x^{1w}, \Delta\tau, Q^{(2)}), x^{1w} \in \Gamma_{1w} \subset \Gamma_{12},
\end{aligned}
\tag{18}
$$

$$
\sum_{v=1}^{V_m} A_v^{(m)}(x^{mw}, \Delta\tau, U_\tau^{(m)}) d_{jv}^{mK} = f_{j\Gamma}^{(1)}(x, \tau) - b_j^{mK}(x^{mw}, \Delta\tau, U^{(m)}), x^{mw} \in \Gamma_{mw} \subset \partial\Omega^{(1)}, \tag{19}
$$

$$
\begin{aligned}
&\sum_{v=1}^{V_m} A_v^{(m)}(x^{mw}, \Delta\tau, Q_{\tau l}^{(m)}) d_{jv}^{mK} - \sum_{v=1}^{V_m} A_v^{(m)}(x^{mw}, \Delta\tau, Q_{\tau j}^{(m)}) d_{lv}^{mK} \\
&= f_{c\Gamma}^{(2)}(x^{mw}, \Delta\tau) - b_j^{mK}(x^{mw}, \Delta\tau, Q_l^{(m)}) + b_l^{mK}(x^{mw}, \Delta\tau, Q_j^{(m)}), x^{mw} \in \Gamma_{mw} \subset \partial\Omega^{(2)},
\end{aligned}
\tag{20}
$$

$$
\sum_{v=1}^{V_m} A_v^{(m)}(x^{mw}, \Delta\tau, Q_\tau^{(m)}) d_{0v}^{mK} = f_{0\Gamma}^{(2)}(x^{mw}, \Delta\tau) - b_0^{mK}(x^{mw}, \Delta\tau, Q^{(m)}). \tag{21}
$$

Collocation points are selected in the middle of each boundary element $\Gamma_{mw} = \partial G_{mw} \cap \partial\Omega_m, w = 1, \ldots, V_m$. After finding unknown constants d_{jv}^{mK} from the system (17), (19), (20) for vector value or from the system (18), (19), (21) for scalar value, all zones Ω_m are considered as completely independent domains. Then, the solutions of SLAE are substituted in (9) to computate the desired values at internal points at time $K\,\Delta\tau$. These values are used to form the right side of the SLAE in the next $(K + 1)$-th time step:

$$u_j^{(mK)}(x^{mz}, \Delta\tau) = \sum_{v=1}^{V_m} A_v^{(m)}(x^{mz}, \Delta\tau, U_\tau^{(m)}) d_{jv}^{mK} + b^{mK}(x^{mz}, \Delta\tau, U^{(m)}), \;\; x^{mz} \in \Omega_m.$$

5 The Numerical Investigations

The developed approach has been tested to find change pressure $p^{(m)}(x,\tau) = u_0^{(m)}(x,\tau)$ in a piecewise-homogeneous reservoir.

In oil and gas industry, what matters the most is the heterogeneity of the filtration-capacitive properties, especially permeability, since it determines the ratio of oil and gas flows to the wells, and, therefore, impacts the development of the deposit. The model of the homogeneous unlimited medium does not always adequately describe the real environment and the processes occurring in it while extracting oil and gas. In order to evaluate the influence of the hydro and piezoelectricity coefficients of different zones, it is necessary to consider an inhomogeneous reservoir model that takes into account the spatial change of its structural-facial and lithologic-physical characteristics. At the same time, these coefficients of the part of the reservoir near the well (near-well bore zone), as a rule, are different from the coefficients in the rest of the reservoir. This change is due to pollution (colmatation) of pores while well drilling and operation or purposeful use of methods of intensification of well productivity, products of corrosion of equipment, precipitates formed in the interaction of reservoir water and impregnated solutions. To assess the effect of such a change in the coefficients in near-well bore zone as a result of colmatation or intensification of the pressure or the flow of the well, it is also necessary to consider the model of a piecewise-homogeneous reservoir [10].

The reservoir of thickness h = 18 m is modeled by an elliptic domain with semiaxis a = 200 m and b = 150 m, and a center at the beginning of the Cartesian coordinate system. The study of the influence of zonal inhomogeneity was carried out for two typical reservoir models: with vertical contact (VC) and with curvilinear contact (CC). The research of colmatation was conducted for one model (CZ) (see Fig. 1).

The reservoir contains a central well of radius r_c = 0.1 m with center coordinates $x_{1c1} = -175,\ x_{2c1} = 0$ and two other wells, located near the outer contour, radii r_c with center coordinates $x_{1c2} = -175,\ x_{2c2} = 0,\ x_{1c3} = -175,\ x_{2c3} = 0$. In the case of VC, the central well is located on the interface of media. In the cases of CC and CZ, a zone Ω_2 with other physical characteristics is located around the central well, the interface of the media Γ_{12} is an ellipse with semiaxis $a_2 = 50$ m and b_2 = 37.5 m, and a circle of radius r_2 = 1 m, respectively.

A software of the proposed numerically-analytical approach was implemented and a ranges of studies of the influence of physical characteristics of components on the pressure change was carried out. The thickness of the near-boundary domains and of extended initial condition domains were chosen to be equal to 20 m (for external contour), 0.05 m (for wells) and 20 m (for interface of the media). The time interval (240 h) was evenly divided into 40 intervals with a constant time step (6 h) for the case of VC and on 100 intervals (2.4 h) for cases of CC and CZ.

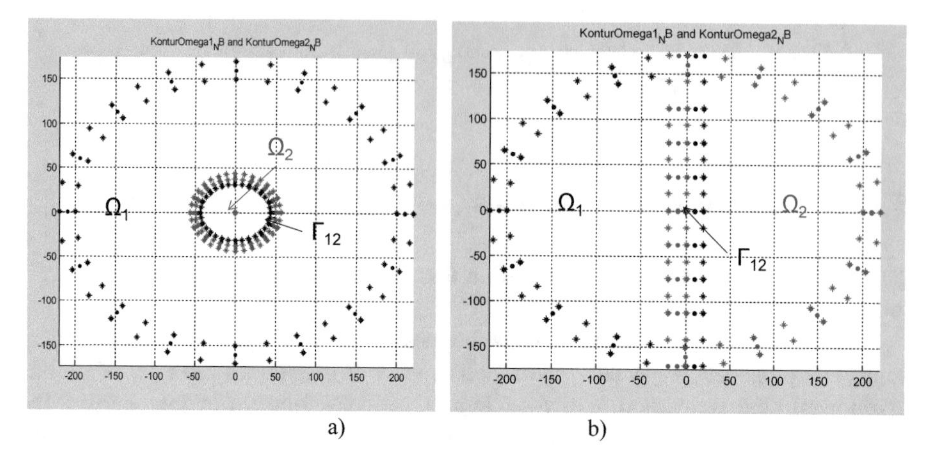

Fig. 1. Typical reservoir models with near-boundary elements: curvilinear (a) and vertical contacts (b)

The outer boundary is evenly discretized into 16 boundary elements, the central well contour – into 4 (in the case of VC) or 2 (in the cases of CC and CZ) elements, the contours of other two wells – into 2 elements, the interface of the media – into 8 elements. The SLAE consisted of 36 (for one well) or 40 (for three wells) equations in the VC model and of 34 (for one well) or 38 (for three wells) equations in the models of CC and CZ.

Domains Ω_m and Ω_{mp} are discretized into 16 and 18 internal elements respectively, it is clear that this quantity had no influence on the matrix dimension, since the integrals from the initial condition are included in the right side of the SLAE.

Numerical values of physical characteristics of a piecewise-homogeneous medium are selected as follows:

$$\begin{aligned} k_1 &= 0.16 \cdot 10^{-12}\,\mathrm{m}^2, \mu_1 = 0.707 \cdot 10^{-3}\mathrm{Pa \cdot s}, \beta_1^* = 7.07 \cdot 10^{-10}\mathrm{Pa}^{-1}, \kappa_1 = 0.32\,\mathrm{m}^2/\mathrm{s} \\ k_2 &= 0.08 \cdot 10^{-12}\,\mathrm{m}^2, \mu_2 = 1.232 \cdot 10^{-3}\mathrm{Pa \cdot s}, \beta_2^* = 4.06 \cdot 10^{-10}\mathrm{Pa}^{-1}, \kappa_2 = 0.16\,\mathrm{m}^2/\mathrm{s} \end{aligned} \quad (22)$$

where k_m is a coefficient of permeability; μ_m is a dynamic coefficient of liquid viscosity; β_m^* is a coefficient of elastic capacity of the saturated reservoir; $\kappa_m = a_m = k_m/(\mu_m\beta_m^*)$ is a piezoelectricity coefficient; $\varepsilon_m = \lambda_m = k_m h/\mu_m$ is a coefficient of hydraulic conductivity of the formation.

For the **VC model** on the outer boundary pressure and on the contours of well flow depended on time:

$$f_{0\Gamma}^{(1)}(x,\tau) = p_z^{(m)}(x,\tau) = \frac{P_z - 0.001k(\Delta\tau)\cdot 10^6}{86400}; P_z = 20 \cdot 10^6; \quad (23)$$

$f_{0\Gamma}^{(2)}(x,\tau) = q_{cl}^{(m)}(x,\tau) = Q^{(m)}(\tau) = \frac{Q_{cl}-0.05k}{2\pi R_c \cdot 86400}$ is the flow of the l-th well, l = 1, …, L,

$$Q_{c1} = 0.347 \cdot 10^{-3}\text{m}^{-3}/\text{s}, Q_{c2} = Q_{c3} = 0.324 \cdot 10^{-3}\text{m}^{-3}/\text{s}.$$

Initial pressure distribution in the reservoir was selected by constant and equal to the outer contour pressure:

$$u_{00}^{(m)}(x) = p_0^{(m)}(x) = P_z.$$

For **models of CC and CZ** pressure on the outer boundary and on the contours of wells depended on time by formulas: (8) and the following one:

$$p_{cl}^{(m)}(x,\tau) = \frac{P_{0l}^{(m)} - 0.05k(\Delta\tau)\cdot 10^6}{86400}, P_{01}^{(2)} == 16\cdot 10^6, P_{02}^{(1)} = P_{03}^{(1)} = 17\cdot 10^6.$$

The initial pressure distribution in the reservoir, taking into account the consistency of the initial and boundary conditions, was modeled as a function:

$$p_0^{(m)}(x) = P_z - \frac{(P_z - P_{0l}^{(m)})(r - r_{el})}{(r_c - r_{el})},$$

where r, r_{el} are the distances in polar coordinates to the point x and to the corresponding one on the outer ellipse.

Since the errors occurring during the application of NBEM only due to procedures of approximation, discretization and numerical integration [7–9], the accuracy of satisfaction of boundary conditions and ideal contact conditions at different moments of time was firstly investigated, a part of the obtained results is shown in Fig. 2 for the CZ model at time $\Delta\tau = 8640$ c (because of the symmetry of the object, a quarter of the outer boundary and a half of the central well contour are given):

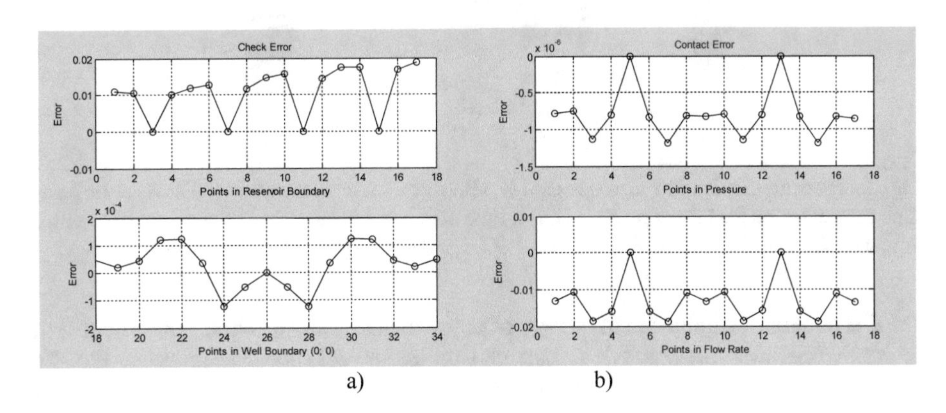

Fig. 2. Relative errors of satisfaction of the boundary conditions (a) and ideal contact conditions (b) for the model CZ

Note that there are no errors in the points of the collocation, the greatest ones are observed when we are approaching the ends of the boundary elements. In the middle of the domain, the errors are less due to the implementation of the maximum principle.

The pressure distribution into the reservoir with three wells at different time moments, as well as the interface of the media (vertical or curvilinear contact) and the location of the wells are shown in Fig. 3.

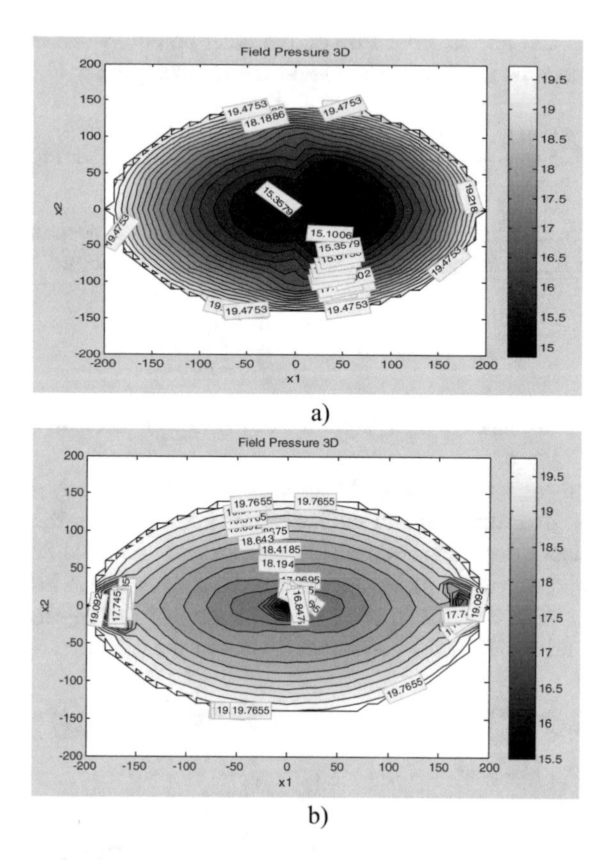

Fig. 3. Pressure distribution in the reservoir with three wells, received by NBEM, at different time moments: vertical contact, 60 h with a time step 6 h (a); curvilinear contact, 240 h with a time step 2.4 h (b).

The cumulative influence of change of physical characteristics in Ω_1 on the value of reservoir pressure was studied. A part of studies for VC model is given in Fig. 4a, where the curves without symbols correspond to a piecewise-homogeneous reservoir with values (22), the curves with the characters "o" and "+" correspond to the following values: $k_1 = 0.12 \cdot 10^{-12}\,\mathrm{m}^2, \mu_1 = 0.898 \cdot 10^{-3}\mathrm{Pa \cdot s}, \beta_1^* = 5.56 \cdot 10^{-10}\mathrm{Pa}^{-1}$, $\kappa_1 = 0.24\,\mathrm{m}^2/\mathrm{s}$ and $k_1 = 0.04 \cdot 10^{-12}\,\mathrm{m}^2, \mu_1 = 1.956 \cdot 10^{-3}\mathrm{Pa \cdot s}, \beta_1^* = 2.55 \cdot 10^{-10}$ $\mathrm{Pa}^{-1}, \kappa_1 = 0.08\,\mathrm{m}^2/\mathrm{s}$ (characteristics in Ω_2 did not change), respectively.

In Fig. 4b, the pressure distribution in the circle Ω_2 (model of CZ) is shown, when the piezoelectricity coefficient has been changed. Solid line corresponds to a piecewise-homogeneous reservoir with values (7). Dotted line corresponds to a homogeneous medium with values (7), as in Ω_1. Line with the symbols "o" represents the case when physical characteristics in the domain Ω_2 are one and a half times smaller than in the domain Ω_1.

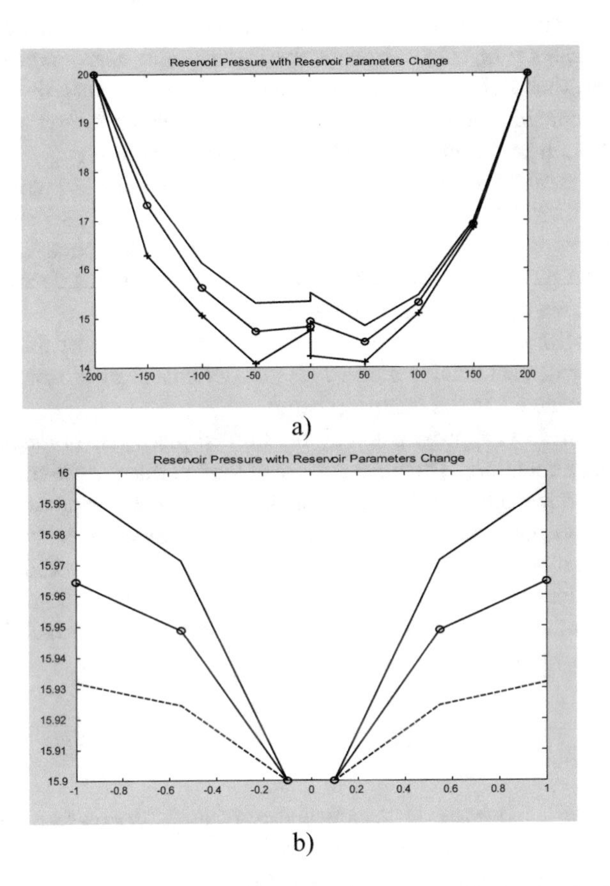

Fig. 4. Influence of piezoelectricity coefficients on the pressure distribution in the time of 48 h at the interior points of a piecewise-homogeneous reservoir with a central well (a) and in the domain Ω_2 (b)

As we can see, decreasing of the piezoelectricity coefficient in Ω_1 leads to lowering of the pressure at the points belonging to Ω_1, Ω_2. A similar conclusion can be drawn to a homogeneous media, but in such a reservoir pressure decreases more sharply. Decreasing of the piezoelectricity coefficient in Ω_2 leads to lowering of the pressure at the points belonging to Ω_2.

6 Conclusion

An indirect near-boundary element method has a number of undeniable advantages for the modeling of non-stationary processes in piecewise-homogeneous domains, since it precisely satisfies the original equations of the model and allows to simulate the complex surface of geological objects located in the Earth's crust. It requires discretization of the near-boundary domains and extended initial the domains only. Also, it gives high accuracy of calculations at the internal and near-boundary points. The ability to choose time step, the number of near-boundary elements, the thickness of the near-boundary domains and the extended initial domains allowed to control the error of satisfaction of boundary conditions and ideal contact conditions.

Integral representations of the initial differential equations are written through the convolution of their fundamental singular solutions with the intensities of fictitious sources distributed within the external boundary domain. The functions of intensity do not have a certain physical meaning, but when they are found, the desired values inside the reservoir can be obtained using simple integration.

The use of NBEM does not require a preliminary analysis of the singularity (the principal value) when calculating derivatives on coordinates and normal of FS, since all integrals are considered in the Riemann sense.

Real environment is approximated by the limited piecewise-homogeneous domain with a curvilinear boundary. The ideal contact at the interface of the media, the number of wells, the impermeability of the outer contour, the pressure depression, and the flow of wells was taken into account in the mathematical model of pressure change. The software of the proposed numerically-analytical approach was carried out. A range of studies of the influence of physical and time parameters on the pressure distribution in the near-well bore zone was provided. It allows to evaluate the possibility of working intensification of wells.

References

1. Beer, G., Smith, I., Duenser, C.: The Boundary Element Method with Programming: for Engineers and Scientists. Springer, Wienn (2008)
2. Banerjee, P.K., Butterfield, R.: Boundary Element Method in Engineering Science. McGraw-Hill Inc., London (1984)
3. Brebbia, C.A., Telles, J.C.F., Wrobel, L.C.: Boundary Element Techniques, Theory and Application in Engineering. Springer, Heidelberg (1984)
4. Zhang, Y., Qu, W., Chen, J.: A new regularized BEM for 3D potential problems. Sci. Sin. Phys. Mech. Astron. **43**, 297 (2013)
5. Qu, W., Chen, W., Fu, Z.: Solutions of 2D and 3D non-homogeneous potential problems by using a boundary element-collocation method. Eng. Anal. Bound. Elem. **6**, 2–9 (2015)
6. Integrated engineering software about boundary-element method. http://www.integratedsoft.com/technology/BEM.aspx
7. Zhuravchak, L.M., Zabrodska, N.V.: Nonstationary thermal fields in inhomogeneous materials with nonlinear behavior of the components. Physicochem. Mech. Mater. (Mater. Sci.) **1**, 33–41 (2010)

8. Zhuravchak, L., Kruk, O.: Solving 3D problems of potential theory in piecewise-homogeneous media by using indirect boundary and near-boundary element methods. JCPEE **6**(2), 117–127 (2016)
9. Zhuravchak, L., Struk, A., Struk, E.: Modeling of nonstationary process of reservoir pressure change in piecewise-homogeneous medium with nonlinear behaviour of regions materials. Int. J. Adv. Res. Comput. Eng. Technol. **5**(5), 1439–1449 (2016)
10. Boyko, V.S.: Underground Hydrogasdynamics. Apriori, Lviv (2007). (in Ukrainian)
11. Zhuravchak, L.: Computation of pressure change in piecewise homogeneous reservoir for elastic regime by indirect near-boundary element method. In: Proceedings of International Scientific Conference "Computer sciences and information technologies" (CSIT-2019), vol. 1, pp. 141–144. IEEE (2019)

Modified Asymptotic Method of Studying the Mathematical Model of Nonlinear Oscillations Under the Impact of a Moving Environment

Petro Pukach[1(✉)], Volodymyr Il'kiv[2], Zinovii Nytrebych[2], Myroslava Vovk[2], and Pavlo Pukach[3]

[1] Department of Computational Mathematics and Programming, Lviv Polytechnic National University, Lviv, Ukraine
ppukach@gmail.com

[2] Department of Mathematics, Lviv Polytechnic National University, Lviv, Ukraine
ilkivv@i.ua, znytrebych@gmail.com, mira.i.kopych@gmail.com

[3] Department of Artificial Intelligence Systems, Lviv Polytechnic National University, Lviv, Ukraine
pavlopukach@gmail.com

Abstract. Wave theory of movement is used to study the mathematical model of a physical system which describes oscillations of a one-dimensional elastic body under the impact of a moving continuous flow of a homogeneous environment. This model accounts for nonlinear elastic properties of the body at transverse oscillations, as well as environment density and movement velocity. Oscillation amplitude and frequency variation laws in nonresonant modes and under the impact of harmonic perturbation are obtained. Variation laws of the aforesaid parameters are defined by geometrical characteristics of the elastic body, physical and mechanical properties of the material, the velocity of the moving environment, the angular velocity of elastic body rotation, and external factors.

Keywords: Mathematical model · Physical system · Dynamical process · Nonlinear oscillations · Wave theory

1 Introduction. Problem Topicality. Literature Review

Nonlinear oscillation systems and mathematical models of nonlinear oscillations have been studied intensively in recent years [1–10]. It is worth noting that these studies are carried out both for the systems with a finite number of freedom degrees [11, 12] and for those with distributed parameters [13]. Qualitative [14–17] and analytical-numerical approaches [18–25] are used for this purpose. Various modifications of asymptotic methods in nonlinear mechanics [26–29] possess a special role among the analytical approximation methods of analyzing nonlinear systems with distributed parameters.

N. Shakhovska and M. O. Medykovskyy (Eds.): CCSIT 2019, AISC 1080, pp. 78–89, 2020.
https://doi.org/10.1007/978-3-030-33695-0_7

With the help of non-linear equations (with weak and strong nonlinearity), mathematical models of longitudinal and bending oscillations of one-dimensional systems (rods, rolls, beams, etc.) are described under the elasticity law close to linear or elasticity law very different from linear. At the same time, the use of some these approaches resulted in resolving a great number of important applications [30]. In particular, various combinations of asymptotic and numerical methods of studying nonlinear oscillation systems are widely used in computer sciences, MEMS technologies [31], telecommunications [32] etc.

Nonlinear oscillation systems were researched using the combination of the principle of single-frequency of oscillations and asymptotic methods of nonlinear mechanics in a number of studies [33, 34]. The study [33] suggests a methodology of researching dynamic processes in systems with distributed parameters for nonlinear hyperbolic equations describing oscillatory processes on the basis of excitation method. In [34], unsteady longitudinal and transverse oscillations of nonlinearly elastic rods and oscillations of rods under the impact of central and side loading etc. are studied with the help of asymptotic methods. The results obtained in the two aforesaid studies consist in the generalization of the single-frequency method for quasilinear partial differential equations. Wave theory methods of researching dynamic processes in the case of different environments and systems are studied combining with asymptotic methods of nonlinear mechanics. To use main ideas of wave theory makes sense in such applications, since classical methods of integration of the partial differential equations cannot be realized. Primarily it concerns to the problems describing dynamic processes of longitudinally-moving environments. The longitudinal component of the movement velocity of the environment influents seriously on the qualitative side of the process, generating an unstable oscillatory process. In [33, 34], a research methodology for certain classes of the aforesaid type systems is proposed, in particular, for elastic bodies rotating around a fixed axis with a constant angular speed and axial motion of the environment. Nonlinear oscillation systems were researched using the combination of the principle of single-frequency of oscillations and asymptotic methods of nonlinear mechanics in a number of studies [33, 34]. The study [33] suggests a methodology of researching dynamic processes in systems with distributed parameters for nonlinear hyperbolic equations describing oscillatory processes on the basis of excitation method. In [34], unsteady longitudinal and transverse oscillations of nonlinearly elastic rods and oscillations of rods under the impact of central and side loading etc. are studied with the help of asymptotic methods. The results obtained in the two aforesaid studies consist in the generalization of the single-frequency method for quasilinear partial differential equations.

In this study, methods of researching dynamic processes in different environments and systems on the basis of wave theory in combination with asymptotic methods of nonlinear mechanics are described. The main ideas of wave theory are widely used in such applications, where classical methods of integration of partial differential equations cannot be used. This primarily applies to problems, which describe dynamic processes of longitudinally-moving environments. The longitudinal component of the movement velocity of the environment affects not only quantitative characteristics of oscillatory systems, but can also have a serious impact on the qualitative side of the process, i.e. lead to an unstable oscillatory process. In [33, 34], a research methodology for certain classes of the aforesaid type of systems is developed, in particular for elastic bodies rotating around an immovable axis with a constant angular velocity and along which the environment is moving. The methodology is based on the key idea of asymptotic integration of partial differential equations, which combines the main ideas of wave theory of movement, and the principle of single-frequency of oscillations in nonlinear systems. Dynamic (in particular, resonant) processes in elastic bodies, along which the flow is moving, are insufficiently studied primarily because of the absence of a device to analyze at least their linear mathematical models. Since mathematical models of such systems are widely used to analyze technological oscillatory systems, various approaches (numerical and analytical) have been developed in recent decades to study linear and nonlinear models of the aforesaid systems. To partially solve it, this study suggests an approach whose main idea lies in the creation of a methodology of research that combines the advantages of asymptotic approach and wave theory of movement.

2 Nonresonant Oscillations of an Elastic Body, Along Which a Continuous Flow of a Homogeneous Environment Is Moving

To study the impact of a moving homogeneous environment flow on nonlinear oscillations of an elastic body, it is necessary to build a solution for the excited boundary value problem first [33]

$$u_{tt} + \beta^2 u_{xx} + \alpha^2 u_{xxxx} = -\varepsilon F(u, \theta, u_t, u_{xt}, u_{xx}, u_{xxx}, u_{xxxx}), \tag{1}$$

where $F(u, \theta, u_t, u_{xt}, u_{xx}, u_{xxx}, u_{xxxx})$ – analytical 2π- periodic $\theta = \nu t$ function, with ν – external perturbation frequency; $\beta^2 = \frac{N - m_1 V^2}{m + m_1}$, α^2 – some positive constant, which depends on physical and mechanical parameters of the body. In Eq. (1), $u(x, t)$ – elastic body movement at an arbitrary point of time with coordinate x; m – mass of unit length of elastic body; m_1 – mass of unit length of the conditional material line of a homogeneous environment flow moving along the body; E – elastic modulus of body material; I – moment of inertia of body cross section; $q(u, x, t)$ – intensity of the resultant of external forces affecting a conditionally selected body element; V – constant movement velocity of homogeneous environment along the elastic body; N – compression force. In the right part of Eq. (1), the function

$$F(u, \theta, u_t, u_{xt}, u_{xx}, u_{xxx}, u_{xxxx}) = -\frac{EI}{m + m_1}\left((u_{xx})^3\right)_{xx} - \frac{m_1}{m + m_1} 2V u_{xt} + \frac{m_1}{m + m_1} q(u, x, t)$$

describes nonlinear constituents of the restoring force, the resistance force, and other forces, whose maximum value is considerably lower than the restoring force value, which is shown by a small parameter ε. The aforesaid function is decomposed in a row by degrees of the small parameter ε.

For Eq. (1), let us consider boundary conditions that correspond to pin-edge fixed ends of the elastic body with the length l, i.e.

$$u(0,t) = u(l,t) = 0, \ u_{xx}(0,t) = u_{xx}(l,t) = 0. \quad (2)$$

Problems (1) and (2) will be solved with the following additional simplifications of the right side (1) in the mathematical model of oscillations: the maximum value of inertia forces of the continuous flow of the moving environment is small compared to the maximum value of magnitude $\alpha^2 u_{xxxx}$. Such simplification allows applying general ideas of excitation methods to build a solution of the aforesaid boundary value problem. In the first asymptotic approximation, the unifrequency process of the elastic body is described with the following dependence

$$u(x,t) = a(\cos(kx+\psi) - \cos(kx-\psi)) + \mu u_1(a,x,\psi,\theta), \ k = 0,1,2,\ldots, \quad (3)$$

where $\psi = \omega t + \varphi$, ω, μ – frequencies, parameters a and φ are unknown time functions, $u_1(a,x,\psi,\theta)$ – 2π - periodic ψ and θ function, which is defined in such a way that asymptotic presentation of solution within the accuracy of second order infinitesimals can satisfy the original equation and the boundary conditions. Besides, the aforesaid function must satisfy the boundary conditions associated with (2), i.e.

$$u_1(a,x,\psi,\theta)|_{x=0} = u_1(a,x,\psi,\theta)|_{x=l}. \quad (4)$$

Unlike linear ones, nonlinear oscillations have a lot of peculiarities. Technically, the impact of nonlinear forces is manifested in time dependence of defining parameters of the dynamic process. Therefore, parameters a and ψ in asymptotic presentation of solution (3) are considered time-varying magnitudes and they account for the impact of nonlinear periodic forces, as well as the homogeneous environment flow movement along the body. The unknown variation laws for parameters a and ψ are found as laws defined by ordinary differential equations [34]

$$\frac{da}{dt} = \varepsilon A_1(a) + \ldots, \quad \frac{d\psi}{dt} = \omega_k + \varepsilon B_1(a) + \ldots, \quad (5)$$

where ω_k – self-oscillation frequency. Functions $A_1(a)$ and $B_1(a)$ are found in such a way that asymptotic image (3) with regard to (5) with a satisfactory accuracy can make a solution for the original problem. Finding variation laws for parameters a and ψ for the general case partially solves the set problem. Therefore, the main focus shall be on defining variation laws for these elastic body oscillation parameters. Since the elastic body is subject to the impact of external periodic excitation with frequency ν, its impact on the elastic body dynamic process depends greatly on the ratio between the aforesaid magnitude and self-oscillation frequency ω_k. If between the aforesaid frequencies there is a relation of the type $p\nu \approx r\omega_k$ (p and r – coprimes), this case of oscillations is resonant and is characterized by a considerable growth of oscillation amplitude. A simpler case of elastic body oscillations is nonresonant, i.e. $p\nu \neq r\omega_k$. The latter is considered in Sect. 2. The problem of building an approximate solution to problem (1) and (2) consists in finding functions $A_1(a)$, $B_1(a)$, and $u_1(a,x,\psi,\theta)$. For that, let us differentiate relation (3) sequentially by independent variables. Having considered (4) and (5), we shall get the following

$$u_t = \varepsilon A_1(a)(\cos(kx+\psi) - \cos(kx-\psi)) - a(\omega + \varepsilon B_1(a))(\sin(kx+\psi) + \sin(kx-\psi)) + \varepsilon\left(\omega_k(u_1(a,x,\psi,\theta))_\psi + \nu(u_1(a,x,\psi,\theta))_\theta\right),$$

$$u_{tt} = -a\omega_k^2(\sin(kx+\psi) + \sin(kx-\psi)) - 2\varepsilon(A_1(a) + aB_1(a))(\cos(kx+\psi) - \cos(kx-\psi)) + \varepsilon\left(\omega_k^2(u_1(a,x,\psi,\theta))_{\psi\psi} + \nu^2(u_1(a,x,\psi,\theta))_{\theta\theta} + 2\omega_k\nu(u_1(a,x,\psi,\theta))_{\psi\theta}\right),$$

$$\begin{aligned} u_x &= -ak(\sin(kx+\psi) - \sin(kx-\psi)) + \mu(u_1)_x, \\ u_{xx} &= -ak^2(\cos(kx+\psi) - \cos(kx-\psi)) + \mu(u_1)_{xx}, \\ u_{xxx} &= ak^3(\sin(kx+\psi) + \sin(kx-\psi)) + \varepsilon(u_1)_{xxx}, \\ u_{xxxx} &= ak^4(\cos(kx+\psi) - \cos(kx-\psi)) + \varepsilon(u_1)_{xxxx}. \end{aligned}$$

The aforesaid equalities after simple transformations allow obtaining relations which link the unknown functions $A_1(a)$, $B_1(a)$, and $u_1(a,x,\psi,\theta)$ in the following form

$$\begin{aligned} &\omega_k^2(u_1(a,x,\psi,\theta))_{\psi\psi} + \nu^2(u_1(a,x,\psi,\theta))_{\theta\theta} + 2\omega_k\nu(u_1(a,x,\psi,\theta))_{\psi\theta} + \alpha^2(u_1)_{xxxx} \\ &= 4a\omega_k A_1(a)\sin\frac{k\pi x}{l}\sin\psi + a\omega_k B_1(a)\sin\frac{k\pi x}{l}\cos\psi - V^2 a\left(\frac{k\pi}{l}\right)^2\sin\frac{k\pi x}{l}\cos\psi \\ &+ 2Va\omega_k\frac{k\pi}{l}\cos\frac{k\pi x}{l}\sin\psi + F(a,\psi,x,\theta). \end{aligned} \tag{6}$$

To define the target functions $A_1(a)$ and $B_1(a)$ from (6), let us impose additional conditions on $u_1(a,x,\psi,\theta)$ – the condition that there are no summands in its development which are proportional to the main wave mode and its time derivative, i.e.

$$\int_0^{2\pi} (\cos(kx+\psi) - \cos(kx-\psi))u_1(a,x,\psi,\theta)d\psi = 0,$$

$$\int_0^{2\pi} (\sin(kx+\psi) + \sin(kx-\psi))u_1(a,x,\psi,\theta)d\psi = 0.$$

These conditions are equivalent to the dynamic process amplitude parameter equaling the fundamental harmonic amplitude. Besides, taking into account that in the nonresonant case small periodic oscillations have impact only on the form of oscillations (not the amplitude or frequency variation laws), we shall obtain the unknown functions $A_1(a)$ and $B_1(a)$ in the following form

$$A_1(a) = -\frac{1}{p_1}\frac{1}{4\omega_k\pi^2}\int\limits_0^l\int\limits_0^{2\pi}\int\limits_0^{2\pi} F(a,\psi,x,\theta)\sin\frac{k\pi x}{l}\sin\psi dx d\psi d\theta,$$

$$B_1(a) = -\frac{1}{p_1}\frac{1}{4\omega_k\pi^2 a}\int\limits_0^l\int\limits_0^{2\pi}\int\limits_0^{2\pi} F(a,\psi,x,\theta)\sin\frac{k\pi x}{l}\cos\psi dx d\psi d\theta,$$

$$p_1 = \int\limits_0^l\left(\sin\frac{k\pi x}{l}\right)^2 dx = \frac{l}{2}.$$

Assuming that the elastic body material satisfies the nonlinear technical elasticity law [35, 36], the right parts of Eq. (5) acquire the following form

$$A_1(a) = 0, \quad \varepsilon B_1(a) = -\frac{3}{32}\left(\frac{\pi a}{l\omega_1}\right)^2\varepsilon + \left(\frac{\pi}{l}\right)^2\frac{m_1}{m+m_1}\frac{V^2}{8\omega_1}. \tag{7}$$

In Fig. 1, the dependence of nonlinear elastic body oscillations on the environment continuous flow movement amplitude and velocity is shown on the basis of relations (7).

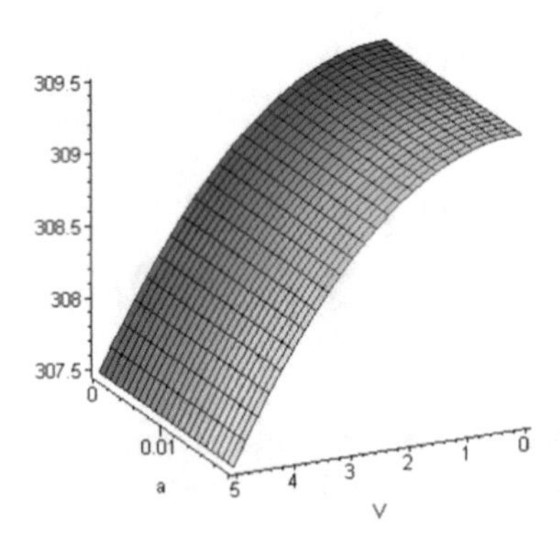

Fig. 1. Dependence of the frequency of linear oscillations of an elastic body on amplitude a and continuous stream of homogeneous environment relative velocity V at $m_1 = 0,6$ kg/m

3 Resonant Oscillations of an Elastic Body, Along Which the Environment is Moving

The resonant case is a more complicated case of nonlinear oscillations of an elastic body, along which a continuous flow of the environment is moving. Resonant oscillations almost always have a negative impact on physical systems, because with resonance, oscillation amplitudes increase greatly in these systems, i.e. dynamic loads increase. This leads to losses in the operation resource of technological systems.

In the case of nonlinear resonant transverse oscillations of an elastic body, along which a continuous flow of homogeneous environment is moving, physical-mechanical and cinematic characteristics of the body affect not only the amplitude, but also the resonant frequency. To discover their impact on the resonant value of amplitude, as well as in the nonresonant case, the solution of the set problem is to be found in the form of an asymptotic image (3) with the only difference that in the resonance case, the amplitude of oscillations of an elastic body relies greatly on the difference of phases of self-oscillations and forced oscillations (parameter $\varphi = \psi - \theta$). Therefore, in the aforesaid asymptotic presentation, the amplitude parameter is defined by more complicated relations, namely

$$\frac{da}{dt} = \mu A_1(a, \phi) + \ldots, \frac{d\psi}{dt} = \omega - \nu + \varepsilon B_1(a, \varphi) + \ldots . \tag{8}$$

The unknown relations (8) of functions $A_1(a, \varphi)$ and $B_1(a, \varphi)$ are found in such a way that asymptotic presentation (3) with the account of (8) can satisfy, with the accuracy considered, the original equation. For this, by means of differentiating (8), similarly to how it was done in other studies [33, 34], it is possible to obtain a differential equation which links the unknown functions in the following way

$$\begin{aligned}
\omega^2(u_1(a,x,\psi,\theta))_{\psi\psi} + \nu^2(u_1(a,x,\psi,\theta))_{\theta\theta} + 2\omega\nu(u_1(a,x,\psi,\theta))_{\psi\theta} + \alpha^2\left(\frac{k\pi}{l}\right)^2(u_1)_{xxxx} \\
= aV^2\left(\frac{k\pi}{l}\right)^2 \sin\frac{k\pi x}{l}\cos\psi - 2V\frac{k\pi}{l}\cos\frac{k\pi x}{l}\cos\psi + F(x,a,\psi,\theta) \\
+ \mu \sin\frac{k\pi x}{l}\left(\left(-\frac{\partial A_1(a,\varphi)}{\partial\varphi}(\omega-\nu) + 2a\omega B_1(a,\varphi)\right)\cos\psi\right. \\
\left. + \left(\frac{\partial B_1(a,\varphi)}{\partial\varphi}(\omega-\nu) + 2\omega A_1(a,\varphi)\right)\sin\psi\right).
\end{aligned}$$

In the resonance case, for the first problem solution approximation, we have a system of differential equations which links the unknown functions in the form

$$(\omega-\nu)\frac{\partial^2 a}{\partial t\partial\varphi} - 2a\omega\frac{\partial\varphi}{\partial t} = \frac{1}{p_1}\frac{1}{4\pi^2}\sum_s e^{is\varphi}\int_0^l\int_0^{2\pi}\int_0^{2\pi} F(a,\psi,x,\theta)\sin\frac{k\pi x}{l}e^{-is\varphi}\cos\psi dx d\psi d\theta$$

$$a(\omega-\nu)\frac{\partial^2\varphi}{\partial t\partial\varphi} - 2a\frac{\partial a}{\partial\varphi} + V^2 a\left(\frac{\pi}{l}\right)^2 =$$

$$= \frac{1}{p_1}\frac{1}{4\pi^2}\sum_s e^{is\varphi}\int_0^l\int_0^{2\pi}\int_0^{2\pi} F(a,\psi,x,\theta)\sin\frac{k\pi x}{l}e^{-is\varphi}\cos\psi dx d\psi d\theta.$$

Below, on the basis of general dependences obtained in Sect. 3, let us consider transverse oscillations of an elastic body, along which a continuous flow of the environment is moving at a steady velocity, under the impact of periodic excitations of harmonic character. This case is the most interesting in terms of theory and practice, and most important for mathematical modelling of nonlinear oscillations of moving environments.

4 Transverse Oscillations of an Elastic Body, Along Which a Continuous Stream of Homogeneous Environment is Moving Under the Impact of Harmonic Excitation

In this case the differential equation of movement of a system of bending oscillations of an elastic body is the following

$$\begin{aligned} u_{tt} + \alpha^2 u_{xxxx} = & -\frac{m_1}{m+m_1}\left(V^2 u_{xx} + 2Vu_{xt}\right) \\ & - \varepsilon\frac{3EI}{m+m_1}\left(u_{xx}u_{xxxx} + 2(u_{xxx})^2\right)_{xx} + \varepsilon\frac{H}{m+m_1}\sin \nu t, \end{aligned} \tag{9}$$

where H – amplitude of external periodic perturbaions. If we consider that boundary conditions for Eq. (9) correspond to pin-edge fixed ends of elastic body, the unifrequency oscillatory process in the mode approximate to external perturbation frequency can be described using relation (3).

Taking into account the internal friction force R, which is proportional to a degree of velocity, i.e. $R = \gamma(u_t)^s$, $\gamma > 0$ – constant, as well as elastic body compression (tension), nonresonant and resonant oscillations shall be described by the following relations respectively

$$\frac{da}{dt} = -\gamma\frac{m_1}{m+m_1}\left(\overline{\omega_k}\right)^{s-1}a^s, \quad \frac{d\psi}{dt} = \overline{\omega_k} - \varepsilon\left(\frac{3}{32}\left(\frac{\pi a}{l\overline{\omega_k}}\right)^2 + \left(\frac{\pi}{l}\right)^2\frac{V^2}{8\overline{\omega_k}}\right)$$

or

$$\frac{da}{dt} = -\frac{\beta}{m+m_1}\left(\overline{\omega_k}\right)^{s-1}a^s + \frac{2\mu H}{\pi(\overline{\omega_k} + \nu(t))}\cos\varphi,$$

$$\frac{d\varphi}{dt} = \overline{\omega_k} - \nu - \left(\frac{\pi}{l}\right)^2\frac{V^2}{8\overline{\omega_k}} - \mu\left(\frac{2H}{\pi(\overline{\omega_k} + \nu(t))a}\sin\varphi + \frac{3}{32}\left(\frac{\pi a}{l\overline{\omega_k}}\right)^2\right),$$

where $\overline{\omega_k} = \frac{k\pi}{l}\sqrt{\left(\frac{k\pi}{l}\right)^2\frac{EI}{m+m_1} \mp \frac{N}{m+m_1}}$.

The " + " sign corresponds to body tension by axial force, whereas the "–" sign – to body compression by axial force. Figure 2 shows dependence of the frequency of nonlinear oscillations of a previously stretched (compressed) elastic body on some parameters.

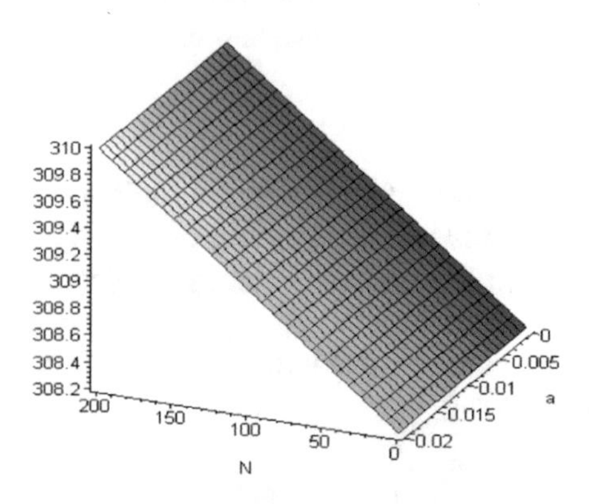

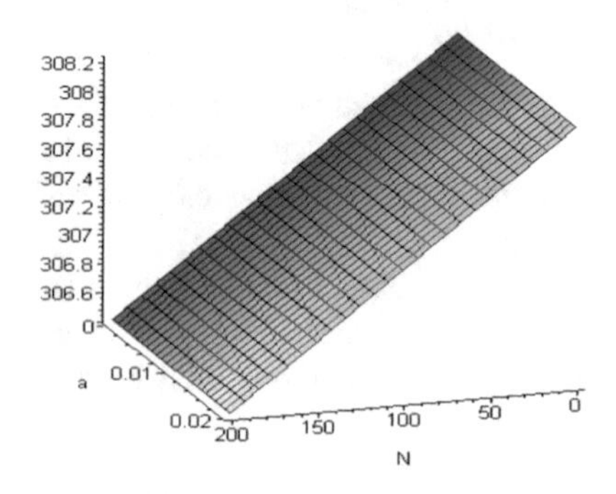

Fig. 2. Dependence of the frequency of nonlinear oscillations of an elastic body on the relative velocity of moving environment V and the compression (tension) force

The differential relations obtained above also allow obtaining relations for defining the amplitude of stationary resonant oscillations

$$\frac{\beta}{m+m_1}(\overline{\omega_k})^{s-1}a^s - \frac{2\mu H}{\pi(\overline{\omega_k}+\nu(t))}\cos\varphi = 0,$$

$$\overline{\omega_k} - \nu(t) - \left(\frac{\pi}{l}\right)^2\frac{V^2}{8\overline{\omega_k}} = \mu\left(\frac{2H}{\pi(\overline{\omega_k}+\nu(t))a}\sin\varphi + \frac{3}{32}\left(\frac{\pi a}{l\overline{\omega_k}}\right)^2\right)$$

and the resonance curve

$$\left(\frac{2\mu H}{\pi(\overline{\omega_k}+\nu(t))}\right)^2 = \left(\frac{\beta}{m+m_1}(\overline{\omega_k})^{s-1}a^s\right)^2 + a^2\left(\overline{\omega_k} - \nu(t) - \left(\frac{\pi}{l}\right)^2\frac{V^2}{8\overline{\omega_k}} - \frac{3\mu}{32}\left(\frac{\pi a}{l\overline{\omega_k}}\right)^2\right)^2.$$

On the basis of these relations, a curve is built (Fig. 3) to show the dependence of amplitude of elastic body resonant oscillations on the imbalance $\alpha = \omega - \nu$ of frequencies of self-oscillations and forced oscillations.

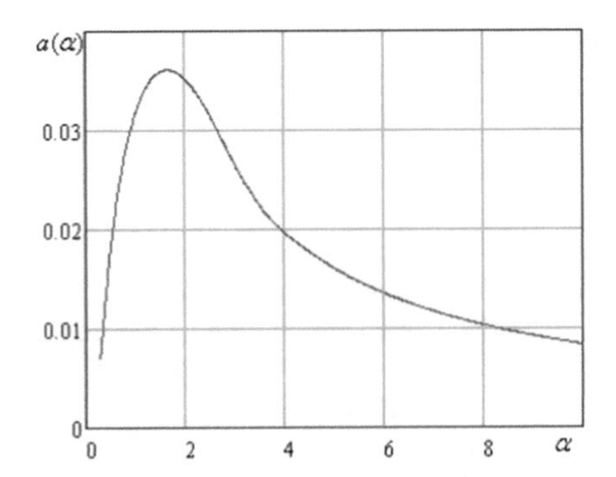

Fig. 3. Dependence of the amplitude of elastic body oscillations on frequencies imbalance at $V = 4\,\mathrm{m/s}$

5 Conclusions

In this paper, a modified approach to the study of mathematical models of nonlinear oscillations of elastic one-dimensional bodies subject to the impact of a moving environment is suggested. The environment is moving along longitudinal oscillations. The research methods are based on the assumption that environment velocity is constant and relatively small. Modification of theoretical approaches to analyzing dynamic processes in such mathematical models consists in combining wave theory of movement and asymptotic methods of nonlinear mechanics. Based on this combination, a new asymptotic method for studying moving systems is created. The effectiveness of applying this approach to the physical systems described in the paper is demonstrated. In particular, the main defining relations are obtained, which allow analyzing amplitude and frequency characteristics of a dynamic process.

As far as practical application to technological systems modelling is concerned, the results obtained in this paper show that the impact of elastic body material nonlinear elastic characteristics and of environment continuous flow movement along the elastic body is manifested in the dependence of body's on its movement amplitude and velocity. At that, self-resonant frequency of nonlinear oscillations of an elastic body is smaller for higher values of continuous environment flow relative velocity; decreases with elastic body's length increase; is smaller for a continuous flow of the environment with a smaller specific weight.

The aforesaid facts are highly important in the research of resonant oscillations of an elastic body. The relations obtained in the paper show that for higher movement velocities of the continuous flow of a homogeneous environment, resonant amplitude values are higher (on condition of relatively low environment movement velocities); for a smaller mass per unit length of the continuous flow of a homogeneous environment, the resonant amplitude value is smaller.

The methodology developed in this paper, which is based on the combination of wave theory of movement and asymptotic methods of nonlinear mechanics, can also be applied to other cases of mathematical modelling of oscillatory processes in complex-structure dynamic systems.

References

1. Magrab, E.B.: An Engineer's Guide to Mathematica. Wiley, Hoboken (2014)
2. Jones, D.I.G.: Handbook of Viscoelastic Vibration Damping. Wiley, Hoboken (2001)
3. Sobotka, Z.: Theory of Plasticity and Limit Design of Plates. Elsevier, Amsterdam (1989)
4. Chen, L.-Q., Yang, X.-D., Cheng, C.-J.: Dynamic stability of an axially moving viscoelastic beam. Eur. J. Mech. A/Solids **23**, 659–666 (2004)
5. Hatami, S., Azhari, M., Saadatpour, M.M.: Free vibration of moving laminated composite plates. Compos. Struct. **80**, 609–620 (2007)
6. Banichuk, N., Jeronen, J., Neittaanmaki, P., Tuovinen, T.: Static instability analysis for traveling membranes and plates interacting with axially moving ideal fluid. J. Fluids Struct. **26**, 274–291 (2010)
7. Czaban, A., Szafraniec, A., Levoniuk, V.: Mathematical modelling of transient processes in power systems considering effect of high-voltage circuit breakers. Przeglad Elektrotechniczny **95**(1), 49–52 (2019)
8. Mockersturm, E.M., Guo, J.: Nonlinear vibration of parametrically excited, visco-elastic, axially moving strings. J. Appl. Mech. ASME **72**, 374–380 (2005)
9. Kuttler, K.L., Renard, Y., Shillor, M.: Models and simulations of dynamic frictional contact. Comput. Methods Appl. Mech. Engrg. **177**, 259–272 (1999)
10. Lim, C.W., Li, C., Yu, J.-L.: Dynamic behaviour of axially moving nanobeams based on non-local elasticity approach. Acta Mech. Sinica **26**, 755–765 (2010)
11. Wickert, J.A., Mote Jr., C.D.: Classical vibration analysis of axially-moving continua. J. Appl. Mech. ASME **57**, 738–744 (1990)
12. Pukach, P.Ya., Kuzio, I.V.: Nonlinear transverse vibrations of semiinfinite cable with consideration paid to resistance. Naukovyi Visnyk Natsionalnoho Hirnychoho Universytetu, no. 3, pp. 82–86 (2013)
13. Pukach, P., Il'kiv, V., Nytrebych, Z., Vovk, M., Pukach, P.: On the asymptotic methods of the mathematical models of strongly nonlinear physical systems. In: Advances in Intelligent Systems and Computing, vol. 689, pp. 421–433 (2018)
14. Lavrenyuk, S.P., Pukach, P.Ya.: Mixed problem for a nonlinear hyperbolic equation in a domain unbounded with respect to space variables. Ukrainian Math. J. **59**, no. 11, pp. 1708–1718 (2007)
15. Buhrii, O.M.: Visco-plastic, newtonian, and dilatant fluids: stokes equations with variable exponent of nonlinearity. Matematychni Studii, vol. 49, no. 2, pp. 165–180 (2018)
16. Nytrebych, Z., Malanchuk, O., Il'kiv, V., Pukach, P.: On the solvability of two-point in time problem for PDE. Italian J. Pure Appl. Math. **38**, 715–726 (2017)
17. Pukach, P.: Investigation of bending vibrations in Voigt-Kelvin bars with regard for non-linear resistance forces. J. Math. Sci. **215**(1), 71–78 (2016)
18. Gurtin, M.E., Murdoch, A.I.: A continuum theory of elastic material surfaces. Arch. Ration. Mech. Anal. **57**, 291–323 (1975)

19. Lam, D.C.C., Yang, F., Chong, A.C.M., Wang, J., Tong, P.: Experiments and theory in strain gradient elasticity. J. Mech. Phys. Solids **51**, 1477–1508 (2003)
20. Papargyri-Beskou, S., Tsepoura, K.G., Polyzos, D., Beskos, D.E.: Bending and stability analysis of gradient elastic beams. Int. J. Solids Struct. **40**, 385–400 (2003)
21. Vardoulakis, I., Sulem, J.: Bifurcation Analysis in Geomechanics. Blackie/ Chapman and Hall, London (1995)
22. Gao, X.-L., Park, S.K.: Variational formulation of a modified couple stress theory and its application to a simple shear problem. Zeitschrift furangewandte Mathematik und Physik **59**, 904–917 (2008)
23. Ma, H.M., Gao, X.-L., Reddy, J.N.: A non-classical Reddy-Levinson beam model based on a modified couple stress theory. Int. J. Multiscale Comput. Eng. **8**, 167–180 (2010)
24. Belmas, I.V., Kolosov, D.L., Kolosov, A.L., Onyshchenko, S.V.: Stress-strain state of rubber-cable tractive element of tubular shape. Naukovyi Visnyk Natsionalnoho Hirnychoho Universytetu, vol. 2, pp. 60–69 (2018)
25. Mahmoodi, S.N., Jalili, N.: Non-linear vibrations and frequency response analysis of piezoelectrically driven microcantilevers. Int. J. Non-Linear Mech. **42**, 577–587 (2007)
26. Nayfeh, A.H.: Introduction to Perturbation Techniques. Wiley, New York (1981)
27. Nayfeh, A.H., Mook, D.T.: Non-Linear Oscillations. Wiley, New York (1979)
28. Pain, H.J.: The Physics of Vibration and Waves, 6th edn. Wiley, New York (2005)
29. Chen, L.-Q., Chen, H.: Asymptotic analysis of nonlinear vibration of axially accelerating visco-elastic strings with the standard linear solid model. J. Eng. Math. **67**, 205–218 (2010)
30. Bayat, M., Barari, A., Shahidi, M.: Dynamic response of axially loaded Euler-Bernoulli beams. Mechanika **17**(2), 172–177 (2011)
31. Teslyuk, V.M.: Models and Information Technologies of Micro-electromechanical Systems Synthesis. Vezha and Ko, Lviv (2008)
32. Nytrebych, Z., Il'kiv, V., Pukach, P., Malanchuk, O.: On nontrivial solutions of homogeneous Dirichlet problem for partial differential equations in a layer. Kragujevac J. Mathem. **42**(2), 193–207 (2018)
33. Pukach, P.Ya., Kuzio, I.V., Nytrebych, Z.M., Ilkiv, V.S.: Analytical methods for determining the effect of the dynamic process on the nonlinear flexural vibrations and the strength of compressed shaft. Naukovyi Visnyk Natsionalnoho Hirnychoho Universytetu **5**, 69–76 (2017)
34. Pukach, P.Ya., Kuzio, I.V., Nytrebych, Z.M., Ilkiv, V.S.: Asymptotic method for investigating resonant regimes of non–linear bending vibrations of elastic shaft. Naukovyi Visnyk Natsionalnoho Hirnychoho Universytetu **1**, 68–73 (2018)
35. Kauderer, H.: Nonlinear Mechanics. Izdatelstvo Inostrannoy Literatury, Moscow (1961). (in Russian)
36. Pukach, P., Nytrebych, Z., Ilkiv, V., Vovk, M., Pukach, Yu.: On the mathematical model of nonlinear oscillations under the impact of a moving environment. In: Proceedings of International scientific conference Computer sciences and information technologies (CSIT-2019), vol. 1, pp. 71–74 (2019)

Solving Systems of Nonlinear Equations on Multi-core Processors

Lesia Mochurad(✉) and Nataliya Boyko(✉)

Lviv Polytechnic National University, Lviv 79013, Ukraine
{lesia.i.mochurad, nataliya.i.boyko}@lpnu.ua

Abstract. The paper proposes an approach to solving multidimensional systems of nonlinear equations based on the use of the OpenMP parallelization mechanism and the multicore architecture of modern computers. The software product, which performs the main function - the parallelization of the numerical solution of multidimensional SNE by the Newton method, is developed. The analysis of the speed and efficiency of calculations with different number of processor cores is carried out. As a result, appropriate estimates of the acceleration and efficiency coefficients were obtained. The proposed method is easily scaled to a different number of processor cores. A number of numerical experiments were conducted. The obtained results also indicate the possibility of further optimization of the computational process by developing the multi-core architecture of modern computers.

Keywords: Newton's method · OpenMP parallel computing technology · Multithreading · Finite difference

1 Introduction

The emergence of multi-core processors was the beginning of a new era of desktop computing. Companies that are processor manufacturers have virtually come to the forefront of increasing computing productivity. An outdated way to increase productivity by increasing clock speeds leads to increased power consumption. The transition to all of the more subtle technologies of manufacturing chips exacerbates the problem of leakage of electric current and the heating of the processor itself. Therefore, in the first decade of the XXI century. there was a need for the development of methods that increase the processor's performance as a result of an increase in the number of computing cores in the same package and program code instructions executed in one cycle. To solve this problem, it was proposed to implement parallel execution at the level of instructions or streams [1].

Nowadays a large number of applied problems is reduced to solving systems of nonlinear equations (SNE) [2, 3]. However, with the advent of technology and scientific progress, the problem of solving multidimensional SNE without time consuming costs. Therefore, the problem of the implication of numerical methods for solving SNE with the modern trends in the development of computer technology, which, in turn, will enable the development of algorithms aimed at the rapid and effective solution of such systems.

N. Shakhovska and M. O. Medykovskyy (Eds.): CCSIT 2019, AISC 1080, pp. 90–106, 2020.
https://doi.org/10.1007/978-3-030-33695-0_8

The object of study is the study of parallel algorithms for numerical solving of multidimensional SNE.
The subject of study is OpenMP's parallel programming technology for current trends in multi-core processors.
The purpose of the work is to create a software product that will enable you to quickly and efficiently deploy multidimensional SNE without time consuming costs.

2 Review of the Literature

An urgent problem of modern science is the study of phenomena and processes of different nature. At the same time, studies are often carried out using numerical experiments on mathematical models. Many of the calculational problems that arise in mathematical modeling are reduced to solving systems of nonlinear equations of high order. When solving some problems, there is a need to solve SNE faster than the real-time process. Moreover, their solution requires multivariate calculations and significant computational resources [4, 11, 12]. Quite often there is a need to solve problems for which the initial data are given approximate. Therefore, for solving such problems and assessing the authenticity of solutions it is expedient to use high-performance computers with parallel computing, including multi-core computers, and appropriate algorithms that take into account the architectural and computational characteristics of these computers.

Problems of solving **nonlinear** equations are constantly arising in practice. In particular, they are used in the manufacture of furniture, the organization of interior rooms, etc. [5–10]. For example, you make a bookshelf for carrying books, the height of which varies from 18½ to 26 cm, a length of 60 cm (Fig. 1). The material is a tree with a Young module. You want to find the maximum vertical deviation of the shelf. The vertical deviation of the shelf will be determined by a nonlinear equation:

$$v(x) = 0.42493 \times 10^{-4}x^3 - 0.13533 \times 10^{-8}x^5 - 0.66722 \times 10^{-6}x^4 - 0.018507x, \quad (1)$$

where x is the position along the length of the beam. So, you need to find the maximum deviation.

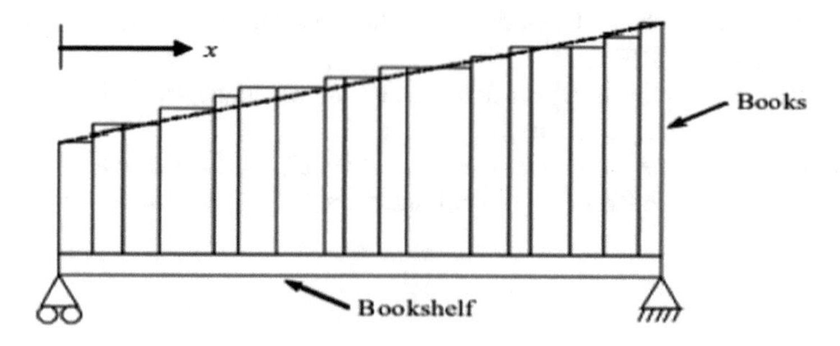

Fig. 1. A loaded bookshelf.

We obtain SNE for describing a large number of shelves interconnected by certain criteria.

Nonlinear equations are used in electrical engineering [6, 15]. Thermistors are temperature meters, which operate on the principle: the material exhibits a change in the electrical resistance with temperature change (Fig. 2). That is, measuring the resistance of the thermistor material can determine the temperature.

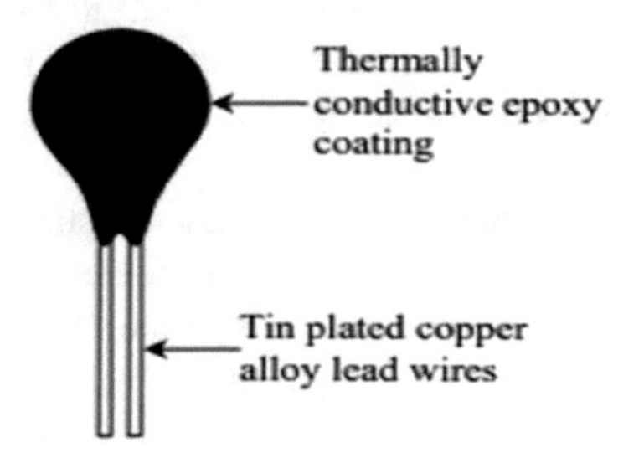

Fig. 2. A typical thermistor.

The connection between the resistance of the thermistor and its temperature is given by a nonlinear equation:

$$\frac{1}{T} = 1.129241 \times 10^{-3} + 2.341077 \times 10^{-4} \ln(R) + 8.775468 \times 10^{-8} \{\ln(R)\}^3, \quad (2)$$

where T is in kelvin, and R is in ohm. From the given equation we find the range of values of the resistance of the thermistor.

Nonlinear equations are used in mechanics [7, 14]. For example, before shrinking in a steel knife should be a chilled shaft (part of the shaft or axis that comes in contact with the winch and directly perceives the load from the latter, Fig. 3).

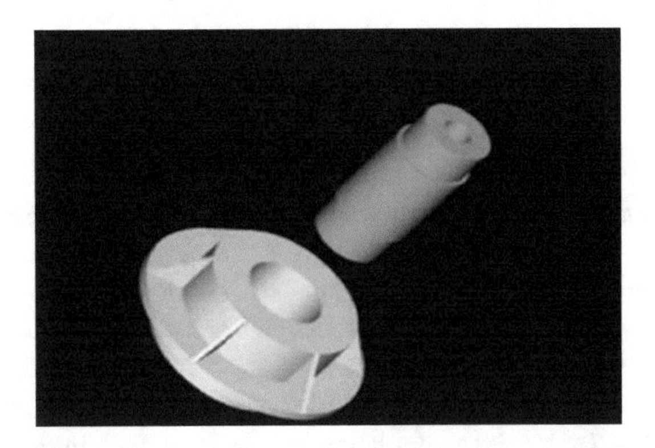

Fig. 3. The column that will slip through the hub after sealing.

The equation that gives the temperature to which the chuck must be cooled is given as follows to obtain the desired restriction:

$$f(T_f) = -0.50598 \times 10^{-10} T_f^3 + 0.38292 \times 10^{-7} T_f^2 + 0.74363 \times 10^{-4} T_f + 0.88318 \times 10^{-2} = 0. \quad (3)$$

You need to find the root of the Tf equations to find the temperature to which the shaft must be cooled.

3 Statement of the Problem

Let us consider the problem of solving a system n of nonlinear equations with n unknowns

$$\begin{aligned} f_1(x_1, x_2, \ldots, x_n) &= 0, \\ f_2(x_1, x_2, \ldots, x_n) &= 0, \\ \ldots\ldots\ldots\ldots\ldots\ldots& \\ f_n(x_1, x_2, \ldots, x_n) &= 0. \end{aligned} \quad (4)$$

We introduce the following vectors $x = (x_1, x_2, \ldots, x_n)^T$ and $f(x) = (f_1(x), f_2(x), \ldots, f_n(x))^T$. In these notation, our problem (4) takes on the form

$$f(x) = 0. \quad (5)$$

Moreover, (5) can be some approximation to the exact system of nonlinear equations, and for these vector-valued functions inequality

$$\|f(u) - \varphi(u)\| \leq \delta \tag{6}$$

on any n measurable vector u. To solve the problem (5), the initial approximation is given $x^{(0)}$ and the area is determined $D = \{a_i \leq x_i \leq b_i \ (i = 1, 2, \ldots, n)\}$, in which the solution is sought, and the required precision ε letting approximation to the solution of the system. In this case, the initial approximation belongs to a definite region $x^{(0)} \in D$. The lower index in the formulas indicates the component numbers, and the upper ones are the numbers of iterations. Let $x^{(k)}$ - another approximation to the solution x^* of the system (5). Then a Taylor series at a point $x^{(k)}$ can be written in the form

$$f(x^{(k)} + p) = f(x^{(k)}) + J(x^{(k)})p + \ldots, \tag{7}$$

where $J(x^{(k)}) = \nabla f(x^{(k)})$ - Jacobi matrix of the first derivatives in the point $x^{(k)}$ that is

$$\left(J\left(x^{(k)}\right)\right)_{ij} = \left(\frac{\partial f_i(x)}{\partial x_j}\right)_{x=x^{(k)}}. \tag{8}$$

We will use the notation $J^{(k)} = J(x^{(k)})$ for the Jacobi matrix and $f^{(k)} = f(x^{(k)})$ for the value vector of the function.

If we consider only the first two terms of a Taylor series, then we obtain an approximation of a nonlinear function

$$f(x^{(k)} + p) \approx y = f^{(k)} + J^{(k)}p. \tag{9}$$

We will then use the zero of our linear approximation to determine the next approximation $x^{(k+1)}$ to the solution. We put $y = 0$, we obtain a system of linear equations

$$J^{(k)}p = -f^{(k)}. \tag{10}$$

Accordingly, this system of equations can be solved by the Gauss method. At the same time, we will obtain a new approximation to the solution by the formula $x^{(k+1)} = x^{(k)} + p$, that is

$$x^{(k+1)} = x^{(k)} - \left(J^{(k)}\right)^{-1} f^{(k)}. \tag{11}$$

Formula (6) is a formula of Newton's method [8]. The latter has a quadratic convergence rate, which is a significant advantage of its use. However, each iteration requires the deployment of SOLA, which is a labor-intensive procedure. In addition, Newton's method requires the computation of all the first partial derivatives of nonlinear functions. Note that derivatives can be approximated using numerical differentiation.

It takes a long time to solve a SOLA or to find an inverse matrix. However, this time can be significantly reduced if, when computing the elements of the Jacobi matrix and solving SOLA, or finding the inverse matrix of execution of all arithmetic operations, Parallel using the modern OpenMP parallel programming technology and such a property as a lot of flow [9]. This approach is as much as possible aimed at supporting the latest development of multi-core processors.

4 Methods of Solving

It is suggested to use the OpenMP parallel programming technology [10, 16] to parallelize programs on multi-core personal computers in order to increase their productivity when solving multidimensional SNAs. Program parallelism is performed in order to more efficient computer time for their execution. OpenMP can be seen as a high-level add-on over Pthreads (or similar thread libraries). OpenMP implements parallel computing using multithreading, in which the master flow creates a set of slave streams and the task is distributed among them. It is assumed that the threads are executed in parallel on a machine with several cores (the number of processor cores does not necessarily have to be larger or equal to the number of threads).

The tasks performed by the threads in parallel, as well as the data needed to perform these tasks, are described by special instructions of the preprocessor of the corresponding language - pragm. The number of streams created can be regulated both by the program itself by calling library procedures, and from the outside, using environment variables.

Advantages of this technology [11, 12]:

- due to the idea of incremental paralleling, OpenMP is ideally suited to developers who want to quickly parse their computing programs with large parallel cycles. The developer does not create a new parallel program, but simply adds sequentially to the text of the program OpenMP-directive;
- at the same time, OpenMP is a flexible mechanism that gives developers great control over the behavior of a parallel program;
- it is assumed that the OpenMP program on a single-processor platform can be used as a sequential program, that is, there is no need to maintain a consistent and parallel version. OpenMP directives are simply ignored by a sequential compiler, and stubs can be used to call OpenMP procedures (stubs);
- support for so-called "orphan" (detached) directives, that is, directives for synchronization and division of work may not be included directly in the lexical context of the parallel region.

The paper proposes a parallel algorithm for the Newton's method. On each iteration of which the following macro operations are performed:

1. the calculation of each OpenMP-process part of the vector-function component corresponding to its logical number;
2. calculate by each OpenMP-process $J^{(k+1)}$: the corresponding number of rows of the Jacobi matrix
3. solving the resulting SOLA using an appropriate parallel algorithm of the Gauss method;

4. using the resulting SOLA solution, calculate by each OpenMP process the corresponding logical number of the next approximation component (4) to the SNE solution
5. verification of conditions for termination of the iterative process by the formula $\left\|f^{(k+1)}\right\| \leq \frac{\varepsilon}{\left\|\left(J^{(k+1)}\right)^{-1}\right\|}$ which is an assessment of the quality of the approximate solution, which ensures the implementation of inequality $\left\|x^{(k+1)} - x^*\right\| \leq \varepsilon$ where x^* is the exact solution of SNE.

OpenMP is supported by virtually all manufacturers of large computing systems, easily integrated into the system software of the multi-core computer. The OpenMP technology allows users to work with single text for a serial and parallel program by performing small insertions in the program. In addition, since OpenMP is based on "incremental" programming, then the user does not need to write a parallel program at once, it can create it consistently. This greatly simplifies the process of creating a parallel program.

5 Numerical Experiments

To verify the validity of the proposed algorithm proposed in the work, without decreasing generality, consider the numerical solution of the system of nonlinear equations of the form:

$$\sum_{j=1}^{n} x_j - 0.5(3n+1) + 2x_i^2 - 2\left(1 + 2\frac{i}{n} + \left(\frac{i}{n}\right)^2\right) = 0, \quad i = 1, 2, \ldots, n \qquad (12)$$

at a given initial approximation $x^{(0)} = 0.5$ in the region $D = \{-1000 \leq x_i \leq 1000\}$.

The realization of solving a system of nonlinear equations is divided into several stages.

At the first stage, the class-equation and the class-system of equations were constructed, which became the basis for creating the future functional. Also, the numerical computation of Jacobi matrix elements is realized. For the approximate computation of partial derivatives, the formula of the central difference derivative is chosen:

$$u'(x) \approx \frac{u(x+h) - u(x-h)}{2h}. \qquad (13)$$

The main functions are: derivativeAdjective (finding the partial derivative of the studied system), functionToDerivative (calculating the value of the derivative at the point), checkingSystem (implementation of the studied system), matrixJacobi (Jacobi matrix calculus).

The most important and time-consuming part of the work was the implementation of the very algorithm of the Newton method. The main functions are Det (Definition calculus), EquationInPoint (finding the next approximation), Gaus (Gaussian method realization), InverseMatrix (finding the inverse matrix), Newton (Newton method implementation), SolutionOfTheSystem (search for solving the studied system).

The following is a snippet of implementation of the main functions:

- checkingSystem (): the function of the task of the researched system;

```
   //initialize coefficients, powers, doubleCoefIndexes
#pragma omp parallel for
   for (int i = 0; i < size; i++)
   {
      coeffcients.push_back(vector<double>());
      powers.push_back(vector<double>());
      doubleCoefIndexes.push_back(i + 1);
   }
   //set coefficients and powers
#pragma omp parallel for
   for (size_t i = 0; i < size; i++)
   {
      sum = 0;
      for (size_t j =0; j < size; j++)
      {
            sum += -0.5 * (3 * size +1) -2* (1+2* (double(i + 1) / 
            size)
                + std::pow(i + 1, 2.0) / std::pow(size, 2.0));
            if (i == j)
            {
                     coefficients[i].push_back(2 * size);
                     powers[i].push_back(2);
   }
   else
   {
                      coefficients[i].push_back(1);
                      powers[i].push_back(1);
      }
   }
   coefficients[i].push back(sum);
```

- derivativeAdjective (): partial calculation function;

```
double   Equation::derivativeAdjective(size_t   adjectivelndex,   double
point)
{
        double h = 0.000000001; //step to calculate derivative
                //approximately calculate the first derivative //central
                derivative
        if  (adjectivelndex  +  1  ==  doubleCoefIndex)  //Only  for  check-
        ingSystem() method
        {
            return
            (functionTo Derevative Double(coefficients[adjectivelndex],
            (point + h)) -
                functionToDerevativeDouble(coefficients[adjectivelndex],
                (point - h))) / (2 * h);
        }
        double a = coefficients[adjectivelndex];
        double b = powers[adjectivelndex];
        double c = point + h;
        return  (functionToDerevative(coefficients[adjectivelndex],  pow-
        ers[adjectivelndex], (point + h)) -
           functionToDerevative(coefficients[adjectivelndex],        pow-
           ers[adjectivelndex], (point - h))) / (2 * h)
}
```

- EquationInPoint (): the function of finding the next approximation;

```
vector<double> Newton::EquationlnPoint(EquationSystem e, vector<double>
points)
{
    vector<double> FxO;
    for (int k = 0; k < e.getEquationsCount(); k++)
    {
        double result = 0;
        vector<double> coefs = e.getEquations()[k].getCoefficients();
        vector<double> pows = e.getEquations()[k].getPowers();
        int countCoefs = e.getEquations()[k].getCoefficientsCount() - 1;
        for (int i = 0; i < countCoefs; i++)
        {
                result += pow(points[i], pows[i]) * coefs[i];
                if (k + 1 == e.getEquations()[k].getDoubleCoefIndex())
                {
                   result += e.getEquations()[k].getDoubleCoefIndex();
                }
        }
        FxO.push_back(result + coefs[countCoefs]);
     }
     return FxO;
  }
```

- Gaus (): paralleling the Gauss method for solving the SOLA;

```
    double* Newton::Gaus(double **brr, int row, int col, double*det,
    bool&mark, doul
    {
         double**arr = new double*[row];
         //initialize array
    #pragma omp parallel for
         for (int i = 0; i < row; i++)
         {
             arr[i] = new double[col];
             //copy array
    #pragma omp parallel for
    for (int j = 0; j < col; j++)
    {
           arr[i] [j] = brr[i][j];
    }
}
     //Print(arr, row, col);
     int p = 0;
     for (int j = 0; j < col - 1; j++)
     //calculate matrix on the next iteration
#pragma omp parallel for
     for (int i = j +1; i < row; i++)
     {
         double m = - (arr[i][j] / arr[j][j]);
         double secondnumb = arr[j][j];
         //cout « m«endl;
         for (int k = j; k < col; k++)
         {
                arr[i][k] = (arr[j][k] * m + arr[i][k])*secondnumb;
         }
     }
     //Print(arr, row, col);
     *det = Det(arr, row, p);
     double*results = new double[row];
     //initialize array
#pragma omp parallel for
     for (int i = 0; i < row; i++)
     {
     results[i] = 0;
     }
     //calculate final results
#pragma omp parallel for
     for (int i = row - 1; i >= 0; i—)
     {
         double sum = 0;
         for (int j = col - 2; j > i; j—)
         {
                sum += arr[i][j] * results[j];
         }
```

- InverseMatrix (): the parallel finding of the inverse matrix;

```
double** Newton::InverseMatrix(vector<vector<double» arr, int size, dou-
ble*det, bool& mark, double epsilon)
{
     double**inverse = new double*[size];
     //initialize array
#pragma omp parallel for
     for (int i = 0; i < size; i++)
     {
               inverse[i] = new double[size];
     }
     double**matrix = new double*[size];
     //initialize array
#pragma omp parallel for
     for (int j = 0; j < size; j++)
     {
               matrix[j] = new double[size + 1];
     }
     //copy matrix
#pragma omp parallel for
     for (int i = 0; i < size; i++)
     {
#pragma omp parallel for
     for (int j = 0; j < size; j++)
          {
               matrix[i][j] = arr[i] [ j] ;
          }
          matrix[i][size] = 0;
     }
     matrix[0][size] = 1;
     //calculate results
#pragma omp parallel for
     for (int i = 0; i < size; i++)
     {
          double*results = Gaus(matrix, size, size +, det, mark, epsi-
          lon);
```

• SolutionOfTheSystem (): parallelization of the SNE solution (Newton's method);

```
Result   Newton::SolutionOfTheSystem(EquationSystem   e,   vector<double>
points, double epsilon)
{
     Result result;
     //EquationSystem efcoefs, pows);
     bool is_first = true;
     vector<double> multi;
     vector<double> r;
     int ii = 0;
     while (true)
     {
          vector<vector<double>> jacobi = e.matrixJacobi(points);
          //this->Print(jacobi);
          vector<double> FxO = this->EquationInPoint(e, points);
          /*for (int i = 0; i < FxO.size (); i++)
          {
          cout « "FxO: " « FxO [i] « ""
          }
          cout « endl;*/
          bool mark = false;
          double*det = new double;
          if (is_first)
          {
               double**inverse  =  InverseMatrix(jacobi,  jacobi.size(),
               det, mark, epsilon);
               if (*det == 0)
               {
                    result.is_solution = false;
                    return result;
               }
               vector   <   vector<double>>   inv(jacobi.size(),   vec-
               tor<double>(jacobi.size()));
               //copy inverse
 #pragma omp parallel
               for (int i = 0; i < jacobi.size(); i++)
               {
                    for (int i = 0; j < jacobi.size(); i++)
                    else
               {
                    double**br = new double *[jacobi.size ()] ;
                    for (int i = 0; i < jacobi.size () ; i++)
                    {
                         br[i] = new double[jacobi.size() + 1];
                         //copy jacobi
 #pragma omp parallel for
               for (int j = 0; j < jacobi.size(); j++)
               {
               br[i][j] = jacobi[i][j];
               }
                    br[i][jacobi.size()] = Fx0[i] * (-1);
               }
               double* gaus = Gaus(br, jacobi.size(), jacobi.size() + 1,
               dec, mark, epsilon);
               for (int i = 0; i < r.size(); i++)
               {
```

```
                r[i] += gaus[i];
                cout « "delta x: " « gaus[i] « ""
            }
            cout « endl « "R:" « endl;
            for (int i = 0; i < r.size(); i++)
            {
            cout « r[i] « ""
            }
            cout « endl;
            vector<double> mm(r.size());
            //copy gaus
#pragma omp parallel for
            for (int i = 0; i < r.size(); i++)
            {
            mm[i] = gaus [i] ;
            }
            multi = mm;
            //delete br
#pragma omp parallel for
            for (int i = 0; i < jacobi.size(); i++)
            double max = Max(multi);
            if (max > epsilon)
            {
                //points = r;
                //copy results]
#pragma omp parallel for
            for (int i = 0; i < r.size(); i++)
            {
                points[i] = r[i];
                }
                continue;
        }
        result.matrix = r;
        result.is_solution = true;
        break;
```

Below we give the results of computing the elements of the Jacobi matrix of the studied system with the number of equations in the system - 8.

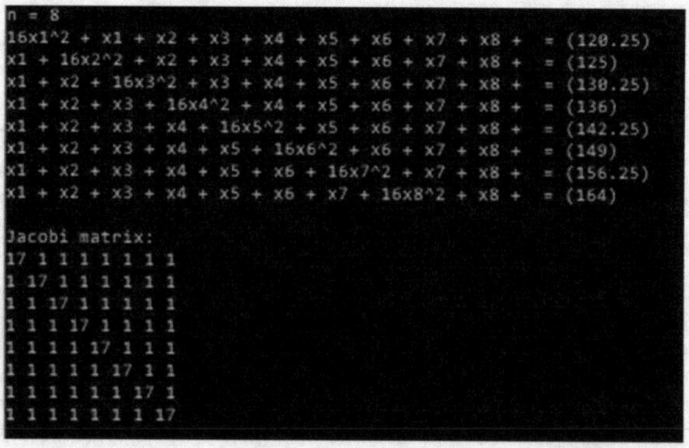

```
n = 8
16x1^2 + x1 + x2 + x3 + x4 + x5 + x6 + x7 + x8 +   = (120.25)
x1 + 16x2^2 + x2 + x3 + x4 + x5 + x6 + x7 + x8 +   = (125)
x1 + x2 + 16x3^2 + x3 + x4 + x5 + x6 + x7 + x8 +   = (130.25)
x1 + x2 + x3 + 16x4^2 + x4 + x5 + x6 + x7 + x8 +   = (136)
x1 + x2 + x3 + x4 + 16x5^2 + x5 + x6 + x7 + x8 +   = (142.25)
x1 + x2 + x3 + x4 + x5 + 16x6^2 + x6 + x7 + x8 +   = (149)
x1 + x2 + x3 + x4 + x5 + x6 + 16x7^2 + x7 + x8 +   = (156.25)
x1 + x2 + x3 + x4 + x5 + x6 + x7 + 16x8^2 + x8 +   = (164)

Jacobi matrix:
17 1 1 1 1 1 1 1
1 17 1 1 1 1 1 1
1 1 17 1 1 1 1 1
1 1 1 17 1 1 1 1
1 1 1 1 17 1 1 1
1 1 1 1 1 17 1 1
1 1 1 1 1 1 17 1
1 1 1 1 1 1 1 17
```

The results of testing the program demonstrating the implementation of the proposed algorithm for the Newton method for numerical solution of the study system with the number of equations in the system - 8:

```
delta x: -1.92774 delta x: -1.82385 delta x: -1.73578 delta x: -1.66361 delta x:
 -1.60739 delta x: -1.56719 delta x: -1.54304 delta x: -1.53499
R:
3.1009 3.00167 2.91786 2.84941 2.79626 2.75834 2.7356 2.72803
delta x: -0.580872 delta x: -0.534069 delta x: -0.494748 delta x: -0.462788 delt
a x: -0.438075 delta x: -0.420507 delta x: -0.410001 delta x: -0.406506
R:
2.52003 2.4676 2.42311 2.38662 2.35818 2.33783 2.3256 2.32152
delta x: -0.0694433 delta x: -0.0597527 delta x: -0.0519949 delta x: -0.0459599
delta x: -0.0414666 delta x: -0.0383668 delta x: -0.0365515 delta x: -0.0359539

R:
2.45059 2.40785 2.37112 2.34067 2.31671 2.29946 2.28905 2.28557
delta x: -0.00176526 delta x: -0.00141084 delta x: -0.00114765 delta x: -0.00095
5733 delta x: -0.000820215 delta x: -0.000730429 delta x: -0.000679268 delta x:
-0.000662648
R:
2.44882 2.40644 2.36997 2.33971 2.31589 2.29873 2.28837 2.28491
Count of iterations: 5

Result:
2.44882 2.40644 2.36997 2.33971 2.31589 2.29873 2.28837 2.28491
Runtime: 0.25 s
Press any key to continue . . .
```

6 Analysis of the Results

In the course of an experimental study, the solving of the considered system was found by executing the program algorithm in parallel on 16, 4, 2, and 1-core computers.

Table 1 shows the execution time of the program (in seconds) for numerical solving SNE with the number of equations n on one-, two-, four- and sixteenth-nuclear processors.

After analyzing the results obtained, one can conclude that with an increase in the number of nuclei twice, the time for which the program algorithm finds the solution of the investigated system decreases practically by half, with a different number of equations in the system (we are talking about multidimensional SNE).

Table 1. Time to execute a program that implements a parallel algorithm

	Number of cores			
n	1	2	4	16
200	129.459	66.484	39.343	15.329
300	208.735	105.172	57.684	20.509
500	1601.003	809.759	411.062	138.428
1000	3102.451	1612.439	848.392	241.469

We will calculate the acceleration coefficient by the formula $S_p(n) = T_1(n)/T_p(n)$ size n it is used to parameterize the computational complexity of a solvable problem and can be understood, for example, as the number of input data of the problem [13].

Table 2 shows the corresponding values, and in Fig. 4 shows the graph of the change in the acceleration coefficient, depending on the number of computer cores.

Table 2. The value of the acceleration factor

n	Sp(n)
1	1
2	1.947220384
4	3.290521821
16	6.069623517

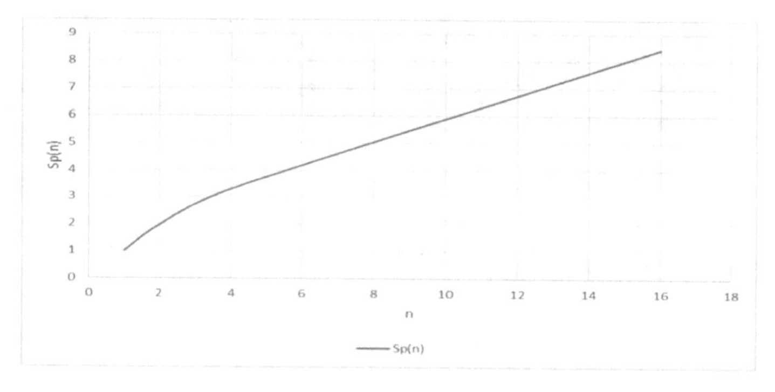

Fig. 4. Changing the acceleration factor depending on the architecture of the computer.

The results shown in Fig. 4, means that the developed parallel algorithm provides normal scalability, that is, the time of execution of the task is proportional to decrease with the growth of the number of processor cores. The greatest acceleration is achieved when using a 16-core processor.

In the work, the coefficient of efficiency is calculated by the formula $E_p(n) = S_p(n)/p$. Table 3 shows the corresponding values, and in Fig. 5 shows a graph of changing the efficiency factor, depending on the number of computer cores.

Table 3. The value of the efficiency factor

n	Ep(n)
1	1
2	0.973610192
4	0.822630455
16	0.37935147

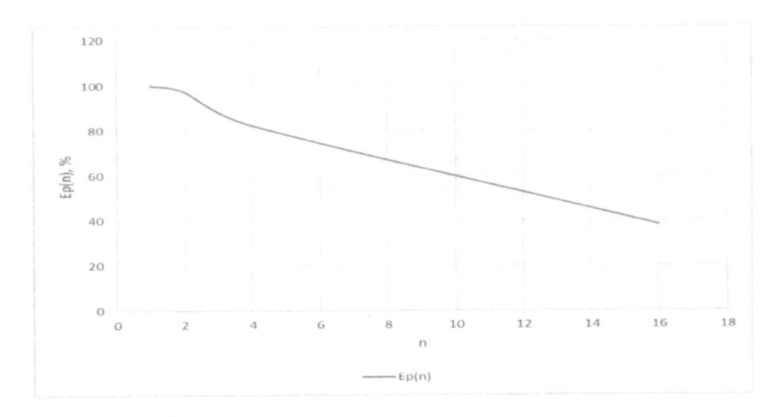

Fig. 5. Change the efficiency factor depending on the architecture of the computer.

On Fig. 5. Represents the value of the efficiency coefficient in percentages obtained by solving a given thousand SNRs in a thousandth order using the parallel algorithm of the Newton method on 16, 4, 2, and 1-core computers.

7 Conclusion

In the work, a software product was developed that allows solving multidimensional SNE, to evaluate the acceleration and efficiency of program execution using different number of processor cores. The developed algorithm leads to a significant reduction in the decision time. The results obtained are achieved by taking into account the architecture of the processor, which gives the opportunity to have several levels of parallelism. Newton's paralleling of the method reduces the execution time of the program, and therefore is effective. The proposed method is easily scaled to a different number of processes.

The research shows how real and important is the possibility of efficient use of multi-core personal computer for solving problems of large volumes. The time for solving a task on multi-core personal computers is greatly reduced when the program is automatically parsed using the OpenMP interface, which implements the parallel execution of both the cycles and the functional blocks of the program code.

References

1. Krste, A., et al.: The landscape of parallel computing research: a view from Berkeley. University of California, Berkeley. Technical report № UCB/EECS-2006-183, 56 p. (2006)
2. Yakovlev, M.F., Gerasymova, T.O., Nesterenko, A.N.: Characteristic feature of the solving both of non-linear systems and systems of ordinary differential equations on parallel computer. In: Proceedings of International Symposium "Optimization Problems of Computations" (OPC - XXXV). V.M. Glushkov Institute of cybernetics of NAS of Ukraine, Kyiv (2009), vol. 2, pp. 435–439 (2009)

3. Yakovlev, M.V.F., Nesterenko, A.N., Brusnikin, V.N.: Problems of the efficient solving of non-linear systems on multi-processor MIMD-architecture computers. Math. Mach. Syst. **4**, 12–17 (2014)
4. Khymych, A.N., Molchanov, Y.N., Popov, A.V., et al.: Parallel Algorithms for Solving Problems of Computational Mathematics, 248 p. Scientific Opinion, Kiev (2008)
5. Autar, K.: Nonlinear equations. In: Newton-Raphson Method-More Examples, Civil Engineering, 7 August, 4 p. (2009)
6. Autar, K.: Nonlinear equations. In: Newton-Raphson Method-More Examples, Electrical Engineering, 7 August, 4 p. (2009)
7. Autar, K.: Nonlinear equations. In: Newton-Raphson Method-More Examples, Mechanical Engineering, 7 August, 3 p. (2009)
8. Kakhaner, D., Mouler, K., Nэsh, S.: Numerical Methods and Software, p. 575. Mir publishing house, Moscow (1998)
9. Mochurad, L.I.: Method of reduction model for calculation of electrostatic fields of electronic optics systems. Sci. J. Radioelektronika Inf. Manage. **1**(48), 29–40 (2019). (In Ukrainian)
10. Voss, M.: OpenMP Share Memory Parallel Programming, Toronto, Canada (2003)
11. Chapman, B., Jost, G.: Ruud van der Pas: Using OpenMP: Portable Shared Memory Parallel Programming (Scientific and Engineering Computation). The MIT Press, Cambridge (2008)
12. Chandra, R., Menon, R., Dagum, L., Kohr, D., Maydan, D., McDonald, J.: Parallel Programming in OpenMP. Morgan Kaufinann Publishers, San Francisco (2000)
13. Ananth, G., Anshul, G., George, K., Vipin, K.: Introduction to Parallel Computing, 856 p. Addison Wesley (2003). ISBN-0-201-64865-2
14. Boyko, N.: A look trough methods of intellectual data analysis and their applying in informational systems. In: Scientific and Technical Conference Computer Sciences and Information Technologies (CSIT), 2016 XIth International, pp. 183–185. IEEE (2016)
15. Boyko, N.: Advanced technologies of big data research in distributed information systems. In: Radio Electronics, Computer Science, Control, vol. 4, pp. 66–77. Zaporizhzhya National Technical University, Zaporizhzhya (2017)
16. Boyko, N.: Machine learning on data lake. Monograph, p. 189, LAP Lambert Academic Publishing (2018)

Mathematical Modelling of Spatial Deformation Process of Soil Massif with Free Surface

Anatoliy Vlasyuk[1], Nataliia Zhukovska[2](✉), Viktor Zhukovskyy[2], and Rajab Hesham[2]

[1] National University of Ostroh Academy, Ostroh, Ukraine
[2] The National University of Water and Environmental Engineering, Rivne, Ukraine
n.a.zhukovska@gmail.com, v.v.zhukovskyy@nuwm.edu.ua

Abstract. The study of deformation processes of soil massifs with free surface under mass and heat transfer is important in the design, construction and operation of buildings. The article presents a mathematical model of the deformation state problem of the soil massif under mass and heat transfer and the present free surface in the three-dimensional case. The basic equations, boundary conditions and conditions of congruence for displacements, strains, stresses, and also additional functions are given. For computer modelling of the set boundary value problem, a software package for the capabilities of the Microsoft Visual Studio 2017 framework for Windows Desktop in the C# programming language was created. It is shown that the presence of a free surface that breaks the area of the investigated soil in the area of water-saturated soil and the area of the soil in its natural state changes the distributions of the displacements of the soil mass in these areas.

Keywords: Mathematical model · Soil massif · Stress-strain state · Mass and heat transfer · Free surface

1 Introduction

Modeling and research of deformation processes of soil massifs are the important problems in the construction and subsequent operation of various objects. The most common way of studying such problems is the mathematical modeling. Modern methods of mathematical modeling allow to predict the influence of various types of physical and chemical factors on the stress-strain state of the soil massif. In our time, this is an urgent topic due to various kinds of soil contamination and violation of their bearing capacity, which arise in the conditions of intensive development of chemical industry and construction.

In particular, under the influence of processes of filtration of groundwater, mass transfer, heat transfer, and other factors the stress-strain state (SSS) of soil massifs can change considerably, and consequently, it can lead to the earth's settling and the emergence of emergency situations.

N. Shakhovska and M. O. Medykovskyy (Eds.): CCSIT 2019, AISC 1080, pp. 107–120, 2020.
https://doi.org/10.1007/978-3-030-33695-0_9

Also important for the modeling and research of deformation processes in soil environments is the consideration of the availability of free surface of groundwater. The change position of the groundwater level may lead to the change soil bases of the objects and eventually may lead to an accident of these structures.

Therefore, the consideration of the presence of free surface in soils and the processes of filtration of saline solutions in non-isothermal conditions, heat transfer and mass transfer is more real and adequately reflects the deformation processes studied in such environments, in particular, in the three-dimensional case.

The purpose of this article is to formulate a spatial problem and construct a mathematical model of the stress-strain state of a soil massif with a free surface under the influence of mass and heat transfer.

2 Related Works

Numerous mathematical models have been developed for simulation of the heat and mass transfer in different types of media according to the use of Darcy's, Fick's, and Fourier's laws [8, 12]. Also, a set of mathematical models took into account the influence of the Cauchy law. Research of deformation and filtration processes of porous media under the influence of various factors carried out by many scientists. In particular, in the works [1, 2, 6, 11]. The recent researches created models for fractional-order plasticity of state-dependent behaviour of granular soils without using plastic potential described [9], large plastic deformations of cohesive soils using sequential limit analysis [3] and proposed system approach for monitoring of World Heritage sites placed on the active landslides [10]. This allows developing earlier and more reliable prediction of the building structures strain-stress state evolution, the intensity of the CLLS landslide activity, carrying out correlation among geliogenous and lithogenous factors, etc. At the same time mathematical expectation, correlation function and the continuum damage parameter variance become important for the reliability analysis for most engineering structures as well as for the stability of soil bases. This analysis is performed on the basis of static and dynamic stress-strain states in [7].

However, in all the aforementioned works, the effect of heat and mass transfer during the filtration of saline solutions under non-isothermal conditions on deformation processes in soil media is not taken into account [14].

3 Formulation of the Problem

Consider a soil massif that has the shape of a rectangular parallelepiped in an area Ω with a free surface Γ_0 (see Fig. 1).

Figure 1 introduces the following notation: Ω_1 is an area of water-saturated soil massif; Ω_2 is an area of the soil massif in the natural state, $\Omega = \Omega_1 \cup \Omega_2$. The ground array which is elastically deformable within a linear theory of elasticity with different elastic parameters (Lame coefficients) $\lambda(c, T)$, $\mu(c, T)$ is considered. These parameters depend from salt concentration and temperature. The gravity acts on the soil, and in the case of a water-saturated layer of soil, the archimedean and filtration forces acts.

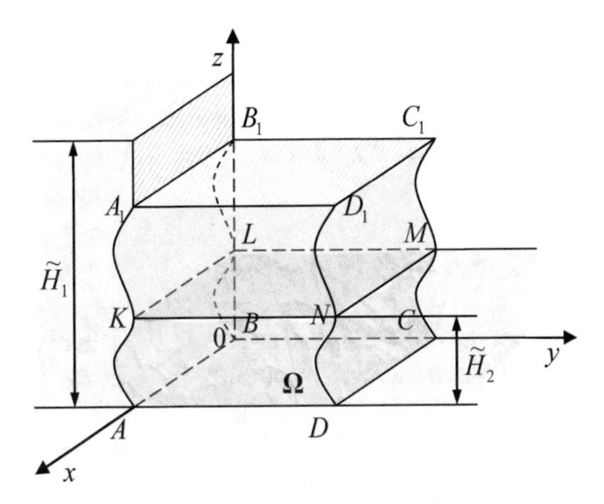

Fig. 1. Scheme of a soil massif with a free surface.

Let us mark the plane ABCD in Fig. 1 as Γ_1, $A_1B_1C_1D_1$ as Γ_2, ABA_1B_1, CDC_1D_1 and BCB_1C_1, ADA_1D_1 as Γ_3, Γ_4 and Γ_5, Γ_6 respectively. Planes AKLB and KA_1B_1L as Γ_3^1 and Γ_3^2, moreover $\Gamma_3 = \Gamma_3^1 \cup \Gamma_3^2$. Planes DNMC and ND_1C_1M let us mark as Γ_4^1 and Γ_4^2 whereas, $\Gamma_4 = \Gamma_4^1 \cup \Gamma_4^2$. Planes BCML, LMC_1D_1 and ADNK, KND_1A_1 let's mark as Γ_5^1, Γ_5^2 and Γ_6^1, Γ_6^2 respectively, when $\Gamma_5 = \Gamma_5^1 \cup \Gamma_5^2$ and $\Gamma_6 = \Gamma_6^1 \cup \Gamma_6^2$.

Planes Γ_1 and Γ_2 are considered as insulated and impenetrable, where as Γ_3, Γ_4, Γ_5, Γ_6 are drainaged. The free surface of groundwater Γ_0 is considered to be stationary.

Piezometric pressure value $\tilde{H}_1$ and temperature $\tilde{T}_1$ and salt concentration $\tilde{C}_1$ in an aquatic pool on the borders of a soil massif $\Gamma_3^1 \cup \Gamma_5^1$, as well as the value of piezometric pressure $\tilde{H}_2$, temperature $\tilde{T}_2$ and the condition of rapid removal of salts in the water basin on the borders of the soil massif $\Gamma_4^1 \cup \Gamma_6^1$ accordingly, whereas $\tilde{H}_1 > \tilde{H}_2$, $\tilde{T}_1 > \tilde{T}_2$ are specified. The transfer of water-dissolved substances and heat to the filtration stream is carried out as a result of the pressure differences. The processes of filtration of substances dissolved in water and heat are occurred according to the laws of Darcy, Fica and Fourier in this case.

It is assumed that there is no displacement of the lower one (Γ_1) and (or) upper (Γ_2) areas of the soil massif.

Thus, in the investigated area of the soil massif Ω it is necessary to calculate:

(1) the distribution of the displacement vector for a soil mass in a water-saturated and natural state $\mathbf{u_i}(\mathbf{X}) = (U_i(\mathbf{X}),\ V_i(\mathbf{X}),\ W_i(\mathbf{X}))$, $\mathbf{X} \in \Omega_i$, $i = \overline{1,\ 2}$;

(2) the distribution of normal components of deformation $\varepsilon_x^{(i)}$, $\varepsilon_y^{(i)}$, $\varepsilon_z^{(i)}$ and tangent components of deformation $\varepsilon_{xy}^{(i)}$, $\varepsilon_{xz}^{(i)}$, $\varepsilon_{yz}^{(i)}$, $i = \overline{1,\ 2}$ for a soil massif located in water-saturated and natural conditions;

(3) the distribution of normal stress components $\sigma_x^{(i)}$, $\sigma_y^{(i)}$, $\sigma_z^{(i)}$ and tangential constituent stresses $\tau_{xy}^{(i)}$, $\tau_{xz}^{(i)}$, $\tau_{yz}^{(i)}$, $i = \overline{1,\ 2}$ for a soil massif located in water-saturated and natural conditions;

(4) distribution of piezometric pressure $h(\mathbf{X}, t)$, $\mathbf{X} \in \Omega_1$, $t > 0$ in the water-saturated area of the soil massif;

(5) distribution of the concentration of saline solutions $c(\mathbf{X}, t)$, $\mathbf{X} \in \Omega_1$, $t > 0$ in the water-saturated area of the soil massif;
temperature distribution $T_i(\mathbf{X}, t)$, $\mathbf{X} \in \Omega_i$, $i = \overline{1, 2}$, $t > 0$ for a soil massif located in water-saturated and natural conditions.

4 Mathematical Model of the Problem

The mathematical model of the deformations processes problem of the influence of mass and heat transfer of the soil massif with a free surface in generally accepted notation can be described as the next three-dimensional boundary value problem [12, 13]:

the system of Lamé equations describing SSS for displacements under the mass and heat transfer in water-saturated and natural state of soil massif

$$
\begin{aligned}
&\mu_i \Delta U_i + (\lambda_i + \mu_i)\frac{\partial}{\partial x}\left(\frac{\partial U_i}{\partial x} + \frac{\partial V_i}{\partial y} + \frac{\partial W_i}{\partial z}\right) + \frac{\partial \lambda_i}{\partial x}\left(\frac{\partial U_i}{\partial x} + \frac{\partial V_i}{\partial y} + \frac{\partial W_i}{\partial z}\right)\\
&\quad + 2\frac{\partial \mu_i}{\partial x}\frac{\partial U_i}{\partial x} + \frac{\partial \mu_i}{\partial y}\left(\frac{\partial U_i}{\partial y} + \frac{\partial V_i}{\partial x}\right) + \frac{\partial \mu_i}{\partial z}\left(\frac{\partial U_i}{\partial z} + \frac{\partial W_i}{\partial x}\right)\\
&\quad - \left(\left(3\frac{\partial \lambda_i}{\partial x} + 2\frac{\partial \mu_i}{\partial x}\right)T_i + (3\lambda_i + 2\mu_i)\frac{\partial T_i}{\partial x}\right)\alpha_T^{(i)} + X_i = 0,\\
&\mu_i \Delta V_i + (\lambda_i + \mu_i)\frac{\partial}{\partial y}\left(\frac{\partial U_i}{\partial x} + \frac{\partial V_i}{\partial y} + \frac{\partial W_i}{\partial z}\right) + \frac{\partial \lambda_i}{\partial y}\left(\frac{\partial U_i}{\partial x} + \frac{\partial V_i}{\partial y} + \frac{\partial W_i}{\partial z}\right)\\
&\quad + 2\frac{\partial \mu_i}{\partial y}\frac{\partial V_i}{\partial y} + \frac{\partial \mu_i}{\partial x}\left(\frac{\partial U_i}{\partial y} + \frac{\partial V_i}{\partial x}\right) + \frac{\partial \mu_i}{\partial z}\left(\frac{\partial V_i}{\partial z} + \frac{\partial W_i}{\partial y}\right)\\
&\quad - \left(\left(3\frac{\partial \lambda_i}{\partial y} + 2\frac{\partial \mu_i}{\partial y}\right)T_i + (3\lambda_i + 2\mu_i)\frac{\partial T_i}{\partial y}\right)\alpha_T^{(i)} + Y_i = 0,\\
&\mu_i \Delta W_i + (\lambda_i + \mu_i)\frac{\partial}{\partial z}\left(\frac{\partial U_i}{\partial x} + \frac{\partial V_i}{\partial y} + \frac{\partial W_i}{\partial z}\right) + \frac{\partial \lambda_i}{\partial z}\left(\frac{\partial U_i}{\partial x} + \frac{\partial V_i}{\partial y} + \frac{\partial W_i}{\partial z}\right)\\
&\quad + 2\frac{\partial \mu_i}{\partial z}\frac{\partial W_i}{\partial z} + \frac{\partial \mu_i}{\partial x}\left(\frac{\partial U_i}{\partial z} + \frac{\partial W_i}{\partial x}\right) + \frac{\partial \mu_i}{\partial y}\left(\frac{\partial V_i}{\partial z} + \frac{\partial W_i}{\partial y}\right)\\
&\quad - \left(\left(3\frac{\partial \lambda_i}{\partial z} + 2\frac{\partial \mu_i}{\partial z}\right)T_i + (3\lambda_i + 2\mu_i)\frac{\partial T_i}{\partial z}\right)\alpha_T^{(i)} + Z_i = 0, \quad \mathbf{X} \in \Omega,
\end{aligned} \tag{1}
$$

where $\lambda_i = \lambda_1(c, T_1)$, $\mu_i = \mu_1(c, T_1)$ at $\mathbf{X} \in \Omega_1$, $i = 1$ and $\lambda_i = \lambda_2(T_2)$, $\mu_i = \mu_2(T_2)$ at $\mathbf{X} \in \Omega_2$, $i = 2$ and the components of mass forces X, Y, Z are calculated by the formulas

$$
\begin{aligned}
&X_i = \begin{cases} \frac{dp_1}{dx}, & \mathbf{X} \in \Omega_1,\ i = 1,\\ 0, & \mathbf{X} \in \Omega_2,\ i = 2,\end{cases} \quad Y_i = \begin{cases} \frac{dp_2}{dy}, & \mathbf{X} \in \Omega_1,\ i = 1,\\ 0, & \mathbf{X} \in \Omega_2,\ i = 2,\end{cases}\\
&Z_i = \begin{cases} \gamma_{zv} + \frac{dp_3}{dy}, & \mathbf{X} \in \Omega_1,\ i = 1,\\ \gamma_{pr}, & \mathbf{X} \in \Omega_2,\ i = 2;\end{cases}
\end{aligned} \tag{2}
$$

the equation of convective diffusion in the presence of mass and heat transfer for a water-saturated area of a soil massif

$$\nabla \cdot (\mathbf{D}(c, T_1)\nabla c) - \upsilon\nabla c - \gamma(c - C_m) + \nabla \cdot (\mathbf{D}_{\mathbf{T}}^{(\mathbf{1})}\nabla T_1) = n_p \frac{\partial c}{\partial t}, \mathbf{X} \in \mathbf{\Omega}_1, t > 0, \quad (3)$$

the equation of convective heat transfer in both areas of the soil massif

$$\nabla \cdot (\boldsymbol{\lambda}_{\mathbf{T}}^{(\mathbf{i})}\nabla T_i) - \rho c_\rho \bar{\upsilon}\nabla T_i = c_T^{(i)} \frac{\partial T_i}{\partial t}, \mathbf{X} \in \Omega_i, i = \overline{1, 2}, t > 0, \quad (4)$$

where $\bar{\upsilon} = \upsilon$ for $\mathbf{X} \in \Omega_1$ and $\bar{\upsilon} = \mathbf{0}$ for $\mathbf{X} \in \Omega_2$;
the generalized equation of filtration of saline solutions in non-isothermal conditions and the equation of continuity of a process in a water-saturated area of a soil massif

$$\upsilon = -\mathbf{K}(c, T_1)\nabla h + \mathbf{v_c}\nabla c + \mathbf{v_T}\nabla T_1, \quad div\, \upsilon + \frac{\partial n_p}{\partial t} = 0, \mathbf{X} \in \Omega_1, t > 0, \quad (5)$$

the specific flow of dissolved salts in the case of non-isothermal conditions in a water-saturated soil massif

$$\mathbf{q_c} = \upsilon c - \mathbf{D}(c, T_1)\nabla c - \mathbf{D}_{\mathbf{T}}^{(\mathbf{1})}\nabla T_1, \mathbf{X} \in \Omega_1, t > 0, \quad (6)$$

the generalized Fourier law in the case of convective mass and heat transfer

$$\mathbf{q}_{\mathbf{T}}^{(\mathbf{i})} = \rho c_\rho \bar{\upsilon} T_i - \boldsymbol{\lambda}_{\mathbf{T}}^{(\mathbf{i})}\nabla T_i, \mathbf{X} \in \Omega_i, i = \overline{1, 2}, t > 0, \quad (7)$$

the normal and shear strains

$$\begin{gathered} \varepsilon_x^{(i)} = \frac{\partial U_i}{\partial x}, \quad \varepsilon_y^{(i)} = \frac{\partial V_i}{\partial y}, \quad \varepsilon_z^{(i)} = \frac{\partial W_i}{\partial z}, \\ \varepsilon_{xy}^{(i)} = \frac{1}{2}\left(\frac{\partial U_i}{\partial y} + \frac{\partial V_i}{\partial x}\right), \quad \varepsilon_{xz}^{(i)} = \frac{1}{2}\left(\frac{\partial U_i}{\partial z} + \frac{\partial W_i}{\partial x}\right), \quad \varepsilon_{yz}^{(i)} = \frac{1}{2}\left(\frac{\partial V_i}{\partial z} + \frac{\partial W_i}{\partial y}\right), \\ \mathbf{X} \in \Omega_i, \quad i = \overline{1, 2}, \end{gathered} \quad (8)$$

the normal and shear stresses

$$\begin{aligned} \sigma_x^{(i)} &= \lambda_i \varepsilon_\theta^{(i)} + 2\mu_i \varepsilon_x^{(i)} - (3\lambda_i + 2\mu_i)\alpha_T^{(i)} \bar{T}_i, \sigma_y^{(i)} = \lambda_i \varepsilon_\theta^{(i)} + 2\mu_i \varepsilon_y^{(i)} - (3\lambda_i + 2\mu_i)\alpha_T^{(i)} \bar{T}_i, \\ \sigma_z^{(i)} &= \lambda_i \varepsilon_\theta^{(i)} + 2\mu_i \varepsilon_z^{(i)} - (3\lambda_i + 2\mu_i)\alpha_T^{(i)} \bar{T}_i, \tau_{xy}^{(i)} = 2\mu_i \varepsilon_{xy}^{(i)}, \tau_{xz}^{(i)} = 2\mu_i \varepsilon_{xz}^{(i)}, \tau_{yz}^{(i)} = 2\mu_i \varepsilon_{yz}^{(i)}, \end{aligned} \quad (9)$$

where $\varepsilon_\theta^{(i)} = \varepsilon_x^{(i)} + \varepsilon_y^{(i)} + \varepsilon_z^{(i)}$, $\mathbf{X} \in \Omega_i$, $i = \overline{1, 2}$.

On a free surface Γ_0 (depression curve) the function $h(\mathbf{X}, t)$ described by the following conditions:

$$h(\mathbf{X}, t)|_{\Gamma_0} = z, (\upsilon, \mathbf{n})|_{\Gamma_0} = n_p \frac{\partial \varphi}{\partial t}, \varphi(\mathbf{X}, t) = (z - h(\mathbf{X}, t))|_{\Gamma_0} = 0, t > 0, \quad (10)$$

where the ratio $\varphi(\mathbf{X}, t)$ describes the free surface Γ_0.

For concentrations of salt solutions in non-isothermal conditions on the depression curve, the following boundary condition is fulfilled:

$$\left(\mathbf{D}(c,T_1)\nabla c+\mathbf{D}_{\mathbf{T}}^{(1)}\nabla T_1,\,\mathbf{n}\right)\Big|_{\Gamma_0}=0,t>0. \tag{11}$$

The boundary conditions on the boundaries of the soil massif and the conditions for conjugation of the ideal contact for displacements and temperature, as well as the boundary conditions for the piezometric pressure and the concentration of salts on the boundaries of the water-saturated soils, have the form

$$(\upsilon,\,\mathbf{n})|_{\Gamma_1}=0,h(\mathbf{X},t)|_{\Gamma_3^1\cup\Gamma_5^1}=\tilde{H}_1(\mathbf{X},t),h(\mathbf{X},t)|_{\Gamma_4^1\cup\Gamma_6^1}=\tilde{H}_2(\mathbf{X},t),\mathbf{X}\in\Omega_1,t>0, \tag{12}$$

$$(\mathbf{q_c},\,\mathbf{n})|_{\Gamma_1}=0,c(\mathbf{X},t)|_{\Gamma_3^1\cup\Gamma_5^1}=\tilde{C}_1(\mathbf{X},t),(\mathbf{D}\nabla c+\mathbf{D}_{\mathbf{T}}^{(1)}\nabla T_1,\,\mathbf{n})\Big|_{\Gamma_4^1\cup\Gamma_6^1}=0, \quad \mathbf{X}\in\Omega_1,t>0, \tag{13}$$

$$(\mathbf{q}_{\mathbf{T}}^{(\mathbf{i})},\,\mathbf{n})\Big|_{\Gamma_1\cup\Gamma_2}=0,T_i(\mathbf{X},t)|_{\Gamma_3\cup\Gamma_5}=\tilde{T}_1(\mathbf{X},t),T_i(\mathbf{X},t)|_{\Gamma_4\cup\Gamma_6}=\tilde{T}_2(\mathbf{X},t),\mathbf{X}\in\Omega_i, \quad i=\overline{1,2},t>0, \tag{14}$$

$$\mathbf{u_1}(\mathbf{X})=0,\mathbf{X}\in\Gamma_1,\mathbf{u_2}(\mathbf{X})=0,\mathbf{X}\in\Gamma_2, \tag{15}$$

$$\sigma_n^{(i)}=0,\tau_n^{(i)}=0,\mathbf{X}\in\Gamma_3\cup\Gamma_4\cup\Gamma_5\cup\Gamma_6,i=\overline{1,2}, \tag{16}$$

$$[T_i(\mathbf{X},t)]|_{\Gamma_0}=0,\,[\mathbf{q}_{\mathbf{T}}^{(\mathbf{i})}]\Big|_{\Gamma_0}=0,\mathbf{X}\in\Gamma_0,i=\overline{1,2},t>0, \tag{17}$$

$$[u_n^{(i)}]|_{\Gamma_0}=[u_s^{(i)}]|_{\Gamma_0}=0,[\sigma_n^{(i)}]|_{\Gamma_0}=[\tau_s^{(i)}]|_{\Gamma_0}=0,\mathbf{X}\in\Gamma_0,i=\overline{1,2}, \tag{18}$$

$$h(\mathbf{X},0)=\tilde{H}_0(\mathbf{X}),c(\mathbf{X},0)=\tilde{C}_0(\mathbf{X}),\mathbf{X}\in\Omega_1,T_i(\mathbf{X},0)=\begin{cases}\tilde{T}_0^1(\mathbf{X}), & \mathbf{X}\in\Omega_1,\\ \tilde{T}_0^2(\mathbf{X}), & \mathbf{X}\in\Omega_2.\end{cases} \tag{19}$$

In (1)–(19) uses the following notation: $\mathbf{X}=(x,y,z)$, point of region $\Omega=\Omega_1\cup\Omega_2$; Ω_1, water-saturated soil massif; Ω_2, soil massif at natural state; Γ, border of region Ω; Γ_0, free surface in region Ω; t, time, $t>0$, day; $\mathbf{u_i}=(U_i,V_i,W_i)$, $i=\overline{1,2}$, displacements vector, m; U_i, $i=\overline{1,2}$, displacements along the Ox axis, m; V_i, $i=\overline{1,2}$, displacements along the Oy axis, m; W_i, $i=\overline{1,2}$, displacements along the Oz axis, m; $c(\mathbf{X},t)$, concentration of salt solution, $\frac{g}{l}$; $T_i(\mathbf{X},t)$, $i=\overline{1,2}$, temperature, °C; $h(\mathbf{X},t)$, piezometric head, m; X_i, Y_i and Z_i, $i=\overline{1,2}$, are components of mass force, H; $\varepsilon_x^{(i)}$, $\varepsilon_y^{(i)}$, $\varepsilon_z^{(i)}$ and $\varepsilon_{xy}^{(i)}$, $\varepsilon_{xz}^{(i)}$, $\varepsilon_{yz}^{(i)}$, $i=\overline{1,2}$, the normal and shear strains; $\sigma_x^{(i)}$, $\sigma_y^{(i)}$, $\sigma_z^{(i)}$ and $\tau_{xy}^{(i)}$, $\tau_{xz}^{(i)}$, $\tau_{yz}^{(i)}$, $i=\overline{1,2}$, the normal and shear stresses, Pa; p_1, p_2, p_3, filtration pressure of salt solution, $p_1=\gamma_p(h(\mathbf{X},t)-x)$, $p_2=\gamma_p(h(\mathbf{X},t)-y)$, $p_3=\gamma_p(h(\mathbf{X},t)-z)$, Pa; υ, speed vector of filtration of the salts in nonisothermal conditions, $\frac{\text{m}}{\text{day}}$; $\lambda_1(c, T_1)$ and $\mu_1(c, T_1)$, Lamé coefficients which depend on the salts concentration and temperature, Pa; $\lambda_2(T_2)$ and $\mu_2(T_2)$, Lamé coefficients which depend on the temperature, Pa; $\mathbf{K}(c, T_1)$, filtration tensor which depend on the concentration of filtrating solution and temperature, $\frac{\text{m}}{\text{day}}$; $\mathbf{D}$ and $\mathbf{D_T^{(1)}}$, tensors of convective diffusion and thermodiffusion respectively, $\frac{\text{m}^2}{\text{day}}$; $\boldsymbol{\lambda}_\mathbf{T}^{(\mathbf{i})}$, $i = \overline{1,2}$, effective heat conductivity tensor of humid soil, $\frac{\text{J}}{\text{m}\cdot\text{deg}\cdot\text{day}}$; n_p, soil porosity; c_ρ, pore solution specific heat capacity at a constant pressure, $\frac{\text{J}}{\text{kg}\cdot\text{deg}}$; $c_T^{(i)}$, $i = \overline{1,2}$, soil specific heat capacity at a constant volume, $\frac{\text{J}}{\text{m}^3\cdot\text{deg}}$; $\mathbf{v_c}$, tensor of chemical osmosis, $\frac{\text{m}^{-5}}{\text{kg}\cdot\text{day}}$; $\mathbf{v_T}$, tensor of thermal osmosis, $\frac{\text{m}^2}{\text{kg}\cdot\text{day}}$; ρ, pore solution density, $\frac{\text{kg}}{\text{m}^3}$; γ, rate constant of mass transfer, $\frac{1}{\text{day}}$; C_m, concentration of limiting saturation, $\frac{g}{l}$; γ_{zv}, the proportion of the soil that is in a suspended state, $\frac{\text{Pa}}{\text{m}}$; $\mathbf{n}$, vector of direction cosines of the outer normal to the region boundary; [] is an jump function; Γ_1, Γ_2, Γ_3, Γ_4, Γ_5, Γ_6 are areas $ABCD$, $A_1B_1C_1D_1$, ABA_1B_1, CDC_1D_1, ADA_1D_1, BCB_1C_1 of the region Ω respectively; $\tilde{H}_0(\mathbf{X})$, $\tilde{C}_0(\mathbf{X})$, $\tilde{T}_0^1(\mathbf{X})$, $\tilde{T}_0^2(\mathbf{X})$, $\tilde{H}_1(\mathbf{X}, t)$, $\tilde{H}_2(\mathbf{X}, t)$, $\tilde{C}_1(\mathbf{X}, t)$, $\tilde{T}_1(\mathbf{X}, t)$, $\tilde{T}_2(\mathbf{X}, t)$ are some given functions; $\bar{T}_i = T_i(\mathbf{X}, t) - T_i(\mathbf{X}, 0)$, $i = \overline{1,2}$, the temperature difference at a given time and temperature not stressful state at a given point; $\alpha_T^{(i)}$, $i = \overline{1,2}$, the average coefficient of linear thermal expansion at temperatures (T_0, T) [4], $\alpha_T = \frac{1}{T}\int\limits_0^{\bar{T}} \alpha\, d\bar{T}$, $\alpha = \frac{\Delta l}{lT}$, linear expansion coefficient, Δl, change of linear dimensions of the test sample.

The dependence of Lame's coefficients on the concentration of saline solutions is follow [5]:

$$\lambda(c) = a_3^2 \cdot c^3 + a_2^2 \cdot c^2 + a_1^2 \cdot c + a_0^2,$$

where $a_3^2 = -1798{,}96$, $a_2^2 = 4314{,}732$, $a_1^2 = -2615{,}37$, $a_0^2 = 2545{,}743$;

$$\mu(c) = a_3^3 \cdot c^3 + a_2^3 \cdot c^2 + a_1^3 \cdot c + a_0^3,$$

where $a_3^3 = -1205{,}28$, $a_2^3 = 2880{,}321$, $a_1^3 = -1741{,}92$, $a_0^3 = 1696{,}324$;

$$E(c) = a_3^4 \cdot c^3 + a_2^4 \cdot c^2 + a_1^4 \cdot c + a_0^4,$$

where $a_3^4 = -0{,}000393$, $a_2^4 = 0{,}1878866$, $a_1^4 = -22{,}70202$, $a_0^4 = 4410{,}552$, $c \in [0, 1]$ is a dimensionless value.

The dependences of the filtration coefficient on the concentration of saline solution and temperature are follow:

$$k_1(c) = a_5^1 \cdot c^5 + a_4^1 \cdot c^4 + a_3^1 \cdot c^3 + a_2^1 \cdot c^2 + a_1^1 \cdot c + a_0^1,$$

where $a_5^1 = 5,9404 \cdot 10^{-2}$, $a_4^1 = -1,6703 \cdot 10^{-1}$, $a_3^1 = 1,7051 \cdot 10^{-1}$, $a_2^1 = -7,4311 \cdot 10^{-2}$, $a_1^1 = 1,0563 \cdot 10^{-2}$, $a_0^1 = 1,0054 \cdot 10^{-3}$;

$$k_2(T) = b_5 \cdot T^5 + b_4 \cdot T^4 + b_3 \cdot T^3 + b_2 \cdot T^2 + b_1 \cdot T + b_0,$$

where $b_5 = 1,4154 \cdot 10^{-2}$, $b_4 = -2,6097 \cdot 10^{-2}$, $b_3 = 1,0819 \cdot 10^{-2}$, $b_2 = 1,2844 \cdot 10^{-4}$, $b_1 = 1,0404 \cdot 10^{-2}$, $b_0 = 3,0925 \cdot 10^{-3}$, $T \in [0,1]$ is a dimensionless value.

The dependence of the coefficient of filtration on the concentration of salt and temperature is follow:

$$k(c,T) = \frac{1}{k_0} k_1(c) \cdot k_2(T), \quad k_0 = 10^{-4} \frac{м}{доба},$$

where k_0 is the filtration rate for pure water at a temperature 20 °C.

5 Numerical Solution of the Problem

Numerical methods of mathematical physics, in particular, the method of finite differences, the Gauss-Seidel method were be used for the numerical solution of the set boundary value problem.

For approximate the Eq. (2) we use a pattern of the “box” (see Fig. 2).

To find the displacement values $U_i(\mathbf{X})$, $V_i(\mathbf{X})$ and $W_i(\mathbf{X})$, $i = \overline{1,\ 2}$, the Gauss-Seidel iteration method was used. To do this, the given equations are derived from equations of equilibrium in the following form:

$$\begin{aligned}
U^{(z+1)}_{i,j_1,j_2,j_3} &= \bar{A}_1 U^{(z)}_{i,j_1+1,j_2,j_3} + \bar{B}_1 U^{(z+1)}_{i,j_1-1,j,j_3} + \bar{C}_1 U^{(z)}_{i,j_1,j_2+1,j_3} + \bar{D}_1 U^{(z+1)}_{i,j_1,j_2-1,j_3} \\
&\quad + \bar{E}_1 U^{(z)}_{i,j_1,j_2,j_3+1} + \bar{G}_1 U^{(z+1)}_{i,j_1,j_2,j_3-1} + \bar{F}_1 (V^{(z)}_{i,j_1,j_2,j_3}, W^{(z)}_{i,j_1,j_2,j_3}, T^{(s)}_{i,j_1,j_2,j_3}), \\
V^{(z+1)}_{i,j_1,j_2,j_3} &= \bar{A}_2 V^{(z)}_{i,j_1+1,j_2,j_3} + \bar{B}_2 V^{(z+1)}_{i,j_1-1,j_2,j_3} + \bar{C}_2 V^{(z)}_{i,j_1,j_2+1,j_3} + \bar{D}_2 V^{(z+1)}_{i,j_1,j_2-1,j_3} \\
&\quad + \bar{E}_2 V^{(z)}_{i,j_1,j_2,j_3+1} + \bar{G}_2 V^{(z+1)}_{i,j_1,j_2,j_3-1} + \bar{F}_2 (U^{(z+1)}_{i,j_1,j_2,j_3}, W^{(z)}_{i,j_1,j_2,j_3}, T^{(s)}_{i,j_1,j_2,j_3}), \\
W^{(z+1)}_{i,j_1,j_2,j_3} &= \bar{A}_3 W^{(z)}_{i,j_1+1,j_2,j_3} + \bar{B}_3 W^{(z+1)}_{i,j_1-1,j_2,j_3} + \bar{C}_3 W^{(z)}_{i,j_1,j_2+1,j_3} + \bar{D}_3 W^{(z+1)}_{i,j_1,j_2-1,j_3} \\
&\quad + \bar{E}_3 W^{(z)}_{i,j_1,j_2,j_3+1} + \bar{G}_3 W^{(z+1)}_{i,j_1,j_2,j_3-1} + \bar{F}_3 (U^{(z+1)}_{i,j_1,j_2,j_3}, V^{(z+1)}_{i,j_1,j_2,j_3}, T^{(s)}_{i,j_1,j_2,j_3}),
\end{aligned} \tag{20}$$

where $\bar{A}_{i_3}$, $\bar{B}_{i_3}$, $\bar{C}_{i_3}$, $\bar{D}_{i_3}$, $\bar{E}_{i_3}$, $\bar{G}_{i_3}$, $\bar{F}_{i_3}$ are some known functions.

The iterations by the formulas (20) are carried out until the difference of values does not become less than a given precision ε

$$\left| U^{(z+1)}_{i,j_1,j_2,j_3} - U^{(z)}_{i,j_1,j_2,j_3} \right| \le \varepsilon, \ \left| V^{(z+1)}_{i,j_1,j_2,j_3} - V^{(z)}_{i,j_1,j_2,j_3} \right| \le \varepsilon, \ \left| W^{(z+1)}_{i,j_1,j_2,j_3} - W^{(z)}_{i,j_1,j_2,j_3} \right| \le \varepsilon$$

where $j_1 = \overline{1, m_1 - 1}$, $j_2 = \overline{1, m_2 - 1}$, $j_3 = \overline{1, m_3 - 1}$, $i = \overline{1,\ 2}$, z is an iteration number.

We approximate the boundary conditions (16) for the lateral planes of the soil massif $\Gamma_3 \cup \Gamma_4 \cup \Gamma_5 \cup \Gamma_6$.

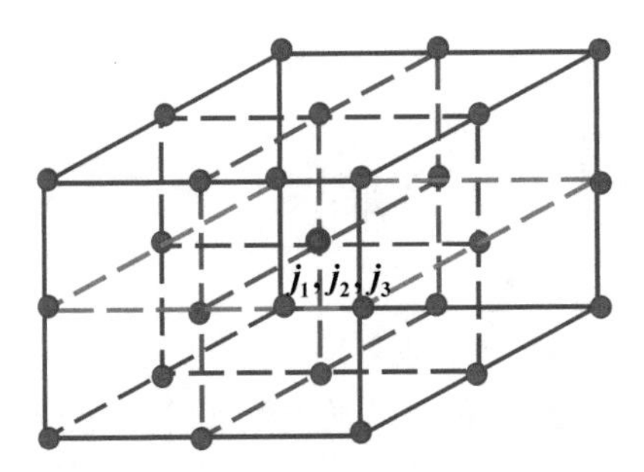

Fig. 2. A pattern of "box"

For conditions $\sigma_n^{(i)} = 0$, $\tau_n^{(i)} = 0$, on the plane Γ_4 we obtain the following:

$$\sigma_n^{(i)} = \sigma_y^{(i)} = 0,\ \tau_n^{(i)} = \sqrt{(\tau_{xy}^{(i)})^2 + (\tau_{zy}^{(i)})^2} = 0,\ i = \overline{1,\ 2}.$$

Then we have

$$\sigma_y^{(i)} = \lambda_i \varepsilon_\theta^{(i)} + 2\mu_i \varepsilon_y^{(i)} - (3\lambda_i + 2\mu_i)\alpha_T^{(i)} \bar{T}_i = 0,$$

$$\tau_n^{(i)} = \sqrt{(2\mu_i \varepsilon_{xy}^{(i)})^2 + (2\mu_i \varepsilon_{zy}^{(i)})^2} = \sqrt{\mu_i^2 \left(\frac{\partial U_i}{\partial y} + \frac{\partial V_i}{\partial x}\right)^2 + \mu_i^2 \left(\frac{\partial V_i}{\partial z} + \frac{\partial W_i}{\partial y}\right)^2} = 0,$$
$i = \overline{1,\ 2}$.

Thus, we obtain the values of the displacements on the lateral plane Γ_4

$$U_{i,j_1,m_2,j_3} = U_{i,j_1,m_2-1,j_3} - \frac{h_2}{2h_1}\left(V_{i,j_1+1,m_2,j_3} - V_{i,j_1-1,m_2,j_3}\right),$$

$$V_{i,j_1,m_2,j_3} = V_{i,j_1,m_2-1,j_3} - \frac{h_2}{\lambda_{i,j_1,m_2,j_3} + 2\mu_{i,j_1,m_2,j_3}} \times \left(\frac{\lambda_{i,j_1,m_2,j_3}}{2}\left(\frac{U_{i,j_1+1,m_2,j_3} - U_{i,j_1-1,m_2,j_3}}{h_1} + \frac{W_{i,j_1,m_2,j_3+1} - W_{i,j_1,m_2,j_3-1}}{h_3}\right) + \theta\right),$$

$$W_{i,j_1,m_2,j_3} = W_{i,j_1,m_2-1,j_3} - \frac{h_2}{2h_3}\left(V_{i,j_1,m_2,j_3+1} - V_{i,j_1,m_2,j_3-1}\right),$$

where $\theta = \left(3\lambda_{i,j_1,m_2,j_3} + 2\mu_{i,j_1,m_2,j_3}\right)\alpha_T^{(i)} \bar{T}_{i,j_1,m_2,j_3}^{(s)}$, $j_1 = \overline{1, m_1 - 1}$, $j_3 = \overline{1, m_3 - 1}$, $i = \overline{1,\ 2}$, $s = \overline{0, n_1}$.

Similarly, for the lateral plane Γ_3 we have

$$U_{i,j_1,0,j_3} = U_{i,j_1,1,j_3} + \frac{h_2}{2h_1}\left(V_{i,j_1+1,0,j_3} - V_{i,j_1-1,0,j_3}\right),$$

$$V_{i,j_1,0,j_3} = V_{i,j_1,1,j_3} + \frac{h_2}{\lambda_{i,j_1,0,j_3} + 2\mu_{i,j_1,0,j_3}} \times \left(\frac{\lambda_{i,j_1,0,j_3}}{2}\left(\frac{U_{i,j_1+1,0,j_3} - U_{i,j_1-1,0,j_3}}{h_1} + \frac{W_{i,j_1,0,j_3+1} - W_{i,j_1,0,j_3-1}}{h_3}\right) - \theta'\right),$$

$$W_{i,j_1,0,j_3} = W_{i,j_1,1,j_3} + \frac{h_2}{2h_3}\left(V_{i,j_1,0,j_3+1} - V_{i,j_1,0,j_3-1}\right),$$

where $\theta' = \left(3\lambda_{i,j_1,0,j_3} + 2\mu_{i,j_1,0,j_3}\right)\alpha_T^{(i)}\bar{T}^{(s)}_{i,j_1,0,j_3}$ $j_1 = \overline{1, m_1 - 1}$, $j_3 = \overline{1, m_3 - 1}$, $i = \overline{1,\ 2}$, $s = \overline{0, n_1}$.

For the lateral plane Γ_5 we have

$$U_{i,0,j_2,j_3} = U_{i,1,j_2,j_3} + \frac{h_1}{\lambda_{i,0,j_2,j_3} + 2\mu_{i,0,j_2,j_3}} \times \left(\frac{\lambda_{i,0,j_2,j_3}}{2}\left(\frac{V_{i,0,j_2+1,j_3} - V_{i,0,j_2-1,j_3}}{h_2} + \frac{W_{i,0,j_2,j_3+1} - W_{i,0,j_2,j_3-1}}{h_3}\right) - \theta''\right),$$

$$V_{i,0,j_2,j_3} = V_{i,1,j_2,j_3} + \frac{h_1}{2h_2}\left(U_{i,0,j_2+1,j_3} - U_{i,0,j_2-1,j_3}\right),$$

$$W_{i,0,j_2,j_3} = W_{i,1,j_2,j_3} + \frac{h_1}{2h_3}\left(U_{i,0,j_2,j_3+1} - U_{i,0,j_2,j_3-1}\right),$$

where $\theta'' = \left(3\lambda_{i,0,j_2,j_3} + 2\mu_{i,0,j_2,j_3}\right)\alpha_T^{(i)}\bar{T}^{(s)}_{i,0,j_2,j_3}$, $j_2 = \overline{1, m_2 - 1}$, $j_3 = \overline{1, m_3 - 1}$, $i = \overline{1,\ 2}$, $s = \overline{0, n_1}$.

From the conjugation conditions (18) $[u_n^{(i)}]\Big|_{\Gamma_0} = [u_s^{(i)}]\Big|_{\Gamma_0} = 0$, taking into account that $U_{1,j_1,j_2,,m_3^*} = U_{2,j_1,j_2,m_3^*} = 0$, $V_{1,j_1,j_2,,m_3^*} = V_{2,j_1,j_2,m_3^*} = 0$, $W_{1,j_1,j_2,,m_3^*} = W_{2,j_1,j_2,m_3^*} = 0$, we obtain the values of displacements on the free surface Γ_0

$$U_{1,j_1,j_2,m_3^*} = U_{1,j_1,j_2,m_3^*-1} - \frac{h_3}{2h_1}\left(W_{1,j_1+1,j_2,m_3^*} - W_{1,j_1-1,j_2,m_3^*}\right),$$

$$V_{1,j_1,j_2,m_3^*} = V_{1,j_1,j_2,m_3^*-1} - \frac{h_3}{2h_2}\left(W_{1,j_1,j_2+1,m_3^*} - W_{1,j_1,j_2-1,m_3^*}\right),$$

$$
\begin{aligned}
W_{1,j_1,j_2,m_3^*} = {} & \frac{1}{\left(\frac{(\lambda_{1,j_1,j_2,m_3^*}+2\mu_{1,j_1,j_2,m_3^*})}{h_3} - \frac{(\lambda_{2,j_1,j_2,m_3^*}+2\mu_{2,j_1,j_2,m_3^*})}{h_3}\right)} \\
& \times \left(\frac{(\lambda_{1,j_1,j_2,m_3^*}+2\mu_{1,j_1,j_2,m_3^*})}{h_3} W_{1,j_1,j_2,m_3^*-1} - \frac{(\lambda_{2,j_1,j_2,m_3^*}+2\mu_{2,j_1,j_2,m_3^*})}{h_3} W_{2,j_1,j_2,m_3^*-1} \right. \\
& - \frac{\lambda_{1,j_1,j_2,m_3^*}}{2}\left(\frac{U_{1,j_1+1,j_2,m_3^*}-U_{1,j_1-1,j_2,m_3^*}}{h_1} + \frac{V_{1,j_1,j_2,m_3^*}-V_{1,j_1,j_2-1,m_3^*}}{h_2}\right) \\
& + (3\lambda_{1,j_1,j_2,m_3^*}+2\mu_{1,j_1,j_2,m_3^*})\alpha_T^{(1)}\overline{T}^{(s)}_{1,j_1,j_2,m_3^*} \\
& + \frac{\lambda_{2,j_1,j_2,m_3^*}}{2}\left(\frac{U_{2,j_1+1,j_2,m_3^*}-U_{2,j_1-1,j_2,m_3^*}}{h_1} + \frac{V_{2,j_1,j_2,m_3^*}-V_{2,j_1,j_2-1,m_3^*}}{h_2}\right) \\
& \left. - (3\lambda_{2,j_1,j_2,m_3^*}+2\mu_{2,j_1,j_2,m_3^*})\alpha_T^{(2)}\overline{T}^{(s)}_{2,j_1,j_2,m_3^*} \right),
\end{aligned}
$$

where $j_1 = \overline{1, m_1 - 1}$, $j_2 = \overline{1, m_2 - 1}$.

The study of the finite difference method computational error for the mathematical model described above will be conducted the next stage of our research.

6 Computer Modelling and Results of Numerical Experiments

For computer modelling of the set boundary value problem, a software package for the capabilities of the Microsoft Visual Studio 2017 framework for Windows Desktop in the C# programming language was created.

As an example, the spatial stress-strain state in a water-saturated soil massif in the region $\Omega = \{\mathbf{X} = (x, y, z) : \; 0 \le x \le l_1, \; 0 \le y \le l_2, \; 0 \le z \le l_3\}$, which has the shape of a rectangular parallelepiped of $l_1 = 10\,\text{m}$ length, $l_2 = 10\,\text{m}$ thickness and $l_3 = 10\,\text{m}$ height. The free surface is at the depth $z = 5\,\text{m}$.

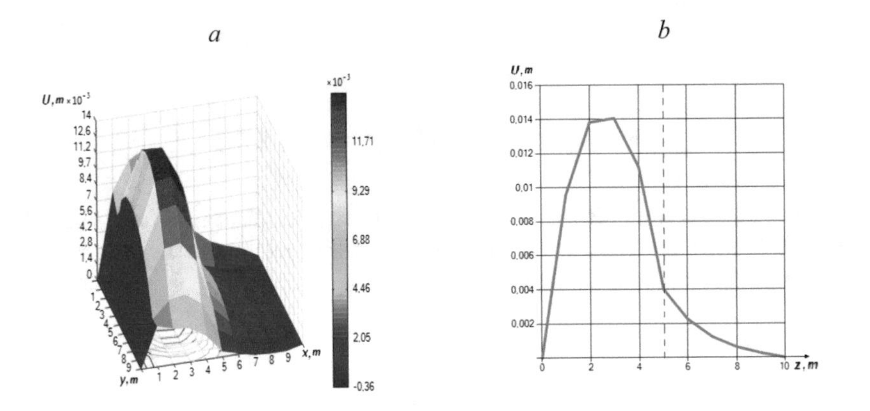

Fig. 3. Distributions of displacements $U(\mathbf{X})$

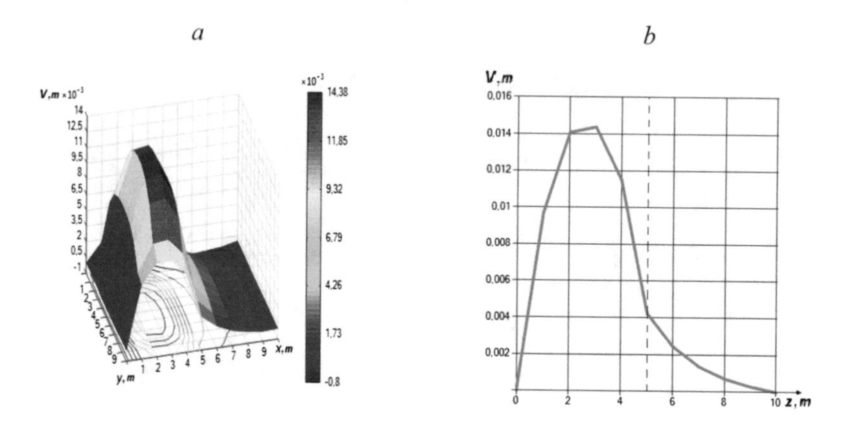

Fig. 4. Distributions of displacements $V(\mathbf{X})$

A series of numerical experiments was conducted, the results of which are presented in the form of graphs at $t = 1080$ days.

In Figs. 2, 3 and 4 shows the displacements graphs $U(\mathbf{X})$, $V(\mathbf{X})$ and $W(\mathbf{X})$ in the section of the plane xOz at $y = 5\,\text{m}$ (a) and in the sections of the planes xOz and yOz at $x = 5\,\text{m}$ and $y = 5\,\text{m}$ (b), taking into account the mass and heat transfer and the present free surface.

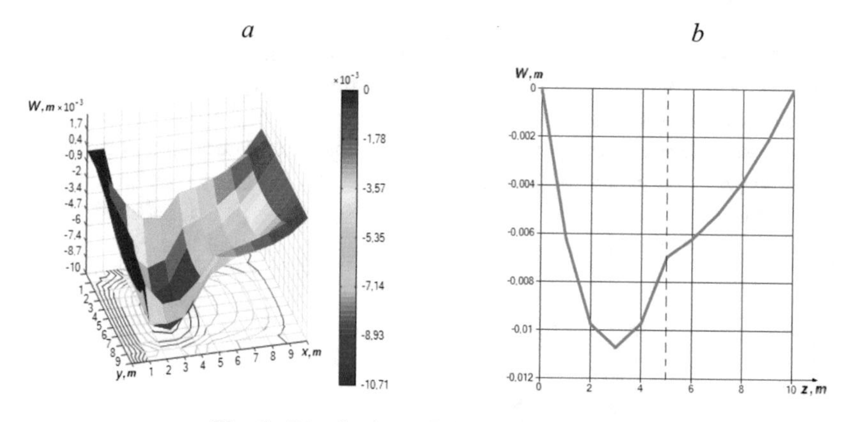

Fig. 5. Distributions of displacements $W(\mathbf{X})$

As can be seen from Figs. 3, 4 and 5, the values of displacements are much larger in the water-saturated area of the soil mass in comparison with the values of displacement in the soil area in the natural state.

7 Conclusions

The task setting are formulated and the mathematical modeling of the spatial problem of the stress-strain state of the soil massif under mass and heat transfer and present of a free surface in it are carried out in this. The main differential equations, boundary conditions and conjugation conditions for the desired functions in the studied region, as well as the dependences of the Lame coefficients, Young's modulus on the concentration of saline solutions and temperature are presented as well. Graphs of distributions of displacements are presented. It is shown that the presence of a free surface that breaks the area of the investigated soil in the area of water-saturated soil and the area of the soil in its natural state changes the distributions of the displacements of the soil mass in these areas.

References

1. Hetnarski, R.B.: Encyclopedia of Thermal Stresses. Springer, Dordrecht (2014)
2. Jafari, M.: Deformation modelling of structures enriched by inter-element continuity condition based on multi-sensor data fusion. Appl. Math. Model. **40**(21–22), 9316–9326 (2016). https://doi.org/10.1016/j.apm.2016.06.010
3. Kong, D., Martin, C.M., Byrne, B.W.: Modelling large plastic deformations of cohesive soils using sequential limit analysis. Int. J. Numer. Anal. Methods Geomech. **41**(18), 1781–1806 (2017). https://doi.org/10.1002/nag.2700
4. Kovalenko, A.D.: Thermoelasticity. K.: High School, 216 p. (1975). (in Russian)
5. Kuzlo, M.T., Filatova, I.A.: Investigation of the influence of salt solutions concentration on deformation characteristics of soils. Hydromelioration and hydrotechnical construction. Collection of scientific works, Rivne: National University of Water and Environmental Engineering, issue 31, pp. 175–182 (2007). (in Ukrainian)
6. Kuzlo, M.T., Moshynskyi, V.S., Martyniuk, P.M.: Mathematical modelling of soil Massif's deformations under its drainage. Int. J. Appl. Math. **31**(6) (2019). https://doi.org/10.12732/ijam.v31i6.5
7. Larin, O., Vodka, O.: A probability approach to the estimation of the process of accumulation of the high-cycle fatigue damage considering the natural aging of a material. Int. J. Damage Mech. **24**(2), 294–310 (2015). https://doi.org/10.1177/1056789514536067
8. Safonyk, A., Martynov, S., Kunytsky, S., Pinchuk, O.: Mathematical modelling of regeneration the filtering media bed of granular filters. Adv. Model. Anal. **73**(2), 72–78 (2018). https://doi.org/10.18280/ama_c.730206
9. Sun, Y., Gao, Y., Zhu, Q.: Fractional order plasticity modelling of state-dependent behaviour of granular soils without using plastic potential. Int. J. Plast. **102**, 53–69 (2018). https://doi.org/10.1016/j.ijplas.2017.12.001
10. Trofymchuk, O., Kaliukh, I., Klymenkov, O.: TXT-tool 2.380-1.1: monitoring and early warning system of the building constructions of the Livadia Palace, Ukraine. In: Sassa, K., Guzzetti, F., Yamagishi, H., Arbanas, Z., Casagli, N., McSaveney, M., Dang, K. (eds.) Landslide Dynamics: ISDR-ICL Landslide Interactive Teaching Tools, vol. 51, pp. 491–508. Springer, Cham (2018)
11. Vlasyuk, A.P., Zhukovska, N.A., Zhukovskyy, V.V., Klos-Witkowska, A., Pazdriy, I., Iat-sykovska, U. (eds.): Mathematical modelling of three-dimensional problem of soil mass stressed-strained state considering mass and heat transfer. In: 2017 9th IEEE International Conference on Intelligent Data Acquisition and Advanced Computing Systems: Technology and Applications (IDAACS) (2017)

12. Vlasyuk, A.P., Zhukovskii, V.V.: Mathematical simulation of the migration of radionuclides in a soil medium under nonisothermal conditions with account for catalytic microparticles and nonlinear processes. J. Eng. Phys. Thermophys. **90**(6), 1386–1398 (2017). https://doi.org/10.1007/s10891-017-1697-4
13. Vlasyuka, A.P., Zhukovskaya, N.A.: Mathematical simulation of the stressed-strained state of the foundation of earth dams with an open surface under the influence of heat and mass transfer in the two-dimensional case. J. Eng. Phys. Thermophys. **88**(2), 329–341 (2015). https://doi.org/10.1007/s10891-015-1197-3
14. Vlasyuk, A.P., Zhukovska, N.A., Zhukovskyy, V.V.: About mathematical modelling of spatial deformation problem of soil massif with free surface. In: Proceedings of International Scientific Conference "Computer Sciences and Information Technologies" (CSIT-2019), vol. 1, pp. 128–131. IEEE (2019)

Searching for Pareto-Optimal Solutions

Igor Kovalenko, Yevhen Davydenko(✉), and Alyona Shved

Petro Mohyla Black Sea National University, Mykolaiv, Ukraine
{ihor.kovalenko,davydenko,avshved}@chmnu.edu.ua

Abstract. The problem of narrowing the Pareto set is considered. The existing approaches aimed at solving the problem of finding a set of Pareto optimal solutions has been analyzed. One approach for solving multi-objective problems based on complex using methods for the construction of the Pareto optimal set and evidence theory has been proposed in this paper. The proposed technique allows to evaluate the obtained set of non-dominated alternatives by methods of the evidence theory to find the best (optimal) solution. The proposed approach allows us to obtain a more formalized procedure for narrowing the Pareto set to obtaining a single optimal solution (a single-element Pareto set). The use of the mathematical apparatus of the evidence theory makes it possible to model uncertainty in expert or decision makers judgments (the strict requirement of the "unambiguous" preference of one alternative over the other is removed). Using the proportional conflict redistribution rules for aggregating group expert assessments makes it possible to process expert evidence generated under conflicting, contradiction expert information. Numerical examples of the proposed methodology for integrated application of evidence theory and methods for Pareto set construction to find optimal solutions are given. The results obtained make it possible to improve the quality and effectiveness of finding optimal solutions.

Keywords: Multi-objective problem · Pareto set · Evidence theory · Optimal solution

1 Introduction

The fundamental Edgeworth-Pareto principle has been established in recent decades. According to this principle the best choice should be made only among the elements of the Pareto set (set of non-dominated alternatives) [1]. It should be noted that the task of forming this set has been solved in papers [1–4, etc.]. However, as the authors of these works indicate, the Pareto set quite often turns out to be wide enough, and the specific choice of solutions (alternatives) within it is not obvious. For this reason, the problem of narrowing the Pareto set arises associated with the choice of a particular Pareto-optimal variant as the "best" choice [5].

In this regard, a number of publications [2–7, etc.] have appeared over recent years, in which a number of approaches for solving this problem are presented.

Among such approaches, the theory of the importance of criteria [8] and the axiomatic approach are most developed. In the framework of the latter a number of axioms are formulated that impose certain requirements on the behavior of the decision

N. Shakhovska and M. O. Medykovskyy (Eds.): CCSIT 2019, AISC 1080, pp. 121–138, 2020.
https://doi.org/10.1007/978-3-030-33695-0_10

maker (DM) in terms of their preferences in the process of choosing alternative solutions [1, 5, 6]. In general, both these approaches can be implemented only on the condition that the additional information ("quanta" of information) [6] of quantitative or qualitative nature about the criteria or decision maker preferences will be obtained.

Most of the existing methods aimed at narrowing the Pareto set, according to the author of [5], cannot be considered strictly justified, since they are based on certain subjective heuristic considerations and/or use information that is extremely difficult to extract from the decision makers. This opens a way for searching the additional approaches to solving the problem of finding optimal solutions on the formulated Pareto set. One such approach can be based on the mathematical apparatus of the Dempster-Shafer evidence theory [9–11].

The aim of this work is to develop an approach for finding optimal solutions, which is based on the complex use of Pareto set construction methods and evidence theory.

2 Pareto Set Construction

Let's use the algorithm proposed in [12] for Pareto set construction. The implementation of this algorithm is determined by the following three conditions:

1. set of possible (admissible) alternative solution $A = \{a_i | i = \overline{1,n}\}$;
2. vector criteria $K = \{K_j | j = \overline{1,m}\}$, $m \geq 2$;
3. preference relations $\underset{A}{\succ}$ over a set A (for example $a_1 \underset{A}{\succ} a_2$).

The first step is a sequential comparison of the first solution a_1 with all the others $a_2,\ldots, a_n$. This comparison is to verify the validity of the relations $a_1 \underset{A}{\succ} a_i$ and $a_i \underset{A}{\succ} a_1$ for each $i = \overline{2,n}$. If (for some i) the relation $a_1 \underset{A}{\succ} a_i$ is satisfied, then the dominant solution a_i is removed from the set A. When the second relation $a_i \underset{A}{\succ} a_1$ is satisfied, the solution a_1 must be removed. If none of the above relations $a_1 \underset{A}{\succ} a_i$ and $a_i \underset{A}{\succ} a_1$ is true, then nothing should be deleted. When decision a_1 was compared with all other decisions a_2, …, a_n, and no $i = \overline{2,n}$ holds the relation $a_i \underset{A}{\succ} a_1$, then the first solution should be remembered as undominated and removed from A.

If, after performing the first step on set A, there is no solution left (that is, all solutions were removed), then the algorithm terminates. And one non-dominated solution a will be stored in memory. Otherwise (if not all solutions were removed), it is necessary to go to the second step of the algorithm, which is similar to the first. First, it is necessary to number the elements of the newly obtained set, and then to conduct a consistent comparison of the first solution of the considered set with all its other elements. Thus, at each step of the algorithm, it is necessary to fix non-dominated solutions that will ultimately construct the Pareto set.

But quite often such a set can be wide enough, and make an optimal choice within it is quite problematic. In this regard, in order to narrow the Pareto set, the approach to reveal information on the relative importance of criteria·was proposed. This approach is implemented through a direct survey of decision makers.

As a result, the decision maker turns out, for example, to the situation when, in order to increase the value of the more important criterion K_i per ω_i units, the decision maker is ready to sacrifice the loss of ω_j units according to K_j (on condition of saving values of all other criteria).

This takes into account the coefficient of relative importance θ_{ij}, that expresses the proportion of the relative amount of loss and increase of K_i compared to K_j [8]:

$$\theta_{ij} = \frac{\omega_i}{\omega_i + \omega_j}, \quad 0 < \theta_{ij} < 1 \tag{1}$$

Taking into account this indicator, the values of the selected criteria are recalculated using the following formulas:

$$\begin{aligned} K_i^* &= \theta_{ij} * K_i + \left(1 - \theta_{ij}\right) * K_j; \\ K_j^* &= \theta_{ij} * K_j + \left(1 - \theta_{ij}\right) * K_i. \end{aligned} \tag{2}$$

Thus obtained K_i^* and K_j^* are used in the procedure of narrowing the Pareto set. To obtain a singleton Pareto set, the considered procedure should be repeated or an additional criterion should be involved in the analysis.

Consider a situation when decision makers find it difficult to determine the relative importance of criteria. In this case, it will be effective to use the Dempster-Shafer theory [9–11, 13, 14]. The mathematical apparatus of this theory allows selecting subgroups from the original set of alternatives and then determining the degree of their preference on a given scale towards all remaining alternatives, instead of comparing the individual alternatives. The aggregation of the results of the expert survey is carried out on the basis of one of combination rules of evidence theory. Based on the obtained combined basic probability assignments, the values of the belief and plausibility measures are determined for all formed subgroups of alternatives, including singleton subgroups.

3 Evidence Theory Basics

Let $\mathrm{A} = \{A_i | i = \overline{1,n}\}$ be a finite set of hypotheses (mutually exclusive and exhaustive) – frame of discernment in Dempster-Shafer notation (Dempster-Shafer theory, DST, evidence theory) [9–11] and group of experts $\mathrm{E} = \{E_j | j = \overline{1,z}\}$. According to Dempster-Shafer's model expert E_j can form a system of subsets $\mathrm{P}_j = \{X_t | t = \overline{1,s}\}$, $s = 2^{\mathrm{A}}$, reflecting his preferences, such as $X_t \subseteq \mathrm{A}$ and satisfying the following conditions:

$$\begin{aligned} &1.\ X_t = \{\varnothing\}; \\ &2.\ X_t = \{A_i\}; \\ &3.\ X_t = \{A_i | i = \overline{1,p}\},\ p < n; \\ &4.\ X_t = \mathrm{A} = \{A_i | i = \overline{1,n}\}. \end{aligned} \tag{3}$$

Condition 1 – expert selected none of the alternatives; condition 2 – the expert choose one alternative ($A_i \in$ A); condition 3 – p alternatives can be satisfied with expert choice; condition 4 – all alternatives are equivalent for expert.

The fulfillment of condition 1 could indicate an unsuccessful formation of a set of evaluated objects (alternatives), an unjustified choice of a method for identifying and (or) analyzing expert information, an unqualified forming an expert group, etc.

The Dempster-Shafer (evidence) theory is based on three main functions ($\forall X \subseteq \mathrm{A}$) [9–11]:

1. the basic probability assignment (*bpa*) or mass function m: $2^{\mathrm{A}} \rightarrow [0, 1]$:

$$\sum_{X_t \subseteq \mathrm{A}} m(X_t) = 1; \;\; m(\varnothing) = 0; \tag{4}$$

2. the belief function *Bel*: $2^{\mathrm{A}} \rightarrow [0, 1]$:

$$Bel(B) = \sum_{X_t \subseteq B} m(X_t) \tag{5}$$

3. the plausibility function *Pl*: $2^{\mathrm{A}} \rightarrow [0, 1]$:

$$Pl(B) = \sum_{X_t \cap B \neq \varnothing} m(X_t) \tag{6}$$

where 2^{A} denotes the power set of the finite and discrete frame A (the set of all subsets of A, including the empty set Ø).

Elements $X \in 2^{\mathrm{A}}$ having $m(X) > 0$ are called focal elements of $m(\cdot)$. The $m(X_t)$ determines the subjective degree of confidence that the best element of the set A is in the subset $X_t \subseteq$ A.

The belief $Bel(B)$ denotes the full degree of support (belief) given to each $B \subseteq$ A, including all subsets $X_t \subseteq B$. The plausibility $Pl(B)$ express the full degree of potential support that can be given to B (how much belief mass potentially supports B).

The value of the $Bel(\cdot)$ and $Pl(\cdot)$ functions determine the upper and lower bounds of the interval, which contains the exact probability value $P(B)$ of the considered subset B: $Bel(B) \leq P(B) \leq Pl(B)$.

The $[Bel(B); Pl(B)]$, $Bel(B) \leq Pl(B)$ is single interval [0, 1] that represents the complete information about the measure of belief in B.

The subsets $X_t \subseteq$ A formed on the basis of frame of discernment can interact with each other in various ways, thus forming the following models of evidence [13, 14]:

1. Equivalent judgments (evidence):

$$X_1 = X_2 = \ldots = X_t = \ldots = X_s. \tag{7}$$

2. Consonant judgments (evidence):

$$X_1 \subset X_2 \subset \ldots \subset X_t \subset \ldots \subset X_s. \tag{8}$$

3. Consistent judgments (evidence):

$$X_1 \cap X_2 \cap \ldots \cap X_t \cap \ldots \cap X_s \neq \varnothing. \tag{9}$$

4. Arbitrary judgments (evidence):

$$X_1 \cap X_2 \cap \ldots \cap X_t \cap \ldots \cap X_s = \varnothing, \text{ but } \exists C : X_i \cap X_j \neq \varnothing. \tag{10}$$

5. Disjoint judgments (evidence):

$$\forall X_i, X_j \subseteq X : X_i \cap X_j = \varnothing. \tag{11}$$

The aggregation of *bpa's* is based on the operation of combination of evidence obtained from different (multiple) sources. It is assumed that each such source of evidence is independent. In the Dempster-Shafer theory, the combination of *bpa's* is made according to the Dempster rule [11, 15]:

$$m_{DS}(X) = \frac{1}{1 - k_{12}} \cdot \sum_{X_1, X_2 \in 2^A,\ X_1 \cap X_2 = X} m_1(X_1) m_2(X_2), \tag{12}$$

where X_1, X_2 are groups of evidence obtained from 1st and 2nd independent sources; k_{12} is a degree of conflict:

$$k_{12} = \sum_{X_1, X_2 \in 2^A,\ X_1 \cap X_2 = \varnothing} m_1(X_1) m_2(X_2). \tag{13}$$

Dempster rule of combination has a significant drawback – information obtained from conflicting sources is completely ignored. This aspect can lead to unsatisfactory combination results (Zadeh's paradox [16]), e.g. Dempster's rule of combination absolutely cannot be applied in situation of extreme height conflict with coefficient $k_{12} = 1$. To eliminate this drawback, a number of alternative combination rules [17–22] were proposed, including the Dempster discounting rule [15]. Let us consider some of them.

4 Alternative Combination Rules in DST

In contrast to the Dempster's rule of combination, Yager [15, 17] does not attribute the combined probabilities for the empty intersections of marginal focal elements to the empty set and does not normalize them, but uses them to reflect the degree of ignorance. The combined *bpa* in the Yager's rule is defined as follows:

$$m_Y(X) = \sum_{X_1, X_2 \in 2^{\mathrm{A}}, X_1 \cap X_2 = X} m_1(X_1) m_2(X_2), \tag{14}$$

provided that $m_Y(\emptyset) = 0$, $\forall\, X \in 2^{\mathrm{A}}$, $X \neq \emptyset$ and $X \neq \mathrm{A}$.

In the case when $X = \mathrm{A}$, the combined *bpa* of the universal set (frame of discernment) is defined in the following way:

$$m_Y(\mathrm{A}) = q(\mathrm{A}) + q(\varnothing) = m_1(\mathrm{A}) m_2(\mathrm{A}) + \sum_{X_1, X_2 \in 2^{\mathrm{A}}, X_1 \cap X_2 = \varnothing} m_1(X_1) m_2(X_2), \tag{15}$$

where $q(\mathrm{A})$ and $q(\emptyset)$ are combined *bpa* of the universal and the null set, respectively.

Inagaki [15, 18] suggests the following rule of combination of probabilities for any non-empty subset $X = X_1 \cap X_2$:

$$m_k^U(X) = [1 + kq(\varnothing)] \cdot q(X), \quad X \neq \mathrm{A}, \varnothing, \tag{16}$$

where $q(X) = \sum_{X_1, X_2 \in 2^{\mathrm{A}}, X_1 \cap X_2 = X} m_1(X_1) m_2(X_2)$; $q(\emptyset)$ is the combined *bpa* over all empty intersections of marginal focal elements; the parameter k is used to normalization:

$$0 \leq k \leq \frac{1}{1 - q(\varnothing) - q(\mathrm{A})}.$$

For $X = \mathrm{A}$ Inagaki the following equation:

$$m_k^U(\mathrm{A}) = [1 + kq(\varnothing)] \cdot q(\mathrm{A}) + [1 + kq(\varnothing) - k] \cdot q(\varnothing), \tag{17}$$

where $q(\mathrm{A})$ is combined *bpa* of the universal set.

Depending on the values of the coefficient k, conflicts on a set of hypotheses may or may not be taken into account: with $k = 0$ Inagaki's combination rule is equal to Yager's rule; with $k = 1/(1 - q(\emptyset))$ Dempster's rule is obtained.

Zhang's combination rule takes into account the degree of intersection of the selected subsets, determined on the basis of various groups of evidence [15, 19], and introduces a measure to their intersection:

$$r(X_1, X_2) = \frac{|X|}{|X_1|\,|X_2|} = \frac{|X_1 \cap X_2|}{|X_1|\,|X_2|}, \tag{18}$$

where $X_1 \cap X_2 = X$; $|\cdot|$ is the cardinality of the corresponding focal elements.

Thus, the value of the combined mass of the probability of the resulting subset is defined as

$$m_Z(X) = k \cdot \sum_{X_1, X_2 \in 2^{A}, X_1 \cap X_2 = X} [r(X_1, X_2) m_1(X_1) m_2(X_2)], \tag{19}$$

where k is a normalization coefficient.

Smets combination rule [15, 20] actually represents an unrated version of the Dempster combination rule.

The combined *bpa* of intersecting focal elements, provided that $\forall\ (X \neq \emptyset) \in 2^{A}$, is defined in the next way:

$$m_S(X) = \sum_{X_1, X_2 \in 2^{A}, X_1 \cap X_2 = X} m_1(X_1) m_2(X_2). \tag{20}$$

The combined *bpa* of the null set can be determined as follows:

$$m_S(\varnothing) \equiv k_{12} = \sum_{X_1, X_2 \in 2^{A}, X_1 \cap X_2 = \varnothing} m_1(X_1) m_2(X_2). \tag{21}$$

If the conflict level is significant, then the Proportional Conflict Redistribution Rule (e.g. PCR5) can be applied to aggregate expert estimates (combine *bpa's*).

The combined *bpa* according to PCR5 rule for two sources of evidence, provided that $\forall (X \neq \emptyset) \in 2^{A}$, $m(\emptyset) = 0$, is defined as follows [15, 21–23]:

$$m_{PCR5}(X) = \sum_{\substack{X_1, X_2 \in 2^{A} \\ X_1 \cap X_2 = X}} m_1(X_1) m_2(X_2) + \sum_{\substack{Y \in 2^{A} \\ Y \cap C = \varnothing}} \left[\frac{m_1(C)^2 \cdot m_2(Y)}{m_1(C) + m_2(Y)} + \frac{m_2(C)^2 \cdot m_1(Y)}{m_2(C) + m_1(Y)} \right]. \tag{22}$$

This rule allows the reallocation of partial conflicting mass to the elements involved in conflict and proportionally to their individual *bpa's*. This property allows to operate correctly with combined *bpa*'s assigned to empty intersections. The PCR5 combination rule allows to process expert judgments in a situation with the total conflict (when mass reaches 1), and the combined basic probability masses for all the subsets selected by the experts, including one-element ones, will be calculated.

5 Examples

As an example, that shows how proposed approach can be used in practical applications, let us consider the problem of selecting the most appropriate dry port location.

Dry ports defined as combination of temporary storage warehouses, auxiliary buildings and structures, roads and railways, sites located outside the territory of the seaport. And they are interconnected by a single technological process and an information system for performing group operations with goods and their temporary storage under customs control. This problem has a multi-criteria nature that requires the formation of a set of factors (criteria) that determine the dry port location, the availability a list of alternatives, as well as the choice of methods for analyzing such criteria, that may include the following:

K_1 – belonging to a specific climatic zone, and the existence of transport corridors;
K_2 – population and gross regional product;
K_3 – industrial output and export/import volume;
K_4 – railway and road traffic volumes;
K_5 – goods properties and nomenclature
K_6 – road and rail network density, the state of transport communications;
K_7 – legal features of customs policy;
K_8 – the presence and implementation of government programs and projects in the transport and logistics complex.

Suppose we have $n = 6$ alternatives (a_1–a_6). For an example, let us consider next Ukrainian seaports: Mykolaiv (a_1), Reni (a_2), Yuzhne (a_3), Ochakov (a_4), Skadovsk (a_5), Izmail (a_6).

5.1 Example 1: Pareto Set Construction

In the first stage, let us consider an example of Pareto set construction based on the described algorithm.

We will use a five-point scale to evaluate alternatives for each criterion. The obtained results of the expert survey are given in Table 1.

Table 1. Evaluation results of experts on all criteria.

Alternatives	Criteria							
	K_1	K_2	K_3	K_4	K_5	K_6	K_7	K_8
a_1	4	3	4	3	1	4	3	5
a_2	5	3	3	3	2	3	4	4
a_3	2	4	2	4	4	2	3	5
a_4	5	3	2	3	2	1	4	3
a_5	3	4	3	4	5	2	4	2
a_6	4	3	5	4	4	4	3	5

Let us perform the following pairwise comparisons according to the considered algorithm:

1. Compare in pairs a_1 and a_2; a_1 and a_3; a_1 and a_4; a_1 and a_5; a_1 and a_6 for each criterion $K_1 \div K_8$:

$$a_1 \underset{K_1}{<} a_2; a_1 \underset{K_2}{=} a_2; a_1 \underset{K_3}{>} a_2; a_1 \underset{K_4}{=} a_2; a_1 \underset{K_5}{<} a_2; a_1 \underset{K_6}{>} a_2; a_1 \underset{K_7}{<} a_2; a_1 \underset{K_8}{>} a_2;$$
$$a_1 \underset{K_1}{>} a_3; a_1 \underset{K_2}{<} a_3; a_1 \underset{K_3}{>} a_3; a_1 \underset{K_4}{<} a_3; a_1 \underset{K_5}{<} a_3; a_1 \underset{K_6}{>} a_3; a_1 \underset{K_7}{=} a_3; a_1 \underset{K_8}{=} a_3;$$
$$a_1 \underset{K_1}{<} a_4; a_1 \underset{K_2}{=} a_4; a_1 \underset{K_3}{>} a_4; a_1 \underset{K_4}{=} a_4; a_1 \underset{K_5}{<} a_4; a_1 \underset{K_6}{>} a_4; a_1 \underset{K_7}{<} a_4; a_1 \underset{K_8}{>} a_4;$$
$$a_1 \underset{K_1}{>} a_5; a_1 \underset{K_2}{<} a_5; a_1 \underset{K_3}{>} a_5; a_1 \underset{K_4}{<} a_5; a_1 \underset{K_5}{<} a_5; a_1 \underset{K_6}{>} a_5; a_1 \underset{K_7}{<} a_5; a_1 \underset{K_8}{>} a_5;$$
$$a_1 \underset{K_1}{=} a_6; a_1 \underset{K_2}{=} a_6; a_1 \underset{K_3}{<} a_6; a_1 \underset{K_4}{<} a_6; a_1 \underset{K_5}{<} a_6; a_1 \underset{K_6}{=} a_6; a_1 \underset{K_7}{=} a_6; a_1 \underset{K_8}{=} a_6.$$

As a result of comparison a_1 and a_6, the relation $a_6 \geq a_1$ is performed. Respectively, a_1 is the dominant alternative and is removed from consideration.

2. Compare in pairs a_2 and a_3; a_2 and a_4; a_2 and a_5; a_2 and a_6 for each criterion $K_1 \div K_8$:

$$a_2 \underset{K_1}{>} a_3; a_2 \underset{K_2}{<} a_3; a_2 \underset{K_3}{>} a_3; a_2 \underset{K_4}{<} a_3; a_2 \underset{K_5}{<} a_3; a_2 \underset{K_6}{>} a_3; a_2 \underset{K_7}{>} a_3; a_2 \underset{K_8}{<} a_3;$$
$$a_2 \underset{K_1}{=} a_4; a_2 \underset{K_2}{=} a_4; a_2 \underset{K_3}{>} a_4; a_2 \underset{K_4}{=} a_4; a_2 \underset{K_5}{=} a_4; a_2 \underset{K_6}{>} a_4; a_2 \underset{K_7}{=} a_4; a_2 \underset{K_8}{>} a_4;$$
$$a_2 \underset{K_1}{>} a_5; a_2 \underset{K_2}{<} a_5; a_2 \underset{K_3}{=} a_5; a_2 \underset{K_4}{<} a_5; a_2 \underset{K_5}{<} a_5; a_2 \underset{K_6}{>} a_5; a_2 \underset{K_7}{=} a_5; a_2 \underset{K_8}{>} a_5;$$
$$a_2 \underset{K_1}{>} a_6; a_2 \underset{K_2}{=} a_6; a_2 \underset{K_3}{<} a_6; a_2 \underset{K_4}{<} a_6; a_2 \underset{K_5}{<} a_6; a_2 \underset{K_6}{<} a_6; a_2 \underset{K_7}{>} a_6; a_2 \underset{K_8}{<} a_6.$$

As a result of comparison a_2 and a_4 the relation $a_2 \geq a_4$ is performed. So a_4 is the dominant alternative and it is not considered further.

3. Compare in pairs a_3 and a_5; a_3 and a_6 for each criterion $K_1 \div K_8$:

$$a_3 \underset{K_1}{<} a_5; a_3 \underset{K_2}{=} a_5; a_3 \underset{K_3}{<} a_5; a_3 \underset{K_4}{=} a_5; a_3 \underset{K_5}{<} a_5; a_3 \underset{K_6}{=} a_5; a_3 \underset{K_7}{<} a_5; a_3 \underset{K_8}{>} a_5;$$

$$a_3 \underset{K_1}{<} a_6; a_3 \underset{K_2}{>} a_6; a_3 \underset{K_3}{<} a_6; a_3 \underset{K_4}{=} a_6; a_3 \underset{K_5}{=} a_6; a_3 \underset{K_6}{<} a_6; a_3 \underset{K_7}{=} a_6; a_3 \underset{K_8}{=} a_6.$$

In this case, the pairs a_3, a_5, and a_3, a_6 are incomparable by relation $\geq$. Respectively, a_3 is an undominated alternative and therefore is included in the Pareto set P = $\{a_3\}$.

4. Compare in pairs a_5 and a_6 for each criterion $K_1 \div K_8$:

$$a_5 \underset{K_1}{<} a_6; a_5 \underset{K_2}{>} a_6; a_5 \underset{K_3}{<} a_6; a_5 \underset{K_4}{=} a_6; a_5 \underset{K_5}{>} a_6; a_5 \underset{K_6}{<} a_6; a_5 \underset{K_7}{>} a_6; a_5 \underset{K_8}{<} a_6.$$

The alternatives a_5 and a_6 are incomparable, so both are included in the Pareto set P = $\{a_5, a_6\}$.

Thus, the resulting Pareto set has the form P = $\{a_3, a_5, a_6\}$. Therefore, the final optimal choice should be made precisely among these alternatives, Table 2.

Table 2. Pareto set.

Alternatives	Criteria							
	K_1	K_2	K_3	K_4	K_5	K_6	K_7	K_8
a_3	2	4	2	4	4 (3)	2 (3)	3	5
a_5	3	4	3	4	5 (3.5)	2 (3.5)	4	2
a_6	4	3	5	4	4	4	3	5

However, this requires possessing information on the preferences of decision makers regarding the considered criteria $K_1 \div K_8$. Let the decision maker is ready to sacrifice the criterion K_5 by the value of $K_6^* = K_5^* = 0.5$ points. The obtained values of criteria are given in brackets in Table 2.

The values of criteria K_5 and K_6 for alternative a_6 remained unchanged, since they were the same in terms of preference from the very beginning.

So, let us make the pairwise comparison of alternatives a_3, a_5 and a_6, considering new values of criteria K_5 and K_6.

1. Compare in pairs a_3 and a_5; a_3 and a_6 for each criterion $K_1 \div K_8$:

$$a_3 \underset{K_1}{<} a_5; a_3 \underset{K_2}{=} a_5; a_3 \underset{K_3}{<} a_5; a_3 \underset{K_4}{=} a_5; a_3 \underset{K_5}{<} a_5; a_3 \underset{K_6}{<} a_5; a_3 \underset{K_7}{=} a_5; a_3 \underset{K_8}{<} a_5;$$

$$a_3 \underset{K_1}{<} a_6; a_3 \underset{K_2}{>} a_6; a_3 \underset{K_3}{<} a_6; a_3 \underset{K_4}{=} a_6; a_3 \underset{K_5}{<} a_6; a_3 \underset{K_6}{<} a_6; a_3 \underset{K_6}{<} a_6; a_3 \underset{K_8}{=} a_6.$$

The alternative a_3 is dominated. It is removed from the resulting Pareto set P = $\{a_3, a_5, a_6\} \Rightarrow$ P = $\{a_5, a_6\}$.

2. Compare in pairs a_5 and a_6 for each criterion $K_1 \div K_8$:

$$a_5 \underset{K_1}{<} a_6; a_5 \underset{K_2}{>} a_6; a_5 \underset{K_3}{<} a_6; a_5 \underset{K_4}{=} a_6; a_5 \underset{K_5}{<} a_6; a_5 \underset{K_6}{<} a_6; a_5 \underset{K_7}{>} a_6; a_5 \underset{K_8}{<} a_6.$$

The alternatives a_5 and a_6 do not dominate one another, and therefore remain in the composition of the set P.

The considered procedure of identifying the relative importance of criteria allows narrowing down the Pareto set and, thus, reducing the number of possible solutions.

5.2 Example 2: Selection from Among the Pareto Solutions

Example 2.1: Equivalent Expert Judgments (Evidences). Let us consider a set of alternatives (frame of discernment) P = $\{a_3, a_5, a_6\}$ formed earlier and the expert group E = $\{E_j | j = \overline{1, z}\}$, z = 2. In Dempster-Shafer model the frame of discernment is a set of exclusive and exhaustive elements.

As a result of the expert survey, a system of subsets $\{Y_1, Y_2\}$ was formed, where $Y_j = \{X_i^j | i = \overline{1, d}\}$, $d \le 2^{|P|} - 1$, $j = \overline{1, z}$, reflects the experts choice. Any subset X_i^j can be formed based on rules (3).

For each system $Y_j = \{X_i^j | i = \overline{1,d}\}$, a vector $B_j = \{b_i^j | i = \overline{1,d}\}$ is generated by expert E_j, that contains numerical values of the degrees of superiority b_i^j in the terms of a given preference scale ($X_i^j \succeq P$, $X_i^j \subseteq Y_j$). To perform the pairwise comparisons nine-point preference scale was used: the value 1 means that the alternatives are the same or cannot be compared, and the value 9 means the absolute degree of superiority. For each formed system $Y_j = \{X_i^j | i = \overline{1,d}\}$ the vector $M_j = \{m_i^j | i = \overline{1,d+1}\}$ will be received, whose elements are calculated in accordance with [24]:

$$m_i^j(X_i^j) = \frac{b_i^j}{\sum_{i=1}^{d} b_i^j + \sqrt{d}}, \tag{23}$$

$$m_{d+1}^j(P) = \frac{\sqrt{d}}{\sum_{i=1}^{d} b_i^j + \sqrt{d}}, \tag{24}$$

where d is a total number of selected groups X_i^j; b_i^j is a degree of preference of X_i^j appointed by the expert E_j; $m_i^j(X_i^j)$ is the basic probability assignments of X_i^j; $m_{d+1}^j(P)$ is the basic probability assignments of frame of discernment P. The value corresponding to $m_{d+1}^j(P)$ reflects the degree of total ignorance of E_j.

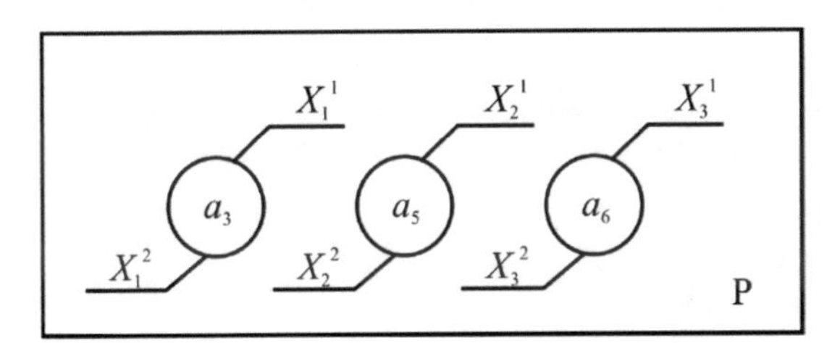

Fig. 1. Formed focal elements (equivalent evidence).

Figure 1 shows the focal elements formed by experts E_1 and E_2. Expert judgments satisfy condition (7) and they are considered equivalent.

The obtained results are given in Table 3.

Table 3. Sources of evidence (Example 2.1).

Expert E_j	X_i^j	b_i^j	$m_i^j(X_i^j)$
E_1	$\{a_3\}$	5	0.3
	$\{a_5\}$	7	0.42
	$\{a_6\}$	3	0.18
	$\{a_3, a_5, a_6\}$	–	0.1
E_2	$\{a_3\}$	7	0.38
	$\{a_5\}$	4	0.21
	$\{a_6\}$	6	0.32
	$\{a_3, a_5, a_6\}$	–	0.09

Table 4 presents the resulting subsets constructed on the basis of the intersection of the subsets formed by experts.

Table 4. Intersections of focal elements (Example 2.1).

Experts' focal		E_1			
elements X_i^j		$\{a_3\}$	$\{a_5\}$	$\{a_6\}$	$\{a_3, a_5, a_6\}$
E_2	$\{a_3\}$	$\{a_3\}$	Ø	Ø	$\{a_3\}$
	$\{a_5\}$	Ø	$\{a_5\}$	Ø	$\{a_5\}$
	{a6}	Ø	Ø	$\{a_6\}$	$\{a_6\}$
	$\{a_3, a_5, a_6\}$	$\{a_3\}$	$\{a_5\}$	$\{a_6\}$	$\{a_3, a_5, a_6\}$

The combined basic probability assignments for the considered subgroups of alternatives were obtained on the basis of the Dempster's rule of combination (12):

$m_{12}(\{a_3\}) = 0.406;$
$m_{12}(\{a_5\}) = 0.333;$
$m_{12}(\{a_5, a_6\}) = 0.240;$
$m_{12}(\{a_3, a_5, a_6\}) = 0.02.$

The coefficient of conflict calculated based on (13) is 0.56.

Given the results obtained, define the values of the belief (5) and plausibility (6) functions for each of the original alternatives a_3, a_5, a_6:

$$a_3 : \begin{cases} Bel(\{a_3\}) = m_{12}(\{a_3\}) = 0.406; \\ Pl(\{a_3\}) = m_{12}(\{a_3\}) + m_{12}(P) = 0.426. \end{cases}$$

$$a_5 : \begin{cases} Bel(\{a_5\}) = m_{12}(\{a_5\}) = 0.333; \\ Pl(\{a_5\}) = 0.353. \end{cases}$$

$$a_6 : \begin{cases} Bel(\{a_6\}) = m(\{a_6\}) = 0.240; \\ Pl(\{a_6\}) = 0.260. \end{cases}$$

The results of the calculations indicate that a_3 has the highest values of belief and plausibility functions. In some situations, it is impossible to unambiguously determine the resulting ranking of alternatives, since the resulting intervals can be nested.

To determine the degree of superiority of the original alternatives, let use the coefficient of pessimism:

$$K = \gamma \cdot bel(\{X_i^{12}\}) + (1 - \gamma) \cdot pl(\{X_i^{12}\}), \tag{25}$$

where $\gamma \in [0;\ 1]$ is a coefficient of pessimism; $\{X_i^{12}\}$ is the i-th group of resulting focal elements (alternatives), $i = \overline{1, r}$; r is a number of resulting focal elements.

As a result of the analysis, the following ranking of alternatives has been obtained: $a_3 \succ a_5 \succ a_6$.

Let's compare the results of solving a choice problem using the combination rules (14)–(22). The results are presented in Table 5.

Table 5. Combination results by different rules (Example 2.1, γ = 0.6).

Alternatives	$[Bel(\cdot);\ Pl(\cdot)]$ interval	Coefficient of pessimism K	$[Bel(\cdot);\ Pl(\cdot)]$ interval	Coefficient of pessimism K
	Yager rule		Inagaki rule	
a_3	[0.179; 0.747]	0.4	[0.411; 0.420]	0.41
a_5	[0.147; 0.715]	0.37	[0.337; 0.346]	0.34
a_6	[0.106; 0.674]	0.33	[0.243; 0.252]	0.25
	Smets rule		Zhang rule	
a_3	[0.179; 0.188]	0.18	[0.424; 0.433]	0.43
a_5	[0.147; 0.156]	0.15	[0.337; 0.346]	0.34
a_6	[0.106; 0.115]	0.11	[0.230; 0.240]	0.23

The results presented in Table 5 show that the best choice is a_3. According to data obtained, Yuzhne sea port is the most suitable option for placing a dry port.

When using different combination rules, the differences between the results are associated with the use of different approaches when combining the basic probability assignments of the original focal elements, mainly in dealing with the combined basic probability assignments for the empty intersections of the original focal elements.

Example 2.2: Arbitrary Expert Judgments (Evidence). Let us consider a frame of discernment P = $\{a_3,\ a_5,\ a_6\}$ and the expert group E $= \{E_j | j = \overline{1, z}\}$, z = 2. In the process of identifying expert preferences the expert E_1 forms vectors $Y_1 = \{\{a_3\}, \{a_5, a_6\}\}$ and $B_1 = \{5,\ 6\}$, and the expert E_2 forms vectors $Y_2 = \{a_5\}$, and $B_2 = \{4\}$ corresponding.

Figure 2 shows the focal elements formed by experts E_1 and E_2. Expert judgments satisfy condition (10) and they are considered arbitrary.

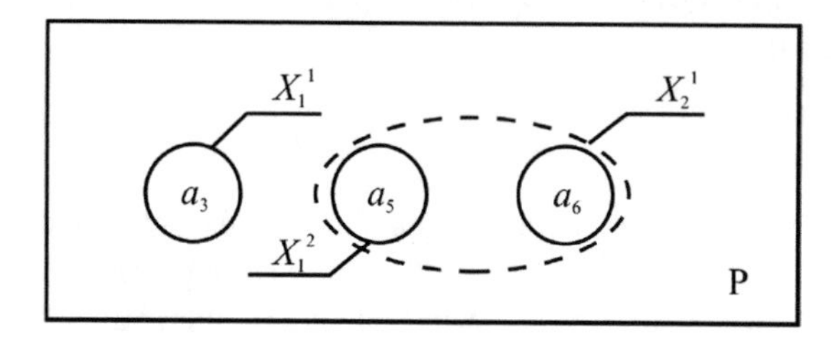

Fig. 2. Formed focal elements (arbitrary evidence).

Using expressions (23)–(24) we have:

E_1: $m_1(\{a_3\}) = 0.4$; $m_1(\{a_5, a_6\}) = 0.48$; $m_1(\{a_3, a_5, a_6\}) = 0.12$.
E_2: $m_2(\{a_5\}) = 0.8$; $m_2(\{a_3, a_5, a_6\}) = 0.2$.

All possible intersections of the elements of the considered groups of alternatives are presented in Table 6.

Table 6. Intersections of focal elements (Example 2.2).

Experts' focal elements X_i^j		E_1		
		$\{a_3\}$	$\{a_5, a_6\}$	$\{a_3, a_5, a_6\}$
E_2	$\{a_5\}$	$\varnothing$	$\{a_5\}$	$\{a_5\}$
	$\{a_3, a_5, a_6\}$	$\{a_3\}$	$\{a_5, a_6\}$	$\{a_3, a_5, a_6\}$

Let's compare the results of solving a choice problem using the combination rules (12)–(22). The results are presented in Table 7.

Table 7. Combination results by different rules (Example 2.2, $\gamma = 0.6$).

Alternatives	$[Bel(\cdot); Pl(\cdot)]$ interval	Coeff. K	$[Bel(\cdot); Pl(\cdot)]$ interval	Coeff. K	$[Bel(\cdot); Pl(\cdot)]$ interval	Coeff. K
	Dempster rule		Yager rule		Inagaki rule	
a_3	[0.118; 0.153]	0.13	[0.08; 0.424]	0.23	[0.119; 0.143]	0.13
a_5	[0.706; 0.882]	0.78	[0.48; 0.920]	0.66	[0.714; 0.881]	0.78
	Smets rule		Zhang rule			
a_3	[0.08; 0.104]	0.09	[0.092; 0.119]	0.10		
a_5	[0.48; 0.600]	0.53	[0.771; 0.908]	0.83		

The coefficient of conflict calculated using (13) is 0.32.

From the results above it can be seen that the alternative a_5 has the highest values of the belief and plausibility functions, therefore, it is optimal. The following ranking of alternatives has been obtained: $a_5 \succ a_3$. According to data obtained, Skadovsk sea port is the most suitable option for placing a dry port.

Example 2.3: Conflicting Expert Judgments (Evidence). Let us consider a frame of discernment P = $\{a_3, a_5, a_6\}$ and the expert group E = $\{E_j | j = \overline{1, z}\}$, z = 2.

Figure 3 shows the focal elements formed by experts E_1 and E_2. Expert judgments satisfy condition (11) and they are considered disjoint.

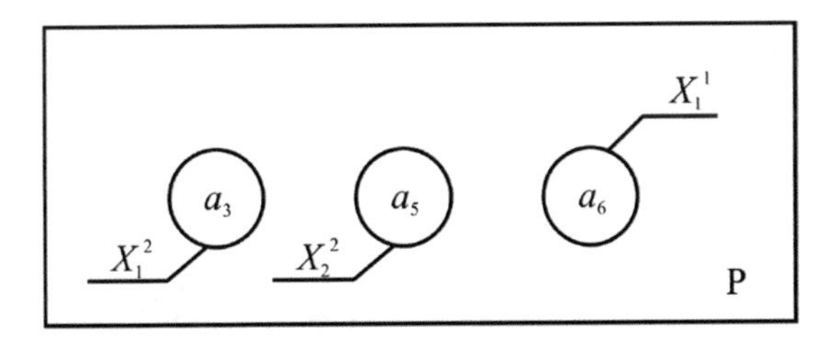

Fig. 3. Formed focal elements (disjoint evidence).

The results of expert survey are given in Table 8.

Table 8. Sources of evidence (Example 2.3).

Expert E_j	X_i^j	b_i^j	$m_i^j(X_i^j)$
E_1	$\{a_6\}$	9	0.9
	$\{a_3, a_5, a_6\}$	–	0.1
E_2	$\{a_3\}$	9	0.79
	$\{a_5\}$	1	0.09
	$\{a_3, a_5, a_6\}$	–	0.12

All possible intersections of the elements of the considered groups of alternatives are presented in Table 9.

Table 9. Intersections of focal elements (Example 2.3).

Experts' focal elements X_i^j		E_1	
		$\{a_6\}$	$\{a_3, a_5, a_6\}$
E_2	$\{a_3\}$	Ø	$\{a_3\}$
	$\{a_5\}$	Ø	$\{a_5\}$
	$\{a_3, a_5, a_6\}$	$\{a_6\}$	$\{a_3, a_5, a_6\}$

As can be seen from Table 9, there are 2 local conflicts in the model: $X_1^1 \cap X_1^2 = \varnothing$, $X_1^1 \cap X_2^2 = \varnothing$.

The conflict coefficient (13) is 0.792. In such a situation, using combination rules of evidence theory (e.g. Dempster rule) is inexpedient because the level of conflict is too large.

The elements of the frame of discernment do not overlap each other; therefore, the conditions for exclusion and exhaustion are met. In this case, the Dempster-Shafer model is applicable, but it is recommended to use one of the proportional conflict redistribution rules (PCR rules). Let us use PCR5 rule (22):

$m_{12}(\{a_3\}) = 0.41;$
$m_{12}(\{a_5\}) = 0.02;$
$m_{12}(\{a_6\}) = 0.56;$
$m_{12}(\{a_3, a_5, a_6\}) = 0.01.$

Given the results obtained, define the values of the belief (5) and plausibility (6) functions for each of the original alternatives a_3, a_5, a_6:

$$a_3 : \begin{cases} Bel(\{a_3\}) = m_{12}(\{a_3\}) = 0.41; \\ Pl(\{a_3\}) = 0.42. \end{cases}$$

$$a_5 : \begin{cases} Bel(\{a_5\}) = m_{12}(\{a_5\}) = 0.02; \\ Pl(\{a_5\}) = 0.03. \end{cases}$$

$$a_6 : \begin{cases} Bel(\{a_6\}) = 0.56; \\ Pl(\{a_6\}) = 0.57 \end{cases}$$

The following ranking of alternatives has been obtained: $a_6 \succ a_3 \succ a_5$. According to data obtained, Izmail sea port is the most suitable option for placing a dry port.

6 Conclusion

The approach based on the integrated use of the Pareto set formation procedure and the evidence theory for finding optimal solution has been proposed in the paper. Using the mathematical apparatus of evidence theory is primarily due to the fact that the initial set consists of exhaustive and mutually exclusive elements. As additional information, in contrast to the existing approaches, this approach uses the information (evidence) on the formation of subgroups of solutions that make up the Pareto set.

It can be considered as the "equivalent" to decision makers or experts' preferences. However, according to the authors, such a formalized procedure is more realistic in practice, since the requirement from the person of the "unambiguous" preference of one variant over the other is quite "strict".

Acknowledgment. This research was partially supported by the state research project "Development of information and communication decision support technologies for strategic decision-making with multiple criteria and uncertainty for military-civilian use" (research project no. 0117U007144, financed by the Government of Ukraine).

References

1. Nogin, V.D.: Priniatie reshenii v mnogokriterialnoi srede: kolichestvennyi podkhod, 2nd edn. Fizmatlit, Moscow (2005). (in Russian)
2. Bogdanova, A.V., Nogin, V.D.: Reduction of the Pareto set based on some compound information on the relative importance of criteria. Vestnik of Saint Petersburg University. Applied Mathematics. Computer Science. Control Processes, vol. 2, pp. 1–16 (2007). (in Russian)
3. Nogin, V.D., Volkova, N.A.: Evolution of the Edgeworth-Pareto principle. Tavricheskiĭ Vestnik Informatiki i Matematiki **1**, 21–33 (2006). (in Russian)
4. Zakharov, A.O.: Suzhenie mnozhestva Pareto na osnove vzaimozavisimoi informatsii zamknutogo tipa. Iskusstvennyi Intellect i Priniatie Reshenii **1**, 67–81 (2011). (in Russian)
5. Nogin, V.D.: Problema suzheniia mnozhestva Pareto: podkhody k resheniiu. Iskusstvennyi Intellect i Priniatie Reshenii **1**, 98–112 (2008). (in Russian)
6. Nogin, V.D.: Algoritm suzheniia mnozhestva Pareto na osnove proizvolnogo konechnogo nabora "kvantov" informatsii. Iskusstvennyi Intellect i Priniatie Reshenii **1**, 63–69 (2013). (in Russian)
7. Nogin, V.D.: Suzhenie mnozhestva Pareto na osnove nechetkoi informatsii. Int. J. Inf. Technol. Knowl. **6**(2), 157–168 (2012). (in Russian)
8. Podinovskii, V.V.: Vvedenie v teoriiu vazhnosti kriteriev v mnogokriterialnykh zadachakh priniatiia reshenii. Fizmatlit, Moscow (2007). (in Russian)
9. Shafer, G.A.: A Mathematical Theory of Evidence. Princeton University Press, Princeton (1976)
10. Beynon, M.J., Curry, B., Morgan, P.: The Dempster-Shafer theory of evidence: an alternative approach to multicriteria decision modeling. Omega **28**(1), 37–50 (2000)
11. Dempster, A.P.: A generalization of Bayesian inference. J. Roy. Stat. Soc. **B**(30), 205–247 (1968)
12. Nogin, V.D.: Priniatie reshenii pri mnogikh kriteriiakh. IUTAS, SPb (2007). (in Russian)
13. Sentz, K., Ferson, S.: Combination of evidence in Dempster–Shafer theory. Sandia National Laboratories, Albuquerque, New Mexico (2002)
14. Shved, A., Davydenko, Ye.: The analysis of uncertainty measures with various types of evidence. In: 1st International Conference on Data Stream Mining & Processing, Lviv, Ukraine, pp. 61–64. IEEE (2016). https://doi.org/10.1109/dsmp.2016.7583508
15. Uzga-Rebrovs, O.: Upravlenie neopredelennostiami, vol. 3. Izdevnieciba, Rezekne (2010). (in Russian)
16. Zadeh, L.A.: Review of Shafer's "A mathematical theory of evidence". AI Mag. **5**(3), 81–83 (1984)
17. Yager, R.R.: On the Dempster-Shafer framework and new combination rules. Inf. Sci. **41**(2), 93–137 (1987)
18. Inagaki, T.: Interdepence between safety-control policy and multiple-sensor schemes via Dempster-Shafer theory. Trans. Reliab. **40**(2), 182–188 (1991)

19. Zhang, L.: Representation, independence and combination of evidence in the Dempster-Shafer theory of evidence. In: Yager, R.R., Kacprzyk, J., Fedrizzi, M. (eds.) Advances in the Dempster-Shafer Theory of Evidence, pp. 51–69. Wiley, New York (1994)
20. Smets, Ph.: The combination of evidence in the transferable belief model. Pattern Anal. Mach. Intell. **12**, 447–458 (1990)
21. Smarandache, F., Dezert, J.: Advances and Applications of DSmT for Information Fusion, vol. 1. American Research Press, Rehoboth (2004)
22. Smarandache, F., Dezert, J.: Advances and Applications of DSmT for Information Fusion, vol. 2. American Research Press, Rehoboth (2006)
23. Kovalenko, I., Davydenko, Ye., Shved, A.: Development of the procedure for integrated application of scenario prediction methods. Eastern-Eur. J. Enterp. Technol. **2/4**(98), 31–37 (2019). https://doi.org/10.15587/1729-4061.2019.163871
24. Beynon, M.J.: DS/AHP method: a mathematical analysis, including an understanding of uncertainty. Eur. J. Oper. Res. **140**, 148–164 (2002)

Logical-Structural Models of Verbal, Formal and Machine-Interpreted Knowledge Representation in Integrative Scientific Medicine

Serhii Lupenko[1(✉)], Oleksandra Orobchuk[1], and Mingtang Xu[2]

[1] Ternopil Ivan Puluj National Technical University, Ternopil 46000, Ukraine
lupenko.san@gmail.com, orobchuko@gmail.com
[2] Beijing Medical Research Institute “Kundawell”, Beijing 100010, China
mingtangxu@126.com

Abstract. The paper is devoted to the development of logical-structural models of presentation of knowledge in the field of Integrative Scientific Medicine at verbal, formal and machine-interpreted levels. The logical and structural models of knowledge representation are the basis of the onto-oriented knowledge base of Integrative scientific medicine, reflect the structure of its theory and provide unification, standardization of the technology of presenting information (data and knowledge) as a result of multidisciplinary, interdisciplinary and transdisciplinary studies of national and traditional medical trends that claim for joining the Integrative Scientific Medicine. On the basis of developed models using the OWL language and the Protégé environment, a computer ontology of the special scientific theory of Chinese Image Medicine was created that is used for an onto-oriented expert system, an electronic multimedia education system and a system for professional activities in the field of Chinese Image Medicine as a promising component of Integrative Scientific Medicine.

Keywords: Logical-structural models of knowledge representation · Ontology · Axiomatic-deductive strategy · Onto-oriented informational systems · E-learning systems · Integrative Medicine · Chinese Image Medicine

1 Introduction

The present stage of development of mankind is characterized by incredibly rapid pace of formation of a new world order, globalization processes in all spheres of social life and realization of synthetic (integrative) tendencies in science, philosophy, religion and art. It also fully applies to the medical field, one of the important innovative directions of which is the formation of Integrative Scientific Medicine, which is based on scientific principles and synthesizes the best achievements of the conventional (western) and nonconventional (alternative, complementary, folk) medicine [1–5]. In spite of the active development of Integrative Medicine in America, China and Europe, the existence of many international periodical scientific journals devoted to the development of Integrative Medicine and the holding of several World Congresses of Integrative

N. Shakhovska and M. O. Medykovskyy (Eds.): CCSIT 2019, AISC 1080, pp. 139–153, 2020.
https://doi.org/10.1007/978-3-030-33695-0_11

Medicine, Integrative Scientific Medicine is not a formed theoretical and applied direction, its formation is only in the initial stage. Solving the following problems are the necessary conditions for the formation of Integrative Scientific Medicine (ISM):

1. Development of a holistic methodology and carrying out a complex of multidisciplinary, interdisciplinary and transdisciplinary studies of selected nonconventional (alternative, complementary, folk) medical areas that claim to be a part of ISM that meets the requirements of evidence-based medicine and is consistent with the WHO's alternative medicine strategy [6].
2. Creation of the scientific theory of Integrative Medicine, which would assume the role of a meta-interpreter of the existing traditional theories of all (conventional and nonconventional) medical areas that claim to be part of it and would provide a basis for their mutual harmonization and coordination.
3. Creation of special scientific theories of individual medical areas, which claim to be part of the ISM. These special theories should be coordinated among themselves at the level of the general theory of ISM, and to act as interpreters of the existing traditional (pre-scientific) theories of the corresponding medical directions.
4. Creation of the developed infrastructure of research institutes, educational establishments, associations of specialists of ISM that would enable the development and implementation of ISM both at the national and international levels and would meet national and international standards for the provision of medical services to the population.

Solution of these tasks is impossible without the development of an integrated information and analytical environment of ISM, which includes software and tools for supporting the implementation of multidisciplinary, interdisciplinary and transdisciplinary studies of certain nonconventional medical areas, which claim to be a part of ISM, program means of analysis, systematization, comparison of the results of diagnostic and therapeutic activities within different (conventional and nonconventional) medical areas, as well as expert support system for diagnostic and therapeutic decision-making and e-learning system.

The intellectualized core of such an integrated information and analytical environment should be the multi-level onto-oriented knowledge base of ISM, which in formal and machine-interpreted form would reflect the theory of ISM, and would provide access to knowledge for information systems supporting the implementation of multidisciplinary, interdisciplinary and transdisciplinary research separate nonconventional medical directions, expert system for diagnostic and therapeutic decision-making support and e-learning systems in the field of ISM.

2 State of Study Problem

The vast majority of modern theories, models and methods of conventional (western) medicine are based on scientific theoretical and experimental principles, they generally meet the requirements of evidence-based medicine and are equipped with modern information-analytical and instrumental diagnostic-therapeutic means. As far as folk and tradition (alternative, complementary) medical areas are concerned, it is a good example to carry out large-scale clinical research and development of relevant information and analytical tools (ontologies, expert systems, grid systems) for Traditional Chinese Medicine, as reflected in a large number of scientific publications [7–14]. There are also similar works for a number of other popular medical areas, including Indian, Tibetan and African alternative medical traditions [15–17]. However, despite the clinical research and information and analytical tools developed for nonconventional medical areas, we cannot yet talk of synthesis, coordination and unification of the theoretical foundations, experimental and clinical studies and the relevant informational and technological means of non-conventional medicine and western medicine within the framework of the paradigm of ISM.

A promising approach to organizing a scientific theory and constructing an onto-oriented information and analytical environment for an nonconventional medical direction within the framework of Integrative Scientific Medicine is an approach to Chinese Image Medicine (CIM), the research program of which was developed in 2016 [18]. In papers [19–24] a number of important results were obtained for the emergence of the theory of Chinese Image Medicine as a scientific theory that exists in three interrelated forms, namely verbal, formal, and machine-interpreted levels. The development of the scientific theory of the CIM is proposed in accordance with its traditional theory and with the theory of Integrative scientific medicine. The research methodology of the CIM in the framework of classical and nonclassical scientific rationality was substantiated, the axiomatic-deductive strategy of the organization of traditional and scientific theories of the CIM was substantiated, satisfying the requirements of semantic quality and providing clear guidelines in its development strategy. The generalized architectures of all information systems, which are part of the integrated onto-oriented information-analytical environment of scientific researches, professional healing and e-learning of Chinese Image Medicine, have been developed.

Based on the results of works [19–24], it is appropriate to disseminate and develop them in relation to the needs of ISM, which, as its components, includes a special scientific and traditional theory of CIM. In particular, the purpose of this work is to develop logical-structural models of presentation of knowledge of ISM based on axiomatic-deductive strategy and an ontological approach to the organization of knowledge that will provide a high level of their semantic quality, reflect the separate structure of the theory of ISM and will be adequate for its consistent representation on the verbal, formal and machine-interpreted levels.

3 Results

Structure of the Integrative Scientific Medicine Theory. Based on the results of works [19–21], the theory of ISM is desirable to be presented in two parts of it, namely, the general theory of ISM and a set of special theories of individual medical areas (conventional or nonconventional) that claim to enter the composition of ISM. The general theory of ISM is currently in the stage of its active formation and development and should play the role of meta-theory and meta-language of the corresponding individual medical direction within the framework of Integrative Medicine. The special scientific theory of the nonconventional medical direction reflects the specific and unique knowledge that is characteristic of it purely and which is based on the principles of the general theory of ISM, as in its meta-theory.

In order to historically preserve and comprehensively study traditional and folk forms of medical knowledge, the traditional historical (nonscientific) theories of separate traditional and alternative medical theories should be included in the theory of ISM, which claim to enter into ISM. Traditional theory of nonconventional medical direction is given in terms and concepts that are characteristic of culture and philosophy of the corresponding historical epoch of the formation of this medical direction, and also reflects the terminology and conceptual apparatus of the understanding of traditional theory by modern specialists practicing the corresponding direction of nonconventional medicine. Special scientific theory of a separate nonconventional medical field, in conjunction with the general theory of ISM, will play the role of interpreter of the corresponding traditional theory, that is, they must have significantly more expressive language and conceptual means in comparison with the corresponding means of traditional theory. This indicates the highest priority of the scientific theory of the nonconventional medical trend in comparison with its traditional version, which consists of the possibility of an adequate rational interpretation of all elements (concepts, relations between concepts) of the terminological and conceptual apparatus, all statements and inferences of the traditional theory of nonconventional medical direction, its means scientific theory.

According to works [19–21] separate structure of the general theory of ISM and special theories of individual medical directions will be presented as in Table 1.

Table 1. Separate structure of general and special theories of Integrative Scientific Medicine.

Sections of general theory of integrative medicine	Sections of the special theory of the special medical direction
The General Theory of reality and human in Integrative Medicine	Special theory of reality and human in a separate medical direction
General theory of health and diseases in Integrative medicine	Special theory of health and diseases in a separate medical direction
General theory and diagnostic technologies in Integrative Medicine	Special theory and diagnostic technologies in a separate medical direction
General theory and technology of therapy in Integrative medicine	Special theory and technology of therapy in a separate medical direction
General theory and technology of training, development of a specialist in Integrative Medicine	Special theory and technology of training, development of specialist of a separate medical direction

The section “Theory (general and special) of reality and human” describes the basic concepts and ideas of Integrative Medicine and/or its corresponding medical component (for example CIM) and serves as an ontological, epistemological, axiological, and praxological philosophical foundation for the rest of their sections. The section “Theory of health and diseases” describes the basic concepts of health and disease; diagnostic standards of health and diseases for their evaluation by different diagnostic methods; classification and definition of types of diseases. The section "Theory and technology of diagnosing" describes and formalizes theoretical foundations, methods and means of obtaining diagnostic medical information, as well as methods for its interpretation. The section "Theory and therapy technologies" describes and formalizes theoretical foundations, methods and means of conducting therapeutic procedures, as well as their interrelations with the corresponding diagnostic information. The section "Theory and technology of education, development of a medical specialist" describes educational theoretically and practically oriented content, as well as technologies for its implementation into the educational process for the preparation and improvement of qualifications of medical specialists of ISM. A similar division into sections takes place also for the traditional theories of certain nonconventional medical areas.

Thus, taking into account the division of the theory of ISM into the general and the totality of special theories, the inclusion in it of a set of traditional theories of certain nonconventional medical directions, as well as the division of general, special and traditional theories into the relevant sections, the structure of the theory of a separate medical nonconventional direction can be presented as schema shown in Fig. 1.

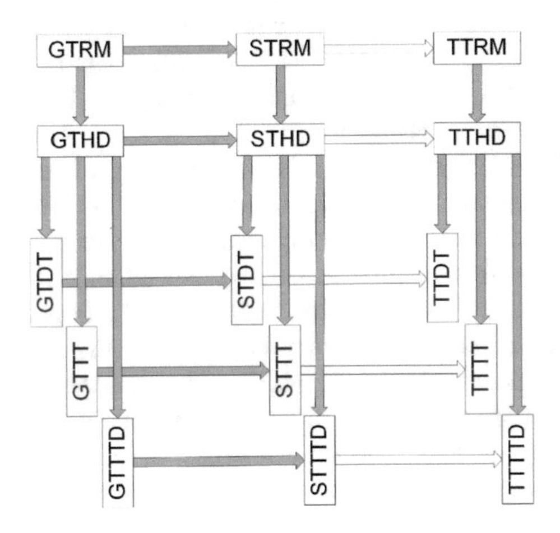

Fig. 1. Structure of the theory of a separate medical nonconventional direction as a component of Integrative Scientific Medicine.

In Fig. 1, the following notation is made: GTRM (STRM, TTRM) - general (special, traditional) theory of reality and human; GTHD (STHD, TTHD) - general (special, traditional) theory of health and diseases; GTDT (STDT, TTDT) - general (special, traditional) theory and diagnostic technologies; GTTT (STTT, TTTT) - general (special, traditional) theory and technology of therapy; GTTTD (STTTD, TTTTD) - general (special, traditional) theory and technology of training, development of a medical specialist. The filled arrows in the structure, depicted in Fig. 1, reflect the procedure for the taxonomic distribution of the terminology-conceptual apparatus and the system statements of the sections of the higher level of universality (abstraction) to the corresponding sections of the lower level of generality (abstraction). The cavity arrows of this structure reflect the fact of the interpretation of traditional theories of certain nonconventional medical areas by the logical-semantic means of the general and corresponding special scientific theories of Integrative Medicine.

4 Logical-Structural Models of Knowledge Representation in Integrative Scientific Medicine

To ensure the high semantic quality of the theory of ISM, we apply the axiomatic-deductive strategy of knowledge organization, which was developed in the works [25, 26], to construct the theory of ISM. According to the axiomatic-deductive strategy of organization of knowledge, the semantic space of the theory of ISM, which includes general, special and traditional theories, is a complex heterogeneous system, the fundamental components of which are the set of terms-concepts that define the terminology-conceptual apparatus (glossary) of ISM, the set of relations between these concepts, a set of statements and reasoning (proofs). Considering that the statements of the ISM theory can be regarded as certain functions in a set of its terms-concepts, that actualize, reflect explicit and implicit relations between the terms-concepts of ISM, and reasonings are certain relations between its statements, it can be stated that the content of the theory of ISM is completely determined by its terminological-conceptual apparatus and the system of statements.

Since the result of the axiomatic-deductive strategy of organizing the semantic space of ISM is its logical-semantic core, which "thread" all the other elements of this semantic space, then all these other components of the semantic space can be conventionally called the periphery of the semantic space of the ISM theory. The periphery of the semantic space of ISM contains concepts and statements that perform an explanatory, supportive role in the components of the logical-semantic core of the content space of the theory of ISM. In this case, the semantic space of the theory of ISM can be presented as the union of its logical-semantic core and periphery.

A fundamental generalized formalized structure, which describes the logical-semantic core of the theory of ISM and sets its knowledge base, is the two:

$$LSKor = \langle CLSKor,\ SLSKor \rangle, \tag{1}$$

where *CLSKor* is a formalized structure that describes the conceptual system (conceptual space) of the logical-semantic core of the theory of ISM, namely, it presents it as a set of concepts, operations on concepts and relationships between concepts, providing a correct formalized description of the terminological and conceptual apparatus of the theory of ISM; formalized structure *SLSKor* describes the system of statements of the logical-semantic core of the theory of ISM, namely, it presents it as a system of axiomatic statements (axioms), rules of proofing and derivative statements (theorems).

The set of elements and connections of the formalized structure *SLSKor* is given on the elements and connections of the structure *CLSKor*, which can be conditionally applied as follows:

$$SLSKor = F[CLSKor], \tag{2}$$

where $F[\cdot]$ - describes a set of transformations that take place between elements and connections of structures *SLSKor* and *CLSKor*.

Taking into account the foregoing, concerning the general structure of the theory of ISM, we will write in more detail the components *CLSKor* and *SLSKor*.

Conceptual system *CLSKor* will be presented as a relational system:

$$CLSKor = \left\langle \left\{ \mathbf{C}_{meta},\ \left\{ \mathbf{C}_{S_n},\ n = \overline{1,N} \right\},\ \left\{ \mathbf{C}_{T_n},\ n = \overline{1,N} \right\} \right\},\ \mathbf{\Psi} \right\rangle. \tag{3}$$

In expression (3) $\mathbf{C}_{meta}$ is a conceptual system of the upper level, which reflects the terminological-conceptual apparatus (concepts, relations and operations on concepts) of the general theory of ISM, which is common to all medical areas that are part of it.

A set $\left\{ \mathbf{C}_{S_n},\ n = \overline{1,N} \right\}$ is a set of conceptual systems of individual medical areas that are part of Integrative scientific medicine, and which describe the special scientific theories of the relevant conventional and unconventional medical areas.

A set $\left\{ \mathbf{C}_{T_n},\ n = \overline{1,N} \right\}$ is a set of conceptual systems of individual medical areas that are part of ISM, and which describe the traditional scientific theories of relevant conventional and nonconventional medical areas.

$\mathbf{\Psi}$ is a set of relations (the function of interpretation and the taxonomic relation) $\mathbf{\Psi} = \left\{ \left\{ f_I^C(\cdot),\ \psi_{Tax}^C \right\} \right\}$, which are given on the concepts of systems $\mathbf{C}_{meta}$, $\left\{ \mathbf{C}_{S_n},\ n = \overline{1,N} \right\}$, $\left\{ \mathbf{C}_{T_n},\ n = \overline{1,N} \right\}$, establishing the relationships between these conceptual systems. Interpretation function $f_I^C(\cdot)$ is given on the set of all concepts of the traditional theory of the corresponding medical direction (that is, on the set of concepts from $\mathbf{C}_{T_n}$) and gains values on the set of vectors of concepts of general and special scientific theories of the corresponding medical direction (that is, from the set of vectors, the elements of which are concepts from $\mathbf{C}_{meta}$ and $\mathbf{C}_{S_n}$).

System of statements *SLSKor* will submit as a relational system:

$$SLSKor = \left\langle \left\{ \mathbf{S}_{meta},\ \left\{ \mathbf{S}_{S_n},\ n = \overline{1,N} \right\},\ \left\{ \mathbf{S}_{T_n},\ n = \overline{1,N} \right\} \right\},\ \mathbf{\Xi} \right\rangle. \tag{4}$$

In the expression (4) $\mathbf{S}_{meta}$ is a system of statements of the logical-semantic core of the general theory of ISM, which are common to all medical areas that are part of it.

A set $\left\{ \mathbf{S}_{S_n},\ n = \overline{1,N} \right\}$ is a set of systems of statements of logical-semantic core of special scientific theories of certain conventional and nonconventional medical areas that are part of ISM.

A set $\left\{ \mathbf{S}_{T_n},\ n = \overline{1,N} \right\}$ is a set of systems of statements of the logical-semantic core of the traditional theories of certain conventional and nonconventional medical areas that are part of ISM.

$\mathbf{\Xi}$ is a set of relations (the function of interpretation and the taxonomic relation) $\mathbf{\Xi} = \left\{ f_I^S(\cdot),\ \psi_{Tax}^S \right\}$, which are given on the elements (statements) of systems $\mathbf{S}_{meta}$, $\{\mathbf{S}_{S_n},\ n = \overline{1,N}\}$, $\{\mathbf{S}_{T_n},\ n = \overline{1,N}\}$, establishing the relationships between them. Interpretation function $f_I^S(\cdot)$ is given on the set of all assertions of the traditional theory of the corresponding medical direction (that is, on the set of statements from $\mathbf{S}_{T_n}$) and gains values from the set of statements of general and special scientific theories of the corresponding medical direction (that is, from the set of statements from $\mathbf{S}_{meta}$ and $\mathbf{S}_{S_n}$).

Taking into account the relationship between systems *CLSKor* and *SLSKor*, which is conventionally depicted in Expression (2), we will detail it by writing the following set of relations:

$$\mathbf{S}_{meta} = F[\mathbf{C}_{meta}],\ \mathbf{S}_{S_n} = F[\mathbf{C}_{S_n}],\ \mathbf{S}_{T_n} = F[\mathbf{C}_{T_n}],\ n = \overline{1,N}. \quad (5)$$

Summarizing the above results, the logical-semantic core of the theory of ISM can be presented as the following graph of formalized structures that describe it (see Fig. 2).

For the convenience of further presentation of the material, we will denote the conceptual system $\mathbf{C}_{meta}$ as $\mathbf{C}_1$, the conceptual system $\mathbf{C}_{S_n}$ we will denote as $\mathbf{C}_{n+1}$, the conceptual system $\mathbf{C}_{T_n}$ denoted as $\mathbf{C}_{N+n+1}$ ($n = \overline{1,N}$), the system of statements $\mathbf{S}_{meta}$ we denote as $\mathbf{S}_1$, the system of statements $\mathbf{S}_{S_n}$ is denoted as $\mathbf{S}_{n+1}$, and the system of statements $\mathbf{S}_{T_n}$ is denoted as $\mathbf{S}_{N+n+1}$ ($n = \overline{1,N}$). Thus, taking into account the notation introduced by us, the conceptual system *CLSKor* of ISM in a more compact and uniform form will be presented as follows:

$$CLSKor = \left\langle \left\{ \mathbf{C}_n,\ n = \overline{1, 2 \cdot N + 1} \right\},\ \mathbf{\Psi} \right\rangle, \quad (6)$$

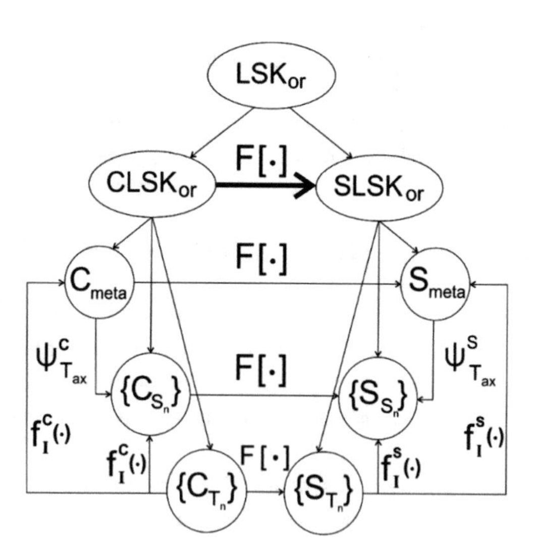

Fig. 2. A graph of formalized structures of the logical-semantic core of the theory of Integrative Scientific Medicine.

the system of statements *SLSKor* ISM in a more compact and uniform form will be presented as follows:

$$SLSKor = \left\langle \left\{ \mathbf{S}_n,\ n = \overline{1, 2 \cdot N + 1} \right\},\ \Xi \right\rangle. \quad (7)$$

We will analyze the elements of the formalized structures (6) and (7), namely, taking into account the separate structure of the theory of ISM (5 sections), components and, presented in the form of the following relational systems:

$$\mathbf{C}_n = \left\langle \left\{ \mathbf{C}_n^j,\ j = \overline{1, 5} \right\},\ \psi_{Tax}^C \right\rangle, \quad (8)$$

$$\mathbf{S}_n = \left\langle \left\{ \mathbf{S}_n^j,\ j = \overline{1, 5} \right\},\ \psi_{Tax}^S \right\rangle, \quad (9)$$

where $\mathbf{C}_n^j$- conceptual system *j*-th section of the general theory $(n = 1)$, special ones $\left(n = \overline{2, N+1}\right)$ scientific theories, as well as traditional theories $\left(n = \overline{N+2, 2 \cdot N}\right)$ separate medical directions as components of ISM; $\mathbf{S}_n^j$- system of statements *j*-th section of the general theory $(n = 1)$, special ones $\left(n = \overline{2, N+1}\right)$ scientific theories, as well as traditional theories $\left(n = \overline{N+2, 2 \cdot N}\right)$ separate medical directions as components of ISM.

We now summarize the above designations in the form of Table 2.

Table 2. Designation of conceptual systems and systems of statements of the sections of the theory of Integrative Scientific Medicine.

Theory section					
The name of the theory	1	2	3	4	5
General scientific theory	$\mathbf{C}_1^1, \mathbf{S}_1^1$	$\mathbf{C}_1^2, \mathbf{S}_1^2$	$\mathbf{C}_1^3, \mathbf{S}_1^3$	$\mathbf{C}_1^4, \mathbf{S}_1^4$	$\mathbf{C}_1^5, \mathbf{S}_1^5$
Special scientific theories	$\mathbf{C}_2^1, \ldots, \mathbf{C}_{N+1}^1$ $\mathbf{S}_2^1, \ldots, \mathbf{S}_{N+1}^1$	$\mathbf{C}_2^2, \ldots, \mathbf{C}_{N+1}^2$ $\mathbf{S}_2^2, \ldots, \mathbf{S}_{N+1}^2$	$\mathbf{C}_2^3, \ldots, \mathbf{C}_{N+1}^3$ $\mathbf{S}_2^3, \ldots, \mathbf{S}_{N+1}^3$	$\mathbf{C}_2^4, \ldots, \mathbf{C}_{N+1}^4$ $\mathbf{S}_2^4, \ldots, \mathbf{S}_{N+1}^4$	$\mathbf{C}_2^5, \ldots, \mathbf{C}_{N+1}^5$ $\mathbf{S}_2^5, \ldots, \mathbf{S}_{N+1}^5$
Traditional theories	$\mathbf{C}_{N+2}^1, \ldots, \mathbf{C}_{2N+1}^1$ $\mathbf{S}_{N+2}^1, \ldots, \mathbf{S}_{2N+1}^1$	$\mathbf{C}_{N+2}^2, \ldots, \mathbf{C}_{2N+1}^2$ $\mathbf{S}_{N+2}^2, \ldots, \mathbf{S}_{2N+1}^2$	$\mathbf{C}_{N+2}^3, \ldots, \mathbf{C}_{2N+1}^3$ $\mathbf{S}_{N+2}^3, \ldots, \mathbf{S}_{2N+1}^3$	$\mathbf{C}_{N+2}^4, \ldots, \mathbf{C}_{2N+1}^4$ $\mathbf{S}_{N+2}^4, \ldots, \mathbf{S}_{2N+1}^4$	$\mathbf{C}_{N+2}^5, \ldots, \mathbf{C}_{2N+1}^5$ $\mathbf{S}_{N+2}^5, \ldots, \mathbf{S}_{2N+1}^5$

According to [21, 24], the development of the theory of ISM is appropriate to carry out on three interrelated levels: verbal, formal, and machine-interpreted. Therefore, we designate the above structures $\mathbf{C}_n^j$ and $\mathbf{S}_n^j$ for these forms as follows:

(1) $\mathbf{C}V_n^j$ та $\mathbf{S}V_n^j$ - conceptual system and system of statements *j*-th section of the general theory $(n = 1)$, special ones $\left(n = \overline{2, N+1}\right)$ scientific theories, as well as traditional theories $\left(n = \overline{N+2, 2 \cdot N+1}\right)$ separate medical directions as components of ISM in verbal form (in natural language);

(2) $\mathbf{C}F_n^j$ та $\mathbf{S}F_n^j$ - conceptual systems and statements systems *j*-th section of the general theory ($n = 1$), special ones $\left(n = \overline{2, N+1}\right)$ scientific theories, as well as traditional theories $\left(n = \overline{N+2, 2 \cdot N+1}\right)$ separate medical directions as components of ISM as a formal axiomatic system;
(3) $\mathbf{C}M_n^j$ та $\mathbf{S}M_n^j$ - conceptual systems and statements systems *j*-th section of the general theory ($n = 1$), special ones $\left(n = \overline{2, N+1}\right)$ scientific theories, as well as traditional theories $\left(n = \overline{N+2, 2 \cdot N+1}\right)$ separate medical directions as components of ISM in machine-interpretated form.

Taking into account the formalized apparatus of the axiomatic-deductive strategy of knowledge organization, described in the papers [25, 26], components $\mathbf{C}_n^j$ and $\mathbf{S}_n^j$ in three forms of their presentation are presented as corresponding axiomatic systems. Let's consider in more detail only axiomatic systems $\mathbf{C}F_n^j$ and $\mathbf{S}F_n^j$.

Conceptual system $\mathbf{C}F_n^j$ *j*-th section of the corresponding n-th (general, special, traditional) theory as a formal system, we will submit the following:

$$\mathbf{C}F_n^j = \left\{\mathbf{AlC}F_n,\ \mathbf{TC}F_n^j,\ \mathbf{AC}F_n^j,\ \mathbf{RulsC}F\right\}, \tag{10}$$

where: $\mathbf{AlC}F_n$ - the alphabet of an artificial language, which is a finite set of symbols, of which the correctly constructed formulas of descriptive logic are formed, which describes the (general, special, traditional) theory of ISM;

$\mathbf{TC}F_n^j = \left\{CF_n^j(k),\ k = \overline{1, K}\right\}$ - the set of all correctly formulated formulas formed from elements of the alphabet $\mathbf{AlC}F_n$, which mutually unambiguously correspond to the terminology of the terminological-conceptual apparatus of the j-th section of the corresponding n-th (general, special, traditional) theory of ISM;

$\mathbf{AC}F_n^j = \left\{CF_n^j(k),\ k = \overline{1, K_1}\right\}$ - the set of axioms of the formal system, which corresponds to the set of the names of base (atomic) concepts in the natural language. Set $\mathbf{AC}F_n^j$ is a subset $\mathbf{TC}F_n^j$ $\left(\mathbf{AC}F_n^j \subset \mathbf{TC}F_n^j\right)$;

$\mathbf{RulsC}F = \{RCF_1,\ RCF_2,\ \ldots, RCF_P\}$ is the set of formally-logical rules (operations) of the set $\mathbf{AC}F_n^j$ of derivatives of correctly constructed formulas $\mathbf{DC}F_n^j = \left\{CF_n^j(k),\ k = \overline{K_1+1, K}\right\}$ in the framework of the j-th section of the n-th theory of ISM.

From the foregoing follows the following relation:

$$\mathbf{TC}F_n^j = \mathbf{AC}F_n^j \cup \mathbf{DC}F_n^j,\ n = \overline{1, 2N+1},\ j = 5. \tag{11}$$

System of statements $\mathbf{SF}_n^j$ *j*-th section of the corresponding n-th (general, special, traditional) theory in the form of a formal system, we will submit as a four:

$$\mathbf{SF}_n^j = \left\{ \mathbf{AlS}F_n,\ \mathbf{TS}F_n^j,\ \mathbf{AS}F_n^j,\ \mathbf{Ruls}\mathbf{S}F \right\}, \tag{12}$$

where: $\mathbf{AlS}F_n$ is the alphabet of the formal language of the logic of predicates of the first order;

$\mathbf{TS}F_n^j = \left\{ SF_n^j(l),\ l = \overline{1, L} \right\}$ - the set of all correctly constructed formulas in the logic of predicates of the first order, formed from elements of the alphabet $\mathbf{AlS}F_n$, which mutually unambiguously correspond to the statement in the natural language of j-th section of the corresponding n-th (general, special, traditional) theory of ISM;

$\mathbf{AS}F_n^j = \left\{ SF_n^j(l),\ l = \overline{1, L_1} \right\}$ - the set of axioms of the formal system that corresponds to the set $\mathbf{AS}V_n^j$ axiomatic statements in the natural language. Set $\mathbf{AS}F_n^j$ is a subset $\mathbf{TS}F_n^j$ ($\mathbf{AS}F_n^j \subset \mathbf{TS}F_n^j$);

$\mathbf{Ruls}\mathbf{S}F = \{RSF_1,\ RSF_2,\ \ldots, RSF_G\}$ is the set of formally-logical rules (operations) of the set $\mathbf{AS}F_n^j$ of derivatives of correctly constructed formulas $\mathbf{DS}F_n^j = \left\{ SF_n^j(k),\ k = \overline{L_1 + 1, L} \right\}$ within *j*-th section *n*-th theory of ISM.

From the foregoing follows the following relation:

$$\mathbf{TS}F_n^j = \mathbf{AS}F_n^j \cup \mathbf{DS}F_n^j,\ n = \overline{1, 2N+1},\ j = 5. \tag{13}$$

According to formulas (2) and (5), the axiomatic systems of true statements $\mathbf{S}V_n^j$, $\mathbf{S}F_n^j$, $\mathbf{S}M_n^j$ are based on the corresponding axiomatic systems $\mathbf{C}V_n^j$, $\mathbf{C}F_n^j$, $\mathbf{C}M_n^j$ of conceptual system of *j*-th section *n*-th theory of ISM.

In particular, axiomatic statements $\mathbf{AS}F_n^j = \left\{ SF_n^j(l),\ l = \overline{1, L_1} \right\}$ can be regarded as functions on atomic concepts $\mathbf{AC}F_n^j = \left\{ CF_n^j(k),\ k = \overline{1, K_1} \right\}$ *j*-th section *n*-th theory of ISM, namely, as identical true predicates:

$$SF_n^j(l), = f_l\left(CF_n^j(k),\ k = \overline{1, K_1} \right),\ l = \overline{1, L_1}. \tag{14}$$

The alphabet $\mathbf{AlS}F_n$ contains alphabet $\mathbf{AlC}F_n$ as its subset.

Thus, summing up the given results, the logical-semantic core of the theory of ISM is given by a matrix of size $(2N+1) \cdot 5$, where each of its elements is a set of verbal, formal and machine-interpreted, isomorphic with each other axiomatic systems:

$$\mathbf{C}V_n^j,\ \mathbf{C}F_n^j,\ \mathbf{C}M_n^j,\ \mathbf{S}V_n^j,\ \mathbf{S}F_n^j,\ \mathbf{S}M_n^j,\ n = \overline{1, 2N+1},\ j = 5. \tag{15}$$

as well as interpretation functions $f_I^O(\cdot), f_I^S(\cdot)$ and taxonomic relationships ψ_{Tax}^O, ψ_{Tax}^S.

A Fragment of the Computer Ontology of Chinese Image Medicine as a Component of Integrative Scientific Medicine. Based on developed models using the OWL language and the Protégé environment, a computer ontology of the special scientific theory of Chinese Image Medicine was created that is used for an onto-oriented expert system, electronic multimedia learning system and system for professional activities in the field of Chinese Image Medicine.

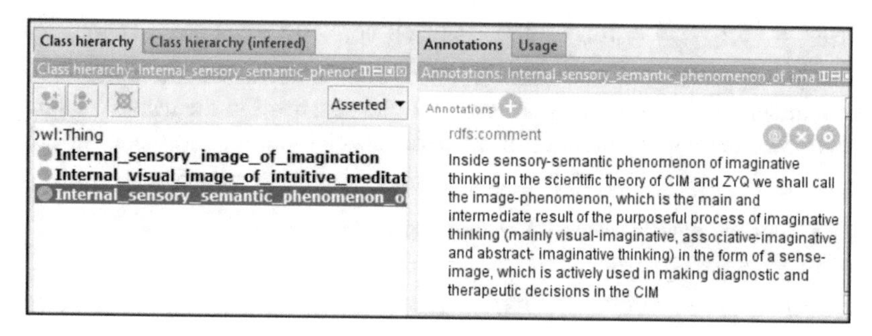

Fig. 3. A fragment of the glossary, which contains scientific interpretations of the traditional notions of CIM.

In particular, Fig. 3 shows a fragment of the glossary of the traditional CIM theory, which contains scientific interpretations of its traditional notions.

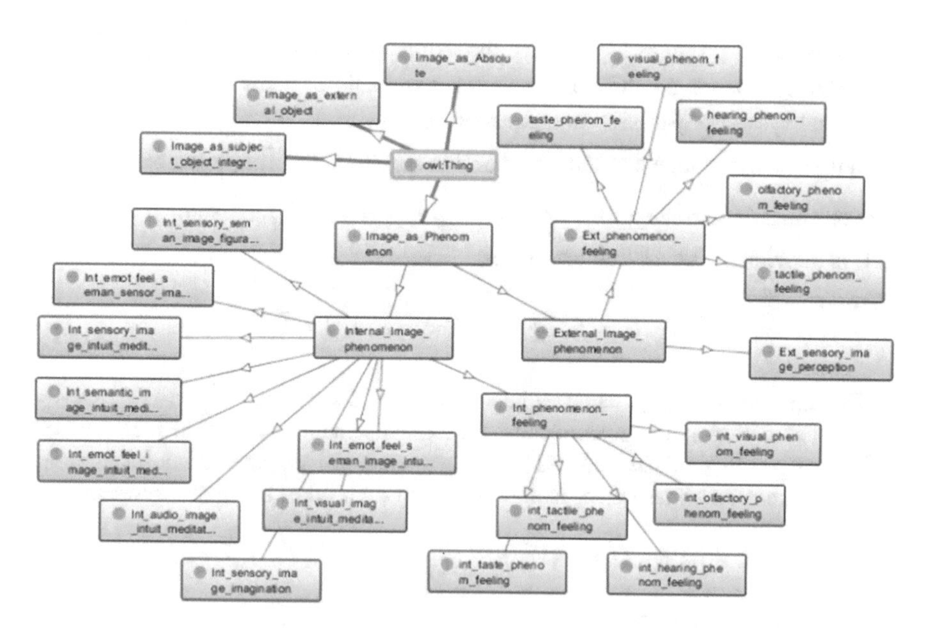

Fig. 4. Taxonomy of concepts that comprehensively cover the fundamental notion of traditional theory of CIM - the concept of "Image".

Figure 4 provides a taxonomy of concepts that comprehensively cover the fundamental notion of the traditional theory of CIM - the concept of "Image", and on the basis of which it is interpreted by means of a special scientific theory of CIM.

5 Conclusions

The paper has developed the logic-structural models of presentation of the knowledge of Integrative Scientific Medicine, which correspond to the axiomatic-deductive strategy for knowledge organization at the verbal, formal and machine-interpreted levels, satisfy the requirements of the logical-semantic quality of scientific theories, provide unification, standardization of the technology of presentation of information (data and knowledge) in the field of Integrative Scientific Medicine and enable to overcome the problem of semantic heterogeneity of poorly structured and difficult formalized knowledge in the field of nonconventional medical areas, eliminate subjective factors, polysemantics, obscurity of concepts and images that are used explicitly or implicitly by specialists of folk and traditional medical directions in the process of diagnostic and therapeutic decision making.

The developed logic-structural models reflect the structure of the theory of Integrative Scientific Medicine and have significant potential for a consistent unified representation of knowledge as a result of multidisciplinary, interdisciplinary and transdisciplinary studies of folk and traditional medical areas that are claiming to become part of Integrative Scientific Medicine.

The developed models for organization of a special scientific theory of Chinese Image Medicine as a component of Integrative Scientific Medicine have been applied, and relevant ontologies for onto-oriented expert systems, electronic multimedia education systems and the system for the professional activities of CIM-therapists have been created.

The obtained results are the basis for further development of the integrated information and analytical environment of Integrative Scientific Medicine, which includes software and tools for supporting the implementation of multidisciplinary, interdisciplinary and transdisciplinary studies of certain nonconventional (folk) medical areas, expert systems for the acceptance of diagnostic and therapeutic decisions and e-learning systems in the field of Integrative Scientific Medicine.

References

1. Micozzi, M.S.: Fundamentals of Complementary, Alternative and Integrative Medicine. Copyright: Elsevier, Amsterdam (2019)
2. Fan, D.: Holistic integrative medicine: toward a new era of medical advancement. Front. Med. **11**(1), 152 (2017)
3. Guarneri, E., Horrigan, B., Pechura, C.: The efficacy and cost effectiveness of integrative medicine: a review of the medical and corporate literature. J. Sci. Healing **5**, 308–312 (2010)
4. Maizes, V., Rakel, D., Niemiec, C.: Integrative medicine and patient-centered care. J. Sci. Healing **5**(5), 277–289 (2009)

5. Horrigan, B.: What is integrative medicine? http://www.bravewell.org/integrative_medicine/what_is_IM. Accessed 23 Nov 2016
6. WHO strategy for traditional medicine for 2014–2023. http://www.who.int/medicines/publications/traditional/trm_strategy14_23/ru/. Accessed 20 Nov 2018
7. Wang, H.: A computerized diagnostic model based on naive Bayesian classifier in traditional Chinese medicine. In: Proceedings of the 1st International Conference on BioMedical Engineering and Informatics (BMEI 2008), pp. 474–477 (2008)
8. Huang, M.-J., Chen, M.-Y.: Integrated design of the intelligent web-based Chinese Medical Diagnostic System (CMDS) – systematic development for digestive health. Expert Syst. Appl. **32**(2), 658–673 (2007)
9. Mao, Y., Yin, A.: Ontology modeling and development for Traditional Chinese Medicine. In: Proceedings of the 2nd International Conference on Biomedical Engineering and Informatics (BMEI 2009), pp. 1–5 (2009)
10. Silva, P., et al.: An expert system for supporting Traditional Chinese Medicine diagnosis and treatment. Procedia Technol. **16**, 1487–1492 (2014)
11. Wang, Y., Zhonghua, Y., Jiang, Y., Liu, Y., Chen, L., Liu, Y.: A framework and its empirical study of automatic diagnosis of traditional Chinese medicine utilizing raw free-text clinical records. J. Biomed. Inform. **45**(2), 210–223 (2012)
12. Wang, X., Qu, H., Liu, P., Cheng, Y.: A self-learning expert system for diagnosis in traditional Chinese medicine. Expert Syst. Appl. **26**(4), 557–566 (2004)
13. Lukman, S., He, Y., Hui, S.C.: Computational methods for traditional Chinese medicine: a survey. Comput. Methods Programs Biomed. **88**, 283–294 (2007)
14. Chen, H., Wang, Y., Wang, H., et al.: Towards a semantic web of relational databases: a practical semantic toolkit and an in-use case from traditional Chinese medicine. In: Proceeding of the 5th International Conference on The Semantic Web (ISWC 2006), pp. 750–763 (2006)
15. Atemezing, G., Pavón, J.: An ontology for African traditional medicine. In: Corchado, J.M., Rodríguez, S., Llinas, J., Molina, J.M. (eds.) International Symposium on Distributed Computing and Artificial Intelligence 2008 (DCAI 2008). Advances in Soft Computing, vol. 50. Springer, Heidelberg (2009)
16. Gayathria, M., Jagadeesh Kannan, R.: Ontology based Indian medical system. Mater. Today Proc. **5**(1), 1974–1979 (2018)
17. Zhu, X., et al.: Research on classification of Tibetan medical syndrome in chronic atrophic gastritis. Appl. Sci. **9**, 1664 (2019)
18. International program of scientific research in Chinese Image Medicine and Zhong Yuan Qigong for 2017–2023. https://kundawell.com/ru/mezhdunarodnaya-programma-nauchnykh-issledovanij-kitajskoj-imidzh-meditsiny-i-chzhun-yuan-tsigun-na-2017-2023-god. Accessed 15 Nov 2018
19. Lupenko, S., Orobchuk, O., Vakulenko, D., Sverstyuk, A., Horkunenko, A.: Integrated onto-based information analytical environment of scientific research, professional healing and e-learning of Chinese Image Medicine. Inf. Sys. Net. J., 10–19 (2017)
20. Lupenko, S., Pavlyshyn, A., Orobchuk, O.: Conceptual fundamentals for ontological simulation of Chinese Image Medicine as a promising component of integrative medcine. Sci. Edu. New. Dim. J. **15**, 28–32 (2017)
21. Lupenko, S., Pasichnyk, V., Kunanets, N., Orobchuk, O., Xu, M.: The axiomatic-deductive strategy of knowledge organization in onto-based e-learning systems for Chinese Image Medicine. In: Proceedings of the 1st International Workshop on Informatics & Data-Driven Medicine, vol. 2255, pp. 126–134 (2018)

22. Lupenko, S., Orobchuk, O., Zahorodna, N.: Formation of the onto-oriented electronic educational environment as a direction the formation of integrated medicine using the example of CIM. In: Actual Scientific Research in the Modern World: Collection of Scientific Papers of the XXIII International Scientific Conference, vol. 12(32), pp. 56–61 (2017)
23. Lupenko, S., Orobchuk, O., Pomazkina, T., Xu, M.: Conceptual, formal and software-information fundamentals of ontological modeling of Chinese Image Medicine as an element of integrative medicine. Wor. Sci. **1** (2017)
24. Lupenko, S., Orobchuk, O., Xu, M.: The ontology as the core of integrated information environment of Chinese Image Medicine. In: Advances in Computer Science for Engineering and Education II, ICCSEEA 2019. Advances in Intelligent Systems and Computing, vol. 938, pp. 471–481. Springer, Cham (2020)
25. Lupenko, S.: Axiomatic-deductive strategy of the organization of the content of academic discipline in the field of information technologies using the ontological approach. In: IEEE 13th International Scientific and Technical Conference on Computer Sciences and Information Technologies (CSIT), Lviv, vol. 1, pp. 387–390 (2018)
26. Lupenko, S.: Organization of the content of academic discipline in the field of information technologies using ontological approach. In: Proceeding of the International Conference on CSIT. Advances in Intelligent Systems and Computing III, Lviv, pp. 312–327 (2018)

Approach for Creating Reference Signals for Detecting Defects in Diagnosing of Composite Materials

Artur Zaporozhets[1(✉)], Volodymyr Eremenko[2], Volodymyr Isaenko[3], and Kateryna Babikova[3]

[1] Institute of Engineering Thermophysics of NAS of Ukraine, Kiev, Ukraine
a.o.zaporozhets@nas.gov.ua
[2] Igor Sikorsky Kyiv Polytechnical Institute, Kiev, Ukraine
[3] National Aviation University, Kiev, Ukraine

Abstract. The article describes the approach to the formation of a simulation model of information signals, which are typical for objects with different types of defects. The dispersive analysis of the signal spectrum components in the bases of the discrete Hartley transform and the discrete cosine transform is carried out. The analysis of the form of the reconstructed information signal is carried out depending on the number of coefficients of the spectral alignment in Hartley bases and cosine functions. The basis of orthogonal functions of a discrete argument is obtained, which can be used for the spectral transformation of information signals of a flaw detector. A function was obtained approximating the distribution of the values of each of the coefficients of the spectral decomposition depending on the degree of damage (defectiveness) of the sample under study. It proved the need to use splines in the approximation of equations. The advantage of the splines is obtaining reliable results even for small degrees of interpolation equations, moreover, the Runge phenomenon does not arise, which occurs when using high-order polynomial interpolation.

Keywords: Diagnostic signal · Information signal · Composite material · Dispersion analysis · Low-speed impact method · Diagnosing · Equation approximation · Splines

1 Introduction

Products made of composite materials, in contrast to products made of metals, are formed from primary raw materials simultaneously with the formation of the materials. Due to the complexity of their manufacturing technology, it becomes impossible to build a priori models describing the definitions of informative parameters of controlled objects, and ignorance of the laws of the probability distribution of changes does not allow to form the corresponding decision rule [1, 2].

In tasks of non-destructive diagnostics of composite materials, as well as in the case of using neural networks as the core of the classifier, the presence of an adequate simulation model of information signals characteristic of objects with different types of defects or damage's degrees has a great importance, since it allows solving several problems simultaneously [3].

N. Shakhovska and M. O. Medykovskyy (Eds.): CCSIT 2019, AISC 1080, pp. 154–172, 2020.
https://doi.org/10.1007/978-3-030-33695-0_12

First, the existence of such model allows you to build a library of information signals that characterize possible defects in composites and therefore can be used to train and configure the information and diagnostic system as a whole or in a particular case of a neural network classifier without physically manufacturing such samples [4]. Secondly, a simulation model of the information signal can be used to verify the accuracy of diagnosis and classification, justify the choice of the most successful architecture and type of neural network classifier, select the threshold sensitivity of the system, validate the information and diagnostic system and, if necessary, adjust its parameters, etc. [5]

The developed methods and systems for diagnosing products made of composite materials most often use the parameters of information signals as the main diagnostic features, the registration of which causes the least complication, namely amplitude, pulse duration, signal phase, and the like [6, 7]. However, the shape of the information signal, i.e. the function of changing it over time provides much more information about the technical state of the research sample and therefore provides more opportunities for its diagnosis [8].

Analysis of the information signal form allows to get a greater number of diagnostic signs, perform object diagnostics under the condition of a limited amount of information, and provides high noise immunity of the system, and modern computing systems, signal acquisition, and processing devices allow to implement high complexity analysis algorithms, thereby increasing the accuracy of control [9–11].

Since at present there is no single universal physical method for diagnosing composite materials that would identify all possible types of defects, the method of modeling reference signals was studied using a low-speed impact method, which allows determining the largest number of types of possible defects in composites.

2 Research Methods

2.1 Selection of the Basis for the Presentation of the Information Signal

The information signal of the sensor $X(t)$ will be considered as a function of the discrete argument [4], that is, a vector which elements are obtained as a result of uniform sampling of the information signal of the sensor:

$$X(\mathrm{Z}) = (X_0, X_1, \ldots . X_j), \tag{1}$$

where $X_j = X(z_j)$, $Z = \{z_0, z_1, \ldots, z_j\}$ are the area of signal detection $X(\mathrm{Z})$; $j \in \overline{0, N-1}$, N is a number of discrete signal samples $X(\mathrm{Z})$.

The task of synthesizing an information signal model with given parameters is most adequately solved by the representation of a signal in the spectral region. And, if in the case of a continuous periodic signal in many cases, the trigonometric Fourier transform takes precedence, then for pulsed signals the problem of choosing the appropriate spectral basis arises, which provides the minimum number of informative spectral components. In addition to ensuring the minimum number of spectral components during choosing an orthogonal basis, an important aspect is also the choice of such a basis in which the components of the spectrum of the information signal are most dependent on changes in the degree of damage of the test object.

In modern informational diagnostic systems, the results of primary measurements are discrete samples of pulsed analog information S_k, according to which, during further processing, informative parameters are determined, such as the amplitude of pulses, their duration, and shape.

Since the pulse information signals obtained during the diagnostics of products made of composite materials using the low-speed impact method have a complex shape, the application of the most common orthogonal transformations (Fourier, Hartley, Haar, cosine and sine transformations, etc.) is made difficult by the large number of spectral decomposition components, which undergo significant changes when the degree of damage to the controlled object changes [12].

Dispersion analysis showed that the number of coefficients of spectral decomposition, which are characterized by a significance coefficient η_x (describes the degree of dependence of the change of the corresponding coefficient on the damage to the object) with a value of more than 0.7, is from 20 to 36 coefficients depending on the chosen basis. Figures 1 and 2 illustrate the results of analysis of variance in the case of using the discrete Hartley transform (DHT) and discrete cosine transform (DCT) [13, 14].

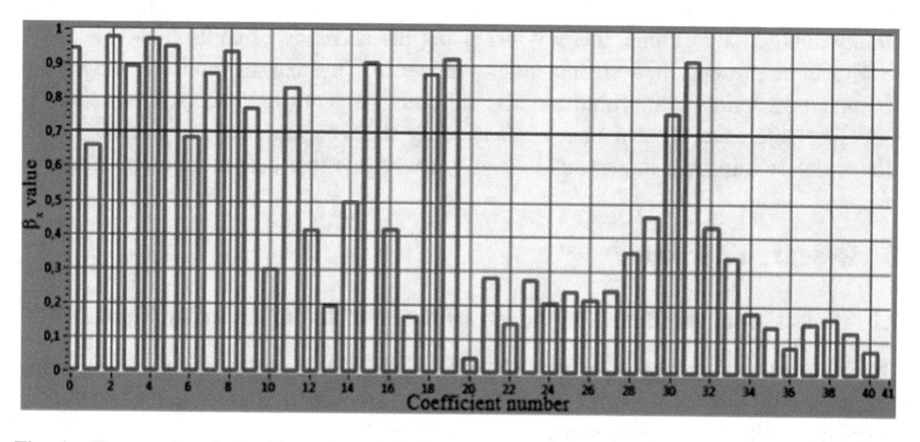

Fig. 1. The results of the dispersion analysis of the components of the signal spectrum in the DHT basis

Figures 3, 4 and 5 illustrate the dependence of the form of the reconstructed information signal on the selected number of spectral decomposition coefficients in Hartley bases and cosine functions. Similar results are characteristic in the case of the use of discrete Fourier transforms, Haar and discrete sine transforms. The spectral alignment is performed by a signal that is obtained by averaging 500 realizations for each of the sample areas studied.

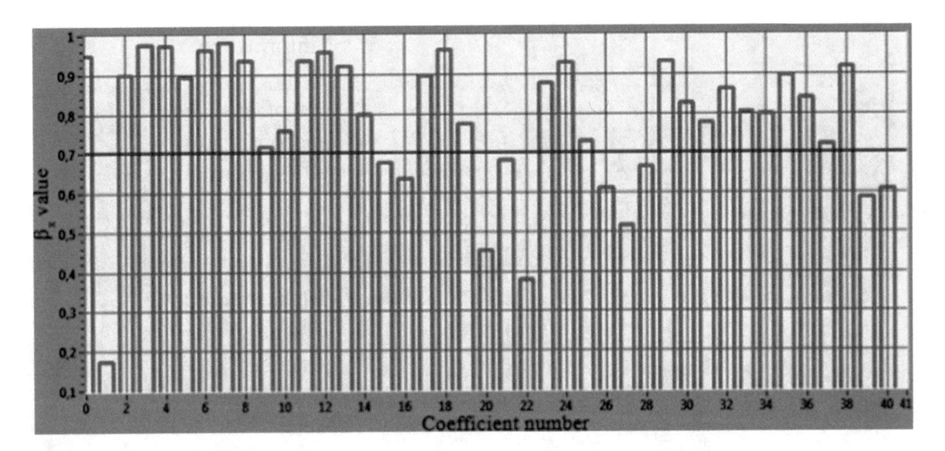

Fig. 2. The results of the dispersion analysis of the components of the signal spectrum in the DCT basis

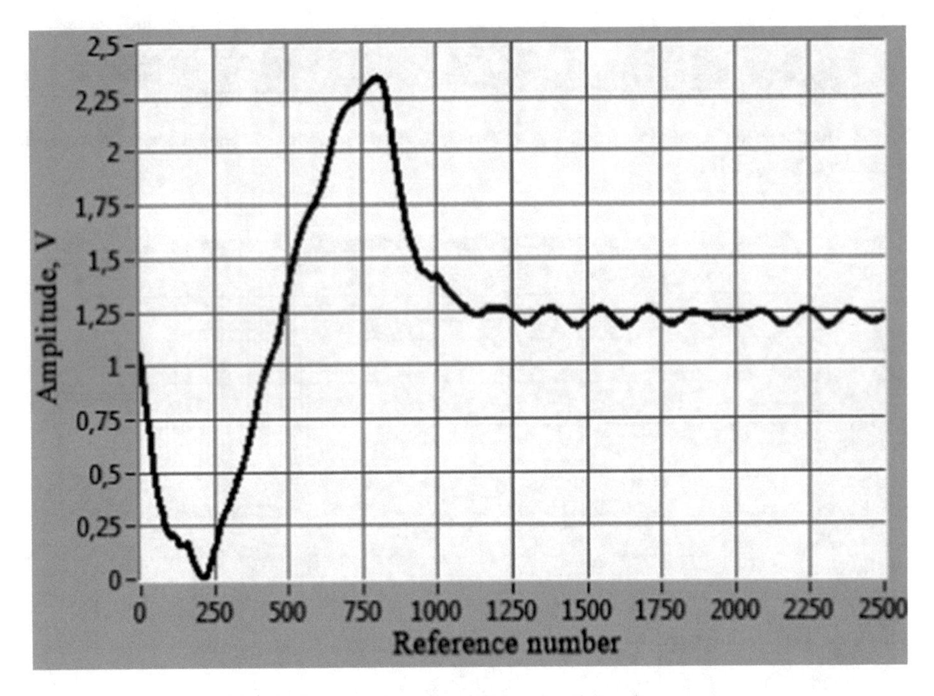

Fig. 3. Information signal from the defect-free area

As can be seen from Figs. 4, 5, 6 and 7 for reliable restoration of signals in the specified bases of orthogonal functions, it is necessary to have at least 30 spectral components, the presence of a smaller number of components leads to significant distortions of the information signal and, as a consequence, the loss of some diagnostic information about the object of study. The need to take into account a large number of

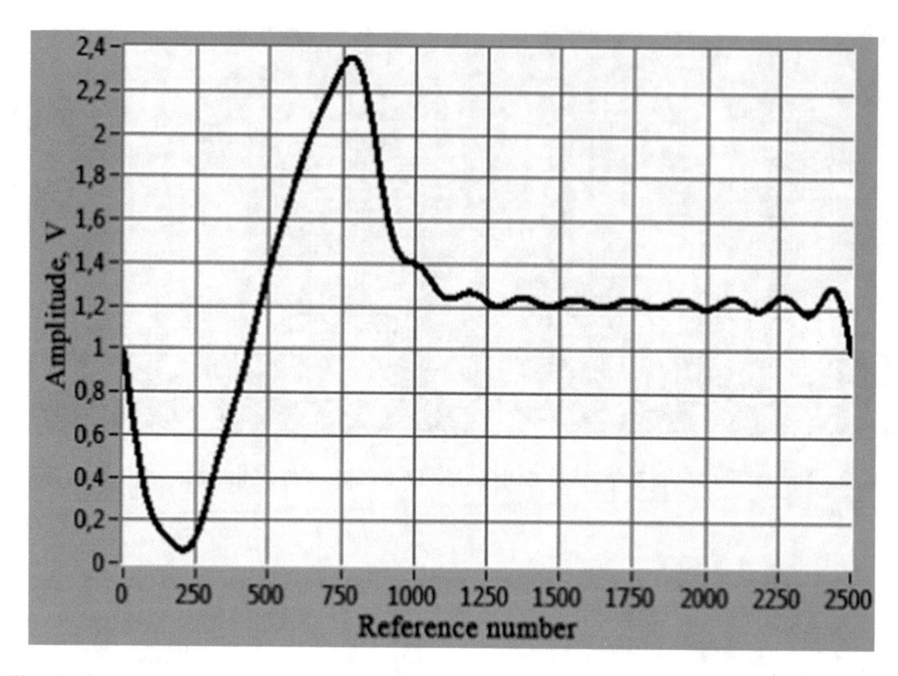

Fig. 4. Reconstructed signal from the defect-free region with 15 spectral decomposition coefficients using DHT

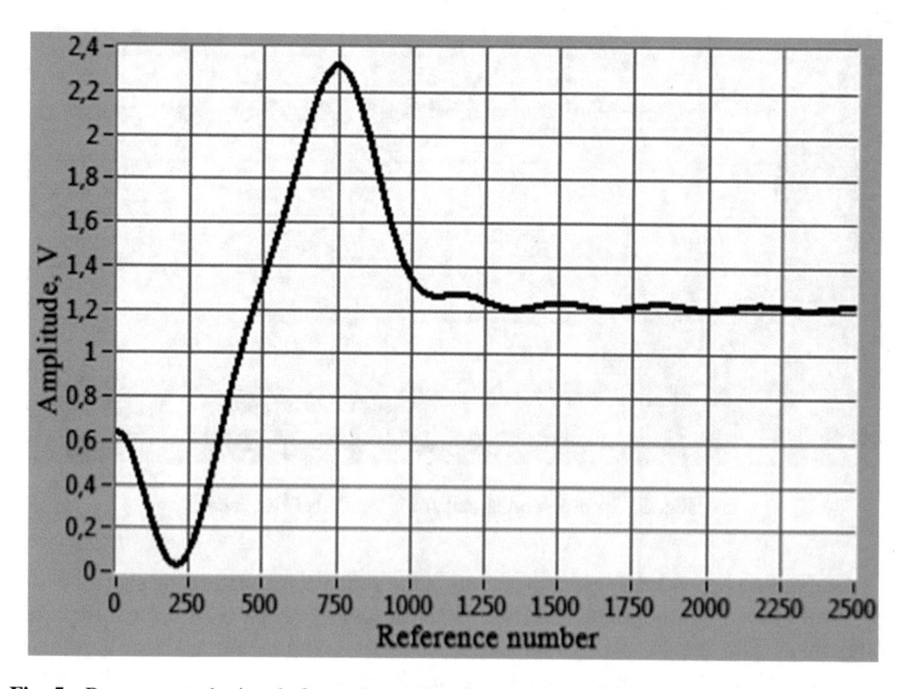

Fig. 5. Reconstructed signal from the defect-free region with 15 spectral decomposition coefficients using DCT

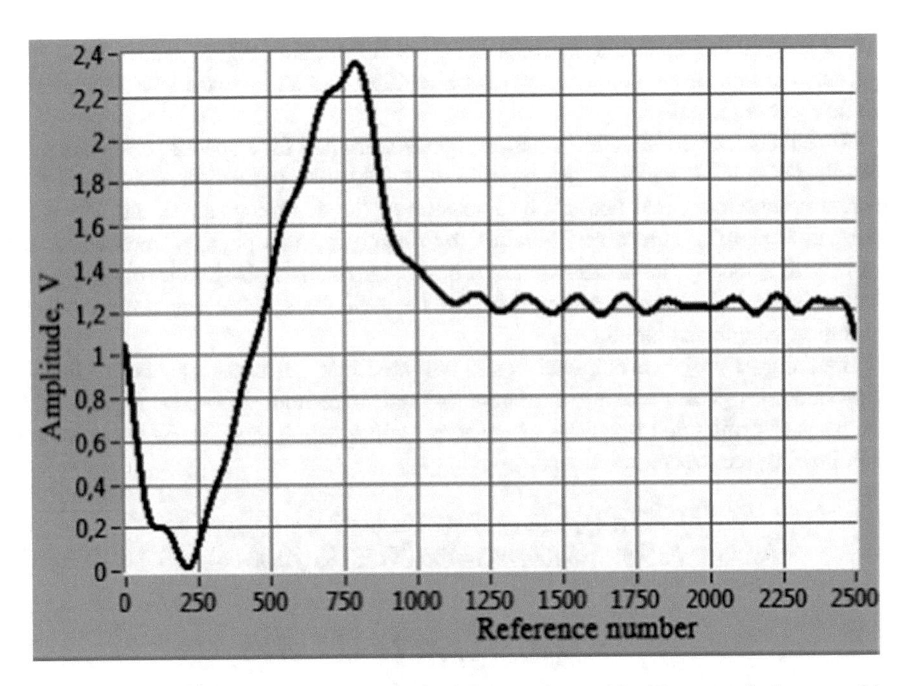

Fig. 6. Reconstructed signal from the defect-free region with 30 spectral decomposition coefficients using DHT

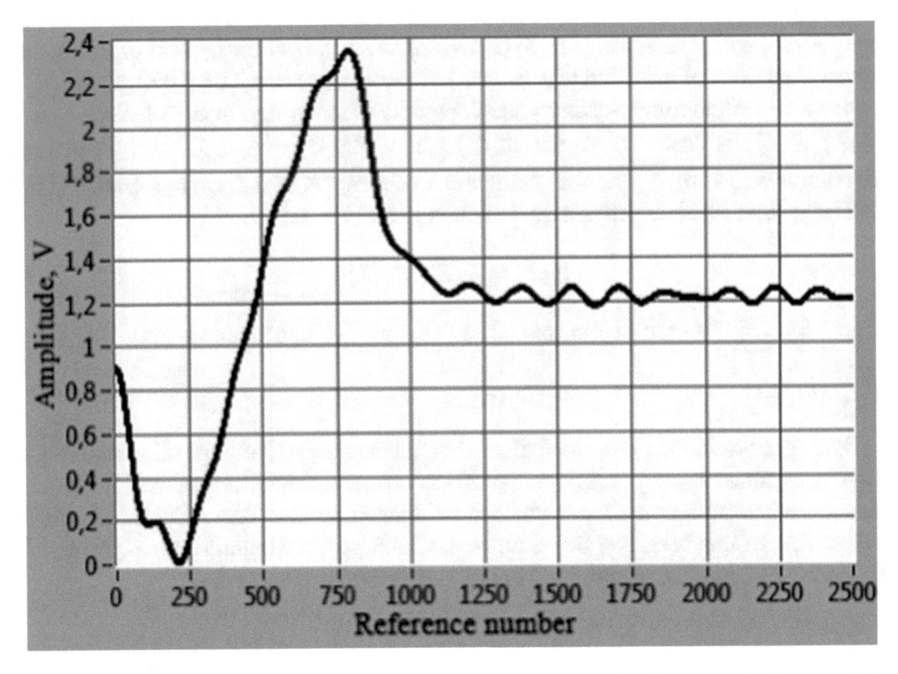

Fig. 7. Reconstructed signal from the defect-free region with 30 spectral decomposition coefficients using DCT

components of the spectrum when building an information signal model leads to a significant complication of the simulation algorithm and an increase in computational and time costs [15, 16].

Since the set of possible information signals is limited for a specific composite and a physical diagnostic method, the output in such a situation (reduction of the number of spectral components, and hence the dimension of the feature space) is the use of the information signals themselves, obtained by diagnosing this physical method, as the basis. With selecting the necessary set of basic signals, it is possible to build on their own basis an orthogonal basis, which will be used for spectral transformation and restoration of information signals.

Obtaining a proper orthogonal basis, provided that the basis signals are linearly independent, is possible using the Gram-Schmidt algorithm. The construction of its own basis of orthogonal functions of a discrete argument, in this case, is performed by the following recurrence relations:

$$\begin{gathered} h_1(Z) = X_1(Z), h_2(Z) = X_2(Z) - \langle X_2(Z), g_1(Z)\rangle g_1(Z), \\ h_3(Z) = X_3(Z) - \langle X_3(Z), g_1(Z)\rangle g_1(Z) - \langle X_3(Z), g_2(Z)\rangle g_2(Z), \\ \vdots \\ h_n(Z) = X_n(Z) - \sum_{k=1}^{n-1} \langle X_n(Z), g_k(Z)\rangle g_k(Z), \end{gathered} \tag{2}$$

where $X_1(Z), X_2(Z), \ldots, X_n(Z)$ are linearly independent vectors (signals) are constructed using the discretization of information signals from the space U; n is the number of components of the desired basis (the dimension of the subspace U^* of the basis signals enters the space U); $h_1(Z), h_2(Z), \ldots, h_n(Z)$ is the system of orthogonal vectors (functions of the discrete argument) of the new basis; $g_1(Z), g_2(Z), \ldots, g_n(Z)$ is a system of orthonormal vectors (functions of a discrete argument) of a new basis; $\langle X_n(Z), g_k(Z)\rangle$ is scalar of vectors $X_n(Z)$ and $g_k(Z)$; $Z = \{z_0, z_1, \ldots, z_j\}$ is a signal detection area; $j \in \overline{0, N-1}$, N is the number of samples of a discrete signal.

The system of basic orthonormal vectors is defined as:

$$g_i(Z) = h_i(Z)/\|h_i(Z)\|, \tag{3}$$

where $\|h_i(Z)\|$ is the L_2-norm of a vector in Euclidean space, $\|h_i(Z)\| = \sqrt{\sum_{j=0}^{N-1} \left[h_{i,j}(Z)\right]^2}$.

After performing the described algorithm, a new basis of orthogonal functions of a discrete argument $\{g_1(Z), g_2(Z), \ldots, g_n(Z)\}$ is obtained, which can be used for spectral transformation of information signals of a flaw detector. The number of spectral components of this basis can be minimized, which greatly simplifies the algorithm for processing information signals and building their simulation model.

The spectral conversion of the signal [17] according to this basis is performed according to the equation:

$$a_{i,j} = \langle X_i(Z), g_j(Z) \rangle, i = \overline{1,L}, j = \overline{1,n} \quad (4)$$

where $a_{i,j}$ is the j-th coefficient of the spectral decomposition of the i-th realization of the information signal; L is the dimension of the sample information signals.

Signal recovery is performed as follows:

$$\overset{*}{X}_i(Z) = \sum_{j=1}^{n} a_{i,j} g_j(Z), i = \overline{1,L} \quad (5)$$

where $\overset{*}{X}_i(Z)$ is the restored i-th implementation of the information signal of the flaw detector by spectral components $j = \overline{1,n}$.

Thus, from a space of dimension U (where U is the set of all possible signals characteristic of each type of defect), a subspace of dimension U^* is selected (U^* is the set of signals chosen to build its own basis), which makes it possible to approximate with a given accuracy any signal from the U-space [18]. Analytically this is described by the expression:

$$\left| X_i(Z) - \overset{*}{X}_i(Z) \right| \leq \alpha \quad (6)$$

where $X_i(Z)$ is the output information signal from the U-space; $\overset{*}{X}_i(Z)$ is an approximated signal; α is the permissible absolute error (discrepancy) between the signals.

Figures 8, 9, 10, 11, 12 and 13 show the corresponding realization of the output information signals, their spectral schedule in the constructed orthogonal basis and the reconstructed signals by the inverse transformation of their spectrum. The values of the coefficients of the spectral decomposition [19] for each type of area (defect-free or defective) are presented in Table 1.

Table 1. The values of the coefficients of the spectral decomposition of information signals

Area type	Coefficient number				
	0	1	2	3	4
No defect	37.442	−0.036	0.013	−0.009	−0.006
2.3 kJ energy damage	10.318	14.952	0.007	0.006	−0.005
2.8 kJ energy damage	9.687	12.861	2.311	0.005	−0.004
3.2 kJ energy damage	3.059	9.180	−2.581	5.496	−0.003
5.1 kJ energy damage	−0.140	2.547	−1.659	3.820	4.023

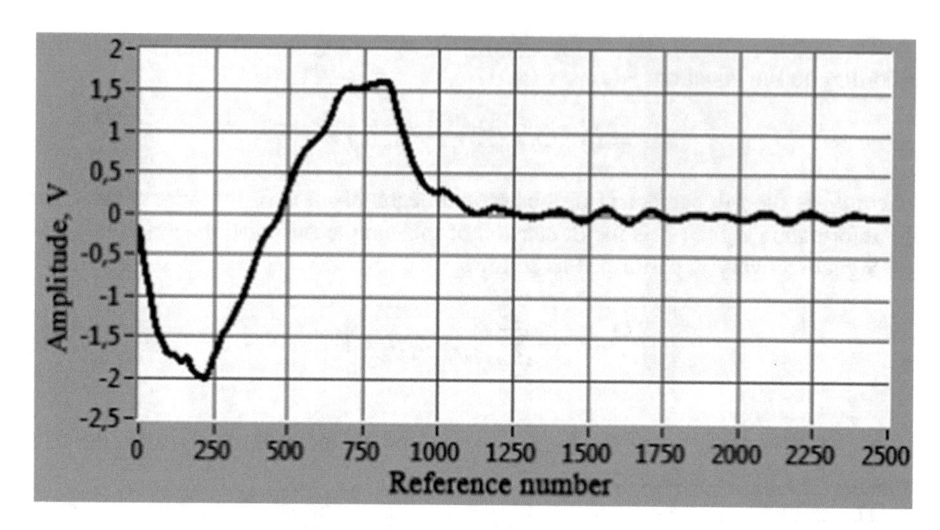

Fig. 8. Output information signals from the defect-free area

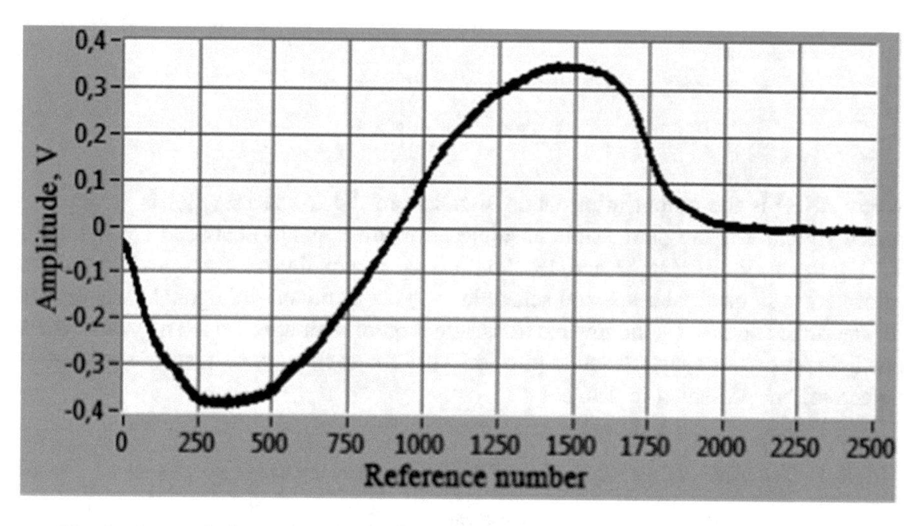

Fig. 9. Output information signals from the area with a damaging impact of 3.24 kJ

Such an approach makes it possible to significantly reduce the number of spectral decomposition coefficients for analyzing and modeling information signals of a flaw detector. In this problem, the dimension of the subspace is reduced to $U^{*} = 5$.

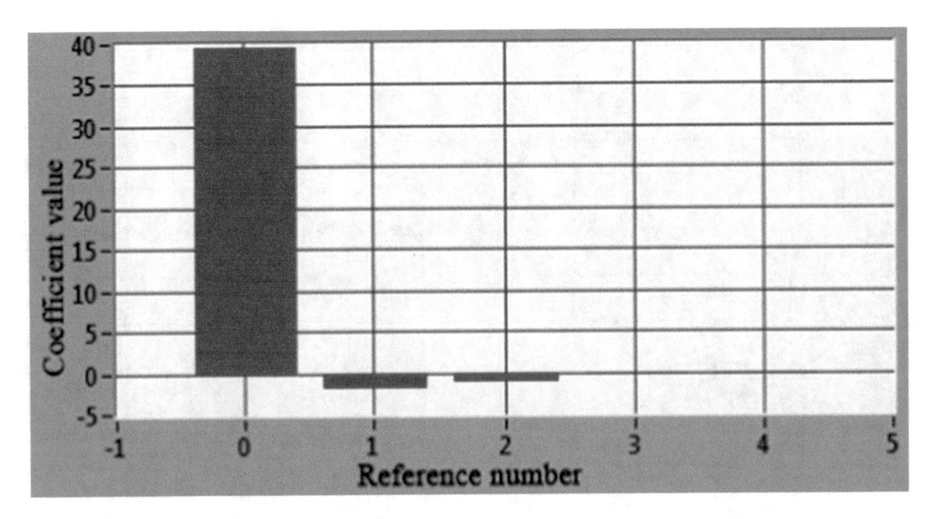

Fig. 10. Spectral alignment of the information signal from the defect-free area

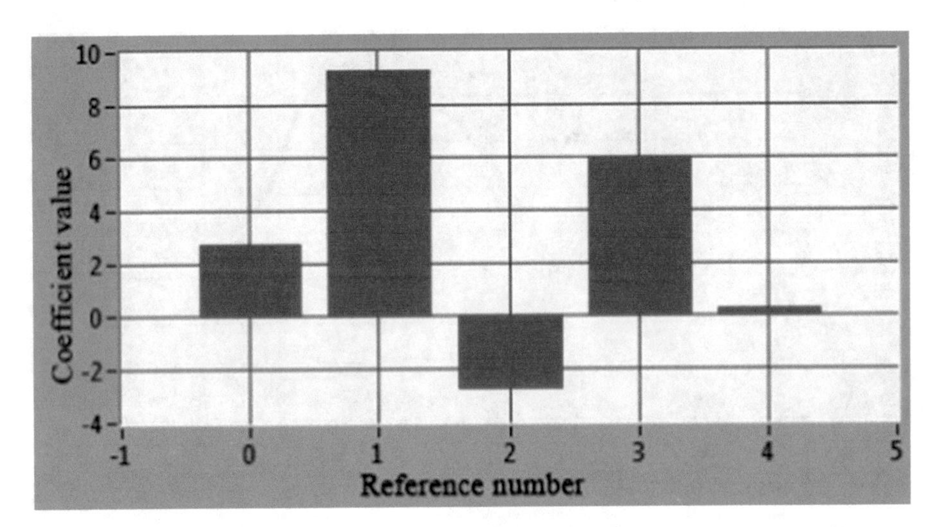

Fig. 11. Spectral alignment of the information signal of the area with a damaging impact of 3.24 kJ

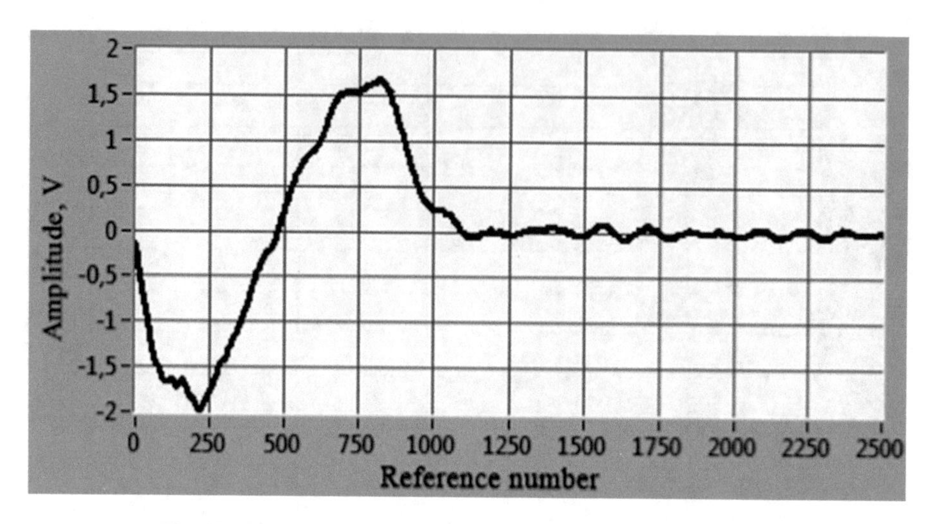

Fig. 12. Recovered information signals from the defect-free area

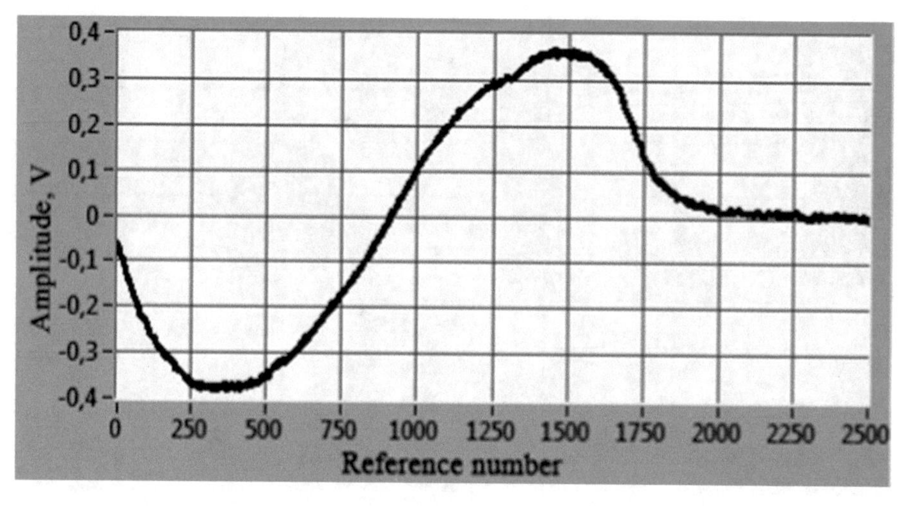

Fig. 13. Recovered information signals from the area with a damaging impact of 3.24 kJ

2.2 Construction and Study of Approximation Equations

To determine the values of the corresponding coefficients, the decomposition characteristic of information signals describing the various defects of the samples under study, it is necessary to obtain a function that approximates the distribution of the values of each of the spectral decomposition coefficients depending on the degree of damage (defectiveness) of the sample under study [20]. Such a function can be determined by interpolating known values of decomposition coefficients, for example, by power polynomials or splines [21]. Further, for each spectral component, it is necessary to select the desired damage degree (defect size) x of the controlled area, determine the value of the spectral components by the established functional dependencies and perform the inverse transformation.

Interpolation using splines is more efficient than interpolation by polynomials [22], since it gives reliable results even for low degrees of interpolation equations, and the Runge phenomenon that occurs when using high-order polynomial interpolation does not occur. The main advantages of spline interpolation are stability and complexity. Systems of linear equations that need to be solved to construct splines are well conditioned, which allows to obtain the coefficients of polynomials with high accuracy. As a result, even for very large N, the computational scheme does not lose stability. Building a table of spline coefficients requires $O(N)$ operations, and calculating the spline value at a given point is $O(\log_2 N)$.

There are many types of interpolation splines [23]. The paper proposes and studies a method for constructing approximations of dependencies of scheduling coefficients on the degree of impact damage using Hermite cubic splines and quadratic splines [24, 25].

Hermite's cubic spline is defined by the following equation:

$$H_b(x) = \sum_{j=0}^{b/2} \frac{(-1)^j}{2^j} \cdot \frac{b!}{j!(b-2j)!} x^{b-2j}, \tag{7}$$

where b is degree of Hermite polynomial.

Hermite polynomials form a complete orthogonal system on the interval $(-\infty, \infty)$ with the weight function $e^{-\frac{x^2}{2}}$:

$$\int_{-\infty}^{\infty} H_b(x) H_m(x) e^{-x^2/2} dx = b!\sqrt{2\pi}\delta_{bm}, \tag{8}$$

where δ_{bm} is Kronecker symbol.

An important consequence of the orthogonality of Hermite polynomials is the possibility of scheduling various functions in series according to Hermite polynomials. For any integral integer p, the equation is true:

$$\frac{x^p}{p!} = \sum_{k=0}^{k \le p/2} \frac{1}{2^k} \cdot \frac{1}{k!(p-2k)!} H_{p-2k}(x). \tag{9}$$

Hermite splines have a continuous first derivative, but the second derivative has a discontinuity in them. This interpolation method uses two control points and two direction vectors. According to this method, the interpolation on the interval (x_k, x_{k+1}), where $k = \overline{1, Q-1}$ (Q is the number of specified points on the interpolation interval that divide the entire interval into a specified number of segments), is given by the formula:

$$\begin{gathered} P(x) = h_{00}(t)p_0 + h_{10}(t)hq_0 + h_{01}(t)p_1 + h_{11}(t)hq_1, \\ h = x_{k+1} - x_k, \quad t = (x - x_k)/h, \end{gathered} \tag{10}$$

where p_0 is initial point at $t = x_k$; p_1 is final point at $t = x_{k+1}$; q_0 and q_1 are respectively the initial (at $t = x_k$) and final (at $t = x_{k+1}$) vectors; $h_{00}(t) - h_{11}(t)$ are base Hermite polynomials: $h_{00}(t) = (1-t)^2(1+2t)$, $h_{01}(t) = t^2(3-2t)$, $h_{10}(t) = t(1-t)^2$, $h_{11}(t) = t^2(t-1)$.

There are such symmetry properties of polynomials:

- $h_{00}(t) + h_{01}(t) = 1$ – symmetry about the y = 1/2;
- $h_{00}(t) = h_{01}(1-t)$ – symmetry about the x = 1/2;
- $h_{10}(t) = -h_{11}(1-t)$ – symmetry with respect to the point (0, 1/2).

The obtained interpolation functions based on cubic Hermite splines for the first two spectral components, depending on the kinetic energy of the damaging impact, are presented in Figs. 14 and 15.

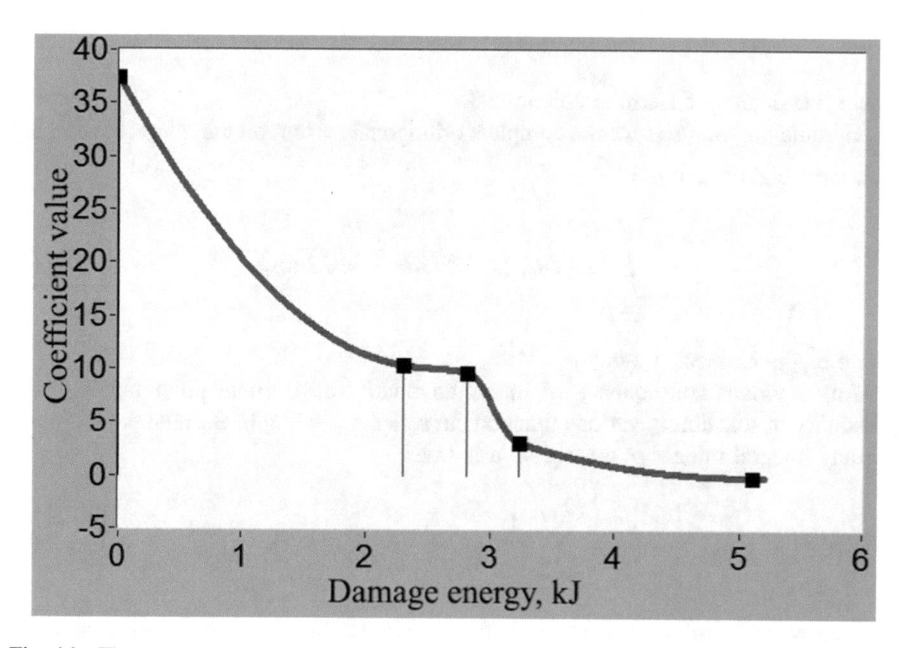

Fig. 14. The approximation of Hermite's splines for the first component of the spectral decomposition of the information signal

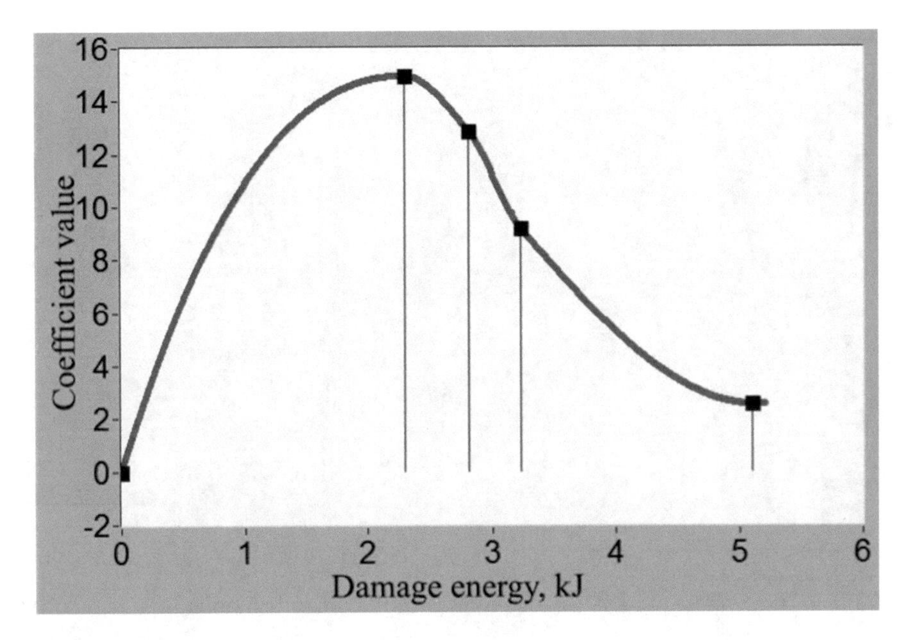

Fig. 15. The approximation of Hermite's splines for the second component of the spectral decomposition of the information signal

Interpolation of a set of points (x_k, y_k) for $k = 1, \ldots, Q$ using quadratic splines is carried out for each interval, and the parameters for one point in different intervals are chosen the same. The interpolation spline will be obtained continuously differentiated by (x_1, x_Q). There are several ways to define parameters. The simplest of them is the following:

$$P_i(x) = y_i + w_i(x - x_i) + \frac{w_{i+1} - w_i}{2(x_{i+1} - x_i)}(x - x_i)^2, \tag{11}$$

The coefficients of this polynomial can be found by choosing the value of w_0 and using the recurrence relation:

$$w_{i+1} = -w_i + 2\frac{y_{i+1} - y_i}{x_{i+1} - x_i}. \tag{12}$$

The coefficients w_i are determined to an approximate degree. Since only two points are used to calculate the next point of the curve (function) (instead of three), this method is prone to serious oscillation effects when the signal changes abruptly. Due to the presence of such effects, this method may not be used for all tasks.

The form of the obtained interpolation functions for the first two spectral components depending on the kinetic energy of the damaging impact using quadratic splines is shown in Figs. 16 and 17.

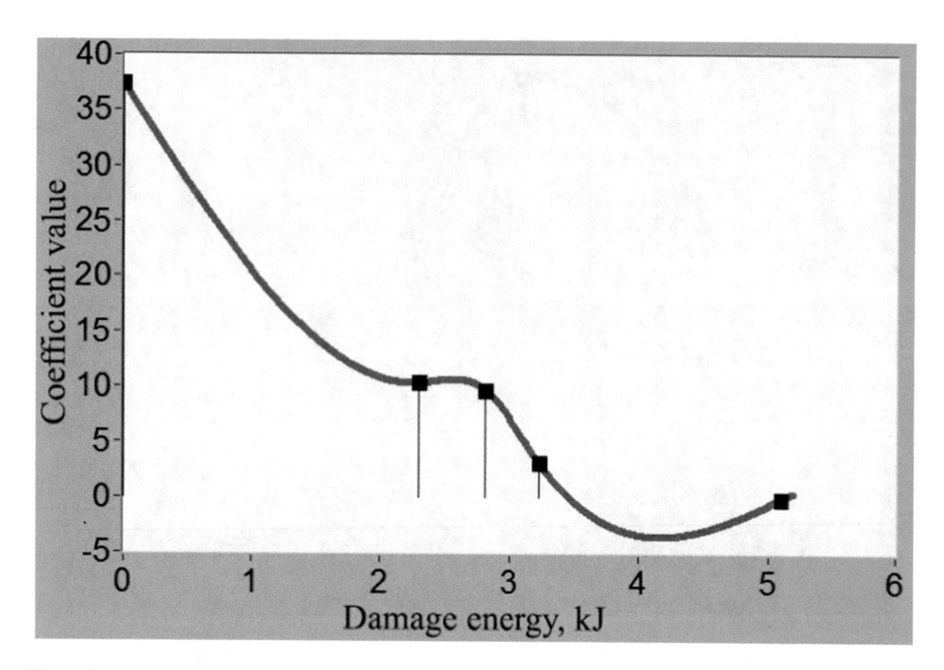

Fig. 16. Approximation by quadratic splines of the first component of the spectral decomposition of the information signal of the flaw detector

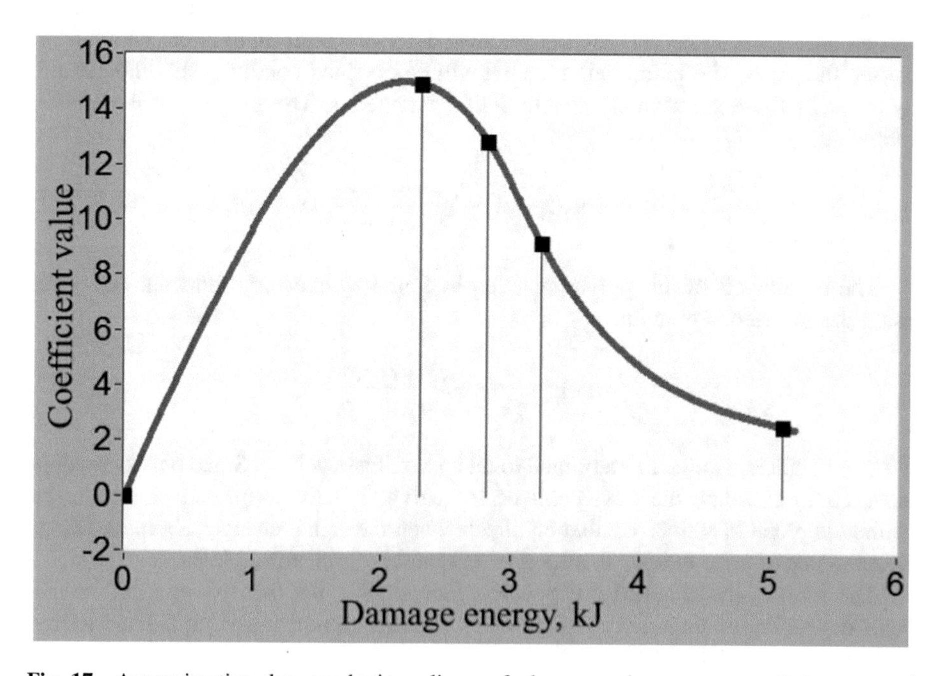

Fig. 17. Approximation by quadratic splines of the second component of the spectral decomposition of the information signal of the flaw detector

To assess the effectiveness of the considered interpolation equations, the information signal received from the site with a damaging impact of 2.81 kJ was compared with the simulated signal corresponds to the same area. Figures 18 and 19 shows the real signals from the damaged area with an energy of 2.81 kJ – curve S_1, as well as the simulated signal using Hermite cubic splines (Fig. 18) and quadratic splines (Fig. 19).

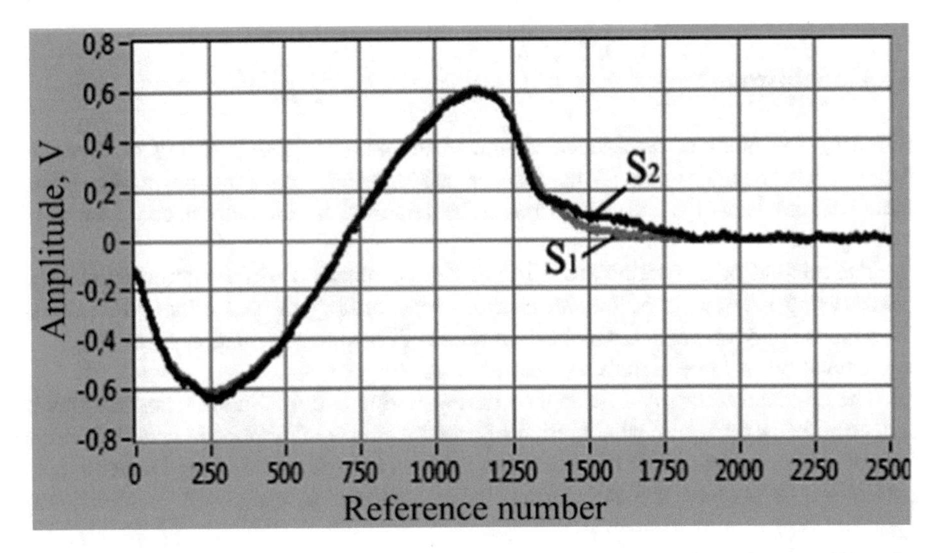

Fig. 18. Comparison of a real signal with damage with an energy of 2.81 kJ (curve S_1) and a simulated signal using Hermite cubic splines (curve S_2)

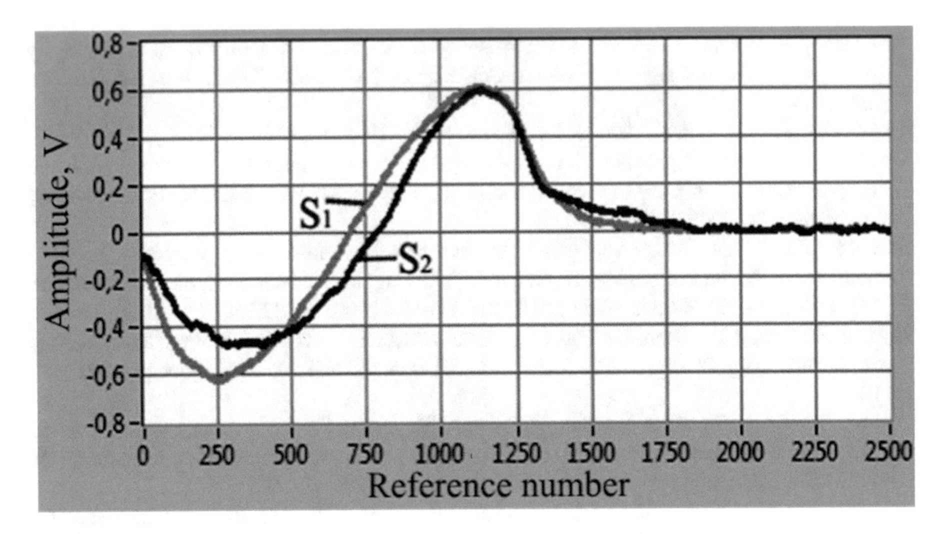

Fig. 19. Comparison of a real signal with damage with an energy of 2.81 kJ (curve S_1) and a simulated signal using quadratic splines (curve S_2)

The numerical estimate of the disagreement between the simulated and real signals was carried out by calculating the RMSE (root mean square error), the value of which is $2.5 \cdot 10^{-3}$ for Hermite's cubic splines and $4.0 \cdot 10^{-3}$ for quadratic splines.

Thus, using interpolation functions, it is possible to obtain the values of spectral components for a given level of damage to the sample zone, and using the inverse transform to obtain a simulated information signal [26–28].

3 Conclusions

Based on the obtained experimental signals for non-destructive testing of products made of composite materials, approaches are developed to construct a simulation model of signals, which takes into account the deterministic and random components of real signals.

The method of simulation modeling of signals obtained while controlling cellular panels using the low-speed impact method using orthogonal and orthonormal transformations is investigated. On its basis, a series of computer simulation experiments on the simulation of these signals was carried out.

The obtained results can be applied during testing and training diagnostic systems for recognizing the technical condition of products made of composite materials about the possible range of defects of a particular material and the nature of their development and allow to analyze the transformation of information signals in real technical systems.

References

1. Lee, M., Thomas, C.E., Wildes, D.G.: Prospects for in-process diagnosis of metal cutting by monitoring vibration signals. J. Mater. Sci. **22**(11), 3821–3830 (1987). https://doi.org/10.1007/BF01133328
2. Widolo, A., Kim, E.Y., Son, J.-D., Yang, B.-S., Tan, A.C.C., Gu, D.-S., Choi, B.-K., Mathew, J.: Fault diagnosis of low speed bearing based on relevance vector machine and support vector machine. Expert Syst. Appl. **36**(2, Part 2), 7252–7261 (2009). https://doi.org/10.1016/j.eswa.2008.09.033
3. Peng, Z.K., Chu, F.L.: Application of the wavelet transform in machine condition monitoring and fault diagnostics: a review with bibliography. Mech. Syst. Signal Process. **18**(2), 199–221 (2004). https://doi.org/10.1016/S0888-3270(03)00075-X
4. Yan, R., Gao, R.X., Chen, X.: Wavelets for fault diagnosis of rotary machines: a review with applications. Sig. Process. **96**(Part A), 1–15 (2014). https://doi.org/10.1016/j.sigpro.2013.04.015
5. Sun, W., Shao, S., Zhao, R., Yan, R., Zhang, X., Chen, X.: A sparse auto-encoder-based deep neural network approach for induction motor faults classification. Measurement **89**, 171–178 (2016). https://doi.org/10.1016/j.measurement.2016.04.007
6. Zaporozhets, A., Eremenko, V., Serhiienko, R., Ivanov, S.: Methods and hardware for diagnosing thermal power equipment based on smart grid technology. In: Advances in Intelligent Systems and Computing III, vol. 871, pp. 476–492 (2019). https://doi.org/10.1007/978-3-030-01069-0_34

7. Zaporozhets, A.A., Eremenko, V.S., Serhiienko, R.V., Ivanov, S.A.: Development of an intelligent system for diagnosing the technical condition of the heat power equipment. In: 2018 IEEE 13th International Scientific and Technical Conference on Computer Sciences and Information Technologies (CSIT), Lviv, Ukraine, 11–14 September 2018. https://doi.org/10.1109/stc-csit.2018.8526742
8. Ali, Y.H., Rahman, R.A., Hamzah, R.I.R.: Acoustic emission signal analysis and artificial intelligence techniques in machine condition monitoring and fault diagnosis: a review. Jurnal Teknologi **69**(2), 121–126 (2014)
9. Sikdar, S., Kudela, P., Radzienski, M., Kundu, A., Ostachowicz, W.: Online detection of barely visible low-speed impact damage in 3D-core sandwich composite structure. Compos. Struct. **185**, 646–655 (2018). https://doi.org/10.1016/j.compstruct.2017.11.067
10. Babak, V., Mokiychuk, V., Zaporozhets, A., Redko, O.: Improving the efficiency of fuel combustion with regard to the uncertainty of measuring oxygen concentration. Eastern-Eur. J. Enterp. Technol. **6**(8), 54–59 (2016). https://doi.org/10.15587/1729-4061.2016.85408
11. Wu, Z., Huang, N.E.: Ensemble empirical mode decomposition: a noise-assisted data analysis method. Adv. Adapt. Data Anal. **01**(2), 1–41 (2009). https://doi.org/10.1142/S1793536909000047
12. Song, Y., Fellouris, G.: Asymptotically optimal, sequential, multiple testing procedures with prior information on the number of signals. Electron. J. Stat. **11**(1), 338–363 (2017). https://doi.org/10.1214/17-EJS1223
13. Hsue, W.-L., Chang, W.-C.: Real discrete fractional Fourier, Hartley, generalized Fourier and generalized Hartley transforms with many parameters. IEEE Trans. Circuits Syst. I Regul. Pap. **62**(10), 2594–2605 (2015). https://doi.org/10.1109/TCSI.2015.2468996
14. Hsue, W.-L., Chang, W.-C.: Multiple-parameter real discrete fractional Fourier and Hartley transforms. In: 2014 19th International Conference on Digital Processing, Hong Kong, China, 20–23 August 2014. https://doi.org/10.1109/icdsp.2014.6900753
15. Zaporozhets, A.: Analysis of control system of fuel combustion in boilers with oxygen sensor. Periodica Polytech. Mech. Eng. (2019). https://doi.org/10.3311/ppme.12572
16. Dertimanis, V.K., Spiridonakos, M.D., Chatzi, E.N.: Data-driven uncertainty quantification of structural systems via B-spline expansion. Comput. Struct. **207**, 245–257 (2018). https://doi.org/10.1016/j.compstruc.2017.03.006
17. Andrews, R.W., Reed, A.P., Cicak, K., Teufel, J.D., Lehnert, K.W.: Quantum-enabled temporal and spectral mode conversion of microwave signals. Nat. Commun. **6**, 10021 (2015). https://doi.org/10.1038/ncomms10021
18. Jung, Y., Cho, H., Lee, I.: MPP-based approximated DRM (ADRM) using simplified bivariate approximation with linear regression. Struct. Multi. Optim. **59**(5), 1761–1773 (2019). https://doi.org/10.1007/s00158-018-2160-7
19. Qu, Y., Wang, W., Guo, R., Ayhan, B., Kwan, C., Vance, S., Qi, H.: Hyperspectral anomaly detection through spectral unmixing and dictionary-based low-rank decomposition. IEEE Trans. Geosci. Remote Sens. **56**(8), 4391–4405 (2018). https://doi.org/10.1109/TGRS.2018.2818159
20. Zaporozhets, A.O., Redko, O.O., Babak, V.P., Eremenko, V.S., Mokiychuk, V.M.: Method of indirect measurement of oxygen concentration in the air. Naukovyi Visnyk Natsionalnoho Hirnychoho Universytetu (5), 105–114 (2018). https://doi.org/10.29202/nvngu/2018-5/14
21. Babak, S., Babak, V., Zaporozhets, A., Sverdlova, A.: Method of statistical spline functions for solving problems of data approximation and prediction of objects state. In: CEUR Workshop Proceedings, vol. 2353, pp. 810–821 (2019). http://ceur-ws.org/Vol-2353/paper64.pdf

22. Ali, A., Khan, K., Haq, F., Shah, S.I.A.: A computational modeling based on trigonometric cubic B-spline functions for the approximate solution of a second order partial integro-differential equation. In: New Knowledge in Information Systems and Technologies. Advances in Intelligent Systems and Computing, WorldCIST 2019, vol. 930, pp. 844–854 (2019). https://doi.org/10.1007/978-3-030-16181-1_79
23. Eremenko, V., Zaporozhets, A., Isaenko, V., Babikova, K.: Application of wavelet transform for determining diagnostic signs. In: CEUR Workshop Proceedings, vol. 2387, pp. 202–214 (2019). http://ceur-ws.org/Vol-2387/20190202.pdf
24. Han, X., Guo, X.: Cubic Hermite interpolation with minimal derivative oscillation. J. Comput. Appl. Math. **331**, 82–87 (2018). https://doi.org/10.1016/j.cam.2017.09.049
25. Meshram, S.G., Powar, P.L., Meshram, C.: Comparison of cubic, quadratic, and quintic splines for soil erosion modeling. Appl. Water Sci. **8**, 173 (2018). https://doi.org/10.1007/s13201-018-0807-6
26. Zaporozhets, A.: Development of software for fuel combustion control system based on frequency regulator. In: CEUR Workshop Proceedings, vol. 2387, pp. 223–230 (2019). http://ceur-ws.org/Vol-2387/20190223.pdf
27. Brajovic, M., Orovic, I., Dakovic, M., Stankovic, S.: On the parameterization of Hermite transform with application to the compression of QRS complexes. Sig. Process. **131**, 113–119 (2017). https://doi.org/10.1016/j.sigpro.2016.08.007
28. Zaporozhets, A.O., Eremenko, V.S., Isaenko, V.M., Babikova, K.O.: Methods for creating reference signals for the diagnosis of composite materials. In: Proceedings of International Scientific Conference Computer Sciences and Information Technologies (CSIT-2019), vol. 1, pp. 84–87 (2019)

Modeling of Lightning Flashes in Thunderstorm Front by Constructive Production of Fractal Time Series

Viktor Shynkarenko(✉), Kostiantyn Lytvynenko(✉), Robert Chyhir(✉), and Iryna Nikitina(✉)

Dnipro National University of Railway Transport named after academician V. Lazaryan, Dnipro, Ukraine
shinkarenko_vi@ua.fm, kostall1l1973@gmail.com, robertchigir@ukr.net, irinasansieva@gmail.com

Abstract. Using the tools of structural-synthesizing modeling, a set of constructors was developed. Implementing parametric multi-character constructors allows to form fractal sequences of characters. Constructor-converter from the character string to time series creates fractal time series, which determine the location, magnitude and decay rate of lightning discharges. Model video images of lightnings in the thunderstorm front are formed in accordance with the implementation of the constructor-assembler. All constructors are developed on the basis of the generalized constructor that was previously presented and repeatedly tested. The model adequacy of the model is confirmed by comparing the video image of the model with the image, what was obtained by NASA satellite. This approach can be the basis for solving the dynamic problems on lightning protection of engineering constructions and civil objects, and development of strategy of aircraft behavior in order to mitigate the risks of lightning strokes in the conditions of movement in the thunderstorm front.

Keywords: L-system · Constrictive-synthesizing modeling · Fractal · Lightning activity · Lightning flash · Thunderstorm front · Time series

1 Introduction

Standard monitoring and forecasting of hazardous thunderstorm phenomena are carried out in all countries on the basis of a unified program and regulatory documents. Such monitoring includes:

– regular monitoring of qualitative and quantitative indicators of the atmospheric thunderstorm state;
– collection, processing and storage of observations of thunderstorm phenomena;
– creation and maintenance of observational databases.

It becomes possible to conduct modeling experiments in order to develop adequate quantitative predictive models of lightning activity of thunderstorm fronts.

The study of patterns of spatial distribution of thunderstorms is the relevant and practically important problem for solving both the essential tasks of atmospheric

N. Shakhovska and M. O. Medykovskyy (Eds.): CCSIT 2019, AISC 1080, pp. 173–185, 2020.
https://doi.org/10.1007/978-3-030-33695-0_13

electricity and lightning protection of engineering constructions and thunderstorm fire risk of forest areas. One of the sources of data on the spatial distribution of thunderstorms is WWLLN (World Wide Lightning Location Network) [1].

Lightning monitoring was performed by satellites using detectors OTD (Optical Transient Detector) and LIS (Lightning Imaging Sensor). They are recording short bursts of infrared radiation, which arise from the lightning discharge and can be seen from space even in daytime under the clouds.

The main directions of modeling and studying of lightning activity are associated with the study of spreading of currents from clouds to the ground [2], impact of lightning on electrical systems [3].

2 State of Problem

Research on regional and global lightning activity and the global electrical circuit is summarized in scientific research [4]. This area of activity has greatly expanded through observations of lightning by satellite and through increased using of the natural resonances of the Earth–ionosphere cavity.

The complex relationships between lightning and rain yields in convective storms have been studied extensively, with different relationships derived in varying geographical locations and atmospheric conditions [5].

Regional behavior of lightning activity is studied for a long time [6].

The effect of solar variability parameters and meteorological parameters on total lightning flashes and convective rain in two selected regions is studied in the article [7], isolation of zones of the lightning activity in specific geographic areas [8], and their impact on breaking-out of fires [9]. Much lesser number of works deals with the problem of modeling of lightning in the thunderstorm front, which is primarily due to its complexity. Typically, such works are limited to isolation of compact zones (clusters) of lightning formation [10].

In this time the modeling of lightning activity on the base of satellite monitoring of flash rate from convective cell can help severe weather forecasts, improve tornado forecasters and their impending threat to the public.

This paper refers to modeling of lightning activity in the thunderstorm front based on the generated fractal time series which determine the time, coordinates and duration of flashes, and comparison of the model video images to video images received from the satellite.

In the constructive approach to the formation of objects of different nature there is an opportunity to understand the constructive notion of the real technical or nature object.

3 The Main Material

3.1 Constructive Modeling Tools

Generalized Constructor. The basis of constructive-synthesizing modeling is the concept of generalized constructive-synthesizing structure [11–13], or generalized constructor (GC):

$$C = \langle M, \Sigma, \Lambda \rangle, \quad (1)$$

where M – heterogeneous replenishable carrier, Σ – signature of relations and relevant operations, such as linking, substitution, and inference, over attributes, Λ – set of statements of the information support of construction (ISC) including: ontology, purpose, rules, restrictions, terms of starting and completion of construction.

Peculiarities of the constructive-synthesizing modeling are as follows [11–13]: attributiveness of elements and operations, replenishable carrier, model of performer in the form of its basic algorithms, relation of operations to the algorithms of their implementation. Main provisions ontological support of constructive-synthesizing modeling developed in [14].

Ontology of generalized constructor in its informal representation is given in [11, 12]; below we provide its part required for the subsequent presentation.

Signature Σ comprises sets of operations: Ξ – linking, Θ – substitution and inference, Φ – operations over attributes. The signature also contains the relations of substitution "$\rightarrow$".

Operations of linking of constructor elements combine the individual elements into constructions or parts thereof (intermediate forms).

Under the form ${}_{w_l}l$ with the set of attributes w_l we understand:

- ${}_{w_l}l = {}_{w_0}\otimes\left({}_{w_1}m_1,\ {}_{w_2}m_2,\ \ldots,\ {}_{w_k}m_k\right)$ for $\forall_{w_i}m_i \in M$;
- ${}_{w_l}l = {}_{w_j}m_j$, if $l = {}_{w_0}\otimes\left(\varepsilon,\ \ldots,\varepsilon,\ {}_{w_j}m_j,\ \varepsilon,\ \ldots,\ \varepsilon\right)$;
- ${}_{w_l}l = {}_{w_0}\otimes\left({}_{w_1}l_1,\ {}_{w_2}l_2,\ \ldots,\ {}_{w_k}l_k\right)$,

where ${}_{w_1}l_1,\ {}_{w_2}l_2,\ \ldots,\ {}_{w_k}l_k$ – forms, ${}_{w_0}\otimes$ – any linking operation of Ξ with attribute w_o, ε – empty element.

The substitution relation is ${}_{w_i}l_i \rightarrow {}_{w_j}l_j$.

Let it be $s = \left\langle {}_{w_1}l_1 \rightarrow {}_{w_2}l_2,\ {}_{w_3}l_3 \rightarrow {}_{w_4}l_4,\ \ldots,\ {}_{w_m}l_m \rightarrow {}_{w_{m+1}}l_{m+1}\right\rangle$ – sequence of substitution relations or $s = \varepsilon$, and $g = \left\langle \oplus_1(w_{1,1},\ w_{2,1},\ \ldots,w_{k_1,1}),\ \oplus_2(w_{1,2},\ w_{2,2},\ \ldots,\ w_{k_2,2},),\ \ldots,\ \oplus_n(w_{1,n},\ w_{2,n},\ \ldots,w_{k_n,n})\right\rangle$ – sequence of operations over attributes. Substitution rule is $\psi : \langle s,\ g\rangle$. Here $\oplus$ is a any operation over attributes ($\oplus \in \Phi$).

A set of substitution rules is $\Psi = \{\psi_i : \langle s_i,\ g_i\rangle\}$.

Suppose the specified form ${}_{w_l}l = \otimes\left({}_{w_1}l_1,\ {}_{w_2}l_2,\ \ldots,\ {}_{w_h}l_h,\ \ldots,\ {}_{w_k}l_k\right)$ and relation of substitution ${}_{w_h}l_h \rightarrow {}_{w_q}l_q$ is such that ${}_{w_h}l_h \prec {}_{w_l}l$ (relation $\prec$ – contains), then the result of ${}_{w_l^*}l^*$ trinary operation of substitution $\Rightarrow \left({}_{w_h}l_h, {}_{w_q}l_q,\ {}_{w_l}l\right)$ will be the form ${}_{w_l^*}l^* = \otimes\left({}_{w_1}l_1,\ {}_{w_2}l_2,\ \ldots,\ {}_{w_q}l_q,\ \ldots,\ {}_{w_k}l_k\right)$ where $\Rightarrow \in \Theta$.

Binary operation of partial inference ${}_{w_l^*}l^* = {}_{v_p}|\!\Rightarrow(\Psi,\, {}_{w_l}l)$ $(|\!\Rightarrow\, \in \Theta)$ consists in:

- choice of one of available substitution rules $\psi_r : \langle s_r, g_r \rangle$ with the relations of substitution s_r;
- performance of substitution operations on its base;
- performance of operations over attributes g_r.

Operation of full inference $(||\!\Rightarrow\, \in \Theta)$ consists in stepwise conversion of forms starting with the initial form and ending with the construction satisfying the condition of inference completion which involves the cyclic performance of partial inference operations. It is a binary operation ${}_{\Delta, w_l^*}l^* = ||\!\Rightarrow(\Psi,\, {}_{w_l}l)$ where ${}_{w_l}l \in M$.

The resulting constructions of full inference operations belong to $\Omega(C)$.

With a view to forming the constructions, a number of clarifying transformations are carried out:

- specialization determines the subject area: semantic nature of the carrier, finite set of operations and their semantics, attributes of operations, order of their performance and limitations of substitution rules;
- interpretation binds of signature operations with their execution algorithms, thus connecting the information model of means of constructions' formation and performer model, which generate the constructive system;
- concretization of the constructor expands axiomatics with a set of substitution rules, assigning of specific sets of nonterminal and terminal characters with their attributes and, where appropriate, the attribute values;
- realization, which essence is formation of a set of constructions using carrier elements.

Specialization of generalized constructor on the basis of constructive-synthesizing approach and L-systems [15] can be considered as

$$C = \langle M, \Sigma, \Lambda \rangle \;{}_{S}\!\mapsto C_L = \langle M_L, \Sigma_L, \Lambda_L \rangle, \tag{2}$$

where M_L includes the character terminals, as well as intermediate forms and multi-character constructions, Σ_L comprises a single operation, i.e. concatenation of characters and character strings (as a rule, the sign of operation between operands is omitted), Λ_L – information support includes the basics of constructive-synthesizing modeling and peculiar features of L-systems $\Lambda_L = \Lambda \cup \Lambda_1$.

Ontology of ISC Λ_1 includes the above designations and their semantics, notions "character", "concatenation", "character string" and other well-known concepts of multi-character processing, as well as provisions given below.

Partial inference operation $|\!\Rightarrow(\Psi, {}_{w_l}l)$ is clarified: all permitted operations of substitution of Ψ applicable to terminals of the form ${}_{w_l}l$ are performed, with looking through from left to right, except the recursion.

Initial conditions are given as a character string (axiom).

A set of non-terminal is empty.

The constructive system allows to generate the certain set of constructions (possibly, one) or to perform the check of attribution of specified construction to the above set.

In some cases, it is necessary to form two or more distinct sets of constructions similarly, where the sets of constructions being formed are different, and the processes of their formation have little variability.

In such cases it is advisable to apply parametric constructors. Suppose the family of constructions is a set of constructions characterized by the limited number of provisions of ISC. When determining the family we specify in parentheses the constructor parameters (variable of ISC elements within the family are listed).

Parametric Multi-character Creator Constructors. Concretization of C_L to the level of the family of parametric multi-character constructors gives

$$C_L = \langle M_L, \Sigma_L, \Lambda_L \rangle_K \mapsto C_{MS}(\mathrm{B}, \mathrm{P}, \mathrm{n}) = \langle M_{MS}, \Sigma_{MS}, \Lambda_{MS} \rangle \tag{3}$$

where B – initial character string (axiom), P – set of substitution rules, n – minimum number of terminals f in output string, $M_{MS} \supset \{f, x, y, p, m, d, u, +, -, /, \backslash\}$, $\Sigma_{MS} = \Xi_{MS} = = \{\circ\}$, $\circ$ – concatenation operation, $\Lambda_{MS} = \Lambda_L \cup \Lambda_2$. Ontological component Λ_2 includes the above terms and their semantics, as well as provisions below:

- purpose of construction – formation of string of fractal structure;
- substitution rules are set by the parameter P;
- limitations – there are no operations over attributes;
- initial conditions – the axiom is specified by B;
- termination condition – number of terminals f in output string $\geq$ n.

As a result of interpretation, we form the constructive system as a set of two models: constructor and internal performer

$$\begin{aligned} &\langle C_{MS}(\mathrm{B}, \mathrm{P}, \mathrm{n}) = \langle M_{MS}, \Sigma_{MS}, \Lambda_{MS} \rangle, C_A = \langle M_A, \Sigma_A, \Lambda_A \rangle \rangle_I \mapsto \\ &C_{A,MS}(\mathrm{B}, \mathrm{P}, \mathrm{n}) = \left\langle M_{A,MS}, \Sigma_{A,MS}, \Lambda_{A,MS} \right\rangle, \end{aligned} \tag{4}$$

where C_A – model of the performer in the constructor form capable of executing the basic and constructed algorithms; M_A – a set of basic and constructed algorithms; $\Sigma_A = \{\cdot, :\}$ includes operations of sequential and conditional algorithms execution; ISC Λ_A is given in [13]; $M_{A,MS} = \langle M_{MS}, M_A \rangle$, $\Sigma_{A,MS} = \langle \Sigma_{MS}, \Sigma_A \rangle$, $\Lambda_{A,MS} = \Lambda_{MS} \cup \Lambda_A \cup$

$$\{(A_1^0 \,|_{A_i, A_j}^{A_i \cdot A_j} \leftarrow\!\!\rfloor \cdot), (A_2^0 \,|_{Z_1, Z_2, A_i}^{:A_i} \leftarrow\!\!\rfloor :), (A_3^0 \,|_{l_i, l_j}^{l_i \circ l_j} \leftarrow\!\!\rfloor \circ); (A_4 \,|_{l_h, l_q, l_i}^{l_j} \leftarrow\!\!\rfloor \Rightarrow); (A_5 \,|_{l_i, \Psi}^{l_j} \leftarrow\!\!\rfloor |\!\Rightarrow);$$

$(A_6 \,|_{l_i, \Psi}^{l_j} \leftarrow\!\!\rfloor ||\!\Rightarrow)\}$..

Algorithms M_A:

- performing an algorithm composition compilation $A_1^0 \Big|_{A_i, A_j}^{A_i \cdot A_j}$ ($A|_X^Y$ – an algorithm over data from an input set X with result values from a set Y, A^0 – generating an algorithm), $A_i, A_j \in \Omega(C_{A,MS})$, $A_i \cdot A_j$ – sequential execution of the algorithm A_j after the algorithm A_i;

- conditional execution $A_2^0|_{Z_1,Z_2,A_i}^{:A_i}$, which consists in executing the algorithm A_i under condition $Z_1 \supseteq Z_2$;
- concatenations of chain of symbols $A_3^0|_{l_i,l_j}^{l_i \circ l_j}$, $l_i, l_j \in M_{MS}$;
- executing the substitution operation $\{A_4|_{l_h,l_q,l_i}^{l_j}$, l_i, l_j, $l_h, l_q \in M_{MS}$, l_i, l_j – the current form in which the substitution operation is performed before and after it is executed, l_h, l_q – the chains in the left and right part of the substitution relation, according to which is executed;
- performing partial and complete output operations $A_5|_{l_i,\Psi}^{l_j}$, $A_6|_{l_i,\Psi}^{l_j}$, $\Psi \subset \Lambda_{MS}$ – a set of rules of substitution.

Constructor-Converter from the Character String to Time Series. The family of such constructors

$$C_{TS}(\Omega_i(C_{MS}), M_x, dM_x, \mathrm{D}_x, d\mathrm{D}_x m) = \langle M_{TS}, \Sigma_{TS}, \Lambda_{TS} \rangle \tag{5}$$

where $\Omega_i(C_{A,MS})$ – strings obtained as a result of implementation of the constructor $C_{A,MS}$; M_x – initial value of mathematical expectation of the time series value, dM_x – its increment (%), D_x – initial value of dispersion of the time series, dD_x – its increment (%), m – number of time series points, M_{TS} includes a set of terminals $T = \{f, v, x, y, p, m,), \ d, u, +, -, /, \backslash\}$ nonterminals $N = \{A\}$, $\Sigma_{TS} = \Xi_{TS} \cup \Phi_{TS}$, $\Xi_{TS} = \{\circ, f\}$, $\Phi_{TS} = \{\wedge, +, -, \times, :, /, \backslash\}$, $\Lambda_{TS} = \Lambda_L \cup \Lambda_3$.

Let's introduce the operations over attributes:

- addition, subtraction, multiplication and division, accordingly, $+(c, a, b)$, $-(c, a, b)$, $\times(c, a, b)$ and $:(c, a, b)$ with operands a, b and result c;
- generation of the random normally distributed number $\wedge(c, a, b)$ with the mathematical expectation a and dispersion b.

ISC Λ_3 includes the above terms, definitions and their semantics, as well as provisions below:

- ontology complemented by known concepts allowing to operate with the time series and the real numbers;
- $''\circ''$ – relation between adjacent array elements;
- purpose of construction is formation of the time series $v(t)$;
- rules of substitution:

$$\begin{aligned}
&\{\langle\langle A \to fA,\ B \to z_i \circ B\rangle, \langle\wedge(v, tM_x, tD_x),\ +(t, t, dt),\ +(i, i, 1)\rangle\rangle,\\
&\langle\langle A \to +A\rangle, \langle\times(\mathrm{qM}, M_x, dM_x), :(\mathrm{qM}, \mathrm{qM}, 100),\ +(\mathrm{t}M_x, \mathrm{t}M_x, qM)\rangle\rangle,\\
&\langle\langle A \to -A\rangle, \langle\times(\mathrm{qM}, M_x, dM_x), :(\mathrm{qM}, \mathrm{qM}, 100), -(\mathrm{t}M_x, \mathrm{t}M_x, qM)\rangle\rangle,\\
&\langle\langle A \to /A\rangle, \langle\times(\mathrm{q}D, D_x, dD_x), :(\mathrm{q}D, \mathrm{q}D, 100),\ +(\mathrm{t}D_x, \mathrm{t}D_x, qD)\rangle\rangle,\\
&\langle\langle A \to \backslash A\rangle, \langle\times(\mathrm{q}D, D_x, dD_x), :(\mathrm{q}D, \mathrm{q}D, 100), -(\mathrm{t}D_x, \mathrm{t}D_x, qD)\rangle\rangle,\\
&\langle\langle A \to xA\rangle, \langle\varepsilon\rangle\rangle, \langle\langle A \to yA\rangle, \langle\varepsilon\rangle\rangle,\\
&\langle\langle A \to pA\rangle, \langle\varepsilon\rangle\rangle, \langle\langle A \to mA\rangle, \langle\varepsilon\rangle\rangle,\\
&\langle\langle A \to dA\rangle, \langle\varepsilon\rangle\rangle, \langle\langle A \to uA\rangle, \langle\varepsilon\rangle\rangle,\\
&\langle\langle A \to \varepsilon\rangle, \langle\varepsilon\rangle\rangle\};
\end{aligned}$$

- operations over attributes:
  - addition, subtraction, multiplication and division, accordingly, $+(c, a, b)$, $-(c, a, b)$, $\times(c, a, b)$ and $:(c, a, b)$ with operands a, b and result c;
  - generation of the random normally distributed number $\wedge(c, a, b)$ with the mathematical expectation a and dispersion b.
- limitation – the rule $\langle A \rightarrow \varepsilon, \langle \varepsilon \rangle \rangle$ is observed if the other ones are not applicable;
- initial conditions – initial terminals A (for multi-character construction), B (for creating time series construction), multi-character construction $\Omega_i(C_{MS})$; initial time $t = 0$, its step $dt = 0.04$ s, current value $tM_x = M_x$, $tD_x = D_x$, $i = 1$;
- termination condition – observance of the empty rule.

Let's define the constructive system by interpreting C_{TS}:

$$\langle C_{TS}(\Omega_i(\mathrm{C}_{MS}), M_x, \mathrm{d}M_x, D_x, \mathrm{m}) = \langle M_{TS}, \Sigma_{TS}, \Lambda_{TS} \rangle, C_B = \langle M_B, \Sigma_B, \Lambda_B \rangle \rangle_I \mapsto$$
$$C_{B,TS}(\Omega_i(\mathrm{C}_{A,MS}), M_x, \mathrm{d}M_x, \mathrm{D}_x, \mathrm{d}D_x, \mathrm{m}) = \left\langle M_{B,TS}, \Sigma_{B,TS}, \Lambda_{B,TS} \right\rangle$$

where $M_B \supset M_A$ and supplemented by algorithms that perform operations on attributes, $\Sigma_B = \Sigma_A$, $\Lambda_B = \Lambda_A$.

Constructor-Assembler. Allows to create a constructive process in the form of a video of the formation of the flashes of lightning based on several time series. Constructor

$$C_{VL}(\mathrm{m}, n, Z_1, Z_2, \ldots, Z_{3*n}) = \langle M_{VL}, \Sigma_{VL}, \Lambda_{VL} \rangle \quad (6)$$

where m – number of time series points, M_{VL} includes a set of terminals $T = \{z_{ij}\}$, ($Z_i = [z_{i1}, z_{i2}, \ldots, z_{im}]$) and computer windows with a picture on it, nonterminals $N = \{A, B\}$, $\Sigma_{VL} = \Xi_{VL} \cup \Phi_{VL}$, $\Xi_{VL} = \{\circ, \bullet\}$, $\Phi_{VL} = \{+, >, \vee\}$, $\Lambda_{VL} = \Lambda_L \cup \Lambda_4$.

ISC Λ_4 includes the above terms, definitions and their semantics, as well as provisions below:

- ontology complemented by known concepts allowing to operate with the time series, the real numbers and computer windows;
- $''\circ''$ – relation between adjacent array elements;
- purpose of construction is video of lightning flashes on computer window;
- rules of substitution:
  $\{\langle\langle A \rightarrow z_{i,j} \circ z_{i+1,j} \circ z_{i+2,j} \circ A,\ B \rightarrow \bullet(z_{i,j} \circ z_{i+1,j} \circ z_{i+2,j})B\rangle,$
  $\langle +(i, i, 3),\ >(q, i, n),\ \vee(q, +(i, 0, 1)),\ \vee(q, +(j, j, 1)),\rangle\rangle,$
  $\langle\langle A \rightarrow \varepsilon\rangle, \langle\varepsilon\rangle\rangle\}$;
- operations over attributes:
  - addition $+(c, a, b)$, with operands a, b and result c;
  - comparison $>(c, a, b)$, if $a > b$ result $c = true$, else – $c = false$;
  - $\vee(c, a)$ – run operations a if $c = true$;
- operation $\bullet(a, b, c)$ – visualization flash of lightning on window in such position: distance along given curve a, distance from given curve b, with imitation flash power c;

- limitation – the rule $\langle A \to \varepsilon, \langle\varepsilon\rangle\rangle$ is observed if the other ones are not applicable;
- initial conditions – initial terminals A (for time series construction), B (for creating video), initial time $t = 0$, its step $dt = 0.04$ s, $i = 1$, $j = 1$; given a curve (curves) of thunderstorm front;
- termination condition – observance of the empty rule when $i = 1, j > m$.

Let's define the constructive system by interpreting C_{VL}:

$$\langle C_{VL}(\mathrm{m}, n, Z_1, Z_2, \ldots, Z_{3*n}) = \langle M_{VL}, \Sigma_{VL}, \Lambda_{VL}\rangle, C_D = \langle M_D, \Sigma_D, \Lambda_D\rangle\rangle_I \mapsto C_{D,VL}(\mathrm{m}, n, Z_1, Z_2, \ldots, Z_{3*n}) = \langle M_{D,VL}, \Sigma_{D,VL}, \Lambda_{D,VL}\rangle,$$

where $M_D \supset M_A$ and supplemented by algorithms that perform operations on attributes and operation $\bullet(a, b, c)$, $\Sigma_D = \Sigma_A$, $\Lambda_D = \Lambda_A$.

3.2 Modeling of Lightning Activity in Thunderstorm Front

NASA and National Oceanic and Atmospheric Administration (NOAA) regularly release images from the Geostationary Lightning Mapper (GLM) instrument onboard the GOES-16 satellite. This is the new step in weather-watching capability, because it might see lightning activity above the clouds [16].

Based on the above constructors, the program modeling the lightning activity was developed.

Testing of the developed methods for modeling lightning flashes is performed by comparing with real data obtained from the satellite. Figure 1 represents: frames of satellite video image [16] (first column), the same with the removal of background (relief and cloudiness, second column), with the flashes brought to the regular form (third column).

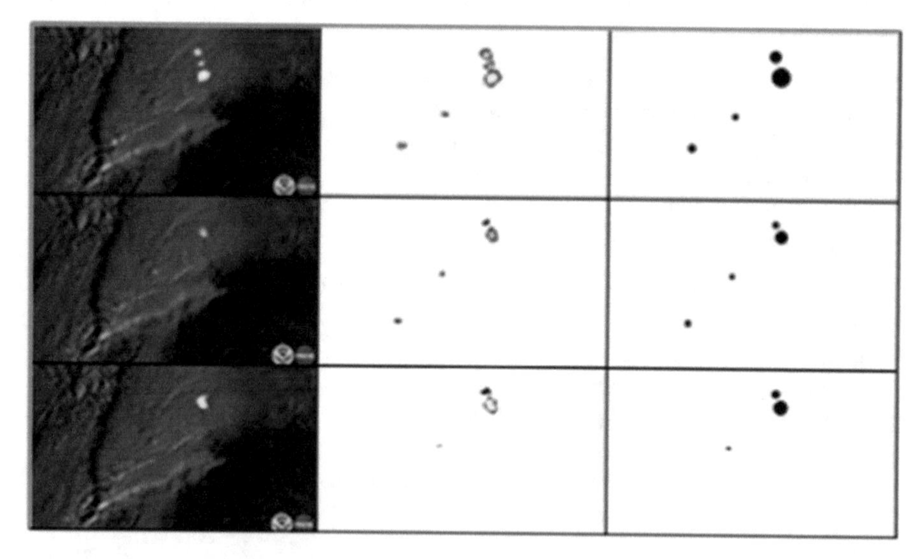

Fig. 1. Separate frames of the original video and flashes of lightning highlighted on them

For this purpose, the original video was converted as follows:

- video of the thunderstorm was divided into frames in the format RRGGBB;
- the blue channel of the images was deleted;
- the image background was removed (only flashes of lightning remain). Each pixel of the i-th frame is compared with pixel at the same position of $(i-1)$-th frame. If similarity condition $\frac{|C_{green,i} - C_{green,i-1}|}{C_{\max}} + \frac{|C_{red,i} - C_{red,i-1}|}{C_{\max}} \leq 0,1$ is doesn't true, pixel's color changes into white (here $C_{green,i}$, $C_{red,i}$ – values of green and red pixel's channels on appropriate frames, and $C_{\max}$ – the maximum possible value for each color channel);
- recursive pixel bypassing for extract highlight lightning flashes;
- smoothing out flashes, noise removal and contour and color conversion of flashes are performed;
- as a result, there is two videos of only flashes of lightning produced: a natural form and a model one having form of a circle.

A similar result was obtained as a result of modeling. The model of thunderstorm front is specified by Bezier curve. The pair of constructors $C_{A,MS}(f, \{f \to f+f\}, 200) \,_R\!\mapsto \Omega_1(C_{A,MS})$ and $C_{A,TS}(\Omega_1(C_{A,MS}), 13, 50, 20, 10, 200) \,_R\!\mapsto \Omega_1(C_{A,TS})$ forms the time series of the lightning discharge position along the curve $u_S(t) = \Omega_1(C_{A,TS})$ (Fig. 2). The second pair $C_{A,MS}(yxyxy, \{x \to +xf, y \to -yf\}, 200) \,_R\!\mapsto \Omega_2(C_{A,MS})$ and $C_{A,TS}(\Omega_2(C_{A,MS}), 9, 5, 20, 10, 200) \,_R\!\mapsto \Omega_2(C_{A,TS})$ – distance from the curve $u_L(t) = \Omega_2(C_{A,TS})$. The third pair $C_{A,MS}(--\backslash\backslash\backslash\backslash\backslash\backslash ppmyuumxmxxmpmmxmpxmff, \{y \to -yf, x \to +xf, f \to ff, p \to +p, m \to -m, d \to \backslash\backslash d, u \to /u\}, 200) \,_R\!\mapsto \Omega_3(C_{A,MS})$ and $C_{A,TS}(\Omega_3(C_{A,MS}), 6, 30, 30, 15, 200) \,_R\!\mapsto \Omega_3(C_{A,TS})$ – value of the discharge $u_R(t) = \Omega_3(C_{A,TS})$. Constructed three time series (Fig. 3).

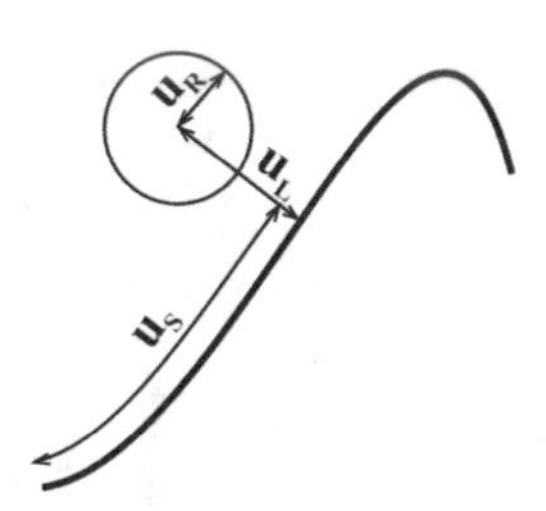

Fig. 2. Determination of the position and strength (radius of the circle) of the lightning flash

According to the constructed three time series constructor-assembler C_{VL} produced model video $C_{D,VL}(200, 1, u_S(t), u_L(t), u_R(t))$. For comparison, Fig. 4 shows the lightning flashes of all frames of satellite (Fig. 4a) and model (Fig. 4b) lightning flashes.

Analysis of other satellite videos [17, 18] shows that the form of a thunderstorm cannot always be given by a single Bezier curve. Therefore, we in the constructor-assembler C_{VL} provide the possibility of assigning several corresponding series of time series for them.

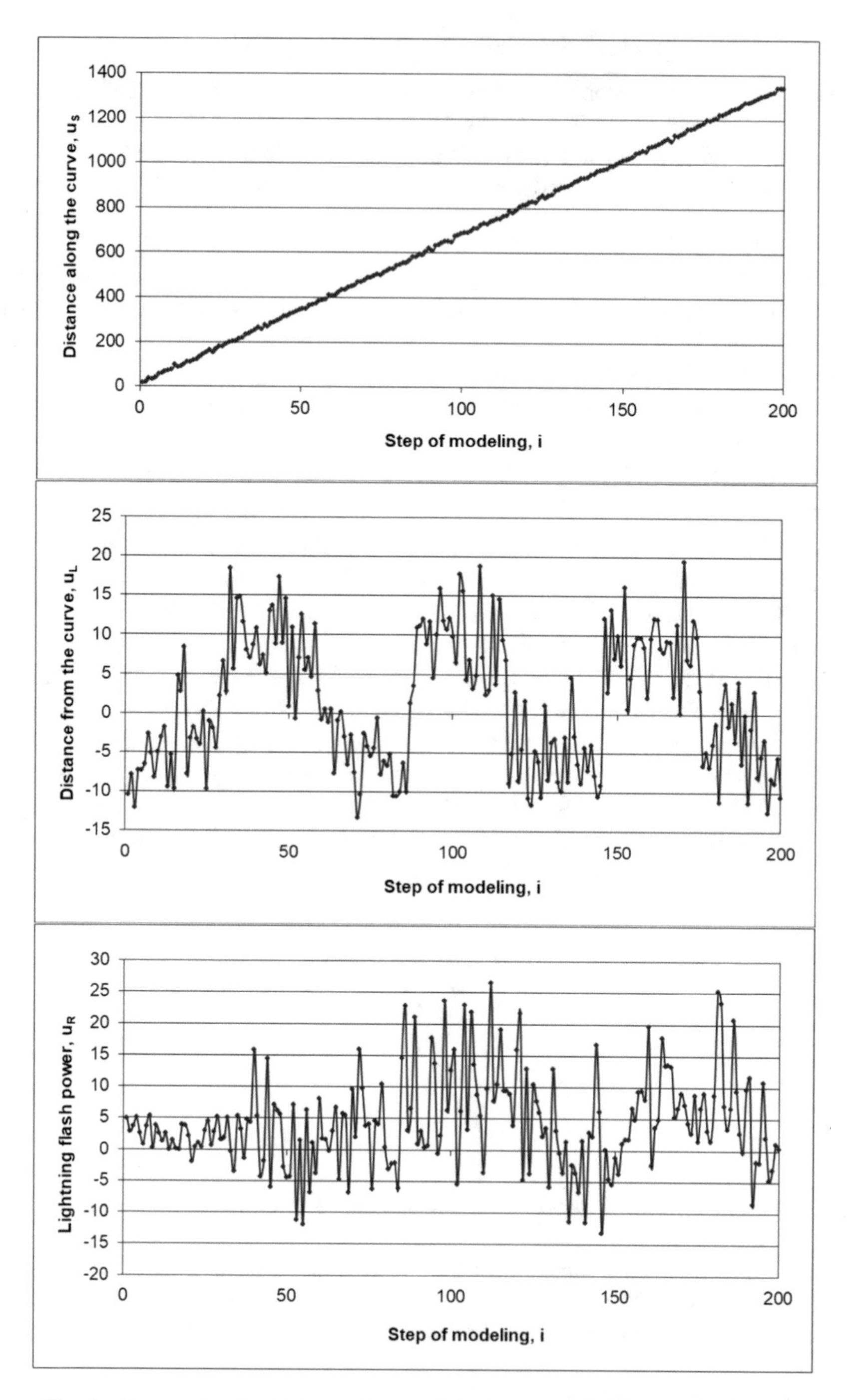

Fig. 3. Time series for distance along and from curve and lightning flash power

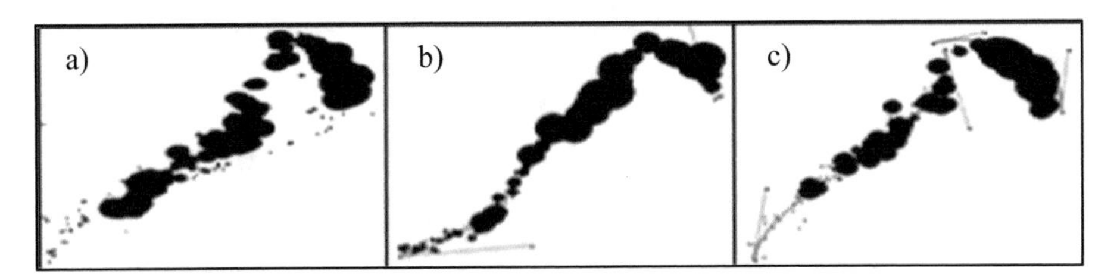

Fig. 4. Frames with all discharges of satellite and model lightning flashes

The model presented above was clarified as follows. We have two Bezier curves: long and short as an Fig. 5. Along the long curve we model two series of flashes of lightning (big and small):

- big flashes:
  - $C_{A,MS}(f,\{f \to f+f-\},50)\ {}_R\!\mapsto \Omega_4(C_{A,MS})$;
  - $C_{B,TS}(\Omega_4(C_{A,MS}),0,12,150,10,50)\ {}_R\!\mapsto \Omega_4(C_{B,TS}) = u_{S,1}(t)$;
  - $C_{A,MS}(f,\{f \to f+f-f\},50)\ {}_R\!\mapsto \Omega_5(C_{A,MS})$;
  - $C_{B,TS}(\Omega_5(C_{A,MS}),8,12,150,10,50)\ {}_R\!\mapsto \Omega_5(C_{B,TS}) == u_{L,1}(t)$;
  - $C_{A,MS}(zzxzzzzy,\{z \to zf, y \to yf-----f, x \to xf+++++f\},50)$ $_R\!\mapsto \Omega_6(C_{A,MS})$;
  - $C_{B,TS}(\Omega_6(C_{A,MS}),0,12,150,10,50)\ {}_R\!\mapsto \Omega_6(C_{B,TS}) = u_{R,1}(t)$;
- small flashes:
  - $C_{A,MS}(f,\{f \to f+f-\},50)\ {}_R\!\mapsto \Omega_7(C_{A,MS})$;
  - $C_{B,TS}(\Omega_7(C_{A,MS}),0,12,150,10,50)\ {}_R\!\mapsto \Omega_7(C_{B,TS}) = u_{S,2}(t)$;
  - $C_{A,MS}(f,\{f \to f+f-f\},50)\ {}_R\!\mapsto \Omega_8(C_{A,MS})$;
  - $C_{B,TS}(\Omega_8(C_{A,MS}),5,12,8,10,50)\ {}_R\!\mapsto \Omega_8(C_{B,TS}) = u_{L,2}(t)$;
  - $C_{A,MS}(f,\{f \to f+f-\},50)\ {}_R\!\mapsto \Omega_9(C_{A,MS})$;
  - $C_{B,TS}(\Omega_9(C_{A,MS}),1,12,0.1,10,50)\ {}_R\!\mapsto \Omega_9(C_{B,TS}) = u_{R,2}(t)$.

Along the shot curve we model one series of flashes of lightning:

- $C_{A,MS}(f,\{f \to f+f-\},50)\ {}_R\!\mapsto \Omega_{10}(C_{A,MS})$;
- $C_{B,TS}(\Omega_{10}(C_{A,MS}),15,12,30,10,50)\ {}_R\!\mapsto \Omega_{10}(C_{B,TS}) = u_{S,3}(t)$;
- $C_{A,MS}(f,\{f \to f+f-\},50)\ {}_R\!\mapsto \Omega_{11}(C_{A,MS})$;
- $C_{B,TS}(\Omega_{11}(C_{A,MS}),15,12,30,10,50)\ {}_R\!\mapsto \Omega_{11}(C_{B,TS}) = u_{L,3}(t)$;
- $C_{A,MS}(f,\{f \to f+f-\},50)\ {}_R\!\mapsto \Omega_{12}(C_{A,MS})$;
- $C_{B,TS}(\Omega_{12}(C_{A,MS}),15,12,30,10,50)\ {}_R\!\mapsto \Omega_{12}(C_{B,TS}) = u_{R,3}(t)$.

According to the constructed time series constructor-assembler C_{VL} produced model video $C_{D,VL}(50,3,u_{S,1}(t),u_{L,1}(t),u_{R,1}(t),u_{S,2}(t),u_{L,2}(t),u_{R,2}(t),u_{S,3}(t),u_{L,3}(t),$ $u_{R,3}(t))$. On Fig. 5 shows the video frames created by the constructor C_{VL}.

These frames to some extent correspond to the frames from the satellite video (Fig. 1, column 3). Exact coincidence is not expected since the real and model processes are stochastic. However, all flashes of lightning from satellite video (Fig. 4a) and from video of improved model (Fig. 4c) correlate quite well.

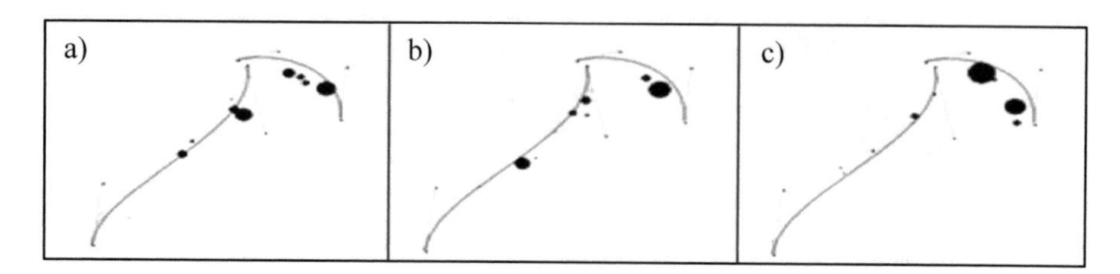

Fig. 5. Video frames of lightning flashes by improved model

4 Conclusions

Usage of modeling in the formation of lightning discharges based on the constructive-synthesizing approach allows obtaining the realistic description of the thunderstorm front lightning activity. This approach can be the basis for solving the dynamic problem on lightning protection of engineering constructions and civil objects, and development of strategy of aircraft behavior in order to mitigate the risks of lightning strokes in the conditions of movement in the thunderstorm front.

References

1. Rodger, C.J., Werner, S., Brundell, J.B., Lay, E.H.: Detection efficiency of the VLF World-Wide Lightning Location Network (WWLLN): initial case study. Ann. Geophys. **24**, 3197–3214 (2006)
2. Kraaij, T., Cowling, R.M., van Wilgen, B.W.: Lightning and fire weather in eastern coastal fynbos shrublands: seasonality and long-term trends. Int. J. Wildland Fire **22**, 288–295 (2013). https://doi.org/10.1071/WF11167
3. Pack, S., Piantini, A.: Lightning research and lightning protection technology. Electr. Power Syst. Res. **113**, 1–2 (2014)
4. Williams, E.R.: Lightning and climate: a review. Atmos. Res. **76**(1–4), 272–287 (2005). https://doi.org/10.1016/j.atmosres.2004.11.014
5. Yair, Y., Lynn, B., Price, C., Kotroni, V., Lagouvardos, K., Morin, E., Mugnai, A., Llasat, M.d.C.: Predicting the potential for lightning activity in Mediterranean storms based on the Weather Research and Forecasting (WRF) model dynamic and microphysical fields. J. Geophys. Res. **115** (2010). D04205. https://doi.org/10.1029/2008JD010868
6. Sátoria, G., Williams, E., Lempergeraams, I.: Variability of global lightning activity on the ENSO time scale. Atmos. Res. **91**(2–4), 500–507 (2009). https://doi.org/10.1016/j.atmosres.2008.06.014
7. Devendraa, S., RameshKumara, P., Kulkarnia, M.N., Singhb, R.P., Singhc, A.K.: Lightning, convective rain and solar activity — over the South/Southeast Asia. Atmos. Res. **120–121**, 99–111 (2013). https://doi.org/10.1016/j.atmosres.2012.07.026
8. Galanaki, E., Kotroni, V., Lagouvardos, K., Argiriou, A.: A ten-year analysis of cloud-to-ground lightning activity over the Eastern Mediterranean region. Atmos. Res. **166**, 213–222 (2015)
9. Ahrens, M.: Lightning fires and lightning strikes. In: National Fire Protection Association, Quincy, 31 p. (2013)

10. Fuchs, B.R., Bruning, E.C., Rutledge, S.A., Carey, L.D., Krehbiel, P.R., Rison, W.: Climatological analyses of LMA data with an open-source lightning flash-clustering algorithm. J. Geophys. Res. Atmos. **121**(14), 8625–8648 (2016)
11. Shynkarenko, V.I., Ilman, V.M.: Constructive-synthesizing structures and their grammatical interpretations. Part I. Generalized formal constructive-synthesizing structure. Cybern. Syst. Anal. **50**(5), 655–662 (2014). https://doi.org/10.1007/s10559-014-9655-z
12. Shynkarenko, V.I., Ilman, V.M.: Constructive-synthesizing structures and their grammatical interpretations. Part II. Refining transformations. Cybern. Syst. Anal. **50**(6), 829–841 (2014). https://doi.org/10.1007/s10559-014-9674-9
13. Shynkarenko, V.I., Ilman, V.M., Skalozub, V.V.: Structural models of algorithms in problems of applied programming. I. Formal algorithmic structures. Cybern. Syst. Anal. **45**(3), 329–339 (2009). https://doi.org/10.1007/s10559-009-9118-0
14. Skalozub, V., Illman, V., Shynkarenko, V.: Development of ontological support of constructive-synthesizing modeling of information systems. Eastern-Eur. J. Enterp. Technol. **6/4**(90), 58–69 (2017). https://doi.org/10.15587/1729-4061.2017.119497
15. Lindenmayer, A.: Mathematical models for cellular interaction in development. J. Theor. Biol. **18**, 280–315 (1968)
16. First Images from GOES-16 Lightning Mapper. https://www.americaspace.com/2017/03/07/goes-16-satellite-returns-first-lightning-mapping-images-like-never-seen-before/. Accessed 11 July 2018
17. Regional and Mesoscale Meteorology Branch. http://rammb.cira.colostate.edu/ramsdis/online/loop.asp?data_folder=loop_of_the_day/goes-16/20190116000000&number_of_images_to_display=120&loop_speed_ms=100. Accessed 11 July 2018
18. Regional and Mesoscale Meteorology Branch. http://rammb.cira.colostate.edu/ramsdis/online/loop.asp?data_folder=loop_of_the_day/goes-16/20180815000000&number_of_images_to_display=100&loop_speed_ms=150. Accessed 11 July 2018

Research and Development of Models and Program for Optimal Product Line Control

Taisa Borovska[1(✉)], Dmitry Grishin[1], Irina Kolesnik[1], Victor Severilov[1], Ivan Stanislavsky[1], and Tetiana Shestakevych[2]

[1] Vinnytsia National Technical University, Vinnytsia, Ukraine
taisaborovska@gmail.com, dmitriygrishin2@gmail.com, iraskolesnyk@gmail.com, severilovvictor0@gmail.com, stanislavskyi.ivan@gmail.com

[2] Lviv Polytechnic National University, Lviv, Ukraine
tetiana.v.shestakevych@lpnu.ua

Abstract. The article is devoted to the development of mathematical models and program for methods of optimal control of product lines. The classes of the lines are analyzed, the mathematical model of the line as a control object for the manufacturer, retailer and customer are proposed. The analysis of previous developments was carried out: market models with asymmetric information structure, models of manufacturers of the production segment and alternative simulation models of the product line. The study of the dynamics and steady state of the product line was carried out. To study the dynamics of the product line, a simulation model of «producers, product lines, consumers» was used, in which the choice of consumers is simulated in the samples. The problem of optimal aggregation of a multidimensional nonlinear, stochastic and non-stationary object «product line» has been set and solved. Optimal control program has been developed on the basis of optimal aggregation. Examples of modeling are given.

Keywords: Optimal aggregation · Product line · Production function · Demand function · Information technology

1 Introduction

A typical production cycle, oriented towards the mass consumer nominally consists of the following stages: production, logistics, bringing a product or service to the final consumer, and the last is recycling of production and consumption waste. In all the ages, prehistoric, industrial, and post-industrial, it was the last stage that caused the change of civilizations. Science and practice confirmed the fundamental nature and the triunity of the elements of the cycle «production, consumption, recycling». From the standpoint of fundamental science (availability of theory, mathematical models and methods for solving the applied problems), the process of bringing the product of production to the final consumer is the most scientifically unsecured. Let us compare the mathematical models of complex technical objects brought to practice: vehicles, bioreactors and market

N. Shakhovska and M. O. Medykovskyy (Eds.): CCSIT 2019, AISC 1080, pp. 186–201, 2020.
https://doi.org/10.1007/978-3-030-33695-0_14

models, product lines based on efficient information technologies, information uncertainty management systems, and cascade project management technologies. An essential component of design and production systems is financial management, taking into account the state of the markets for raw materials and products. The problems in these technical systems are the quality of the software for automated control systems. Modern trading enterprises are substantially globalized, they use efficient customer service technologies, and are integrated with supplier enterprises. This area has developed its own empirical models and methods for managing the production of services. Intensive studies are being conducted, the search for improving management efficiency in the context of global competition, «marketing wars» and «consumer whims».

We did not find mathematical models that are adequate to modern requirements for the specific features of the retail. The question arises: «Is it possible to create models of the level of models of aerodynamics, metallurgy, chemistry in this area?» Retail is an «interface» that is based on consumer psychology. In the classical model, the consumer chooses a certain product on the set price. In today's retail, the consumer chooses from a variety of products of a certain class with different prices and values in the product line. In the search for the key «mathematical models of the product line» millions of sources with empirical models are found. At the same time, authors could not find the appropriate mathematical models. The lack of mathematical models of the market based on product lines is due to the significant complexity of retail processes and the absence of reliable and complete statistical data. However, modern mathematical modeling packages reduce restrictions on the development of adequate and adaptive simulation models of modern production, consumption, and retail, i.e. «virtual reality» which complements the incomplete statistics of «real reality». This paper presents mathematical models of statics and dynamics of the product line as a control object. The possibility of obtaining non-trivial results is due to a certain amount of research in the development of computationally efficient consumer choice models in the product line and market models with a dynamic information structure, as well as using the methodology of optimal aggregation for the formulation and solution of the problem of optimal control of the product line [1–3]. The initial prototype was a market model with an «asymmetric information structure,» in essence, a model of consumer choice between a quality and a low-quality product [4–8].

The aim of the work is to develop a model and program for solving the problem of optimal aggregation of a product line, which allows to reduce the multidimensional nonlinear programming problem to a system of one-dimensional optimization problems and develop an appropriate system for managing the product line state.

To achieve this goal, it is necessary to solve the following tasks: development of a simulation model of the dynamics of the product line and research on «virtual reality»; modification of the consumer choice model on the product line; development of a binary operator for optimal aggregation of products of the line.

The practical use of the research results is the possibility of embedding in an automated production management system (CAM) as an automated decision-making system (ASPR) for analyst. The results of the research will be useful for methodological support of relevant academic disciplines. It can be also relevant as a stand for researching a full-scale object if for manufacturers such object is difficult to conduct experiments on.

Today, the concept of modeling for large systems is changing, instead of «model as a mapping of object properties, essential for the researcher» the «model as a means for searching and experimental verification of the design solutions» is used. On the basis of this concept, a «metamodel» was developed and investigated, which is a model of joint development of an enterprise and of its imitation model [3]. The topic of the article has an interdisciplinary nature, i.e. models and methods should be built on the basis of production technologies of marketing, sociopsychology of an individual, groups, and socium, system dynamics, information systems, production technologies, and just finance. Searching for analogs of optimal aggregation methods, similarities were found by search keys, but not by content [7–10].

2 Analysis and Development of Mathematical Models and Product Line Modeling Program

In the practice of production and retail, there are often problems associated with the transfer and expansion of production, problems arise with the expansion of retail chains [11–13]. Figure 1 presents: the functional structure of modern production, the proposed means to solve or eliminate problems. There are two blocks: production and retail – the objects of development and research. The environment for these blocks – competitors and users. On the right, the resource model of the system is presented: the «production» block is closed by two resource feedbacks (RFB): production and sales costs are reimbursed via «retail»; through «ecology» – production and consumption wastes are processed into safe or useful components. A separate production system today operates in the environment of other production systems in conditions of limited markets for products and resources. Therefore, it is impossible to build current management and development without taking into account the active environment [3, 14, 15]. In particular, the solution of the optimal control problem should be carried out online based on the current state and forecasts. This requires changes in the structure and functions of control systems. This cannot be done on the basis of classical optimization methods and Gaussian statistics. In this regard, the product line is a complex management object in all aspects [5, 12, 16–19].

Figure 2 presents the system of author's models (the results of the development are presented in master's works, in PhD theses, in the doctoral dissertation). These models evolved in accordance with the development of software and hardware. At the first stages of work, the models reproduced and improved the well-known prototypes, and then their own developments were carried out as they gained experience in building mathematical models and software design.

Product Line Model. Today, the main tool for analyzing and synthesizing new models and program for new tasks is modeling. Improving the efficiency and reducing the cost of software and hardware makes it advantageous to use models for «instrumentless measurements», computational experiments, and obtaining «statistics of virtual reality». The model of the product line shown in Fig. 3, is the first «working tool» of the development.

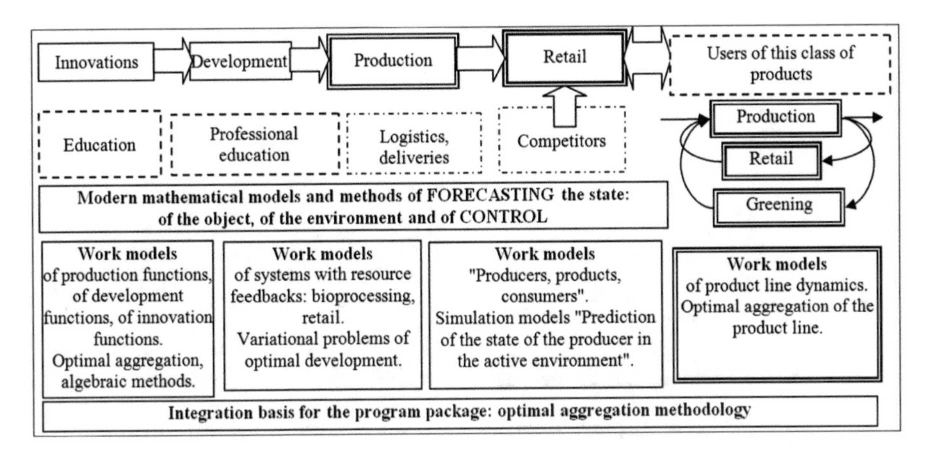

Fig. 1. Modern enterprise: the required mathematical and software support

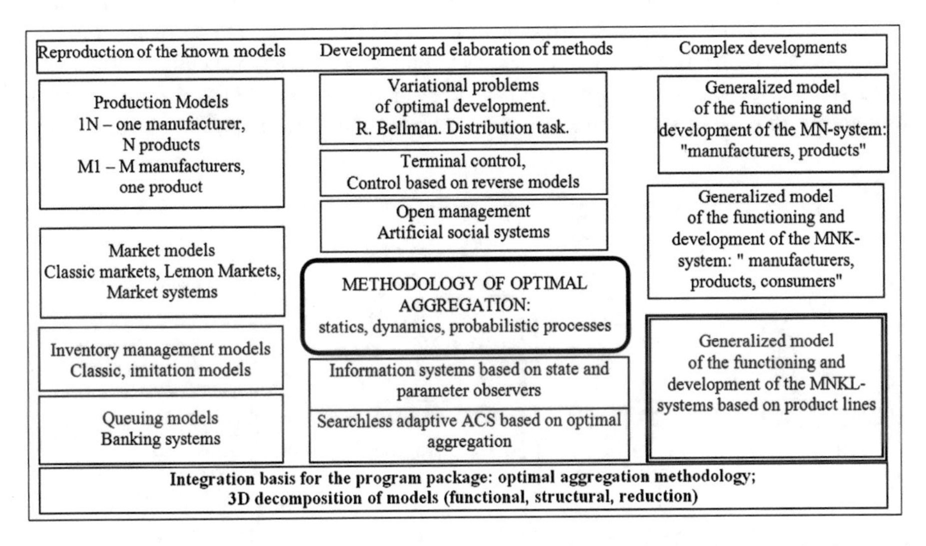

Fig. 2. Scheme of the project «modeling and management of modern production and development of socio-technical-ecological systems (STES)»

Figures 2 and 3 are given as the examples of elements of the information technology «designing of new models for new tasks». The term «design» refers to the development that does not have the direct prototypes of a mathematical model for a certain system of production of goods or services. This is possible on condition of the combination of analytical and numerical methods based on the use of the abilities of mathematical packages that rely on the basic computer resources. An example is the SimuLink package, which takes 80% the operation system resources.

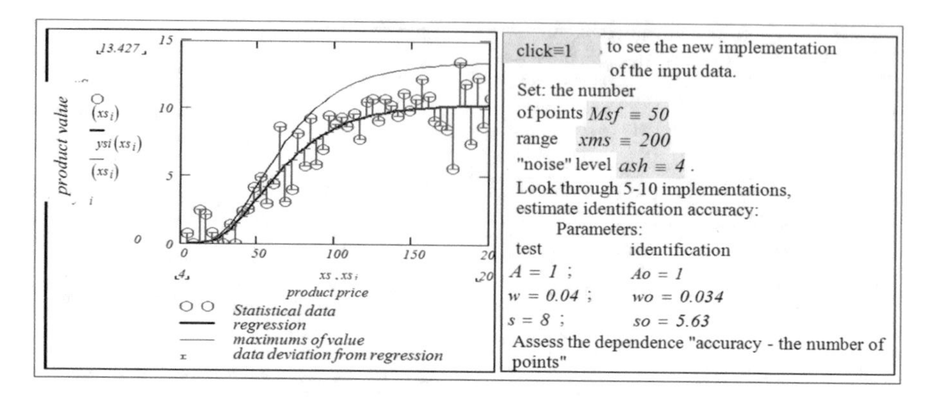

Fig. 3. Regression model of the product line. Example

Alternative Consumer Choice Models on the Product Line. According to the models and modeling results that are presented in this article, no analogs were found. Several alternatives have been brought to the level of appropriate software. Figure 4 shows an alternative to the model of product choice by a consumer with a certain level of income based on fuzzy logic. We use fuzzy logic as a substitute for the theory of probability in case of the absence of the necessary statistical data.

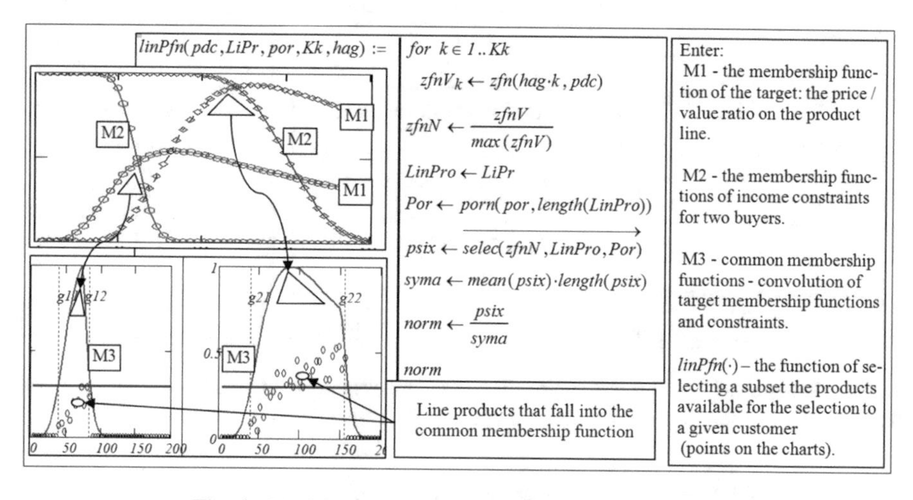

Fig. 4. Model of customer choice based on fuzzy logic

Working model «regular choice of the buyer with training». The model is focused on products of regular consumption: bread, beer, kvass, cheese. We use samples of 100–2000 people in the main program to simulate the choice of customers and the number of price positions in the range of 30–100 products, from which certain product is randomly selected. We believe that it is possible to arrange products according to the complex indicator «value». In the simulation program «true value estimates» are introduced. We believe that the purchase and consumption of a certain product causes a change in the value estimate. The source of the change in the value estimate can be information exchange in the Internet and society. The following is the technology for developing a module of selection imitation on a product line for consumer sample. The characteristic feature of the program is embedded cycles of a large dimension. For testing, we take the system «3 customers, 3 products.» We specify the current matrix of the normalized *Mu* product ratings, the vector of «true» product ratings by *Muis* consumers, and the matrix of current consumer ratings for *dMu*. Figure 5 presents the module «revaluation of values» $dmu(vyb, bolv, nvz)$.

We set the matrix of "true values" of products and the matrix of the current choice

$$Nvz = \begin{pmatrix} 0.2 & 0.2 & 0.2 \\ 0.2 & 0.2 & 0.2 \\ 0.2 & 0.2 & 0.2 \end{pmatrix} \quad Bolv := \begin{pmatrix} 0.75 & 0.75 & 0.75 \\ 0.15 & 0.15 & 0.15 \\ 0.10 & 0.10 & 0.10 \end{pmatrix} \quad Vyb := \begin{pmatrix} 0 & 1 & 0 \\ 1 & 0 & 0 \\ 0 & 0 & 1 \end{pmatrix}$$

We make a module designed for vectorization of calculations

$$dmu(vyb, bolv, nvz) := \begin{array}{|l} kys \leftarrow vyb \cdot nvz \\ qq \leftarrow vyb \cdot bolv + rnd(kys) - 0.5 \cdot kys \\ qq \end{array}$$

We test the module (it is clear why all the parameters are set by the matrices of the same dimension):

$$dMU := \overrightarrow{dmu(Vyb, Bolv, Nvz)} \qquad dMU = \begin{pmatrix} 0 & 0.665 & 0 \\ 0.117 & 0 & 0 \\ 0 & 0 & 0.143 \end{pmatrix}$$

Fig. 5. Model of the customer choice with training

Module «correction of product values assessments». We compile two versions of the module, in which two embedded cycles are replaced by two operations of computation vectorization. We will pay attention to the structures of the above expressions: the vectorized user function *zmin()* has as the parameter the vectorized user function *dmu()*. The results of calculations for test data proofs the expressions are syntactically correct. In the information aspect, it means the integration of the successive stages of input data processing into one stage. In the line modeling program, these modules process high-dimensional matrices.

$$\overrightarrow{zmin\left(Mu, \overrightarrow{dmu(Vyb, Bolv, Nvz)}\right)} = \begin{pmatrix} 0.6 & 0.765 & 0.1 \\ 0.164 & 0.33 & 0.3 \\ 0.1 & 0.34 & 0.096 \end{pmatrix};$$

$$\overrightarrow{zmip\left(Mu, \overrightarrow{dmu(Vyb, Bolv, Nvz)}, \alpha l\right)} = \begin{pmatrix} 0.6 & 0.33 & 0.1 \\ 0.3 & 0.33 & 0.3 \\ 0.1 & 0.34 & 0.6 \end{pmatrix}.$$

Based on these modules, a module *nrmlz2(Mu, Vyb, Bolv, Nvz, Al)* is realized. This module uses the matrix of the current state of product ratings by customers *Mu*, the current matrix of customer choice *Vyb* and parameters: «true product ratings» *Bolv* and uncertainties of product ratings by the customers *Nvz* (factors of subjectivity, non-qualification, and customers «training» rates *Al*). The module returns the matrix of the subsequent state of product ratings by the customers.

Development and Testing of Alternative Models of Product Line Dynamics. Due to the lack of direct analogs, we developed two alternative versions of models and software. And we obtain two structurally different (as in non-identical redundancy) interconnected and complementary models. We use in parallel two alternative models: *a simulation model of the system*; *integrated model of the system.*

The simulation model reproduces the behavior of each «element» of a distributed system: «producer», «product», «consumer». Terms in quotes – only labels, specified in accordance with the specific characteristic of the task – banking systems, construction, Internet, air travel, auto auto-service. In particular, «production» is the production of information and material products, as well as services, in particular, economy and finance. Logistics (supply of resources and products) and retail (retail to the final consumer) must be referred to the production and it is not only natural but necessary. In the modern world, logistics has a planetary scale – the discovery of effective communication allows the efficient producer to expand sales in the former remote region and at the same time «kill» the local production [12, 13].

The integrated model reproduces the behavior of such objects as the «producer system», «consumers system», and «products system». In the «products» class, various product systems (taxa) can be distinguished, for example, «cars and related products and services», «product line of mobile phones», «product line of business intelligence programs», «line of coffee products», «coffee and related products, modeling manuals [2, 3].

Figure 6 presents the alternative programs. All the «generating mechanisms» are implemented in the subprograms. Parameterization of the software allows conducting research «what will happen if» and «risk analysis».

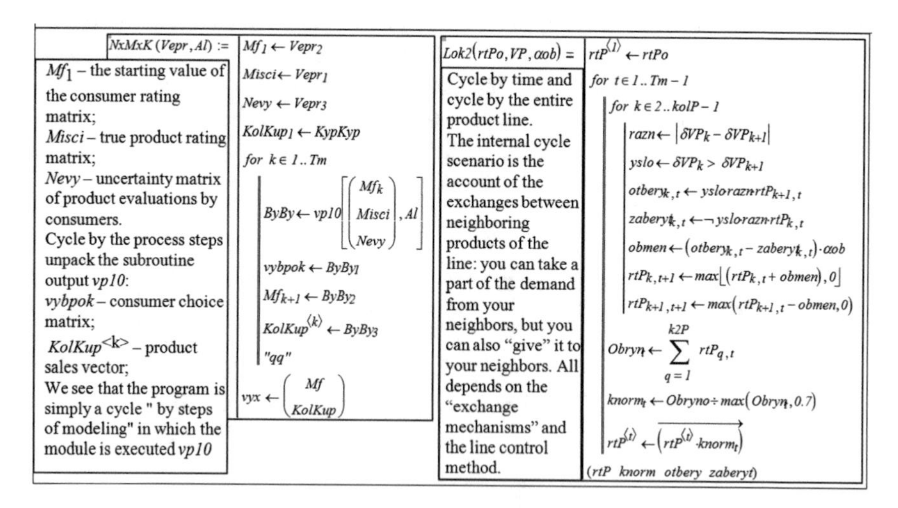

Fig. 6. Texts of programs for the alternative models of the product line dynamics

Figure 7 presents the examples of simulation results for the alternative models of the dynamics of the product line on three-dimensional graphs – the rate of sales for each price. On the right – the evolution of the demand distribution on the product line. One can see the general in the processes and the differences: in the model M1, the random choice of each consumer is taken into account; in the M2 model, consumers are integrated into the consumption process. We see the convergence of the processes (both deterministic and stochastic) to the equilibrium of supply and demand. However, on the right, we see the convergence of the demand distribution of the product line to a substantially non-smooth profile.

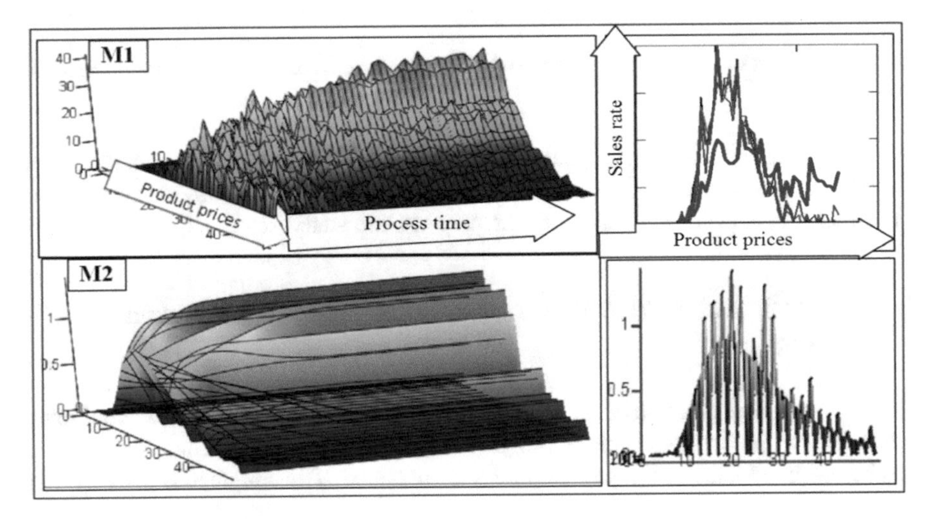

Fig. 7. Modeling of the dynamics of the product line using the models M1 and M2

Alternative models allow to see the general properties of the product line market. Let us consider in details the dynamics of the demand profile of the product line. Figure 8 presents the graphs: estimates of the value, current demand, and consumer distribution by the incomes. It should be noted that the size of consumer income determines the structure of the consumption – the distribution of consumer spending among the product groups – food, clothing, furniture, and transport. These issues were considered in [3]. We will compare the simulation results for the average income of 100 and 70 conventional units (c.u). The distribution of demand reflects the distribution of incomes; it can be quantified (as in Fig. 3). Research on models of the line dynamics allowed us to make some generalizations about the laws of the «market of the products line».

Figure 9 shows a visual presentation of the generalization. In the upper part – the state of the line for three moments of time. The «generating mechanisms» of the demand distribution are superimposed on the last graph, these are the «price, value» function of the line and the function of buyers' distribution by incomes, as well as the regression graph of the demand distribution. In the lower part, the statement that the steady-state distribution of demand rates is a mapping of the distribution of buyers by income and the function «price, value» for the product line. A prospective continuation

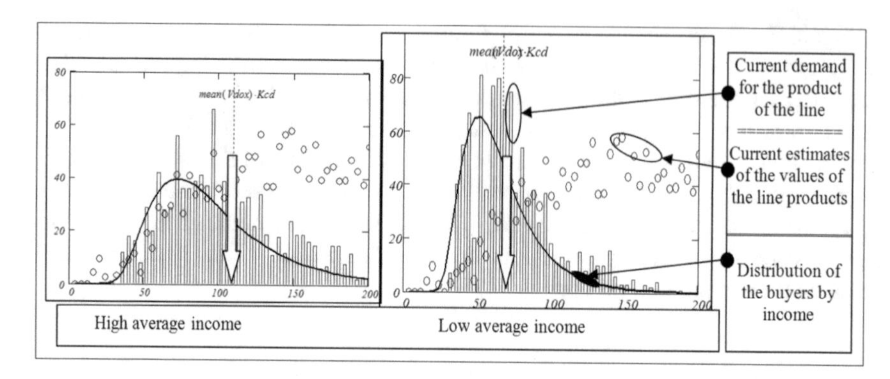

Fig. 8. Analysis of simulation results: the relationship of the distribution of income, demand and value estimates

of the study is the obtaining of a parameterized user function as the reflection of the two functions: frequency distribution and rank distribution. Such a mapping is necessary to construct a system for the control of the product line state.

3 Analysis and Development of Product Line Management Systems

Based on the consideration of the logic of the functioning processes of the penultimate stage of the production cycle - bringing the products of production to the user, mathematical models of the product line were obtained. The last stage is the recycling

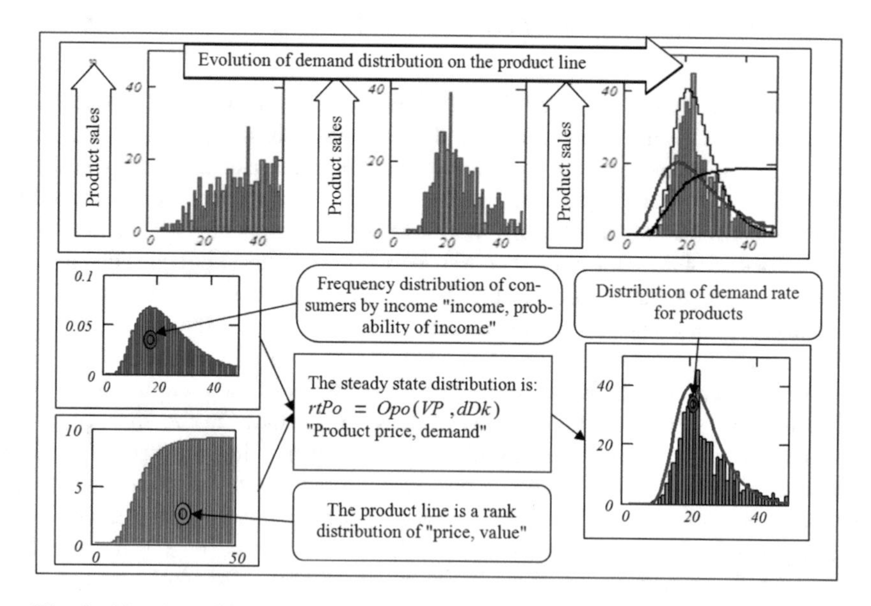

Fig. 9. Mapping of the resources distribution and values in demand rates distribution

of food products, cars, clothes, TVs, video recorders and megatons of various packages for millions of products of various products. The objectives of the last stage are presented in Fig. 1 as a production structure with two resource feedbacks (RFB): returning costs as a result of product sales and greening - processing production and consumption wastes into safe and healthy products. The task of managing only the product line allows us to solve only local retail tasks.

The solution of the extended management problem, taking into account financial and environmental resource feedback (RFB), is the subject of the next research. In this article, on the basis of modeling the dynamics of a product line, we create a statement of the task of the product line management. In analogs and prototypes only high-quality, empirical settings of the line management are given. It should be noted that the task of the line management is essentially stochastic, non-stationary and non-linear. Therefore, it is impossible to obtain an effective solution based on known control methods. To formulate the problem in control theory, it is necessary to specify or obtain: – a model of the object dynamics; – controllable object state variables, – control variables, – control requirements, – controller dynamics model, – model of the closed system dynamics.

To solve the problem of control synthesis, it is necessary to find the parameters of the regulator control law, which can be found analytically or by search methods. Today, the mass products are designed and manufactured as the distributed and simultaneously complete systems, i.e. product lines. The product line is a multi-dimensional, stochastic, non-stationary, essentially non-linear object. Elements of the product line form the parallel structure. The output of this structure is the total sale of all the products, and a radical difference from other parallel structures is a significant mutual influence of the neighboring (by prices) products of the line. A product with a higher value can «pick up» the demand from related products.

Models of large and complex systems are designated to give the manager the understanding of this system and the opportunity to experiment. In the line control models, one more feature of the line should be shown: the evaluation by a buyer of a certain product of the line by means of information exchange becomes available to

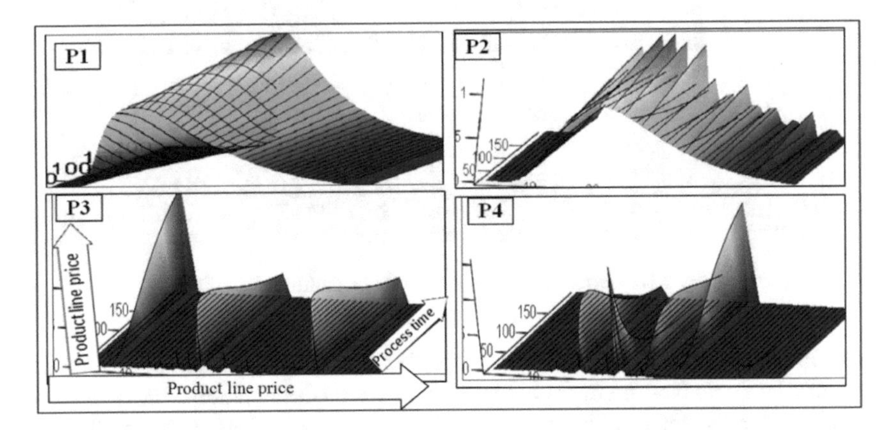

Fig. 10. Examples of the results of modeling the dynamics of the product line

other buyers. Such information exchange leads to a significant change of the processes in the «producers, products, consumers» systems. Figure 10 presents several examples of the line dynamics at certain sets of the line model parameters: learning speed, disturbance level, etc.

Note that the results obtained on the model of a complex for modeling system can not necessarily be reproduced in the real systems. However, these examples may be helpful in finding innovative management practices. Let us analyze the examples in Fig. 10. Example P1: the line evolves without local disturbances of the initial profile, but the demand for cheap products is falling, for products with the maximum demand – is growing, the demand for expensive products with low demand is stable. Example P2: small disturbances, slow customer training for effective selection, the result – the initial profile is «destroyed» by the local competition of adjacent products of the line. Examples of P3 and P4: large disturbances and rapid customer training, first there are 8–15 products out of 50, then 2–5 products and manufacturers remain. Further, the only product that takes away demand from the remaining products appears. The graphs P3 and P4 are two implementations of the random process of the line functioning. We give these examples of modeling because in the world the similar processes take place: the markets for clothing and footwear, liners and turbines, smartphones and cars have one or two leaders in production and retail. The «generating mechanisms» of real processes are much more complicated than in this model, but the results are qualitatively similar: the number of market participants is decreasing.

Figure 11 shows a detailed example of the dynamics of a product line – two projections of a 3D graphic. The dots mark the moments of the demand zeroing for the corresponding product of the line.

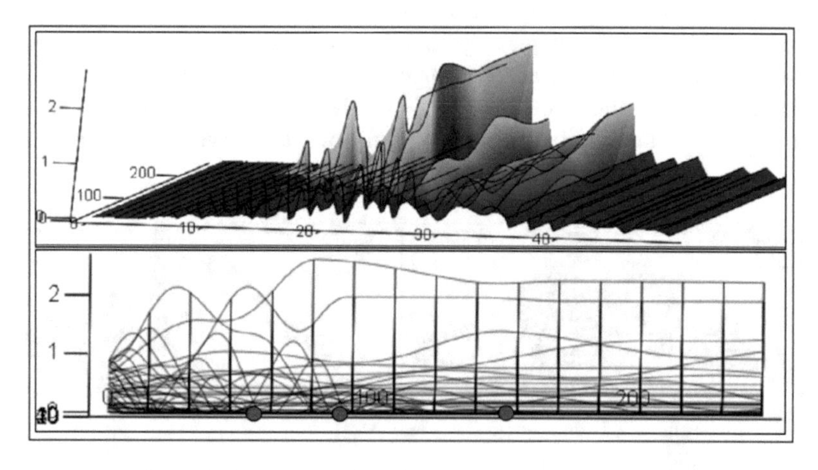

Fig. 11. Dynamics of the line without control. Detailing

Local Product Line Control. We will perform the formalization of relationships between related products on the line. Figure 12 presents the situation of the relationship of demand for the three adjacent elements and the block diagram of the line local control. The block diagram of local control is a standard automatic control system (ACS), with the exception of the «value increment controller» block. The functions of this block are the selection and implementation of ways to increase the value of a product line without changing the sale price. Examples of the simplest methods are elongation of the tube «20% more for the same price», as well as increase the capacity of the smartphone battery, «protective film included».

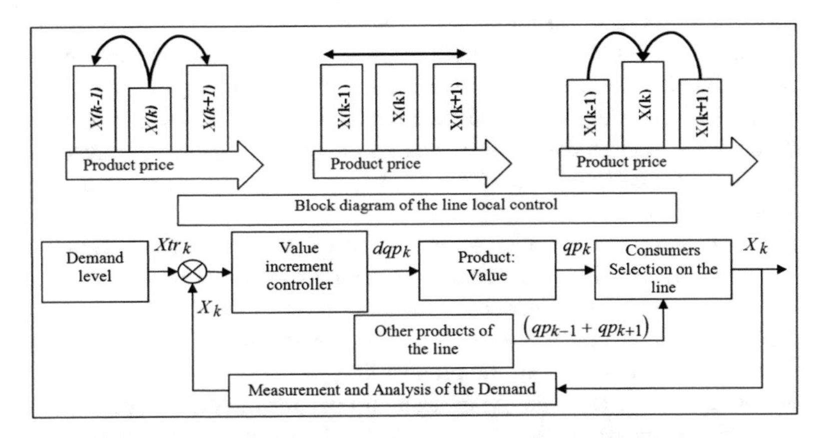

Fig. 12. Development of a local product line control system

Global Product Line Control. The goal of global management is to maximize the effectiveness of the line as a «cost, income» converter. This is the task of non-linear programming: the distribution of a limited resource. Figure 13 shows the scheme for the problem of optimal aggregation of a parallel structure, in which the subsystems are connected only through the limitation of the total resource. In the upper part of Fig. 13 – the initial system, in the lower part – the scheme of the optimally aggregated system. The peculiarity of optimal aggregation methods is that the solution to the optimization problem is an optimal equivalent function of the «costs, output» class [3].

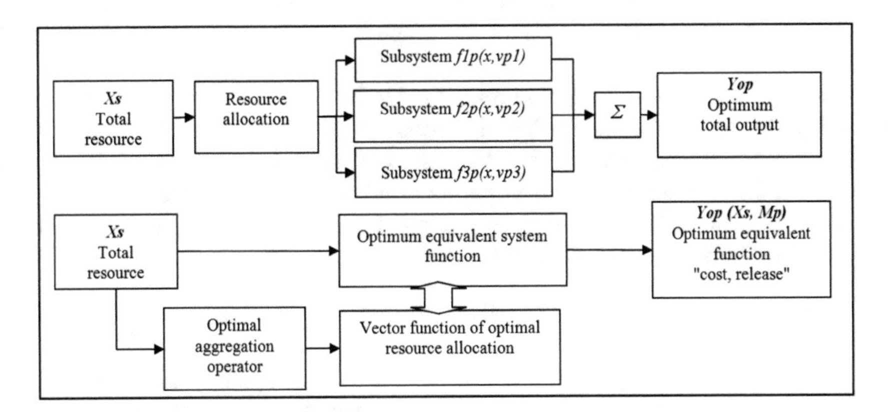

Fig. 13. Optimal aggregation of the structure with parallel functioning elements

Main advantages of the optimal aggregation method are: algebraic method, searchless, computational costs grow no more than linearly with an increase in the dimensionality of the optimization object. The consequence of this advantage is the ability to perform testing on the models of small dimension (for the transition from the dimensionality 4 to dimensionality 400, you are to change the dimensionalities of the vectors and matrices in the source data).

A binary operator of optimal aggregation of a product line is a new scientific result – an extension of a binary operator for the parallel structures, where the elements are connected only through a common resource. The essence of the expansion is taking into account the demand constraint, which is determined by the demand function for the product line. Sources of demand functions are sales statistics and (or) analytics obtained on simulation models, constructed on the base of «generating mechanisms».

Consider the formal definition of the operator of the optimal aggregation of parallel resource structures.

The carrier of the algebra is the set of generalized admissible production functions (PF) – non-strictly positive and non-strictly monotonic functions of the variable x: $0 \leq x \leq R$. Interpretation: functions of the class «costs, output», where «costs» – the cost of the actual production, delivery, storage and sale to the consumer; «output» – the rate of the given product sales of the product line.

A set of operations is a binary aggregation operator of two production functions (FP) $f1, f2$ with a constant parameter α: $0 \leq \alpha \leq 1$, giving a set of α-functions:

$$f\alpha(f1, f2, \alpha, x) := f1(\alpha \times x) + f2[(1 - \alpha) \times x].$$

Optimization. Let a finite set of values $\alpha 1, \ldots, \alpha N$, be given, uniformly covering the range of definition and the corresponding set of α-functions: $(f1, f2, \ldots, fN)$, then

$$FopN(f1, f2, \ldots, fN) \rightarrow \max(f1, f2, \ldots, fN),$$

with $N \rightarrow \alpha$ converges to the optimal production function of the system of two production elements operating in parallel, and the hodograph of the maxima for x: $0 \leq x \leq R$ determines the optimal resource distribution function $\alpha_{op}(x)$. α-functions – a tool for determining and analyzing the properties of the operator of the optimal aggregation.

The binary optimal aggregation operator $f2o(f1, f2)$ of two production functions (PF) $f1, f2$ *takes* two discretized PFs, represented by matrices with a fixed number of rows and columns representing the function values, and the values of the vector function of the optimal resource distribution in previous aggregations; and *returns* an object of the same class – a discretized optimal production function represented by a matrix of the same structure. Operator – associative and commutative.

$$f2o(f1, f2o(f2, f3)) = f2o(f2, f2o(f3, f1)).$$

Similarly to the $f2o$ operator, we form the operator of optimal aggregation of the $f2opl$ product line. The construction operations are presented visually on a test example (Fig. 14). Functions $f1p(x, vP1)$, $f2p(x, vP2)$, we take from the class of S-functions:

$$Fvv(x, A, w, s) = A \times (1 - e^{-w \cdot x})s.$$

We set the basic matrix of parameters for the functions of production and development of *MPko*: the first and second columns of which are the functions of production and development of the first subsystem, and the third and fourth columns are the functions of production and development of the second subsystem:

$$MPko = \begin{pmatrix} vp1_1 & vpr1_1 & vp2_1 & vpr2_1 \\ vp1_2 & vpr1_2 & vp2_2 & vpr2_2 \\ vp1_3 & vpr1_3 & vp2_3 & vpr2_3 \end{pmatrix}; \; MPko := \begin{pmatrix} 30 & 40 & 50 & 60 \\ 0.04 & 0.08 & 0.06 & 0.07 \\ 5 & 12 & 27 & 42 \end{pmatrix}.$$

In the variable «production costs» we take into account the costs of production itself, the costs of logistics, sales, and advertising. The variable «release» is «the return of costs» to the «production, retail» system. This is also a «buyer received product» event. We introduce the notations:

$$vp1 := MPko^{\langle 1 \rangle}; \; vp1 := MPko^{\langle 1 \rangle}; \; vp1 := MPko^{\langle 1 \rangle}; \; vp1 := MPko^{\langle 1 \rangle};$$

$$f1p(x,\, vp1) := Fvv(x,\, vp1_1,\, vp1_2,\, vp1_3); \; f2p(x,\, vp2) := Fvv(x,\, vp2_1,\, vp2_2,\, vp2_3);$$
$$f3p(x,\, vp3) := Fvv(x,\, vp3_1,\, vp3_2,\, vp3_3); \; f4p(x,\, vp4) := Fvv(x,\, vp4_1,\, vp4_2,\, vp4_3).$$

We modify the production functions taking into account the state of the market and consumers, we introduce the vector of market parameters. We enter the vectors of parameters of the technological process of the production *Vp* and «market» *Vm*: *fp* (*x*, *Vp*, *Vm*), where

$$Vp = \begin{pmatrix} \text{"productive capacity"} \\ \text{"maximum efficiency"} \\ \text{"starting costs"} \end{pmatrix}; \; Vm = \begin{pmatrix} \text{"resource price"} \\ \text{"product price"} \\ \text{"demand function"} \end{pmatrix}.$$

In the basic works [1–3], for the problems of optimal aggregation, a unified system of parameters for the resource functions was chosen, we follow this system. Figure 14 shows the logic for solving the problem of optimal aggregation of the line (solution - a trivial software module). We build graphs of the test system. We set the demand constraints on the graphs on the coordinate axis «output» (squares) and determine the constraints on the cost of production for each product of the line. Thus, we obtained a working model of the dynamics of a product line with local and global optimal control.

Designations in Fig. 14: *Fopl* – optimal equivalent function of the «production» line (OEPF); *Fop* – OEPF systems «production of the line products»; *f2opl* – binary operators for optimal aggregation of the line; *fpl*1–*fpl*4 – functions of the «production» of the line elements taking into account demand constraints.

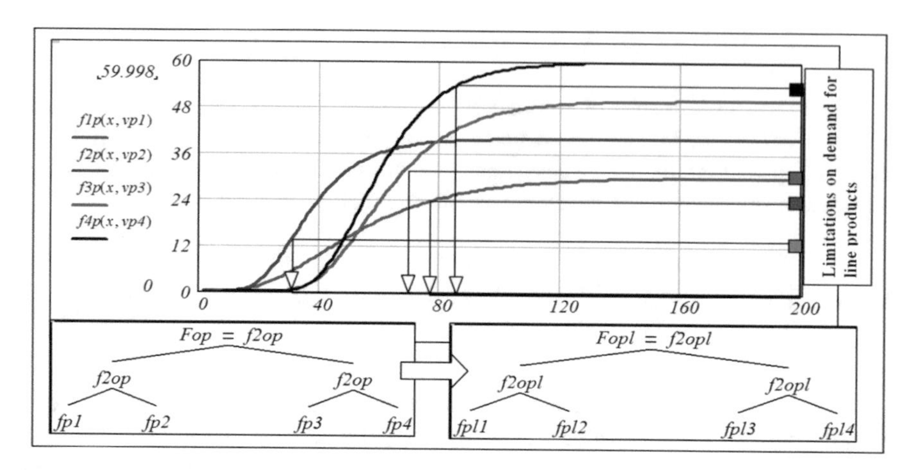

Fig. 14. The logic of solving the problem of global control of a product line based on the method of optimal aggregation

4 Conclusion and Future Developments

In this paper, authors set the task of studying the object, relevant both for science and practice, a product line, in order to formulate and solve the problems of managing this complex, interdisciplinary object which has no clear boundaries. The relevance of the product line control was reflected only in the number of publications of empirical and descriptive nature. The tasks of managing a product line cannot be solved within the framework of known models and methods. The substitute for the necessary missing information about the properties of the line as the object of control was the results of modeling on simulation models of the systems «manufacturers, products, consumers». To solve the problems, stated in the research, authors used the methodology of optimal aggregation and resource approach, where the product line is considered as a key element of the contour of the resource feedback «production, retail, greening». Violations in the functioning of such a system lead to social, economic and environmental crises. Research and developments of this article provide the opportunity to solve global problems. Authors developed a mathematical model of systems with distributed and price-ranked products of the same class. Also, two alternative models of the product line were developed and investigated, which ensured the mutual control of models in the absence of field data and close analogs. To solve the management problem, the authors proposed a method of division into local control of related products, and a global control of the line as a whole. In addition, the task of global line control, based on optimal aggregation, was set and solved.

The prospects for further research are expansion according the directions: elaboration of new models, development of modeling program and models research.

References

1. Borovska, T.M., Vernigora, I.V., Grishin, D.I., Severilov, V.A., Gromaszek, K., Aizhanova, A.: Adaptive production control system based on optimal aggregation methods. In: Proceedings of the SPIE 10808, Photonics Applications in Astronomy, Communications, Industry, and High-Energy Physics Experiments 2018, 108086O, 1 October 2018). https://doi.org/10.1117/12.2501520
2. Borovska, T., Vernigora, I., Kolesnyk, I., Kushnir, A.: Control of multi-channel multiphase queuing system based on optimal aggregation methodology. In: Proceedings of the 13th International Scientific and Technical Conference «Computer science and information technologies» CSIT 2018, Lviv, Ukraine, 11–14 September 2018, vol. 1, pp. 259–265. Publishing House «Vezha and Ko», Lviv (2018). ISBN 978-1-5386-6463-6
3. Borovska, T.M.: Mathematical models of functioning and development of production systems based on optimal aggregation methodology. VNTU, Vinnitsa, Ukraine, 308. ISBN 978–966–641–731–5
4. Krishnamurthi, P.: Product lining and price lining (2007). http://fmcg-marketing.blogspot.com/2007/10/product-lining-and-price-lining.html. Accessed 20 Mar 2016
5. Hanks, G.: Examples of a product line extension (n.d.). http://smallbusiness.chron.com/examples-product-line-extension-69425.html. Accessed 20 Mar 2016
6. Learn marketing: Product strategy (n.d.). http://www.learnmarketing.net/productobjectives.htm. Accessed 20 Mar 2016
7. Wilsom, O.L., Norton, A.J.: Optimal entry timing for a product line extension. Market Sci. **8** (1), 1–17 (1989). JSTOR 184099. https://www.jstor.org/stable/184099
8. Quinonez, N.: 5 Product line pricing strategies you need to know (2014). https://blog.udemy.com/product-line-pricing/. Accessed 20 Mar 2016
9. Xinxin, L.: Self-selection and information role of online product reviews. Inf. Syst. Res. **19** (4), 56–64 (2007)
10. Neubauer, J., Steffen, B., Margaria, T.: Higher-order process modeling: product-lining, variability modeling and beyond (2013)
11. Mukha, Ap.A.: Control of the process of complex engineering systems and processes development. Characteristic features of FMEA-analysis application. Mathematical machine and systems, no. 2, pp. 168–176 (2012). ISSN 1028-9763
12. Rüttimann, B.: Introduction to Modern Manufacturing Theory, p. 149. Springer International Publishing AG 2018 (2016). https://doi.org/10.1007/978-3-319-58601-4
13. Nersessian, N.J., Chandrasekharan, S.: Hybrid analogies in conceptual innovation in science. Cogn. Syst. Res. **10**, 178–188 (2009). https://doi.org/10.1016/j.cogsys.2008.09.009
14. Burkov, V.N., Novikov, D.A.: Introduction to the theory of active systems, 125 p. ICP RAS (1996)
15. Opoitsev, V.I.: Equilibrium and stability in models of collective behavior, 245 p. World, Moscow, USSR (1977)
16. Forrester, J.W.: Basics of Cybernetics Enterprises (Industrial Dynamics), 340 p. Progress, Moscow, USSR (1971)
17. Leontiev, V.: Theoretical assumptions and nonobservable facts. Economy, ideology, politics, USA, no. 9, p. 15 (1972)
18. Bellman, R., Gliksberg, I., Gross, O.: Certain Problems of Mathematical Control Theory, 233 p. Publishing House of Foreign Literature (1962)
19. Weijia, D., Ginger, Z., Jungmin, L.: Optimal aggregation of consumer ratings. NBER Working Paper No. 18567, pp. 12–23

Properties of Logical Functions Implemented by One Generalized Neural Element over the Galois Field

Fedir Geche[1], Oksana Mulesa[1], Anatoliy Batyuk[2](✉), and Veronika Voloshchuk[1]

[1] Departament of Cybernetics and Applied Mathematics, Uzhhorod National University, sq.Narodna, 3, Uzhhorod 88000, Ukraine
{fedir.geche, oksana.mulesa, veronika.smolanka}@uzhnu.edu.ua
[2] ACS Department, Lviv Polytechnic National University, Lviv, Ukraine
abatyuk@gmail.com

Abstract. We introduce the concept of multivalued neural element over Galois field relative to any system of characters of group on which we define logical functions. We deduced the criteria of realization of logical functions on such neural elements. Based on spectral attributes of logical functions the method of synthesis of generalized neural elements over Galois field was developed. Invariant operations over logical functions, which are realizable by one generalized neural element relative to the system of characters where no one of them have common parent, were described.

Keywords: Galois field · Spectrum of logical function · Vector of the structure · Basis of a vectors space · Character of a group · Synthesis · Invariant operations

1 Introduction

Finitesimal fields, Abelian group and theirs characters over different fields are effectively used in theory of logical functions, machines, and in the coding theory [1–3]. Main methods in spectral analysis cannot be successfully applied for research of attributes of multivalued neural functions and synthesis multivalued neural elements over finitesimal Galois field [4, 5].

In this paper we introduce a concept of generalized neural element relative to any system of characters over Galois field, define criteria of realizable logical functions by one generalized element and develop method of synthesis such elements. We described invariant operations over logical functions, which are realizable by one generalized neural element relative to the given system of characters.

N. Shakhovska and M. O. Medykovskyy (Eds.): CCSIT 2019, AISC 1080, pp. 202–213, 2020.
https://doi.org/10.1007/978-3-030-33695-0_15

2 Criteria of Realization Logical Functions by One Generalized Neural Element

Let's assume $F = GF(p^m)$ is the Galois field which contains cyclic group $H_k = \langle a | a^k = 1 \rangle$ and $G_n = H_k \otimes H_k \otimes \ldots \otimes H_k$ is the direct product of n cyclic groups H_k.

Logical function of n variables over field F means unambiguous reflection like $f : G_n \to H_k$. Obviously, if $k = 2$, then we have boolean function, and in case $k > 2$ function k- significant logic.

To introduce concept of generalized neural element (GNE) and to define criteria of feasibility logical functions by one GNE is necessary, that group of characters $\chi(G_n)$ [6, 7] over field F forms the basis of the vector space $V_F^n = \{f | f : G_n \to F\}$. Considering orthogonal property of characters [7] we can confirm that $\chi(G_n)$ is the basis of space V_F^n, if ranks of groups $\chi(G_n)$ and G_n are equal, namely when k is a divisor of rank of cyclic group $F^* = GF(p^m) \backslash \{0\}$ So, spectral analysis logical functions over field $F = GF(p^m)$ is possible, if $p^m - 1$ divides exactly by k.

Let's find analytical form of characters of group G_n over the field $F = GF(p^m)$. Let's assume k is a positive integer number which is a divisor of number $p^m - 1$ and $k \geq 2$. Let's denote by ε a primitive element of field $F = GF(p^m)$ and by H_k a cyclic group of rank k, to wit $H_k = \langle a | a^k = 1 \rangle$. Then for any element $h \in H_k$ exists such number $j \in \{0, 1, \ldots, k-1\}$, that $h = a^j$. As you know [7] characters χ_t of group H_k over the field $F = GF(p^m)$ can be defined as:

$$\chi_t(a^j) = \sigma^{tj} \tag{1}$$

where $\sigma = \varepsilon^{(p^m - 1)/k}$, $t, j \in \{0, 1, \ldots, k-1\}$.

With that group G_n is a direct product of n cyclic groups H_k, follows: for any $\mathbf{g} \in G_n$ exist such numbers $j_t \in \{0, 1, \ldots, k-1\}$, $t = 1, 2, \ldots, n$, that $\mathbf{g} = (a^{j_1}, \ldots, a^{j_n})$.

Considering multiplicative property of characters [6, 7] from (1) we obtain: all the characters of group G_n deflate by functions

$$\chi_t(\mathbf{g}) = \sigma^{t_1 j_1 + t_2 j_2 + \ldots + t_n j_n}, \tag{2}$$

where $t = t_1 k^{n-1} + t_2 k^{n-2} + \ldots + t_n (t_1, t_2, \ldots, t_n \in \{0, 1, 2, \ldots, k-1\})$.

We can build all the elements of group $\chi(G_n)$ by formula (2). Since ranks of groups $\chi(G_n)$ and G_n are equal to $k^n - 1$, then any logical function $f : G_n \to H_k$ definitely can be defined as

$$f(\mathbf{g}) = \sum_{t=0}^{k^n - 1} s_t \chi_t(\mathbf{g}). \tag{3}$$

Vector $S_f = (s_0, s_1, \ldots, s_{k^n - 1})$ is calling the spectrum of logical function f in the system of basic functions $\chi(G_n)$. Based on formula (3) and considering the orthogonally of characters succeed with a formula for definition of spectral coefficients s_t:

$$s_t = |G_n|^{-1} \sum_{\mathbf{g} \in G_n} f(\mathbf{g}) \chi_t^{-1}(\mathbf{g}), \tag{4}$$

where $|G_n|$ is the rank of the group G_n, and $t = 0, 1, 2, \ldots, k^n - 1$.

Should be noted, when $k = 2$, then by formula (4) we get spectral coefficients of logical function f in the system of basic functions of Walsh–Hadamard [8], in case $k > 2$ – spectral coefficients of logical function of Vilenkin–Krestenson [9] over the Galois field.

Remark. To find spectral coefficients s_t of logical functions $f : G_n \to H_k$ we can use fast algorithms which are based on well known theorem from group representation theory [6] and based on theorem of matrix factorization [10, 11].

Let's assume ε is a primitive element of the field $F = GF(p^m)$. We define function on set $F \backslash \{0\}$ by: $\forall \xi \in F \backslash \{0\}$ $\text{Fsign}\,\xi = \sigma^j$,

if $\frac{j(p^m - 1)}{k} \leq \deg\xi < \frac{(j+1)(p^m - 1)}{k}$, where $\deg\xi$ – degree of element ξ $\left(\xi = \varepsilon^{\deg\xi}\right)$, $j \in \{0, 1, \ldots, k-1\}$.

Let's create set $\chi = \left\{\chi_{i_1}, \chi_{i_2}, \ldots, \chi_{i_t}\right\}$ of different elements of group of characters $X(G_n)$ over the field F, except the main element χ_0, and define concept of generalized neural element relatively to system χ.

Generalized neural element (GNE) over the field $F = GF(p^m)$ relatively to system of characters $\chi = \left\{\chi_{i_1}, \chi_{i_2}, \ldots, \chi_{i_t}\right\}$ is called logical device with $t+1$ inputs $\chi_{i_1}, \chi_{i_2}, \ldots, \chi_{i_t}$ $\chi_0 = 1$ $(t \geq 1)$ and one output, which accepts values from set H_k. Each input is associated with particular element ω_i of the field F. Then the output signal can be find as follows: each input signal should be multiplied by particular element ω_i, then summarize those products. As result, we obtain $\text{Fsign}\,\xi$.

Mathematical model of generalized neural element relatively to the system of characters $\chi = \left\{\chi_{i_1}, \chi_{i_2}, \ldots, \chi_{i_t}\right\}$ is defined so:

$$f(\mathbf{g}) = \text{Fsign}\left(\sum_{j=1}^{t} \omega_j \chi_{i_j}(\mathbf{g}) + \omega_0\right), \tag{5}$$

where $\mathbf{w}_\chi = (\omega_1, \ldots, \omega_t; \omega_0)$ – vector of the structure of generalized neural element and $\mathbf{g} \in G_n$.

Let $\mathrm{w}_\chi(\mathbf{g}) = \omega_1 \chi_{i_1}(\mathbf{g}) + \ldots + \omega_t \chi_t(\mathbf{g}) + \omega_0$.

Theorem 1. Logical function $f : G_n \to H_k$ can be realizabled by one generalized neural element relative to the system of characters $\chi = \left\{\chi_{i_1}, \chi_{i_2}, \ldots, \chi_{i_t}\right\}$ over the field $F = GF(p^m)$ with vector of structure $\mathbf{w}_\chi = (\omega_1, \ldots, \omega_t; \omega_0)$ then and only then when exists such function $r : G_n \to F \backslash \{0\}$, that

$$\forall \mathbf{g} \in G_n \quad r(\mathbf{g}) f(\mathbf{g}) = \mathrm{w}_\chi(\mathbf{g}) \tag{6}$$

and

$$0 \le \deg r(\mathbf{g}) < \frac{p^m - 1}{k}. \tag{7}$$

Necessity. Let function $f : G_n \to H_k$ can be realizabled by one generalized neural element relative to the system of characters $\chi = \{\chi_{i_1}, \chi_{i_2}, \ldots, \chi_{i_t}\}$ over the field $F = GF(p^m)$ with vector of structure $\mathbf{w}_\chi = (\omega_1, \ldots, \omega_t; \omega_0)$, to wit

$$\forall \mathbf{g} \in G_n \; f(\mathbf{g}) = \; \mathrm{Fsign}\, \mathrm{w}_\chi(\mathbf{g}). \tag{8}$$

Let's build function $r(\mathbf{g})$ in the way: for all the $\mathbf{g} \in G_n$

$$\deg r(\mathbf{g}) = \deg \mathrm{w}_\chi(\mathbf{g}) - \deg f(\mathbf{g}). \tag{9}$$

Then $\deg \mathrm{w}_\chi(\mathbf{g}) = \deg r(\mathbf{g}) + \deg f(\mathbf{g})$. That's why for any $\mathbf{g} \in G_n \mathrm{w}_\chi(\mathbf{g}) = r(\mathbf{g}) f(\mathbf{g})$. Suppose, that on any fixed $\mathbf{g} \in G_n$ function f is equal to σ^j. Then from equality (8) and definition of function $\mathrm{Fsign}\, \xi$ follows that

$$\frac{j(p^m - 1)}{k} \le \deg \mathrm{w}_\chi(\mathbf{g}) < \frac{(j+1)(p^m - 1)}{k}. \tag{10}$$

Inequality (10) could represented as:

$$\frac{j(p^m - 1)}{k} - \deg f(\mathbf{g}) \le \deg \mathrm{w}_\chi(\mathbf{g}) - \deg f(\mathbf{g}) < \frac{(j+1)(p^m - 1)}{k} - \deg f(\mathbf{g}).$$

From the last inequality considering (9) and $\deg f(\mathbf{g}) = \frac{j(p^m-1)}{k}$, $\left(f(\mathbf{g}) = \sigma^j = \varepsilon^{j(p^m-1)/k}\right)$, we have directly $0 \le \deg r(\mathbf{g}) < \frac{p^m-1}{k}$.

Sufficiency. Let's assume for the function $f : G_n \to H_k$ exists such function $r : G_n \to F \backslash \{0\}$, which satisfies the conditions (6), (7). We will show that function f can be realizabled by one GNE relatively to the system of characters $\chi = \{\chi_{i_1}, \ldots, \chi_{i_t}\}$ over the field F with vector of the structure $\mathbf{w}_\chi = (\omega_1, \ldots, \omega_t; \omega_0)$. From (6) and (7) follows that $\deg r(\mathbf{g}) = \deg \mathrm{w}_\chi(\mathbf{g}) - \deg f(\mathbf{g})$ and

$$0 \le \deg \mathrm{w}_\chi(\mathbf{g}) - \deg f(\mathbf{g}) < \frac{p^m - 1}{q}. \tag{11}$$

Let $\mathbf{g} \in G_n$ and $f(\mathbf{g}) = \sigma^j$. Inequality (11) considering equality $\deg f(\mathbf{g}) = \frac{j(p^m-1)}{k}$ can be represented as: $\frac{j(p^m-1)}{k} - \deg \sigma^j \le \deg \mathrm{w}_\chi(\mathbf{g}) - \deg \sigma^j < \frac{(j+1)(p^m-1)}{k} - \deg \sigma^j$. Hence for all the $\mathbf{g} \in G_n$ $\frac{j(p^m-1)}{k} \le \deg \mathrm{w}_\chi(\mathbf{g}) < \frac{(j+1)(p^m-1)}{k}$ and by definition of the function $\mathrm{Fsign}\, \xi$ we have: $f(\mathbf{g}) \; = \; \mathrm{Fsign}\, \mathrm{w}_\chi(\mathbf{g})$.

So, function f can be realizabled by one GNE relativelly to the system of characters $\chi = \{\chi_{i_1}, \chi_{i_2}, \ldots, \chi_{i_t}\}$ with vector of the structure $\mathbf{w}_\chi = (\omega_1, \ldots, \omega_t; \omega_0)$.

Often to solve number of practical tasks of image recognition, diagnostic, building of neural networks, etc. is necessary to realize partially determined logical functions by one neural element with different activation function. So, development of methods of synthesis GNE which realize partially determined logical functions is practically important task to build logical schemas in neural basis.

Let logical function $f : G_n \rightarrow H_k$ is not defined on some elements of group G_n. Let's denote $D_n \subset G_n$ set of such elements where function f is determined.

Partially determined logical function $f : G_n \rightarrow H_k$ can be realizabled by one GNE relatively to the system of characters $\chi = \{\chi_{i_1}, \chi_{i_2}, \ldots, \chi_{i_t}\}$ over the field F, if exists such $t+1$-dimensional vector $\mathbf{w}_\chi = (\omega_1, \ldots, \omega_t; \omega_0)$, that for all the $\mathbf{g} \in D_n$ $f(\mathbf{g}) = \mathrm{Fsign}\,\mathrm{w}_\chi(\mathbf{g})$.

Theorem 2. Partially defined logical function $f : G_n \rightarrow H_k$ can be realized by one GNE relatively to the system of characters $\chi = \{\chi_{i_1}, \chi_{i_2}, \ldots, \chi_{i_t}\}$ over the field $F = GF(p^m)$ with vector of the structure $\mathbf{w}_\chi = (\omega_1, \ldots, \omega_t; \omega_0)$ then and only then when exist such fully defined functions $r : G_n \rightarrow F \backslash \{0\}$, $h : G_n \rightarrow H_k$, that restriction h on the D_n matches with f ($f = h|D_n$) and next conditions are fulfilled

$\forall \mathbf{g} \in D_n \;\; r(\mathbf{g})f(\mathbf{g}) = \mathrm{w}_\chi(\mathbf{g})$ and $\forall \mathbf{g} \in D_n \;\; 0 \le \deg r(\mathbf{g}) < \frac{p^m - 1}{k}$.

This theorem is proving in the same way as Theorem 1.

Theorem 1 (Theorem 2) designates logical functions (partially determined logical functions) as realizable by one GNE relatively to the system of characters $\chi = \{\chi_{i_1}, \chi_{i_2}, \ldots, \chi_{i_t}\}$ over the Galois field, but it's difficult to apply in practice for finding a vector of a structure $\mathbf{w}_\chi = (\omega_1, \ldots, \omega_t; \omega_0)$. In the next section we are going to present a practically suitable method of synthesis GNE relatively to system of characters $\chi = \{\chi_{i_1}, \chi_{i_2}, \ldots, \chi_{i_t}\}$ over the field $F = GF(p^m)$.

3 Synthesis of Generalized Neural Elements over the Galois Field

Let $f : G_n \rightarrow H_k$ is any logical function, $F = GF(p^m)$ and $(p^m - 1)$ divided into k. The question arises whether the function f is realizable by one GNE related to the system of characters $\chi = \{\chi_{i_1}, \chi_{i_2}, \ldots, \chi_{i_t}\}$ over the field $F = GF(p^m)$. And if yes the how to find the vector of the structure $\mathbf{w}_\chi = (\omega_1, \ldots, \omega_t; \omega_0)$ of particular GNE?

Let $\chi^*(G_n) = \{\chi_0, \chi_{i_1}, \chi_{i_2}, \ldots, \chi_{i_t}\} \subset \chi(G_n)$.

Theorem 3. Logical function $f : G_n \rightarrow H_k$ is realizable by one generalized neural element relatively to the system of characters $\chi = \{\chi_{i_1}, \chi_{i_2}, \ldots, \chi_{i_t}\}$ over the field $F = GF(p^m)$ with vector of the structure $\mathbf{w}_\chi = (\omega_1, \ldots, \omega_t; \omega_0)$ then and only then when exists such function $r : G_n \rightarrow F \backslash \{0\}$, that

$$0 \le \deg r(\mathbf{x}) < \frac{p^m - 1}{k}, \text{ and } \left(r(\mathbf{x})f(\mathbf{x}), \chi^{-1}(\mathbf{x})\right) = 0,$$

for all the $\chi \in X(G_n) \backslash X^*(G_n)$, where $(\mathbf{a}, \mathbf{b})$ is a scalar product of vectors $\mathbf{a}$ and $\mathbf{b}$ over the field F.

Necessity. Let logical function $f : G_n \to H_k$ is realizable by one GNE related to the system of characters $\chi = \{\chi_{i_1}, \chi_{i_2}, \ldots, \chi_{i_t}\}$ over the field F with vector of the structure $\mathbf{w}_\chi = (\omega_1, \ldots, \omega_t; \omega_0)$. Then by theorem 1 there exists such function $r : G_n \to F \backslash \{0\}$, that $\forall \mathbf{x} \in G_n\ 0 \le \deg r(\mathbf{x}) < \frac{p^m - 1}{k}$ and

$$r(\mathbf{x})f(\mathbf{x}) = \omega_0 \chi_0(\mathbf{x}) + \omega_1 \chi_{i_1}(\mathbf{x}) + \ldots + \omega_t \chi_{i_t}(\mathbf{x}). \tag{12}$$

Let's decompose the function $r(\mathbf{x})f(\mathbf{x}) \in V_F^n$ by the basis of the space V_F^n, which consists of the characters of the group G_n, to wit

$$r(\mathbf{x})f(\mathbf{x}) = \sum_{\chi \in X^*(G_n)} s_\chi \chi(\mathbf{x}) + \sum_{\chi \in X(G_n) \backslash X^*(G_n)} s_\chi \chi(\mathbf{x}). \tag{13}$$

From (12) and (13) follows that if $\chi \in X(G_n) \backslash X^*(G_n)$ then $s_\chi = 0$. Then from the equality $|G_n| s_\chi = (r(\mathbf{x})f(\mathbf{x}), \chi^{-1}(\mathbf{x}))$ follows:

$$\left(r(\mathbf{x})f(\mathbf{x}), \chi^{-1}(\mathbf{x}) \right) = 0$$

for all the $\chi \notin X^*(G_n)$. So, necessity is proved.

Sufficiency is clear. If logical function $f : G_n \to H_q$ satisfies conditions of Theorem 3 then coordinates of vector of the structure $\mathbf{w}_\chi = (\omega_1, \ldots, \omega_t; \omega_0)$ GNE which relize function f relatively to the system of characters $\chi = \{\chi_{i_1}, \chi_{i_2}, \ldots, \chi_{i_t}\}$ over the field $F = GF(p^m)$ can be find by formulas:

$$\begin{aligned} \omega_0 &= |G_n|^{-1} \left(r(\mathbf{x})f(\mathbf{x}), \chi_0^{-1}(\mathbf{x}) \right), \\ \omega_1 &= |G_n|^{-1} \left(r(\mathbf{x})f(\mathbf{x}), \chi_{i_1}^{-1}(\mathbf{x}) \right), \\ &\cdots\cdots\cdots\cdots\cdots\cdots \\ \omega_t &= |G_n|^{-1} \left(r(\mathbf{x})f(\mathbf{x}), \chi_{i_t}^{-1}(\mathbf{x}) \right). \end{aligned} \tag{14}$$

4 Invariant Operations over Generalized Logical Neurofunctions

In the study of classes of generalized logical functions $f : G_n \to H_k$, which can be realized by one GNE relative to different systems of chareacters $\chi = \{\chi_{i_1}, \chi_{i_2}, \ldots, \chi_{i_t}\}$ over the field $F = GF(p^m)$ it's important to find out transformations over those functions which retain the property their realizibility by one generalized neural element.

Let $f : G_n \to H_k$ $(f = f(x_1, \ldots, x_n), x_r = \chi_{k^{n-r}}; r = 1, \ldots, n)$, $\chi_q \in \chi(G_n)$. Obviously $q \in \{0, 1, 2, \ldots, k^n - 1\}$ and the number q can definitely be written as $q = q_1 k^{n-1} + q_2 k^{n-2} + \ldots + q_n$, where $q_1, q_2, \ldots, q_n \in \{0, 1, 2, \ldots, k - 1\}$. Let's build the vector $(q_1^{`}, \ldots, q_n^{`})$ for the number q in the next way: if $q_i \neq 0$ then $q_i^{`} = 1$; if $q_i = 0$

then $q_i^{\cdot} = 0$. For a character $\chi_q \in \chi(G_n)$ $(q \neq 0)$ we will build a set $q(\chi_q) = \bigcup_{t=1}^{n} \{q_t^{\cdot} \cdot x_t\}$. Elements of set $q(\chi_q)$ generate character χ_q.

In the future, studying the realizability of the logical function $f : G_n \to H_k$ by one GNE relative to the system of characters $\chi = \{\chi_{i_1}, \ldots, \chi_{i_t}\} \subset X(G_n)$ we will use next denotation: $f_\chi(\mathbf{g}) = f_\chi(\chi_{i_1}(\mathbf{g}), \ldots, \chi_{i_t}(\mathbf{g}))$.

The system of characters $\chi = \{\chi_{i_1}, \ldots, \chi_{i_t}\} \subset X(G_n) \backslash \{\chi_0\}$, which elements satisfy the condition: $q(\chi_{i_r}) \cap q(\chi_{i_t}) = \varnothing$, if r ≠ t, we will denote as χ_*. Let $X_*(G_n) = \chi_* \cup \{\chi_0\}$.

Theorem 4. If logical function $f : G_n \to H_k$ is realizable by one GNE related to the system of characters $\chi_* = \left\{\chi_{i_1}, \ldots, \chi_{i_j}, \ldots, \chi_{i_t}\right\}$ over the field $F = GF(p^m)$ with vector of the structure $\mathbf{w}_{\chi_*} = (\omega_1, \ldots, \omega_j, \ldots, \omega_t; \omega_0)$, then function $f_{\chi_*}^{(1)}\left(\chi_{i_1}, \ldots, \chi_{i_j}, \ldots, \chi_{i_t}\right) = f_{\chi_*}\left(\chi_{i_1}, \ldots, \xi\chi_{i_j}, \ldots, \chi_{i_t}\right)$, where $\xi \in H_k$, is also realizable by one GNE over the field F with vector of the structure $\mathbf{w}_{\chi_*}^{(1)} = (\omega_1, \ldots, \xi\omega_j, \ldots, \omega_n; \omega_0)$.

Prooving. We have, that function $f : G_n \to H_k$ is realizable by one GNE related to the system of characters $\chi_* = \left\{\chi_{i_1}, \ldots, \chi_{i_j}, \ldots, \chi_{i_t}\right\}$ over $F = GF(p^m)$ with vector of the structure $\mathbf{w}_{\gamma_*} = (\omega_1, \ldots, \omega_j, \ldots, \omega_n; \omega_0)$. Then, on the basis of Theorem 3, is exists a such function $r : G_n \to F\backslash\{0\}$, that $\left(r(\mathbf{x})f_{\chi_*}(\mathbf{x}), \chi^{-1}(\mathbf{x})\right) = 0$, for all $\chi \in X(G_n)\backslash X_*(G_n)$ and $0 \leq \deg r(\mathbf{x}) < \frac{p^m - 1}{k}$. Let $\xi \in H_k$.

Let define the function $r_1(x_1, \ldots, x_j, \ldots, x_n)$ in this way:

$$r_1(x_1, \ldots, x_j, \ldots, x_n) = r(x_1, \ldots, \xi x_j, \ldots, x_n).$$

The element $\mathbf{x}' = (x_1, \ldots, x_{j-1}, \xi x_j, x_{j+1}, \ldots, x_n)$ of the group G_n can be written in this way: $\mathbf{x}' = (1, \ldots, 1, \xi, 1, \ldots, 1) \circ (x_1, \ldots, x_{j-1}, x_j, x_{j+1}, \ldots, x_n)$, where $\circ$ - is the symbol of the operation of coordinate multiplication of vectors. Characters $\chi \in X(G_n)$ are multiplicative functions, that defined on the group G_n, to wit $\chi(\mathbf{x}') = \chi(1, \ldots, 1, \xi_j, 1, \ldots, 1)\chi(\mathbf{x})$. So,

$$\left(r(\mathbf{x})f_{\chi_*}(\mathbf{x}), \chi^{-1}(\mathbf{x})\right) = \left(r(\mathbf{x}')f_{\chi_*}(\mathbf{x}'), \chi^{-1}(\mathbf{x}')\right) = \chi^{-1}(1, \ldots, 1, \xi, 1, \ldots, 1)\left(r_1(\mathbf{x})f_{\chi_*}^{(1)}(\mathbf{x}), \chi^{-1}(\mathbf{x})\right) = 0$$

for all $\chi \notin X_*(G_n)$.

Then, given that $\forall \mathbf{x} \in G_n$ $0 \leq \deg r_1(\mathbf{x}) < \frac{p^m - 1}{k}$, the function $f_{\chi_*}^{(1)}$ is realizable by one GNE related to $\chi_* = \left\{\chi_{i_1}, \ldots, \chi_{i_j}, \ldots, \chi_{i_t}\right\}$ over the field F.

Given, that $\mathbf{w}_{\chi_*} = (\omega_1, \ldots, \omega_j, \ldots, \omega_t; \omega_0)$ – is vector of structure of GNE related to the system of characters $\chi_* = \left\{\chi_{i_1}, \ldots, \chi_{i_j}, \ldots, \chi_{i_t}\right\}$ over the field $F = GF(p^m)$, that realize function $f(x_1, \ldots, x_j, \ldots, x_n)$, and $\mathbf{w}' = \left(\omega_1', \ldots, \omega_j', \ldots, \omega_n'; \omega_0'\right)$ is vector of structure of GNE, related to the system χ_*, that realize function $f_{\chi_*}^{(1)}(x_1,\ldots,x_j,\ldots,x_n)$. Then, based on (14) and $q(\chi_{i_r})\cap q(\chi_{i_s})=\varnothing$, *if* r ≠ s, we have:

$$\omega_0' = |G_n|^{-1}\left(r_1(\mathbf{x})f_{\chi_*}^{(1)}(\mathbf{x}),\chi_0^{-1}(\mathbf{x})\right) = |G_n|^{-1}\left(r(\mathbf{x}')f_{\chi_*}(\mathbf{x}'),\chi_0^{-1}(\mathbf{x}')\right) = \omega_0$$

if $s \neq j$, then

$$\omega_s' = |G_n|^{-1}\left(r_1(\mathbf{x})f_{\chi_*}^{(1)}(\mathbf{x}),\chi_{i_s}^{-1}(\mathbf{x})\right) = |G_n|^{-1}\left(r(\mathbf{x}')f_{\chi_*}(\mathbf{x}'),\chi_{i_s}^{-1}(\mathbf{x}')\right) = \omega_s$$

and

$$\omega_j' = |G_n|^{-1}\left(r_1(\mathbf{x})f_{\chi_*}^{(1)}(\mathbf{x}),\chi_{i_j}^{-1}(\mathbf{x})\right) = |G_n|^{-1}\xi\left(r(\mathbf{x}')f_{\chi_*}(\mathbf{x}'),\chi_{i_j}^{-1}(\mathbf{x}')\right) = \xi\omega_j.$$

So, GNE related to the system of characters $\chi_* = \left\{\chi_{i_1},\ldots,\chi_{i_j},\ldots,\chi_{i_t}\right\}$ with vector of structure $\mathbf{w}_{\chi_*}^{(1)} = \left(\omega_1,\ldots,\xi\omega_j,\ldots,\omega_n;\omega_0\right)$ is realize function $f_{\chi_*}^{(1)}$. The theorem is proved.

Theorem 5. If logical function $f: G_n \to H_k$ is realizable by one GNE related to the system of characters $\chi_* = \left\{\chi_{i_1},\ldots,\chi_{i_j},\ldots,\chi_{i_r},\ldots,\chi_{i_t}\right\}$ over the field $F = GF(p^m)$ with vector of the structure $\mathbf{w}_{\chi_*} = \left(\omega_1,\ldots,\omega_j,\ldots,\omega_r,\ldots,\omega_t;\omega_0\right)$ is implementable by one GNE related to the system of characters, then function $f_{\chi_*}^{(2)}\left(\chi_{i_1},\ldots,\chi_{i_j},\ldots,\chi_{i_r},\ldots,\chi_{i_t}\right) = f_{\chi_*}\left(\chi_{i_1},\ldots,\chi_{i_r},\ldots,\chi_{i_j},\ldots,\chi_{i_t}\right)$ is also realizable by one GNE with vector of the structure $\mathbf{w}_{\chi_*}^{(2)} = (\omega_1,\ldots,\omega_r,\ldots,\omega_j,\ldots,\omega_t;\omega_0)$.

Prooving. We have, that the function $f: G_n \to H_k$ is realizable by one GNE related the system of characters $\chi_* = \left\{\chi_{i_1},\ldots,\chi_{i_j},\ldots,\chi_{i_r},\ldots,\chi_{i_t}\right\}$ over the field $F = GF(p^m)$ with vector of the structure $\mathbf{w}_{\chi_*} = \left(\omega_1,\ldots,\omega_j,\ldots,\omega_r,\ldots,\omega_t;\omega_0\right)$. Then, based on the Theorem 3, is exists such function $r: G_n \to F\backslash\{0\}$, that

$$0 \leq \deg r(\mathbf{x}) < \frac{p^m - 1}{k} \tag{15}$$

and

$$\left(r(\mathbf{x})f_{\chi_*}(\mathbf{x}),\chi^{-1}(\mathbf{x})\right) = 0 \tag{16}$$

for all $\chi \in X(G_n)\backslash X_*(G_n)$. Let $\mathbf{x} = \left(x_1,\ldots,x_i,\ldots,x_j,\ldots,x_n\right)$, $\mathbf{x}' = \left(x_1,\ldots,x_j,\ldots,x_i,\ldots,x_n\right) \in G_n$, $r_2(\mathbf{x}) = r(\mathbf{x}')$ i $\tilde{\chi}(\mathbf{x}') = \chi(\mathbf{x})$. It is obviously that $\max_{\mathbf{x}\in G_n}\{\deg r_2(\mathbf{x})\} = \max_{\mathbf{x}\in G_n}\{\deg r(\mathbf{x})\}$.

A function $r_2(\mathbf{x})$ satisfies inequality (15) and takes values from the set $F\backslash\{0\}$. We have that $q(\chi_{i_r})\cap q(\chi_{i_j})=\varnothing$, *if* r ≠ j, so, the system of Eq. (16) can be written by such way:

$$\left(r_2(\mathbf{x})f_{\chi_*}^{(2)}(\mathbf{x}),\chi^{-1}(\mathbf{x})\right)=\left(r(\mathbf{x}')f_{\chi_*}(\mathbf{x}'),\tilde{\chi}^{-1}(\mathbf{x}')\right)=0,$$

for all $\chi\in X(G_n)\backslash X_*(G_n)$. So, based on inequality $0\leq\deg r_2(\mathbf{x})<\frac{p^m-1}{k}$, we have realizable of the function $f_{\chi_*}^{(2)}(\mathbf{x})$ by one GNE related to the system of characters $\chi_*=\left\{\chi_{i_1},\ldots,\chi_{i_j},\ldots,\chi_{i_r},\ldots,\chi_{i_t}\right\}$. Let $\mathbf{w}=\left(\omega_1,\ldots,\omega_j,\ldots,\omega_r,\ldots,\omega_n;\omega_0\right)$ is the vector of structure GNE, that realize the function $f(\mathbf{x})$. We denote $\mathbf{w}_{\chi_*}^{(2)}=\left(\omega_1',\ldots,\omega_j',\ldots,\omega_r',\ldots,\omega_n';\omega_0'\right)$ - such vector of the structure GNE related to the system of characters $\chi_*=\left\{\chi_{i_1},\ldots,\chi_{i_j},\ldots,\chi_{i_r},\ldots,\chi_{i_t}\right\}$, that realize the function $f_{\chi_*}^{(2)}(\mathbf{x})$. So, based on (14) we have:

$$\omega_0'=|G_n|^{-1}\left(r_2(\mathbf{x})f_{\chi_*}^{(2)}(\mathbf{x}),\chi_0^{-1}(\mathbf{x})\right)=|G_n|^{-1}\left(r(\mathbf{x}')f_{\chi_*}(\mathbf{x}'),\chi_0^{-1}(\mathbf{x}')\right)=\omega_0.$$

If $s\neq r$ and $s\neq j$, then

$$\omega_s'=|G_n|^{-1}\left(r_2(\mathbf{x})f_{\chi_*}^{(2)}(\mathbf{x}),\chi_{i_s}^{-1}(\mathbf{x})\right)=|G_n|^{-1}\left(r(\mathbf{x}')f_{\chi_*}(\mathbf{x}'),\chi_{i_s}^{-1}(\mathbf{x}')\right)=\omega_s.$$

If $s=r$, then

$$\omega_r'=|G_n|^{-1}\left(r_2(\mathbf{x})f_{\chi_*}^{(2)}(\mathbf{x}),\chi_{i_r}^{-1}(\mathbf{x})\right)=|G_n|^{-1}\left(r(\mathbf{x}')f_{\chi_*}(\mathbf{x}'),\tilde{\chi}_{i_j}^{-1}(\mathbf{x}')\right)=\omega_j,$$

and, $\omega_j'=\omega_r$. So, the function $f_{\chi_*}^{(2)}(\mathbf{x})$ is realizable by one GNE related the system of characters $\chi_*=\left\{\chi_{i_1},\ldots,\chi_{i_j},\ldots,\chi_{i_r},\ldots,\chi_{i_t}\right\}$ over the field $F=GF(p^m)$ with vector of the structure $\mathbf{w}_{\chi_*}^{(2)}=\left(\omega_1,\ldots,\omega_r,\ldots,\omega_j,\ldots,\omega_n;\omega_0\right)$. The theorem is proved.

Theorem 6. If logical function $f:G_n\to H_k$ is realizable by one GNE related to the system of characters $\chi_*=\left\{\chi_{i_1},\ldots,\chi_{i_t}\right\}$ over the field $F=GF(p^m)$ with vector of the structure $\mathbf{w}_{\chi_*}=(\omega_1,\ldots,\omega_t;\omega_0)$, then function $f_{\chi_*}^{(3)}\left(\chi_{i_1},\ldots,\chi_{i_t}\right)=\xi f_{\chi_*}\left(\xi^{-1}\chi_{i_1},\ldots,\xi^{-1}\chi_{i_t}\right)$, where $\xi\in H_k$, is also realizable by one GNE over the field F with vector of the structure $\mathbf{w}_{\chi_*}^{(3)}=(\omega_1,\ldots,\omega_n;\xi\cdot\omega_0)$.

Prooving. First, we show that the theorem is true for the function $f_{\chi_*}^{\prime(3)}\left(\chi_{i_1},\ldots,\chi_{i_t}\right)=\xi f\left(\chi_{i_1},\ldots,\chi_{i_t}\right)$. Let the function $f:G_n\to H_k$ is realized by one GNE related the system of characters $\chi_*=\left\{\chi_{i_1},\ldots,\chi_{i_t}\right\}$ over the field $F=GF(p^m)$ with vector of the structure $\mathbf{w}_{\chi_*}=(\omega_1,\ldots,\omega_t;\omega_0)$ and the function $r(\mathbf{x})$ is satisfied conditions of the Theorem 3. If we put that $r_3(\mathbf{x})=r(\mathbf{x})$, so we have:

$$\left(r_3(\mathbf{x})f'^{(3)}_{\chi_*}(\mathbf{x}), \chi^{-1}(\mathbf{x})\right) = \xi\left(r(\mathbf{x})f_{\chi_*}(\mathbf{x}), \chi^{-1}(\mathbf{x})\right) = 0,$$

for all $\chi \in X(G_n)\backslash X_*(G_n)$. So $f'^{(3)}_{\chi_*}(\mathbf{x})$ is a neurofunction over F. Next, we are going to finding the vector of the structure $\mathbf{w}'^{(3)}_{\chi_*} = (\omega'_1, \ldots, \omega'_n; \omega'_0)$ GNE, that is realize the function $f'^{(3)}_{\chi_*}(\mathbf{x})$. Based on (14) we have:

$$\omega'_0 = |G_n|^{-1}\left(r_3(\mathbf{x})f^{(3)}_{\chi_*}(\mathbf{x}), \chi_0^{-1}(\mathbf{x})\right) = |G_n|^{-1}\cdot\xi\left(r(\mathbf{x})f(\mathbf{x}), \chi_0^{-1}(\mathbf{x})\right) = \xi\omega_0,$$

and for $1 \leq j \leq n$

$$\omega'_j = |G_n|^{-1}\left(r_3(\mathbf{x})f'^{(3)}_{\chi_*}(\mathbf{x}), \chi_{i_j}^{-1}(\mathbf{x})\right) = |G_n|^{-1}\cdot\xi\left(r(\mathbf{x})f_{\chi_*}(\mathbf{x}), \chi_{i_j}^{-1}(\mathbf{x})\right) = \xi\omega_j,$$

So, the function $f'^{(3)}_{\chi_*}(\mathbf{x})$ is realized by one GNE related the system of characters $\chi_* = \{\chi_{i_1}, \ldots, \chi_{i_t}\}$ over the field F with vector of the structure $\mathbf{w}'^{(3)}_{\chi_*} = \xi(\omega_1, \ldots, \omega_t; \omega_0)$. From this and from the Theorem 4 is following, that the function $f^{(3)}_{\chi_*}(\chi_{i_1}, \ldots, \chi_{i_t}) = \xi f_{\chi_*}(\xi^{-1}\chi_{i_1}, \ldots, \xi^{-1}\chi_{i_t})$ is realized by one GNE over the field F with vector of the structure $\mathbf{w}^{(3)}_{\chi_*} = (\omega_1, \ldots, \omega_n; \xi \cdot \omega_0)$. The theorem is proved.

Theorem 7. If logical function $f : G_n \to H_k$ is realizable by one GNE related to the system of characters $\chi_* = \left\{\chi_{i_1}, \ldots, \chi_{i_j}, \ldots, \chi_{i_t}\right\}$ over the field $F = GF(p^m)$ with vector of the structure $\mathbf{w}_{\chi_*} = (\omega_1, \ldots, \omega_{j-1}, \omega_j, \omega_{j+1}, \ldots, \omega_t; \omega_0)$, then function $f^{(4)}_{\chi_*}(\chi_{i_1}, \ldots, \chi_{i_{j-1}}, \chi_{i_j}, \chi_{i_{j+1}}, \ldots, \chi_{i_t}) = \chi_{i_j} f\left(\chi_{i_j}^{-1}\chi_{i_1}, \ldots, \chi_{i_j}^{-1}\chi_{i_{j-1}}, \chi_{i_j}^{-1}, \chi_{i_j}^{-1}\chi_{i_{j+1}}, \ldots, \chi_{i_j}^{-1}\chi_{i_t}\right)$, is also realizable by one GNE over the field F with vector of the structure $\mathbf{w}^{(4)}_{\chi_*} = (\omega_1, \ldots, \omega_{j-1}, \cdot\omega_0, .\omega_{j+1}, \ldots, \omega_t; \omega_j)$.

Proving. Let $\mathbf{x} = (x_1, \ldots, x_n) \in G_n$ and $\mathbf{x}^j = (\chi_{i_j}^{-1}(\mathbf{x})\chi_{i_1}(\mathbf{x}), \ldots, \chi_{i_j}^{-1}(\mathbf{x})\chi_{i_{j-1}}(\mathbf{x}), \chi_{i_j}^{-1}(\mathbf{x}), \chi_{i_j}^{-1}(\mathbf{x})\chi_{i_{j+1}}(\mathbf{x}), \ldots, \chi_{i_j}^{-1}(\mathbf{x})\chi_{i_n}(\mathbf{x}))$ i $r_4(x) = r(x^j)$. From that the function $f : G_n \to H_k$ is realized by one GNE related the system of characters $\chi_* = \left\{\chi_{i_1}, \ldots, \chi_{i_j}, \ldots, \chi_{i_t}\right\}$ over the field $F = GF(p^m)$ and that the multiplication of two characters of the group G_n also is a character, we have:

$$\begin{aligned}(r_4(\mathbf{x})f^{(4)}_{\chi_*}(\mathbf{x}), \chi^{-1}(\mathbf{x})) &= (r(\mathbf{x}^j)\chi_{i_j}(\mathbf{x})f_{\chi_*}(\mathbf{x}^j), \chi^{-1}(\mathbf{x})) \\ = (r(\mathbf{x}^j)f_{\chi_*}(\mathbf{x}^j), \chi_{i_j}^{-1}(\mathbf{x})\chi^{-1}(\mathbf{x})) &= (r(\mathbf{x}^j)f_{\chi_*}(\mathbf{x}^j), \chi^{-1}(\mathbf{x}^j)) = 0\end{aligned}$$

for all $\chi \in X(G_n)\backslash X_*(G_n)$.

Taking into account, that $\max\deg r_4(\mathbf{x})\,(\min\deg r_4(\mathbf{x}))$ and $\max\deg r(\mathbf{x})$ $(\min\deg r(\mathbf{x}))$ in the group G_n are the same, is following that the function $f_{\chi_*}^{(4)}$ is realizable by one GNE related to the system of characters $\chi_* = \left\{\chi_{i_1},\ldots,\chi_{i_j},\ldots,\chi_{i_t}\right\}$ over the field $F = GF(p^m)$.

We have, that GNE with vector of the structure $\mathbf{w}_{\chi_*} = (\omega_1,\ldots,\omega_{j-1},\omega_j,$ $\omega_{j+1},\ldots,\omega_t;\omega_0)$ is realize the function $f_{\chi_*}(\mathbf{x})$. Let the GNE with vector of the structure $\mathbf{w}'_{\chi_*} = (\omega'_1,\ldots,\omega'_{j-1},\omega'_j,\omega'_{j+1},\ldots,\omega'_t;\omega'_0)$ is realizes the function $f_{\chi_*}^{(4)}(\mathbf{x})$. Express the coordinates of the vector $\mathbf{w}'_{\chi_*}$ with the coordinates of the vector $\mathbf{w}_{\chi_*}$

$$\begin{aligned}
\omega'_0 &= |G_n|^{-1}\left(r_4(\mathbf{x})f_{\chi_*}^{(4)}(\mathbf{x}),\chi_0^{-1}(\mathbf{x})\right) = |G_n|^{-1}\left(r(\mathbf{x}^j)\chi_{i_j}(\mathbf{x})f_{\chi_*}(\mathbf{x}^j),\chi_0^{-1}(\mathbf{x})\right)\\
&= |G_n|^{-1}\left(r(\mathbf{x}^j)f_{\chi_*}(\mathbf{x}^j),\chi_{i_j}^{-1}(\mathbf{x}^j)\right) = \omega_j,\\
\omega'_j &= |G_n|^{-1}\left(r_4(\mathbf{x})f_{\chi_*}^{(4)}(\mathbf{x}),\chi_{i_j}^{-1}(\mathbf{x})\right) = |G_n|^{-1}\left(r(\mathbf{x}^j)\chi_{i_j}(\mathbf{x})f_{\chi_*}(\mathbf{x}^j),\chi_{i_j}^{-1}(\mathbf{x})\right)\\
&= |G_n|^{-1}(r(\mathbf{x}^j)f_{\chi_*}(\mathbf{x}^j),\chi_0^{-1}(\mathbf{x}^j)) = \omega_0,
\end{aligned}$$

If $r \neq 0$ and $r \neq j$, then

$$\begin{aligned}
\omega'_r &= |G_n|^{-1}\left(r_4(\mathbf{x})f_{\chi_*}^{(4)}(\mathbf{x}),\chi_{i_r}^{-1}(\mathbf{x})\right) = |G_n|^{-1}\left(r(\mathbf{x}^j)\chi_{i_j}(\mathbf{x})f_{\chi_*}(\mathbf{x}^j),\chi_{i_r}^{-1}(\mathbf{x})\right)\\
&= |G_n|^{-1}\left(r(\mathbf{x}^j)\chi_{i_j}(\mathbf{x})f_{\chi_*}(\mathbf{x}^j),\chi_{i_r}^{-1}(\mathbf{x})\right) = |G_n|^{-1}\left(r(\mathbf{x}^j)f_{\chi_*}(\mathbf{x}^j),\chi_{i_r}^{-1}(\mathbf{x})\chi_{i_j}^{-1}(\mathbf{x})\right)\\
&= |G_n|^{-1}\left(r(\mathbf{x}^j)f_{\chi_*}(\mathbf{x}^j),\chi_{i_r}^{-1}(\mathbf{x}^j)\right) = \omega_r.
\end{aligned}$$

So, the function $f_{\chi_*}^{(4)}\left(\chi_{i_1},\ldots,\chi_{i_{j-1}},\chi_{i_j},\chi_{i_{j+1}},\ldots,\chi_{i_t}\right)$ is realizable by one GNE related the system of characters $\chi_* = \left\{\chi_{i_1},\ldots,\chi_{i_j},\ldots,\chi_{i_t}\right\}$ over the field $F = GF(p^m)$ with vector of structure $\mathbf{w}_{\chi_*}^{(4)} = \left(\omega_1,\ldots,\omega_{j-1},\omega_0,\omega_{j+1},\ldots,\omega_t;\omega_j\right)$. The theorem is proved.

5 Summary and Conclusion

In the paper were shown criteria's of realizability logical functions by one generalized neural element relative to the given system of characters and developed an effective method of synthesis such neural elements over the Galois field. Was described invariant operations over logical functions, which are realizable relative to some systems of characters.

References

1. Berlekemp, E.: Algebraic Coding Theory, p. 477. Mir, Moscow (1971)
2. Kuzmin, I.V., Kedrus, V.A.: Fundamentals of Information Theory and Coding, p. 278. Vischa shkola, Kiev (1977)

3. Clark, J., Kane, J.: Coding with Error Correction in Digital Communication Systems, p. 391. Radio i sviaz, Moscow (1987)
4. Geche, F.E.: Neural elements over finite fields. Inf. Technol. Syst. **1**(1/2), 100–104 (1998)
5. Geche, F.E., Batyuk, A.Y., Buchok, V.Y.: Invariant operations on discrete neural functions over galois field. In: IEEE Firs International Conference on Data Stream Mining & Processing (DSMP), Lviv, Ukraine, pp. 112–116, 23–27 August 2016
6. Curtis, C., Rainer, I.: Theory of Representations of Finite Groups and Associative Algebras, p. 667. Nauka, Moscow (1969)
7. Warden, V.: Algebra, p. 623. Nauka, Moscow (1979)
8. Zalmanzon, L.A.: The Fourier, Walsh, Haar transformation and their application, p. 493. Nauka, Moscow (1989)
9. Labunets, V.G., Sitnikov, O.P.: Harmonic analysis of Boolean functions and functions of valued logic over finite fields. Izv. Academy of Sciences of the USSR. Ser.: Technical cybernetics, Moscow, no. 1, pp. 141–148 (1975)
10. Yaroslavsky, L.P.: Introduction to Digital Image Processing, p. 312. Sovetskoe radio, Moscow (1979)
11. Geche, F., Mulesa, O., Voloshchuk, V., Batyuk, A.: Generalized logical neural functions over the galois field and their properties. In: Proceedings of International Scientific Conference Computer Sciences and Information Technologies (CSIT-2019), vol. 1, pp. 21–24 (2019)

Suitable Site Selection Using Two-Stage GIS-Based Fuzzy Multi-criteria Decision Analysis

Svitlana Kuznichenko[1(✉)], Iryna Buchynska[1], Ludmila Kovalenko[1], and Yurii Gunchenko[2]

[1] Odessa State Environmental University, Odessa, Ukraine
skuznichenko@gmail.com, buchinskayaira@gmail.com, l.b.kovalenko@ukr.net
[2] Odessa I.I.Mechnikov National University, Odessa, Ukraine
7996445@gmail.com

Abstract. The paper proposes a methodology for two-stage fuzzy multi-criteria decision analysis in a raster-based geographical information system (GIS) to determine the suitable locations for territorial objects. Recommendations about the stages of choosing alternatives for spatial and non-spatial constraints are given. It is shown that the fuzzyfication of criteria, that is, the conversion of their attribute values into a fuzzy set, based on expert evaluation of a fuzzy membership function, allows screening alternatives by determining thresholds of alpha-cut of fuzzy sets for each criterion, followed by combining criteria attributes using any aggregation operators: minimum, maximum, weighted sum, OWA operator Jager. Adding to the procedure of multicriteria analysis of the additional stage of filtration of alternatives gives the opportunity to reduce the number of alternatives, and in the future and the processing time of the criteria layers by aggregator operators. The proposed algorithm for screening alternatives can be performed in a GIS environment using Fuzzy Membership, Overlay and raster calculators tools.

Keywords: Geographic information system · Multiple-Criteria decision analysis · Fuzzy set theory · Alpha-cut · Site selection

1 Introduction

Spatial problems, in particular the problem of suitable site selection, according to their nature, are always multi-criteria [1] and require taking into consideration of the number of economic, ecological, social and other factors, which allow to assess the suitability of the territory.

To solve the facility location problem, various combinatorial methods, methods using network models, numerical methods, simulation modeling, etc. are often used [2–4]. The presence of spatial factors determines the use of methods based on GIS technologies [5]. GIS capabilities for generating alternatives and choosing the best solution are usually based on surface analysis, proximity analysis, and overlay analysis [6, 7].

N. Shakhovska and M. O. Medykovskyy (Eds.): CCSIT 2019, AISC 1080, pp. 214–230, 2020.
https://doi.org/10.1007/978-3-030-33695-0_16

The disadvantages of most of the aforementioned methods are the requirement for crisp information, but in practice the problems of suitable site selection are poorly structured [8], that is, those requiring the use of unformalized (fuzzy) knowledge based on expert experience. Therefore, algorithms for solving the suitable location problem using fuzzy information on the basis of GIS technologies represent practical and theoretical interest. A promising approach that allows the most adequate description of this process is the mathematical apparatus of the theory of fuzzy sets [9–13].

2 The Proposed Fuzzy Multi-criteria Decision Analysis Approach in GIS

For the last several decades, GIS has been used in conjunction with other systems and methods, such as Decision Support Systems (DSS) and Multicriteria Decision Analysis (MCDA) [14–17]. The combination of GIS and MCDA tools gives a synergistic effect and helps to increase the efficiency and quality of spatial analysis when selecting the optimal location of objects. At an elementary level, the combination of GIS-MCDA can be considered as a process that converts and combines geographic data and assessment judgments, that is, the benefits of the decision maker (DM) to obtain information for decision-making.

Let us consider a formal description of the procedure for multi-criteria decision analysis in a geographic context. Selection of suitable places is carried out by spatial analysis using GIS, based on criteria that take into account various factors of influence: nature protection requirements, features of terrain, landscape morphology, socio-economic factors, etc. To do this, perform the procedure of decomposing a set of objects that belong to the investigated territory and affect the decision making, after which a map K is received, which is a set of thematic layers of criteria K_i [18]:

$$K = \{K_i\}, \quad i = \overline{1,n}. \tag{1}$$

Schematically, the process of decomposition of the set of objects O on thematic layers of criteria is shown in Fig. 1.

For spatial modeling in order to select a suitable spatial object location, we will use a raster data model. Therefore, all the received thematic layers of objects should be presented as a set of cells (pixels) in a raster model of GIS, which has the form of a two-dimensional discrete rectangular grid with mx × my cells, where $\Delta x = \Delta y = \Delta r$ – cell sizes.

Each cell is an alternative, which is described by its spatial data (geographic coordinates) and attributive data (criteria values). We will write the set of alternatives A, which are evaluated according to the criteria C_j:

$$A = \left\{a_{ij} \middle| \, i = \overline{1,m}, \quad j = \overline{1,n}\right\} \tag{2}$$

where a_{ij} is the value of the attribute of the alternative, that is, the value of the attribute according to the j-th criterion and the i-th alternative; n is the inumber of criteria; $m = m_x{\cdot}m_y$ is the number of alternatives.

It is important to choose such a procedure for rasterize vector layer criteria, which will get a set of cells whose attributes bear the content information about the value of the function of the effect of the objects of the layer. For example, attributes can be derived from vector maps that contain point objects of observation points by the value of some factor using different methods of interpolation. Often, to obtain spatial relationships between objects, different distance metrics are used: Euclidean, Manhattan, Chebyshev metrics, etc.

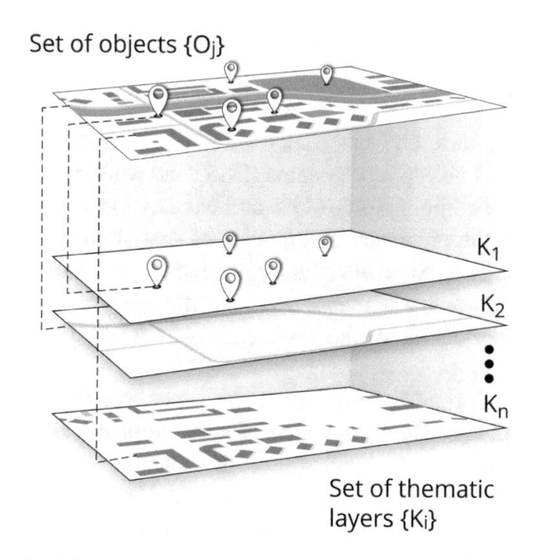

Fig. 1. Scheme of decomposition of object in thematic layers.

DM's preferences to evaluation the criterion are determined by assigning the weight of the criteria w_j, where $j = 1, 2,\ldots, n$. We will assume that the DM's preferences are spatially homogeneous, that is, each criterion C_k is assigned one weight w_k. Thus, the matrix of decision-making will have the form shown in Table 1 [16].

Table 1. Matrix of decision making.

Alternatives	Spatial coordinates		Criteria/attributes C_j			
A_i	X_i	Y_i	C_1	C_2	...	C_n
A_1	x_1	y_1	a_{11}	a_{12}	...	a_{1n}
A_2	x_2	y_2	a_{21}	a_{22}	...	a_{2n}
A_3	x_3	y_3	a_{31}	a_{32}	...	a_{3n}
...	...	...	...	...	...	...
A_m	x_m	y_m	a_{m1}	a_{m2}	...	a_{mn}
Weight, w_j			w_1	w_2	...	w_n

To select suitable sites locations of objects, it is advisable to apply a procedure consisting of two phases: macroanalysis (site screening) and microanalysis (site evaluation) [19]. The two-stage selection approach suggests that for those alternatives that were tested in the first stage for compliance with minimum requirements, in the second stage a more detailed analysis by the MCDA methods is carried out. Preliminary screening of alternatives can be made taking into account restrictions: for attribute values (non-spatial constraints) or for location (spatial constraints). Constraints may be represented by raster layers in which attributes of cells with ineligible alternatives have a value of 0, and with acceptable alternatives – value 1. Using the constraint layer as a conjunctive filter, one can determine the set of possible alternatives.

The general diagram of the GIS-MCDA site selection process is shown in Fig. 2.

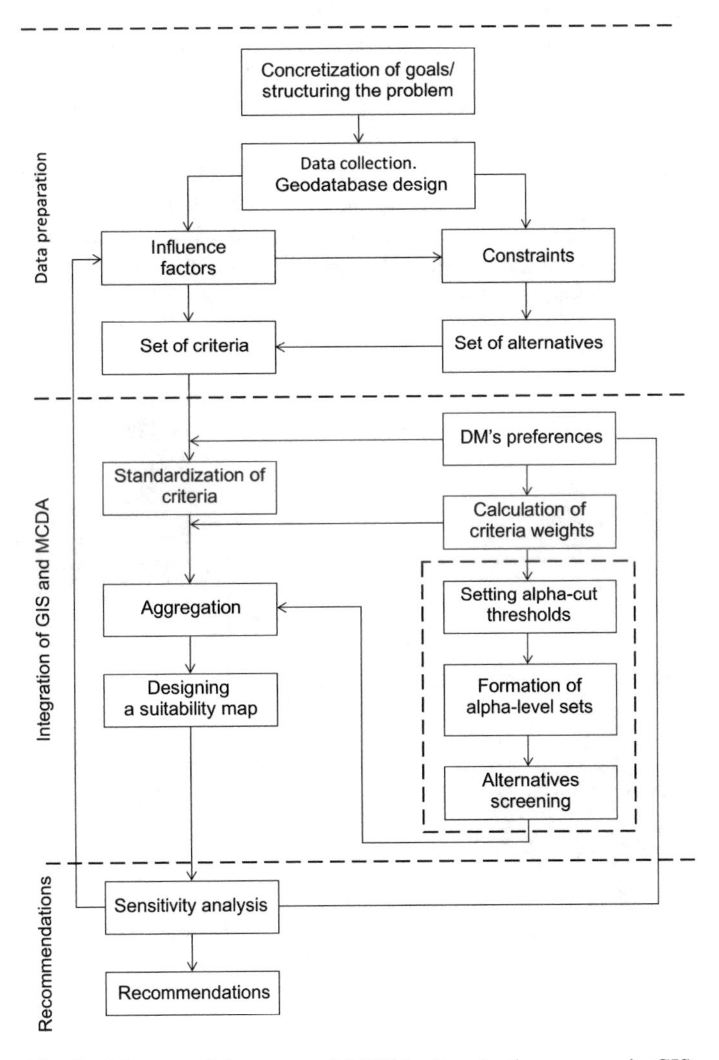

Fig. 2. Diagram of the proposed MCDA site selection process in GIS.

2.1 Standardization of Criteria

Criterion layers generally has different ranges or scale values of attributes, so they require the transformation to comparable units. In addition, the criteria can be not only quantitative but also qualitative. Procedures for transforming raw data to comparable units are called scaling methods or standardization. The standardization procedure allows you to scale the attributes in the scale [0, 1]. The approach to scaling attributes based on fuzzy logic methods is based on the transformation of the values of the j-th layer attributes in the degree of membership to the fuzzy set $B_j \subseteq A$:

$$B_j = \left\{ \left(a, \mu_b^j(a) \right) | a \in A \right\}, \quad \mu_b^j(a) : a \to [0, 1] \tag{3}$$

where a is the attribute value, A is the attributes values set.

Membership function $\mu_b(a)$ specifies the degree of membership of the attribute a to the fuzzy set B_j. The bigger the value $\mu_b(a)$, to a greater extent the attribute corresponds to the properties of the fuzzy set. As a rule, membership function is built with the participation of an expert or a group of experts.

Let us take a look on example of implementation of the proposed fuzzy spatial information processing model on an example of fuzzification of a vector layer that contains objects of a traffic network on the southern territory of the Odessa Region. Figure 3 shows phases of the fuzzification process. The vector layer with linear objects of the traffic network was transformed into a raster layer of the Euclidean distances. For

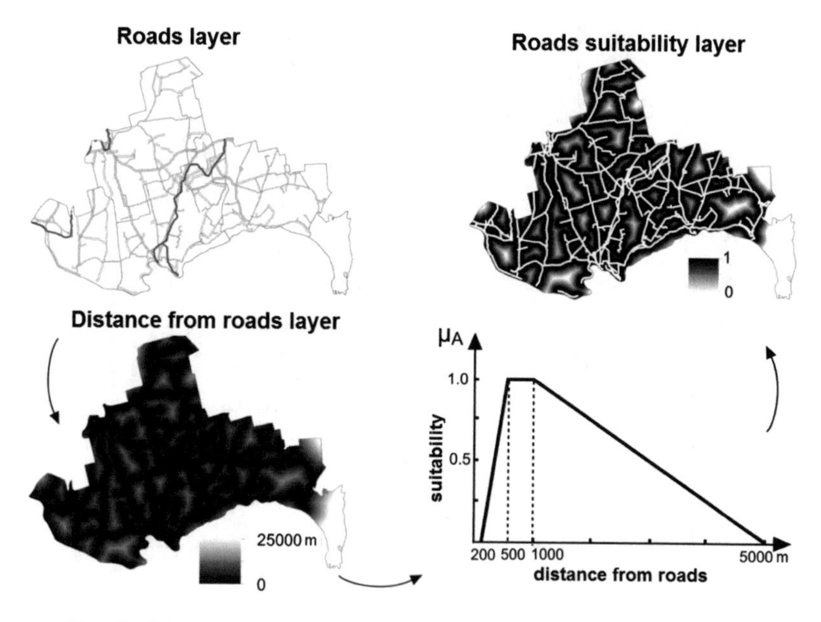

Fig. 3. Scheme of the fuzzification process of distance from road layer.

this purpose the tool of geo-processing of ArcGIS ArcToolbox Euclidean Distance data set was applied.

The raster that considers the degree of appropriateness of each cell under the "Distance from rail and motor roads" criteria was constructed in accordance with the proposed by experts piecewise-linear trapezoid-shaped membership function $\mu_b(a)$, which was defined in terms of an appropriate analytical expression. The Fuzzy Membership tool can be used to construct the raster of appropriateness, but if the applied membership function is not included to the set of this tool, Reclassify and Divide reclassification tools may be used.

2.2 Calculation of Criteria Weights

Use of GIS-MCDA results in assigning criteria weights to indicate their relative importance. In the case of n criteria, the multitude of weights is determined as follows:

$$W = \left\{ w_i \middle| \sum w = 1,\ i = \overline{1, n} \right\} \tag{4}$$

The simplest way to evaluate the importance of the criteria is ranking method, i.e. the ranking of the criteria by the expert in order of importance (the most important – 1, the second important – 2, etc.). Once the rating is established, the weights of the criteria are calculated according to the equation [16]:

$$w_i = \frac{n - r_i + 1}{\sum_{i=1}^{n} (n - r_i + 1)} \tag{5}$$

where w_i is normalized weight for the *i*th criteria, n is the amount of criteria ($j = 1, 2, \ldots, n$), and r_i is the rank position of the criteria.

The criteria weights can be found by direct assessment by experts on the basis of a certain scale, for example, from 0 to 100. In this case, the standard weight of the criteria is calculated as follows [16]:

$$w_i = \frac{w'_i}{\sum w'} \tag{5}$$

where w_i is normalized weight for the *i*th criteria, and w'_i is the assessed value for the *i*th criteria.

Fuzzy AHP technique. For calculation of the normalized weight of the criteria multicriterial technique AHP (Analytic hierarchy process) [20–24] is often used, which is based on a pair-wise comparison. Using the pair-wise comparison method (PCM), you can compare the criteria with each other and calculate their relative importance for the goal. The result is a pair-wise comparison matrix based on the formula (6).

$$A = [r_{ij}] = \begin{bmatrix} 1 & r_{12} & \dots & r_{1n} \\ 1/r_{12} & 1 & \dots & r_{2n} \\ \dots & \dots & \dots & \dots \\ 1/r_{1n} & 1/r_{2n} & \dots & 1 \end{bmatrix} \quad (6)$$

where r_{ij} are numbers that represent the relative importance of the i-th element in comparison with the j-th in relation to the goal.

If, according to some criteria, it is possible to obtain objective quantitative estimates of elements, then the relation of these estimates is taken as a priority. When evaluating criteria on the basis of subjective judgments of experts, the 9-point scale of relative importance Saaty [20] is used (Table 2).

Table 2. The scale of relative importance Saaty and corresponding fuzzy number.

Saaty scale	Definition	Unfuzzy triangular scale
1	no benefit	(1, 1, 1)
3	weak benefit	(2, 3, 4)
5	essential benefit	(4, 5, 6)
7	clear benefit	(6, 7, 8)
9	absolute benefit	(9, 9, 9)
2	intermediate values between adjacent values of the scale	(1, 2, 3)
4		(3, 4, 5)
6		(5, 6, 7)
8		(7, 8, 9)

At the next stage, there are eigenvalues and eigenvector of the matrix and a vector of local priorities is formed.

To control the consistency of expert assessments, two related characteristics are introduced - the Consistency Index (C.I.) and the Consistency Ratio (C.R.):

$$C.I. = \frac{\lambda_{\max} - n}{n - 1} \quad (7)$$

where n is the number of criteria and λ_{max} is the biggest eigenvalue.

$$C.R. = \frac{C.I.}{R.I.} \quad (8)$$

where R.I. is the Random Inconsistency index that is dependent on the sample size (Table 3).

A reasonable level of consistency in the pair-wise comparisons is assumed if C. R. < 0.10, while C.R. $\geq$ 0.10 indicates inconsistent judgments.

Table 3. Value of random index (R.I.) depending on the rank of the matrix.

n	1	2	3	4	5	6	7	8	9	10
R.I.	0.00	0.00	0.52	0.89	1.11	1.25	1.35	1.40	1.45	1.49

The AHP method may only be used for mutually independent criteria. The disadvantages of the method include the possibility of processing only crisp assessed values of experts, which complicates its use in solving issues that are characterized by uncertainty and insufficiency of information.

In [25–27], modified fuzzy version FAHP was proposed. In this method, paired comparisons of criteria are carried out in terms of linguistic variables represented by triangular numbers (see Table 2).

During the first stage, the expert transforms a crisp matrix of pair comparisons A (after checking the consistency of the assessed values, C.R. < 0.10) into a fuzzy matrix $\tilde{A}$ using a scale with triangular fuzzy numbers:

$$\tilde{A} = \begin{bmatrix} \tilde{a}_{11} & \tilde{a}_{12} & \cdots & \tilde{a}_{1n} \\ \tilde{a}_{21} & \tilde{a}_{22} & \cdots & \tilde{a}_{2n} \\ \cdots & \cdots & \cdots & \cdots \\ \tilde{a}_{n1} & \tilde{a}_{n2} & \cdots & \tilde{a}_{nn} \end{bmatrix} \tag{9}$$

where $\tilde{a}_{ij}$ is the result of comparing the ith criteria to the jth criteria, expressed in terms of the fuzzy triangular scale.

The fuzzy weights of each criterion can be found using vector summation according to the equation:

$$\tilde{w}_i = \tilde{r}_i \otimes (\tilde{r}_1 \oplus \tilde{r}_2 \oplus \ldots \oplus \tilde{r}_n)^{-1} = (lw_i, mw_i, uw_i) \tag{10}$$

The geometric mean of fuzzy comparison values of each criterion is calculated as [25]:

$$\tilde{r}_i = \left(\prod_{j=1}^{n} \tilde{a}_{ij} \right)^{\frac{1}{n}}, i = 1, 2, \ldots, n \tag{11}$$

Defuzzification of the fuzzy weight is carried out using the equation:

$$M_i = \frac{lw_i + mw_i + uw_i}{3} \tag{12}$$

where M_i is a crisp number that needs to be standardized:

$$w_i = \frac{M_i}{\sum_{i=1}^{n} M_i} \tag{13}$$

2.3 Formation of Alpha-Level Sets

After standardizing attributes of criteria, DM can perform an additional filtering (screening) alternatives, applying non-spatial constraints to attribute values. The following method is proposed for this purpose. Expert evaluations of alternatives according to the criteria are presented as fuzzy sets expressed by membership functions:

$$\tilde{C}_j = \{\mu_j(a_i)/a_i\}, \quad \mu_j(a_i) \in [0,1], \quad j = \overline{1,n}. \tag{14}$$

Next, we will perform a ranking of criteria C_j by importance and number them in the order of decreasing the weight of the criteria w_j.

Let us set a threshold $\alpha_j \in (0,1]$ and α-cut of the membership function $\mu_j(a)$ of the following type:

$$A_j = \{a | a \in A_{j-1},\ \mu_j(a) \geq \alpha_j\}, \quad A_0 \equiv A, \quad j \leq n. \tag{15}$$

An α-cut threshold for fuzzy set A defines a minimum truth membership level for a fuzzy set. All membership values below the α-cut are considered equivalent to zero. Figure 4 shows the example of the α-cut threshold applied to fuzzy set.

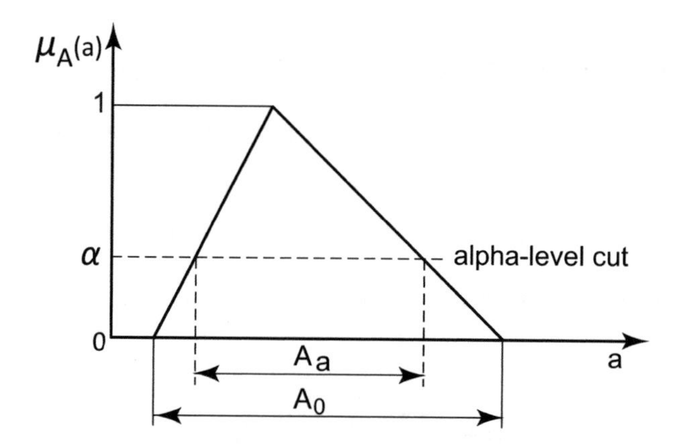

Fig. 4. Alpha-cut threshold applied to fuzzy set.

The calculation is repeated until the last iteration of the set A_n contain only alternatives that are considered by experts. DM can change the set A_n by varying the weights of the criteria wj or thresholds α_j.

If the criteria are equivalent by importance, then for each criterion C_j, separate sets of α-levels A_j are calculated at given thresholds α_j, and then the set is built of the following type:

$$A^* = \cap_{j=1}^{n} A_j \tag{16}$$

For the set of alternatives A^*, a convolution of the criteria is performed. To do this, the GIS environment usually uses various aggregation operators: minimum, maximum, average, weighed sum, OWA operator [28–31].

2.4 Aggregation

The combining rule (aggregation) integrates data and information about alternatives (criteria attributes values) and the preferences of DMs (criteria weight) in the overall assessment of alternatives.

One of the simplest compensative aggregation operators implemented in the GIS is the weighted linear combination (WLC) [30]:

$$\mu(a_i^*) = \sum_{j=1}^{n} w_j \mu_j(a_i) \tag{17}$$

where $\mu_j(a_i)$ is membership function of ith alternative to jth criteria, and w_j is normalized weight of jth criteria.

The WLC method is compensatory, i.e. it allows compensating the poor suitability of one criterion by the good suitability of another criterion.

An example of using the proposed algorithm for screening alternatives for the three layers of criteria $C = \{C_1, C_2, C_3\}$ is presented in Fig. 5.

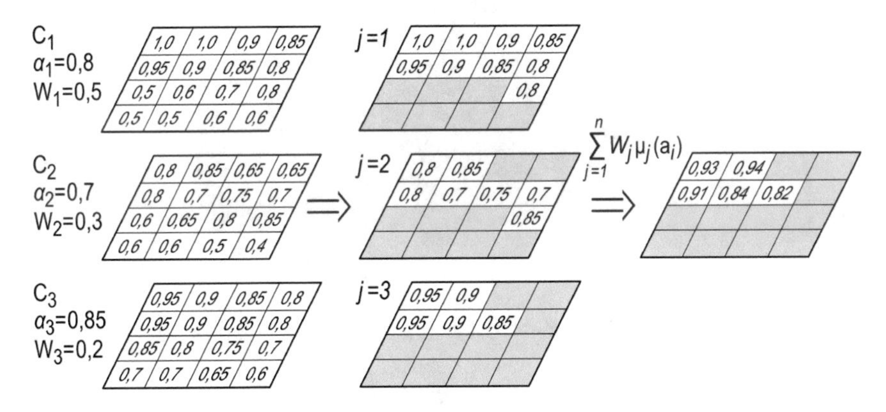

Fig. 5. An example of implementing an algorithm for screening alternatives for three criteria.

3 Results of the Experimental Study

To illustrate the proposed algorithm for screening alternatives for fuzzy sets of α-level, we use the data of the multicriterial decision-making model on the location of solid waste (SW) landfills site in the south of the Odessa region proposed by the authors in [32]. The model takes into account construction norms, physical, environmental and socio-economic factors for the location of solid waste landfill site. One of the sites found is located in Izmail Raion in southeast from Suvorovo village (45.5692N, 29.0088E).

Let us consider the implementation of the algorithm for the three criteria of the model: C_1 – Distance from rail and motor roads; C_2 – Distance from city limits; C_3 – Distance from settlements.

The following vector layers were used to calculate the raster layers of the C_1, C_2, and C_3 criteria, which were downloaded from the cartographic web service OpenStreetMap: V_1 – Rail and motor roads; V_2 – Cities; V_3 – Settlements shown in Fig. 6

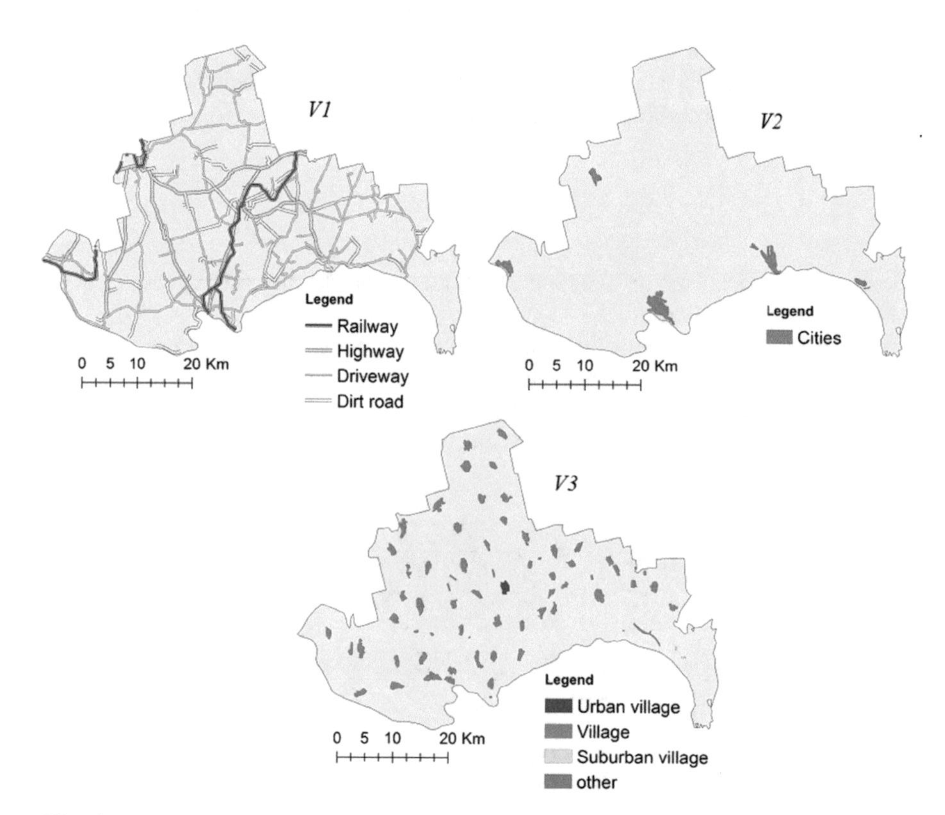

Fig. 6. Vector layers of criteria: V_1 – Rail and motor roads; V_2 – Cities; V_3 – Settlements.

To obtain the raster layers of the criteria and calculate the distance between the objects used the Euclidean metric, the value of which between two point objects O_1 (x_1, y_1) and O_2 (x_2, y_2) is calculated by equation:

$$ED(O_1, O_2) = \sqrt{(x_1 - x_2)^2 + (y_1 - y_2)^2} \tag{18}$$

In the case of a raster data model, the distance from any cell of the raster to the object O_i will be equal to the minimum distance from this cell to each cell that covers the object being investigated. The raster layers of the C_1, C_2, and C_3 criteria are shown in Fig. 7.

The attributes of the criteria are standardized by an expert assessment of their fuzzy membership functions. To standardize the criteria, piecewise-linear membership functions were used, whose form is shown in Fig. 8, 9, 10.

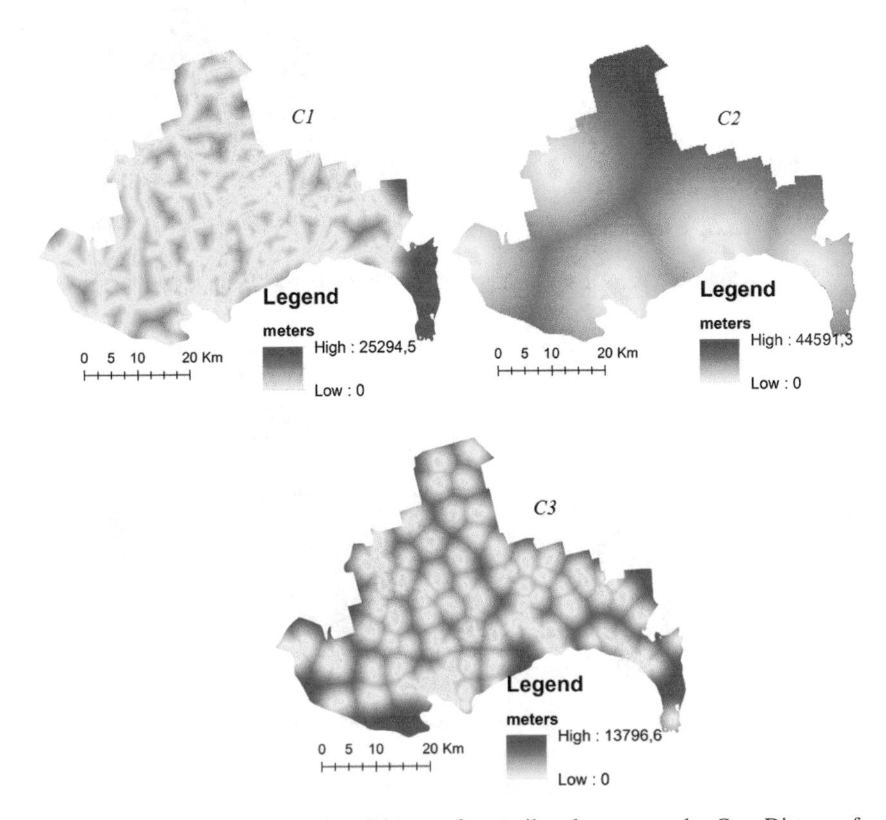

Fig. 7. Raster layers of criteria: C_1 – Distance from rail and motor roads; C_2 – Distance from city limits; C_3 – Distance from settlements.

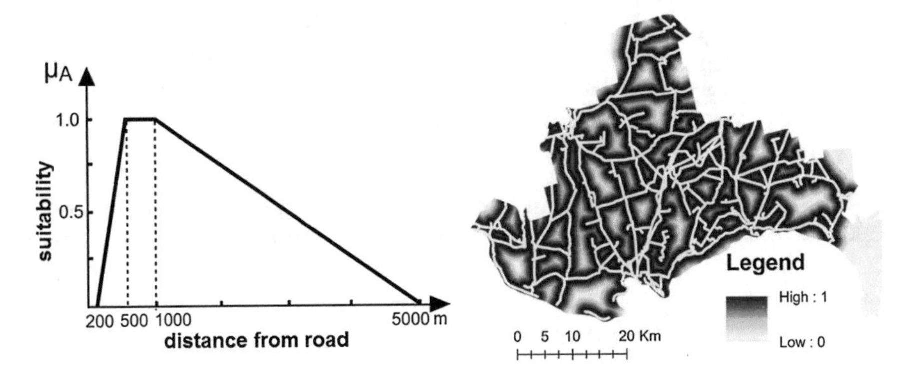

Fig. 8. Fuzzy membership functions and roads suitability layer.

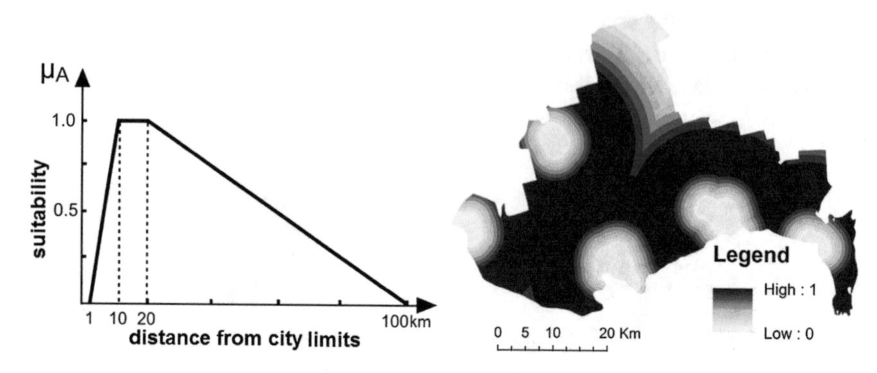

Fig. 9. Fuzzy membership functions and city suitability layer.

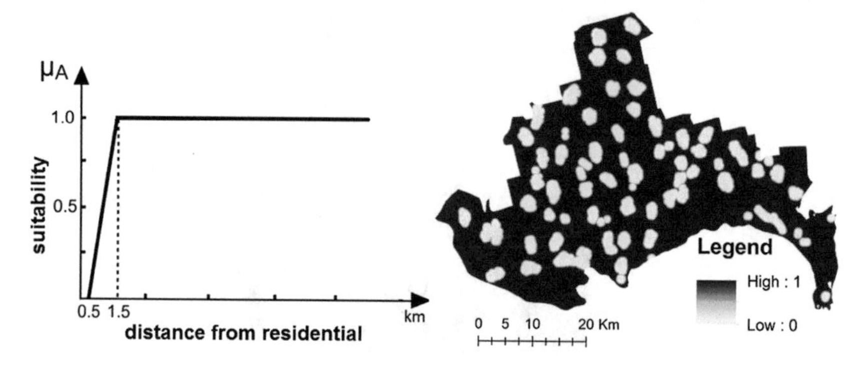

Fig. 10. Fuzzy membership functions and settlement suitability layer.

By applying Eqs. (9)–(13), we calculated weight coefficients of criteria. Table 4 gives a matrix of pairwise comparisons for three criteria C_1, C_2, C_3.

Table 4. Matrix of paired comparisons for criteria.

Criteria	C_1	C_2	C_3
C_1: Distance from road and rail	(1, 1, 1)	(1, 2, 3)	(2, 3, 4)
C_2: Distance from the city limits	(1/3, 1/2, 1)	(1, 1, 1)	(1, 2, 3)
C_3: Distance from settlements	(1/4, 1/3,1/2)	(1/3, 1/2, 1)	(1, 1, 1)

At the next step, the geometric mean of fuzzy comparison values of each criterion is calculated by Eq. (11). For example, geometric mean of fuzzy comparison values of „Distance from rail and motor roads criterion is calculated as:

$$\tilde{r}_1 = \left[(1 \cdot 1 \cdot 2)^{\frac{1}{3}}; (1 \cdot 2 \cdot 3)^{\frac{1}{3}}; (1 \cdot 3 \cdot 4)^{\frac{1}{3}}\right] = [1.260; 1.817; 2.289] \tag{19}$$

The geometric means of fuzzy comparison values of all criteria are shown in Table 5. In addition, the total values and the reverse values are also presented. In the last row of Table 5, since the fuzzy triangular number should be in increasing order, the order of the numbers is changed.

Table 5. Geometric means of fuzzy comparison value.

Criteria	$\tilde{r}_i$		
C_1: Distance from rail and motor roads	1.260	1.817	2.289
C_2: Distance from the city limits	0.693	1.000	1.442
C_3: Distance from settlements	0.437	0.550	0.794
Total	2.390	3.367	4.525
Revers	0.418	0.297	0.221
Increasing order	0.221	0.297	0.418

In the next step, the fuzzy weight of „Distance from rail and motor roads criterion is found by the help of Eq. (10) as:

$$\tilde{w}_1 = [(0.260 \cdot 0.221); (1.817 \cdot 0.297); (2.289 \cdot 0.418)] = [0.278; 0.540; 0.958] \quad (20)$$

The relative fuzzy weights of each criterion $\tilde{w_i}$ (10), as well as a definite weight of each criterion M_i (12) and normalized weights w_i (13) are given in Table 6.

Table 6. Relative fuzzy weights and normalized relative weights of each criterion.

Criteria	$\tilde{w}_i$			M_i	w_i
C_1: Distance from rail and motor roads	0.278	0.540	0.958	0.592	0.519
C_2: Distance from the city limits	0.153	0.297	0.603	0.351	0.308
C_3: Distance from settlements	0.097	0.163	0.332	0.197	0.173

As an aggregation operator, we use the weighted linear combination (16).

In Fig. 11, alternative models of suitability for the placement of a SW landfill for different threshold values are given (α_1, α_2, α_3).

Characteristics of the implementation of models in the GIS environment are presented in Table 7.

Table 7. Characteristics of implementation of models in GIS.

The values of the thresholds α-cuts (α_1, α_2, α_3)	Number of alternatives (raster cells with value)	Time to perform a weighted sum operator
(0, 0, 0)	6900583	4 870 ms
(0.2, 0.35, 0.3)	4084035	4 800 ms
(0.3, 0.5, 0.45)	3540798	4 760 ms
(0.8, 0.85, 0.9)	1079139	4 580 ms

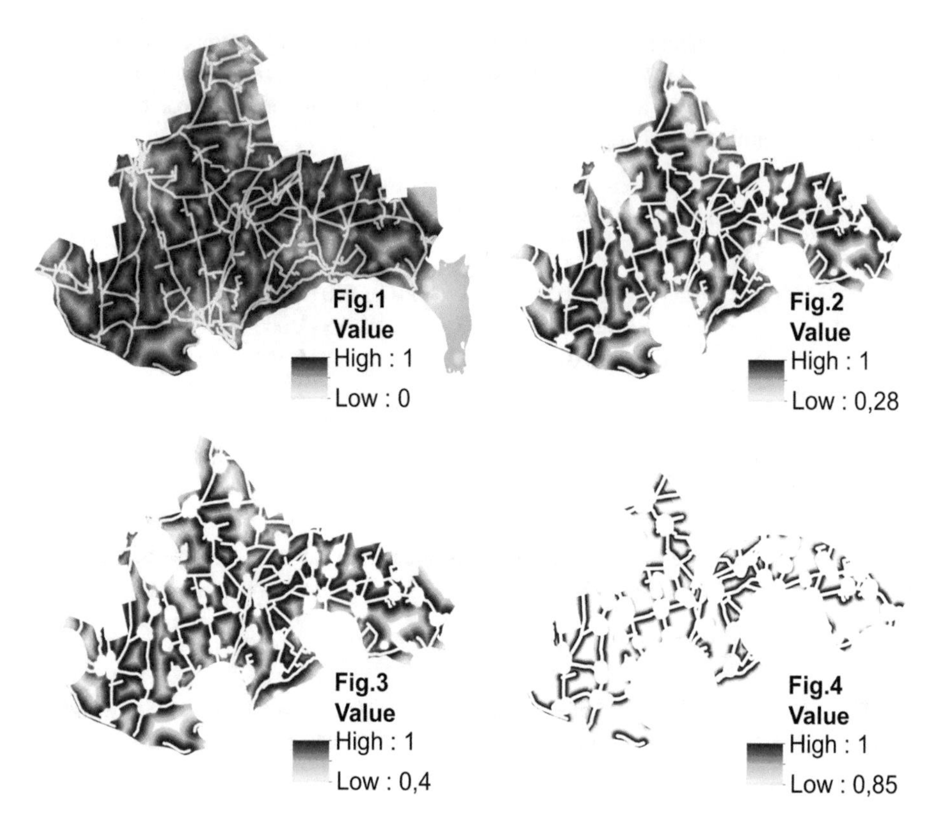

Fig. 11. Combined suitability maps for the placement of a solid waste landfill for different values of α-thresholds of fuzzy sets.

Combined maps of suitability for placement of a landfill of solid waste, shown in Fig. 11, are constructed for four different set of thresholds α-cut of fuzzy sets of criteria. The higher the threshold value, the less the number of alternatives remains for further analysis and the faster the weighted sum operator performs (the operator's execution time is calculated for the ArcGIS 10.5 environment and computer system containing OS Windows 10, Intel Core i3-7100 Kabylake, RAM 8 Gb).

4 Conclusions

Using in multi-criteria analysis of a raster data model allows you to display continuous surfaces, analyze them and perform overlays using complex data sets. In the study of large areas, the sets of raster data can be large enough, which leads to a significant increase in data volumes and a decrease in the processing speed. Adding to the procedure of multicriteria analysis of the additional stage of filtering alternatives, by specifying on the basis of the advantages of the DM thresholds of the level α_j by which the fuzzy set of α-levels is constructed in accordance with (15), enables to reduce the number of alternatives, and in the subsequent and the processing time of the criteria layers by the aggregator operators. The proposed algorithm for screening alternatives can be quite simply performed in a GIS environment using Fuzzy Membership, Overlay and Raster calculators tools.

Alpha cuts play a crucial role in many fuzzy models by removing unnecessary noise and specifying a degree of confidence necessary in the model to effect a correct outcome. Thresholds must be used with care, however, in that very high alpha cuts (higher than the crossover point) can have serious deleterious effects on a model's performance.

Uncertainty of DM preferences is one aspect of the manifestation of fuzzy information in MCDA. This information often refers to the relative importance (weights) of the criteria. As reasons for the fuzziness of the criteria weights, one can note their expression in verbal form, also the difficulty or inability to accurately and unambiguously evaluate the value of the corresponding parameters. An advantage of the MCDA approaches presented in the work is the use of the apparatus of the theory of fuzzy sets to formalize uncertain qualitative concepts that have a verbal form of expression. A modification of the AHP method is given in the paper, which makes it possible to present and process fuzzy expert judgments about the criteria weights. Fuzzification of criteria attributes allows you to get a more informative map of suitability by determining the rank of suitability of alternatives.

References

1. Chakhar, S., Mousseau, V.: Spatial multicriteria decision making. In: Shekhar, S., Xiong, H. (eds) Encyclopedia of GIS, pp. 747–753. Springer, New York (2008)
2. Current, J., Daskin, M. S., Schilling, D.: Discrete network location models. In: Drezner, Z., Hamacher, H., (eds.) Facility Location Theory: Applications and Methods, pp. 83–120. Springer, Berlin (2001)
3. Daskin, M.S.: What you should know about location modeling. Naval Res. Logistics **55**, 283–294 (2008)
4. ReVelle, C., Swain, R.: Central facilities location. Geog. Anal. **2**, 30–34 (1970)
5. Church, R.L.: Location modelling and GIS. In: Goodchild et al., (eds.) Handbook on GIS Cambridge, GeoInformation International, UK, pp. 293–303 (1997)
6. Church, R.L.: Geographical information systems and location science. Comput. Oper. Res. **29**, 541–562 (2002)
7. Malczewski, J.: GIS-based land-use suitability ana-lysis: a critical overview. Prog. Plann. **62**, 3–6 (2004)
8. Simon, H.: The structure of Ill-structured problems. Artif. Intell. **4**, 181–202 (1973)
9. Zadeh, L.: Fuzzy sets. Inf. Control **8**(3), 338–353 (1965)
10. Petry, F.E., Robinson, V.B., Cobb, M.A. (eds.): Fuzzy Modeling with Spatial Information for Geographic Problems. Springer, Berlin (2005)
11. Leung, Y.: Fuzzy sets approach to spatial analysis and planning: a nontechnical evaluation. Geogr. Ann. **65**, 65–75 (1983)
12. Charnpratheep, K., Zhou, Q., Garner, B.: Preliminary landfill site screening using fuzzy geographical information systems. Waste Manag. Res. **15**(2), 197–215 (1997). https://doi.org/10.1177/0734242X9701500207

13. Morris, A., Jankowski P.: Spatial decision making using fuzzy GIS. In: Petry, F., Robinson, V., Cobb, M. (eds.) Fuzzy Modeling with Spatial Information for Geographic Problems, pp. 275–298. Springer, Berlin (2005)
14. Chakhar, S., Martel, J.: Enhancing geographical information systems capabilities with multicriteria evaluation functions. J. Geogr. Inf. Decis. Anal. **7**(2), 69–71 (2003)
15. Malczewski, J.: GIS–based multicriteria decision analysis: a survey of the literature. Int. J. Geogr. Inf. Sci. **20**(7), 703–726 (2006)
16. Malczewski, J., Rinner, C.: Multicriteria Decision Analysis in Geographic Information Science, 331 p. Springer, New York (2015)
17. Lidouh, K.: On themotivation behind MCDA and GIS integration. Int. J. Multicriteria Decis. Making **3**(2/3), 101–113 (2013)
18. Kuznichenko, S., Gunchenko, Yu., Buchynska, I.: Fuzzy model of geospatial data processing in multi-criteria suitability analysis. Collection of scientific works of the Military Institute of Kyiv National Taras Shevchenko University, vol. 61, pp. 90–103 (2018)
19. Rikalovic, A., Cosic, I., Lazarevic, D.: GIS Based Multi-Criteria Analysis for Industrial Site Selection. Procedia Eng. **69**, 1054–1063 (2014)
20. Saaty, T.: The Analytic Hierarchy Process: Planning, Priority Setting, Resources Allocation, p. 287. McGraw, New York (1980)
21. Saaty, T.: Fundamentals of the Analytic Hierarchy Process. RWS Publications, Pittsburgh (2000)
22. Saaty, T.: Highlights and critical points in the theory and application of the analytic hierarchy process. Eur. J. Oper. Res. **74**, 426–447 (1994)
23. Zahedi, F.: The analytical hierarchy process – a survey of the method and its applications. Interfaces **16**, 96–108 (1986)
24. Vargas, L.: An overview of the analytic hierarchy process and its applications. Eur. J. Oper. Res. **48**, 2–8 (1990)
25. Buckley, J.: Fuzzy hierarchical analysis. Fuzzy Sets Syst. **17**(1), 233–247 (1985)
26. Vahidnia, M., Alesheikh, A., Alimohammadi, A., Bassiri, A.: Fuzzy analytical hierarchy process in GIS application. Int. Arch. Photogrammetry Remote Sens. Spatial Inf. Sci. Beijing **37**, 593–596 (2008)
27. Ziaei, M., Hajizade, F.: Fuzzy analytical hierarchy process (FAHP): a GIS-based multicriteria evaluation/selection analysis. In: 19th International Conference on Geoinformatics, vol.19, no. 11, pp. 24–26 (2011)
28. Yager, R.: On ordered weighted averaging aggregation operators in multicriteria decision making. IEEE Trans. Syst. Man Cybern. **18**, 183–190 (1988)
29. Malczewski, J.: GIS and Multicriteria Decision Analysis, p. 392. Wiley, New York (1999)
30. Malczewski, J.: On the use of weighted linear combination method in GIS. In: Common and best practice approaches. Transactions in GIS, vol. 4, no. 1, pp. 5–22 (2000). https://doi.org/10.1111/1467-9671.00035
31. Malczewski, J.: Ordered weighted averaging with fuzzy quantifiers: GIS-based multicriteria evaluation for land-use suitability analysis. Int. J. Appl. Earth Obs. Geoinf. **8**(4), 270–277 (2006). https://doi.org/10.1016/j.jag.2006.01.003
32. Kuznichenko, S., Kovalenko, L., Buchynska, I., Gunchenko, Y.: Development of a multi-criteria model for making decisions on the location of solid waste landfills. Eastern-Eur. J. Enterp. Technol. **2**(3(92)), 21–31 (2018). https://doi.org/10.15587/1729-4061.2018.129287

Quadratic Optimization Models and Convex Extensions on Permutation Matrix Set

Oksana Pichugina(✉) and Sergiy Yakovlev

National Aerospace University "Kharkiv Aviation Institute", 17 Chkalova Street, Kharkiv 61070, Ukraine
oksanapichugina1@gmail.com, svsyak7@gmail.com

Abstract. A new approach to the construction of lower bounds of quadratic function the permutation matrix set, based on the utilization of functional representations and convex extensions, is offered. Several quadratic functional representations of the are formed. A family of one-parametric convex quadratic extensions of a quadratic function from the set onto the Euclidean space is formed. The results can be applied in approximate and exact methods of quadratic optimization on the permutation matrix set.

Keywords: Permutation matrix set · Euclidean combinatorial set · Unconstrained quadratic optimization · Convex extension · Continuous functional representation

1 Introduction

The study of quadratic problems of combinatorial optimization known by a wide range of real-world applications in such areas as optimal planning, management, design, info-communications, etc. Among areas of engineering applications are problems of finding an optimal topology of an info-communication network, balancing problems for a system of discretely distributed masses, community detection, clustering, and many others [13,15,17,27,33,35,43,46,56]. Different graph problems are reduced to Boolean quadratic problems [1,25,34,44–46]. Two more theoretical applications of the class are the unconstrained binary quadratic problems (UBQP) and the Quadratic Assignment Problem (QAP) [8,16] having various real-world applications in Facility Layout, Electronics, Electricity, Economics, Scheduling, Supply Chains, Image processing, Manufacturing lines, Molecular conformations in chemistry, Bandwidth minimization of a graph, Keyboard and DNA MicroArray Layout [8,10,16,20,21,26,32,40], Graph Theory [1,5,25]. In addition, many well-known problems of combinatorial optimization are writable as QAP, e.g., Traveling salesman problem, Maximum cut problem, etc. [11,24,30,31].

Note, that majority of combinatorial optimization problems are NP-hard [43, 44,46]. However, in some cases it is possible to single out special classes of the

N. Shakhovska and M. O. Medykovskyy (Eds.): CCSIT 2019, AISC 1080, pp. 231–246, 2020.
https://doi.org/10.1007/978-3-030-33695-0_17

problems, which polynomially solvable [2,8,9,12,19,28,29,46]. For instance, this concerns the linear assignment problem (LAP) [9], which is solved on a set of permutation matrices [7]. At the same time, QAP is solved on the same combinatorial set and belongs to NP-hard class of problems.

On the other hand, various classes of mathematical models of quadratic problems can be formulated in a graph-theoretic way [15,22,23], as well as by discrete or continuous formulations [8,13,36,42,54]. In the first case, a search is conducted on geometric graphs with a finite point configuration as a node-set. In the second one, a reformulation of original combinatorial problem as an optimization problem on a set of vectors is required.

With this regards, of particular interest are properties of problems of quadratic combinatorial optimization after their mapping into Euclidean space. Combinatorial sets allowing bijective mapping into $\mathbb{R}^n$ are called Euclidean combinatorial sets (e-sets), and those images are sets of euclidean combinatorial configurations ($\mathcal{C}$-sets) [39]. An important class of $\mathcal{C}$-sets is *vertex-located sets* (*VLSs*) coinciding with vertex sets of their convex hulls [48]. Note that there exists a strong connection between optimization problems on any finite set in $\mathbb{R}^n$ and the ones on VLSs. Indeed, on the one hand, an arbitrary finite set $E \subset \mathbb{R}^n$ are representable as a collection of finitely many its vertex-located subsets [53]. On the other hand, a lifting of E onto $\mathbb{R}^{n'}$, where $n' > n$ allows to move to a consideration of a VLS $E' \subset \mathbb{R}^{n'}$ instead of E [53]. A special feature of optimization problems on VLSs is an existence of convex extensions of functions defined on such sets [48]. Some methods for optimization of linear, quadratic, and convex functions on some classes of VLSs were considered in [36,38,39,45,47,49–51] and for multi-objective formulation – in [22,23,41]. Modern studies of combinatorial sets are associated with a concept of combinatorial configuration [3,18,39]. In particular, optimization models and methods on combinatorial configuration sets and their images in Euclidean space are presented in the literature, e.g., in [22,23,36,38,39,48–53].

Let us consider the following constrained quadratic Boolean problem:

$$f(\mathcal{X}) = \sum_{i,j,k,l=1}^{n} a_{ijkl} x_{ij} x_{kl} + \sum_{i,j=1}^{n} c_{ij} x_{ij} \rightarrow \min_{\Pi_n}, \tag{1}$$

where

$$\Pi_n = \left\{ \mathcal{X} = [x_{ij}]_{n\times n} : x_{ij} \in \{0,1\}\,,\ i,j \in J_n, \mathcal{X}\mathbf{e} = \mathcal{X}^T\mathbf{e} = \mathbf{e} \right\}, \tag{2}$$

$J_n = \{1, ..., n\}\,,\ \mathbf{e} = (1, ..., 1)^T$.

One can assume that the matrix $A = [a_{ijkl}]_{n\times n\times n\times n}$ has a property:

$$a_{ijkl} = a_{klij},\ i,j,k,l \in J_n. \tag{3}$$

(2) defines a set of permutation matrices [7], hence the problem (1)–(3) is the general QAP [8,16].

The problem is intensively studied due to a variety of its applications. A detailed review of exact and approximate methods for solving QAP is given in [8].

Let us highlight exact methods based on the Branches and Bound paradigm. For Π_n, branching is performed based on two-levelness of the set, while obtaining lower bounds on objective function, as a rule, is connected with solving polyhedral relaxation optimization problem requiring a convexity of the objective function. Thus, to apply this solution scheme, preliminary convexification of the quadratic objective function from the set is needed. One way to accomplish this is constructing a convex extension of $f(x)$, and some approaches to deal with this can be found in [5,6,44,55]. For Π_n as a Boolean set, the convexification is typically conducted using the following expression $x^2 - x = 0$ of $x \in \{0,1\}$-condition [44]. An issue is that the low bound can be rather rough, which leads to a situation when an exact solution can not be found in reasonable time.

This paper is dedicated to deriving new properties of the permutation matrices' set and quadratic functions defined on it, including convexification techniques, making it possible offering new low bounds of $f(x)$ and increasing an efficiency of optimization methods to solving QAP and other polynomial problems over the set of permutation matrices.

2 Convex Extensions and Continuous Functional Representation

Function $\mathcal{F}(x)$ is called an *extension* of function $f(x)$ defined on E, if $\mathcal{F}(x)$ is defined on $E' \supset E$ and $\mathcal{F}(x) \underset{E}{=} f(x)$ (that is $\mathcal{F}(x)$ and $f(x)$ coincide with E) [39].

If $E' = \mathbb{R}^n$, then such a function is called an extension from E [39].

Extensions of functions can be continuous, convex, nonlinear, differentiable, etc. For instance, if $\mathcal{F}(x)$ is an extension of $f(x)$, which is convex on a convex set $E' \supset E$, then $\mathcal{F}(x)$ is *a convex extension* of $f(x)$ ($f.CE$, CE) from E onto E' [48].

Let us introduce a concept of a parametric extension as a generalization of a concept of a function extension.

A family of functions $\mathcal{F}(x,\lambda) : E' \times \Lambda \to \mathbb{R}^1$, where $E' \supset E$ and $\Lambda \subseteq \mathbb{R}^k$ is called a *k-parametric extension* of $f(x) : E \to \mathbb{R}^1$, if $\mathcal{F}(x,\lambda)$ is an extension of $f(x)$ from E onto E' $\forall \lambda \in \Lambda$.

Regarding existence convex sets, let us consider the following class.

A set E is called *vertex located* (VLS) if its points and only they are the vertices of its convex hull – a combinatorial polytope $P = conv\, E$ [39,48]: $E = vert\ conv\ E$.

Particular interest in vertex located sets (VLSs) is caused by the following of their common feature that any function defined on a VLS E has a convex extension at least onto the polytope P [48].

A representation of a set $E \subset \mathbb{R}^n$ by analytic dependencies:

$$f_j(x) = 0,\ j \in J_{m'}, \tag{4}$$

$$f_j(x) \leq 0,\ j \in J_m \backslash J_{m'}, \tag{5}$$

where all functions involved are continuous, is called *a continuous functional representation* (an f-representation) of E [36,37,39].

With regard to functions involved in (4), (5), such representations can be linear and nonlinear, polynomial or differentiable, convex or concave, etc.

The relations (4), (5) can be rewritten as:

$$E = \bigcap_{i \in J_m} M_i. \tag{6}$$

where

$$M_j = \{x \in \mathbb{R}^n : f_j(x) = 0, \ j \in J_{m'}; f_j(x) \leq 0, \ j \in J_m \backslash J_{m'}\}, \tag{7}$$

thus providing a geometric meaning to any f-representation and their connection with real varieties.

A functional representation of E is called:

- *strict*, if $m = m'$;
- *irredundant*, if the exclusion of any of the constraints (4), (5) leads to a violation of the condition (6);
- *bounded*, if, among the sets (7), bounded sets are presented.

In this paper, we will use parametric convex extensions as a tool for obtaining lower bound on $f^* = f(\mathcal{X}^*)$, where $\langle \mathcal{X}^*, X^* \rangle$ is a solution of QAP. In turn, f-representations of Π will be formed in order to construct f. CEs. It is due to a close connection of f-CEs with f-representation of feasible sets. Indeed, $\forall j \in J_{m'}, \forall \alpha \in \mathbb{R}^1$ if $f_j : \mathbb{R}^n \to \mathbb{R}^1$, then $f(x) + \alpha f_j(x)$ is an extension of $f(x)$.

3 Π_n: Properties and Functional Representations

Let us list some algebraic topological and topological metric peculiarities of Π_n and explore new ones, which will be used later in constructing f-representations of Π_n.

It is known [7] that a convex hull of Π_n is a Birkgoff polytope $D_n = conv\, \Pi_n$, which H-presentation is:

$$D_n = \left\{ \mathcal{X} \in \mathbb{R}^{n \times n} : \mathcal{X} \geq \mathbf{0}, \ \mathcal{X}\mathbf{e} = \mathcal{X}^T\mathbf{e} = \mathbf{e} \right\}, \tag{8}$$

where $\mathbf{0}$ – is a relevant matrix of zeros.

A highly important property of Π_n and D_n is that $\Pi_n = vert D_n$ [7] implying a vertex locality of Π_n. Respectively, any function $f : \Pi_n \to \mathbb{R}^1$ can be extended in a convex way at least onto D_n. An issue is how exactly the CE $\mathcal{F}(x)$ looks like depending on $f(x)$. Here, we present some ways to solve this problem of finding $\mathcal{F}(x)$ for quadratic $f(x)$ based on properties of Π_n and the behavior of such $f(x)$ on it.

We denote the Boolean $\mathcal{C}$-set [39] in $\mathbb{R}^n$ as B_n and single out the Boolean permutation $\mathcal{C}$-set induced by $\{0^{n-m}, 1^m\}$ from it [39]:

$$B_n(m) = \left\{ x \in B_n : \sum_{i=1}^{n} x_i = m \right\}.$$

We will also use the following notations:

- for vectors of a-s: $\mathbf{a} = (a, ..., a)^T \in \mathbb{R}^n$, $\forall a \in \mathbb{R}^1$;
- for rows and columns of a matrix $\mathcal{X}$ of an order n:

$$\mathbf{x}_i = (x_{ij})_j = (x_{ij})_{j \in J_n}, \; i \in J_n; \; \mathbf{x}_j^{'} = (x_{ij})_i = (x_{ij})_{i \in J_n}, \; j \in J_n;$$

- if $M \in \mathbb{R}^k$, then $M^m \in \mathbb{R}^{m \times k} : M^m = \overset{m}{\underset{i=1}{\otimes}} M$;

$$\mathbf{B}_n(m) = B_n^n(m); \tag{9}$$

$$\mathbf{B}_n^{'}(m) = \overset{n}{\underset{j=1}{\otimes}} B_n^{'}(m). \tag{10}$$

Proposition 1. *The set Π_n can be represented as an intersection:*

$$\Pi_n = \boldsymbol{B}_n(1) \cap \boldsymbol{B}_n^{'}(1).$$

Proposition 2. *If $n > 2$, $B_n(1)$ has the following strict quadratic f-representation:*

$$x_i^2 - x_i = 0, \; i \in J_n; \tag{11}$$

$$\sum_{i=1}^{n} x_i^2 - 2/n \sum_{i=1}^{n} x_i - 1 + 2/n = 0. \tag{12}$$

Proof. (11) defines B_n. To single out $B_n(1)$ from it, equation

$$x^T \mathbf{e} = 1 \tag{13}$$

can be added. Another way is to add an equation of a circumsphere for $B_n(1)$ of a minimal radius. The equation is:

$$\sum_{i=1}^{n} (x_i - 1/n)^2 = 1 - 1/n \tag{14}$$

writable as (12). Note that a center of this hypersphere lies on the hyperplane (13). It is easy to see that no other points of B_n are on the sphere. It completes the proof.

Proposition 3. *The set Π_n is inscribed in the minimal radius hypersphere $S^{\min} = S_{\mathbb{R}^{\min}}\left((\boldsymbol{a}^{\min})^n\right)$:*

$$\sum_{i,j=1}^{n} (x_{ij} - 1/n)^2 = n - 1, \tag{15}$$

centered at $(\boldsymbol{a}^{\min})^n$, where $a^{\min} = 1/n$, with a radius $R^{\min} = \sqrt{n-1}$.

Proof. To prove this, we apply (14) to the rows and columns of $\mathcal{X} \in \Pi_n$:

$$\mathbf{x}_i \in S^i_{r\min}\left(\mathbf{a}^{\min}\right) : \sum_{j=1}^{n} \left(x_{ij} - \frac{1}{n}\right)^2 = 1 - \frac{1}{n},\ i \in J_n, \tag{16}$$

$$\mathbf{x}^{'}_j \in S^j_{r\min}\left(\mathbf{a}'^{\min}\right) : \sum_{i=1}^{n} \left(x_{ij} - \frac{1}{n}\right)^2 = 1 - \frac{1}{n},\ j \in J_n. \tag{17}$$

Then add (16) for $j \in J_n$ and (17) for $i \in J_n$ obtaining identical equation (15). According to [38] and (9), it defines the minimum circumsphere for $\mathbf{B}_n(1)$. Similarly, by (10), Eq. (15) describes the minimum circumsphere for $\mathbf{B}'_n(1)$. The fact that we got (15) in both cases justifies that it is the minimum circumsphere equation for Π_n as well.

Remark 1. It should be noted that, in contrast to $B_n(1)$, adding to (15) the Boolean constraints:

$$x_{ij}^2 - x_{ij} = 0,\ i, j \in J_n, \tag{18}$$

where each of them is an equation of 0-circumsphere of minimum radius for B_1, does not yield a strict f-representation of Π_n, since (18) defines B_{n^2}, and (15), (18) – $B_{n^2} \cap S_{\mathbb{R}\min}\left((\mathbf{a}^{\min})^n\right) = B_{n^2}(n)$, which cardinality is $|B_{n^2}(n)| = C^n_{n^2}$.

Let us list other f-representations of Π_n.

Π_n is representable as an intersection of D_n and S^{min}:

$$\Pi_n = D_n \cap S^{min}. \tag{19}$$

In analytic form. it is a quadratic, non-strict, convex, bounded representation, called a *polyhedral-spherical* representation [37,39] of Π_n. It consists of n^2 inequalities, $2n$ equations of the system (8), and the collection (15). It is redundant due to the redundancy of (8). The representation (19) is of our interest because it allows considering continuous relaxations of QAP such as:

- traditional polyhedral relaxation [5]: $f(\mathcal{X}) \to \min,\ \mathcal{X} \in D_n$;
- spherical relaxation: $f(\mathcal{X}) \to \min,\ \mathcal{X} \in S_{\mathbb{R}\min}\left((\mathbf{a}^{\min})^n\right)$.

In many cases, the latter allows obtaining more accurate lower bounds on function (1), since, in practical problems, it is quite often that the minimizer to the unconstrained problem (1) is within an interior of the polytope D_n and, respectively, is inside of a ball bounded by a sphere (15).

Another type of f-representations is obtained as a result of replacement of linear constraints by quadratic ones in the strict representation of Π_n given by (18),

$$\sum_{i=1}^{n} x_{ij} = 1,\ j \in J_n,\ \sum_{j=1}^{n} x_{ij} = 1,\ i \in J_n, \tag{20}$$

(further referred to as **R1**).

Proposition 4. *For $n > 2$, the set Π_n has the following strict quadratic f-representation:* (18),

$$\sum_{j=1}^{n} x_{ij}^2 - 2/n \sum_{j=1}^{n} x_{ij} - 1 + 2/n = 0, \; i \in J_n, \tag{21}$$

$$\sum_{i=1}^{n} x_{ij}^2 - 2/n \sum_{i=1}^{n} x_{ij} - 1 + 2/n = 0, \; i \in J_n. \tag{22}$$

Proof. The conditions of inscribing Π_n in the hyperspheres (16), (17) are able to replace the linear constraints (20) with these quadratic ones. As a result, the f-representation (16)–(18) (further referred to as **R2**) can be written in an equivalent form (18), (21), (22).

Note that both representations R1 and R2 are strict, quadratic, convex, and unbounded. However, they are easily transformed into bounded by adding Eq. (15) of a circumsphere. To (15), (18), (21), (22) we refer to as **R3**.

4 Convex Extensions of Quadratic Functions from Π_n

Using the strict f-representation R3, let us construct $\mathcal{F}$.CE onto Euclidean space $\mathbb{R}^{n^2}$ of a quadratic function $f : \Pi_n \to \mathbb{R}^{n^2}$ given by (1). This function can be represented as a linear and a quadratic form:

$$f(\mathcal{X}) = f_1(\mathcal{X}) + f_2(\mathcal{X}) \to \min_{\Pi_n}, \tag{23}$$

where

$$f_1(\mathcal{X}) = \sum_{i,j=1}^{n} c_{ij} x_{ij}; \; f_2(\mathcal{X}) = \sum_{i,j,k,l=1}^{n} a_{ijkl} x_{ij} x_{kl}. \tag{24}$$

In turn, by (3), $f_2(\mathcal{X})$ can be rewritten as follows:

$$f_2(\mathcal{X}) = f_2'(\mathcal{X}) + f_2''(\mathcal{X}), \tag{25}$$

where

$$f_2'(\mathcal{X}) = \sum_{i,j=1}^{n} a_{ijij} x_{ij}^2, \; f_2''(\mathcal{X}) = 2 \sum_{(i,j) \prec (k,l)} a_{ijkl} x_{ij} x_{kl}, \tag{26}$$

and a lexicographic ordering of $\prec$ for a pair of indices $(i,j), (k,l)$ means that $(i,j) \prec (k,l)$ if and only if either $i < k$ or $i = k$ and $j < l$.

Let us construct CEs of functions (24) separately. For the term $f_2'(\mathcal{X})$, a f.CE from B_{n^2} will be formed by (18):

$$f_2'(\mathcal{X}) = \sum_{i,j=1}^{n} a_{ijij} x_{ij}^2 \underset{B_{n^2}}{=} \sum_{i,j=1}^{n} a_{ijij} x_{ij} = \mathcal{F}_2'(\mathcal{X}). \tag{27}$$

Another term in (26) is $f_2''(\mathcal{X})$ involving cross quadratic terms:

$$f_{ijkl} = 2a_{ijkl}x_{ij}x_{kl}, (i,j) \prec (k,l), \ a_{ijkl} \neq 0. \tag{28}$$

In the notation (28), function $f_2''(\mathcal{X})$ becomes:

$$f_2''(\mathcal{X}) = \sum_{(i,j)\prec(k,l), a_{ijkl}\neq 0} f_{ijkl}. \tag{29}$$

To each term (28) if (29), apply a transformation:

$$f_{ijkl} = 2\left|a_{ijkl}\right| \left((x_{ij} \pm x_{kl})^2 - x_{ij}^2 - x_{kl}^2 \right) \tag{30}$$

and construct a CE based on of applying the strict representation R3. We replace a non-convex component of (30):

$$- x_{ij}^2 - x_{kl}^2 \tag{31}$$

by its CE. We offer some ways depending on which component of R3 is used.

4.1 Method 1: f_{ijkl}.CE from B_{n^2}

To the expressions (31), let us apply the substitution (18) thus constructing a CE $\mathcal{F}_{ijkl}$ of each summands (28) from B_{n^2}:

$$\begin{aligned} f_{ijkl} &= \left|a_{ijkl}\right| \left((x_{ij} \pm x_{kl})^2 - x_{ij}^2 - x_{kl}^2 \right) \\ &\underset{B_{n^2}}{=} \left|a_{ijkl}\right| \left((x_{ij} \pm x_{kl})^2 - x_{ij} - x_{kl} \right) = \mathcal{F}_{ijkl}. \end{aligned} \tag{32}$$

Substituting the functions (32) into (29) we obtain a $f_2''(\mathcal{X})$.CE:

$$\begin{aligned} \mathcal{F}_2''(\mathcal{X}) &\underset{B^{n^2}}{=} \sum\nolimits_{(i,j)\prec(k,l)} \mathcal{F}_{ijkl} = \sum\nolimits_{(i,j)\prec(k,l)} \left|a_{ijkl}\right| \left((x_{ij} \pm x_{kl})^2 - x_{ij} - x_{kl} \right) \\ &= f_2''(\mathcal{X}) + 2\sum\nolimits_{(i,j)\prec(k,l)} \left|a_{ijkl}\right| \left(x_{ij}^2 - x_{ij}\right) = f_2''(\mathcal{X}) + \sum\nolimits_{i,j=1}^{n} \left(x_{ij}^2 - x_{ij}\right) A_{ij}; \end{aligned} \tag{33}$$

where

$$A_{ij} = \sum_{k,l:(i,j)\neq(k,l)} \left|a_{ijkl}\right|, \ i,j \in J_n. \tag{34}$$

4.2 Method 2: f_{ijkl}.CE from a Π_n-circumsphere S^{min}

In the expressions (30), we consider several cases depending on a i,j,k,l-combination: (a) $i = k, \ j < l$, (b) $i < k, \ j = l$, (c) $i < k, \ j < l$. For each of them, the nonconvex function (31) in (30) will be replaced by convex one, using the equations of the minimum radius circumspheres for $\mathcal{X} \in \Pi_n$ (see (14)) and its components (see (16), (17)). As a result, we come to the following cases.

Case (a) From (22), we get

$$-x_{ij}^2 - x_{kl}^2 = -x_{ij}^2 - x_{il}^2 \underset{x_i \in B_n(1)}{=} \sum_{j' \neq j,l} x_{ij'}^2 - 2/n \sum_{j'=1} x_{ij'} - 1 + 2/n; \tag{35}$$

Case (b) From (21), we obtain

$$-x_{ij}^2 - x_{kl}^2 = -x_{ij}^2 - x_{kj}^2 \underset{x'_j \in B_n(1)}{=} \sum_{i' \neq i,k} x_{i'j}^2 - 2/n \sum_{i'=1} x_{i'j} - 1 + 2/n; \tag{36}$$

Case (c) From (14), we have

$$-x_{ij}^2 - x_{kl}^2 \underset{\mathcal{X} \in \mathbf{B}_n(1)}{=} \sum_{i' \neq i,k; j' \neq j,l} x_{i'j'}^2 - 2/n \sum_{i' \neq i,k; j' \neq j,l} x_{i'j'} - n + 2. \tag{37}$$

Substituting (35)–(37) into (30), we get three types of CEs:

$$\begin{aligned} f_{ijil} &\underset{\mathbf{x}_i \in S^i_{rmin}(\mathbf{a}^{\min})}{=} |a_{ijil}| \left((x_{ij} \pm x_{il})^2 + \sum_{j' \neq j,l} x_{ij'}^2 - 2/n \sum_{j'=1} x_{ij'} + 2/n - 1 \right) \\ &= |a_{ijil}| \left(2x_{ij}x_{il} + \sum\nolimits_{j'=1}^{n} x_{ij'}^2 - 2/n \sum\nolimits_{j'=1} x_{ij'} + 2/n - 1 \right) \\ &= f_{ijil} + |a_{ijil}| \left(\sum\nolimits_{j'=1}^{n} x_{ij'}^2 - 2/n \sum\nolimits_{j'=1} x_{ij'} + 2/n - 1 \right) = \mathcal{F}^1_{ijil}. \end{aligned} \tag{38}$$

$$\begin{aligned} f_{ijkj} &\underset{\mathbf{x}'_j \in S'^j_{rmin}(\mathbf{a}^{\min})}{=} |a_{ijkj}| \left((x_{ij} \pm x_{kj})^2 + \sum_{i' \neq i,k} x_{i'j}^2 - 2/n \sum_{i'=1} x_{i'j} + 2/n - 1 \right) \\ &= |a_{ijkj}| \left(2x_{ij}x_{kj} + \sum\nolimits_{i'=1}^{n} x_{i'j}^2 - 2/n \sum\nolimits_{i'=1} x_{i'j} + 2/n - 1 \right) \\ &= f_{ijkj} + |a_{ijkj}| \left(\sum\nolimits_{i'=1}^{n} x_{i'j}^2 - 2/n \sum\nolimits_{i'=1} x_{i'j} + 2/n - 1 \right) = \mathcal{F}^2_{ijkj}. \end{aligned} \tag{39}$$

$$\begin{aligned} &f_{ijkl} \underset{\mathcal{X} \in S_{\mathbb{R}\min}((\mathbf{a}^{\min})^n)}{=} \\ &|a_{ijkl}| \left((x_{ij} \pm x_{kl})^2 + \sum\nolimits_{i' \neq i,k;\ j' \neq j,l} x_{i'j'}^2 - 2/n \sum\nolimits_{i',j'=1}^{n} x_{i'j'} + 2 - n \right) \\ &\qquad = |a_{ijkl}| \left(2x_{ij}x_{kl} + \sum\nolimits_{i',j'=1}^{n} x_{i'j'}^2 - 2/n \sum\nolimits_{i',j'=1}^{n} x_{i'j'} + 2 - n \right) \\ &\qquad = f_{ijkl} + |a_{ijkl}| \left(\sum\nolimits_{i',j'=1}^{n} x_{i'j'}^2 - 2/n \sum\nolimits_{i',j'=1}^{n} x_{i'j'} + 2 - n \right) = \mathcal{F}^3_{ijkl}. \end{aligned} \tag{40}$$

In the expression (29) for $f_2''(\mathcal{X})$, we single out terms $f_2^1(\mathcal{X})$, $f_2^2(\mathcal{X})$, $f_2^3(\mathcal{X})$ corresponding for each pair x_{ij}, x_{kl} to the cases (a)–(c). Let us apply the corresponding way (38)–(40) of convexification in order to form three components of $\mathcal{F}_2''(\mathcal{X})$ – a CE of $f_2''(\mathcal{X})$ from a set CS:

$$CS = \left\{ \mathcal{X} \in S_{\mathbb{R}^{\min}}\left((\mathbf{a}^{\min})^n\right) : \mathbf{x}_i \in S^i_{rmin}\left(\mathbf{a}^{\min}\right),\ \mathbf{x}_j^{'} \in S^j_{rmin}\left(\mathbf{a}'^{\min}\right) \right\}_{i,j}, \quad (41)$$

formed as an intersection of hypersphere (15) with surfaces (16), (17).

Let us denote these three $\mathcal{F}_2^1(\mathcal{X})$, $\mathcal{F}_2^2(\mathcal{X})$, $\mathcal{F}_2^3(\mathcal{X})$, respectively. Correspondingly,

$$\begin{aligned} f_2''(\mathcal{X}) &= \sum_{\substack{(i,j) \prec (k,l), \\ a_{ijkl} \neq 0}} f_{ijkl} = \sum_{i=1}^{n} \sum_{j<l} f_{ijkl} + \sum_{j=1}^{n} \sum_{i<k} f_{ijkl} + \sum_{i<k, j<l} f_{ijkl} \\ &= f_2^1(\mathcal{X}) + f_2^2(\mathcal{X}) + f_2^3(\mathcal{X}) \underset{CS}{=} \mathcal{F}_2^1(\mathcal{X}) + \mathcal{F}_2^2(\mathcal{X}) + \mathcal{F}_2^3(\mathcal{X}) = \mathcal{F}_2^{'''}(\mathcal{X}), \end{aligned}$$

$$\mathcal{F}_2^1(\mathcal{X}) = \sum_{i=1}^{n} \sum_{j<l} \mathcal{F}^1_{ijil},\ \mathcal{F}_2^2(\mathcal{X}) = \sum_{j=1}^{n} \sum_{i<k} \mathcal{F}^2_{ijkj},\ \mathcal{F}_2^3(\mathcal{X}) = \sum_{i<k, j<l} \mathcal{F}^3_{ijkl}. \quad (42)$$

Introduce notation for power sums for components of columns and rows of $\mathcal{X}$.

$$s_i^m = \sum_{j=1}^{n} x_{ij}^m,\ i \in J_n;\ s_j^{'m} = \sum_{i=1}^{n} x_{ij}^m,\ j \in J_n;\ S^m = \sum_{i,j=1}^{n} x_{ij}^m.$$

Now, formula (38)–(42) can be rewritten as follows:

$$\begin{aligned} \mathcal{F}_2^1(\mathcal{X}) &= \sum_{i=1}^{n} \sum_{j<l} \mathcal{F}^1_{ijil} \\ &= \sum_{i=1}^{n} \sum_{j<l} \left(f_{ijil} + |a_{ijil}| \left(\sum_{j'=1}^{n} x_{ij'}^2 - 2/n \sum_{j'=1} x_{ij'} + 2/n - 1 \right) \right) \\ &= f_2^1(\mathcal{X}) + \sum_{i=1}^{n} \left(s_i^2 - 2/n s_i + 2/n - 1 \right) \sum_{j<j'} |a_{ijil}|; \end{aligned} \quad (43)$$

$$\begin{aligned} \mathcal{F}_2^2(\mathcal{X}) &= \sum_{j=1}^{n} \sum_{i<i'} \mathcal{F}^2_{ijkj} \\ &= \sum_{j=1}^{n} \sum_{i<k} \left(f_{ijkj} + |a_{ijkj}| \left(\sum_{i'=1}^{n} x_{i'j}^2 - 2/n \sum_{i'=1}^{n} x_{i'j} + 2/n - 1 \right) \right) \\ &= f_2^2(\mathcal{X}) + \sum_{j=1}^{n} \left(s_j^{'2} - 2/n s_j^{'} + 2/n - 1 \right) \sum_{i<i'} |a_{ijkj}|; \end{aligned} \quad (44)$$

$$\begin{aligned}\mathcal{F}_2^3(\mathcal{X}) &= \sum_{i<k,j<l} \mathcal{F}_{ijkl}^3 \\ &= \sum_{i<k,j<l} \left(f_{ijkl} + |a_{ijkl}| \left(\sum_{i',j'=1}^{n} x_{i'j'}^2 - 2/n \sum_{i',j'=1} x_{i'j'} + 2 - n \right) \right) \\ &= f_2^3(\mathcal{X}) + \left(S^2 - 2/nS^1 + 2 - n\right) \sum_{i<k,j<l} |a_{ijkl}| \,.\end{aligned} \tag{45}$$

Applying (43)–(45) to (42), we get:

$$\begin{aligned}\mathcal{F}_2^{'''}(\mathcal{X}) &= \left(f_2^1(\mathcal{X}) + f_2^2(\mathcal{X}) + f_2^3(\mathcal{X})\right) + \sum_{i=1}^{n} \left(s_i^2 - 2/ns_i + 2/n - 1\right) \sum_{j<j'} |a_{ijil}| \\ &+ \sum_{j=1}^{n} \left(s_j^{'2} - 2/ns_j^{'} + 2/n - 1\right) \sum_{i<i'} |a_{ijkj}| + \left(S^2 - 2/nS^1 + 2 - n\right) \sum_{i<k,j<l} |a_{ijkl}| \,.\end{aligned} \tag{46}$$

In notations $\overline{a}_i = \sum_{j<l} |a_{ijil}|$, $\overline{a}_j^{'} = \sum_{i<k} |a_{ijkj}|$, $A = \sum_{i<k,j<l} |a_{ijkl}|$, the expression (46) is rewritten in the following way:

$$\begin{aligned}\mathcal{F}_2^{'''}(\mathcal{X}) &\underset{CS}{=} f_2''(\mathcal{X}) + \sum_{i=1}^{n} \overline{a}_i \left(s_i^2 - 2/ns_i + 2/n - 1\right) \\ &+ \sum_{j=1}^{n} \overline{a}_j^{'} \left(s_j^{'2} - 2/ns_j^{'} + 2/n - 1\right) + A\left(S^2 - 2/nS^1 - n + 2\right).\end{aligned} \tag{47}$$

Since Π_n is formed at the intersection of B_{n^2} with CS (see (41)), to obtain a CE (1) from Π_n only we will combine in a resulting f.CE all components of R3. For that, a convex linear combination of $f''(\mathcal{X})$.CEs in the forms (33), (34), and (47) from sets B_{n^2} and CS onto $\mathbb{R}^{n^2}$ will be used resulting in

$$\forall \lambda \in (0,1) \quad f_2''(\mathcal{X}) \underset{\Pi_n}{=} \lambda \mathcal{F}_2''(\mathcal{X}) + (1-\lambda)\, \mathcal{F}_2^{'''}(\mathcal{X}) = \mathcal{F}_2''(\mathcal{X}, \lambda).$$

Similarly, coming back to the original function (23), a one-parametric f.CEs is obtained. Namely, by (23), (25):

$$\begin{aligned}\mathcal{F}(\mathcal{X}, \lambda) &= f_1(\mathcal{X}) + \mathcal{F}_2(\mathcal{X}, \lambda) = f_1(\mathcal{X}) + \mathcal{F}_2^{'}(\mathcal{X}) + \mathcal{F}_2''(\mathcal{X}, \lambda) \\ &= f_1(\mathcal{X}) + \mathcal{F}_2^{'}(\mathcal{X}) + \lambda \mathcal{F}_2''(\mathcal{X}) + (1-\lambda)\, \mathcal{F}_2^{'''}(\mathcal{X}),\end{aligned} \tag{48}$$

where $\lambda \in (0,1)$, $f_1(\mathcal{X})$, $\mathcal{F}_2^{'}(\mathcal{X})$, $\mathcal{F}_2''(\mathcal{X})$, $\mathcal{F}_2^{'''}(\mathcal{X})$ – are linear and quadratic functions given by the expressions (24), (27), (33), (47), respectively.

For a fixed λ, (48) is a quadratic f.CE, which can be represented in the form (1):

$$\mathcal{F}(\mathcal{X}, \lambda) = \sum_{i,j,k,l=1}^{n} a_{ijkl}^{'}(\lambda) x_{ij} x_{kl} + \sum_{i,j=1}^{n} c_{ij}^{'}(\lambda) x_{ij} + d'(\lambda), \lambda \in (0,1)\,. \tag{49}$$

This means, for the quadratic function (1), a family (49) of f.CEs was found in the same class of quadratic functions. In such a way, we justified that for a fixed $\lambda \in (0,1)$, the problem (1) is equivalent to the following combinatorial optimization problem $\mathcal{F}(\mathcal{X},\lambda) \to \min\limits_{\Pi_n}$ operating with a convex objective function.

In order to find a lower bound z^l on z^*, we offer the following scheme: randomly generate a finite set $\Lambda = \{\lambda_i\}_{i\in I} \subset [0,1]$, solve relaxation problems

$$z^{l1}(\lambda_i) = \min_{\mathcal{X}\in D_n} \mathcal{F}(\mathcal{X},\lambda),\ z^{l2}(\lambda_i) = \min_{\mathcal{X}\in S^{min}} \mathcal{F}(\mathcal{X},\lambda),\ i \in I.$$

getting a set of lower bounds on f^* and taking their minimum as a lower bound. Obtaining $z^{l1}(\lambda_i)$ is a convex optimization problem over the Birkhoff polytope, which can be effectively solved using the conditional gradient and projection methods [4] due to possibility effectively solve auxiliary linear problems over D_n. The problem of finding $z^{l2}(\lambda_i)$ can be solved explicitly [14] as a problem of optimizing a convex quadratic function on a hypersphere. Respectively, $z^l(\Lambda) = \min\limits_{i\in I}\{z^{l1}(\lambda_i), z^{l2}(\lambda_i)\}$ will be a lower bound on z^l. To improve this bound, $z^l([0,1])$ can be chosen instead of $z^l(\Lambda)$. It assumes solving two quadratic parametric optimization problems of getting $z^{l1} = \min\limits_{\lambda[0,1]}\{z^{l1}(\lambda)$ and $z^{l2} = \min\limits_{\lambda[0,1]}\{z^{l2}(\lambda)$ and taking a minimum of them as $z^l - z^l = \min\{z^{l1}, z^{l2}\}$. The method of getting $z^l(\Lambda)$ and z^l is universal for the whole class of quadratic functions and QAPs. However, the accuracy of estimating z^* by $z^l(\Lambda)$ depends on a choice of Λ, while the accuracy of estimating z^* by z^l depends on exactly or approximately z^l.

5 Conclusion

The paper is dedicated to studying the set Π_n of permutation matrices and behavior of quadratic functions on it. For this purpose, a notion of a parametric extension of functions is introduced, new algebraic topological properties of Π_n are derived, continuous functional representations of Π_n are obtained. Based on utilizing these functional representations, a one-parametric convex extension of a quadratic function f from Π_n onto $\mathbb{R}^{n^2}$ is formed and is used for obtaining new lower bounds f than currently known ones based on combining solutions of polyhedral and spherical relaxations of QAP. Expectedly, the reduction of a search domain, when QAP is solved, will lead to solving numerous real-world problems modeled as QAPs in more effective way and in reasonable time.

References

1. Armbruster, M., Fügenschuh, M., Helmberg, C., Martin, A.: A comparative study of linear and semidefinite branch-and-cut methods for solving the minimum graph bisection problem. In: Lodi, A., Panconesi, A., Rinaldi, G. (eds.) Integer Programming and Combinatorial Optimization, pp. 112–124. Springer, Heidelberg (2008). https://doi.org/10.1007/978-3-540-68891-4_8

2. Bachem, A., Euler, R.: Recent trends in combinatorial optimization. OR Spektrum **6**, 1–21 (1984). https://doi.org/10.1007/BF01721246
3. Berge, C.: Principes de combinatoire. Dunod, Paris (1968)
4. Bertsekas, D.P.: Nonlinear Programming. Athena Scientific, Belmont (1999)
5. Billionnet, A., Elloumi, S., Plateau, M.-C.: Improving the performance of standard solvers for quadratic 0-1 programs by a tight convex reformulation: the QCR method. Discrete Appl. Math. **157**, 1185–1197 (2009). https://doi.org/10.1016/j.dam.2007.12.007
6. Billionnet, A., Jarray, F., Tlig, G., Zagrouba, E.: Reconstructing convex matrices by integer programming approaches. J. Math. Model. Algor. **12**, 329–343 (2012). https://doi.org/10.1007/s10852-012-9193-5
7. Brualdi, R.A.: Combinatorial Matrix Classes. Cambridge University Press, Cambridge (2006)
8. Burkard, R.E.: Quadratic assignment problems. In: Pardalos, P.M., Du, D.-Z., Graham, R.L. (eds.) Handbook of Combinatorial Optimization, pp. 2741–2814. Springer, New York (2013). https://doi.org/10.1007/978-1-4419-7997-1_22
9. Burkard, R.E., Çela, E.: Linear assignment problems and extensions. In: Du, D.-Z., Pardalos, P.M. (eds.) Handbook of Combinatorial Optimization, pp. 75–149. Springer, New York (1999). https://doi.org/10.1007/978-1-4757-3023-4_2
10. Cela, E.: The Quadratic Assignment Problem: Theory and Algorithms. Springer, New York (2010)
11. Cook, W.J., Cunningham, W.H., Pulleyblank, W.R., Schrijver, A.: Combinatorial Optimization. Wiley, New York (1998)
12. Crama, Y., Spieksma, F.C.R.: Scheduling jobs of equal length: complexity, facets and computational results. In: Balas, E., Clausen, J. (eds.) Integer Programming and Combinatorial Optimization, pp. 277–291. Springer, Heidelberg (1995). https://doi.org/10.1007/3-540-59408-6_58
13. Colbourn, C.J., Dinitz, J.H. (eds.): Handbook of Combinatorial Designs. Chapman and Hall, CRC Press, New York (2006)
14. Dahl, J.: Convex optimization in signal processing and communications (2003)
15. Farzad, B., Pichugina, O., Koliechkina, L.: Multi-layer community detection. In: 2018 International Conference on Control, Artificial Intelligence, Robotics Optimization (ICCAIRO), pp. 133–140 (2018). https://doi.org/10.1109/ICCAIRO.2018.00030
16. Floudas, C.A., Pardalos, P.M., Adjiman, C.S., Esposito, W.R., Gümüş, Z.H., Harding, S.T., Klepeis, J.L., Meyer, C.A., Schweiger, C.A.: Quadratic programming problems. In: Handbook of Test Problems in Local and Global Optimization, pp. 5–19. Springer, New York (1999)
17. Stoyan, Yu.G., Sokolovskii, V.Z., Yakovlev, S.V.: Method of balancing rotating discretely distributed masses. Energomashinostroenie **2**, 4–5 (1982)
18. Hulianytskyi, L., Riasna, I.: Formalization and classification of combinatorial optimization problems. In: Optimization Methods and Applications, pp. 239–250. Springer, Cham (2017). https://doi.org/10.1007/978-3-319-68640-0_11
19. Kabadi, S.N.: Polynomially solvable cases of the TSP. In: Gutin, G., Punnen, A.P. (eds.) The Traveling Salesman Problem and Its Variations, pp. 489–583. Springer, New York (2007). https://doi.org/10.1007/0-306-48213-4_11
20. Kaibel, V.: Polyhedral methods for the QAP. In: Pardalos, P.M., Pitsoulis, L.S. (eds.) Nonlinear Assignment Problems, pp. 109–141. Springer, New York (2000). https://doi.org/10.1007/978-1-4757-3155-2_6

21. Kammerdiner, A., Gevezes, T., Pasiliao, E., Pitsoulis, L., Pardalos, P.M.: Quadratic assignment problem. In: Gass, S.I., Fu, M.C. (eds.) Encyclopedia of Operations Research and Management Science, pp. 1193–1207. Springer, New York (2013). https://doi.org/10.1007/978-1-4419-1153-7_1152
22. Koliechkina, L.M., Dvirna, O.A.: Solving extremum problems with linear fractional objective functions on the combinatorial configuration of permutations under multicriteriality. Cybern. Syst. Anal. **53**, 590–599 (2017). https://doi.org/10.1007/s10559-017-9961-3
23. Koliechkina, L., Pichugina, O.: A horizontal method of localizing values of a linear function in permutation-based optimization. In: Le Thi, H.A., Le, H.M., Pham Dinh, T. (eds.) Optimization of Complex Systems: Theory, Models, Algorithms and Applications, pp. 355–364. Springer, Cham (2019)
24. Korte, B., Vygen, J.: Combinatorial Optimization: Theory and Algorithms. Springer, Heidelberg (2012)
25. Krislock, N., Malick, J., Roupin, F.: Computational results of a semidefinite branch-and-bound algorithm for k-cluster. Comput. Oper. Res. **66**, 153–159 (2016). https://doi.org/10.1016/j.cor.2015.07.008
26. Lawler, E.L.: The quadratic assignment problem. Manage. Sci. **9**, 586–599 (1963)
27. Mashtalir, V.P., Yakovlev, S.V.: Point-set methods of clusterization of standard information. Cybern. Syst. Anal. **37**(3), 295–307 (2001). https://doi.org/10.1023/A:1011985908177
28. Miller, A.J., Nemhauser, G.L., Savelsbergh, M.W.P.: Facets, algorithms, and polyhedral characterizations for a multi-item production planning model with setup times. In: Aardal, K., Gerards, B. (eds.) Integer Programming and Combinatorial Optimization, pp. 318–332. Springer, Heidelberg (2001). https://doi.org/10.1007/3-540-45535-3_25
29. Nakamura, D., Tamura, A.: The generalized stable set problem for claw-free bidirected graphs. In: Bixby, R.E., Boyd, E.A., Ríos-Mercado, R.Z. (eds.) Integer Programming and Combinatorial Optimization, pp. 69–83. Springer, Heidelberg (1998). https://doi.org/10.1007/3-540-69346-7_6
30. Papadimitriou, C.H., Steiglitz, K.: Combinatorial Optimization: Algorithms and Complexity. Dover Publications, Mineola (2013)
31. Schrijver, A.: Combinatorial Optimization: Polyhedra and Efficiency. Springer, Heidelberg (2002)
32. Pardalos, P.M., Wolkowicz, H.: Quadratic Assignment and Related Problems: DIMACS Workshop, 20–21 May 1993. American Mathematical Soc. (1994)
33. Pichugina, O.: Placement problems in chip design: modeling and optimization. In: 2017 4th International Scientific-Practical Conference Problems of Infocommunications. Science and Technology (PIC S&T), pp. 465–473 (2017). https://doi.org/10.1109/INFOCOMMST.2017.8246440
34. Pichugina, O., Farzad, B.: A human communication network model. In: CEUR Workshop Proceedings, KNU, Kyiv, pp. 33–40 (2016)
35. Pichugina, O., Kartashov, O.: Signed permutation polytope packing in VLSI design. In: 2019 IEEE 15th International Conference on the Experience of Designing and Application of CAD Systems (CADSM) Conference Proceedings, Lviv, pp. 4/50–4/55 (2019). https://doi.org/10.1109/CADSM.2019.8779353
36. Pichugina, O.S., Yakovlev, S.V.: Continuous representations and functional extensions in combinatorial optimization. Cybern. Syst. Anal. **52**(6), 921–930 (2016). https://doi.org/10.1007/s10559-016-9894-2

37. Pichugina, O.S., Yakovlev, S.V.: Functional and analytic representations of the general permutation. Eastern-Eur. J. Enterp. Technol. **79**, 27–38 (2016). https://doi.org/10.15587/1729-4061.2016.58550
38. Pichugina, O., Yakovlev, S.: Optimization on polyhedral-spherical sets: theory and applications. In: 2017 IEEE 1st Ukraine Conference on Electrical and Computer Engineering, UKRCON 2017 - Proceedings, KPI, Kiev, pp. 1167–1174 (2017). https://doi.org/10.1109/UKRCON.2017.8100436
39. Pichugina, O., Yakovlev, S.: Euclidean combinatorial configurations: continuous representations and convex extensions. In: Lytvynenko, V., Babichev, S., Wójcik, W., Vynokurova, O., Vyshemyrskaya, S., Radetskaya, S. (eds.) Lecture Notes in Computational Intelligence and Decision Making, pp. 65–80. Springer, Cham (2019). https://doi.org/10.1007/978-3-030-26474-1_5
40. Pitsoulis, L., Pardalos, P.M.: Quadratic assignment problem. In: Floudas, C.A., Pardalos, P.M. (eds.) Encyclopedia of Optimization, pp. 2075–2107. Springer, New York (2001). https://doi.org/10.1007/0-306-48332-7_405
41. Semenova, N.V., Kolechkina, L.N., Nagornaya, A.N.: One approach to solving vector problems with fractionally linear functions of the criteria on the combinatorial set of arrangements. J. Autom. Inf. Sci. **42**, 67–80 (2010). https://doi.org/10.1615/JAutomatInfScien.v42.i2.50
42. Sergienko, I.V., Hulianytskyi, L.F., Sirenko, S.I.: Classification of applied methods of combinatorial optimization. Cybern. Syst. Anal. **45**, 732 (2009). https://doi.org/10.1007/s10559-009-9134-0
43. Sergienko, I.V., Shylo, V.P.: Modern approaches to solving complex discrete optimization problems. J. Autom. Inf. Sci. **48**, 15–24 (2016). https://doi.org/10.1615/JAutomatInfScien.v48.i1.30
44. Sherali, H.D., Adams, W.P.: A Reformulation-Linearization Technique for Solving Discrete and Continuous Nonconvex Problems. Kluwer Academic Publishers, Dordrecht (1999)
45. Shor, N.Z., Stetsyuk, P.I.: Lagrangian bounds in multiextremal polynomial and discrete optimization problems. J. Global Optim. **23**, 1–41 (2002). https://doi.org/10.1023/A:1014004625997
46. Stetsyuk, P.I.: Problem statements for k-node shortest path and k-node shortest cycle in a complete graph. Cybern. Syst. Anal. **52**, 71–75 (2016). https://doi.org/10.1007/s10559-016-9801-x
47. Yakovlev, S.V.: Bounds on the minimum of convex functions on Euclidean combinatorial sets. Cybernetics **25**, 385–391 (1989). https://doi.org/10.1007/BF01069996
48. Yakovlev, S.V.: The theory of convex continuations of functions on vertices of convex polyhedra. Comp. Math. Math. Phys. **34**, 1112–1119 (1994)
49. Yakovlev, S.V., Grebennik, I.V.: Localization of solutions of some problems of nonlinear integer optimization. Cybern. Syst. Anal. **29**, 727–734 (1993). https://doi.org/10.1007/BF01125802
50. Yakovlev, S., Pichugina, O.: On constrained optimization of polynomials on permutation set. In: Proceedings of the Second International Workshop on Computer Modeling and Intelligent Systems (CMIS-2019), CEUR Vol-2353 urn:nbn:de:0074-2353-0, Zaporizhzhia, Ukraine, pp. 570–580 (2019)
51. Yakovlev, S.V., Valuiskaya, O.A.: Optimization of linear functions at the vertices of a permutation polyhedron with additional linear constraints. Ukr. Math. J. **53**, 1535–1545 (2001). https://doi.org/10.1023/A:1014374926840

52. Yakovlev, S., Pichugina, O., Yarovaya, O.: On optimization problems on the polyhedral-spherical configurations with their properties. In: 2018 IEEE First International Conference on System Analysis Intelligent Computing (SAIC), pp. 94–100 (2018). https://doi.org/10.1109/SAIC.2018.8516801
53. Yakovlev, S., Pichugina, O., Yarovaya, O.: Polyhedral-spherical configurations in discrete optimization problems. J. Autom. Inf. Sci. **51**, 26–40 (2019). https://doi.org/10.1615/JAutomatInfScien.v51.i1.30
54. Yemelichev, V.A., Kovalev, M.M., Kravtsov, M.K.: Polytopes, Graphs and Optimisation. Cambridge University Press, Cambridge (1984)
55. Xia, Y., Gharibi, W.: On improving convex quadratic programming relaxation for the quadratic assignment problem. J. Comb. Optim. **30**, 647–667 (2013)
56. Zgurovsky, M.Z., Pavlov, A.A.: Combinatorial Optimization Problems in Planning and Decision Making: Theory and Applications. Springer, Cham (2019)

The Ateb-Gabor Filter for Fingerprinting

Mariya Nazarkevych[1](✉), Mykola Logoyda[1], Oksana Troyan[2], Yaroslav Vozniy[1], and Zoreslava Shpak[2]

[1] Publishing Information Technology Department, Institute of Computer Science and Information Technologies, Lviv Polytechnic National University, 12 Bandery str., Lviv 79013, Ukraine
mariia.a.nazarkevych@lpnu.ua

[2] Department of Automated Control Systems, Institute of Computer Science and Information Technologies, Lviv Polytechnic National University, 12 Bandery str., Lviv 79013, Ukraine

Abstract. In biometric protection systems, a lot of time is spent on recognition processes. The quality of recognition also remains unsatisfactory. A new Ateb-Gabor filtration method is proposed that extends the classic filtration methods. Applying the Ateb-Gabor filter fully utilizes the Gabor filter and uses the apparatus of Ateb functions. These functions extend the capabilities of trigonometry and build on accurate solutions of differential equations with significant second order nonlinearity. This approach allows the intensity of both the entire image and certain predefined portions to be altered, allowing for more accurate outlines in biometric images. The functions used depend on two rational parameters m and n, the change of which leads to the change of certain areas of the image. Fingerprints were filtered with a developed filter, showing the effectiveness of its use. Sketches of filtered biometric images have been developed.

Keywords: Image processing · fingerprinting · Ateb-Gabor filter

1 Introduction

From the perspective of biologists, the inner folds are very useful, they enhance the tactile sensitivity of the skin, which improves the perception of a person and increases his/her chances of survival. The papillary pattern can not be changed and can not be eliminated. This is proved by physicians, even after burns or surgery, it is still recovering.

In the past people certified important documents with the imprint of their fingers. Papillary lines are formed in the embryo in 3–4 months of intrauterine development. At this stage, the derma, which is a layer and connects the growth skin very quickly, and has a higher growth rate than close tissue. As a result, folds are formed which are "pushed out" in the upper layers of the epidermis. Their drawing depends on many factors, among which one of the most important

N. Shakhovska and M. O. Medykovskyy (Eds.): CCSIT 2019, AISC 1080, pp. 247–255, 2020.
https://doi.org/10.1007/978-3-030-33695-0_18

factors is heredity and the occasional touch of the hands of the fetus with other parts of the body. A little later the intensive growth of the external epidermis begins. Thus, the unique drawing is fixed for the whole life of a person. When the embryo is six months, it has already clear fingerprints that will accompany a person from birth to death.

2 The Analysis of Fingerprinting Research

When creating automated biometric identification systems, one of the most labor intensive is image processing. Such systems can be used on restricted sites, voting systems, electronic payments, mobile phones, crossing state borders, investigating crimes, etc.

There is an urgent need to create new remedies, as recent NYU scientists have used a neural network to fake human fingerprints. The results of the work were published at a security conference in Los Angeles [1].

Generally, a biometric fingerprint authentication system does not scan the whole fingerprint. This system only needs a small snippet that compares with the snippet in the database. Therefore, using machine learning, researchers have artificially created fingerprints. Some of them were generally rectangular, or had sharp angles. Our design remembers the whole fingerprint, not the snippet.

After receiving the fingerprint image, it must be processed. The main steps of the image processing process are the following: troubleshooting and enhancing clarity, identifying key features, and recognition itself. To this end, adaptive matching and filtering are performed [2]. And Adaptive Threshold Segmentation [3]. Despite the possible breaks and heterogeneity of individual papillaries, you can determine their direction. This filtering is applied to every image pixel. Since one of the most difficult tasks of fingerprint image processing is to get a clear image for recognition, there are several ways to solve it. Most of these methods use adaptive recognition of individual parts of the image. Initially, the image is divided into squares that determine the characteristics of the papillaries and their orientation. This methodology allows to select the necessary imprint even from a few prints that are left at different time intervals. The methodology is complex and requires high-cost equipment [4].

A new type of filtration of biometric images, which made it possible to more effectively perform identification in fingerprinting, has been considered.

3 Mathematical Apparatus of Ateb-Functions

The mathematical apparatus of the Ateb-functions allows to obtain analytic solutions of the system of differential equations that describe essentially non-linear processes of a system with a single degree of freedom [5] Ateb-functions are a reversal to the Beta-functions. An incomplete Beta-functions defined by equality

$$B_x(p,q) = \int_0^x (1-t)^{q-1} dt, \tag{1}$$

where p and q are the real numbers. If

$$p = \frac{1}{n+1}, \qquad q = \frac{1}{m+1}, \tag{2}$$

where m and n are determined by the formulas (2). If $p > 0, q > 0$, then the *Beta*-function is defined and continuous. *Ateb*-functions constructed for values (2) are periodic and describe oscillatory motion [6].

4 Construction of Mathematical Models of the Class of Periodic Ateb-Functions

Periodical Ateb-function $u = ca(m, n, \omega)$ is constructed as follows. Let's consider the expression:

$$-\frac{m+1}{2} \int_{1}^{-1 \leq u \leq 1} \left(1 - \bar{u}^{m+1}\right)^{-\frac{n}{n+1}} d\bar{u} = \omega \tag{3}$$

The dependence u of ω for the integral (3) is a function of n and m, which are called the cosine of the Ateb-function and is marked.

$$u = ca(m, n, \omega) \tag{4}$$

Let's consider the Ateb-Gabor filtration on condition that one of the parameters m takes a unit value. It should be noted that the function $u = ca(m, n, \omega)$ is symmetric according to the proofs of [6]. Thus, at the value of the parameter $n = 1$, and m takes different values, so we obtain the same results as if case when $m = 1$, and n takes different values. Let suppose $m = 1$, and n - takes any rational number. According to formula (2), $Ateb - ca$ is continuous and defined. The method of numerical representation of the cosine of the Ateb-function is based on the calculation of defined integral (3), the search of zeros of the function, and also finding the value of the Taylor series with a given accuracy of order 10^{-10}. To achieve such indicators, it is enough to perform calculations of seven members of the Taylor series. The introduction of Ateb-functions provides a large variety of options depending on the parameters that are rational numbers.

Thus, the set of functions in the process of changing n parameter in the case that $m = 1$ is shown in Fig. 1.

We choose a class of periodic Ateb functions for filtering, since their characteristics are repeated every period.

5 The Ateb-Gabor Filter

Based on the *Ateb*-Gabor filter [7], images can be filtered with a large number of crests. This can provide better characteristics than the well-known Gabor filter. In this way, filtration with a large spectrum of curves and a larger set of control

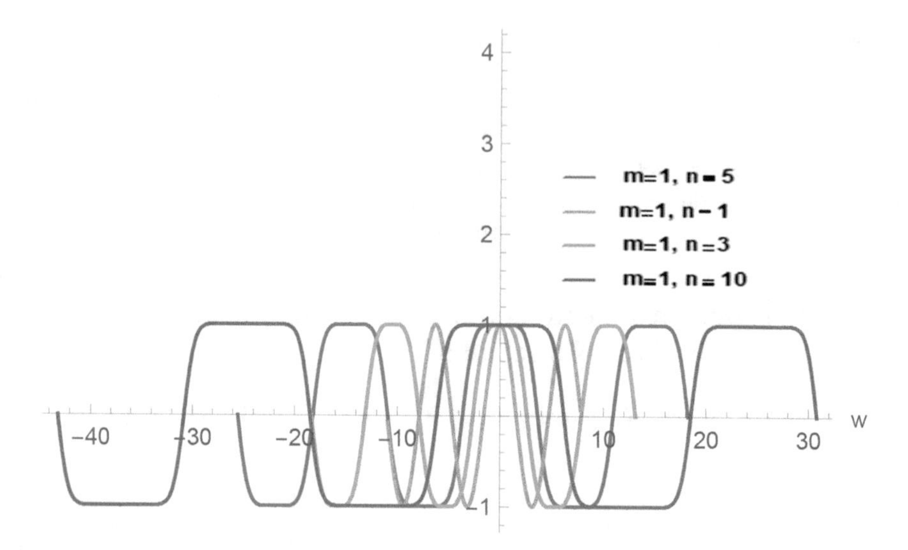

Fig. 1. Graphic Ateb-function $u = ca(m, n, \omega)$ changing n parameter in the case that m = 1

parameters can be realized. In particular, four parameters for the *Ateb*-Gabor filter - m, n, σ, θ, as opposed to two for the previously known Gabor filter - σ, θ. Two-dimensional Ateb-Gabor filtering [7] is performed using the formula:

$$AtebG(x, y, \lambda, \theta, \psi, \sigma, \xi, m, n) = \exp\Big(\frac{-\acute{x}^2 + \psi \cdot \acute{y}^2}{2\sigma^2}\Big) \cdot Atebca\Big(\frac{2\Pi\acute{x}}{\lambda} + \xi\Big) \quad (5)$$

$$\begin{cases} \acute{x} = x \cdot \cos(\theta) + y \cdot \sin(\theta) \\ \acute{y} = -x \cdot \sin(\theta) + y \cdot \cos(\theta), \end{cases}$$

where x, y - is the intensity at the point of the input image; σ - standard deviation of the Gaussian kernel; λ - is the wavelength of the cosine multiplier; θ - is the orientation of the normal parallel bands; ξ - is the phase shift; ψ - is the compression coefficient; m, n - real numbers with formula (2).

As we can notice from Fig. 1, $u = ca(m, n, \omega)$ changes own size and shape while increasing n parameter. The form of this function increases with the increase of n in the direction of expansion by the argument ω.

With the help of Fig. 2, we can notice how the frequency response of the Ateb-Gabor filtering is changed at the rational parameter n is changed.

Let's show a change in the parameter n from 0.1 to 1 at a constant $m = 1$ on the Fig. 3.

A new method of image filtering has been considered. The complexity of recognizing the biometric images generates different ways to search for the approximate methods. For the efficient image processing, filtering has been used, including the Gabor filter method, which allows to modify the images with brighter

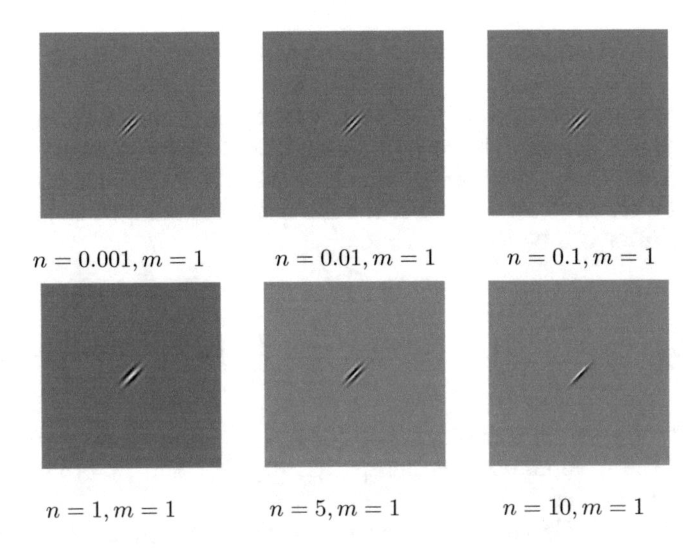

Fig. 2. The frequency characteristics of Ateb-Gabor filtration $u = ca(m, n, \omega)$ at changing of n parameter and constant parameter $m = 1$

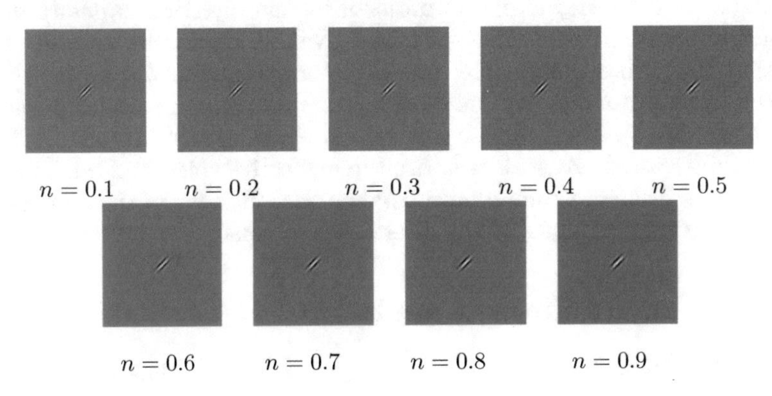

Fig. 3. The frequency characteristics of Ateb-Gabor filtration at constant parameter $m = 1$ and changing of n parameter within from 0.1 to 1.

contours. The introduction of the Ateb-Gabor filter will provide more options to change the accuracy and brightness of the image than when using the Gabor filter. Ateb-functions are a generalization of elementary trigonometry, and, accordingly, have greater functionality for harmonic functions. The Ateb-Gabor filter allows to change the intensity of the entire image, as well as the intensity in certain ranges. The influence of the parameters m, n of the functions of Ateb, as well as the influence of the amplitude of the function, the frequency of oscillations on the Ateb-Gabor filter have been regarded. These dependencies have been analyzed. Fingerprints have been filtered by the developed filter, and it shows the effectiveness of its usage. Sketches for biometric images have been developed.

The image processing of the *Ateb*-Gabor filter is achieved by averaging the image values in a specific area at each point. The filtration pattern has v and w pixels according to the rows and columns, as well as the current pixels and i, j, which will be changed as a result of the filtering. Accordingly, the usage of the *Ateb*-Gabor filter will be the following,

$$AtebI\grave{}(\grave{x}, \grave{y}) = \frac{1}{\sigma^2} \sum_{i=1}^{n} \sum_{j=1}^{m} I\Big(x - \frac{l}{2} + i, y - \frac{k}{2} + j\Big) AtebG\Big(x, y, \lambda, \theta, \psi, \sigma, \xi, m, n\Big) \tag{6}$$

where $AtebI\grave{}(\grave{x}, \grave{y})$ is the intensity converted image at the point (x, y). $I(i, j)$ - is the value of the Ateb-Gabor function $x \in [1, v], y \in [1, w]$, l, k - the value of the filter in the flow point.

To use the Ateb-Gabor filter, it is necessary to have values from the formula 6 of the filter direction; orientation of the normal parallel bands θ; deviation of the Gaussian nucleus σ [7]. The frequency response of the filter is determined from the local frequency ω of the projections, the direction is determined by the local orientation. The values of σ are given in the implementation of the algorithm. The more these values are, the more the noise will be noise-proof. However, more distortions will be introduced, creating non-existent projections and depressions [8]. If we select σ low values, the filter will not distort, but its filtering ability will decrease. It will result in ineffective noise removal. Therefore, when selecting values, a compromise between the efficiency of the filter and the absence of distortions made by the filter has to be made.

6 The Study of Finger Papules

In the study of finger papules, there are such peculiarities. For each fingerprint two types of attributes - global and local - can be defined. Global signs can be seen with the naked eye. Local signs are the features of the direction of the papillary lines that are unique to each imprint [9]. The secretion is due to the fact that the lines of fingerprints are not linear. They are often broken, branched, change direction and have breaks (see Fig. 4). The points in which the lines end, branch, or change the direction are called minucius [10]. These

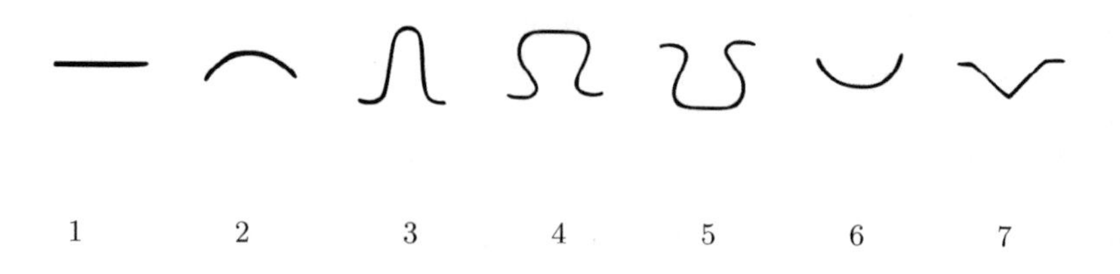

Fig. 4. Classification of crests [12]

points provide unique fingerprint information when identifying a person. Each footprint contains up to 70 minucia [11].

7 Database of Minusculuses

The shape of the edge of the minuscle occurs due to the difference in the height of the crests. It is possible to classify crests according to the scheme of Chatterjee [12].

There are typical variants of the crests: straight (Fig. 4.1), convex crest (Fig. 4.2), peak (Fig. 4.3), which is more convex. The shape of the crest is shown in Fig. 4.4 has the name of the table, and in Fig. 4.5 there is a pocket, which is a table display. In Fig. 4.6 there is a concave crest, and in Fig. 4.7 - an angle

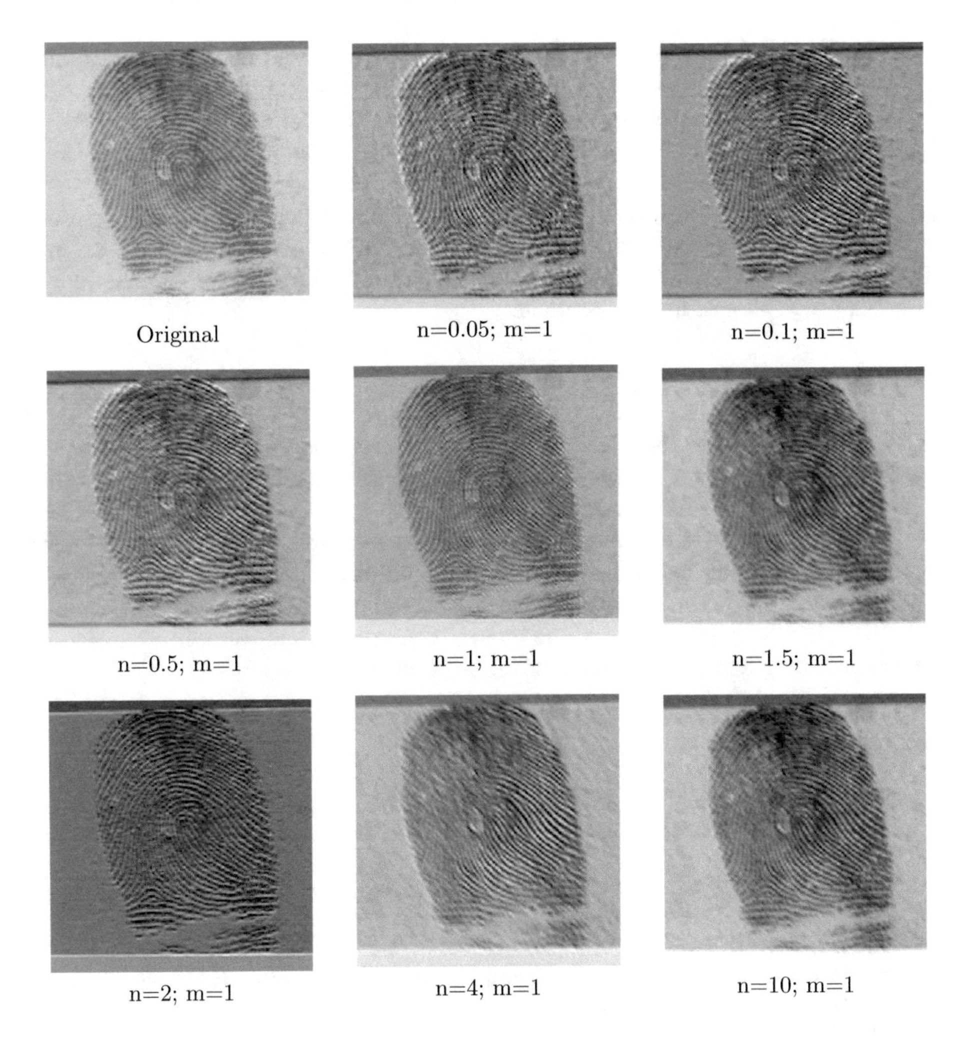

Fig. 5. The construction of a two-dimensional Ateb-Gabor filter

crest. A typical imprint has 150 variants of crests, and a crest of 5 mm in length contains approximately 10 modifications [13].

8 The Implementation of *Ateb*-Gabor Filtration

The relationship between the frequency and width of the *Ateb*-Gabor filter has been determined. It allowed the filters to automatically find the edges of objects with different frequencies, sizes and directions. The method of removing the average component of the *Ateb*-Gabor filter has been considered. It allows to reduce the magnitude of the average filter to zero without deforming the filter [14–16]. When processing images, a technique was used that was presented in [17].

The results of the *Ateb*-Gabor filter, which filters the fingerprints, is shown in Fig. 5.

The differential equations with degree nonlinearity have been solved. Graphic solutions with various parameters m and n, which are periodic Ateb-functions, have been shown. When applying the mathematical apparatus of the Ateb-functions to filtering the images, the control effects to change the gradation characteristics of the image can be significantly expanded. The program is a prototype of the working version without optimizing the algorithm over the program time. The algorithm can be accelerated by using a wavelet transform for the Ateb-Gabor filter when performing a convolutional operation, as well as by parallel processing of image segments.

9 Conclusions

The new Ateb-Gabor filtration method has been considered. The new filtering method has wider properties, since it has been based on the theory of the Ateb-function, which extends elementary trigonometry.

The developed Ateb-Gabor filter allows you to adjust the intensity of both the entire image and certain predefined parts, allowing you to clearly highlight the contours. Used Ateb functions that are sensitive to two rational parameters m and n change certain areas of the image. The fingerprints were filtered with a developed filter and the efficiency of its use is shown.

References

1. Hammad, M., Wang, K.: Parallel score ECG fusion and fingerprint for human authentication based on convolution neural network. Comput. Secur. **81**, 107–122 (2019)
2. Sharma, R.P., Dey, S.: Two-stage quality adaptive fingerprint image enhancement using Fuzzy C-means clustering based fingerprint quality analysis. Image Vis. Comput. **83**, 1–16 (2019)
3. Sundaresan, V., Zamboni, G., Le Heron, C., Rothwell, P.M., Husain, M., Battaglini, M., De Stefano, N., Jenkinson, M., Griffanti, L.: Automated lesion segmentation with BIANCA: impact of population-level features, classification algorithm and locally adaptive thresholding. NeuroImage, 116056 (2019)

4. Varetskyy, Y., Rusyn, B., Molga, A., Ignatovych, A.: A new method of fingerprint key protection of grid credential. In: Advances in Intelligent and Soft Computing, vol. 84, pp. 99–103 (2010)
5. Rosenberg, R.M.: The Ateb ()-functions and their properties. Q. Appl. Math. **21**(1), 37–47 (1963)
6. Nazarkevych, M., Kynash, Y., Oliarnyk, R., Klyujnyk, I., Nazarkevych, H.: Application perfected wave tracing algorithm. In: 2017 IEEE First Ukraine Conference on Electrical and Computer Engineering (UKRCON), pp. 1011–1014. IEEE, May 2017
7. Nazarkevych, M., Oliarnyk, R., Dmytruk, S.: An images filtration using the Ateb-Gabor method. In: 2017 12th International Scientific and Technical Conference on Computer Sciences and Information Technologies (CSIT), vol. 1, pp. 208–211. IEEE, September 2017
8. Martsyshyn, R., Medykovskyy, M., Sikora, L., Miyushkovych, Y., Lysa, N., Yakymchuk, B.: Technology of speaker recognition of multimodal interfaces automated systems under stress. In: 2013 12th International Conference on the Experience of Designing and Application of CAD Systems in Microelectronics (CADSM), pp. 447–448. IEEE, February 2013
9. Huckemann, S., Hotz, T., Munk, A.: Global models for the orientation field of fingerprints: an approach based on quadratic differentials. IEEE Trans. Pattern Anal. Mach. Intell. **30**(9), 1507–1519 (2008)
10. Krill-Burger, J.M., Lyons, M.A., Kelly, L.A., Sciulli, C.M., Petrosko, P., Chandran, U.R., LaFramboise, W.A.: Renal cell neoplasms contain shared tumor type-specific copy number variations. Am. J. Pathol. **180**(6), 2427–2439 (2012)
11. Jain, A.K., Feng, J., Nandakumar, K.: Fingerprint matching. Computer **43**(2), 36–44 (2010)
12. Chatterjee, A., Mandal, S., Rahaman, G.A., Arif, A.S.M.: Fingerprint identification and verification system by minutiae extraction using artificial neural network. JCIT **1**(1), 12–16 (2010)
13. Jain, A.K., Chen, Y., Demirkus, M.: Pores and ridges: high-resolution fingerprint matching using level 3 features. IEEE Trans. Model Anal. Mach. Intell. **29**(1), 15–27 (2007)
14. Nazarkevych, M., Klyujnyk, I., Nazarkevych, H.: Investigation the Ateb-Gabor filter in biometric security systems. In: 2018 IEEE Second International Conference on Data Stream Mining & Processing (DSMP), pp. 580–583. IEEE, August 2018
15. Nazarkevych, M., Buriachok, V., Lotoshynska, N., Dmytryk, S.: Research of Ateb-Gabor filter in biometric protection systems. In: 2018 IEEE 13th International Scientific and Technical Conference on Computer Sciences and Information Technologies (CSIT), vol. 1, pp. 310–313. IEEE, September 2018
16. Nazarkevych, M., Klyujnyk, I., Maslanych, I., Havrysh, B., Nazarkevych, H.: Image filtration using the Ateb-Gabor filter in the biometric security systems. In: 2018 XIV-th International Conference on Perspective Technologies and Methods in MEMS Design (MEMSTECH), pp. 276–279. IEEE, April 2018
17. Lytvyn, V., Peleshchak, I., Vysotska, V., Peleshchak, R.: Satellite spectral information recognition based on the synthesis of modified dynamic neural networks and holographic data processing techniques. In: 2018 IEEE 13th International Scientific and Technical Conference on Computer Sciences and Information Technologies (CSIT), Lviv, pp. 330–334 (2018)

Project Management

Methodical Approach to Assessing the Readiness Level of Technologies for the Transfer

Nataliya Chukhray(✉), Nataliya Shakhovska(✉), Oleksandra Mrykhina, Myroslava Bublyk, and Lidiya Lisovska

Lviv Polytechnic National University, Lviv 79013, Ukraine
Natalia.i.chuhraj@lp.edu.ua, {Nataliya.b.shakhovska, Oleksandra.b.mrykhina}@lpnu.ua, my.bublyk@gmail.com, Lida_lissovcka@ukr.net

Abstract. The author's method of assessing the readiness level of the research and development for commercialization in the market named TAPDS is developed. The TAPDS method thoroughly takes into account the market component and the readiness of consumers to acquire the research and development. The author's TAPDS methodology, in contrast to the existing methodological approaches, will enable the evaluators more thoroughly to take into account the market component and readiness of consumers to acquire the researched R&D products. Unlike existing ones, the value of the proposed TAPDS method is to provide a comprehensive assessment of the technology's readiness for the transfer. The complexity of the assessment is ensured by a specially developed formalized toolkit, which includes the evaluation of technology at the analytical, technological and patent levels of its readiness, as well as the level of readiness of demand for technology and the impact of society on its development. The method of hierarchy analysis is used for expert evaluation comparison. The clustering analysis is provided using k-means and c-means algorithms. The discriminant analysis is used for finding parameters influents for level of readiness and complexity reduction.

Keywords: Technology transfer · Readiness level · Clustering · Complexity reduction

1 Introduction

Among the contemporary factors in the development of the world economy, one of the leading ones is the knowledge that transforms into skills and innovative technologies. Knowledge becomes the basis for increasing the effectiveness of interaction of participants in innovative infrastructures of countries. An important tool for such interaction is the transfer of technology. The time of developing, mastering and generating market effects depends on the efficiency of technology transfer. However, the complexity of today's market demand for technology transfer is growing faster than developing appropriate methods to improve its efficiency.

N. Shakhovska and M. O. Medykovskyy (Eds.): CCSIT 2019, AISC 1080, pp. 259–282, 2020.
https://doi.org/10.1007/978-3-030-33695-0_19

One of the main components of a successful technology transfer process is to assess the level of technology readiness for the transfer. In the world, scientists and practitioners have developed methods and models to make technology ready for transfer, but for the most part, their developments are fragmentary in nature, aimed at evaluating specific technologies or using a narrow circle of business entities.

Current approaches to determining the level of technology readiness for transfer can be divided into two groups:

- approaches that focus on assessing the level of technology development as such (internal-oriented approach);
- approaches oriented to assessing the market susceptibility of technologies (external-oriented approach).

The first group of approaches is aimed at establishing the technological level of technology readiness. At the same time, other factors affecting the readiness of technology for transfer are mostly evaluated indirectly. However, a high level of technological innovation and technology readiness will not always mean its efficient transfer and commercialization. Market characteristics, elasticity and demand trends, target audience behavior, and other factors often drastically change the technology transfer vector.

There are developments in assessing the level of technology readiness for transfer, which partly aggregate the positions of these approaches. However, they have the following drawbacks:

- there is no correlation between the levels of technology and market readiness of technologies;
- the level of market readiness of the technology is determined by enlargement: it does not differentiate the marketing readiness of the technology, as analytical, and, in fact, dictated by demand (due to the consumer's receptivity);
- not developed formalized apparatus for determining the level of readiness of technologies for the transfer, the results of evaluation on the available approaches are mostly averaged;
- valid approaches do not always take into account social aspects when evaluating technologies (social corporate responsibility, etc.).

So far, a methodological approach has not been created that would allow to determine the level of readiness of technologies based on the aggregation of heterogeneous aggregate data on its development. The need for an integrated approach is determined by the current level and rates of market development.

Creating of such a methodological approach can become a major lever during: selecting and justifying investment decisions on technology to be transferred; making decision on further technology development; establishment of a strategy for the technology promotion and development; pricing on technology; modelling effects from market technology launch, etc. It is also important to take into account the expert opinion regarding the technology readiness.

2 State of Art

The results of the literature analysis indicate that the problems of assessing the technologies readiness for the transfer were not given due importance by the international community of scientists and practitioners. The experts focus mostly on developing the fundamentals of technology evaluation for commercialization, taking into account the specific features of the countries [1, 2]. Alternatively, they are based on the specifics of the activities of particular universities [3], at which these technologies were investigated.

Scientists [4, 5] propose methods and models for the transfer of university technologies. Some issues related to the readiness of technologies for the transfer are upgraded [6]. A group of scientists - [7, 8] and others investigated problems of technology transfer evaluation.

The question of assessing the level of technological readiness for the transfer is partially considered from the standpoint of academic entrepreneurship in the work [9, 10]; from the standpoint of technological entrepreneurship - [11, 12].

From the point of view of the interaction of universities and industry, scientists [13] investigate technology transfer. Approaches to the measurement of individual subsystems of technology transfer are shown in the study [14].

For the evaluation of innovative technologies, the practice of using common approaches (manuals) is widespread in the world. Among them, the most popular ones regarding deciding when to make technology transfer are “Oslo Manual” (2012), “Frascati Manual” (2002), “Canberra’s Guide” (1995), etc. In recent years, models of technology evaluation have been spread, in particular: NABC model, balanced scorecard model, value researcher model, technology broker model, production model for production technology S. Muege (2015), etc.

Despite the fact that there are significant issues in assessing the availability of technology for the transfer, so far, the methods and models that answer the above questions are fragmented. The degree of justification of the methodological provision for assessing the level of technology readiness for the transfer depends on the speed of its implementation and the effectiveness of technological development of both market actors and the state as a whole.

Particularly important is the consideration of this problem from the point of view of increasing the effectiveness of interaction between R&D participants. This is in particular confirmed by statistical data, which in recent years indicate a significant elevation of the role of such interaction. According to R&D Trends Forecast [26], the positions of the integral indicator Expectations for 2018 cooperation efforts relative to 2017, in particular, Participation in R&D alliances increased by 50%. The contracts/ grants position with academia increased by 37% in the period under review. At the same time, these indicators are the highest values in the framework of the integral index.

One of the most popular and widely used method in the world is the approach of the National Agency for the Exploration of the Space (NASA). This organization has developed and successfully used the Technological Readiness Level (TRL) Assessment Methodology, starting in 1974. In 2013, the TRL was standardized in the form of ISO 16290: 2013.

Despite the widespread approach of NASA [15] to assess the readiness of technology for their transfer, it is not without flaws. The first disadvantage is that among the 9 levels of technology readiness, not enough attention was paid to studying and analyzing markets and the demand for a product. More attention is paid to the process of product entry into the market, rather than the study of market readiness to perceive technology and the willingness of the consumer to pay for it.

The Readiness Development Model, published on the site of the National Academy of Sciences of Ukraine [16], also covers 9 levels of readiness. However, there is also no proper level of research on potential markets and their peculiarities in terms of readiness to accept innovation (technology).

In the vast majority of sources available to authors [17–19] the level of marketing readiness is considered in terms of micro level. More attention is paid to the development of the business plan [15–17], the development of procedures for the market entry of both pioneer and finished product [19, 20], etc.

The site of the independent, global open source community OW2 [17], which aims to promote the development of open source software, business applications, cloud computing platforms, offers a flexible, component-based approach to determining the marketing readiness of technologies for the transfer. The OW2 Market Readiness Levels from 3 to 9 consider OW2 Market Readiness Levels as a sequence of business situations that reflect the state of the project from its initial market readiness (development) to the final readiness (market leader) (Table 1).

Separately, the technology readiness scale and its relevance to the NASA TRL classification is presented. In this marketing activity case, more attention is paid to the internal levels of marketing (market) readiness. Attention to the external components of market readiness is focused only on OW2 MRL 5 – Market entry, OW2 MRL 6 - Market Expansion and OW2 MRL 9 - The market leader, which is the market entry, market expansion and market recognition of the product. Most researchers [17–20] adhere to such a simplified interpretation of marketing readiness. In general, all this marketing readiness should be called internal marketing readiness, which refers to the level of developer, i.e., the micro level. The level of market readiness at the macro level (external component) – describes the readiness of the market to perceive technology, solvency of the consumer of the corresponding segment, the presence of analogues-substitutes, the assessment of added value created for the consumer, etc.

Darmani and Jullien [19] are investing market readiness level (MRL) in the need to measure the maturity of the demand for this technology on the market. The MRL is described at twelve levels, from the definition of the existence of an unmet need in the product until full commercialization, along with the scale of the process. At each of the 12 levels, the following market parameters that influence the development and deployment of certain technologies are explored, examples are given for energy saving technologies.

First, various potential possibilities of application of the proposed technologies are analyzed. Further, after receiving research results, REEEM Roadmap for Technology and Innovation is used to evaluate the MRL. The approach consistently assesses whether there is a chain of value creation technology, supply chain and customer chain.

The competition is evaluated on the market and the market is susceptible to future technologies, which the authors call potential technology deployment. At each level, the market is investigated as to the existence of the need for proposed technologies, for the full commercialization and successful introduction of technology to the market.

Table 1. Comparison of used levels of readiness to the levels of technological readiness of NASA

Readiness levels	NASA TRL [15]	IRT [16]	OW2 MRL [17]	CRL [20]
9	Actual system proven in operational environment	Fully articulated business with appropriate infrastructure and staffing	Established Player	–
8	System complete and qualified	Capability to transition to full production and distribution	Actively Competitive	–
7	System prototype demonstration in operational environment	Capability to support limited production; full business team in place	Business Build-up	–
6	Technology demonstrated in relevant environment	Capability to support development and design with a market-driven business team	Market Broadening	Distribution and product sale in the open market is organized
5	Technology validated in relevant environment	Capability to support project engineering development and design	Market Opening	Test marketing is completed and readiness to enter the various sales markets is demonstrated
4	Technology validated in laboratory	Capability to work limited-scope programs with project teams	Usefulness Demonstrated	The second phase of prototype testing by the customers is finished and priority plan for sales market development is made
3	Experimental proof of concept	Experimental evidence of business opportunity	Fledgeling Usefulness	The first phase of prototype testing is carried out by the customers
2	Technology concept formulated	Paper studies produced	Development	Pre-planning of main features of the product and production volume is completed
1	Basic principles observed	Inventor or team with a dream	Research	The concept of the projected product is studied, explained and improved

Source: [15–19]

In [20], authors highlight the level of consumer readiness (CRL), which defines the level of knowledge about consumers and their needs for the proposed technology. This CRL consists of six levels that begin with the identification of the consumer and his/ her needs and end with the integration of the consumer into the development of technology. In addition, in the paper [20], authors highlight the level of readiness of society (SRL). This SRL consists of five levels that define the level of knowledge about the interests of stakeholders when launching the proposed technology. It also assesses the impact of technology on society, from the selection of stakeholders to their involvement in the development process.

In order to further study the issues of assessing the availability of technologies for the transfer, it is necessary to determine the categorical framework of the basic concept "technology".

The study of literary sources on the subject of "technology" has shown poly-aspectance of approaches. In particular, the Merriam-Webster dictionary interprets technology as: the practical application of knowledge in one or another field; way of performing the task, in particular using technical processes, methods or knowledge; specialized aspects of a specific field of activity [21]. The Oxford Dictionary provides an understanding of technology as the application of scientific knowledge for practical purposes, in particular in industry [22], Cambridge Dictionary - as a study and knowledge of practical, in particular, industrial, the use of scientific discoveries [23]. The American Heritage Dictionary provides several definitions of the term "technology":

- the use of science, in particular for industrial or commercial purposes;
- scientific method and material used to achieve commercial or industrial purpose;
- the core of knowledge available to society, obtained on the basis of the invention of the means obtained during the development of art and skills, as well as extraction or collection of materials [24].

R. Rhodes and co-authors have collected a number of essays on technologies that determine the development of the twentieth century. The book contains a brief overview of technology judgments from more than a hundred leading scholars, critics, historians and other professionals. Among them one of the most generalized definitions of technology is: "technology is the application of science, engineering and industrial organization for the creation of a world built by people" [25, p. 19].

Consequently, in this paper we will use the traditional generalized understanding of the concept of "technology", which can be both a scientific development, and a technological product, which is the result of a set of scientific and technical decisions. Based on this, technologies cannot be directly transmitted to consumers, as other categories in the B2B, B2G, etc. fields.

3 Materials and Methods

When forming the research goal, hypotheses were put forward.

Hypothesis 1. The level of readiness of R&D products depends, in particular, on the awareness of potential consumers and the potential benefits of using this development in real terms. The verification of this hypothesis will be carried out through a comparative analysis of the results of the assessment of scientific developments by the proposed author's methodology and methodology of NASA.

Hypothesis 2. The proposed methodology for assessing the readiness level of R&D products is universal and does not depend on technological complexity, the type of potential market and development categories.

In order to justify these hypotheses, NASA's methodological approach is used. Compared to others, this approach is characterized by a significant level of complexity, which contributes to the adequacy of the evaluations received. NASA's approach allows us to consistently assess the technical maturity of different types of technology. However, the disadvantage of this approach is the lack of developed tools for formalizing assessments, which leads to the groundlessness of its findings.

In addition, NASA (TRL, MRL, IPRL) technology assessment components provide a rather internal view of the technology's readiness for the transfer. In particular, NASA's MRLs receive a generalized assessment of technology's marketing readiness; do not take into account many factors that have a direct impact on its change.

Therefore, the purpose of this study is to develop the existing NASA methodology by combining the assessment of the transferability of R&D products from the standpoint of their internal potential (internal factors of readiness for commercialization), taking into account the level of market development and the availability of consumer needs on it (external factors of readiness for commercialization).

Thus, the proposed methodological approach is based on the evaluation of five components of R&D products of readiness for commercialization, namely:

- the technological readiness level (**T**RL),
- the level of analytical readiness (**A**RL),
- the patent level (**P**RL),
- the demand readiness level (**D**RL),
- the society impact level (**S**IL).

Using the initial letters of the abbreviation of the names of the components of the author's method was called **TAPDS**.

The application of such an integrated methodological approach to assessing the readiness of R&D products for commercialization will make it possible to more clearly identify the integrated level of development readiness, which in turn will form the basis for making more informed management decisions regarding the prospects and forms of commercialization of the development.

Thus, the technological readiness level TRL is estimated using the NASA division at the levels given in Table 1.

In order to assess the level of patent availability of the technology, PRL is applied by the NASA approach (IPRL) to five levels [15, 16]:

- technical solutions make know-how; provision is made for submission or submission of applications for obtaining security documents for industrial property rights;
- obtained and supported by Ukraine security documents on industrial property objects; an international patent application (s) for obtaining a patent (for the PCT system, etc.) has been filed;
- an application (s) for obtaining a security document in a foreign (s) country (s) is filed under the national procedure;
- received and maintained patent (s) for invention in a foreign (s) country (s).

It is expedient to extend the NASA MRL to the level of analytical readiness of the (ARL) and the demand readiness level (DRL). These components will allow managers to take into account both the external and internal characteristics of the market readiness of the technology to the transfer.

One of the most important questions of the initial stages of technology development is the rationale for its concept, accessible to the understanding of a wide audience. The installation of ARL focuses on internal development characteristics in the context of the economic justification of its readiness for transfer. Depending on the depth of understanding of the organizational and economic aspects of the possibilities of commercialization of the development and substantiation of the business model in accordance with the methodology, the level of ARL is identified.

The definition of ARL should be guided by marketing information, taking into account the completeness and quality of the analytical data on technology. ARL substantiates the entire R&D products process for this technology. The highest level of ARL is considered to be a business plan for introducing innovations into the market. The evaluation of technology for ARL involves a thorough evaluation of each of its levels for establishing an integral (aggregate) APR.

The assessment of the ARL can be carried out by: employees of analytical units of technology developers, independent experts, etc.

The ARL indicator group includes six levels. The methodological approach assumes that the transition to the next level of analytical maturity occurs after the implementation of the previous one.

Applying this approach to the ARL level assessment will reduce the risk level of technology implementation in the specific conditions of the future consumer and innovation entrepreneurship in general; make a positive contribution to the overall success of the development and implementation of advanced breakthrough technologies. It is concluded that attracting potential consumers to assess the readiness of technology for transfer in the aspect of the formation of a business plan is necessary at relatively early levels of marketing readiness.

In the author's technique, it is suggested to evaluate the level of the demand readiness level, which enables to determine the level of technology value for consumers, named DRL.

R&D result becomes a product when it brings consumer value: appropriate benefits and costs. Such a duality of R&D product is explained by the work it has invested in, which also has duality. Benefits and costs are two categories of economic evaluation of technologies that are mediated by mutual influence. A cost-benefit analysis is a process businesses use to analyze decisions. The business or analyst sums the benefits of a technology or action and then subtracts the costs associated with taking that technology or R&D product. Therefore, it is expedient to determine the DRL, designed to reflect the external characteristics of the technology from the standpoint of its transfer readiness.

The lifetime of technology in the market is determined not only by the producers of the technology that compete with each other, but also by the characteristics of the promise of this technology from the consumer's perspective, which are formed on the basis of the incorporation into the technology of consumer value. To find out the level of technology compliance with the requirements of consumers, it is necessary to distinguish between the nature of the need, which is either implemented or predictable, and is inherent. The consumer deals with a set of choices of goods (technologies) and chooses which one will satisfy his specific need. The choice will be determined by a list of consumer needs, according to which he evaluates each product. In view of this, consumer value is a measure of utility, and a set of consumer values determines the quality of products.

The value that a developer puts in technology determines the consumer value of technology in the form of a finished product and dictates the choice of the method of its valuation. Demand value is the basis for determining the value of technology, which in the future becomes the basis for pricing it. Sources of obtaining utility for the consumer are the basis for the implementation of demand value of goods.

DRL describes the readiness of the market to perceive technology, solvency of the consumer of the corresponding segment, the presence of analogues-substitutes, the assessment of added value created for the consumer, etc. The assessment of the market in terms of its readiness for technology transfer involves a thorough evaluation of each of the stages by the levels of the approach. In order to establish an integral (aggregated) indicator of the DRL for the technology transfer it is expedient to aggregate the obtained estimates.

The requirement readiness level DRL includes indicators that describe 4 demand states, that is, 4 levels. The methodological approach implies that the transition to the next level of demand readiness occurs after the achievement of the indicators of the previous one. DRL should reflect the level of market demand for the proposed technology, that is, the level of consumer interest in a particular technology. Here it is necessary to determine the level of readiness of the demand to adopt the technology and level of solvency of the consumer of the corresponding segment, the presence of analogues-substitutes, value added assessment.

The society impact level (SIL) also has 4 levels that describe the level of R&D impact on society. It deeper reveals the level of readiness of technology to transfer through the allocation of R&D products impact on society itself, changing its levels of development under the influence of R&D products. In the study of the SIL of R&D, the indicators of influence are not only scientific and technological advances in industrial markets and their transformation, but also changes in the level of households. According to the authors of work [18], the impact on households is deeper, affects the mental readiness of R&D products.

In contrast to NASA's MRL, the proposed DRL and SIL allow for the variability of both the operating environment of the technology and its external marketing environment to be taken into account, and the interrelations between them.

The results of the study indicate that the methods for determining the TRL and IPRL NASA meet the current needs of feasibility study. Each of the levels of technology readiness for the transfer is defined in nine stages, except IPRL, which consists of four.

In order to assess the market demand for technology, we believe that it is necessary to conduct an expert evaluation of the opinion of leading experts, which should include experts who directly make decisions on the introduction of new technologies or equipment in the company, the company. To assess the level of interest in a particular technology, readiness to accept technology, level of acceptance of prices and readiness, etc., an online survey should be conducted along with panel interviews with experts. This will help the end user to understand the new technology.

In order to substantiate the integral assessment of technology readiness for the transfer, it is proposed to use a formalized apparatus on the principles of complexity. An integrated approach to assessing the availability of technologies for a transfer provides the following functions:

1. orienting the assessment to the target result, rather than the interim results (this makes it possible to assess the readiness of the technology in the context of its market competitiveness);
2. promotes thorough and comprehensive research of technology at all stages of its development;
3. leads to the successful preparation of technology through the correction of structural and functional interrelations between the stages of technology preparation and management, taking into account their hierarchy;
4. facilitates the efficient transfer of technologies.

4 Results

Approval of the methodology was carried out in two stages.

Stage 1

Component indicators of technology readiness for the transfer are uneven within the integral index of such an assessment, which requires their normalization. At the beginning, in order to determine the values of the weights, a set of experts' interviews were conducted that consisted of potential customers of the research and development. The survey data was processed using the method of hierarchy analysis.

To determine the values of the weights, a multi-choice problem has been formed: we have a global goal and five components of R&D products of readiness for the transfer: TRL, ARL, PRL, DRL, SIL. Consequently, it is necessary to find the specific weights (w_{ij}) of the indicated components (X) in the integral index of the process of assessing the readiness level of the technology to the transfer (as goal) (L). Weights are determined by the empirical matrix of pairwise comparisons given by the expert.

In order to form the matrices of pair comparisons, the Saati scale is used in [27]. The matrix of pair comparisons is formed based on comparing the elements in the table with each other in terms of decision makers and is inversely symmetric.

To assess the reliability of the presented source data, an index of information consistency K is calculated. To do this, the value of L is calculated:

$$L = \sum_{i=1}^{n} w_{i1}X_1 + \sum_{i=1}^{n} w_{i2}X_2 + \ldots + \sum_{i=1}^{n} w_{in}X_n = \sum_{j=1}^{n}\sum_{i=1}^{n} w_{ij}X_j \tag{1}$$

After L calculation, the index of consistency is calculated. This index is reflecting the degree of accuracy of expert information.

In the paper, we propose to evaluate the readiness of 16 R&D products. The scope of the expert group is 17 people. This expert group consists of specialists of different domain area, particularly10 scientists, 2 patent scientists, 3 marketers, research director and her deputy (2). That is why we should calculate the weight of the components of assessing the readiness of the technology for the transfer. Table 2 shows the results of determining the weight of five components of R&D products in the process of evaluating their readiness for the transfer.

Table 2. Specific weight of the components of assessing the readiness of the technology for the transfer

Evaluation components	Values of scales, points	Normalization of weighting coefficients, %
TRL	51,35	32,1
ARL	33,2	20,7
PRL	23,6	14,7
DRL	38,8	24,3
SIL	13,05	8,2
Total	160	100

Based on the results of the evaluation, TRL is recognized by the experts as the most important; it makes up 32.1% of the total number of evaluations received. DRL is characterized by the following important value - 24.3%. The smallest weight value is inherent in SIL - 8.2%.

Stage 2

The next step was to conduct a quantitative survey during personal meetings with the respondents - the leaders of the research groups of the Lviv Polytechnic, who are the developers of the researched R&D products. Polling place was research laboratories of Lviv Polytechnic, polling time was 2018-2019. To make sure that our sample of that population is representative it was used formula for the case when the population is known by researchers:

$$n = \frac{t^2 \times V^2 \times N}{Vx^2 \times N + t^2 \times V^2}, \tag{2}$$

where n is size of training dataset, N is size of whole dataset, V_x is level of accuracy, t is the level of reliability, V is dispersion of the studied feature.

The sample of n = 16 developments are representative under the conditions, given the population N = 85 developments, confidence level 95%, which corresponds to a table-valued t-test, equal to 2,015, given coefficient of variation of the signs V_x = 0,33 (maximum coefficient if variation, ensuring the uniformity of signs) and the level of Confidence Interval of survey results V = 0,15.

Therefore, according to the calculation below, the sample size is 16 respondents. The tool of research was the questionnaire, which consisted of 12 closed issues of a substantive nature and 4 questions of a personal character.

Table 3. Resulted table of evaluation of levels of readiness of R&D products by the TAPDS method

№	Name of R&D products	TRL	ARL	PRL	DRL	SIL	Level of readiness
1	Packing equipment	0,111	0,167	0,200	0,500	0,750	**0,287**
2	Planner for military action	0,222	0,167	0,200	0,250	0,500	**0,239**
3	On-Board Equipment of the Signal Processing for Multi-Spectral Scanner MSU-M of the Perspective Spacecraft Sich-2 M	0,556	0,333	0,200	0,250	0,250	**0,357**
4	Device for Determination of Optical Duration of GNSS Measurement in Monitoring of Deformation of Engineering Structures	0,111	0,167	0,200	0,250	0,500	**0,204**
5	Biocides for the protection of petroleum products and materials from biological influences	0,111	0,167	0,200	0,250	0,500	**0,204**
6	Growth regulator and biocide for protection against phytopathogenic microflora in phytochemical treatment of oil contaminated soils	0,111	0,167	0,200	0,500	0,500	**0,266**
7	Ensuring Technological Strength of Welded Joints with Armored Steel of ARMSTAL 500-Type	0,667	0,167	0,200	0,500	0,250	**0,419**
8	Information Technology for Personalized Medical Information Processing	0,778	0,167	0,200	0,750	0,500	**0,538**
9	Lactic fermented beverages based on the microbiota "Tibetan fungus"	0,667	0,167	0,800	0,250	0,250	**0,442**
10	Fundamentals of technology for the production of copper nitrate (copper nitrate) from copper waste	0,444	0,167	0,400	0,750	0,250	**0,439**
11	Automated System of Geodetic Monitoring	0,778	0,167	0,600	0,500	0,250	**0,511**
12	Personal passive Optically stimulated luminescence dosimetry means of ionizing radiation	0,333	0,333	0,200	0,250	0,750	**0,330**
13	Structural and finishing materials of new generation based on nanomodified cementing systems	0,444	0,333	0,200	0,500	0,750	**0,427**
14	Seismic Vibration Sensor of Special Purpose	0,444	0,167	0,400	0,250	0,500	**0,337**
15	Granular hydrogel materials for controlled release of drugs	0,556	0,167	0,200	0,250	0,750	**0,366**
16	Hydrogel bandages for the treatment of damaged skin	0,667	0,167	1,000	0,750	0,500	**0,616**
Average		**0,438**	**0,198**	**0,338**	**0,422**	**0,484**	**0,374**

Each of the components of R&D products readiness for commercialization has the following levels:

- The technological readiness level (**T**RL) – 9;
- The level of analytical readiness (**A**RL) – 6;
- The patent level (**P**RL) – 5;
- The demand readiness level (**D**RL) – 4;
- The society impact level (**S**IL) – 4.

The 16 R&D products were analyzed. The results of evaluated is given in Table 3. Normalizes as feature scaling, min - max, or unity-based normalization typically used to bring all values into the range [0,1]. However, this can be generalized to restrict the range of values in the dataset between any arbitrary points a and b, using:

$$X' = a + (x - x_{min})(b - a)/(x_{max} - x_{min}). \tag{3}$$

The results of calculating R&D product readiness values for the transfer (Table 3) show that the highest values among them are SIL - 0,484 and TRL - 0,438. Values of DRL (0,422) are closed to the meaning of them. Significantly, lower values of ARL and PRL are 0.198 and 0.338, respectively. In general, the evaluation practice has shown that in most cases the TRL, DRL and SIL have a dominant influence on the overall assessment of the level of readiness of R&D products. The advantage of evaluating these constituents, as well as PRL, is that they can be determined quickly. The obtained values do not substantiate the necessary and sufficient level of readiness of R&D products to the transfer; in particular, they do not reflect the features and characteristics of the development of R&D products in a particular market segment. The solution to this collision is the careful evaluation of the ARL, based on the research of the prospects for the development of R&D products, taking into account the specifics of the market, that is, the analytical determination of their transferability. Attention to the ARL appraisers will increase the level of accuracy of the justification of the level of readiness of R&D products for the transfer into the business environment, based on which the commercialization of R&D products will be intensified.

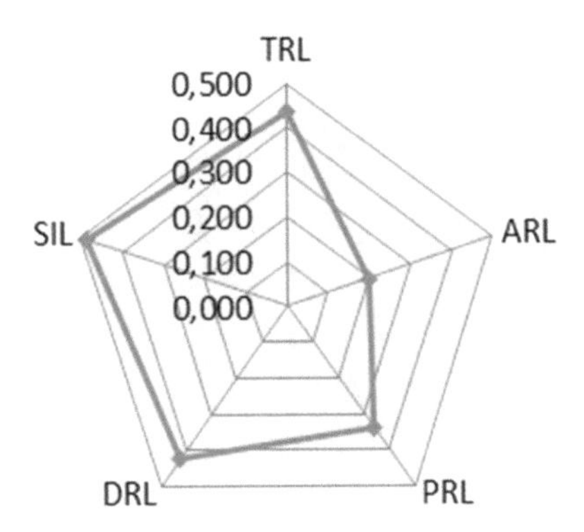

Fig. 1. The petaldiagram of the average levels of readiness of the researched R&D products

The conducted researches prove that according to the average indicators, the level of readiness of the investigated R&D products can be represented by a petaldiagram (Fig. 1), where TRL is of greatest importance.

Let us provide clustering analysis using language R and different clustering approaches. We have 16 instances and want to create clusters from similar objects.

The basic idea behind k-means clustering consists of defining clusters so that the total intra-cluster variation (known as total within-cluster variation) is minimized.

There are several k-means algorithms available. The standard algorithm defines the total within-cluster variation as the sum of squared distances Euclidean distances between items and the corresponding centroid:

$$W(C_k) = \sum x_i \in C_k (x_i - \mu_k)^2, \tag{4}$$

where x_i design a data point belonging to the cluster $C_{k,}$, μ_k is the mean value of the points assigned to the cluster C_k. Each observation (x_i) is assigned to a given cluster such that the sum of squares (SS) distance of the observation to their assigned cluster centers μ_k is a minimum. The total within-cluster variation is defined as follow:

$$tot.withinss = \sum_{k=1}^{k} W(C_k) = \sum_{k=1}^{k} \sum_{x_i \in C_k} (x_i - \mu_k)^2. \tag{5}$$

The *total within-cluster sum of square* measures the compactness (i.e. *goodness*) of the clustering and we want it to be as small as possible.

The first step when using k-means clustering is to indicate the number of clusters k that will be generated in the final solution. We propose to compute a clustering algorithm of interest using different values of clusters k. Next, the *wss* (within sum of square) is drawn according to the number of clusters. The location of a bend (knee) in the plot is generally considered as an indicator of the appropriate number of clusters.

By default, the **R** software uses 10 as the default value for the maximum number of iterations (Fig. 2).

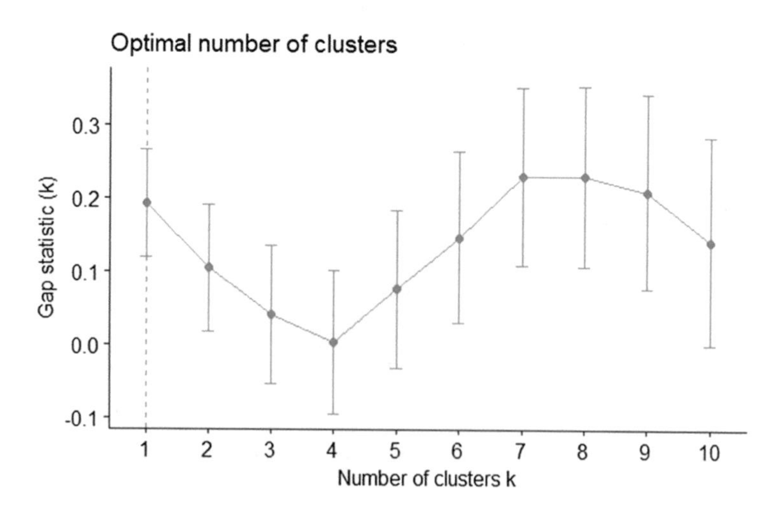

Fig. 2. Optimal number of clusters

Let us create 4 clusters. The result of k-means looks like this (Fig. 3):

```
K-means clustering with 4 clusters of sizes 4, 3, 3, 6
Cluster means:
          V1         V2        V3        V4         V5        V6
1 -1.2036286 -1.910283 -3.356439 -1.168461 -0.4904146 -1.218457
2 -0.7218424 -1.588728 -2.900823 -3.819132 -0.6538862 -1.352350
3 -3.8167128 -2.488915 -1.766092 -1.998152 -0.9241962 -4.605170
4 -0.5269124 -2.782902 -0.906862 -1.350218 -1.2707698 -0.898685
Clustering vector:
 [1] 1 1 2 3 3 3 4 1 4 4 4 2 1 4 2 4
Within cluster sum of squares by cluster:
[1]  7.028439  4.139010  1.618399 19.643758
 (between_SS / total_SS =  73.8 %)
```

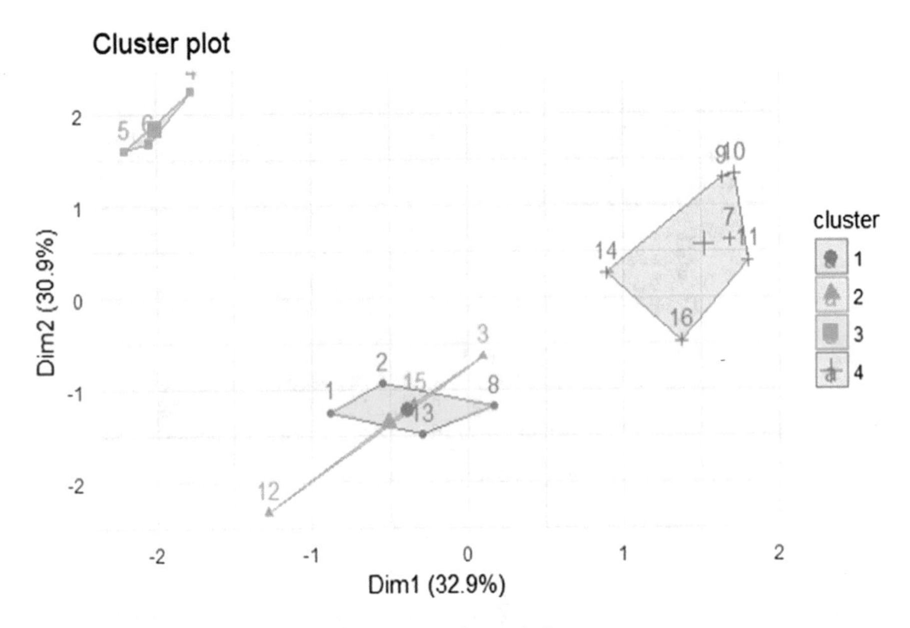

Fig. 3. The clusters' shape

Fuzzy c-means (FCM) is a method of clustering which allows one piece of data to belong to two or more clusters. This method is frequently used in pattern recognition. It is based on minimization of the following objective function:

$$J_m = \sum_{i=1}^{N} \sum_{j=1}^{C} u_{ij}^m \left\| x_i - c_j \right\|^2, 1 \leq m \leq \infty \quad (6)$$

where m is any real number greater than 1, u_{ij} is the degree of membership of x_i in the cluster j, x_i is the ith of d-dimensional measured data, c_j is the d-dimension center of the cluster, and $\|*\|$ is any norm expressing the similarity between any measured data and the center.

Fuzzy partitioning is carried out through an iterative optimization of the objective function shown above, with the update of membership u_{ij} and the cluster centers c_j by:

$$u_{ij} = \frac{1}{\sum_{k=1}^{C}\left(\frac{\|x_i - c_j\|}{\|x_i - c_k\|}\right)^{\frac{2}{m-1}}},\ c_j = \frac{\sum_{i=1}^{N} u_{ij}^m \cdot x_i}{\sum_{i=1}^{N} u_{ij}^m} \tag{7}$$

This iteration will stop when $max_{ij}\left\{\left|u_{ij}^{(k+1)} - u_{ij}^{(k)}\right|\right\} < \varepsilon$, where ε is a termination criterion between 0 and 1, whereas k are the iteration steps.

This procedure converges to a local minimum or a saddle point of J_m.

```
Fuzzy c-means clustering with 4 clusters
Memberships:
               1           2           3           4
 [1,] 0.066852661 0.680993885 0.144388444 0.107765010
 [2,] 0.040084749 0.634488764 0.242865294 0.082561193
 [3,] 0.034816589 0.103690545 0.788096594 0.073396272
 [4,] 0.911394179 0.030261533 0.033045113 0.025299176
 [5,] 0.992863472 0.002593118 0.002454238 0.002089173
 [6,] 0.914499866 0.032766787 0.026164947 0.026568400
 [7,] 0.065332079 0.397206079 0.157745894 0.379715949
 [8,] 0.047852709 0.704069585 0.119041709 0.129035997
 [9,] 0.112133673 0.186211460 0.324853298 0.376801569
[10,] 0.058355325 0.162966091 0.086119323 0.692559261
[11,] 0.005140135 0.019086958 0.014041735 0.961731173
[12,] 0.065242910 0.171358351 0.674275185 0.089123554
[13,] 0.050943908 0.541952165 0.175205774 0.231898153
[14,] 0.060630740 0.168195816 0.241968645 0.529204798
[15,] 0.025024673 0.093098170 0.823463759 0.058413398
[16,] 0.049984415 0.143818288 0.098691555 0.707505742
Closest hard clustering:
 [1] 2 2 3 1 1 1 2 2 4 4 4 3 2 4 3 4
```

The key operation in hierarchical agglomerative clustering is to repeatedly combine the two nearest clusters into a larger cluster. Let us have a look at its working:

1. It starts by calculating the distance between every pair of observation points and store it in a distance matrix.
2. It then puts every point in its own cluster.
3. Then it starts merging the closest pairs of points based on the distances from the distance matrix and as a result the amount of clusters goes down by 1.
4. Then it recomputes the distance between the new cluster and the old ones and stores them in a new distance matrix.

5. Lastly, it repeats steps 2 and 3 until all the clusters are merged into one single cluster.
6. One of the common linkage methods is complete-linkage, which calculates the maximum distance between clusters before merging (Fig. 4).

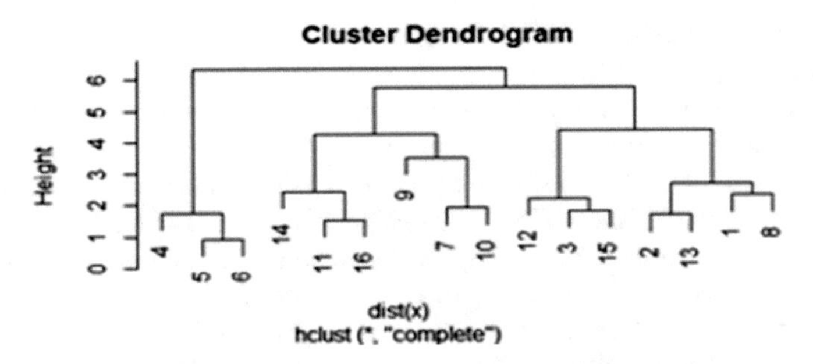

Fig. 4. Result of hierarchical clustering

Let us analyze the importance of each characteristics for each cluster. For this purpose, we use PCA. Figure 5 presents result of analysis for five variables (without Level of readiness).

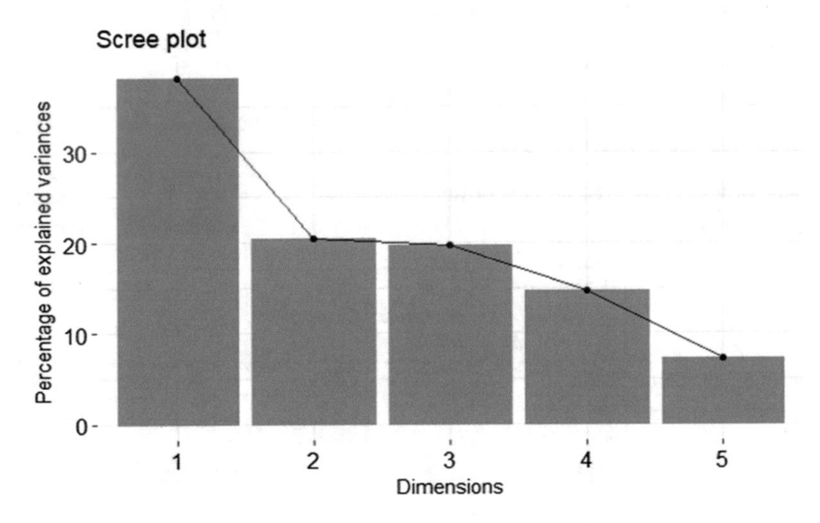

Fig. 5. PCA analysis

As you can see, the variable 1 is most important for result. Let us compare given method with well known. Comparative data on the readiness level of R&D products by the NASA method and the author's TAPDS method is given in Table 4.

The comparison shows the importance of considering the readiness of the technology to take into account the levels of its demand readiness and society impact for commercialization. The higher the DRL, the higher the probability of the success of the commercialization process. The level of readiness of R&D products depends, in particular, on the awareness of potential consumers and the potential benefits of using this development in real terms. Thus, hypothesis 1 is confirmed by comparative analysis of the results of the assessment of R&D products by the proposed author's methodology and methodology of NASA.

Listed on Fig. 6 levels of technology readiness for transfer to each R&D product point to a significant imbalance in most R&D products. None of the constructed pentagons has the correct form with the maximum value of the readiness levels of R&D products for the transfer, most of them have high values at one of the levels of readiness and very low on the other. Such irregular pentagon shapes may point to future problems with their commercialization process. The obtained results of the levels of readiness of R&D products for the transfer point also to the success of the transfer process itself and to the way of commercialization of R&D products by enhancing those components of readiness levels with low of level means.

It is worth noticing that the results of the research do not confirm the uniqueness of the hypothesis 2. It has not been proved that the proposed methodology for assessing the readiness level of R&D products is universal and does not depend on the technological complexity and category of development. During the study, the limited use of this technique to assess the readiness level of scientific and technological services, as scientific and technical services as a scientific product have a number of specific characteristics that do not take into account the proposed TAPDS methodology.

Table 4. Comparative data on the readiness level of R&D products by the NASA method and the author's TAPDS method

№	Name of R&D products	The features of R&D products			Level of readiness by method, points		Deviation level of readiness, points
		*Complexity of technology**	*Type of potential Market***	*CEA Groups****	*NASA*	*TAPDS*	
1	Packing equipment	High	A, B,	C28	0,581	0,287	−0,294
2	Planner for military action	Medium	A, C	O84	0,42	0,239	−0,181
3	On-Board equipment of the signal processing for Multi-spectral scanner MSU-M of the perspective spacecraft Sich-2M	High	A, B	O84	0,111	0,357	0,246

(*continued*)

Table 4. (*continued*)

№	Name of R&D products	The features of R&D products			Level of readiness by method, points		Deviation level of readiness, points
		*Complexity of technology**	*Type of potential Market***	*CEA Groups****	*NASA*	*TAPDS*	
4	Device for determination of optical duration of GNSS measurement in monitoring of deformation of engineering structures	High	A, B, C	F43	0,022	0,204	0,182
5	Biocides for the protection of petroleum products and materials from biological influences	Medium	A, B, C	C19, C20	0,022	0,204	0,182
6	Growth regulator and biocide for protection against phytopathogenic microflora in phytochemical treatment of oil contaminated soils	Medium	A, B, C	C20	0,022	0,266	0,244
7	Ensuring technological strength of welded joints with armored steel of ARMSTAL 500-type	Medium	C	O84	0,69	0,419	−0,271
8	Information technology for personalized medical information processing	High	A, B, C	Q86, J 62	0,78	0,538	−0,242
9	Lactic fermented beverages based on the microbiota "Tibetan fungus"	Medium	A, B, C	C10, A01	0,678	0,442	−0,236

(*continued*)

Table 4. (*continued*)

№	Name of R&D products	The features of R&D products			Level of readiness by method, points		Deviation level of readiness, points
		*Complexity of technology**	*Type of potential Market***	*CEA Groups****	*NASA*	*TAPDS*	
10	Fundamentals of technology for the production of copper nitrate (copper nitrate) from copper waste	Medium	A, B, C	A01, C20, C14, E38	0,46	0,439	−0,021
11	Automated System of Geodetic Monitoring	High	B, C	M71	0,723	0,511	−0,212
12	Personal passive Optically stimulated luminescence dosimetry means of ionizing radiation	High	A, C	O84	0,371	0,330	−0,041
13	Structural and finishing materials of new generation based on nanomodified cementing systems	Medium	A	F43	0,422	0,427	0,005
14	Seismic vibration sensor of special purpose	Medium	A, C	O84	0,11	0,337	0,227
15	Granular hydrogel materials for controlled release of drugs	Medium	A, C	C21	0,721	0,366	−0,355
16	Hydrogel bandages for the treatment of damaged skin	High	A, B, C	Q86, J 62, O84	0,679	0,616	−0,063

* *The complexity of technology indicates the number of developments that make up this technology (in particular, the number of patents, or the number of developments subject to patenting within the same technology, etc.)*

***The market type is defined on the basis of the subjects between which it is planned to conclude agreements on technology transfer: A - market of manufacturers; B - the market of intermediate sellers; C - market of state institutions.*

****The choice of the Classification of Economic Activities (from A to U) group is carried out in accordance with [28].*

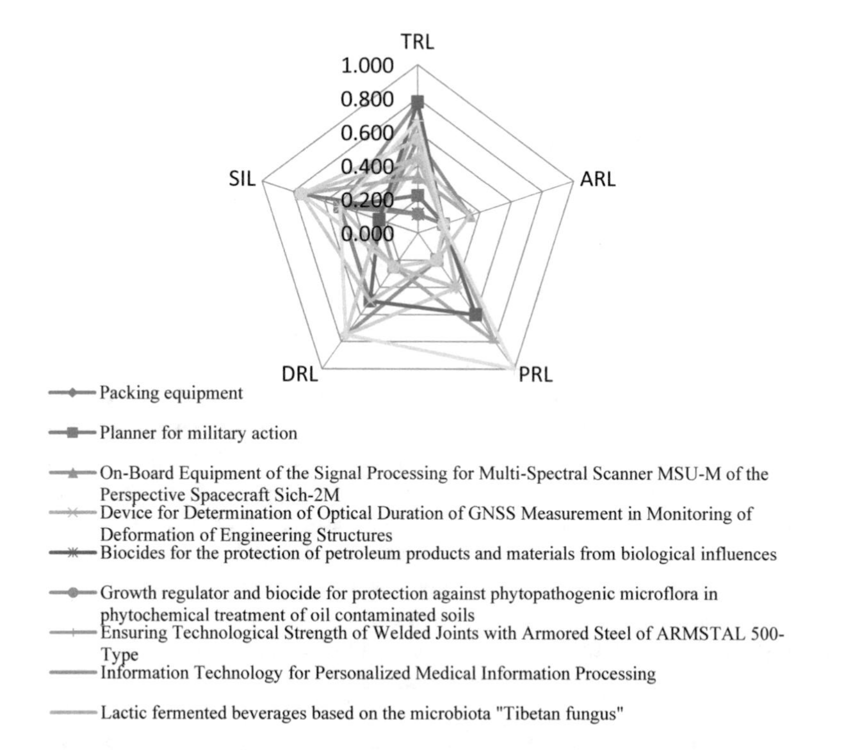

Fig. 6. The petaldiagram of the each levels of readiness of the each evaluated R&D products

5 Conclusions

The synthesis of the results obtained with the author's TAPDS method makes it possible to draw conclusions about the level of readiness of R&D products for commercialization on the market. The conducted studies prove the usefulness of the author's TAPDS methodology, the use of which, in contrast to the existing methodological approaches, will enable the evaluators more thoroughly to take into account the market component and readiness of consumers to acquire the researched R&D products.

Unlike existing ones, the value of the proposed TAPDS method is to provide a comprehensive assessment of the technology's readiness for the transfer. The complexity of the assessment is ensured by a specially developed formalized toolkit, which includes the evaluation of technology at the analytical, technological and patent levels of its readiness, as well as the level of readiness of demand for technology and the impact of society on its development. The indicated levels of technology readiness for the transfer focus the evaluators on the target result (it can be both commercialization and market technology, and the choice of scenarios for social technology transfer, etc.). At the same time, each of the stages of evaluation of the specified levels allows you to trace the peculiarities of the development of technology in one or another context, which contributes to taking into account their impact on the target result. In particular, the objectivity of this is ensured by the author's approach to the valuation of technology readiness at the appropriate stages of the assessment levels.

The proposed TAPDS method provides reasonable estimation results, since they are based on the interconnection of the structural-functional interaction between the stages of technology preparation and the transfer. This contributes to a comprehensive study and consideration of aspects of passing the stages of technology readiness to the transfer.

An important characteristic of the proposed TAPDS method is the relevance of the method to modern trends of innovation development: thanks to the concept of open innovation and other events and phenomena caused by it, the technologies are characterized by a high level of duality of the work embedded in them. This imposes new requirements for evaluating such technologies. It is important not only to evaluate the readiness of the technology for the transfer (mainly NASA-oriented approaches and analogues to it), but to take into account the peculiarities of the development of value added from this technology, which may affect the further market convergence of technology. Accordingly, taking into account this in the TAPDS (in the part of ARL, DRL and SIL) will contribute to an objective assessment of the market value of technology, which will be determined based on its level of development.

The TAPDS method allows analysts to take into account the variability of both the internal (operational) environment of the technology and its external environment (market development) and coordinate them in order to effectively justify the level of technology readiness for the transfer. The method is the basis for the further formation of a business model of technology development.

Acknowledgment. The research was supported by Ministry of Education and Science of Ukraine in the frame of the state budget theme TRANSFER (Project "Evaluation of technology value and readiness to transfer from universities to the business environment" (2019-2021) state number registration 0119U002251).

References

1. Gilsing, V., Bekkers, R., Freitas, I.M.B., Van der Steen, M.: Differences in technology transfer between science-based and development-based industries: transfer mechanisms and barriers. Technovation **31**(12), 638–647 (2011)
2. Hülsbeck, M., Lehmann, E.E., Starnecker, A.: Performance of technology transfer ofces in Germany. J. Technol. Transfer **38**(3), 199–215 (2013)
3. Chapple, W., Lockett, A., Siegel, D., Wright, M.: Assessing the relative performance of U.K. University technology transfer offices: parametric and non-parametric evidence, Rensselaer Working Papers in Economics from Rensselaer Polytechnic Institute, Department of Economics, 38 p. (2005)
4. Hess, S., Siegwart, R.Y.: University technology incubator: technology transfer of early stage technologies in cross-border collaboration with industry. Bus. Manag. Res. **2**(2) (2013). https://doi.org/10.5430/bmr.v2n2p22 URL: http://dx.doi.org/10.5430/bmr.v2n2p22
5. Bradley, S., Hayter, C., Link, A.: Models and methods of university technology transfer. Found. Trends Entrepreneurship **9**(6), 571–650 (2013)
6. Tyre, M.J., Orlikowski, W.J.: Windows of opportunity: temporal patterns of technological adaptation in organizations. Organ. Sci. **5**(1), 98–118 (1994)

7. Kumar, S., Luthra, S., Haleem, A., Mangla, S.K., Garg, D.: Identification and evaluation of critical factors to technology transfer using AHP approach. Int. Strateg. Manag. Rev. **3**, 24–42 (2015). https://doi.org/10.1016/j.ism.2015.09.001
8. Siegel, D.S., Waldman, D.A., Atwater, L.E., Link, A.N.: Toward a model of the effective transfer of scientific knowledge from academics to practitioners: quantitative evidence from the commercialization of university technologies. J. Eng. Technol. Manag. **21**(1–2), 115–142 (2004). https://doi.org/10.1016/j.jengtecman.2003.12.006
9. Chukhray, N.I., Mrykhina, O.B.: Theoretical and methodological basis for technology transfer from universities to the business environment. Probl. Perspect. Manag. **16**(1), 399–416 (2018). https://doi.org/10.21511/ppm.16(1).2018.38
10. Litan, R.E., Cook-Deegan, R.: Universities and economic growth: the importance of academic entrepreneurship. In The Kauffman Task Force on Law, Innovation, and Growth (ed.), Rules for growth: Promoting innovation and growth through legal reform. Ewing Marion Kauffman Foundation, Kansas City (2011)
11. Fleming, L., Woodward, Y., Golden, J.: Science and technology entrepreneurship for greater societal benefits: ideas for curriculum innovation. In: Libecap, G.D., Thursby, M., Hoskinson, S. (eds.) Advances in the Study of Entrepreneurship, Innovation and Economic Growth. Spanning boundaries and disciplines: University technology commercialization in the idea age, vol. 21. Emerald Books (2010)
12. Mosey, S., Muerrero, G., Greenman, A.: Technology entrepreneurship research opportunities: insights from across Europe. J. Technol. Transfer **42**(1), 1–9 (2017). https://doi.org/10.1007/s10961-015-9462-3
13. Albats, E., Fiegenbaum, I., Cunningham, J.A.: A micro level study of university industry collaborative lifecycle key performance indicators. J. Technol. Transfer (2017). https://doi.org/10.1007/s10961-017-9555-2
14. Autio, E., Laamanen, T.: Measurement and evaluation of technology transfer: review of technology transfer mechanisms and indicators. Int. J. Technol. Manag. **10**(7–8), 643–664 (1995)
15. NASA (2017). Technology Readiness Level. https://www.nasa.gov/directorates/heo/scan/engineering/technology/txt_accordion1.html
16. Levels of development readiness. http://www.nas.gov.ua/RDOutput/EN/book2017/Pages/iprnav.aspx
17. The OW2 Market Readiness Levels. https://oscar.ow2.org/view/MRL/
18. Bublyk, M., Shpak, N, Rybytska, O.: Social minima and their role in the formation of household welfare in Ukraine. Sci. Bull. Polissia **1**(9), 1, 63–71 (2017)
19. Innovation Readiness Level Report Energy Storage Technologies Authors This report is led and developed by InnoEnergy (Anna Darmani, Celine Jullien). http://www.reeem.org/wp-content/uploads/2017/09/REEEM-D2.2a.pdf
20. Brutyan, M.M.: Conceptual aspects of marketing readiness level assessment model. Global J. Eng. Sci. Res. Manag. **4**(10), 34–44 (2017)
21. Merriam-Webster Dictionary: Definition of Technology (2019). https://www.merriam-webster.com/dictionary/technology. Accessed 20 April 2019
22. Oxford dictionaries: Definition of Technology (2019). https://en.oxforddictionaries.com/definition/technology. Accessed 20 April 2019
23. Cambridge dictionary: Definition of Technology (2019). http://dictionary.cambridge.org/dictionary/english/technology. Accessed 20 April 2019

24. American Heritage Dictionary: Definition of Technology (2019). https://ahdictionary.com/word/search.html?q=technology. Accessed 20 April 2019
25. Rhodes, R.: Visions Of Technology: A Century Of Vital Debate About Machines Systems And The Human World. Simon & Schuster, New York (2000)
26. Results from the Innovation Research Interchange's Annual Survey. R&D Trends Forecast, Res.-Technol. Manag. J. 61(1), 23–34 (2016). https://doi.org/10.1080/08956308.2018.1399021
27. Chukhray, N., Shakhovska, N., Mrykhina, O., Bublyk, M., Lisovska, L.: Consumer aspects in assessing the suitability of technologies for the transfer. In: Proceedings of International scientific conference "Computer sciences and information technologies" (CSIT-2019), vol. 3 pp. 142–147. IEEE (2019)
28. Classification of types of economic activity DK 009: 2010. https://zakon.rada.gov.ua/rada/show/vb457609-10. Accessed 20 April 2019

Development of Project Managers' Creative Potential: Determination of Components and Results of Research

Anastasiia Voitushenko(✉) and Sergiy Bushuyev(✉)

Kyiv National University of Construction and Architecture, Kiev, Ukraine
voityshenko@gmail.com, sbushuyev@ukr.net

Abstract. The article gives a description of the creative potential, problems of its formation, development and effective use in the formation and implementation of the strategy of innovation development of Ukrainian enterprises. Earlier it was believed that the development of creative potential is a task of psychology, but nowadays more and more scientists are paying attention to the development of creative potential from the point of view of competence in project management. This is due to the fact that the project management industry is developing and there are new characteristics that affect the work of the team and the enterprise as a whole. Our study showed that only 11 competencies were identified by experts as a set of competences that impacted the development of creative potential. Of course, in some cases, there may be more criteria, but in most cases 11 competencies will suffice. We also investigated the priority of these competencies (criteria), the ways of their development, and provided methodological recommendations on how to improve them.

Keywords: Creativity · Creative potential · Competences · Matrix of ranking competences

1 Introduction

The development of innovative projects is a key trend in globalization and requires research related to the formalization of knowledge about project management, portfolios and programs based on competencies.

In today's economy, innovation plays a huge role. Without the use of innovation, it is virtually impossible to create competitive products that have a high degree of knowledge and novelty.

Thus, innovations are an effective means of competition because they lead to the creation of new needs, to lower the cost of production, to the inflow of investments, to improve the image (rating) of the manufacturer of new products, to the opening and capture of new markets, including foreign markets.

The innovative project aims to develop theoretical and practical issues related to the creation, distribution and use of new technologies and new products.

N. Shakhovska and M. O. Medykovskyy (Eds.): CCSIT 2019, AISC 1080, pp. 283–292, 2020.
https://doi.org/10.1007/978-3-030-33695-0_20

The concept of creativity is used to detail the impact of innovative solutions on the economic system. The categories of innovation and creativity are similar, but not identical. There is a connection between them, though each category is one in itself.

The theme of the development of creative management is extremely relevant for the contemporary development of Ukrainian society, which is of great importance.

Today, personal competences occupy a leading place in the practical activities of organizations. It is common knowledge that it is from the competencies that managers have in determining the success of business in general. Therefore, it is extremely important to study the problems of competences, levels of development of creative potential in the practical activities of organizations.

To achieve an organizational goal, it is necessary that managers have specific skills and skills and have the opportunity to use them practically. One of the most common perspectives on this problem is based on the definition of managerial organizational capabilities and competencies as the most influential factors in its successful long-term development. That is, the company must create the necessary conditions not only for the development and upgrading of the personnel, but also for the effective realization of their creative potential, since the modern strategy of human resources management is both in determining the ways of developing these competencies of all personnel of the organization and each of the project managers separately.

To date, the main problem is that not developed a universal method for evaluating and selecting competences that influence the formation of creative potential. Although the universal method of evaluation is not defined, however, the most effective methods for assessing competences are proposed in this article, and a mathematical model for forming the required set of competences and determining their priority is developed.

The methodological basis is the conceptual provisions on the essence of the competence approach in the management of personnel of the organization, specific models of assessment and development of competences, the system of general scientific and special methods.

2 Theoretical Studies

The analysis of the competencies of the enterprise and their role in ensuring its successful functioning has been reflected in the work of many scholars and practitioners in the field of strategic management and human resources management. The last few years have been marked by an active search for the methodological foundations for identifying and assessing the competencies of the enterprise and managing the process of their formation [1–4].

Teamwork is one of the skills that today are highly valued in the professional environment of the project, driven by the implementation of various personal and interpersonal skills. At present, the effectiveness of teamwork is estimated on the basis of goal-setting, interpersonal, competency and role-based approaches.

Competence approach in assessing professional qualifications is based on the appropriate set of criteria, which allows forming the assessment of the professional level of candidates, depending on the specificity of the project. It allows to form assessments of candidates by their ability to apply the creative knowledge and skills of their professional activity, which manifests itself in a wide range of research and production tasks [5].

Competences are key elements in the following cases: high-quality job descriptions, development of methods and criteria for selecting employees; personnel productivity management; development of methods and tools for personnel assessment; description of the processes of assessment of individual competence and results; development of personnel development programs. In general, the use of competence modeling as a method for assessment, training and competency development is a common practice in Human Resource Management a science that studies the recruitment, distribution, development, leadership, retention and dismissal of staff.

It is important for the project team to identify the competences that affect the success of the project. Building appropriate models of project team members' behavior to predict their impact on successful project implementation can serve as the basis for building a project management strategy for each project based on specific situations.

The project team is one of the main concepts of project management. This is a group of employees who are directly involved in the implementation of the project and subordinate to the head of the latter; the main element of its structure, as it is the project team that implements its design. This group is created for the period of project implementation and after its completion is dissolved [6].

In modern design management there are two main approaches to team formation. The first is based on the consolidation and development of teams that were formed naturally (team building). The second is focused on the competence of the project and the distribution of roles in the team [7].

The task of this approach is to increase the professionalism of the team through the combination of the roles of all participants and the creation of conditions for non-conflict intra-team interaction.

Consequently, when forming a team, it is necessary to take into account the peculiarities of the competence providing of the project's work and the role interaction of the participants, which contributes to the achievement of the objective of the project and to increase the effectiveness of its implementation.

Therefore, ICB 4 represents important responsibilities for managing projects, programs and portfolios. Elements are classified into three groups: personal and social (people), contextual (perspective) and technical (practice). In ICB 4, each element of competence includes several key competences. They describe the main aspects of the element of this competence and are written in such a way that it is possible to assess the competence of these indicators. To facilitate evaluation, each indicator contains, in addition to the description, a number of specific and empirical measures that indicate the necessary or possible actions.

However, for a modern manager, it is important not so much to have the amount of knowledge, how to creatively think, to be able to independently learn new realities and solve new problems arising in management activities.

To develop creativity, you need the following elements:

1. competence (knowledge, skills, experience);
2. creative thinking, flexibility and persistence in finding a solution;
3. motivation is internal (personal interest in solving the problem) and external (material incentives and career advancement).

ICB 4 defines the concept of "individual competence" as "the application of knowledge, skills and capabilities in order to achieve the expected results" [8].

The competences defined in the standard of competence ICB 4.0, necessitates selection of the most important competences required for the formation of creative potential of the project team.

Different levels of competence include:

- knowledge: To demonstrate the memory of the materials learned by means of reference to facts, terms, basic concepts and answers;
- understanding: To demonstrate understanding of facts and ideas by organizing, comparing, translating, interpreting through the description and formulation of basic ideas;
- application: Using the obtained knowledge to solve problems in new situations by applying facts, methods and rules;
- analysis: Study and break down the information into parts, identifying motives or reasons, draw conclusions and find evidence in support of generalization;
- synthesis: Construct a structure or template from a variety of elements and acts of putting the parts together to form as a whole; collecting information together in a different way, combining elements in a new model or offering alternative solutions;
- assessment: Express and defend opinions through judgments about information, reasonableness of ideas or quality of work based on a set of criteria.

When forming an assessment model to determine the competence of project team members, evaluation methods such as professional testing, case analysis, case studies are used.

Creativity is the ability to find and identify a problem; generate a large number of ideas; produce incompatible problems (that is, have the flexibility of thinking); find original answers, non-standard solutions; perfect the object by adding individual details; see new features in the site, opportunities for its new use (i.e., analyze and synthesize elements problems).

Under the notion of "creative potential" we will understand: the ability to produce a large number of ideas (productivity), the ability to improve and develop ideas (flexibility), the ability to analyze and solve problems, the ability to unusual way of solving problems (originality), internal motivation manager and his professional competence (knowledge and skills that a manager has in his field of activity) [9].

3 Study Setup

The research was conducted on the example of the teaching staff of the Department of Project Management of the Kyiv National University of Construction and Architecture. The research is devoted to the definition and selection of competences that influence the formation of creative potential of the team of managers. In the course of research, a number of methods have been used that are consistent with the objectives of the study, namely: theoretical (analysis of scientific literature); empirical methods (questionnaires, observations, interviews, surveys and experimental methods (processing of research results in order to systematize research results).

The study was conducted as an independent questionnaire for experts in project management and questionnaires for Associate Professors and Professors of the department of project management KNUCA.

Initially, experts had to choose the competencies that influence the formation of the creative potential of the team of managers. In the next step, they had to rate the priority of each of the selected criteria.

Responsible experts in project management, as a rule, were associate professors (73%), doctor of technical sciences (25%) and postgraduate students (2%).

4 Findings

Of the 28 competences of the ICB 4.0 standard by means of peer review, 11 competences for the project manager were selected that influence the creative potential of the project manager.

Within the scope of competence of management of innovative projects and programs, we distinguish the following criteria which the project manager must meet:

1. Strategy: Strategic perception. Ability to perceive the strategic elements of the program/project and align them according to priorities for proper application.
2. Power and interests: Assessing personal ambitions and other people's interests and their potential impact on flexible work; Estimation of the informal influence of flexible teams and its potential impact on project work.
3. Culture and Values: Assessing the culture and value of society and its implications for flexible activities. Building a bridge between flexible activities with formal culture and corporate values of the organization. Assessing the informal culture and values of the organization and their implications for flexible activities.
4. Self-reflection and self-management: The ability to recognize, reflect and understand their own emotions, behavior, values. Ability to set personal goals, check and regulate the process, as well as perform work depending on the situation and own resources.
5. Personal integrity and reliability: The ability to assume responsibility for their own decisions and actions, demonstrating personal reliability and integrity, and thoroughly performing the tasks.
6. Personal communication: Timely exchange of information with all interested parties, communication with and within the virtual teams.
7. Relationships and engagement: The ability to create strong relationships due to social competences. Initiation and development of professional relations, demonstration of empathy.
8. Leadership: management of individuals or groups of people. Provide direction, coaching and mentoring to guide and improve the work of individuals and teams.
9. Teamwork: working together to solve a task. Choice and team building, its support, training and development.
10. Resourcefulness: The ability to use different techniques and ways of thinking, for analysis and the search for new (alternative) solutions in difficult situations. It requires action by original and creative methods and stimulating the creativity of individuals.

11. Resource orientation: The ability to evaluate all decisions and actions and their impact on the organization's goals and project success. Create and maintain a healthy, safe and productive flexible job.

The set of competencies required by the manager of the innovation project and that influencing the formation of creative potential can be displayed on Fig. 1.

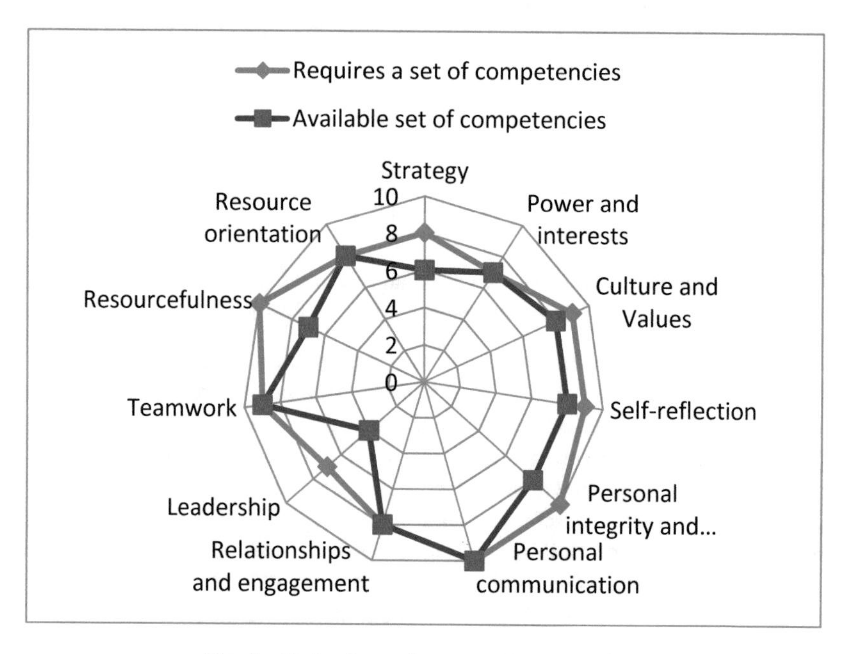

Fig. 1. Defined set of manager competencies.

After identifying the competencies that influence the formation of creative potential, the main problem remains, which is to attempt to compare competencies of project team members whose application is aimed at fulfilling various tasks in an innovation project.

To determine the priority of competencies for project management, we use a mathematical model based on the multi-criteria approach and expert evaluation. As criteria for evaluating the functions performed by team members selected competencies.

Twelve experts (m = 12) were selected, which were involved in assessing the degree of influence of eleven (n = 11) criteria for the formation of creative potential.

Determination of weighting factors involves: determining the degree of importance of the parameters, assigning them various ranks; validation of expert assessments for further use. The result of the expert ranking is shown in Table 1.

Table 1. The result of expert rankings

Criteria aj = $\overline{1, 11}$	Experts i = $\overline{1, 12}$												$\sum Rij$	Δ	Δ^2
	1	*2*	*3*	*4*	*5*	*6*	*7*	*8*	*9*	*10*	*11*	*m = 12*			
1	9	10	11	11	8	11	8	8	11	7	11	11	116	46	2116
2	10	8	10	10	9	8	11	10	5	10	10	7	108	38	1444
3	2	9	9	5	10	9	9	11	4	7	3	10	88	18	324
4	6	11	7	9	11	10	10	7	10	8	7	9	105	35	1225
5	4	5	5	8	4	4	1	6	2	9	9	8	65	–5	25
6	3	7	6	7	7	7	2	4	6	5	5	3	62	–8	64
7	5	6	8	6	6	3	3	4	3	2	5	5	56	–14	196
8	8	4	1	1	3	6	5	1	7	4	8	6	54	–16	256
9	1	3	4	2	2	2	6	6	8	1	1	2	38	–32	1024
10	1	1	2	3	1	1	4	2	1	3	2	1	22	–48	2304
n = 11	7	2	3	4	5	5	7	3	9	6	4	1	56	–14	196
	56	66	66	66	66	66	66	62	66	62	65	63	770		9174

Now, according to Table 1, we will calculate the sum of the ranks *Rij*, which got the j factor after a survey of all m experts. These amounts will be the main indicators of the influence of the criteria. From the example above, it is obvious that the 10th factor has the greatest influence on the formation of creative potential. The following are the factors 6, 8, 11, 7, 9, 5, 3, 4, 2, 1.

Determine the relative degree of influence of each of the factors on the formation of creative potential and verify the consistency of expert opinions.

After obtaining the results of the examination, we check the hypothesis of consistency in minds with the help of Kendall's coefficient of concordance [10]. The sum of ranks for each criterion is determined by:

$$Ri = \sum (m, j = 1)\, Rij \tag{1}$$

The average sum of ranks (T) is calculated by the formula:

$$T = Rij/n, \tag{2}$$

We calculate the square of deviations for each parameter (Δ^2) and the total sum of squares of deviations.

$$S = \sum (n, i = 1)\, \Delta^\wedge 2. \tag{3}$$

Determine the coefficient of consistency (concordation). The values of this coefficient are in the range from 0 to 1.

$$W = D/Dmax = (12 * S)/(m^{\wedge}2\ (n^{\wedge}3 - n)), \tag{4}$$

If the coefficient of concordance is zero or close to it, this means a complete inconsistency of expert opinions. With the approximation of the coefficient of concordance to one, one can speak of the unity of expert opinions. Further work with a group of experts is only appropriate if the coefficient of concordation is greater than or equal to 0.4. In our case, the coefficient of concordance $W = 0{,}57$.

The Fig. 2 shows the precedence of the criteria, which is determined by the expert assessment. Competences are indicated according to their ordinal number in the Table 1.

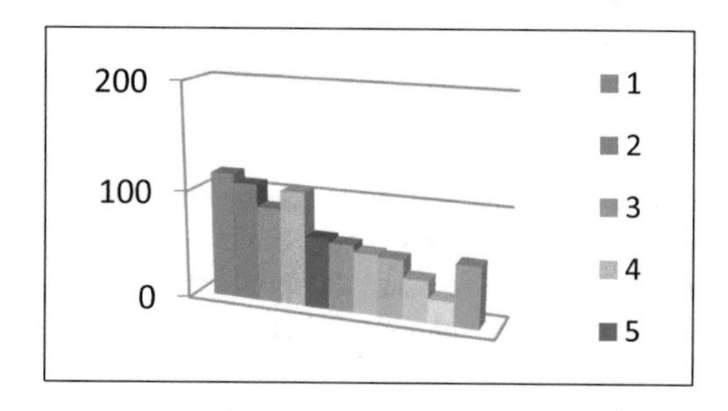

Fig. 2. Priority of competencies (criteria)

An assessment of the team's competences gives the organization's management the information it needs to make informed decisions about raising, transferring, or training employees. Although, to date, there is no universal method for assessing the creative potential for each member of the team, but based on the mathematical model considered, the competencies needed can be identified and evaluated as efficiently as possible. This will allow you to see in which competencies managers lack knowledge and skills and to select appropriate methods for developing competencies.

5 Discussion and Conclusions

The examples presented show the lack of competence of the project manager on the criteria - strategy, culture and values, self-reflection and self-management, personal integrity and reliability, leadership, resourcefulness. In some cases, there is knowledge and skills to apply them.

There are many different approaches to developing competencies that shape the creative potential of project managers.

The first method used by enterprises to increase the competence of employees is training. Their goal is mainly to increase the level of knowledge and skills in the field of logistics. Another tool is mentoring and/or consulting, the main purpose of which is to provide counseling or suggestion solutions for each employee in a particular situation, mentor with a great professional background.

Simulation and games. Facilitates the development of competences through situational simulation games (desktop or computer games) that provide information on the interaction and behavior of people depicted in a specific environment. Often, simulation games and other forms of game teaching are a combination of approaches that stimulate, for example, self-development in combination with partner development, coaching in the educational environment. It can also be useful in combining approaches based on prior experience of the development stage of individuals who are in an organization in certain circumstances.

Development among partners. An example of such development can be discussion with colleagues of the results of joint activities. In this case, information on the effectiveness of their own activities and ways to increase its effectiveness may be requested. Training partners from different fields of activity could help to see the situation in a different angle and become a major part of the mutually beneficial development of both partners. For example, through questions to partners through mutual understanding of existing conditions [11].

One of the popular methods is coaching, which aims to change behavior (especially skills change) through the most effective use of resources available at the disposal of the enterprise in their daily professional activities, integrated into strategic decisions of the company coaching helps to release the potential of the individual to maximize his achievements and improving performance. Coaching is more of a help in the learning process than learning [12, 13]:

- The main task of coaching is the development of human potential through targeted changes in accordance with the client's expectations;
- Coaching is a long-term relationship between coach and client that helps the client to achieve extraordinary results in their life, career, business or organization

The peculiarities of the implementation of a competence approach in the management of human resources at domestic enterprises are investigated. The analysis and evaluation of key competences has been carried out. During the evaluation, the importance of competences was determined, their ranking was conducted; the priority of competences that influence the development of creative potential of the team of managers is determined.

Consequently, we investigated the peculiarities of introducing a competent approach in the management of human resources at domestic enterprises; we found problems with the formation of a team of managers and remoteness in enterprises of the system for assessing the creative potential of project participants. We have analyzed and evaluated key competencies; conducted their ranking and, as a result, determined the priority of the competences that influence the development of creative potential of the team of executives.

The model of competencies of the project manager was suggested, which allows to define the structure and necessary level of competence for successful management of innovative projects.

An assessment of the team's competencies provides the organization with the information necessary to make informed decisions about raising, transferring, or training employees. Although, to date, there is no universal method for assessing the creative potential for each member of the team, but based on the mathematical model considered, the competencies needed can be identified and evaluated as efficiently as possible. This will allow you to see in which competencies managers lack knowledge and skills and to select appropriate methods for developing competencies.

References

1. Bushuyev, S., Verenych, O.: The blended mental space: mobility and flexibility as characteristics of project/program success. In: 2018 IEEE 13th International Scientific and Technical Conference on Computer Sciences and Information Technologies, CSIT (2018)
2. Bushuyev, S., Verenych, O.: Organizational maturity and project: Program and portfolio success. In: 2018 Developing Organizational Maturity for Effective Project Management (2018)
3. Belyakova, G., Sumina, Y.: Key competences as the basis for sustainable competitive advantage of an enterprise. Electron. J. "Res. Russia" http://zhurnal.ape.relarn.ru/articles/2005/104.pdf
4. Verba, V., Grebeshkova, O.: Problems of identification of competencies of the enterprise. Problems of science. Kyiv, vol. 7, pp. 23–28 (2004)
5. Bushuyev, S.D., Yaroshenko, R.F.: Vector model of bifurcation points in programs of organizations development. Management of complex systems development, vol 12, pp. 26–29
6. Bushuyev, S., Murzabekova, A., Murzabekova, S., Khusainova, M.: Develop breakthrough competence of project managers based on entrepreneurship energy. In: Proceedings of the 12th International Scientific and Technical Conference on Computer Sciences and Information Technologies, CSIT (2017)
7. Bushuyev, S., Wagner, R.: IPMA Delta and IPMA Organisational Competence Baseline (OCB): new approaches in the field of project management maturity. Int. J. Managing Projects Bus. **7**(2), 302–310 (2014)
8. ICB 4 – Standard. http://www.ipma.world/certification/competence/ipma-competence-baseline
9. Voitushenko, A.: Conceptual model for the production of innovative products based on the creative potential of the commands of project. Manag. Dev. Complex Syst. **33**, 31–36 (2018)
10. Beshelev, S.D., Gurvich, F.G.: Mathematical-statistical methods of expert evaluations. Statistics **2**, 263 (1980)
11. Bushuyev, S.D., Bushuiev, D.A., Yaroshenko, R.F.: Breakthrough competences in the management of innovative projects and programs. Bulletin of the National Technical University "KhPI", vol. 1, pp. 3–9 (2018)
12. Bushuyev, S.D., Bushuiev, D.A., Bushuyeva, N.S., Kozyr, B.Y.: Information technologies for project management competences development on the basis of global trends. Inf. Technol. Learn. Tools **68**(6), 218–234 (2018)
13. Voitushenko, A., Bushuev, S.: Determination of competences that take affect the formation of creative capabilities of team of managers. In Proceedings International Scientific Conference "Computer Sciences and Information Technologies", (CSIT-2019), vol. 3, pp. 122–126 (2019)

Content Management Method of Complex Technical System Development Projects

D. N. Kritskiy(✉), E. A. Druzhinin, A. V. Karatanov, and O. S. Kritskaya

National Aerospace University «KhAI», Kharkiv, Ukraine
d.krickiy@khai.edu

Abstract. The article covers unmanned aerial vehicle development project content. To achieve this goal the project stages table was suggested, including technical solutions volumes indicators obtain during their implementation. Further development of the method of mastered volume due to the introduction of such indicators as planned decisions, actual decisions, confirmed decisions; on the basis of this approach, a system of inequalities is proposed, which allows assessing the feasibility of the project to create unmanned aircraft; presented the results obtained during the implementation of the project in the Research Institute of Aircraft Flight Modes of the National Aerospace University "Kharkiv Aviation Institute".

Keywords: Complex technical systems · Earned value method · Content management · System of inequalities of project indices management

1 Introduction

Formulation of the problem. Complex technical systems (CTS) development projects represent a multilevel work package with branching structure and iterative progress of works [1]. Most frequently at early stages of development the result is formed in general view and provides several alternatives for implementation. This is attributed to a big number of elements in the product structure, and layout complexity. Such development is carried out nowadays only in conditions of computer support for all stages of development and complete integration of data about intermediate results. The unmanned aircraft vehicle (UAV), as an example of CTS, is a special robotized type of aircraft vehicles (AV). The main problem when developing UAVs is the absence in Ukraine and other countries of a normative base, which controls the order of its development and integration into airspace [1, 2]. The normative base, applied for AVs, cannot be always adapted for project needs, which is first of all connected to the fact that it has a lot of works intended for reiterated testing of UAV. For instance, in case of small UAVs it is cheaper to carry out testing on several samples than to develop and implement UAV parts models.

In such conditions there appear problems with managing the project content as well as the problems, connected with the quality monitoring and resource allocation management. At present design departments of Ukraine develop unmanned aircraft vehicles

N. Shakhovska and M. O. Medykovskyy (Eds.): CCSIT 2019, AISC 1080, pp. 293–303, 2020.
https://doi.org/10.1007/978-3-030-33695-0_21

(UAVs) from the "faster and cheaper" position, which in the future can lead to the use in the airspace of products that do not correspond to the manned aviation safety standards. Moreover, when managing the content of UAVs development, we should consider the special feature of the project, related to the profitability calculation: the development and production costs are repaid only at the operation stage.

Thus, there appears a contradiction between the development of high-quality product for its safe operation in the air space together with manned aviation and reduction of time-frame and UAVS developmentproject costs. This contradiction is solved owing to the use of up-to-date models and methods of project management to ensure the operation profitability.

Analysis of Recent Research and Publications. Having analyzed the established models and methods of project content management [3–9] we should note that they do not allow to effectively manage the content of development projects of such CTS category as they do not take into consideration the special features of UAVs. The carried out research is based on the use of:

system analysis method used to study the special features of complex projects and the processes of their indices assessment[2, 7];

CTS rational structure design method used to form the project hierarchy [6, 11];

The decision tree method used to classify the project alternatives[10].

The Aim of the Study. The content management method considered in this work can be used to evaluate the actual course of the UAVs development project and to forecast the possibility of getting qualitative and financially sound result. To do so the earned value method has been modified. The use of decision-making theory method has been proposed to choose the rational structure of works for project completion out of alternatives set.

2 The CTS Development Project Content Management with the Use of Earned Value Method

The CTS development project structure is closely related to the structure of the developed model of equipment's future appearance. Therefore, the content management processes are interrelated with the works of the project and its product quality assessment. The project alternative products quality assessment is necessary to choose the realizable UAVs models out of proposed ones. At every stage of UAVs development a set of decisions is formed: from innovative to classical. The choice of classical decisions provides the reduction of the number of experimental works in the project, necessary to confirm the results quality. At the same time the innovative decisions (referring either to the UAVs features or new components) are always present in the process of new UAV development. Such decisions are planned and checked on realizability several times after the main stages of development are carried out.

To monitor the project the use of the method of earned value is proposed. Traditionally the earned value method has been used to control the project costs. In this work we propose to use parameters (Fig. 1): the number of planned decisions (R_0), the number of confirmed decisions (R_e) and number of actually taken decisions (R_f) to monitor the project content in the earned value method [12, 13].

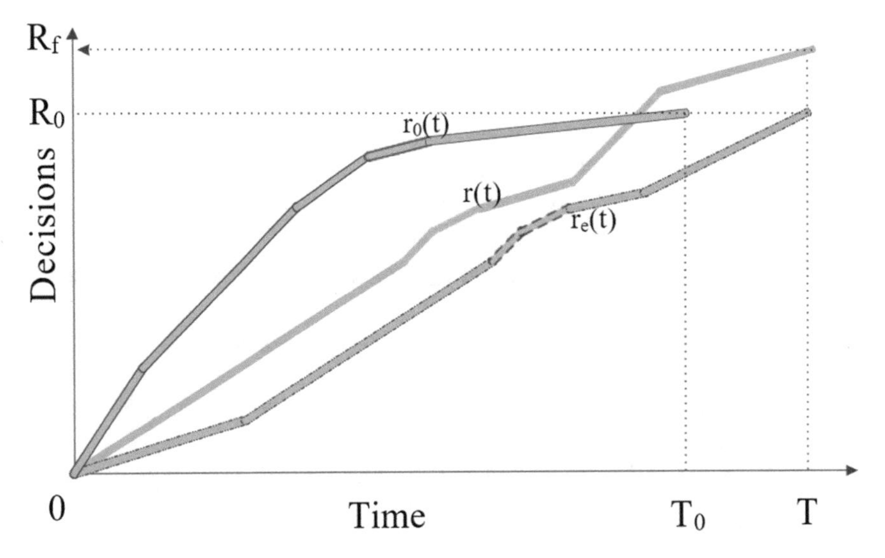

Fig. 1. The decision-making dynamics indices

The planned total decisions R_0 are the number of hypotheses, which are planned to be realized in the course of the project; the actually taken decisions R_f are the hypotheses, which have been actually approved in the course of carrying out the project at the management level and which have an impact on the goal achievement result of the project; the confirmed decisions R_e are experimentally proved hypotheses.

Every operation and the project in whole are described by the following variables:

- R_0 is the planned total decisions of the project;
- T_0 is the planned project completion time;
- X_0 is the total value of works on the project;
- $r_0(t)$ is the planned decision making dynamics;
- r(t) is the actual decision making dynamics;
- $r_e(t)$ is the dynamics of decisions experimental verification;
- $x_0(t)$ is the planned value of works dynamics;
- x(t) is the actual value of works dynamics;
- $x_e(t)$ is the earned value of works;
- T is the project's actual completion time;
- R_f is the actually taken decisions during the project.

Earned value derived indices:

- $\Delta r_0(t) = r_0(t) - r(t)$ is the difference between planned and actually approved decisions;
- $\Delta r(t) = r_0(t) - r_e(t)$ is the difference between planned and experimentally proven decisions;
- $\Delta r_e(t) = r(t) - r_e(t)$, $\Delta r_e(t) \geq 0$ is the difference between actually approved and experimentally proven decisions;
- $\Delta x_0(t) = x_0(t) - x(t)$ is the difference between planned and actual value of works;

- $\Delta x(t) = x_0(t) - x_e(t)$ is the difference between planned and earned value of works;
- $\Delta x_e(t) = x(t) - x_e(t)$, $\Delta x_e(t) \geq 0$ is the difference between actually approved and earned value of works;
- $\alpha_{r(t)} = r_e(t) / r_0(t)$ is the index of approved decisions value;
- $\beta_{r(t)} = r_e(t) / r(t)$ is the index of approved decisions dynamics;
- $\alpha_{x(t)} = x_e(t) / x_0(t)$ is the index of works earned value;
- $\beta_{x(t)} = x_e(t) / x(t)$ is the value dynamics index;
- $\tau_{0r}(t) = t - r_0^{-1}(r_e(t))$ is the current delay (from the plan), which is determined from the condition: $r_0(t - \tau_{0r}(t)) = r_e(t)$;
- $\tau_r(t) = t - r^{-1}(r_e(t))$ is the current delay in decisions, which is determined from the condition: $r(t - \tau_r(t)) = r_e(t)$
- $\tau_{0x}(t) = t - x_0^{-1}(x_e(t))$ is the current delay (from the plan), which is determined from the condition: $x_0(t - \tau_{0x}(t)) = x_e(t)$;
- $\tau_x(t) = t - x^{-1}(x_e(t))$ is the current delay in decision which is determined from the condition: $x(t - \tau_x(t)) = x_e(t)$;
- $e_0 = X_0 / R_0$ is the planned project efficiency on the whole;
- $e_0(t) = x_0(t) / r_0(t)$ is planned efficiency of decisions usage at the moment of time t;
- $e = X / R_f$ is the actual project efficiency on the whole;
- $e(t) = x_e(t) / r(t)$ is the actual efficiency of decisions usage at the moment of time t.

The analysis of the given above indices will help the project manager to monitor if the project is realized successfully. The assessment is carried out using the analysis of the ration of approved decisions to the general number of decisions in the project. Thus, the project is considered to be complete (the project goal is reached) as soon as the value of experimentally proven decisions coincides with the total value of decisions. The project duration is herein the main indicator, playing the role of efficiency and constraints criterion components.

With such representation of parameters describing the project it is reasonable to present the project risk index at the fixed moments of time t with such a formula:

$$R(t) = 1 - \frac{R_e(t)}{R_f(t)}$$

Such representation of project risks allows us to track the dynamics of product quality change (the more taken decisions are approved, the higher is the quality of the obtained product).

3 The Conditions of Project's Successful Implementation

The successful implementation of UAVs development project depends on a range of conditions, which are real constraints and form a system of inequalities, which can be called the system of inequalities of project indices management.

Let's consider these conditions, which ensure the success of project implementation.

The first condition is the obtaining of the project product of the necessary quality level. For projects of developing CTS for instance UAVs this conditions are shown in meeting the requirements of access to the air space to carry out aviation works.

To meet the requirements of Ukraine's Air Law and Aviation rules it is necessary for the flight to be controlled and managed by the operator from a special control station, positioned out of the apparatus, for the UAV to possess airworthiness property and its airworthiness should be maintained in the course of service life. Meeting these requirements is provided by the corresponding structure and by presence of certain components in its composition. According to the ideas of V.F. Bolchovitnikov the mass balance equation can be used to manage development quality as it shows the access conditions of UAVs into the air space.

At different stages and phases of the project members of this equation are determined in different ways: statistically, by weight equation, drawings calculation, direct weighing. In so doing some members are known from the very beginning and in the course of development do not change. This is, for instance, payload weight, mass of radar transponder and of other equipment, which has to be installed on the UAV. The calculations using this equation do not require any special knowledge, and the input data for it (components masses) are determined by specialists–designers.

The second condition of project implementation is the condition of return of funds, invested into development and production at the expense of profits, received at operation. This condition can be written in such a way:

$$P \geq C_d + C_p + C_o,$$

where P is the profit from operation;

C_d is the development costs;
C_p is the production costs;
C_o is the operation costs.

Let's consider all the members of this equation in succession. The operation profits P depend on the period of work t_{ok} in the course of which effective work has been carried out, on the cost of payment for one hour of this work (α) and on the quantity of product samples (n), which are used in the effective work implementation. Therefore, the profits from operation in the assumption that all the n samples in the course of period t_{ok}.

t_{ok} will look as follows:

$$t_{ok} = \frac{1}{n(\alpha - b)}(C_d + C_p + C_o) + \frac{C_1}{\alpha - b}$$

From the equation we can see that the funds return period t_{ok} consists of two parts – dependent and independent on the quantity of produced samples n.

It should be considered that absolutely all further costs depend on development quality, the cost of sample production depends on preproduction quality, and operation costs depend on operation preparation quality.

The third condition of project implementation is the resource component. The value t_{ok} from the point of view of features is of significant importance as it determines the equipment's needed resource. For developed and produced products to be able to repay costs and bring profit its technical resource (T) should be greater than the funds return period. In doing so, in operation we need to have technical unpaid flights. Unpaid flights are also possible due to failures, thus the technical resource in relation to the funds return period should be determined with the coefficient significantly greater than one:

$$T = C_{res} \times t_{ok},$$

where C_{res} is the coefficient for technical resource determination, $C_{res} >> 1$.

The technical resource of aviation equipment product is its most important technical feature, which determines time without failures, serviceability and safety.

The technical resource of aviation equipment product is product operating time in hours or in the number of applications before the limiting state occurs when the further operation is impossible under safety conditions.

The resource is determined based on calculations and tests of failure allowable probability.

The resource is set to the type of the product, and for certain samples it is written off in operation within the established service life.

The following resources are set: assigned, warranty resource, overhaul life.

Inequality $T > t_{ok}$ or equality $T = C_{res} \times t_{ok}$ is the most important condition of CTS (e.g. UAV) development project implementation. In the course of assigned resource not only repay should occur, but also profit should be made, which justifies the funds invested into development, production and operation. Wherein repair works, with the help of which it is possible to set and reach greater values of assigned resource increase the operation cost. Malfunctions, which occur in the course of the warranty resource, are eliminated at the expense of the producer, which makes the production more expensive; moreover, the operation cost is raised due to losses caused by backlog.

After reaching the funds return period the operation can be continued to obtain profit during several overhaul life resources to achieve the assigned resource, when the operation can be terminated.

The fourth condition of project implementation is in the sufficiency of the necessary equipment and personnel with the right qualifications. All the projects of CTS development are carried out in the conditions of tight constraints in terms and costs with high market requirements to the project product quality. The cost and project completion time indices are considered only if the proper project product quality is achieved.

The work on carrying out the project is characterized with labor-intensiveness, duration and cost. These works are complicated and are carried out in a parallel-serial way. Therefore, labor intensiveness is a stable characteristic. The duration of the project depends on logical structure of works and possibility to carry out simultaneous operation. The cost of works is unstable value, it depends on salaries level and energy supply costs. From the point of view of well-timed market launch of project product, on which other works and projects implementation can depend, the completion time of the project is very important.

Therefore, when realizing the project it is necessary to prioritize in the following way: the project product quality, the project completion time, cost. Thus, the most important requirement is set to the quality of project product, some variations are possible with the project completion time and even greater variations are allowed with the cost.

Therefore, it is necessary to have the corresponding production facilities, scientific and technical ground work and necessary materials as well as bought-in components.

The availability of corresponding material and bought-in components means that standardized aircraft materials and product's aircraft components are also available, or that there are materials and components suppliers, with whom there is the protocol for endorsement of production facilities usage in the development of the product.

The described above conditions are written down as a system of inequities and equations:

$$\begin{cases} \sum_{i=1}^{n} \overline{m_i} = 1; \\ \sum_{i=1}^{3} C_i = P|t = t_{ok} \\ T = C_{res} \times t_{ok}, C_{res} > 1; \\ F_{ij}^{ent} \geq F_{ij}, \end{cases}$$

where F_j^{ent} are the enterprise's funds (equipment, personnel, etc.);

F_{ij} are funds necessary to realize the project (equipment, personnel, etc.).

The availability of corresponding production facilities, scientific and technical groundwork, materials and bought-in components make up the fourth condition of successful implementation of CTS development project. If this condition is not met, then the project of these conditions development is also required. If this condition is not met, it is impossible to carry out the CTS development project (for instance, there is no production area, where this project can be carried out or there are no bought-in components and materials).

4 The Results Obtained After Implementation of Research Results

The developed model of project implementation conditions and updated method of earned value have been implemented in the Inter-branch Research Institute of Aircraft Flight Modes Physical Modeling Problems of the National Aerospace University "Kharkiv Aviation Institute". As the result of their implementation the development period is reduced up to 21% and if the developed model and method at the development stage are used the project timeline is reduced up to 7%.

The data on planned decisions, development stages deadlines and planned value of experimentally proven hypotheses in the course of UAVs development are shown in Table 1. The base period is equal to 30 months.

Table 1. The planned data when carrying out the project

Stage name	Project deadlines				Cost, %		Decisions made %		Decisions approved, %	
	Time		%							
	Months	Scope	Project duration	Total	Stage cost, %	Total cost %	Stage %	Total %	Stage	Total
	0	0	0	0	0	0	0	0	0.00	0.00
Research work Marketing	1	1	3.33	3.33	0.5	0.5	10	10	0.00	0.00
Research work finding ways and justifying the development possibility	2	3	6.67	10.00	2	2.5	25	35	0.00	0.00
Pilot project (technical and economic study)	2	5	6.67	16.67	2.5	5	20	55	0.00	0.00
Filling of an application to SAA	1	6	3.33	20.00	0,5	5.5	5	60	0.00	0.00
Schematic design	2	8	6.67	26.67	3.5	9	18	78	6.00	6.00
Experimental model	0.25	8.25	0.83	27.50	0.5	9.5	2	80	1.00	7.00
Technological Process	2	10.25	6.67	34.17	7.5	17	15	95	5.00	12.00
Design Documentation	2	12.25	6.67	40.83	7.5	24.5	5	100	16.00	28.00
Preproduction prototype	1	13.25	3.33	44.17	7.5	32	0	100	2.00	30.00
Prototype production	6	19.25	20.00	64.17	33	65	5	105	24.00	54.00
Preparation for testing	1	20.25	3.33	67.50	7.5	72.5	0	105	6.00	60.00
Certification production test. (flight developmental testing)	6	26.25	20.00	87.50	18	90.5	15	120	40.00	100.00
Design Documents Update	0.25	26.5	0.83	88.33	1.5	92	0	120	0.00	100.00
Certification State test	2	28.5	6.67	95.00	7	99	0	120	0.00	100.00
Documentation final adjustment consideration	1	29.5	3.33	98.33	0.5	99.5	0	120	0.00	100.00
Issuance of a certificate of type	0.5	30	1.67	100.00	0.5	100	0	120	0.00	100.00

Figure 2 shows the difference of planned (Fig. 2a) from actual indices (Fig. 2b).

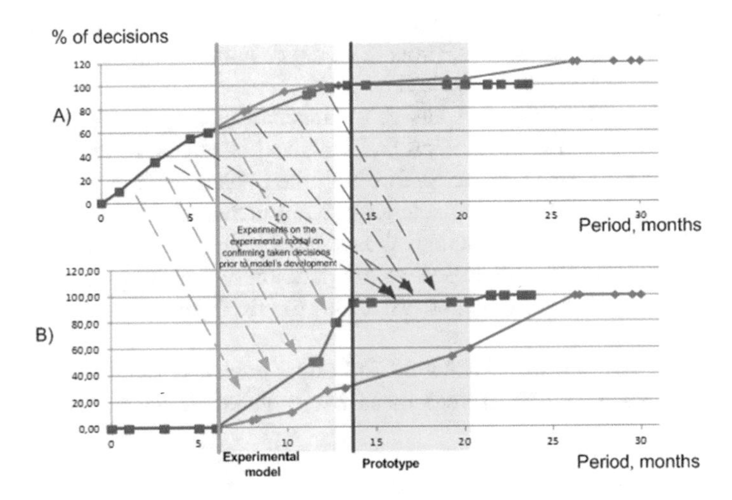

Fig. 2. The ratio of taken decisions to the project implementation deadlines

Table 2. The actual data of project implementation

Stage name	Duration and deadline				Cost %		Decisions made %		Proven by experiments	
	Duration, months		Duration, %							
	Stage	From the beginning	Stage	From the beginning	Stage	Total	Stage	Total	Stage	Total
	0.00	0.00	0.00	0.00	0.00	0.00	0.00	0.00	0.00	0.00
Research work Marketing	1.00	1.00	3.33	3.33	0.50	0.50	10.00	10.00	0.00	0.00
Research work finding ways and justifying the development possibility	2.00	3.00	6.67	10.00	2.00	2.50	25.00	35.00	0.00	0.00
Pilot project (technical and economic study)	2.00	5.00	6.67	16.67	2.50	5.00	20.00	55.00	0.00	0.00
Filling of an application to SAA, the development of projects certification basis	1.00	6.00	3.33	20.00	0.50	5.50	5.00	60.00	0.00	0.00
Schematic design	5.50	11.50	18.33	38.33	24.00	39.50	32.00	92.00	50.00	50.00
Experimental model	0.25	11.75	0.83	39.17	0.50	40.00	2.00	94.00	0.00	50.00
Technological Process	1.00	12.75	3.33	42.50	3.50	43.50	4.00	98.00	30.00	80.00
Design Documentation	1.00	13.75	3.33	45.83	3.50	57.00	2.00	100.00	15.00	95.00
Preproduction prototype	1.00	14.75	3.33	49.17	7.50	64.50	0.00	100.00	0.00	95.00
Prototype production	4.50	19.25	15.00	64.17	20.00	84.50	0.00	100.00	0.00	95.00
Preparation for testing	1.00	20.25	3.33	67.50	7.50	92.00	0.00	100.00	0.00	95.00
Certification production test. (flight developmental testing)	1.25	21.50	4.17	71.67	4.00	96.00	0.00	100.00	5.00	100.00
Design Documents Update	0.00	21.50	0.00	71.67	0.00	96.00	0.00	100.00	0.00	100.00
Certification State test	0.75	22.25	2.50	74.17	3.00	99.00	0.00	100.00	0.00	100.00
Consideration of document final adjustment, validation of projects certification basis, requirements specification, design documentation	1.00	23.25	3.33	77.50	0.50	99.50	0.00	100.00	0.00	100.00
Issuance of a certificate of type	0.50	23.75	1.67	79.17	0.50	100.00	0.00	100.00	0.00	100.00

The areas, where the approved decisions are confirmed, are highlighted (the first area is the approval of taken decisions with testing of the prototype, the second area is the confirmation of the taken decisions by testing on the prototype).

The actual data, obtained in the course of the development project of commercial UAVs in PMPRI are shown in Table 2.

As can be seen from Tables 1 and 2, the mismatch occurs at the stage of "Schematic design". The 2 months duration and technical solutions volume equal to 18% from total result were assumed, but in fact both the implementation time (5.5 months) and volume of confirmed technical solutions (32%) increased. This led to the fact that in the subsequent stages (Technological Process, Design Documentation, Prototype production) it became possible to reduce the duration by lowering the number of the resulting product quality confirmation tasks. Due to this, the duration of the project was reduced by 6.5 months.

At the same time, the tables show the indicators proposed according to the content management method for CTS development, based on earned value method.

5 Conclusions

Project content management method for CTS development, based on earned value method, allows to analyze current project decisions at all stages and phases of project implementation by finding discrepancies in the project at the earliest stages.

The proposed approach allows to use the advantages of project management methods when managing the development of complex equipment taking its special features into consideration. It makes it possible to develop a set of tools, intended for project managers to monitor and support the decision-making process to confirm that the right decision about development has been made.

The obtained theoretical regulations and dependencies can be used at aviation enterprises. In the future it is also necessary to consider which works specifically and their extent should be carried out when iteration occurs when developing commercial UAVs.

References

1. Karimov, A.H.: Features of designing unmanned aircraft systems of the new party and special. In: Proceedings of MAI, 47. https://www.mai.ru/science/trudy (2011)
2. Chenarani, A., Druzhinin, E.A., Kritskiy, D.N.: Simulating the impact of activity uncertainties and risk combinations in R&D projects. J. Eng. Sci. Technol. Rev. **10**(4), 1–9 (2017)
3. Mazur, I.I., Shapiro, V.D.: Project Management, 6th edn, p. 960. Omega-L, Moscow (2010)
4. Guide to the Body of Knowledge Project Management American National Standard ANSI: Project Management Institute (2013)
5. P2M: A guidebook of Project and Program Managements for Enterprise Innovation, vol. 1, Revision 3. Project Management Assosiation of Japan (PMJA) (2005)

6. Shakhovska, N., Vovk, O., Hasko, R., Kryvenchuk, Y.: The method of big data processing for distance educational system. In: Conference on Computer Science and Information Technologies, pp. 461–473. Springer, Cham, September 2017
7. Shakhovska, N., Vovk, O., Kryvenchuk, Y.: Uncertainty reduction in Big data catalogue for information product quality evaluation. Eastern Eur. J. Enterp. Technol. **1**(2–91), 12–20 (2018)
8. Pavlov, A.N.: Project Management based on the standard PMI PMBOK. Statement of the methodology and experience. BINOM. Laboratory of Knowledge, Moskow (2012)
9. Ayupov, A.I., Plyaskota, S.I.: Testing and quality requirements. Modern approaches to aircraft building. In: Management of development of large-scale systems: Proceedings IV International, pp. 179–180 (2010)
10. Popov, V.L.: Management of innovative projects, p. 336. INFRA-M, Moskow (2009)
11. Kritskiy, D.N., Druzhynin, Y.A., Pohudina, O.K. Krytska, O.S.: Decision making by the analysis of project risks based on the FMEA method. In: Computer Sciences and Information Technologies, CSIT 2018, no. CFP18D36, pp. 187–190 (2018)
12. Kritsky, D.N., Druzhinin, E.A., Pogudina, O.K. Kritskaya, O.S.: A method for assessing the impact of technical risks on the aerospace product development projects. In: Advances in Intelligent Systems and Computing, vol. 871, pp. 504–521. Springer, Switzerland (2019)
13. Kolosov, E.V., Novikov, D.A., Tsvetkov, A.V.: Earned value in operational project management. NIC Apost- REF, Moskow (2000)

Model of Forming and Analysis of Energy Saving Projects Portfolio at Metallurgical Enterprises

Kiyko Sergey[1(✉)], Druzhinin Evgeniy[1,2(✉)], Prokhorov Oleksandr[1,2(✉)], and Kritsky Dmitry[2(✉)]

[1] PJSC "Electrometallurgical Works "Dniprospetsstal" named after A.M. Kuzmin, Zaporozhye, Ukraine
kiyko@dss.com.ua, druzhininl05@gmail.com
[2] National Aerospace University, Kharkiv Aviation Institute (KHAI), Kharkiv, Ukraine
{o.prokhorov,d.krickiy}@khai.edu

Abstract. The factors influencing the energy efficiency level of a metallurgical enterprise have been determined. They are necessary to develop a theoretical and methodological approach for managing the energy consumption mechanism of a metallurgical enterprise in the new business environment and calculating energy efficiency indicators. A system model for the formation of projects portfolio in accordance with the energy efficiency strategy of a metallurgical enterprise is presented, using methods of set theory, systemic and multi-criteria analysis, agent-based simulation modeling, and methods for analyzing risks and uncertainties.

Keywords: Energy efficiency · Metallurgical enterprises · Projects portfolio · Energy saving management

1 Introduction

The development of organizational management systems, management accounting mechanisms and the elimination of energy losses is the main task of energy saving projects at metallurgical enterprises. Success criteria for energy saving projects include: efficiency; operating costs, losses, etc. It is a complex task to objectively estimate each energy resource share in the total flow and determine the energy intensity of an individual production plant, the entire enterprise, etc. Therefore, the basic concept of this work is to create methodological basis for managing projects and energy saving programs of metallurgical enterprises based on the various types of resources management, both internal and external.

The implementation of the company's energy conservation program involves the creation of a management system that is formed through the mutual coordination of the three most complex categories: the control object, the management mechanism and the management organization.

N. Shakhovska and M. O. Medykovskyy (Eds.): CCSIT 2019, AISC 1080, pp. 304–314, 2020.
https://doi.org/10.1007/978-3-030-33695-0_22

With mutual coordination of these categories, the interpretation of the energy saving management will be as follows: in the management system of the enterprise energy economy, energy management activities of the facility (enterprise) management are carried out under systematic influences from a specially formed mechanism, which is activated by an appropriate management organization. This mechanism is the management of the energy saving projects portfolio.

The methodology for managing a portfolio of energy saving projects at metallurgical enterprises, based on a systems approach, will reveal the integrity of the mechanisms of the enterprise, identify its connections and bring them together into a single dynamic complex. A systems approach to energy management is based on the fact that the specifics of enterprise management are not limited to the features of its constituent elements, but is manifested in the nature of the connections and relationships between certain elements and the factors affecting their functioning. It is based on the fact that all elements of enterprise management are considered as one whole in interrelation with each other and with their characteristic dynamic process of interaction with the external environment.

2 Review of the Recent Papers

The works of many scientists are devoted to the study of energy saving reserves and raising the level of enterprises production processes energy efficiency: Bunse [1], Wang [2], Brunke [3] etc. The researchers consider approaches and peculiarities of these problems possible solutions, such as energy survey, monitoring and planning. They emphasize, that the purpose-oriented energy saving programs are implemented considering the peculiarities of the particular enterprise, and the energy supplies exhaustion arrangements for accounting are considered tactical choice.

The authors in [4] considered the theoretical and practical aspects of operational planning and power management in an industrial enterprise. The possibilities of mathematical modeling and optimization of technological regimes are presented. In this example of a large metallurgical plant sinter production, calculations of the optimal-compromise mode are performed. It is shown that as a result of the coordination task, the shift productivity of the sinter plant will increase, and the power consumption will decrease. This article [5] considers scientific and methodological basis of the mining and metallurgical enterprises energy management from the formation of mathematical models of energy consumption to the operational management of power modes. In [6] presented a conceptual framework for managing projects portfolios in organization and formalization of value-oriented development portfolio creation methodological bases for metallurgical enterprises. The article [7] proposes a comprehensive algorithm for evaluating the prospects for the implementation of innovative projects at metallurgical enterprises in Ukraine, based on a set of qualitative and quantitative indicators to take into account the effectiveness of innovative activities of an enterprise. The result of paper [8] is a methodological and organizational basis for creating effective systems and technologies for managing the project programs and portfolios. A balanced system of parametric single indicators of innovation is presented – the risks, personnel, quality, innovation, resources, and performers, which allows getting a comprehensive idea of any project already in the initial stages.

3 Energy Saving Factors that Determine the Efficiency of Metallurgical Enterprises

It is important to structure the supporting energy saving factors that determine the efficiency of the metallurgical enterprises (Fig. 1), as well as to identify the dependencies between the factors that ultimately will have an impact on the structuring of the energy saving program. Between the factors influencing the level of energy efficiency of the enterprise, there are correlation and regression relationships, interaction and multiplicity of cause-effect relationships.

It is important to understand that the basis of an effective energy strategy of an enterprise is an individual set of factors that should be in the field of view of the energy management of an enterprise, because over time, under the influence of certain factors, the potential for energy saving is formed, and it must be implemented in a timely manner to improve competitiveness and social economic development of the enterprise, otherwise the growth of the energy saving potential is characterized by a decrease in the efficiency of energy consumption management. It should be noted that the basis of market management of energy consumption of the metallurgical enterprise is planning, but at the same time the center of gravity shifts towards strategic planning, justification of the priorities of the energy policy and the most important areas of economic development.

A comprehensive energy conservation program for a metallurgical enterprise should cover such areas as: developing its own energy base; development and improvement of the energy accounting system; improving the efficiency of energy use by consumers, reducing losses during transportation of energy resources; full involvement in the fuel balance of the enterprise of secondary energy resources; energy consumption monitoring; modernization of technological processes and units; improving the energy balance of the enterprise; involvement of staff in energy saving activities.

Analysis of the structure of the energy saving program at metallurgical enterprises shows that the main problems that energy saving projects solve are inefficient use (significant losses) of energy resources (natural resources, heat energy, electricity), control over the formation of costs and results of improvements in energy consumption.

4 Conceptual Approach of Energy Saving Project Portfolios Integrated Management at Metallurgical Enterprises

Using a systematic approach, an energy management strategy was formulated, which is based on a conceptual approach to managing prospective energy consumption of a metallurgical enterprise, on an integrated system of requirements for participants in the

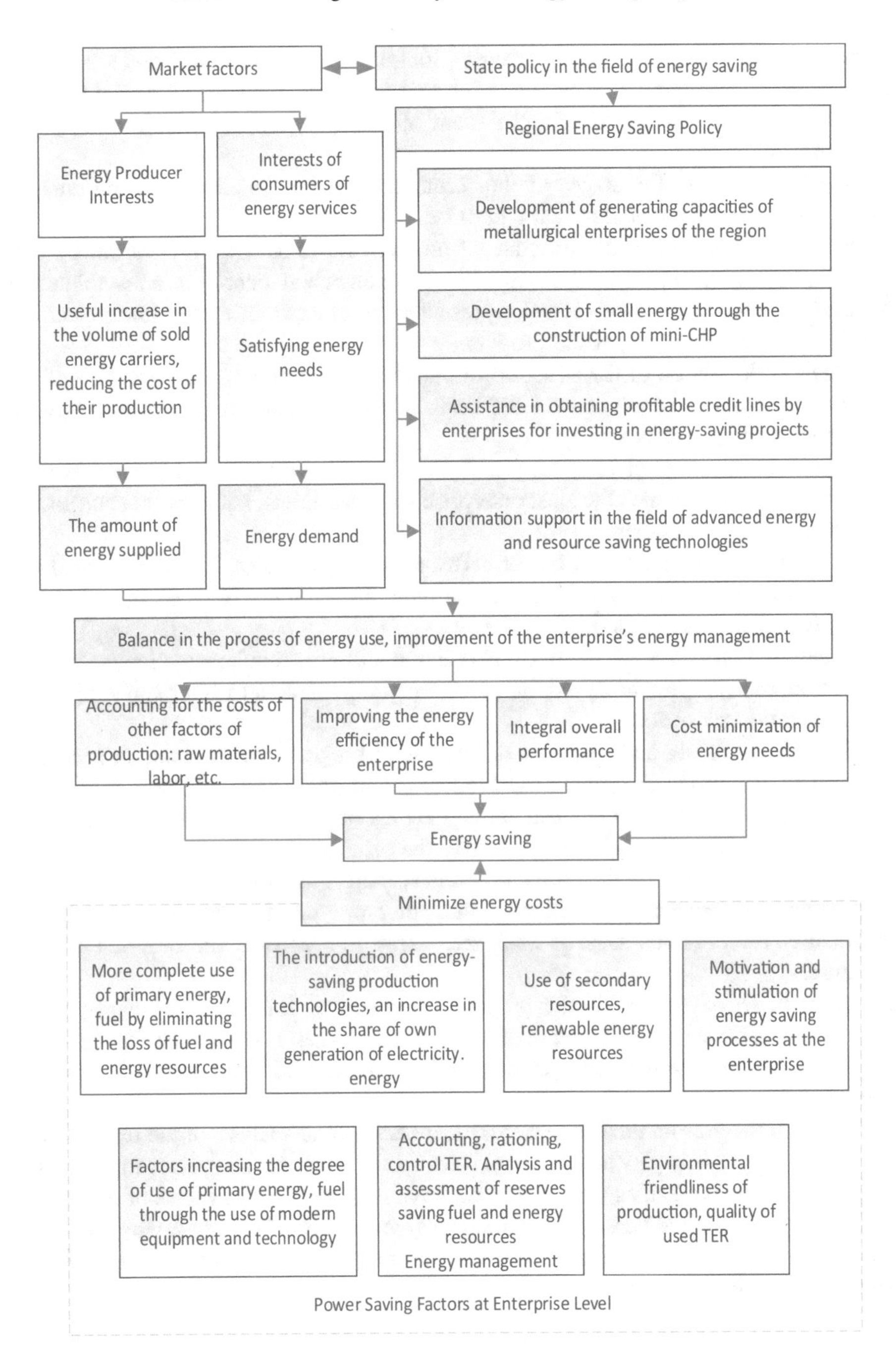

Fig. 1. Energy saving factors that determine the efficiency of metallurgical production

markets of metallurgical goods in ensuring long-term and sustainable development of the enterprise's economy, efficient use of energy resources of the metallurgical industry power consumption, using a unified system of goals, objectives, principles and performance indicators.

We will consider the stages of the formation and implementation of an energy-saving program for a metallurgical enterprise.

Stage 1. Formation of the structure of program stakeholders. To create this structure, it is necessary to analyze the program environment and identify its stakeholders. It is also advisable to allocate several representatives of each group for further participation in the development of the program.

Stage 2. Formation of the structure of values of the program. The structure of the program's values is developed on the basis of data obtained from the stakeholders identified in the previous step.

Stage 3. Formation of temporary and financial reserves of the program. Possible financial sources for a metallurgical enterprise are: own funds; bank loans; state and/or municipal budget subsidies.

Stage 4. Development of a list of software limitations. The list of program limitations is formed during the analysis of the program resources in terms of their sufficiency for its implementation.

Stage 5. Formation of a set of alternative program implementation scenarios. Project alternatives are analyzed in order to group them into alternative program implementation scenarios.

Stage 6. Modeling and selection of alternative program implementation scenarios that satisfy the program limitations. Modeling and analysis of alternative program implementation scenarios developed at the previous step are carried out with the aim of identifying a subset of scenarios that satisfy the program's limitations.

Stage 7. Evaluation of alternative program scenarios in terms of meeting the values of program stakeholders. The subset of alternative program implementation scenarios obtained at the previous stage is analyzed taking into account the program's value structure.

Stage 8. Development of the program architecture. Based on the high-priority program scenario selected at the previous stage, a roadmap is developed for achieving the total program value, and then transformed into a more detailed program architecture.

As can be seen from the above steps, the modeling and analysis of the feasibility of energy saving projects play an important role. The basis of this analysis is to justify the possibility of the project, as well as the degree of influence on the values of the energy saving program. In addition, it determines the availability of energy-saving technologies and equipment, that are necessary for the project, the possibility of their development and effective operation in specific conditions, etc.

Figure 2 shows the structural synthesis model of the portfolio of energy-saving projects. At each point in time, the company runs many projects that are in various phases of its implementation. In this regard, the assessment of the effectiveness of the enterprise is a rather difficult task.

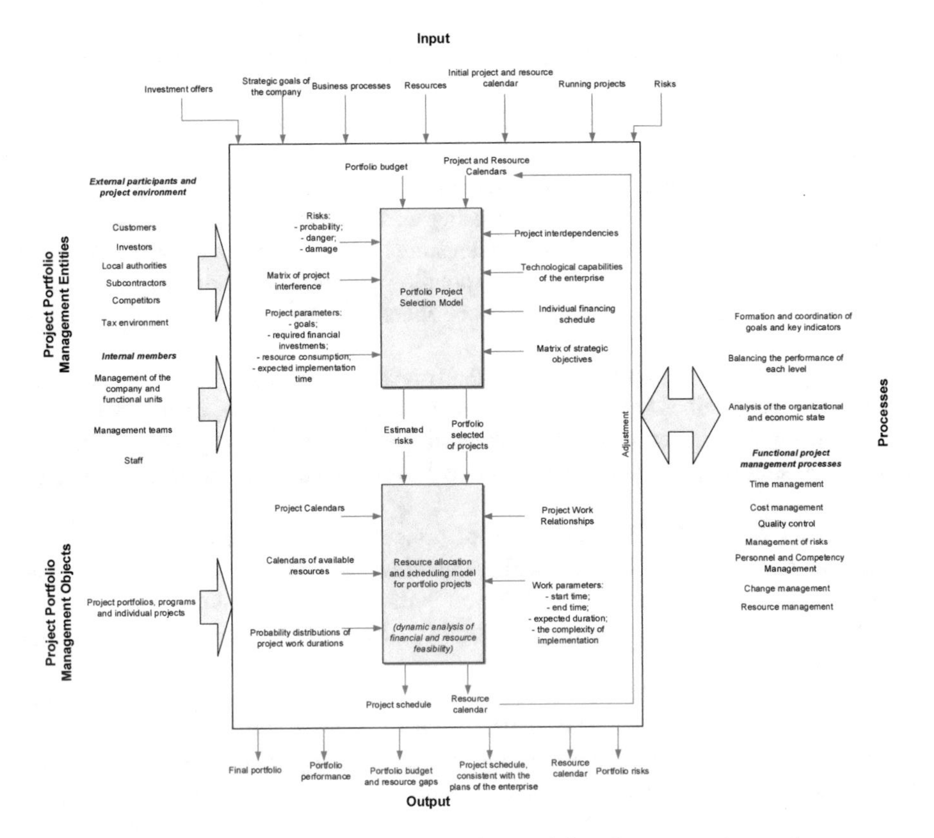

Fig. 2. Model of structural synthesis of a portfolio of energy saving projects

Energy saving projects planned for implementation at the enterprise are determined by a variety of parameters or attributes. The portfolio can be made as individual projects of the energy saving program, and projects to improve current activities. Thus, the portfolio at each point in time is a set of projects, some of which are aimed at implementing the program, and some are independent projects.

Further, one should take into account the typology of interdependencies of project operations: common operations for several projects; shared resources for multiple projects; dependence of the income received by the project from the income received by other projects. It should be borne in mind that the process of forming the final portfolio of projects will also be determined by a set of actions to prevent and eliminate undesirable manifestations of risk factors.

Thus, the overall task of building a portfolio of projects can be represented as follows.

At the first stage it is necessary to form a preliminary portfolio of projects that meets the requirements of the enterprise and the external environment. The task of forming an effective portfolio of projects is to determine such a procedure for selecting projects, allowing to take into account both the influence of the external environment and the risks of the internal environment of the enterprise, and for the enterprise to ensure maximum guaranteed energy efficiency. For this, we will use the project selection model, the main points of which will be discussed in the following publications.

If, however, the probabilistic characteristics of project flows to a portfolio and their resource requirements are known, then using an agent-based simulation model of a project portfolio [9], one can obtain an interesting estimate of the value brought by the adopted strategic decision and the corresponding program. In addition, the risks of program and project implementation can be identified and analyzed in a simulation model.

Using this model allows you to consistently analyze energy saving projects in order to identify opportunities for their implementation at the enterprise, coordinate project implementation plans and plans of the enterprise at various planning levels, select the most promising projects for implementation in accordance with an effective energy strategy.

The solution for the main problems of project portfolio management should be carried out on the basis of building and researching a system of mathematical models.

As part of our study, the system of project portfolio management models includes: a model for the formation of a project portfolio; agent simulation model of energy consumption.

Together, they will allow evaluating the effectiveness of selected projects for the implementation of energy saving measures, objectively assess the share of each energy resource in the total flow, determine the energy intensity of a separate production facility, the entire enterprise, and adjust the strategic direction in energy management.

The assessment of prospective power consumption should be considered as the primary stage in the formation of a prospective fuel and energy balance of a metallurgical enterprise, within which all private energy balances are interconnected.

The transformation of the structure of energy consumption should occur as a result of a significant reduction in the overall energy intensity of production and especially of fuel and heat capacity.

By changing exogenous variables within acceptable limits, it is possible to form a certain area of promising options for energy consumption, to consider ways and possibilities of meeting the received needs separately for each option. In the complex studies of the energy consumption system of a metallurgical enterprise, the identification of interconnections, that is, of direct and inverse connections between individual objects and processes, is extremely important.

This increases the degree of adequacy of the model to the described system, reliability and reliability of forecasts for such a model. In the scheme, such relationships are realized in the fuel and energy balance, that is, between the production and consumption of electric and thermal energy, as well as between electricity and other energy sources.

The results obtained in the study and long-term forecasting of energy consumption according to the principles outlined can serve as preliminary basic information at all stages of planning and managing the process of energy consumption of an enterprise.

The metallurgical enterprise project portfolio will be associated with a vector *PPR* with a dimension corresponding to the number of projects $P = \{P_1, P_2, \ldots, P_n\}$ in the set of projects under consideration, the values of which are binary values pp_i, where 1 means that the *i*-th project is included in the portfolio, 0 means that the *i*-th project is not included in a portfolio.

Thus, for example, projects can be defined that under any conditions cannot be excluded from the portfolio. Each project included in the portfolio $i \in P$ is a subject to management and has a number of characteristics that require clarification and formalization.

A set of energy efficiency projects of a metallurgical enterprise, or a project portfolio, is also subject to management and has such parameters as profitability, risk, time of implementation, required resources, etc.

At the same time, the implementation of each project influences the implementation of other projects included in the portfolio, and thereby affects the parameters of the entire portfolio of projects.

Given the absolute importance of the characteristics of each of the projects included in the portfolio, it should be noted that the strategic competitiveness and development of the enterprise depend on the characteristics of the entire portfolio of projects.

Formalized to submit a project in accordance with the directions of improving the energy efficiency of an enterprise can be in the form of a combination of the following components:

$$P_i = \langle X_i, W_i, R_i \rangle,$$

where X_i is the vector of initial characteristics of the *i*-th project; W_i- vector of the characteristics of the attractiveness and feasibility of the project; R_i- cumulative project risk. The vector of the initial characteristics of the project is represented as

$$X_i = \langle C_i, Y_i, S_i, H_i, T_i, R_i, I_i \rangle,$$

where C_i are the goals of the project; Y_i- complex of jobs on the project; S_i- required financial investments in the project; H_i- resource intensity of the project; T_i- the expected time of the project; I_i- vector of mutual influence on other projects in the portfolio. In the interference vector I_i, coefficients are put down, which can take values from 0 to 1, indicating the level of dependence of project i on other portfolio projects.

An important step in the analysis is the grouping of the projects under consideration for selection into the portfolio in the following aspects: from the standpoint of energy efficiency goals, finances, conditions (resources).

For this purpose, indicators of attractiveness and feasibility are used, which together represent the possibility of implementing the project at this enterprise, taking into account the strategic directions of the enterprise's activities, resource, financial and temporary support.

$$W_i = \langle SC_i, E_i, SR_i, HR_i \rangle,$$

where SC_i is the index of compliance with the strategic goals of the enterprise and increasing energy efficiency during the project implementation; E_i- indicators for assessing the economic efficiency of the project; SR_i- financial feasibility of the project; HR_i- resource realizability of the project.

The objectives C_i of the project are formulated as a set of indicators with an indication of their values, which should be achieved as a result of the project $\{K_j^{P_i}\}$. The conformity indicator $SC_j^{Str_i}$ is considered for all strategic objectives in the four projections indicated. If there is no parameter $K_j^{Str_i}$ in the project description, the compliance indicator $SC_j^{Str_i}$ for the strategic goal Str_i is zero. Otherwise, the target values of this indicator in the project $K_t^{P_i}$ and strategy $K_t^{Str_i}$ are compared with the current value (at the time t) of this indicator for the enterprise K_t^E:

$$SC_t^{Str_i} = \frac{K_t^{Str_i} - K_t^E}{K_t^{P_i} - K_t^E}.$$

After determining the conformity assessments for individual parameters, the strategic compliance of the project with respect to the energy efficiency strategy Str_i can be calculated by averaging the estimates for individual indicators:

$$SC^{Str_i} = \frac{1}{N_{KPI}^{Str_i}} \sum_{t=1}^{N_{KPI}^{Str_i}} SC_t^{Str_i},$$

where $N_{KPI}^{Str_i}$ – number of indicators in the strategy description Str_i.

This is suitable if only one strategic goal is assigned to each project. In reality, the situation is possible with several goals. In this case, after determining the conformity assessments for each goal, the strategic fit of the project P_i can be calculated by averaging the estimates SC^{Str_i} for individual indicators. At the same time, it is also possible to take into account the importance of strategic goals by introducing weights that can be obtained by an expert using the method of analyzing hierarchies or pairwise comparisons. If N_i^{Str} strategic objectives are associated with each portfolio project, the compliance index is calculated as follows

$$SC_i = \frac{1}{N_i^{Str}} \sum_{k=1}^{N_i^{Str}} w^{\mathrm{Str}_k} SC^{Str_k},$$

where w^{Str_k} – the importance of a strategic goal, $\sum_k w_k^{\mathrm{Str}_i} = 1$.

Thus, the index of compliance of the project $SC_i \in [0, 1]$ with the energy efficiency strategy is formed, the values of which are interpreted as follows: $SC_i = 1$ - if the project fully complies with the strategy; $SC_i = 0$ - if the project does not match the strategy; $0 < SC_i < 1$ - if the project is partially consistent with the strategy and is associated with the development of the strategic potential of the enterprise.

5 The Results of the Energy Saving Program Implementation at the Enterprise

The developed models allow us to identify, to analyze and to find the promising energy-saving projects in order to choose among them the most viable one. We managed to optimize the portfolio and this allowed us to focus on the most desirable goals.

Approbation of the developed models and computer means in PJSC Dniprospetsstal proved that the effective management of energy efficiency based on the program and portfolio projects management is possible.

The application economic effect calculation of the developed methods for PJSC Dniprospetsstal multi-level planning and power management is given in Table 1.

Table 1. The results of the developed methods application

Parameter	Unit	Base period 2011	Reporting period 2017
Planned total electricity consumption	thousand kWh	592837,5	445023,4
Actual total electricity consumption	thousand kWh	560425,2	421168,0
Consumption deviation	thousand kWh	32412,3	23855,4
Reducing deviations in electricity consumption from request	%		26,4
Declared reduction of deviations from request	%	20	
Increase forecast accuracy	%		9,2

The consumption of electricity by the enterprise in 2017 has made 421168,0 thousand kWh (Fig. 3).

Due to implementation of the energy saving program of in 2017 the enterprise saved 1,7 million kWh and 350 thousand m^3 of gaseous fuel. In comparison with 2011 electricity consumption on steel-melting production has decreased by 30%. Also the costs share of the electricity for steel-melting production has decreased from 71% in 2011 up to 67,8% in 2017.

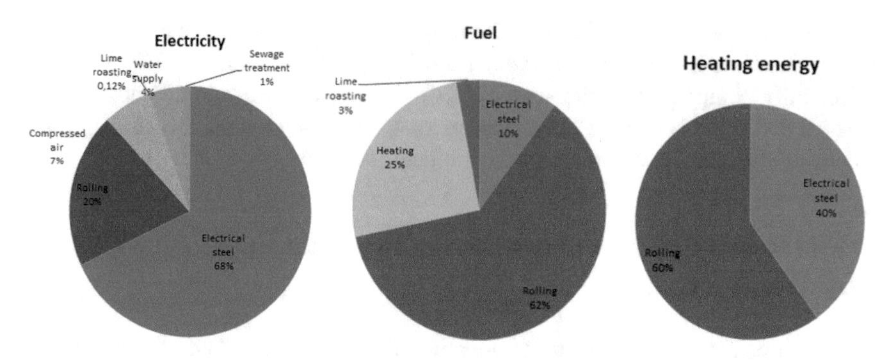

Fig. 3. The balance of energy consumption in PJSC Dniprospetsstal in 2017

References

1. Bunse, K., Vodicka, M., Schönsleben, P., Brülhart, M., Ernst, F.O.: Integrating energy efficiency performance in production management – gap analysis between industrial needs and scientific literature. J. Clean. Prod. **19**(6–7), 667–679 (2011)
2. Wang, H., Zhou, P., Zhou, D.Q.: Scenario-based energy efficiency and productivity in China: a non-radial directional distance function analysis. Energy Econ. **40**, 795–803 (2013)
3. Brunke, J.-C., Johansson, M., Thollander, P.: Empirical investigation of barriers and drivers to the adoption of energy conservation measures, energy management practices and energy services in the Swedish iron and steel industry. J. Clean. Prod. **84**(1), 509–525 (2014)
4. Shemetov, A.N., Oleynikov, V.K.: Adaptive management of power consumption of an industrial facility according to a complex criterion. Glavnyy energetik **4**, 30–37 (2014). (In Russian)
5. Shemetov, A.N., Fedorova, S.V., Kuznetsov, S.V., Lyapin, R.N.: Modern problems and prospects of model forming of energy management at enterprises of mining and metallurgical complex. Elektrotekhnicheskie sistemy i kompleksy [Electrotechnical Syst. Complexes] **4** (33), 41–48 (2016). (In Russian)
6. Molokanova, V., Petrenko, V.: Project-oriented approach to metallurgical enterprises sustainable development management. Metall. Min. Ind. **8**, 28–34 (2016)
7. Kukhta, P.V., Sviderska, S.Y.: Algorithm for evaluation of realisation potential for innovation projects of metallurgical enterprises of Ukraine. Efektyvna ekonomika, vol. 12 (2018). http://www.economy.nayka.com.ua/?op=1&z=6733
8. Vishnevskaya, M., Kozenkov, D., Kaut, O.: Development of methodology for the calculation of the project innovation indicator and its criteria components. Baltic J. Econ. Stud. **3**(5), 61–69 (2017)
9. Kiyko, S., Druzhinin, E., Prokhorov, O.: Managing the energy-saving projects Portfolio at the metallurgical enterprises. In: Shakhovska, N., Medykovskyy, M. (eds.) Advances in Intelligent Systems and Computing III, CSIT 2018, Advances in Intelligent Systems and Computing, vol. 871. Springer, Cham (2018)

Aviation Aircraft Planning System Project Development

Vasyl Lytvyn[1(✉)], Agnieszka Kowalska-Styczen[2], Dmytro Peleshko[3], Taras Rak[3], Viktor Voloshyn[3], Jörg Rainer Noennig[4], Victoria Vysotska[1], Lesia Nykolyshyn[1], and Hanna Pryshchepa[1]

[1] Lviv Polytechnic National University, Lviv, Ukraine
Vasyl.V.Lytvyn@lpnu.ua
[2] Silesian University of Technology, Gliwice, Poland
[3] IT STEP University, Lviv, Ukraine
[4] Technische Universität Dresden, Dresden, Germany

Abstract. Since the airspace is subordinated to the territories of different states, different authorities, such as Eurocontrol in Europe and FAA in America, etc., an acute need for a product that would provide easy, fast, high-quality flight planning, their proper dissemination, formation of necessary documents, etc. arises. Purpose is to cluster all necessary planning flights items, functions, united informational system data. This system is designed to ease flight plan description in Eurocontrol database etc., provide its correct dissemination to all controllers, towers, airports etc. This project will help to omit difficult and long-lasting phone calls, automate and optimize flight plan rendering, provide with high data accuracy. The pilot also can easily check the weather at the airports during the flight. Adding all necessary characteristics of the plane, the system will calculate fuel consumption for all approved flights. Besides, all available and necessary documents will be established into unified document or database. Object study is aviation aircrafts flights planning. Subject study is aviation aircraft planning informational system project development. Novelties are few programs nowadays, which could rapidly, easily and qualitatively schedule aviation aircrafts flights of general-purpose. Many programs are very narrow focused and don't give access to full functionality, others are highly expensive to operate them. So, this informational system project will be multifunctional and of high quality at the same time which cause its enormous demand. Flights planning aviation aircrafts of general-purpose informational system project will have enormous demand all across the globe. However, Europe would be the dominant user since airspace is under vigilant superintendence of Eurocontrol and number of rules, prohibitions and requirements covers these lands. Created application ensures easy and fast itinerary scheduling around the most problematical areas of Europe.

Keywords: Aviation aircraft planning · Machine learning · Project development · Applied methods and procedures for general aviation aircraft design · Project management · Flight planning · Gantt chart · Flight plan · System analysis · Life cycle · Decision making · Aviation aircraft · Risk management

N. Shakhovska and M. O. Medykovskyy (Eds.): CCSIT 2019, AISC 1080, pp. 315–348, 2020.
https://doi.org/10.1007/978-3-030-33695-0_23

1 Introduction

Given the popularity and rapid development of aviation today, there are many problems associated with synchronization of aircraft, the provision of take-off and landing conditions for aircraft, laying and registration of rational routes, calculation of fuel for flights, etc. To solve the problem today, the high development of IT technologies is rapidly helping. Modern IT technology computer systems are able to store data in large volumes, constantly read and update the location of objects, using navigation systems, predict weather conditions and display the necessary data in a user-understandable format. Systems operating in the context of general aviation use the latest technology for convenient and fast data processing. Systems read the coordinates of the aircraft during the flight, in order to follow the schedule and the correctness of the movement. In addition, these systems make inquiries into air control from different countries of the world for registration and reservation of runways for a specific time. In addition, such software controls fuel consumption by analyzing the total weight of luggage. There are many technical features and benefits of these systems:

- Versatility. Systems have the ability to control all business processes of aviation and include a wide range of functions [1];
- Portability. Systems can be developed as a web application or mobile application, making them very user-friendly [2];
- Cross-platform. Any operating systems and browsers can support systems;
- Price availability. The cost of using these systems is small, and the basic set of necessary services in some systems is free [3].

The project, which is being developed, should provide to different users a range of business services in the context of general aviation [4]. Project should solve the problem of scheduling and route planning, calculating fuel load [5], booking runways for landing and departure of aircraft [6]. After the analysis of competitors' systems, you must select the following features and advantages of this system [7]:

- a large database of routes [8];
- data processing in real time [9];
- implementation of optimal and fast algorithms for calculation [10];
- a wide range of services in the Basic (free) version of the program [11];
- portability: the program should be supported by any devices (PC, Tablet, Smartphone). Should also be made a task decomposition, starting with the main modules of the system [12].

The ability to use the system have only authorized users [13]. Each user must have a personal account with your own data [14]. Also features should vary depending on the version of the program and from the payment [15]. The system should be provided into 2 groups of users: pilots and administrators. Events in the system should be divided into 2 types: events with the confirmation and events without confirmation [16]. Event. that are associated with personal data change should be validated automatically. Events, registration, planning must be confirmed by the administrators [17].

2 Analysis of the Systems of Analogues

Software market analysis is an extremely important aspect in the backlog project creation, because it allows you to take a look at existing solutions to problematic issues [18–21]. According to this you can clearly establish the statements about the advantages and disadvantages of systems-analogues in the sphere of air flights [22–25]. Project Manager in this case generates a list of the problems of system analysis that one should consider within the framework of market analysis software [26–29]. In Fig. 1 is a software evaluation metric [30].

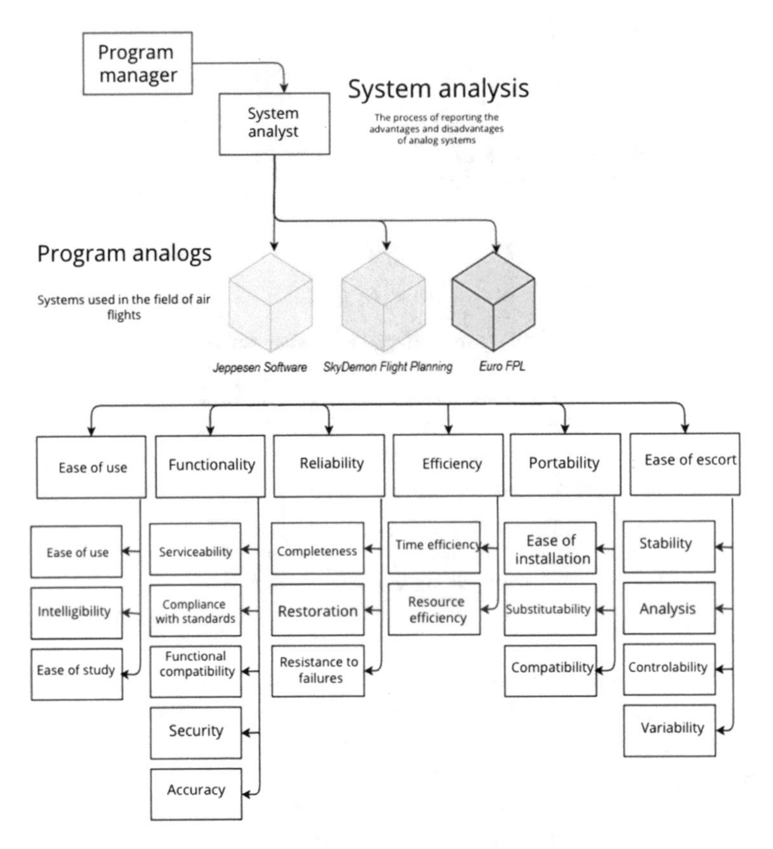

Fig. 1. Software evaluation metric

1. Jeppesen Software is the software that specializes in navigation information, aircraft operations, and management solution for the optimization of decision management crew and fleet, as well as the products and services of the training flight. The airline and the pilots all Control Center networks used in its operations

schema, data, and management tools. In Fig. 2 is displayed a program interface Jeppesen. Software. Advantages: maintenance of the four segments of the market (Commercial, Business, military and General Aviation); quick update navigation data; portability. Disadvantages: high price for the use of the services (Table 1) [31–34].

Fig. 2. Program interface Jeppesen. Software

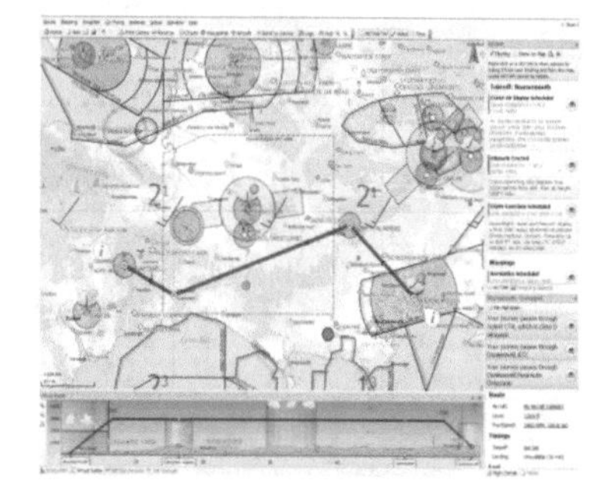

Fig. 3. Program SkyDemon flight planning

Table 1. Comparisons of similar software

Property	1	2	3	4
Intuitive interface	+	+	+	+
A wide range of itineraries	+	–	–	+
Work in real time	+	–	–	–
Portability	–	+	+	+
A quick update	+	+	–	–
The ability to filter routes	+	+	+	+

2. SkyDemon Flight Planning is software for the scheduling air routes, taking into account all possible natural obstacles and shortcomings (Fig. 3). Advantages: the system works in real time; -quick to assimilate; intuitive interface. Disadvantages: limited range of functions (Table 1) [35–39].
3. Euro FPL is software for planning air routes, with access to cloud storage environments (Fig. 4) [40]. Advantages: a large number of routes; notifications about flight tracking; journals navigation in real time around the world; tracking winds in different directions. Disadvantages: limited support service; low volume routes (Table 1).

Fig. 4. The interface of the program Euro FPL

Fig. 5. Interface AerosDB

4. AerosDB has been developed with the help of more than 42 years of experience in the field of aviation and more than 31 years of practical knowledge gained in areas such as air, airline pilot and command and the certification of the airline, as well as an extensive knowledge in it, including aerospace, government, financial and utility companies (Fig. 5). Advantages: multifunctional; the presence of the filter routes; aircraft settings. Disadvantages: complex interface to use (Table 1) [41–47].

3 Algorithmization of the System

According to the UML diagrams algorithmization of the information system for flight planning is shown. Diagram using UML is a diagram that depicts the actors and use cases and their relationship. Use case diagram is a graph that consists of a set of actors, use cases (use) the restricted limit system (rectangle), associations between actors and precedents, relationships among use cases, and of generalization between actors [48–51].

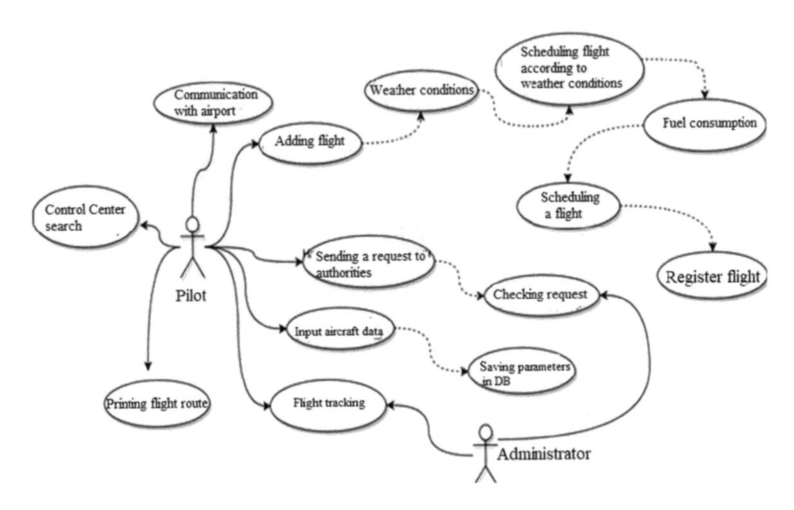

Fig. 6. Chart usage

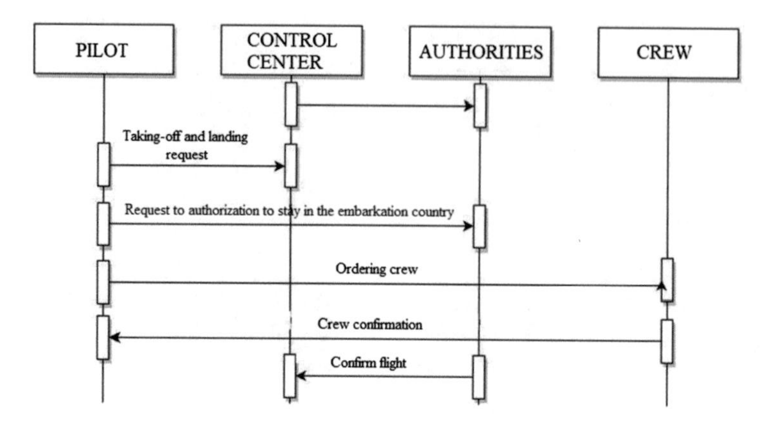

Fig. 7. Chart sequences

Use case diagrams represent the elements of the model use (Fig. 6) [52–55]. Figure 7 is the diagram of the sequences of IS for flight planning [37, 56–63].

Figure 8 shows activity diagram. Figure 9 shows an system of planning state diagram.

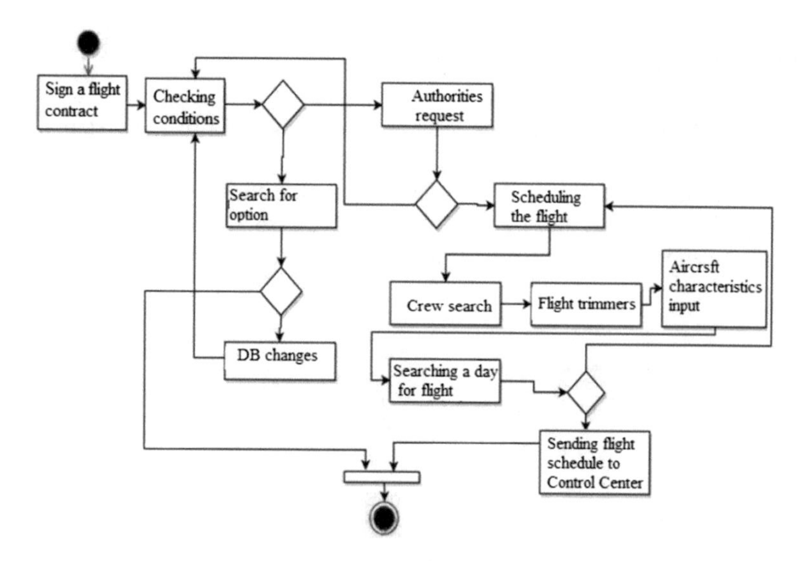

Fig. 8. Activity diagram

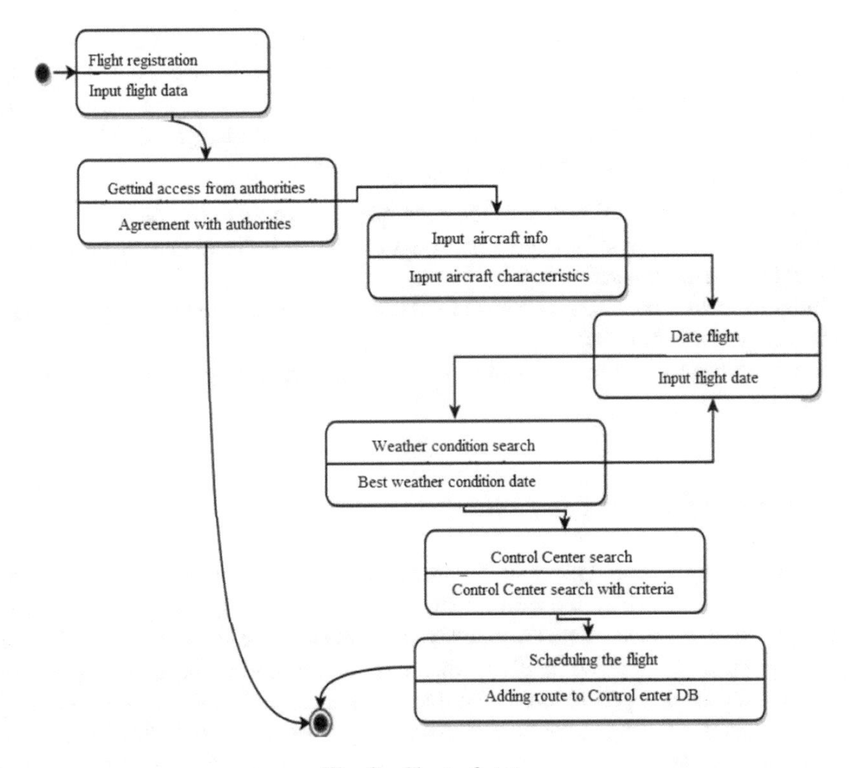

Fig. 9. Chart of states

4 The Purpose of the Project

The purpose of this project of flight aircraft aviation information system planning is effectively cooperation all necessary items for flight planning products, features, materials in a single information system. This system should facilitate the filing of flight plan in the database, operations and ensure its proper distribution: all necessary Control Center, tower, airports, etc., the project will help to avoid difficult and long-lasting phone calls, automates and optimizes flight plan filing, will ensure maximum correctness of submitted data. In addition, the pilot will be able to check easily the weather at airports, as well as for the entire route. Adding all the necessary characteristics of the aircraft, the system will calculate the fuel calculations for all registered. As well as all-existing and required documents will be formed in a single document.

This project will have a great demand around the world, but the largest popularity it will cover in Europe, as well as air space of this territory is under the control of Eurocontrol, precisely in this territory there are too many rules, prohibitions and requirements. And created program is developed to provide an easy and rapid assembly routes on the most problematic territories of Europe.

The main aim of the project is something that every pilot who is the owner of the aircraft will be able to register in the system by himself/herself when registering it must enter your license and then she will verified for authenticity. After registration, the pilot will be granted with great interface and flexibility regarding the planning and calculation of flight selection of ideal conditions. This software is targeted at automation phases of flight planning. Thus, the functional content of the system will save time, human and financial resources which are important in the design of the flight. The program will provide a pilot the following opportunities:

- the possibility of creation a new flight to specify start and end route;
- a pilot can see weather for a week and the system will indicate that the best day will be able to make the safest flight;
- ability to specify the characteristics of the aeroplane relative to what would be hand-picked by the best route;
- the possibility to confirm the landing online and bargain with different government agencies.

5 The Relevance of the System

Air flights occupy a special place in the process of rapid movement between the initial and final points. This way the movement stands out for its speed and the ability to travel over long distances. The air industry in Ukraine, there is quite a long time, but the main progress it achieved in 2012, when our country hosted the European Football Championship. In this period were reconstructed and built airports, to improve travel service for Ukraine football fans from all over Europe. From this point we can conclude that air travel is one of the most important aspects, forming the level of the State. Diploma subject is establishment of information system for planning flights of aircraft, so the main task is the analysis of the design, and search for potential problems in this area. In order to understand how extensive the scope of the design you want to analyze all the statistics related to the flight of aircraft in Ukraine. This formed a statistical indexes of passenger traffic in the largest Ukraine's airports. Statistics is in Fig. 10.

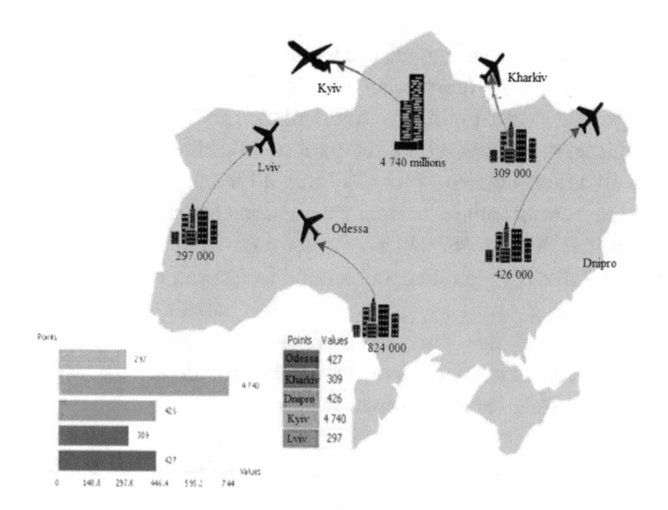

Fig. 10. Statistics traffic airports

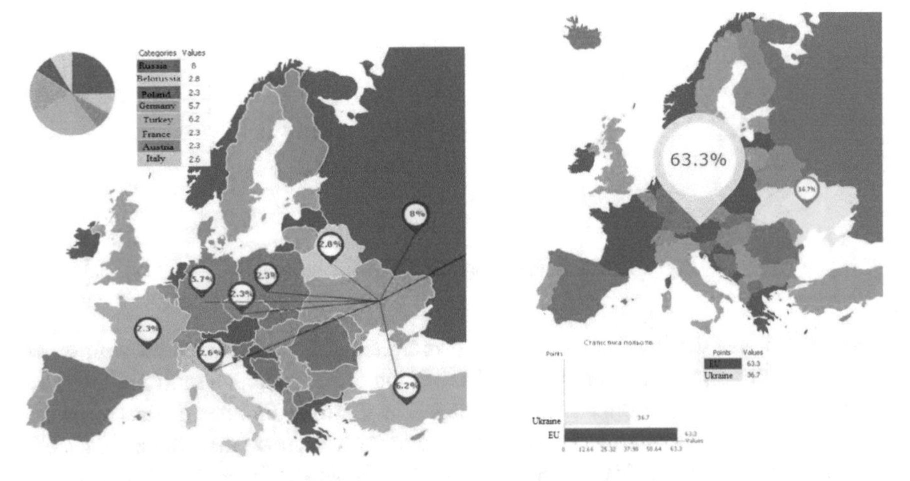

Fig. 11. The most popular directions statistics **Fig. 12.** Infographics of directions of travel

As you can see there are at least five major airports (excluding the occupied territory), and a dozen smaller where passenger traffic does not exceed one hundred thousand. Once the data have been obtained from level of the use of this type of transport should analyze in any point of the world sent more aircraft, and why this kind of transport is so popular. Aircraft are the most comfortable means of travel over long distances, because consumes less time compared with cars or trains. Also in recent years, air travel became more accessible financially. All these aspects form the following the popularity of the use of air transport. Figure 11 makes it clear what countries in the world are the most popular for our residents. Figure 11 clearly explains where exactly the Ukrainians are sent using the airlines. For comprehensive understanding parameters of air travel developed Fig. 12. Besides the small percentage of flights to the United States, you can understand that the main direction is European countries. This leads to the task of the Major diploma project in the automation of this direction of the route.

Functional content provided in IC will reduce the level of accidents occured among air flights. Figure 13 is infographics that shows statistics of accidents causes.

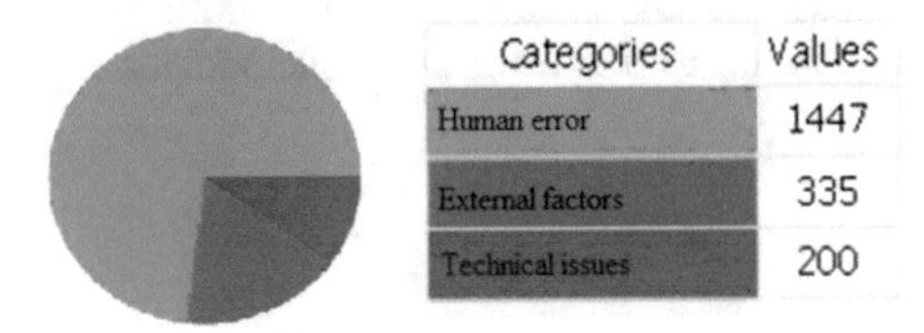

Categories Values
Human error 1447
External factors 335
Technical issues 200

Fig. 13. The statistics of aircraft accidents causes

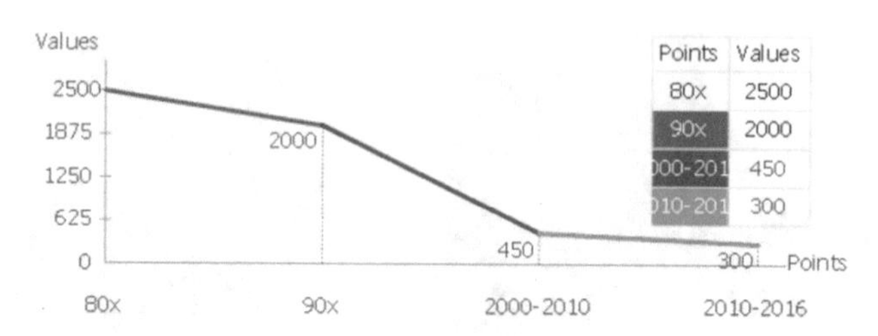

Fig. 14. The statistics of mortality in aircrashes

As you can see the level of accidents of aircraft not as high compared with other means of transportation. For example, world aviation estimates that a million flew have only two crashes. In addition, statistics shows that the total number of deaths in the aircrashes is several hundred per year, while autocrash is more than three thousand. I want to add that with the development of technology also the safety of air operations is increasing. Every year the Experts count less and less air crashes that positively responds to the popularity of this kind of movement. In Fig. 14 clearly can be seen the decline of mortality during the flights. According to modern technology we can see the following result, because the information system should contain a function that will be aimed at ensurement the flight safety.

6 Place of Application

In our modern world, most processes are automated and play an important role in organizations of certain works. The processes that belong to air traffic and flights themselves are automated as they require exact calculations and reliable control of the flight. You can say that the route is one of the basic tasks that require accuracy in calculations because the airlines are investing large sums of money in software development that will solve this problem. This process is important because it concerns the time and money that will be spent for the route calculation and flight as well as on the people' lives that are extremely important for all airlines.

Due to implementation of the goals, the information system can be used in different directions. Expected result from implementation should be the improvement of the processes associated with the air mission control according to automated module. Figure 15 shows the expected results from the implementation of the established IS.

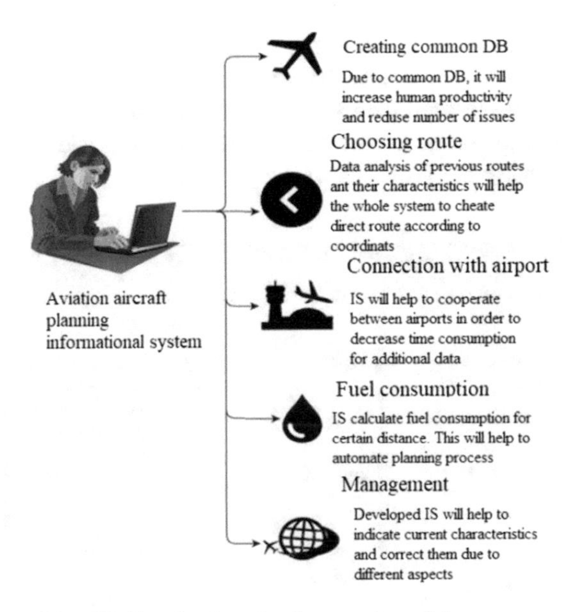

Fig. 15. Results from implementation of the system

7 Description of the Functions and Structure of the System

Requirements formation to system is one of the most important stages of the life cycle of the system [16]. At this point, the systems analyst must research the subject area and identify problem areas. Thus formed the basic framework of the tasks and goals that should be automated in the future. In the Scrum methodology provided an important tool which is backlog project. It is organized by functions describing individual characteristics. Figure 16 is the backlog project. Functions:

- registration of flight/sending all the necessary authorities. A function that is provided for the formation of cooperation between other airports, and sending data on all flights;
- calculations of the fuel for the flight. Important feature is aimed at additional anticipation of controllers that will improve the safety of the flights;
- crew control. A feature that improves the interaction of crew of controllers. This process will launch the exchange of information between the two modules of the system, the status of current flight;
- formation of a single document with all the above mentioned information. Documentation is an important aspect of the process, because due to it you can track the status of those or other air travel;
- ability to manage already registered flight. An important aspect that will make adjustments in flight status in case of external factors. External factors can be: weather conditions, damage to the airports and others;
- ability to create drafts. A function to keep track of completed operations or those that do not have occurred due to certain factors;

- ability to create routes as yourself and receive suggestions automatically and the ability to edit a route on the map. Based on the indicators of the environment, the system will be able to form their own routes that will be safe for travel. Also the Manager himself can create a route on the specified points and create the changes in it. This will increase the flexibility of the product use.

№	Function name	Value	Complexity	Demonstration
1	Flight registration/ sending to all authorities	Obligatory	7,2	by inputting data, information is sent to authorities
2	Fuel consumption	Obligatory	6,2	By inputting current fuel totals, system calculates the distance before flight
3	Crew management	Obligatory	10,6	Client-server architecture will help to run the crew
4	Creating common document with all information listed above	Not obligatory	1,8	All data in the DB can be gathered in one document
5	Opportunity to run existed flight	Obligatory	6,2	Client-server architecture will help constant cooperation between 2 modules and Control Center can run the flight
6	Opportunity to create notes	Not obligatory	3	To create a note, there should be one flight that soon will be listed in general history activity of aircraft
7	Opportunity to create routes both manually and receive invitations automatically and change the route on map	Obligatory	8,8	Route creation can be formed according to start and finish coordinate, besides, previous flights can be chosen from flights' history

Fig. 16. Information system Backlog

One of the important parts of the system is its administrative part, with which you can keep track of all the users of the system and their interaction with the system in General. Further be described the main functions of the administrative part of the system:

- Users Tab or Route Users. Function is the management of the users of the system, an administrator can view personal information about the pilot, his license and all permits that were presented to him during the creating a profile in the system. The administrator login and password available to the pilot, there is the possibility of log in under the login user, you can also edit his profile and the ability to lock in the system.
- Tab Messages;

  (a) Send Messages function in the system between users and administrators;
  (b) Messages is browsing feature of dialogue between users and Administrators;
  (c) Inbox is incoming messages that are received
  (d) Outbox is outgoing messages sent to specific Administrator

- Tab Flight Plans;
  (a) Draft Flight Plans. This tab contains drafts of flights that create users in their personal offices;
  (b) Live Flight Plans is active flight plans that are registered and will be in the nearest future;
  (c) History Flight Plans is saved flight plans, which were made in the past;
- Tab Airport;
  (a) Airports consists documentation;
  (b) Airways is tab that contains data on air routes as well as the documents that you want to confirm with government agencies on the flight plan;
  (c) Waypoints is a tab that contains data on points of the route which will be carried out by a scheduled route;
  (d) SIDSTAR is a tab that contains data on standardized departures and landings;
  (e) NOTAM (Notice to Airmen) is notifications to the air staff that are sent to the judicial authorities, which is sent by means of telecommunication, and contains information about the introduction of the action in a specific State or even the replacement of air navigation facilities and their equipment. Contains information about the dangers, and on-time warnings about which is essential for crew associated with the execution of the operation;
- Tab Aircraft;
  (a) Manage Aircraft is the tab that contains the information about the model airplane (fuel use and the physical characteristics of the aircraft);
  (b) Manage User Aircraft is a tab that contains data for the user plane (fuel use and the physical characteristics of the aircraft);

Figure 17 is an example of a document the calculation of flight.

BRIEFING PACK   NDEMO1 G150

DATE	26-NOV-2017 15:30 Z
STATUS	DRAFT
ROUTE	UKLL - UKBB LVIV INTERNATIONAL - BORYSPIL INTERNATIONAL AIRPORT DCT ADBAN N983 SORON M986 SLV
FLIGHT LEVEL	410
ALT	
DISTANCE	273.6 NM
FLIGHT TIME	00 HR 53 MINS
TRIP FUEL	2005 Pounds
MIN. RAMP FUEL	3455 Pounds
BURN [Trip Fuel]	2155 Pounds
ENDURANCE	0215 HRS
PIC	NYKOLYSHYN
POB	001
UKLL	**METAR 26-Nov-2017 15:00 UTC** **Wind**: True direction = 350 degrees, Speed: 3 m/s

Fig. 17. Example of a document the calculation of flight

One of the most important features included in the system of calculation of flight is the Autoroute. The user must specify the start point of the flight and the end, and then the algorithm of the system will generate the entire route with the best-selected points. Once the route is generated to the user, VALIDATE function becomes available, which allows you to send such powerful institutions as Eurocontrol and Eurokontal incoming documents relative to the aircraft, which will fly the route on which will be flying, information about the people who will be on board during the flight, and the information about the possibility of their stay on the territory of the country, flight plan and other governing documents that would indicate the purity of the flight.

8 Planning the Design Decisions

Project management is a responsible process that requires a Manager making decisions that will achieve the goals of the project by taking the minimum number of resources available. To do this, and created a project plan, where clearly formed the structure of interaction objects system, understanding the concept of workflow helps. In Fig. 18 is the project processes decomposition.

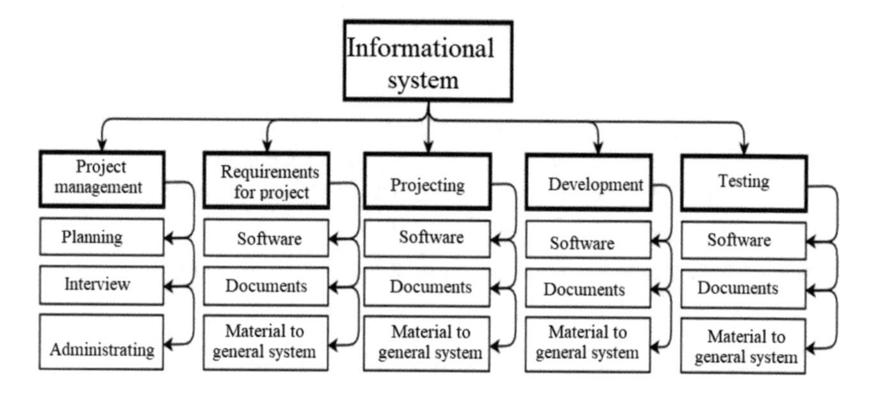

Fig. 18. Project processes structure

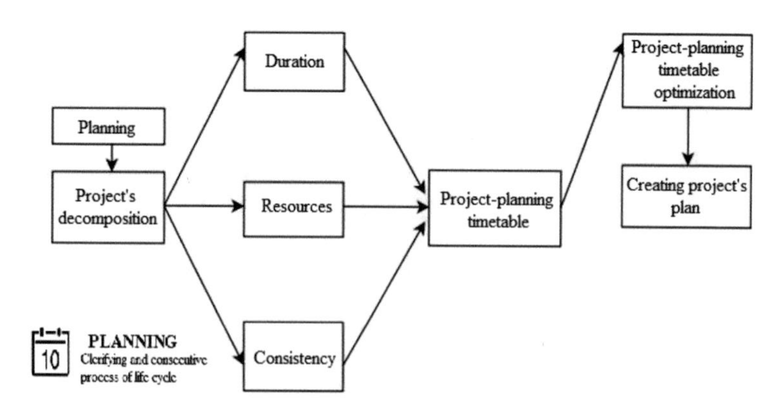

Fig. 19. The planning process structure

On Fig. 19 can be seen processes' blocks that combine into the software creation. The first block is the management of the project, which includes three subitems. Figure 19 is the planning process structure. Decomposition of the project is the cooperation of various kinds of objects, methods and structures that form the concept of activities aimed at the creation of a new product, technology or tool to use. A starting point in creating a new project is the search for improvements in the design. Systems analysis allows you to understand that you can improve upon examining the problematic aspects of the sphere that you are automating. Thus derived from the ordered list of important problems that must be solved by means of the software. After that is formed by a team of developers, which will step by step line up the architecture of a software system. In addition, the program development must be designed by a functional product that will solve all the problems found in the analysis area. A successful result can be considered as the project management, which meet all the business goals and have formed such a software system that meets all the commonly used standards. Figure 20 is the decomposition of the project information system for flight planning.

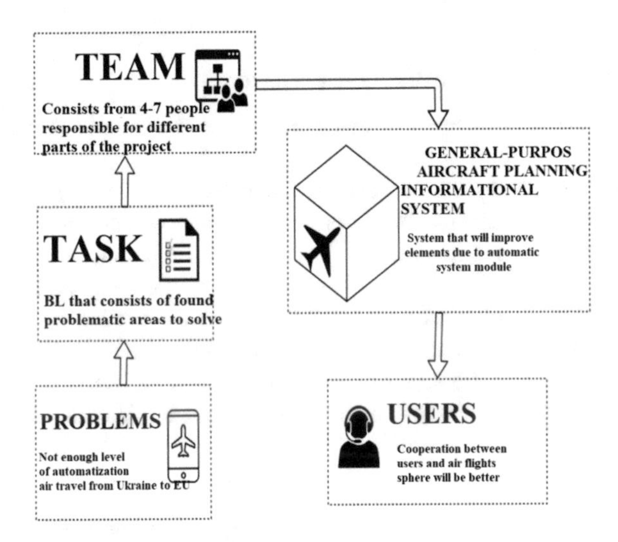

Fig. 20. The decomposition of the project

Decomposition of the project is divided into more specific issues in the management of the project. The first issue is the assessment of the duration of the works. This is an extremely important issue in the project management, as the Head shall examine the detailed scope of the specified requirements to properly distribute the resources available for the development of this system. The question of how to determine the scale of the design, even before the beginning of works. To do this, there is an interesting method that involves the ordering of Fibonacci numbers. We have the numbers: one, two, three, five, eight, thirteen, twenty-one, thirty-four, and so on. These numbers will serve as a kind of assessment of the severity of a given function. An important detail is that the assessment of gravity will exhibit all of the elements of a team project. This method will take a look at the immense performance from all points of view, ranging from project manager finishing tester. In Table 2 is defined the complexity size of implementation. After design team puts up heaviness scores of implementation functions expected in the project, you must determine the average of these estimates and further plan for the resource management development. In Table 2 displayed calculation on the severity of the execution of different features of the project.

Table 2. Complexity size of implementation

Name of the function	PM	SA	BA	DEV	TEST	Sum
1. Register a flight/sending all the necessary authorities	5	5	8	13	5	7.2
2. Calculation of fuel for flight	8	5	8	8	2	6.2
3. Crew management	8	8	8	21	8	10.6
4. Formation of a single document with all the listed above information	2	1	2	2	2	1.8
5. Ability to manage already registered flight	5	5	8	8	5	6.2
6. The ability to create drafts	1	3	3	3	2	3
7. The ability to create routes as yourself and receive suggestions automatically and the ability to edit a route on the map	5	5	8	21	5	8.8

The result is a twenty one, as the maximum possible score in determining the severity of the implementation functions and unit, as estimated the simplest feature amongst the entire list of tasks of the project. Now you can understand how largescale system is, within the maximum and minimum values. The next stage in the planning process is to define the resources for development. A description of the resources IT project is ruled by the same principles as the description of the WBS. However, unlike the structuring and the description of the WBS, you should take into account that the project operates multiple resources simultaneously. So instead of one structure

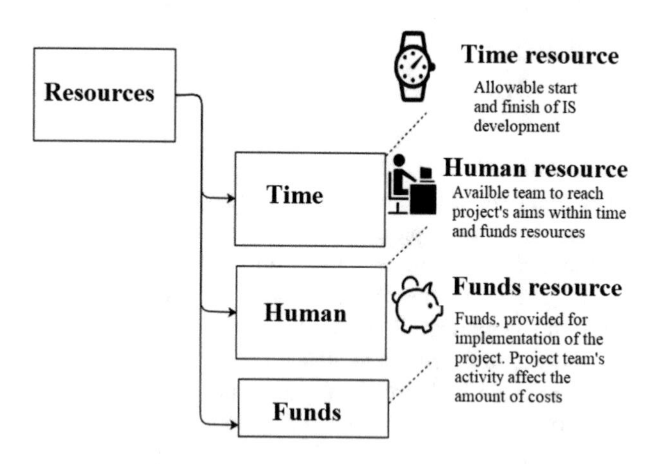

Fig. 21. The system resources structure

decomposition during description and structuring of resources you want to create multiple structures of decomposition (decomposition of the structure is used to describe one type of resources). Types of resources that are used to implement plans IT-project are depicted in Fig. 21.

Receipt of the draft order in IT company creates the need for search and selection of professionals for the realization of the task. IT-enterprises typically have significant personnel reserve, so for most, the best and fastest order execution raises the problem of selection of such experts, statistics work as possible for the customer. For the development of the information system planning of flights the aircraft also formed the team. To do this, consider the selection of people closely related to the system analysis, business process management, software development and software testing. The team of the project development was formed on the basis of checking personal qualities of employees. The stage of establishing the final composition of the team allows you to start a workflow, but before that team must go through several phases of installation, without which the performance of the Group of people would be impossible [4]. Figure 22 is the structure of the temporal phases of establishment of project team.

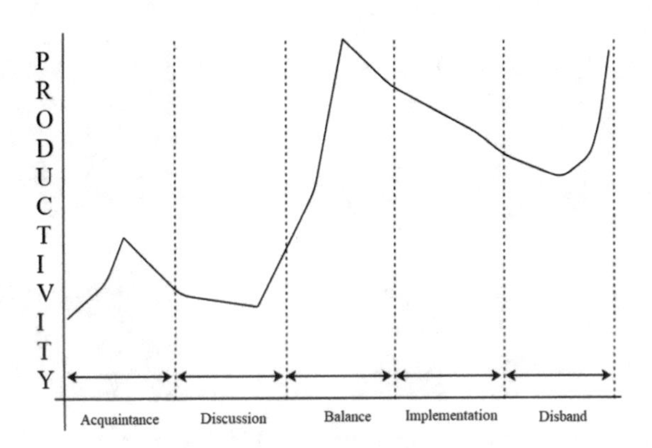

Fig. 22. Creating team phase

As you can see in Fig. 22 has five phases. The first phase is lapping at which people show their best qualities and they are trying to gain insight into the overall concept of interaction. The second stage is discussion and it is characterized by the presence of certain disputes and controversies in the middle of the team. If the team passes, it will led to milestone phase equilibrium, which smoothly moves on stage. A successful phase of execution will end the implementation of the system after that will lead to the disbandment of the interrelated elements of project development. Based on these phases, the project manager should anticipate all aspects of phases of the team creation and choose such people without any problems will cooperation with each other.

Forming a team, smoothly moves on the stage of the formation of the tasks of the task sequence. This is an extremely important step, because the sequence of shapes a specific methodology for the development of an IT product. Project Manager based on subject area development chooses more suitable and more efficient way of doing project management. For the creation of information system of choice was based on the two most popular methods. The first methodology was the Scrum, which is a flexible way of managing a project. The second method is to outline a way to manage smooth completion of the phases and the transition to the next level of development. In order to properly selection a methodology should consider the advantages and disadvantages of these two tools. For example, the Scrum can break the whole process of development on the sprinti that will constantly monitor all work progress, and due to the use of cascade model, a group of developers creates software based on contract requirements [4, 5]. The target of the Manager in this case was the election of such a methodology, that will create a product using the form and time. Figure 23 is an information graphics listing the advantages of the two drugs.

Fig. 23. Scrum and waterfall characteristics

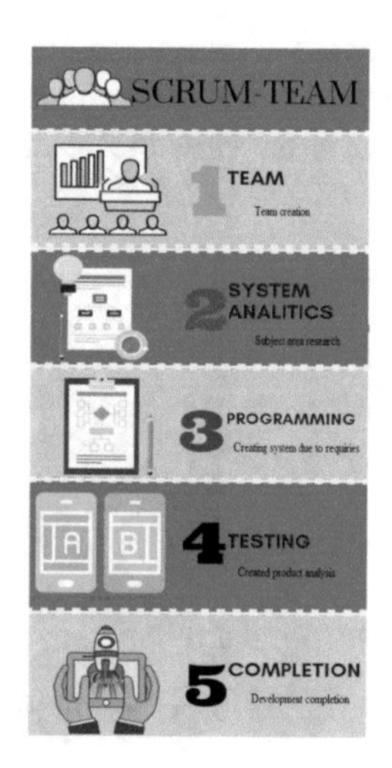

Fig. 24. Sequence of tasks execution

According to statistical indicators and benefits, the choice fell on a flexible design method. Scrum is an extremely flexible methodology that allows you constantly inspect the workflow and engage additional elements such as the customer of the system to the stages of creation. Figure 24 shows the sequence of tasks with Scrum. Figure 25 clearly shows Scrum's workflow.

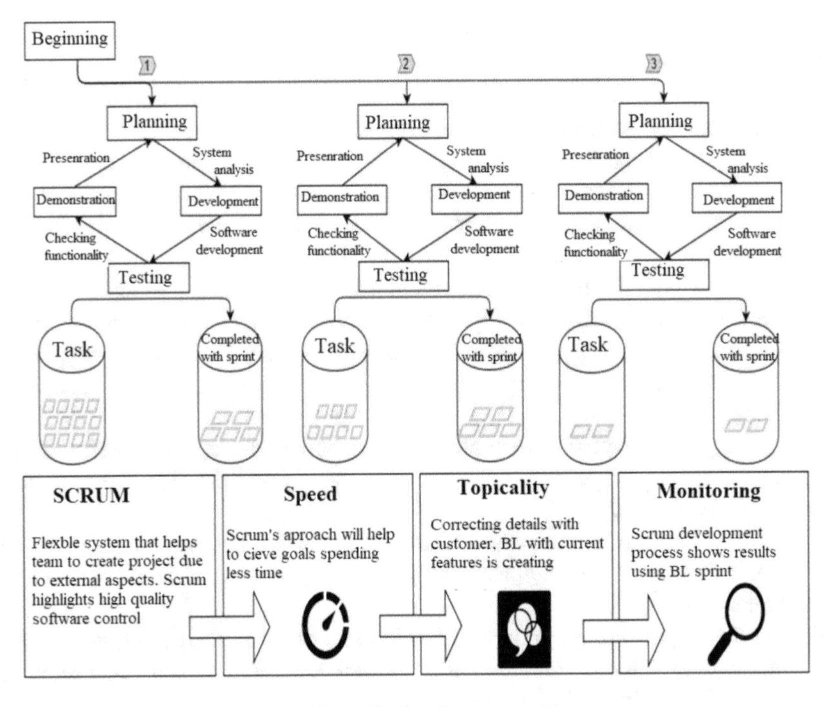

Fig. 25. Life cycle development with scrum

9 Functional Roles and Responsibilities

One of the main tasks in project management is a team that will work on the project [10]. The project manager, as a managerial element, assumes responsibility for the project and for the team, its role is to establish the framework, goals and resources allocated, time for the quality project implementation and objectives. The project manager, together with the project team, establishes the goals and the work that will be required. The project manager is responsible not only for functional requirements such as setting goals, time, jobs, but also for the atmosphere in the team, which also plays an important role in teamwork, solving conflicts inside the team of finding alternatives and also motivating all these aspects play an important role in building a cultural atmosphere. As we all know, the client wants to get good results and well-designed software that satisfies their wishes and the purpose they are settled. Since each good project requires qualitative analysis and changes during the project to achieve the final goal. Therefore, the project manager must respond quickly to all changes made by the client during the development of the project, and to satisfy all of his needs and implement them. Nevertheless, in some cases, when customers make a lot of changes to the project, it can negatively affect the deadline, settled at the beginning, as well as the quality of the project. From the customer's side there are the following roles:

- The main customer is the person who specifies the requirements for the project, the expected results that he is planning to achieve and setting the timeframe for the project. Responsible for resolving general issues and differences arising during the development of the project;

- System administrator is the person responsible for access to all internal systems companies to be involved in the development of the project. The person responsible for verifying the outcome of the project development as well as for verifying that all requirements that were set at the planning stage have been met;
- Controller is the high level aviation specialist who is one of the main elements of air traffic services. The main tasks are to maintain orderly safe collapse, and to ensure that aircraft are at a safe distance from each other and within their area of responsibility;
- The pilot is one of the most important roles of the entire system, both from knowledge and the skills of the pilot depend on the lives of people, in our time the pilot is already fully automated and algorithmic, so the main role of the pilot is to track the parameters of the flight sensors. That is tracking the normal condition of the flight;
- Crew is a group of people who are united in some sort of original structure or a hierarchy in order to perform a collaborative task. From the side of the performer are the following roles;
- The project manager is the person who allocates resources, solves the main issues that arose during the development of the project and controls the whole process of its implementation;
- Project developers are programmers who are involved in the design of the project architecture. Develop all functionalities as well as all requirements that were set at the beginning of the project planning;
- Tester of the project is a set of test workers developed programmer's software; they help to avoid any problems that can be found during the creation of the project. Often, software runs several cycles from software developers to testing. The main project resources are workstations that will be used to develop software by programmers. The performer carries out provision of workstations. The customer must provide the API executor that will be used to obtain flight, route, airborne characteristics, and start and end flight data for a more accurate and detailed flight calculation.

An executor provides the customer with a GIT version control system to monitor the work performed over a period to monitor the stability of the design and control of project execution times.

10 Power Structures

As an information system for flight planning is targeted at practical application, it should function within permitted rights of the field that is automated. In other words, a functional module should be on the legislative bodies, and enable opportunities so that the result was correct in relation to the air code. For example, with a developed product, we plan to fly from the starting point to the destination. In this case, we must reconcile all the details with the authorities of the landing place. This is an important aspect, because not having agreed the details of the entire flight, the laws and general safety of aircraft will be violated. Therefore, IS decides this task through the interaction of various airport points of the world. In the coordination of details with the authorities, the Control Center must send all necessary information about the flight, and also to send messages with information about current flight. For example, type of aircraft, point and time of departure, the height of the aircraft and other no less important aspects. Successful flight completion is possible only when all standards are met, so system capabilities should be expected. The result of the creation is a multifunctional product that does not violate the legislative acts of flights. With IS, you can schedule a flight and reconcile all of the details with the authorities.

11 Hierarchy and Work Schedule

The schedule of work is an extremely important aspect of the life cycle of the project, because it allows you to understand at what stage is the development. The information system for aircraft flight planning is designed using the Agile/Scrum methodology, which allows flexible development to base on external factors. The entire development process is divided into sprints, which are planned to create a certain part of the system and in the subsequent demonstration of the work performed by the customer. Through this approach, both parties will understand the ideals that need to guide the resources of project. Figure 26 shows Gantt chart of the first sprint development stages.

#	Name	Duration	Priority	Start	Finish	Complete
1	System decomposition	4 d	★★★★☆	21.08.2017	24.08.2017	0 %
2	Team creation	1 d	★★★★★	25.08.2017	25.08.2017	0 %
3	Sequence details with customer	1 d	★★★★★	29.08.2017	29.08.2017	0 %
4	Literature revision	1 d	★★★☆☆	30.08.2017	30.08.2017	0 %
5	Subject area analysis	2 d	★★★★★	31.08.2017	01.09.2017	0 %
6	Creating backlog idea	1 d	★★★★☆	04.09.2017	04.09.2017	0 %
7	Creating DB idea	5 d	★★★★★	06.09.2017	12.09.2017	0 %
8	Creating operation module	4 d	★★★★★	13.09.2017	18.09.2017	0 %
9	Creating prototype	2 d	★★★★☆	19.09.2017	20.09.2017	0 %

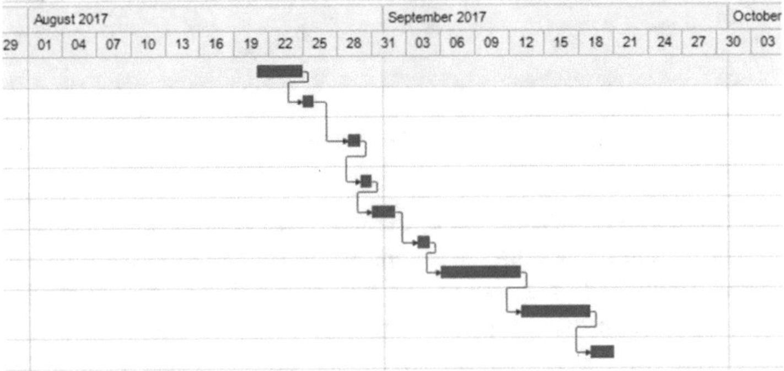

August 2017 September 2017 October
29 01 04 07 10 13 16 19 22 25 28 31 03 06 09 12 15 18 21 24 27 30 03

Fig. 26. The Gantt chart of the first sprint development stages

During the first sprint, the decomposition of the system was developed, which made it possible to understand what goals of this system follows. After that a development team was formed. The team was selected in such a way that the interaction was very effective and the elements of the team understood what needs to be done in the project. Now when the team has been formed, the coordination of all parts with the customer was carried out, which made it possible to understand the main directions of application of the product. The analysis of the subject area is the stage in which the system analyst analyzed the problems of the design sphere and determined the additional functionalities that should be implemented. Combining customer offers and problem issues, the backlog-product is created. Creating DB concept in the first sprint was to choose the technology to create a repository and start the main tables. Creating a functional module in the first sprint is to choose the architecture of the future system, as well as the programming of the main classes of IS. Creating a prototype is an important process that is developed on the basis of the subject area and other equally important factors. It should also be added that the first sprint does not have a testing process, because this stage is relevant at later sprint development (Fig. 27).

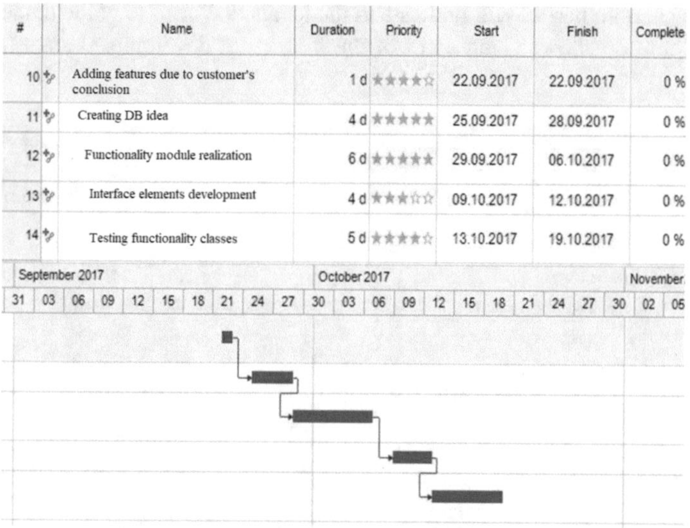

#	Name	Duration	Priority	Start	Finish	Complete
10	Adding features due to customer's conclusion	1 d	★★★★☆	22.09.2017	22.09.2017	0 %
11	Creating DB idea	4 d	★★★★★	25.09.2017	28.09.2017	0 %
12	Functionality module realization	6 d	★★★★★	29.09.2017	06.10.2017	0 %
13	Interface elements development	4 d	★★★☆☆	09.10.2017	12.10.2017	0 %
14	Testing functionality classes	5 d	★★★★☆	13.10.2017	19.10.2017	0 %

Fig. 27. A Gantt chart for the second sprint

After the first sprint demonstration of the work was completed, the customer makes his own adjustments and planning the second sprint. At the second sprint, virtually the same processes occur, only the testing phase is added. Thus, the project design team can continuously monitor the quality of the performed work. This approach will reduce the cost of resources for development, thereby increasing productivity (Fig. 28).

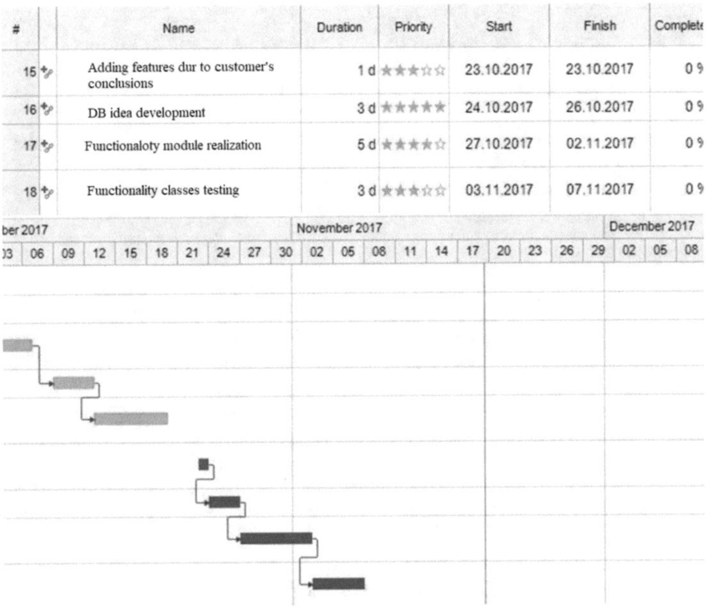

#	Name	Duration	Priority	Start	Finish	Complete
15	Adding features dur to customer's conclusions	1 d	★★★☆☆	23.10.2017	23.10.2017	0 %
16	DB idea development	3 d	★★★★★	24.10.2017	26.10.2017	0 %
17	Functionaloty module realization	5 d	★★★★☆	27.10.2017	02.11.2017	0 %
18	Functionality classes testing	3 d	★★★☆☆	03.11.2017	07.11.2017	0 %

Fig. 28. A Gantt chart for the third sprint

As this part was completed, the sprint became smaller in volume. This tendency suggests the correct planning and coordination of details with the customer, since the direction of development in the right direction from the very beginning that allow spending less project resources. As a result, the team for the third sprint already clearly understands the concept of the future system and is engaged in the accumulation of functional and testing existing product class software. Figure 29 shows the latest Sprint development that counts the latest blocks of the system life cycle tasks.

The features of this phase are the presence of a block with interface testing. At this stage, the interface is examined for compliance with general building standards, namely clarity, intuition, informativeness and other no less important properties of the design of the windows of the program. The latest process in the life cycle of this IS was the creation of documentation and further implementation. The documentation should contain all the necessary information, which will allow the enterprise-client to use its functionality without unnecessary problems. Figure 30 shows Gantt chart with all sprints for development, in Fig. 31 depicts the chronology of all development sprints.

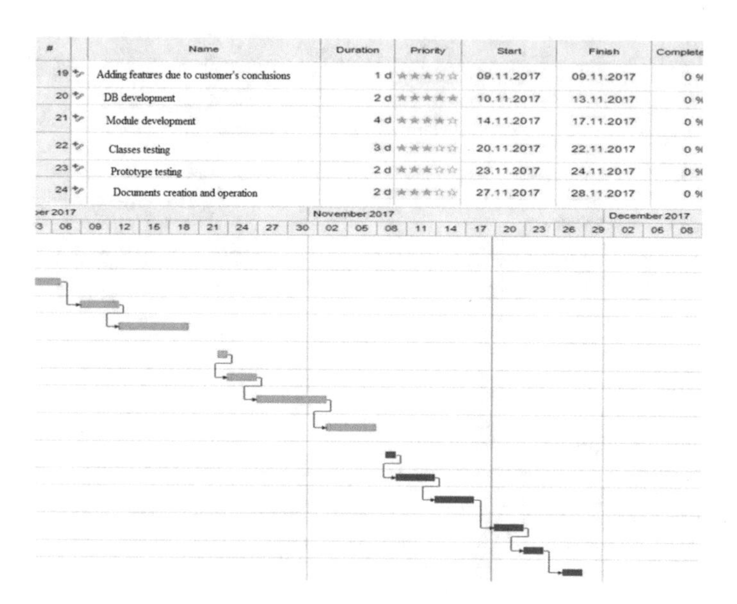

#	Name	Duration	Priority	Start	Finish	Complete
19	Adding features due to customer's conclusions	1 d		09.11.2017	09.11.2017	0 %
20	DB development	2 d		10.11.2017	13.11.2017	0 %
21	Module development	4 d		14.11.2017	17.11.2017	0 %
22	Classes testing	3 d		20.11.2017	22.11.2017	0 %
23	Prototype testing	2 d		23.11.2017	24.11.2017	0 %
24	Documents creation and operation	2 d		27.11.2017	28.11.2017	0 %

Fig. 29. A Gantt chart for the fourth sprint

#	Name	Duration	Priority	Start	Finish	Complete
1	System decomposition (technical task)	4 d		21.08.2017	24.08.2017	100 %
2	Team creation	1 d		25.08.2017	25.08.2017	100 %
3	Details sequence with customer	1 d		29.08.2017	29.08.2017	100 %
4	References revision	1 d		30.08.2017	30.08.2017	100 %
5	Subject area analysis	2 d		31.08.2017	01.09.2017	100 %
6	Backlog idea creation	1 d		04.09.2017	04.09.2017	100 %
7	DB idea development	5 d		06.09.2017	12.09.2017	100 %
8	Functional module creation	4 d		13.09.2017	18.09.2017	100 %
9	Prototype creation	2 d		19.09.2017	20.09.2017	100 %
10	Adding features due to customer's conclusion	1 d		22.09.2017	22.09.2017	100 %
11	DB development	4 d		25.09.2017	28.09.2017	100 %
12	Functional module development	6 d		29.09.2017	06.10.2017	100 %
13	Interface development	4 d		09.10.2017	12.10.2017	100 %
14	Functional classes testing	5 d		13.10.2017	19.10.2017	100 %
15	Adding features due to customer's conclusion	1 d		23.10.2017	23.10.2017	100 %
16	DB development	3 d		24.10.2017	26.10.2017	100 %
17	Functional module development	5 d		27.10.2017	02.11.2017	100 %
18	Functional classes testing	3 d		03.11.2017	07.11.2017	100 %
19	Adding features due to customer's conclusion	1 d		09.11.2017	09.11.2017	100 %
20	DB development	2 d		10.11.2017	13.11.2017	100 %
21	Functional module development	4 d		14.11.2017	17.11.2017	100 %
22	Functional classes testing	3 d		20.11.2017	22.11.2017	100 %
23	Prototype testing	2 d		23.11.2017	24.11.2017	100 %
24	Documents creation and operation	2 d		27.11.2017	28.11.2017	100 %

Fig. 30. Gantt chart of the full life cycle

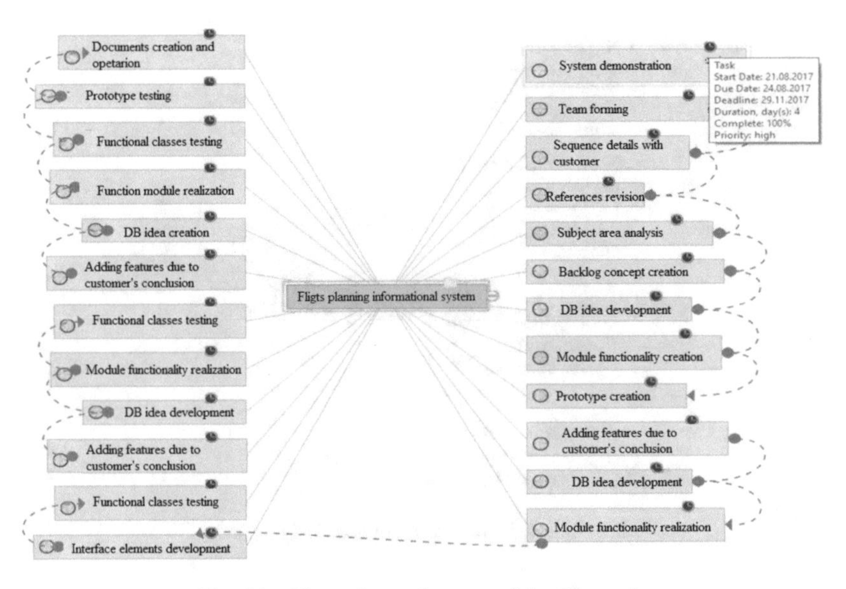

Fig. 31. Chronology of stages of the life cycle

12 Risk Management

Extremely important task in project management is risk management. This task is not separated from most other project management functions.

Fig. 32. Risk management processes

While defining financial needs, calculating the budget, preparing and contracting, monitoring the implementation of the project poses the task of protecting project participants from different types of risks is appearing. Figure 32 is a schematic diagram for risk management processes. The study found several possible risks that could negatively respond to the final concept of the software. All these risks are listed in Fig. 33.

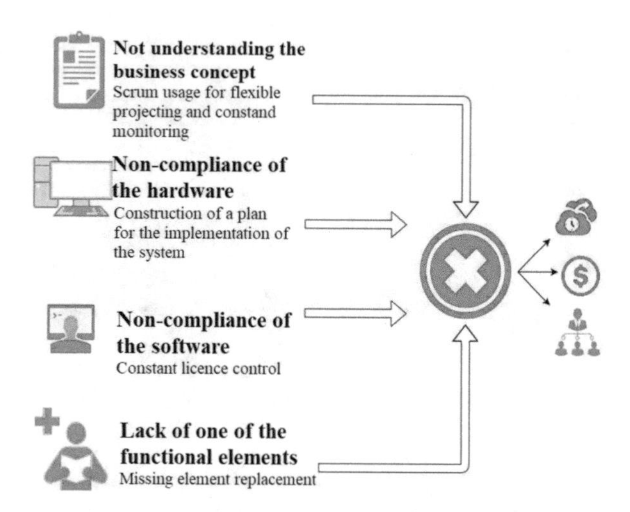

Fig. 33. Diagram of analyzed design risks

Once the risks have been analyzed, plan should be formulated to avoid them. The first risk is not understanding the business concept of the ordered system. This risk can be avoided if the workflow is constructed flexibly. Thus, after each sprint, the customer will be able to specify corrections that should be entered into the structure of the product. The second risk is the non-compliance of the hardware. This applies to various breakdowns, which generally affect the stages of the life cycle. The software should also be fully licensed, since any problems with the work of the creation tools will incur costs. [14] The most important risk when creating airplane planning ICs is the lack of one of the functional elements of the team. Even considering that, development is carried out using a flexible methodology in which processes can be implemented in parallel to each other, the loss of one of the key developers will seriously respond to overall performance. Therefore, the manager's task is to take this aspect into account and plan an action plan (Table 3).

Table 3. Project's risk characteristics

Risk	Probability	Importance	Influence	Overcome strategy
Not understanding the business concept	High	High	Time, funds	Scrum usage for flexible projecting and constant monitoring
Non-compliance of the hardware	Low	Middle	Time, funds	Construction of a plan for the implementation of the system
Non-compliance of the software	Low	Middle	Time, funds	Constant licence control
Lack of one of the functional elements	Middle	High	Time, funds, human	Missing element replacement

13 Description of the Project Task Realization

Sprint planning is an extremely important stage of development (Fig. 34), because it determines what tasks to perform in order to further results demonstration to the customer. Thus, it is necessary to formulate a plan of activity so that each sprint is interrelated and in the general concept led to the creation of a multifunctional product.

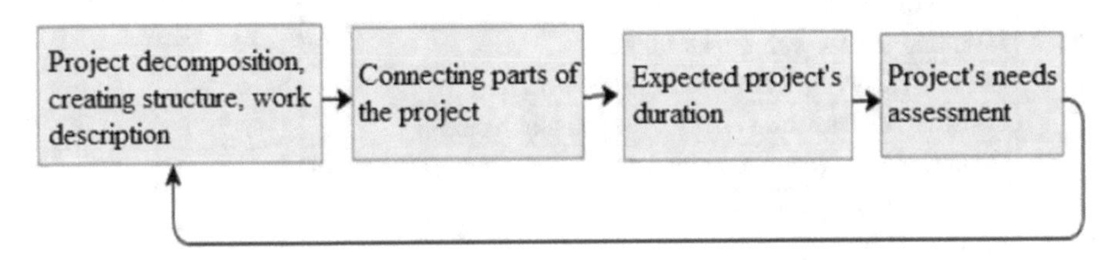

Fig. 34. Scheme of development planning stages

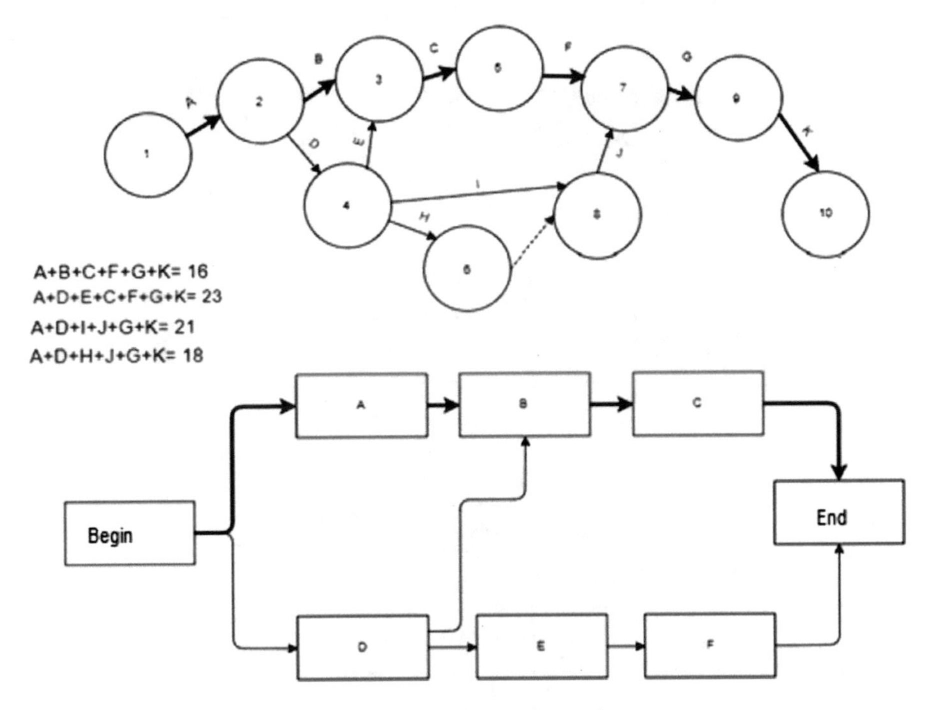

Fig. 35. Planning the first sprint

Network analysis tool serves network graphs. There are different types of network graphs, but the most commonly used arrow graphs. In arrows, each operation is indicated by a letter and represented by an arrow, each operation begins and ends with an event that has a certain number. In Fig. 35 in Table 4 is sprint planning. Figure 36 is the planning of the second sprint with additional symbols. In Fig. 37 there is a workflow for the third sprint of development, which also has additional designations.

Table 4. The sprint planning

Code	Operation description	Accomplishment time
A	Formation of the group of developers	1
B	Matching details with the customer	1
C	Domain analysis (field flight statistics research)	3
D	Construct the structure of the system (authorization)	7
F	Definition of a system requirement (creating a project sketch)	1
H	Prototype development (choice of interface type)	3
I	Iterative testing	6
J	Coordination	1
G	Functional development	4
K	Designing a prototype (creating a template option)	2

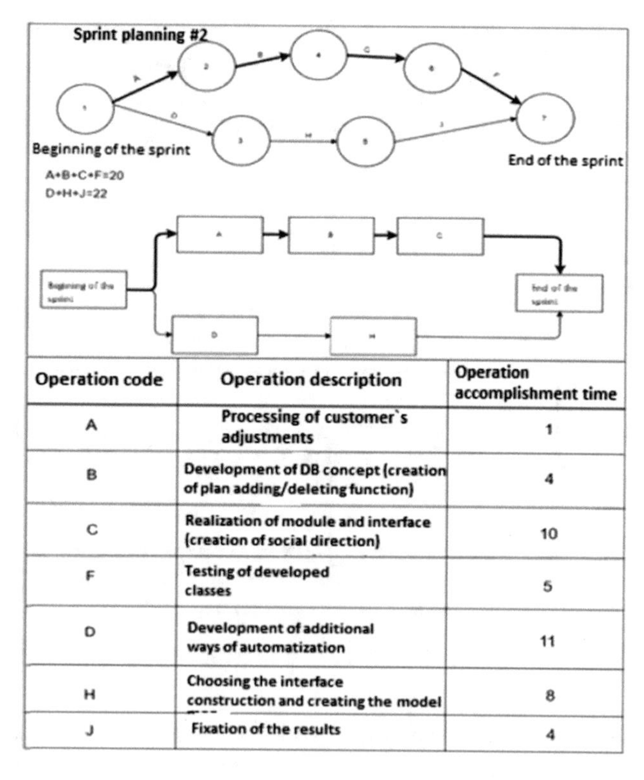

Operation code	Operation description	Operation accomplishment time
A	Processing of customer`s adjustments	1
B	Development of DB concept (creation of plan adding/deleting function)	4
C	Realization of module and interface (creation of social direction)	10
F	Testing of developed classes	5
D	Development of additional ways of automatization	11
H	Choosing the interface construction and creating the model	8
J	Fixation of the results	4

Fig. 36. Planning the second sprint

The system allows the planning of flights using functional tools whose calculation algorithms are based on environmental performance and aircraft characteristics. The software automates the decision-making process by the authorities about the permissions of a flight in one or another country, which saves both time and material resources. By creating this system we will reduce the cost of using various kinds of resources.

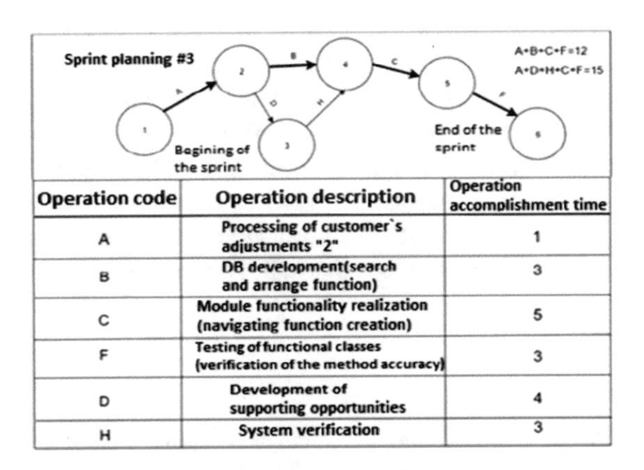

Operation code	Operation description	Operation accomplishment time
A	Processing of customer`s adjustments "2"	1
B	DB development(search and arrange function)	3
C	Module functionality realization (navigating function creation)	5
F	Testing of functional classes (verification of the method accuracy)	3
D	Development of supporting opportunities	4
H	System verification	3

Fig. 37. Planning a third sprint

The system allows to solve several important problems in the directions of flight planning activity, which turn will strengthen the relevance of air travel in the field of travel.

14 Conclusions

As a result of project management, it has been possible to create an automated system that simplifies route planning for any distance, without the use of additional experts. To create the system, you have to choose the right team of developers and form the decomposition of the future product. In addition, a related element of development was the selection of design methodology. To do this, choose the Agile/Scrum design methodology, which allows flexible decision-making.

The entire life cycle of the development was split into sprints, each of which built an action plan. Thus, during project management, there were no additional problems, as the design structure was formed so as not to ignore the possible risks. Having overcome the possible risks, the implementation of the system has spent less resources for development. Therefore, in terms of planning and risk management, the project can be considered to be successful. An important aspect in project management is personnel management. All stages of the development were formed in such a way that each element of the group was closely interconnected and understood the essence and objectives of the project implementation. System analysis allowed correctly investigating the relevance and identifying problem areas of the subject area of design.

This allowed from the very beginning of the development to build the future decomposition of the information system. The correct approach to programming and testing allowed to design a functional module with no errors in use and with a list of tasks. This is an extremely important aspect, because it determines how competitive the system will be compared with its counterparts. The final stage in the development management was the implementation process. The main complications when applying financial indicators of information technology implementation projects are the need to consider a large number of parameters, first, to identify the benefits and costs. To do this, the business analyst relied on the cost of time resources, as well as the cost of human activities that were designed using Gantt charts. Thus, it was easy to track how costly the project came out. The total development lasted seventy-two days, which were divided into four sprints, which is a successful result because the indicator is equivalent to the set terms. Consequently, after conducting the metrics of the project assessment, it can be confident that the information system of aircraft flight planning proved to be a successful software tool that will be relevant to use.

References

1. Harik, R.F., Derigent, W.J., Ris, G.: Computer aided process planning in aircraft manufacturing. Comput.-Aided Des. Appl. **5**(6), 953–962 (2008)
2. Ashford, N.J., Mumayiz, S., Wright, P.H.: Airport Engineering: Planning, Design, and Development of 21st Century Airports. Wiley, Hoboken (2011)
3. Torenbeek, E.: Synthesis of Subsonic Airplane Design: An Introduction to the Preliminary Design of Subsonic General Aviation and Transport Aircraft, with Emphasis on Layout, Aerodynamic Design. Propulsion and Performance. Springer, Dordrecht (2013)
4. Austin, R.: Unmanned Aircraft Systems: UAVS Design, Development and Deployment, vol. 54. Wiley, Chichester (2011)
5. Lypak, H., Rzheuskyi, A., Kunanets, N., Pasichnyk, V.: Formation of a consolidated information resource by means of cloud technologies. In: International Scientific-Practical Conference on Problems of Info Communications Science and Technology (2018)
6. Rzheuskyi, A., Kunanets, N., Stakhiv, M.: Recommendation system: virtual reference. In: 13th International Scientific and Technical Conference on Computer Sciences and Information Technologies (CSIT), pp. 203–206 (2018)
7. Kaminskyi, R., Kunanets, N., Rzheuskyi, A.: Mathematical support for statistical research based on informational technologies. In: CEUR Workshop Proceedings, vol. 2105, pp. 449–452 (2018)
8. Obermaier, J., Hutle, M.: Analyzing the security and privacy of cloud-based video surveillance systems. In: Proceedings of the 2nd ACM International Workshop on IoT Privacy, Trust, and Security, pp. 22–28 (2016)
9. Xu, D., Wang, R., Shi, Y.Q.: Data hiding in encrypted H. 264/AVC video streams by codeword substitution. IEEE Trans. Inf. Forensics Secur. **9**(4), 596–606 (2014)
10. Saxena, M., Sharan, U., Fahmy, S.: Analyzing video services in web 2.0: a global perspective. In: Proceedings of the 18th International Workshop on Network and Operating Systems Support for Digital Audio and Video, pp. 39–44 (2008)

11. Brône, G., Oben, B., Goedemé, T.: Towards a more effective method for analyzing mobile eye-tracking data: integrating gaze data with object recognition algorithms. In: Proceedings of the 1st International Workshop on Pervasive Eye Tracking & Mobile Eye-Based Interaction, pp. 53–56 (2011)
12. Reibman, A.R., Sen, S., Van der Merwe, J.: Analyzing the spatial quality of internet streaming video. In: Proceedings of International Workshop on Video Processing and Quality Metrics for Consumer Electronics (2005)
13. Perniss, P.: Collecting and analyzing sign language data: video requirements and use of annotation software. In: Research Methods in Sign Language Studies, pp. 56–73 (2015)
14. Tran, B.Q.: U.S. Patent No. 8,849,659. U.S. Patent and Trademark Office, Washington, DC (2014)
15. Badawy, W., Gomaa, H.: U.S. Patent No. 9,014,429. U.S. Patent and Trademark Office, Washington, DC (2015)
16. Badawy, W., Gomaa, H.: U.S. Patent No. 8,630,497. U.S. Patent and Trademark Office, Washington, DC (2014)
17. Golan, O., Dudovich, B., Daliyot, S., Horovitz, I., Kiro, S.: U.S. Patent No. 8,885,047. U.S. Patent and Trademark Office, Washington, DC (2014)
18. Chambers, C.A., Gagvani, N., Robertson, P., Shepro, H.E.: U.S. Patent No. 8,204,273. U.S. Patent and Trademark Office, Washington, DC (2012)
19. Maes, S.H.: U.S. Patent No. 7,917,612. U.S. Patent and Trademark Office, Washington, DC (2011)
20. Zdebskyi, P., Vysotska, V., Peleshchak, R., Peleshchak, I., Demchuk, A., Krylyshyn, M.: An application development for recognizing of view in order to control the mouse pointer. In: CEUR Workshop Proceedings, vol. 2386, pp. 55–74 (2019)
21. Rusyn, B., Lytvyn, V., Vysotska, V., Emmerich, M., Pohreliuk, L.: The virtual library system design and development. In: Advances in Intelligent Systems and Computing, vol. 871, pp. 328–349 (2019)
22. Rusyn, B., Lutsyk, O., Lysak, O., Lukeniuk, A., Pohreliuk, L.: Lossless image compression in the remote sensing applications. In: International Conference on Data Stream Mining & Processing (DSMP), pp. 195–198 (2016)
23. Rusyn, B., Vysotska, V., Pohreliuk, L.: Model and architecture for virtual library information system. In: Computer Sciences and Information Technologies, CSIT, pp. 37–41 (2018)
24. Kravets, P.: The control agent with fuzzy logic. In: Perspective Technologies and Methods in MEMS Design, MEMSTECH 2010, pp. 40–41 (2010)
25. Babichev, S., Gozhyj, A., Kornelyuk, A., Litvinenko, V.: Objective clustering inductive technology of gene expression profiles based on SOTA clustering algorithm. Biopolymers and Cell **33**(5), 379–392 (2017)
26. Nazarkevych, M., Klyujnyk, I., Nazarkevych, H.: Investigation the ateb-gabor filter in biometric security systems. In: Data Stream Mining & Processing, pp. 580–583 (2018)
27. Emmerich, M., Lytvyn, V., Yevseyeva, I., Fernandes, V.B., Dosyn, D., Vysotska, V.: Preface: modern machine learning technologies and data science (MoMLeT&DS-2019). In: CEUR Workshop Proceedings, vol. 2386 (2019)
28. Vysotska, V., Burov, Y., Lytvyn, V., Demchuk, A.: Defining author's style for plagiarism detection in academic environment. In: Proceedings of the 2018 IEEE 2nd International Conference on Data Stream Mining and Processing, DSMP 2018, pp. 128–133 (2018)

29. Lytvyn, V., Peleshchak, I., Vysotska, V., Peleshchak, R.: Satellite spectral information recognition based on the synthesis of modified dynamic neural networks and holographic data processing techniques. In: International Scientific and Technical Conference on Computer Sciences and Information Technologies (CSIT), pp. 330–334 (2018)
30. Su, J., Sachenko, A., Lytvyn, V., Vysotska, V., Dosyn, D.: Model of touristic information resources integration according to user needs. In: International Scientific and Technical Conference on Computer Sciences and Information Technologies, pp. 113–116 (2018)
31. Lytvyn, V., Sharonova, N., Hamon, T., Cherednichenko, O., Grabar, N., Kowalska-Styczen, A., Vysotska, V.: Preface: Computational Linguistics and Intelligent Systems (COLINS-2019). In: CEUR Workshop Proceedings, vol. 2362 (2019)
32. Burov, Y., Vysotska, V., Kravets, P.: Ontological approach to plot analysis and modeling. In: CEUR Workshop Proceedings, vol. 2362, pp. 22–31 (2019)
33. Vysotska, V., Lytvyn, V., Burov, Y., Berezin, P., Emmerich, M., Basto Fernandes V.: Development of information system for textual content categorizing based on ontology. In: CEUR Workshop Proceedings, vol. 2362, pp. 53–70 (2019)
34. Lytvyn, V., Vysotska, V., Kuchkovskiy, V., Bobyk, I., Malanchuk, O., Ryshkovets, Y., Pelekh, I., Brodyak, O., Bobrivetc, V., Panasyuk, V.: Development of the system to integrate and generate content considering the cryptocurrent needs of users. Eastern-Eur. J. Enterp. Technol. **1**(2–97), 18–39 (2019)
35. Lytvyn, V., Kuchkovskiy, V., Vysotska, V., Markiv, O., Pabyrivskyy, V.: Architecture of system for content integration and formation based on cryptographic consumer needs. In: Computer Sciences and Information Technologies, CSIT, pp. 391–395 (2018)
36. Lytvyn, V., Vysotska, V., Demchuk, A., Demkiv, I., Ukhanska, O., Hladun, V., Kovalchuk, R., Petruchenko, O., Dzyubyk, L., Sokulska, N.: Design of the architecture of an intelligent system for distributing commercial content in the internet space based on SEO-technologies, neural networks, and machine learning. Eastern-Eur. J. Enterp. Technol. **2**(2–98), 15–34 (2019)
37. Chyrun, L., Gozhyj, A., Yevseyeva, I., Dosyn, D., Tyhonov, V., Zakharchuk, M.: Web content monitoring system development. In: CEUR Workshop Proceedings, vol. 2362, pp. 126–142 (2019)
38. Bisikalo, O., Ivanov, Y., Sholota, V.: Modeling the phenomenological concepts for figurative processing of natural-language constructions. In: CEUR Workshop Proceedings, vol. 2362, pp. 1–11 (2019)
39. Babichev, S., Taif, M.A., Lytvynenko, V., Osypenko, V.: Criterial analysis of gene expression sequences to create the objective clustering inductive technology. In: 2017 IEEE 37th International Conference on Electronics and Nanotechnology, pp. 244–248 (2017)
40. Kazarian, A., Kunanets, N., Pasichnyk, V., Veretennikova, N., Rzheuskyi, A., Leheza, A., Kunanets, O.: complex information e-science system architecture based on cloud computing model. In: CEUR Workshop Proceedings, vol. 2362, pp. 366–377 (2019)
41. Veres, O., Rishnyak, I., Rishniak, H.: Application of methods of machine learning for the recognition of mathematical expressions. In: CEUR Workshop Proceedings, vol. 2362, pp. 378–389 (2019)
42. Lytvyn, V., Vysotska, V., Rusyn, B., Pohreliuk, L., Berezin, P., Naum O.: Textual content categorizing technology development based on ontology. In: CEUR Workshop Proceedings, vol. 2386, pp. 234–254 (2019)
43. Lytvyn, V., Vysotska, V., Dosyn, D., Lozynska, O., Oborska, O.: Methods of building intelligent decision support systems based on adaptive ontology. In: Proceedings of the 2018 IEEE 2nd International Conference on Data Stream Mining and Processing, DSMP 2018, pp. 145–150 (2018)

44. Vysotska, V., Lytvyn, V., Burov, Y., Gozhyj, A., Makara, S.: The consolidated information web-resource about pharmacy networks in city. In: CEUR Workshop Proceedings, pp. 239–255 (2018)
45. Basyuk, T.: The main reasons of attendance falling of internet resource. In: Proceedings of the X-th International Conference on Computer Science and Information Technologies, CSIT 2015, pp. 91–93 (2015)
46. Gozhyj, A., Chyrun, L., Kowalska-Styczen, A., Lozynska, O.: Uniform method of operative content management in web systems. In: CEUR Workshop Proceedings (Computational Linguistics and Intelligent Systems, vol. 2136, pp. 62–77 (2018)
47. Lytvyn, V., Vysotska, V., Rzheuskyi, A.: Technology for the psychological portraits formation of social networks users for the IT specialists recruitment based on big five, NLP and big data analysis. In: CEUR Workshop Proceedings, vol. 2392, pp. 147–171 (2019)
48. Vysotska, V., Burov, Y., Lytvyn, V., Oleshek, O.: Automated monitoring of changes in web resources. In: Lecture Notes in Computational Intelligence and Decision Making, vol. 1020, pp. 348–363 (2020)
49. Demchuk, A., Lytvyn, V., Vysotska, V., Dilai, M.: Methods and means of web content personalization for commercial information products distribution. In: Lecture Notes in Computational Intelligence and Decision Making, vol. 1020, pp. 332–347 (2020)
50. Vysotska, V., Mykhailyshyn, V., Rzheuskyi, A., Semianchuk, S.: System development for video stream data analyzing. In: Lecture Notes in Computational Intelligence and Decision Making, vol. 1020, pp. 315–331 (2020)
51. Lytvynenko, V., Wojcik, W., Fefelov, A., Lurie, I., Savina, N., Voronenko, M., et al.: Hybrid methods of GMDH-neural networks synthesis and training for solving problems of time series forecasting. In: Lecture Notes in Computational Intelligence and Decision Making, vol. 1020, pp. 513–531 (2020)
52. Babichev, S., Durnyak, B., Pikh, I., Senkivskyy, V.: An evaluation of the objective clustering inductive technology effectiveness implemented using density-based and agglomerative hierarchical clustering algorithms. In: Lecture Notes in Computational Intelligence and Decision Making, vol. 1020, pp. 532–553 (2020)
53. Bidyuk, P., Gozhyj, A., Kalinina, I.: Probabilistic inference based on LS-method modifications in decision making problems. In: Lecture Notes in Computational Intelligence and Decision Making, vol. 1020, pp. 422–433 (2020)
54. Chyrun, L., Chyrun, L., Kis, Y., Rybak, L.: Automated information system for connection to the access point with encryption WPA2 enterprise. In: Lecture Notes in Computational Intelligence and Decision Making, vol. 1020, pp. 389–404 (2020)
55. Kis, Y., Chyrun, L., Tsymbaliak, T., Chyrun, L.: Development of system for managers relationship management with customers. In: Lecture Notes in Computational Intelligence and Decision Making, vol. 1020, pp. 405–421 (2020)
56. Chyrun, L., Kowalska-Styczen, A., Burov, Y., Berko, A., Vasevych, A., Pelekh, I., Ryshkovets, Y.: Heterogeneous data with agreed content aggregation system development. In: CEUR Workshop Proceedings, vol. 2386, pp. 35–54 (2019)
57. Chyrun, L., Burov, Y., Rusyn, B., Pohreliuk, L., Oleshek, O., Gozhyj, Bobyk, I.: Web resource changes monitoring system development. In: CEUR Workshop Proceedings, vol. 2386, pp. 255–273 (2019)
58. Veres, O., Rusyn, B., Sachenko, A., Rishnyak, I.: Choosing the method of finding similar images in the reverse search system. In: CEUR Workshop Proceedings, vol. 2136, pp. 99–107 (2018)

59. Mukalov, P., Zelinskyi, O., Levkovych, R., Tarnavskyi, P., Pylyp, A., Shakhovska, N.: Development of system for auto-tagging articles, based on neural network. In: CEUR Workshop Proceedings, vol. 2362, pp. 106–115 (2019)
60. Rzheuskyi, A., Gozhyj, A., Stefanchuk, A., Oborska, O., Chyrun, L., Lozynska, O., Mykich, K., Basyuk, T.: Development of mobile application for choreographic productions creation and visualization. In: CEUR Workshop Proceedings, vol. 2386, pp. 340–358 (2019)
61. Sachenko, S., Rippa, S., Krupka, Y.: Pre-conditions of ontological approaches application for knowledge management in accounting. In: IEEE International Workshop on Intelligent Data Acquisition and Advanced Computing Systems: Technology and Applications, pp. 605–608 (2009)
62. Sachenkom, S., Lendyuk, T., Rippa, S.: Simulation of computer adaptive learning and improved algorithm of pyramidal testing. In: International Conference on Intelligent Data Acquisition and Advanced Computing Systems (IDAACS), vol. 2, pp. 764–770 (2013)
63. Sachenko, S., Lendyuk, T., Rippa, S., Sapojnyk, G.: Fuzzy rules for tests complexity changing for individual learning path construction. In: Intelligent Data Acquisition and Advanced Computing Systems: Technology and Applications, pp. 945–948 (2015)

Artificial Intelligence

Decision Fusion for Partially Occluded Face Recognition Using Common Vector Approach

Mehmet Koc(✉)

Bilecik Seyh Edebali University, Bilecik, Turkey
mehmet.koc@bilecik.edu.tr

Abstract. Partial occlusions in the face image negatively affect the performance of a face recognition system. Modular versions of some methods are used to overcome this problem. Modular Common Vector Approach (MCVA) was successfully applied partial occlusion problem. In this work, we apply some well-known decision fusion methods (product rule, borda count, and majority voting) to the decision stage of MCVA approach to increase the classification performance. A well-known appearance based feature descriptor so called Local Binary Patterns (LBP) is used to extract the facial features. The performance comparisons are conducted on AR face database with several experiments. It is observed that combining the classifier outputs using decision fusion methods increase the classification performance of MCVA.

Keywords: Face recognition · Modular · Common vector · Decision fusion · Partial occlusion · Local binary patterns

1 Introduction

Face recognition is a popular and active research area in pattern recognition and has several application areas such as human computer interaction, security, human identification [1]. Parameters such as, lighting conditions, different facial expressions, and partial occlusion negatively affect the performance of the recognition system. Various methods purposed to solve the partial occlusion problem [2–4]. In [3], Linear Regression Classification (LRC) which is a subspace based classification method is proposed. In the same paper, modular variation of LRC is also proposed to solve the partial occlusion problem. Modular LRC method divides the training images into k equal sized rectangular regions(modules). LRC method is applied these modules separately. Then finally, k classification models obtained using LRC are combined using a distance-based decision fusion method.

Common Vector Approach (CVA) method is firstly proposed to solve a speech recognition problem [5]. Then the method is applied to several pattern recognition problems such as face recognition [6] and feature selection [7]. But the performance of CVA in recognition of partially occluded faces is not at the

N. Shakhovska and M. O. Medykovskyy (Eds.): CCSIT 2019, AISC 1080, pp. 351–360, 2020.
https://doi.org/10.1007/978-3-030-33695-0_24

desired level. In [4], authors purpose a modular variation of CVA to improve the recognition performance at the recognition of partially occluded face images. They improve the recognition performance of CVA at the occluded images using the modular idea. The derivation procedure of MCVA from CVA is the same as the derivation procedure of MLRC from LRC. MCVA compares the decision outputs of the modules to give the final decision. Thus, the final class label is decided using just one module. Combining the module output decisions using appropriate fusion methods may increase the recognition performance.

The performance of the recognition system can be improved by using several decision fusion strategies [8]. In this work, we compare the classification performances of well-known decision fusion methods(product rule, majority voting, and borda count) and the conventional decision rule of MCVA for the recognition of occluded faces.

There are many feature extraction methods in the literature. Local Binary Pattern (LBP) is one of the most well-known appearance based feature extraction method [9]. LBP generates codes for each pixel by comparing the gray level intensity value of center pixel and its neighbors. Then it accumulates the codes in a histogram to generate the feature descriptor. Numerous variations of LBP for different applications area are proposed [13,14]. In this work, we used the conventional LBP as feature descriptor for face images.

The rest of the paper is organized as follows. In Sects. 2 and 3 we briefly give CVA and MCVA approaches respectively. A brief explanation of LBP and LBP based feature extraction from face images are given in Sect. 4. Section 5 represents the classifier output fusion methods that we use in this work. In Sect. 6, we perform a comparative analysis of decision fusion methods and MCVA's conventional decision rule in AR face database. Finally, conclusion of the work is given in Sect. 7.

2 Common Vector Approach

The data in pattern class has two basic property group; the first one is the component which is common to all samples in the class and the second is the group of the properties that are different from each other. After subtracting the differences in the feature vectors of a class, the vector formed by the properties that remain the same for the rest of the class is called *common vector*. The common vector is unique for all samples that are in the same class. In Common Vector Approach (CVA), it is assumed that the feature vector has two components. The first one is the component which is the same for all feature vectors in the class and the second one is the remaining part which contains the variations that are different for each of the feature vector. The first part is common vector and the second part is difference vector. In CVA, there exist two cases, namely insufficient ($n >= m$) and sufficient case ($n < m$). Here n is the dimension of the feature vectors and m is the number of the feature vectors in a class. In face recognition problems we encounter insufficient data case because the dimension of the face space is generally much more higher than the number of face images.

Let assume that $\mathbf{x}_i$ denotes the ith sample from a feature class that has m feature vectors in n dimensional feature space and let $\boldsymbol{\mu}$ denotes the mean of the class. Then covariance matrix of the class can be defined as

$$\Sigma = \sum_{i=1}^{m} (\mathbf{x}_i - \boldsymbol{\mu})(\mathbf{x}_i - \boldsymbol{\mu})^T . \tag{1}$$

It is well-known that a covariance matrix is positive semidefinite, i.e., its eigenvalues are zero or positive real numbers. In insufficient case, the common vector of a class lies in the subspace which is spanned by the eigenvectors corresponding to the zero eigenvalues of the covariance matrix of that class [5]. Therefore, common vector of the class can be found by projecting any sample in the training set to the null space of the covariance matrix. Also the common vector of the class can be calculated by subtracting the variation of any sample in the training set:

$$\mathbf{a}_{com} = P^{\perp}\mathbf{x}_i, i = 1, ..., m \tag{2}$$

where $P^{\perp}$ is the null space projection matrix for the covariance matrix of class. $P^{\perp}$ can be calculated as follows:

$$P^{\perp} = \sum_{i=1}^{n-m+1} \mathbf{u}_i\mathbf{u}_i^T \tag{3}$$

Here $\mathbf{u}_i, i = 1, ...m$ are the eigenvectors corresponding to the zero eigenvalues of the covariance matrix of class. It must be also noted that since the common vector of a class is unique, the common vector $\mathbf{a}_{com}$ in (2) is independent from the index i in the same equation [5].

When someone wants to classify an unknown sample $\mathbf{x}_{test}$ to one of the C classes, he must firstly find the projection of the vector onto the null spaces of the class covariance matrices. Then, the unknown vector is classified to the nearest common vector's class. The classification is done according to the following criterion.

$$C^* = \arg\min_{j} \left\{ \left|\left| \mathbf{a}_{comj} - P_j^{\perp}\mathbf{x}_{test} \right|\right| \right\}, j = 1, ..., C \tag{4}$$

where $\mathbf{a}_{comj}$ and $P_j^{\perp}$ are the common vector and the null space projection matrix of the jth class respectively.

3 Modular Common Vector Approach

A face recognition system generally has low recognition performance when face image has partially occlusion. One may overcome this problem using the modular representation of face image [3,10]. In this section we briefly represent the Modular Common Vector (MCV). Like any modular approach, firstly the face image is divided into several number of rectangular regions(modules, partitions). Then, CVA is applied to these modules of images separately. In Fig. 1, an example face image from AR face database is divided into 9, 16, and 25 modules.

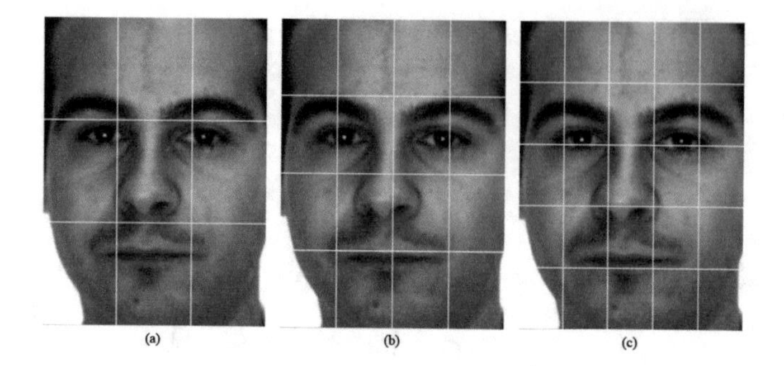

Fig. 1. Example of a nonoccluded face image divided into (a) $3 \times 3 = 9$, (b) $4 \times 4 = 16$, and (c) $5 \times 5 = 25$ partitions.

Let assume that face images have K different modules and let $\mathbf{x}_i^k$ represents the kth module of the ith image. Then the covariance matrix of the kth module is calculated as follows:

$$\Sigma^{(k)} = \sum_{i=1}^{m} \left(\mathbf{x}_i^k - \boldsymbol{\mu}^k\right) \left(\mathbf{x}_i^k - \boldsymbol{\mu}^k\right)^T \tag{5}$$

Here $\boldsymbol{\mu}^k$ is the mean of the kth module. By following the similar procedure in CVA, the null space projection matrix of the kth module is calculated using the eigenvectors corresponding to the zero eigenvalues of the covariance matrix of the kth module as follows:

$$P^{\perp(k)} = \sum_{i=1}^{n-m+1} \mathbf{u}_i^{(k)} \mathbf{u}_i^{(k)T} \tag{6}$$

where $\mathbf{u}_i^{(k)}, i = 1, ..., n - m + 1$ show the eigenvectors corresponding to the zero eigenvalues of kth module's covariance matrix. Then the common vector of the kth module can be calculated as follows:

$$\mathbf{a}_{com}^{(k)} = P^{\perp(k)} \mathbf{x}_i^k, i = 1, ..., m \tag{7}$$

Let $\mathbf{x}_{test}$ is the unknown query that is going to be classified. Firstly, $\mathbf{x}_{test}$ is separated into K modules, i.e., $\mathbf{x}_{test}^k, k = 1, ..., K$. Then CVA is applied to each of K modules separately to find the minimum distance and its corresponding class.

$$d_j^{(k)} = \min_j \left\{ \left|\left| \mathbf{a}_{comj}^{(k)} - P_j^{\perp(k)} \mathbf{x}_{test}^k \right|\right| \right\}, j = 1, ..., C \tag{8}$$

Finally, the minimum distances for each of K modules are compared to find the minimum. The unknown query $\mathbf{x}_{test}$ is classified to the class according to the following criterion.

$$C^* = \arg\min_j \left\{ d_j^{(k)} \right\}, k = 1, ..., K \tag{9}$$

4 Local Binary Patterns

Local binary patterns (LBP) is a well-known and important feature extraction method which captures the local properties of an image. LBP is firstly proposed for texture recognition problems [9]. It is then applied several computer vision and pattern recognition problems [9,11–13]. It is assumed that face and non-face partitions have different local characteristics. LBP operator can be used to detect and to discriminate occlusion in face image.

Original LBP uses 3×3 neighborhood of a pixel. It thresholds the pixels in the neighborhood with center pixel. If the gray level intensity value of the neighborhood pixel is larger than the center pixel it assigns 1 otherwise 0. The results are multiplied by the powers of two and then summed to generate a code for the center pixel. This procedure is repeated for all pixels. Finally, a 256 dimensional histogram is generated using the LBP codes of pixels. LBP operator for circular topology can be given as follows:

$$LBP_{P,R} = \sum_{p=0}^{P-1} s(g_p - g_c)2^p \quad (10)$$

where R is the radius of the circle, P is the number of sampling points, $s(x)$ is the thresholding function which generates 1 for $x \geq 0$ otherwise 0, g_c and g_p are the gray level values of the center pixel and the sampling points respectively. An illustration of LBP code computation for $R = 1$ and $P = 8$ is given in Fig. 2. It should be noted that since some of the sampling points do not correspond to the center of the neighbor pixels, the values of these sampling points are calculated using bilinear interpolation. After tresholding operation binary code 11001101 is generated which is 205 in decimal system. LBP code with eight sampling points generates 256 dimensional histogram. Uniform LBP (uLBP) is proposed in [15] to reduce the dimension of the histogram. LBP code is accepted as uniform if bitwise transitions from 0 to 1 or 1 to 0 is at most two. For example, the LBP code in Fig. 2 is nonuniform since the number of bitwise transitions is four. In uLBP the dimension of the histogram is reduced to 59. 58 of the histogram bins correspond to uniform codes and all of the nonuniform codes are assigned to the last bin.

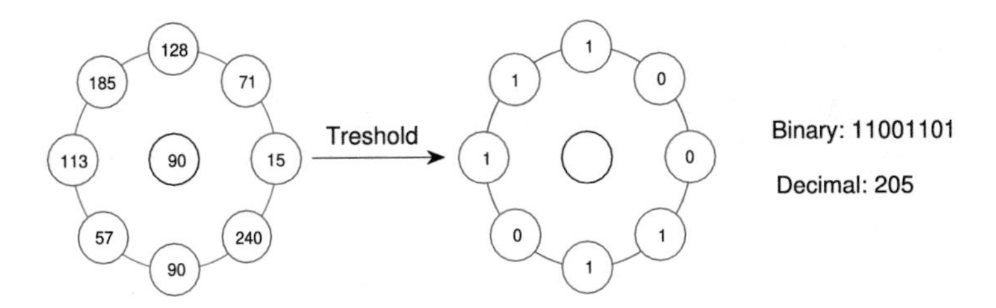

Fig. 2. An illustration of LBP code generation for $R = 1$ and $P = 8$.

5 Decision Fusion

In our previous work [4], we gave several experiments to show the performance of MCVA. Then, in [20], authors replaced the conventional decision rule, namely distance-based evidence fusion [3], with other decision fusion methods (product, majority voting, and borda count) to increase the performance of MCVA. In this section we briefly give some well-known decision fusion methods. These methods do not need additional training [16]. Assume that we obtain K different modules from an image. Then we have K different classifiers with K decision outputs. Assume that the decison of the kth classifier is $d_j^{(k)}, k = 1, ..., K, j = 1, ..., C$.

5.1 Product Rule

In product rule, basically, the outputs of different classifiers are multiplied to get final decision. If there are C classes in a classification problem. The final value for each class is calculated as follows.

$$\mu_j = \prod_{k=1}^{K} d_j^{(k)}, j = 1, ..., C \tag{11}$$

Test sample is classified according to the minimum rule, i.e.,

$$C^* = \arg\min_j \{\mu_j\}, k = 1, ..., K. \tag{12}$$

5.2 Majority Voting Rule

There are three types of majority voting: (i) all classifiers agree, (ii) more than half of the classifiers agree, (iii) majority of the classifiers agree [16]. In the experiments, we used the third version. In majority voting rule the classifier outputs are binary values, i.e., $d_j^{(k)} \in \{0, 1\}$. Unknown test sample is classified to class J which satisfies the following equation.

$$\sum_{k=1}^{K} d_J^{(k)} = \max_{j=1}^{C} \sum_{k=1}^{K} d_j^{(k)} \tag{13}$$

5.3 Borda Count

It is a mapping from individual rankings to a combined ranking to get the most relevant decision [17]. In borda count method, each classifier orders classes. It assigns a level of importance to all of the classes according to their orders. Assume that there are C classes. Then the first-place class receives $C-1$ points, the second-place class receives $C-2$ points, and the last class receives 0 points. This procedure is repeated for all classifiers and the points are summed up. The class with the highest point is the final decision [16]. The main disadvantage of this method is that it treats all classifiers equally by ignoring the characteristics of the classifiers individually [17].

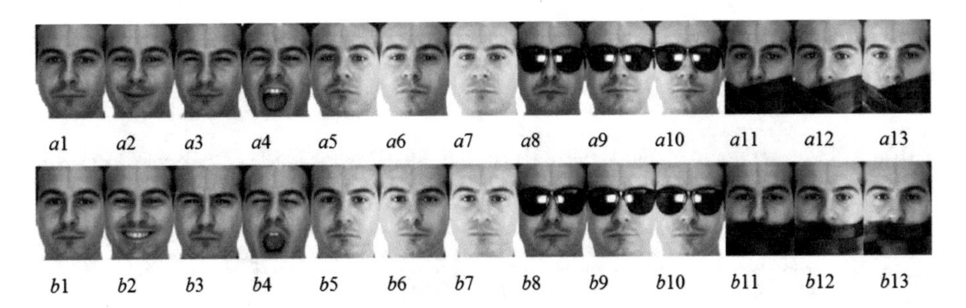

Fig. 3. Images of a subject from AR face database.

6 Experimental Work

In the experimental work, we used a very well-known face database AR [18]. AR face database includes 26 images from 126 different subjects. Images are taken under different lighting conditions(left, right, all sides), face expressions(neutral expression, smile, anger, scream), and occlusion conditions(scarf, sun glass). We used exactly the same subset of AR with the work [19]. This subset includes 30 male and 20 female subjects. The original size of the images are 768×576. After preprocessing operations the final size of the images is 115×87. In Fig. 3 images of a subject from AR face database after the preprocessing operations are shown. In [4,20], the success of MCVA approach was presented with the experiments in AR face database.

In this work, all non-occluded face images (a1, a2, a3, a4, a5, a6, a7, b1, b2, b3, b4, b5, b6, b7) in Fig. 3 are used for training purposes. Training set includes neutral, smile, anger, scream, and images with different lighting conditions.

In the first experiment, we choose a11, a12, a13, b11, b12, b13 in Fig. 3 to form the test set. With this experiment we try to compare the performances of several fusion strategies in scarf-occluded images. We firstly divide face images into 3×3, 4×4, and 5×5 non-overlapping partitions. Then we generate uniform $LBP_{8,1}$ for each partition. CVA is applied to each of the partitions. The distance measures evaluated from each partition are combined using the fusion methods. The results of the experiment are shown in Table 1. In the second part of the experiment we choose neutral, smile, anger, scream images (a1, a2, a3, a4, b1, b2, b3, b4) as the training set and scarf-occluded images as the test set. The results are given in Table 1. MCVA's conventional decision criteria generally has the worst performance except 3×3 partition case in the second part of the experiments. Borda count method takes the lead three out of six experiments. Majority vote, borda count, and product rules' performances increase as the number of partitions increases whereas the performance of conventional MCVA decreases. We can expect these results since when the number of partitions increases, the dimension of partition decreases, correspondingly the information kept by a partition decreases. Since conventional MCVA chooses one partition, its discrimination ability decreases with the increasing partition number. Also, even if the information stored in a partition decreases as the number of partition increases, the decision fusion methods combine the classification ability of CVA at each partition.

Table 1. Comparison of fusion methods on AR face database for scarf-occlusion problem using uniform $LBP_{8,1}$ features in terms of recognition rate (%)

	Part	MCVA	Product	Borda Count	Majority Vote
1st Exp.	3×3	90.00	**95.00**	94.00	87.33
	4×4	81.00	97.67	**99.00**	97.00
	5×5	73.67	97.67	**99.00**	98.33
2nd Exp.	3×3	**81.67**	62.67	59.67	52.22
	4×4	79.33	85.33	**88.67**	82.33
	5×5	76.67	**91.33**	90.67	83.33

Table 2. Comparison of fusion methods on AR face database sun glasses-occlusion problem using uniform $LBP_{8,1}$ features in terms of recognition rates (%)

	Part	MCVA	Product	Borda Count	Majority Vote
1st Exp.	3×3	62.33	**94.33**	84.67	73.67
	4×4	56.33	**96.00**	92.67	86.00
	5×5	51.00	**91.00**	84.33	73.00
2nd Exp.	3×3	48.67	**67.33**	55.00	46.67
	4×4	51.67	**72.00**	68.00	58.00
	5×5	51.00	**88.67**	80.00	70.67

In the second part of the experiments, we compare MCVA's conventional classification criteria with classifier fusion methods using images with sun glasses (a8, a9, a10, b8, b9, b10) in Fig. 3 as test set. We use the same training set with the first part. The experimental outcomes are summarized in Table 2. It is seen that, the product rule has the highest classification rates in all cases.

We again divide all images into 3×3, 4×4, and 5×5 subregions and generate $LBP_{8,1}$ features. Then, we repeat all previous experiments compare the $LBP_{8,1}$ and the uniform $LBP_{8,1}$ features. The results of the experiments are summarized in Table 3 for scarf occlusion, and in Table 4 for sun glass occlusion problems. $LBP_{8,1}$ based feature vector is 256 dimensional whereas uniform $LBP_{8,1}$ is just 59 dimensional. Also it is known that CVA's classification performance increases as the dimension of the feature vector increases [21]. When we compare the result in Tables 1–2 with Tables 3–4 it is clearly seen that in almost all of the experiments $LBP_{8,1}$ features outperform uniform $LBP_{8,1}$ features.

Table 3. Comparison of fusion methods on AR face database for scarf-occlusion problem using $LBP_{8,1}$ features in terms of recognition rate (%)

	Part	MCVA	Product	Borda Count	Majority Vote
1st Exp.	3×3	91.67	**96.33**	95.00	91.33
	4×4	85.33	98.33	**99.33**	98.33
	5×5	80.67	98.33	**99.33**	98.33
2nd Exp.	3×3	**86.00**	68.67	62.67	58.00
	4×4	80.67	87.67	**91.33**	83.67
	5×5	79.67	90.00	**90.67**	85.67

Table 4. Comparison of fusion methods on AR face database sun glasses-occlusion problem using $LBP_{8,1}$ features in terms of recognition rates (%)

	Part	MCVA	Product	Borda Count	Majority Vote
1st Exp.	3×3	67.00	**95.33**	87.67	76.67
	4×4	67.00	**95.33**	94.33	89.00
	5×5	53.67	**91.67**	87.33	76.33
2nd Exp.	3×3	50.00	**69.33**	56.67	46.00
	4×4	53.00	**77.00**	71.00	60.33
	5×5	52.67	**88.67**	80.33	71.33

7 Conclusion

Partial occlusion negatively affects the decision performance of face recognition system. To overcome this problem modular methods are proposed. Modular Common Vector Approach increases the performance of conventional CVA at occluded face images. In this work, we try to make an improvement at the decision level of MCVA using several decision fusion methods. We extract facial properties using LBP instead of pixel gray level intensities. In the experimental work, we perform a comparative analysis of these different decision fusion methods with the conventional MCVA's decision stage, namely distance-based evidence fusion. Also, we compare LBP and uniform LBP features. It is observed that conventional LBP features have better recognition performance than uniform LBP. One of the reason for the superiority is the larger feature vector size of conventional LBP. Because the common vector lies in the null space of the covariance matrix of a class. If the feature vector size is increased with the fixed number of feature vectors, the dimension of the null space of the covariance matrix increase.

When we use LBP features, combining classifier outputs of the partitions using decision fusion methods increase the recognition performance except one case.

References

1. Zhao, W., Chellappa, R., Phillips, P.J., Rosenfeld, A.: Face recognition: a literature survey. ACM Comput. Surv. **35**(4), 399–458 (2003)
2. Azeem, A., Sharif, M., Raza, M., Murtaza, M.: A survey: face recognition techniques under partial occlusion. Int. Arab J. Inf. Technol. **11**(1), 1–10 (2014)
3. Naseem, I., Togneri, R., Bennamoun, M.: Robust regression for face recognition. Pattern Recogn. **45**, 104–118 (2012)
4. Koc, M., Barkana, A.: Modular common vector approach. In: 22nd Signal Processing and Communications Applications Conference (SIU2014), Trabzon, Turkey, pp. 533–535 (2014)
5. Gulmezoglu, M.B., Dzhafarov, V., Barkana, A.: The common vector approach and its relation to principal component analysis. IEEE Trans. Speech Audio Process. **9**(6), 655–662 (2001)
6. Koc, M., Barkana, A., Gerek, O.N.: A fast method for the implementation of common vector approach. Inf. Sci. **180**(20), 4084–4098 (2010)
7. Gunal, S., Edizkan, R.: Subspace based feature selection for pattern recognition. Inf. Sci. **178**(19), 3716–3726 (2008)
8. Kittler, J., Hatef, M., Duin, R.P.W., Matas, J.: On combining classifiers. IEEE Trans. PAMI **20**(3), 226–239 (1998)
9. Ojala, T., Pietikainen, M., Maenpaa, T.: Multiresolution gray-scale and rotation invariant texture classification with local binary patterns. IEEE Trans. PAMI **24**(7), 971–987 (2002)
10. Pentland, A., Moghaddam, B., Starner, T.: View-based and modular eigenspaces for face recognition. In: 1994 Proceedings of IEEE Conference on Computer Vision and Pattern Recognition, Seattle, USA, pp. 84–91 (1994)
11. Ahonen, T., Hadid, A., Pietikainen, M.: Face description with local binary patterns: application to face recognition. IEEE Trans. PAMI **28**(12), 2037–2041 (2006)
12. Shang, J., Chen, C., Liang, H., Tang, H.: Object recognition using rotation invariant local binary pattern of significant bit planes. IET Image Proc. **10**(9), 662–670 (2016)
13. Liu, L., Fieguth, P., Guo, Y., Wang, X., Pietikäinen, M.: Local binary features for texture classification: taxonomy and experimental study. Pattern Recogn. **62**, 135–160 (2017)
14. Kazak, N., Koc, M.: Some variants of spiral LBP in texture recognition. IET Image Proc. **12**(8), 1388–1393 (2018)
15. Mäenpää, T.: The local binary pattern approach to texture analysis - extensions and applications. Ph.D Thesis, University of Oulu (2003)
16. Polikar, R.: Ensemble based systems in decision making. IEEE Circuits Syst. Mag. **6**(3), 21–45 (2006)
17. Mangai, U.G., Samanta, S., Das, S., Chowdhury, P.R.: A survey of decision fusion and feature fusion strategies for pattern classification. IETE Tech. Rew. **27**(4), 293–307 (2010)
18. Martinez, A.M., Benavente R.: The AR face database. CVC Technical report No: 24 (1998)
19. Cevikalp, H., Neamtu, M., Wilkes, M., Barkana, A.: Discriminative common vectors for face recognition. IEEE Trans. PAMI **27**(1), 4–13 (2005)
20. Koc, M.: Different decision fusion methods for modular common vector approach. In: IEEE 14th International Scientific and Technical Conference on Computer Sciences and Information Technologies (CSIT), Ukraine, Lviv, pp. 67–70 (2019)
21. Koc, M., Barkana, A.: Discriminative common vector approach based feature selection in face recognition. Comput. Electr. Eng. **40**(8), 37–50 (2014)

Features of Application of Monte Carlo Method with Markov Chain Algorithms in Bayesian Data Analysis

Peter Bidyuk[1], Yoshio Matsuki[1,2], Aleksandr Gozhyj[3](✉), Volodymyr Beglytsia[3], and Irina Kalinina[3]

[1] National Technical University of Ukraine "Igor Sikorsky Kyiv Polytechnic Institute", kyiv, Ukraine
pbidyuke_00@ukr.net, matsuki@wdc.org.ua
[2] Kyoto University, Kyoto, Japan
[3] Petro Mohyla Black Sea National University, Nikolaev, Ukraine
alex.gozhyj@gmail.com, science@chmnu.edu.ua, irina.kalininal612@gmail.com

Abstract. The article discusses the algorithms of the Monte Carlo method with Markov chains (MCMC). These are the Metropolis-Hastings and Gibbs algorithms. Descriptions and the main features of the algorithms application are given. The MCMC methods are developed to model sets of vectors corresponding to multidimensional probability distributions. The main application of these methods and algorithms in Bayesian data analysis procedures is directed towards study of posterior distributions. The main procedures of Bayesian data analysis are considered and the features of the application of the Metropolis-Hastings and Gibbs algorithms with different types of input data are considered. Examples of application of the algorithms and methods for their evaluation are provided.

Keywords: Monte Carlo method with Markov chains · Bayesian data analysis · Metropolis-Hastings algorithm · Gibbs algorithm

1 Introduction

The use of probabilistic-statistical methods provides significant benefits in the terms of making correct decisions in actually all areas of human activity: in competition for quality improvement and sale of new products it enables to receive high-quality estimates of forecasts, substantiate financial and macroeconomic decisions, solve the problems of image recognition, machine learning and in other tasks. The basis of most probabilistic-statistical methods is the Bayesian methodology of modeling and the formation of a probabilistic inference, which is called the Bayesian analysis.

The notion of methodology means that a number of methods are used to solve the following problems [2–4, 7–10, 12, 15]:

– construction of probabilistic-statistical models of different types (evaluation of structure and parameters) using statistical data and expert estimates;

N. Shakhovska and M. O. Medykovskyy (Eds.): CCSIT 2019, AISC 1080, pp. 361–376, 2020.
https://doi.org/10.1007/978-3-030-33695-0_25

- calculation of the final results on the basis of the created model in accordance to the problem statement: estimates of forecasts, generation of control effects, estimates of variables and parameters at the output of filters, pattern recognition, finding solutions for the management of investigated processes and systems, etc.;
- analysis of correctness of the results obtained by the appropriate sets of statistical quality criteria.

The methodology of Bayesian analysis includes the following methods [1, 3, 4, 7, 8]:

- *Recursive Bayesian estimation: filtering, forecasting, smoothing of variables;*
- *Hidden Markov models* - this is a modification of Bayesian filter, which assumes that the data are discrete; models of state and transition observations are given by probability matrices or conditional probability tables;
- *Optimal recursive Kalman filters*;
- *Particle filters;*
- *Static Bayesian networks (BN);*
- *Dynamic Bayesian networks (DBN);*
- *Markov localization (ML) models;*
- *Bayesian maps;*
- *Bayesian method of data processing* and *decision making* based on hierarchical models;
- *Bayesian regression*, generalized linear models.

The basic algorithms used in simulation of statistical systems are Metropolis-Hastings and Gibbs algorithms [5, 6, 14–17, 20, 23]. They are related to posterior distribution algorithms. In Bayesian analysis methods these algorithms and their modifications are used to find various values numerically using statistical modeling. This procedure is not entirely suitable for a wide statistical choice since the algorithms are iterative and at each stage require performing the calculation of the likelihood function, which requires significant computational resources. This feature significantly slows down the analysis process. Existing modifications of the Metropolis-Hastings and Gibbs algorithms simplify the modeling process, but significantly accelerate it. In the paper proposed the Metropolis-Hastings algorithm, the Gibbs algorithm and the features of their practical application are investigated.

2 Problem Statement

Investigate the features of the Metropolis-Hastings algorithm and the Gibbs algorithm in Bayesian data analysis tasks.

3 Features of Bayesian Data Analysis

Consider the generalized approach based on simulation procedures for calculation of the quantities (moments and quantiles of the posterior and predictive distributions) that arise in Bayesian data analysis. For example, it is necessary to find mean of the posterior distribution

$$E(\theta|\mathbf{y}) = \int \theta h(\theta|\mathbf{y})d\theta,$$

where θ is vector of parameters; $\mathbf{y}$ is vector of observations. In some cases such integral can be found analytically, but in most cases it is not possible. Therefore, it is more common to calculate such integrals using the Monte-Carlo method, based on generating pseudo-random sequences with specified parameters with subsequent averaging the results.

The general idea of the estimation procedure is to correctly generate pseudo-random numbers with a posterior density, $h(\theta|\mathbf{y})$. If it is possible to generate a sequence of parameter values, $\{\theta^{(1)},\ldots,\theta^{(M)}\} \sim h(\theta|\mathbf{y})$, with a posterior density, then when a sufficiently large sample is reached, it is possible to calculate not only the above-mentioned integral, but also other characteristics of the posterior distribution by forming the corresponding estimates. For example, an average of the generated sample is the posterior average calculated by simulating the process, and quantile of the generated distribution is the estimate of quantiles of the posterior distribution. In compliance with the laws of large numbers, that is, when the power of the samples will be sufficiently large, the calculated estimates will be the same up to a posteriori values.

If the model under construction allows the use of natural conjugate prior density, then the posterior density will also belong to the same family as the prior one. It will be known which expression should be used to generate a pseudo-random sequence. If such approach (based on the use of the conjugate density) is impossible, then generating with posterior density will be more difficult task.

The generation of complex posterior density of high dimensionality can be performed using Monte Carlo methods for Markov chains (MCMC) [14, 22]. This class of methods is based on the simulation modeling of properly formed Markov chains, which coincide with the density of the desired type (as a rule this is posterior density).

The main property of the Markov chain is that conditional density of the parameter $\left(\theta^{(j)}\right)$ (this is j-th element of the sequence) is consistent with the specified distribution, but the value of this element depends only on the previous value, $\theta^{(j-1)}$. Denote this *conditional transition density* as follows: $T(\theta^{(j-1)},.|\mathbf{y})$.

The modeling method of the MCMC should ensure generating of such transition density that will coincide with posterior density, starting from the initial condition, θ_0. The convergence here is reached in the sense that for any countable set A from a region

h, the distribution $P(\theta^{(j)} \in A|\mathbf{y}, \theta_0)$ converges to $\int_A h(\theta|\mathbf{y})d\theta$ as $j \to \infty$. Thus, using this approach, the pseudo-random numbers related to posterior density can be obtained using the following recursive computational scheme:

$$\begin{gathered}\theta^{(1)} \sim T(\theta^{(0)}, .|\mathbf{y}) \\ \theta^{(2)} \sim T(\theta^{(1)}, .|\mathbf{y}) \\ \vdots \\ \theta^{(j)} \sim T(\theta^{(j-1)}, .|\mathbf{y}) \\ \vdots\end{gathered}$$

Since the chain generated in this manner must coincide with posterior density, then after a certain number n_0 of iterations, $\theta^{(n_0+1)}, \theta^{(n_0+2)}, \ldots, \theta^{(n_0+M)}$, the sequence can be considered as the values related to the posterior distribution $h(\theta|\mathbf{y})$. The initial n_0 values of the chain, as a rule, are not used, since this is a transient process with rather coarse quantities that are unacceptable for further use. This transitional process is also called the "entry period" (or "burn - in" period). It should be noted that the pseudo-random numbers generated by the Markov chain procedure are correlated to sample of a posteriori density. However, these numbers can be used to calculate sample means and quantiles.

Using large numbers for Markov sequences, it can be shown that the estimates calculated with this method will coincide with the posterior values with a large number of generated items. So, for any integrated function from, θ, for example $g(\theta)$, sample mean of values $\{g(\theta^{(j)})\}$ coincides, under weak regularity conditions, to its mathematical expectation:

$$M^{-1} = \sum_{j=1}^{M} g(\theta^{(j)}) \underset{M \uparrow \infty}{\to} \int_{\theta} g(\theta)h(\theta|\mathbf{y})d\theta \; .$$

The utility of generating pseudo-random sequences such as Markov chains is confirmed by the fact that, as a rule, it is possible to construct a transition density, $T(\theta^{(j-1)}, .|\mathbf{y})$, that coincides with the target density.

4 Metropolis-Hastings Algorithm

Generate a sequence with continuous target density, $h(\theta|\mathbf{y})$, where θ is parameter of vector; $h(\theta|\mathbf{y})$ is continuous density. Suppose that the value of θ is generated in the form of a single block, and a Markov chain is determined by the transition density, $q(\theta', \theta|\mathbf{y})$, where (θ', θ) are any two values from this sample. Let the specified transition density be specified without regard to the target density, so it doesn't coincide with the target density [6, 19].

For example, for a transitive density $q(\theta', \theta|\mathbf{y})$ one can accept a multidimensional normal density with a vector of mean θ' and covariance (dispersion) matrix $\mathbf{V}$, like, $E[\theta'_i(\theta'_l)^T] = \mathbf{V}$. So having as input, $q(\theta', \theta|\mathbf{y})$, it is necessary to build a Markov chain, which coincides with $h(\theta|\mathbf{y})$. This can be done using Metropolis-Hastings method and the algorithm based on it. The idea of the method is to modify the distribution $q(\theta', \theta|\mathbf{y})$ in such a way that the transient density of the modified chain "coincides" with the target density.

In order to determine the sequence of operations when executing the Metropolis-Hastings algorithm, we define the initial value of the chain $\theta^{(0)}$ with the subsequent goal of finding (generate) a sequence of values: $\theta^{(0)}, \theta^{(1)}, \ldots, \theta^{(j-1)}$. The next value of the chain $\theta^{(j)}$ is calculated using the procedure consisting of two steps given below.

- *Proposal step,* i.e. generating a candidate. The candidate for the next value of the chain θ is generated by the distribution $q(\theta^{(j-1)}, \theta|\mathbf{y})$ (*density of supply*) and the following value is calculated:

$$\alpha(\theta^{(j-1)}, \theta|\mathbf{y}) = \min\left\{1, \frac{h(\theta|\mathbf{y})}{h(\theta^{(j-1)}|\mathbf{y})} \frac{q(\theta, \theta^{(j-1)}|\mathbf{y})}{q(\theta^{(j-1)}, \theta|\mathbf{y})}\right\} \tag{1}$$

- *Move step* of the chain with the following value. Take the following chain value based on the analysis of the following condition:

$$\theta^{(j)} = \begin{cases} \theta, & \text{with probability} \quad \alpha(\theta^{(j-1)}, \theta|\mathbf{y}), \\ \theta^{(j-1)} & \text{with probability} \quad 1 - \alpha(\theta^{(j-1)}, \theta|\mathbf{y}). \end{cases}$$

It should be noted that $q(\theta', \theta|\mathbf{y})$ is the density from which the data is generated to simulate a Markov chain. It is called the density for generating candidates or the density that generates proposals for the chain (proposal density). General ways of its setting will be discussed below. We also note that the function $\alpha(\theta^{(j-1)}, \theta|\mathbf{y})$ in (1) can be calculated without knowing the normalizing constant of the posterior density. The value $h(\theta|\mathbf{y})$ is called the *acceptance probability* or *probability of move.*

The theoretical properties of the above algorithm depend largely on the nature (method of determining) the density with which the pseudorandom variable is generated. A typical requirement for this method of generating density is its positive definiteness on the posterior density support set. It follows from these considerations that the chain generated by the Metropolis-Hastings algorithm can go to any point (accept any value) of the reference set in one step.

In addition to this, we note that as a result of the randomization procedure at the stage of filling in the chain (move step), the transition density $T(\theta^{(j-1)}, |\mathbf{y})$ of the generated chain has an interesting form, since it is a mixture of continuous density (away from $(\theta^{(j-1)})$) and a discrete component (to provide some (non-zero) values of the probability of being near $(\theta^{(j-1)})$). The transition density has the following form:

$$T(\theta^{(j-1)}, |\mathbf{y}) = q(\theta^{(j-1)}, \theta|\mathbf{y})\alpha(\theta^{(j-1)}, \theta|\mathbf{y}) + r(\theta^{(j-1)}|\mathbf{y})\delta_{\theta^{(j-1)}},$$

where

$$r(\theta^{(j-1)}|\mathbf{y}) = \int_{\theta} q(\theta^{(j-1)}, \theta|\mathbf{y})\alpha(\theta^{(j-1)}, \theta|\mathbf{y})d\theta \quad (2)$$

$\delta_{\theta^{(j-1)}}$ is a Dirac function for $\theta^{(j-1)}$, which is defined as follows:

$$\delta_\theta(\theta') = \begin{cases} 0, \ \textit{if} \ \theta' \neq \theta, \\ 1, \ \textit{if} \ \int \delta_\theta(\theta')\, d\theta' = 1, \end{cases}$$

or almost this way:

$$\delta_\theta(\theta') = \begin{cases} 0, \ \textit{if} \ \theta' \neq \theta, \\ 1, \ \textit{if} \ \theta' = \theta. \end{cases}$$

It is easy to verify that the integral of the transition density over all possible values of θ is equal to unity, as it should be. Note that the functions $T(\theta^{(j-1)}, \theta|\mathbf{y})$ and $r(\theta^{(j-1)}|\mathbf{y})$ are not calculated in the process of implementing the Metropolis-Hastings algorithm presented above.

Due to the chosen method of determining the assessment $\theta^{(j)}$ and its presentation explicitly in the form of a transition density, according to the Metropolis-Hastings algorithm, the values can be repeated in a generated Markov chain. It is obvious that in order to ensure the efficient movement of the generated values $\theta^{(j-1)}$ over the reference set of the target density, the chain should not often be held at one point within multiple iterations. This behavior of the algorithm can be avoided by proper selection of the proposed density, $q(\theta^{(j-1)}, \theta|\mathbf{y})$.

5 Markov Chain Convergence Control

To monitor the process of estimating parameters when using the Metropolis-Hastings algorithm, one can use a criterion called the autocorrelation period (time) or inefficiency factor, which is calculated for each scalar parameter as follows:

$$IF = 1 + 2\sum_{S=1}^{M}\left(1 - \frac{s}{M}\right)\rho(s)$$

where $\rho(s)$ is the sample autocorrelation function with lags s, which is calculated for the generated Markov chain elements, representing evolution of the estimate of selected parameter: $\theta^{(n_0+1)}, \theta^{(n_0+2)}, \ldots, \theta^{(n_0+M)}$.

The *IF* criterion can be interpreted as an effective sample size (*ESS*), which is calculated as follows:

$$ESS = \frac{M}{IF}. \tag{3}$$

If the generated Markov chain values are independent from each other, then the autocorrelation period is theoretically equal to 1, and *ESS* = *M*. When the inefficiency factor takes large values, then the effective sample size is significantly reduced.

In practice, the effective sample size means how many approximately iterations of the Metropolis-Hastings algorithm need to be performed (after rejecting the initial iterations ("burn - in" period), which we do not take into account in the estimation process) in order for the parameter estimate to stop changing [14]. So the value of *M* depends on the dimension of the vector of parameters and the correctness of functioning of the algorithm. The number of initial iterations that are not taken into account in the process of parameter estimation can range from several thousand to several tens of thousands.

6 An Application Example of the Metropolis-Hastings Algorithm

Consider an example of application of the algorithm on binary data characterizing infection (or absence of infection) of patients after a complex surgery operation (Table 1).

The numbers in the first column, for example 11/87, means that 98 operations were performed on the values of independent variables, $(x_1 = 1, x_2 = 1, x_3 = 1)$; and after these 98 operations, there were 11 cases of patient infection and 87 cases without infection. A total of 251 cases were considered.

So, the main (dependent) variable $y(\cdot)$ is binary and for its formal description we use the probit model.

Table 1. Data on patient infection after surgery

y(1/10)	x_1	x_2	x_3
11/87	1	1	1
1/17	0	1	1
0/2	0	0	1
23/3	1	1	0
28/30	0	1	0
0/9	1	0	0
8/32	0	0	0

This way, the probability of infecting a i-patient can be calculated using the expression:

$$\Pr(y_i = 1|\mathbf{x}_i, \beta) = \Phi(\mathbf{x}_i^T \beta),$$

where $\mathbf{x}_i = (1, x_{i1}, x_{i2}, x_{i3})^T$ is the vector of independent variables; $\beta = (\beta_0, \beta_1, \beta_2, \beta_3)^T$ is the vector of unknown parameters (coefficients) of the model; $\Phi(\cdot)$ is the cumulative function of the standard normal distribution of a random variable (or simply the distribution function). If we accept that the results of the infection $\mathbf{y} = (y_1, y_2, \ldots, y_{251})$ are conditionally independent events, then the likelihood function can be written as follows:

$$\begin{aligned} L(\mathbf{y}|\mathbf{X}, \beta) &= \prod_{i=1}^{251} \Phi(\mathbf{x}_i^T \beta)^{y_i} [1 - \Phi(\mathbf{x}_i^T \beta)]^{(1-y_i)} \\ &= \prod_{i=1}^{251} \begin{cases} \Phi(\mathbf{x}_i^T \beta), & \text{if } y_i = 1, \\ 1 - \Phi(\mathbf{x}_i^T \beta), & \text{if } y_i = 0. \end{cases} \end{aligned}$$

This likelihood function does not involve the adoption of natural conjugate a priori density. Suppose that a priori information about the vector of parameters β can be a multidimensional normal density with zero average for each parameter, and the variance is given by $5\mathbf{I}_4$, where $\mathbf{I}_4$ is the unity matrix of dimensionality 4×4. So, the prior density of the parameter vector β is described by the expression:

$$g(\beta) \propto \exp[-0,5\beta^T (5\mathbf{I}_4)^{-1} \beta],$$

while posterior density can be found in the form:

$$\begin{aligned} h(\beta|\mathbf{y}, \mathbf{X}) &\propto g(\beta) L(\mathbf{y}|\mathbf{X}, \beta) \\ &\propto \exp[-0,5\beta^T (5\mathbf{I}_4)^{-1} \beta] \prod_{i=1}^{251} \Phi(\mathbf{x}_i^T \beta)^{y_i} [1 - \Phi(\mathbf{x}_i^T \beta)]^{(1-y_i)}. \end{aligned}$$

Obviously, this posterior density does not belong to known family of distributions.

If we estimate the vector of parameters β as the average of the corresponding generated Markov chains by this distribution, then we will need to calculate the sample averages and, possibly, their covariance matrix to control the quality of the estimates. A simple way to solve this problem is to generate pseudo-random sequences from this a posteriori distribution, and then use these sequences to calculate the required posterior parameter estimates.

7 Metropolis-Hastings Algorithm Without Latent Data

In order to select the output density proposal distribution, first consider the likelihood function. Let the use the method of maximum likelihood (ML) with Newton-Raphson procedure helped to find such estimates for unknown parameter vector:

$$\begin{aligned}\hat{\beta} &= \arg\max \ln[L(\mathbf{y}|\mathbf{x}, \beta)]\\ &=(-1,093022; 0,607643; 1,197543, -1,904730)^T\end{aligned}$$

and let the symmetric matrix, obtained as a result of the inverse of a Hessian with the opposite sign (the matrix of second derivatives), of the logarithmic function of probability, evaluated at the point, has the following form:

$$\mathbf{V} = \begin{bmatrix} 0,040745 & -0,007038 & -0,039399 & 0,004829 \\ & 0,073101 & -0,006940 & -0,050162 \\ & & 0,062292 & -0,016803 \\ & & & 0,080788 \end{bmatrix}.$$

Now assume that the initial proposed values of the parameter estimates are generated by the random walk random pattern:

$$\beta = \beta^{(j-1)} + \varepsilon^{(j)}, \varepsilon^{(j)} \sim N_4(0, \tau\mathbf{V}),$$

where $\tau-$ is the scale parameter that is used to configure the M-H algorithm so as to generate competing pseudorandom sequences. In this case, the proposed density has the following form:

$$q(\beta^{(j-1)}, \beta|\mathbf{y}) \propto \exp\left[-0,5\left(\beta - \beta^{(j-1)}\right)^T (\tau\mathbf{V})^{-1}\left(\beta - \beta^{(j-1)}\right)\right],$$

where there is no normalizing constant in the denominator.

To generate pseudorandom numbers from a multivariate normal density make use of matrix algebra that any matrix $\mathbf{A}$ can be positively presented uniquely by decomposition into $\mathbf{L}\mathbf{L}^T$ where $\mathbf{L}$ is lower triangular matrix with positive elements on the main diagonal. Assume that such matrix was found, that is, $\tau\mathbf{V} = \mathbf{L}\mathbf{L}^T$. The value of the vector $\varepsilon^{(j)}$ can be calculated from the distribution $N_4(\mathbf{0}, \tau\mathbf{V})$ as follows:

$$\varepsilon^{(j)} = \mathbf{L}\mathbf{z}^{(j)},$$

where $\mathbf{z}^{(j)} \sim N_4(\mathbf{0}, \mathbf{I}_4)$.

Suppose that the $(j-1)-$ algorithm's step is also completed and it is necessary to generate numbers in the next step. Initially, the proposed value β is generated by the equation, $\beta = \beta^{(j-1)} + \varepsilon^{(j)}, \varepsilon^{(j)} \sim N_4(0, \tau \mathbf{V})$, and then the probability of filling in the chain is determined by the expression:

$$\alpha(\beta^{(j-1)}, \beta|\mathbf{y}) = \min\left\{1, \frac{h(\beta|\mathbf{y})}{h(\beta^{(j-1)}|\mathbf{y})} \frac{q(\beta, \beta^{(j-1)}|\mathbf{y})}{q(\beta^{(j-1)}, \beta|\mathbf{y})}\right\},$$

where $\alpha(\beta^{(j-1)}, \beta|\mathbf{y})$ is the conditional probability of selecting a value or for filling in the chain.

Note that the members that contain the proposed density are reduced (i.e., members $q(\beta, \beta^{(j-1)}|\mathbf{y})$ and $q(\beta^{(j-1)}, \beta|\mathbf{y})$. This happens due to the fact that the proposed density in the form of a random step equation is symmetric in relation to its arguments. If a uniform distribution is used, that is $U \sim Uniform(0, 1)$, the iteration will be performed with the following assignment:

$$\beta^{(j)} = \begin{cases} \beta, \ if \ U < \alpha(\beta^{(j-1)}, \ \beta|\mathbf{y}), \\ \beta^{(j-1)}, \ else. \end{cases}$$

Then this process is repeated in order to obtain the sequence, $\{\beta^{(1)}, \beta^{(2)}, \ldots, \beta^{(n_0+M)}\}$.

It is important to understand the role of the parameter setting algorithm. From the expression for the probability of filling in the chain it is clear that for large values the values proposed for filling in the chain will be rather far from the current value and will probably be rejected. If the value is too small, the proposed value will most likely be accepted, but the following values will be close to each other and thus the chain will slowly use the posterior distribution. In both cases, the values of β generated from the posterior distribution will have a significant sequential correlation.

In order to avoid this problem it is advisable to make trial runs of the algorithm with different values of τ. There is a recommendation [6] to use a value of τ that will ensure acceptance of the proposed values at the level of about (30–50)%. Based on these recommendations, the value of the debugging parameter $\tau = 1$ is used for data related to this example. The number of initial iterations of the M-H algorithm that was used to calculate parameter estimates is taken from 100, and 5000 subsequent iterations are used to calculate the estimates.

In order to illustrate the results of calculations (simulation modeling) there are different ways: graphs, tables, and individual numbers. Table 2 below shows the values of the prior and posterior first two moments, the 2.5 (lower) and 97.5 (upper) percentiles of the marginal density for β. Each of these values is calculated by posterior distribution using known formulas; the posterior standard deviation is calculated, this is standard deviation for the generated values, and the posterior percentiles are also calculated based on the generated sequences.

Table 2. Results of applying the M-H algorithm to the parameter estimation using the random step equation for the proposed values

	A priori parameters		A posteriori parameters			
	Middle	Stand deviation	Middle	Stand deviation	The lower percentile	The upper percentile
β_0	0,000	3,162	–1,110	0,224	–1,553	–0,677
β_1	0,000	3,162	0,612	0,254	0,116	1,127
β_2	0,000	3,162	1,198	0,263	0,689	1,725
β_3	0,000	3,162	–1,901	0,275	–2,477	–1,354

As expected, the first and second regressors increase the probability of infection ($\beta_1 > 0$; $\beta_2 > 0$), and the third (antibiotic prophylaxis) reduces likelihood of infection.

Abbreviated version of the proposed density

Pseudorandom numbers that are candidates for the values of a new chain can be generated under a slightly different scheme. If a normal distribution is chosen, the form of the proposed density will be as follows:

$$q(\beta|\mathbf{y}) \propto \exp\left\{-0,5(\beta - \hat{\beta})^T(\tau\mathbf{V})^{-1}(\beta - \hat{\beta})\right\}.$$

This proposing density is similar to that based on the random step equation, except that this distribution is centered around a fixed point (value), $\hat{\beta}$, not the previous value of the chain. This density is called a shortened offering density. The proposed values are generated by the expression:

$$\beta = \hat{\beta} + \varepsilon^{(j)},$$
$$\varepsilon^{(j)} \sim N_4(\mathbf{0}, \tau\mathbf{V}),$$

where $\hat{\beta}$ are the values obtained as a result of application of the MIS to the logarithm of the plausibility function.

The probability of filling in the chain with a new value is determined as follows:

$$\alpha(\beta^{(j-1)}, \beta|\mathbf{y}) = \min\left\{1, \frac{w(\beta|\mathbf{y})}{w(\beta^{(j-1)}|\mathbf{y})}\right\},$$

where

$$w(\beta|\mathbf{y}) = \frac{h(\beta|\mathbf{y}, \mathbf{X})}{q(\beta|\mathbf{y})}; w(\beta^{(j-1)}|\mathbf{y}) = \frac{h(\beta^{(j-1)}|\mathbf{y}, \mathbf{X})}{q(\beta^{(j-1)}|\mathbf{y})}$$

In order for this shortened version of the proposed density to function properly, it is necessary that the function $w(\cdot|\mathbf{y}, \mathbf{X})$ is limited. For a unimodal, absolutely continuous, function, $h(\cdot|\mathbf{y},\ \mathbf{X})$, this condition holds in the case when the tails of the proposed density are thicker than the target density tails. This condition needs to be verified analytically, but it is usually satisfied when $\tau > 1$.

8 Gibbs Algorithm for Generating Markov Chains

One of the simple algorithms for generating Markov chains by the Monte Carlo method is Gibbs algorithm. He was proposed in [13], where the problem of pattern recognition was considered; in his work [19] applied it to solving the problem of missing data imputation; and [11] used it to solve a number of Bayesian data analysis problems. Construction of the Markov chain by the Gibbs algorithm is implemented by generating a set of conditional distributions of this type:

$$\begin{gathered} h(\theta_1|\mathbf{y},\theta_2,\theta_3,\ldots,\theta_p) \\ h(\theta_2|\mathbf{y},\theta_1,\theta_3,\ldots,\theta_p) \\ \vdots \\ h(\theta_p|\mathbf{y},\theta_1,\theta_2,\ldots,\theta_{p-1}) \end{gathered}$$

The number of generated units should be such as to guarantee proper interpretation (i.e. perform the necessary analysis) of each full conditional density. When solving many problems of Bayesian data analysis, the Gibbs generation procedure follows from the structure of a model itself [24].

As a simple example, $\mathbf{y} = (y_1,\ldots,y_n)$, consider the values generated by the linear regression model:

$$\begin{gathered} \{y_i|\beta,\sigma^2\} \sim N(\mathbf{x}_i^T\beta,\sigma^2),\ i = 1,2,\ldots,n, \\ \beta \sim N_k(\beta_0,\mathbf{B}_0), \\ \sigma^2 \sim IG(0,5\,v_0,0,5\,\delta_0). \end{gathered}$$

Where *IG* means an inverse gamma density distribution:

$$p(\sigma^2) \propto \left(\frac{1}{\sigma^2}\right)^{0,5\,v_0+1} \exp\left(-\frac{\delta_0}{2\sigma^2}\right),\ \sigma^2 > 0$$

In this case, when the distributions of β and σ^2 are considered by two separate blocks, the full conditional density β is defined as follows:

$$\begin{gathered} h(\beta|\mathbf{y},\sigma^2) \propto f(\mathbf{y},\beta,\sigma^2) \propto \\ \propto g(\beta)L(\mathbf{y}|\beta,\sigma^2) \propto \\ \propto \exp[-0,5(\beta-\hat{\beta})^T\mathbf{B}_n^{-1}(\beta-\hat{\beta})], \end{gathered}$$

where

$$\hat{\beta} = \mathbf{B}_n\left(\mathbf{B}_0^{-1}\beta_0 + \sigma^{-2}\sum_{i=1}^{n}\mathbf{x}_i y_i\right),$$

and

$$\mathbf{B}_n = \left(\mathbf{B}_0^{-1} + \sigma^{-2}\sum_{i=1}^{n}\mathbf{x}_i\mathbf{x}_i^T\right)^{-1}$$

Vector β has a multidimensional normal full conditional density. It can be shown that the total conditional density of parameter, σ^2, that is, $h(\sigma^2|\mathbf{y}, \beta)$, has an updated inverse gamma distribution:

$$(\sigma^2|\mathbf{y}, \beta) \sim IG\left\{\frac{v_0 + n}{2}, \frac{\delta_0 + \sum_{i=1}^{n}(y_i - \mathbf{x}_i^T\beta)^2}{2}\right\}.$$

Thus, it is possible to perform an analysis of each of the considered full conditional density. Note that the data augmentation process, which will be discussed below, often assists in obtaining a plurality of complete conditional distributions that are subject to analysis.

One execution cycle of the Gibbs algorithm ends with generation of parameter estimates, $\{\theta_k\}_{k=1}^{p}$, from each full conditional distribution, which makes it possible recursively to update estimates of unknown parameters [21]. The Gibbs algorithm, in which each block is implemented in a fixed order, is given below.

9 Gibbs Algorithm for Generating Pseudorandom Values

1. Set initial values of the parameter vector:

$$\theta^{(0)} = (\theta_1^{(0)}, \ldots, \theta_1^{(p)}).$$

2. Repeat for $j = 1, 2, \ldots, n_0 + M$ the following calculations:

- generate $\theta_1^{(j)}$ from, $h(\theta_1|\mathbf{y}, \theta_2^{(j-1)}, \theta_3^{(j-1)}, \ldots, \theta_p^{(j-1)})$;
- generate $\theta_2^{(j)}$ from, $h(\theta_2|\mathbf{y}, \theta_1^{(j)}, \theta_3^{(j-1)}, \ldots, \theta_p^{(j-1)})$;

$$\vdots$$

- generate $\theta_p^{(j)}$ from, $h(\theta_p|\mathbf{y}, \theta_1^{(j)}, \theta_2^{(j)}, \ldots, \theta_{p-1}^{(j)})$;

3. Calculate and save values of vectors estimates

$$\{\theta^{(n_0+1)}, \theta^{(n_0+2)}, \ldots, \theta^{(n_0+M)}\}.$$

In this algorithm, the block of θ_k is generated from the full conditional distribution

$$h(\theta_k|\mathbf{y}, \theta_1^{(j)}, \ldots, \theta_{k-1}^{(j)}, \theta_{k+1}^{(j-1)}, \ldots, \theta_p^{(j-1)}).$$

Here the elements that specify the condition, reflect the fact that when the k *-th* block is achieved, the previous blocks, $(k-1)$, have already been updated. The transition density of the chain (again assuming that h is absolutely continuous) is given by the product of the transition density for each block:

$$T(\theta^{(j-1)}, \theta^{(j)}|\mathbf{y}) = \prod_{k=1}^{p} h(\theta_k|\mathbf{y}, \theta_1^{(j)}, \ldots, \theta_{k-1}^{(j)}, \theta_{k+1}^{(j-1)}, \ldots, \theta_{k+1}^{(j-1)})$$

To illustrate the method by which the blocks are updated, consider the structure of the two blocks. Each block contains one component, (θ_1 or θ_2). Note that within one iteration of the algorithm both components of the algorithm are updated. At the same time each component is updated along its coordinate axis. This property of the algorithm can cause some problems if the components of a parameter vector are highly correlated with each other: as a result, the contours are compressed and the motions along the coordinate axes become very small (the increment of the parameter values becomes less noticeable).

The use of the Gibbs algorithm is based on the assumption that the full conditional distributions used in the process of chain generation have a clear tractable representation. If one or more full conditional distributions take a form that cannot be recognized, then the generation cycle for this parameter block can be performed using the appropriate proposed density and the multi-block algorithm of Metropolis-Hastings. This way of generating the chain is sometimes called "Metropolis inside Gibbs".

10 Conclusions

The study discusses the features of the Metropolis-Hastings and Gibbs algorithms with its application to solving Bayesian data analysis problems. Application of the Monte Carlo methods for Markov chains (MCMC) for generating complex posterior density of large dimension is very promising for solving many practical problems. The algorithms were proposed to model sets of vectors for multidimensional probability distribution, and details of the algorithms are presented that illustrate specific computational procedures required for their implementation. The principal application of these methods and algorithms in Bayesian data analysis procedures is directed towards generating and analysis of posterior distributions for the variables of interest. Examples of the algorithms application were given that illustrate some possibilities for their practical use.

In future studies it is supposed to continue development of new algorithms of the type mentioned above and to extend their application to solving the problems of modeling nonlinear nonstationary process in finances, economy, ecology etc. It is convenient to perform the studies in the frames of specialized decision support systems allowing to combine the results received with different research instruments.

References

1. Andersen, T., Bollerslev, T., Lange, S.: Forecasting financial market volatility: sample frequency visa-vis forecast horizon. J. Empirical Finance **6**(5), 457–477 (1999)
2. Bidyuk, P., Gozhyj, A., Kalinina, I., et al.: The methods Bayesian analysis of the threshold stochastic volatility model. In: Proceedings of the 2018 IEEE 2nd International Conference on Data Stream Mining and Processing, DSMP 2018, Lviv, pp. 70–75 (2018)
3. Bidyuk, P., Gozhyj, A., Kalinina, I., Gozhyj, V.: Analysis of uncertainty types for model building and forecasting dynamic processes. In: Advances in Intelligent Systems and Computing II, vol. 689, pp. 66–78. Springer, Cham (2017)
4. Blasco, A., Sorensen, D., Bidanel, J.P.: Bayesian inference of genetic parameters and selection response for litter size components in pigs. Genetics **149**, 301–306 (1998)
5. Casella, G., George, E.I.: Explaining the Gibbs sampler. Am. Stat. **46**, 167–174 (1992)
6. Chib, S., Greenberg, E.: Understanding the metropolis-hastings algorithm. Am. Stat. **49**, 327–335 (1995)
7. Insua, D., Ruggeri, F., Wiper, M.: Bayesian Analysis of Stochastic Process Models. Wiley, Chichester (2012)
8. Jacquier, E., Polson, N.G., Rossi, P.E.: Bayesian analysis of stochastic volatility models. J. Bus. Econ. Stat. **20**, 69–87 (1994)
9. Jacquier, E., Polson, N.G., Rossi, P.E.: Bayesian analysis of stochastic volatility models with fat-tails and correlated errors. J. Econometrics **122**, 185–212 (2004)
10. Gelman, A., Meng, X.: Applied Bayesian Modeling and Causal Inference from Incomplete-Data Perspectives. Wiley, Chichester (2004)
11. Gelfand, A.E., Smith, A.F.M.: Sampling-based approaches to calculating marginal densities. J. Am. Stat. Assoc. **85**, 398–409 (1990)
12. Gelman, A., Carlin, J.B., Stern, H.S., Rubin, D.B.: Bayesian Data Analysis. A CRC Press Company, Boca Raton (2004)
13. Geman, S., Geman, D.: Stochastic relaxation, gibbs distributions, and the Bayesian restoration of images. IEEE Trans. Pattern Anal. Math. Intell. **6**, 721–741 (1984)
14. Hastings, W.K.: Monte-Carlo sampling methods using Markov chains and their applications. Biometrika **57**(1), 97–109 (1970)
15. Hoff, P.D.: A First Course in Bayesian Statistical Methods. Springer, New York (2009)
16. Kondratenko Yuriy, P., Kondratenko Nina, Y.: Reduced library of the soft computing analytic models for arithmetic operations with asymmetrical fuzzy numbers. Int. J. Comput. Res. Huttington **23**(4), 349–370 (2016)
17. Kondratenko, Y.P., Kozlov, O.V., Korobko, O.V., Topalov, A.M.: Internet of Things approach for automation of the complex industrial systems. In: ICTERI-2017, CEUR Workshop Proceedings Open Access, vol. 1844, pp. 3–18 (2017)
18. Lee, P.: Bayesian Statistics: An Introduction, 2nd edn. Wiley, New York (1997)
19. Tanner, M.A., Wong, W.H.: The calculation of posterior distributions by data augmentation source. J. Am. Stat. Assoc. **82**(398), 528–540 (1987)

20. Metropolis, N., Rosenbluth, A.W., Rosenbluth, M.N., Teller, A.H.: Equations of state calculations by fast computing machines. J. Chem. Phys. **21**(6), 1087–1092 (1953)
21. Raftery, A.E., Lewis, S.: How many iterations in the Gibbs sampler. In: Bernardo, J.M., Berger, J.O., Dawid, A.P., Smith, A.F.M. (eds.) Bayesian Statistics, vol. 4, pp. 763–773. Oxford University Press (1992)
22. Raftery, A.E., Lewis, S.: Comment: one long run with diagnostics: implementation strategies for Markov Chain Monte Carlo. Stat. Sci. **7**, 493–497 (1992)
23. Smith, A.F.M.: Bayesian computational methods. Phil. Trans. R. Soc. Lond. A **337**, 369–386 (1991)
24. Smith, A.F.M., Roberts, G.O.: Bayesian computation via the Gibbs sampler and related Markov chain Monte-Carlo methods (with discussion). J. Roy. Stat. Soc. Ser. B **55**, 3–23 (1993)

Applied Aspects of Implementation of Intelligent Information Technology for Fraud Detection During Mobile Applications Installation

Andrii Yarovyi and Tetiana Polhul(✉)

Vinnytsia National Technical University, Vinnytsia, Ukraine
a.yarovyy@vntu.edu.ua, tanapolg93@gmail.com

Abstract. The intelligent information technology for fraud detection during mobile applications installation is proposed in this paper. The structure of such intelligent information technology of fraud detection is offered in accordance with the tasks which should be solved by it: subsystem for available data analysis; subsystem for intellectual processing of available data; subsystem for developing a database and knowledge base (for detecting fraudsters); classification model building and user classification subsystem; subsystem for users' templates formation; subsystem for the generalized fraudsters fingerprint formation. The proposed intelligent information technology allows the processing of various input data, which in the process gives the opportunity to form a generalized template of fraudster.

Keywords: Fraud detection · Anomaly detection · Mobile application installation · Data mining · Intelligent information technology

1 Introduction

Nowadays there appeared a need to create intelligent information technology for fraud detection. This is due to the emergence of a huge number of new competing products among the billions of users in the mobile application market. To attract most users to their apps, mobile apps developers use the marketing campaigns service. Such a need in marketing campaigns has led to the massive appearance of fraudsters and fraudulent types of mobile applications installation, which can bring companies the required number of "users" and receive appropriate money reward for it [1]. However, it should be noted that such "users" never return to the mobile application because they are fake, we will call them fraudulent ones. Therefore, the creation of intelligent information technology for fraud detection during mobile applications installation is an important task.

2 State of Study Problem

For the time being, there are already known types of mobile applications installation fraud such as mobile hijacking, click spamming, action farms [2–7] and methods and systems for fraud detection during mobile applications installation such as:

N. Shakhovska and M. O. Medykovskyy (Eds.): CCSIT 2019, AISC 1080, pp. 377–386, 2020.
https://doi.org/10.1007/978-3-030-33695-0_26

- Kochava [8] and TCM Attribution Analytics [9], that do not use all user input data and have a limited number of features for fraud detection. This approach has a disadvantage because fraudsters have new characteristics and new data that needs to be updated every time. Also, since these systems do not use all available user data, it reduces its fraud detection efficiency;
- Fraudlogix [10] and Kraken [11]. These systems determine the user's belonging to a blacklist/bots farm according to a certain feature. The disadvantage of such systems is that they do not take into account the appearance of new fraudsters, bots with new parameters and characteristics and do not take into account all user data;
- Protect360 from Appsflyer [12], Adjust [13]. The first of these systems has a huge database of IP addresses and devices that are labeled with fraud. The second system uses not all the set of input data. It should be noted that using more inputs would allow detecting fraudsters more precise. And, as the Appsflyer company says, their fraud label is only a reason for additional expertise, as it is not always correct. Also, it is impossible to find out the reason why a user is a fraudster;
- FraudScore [14] and AppMetrica [15, 16]. Unlike most, they have updates of their algorithms and have a self-learning system based on the neural network. However, all their estimates are based on only the part of input data. But the analysis of only the part of data also causes non-disclosure of some fraudsters.

So, it should be noted that only the last couple uses the intellectual component, and most other specified analog systems perform a user evaluation based on not all but selective input data, so there is an omission of fraudsters having other properties, patterns, behavior, by such systems [17].

Obviously, the reason for the above mentioned disadvantages of the systems is the lack of a single concept of fraud detection based on available data. Also, the disadvantage of existing systems is that they recognize only known types of fraud and cannot recognize new fraudulent patterns. And the ability of the system to adapt is important in the modern world, therefore, it is necessary to create an intellectual information technology that will be able to learn. Therefore, there is a need to create intelligent information technology for fraud detection during mobile applications installation, which would track and identify fraudsters' patterns that are unnoticed to humans.

3 Task of the Research

In the process of development of intelligent information technology for fraud detection during mobile applications installation, the following problems arose:

- detection, analysis and use of available data of various templates, dimensionality, metrics by the methods of classification and scaling;
- intelligent processing of user's data by classification methods, similarity coefficients and fuzzy logic;
- users classification applying a deep neural networks model;

- creation of databases that contain characteristics of fraudsters not only of human, but of various software bots, and knowledge bases with a set of fuzzy rules for fraud detection. In so doing, such databases and knowledge bases should evolve, depending on new data and new fraudsters, in order to create a general fingerprint of a fraudster.

Let us consider the solution of each task for constructing intelligent information technology for fraud detection during mobile applications installation and analyze its efficiency on the test system.

4 Analysis of the Input Data

A detailed analysis of data has shown that data in such systems is heterogeneous, namely, of different templates, metrics, dimensions. Figure 1 [18] provides an example of data used in this class of information technologies. All these data differ not only in metrics, dimensions, and patterns, but also there are both qualitative, quantitative data, and arrays of qualitative and quantitative data, so it is clear that such data analysis is a complicated task.

Usually for the analysis of such data in different fields of science and technology, different methods of scaling are used, for example:

- data normalization, used for example, in the problems of statistical data processing;
- using rule-based systems, built using fuzzy logic;
- normal scaling of measuring devices.

However, how is it possible to perform the scaling of users' data during mobile applications installation that is presented in Fig. 1? It is known that the IP address may look like "127.0.0.1", "192.168.5.1". It is not possible to normalize data of such type. Similarly, it is impractical to use a rule-based system for data of such type, since in order to take into account all of these data, it is necessary to form 4-12 rules, which greatly complicates the analysis process. So, only scaling remains among these three options. According to the authors, the traditional scaling of the whole set of such data is not possible, since the values in all these data are different (number, category of qualitative data, arrays of quantitative or qualitative data). However, it should be noted that the information provided by these data uniquely determines whether this feature characterizes the fraudster or has the properties of the organic user.

Therefore, the method for scaling user's data based on the value of information they carry was proposed in paper [16]. As it is important not only to make a rating of the user but also to indicate the reason why the user was noticed as a fraudster. This is important, for example, when considering lawsuits in which it is necessary to put forward clear arguments. For example, this is one of the most important reasons why self-driving cars are not yet available for sale since such a justification is a mandatory in the researched area [18].

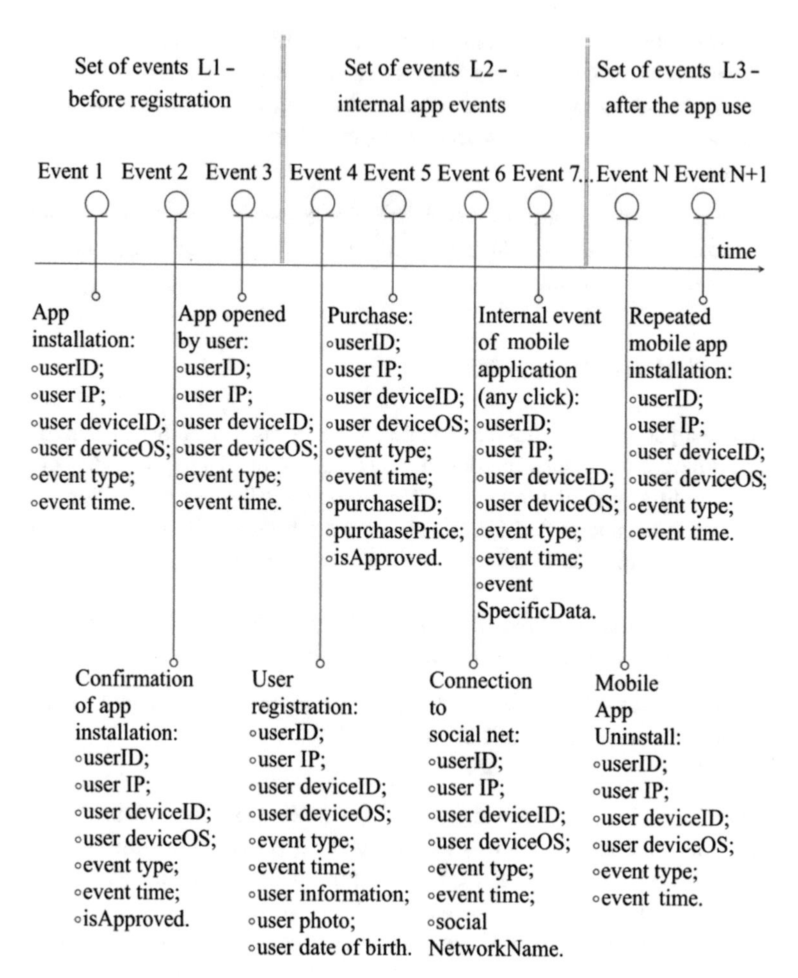

Fig. 1. Example of the input data regarding the user in information technology for fraud detection during mobile applications installation

5 Intelligent Processing of the Available Data

17 scales were developed in the work after analyzing the existing input data and using the proposed scaling method. The peculiarity of each scale is that they allows converting all heterogeneous data into homogeneous with values from 0 to 1. The number of scales was determined both by the number of input data and by the results of expert evaluation considered in the paper [18], on the basis of which a method for overcoming heterogeneity of data in the considered systems was developed. Scales correspond to specific information on the user. For example, the user purchases, information of user connection to social networks, user IP address, time between user internal events, and more other information are important.

Coefficients by which one can determine the value of valuable information for each of the scales, were formed for the presentation of each of these characteristics in normalized form. That is, data analysis using the developed scales is carried out, as a result of which the 17 coefficients are formed for each scale. These coefficients are used in decision-making algorithms for fraud detection and are presented in more detail in [16]. So, the quantitative and qualitative data and their arrays are scaled to homogeneous normalized values without loss of information using the proposed coefficients since the scaling follows the end-point goal, rather than the value of the feature. The peculiarity of each of the proposed coefficients is that each of them can take a value between 0 and 1 and has the same type of data, which allows to convert heterogeneous data to homogeneous data. That is, both quantitative and qualitative data and their arrays are scaled to homogeneous normalized values without loss of information since the scaling is on the final goal. This is an important feature since after converting all available data to homogeneous by such coefficients, it will be possible to submit the obtained homogeneous data to any classification model. Thus, it will be possible to track which of the coefficients has most influenced the attribution of the user to a particular class during users' classification by obtained homogeneous data. This possibility will help to explain why information technology has just taken such a decision.

After analyzing the available data, proposed scales and coefficients, it should be noted that some of the coefficients allow identifying the user unambiguous. Such coefficients allow to form a database of fraudsters and a knowledge base with the characteristics of fraudsters. And some coefficients do not allow to determine the user's class definitely. This kind of data analysis in the process of testing the developed information technology allowed to split the coefficients, using algorithm 1, which is presented in detail in [18, 19], into such groups:

- the first group covers the coefficients that allow a preliminary analysis of the data, namely definitely to identify fraudulent, organic and suspicious users from a set of all the users. For example, the first group includes 13 coefficients, most of which are presented in [4]. There is a coefficient among the coefficients of the first group that, for example, determines whether the purchase was confirmed. If the user has sent a confirmation ID that the mobile store (AppStore, Google PlayMarket, etc.) does not know, then that user is precisely a fraudster. That is, the scales of the first group allow to identify definitely the user's class, which will allow to form the database and knowledge base with known unambiguous users in the future. Examples of scales of the first group are discussed in detail in [16];
- the second group covers the coefficients that make it impossible to do a preliminary analysis. However, one can determine the similarity coefficients of all users with defined users from the generated database by each of the characteristics of the second group, based on databases of fraudsters, organic and suspicious users, and knowledge bases gained from coefficients from the first group. Thus, the second group includes 6 coefficients, most of which are considered in the paper [18]. For example, there is the coefficient that determines the amount of clicks from one device per minute. Since the limit values for this coefficient are not known in advance. It should be noted that during determining the coefficients of the second group it is necessary to check all other features. It is proposed to determine the

values of the coefficients of the second group using formulas to determine the similarity coefficients between users. The model for scaling the coefficients of the second group and the choice of similarity coefficients [20, 21], which is important, is discussed in detail in [17].

So, the next step is to define the classes of the uncertain users. For this purpose, the following actions are performed for each of the determined coefficients of the second group using algorithm 2, proposed in [18]:

1. Determination of similarity coefficients of the uncertain users with organic users from the formed database.
2. Determination of similarity coefficients of uncertain users with fraudsters from the formed database, which forms a set of coefficients, which values range from 0 to 1.
3. Determination of similarity coefficients of uncertain users with suspicious users from the formed database.
4. Combining the resulting sets of coefficients' values into one set of homogeneous values.

After obtaining a set of homogeneous values, which are normalized between zero and one, of all available data received using the proposed scales, we will use the data classification step using the well-known classification model (extreme gradient enhancement XGBoost, random forest, deep neural network, neural networks [22–28], etc.) in order to detect data that definitely identify fraudsters and data which definitely identify organic users.

6 Creation of Developing Knowledge Bases

In order to increase the efficiency, speed and accuracy of information technology for fraud detection, it is not enough to just classify users onto "fraudulent" or "organic" classes; it is still important to use knowledge bases based on the above methods, models and algorithms. In addition, it should be noted that in today's fraud detection systems there are problems, one of which is the lack of intelligent data analysis that prevents such systems from adapting to new types of fraud. In order for information technology to be able to adapt to new types and learn with the use of prediction algorithms, the authors of the work proposed a system with an intellectual component – the formation of a knowledge base that will allow fraudsters to be identified and which will include rules for analyzing anomalies, so that the appearance a new anomaly in the data allowed to create a new rule [29]. That is, such knowledge base can be expanded through the possibility of a new kind of anomaly in data (fraud) emergence. The received set of rules, which in the future, based on the algorithms developed in the work will allow to create a generalized fingerprint of the fraudster, noting even the new and unknown for experts fraudulent properties. The attribution of suspicious users to a class of fraudsters or organic occurs using fuzzy logic. Creation of developing knowledge bases, in turn, will allow determining the reason why the user was labeled as a fraudster. And it also will show when it is necessary to retrain the classification model.

7 Classification of Users Using Deep Neural Networks

The next step in the fraud detection method is the classification of the received sample of users into organic and fraudulent. But the development of the method to overcome heterogeneity of data was necessary since all classification models operate with homogeneous data. Therefore, having obtained the scaled and normalized homogeneous data in the range from 0 to 1 after the stage of overcoming heterogeneous data, any of the known classification models can be used. From the authors' point of view, it is convenient to use the classification model based on deep neural networks (DNN) to solve such a problem. It should be noted that the use of classification models to divide users into two distinct classes is a well-known and widespread approach. In our view, it is enough to use fully connected DNN with three hidden layers to solve such a problem. Training was made in 100 epochs, with dropout.

Let us carry out the training of the classification model on the sample from the developed mobile application [2, 30, 31] and using a method to convert heterogeneous data to homogeneous data. It should be noted that it is necessary to have three sets of data: training, validation, and control for the correct study and evaluation of the system, as well as to avoid re-training of the system. It should also be noted that, above all, the developed information technology must be endowed with predictive ability and well-summarized with the new data for it. The control sample contains 56962 entries, 56657 of which correspond to organic users, and 305 correspond to fraudsters. The samples are labeled, however, the label is used only to further assess the adequacy of the model. Deep neural networks with three hidden layers are selected as a classification model.

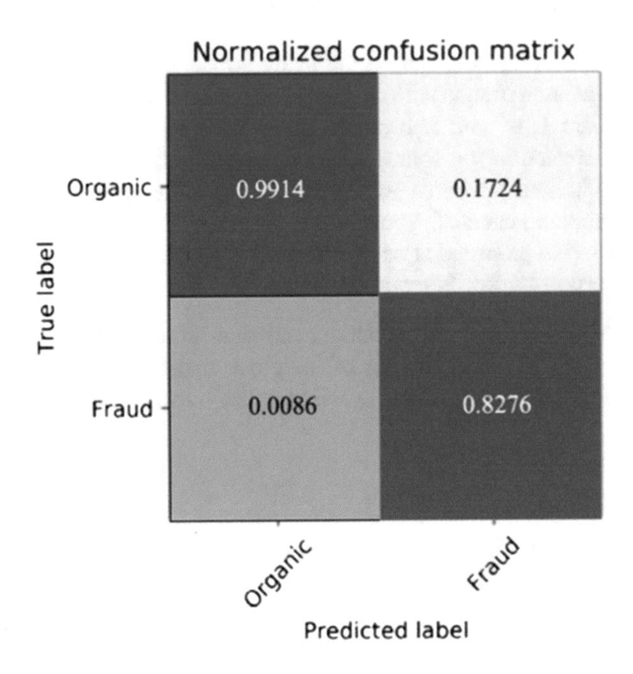

Fig. 2. Normalized confusion matrix of the control sample

The confusion matrix which is given in Fig. 3 shows that 80.43% of fraudsters are correctly classified and 19.57% are classified incorrectly. In general, 99.56% of objects (56639 + 74 = 56713 objects out of 56962 objects) are correctly classified, 0.44% of objects are incorrectly classified.

A generalized structure of intelligent information technology for fraud detection was developed based on the methods, models and algorithms proposed in the paper. It consists of the following subsystems: subsystem of analysis of available data; subsystem of intellectual processing of available data; subsystem of developing a database and knowledge base (for detecting fraudsters); classification model building and user classification subsystem; subsystem of forming users' templates; subsystem of the generalized fraudsters fingerprint formation. It should be noted that this is the basic set of subsystems of the proposed generalized structure of the intelligent information technology for fraud detection and extension for prediction, when and where the next fraudster will come is possible.

8 Conclusions

Thus, in the work, the intellectual information technology for fraud detection during the mobile applications installation is proposed and the following tasks are solved:

- data scaling based on the proposed binary scales to identify, analyze and use all available data of different templates, dimensions, metrics;
- the method of overcoming heterogeneity of data, algorithms 1 and 2 (scaling of the first and second groups) for the intellectual processing of available data on users is proposed;
- the classification of users is carried out using the deep neural networks model;
- databases containing fraudsters' characteristics, not only of humans, but also of various software bots, and knowledge bases with a set of fuzzy rules of fraud detection are created in the course of fraud detection. The main point is that such databases and knowledge bases evolve, depending on new data and new fraudsters, in order to create a general fingerprint of a fraudster;
- the adequacy of the proposed classification model on the sample from the developed mobile application [2, 30, 31] was evaluated.

All the above-considered method, model and algorithms developed for solving the problems highlighted at the beginning of the paper make up intelligent information technology for fraud detection during mobile applications installation.

References

1. Melnykova, N.: The basic approaches to automation of management by enterprise finances. In: 12th International Scientific and Technical Conference on Computer Sciences and Information Technologies (CSIT), Lviv, pp. 288–291 (2017). https://doi.org/10.1109/stc-csit.2017.8098788

2. Yarovyi, A., Polhul, T., Krylyk, L.: Rozrobka metodu vyiavlennia shakhraistva pry instaliuvanni mobilnykh dodatkiv z vykorystanniam intelektualnoho analizu danykh. In: Materialy konferentsiyi XLVII Naukovo-tekhnichna konferentsiya pidrozdiliv Vinnytskoho natsionalnoho tekhnichnoho universytetu, Vinnytsia (2018). http://ir.lib.vntu.edu.ua/bitstream/handle/123456789/22722/079.pdf?sequence=1
3. Our take on mobile fraud detection. http://geeks.jampp.com/data-science/mobile-fraud/
4. Dave, V., Guha, S., Zhang, Y.: ViceROI: Catching Click-Spam in Search Ad Networks. http://www.sysnet.ucsd.edu/~vacha/ccs13.pdf
5. Dave, V., Guha, S., Zhang, Y.: Measuring and fingerprinting click-spam in ad networks. In: Proceedings of the Annual Conference of the ACM Special Interest Group on Data Communication (SIGCOMM), Helsinki, Finland, pp. 175–186, August 2012
6. Yarovyi, A.A., Romanyuk, O.N., Arsenyuk, I.R., Polhul, T.D.: Program applications install fraud detection using data mining. Naukovi pratsi Donetskoho natsionalnoho tekhnichnoho universytetu. Seriya: "Informatyka, kibernetyka ta obchysliuvalna tekhnika", issue 2(25), pp. 126–131 (2017). http://science.donntu.edu.ua/wp-content/uploads/2018/03/ikvt_2017_2_site-1.pdf
7. Polhul, T.D., Yarovyi, A.A.: Vyznachennia shakhraiskykh operatsiy pry vstanovlenni mobilnykh dodatkiv z vykorystanniam intelektualnoho analizu danykh. Suchasni tendentsiyi rozvytku systemnoho prohramuvannia. Tezy dopovidei, Kyiv, pp. 55–56 (2016). http://ccs.nau.edu.ua/wp-content/uploads/2017/12/%D0%A1%D0%A2%D0%A0%D0%A1%D0%9F_2016_07.pdf
8. Polhul, T.D., Yarovyi, A.A.: Vyznachennia shakhraiskykh operatsiy pry instaliatsiyi mobilnykh dodatkiv z vykorystanniam intelektualnoho analizu danykh. Materialy XLVI naukovo-tekhnichnoi konferentsiyi pidrozdiliv VNTU, Vinnytsia (2017)
9. Kochava Uncovers Global Ad Fraud Scam. https://www.kochava.com/
10. TMC Attribution Analytics. https://help.tune.com/marketing-console/attribution-analytics/
11. Fraudlogix: Ad Fraud Solutions for Exchanges, Networks, SSPs & DSPs. https://www.fraudlogix.com/
12. Kraken Antibot. http://kraken.run/
13. AppsFlyer: Measure In-App To Grow Your Mobile Business. https://www.appsflyer.com/
14. Adjust. https://www.adjust.com/
15. FraudScore: FraudScore fights ad fraud using Machine Learning. https://fraudscore.mobi/
16. AppMetrica. https://appmetrica.yandex.ru/
17. Polhul, T., Yarovyi, A.: The input data heterogeneities resolution method during mobile applications installation fraud detection. Visnyk SNU named after V. Dal' – Severodonetsk: SNU named after V. Dal', № 7(248), pp. 60–69 (2018)
18. Polhul, T.: Development of an intelligent system for detecting mobile app install fraud. In: Proceedings of the IRES 156th International Conference, Bangkok, Thailand, 21–22 March, 2019, pp. 25–29
19. Polhul, T.D., Yarovyi, A.A.: Heterogeneous data analysis in intelligent fraud detection systems, № 2, pp. 78–90. Visnyk of Vinnutsia Polytechnic Institute, April 2019. https://doi.org/10.31649/1997-9266-2019-143-2-78-90
20. Polhul, T., Yarovyi, A.: Development of a method for fraud detection in heterogeneous data during installation of mobile applications. East. Eur. J. Enterp. Technol. 1/2(97) (2019). https://doi.org/10.15587/1729-4061.2019.155060
21. Segaran, T.: Programming Collective Intelligence. Building Smart Web 2.0 Applications. O'Reilly Media, Newton (2008). 368 p.

22. Kiulian, A.H., Polhul, T.D., Khazin, M.B.: Matematychna model rekomendatsiynoho servisu na osnovi metodu kolaboratyvnoi filtratsiyi. In: Kompiuterni tekhnolohiyi ta Internet v informatsiynomu suspilstvi, pp. 226–227 (2012). http://ir.lib.vntu.edu.ua/bitstream/handle/123456789/7911/226-227.pdf?sequence=1&isAllowed=y
23. Guido, S., Müller, A.: Introduction to Machine Learning with Python: A Guide for Data Scientists. O'Reilly Media, Newton (2016). 400 p.
24. Yarovyi, A., Ilchenko, R., Arseniuk, I., Shemet, Y., Kotyra, A., Smailova, S.: An intelligent system of neural networking recognition of multicolor spot images of laser beam profile. In: Proceedings of SPIE 10808, Photonics Applications in Astronomy, Communications, Industry, and High-Energy Physics Experiments 2018, vol. 108081, October 2018. https://doi.org/10.1117/12.2501691
25. Géron, A.: Hands-On Machine Learning with Scikit-Learn and TensorFlow: Concepts, Tools, and Techniques to Build Intelligent Systems. Aurélien Géron, O'Reilly Media, Newton (2017). 574 p.
26. Dong, X., Qiu, P., Lü, J., Cao, L., Xu, T.: Mining top-k useful negative sequential patterns via learning. IEEE Trans. Neural Netw. Learn. Syst. https://doi.org/10.1109/tnnls.2018.2886199
27. Kozhemyako, V., Timchenko, L., Yarovyy, A.: Methodological principles of pyramidal and parallel-hierarchical image processing on the base of neural-like network systems. Adv. Electr. Comput. Eng. **8**(2), 54–60 (2008). https://doi.org/10.4316/aece.2008.02010
28. Granik, M., Mesyura, V., Yarovyi, A.: Determining fake statements made by public figures by means of artificial intelligence. In: IEEE 13th International Scientific and Technical Conference on Computer Sciences and Information Technologies (CSIT), Lviv, pp. 424–427 (2018). https://doi.org/10.1109/stc-csit.2018.8526631
29. Polhul, T.D.: Information technology for the construction of intelligent systems for detecting fraud during mobile applications installation. Information Technologies and Computing Engineering, vol. 44, № 1, pp. 4–16, May 2019. https://doi.org/10.31649/1999-9941-2019-44-1-4-16
30. Cielen, D., Meysman, A.D.B., Ali, M.: Introducing Data Science: Big Data, Machine Learning, and More, Using Python Tools. Manning, New york (2016). 320 p.
31. Yarovyi, A.A., Polhul, T.D.: Kompiuterna prohrama "Prohramnyi modul zboru danykh informatsiynoi tekhnolohiyi" vyiavlennia shakhraistva pry instaliuvanni prohramnykh dodatkiv. Cvidotstvo pro reiestratsiu avtorskoho prava na tvir No. 76348. Ministerstvo ekonomichnoho rozvytku i torhivli Ukrainy, Kyiv (2018)
32. Yarovyi, A.A., Polhul, T.D.: Kompiuterna prohrama "Prohramnyi modul vyznachennia skhozhosti korystuvachiv informatsiynoi tekhnolohiyi vyiavlennia shakhraistva pry instaliuvanni prohramnykh dodatkiv". Cvidotstvo pro reiestratsiu avtorskoho prava na tvir No. 76347. Ministerstvo ekonomichnoho rozvytku i torhivli Ukrainy, Kyiv (2018)

MapReduce Hadoop Models for Distributed Neural Network Processing of Big Data Using Cloud Services

Natalia Axak, Mykola Korablyov(✉), and Dmytro Rosinskiy

Kharkiv National University of Radio Electronics, Nauky Ave. 14, Kharkiv 61166, Ukraine
{nataliia.axak,mykola.korablyov, dmytro.rosinskyi}@nure.ua

Abstract. The paper proposes a formalization process of Big Data distributed intelligent processing using Cloud-Fog-Dew architecture. This process provides specialized services, based on continuous support of experts in areas of concern, and advisory support for their actions in diagnostically complex cases. The method of Big Data processing based on neural networks is considered, which is distinguished by dynamic redistribution of work between computers, that allows to uniformly load the computing cluster with different data topologies. Proposed method is less by one order of computational complexity and less time spent. The MapReduce Hadoop models for distributed neural network processing of Big Data were proposed, characterized by the adaptation of data topology to the corresponding architectural computer cluster. This reduces the amount of information transmitted between nodes to increase productivity in solving complex tasks and effectively balancing the load of computing resources with different data topologies. An experimental Hadoop cluster was created to evaluate the performance of developed models for Big Data distributed processing. It allows for the implementation of parallel learning procedures for multilayer neural networks based on "star" and "fully connected graph" data topology with different amounts of input data.

Keywords: Cloud-Fog-Dew architecture · Distributed processing · MapReduce Hadoop · Model · Neural network · Service-oriented system

1 Introduction

Rapid development of embedded intelligent devices and computer networks spawned a wide variety of network applications and services such as [1] Internet of Things (IoT), Internet of Vehicles (IoV), Internet of Everything (IoE), Intelligent Planet, Smart City, Smart Network and Network Services. The popularity of Augmented reality (AR), Unmanned aerial vehicle (UAVs) and others is growing. This raises a relevant question how to meet all the requirements related to the rapid growth of new network applications and services using centralized computing paradigm. Nowaday conditions require to organize a new interaction between people and different systems as a operational response to real-time conversion.

N. Shakhovska and M. O. Medykovskyy (Eds.): CCSIT 2019, AISC 1080, pp. 387–400, 2020.
https://doi.org/10.1007/978-3-030-33695-0_27

Overall development of existing universal computer systems (UCS) can solve many problems of scientific, technological and other kinds [2]. However, there are extremely important connectivity and decision-making tasks that require a quick response to real-time events. To solve such problems today UCS features lacking. All this requires interaction between system and application software developers as well as end-users, and encourages the search and implementation of new solutions for a new class of specialized computer systems (SCS) which are service-oriented systems (SOS) using intelligent information processing. The SCS creation is primarily due to the conflict between the methods of formulating and solving difficultly formalized problems, on the one hand, and the technical capabilities of UCS, on the other.

The use of neural networks (NN) is of practical importance in every domain and has the advantages over traditional mathematical methods when [3]: the problem can not be formalized by traditional mathematical methods; the problem was formalized, but today there is no apparatus for its solution; there is mathematical apparatus corresponding well formalized task, but its implementation in existing computer systems do not satisfy the requirements to obtain solutions in time, size, weight, power and others. However, when the size of NN input increases the neural structure becomes more complex, which, in turn, leads to an increase in training time and a decrease in rate of convergence. At the same time, NN is a high-performance computer, since the algorithms developed for working with NN are parallel.

For decision-making, there is a huge variety of methods with varying levels of difficulty [4, 5]. The peculiarity of decisions is the accumulation and processing of huge amounts of information, which in turn refers to the Big Data concept. The problems of Big Data handling are associated with their manifold difficulties of acquisition, storage, management and analysis, memory size, and calculation speed [6–8]. Therefore, the task of creating intelligent SCS for Big Data distributed processing within a single technology of SOS is certainly relevant.

2 Formalization of Big Data Distributed Intelligent Processing

We proposed distributed intelligent processing of Big Data based on the Cloud-Fog-Dew architecture for personalized SOS (Fig. 1), which solves very important communication and decision-making tasks, and provides the expected operational response to real time conversion during human interaction with different computer systems. These problems are directly related to the methods of handling the growing volumes of data, and to the creation of computing devices that will solve such problems.

To ensure effective access to a web portal that provides a full range of services to a large number of users using all types of sensors and mobile devices, the approach is based on distributed Cloud-Fog-Dew computing (Fig. 1). Examples of issues to manage incidents include [9]: health monitoring, diagnosis and medical treatment and so on. The monitoring mechanism uses “graduation” of the patient, determining the ranges of measured values such as cardiogram, blood pressure, blood sugar level, body temperature, etc., which are divided into the following ranges: critical, high, medium, low.

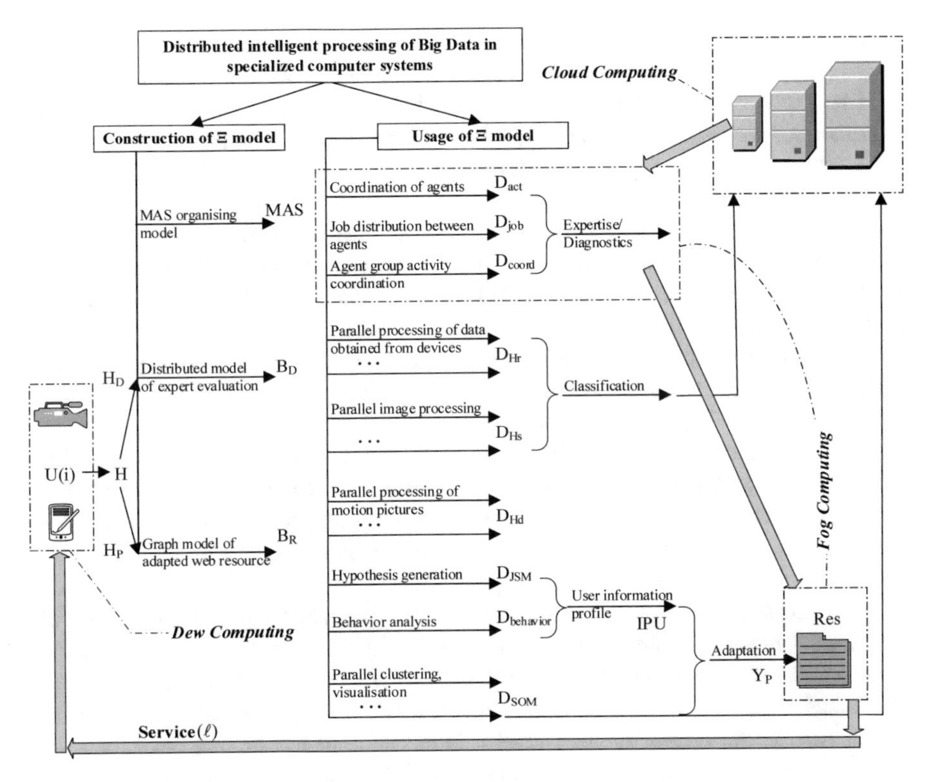

Fig. 1. The process of Big Data distributed intelligent processing in SCS.

In general, the principle of operation of the developed system for remote medical diagnostics and monitoring can be described as follows. A patient (who wishes to consult with a doctor, or needs the services of a nurse, etc.) after registering on medical web-site receives a unique identifier and special application will be installed on his mobile device. To collect information about the patient's condition, various medical and supplementary equipment can be used: diagnostic medical equipment, laboratory medical equipment (including based on biosensors, BioMEMS, LOCs), video surveillance cameras, photocameras, etc. This equipment is connected to the mobile device via wired or wireless digital interfaces. Mobile device sends data to the Fog-server, which in turn, aggregates and converts the readings of the devices into a form suitable for further processing.

Use of Cloud-Fog-Dew architecture allows to store a local copy of the application on user equipment and to work offline. When Internet is available, the state of user equipment is synchronized with the cloud. Slave-server deployed on local computer provides client with the same services as cloud server, synchronizes its own database with cloud servers database, serves only one user and retains only its data. The volume of this server is much smaller than the cloud server and is accessible to the user both with and without Internet connection. With this approach two issues raises. First, not always possible to store numerous data on a local computer according to the cloud data. Second, the user may need to remember local and cloud location of own data [10].

The problems solved by slave-server: (1) registering medical and diagnostic information D_{Hr}; (2) image processing of stationary objects D_{Hst}; (3) image processing of objects in motion D_{Hdn}; (4) generating the preliminary diagnosis using neural network classifier D_{Ξ}.

Slave-server with its associated databases have two functions: first, it provides clients the same services as the cloud Master server; secondly, it synchronizes its database with cloud database servers. It's lightweight web server that serves only one user and retains only its own data.

The main feature of Fog-server is its proximity to the end user, providing mobility of devices for different locations with minimal delay in data processing. It means that data is processed far from the sources of their formation without having to transfer them to the main center: a result of the task D_{Ξ} is available service $Service(H, Y) = Fasaa(\mathrm{T}(H_D, Y_D), \Pi(H_P, Y_P)$. When a patient's critical condition is detected (i.e., when emergency care is needed) data latency or the lack of Internet connection may occur. To solve this problem, the results of processed data from user's computer is transmitted to a mobile device through which the call to the hospital.

Thus, Dew-computing provides for acquisition and primary processing of user data on a local computer, mobile devices and so on. Fog-computing involves constructing models in form of multi-agent system (MAS) organizational models, expert evaluation distributed models, and graph models of adaptation and personalization of the web portal. Also at the Fog-computing level it is expected to use models for expert diagnostics based on the multi-agent approach, user classification and creation of their information profiles, etc. The Cloud-computing level dirernctly processes and stores information.

Let's provide formalization of Big Data distributed intelligent processing in SCS based Cloud-Fog-Dew architecture as follows (Fig. 1). Distributed intelligent processing of Big Data in SCS is the process providing specialized services $Service(\ell)$ which provides for continued support of experts in this subject area and consultative support for their actions in diagnostically difficult cases. Input data can be as follows: $H_D(i)$ which is object image, data from sensors, measuring devices, mobile devices etc.; $H_P(i)$ which describes profile of web service user $U(i)$ $(i = \overline{1, M})$ (i.e., name of browser, its version number, language, platform, built extensions, address, previous page, time zone, time of visit to the page, information about videomonitor etc.). Data is represented in matrix form: $H(i) = \{H_D(i), H_P(i)\}$, $(i = \overline{1, M})$.

Generally, the desired process can be presented as the development of:

- generalized model of problem-oriented computing Ξ, its constituents and their interaction;
- methodological framework for construction and operation of SCS, which provides each user $U(i)$ with effective access to service-oriented senvironment Res for customized $R(n)$ category of user to provide specialized services $Service(\ell)$ and allows to respond quickly to changing the object critical conditions.

The criterion of efficiency is to satisfy to the requirements:

$$k : \forall(D_i \in D_\Xi)[(\tau^a < \tau^{\max})\&(\tau^p < \tau^{\max})\&(\tau^r < \tau^{\max})\&(\varpi > \varpi^{\min})] \Rightarrow \Xi,$$

where D_i is a subtask of general task D_Ξ, τ^a is adapting time of local site, τ^p is processing time of diagnostic indicators, τ^r is response time, ϖ is a degree of relevance of displayed information.

3 MapReduce Models of Big Data Distributed Neural Processing

Processing and creation of Big Data is closely related to programming model MapReduce developed by Google [11]. MapReduce provides a secure, scalable and fault-tolerant computing environment for storing and processing massive data sets. At the same time, it is necessary to substantiate the choice of technology, among other MapReduce solutions for processing Big Data. The main selection criterion is the dimension of data, the scalability of system, the type of computing, the ability to perform tasks automatically to reduce the professional requirements to a specialist. However, the performance of this model is limited by hard configuration strategy. In practice, end users need to set up a simple MapReduce with a lot of parameters, which often leads to performance problems.

MapReduce computing model within map/reduce paradigm can be described by two main phases:

1. Map: (*key1*, *value1*) -> (*key2*, *value2*) is applied to each input pair (*key1*, *value1*) and displays a list of key/value intermediate pairs (*key2*, *value2*). It is posible also to select *Partition/Combain* stage. The purpose of *Partition* stage is the distribution of interim results obtained from Map phase on Reduce-tasks (*key2'*, *reduces_count*) -> (*reduces_id*) where *reduces_count* is the number of nodes which runs convolution operation, *reduces_id* is target node ID. The main objective of this stage is load balancing. Incorrectly implemented *Partition* function may lead to uneven distribution of data between Reduce-nodes.
2. Reduce: (*key2*, *value2*) -> *value3* applies to all intermediate values associated with the same key and creates a list of initial values. Programmers define the application logic using these two primitives. It should be noted that the implementation of Map function is carried independently and in parallel on different computers of cluster.

The most popular implementations of MapReduce model are Hadoop, Mars and Phoenix++. Hadoop is a project with open source managed by Apache Software Foundation and used for reliable, scalable and distributed computing, but can also be used as general purpose file storage able to accommodate petabytes of data [12]. Mars takes advantage of GPU on Nvidia G80. In the Phoenix project, MapReduce model used on computers with shared memory.

Thus, there is a need to develop methods and models for Big Data processing by neural networks that allow to schedule uniform load of computing cluster. Usage of MapReduce Hadoop programming model for neural network processing of Big Data allows to significantly accelerate the calculation with decreasing size of transmitted information between nodes and efficient load balancing of computing resources with different data transfer topologies.

3.1 Parallel Implementation of Multilayer Neural Network Using MapReduce Hadoop Model

Developing efficient parallel and distributed training and operation procedures of neural networks for solving tedious tasks is necessary to consider a large number of interdependent data:

- input parameters: neural network structure, the size of training and test samples, the number of training epochs;
- hardware specifications of distributed computing environment: the number of computers, physical environment topology, bandwidth, latency;
- algorithmic features: decomposition procedures, the nature of information interactions (data transfer topology).

The back-propagation neural network is a multilayer feed-forward network (multilayer perceptron – MLP) which studies the input data using back-propagation of errors [3]. During error propagation MLP continues to adjust the parameters until adapts to all incoming samples. Typical MLP consisting with any number of inputs is shown in Fig. 2.

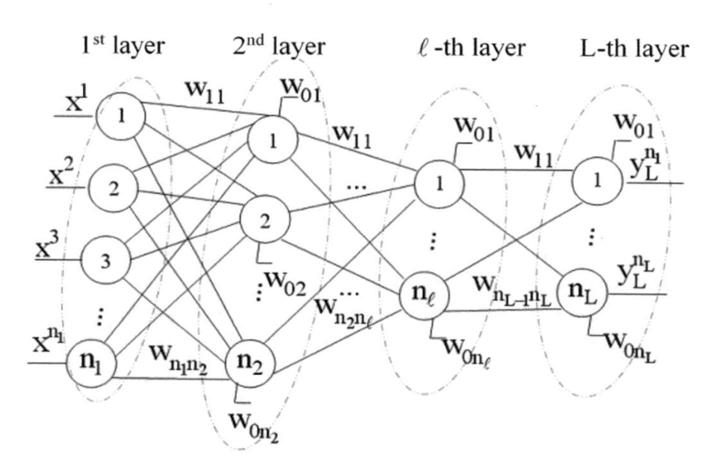

Fig. 2. Architecture of multilayer feedforward neural network.

In general, the MLP is defined by its architecture as follows: $n_1 - n_2 - \ldots - n_L$, where n_1 is number of input layer neurons; n_ℓ is number of neurons in hidden layers; n_L is number of output layer neurons. The study sample $\{(X(1), D(1)), (X(2), D(2)), \ldots, (X(K), D(K))\}$ consists of K examples $X(k) = [x^1(k), \ldots, x^{n_L}(k)]^T$ and values of target indicator $D(k) = [d^1(k), \ldots, d^{n_L}(k)]^T$, $k = \overline{1, K}$. To handle training and test samples U epochs are needed.

Vector of neuron outputs $Y_\ell(k) = [y_\ell^1(k), \ldots, y_\ell^{n_\ell}(k)]^T$ of layers $\ell = \overline{1, L}$ is calculated according to:

$$Y_\ell(k) = f\big(W_\ell(k)Y_{\ell-1}(k) + W_0^T(k)\big), \tag{1}$$

where $f(\bullet)$ is activation function; $W_\ell(k) = (w_{ij}(k))_{i=o,j=1}^{n_{\ell-1},n_\ell}$ is layer's weight matrix $\ell = \overline{2, L}$, $\Sigma_\ell(k) = [\delta_\ell^1(k), \ldots, \delta_\ell^{n_\ell}(k)]^T$ is neuron response error vector of output L-th layer:

$$\Sigma_L(k) = Y_L(k)(1 - Y_L(k))(D(k) - Y_L(k)) \tag{2}$$

and hidden layers $\ell = \overline{L - 1, 2}$:

$$\Sigma_L(k) = Y_\ell(k)(1 - Y_\ell(k))\Sigma_{\ell+1}(k)W_{\ell+1}(k). \tag{3}$$

Weights of layers $\ell = \overline{2, L}$ for neurons $j = \overline{1, n_\ell}$ are set as follows:

$$w_{ij}(k+1) = \eta\delta_\ell^j(k)y_\ell^j(k) + \alpha w_{ij}(k), \tag{4}$$

where η is training speed parameter; α is inertia factor.

Estimation of the current mean-square error E is performed as:

$$E = \frac{1}{2}\sum_{k=1}^{K}(D(k) - Y_L(k))^2. \tag{5}$$

Definition of generalization error during neural network testing is done as:

$$E_g = \frac{1}{K}\sum_{t=1}^{T}\Delta_t, \tag{6}$$

where Δ_t is calculated as:

$$\Delta_t = \begin{cases} 1 \text{ if } D(t) \neq y(t), \\ 0 \text{ if } D(t) = y(t). \end{cases}$$

To implement MLP we use Hadoop cluster with Hadoop Distributed File System (HDFS) which provides high-speed application data access to handle this data. In the Hadoop cluster is a single name node (Namenode) and multiple data nodes (Datanodes) to perform jobs. Namenode manages cluster metadata, Datanodes is an actual processing node. Card functions (mappers) and compression functions (reductors) performed on the data nodes. When the task is sent to a Hadoop cluster, the input data is divided into small pieces of the same size and stored in HDFS.

To reduce MLP's training and operating time, as well as for efficient use of computing resources, the dynamic adjustment of formed part of subtasks is required: if there is a small number of computers, size of subtasks increases, elsewhere decreases. For uniform load of cluster machines, the neurons of each layer MLP are distributed between groups for parallel processing considering cluster architecture ("star" or "fully connected graph" [13]). For each topology, there was developed a model of multilayer neural network parallel implementation based on MapReduce (MRMLP).

3.2 Model of Uniform Distribution of Neural Processing in a "Star" Topology Computing Environment

Each training sample and information defined by format (7) are stored in a single HDFS file. The HDFS catalog can locate any number of files. For each training case the relations (7)–(16) to be performed until the desired accuracy is reached in accordance with (5) and (6). The first time the Map function (Mapping process 1) initializes one entry from HDFS file in the following format:

$$\langle KeyM1_h, X(k), W_{\ell_h}(k), D_h(k)\rangle, \tag{7}$$

where $KeyM1_h$ is h-th launch of Map function, $h = \overline{0, P-1}$; $W_{\ell_h}(k)$ is weight sub-matrix of matrix $W_\ell(k) = \lfloor W_{\ell_1}(k), W_{\ell_2}(k), \ldots, W_{\ell_{p-1}}(k)\rfloor$, with dimension $n_{\ell-1} \times |g_h|$ ($\ell = \overline{0, L-1}$); $D_h(k)$ are groups of target indicators values $D(k) = [D_1(k), D_2(k), \ldots, D_{p-1}(k)]^T$.

The function output provides for calculated g_h-th outputs of neurons using Eq. (1):

$$\langle KeyM1_h, Y_{\ell_h}(k), \{W_{\ell_h}(k), D_h(k)\}\rangle, \tag{8}$$

where $Y_{\ell_h}(k)$ are sets of neurons $Y_\ell(k) = [Y_{\ell_1}(k), Y_{\ell_2}(k), \ldots, Y_{\ell_{p-1}}(k)]^T$ ($\ell = \overline{2, L}$); g_h is increasing sequence calculated according to the uniform distribution of a "star" topology neural processing [13]:

$$g_h = \begin{cases} (h-1)\left(\left[{}^{n_\ell}/_{P-1}\right]+1\right)+\varphi,\ \varphi = \overline{1, \left(\left[{}^{n_\ell}/_{P-1}\right]+1\right)}, & \text{if } 1 \le h \le b; \\ (P-1-h_d)\left(\left[{}^{n_\ell}/_{P-1}\right]+1\right)+(h-P-1+b-1)\left[{}^{n_\ell}/_{P-1}\right]+\varphi,\ \varphi = \overline{1, \left[{}^{n_\ell}/_{P-1}\right]} & \text{if } b \le h \le P-1; \end{cases} \tag{9}$$

whose elements are numbers of neurons in layers $\ell = \overline{2, L}$ are processed by h-th processor; $b = n_\ell \bmod (P-1)$ is a number of computer after which for $h_d = P - 1 - b$ computers, since $(b+1)$-th processor, amount of processed neurons is reduced by one for uniform load.

The Reduce function (Reducing process 2) which is associated with $KeyM1_h$ collects all $Y_{\ell_h}(k)$ results ($h = \overline{0, P-1}$):

$$\langle KeyR2_0, Y_\ell(k), \{Y_{\ell_h}(k)\}\rangle. \tag{10}$$

Next launch of Map function (Mapping process 3) generates local errors for g_h-th sets of output layer neurons according to (2):

$$\langle KeyM3_h, \Sigma_{L_h}(k), \{Y_\ell(k), D_h(k)\}\rangle, \tag{11}$$

where $\Sigma_{L_h}(k)$ are groups of neuron reaction error $\Sigma_L(k) = [\Sigma_{L_1}(k), \Sigma_{L_2}(k), \ldots, \Sigma_{L_{p-1}}(k)]^T$. Accordingly, the Reduce function (Reducing process 4) form a common vector $\Sigma_L(k)$:

$$\langle KeyR4_0, \Sigma_L(k), \{\Sigma_{L_h}(k)\}\rangle. \tag{12}$$

Based (3) the Map function (Mapping process 5) determines local errors $\Sigma_{\ell_h}(k)$ for g_h-th sets of neurons in hidden layers $\ell = \overline{L-1, 2}$ to create vectors $\Sigma_\ell(k)$:

$$\langle KeyM5_h, \Sigma_{\ell_h}(k), \{Y_\ell(k), \Sigma_{\ell+1}(k), W_{\ell+1}(k)\}\rangle, \tag{13}$$

where $\Sigma_\ell(k) = \left[\Sigma_{\ell_1}(k), \Sigma_{\ell_2}(k), \ldots, \Sigma_{\ell_{p-1}}(k)\right]^T$ ($h = \overline{1, P-1}$).

Further the Reduce function (Reducing process 6) forms a general vector $\Sigma_\ell(k)$:

$$\langle KeyR6_0, \Sigma_\ell(k), \{\Sigma_{\ell_h}(k)\}\rangle. \tag{14}$$

Element adjustment of submatrices $W_{\ell_h}(k+1)$ of matrices $W_\ell(k+1)$ of layers $\ell = \overline{2, L}$, with dimension $n_{\ell-1} \times |g_h|$ according to (4) is implemented via Map function (Mapping process 7):

$$\langle KeyM7_h, W_{\ell_h}(k+1), \{Y_\ell(k), W_\ell(k)\}\rangle. \tag{15}$$

Addition of matrices $W_\ell(k+1)$ is implemented via Reduce function (Reducing process 8):

$$\langle KeyR8_0, W_\ell(k+1), \{W_{\ell_h}(k+1)\}\rangle. \tag{16}$$

Calculating the E mean square training error is based on (5).

Graph model of neural processing based on MapReduce in a distributed environment with a "star" topology is shown in Fig. 3.

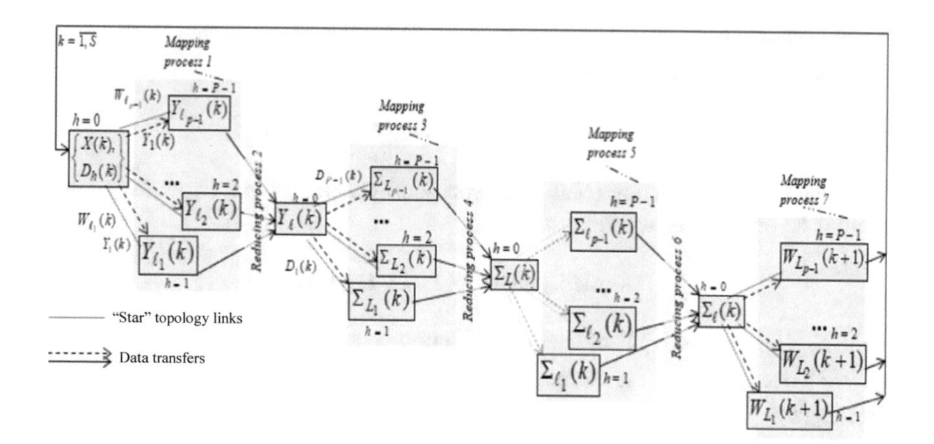

Fig. 3. MapReduce model for parallel neuroprocessing in a distributed environment with "star" topology.

The graph vertices correspond to independent parallel operations, and directed arcs – to the transfers of data between the computers (the results of Map operations are passed as arguments for the Reduce operation).

3.3 Moded of Uniform Distribution of Neural Processing in a "Fully Connected Graph" Topology Computing Environment

Each training sample and information defined by format (17) are stored in a single HDFS file. For each training case the relations (17)–(22) to be performed until the desired accuracy is reached in accordance with (5) and (6). The first time the Map function (Mapping process 1) initializes one entry from HDFS file in the following format:

$$\langle KeyM1_h, X(k), W_{\ell_h}(k), D(k)\rangle, \tag{17}$$

where $KeyM1_h$ is h-th launch of Map function, $h = \overline{0, P-1}$; $W_{\ell_h}(k)$ is submatrix of matrix $W_\ell(k) = \lfloor W_{\ell_0}(k), W_{\ell_1}(k), \ldots, W_{\ell_{p-1}}(k)\rfloor$ $(\ell = \overline{2, L})$, with dimension $n_{\ell-1} \times |r_h|$, for k-th current training sample values of target indicator $D(k) = [D_0(k), D_1(k), \ldots, D_{p-1}(k)]^T$ according to the set of neurons in r_h increasing sequence of uniform distribution of "fully connected graph" neural network topology [13]:

$$r_h = \begin{cases} h(\lceil n_\ell/P \rceil + 1) + \varphi,\ \varphi = \overline{1, (\lceil n_\ell/P \rceil + 1)}, & \text{if } 0 \le h \le b; \\ (P - h_d)(\lceil n_\ell/P \rceil + 1) + (h - P + b)\lceil n_\ell/P \rceil + \varphi,\ \varphi = \overline{1, \lceil n_\ell/P \rceil}, & \text{if } b \le h \le P-1; \end{cases}, \tag{18}$$

whose elements are numbers of neurons in layers $\ell = \overline{2, L}$ are processed by h-th processor; $b = n_\ell \bmod (P)$ is a number of computer after which for $h_d = P - b$

computers, since $(b+1)$ th processor, amount of processed neurons is reduced by one for uniform load.

As result of this function r_h-th neuron outputs will be calculated using Eq. (1):

$$\langle KeyM1_h, Y_{\ell_h}(k), \{W_{\ell_h}(k), D_h(k)\}\rangle, \tag{19}$$

where $Y_{\ell_h}(k)$ are sets of neurons $Y_\ell(k) = \left[Y_{\ell_0}(k), Y_{\ell_1}(k), \ldots, Y_{\ell_{p-1}}(k)\right]^T$, $\ell = \overline{2, L}$, $h = \overline{0, P-1}$.

The Reduce function (Reducing process 2_h $(h = \overline{0, P-1})$) simultaneously measures the partial values of local errors $\Sigma_L(k) = \left[\Sigma_{L_0}(k), \Sigma_{L_1}(k), \ldots, \Sigma_{L_{p-1}}(k)\right]^T$ for r_h-th sets of output layer neurons according to (5):

$$\left\langle \left\{ \begin{array}{l} KeyR2_0, \Sigma_{L_0}(k), \left\{Y_{\ell_h}, D_h(k)\right\}, \\ KeyR2_1, \Sigma_{L_1}(k), \left\{Y_{\ell_h}, D_h(k)\right\}, \\ \ldots, \\ KeyR2_{P-1}, \Sigma_{L_{P-1}}(k), \left\{Y_{\ell_h}, D_h(k)\right\} \end{array} \right\} \right\rangle. \tag{20}$$

The next call of Reduce function (Reducing process 3_h $(h = \overline{0, P-1})$) calculates (using (3)) local errors $\Sigma_\ell(k) = \left[\Sigma_{\ell_0}(k), \Sigma_{\ell_1}(k), \ldots, \Sigma_{\ell_{p-1}}(k)\right]^T$ for r_h-th sets of hidden layers neurons:

$$\left\langle \left\{ \begin{array}{l} KeyR3_0, \Sigma_{\ell_0}(k), \left\{\Sigma_{L_h}(k), Y_{\ell_h}, \Sigma_{\ell+1_h}(k), W_{\ell+1}(k)\right\}, \\ KeyR3_1, \Sigma_{\ell_1}(k), \left\{\Sigma_{L_h}(k), Y_{\ell_h}, \Sigma_{\ell+1_h}(k), W_{\ell+1}(k)\right\}, \\ \ldots, \\ KeyR3_{P-1}, \Sigma_{\ell_{P-1}}(k), \left\{\Sigma_{L_h}(k), Y_{\ell_h}, \Sigma_{\ell+1_h}(k), W_{\ell+1}(k)\right\} \end{array} \right\} \right\rangle \tag{21}$$

The Reduce functions (Reducing process 4_h $(h = \overline{0, P-1})$) form the elements of submatrices $W_{\ell_h}(k+1)$:

$$\left\langle \left\{ \begin{array}{l} KeyR4_0, W_{\ell_0}(k+1), \left\{\Sigma_{L_h}(k), Y_{\ell_h}, \Sigma_{\ell+1_h}(k), W_{\ell+1}(k)\right\}, \\ KeyR4_1, W_{\ell_1}(k+1), \left\{\Sigma_{L_h}(k), Y_{\ell_h}, \Sigma_{\ell+1_h}(k), W_{\ell+1}(k)\right\}, \\ \ldots, \\ KeyR4_{P-1}, W_{\ell_{P-1}}(k+1), \left\{\Sigma_{L_h}(k), Y_{\ell_h}, \Sigma_{\ell+1_h}(k), W_{\ell+1}(k)\right\} \end{array} \right\} \right\rangle.. \tag{22}$$

The MapReduce graph model of parallel neyroprocessing in a “fully connected graph” distributed environment is shown in Fig. 4.

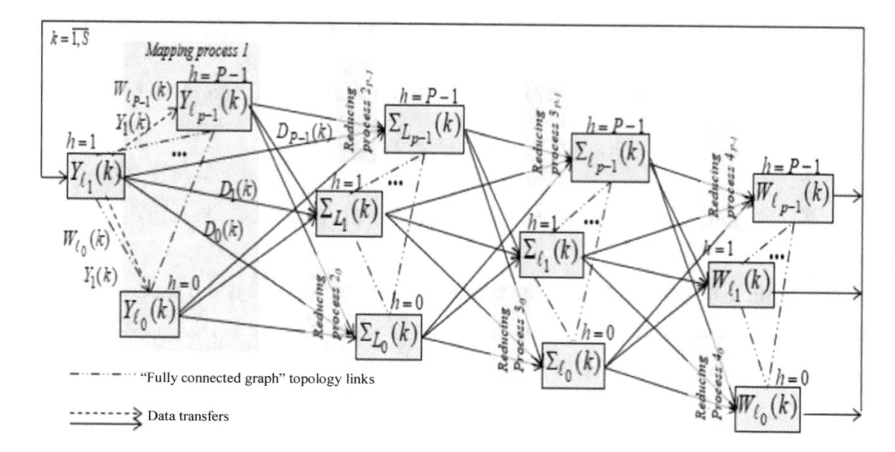

Fig. 4. MapReduce model for parallel data neuroprocessing in a distributed environment with "fully connected graph" topology.

3.4 Simulation of Big Data Distributed Processing

To evaluate the efficiency of algorithms an experimental Hadoop (ver. 2.7.5) cluster was created. The cluster consists of 5 computers (Intel Core 2 Quad CPU Q8200 @2.33 Ghz, Microsoft Windows 10 OS) connected by "star" physical topology: four nodes are Datanodes, one is Namenode. Jobs were managed by YARN. When installing the cluster servers very hard requirements for memory and hard drive are considered. This is what limited the size of processed data.

Testing was conducted to solve the task of classifying hand printed characters using multilayer neural networks. The choice of multilayer neural network structures implemented experimentally as a compromise between achieving the training mean square error values (E < 0.0001) and the maximum possible reduction of generalization error. Generalization error when comparing the desired value and class using MLP was 0.23% of test samples in accordance with (6), which is a good indicator. The results of experiments in which parallel training procedures were implemented in MRMLP with data transfer topologies "star" and "fully connected graph" with different amounts of input data are presented in Table 1.

Table 1. MRMLP parallel training in different topology distributed environment.

Size of training set (samples)	Computational time (seconds)		
	Sequential implementation	Cluster based on topology	
		"star"	"fully connected graph"
650000	923	219	226
1050000	3613	759	892
2460000	5431	925	1052

Usage of MLP parallel training with MapReduce allowed to reduce the computing time with an increase in the training and test samples. The uniform separation of parallel code parts between comuters significantly increased the procedure efficiency.

It was defined that more data is transferred using "fully connected graph" topology; the minimum number of transfers is achieved when using "star" topology. It means that an increase in MLP structure and size of training set for the fastest tasks solving it is expedient to apply the MRMLP training based on a data transfer topology that matches the physical. MRMLP method computational complexity evaluation has confirmed that the proposed method has a smaller one-order computational complexity and less time costs compared to sequential computations.

4 Conclusion

The paper solved the problem of Big Data distributed neural processing based on cloud services and using MapReduce programming model. A formalization of distributed intelligent processing of Big Data using Cloud-Fog-Dew architecture for personalized service-oriented systems is proposed. Use of Cloud-Fog-Dew architecture allows to store a local copy of the application on user equipment and to work offline. To ensure effective access to a web portal that provides a full range of services to a large number of users, an approach based on distributed Cloud-Fog-Dew computing has been implemented.

MapReduce Hadoop model for neural network distributed processing of Big Data characterized by adapting data transfer topology to the appropriate neural network architecture on computing cluster is proposed. The parallel implementation of multi-layer neural network using MapReduce Hadoop model is proposed; basing on this model the uniform distribution models of neural network data processing in the computing environment with "star" and "fully connected graph" topologies are developed. Application of MapReduce Hadoop programming model for Big Data neural network processing can significantly speed up the calculation with decreasing volume of transmitted information between nodes and efficient load balancing of computing resources with different data transfer topologies.

Experimental Hadoop cluster was implemented to estimate the developed model effectiveness. This helped to implement parallel training procedures for multilayer topologies of neural networks with different data transfer topologies and different volumes of input data. The use of parallel training procedures for multilayer neural networks based on MapReduce has reduced the computing time with an increase in volume of the training and test samples.

References

1. Marjani, M.: Big IoT data analytics: architecture, opportunities, and open research chasllenges. IEEE Access **5**, 5247–5261 (2017)
2. Nikolaychuk, Ya.M.: Specialized computer technologies in informatics: monography. In: Nikolaychuk, Ya.M. (ed.) Beskydy, Ternopil, 919 p. (2017). (in ukrainian)

3. Haykin, S.: Neural Networks and Learning Machines, 3rd edn., 937 p. Pearson Education (2009)
4. Bogdanova, V.G.: Multiagent approach to controlling distributed computing in a cluster grid system. J. Comput. Syst. Sci. Int. **53**, 713–722 (2014)
5. Lützenberger, M.: Multi-agent system in practice: when research meets reality. In: Proceedings of the 2016 International Conference on Autonomous Agents & Multiagent Systems. International Foundation for Autonomous Agents and Multiagent Systems, pp. 796–805 (2016)
6. Fernandes, L.M., O'Connor, M., Weaver, V.: Big data, bigger outcomes. J. AHIMA **83**(10), 38–43 (2012)
7. Patibandla, R.S.M.L., Veeranjaneyulu, N.: Survey on clustering algorithms for unstructured data. In: Intelligent Engineering Informatics, pp. 421–429. Springer, Singapore (2018)
8. Shirkhorshidi, A.S.: Big data clustering: a review. In: International Conference on Computational Science and Its Applications, pp. 707–720. Springer, Cham (2014)
9. Chen, H.: Smart health and wellbeing [Trends & Controversies]. IEEE Intell. Syst. **26**(5), 78–90 (2011)
10. Axak, N., Rosinskiy, D., Barkovska O., Novoseltsev, I.: Cloud-fog-dew architecture for personalized service-oriented systems. In: The 9th IEEE International Conference on Dependable Systems, Services and Technologies, DESSERT 2018, Kyiv, pp. 80–84 (2018)
11. Dean, J., Ghemawat, S.: MapReduce: simplified data processing on large clusters. Commun. ACM **51**(1), 107–113 (2008)
12. Ghemawat, S., Gobioff, H., Leung, ST.: The Googlefile system. SIGOPS Oper. Syst. Rev. **37,** 29–43 (2003)
13. Axak, N.G., Lebyodkina, A.Yu.: Method of uniform distribution of parallel operatrions for the multy-layer neural network accellerated learning based on different data topologies. Control. Navig. Commun. Syst. **2**(18), 66–73 (2011). (in Russian)

A New Approach to Image Intensity Transformation Based on Equalizing the Distribution Density of Contrast

Sergei Yelmanov[1(✉)] and Yuriy Romanyshyn[2,3]

[1] Special Design Office of Television Systems, Lviv, Ukraine
sergei.yelmanov@gmail.com
[2] Lviv Polytechnic National University, Lviv, Ukraine
yuriy.romanyshynl@gmail.com
[3] University of Warmia and Mazury, Olsztyn, Poland

Abstract. Intensity transformation is one of the basic approaches to enhance the image. However, known methods of intensity transformation have several disadvantages which significantly limit their use for image processing in automatic mode. In this paper, the problem of improving the efficiency of the intensity transformation of complex images in the automatic mode was considered. A new approach to the intensity transformation was proposed based on equalizing the distribution density of contrast in an image. The distribution of contrast is estimated based on bivariate distribution and brightness increments for pairs of pixels in the image. A new generalized description of the intensity transformation based on the joint distribution of brightness was proposed. It was shown that the traditional definition of histogram equalization is a particular case of the proposed generalized description. A new technique of parameter-free intensity transformation was proposed based on equalizing the distribution density of contrast in an image. The proposed technique provides an effective enhance the contrast of complex images without the appearance of unwanted artifacts has several advantages to the well-known histogram equalization technique. The results of experimental research confirm the effectiveness of the proposed approach to enhance images in automatic mode.

Keywords: Image enhancement · Intensity transformation · Brightness increment · Bivariate · Distribution density · Histogram equalization · Contrast

1 Introduction

The widespread use of video in various applications is one of the major trends in information technology.

Extensive use of video content (video and photo images) requires solving the problem of effectively enhancing source raw images in automatic (parameter-free) mode. The problem of enhancing the images in real-time applications is particularly acute now.

There are a large number of different approaches to enhancing the images [1, 2].

N. Shakhovska and M. O. Medykovskyy (Eds.): CCSIT 2019, AISC 1080, pp. 401–420, 2020.
https://doi.org/10.1007/978-3-030-33695-0_28

However, of particular interest for use in real-time applications are methods of image processing in the spatial domain [3, 4].

Methods of image processing in the spatial domain are based on direct manipulation of pixels in an image and are generally very efficient computationally and simple to implement [1, 2].

Known methods of processing in the spatial domain generally divided into two main groups, such as spatial filtering and intensity transformations [2]. Spatial filtering performs operations in a sliding neighborhood for every pixel in an image. An intensity transformation is the simplest case of processing in the spatial domain and operates on single pixels of an image when the neighborhood has size 1×1 pixel.

Intensity transformation methods are very widely used for enhancing the images and increasing their contrast [5, 6]. Generally, methods of intensity transformation are more efficient computationally and simple to implement, and are very widely used to enhance images in real-time applications.

However, the vast majority of known intensity transformation methods have several significant disadvantages that significantly limits their use in automatic modes, such as excessively high gain the contrast of large extended objects, reducing the contrast of objects with a small size, the emergence of unwanted artifacts in the image, inefficiency of transforming the images with full dynamic range, the need to adjust the parameters of transformation in interactive mode, etc.

In this paper, the problem of improving the efficiency of the intensity transformation of complex images in the automatic mode is considered.

The purpose of this work is to increase the efficiency of transforming the intensity of complex multi-element images.

2 Intensity Transformation

Intensity transformation is the simplest case of processing in the spatial domain when the neighborhood is of size 1×1 pixel (a single pixel) [2].

Intensity transformation is among the simplest of all image processing techniques and is more efficient computationally and simple to implement [1, 2].

In general, the procedure of intensity transformation most often defined as [3]:

$$r = r_{\text{low}} + \left(r_{\text{upp}} - r_{\text{low}}\right) \cdot T(b), \tag{1}$$

$$0 \le b_{\min} \le b \le b_{\max} \le 1, \text{ and } 0 \le r_{\text{low}} < r_{\text{upp}} \le 1, \tag{2}$$

where b is the brightness of a current pixel in the source image B; r is the brightness of a current pixel in the transformed image R; r_{low} and r_{upp} are the lower and upper boundaries of values range in the image R; $T(b)$ is the function of intensity transformation; b_{min} and b_{max} are minimal and maximal brightness of pixels in the image B.

It is usually assumed that the transformed image is normalized to the range [0, 1], that is, $r_{low} = 0$ and $r_{upp} = 1$ [2].

In this case, the transformation (1) takes the form [2]:

$$r = T(b). \tag{3}$$

In (1) it is assumed that the transformation $T(b)$ satisfies the following conditions:

(1) $T(b)$ is well-defined and unambiguous (single-valued) on the interval [0, 1]:

$$T : B \mapsto R,\ T : r = T(b),\ b \in [0,\ 1]; \tag{4}$$

(2) the range of possible values for $T(b)$ coincides with its domain of definition:

$$0 \le T(b) \le 1,\ \forall b | 0 \le b_{\min} \le b \le b_{\max} \le 1; \tag{5}$$

(3) $T(b)$ is a monotonically increasing function in the interval $[b_{\min}, b_{\max}]$ of possible values of brightness in the source image B:

$$\forall b_i, b_j \in B :\ if\ b_i > b_j \Rightarrow T(b_i) > T\left(b_j\right); \tag{6}$$

(4) $T(b)$ takes extreme values in the interval [0, 1] with minimum and maximum values of brightness in the source image:

$$T(b_{\min}) = 0,\ \text{and}\ T(b_{\max}) = 1. \tag{7}$$

Known intensity transformation techniques are usually divided into several main groups, such as basic intensity transformations [1, 2], contrast stretching techniques [1, 6–9], and histogram equalization based techniques [10].

The most well-known technique of intensity transformation is the histogram equalization in which the brightness in the transformed image is distributed evenly.

Another well-known technique is the equalization of sub-histograms of the image [11–15]. In this case, the dynamic range of image separates into several non-intersecting sub-ranges, in each of which then equalizes its sub-histogram.

Histogram equalization and its modifications are very widely used to enhance images and increase their contrast in automatic mode.

Histogram equalization is one of the core procedure in image pre-processing and characterized by high efficiency, low computational cost, simplicity to implement, and is now very popular.

Let us look more detail on the main features of the procedure of histogram equalization.

2.1 Histogram Equalization Technique

Histogram equalization is best known technique of intensity transformation [2, 10–12] and is very widely used to enhance images in automatic mode.

Equalization is a special case of the procedure of histogram specification.

Histogram specification (or histogram matching) is a statistical non-inertial transformation at which the distribution of brightness in the transformed image takes a specified shape and corresponds to the known distribution for the reference image.

In histogram equalization it is assumed that the brightness of the transformed image has an evenly distribution.

In the general case, the procedure of equalizing the density of distribution is defined as:

$$r = T(b) = \Pr(B \le b), \tag{8}$$

where $\Pr(\cdot)$ is probability of an event.

The most well-known definition of histogram equalization has the form:

$$r = HE(b) = F(b) = \int_0^b f(x)\,dx, \tag{9}$$

where $F(b)$ is cumulative distribution function for brightness b in the source image B; $f(b)$ is probability density function of brightness, and:

$$F(b) = \Pr(\{\omega\} \in \Omega : B(\omega) \le b), \tag{10}$$

$$f(b) = \frac{d}{db} F(b), \tag{11}$$

where Ω is the sample space (the total area of the source image); ω is the current image point in the source image.

From (10) it follows that the traditional definition of histogram equalization is based on an analysis of the probability distribution of the brightness of points (pixels) of an image and does not take into account the relationships between them.

However, most of the image characteristics are based on the joint bivariate distribution of the brightness of the pairs of image pixels.

An analysis of the joint distribution of brightness of pixels in an image can significantly increase the efficiency of transforming its intensity.

Consider the set of all pairs of points in the image:

$$S = \{(\omega', \omega'') : \omega' \in \Omega \wedge \omega'' \in \Omega\}, \tag{12}$$

where S is the set of all pairs (ω', ω'') of points (pixels) in the source image.

Based on (12), the bivariate distribution of brightness for the pairs of pixels in the image is generally defined as:

$$F(b_1, b_2) = \Pr((\omega', \omega'') \in S : B(\omega') \le b_1 \text{ and } B(\omega'') \le b_2), \tag{13}$$

$$f(b_1, b_2) = \frac{d^2}{db_1 db_2} F(b_1, b_2), \tag{14}$$

where $F(x_1, x_2)$ is the joint cumulative distribution function (joint CDF) of the brightness of the pairs of image pixels; $f(x_1, x_2)$ is the joint probability density function (joint pdf).

In this case, (10) and (11) are the marginal distribution, which, taking into account (13) and (14), can be defined as:

$$f(b) = \int_0^1 f(x_1, b)\, dx_1 = \int_0^1 f(b, x_2)\, dx_2, \tag{15}$$

$$F(b) = F(1, b) = F(b, 1). \tag{16}$$

Based on (15) and (16), the traditional definition (9) of histogram equalization can be represented in the form:

$$HE(b) = \int_0^1 f(x_1, b)\, dx_1 = \int_0^1 f(b, x_2)\, dx_2, \tag{17}$$

$$HE(b) = F(b, 1) = F(1, b). \tag{18}$$

From of (17) and (18), it follows that the traditional definition (9) of the histogram equalization is based on a marginal distribution and ignores the joint distribution of brightness in an image.

Histogram equalization is the most well-known technique of intensity transformation.

Histogram equalization and its modifications are very widely used to enhance contrast and improve visual perception of images at the stage of their pre-processing.

Histogram equalization is highly efficient, effective computationally, and simple to implement.

However, histogram equalization has several disadvantages that limit its use for image processing in automatic mode, in particular, a reduction in contrast of objects with small sizes, excessively high gain of contrast for large extended objects, and the emergence of unwanted artifacts in the image, etc.

All of these disadvantages are due that histogram equalization is based on an analysis of the marginal distribution of brightness and does not take into account the statistical interrelationships between the brightness values of the pixels in the image.

This is a significant drawback, which considerably reduces the efficiency of image intensity transformation.

This drawback is inherent in the overwhelming majority of known techniques of intensity transformation based on the analysis of the image histogram.

Analyzing the statistical interrelationships between the brightness values of pixels in an image allows to eliminate the existing disadvantages, and to increase the efficiency of transforming the intensity in an image.

This paper proposes a new approach to intensity transformation based on the analysis of the distribution of brightness increments of pairs of pixels in an image.

3 Proposed Approach

This paper addresses the problem of image enhancement by intensity transformation in automatic mode. The purpose of this work is to improve the efficiency of transforming the intensity of complex multi-element images.

In this paper, a new approach to the intensity transformation is proposed based on equalizing the distribution density of contrast for pairs of pixels in an image.

The distribution of contrast is estimated based on an analysis of bivariate distribution and brightness increments for pairs of pixels in the image.

This approach provides an increase in the efficiency of transforming the intensity through a more evenly enhance and redistribution of the contrast of objects in the image without the appearance of unwanted artifacts.

In this paper, we propose a new generalized description of the intensity transformation based on an analysis of the joint distribution of brightness and the magnitude of brightness increments of pairs of pixels in an image.

Based on the proposed approach, a new technique of intensity transformation is proposed based on equalizing the distribution density of contrast in an image.

Assume that $T(b)$ (1) is a statistical non-inertial transformation, which has the form:

$$T(b) = \int_{0}^{b} t(x)\, dx, \tag{19}$$

where $t(b)$ is a distribution density of contrast distribution.

An assessment of distribution density $t(b)$ has the form:

$$t(b) = \alpha \cdot \int_{0}^{b} \int_{b}^{1} \Delta(x, y) \cdot f(x, y) \cdot \xi_b(x, y)\, dxdy, \tag{20}$$

$$\alpha = \left[\int_{0}^{1} t(b)\, db \right]^{-1}, \tag{21}$$

where α is a normalizing factor; $\Delta(x, y)$ is the magnitude of the brightness increment of two pixels with brightness values of x and y; $f(x, y)$ is the density of bivariate probability distribution of brightness; $\xi_b(x, y)$ is the value of distribution density of contrast in the point b from interval $[x, y]$.

In the general case, the distribution density $\xi_b(x, y)$ is defined as:

$$\xi_b(x, y) = \Pr\left(z = b | z \in [x, y]\right). \tag{22}$$

Expressions (1) and (19)–(22) are a generalized description of the procedure of intensity transformation by equalizing the distribution density of contrast.

To demonstrate the proposed approach, assume that the distribution density $\xi_b(x, y)$ is uniformly distributed on the interval (x, y) and in point b is equal to:

$$\xi_b(x,y) = \begin{cases} (x-y)^{-1}, & \text{if } y \leq b \leq x \text{ and } y < x, \\ (y-x)^{-1}, & \text{if } x \leq b \leq y \text{ and } x < y, \\ 0, & \text{otherwise}. \end{cases} \tag{23}$$

Assume that the assessment of the increment $\Delta(x, y)$ of the brightness of two pixels is equal to their contrast:

$$\Delta(x,y) = |C(x,y)|, \tag{24}$$

where $C(x, y)$ is the contrast of the pair of objects in a simple two-element image (the kernel of contrast).

Currently, the most well-known definition is weighted contrast:

$$C^{wei}(x,y) = (x-y)/(x+y). \tag{25}$$

Another well-known definition is relative contrast:

$$C_1^{rel}(x,y) = (x-y)/\max(x,y), \tag{26}$$

$$C_2^{rel}(x,y) = (x-y)/(1-\min(x,y)). \tag{27}$$

In [16] the definition of linear contrast was proposed:

$$C^{lin}(x_1,x_2) = (x_1-x_2)/(x_{\max}-x_{\min}). \tag{28}$$

In the case when the objects in the image are independent, the density $f(x, y)$ of bivariate distribution of brightness is equal to:

$$f(x,y) = f(x)\cdot f(y). \tag{29}$$

Expressions (1), (19)–(23) and (24)–(29) describe the proposed technique of intensity transformation by equalizing the distribution density of contrast.

Let us demonstrate the relationship between the proposed generalized description (19)–(22) for the procedure of intensity transformation and the traditional definition (9) of histogram equalization.

Assume that:

$$\tilde{\Delta}(x,y) = \text{const} = 1. \tag{30}$$

Assume that the distribution density $\xi_b(x, y)$ is equal to:

$$\tilde{\xi}_b(x, y) = \begin{cases} 0, & if\ x \neq b \wedge y \neq b \\ 1/2, & if\ x = b \vee y = b \end{cases} \tag{31}$$

In the case of (30)–(31), the definitions (19) and (20) take the form:

$$\tilde{t}(b) = \frac{1}{2} \cdot \left(\int_0^1 f(x, b)\, dx + \int_0^1 f(b, y)\, dy \right) = f(b), \tag{32}$$

$$\tilde{T}(b) = \int_0^b \tilde{t}(x)\, dx = \int_0^b f(x)\, dx = HE(b). \tag{33}$$

Thereby, the traditional definition (9) for the histogram equalization is a particular case of generalized description (19)–(22) for the procedure of intensity transformation.

4 Research

The research is carried out by quantifying the integral contrast and by expert estimates of the quality for the images that have been previously processed using various techniques of intensity transformation, namely:

(1) percentage linear stretching [1, 2];
(2) sigmoid contrast stretching [7];
(3) adaptive piecewise-linear stretching to mean for r_{mean} = ½ [8];
(4) power law transformation [1, 3, 9] for r_{mean} = ½;
(5) traditional histogram equalization (*HE*) technique [2, 10];
(6) BBHE technique [11–13];
(7) RSMHE technique [12, 14];
(8) DSIHE technique [12, 13, 15];
(9) proposed technique (1), (19)–(29) for linear contrast kernel (28);
(10) proposed technique (1), (19)–(29) for relative contrast kernel (26);
(11) proposed technique (1), (19)–(29) for weighted contrast kernel (25).

Quantifying the integral contrast of images is carried out using the various no-reference metrics, namely:

(1) incomplete integral contrast based on the weighted kernel (25) [16];

$$C_{inc}^{wei} = \int_0^1 \frac{|b - b_0|}{b + b_0} \cdot f(b) db, \tag{34}$$

where b_0 is the value of adaptation level.

(2) incomplete integral contrast based on linear kernel (28) [16];

$$C_{inc}^{lin} = \int_{0}^{1} \frac{|b - b_0|}{b_{\max} - b_{\min}} \cdot f(b)db. \tag{35}$$

(3) complete integral (generalized) contrast based on weighted kernel (25) [17]:

$$C_{gen}^{wei} = \int_{0}^{1} \int_{0}^{1} \frac{|b_i - b_j|}{b_i + b_j} f(b_i)f(b_j)\, db_i db_j. \tag{36}$$

(a) image 1, 750×750×8bit. (b) image 2, 564×845×8bit. (c) image 3, 367×367×8bit.

(d) image 4, 400×400×8bit. (e) image 5, 512×512×8bit. (f) image 6, 732×732×8bit.

Fig. 1. Test images and their histograms.

(a) source image 1. (b) percentage stretching. (c) sigmoid stretching.

(d) piecewise-linear stretching. (e) power law transformation. (f) HE technique.

(g) BBHE technique. (h) RSMHE technique. (i) DSIHE technique.

(j) proposed using (28). (k) proposed using (26). (l) proposed using (25).

Fig. 2. Test image 1 and the results of its processing.

(a) source image 2. (b) percentage stretching. (c) sigmoid stretching.

(d) piecewise-linear stretching. (e) power law transformation. (f) HE technique.

(g) BBHE technique. (h) RSMHE technique. (i) DSIHE technique.

(j) proposed using (28). (k) proposed using (26). (l) proposed using (25).

Fig. 3. Test image 2 and the results of its processing.

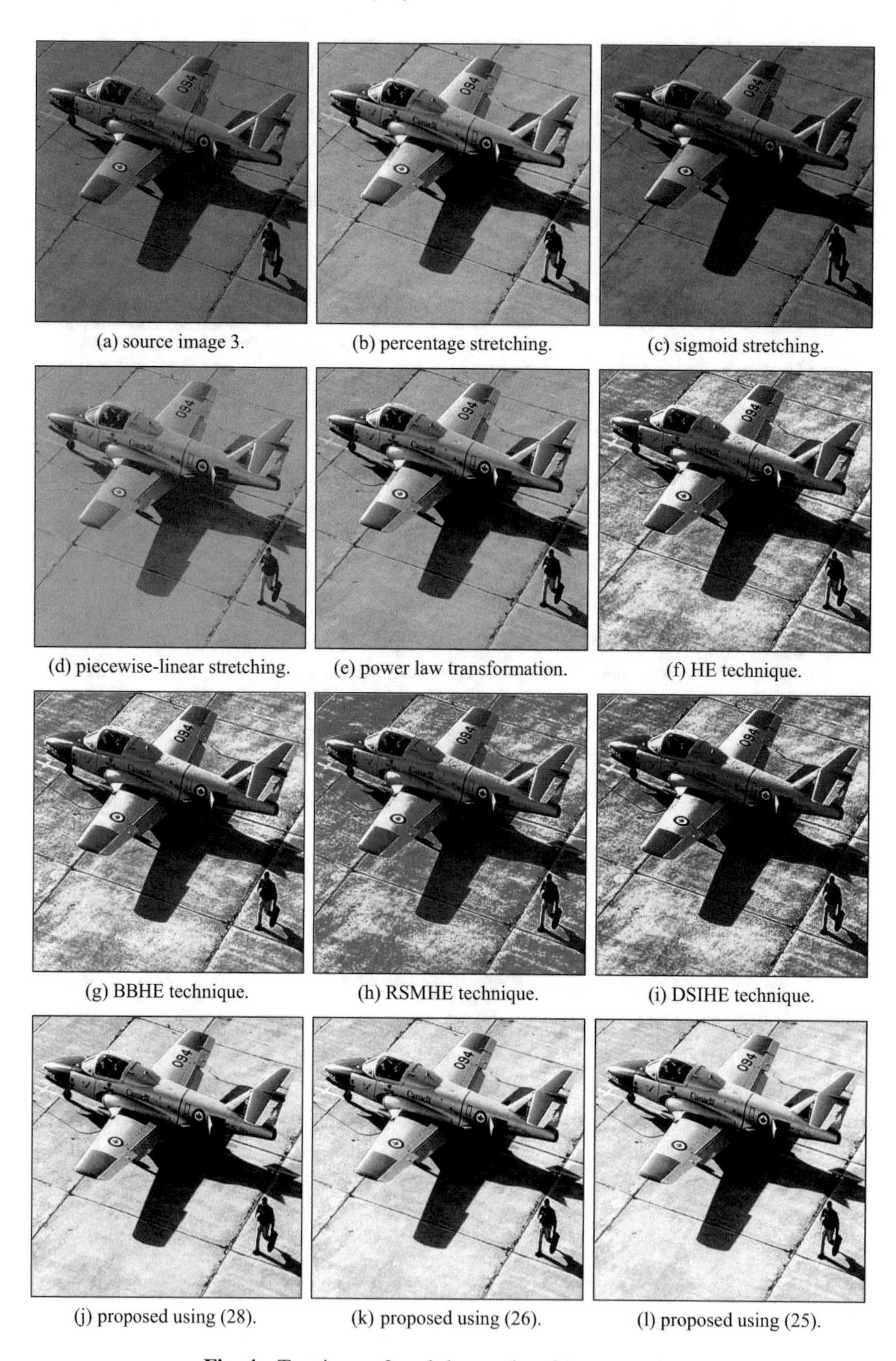

(a) source image 3. (b) percentage stretching. (c) sigmoid stretching.

(d) piecewise-linear stretching. (e) power law transformation. (f) HE technique.

(g) BBHE technique. (h) RSMHE technique. (i) DSIHE technique.

(j) proposed using (28). (k) proposed using (26). (l) proposed using (25).

Fig. 4. Test image 3 and the results of its processing.

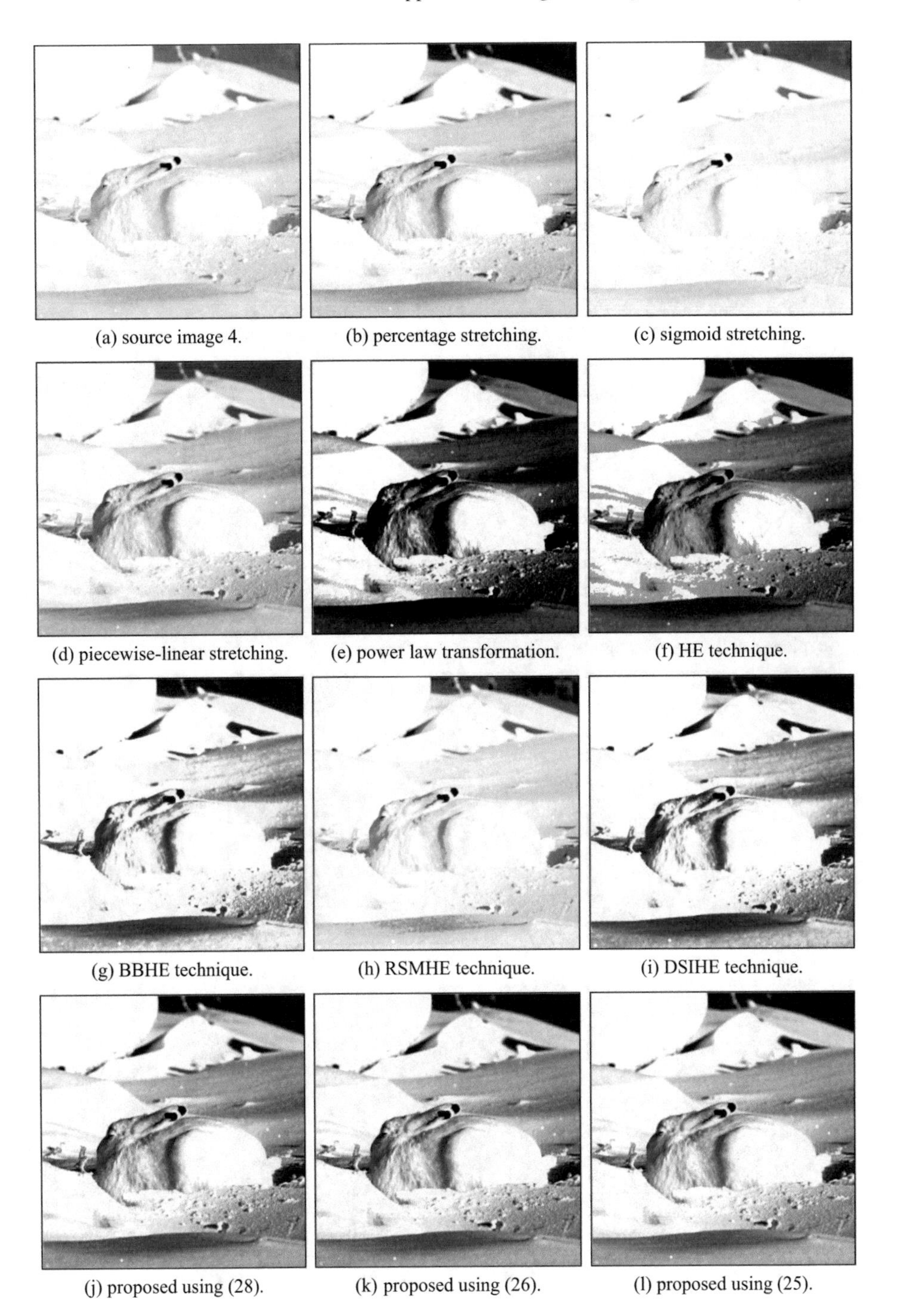

Fig. 5. Test image 4 and the results of its processing.

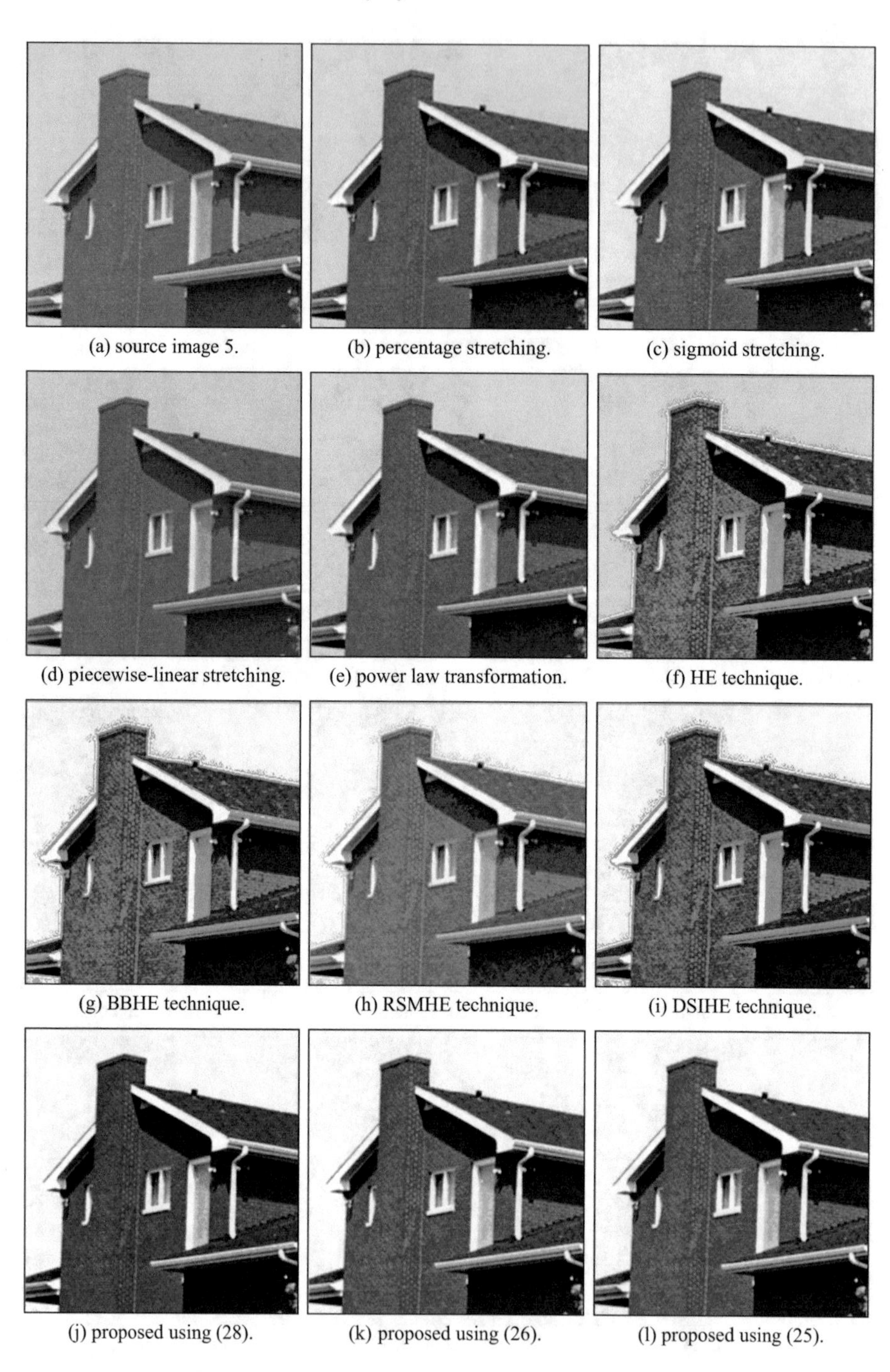

Fig. 6. Test image 5 and the results of its processing.

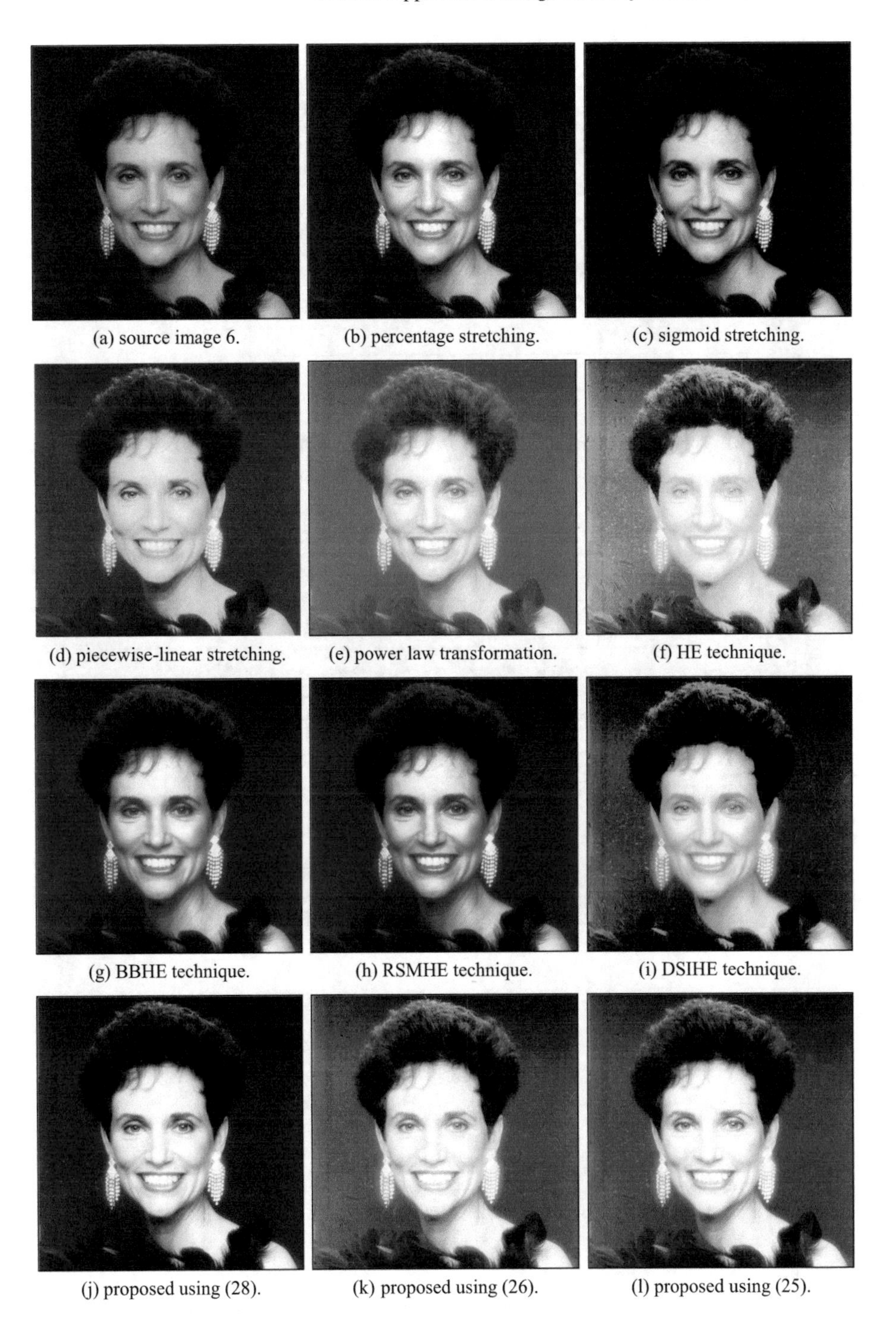

(a) source image 6. (b) percentage stretching. (c) sigmoid stretching.

(d) piecewise-linear stretching. (e) power law transformation. (f) HE technique.

(g) BBHE technique. (h) RSMHE technique. (i) DSIHE technique.

(j) proposed using (28). (k) proposed using (26). (l) proposed using (25).

Fig. 7. Test image 6 and the results of its processing.

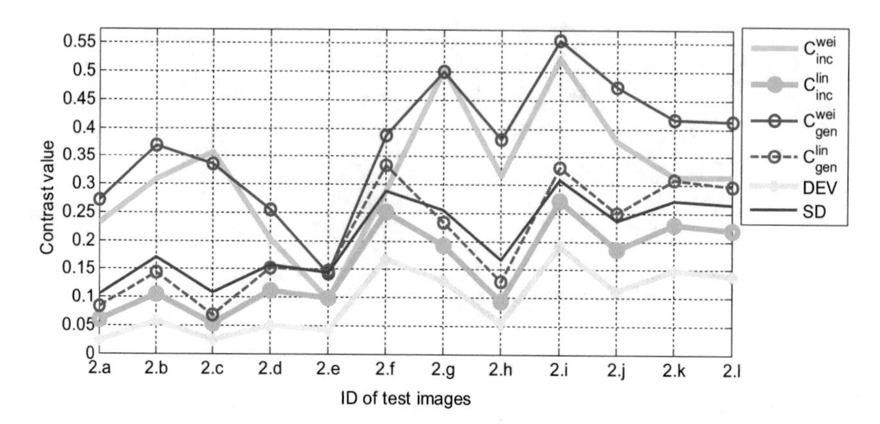

Fig. 8. The results of measuring the integral contrast for the images in Fig. 2.

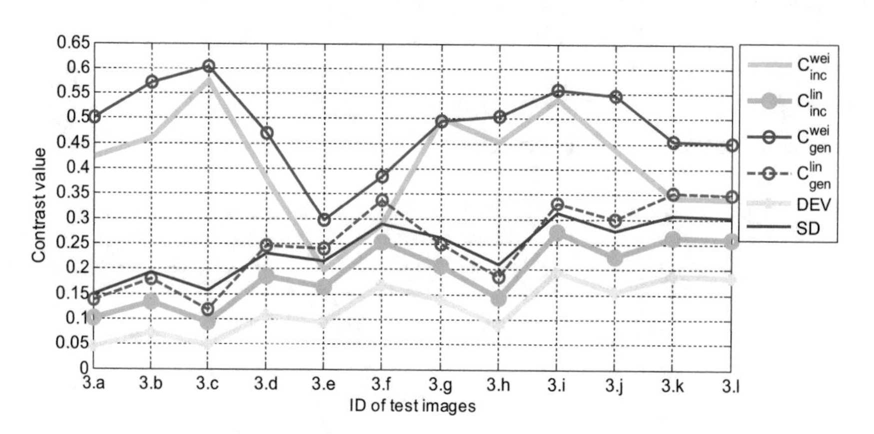

Fig. 9. The results of measuring the integral contrast for the images in Fig. 3.

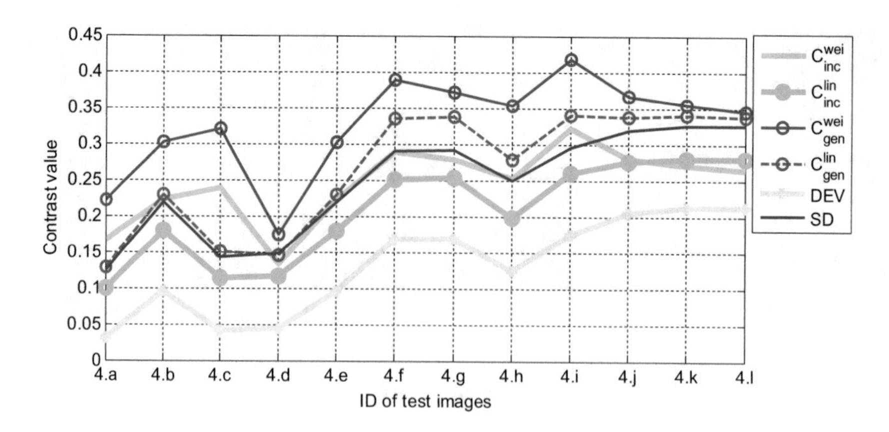

Fig. 10. The results of measuring the integral contrast for the images in Fig. 4.

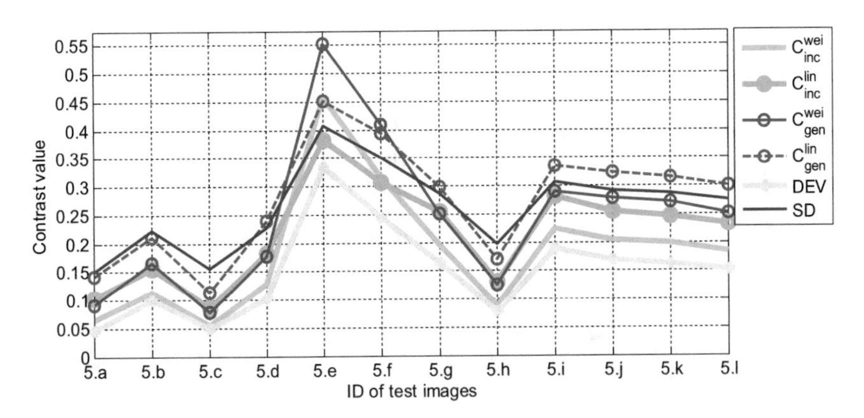

Fig. 11. The results of measuring the integral contrast for the images in Fig. 5.

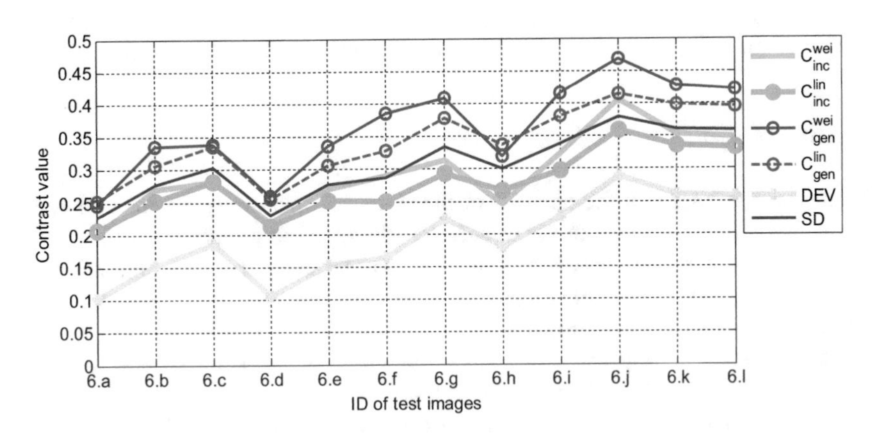

Fig. 12. The results of measuring the integral contrast for the images in Fig. 6.

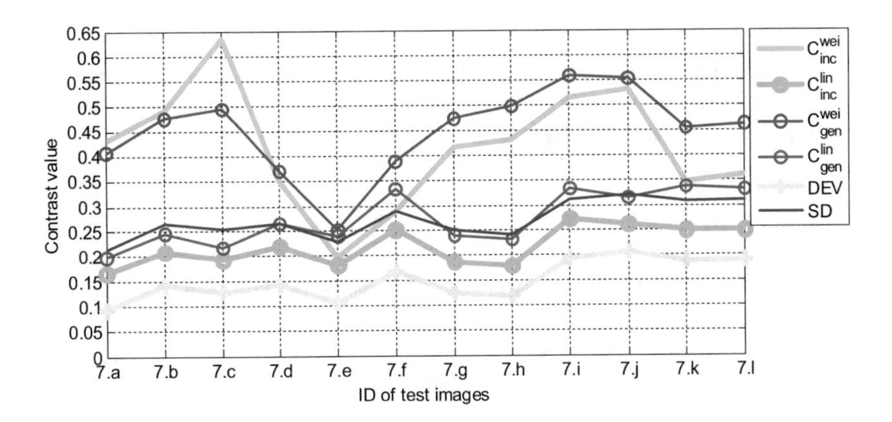

Fig. 13. The results of measuring the integral contrast for the images in Fig. 7.

(4) complete integral contrast based on linear kernel (28) [17]:

$$C_{gen}^{lin} = \int_{0}^{1}\int_{0}^{1} \frac{|b_i - b_j|}{b_{\max} - b_{\min}} f(b_i) f(b_j)\, db_i db_j. \tag{37}$$

(5) mean of squared deviations between values of brightness for all pairs of pixels:

$$DEV = \int_{0}^{1}\int_{0}^{1} (b_i - b_j)^2 f(b_i) f(b_j)\, db_i db_j. \tag{38}$$

(6) standard deviation from the mean (*SD*).

Test images and their histograms are shown in Fig. 1. The results of the processing of test images are presented in Figs. 2, 3, 4, 5, 6 and 7. The results of measuring the integral contrast for the processed images are presented in Figs. 8, 9, 10, 11, 12 and 13.

5 Discussion

The research is based on the analysis of the results of measuring the integral contrast and on the expert estimates of the quality for the transformed images.

Studies show that the results of image processing using known techniques of contrast stretching depends heavily on the brightness distribution and are unsatisfactory for complex images with a full brightness range (Figs. 1b, c, 2b, c, 3c, 4c, 5d and 6b, c).

The efficiency of basic intensity transformations depends on the choice of values of transformation parameters (Figs. 1e, 2e and 4e)

Histogram equalization and its modifications provide most effectively enhance the contrast of complex images. However, histogram equalization based techniques can lead to the emergence of unwanted artifacts in the image, a reduction in contrast for objects with small size, and excessively high increasing contrast of large extended objects (Figs. 1g, h, 2g, h, 4f, g, i, 5f, g, h, i and 6f, i).

The proposed technique of intensity transformation provides an effective increase in contrast without the emergence of unwanted artifacts for all test images and a more evenly distributed the contrast than with traditional histogram equalization.

6 Conclusions

In this paper, the problem of improving the efficiency of the intensity transformation of complex images in the automatic mode was considered.

A new approach to the intensity transformation is proposed based on equalizing the distribution density of contrast for pairs of pixels in an image. The distribution of contrast is estimated based on an analysis of bivariate distribution and brightness increments for pairs of pixels in the image. This approach provides an increase in the efficiency of transforming the intensity through a more evenly enhance and redistribution of the contrast of objects in the image without the appearance of unwanted artifacts. A new generalized description of the intensity transformation based on an analysis of the joint distribution of brightness and the magnitude of brightness increments of pairs of pixels in an image is proposed. The relationship between the proposed generalized description and the traditional definition of histogram equalization is shown. Based on the proposed approach, a new technique of parameter-free intensity transformation is proposed based on equalizing the distribution density of contrast in an image. The proposed technique provides an effective enhance the contrast of complex images without the appearance of unwanted artifacts through a more evenly enhances contrast and its redistribution between low-contrast objects and objects with excessively high contrast in an image and has several advantages compared with the well-known technique of histogram equalization. The results of experimental research using the no-reference metrics of contrast and the expert estimates of the image quality confirm the effectiveness of the proposed approach to enhance images in automatic mode.

References

1. Pratt, W.K.: Digital Image Processing: PIKS Scientific Inside, 4th edn. PixelSoft Inc., Los Altos (2017)
2. Gonzalez, R.C., Woods, R.E.: Digital Image Processing, 3rd edn. Prentice Hall, Upper Saddle River (2010)
3. Gonzalez, R.C., Woods, R.E., Eddins, S.L.: Digital Image Processing Using MATLAB, 2nd edn. Gatesmark Publishing, Knoxville (2009)
4. Rao, Y., Chen, L.: A survey of video enhancement techniques. J. Inf. Hiding Multimed. Signal Process. **3**(1), 71–99 (2012)
5. Kotkar, V.A., Gharde, S.S.: Review of various image contrast enhancement techniques. Int. J. Innov. Res. Sci. Eng. Technol. **2**(7), 2786–2793 (2013)
6. Radha, N., Tech, M.: Comparison of contrast stretching methods of image enhancement techniques for acute leukemia images. Int. J. Eng. Res. Technol. (IJERT) **1**(6), 1–7 (2012)
7. Saruchi, S.: Adaptive sigmoid function to enhance low contrast images. Int. J. Comput. Appl. **55**(4), 45–49 (2012)
8. Yelmanov, S., Romanyshyn, Y.: Image contrast enhancement for smart cameras in wireless/mobile video applications. In: Proceedings of the 2018 IEEE 4th International Symposium on Wireless Systems within the International Conferences on IDAACS-SWS, Lviv, Ukraine, pp. 184–186 (2018)
9. Zhang, D., et al.: Histogram partition based gamma correction for image contrast enhancement. In: 2012 IEEE 16th International Symposium on Consumer Electronics (ISCE), pp. 1–4. IEEE (2012)
10. Hummel, R.A.: Histogram modification techniques. Comput. Graph. Image Process. **4**(3), 209–224 (1975)
11. Kong, N.S.P., Ibrahim, H., Hoo, S.C.: A literature review on histogram equalization and its variations for digital image enhancement. Int. J. Innov. Manag. Technol. **4**(4), 386 (2013)
12. Kaur, M., Kaur, J., Kaur, J.: Survey of contrast enhancement techniques based on histogram equalization, vol. 2, no. 7, p. 136 (2011)

13. Kim, Y.-T.: Contrast enhancement using brightness preserving bi-histogram equalization. IEEE Trans. Consum. Electron. **43**(1), 1–8 (1997)
14. Chen, S.-D., Ramli, A.: Contrast enhancement using recursive mean separate histogram equalization for scalable brightness preservation. IEEE Trans. Consum. Electron. **49**(4), 1301–1309 (2003)
15. Wang, Y., Chen, Q., Zhang, B.: Image enhancement based on equal area dualistic sub-image histogram equalization method. IEEE Trans. Consum. Electron. **45**(1), 68–75 (1999)
16. Yelmanov, S., Romanyshyn, Y.: Rapid no-reference contrast assessment for wireless based smart video applications. In: Proceedings of the 2018 IEEE 4th International Symposium on Wireless Systems within the International Conferences on IDAACS-SWS, Lviv, Ukraine, pp. 171–174 (2018)
17. Yelmanov, S., Romanyshyn, Y.: A new approach to measuring perceived contrast for complex images. In: Shakhovska, N., Medykovskyy, M.O. (eds.) Advances in Intelligent Systems and Computing III. AISC, vol. 871, pp. 85–101. Springer, Cham (2019)

Dynamic Bayesian Networks in the Problem of Localizing the Narcotic Substances Distribution

Volodymyr Lytvynenko[1], Nataliia Savina[2], Jan Krejci[3], Andrey Fefelov[1], Iryna Lurie[1], Mariia Voronenko[1](✉), Ivan Lopushynskyi[1], and Petro Vorona[4]

[1] Kherson National Technical University, Kherson, Ukraine
immun56@gmail.com, faol976@ukr.net, lurieira@gmail.com, mary_voronenko@i.ua, doctordumetaua@meta.ua

[2] National University of Water and Environmental Engineering, Rivne, Ukraine
n.b.savina@nuwm.edu.ua

[3] Jan Evangelista Purkyne University in Usti nad Labem, Ústí nad Labem, Czech Republic
jan.krejci@ujep.cz

[4] Institute of Personnel Training of the State Employment Service of Ukraine, Kyiv, Ukraine
Voron67@ukr.net

Abstract. This paper proposed a methodology for the use of static and dynamic Bayesian networks (BN) in the problems of localizing the distribution of narcotic substances. Methods for constructing the BN structure, their parametric training, validation, sensitivity analysis and "What-if" scenario analysis are considered. A model of dynamic Bayesian networks (DBN) for scenario analysis and prediction of the composition of a narcotic substance has been developed. The model was designed in collaboration with law enforcement officers, as well as forensic experts in the selection and quantification of input and output variables.

Keywords: Narcotic substance · Profiling · Bayesian networks · Dynamic bayesian networks · Structural learning · Parametric learning · Sensitivity analysis · Validation

1 Introduction

1.1 A Subsection Sample

The abuse of illicit drugs today is a serious problem [1]. One of the most effective ways to combat drugs is to prevent their production and distribution. In today's conditions, the illegal circulation of narcotic substances has acquired a huge scope and began to develop into a direct threat to the existence of society. Therefore, the search for various means of dealing with this traffic is not only an urgent task, but such, the success of

N. Shakhovska and M. O. Medykovskyy (Eds.): CCSIT 2019, AISC 1080, pp. 421–438, 2020.
https://doi.org/10.1007/978-3-030-33695-0_29

which determines the future development of both individual social groups and humanity as a whole. Today, quite a lot of various anti-drug software systems have been developed. From a practical point of view, their common drawback is that the analysis process is not automated and the final decision must be made by the operator. The interfaces of such systems are closed, which makes it impossible to dynamically expand their functionality and ability to coordinate actions as part of distributed information systems. However, the socio-economic processes that control law enforcement agencies, like many phenomena occurring in society, are dynamic. This analyzed social phenomenon is characterized not only by the current values but also by the results of previous periods. To represent and analyze social phenomena that change over time, we need to use a model that takes into account the relationship between the values of all parts of the model at different points in time. In this paper, probabilistic graphical models such as dynamic Bayesian networks (DBN) are used to solve this problem.

The proposed software solution is devoid of most of the known drawbacks and has a fundamental difference from the programs used, namely: the analysis process is fully automated (that is, it does not require the presence of an operator). The proposed software solution can find practical application in many fields of applied sciences.

The aim of the work is to develop the static and dynamic Bayesian model in the task of localizing the distribution of narcotic substances.

2 Problem Statement

Drug Profiling System Architecture. Due to the fact that [2], illicit drugs, as a rule, are a complex mixture, in rare cases containing a pure narcotic substance, regardless of whether they are of plant origin, such as heroin, cocaine, cannabis, or are synthetic, such as various types of amphetamines. Since they are produced in artisanal conditions of clandestine laboratories, their chemical composition varies considerably. Along with a narcotic substance, a drug may contain one or more than three different components:

(1) natural ingredients present in raw materials (for example, coca leaf, opium) and used in the production of certain "plant-derived" drugs, such as cocaine or heroin; during the production of the drug, they are extracted together with the narcotic substance and are not completely removed from the composition of the final product;
(2) by-products formed during the production of a drug and related to the method of its manufacture;
(3) diluents that can be added at any stage following the manufacture of a drug chain.

The content of impurities can serve various purposes, in particular, it can contribute to:

(a) identifying specific links between two or more samples;
(b) the classification of material obtained from the various batches seized into groups of related samples, which makes it possible to determine their distribution networks;
(c) identification of the source of the drug sample, including its geographical origin.

These data can be used as an evidence base or as a source of intelligence data in order to identify samples that have a common history [3].

Having a set of well-known statistics on distributors and laboratories producing the drug, which were obtained at different time intervals, it is possible to predict the approximate composition of the narcotic substance.

3 Review of the Literature

In [3] it is shown that illegal drugs and psychotropic substances, as a rule, are a complex mixture, and only in some cases contain a pure narcotic substance. Along with a narcotic or psychotropic substance, samples may contain one or more different types of basic components. Depending on the nature of the analysis of a substance, each component can be characterized using various qualitative or quantitative characteristics. At this point in time, some characteristics are known, while others are unknown. For example, the percentage of the components of a narcotic substance found by chemical analysis is a subset of known characteristics, and the issuing laboratory and the delivery route are a subset of unknowns. The task is to have, with a set of known characteristics of the chemical being studied, determine its unknown characteristics.

In [4–7], the authors suggested using the Bayes network (BN) for solving such types of problems. The main problem here is the choice of the network structure, taking into account the nature of the relationship between the nodes and the a priori definition of the conditional probabilities of the ancestor nodes of the network. However, in the field of forensic science and, in particular, to identify the routes of supply of narcotic substances, the BN is currently not widely spread.

BNs are used in design [8, 9], consumer behavior [10], social behavior [11], support for clinical decision-making [12], system biology [13], ecology [13] and Further. The use of BNs in socio-economic research is widely discussed in [14], where they are one of the mathematical tools for analyzing social behavior, since they allow one to describe, model and predict any empirical data: quantitative, qualitative, and also mixed nature. Bayesian networks make it possible to use both probabilities obtained by analytical or statistical methods and expert estimates, as shown in [15].

A description of DBN successful use for speech recognition is given in [16], the results of predicting the state of patients in the intensive care unit are described in [17]. In [18], modeling of cascade effects in a power system using DBN was described. In work [19], DBN is presented for predicting and diagnosing accidents in atomic reactors. The paper [20] presents the methodology for using DBN in the decision-making process, with daily monitoring and prediction of decompensation in patients with chronic heart failure. [21] presented a methodology for modeling physiological processes in the cardiovascular system as well as glucose-insulin regulation. In [22] developed a methodology for the development and use of DBN for weather forecasting.

The advantage of using BNs is their resistance to incomplete, inaccurate, noisy information, because even in this case, the result will reflect the most likely outcome of events [15].

4 Materials and Methods

Bayesian network (BN) – this is a pair <G, B>, in which the first component of G is a directed acyclic graph corresponding to random variables. A graph is written as a set of conditions of independence: each variable is independent of its parents in G. The second component of the pair, B, is a set of parameters defining the network. It contains the parameters $Q_{x^i|pa(X^i)} = P(x^i|pa(X^i))$ for each possible value of x^i from X^i and $pa(X^i)$ from $Pa(X^i)$, where $Pa(X^i)$ denotes the set of parents of the variable X^i in G. Each variable X^i in the graph G is represented as a vertex. If we consider more than one graph, then we use the notation $Pa^G(X^i)$ to identify the parents X^i in the graph G.

The total joint BN's probability B is calculated by the formula $P_B(X^1, \ldots, X^N) = \prod_{i=1}^{N} P_B(X^i|Pa(X^i))$. From a mathematical point of view, BN is a model for representing probabilistic dependencies, as well as the absence of these dependencies. At the same time, the A → B relationship is causal, when event A causes B to occur, that is, when there is a mechanism whereby the value adopted by A affects the value adopted by B. BN is called causal (causal) when all its connections are causal.

The concept of DBN, first introduced by Dean and Kanazawa [23] in 1988, is an extension of the Bayesian network (BN) [24, 25] to simulate dynamic systems that change over time. In this regard, a DBN allows a probabilistic graphical model to describe the level of uncertainty with a multitude of applications aimed at reducing the complexity of computations and justifying a particular decision in uncertain situations. Analysis of very complex phenomena and decision making in complex situations, where, for example, variables are highly interrelated [26] and/or data are ambiguous, are also parallel achievements in the use of a DBN.

Dynamic Bayesian networks are a generalized model in the state space. The name "dynamic" indicates not the dependence of the structure on time, but only on the dependence on the modeling process.

Hidden Markov model is a simplified DBN without cause-effect relationships. If the state space is continuous, then such a model is called a Kalman filter, in the discrete case - DBN. The structure of the BN remains unchanged in all time slices. Slice is the current state of a DBN at a discrete point in time. A vertex in a network can have a parent only in its own time slice or in the immediately preceding time slice. In other words, a DBN is defined as a first-order Markov process.

In DBN, the state of the system in time is displayed by a set of random variables $X^t = X_1^t \ldots X_n^t$. If the state of the system depends only on the immediately preceding state (i.e. $k = 1$), The system is called the Markov distribution (Markov chains) of the first order with the transition parameter $P(X^t|X^{t-1})$ [27]:

$$P(X^t|X^{t-1}) = \prod_{i-1}^{n} P(X_i^t|P_a(X_i^t)) \tag{1}$$

Given that GeNIe [28, 29] is the implementation domain of the DBN, the design procedure is very similar to the procedure planned for BN, as reported in [27]. Accordingly, we start with static BNs containing variables (nodes) and dependencies (arcs). By determining the conditional probabilities table (CPT) of the nodes, we then quantify the qualitative factors under consideration, which have the most important influence on the cluster that occurs. In each CPT column, the probabilities are normalized to the range.

As a result, the sum of the probabilities in each column must correspond to one. To obtain the aforementioned probability distributions, several methods exist in the literature. For example, if accurate knowledge of the data is not at hand, they can be determined on the basis of suggestions from relevant experts. Another classic method is to use the database of maritime incidents presented, for example, in [28, 29].

The next step is to transform the network into a DBN by equipping the BN with temporary layers and temporary arcs. In this regard, Fig. 1 shows an example of a DBN designed in GeNIe.

The presented network, as an example of a DBN, has one normal node, Z1, two temporary nodes, X3 and Y, and a temporary plate consisting of two-time steps. Straight bold arrows mean normal (static) arcs, while bright arrows with marks on them show temporary arcs.

The time arc with the tag “1” shows the dependence of the child node in time t on the status of the parent node in time. The X3 node with a time arc demonstrates the fact that the state of the X3 node in the previous time step affects its current status. The selection of tags and their orders varies from scenario to scenario.

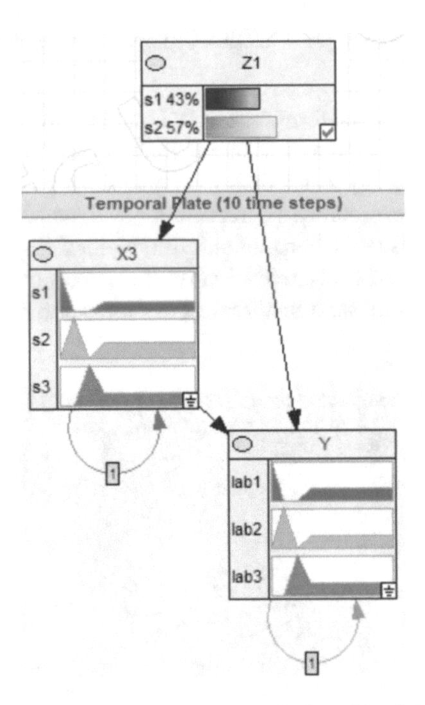

Fig. 1. An example of a DBN designed in GeNIe.

The joint probability density function of a sequence of time steps can be achieved by "unfolding" the network until the edges are clipped (Fig. 2).

Thus, the trend of probability fluctuations for each node in all time steps can be analyzed in detail. In the DBN area, the actual observation of an event is called evidence. By providing such evidence and updating the network, the entire network and node confidence will be changed [27–29].

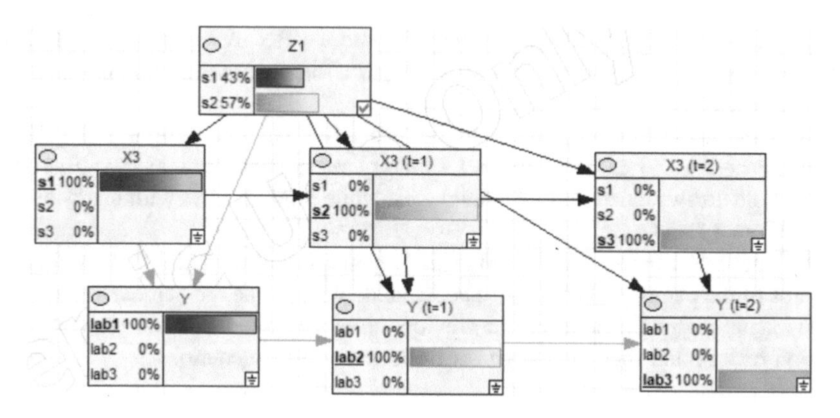

Fig. 2. An example of a DBN deployed in three time steps.

One of the approaches to the formation of BN output is the preliminary discretization of continuous variables. The area of each continuous variable is divided into a certain finite number of sets (Fig. 3). After that, the conditional probability distributions are sampled and, as a result, we obtain a discrete model, with which it is much easier to work [12]. From the existing discretization methods (hierarchical discretization, discretization on the same width of classes, discretization on the same number of points inside the classes) for the existing data set a hierarchical discretization method was chosen.

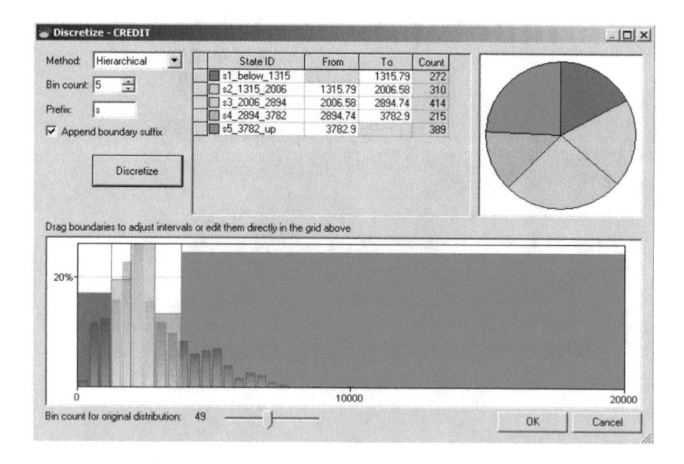

Fig. 3. The process of hierarchical data discretization.

The goal of parametric learning is to find the most likely θ variables that explain the data. Let $D = \{D_1, D_2, \ldots, D_N\}$ be learning data, where $D_l = \{x_1[l], x_2[l], \ldots, x_n[l]\}$ consists of instances of Bayesian network nodes. The learning parameter is quantified by a log-likelihood function, denoted as $L_D(\theta)$. If the data to be trained are complete, the following equations apply:

$$L_D(\theta) = \log\left\{\prod_{l=1}^{N} P(\mathrm{x}_1[l],\ \mathrm{x}_2[l], \ldots, \mathrm{x}_n[l]: \theta)\right\} \tag{2}$$

$$L_D(\theta) = \log\left\{\prod_{i=1}^{n}\prod_{l=1}^{N} P(\mathrm{x}_i[l] | pa(\mathrm{x}_i[l]): \theta)\right\} \tag{3}$$

If all variables are observed, then the simplest and most frequently used method is a statistical estimate that calculates the probability of an event according to the frequency of this event's occurrence in the database. An approach (called maximum likelihood (ML)) uses the formula:

$$P(\mathrm{X_i} = x_k | pa(\mathrm{X_i}) = x_j) = \theta_{\mathrm{i,j,k}} = \frac{\mathrm{N_{i,j,k}}}{\sum_k \mathrm{N_{i,j,k}}}, \tag{4}$$

$\mathrm{N_{i,j,k}}$ - the number of events in the database for which the variable $\mathrm{X_i}$ is in the $\mathrm{x_k}$ state, and its parents are present in the $\mathrm{x_j}$ configuration.

The sensitivity analysis of the Bayesian network allows you to set for each of the network parameters a function expressing the output probability from the point of view of the parameter being studied.

To derive the probability, we will consider the posterior marginal probability of the form $y = p(a|e)$, where a is the value of the variable A and e means available evidence. Each of the network parameters has the form $x = p(b_i|\pi)$, where b_i is the value of the variable B and π is an arbitrary combination of the values of the set of parents $\Pi = pa(B)$ of B.

Denote $p(a|e)(x)$ as a function expressing the a posteriori marginal probability $p(a|e)$ in terms of the parameter x. In the future, we will assume that in a sensitivity analysis, as the parameter $x = p(b_i|\pi)$ changes, each of the probabilities $p(b_j|\pi)$ changes accordingly. The function $y(x)$, obtained as a result of sensitivity analysis, is a quotient of two linear functions in x.

Theorem 1. Let p be a probability function of a Bayesian network over a set of variables V. Let $y = p(a|e)$ and $x = p(b|r)$, as indicated above. We have

$$y = \frac{p(a,e)(x)}{p(e)(x)} = \frac{\alpha x + \beta}{\gamma x + \delta}, \tag{5}$$

where α, β, γ and δ - constants for x.

Proof: the joint probability $p(a|e)$ can be expressed in terms of x by the formula:

$$p(a,e)(x) = \left(\sum_{V:a,e} p(V)\right)(x), \tag{6}$$

where $\sum_{V:a,\ldots,d} p(V)$ denotes the summation over the variables $V \backslash \{A,\ldots, D\}$ with $A, \ldots, D \in V$ with fixed values $a,\ldots,d$ respectively.

The sum $\sum_{V:a,e} p(V)$ in the above equation can be divided into n + 1 individual sums, so the first sum includes only terms with b_1 for B and state π for Π, the second sum includes only conditions with b_2 for B and Π in the state π etc., and the last amount includes the remaining conditions.

Therefore, we can write the formula:

$$\begin{aligned}
p(a,e)(x) &= \left(\sum_j \sum_{V:a,e,b_j,\pi} p(V) + \sum_{V:a,e,\Pi\neq\pi} p(V)\right)(x) \\
&= \sum_j p(b_j|\pi)(x)\frac{\sum_{V:a,e,b_j,\pi} p(V)}{p(b_j|\pi)} + \sum_{V:a,e,\Pi\neq\pi} p(V) \\
&= x\frac{\sum_{V:a,e,b_i,\pi} p(V)}{p(b_i|\pi)} + \sum_{j\neq i}\left(\frac{p(b_j|\pi)}{1-p(b_i|\pi)} - x\frac{p(b_j|\pi)}{1-p(b_i|\pi)}\right) \\
&\quad \cdot \frac{\sum_{V:a,e,b_j,\pi} p(V)}{p(b_j|\pi)} + \sum_{V:a,e,\Pi\neq\pi} p(V) \\
&= x\left(\frac{\sum_{V:a,e,b_i,\pi} p(V)}{p(b_i|\pi)} - \sum_{j\neq i}\frac{\sum_{V:a,e,b_j,\pi} p(V)}{1-p(b_i|\pi)}\right) \\
&\quad + \sum_{j\neq i}\frac{\sum_{V:a,e,b_j,\pi} p(V)}{1-p(b_i|\pi)} + \sum_{V:a,e,\Pi\neq\pi} p(V)
\end{aligned} \tag{7}$$

For the probability $p(e)$, we derive a similar expression, summing up in the above output, over all values of the variable A instead of keeping it fixed to a. As a result of calculating the expressions $p(a|e)(x)$ and $p(e)(x)$, it can be shown that the output $y = p(a|e)$ can be written as a quotient of two functions that are linear in.

From Theorem 1 we find that the function expressing the posterior marginal probability y in terms of one parameter x is characterized by no more than three coefficients. The theorem is easily transferred to n parameters. The function then includes products from all possible combinations of parameters called monomials, both in the numerator and in the denominator. The numerator, as well as the denominator, is characterized by 2^n coefficients, many of which may have a value of zero.

The sensitivity analysis in the GeNie software environment is performed using influence diagrams (Fig. 4). The influence diagram shows the most sensitive parameters for the selected state of the target node Y, sorted from the most sensitive to the least sensitive.

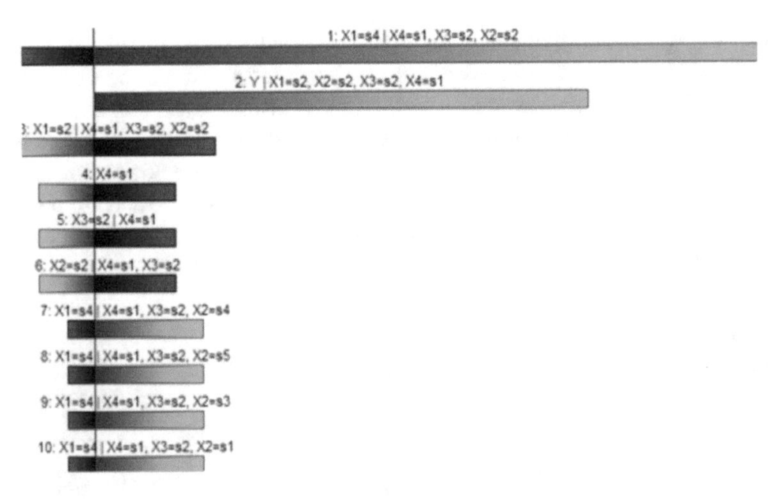

Fig. 4. The sample of influence chart.

Validation of the developed network was carried out according to the algorithm of maximizing the expectation, which was proposed for the first time in 1977 in [30]. The algorithm finds local optimal estimates of the maximum parameter's likelihood. The main idea of the algorithm is that if we knew the values of all nodes, then learning (at some step M) would be simple, since we would have to have all the necessary information.

Therefore, at stage E, calculations of the expected likelihood value, including latent variables, are made as if we were able to observe them. In step M, the maximum likelihood parameter's values are calculated using the maximization of the expected likelihood values obtained in step E. Next, the algorithm performs step E again using the parameters obtained in step M and so on.

Based on the algorithm of maximizing the expectation, a whole series of similar algorithms was developed [31, 32]. For example, the structural algorithm for maximizing the mathematical expectation of the EM combines a standard algorithm for maximizing the mathematical expectation for optimizing parameters, and an algorithm for the structural search for a selection model. This algorithm builds networks based on penalty probabilistic values, which include values obtained using Bayesian information criteria, the principle of minimum description length, and others.

5 Experiments and Results

Figure 5 shows a conceptual model of the proposed BN. According to the presented model, a wide variety of concepts that characterize it can be associated with each narcotic substance. At the same time, the proposed model does not limit the researcher to a rigid unidirectionality of actions, and can be used both to detect causal characteristics (for example, the laboratory and possible delivery can be determined by the known chemical composition and location) and to predict the consequences (for example, if it was found that some laboratory released a batch of substance, then the emergence of what type of substance and in what territory should be expected in the near future). The described properties of the model in Fig. 5 are illustrated by arrows showing the movement of information between the selected blocks. It should be noted that, due to the specifics of the Bayesian networks work, all conclusions of this model, with respect to the information sought, are of a probabilistic nature and are presented in the form of a ranked list (according to the values of probability of fidelity of one or another conclusion). The final decision is made by the researcher.

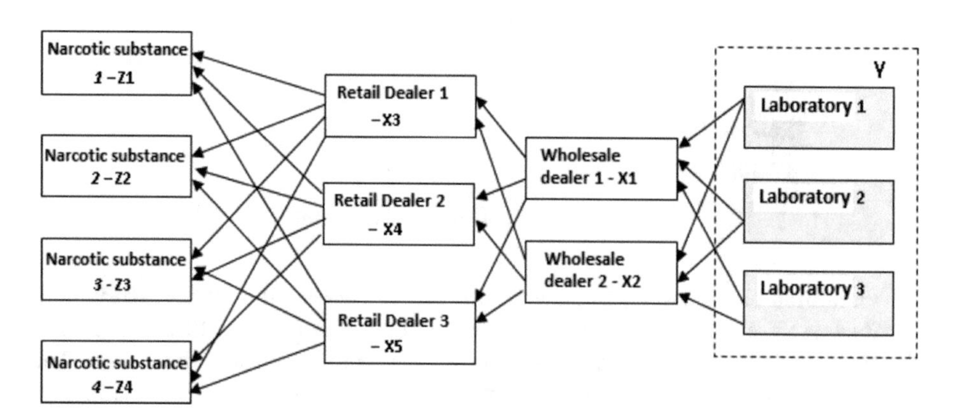

Fig. 5. Conceptual model of the localized distribution of narcotic substances being developed by BN.

For example, an analysis of a drug showed that it, along with other components, contains narcotic substance 3. As can be seen in Fig. 5, this component could have been added by a retailer of 1, 2 or 3, which received goods from a wholesale dealer 1 or 2. Supplies to this the dealer is made in turn from laboratories 1, 2 and 3. The specific laboratory that supplied the goods can be clarified by analyzing other components of the narcotic substance. Thus, having established on the basis of the existing database [1] the nature of the relationship between network nodes, it is possible with a high degree of probability to predict the path of delivery of goods to a given point of sale.

The solution to the problem of building a BN was made using the GeNIe 2.3 Academic software environment. Having a set of known X data (such as the number of components of a narcotic substance, for example), it is necessary to determine the unknown parameter Y (manufacturing laboratory). The main problem here is the choice of the network structure, taking into account the nature of the relationship between the nodes and the prior determination of the conditional probabilities of the ancestor nodes of the network. At the same time, we conduct a sensitivity analysis of the model and re-validation. As the target node was taken node Y, that defines the laboratory producing the drugs.

Let nodes X1, X2, X3, X4, X5 represent dealers, with nodes X1 and X2 symbolizing wholesale dealers, and X3, X4, X5 - retail dealers, respectively. Nodes Z1, Z2, Z3, Z4 - the number of different components of drugs present in the drug. U is a specific laboratory that we want to learn.

Nodes Z1, Z2, Z3, Z4 have two states:

- state s1 - means the absence of this component in the composition of the drug;
- state s2 - means the presence of this component in the composition of the drug.

Nodes X1, X2, X3, X4, X5 each have three states. For example, for node X1, these states will be:

- state s1 means that the drug was not distributed by the wholesale dealer 1 with a probability of 33.3%;
- state s2 - with a probability of 33.3%, we believe that wholesale dealer 1 could either distribute or not distribute the drug;
- s3 state means that the drug was accurately distributed by the wholesale dealer 1 probability of 33.3%.

Node Y has also three states:

- state s1 - the drug is produced in laboratory 1;
- state s1 - drug produced in laboratory 2;
- state s1 - drug produced in the laboratory 3.

After preliminary sampling of the available data set, we obtained the resulting sampled set, that is shown in Fig. 6.

Tree View
Model_наркотики_BS (Mo(
X1
X2
X3
X4
X5
Y
Z1
Z2
Z3
Z4

X1	X2	X3	X4	X5	Y	Z1	Z2	Z3	Z4
s2	s1	s1	s3	s2	lab1	s2	s2	s2	s1
s1	s2	s3	s2	s1	lab3	s1	s2	s2	s1
s2	s3	s3	s3	s2	lab2	s2	s1	s2	s2
s2	s2	s2	s1	s3	lab1	s1	s1	s1	s2
s1	s3	s1	s2	s2	lab2	s1	s2	s2	s1
s2	s2	s2	s2	s3	lab3	s1	s2	s1	s1
s1	s1	s1	s3	s3	lab3	s1	s2	s1	s1
s1	s2	s2	s3	s2	lab2	s1	s2	s2	s2
s2	s1	s1	s1	s2	lab1	s1	s1	s1	s1
s1	s3	s3	s3	s3	lab2	s2	s2	s2	s1
s1	s3	s3	s1	s1	lab3	s2	s2	s2	s1
s3	s1	s3	s2	s2	lab1	s2	s2	s2	s2
s3	s3	s2	s1	s1	lab1	s1	s1	s1	s2
s1	s2	s1	s2	s3	lab3	s1	s1	s2	s1
s3	s2	s2	s2	s3	lab2	s2	s1	s1	s1
s1	s1	s2	s2	s2	lab1	s2	s2	s2	s2
s2	s2	s3	s2	s2	lab3	s2	s2	s1	s1
s3	s3	s3	s1	s2	lab3	s1	s2	s2	s2
s1	s3	s2	s3	s1	lab2	s1	s2	s1	s2
s1	s1	s2	s2	s3	lab3	s2	s2	s2	s2
s2	s2	s2	s2	s1	lab2	s1	s1	s1	s1
s2	s1	s3	s3	s2	lab2	s1	s2	s2	s2
s2	s2	s3	s2	s1	lab3	s2	s2	s1	s2
s3	s1	s1	s2	s3	lab3	s2	s2	s2	s2
s3	s1	s1	s1	s3	lab1	s2	s2	s2	s1
s2	s2	s2	s3	s2	lab1	s2	s1	s2	s1
s1	s3	s3	s3	s1	lab3	s2	s2	s2	s1
s3	s2	s2	s1	s2	lab1	s2	s2	s2	s2
s1	s3	s1	s1	s2	lab3	s2	s2	s2	s1
s3	s1	s2	s1	s3	lab3	s1	s2	s2	s2
s2	s2	s1	s3	s2	lab2	s1	s2	s2	s2
s3	s1	s2	s2	s1	lab2	s1	s1	s1	s2
s2	s1	s1	s2	s1	lab2	s1	s2	s2	s2
s2	s1	s3	s1	s1	lab1	s2	s1	s1	s1
s3	s3	s2	s3	s2	lab1	s2	s2	s2	s2

Fig. 6. The discretized dataset

The Bayesian network model, focused on solving the problem of localizing the distribution of narcotic substances, is shown in Fig. 7.

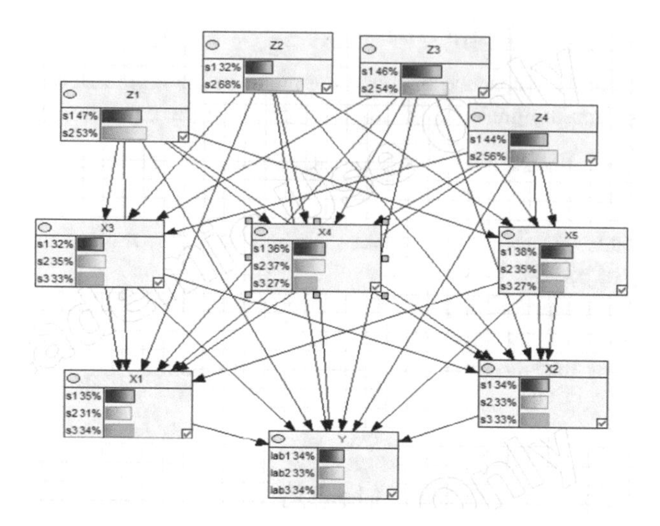

Fig. 7. The static BN model, focused on solving the problem of localizing the distribution of narcotic substances.

After a sensitivity analysis, re-parameterization and cross-validation, the accuracy of the result increased and amounted to more than 61%. The obtained Bayesian model was tested on different data sets using various sampling methods. Considering that the Bayesian network is a probabilistic approach, with the presence of various kinds of uncertainties, the resulting model is adequate to the processes under study. It is shown that validation with subsequent verification allows to ensure the reliability of the model, and significantly increases the accuracy of the obtained model.

At the last stage of the comparative analysis during the testing procedure, it was possible to trace the most probable trends in drug supply routes, which are shown in Fig. 8.

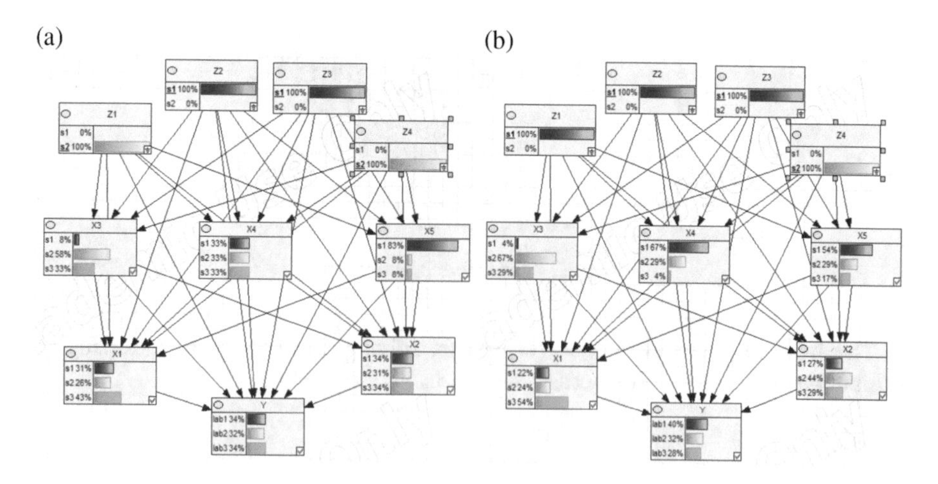

Fig. 8. The results of the comparative analysis: (a) determination of the dealer-supplier of the drug, (b) determination of the laboratory - the manufacturer of the drug.

Figure 8(a) shows that if the Z2 and Z3 components are guaranteed in a narcotic substance, then with 83% certainty, the supply route is carried out through the retail dealer X5. As can be seen from Fig. 8(b), with the presence of components in the narcotic substance Z1, Z2 and Z3 and the absence of component Z4 with a probability of 40%, it can be argued that production was carried out in laboratory 1, and the delivery route with 67% certainty was performed by the turn of the retail dealer X4 and with a probability of 54% turn of the retail dealer X5, who in turn received the goods with a probability of 54% from the wholesale dealer X1.

At the next stage of the study, on the basis of the static BN obtained on the localization of the distribution of narcotic substances, we will build a dynamic BN, which makes it possible to determine the composition of the narcotic substance.

Designed DBN has qualitative and quantitative components. A quality component is a network structure, its conceptual model, extracted from the knowledge of experts. The quantitative component is the conditional probability distribution, which determines the exact relationship between states and variables.

We take the static Bayesian network of 10 nodes (Fig. 6), after re-verification, the accuracy of the initial model reached 61%. Next, we determine the normalized and terminal nodes, and then rebuild the network, taking into account the presence of the time domain. We assume that the nodes Z1, Z2, Z3, Z4 are normalized static, whose values do not change with time when adding time arcs. On nodes, X1–X5 we set temporary arcs and now previous probabilities are taken into account on key variables X1–X5. After learning the parameters, you need to set time periods for all key variables. The resulting dynamic network will look like that shown in Fig. 9.

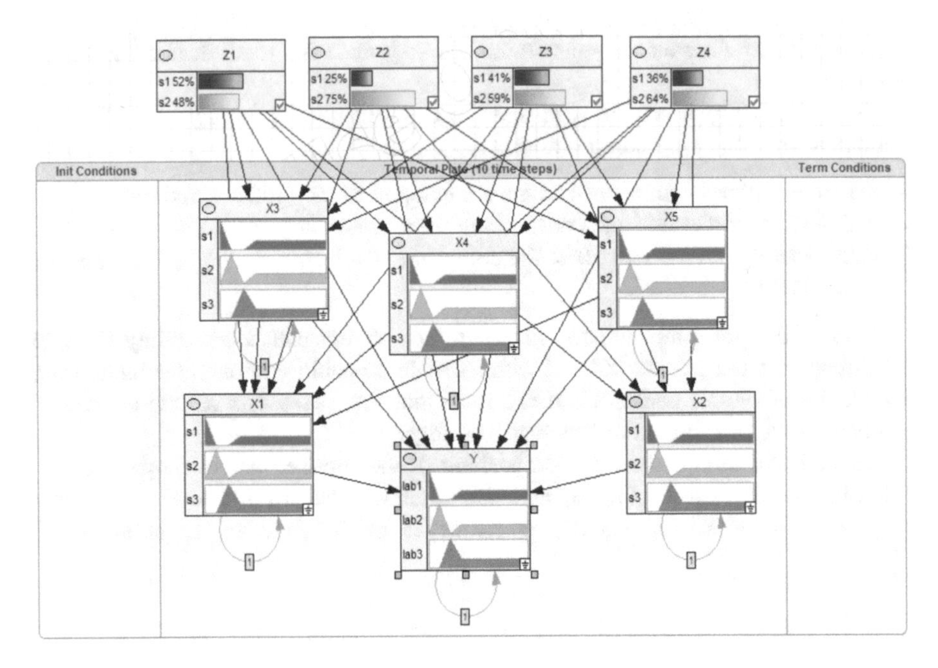

Fig. 9. Dynamic BN for determine the composition of a narcotic substance.

6 Discussion

BNs are interesting for representing knowledge because they allow both top-down and bottom-up, they easily capture the opinions of experts and can be trained in data, updated and personalized. Unlike a system that uses input from a numerical model as an input to a static network to predict the values of key variables, the presented DBN uses the observed values of key variables to predict the future values of the same variables.

Let us analyze the possible scenarios for the supply of narcotic substances, shown in Fig. 10(a)–(c):

Figure 10(a) shows the situation in which the delivery will be made by the wholesale dealer X1 through the retail dealer X3. To find out the approximate composition of the supplied narcotic substance, we set the evidence for the key node X1 wholesale dealer for state s3 - maximum, the key node X3 retail dealer for state s3 maximum.

From Fig. 9(a) it is clear that the drug distributed through the X1 $\rightarrow$ X3 branch has the following composition:

- with confidence at 70% and at 92%, the components of the narcotic substance are components Z2 and Z4, respectively;
- with confidence at 94% and 69%, there are no components Z1 and Z3 in the composition, respectively.

Figure 10(b) shows the situation in which the delivery will be made by the wholesale dealer X2 through the retail dealer X5. To find out the approximate composition of the supplied narcotic substance, we set the evidence for the key node X2 wholesale dealer for state s3 - maximum, the key node X5 retail dealer for state s3 maximum.

From Fig. 10(b) it can be seen that the drug distributed through the X2 $\rightarrow$ X5 branch has the following composition:

- the composition is guaranteed to have a component Z4 (100% certainty);
- with 85% confidence in the composition there is no component Z2;
- with certainty at 85% and 79%, the composition contains components Z1 and Z3, respectively.

Figure 10c shows the situation in which the delivery with a probability of 100% lies through the retail dealer X5 and with a certain probability through the retail dealer X4 and the wholesale dealer X1. It is known that dealers X2 and X3 are not exactly involved in the distribution of this batch of drugs.

To find out the approximate composition of the supplied narcotic substance, we will establish evidence for the key node X5 wholesale dealer for state s3 - maximum, key nodes X4 and X1 for state s2, key nodes X3 and X2 for state s1 - minimum.

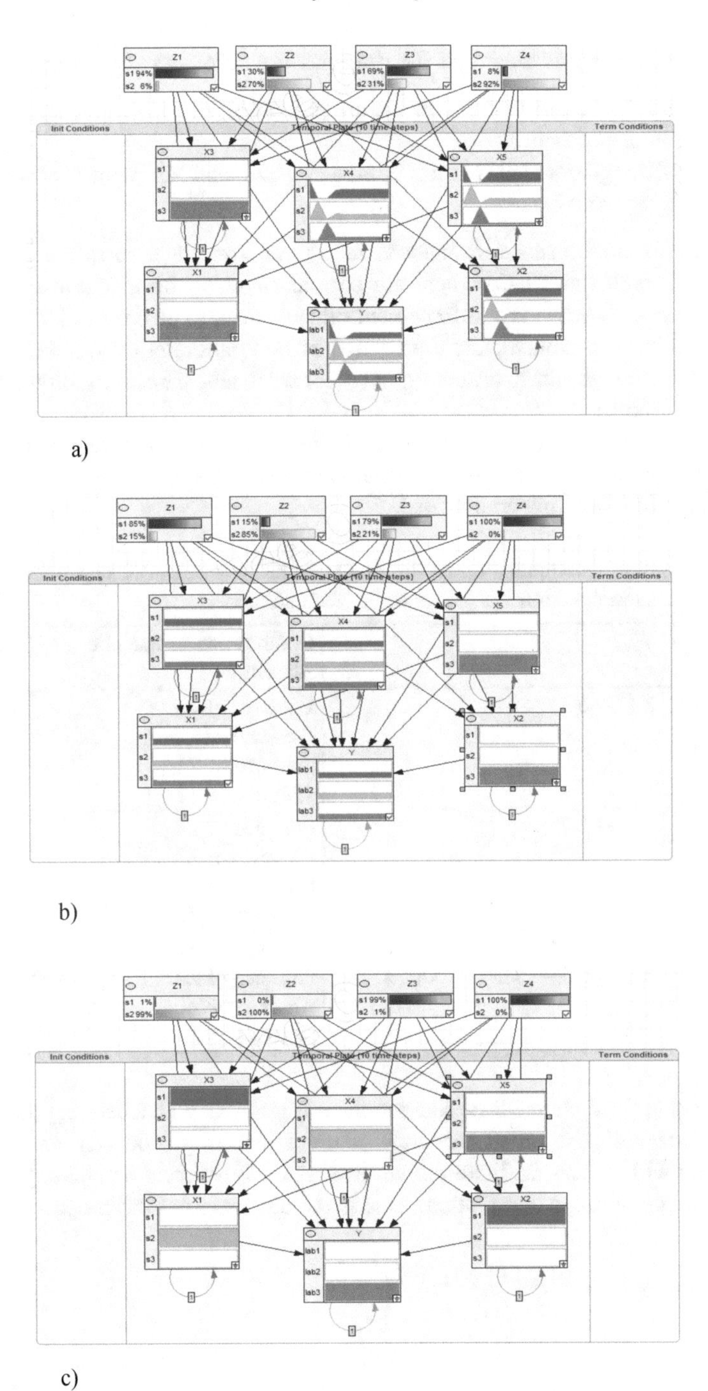

Fig. 10. Determining the composition of a narcotic substance depending on the path of the narcotic substance from the laboratory to the end user.

From Fig. 10(c) it is clear that the drug has the following composition:

- as part of the guaranteed components are present Z3 and Z4 (with confidence 99% and 100%, respectively);
- as part of the guaranteed missing components Z2 and Z1 (with confidence 100% and 99%, respectively).

The results of the scenario analysis in a more structured form are presented in Table 1. It is well known that there are periods of recession in crime and there are periods of recovery when law enforcement officials need to be prepared to increase the number of drugs distributed, the emergence of new laboratories and dealers, etc. In such cases, it is necessary to ensure rapid response, to take care of additional personnel and power reserves.

Previously, in order to achieve the goal of finding out a complete or approximate list of the components that make up a narcotic substance, a test purchase was made. This entailed time and material costs.

Table 1. Results of determining the composition of a narcotic substance depending on the path of the narcotic substance from the laboratory to the end user.

Scenario	Way	Composition of narcotic substance	Probability
Scenario 1	X1 → X3	Presence of Z4 Presence of Z2 Absence of Z1 Absence of Z3	92% 70% 94% 69%
Scenario 2	X2 → X5	Presence of Z4 Presence of Z3 Presence of Z1 Absence of Z2	100% 79% 85% 85%
Scenario 3	May be X1 or X2 → X5 → May be X4	Presence of Z4 Presence of Z2 Absence of Z1 Absence of Z3	92% 70% 94% 69%

Now, using the dynamic Bayesian model, it is possible with a certain probability, to determine the composition of the drug, without incurring monetary and temporary losses. This will greatly facilitate the work of law enforcement agencies to detect the source of narcotic drugs distribution. This is the advantage of the proposed approach.

7 Conclusion

The Bayesian network model for solving the problem of profiling drugs was developed. The proposed model allows for the analysis of a narcotic substance to assess the probability of the production place of the substance and determine the possible ways of its transportation to the place of sale. Dynamic BN is also designed to determine the composition of a specific narcotic substance.

Scope: The end users of the DBN are: law enforcement officials, whose goal is to prevent the spread of narcotic substances in an effort to reduce crime. In the area of offenses, monitoring is an additional opportunity to provide timely information to law enforcement and, in particular, the police.

A further direction of research by the authors is the implementation of this model on real data in accordance with [1].

References

1. Vartuzov, V.V., Shkurdota, S.V., Litvinenko, V.I., Fefelov, A.A.: Computerized technology for compiling profiles of narcotic substances. In: Intertechnical System and Resolution of the Problem of Numerical Object 2012, ISRPN, pp. 39–41 (2012)
2. Shkurdota, S.V., Vartuzov, V.V., Litvinenko, V.I., Fefelov, A.A.: Methods of processing chromatograms for obtaining profiles of narcotic substances. In: Intertechnical System and Resolution of the Problem of Numerical Object 2012, ISRPN, pp. 227–229 (2012)
3. United Nations International Drug Control Program. United Nations, Vienna, New York, p. 27 (2004)
4. Bidyuk, P.I., Terent'yev, O.M., Korshevnyuk, L.O.: Intelektual'nyy analiz slabostrukturovanykh danykh za dopomohoyu bayyesovykh merezh, No. 3/5-HR, p. 85 (2007). (ukr)
5. Zhurovs'kyy, M.Z., Bidyuk, P.I., Terent'ev, O.M.: Systemna metodyka pobudovy bayyesovykh merezh. "Naukovi visti" NTUU "KPI", no. 4, pp. 47–61 (2007). (ukr)
6. Bidyuk, P.I., Terent'yev, O.M., Korshevnyuk, L.O.: Bayyesovskaya set' – instrument intellektual'nogo analiza dannykh. In: Problemy upravleniya i informatiki. K.: IKI NANU-NKAU, no.4, pp. 83–92 (2007). (rus)
7. Terent'ev, O.M., Gasanova L.T.: Bayesian networks in credit scoring. In: Second International Conference on Control and Optimization with Industrial Applications (COIA-1008). Institute of Applied Mathematics BSU, Baku, p. 171 (2008)
8. Andreassen, S., Woldbye, M., Falck, B., Andersen, S.K.: MUNIN - a causal probabilistic network for interpretation of electromyographic findings. In: International Joint Conference on Artificial Intelligence - Proceedings Milan, Italy, pp. 366–372 (1987)
9. Beinlich, I.A., Suermondt, H.J., Chavez, R.M., Cooper, G.F.: The ALARM monitoring system. In: 2nd European Conference on Artificial Intelligence in Medicine- Proceedings, London, England, pp. 247–256 (1989)
10. Castillo, E.F., Gutiérrez, J.M., Hadi, A.S.: Sensitivity analysis in discrete Bayesian networks. In: IEEE Transactions on Systems, Man, and Cybernetics - Part A: Systems and Humans, no. 27(4), pp. 412–423 (1997)
11. Chavez, R.M., Cooper, G.F.: KNET: Integrating hypermedia and normative Bayesian modeling. In: Uncertainty in Artificial Intelligence 4, North-Holland, Amsterdam, pp. 339–349 (1990)
12. Cheeseman, P., Kelly, M., Taylor, W., Freeman, D., Stutz, J.: Bayesian classification. In: AAAI, St. Paul: MN, pp. 607–611 (1988)
13. Cooper, G.F.: Current research directions in the development of expert systems based on belief networks. In: Applied Stochastic Models and Data Analysis, no. 5, pp. 39–52 (1989)
14. Cheng, J., Druzdzel, M.: AIS-BN: an adaptive importance sampling algorithm for evidential reasoning in large bayesian networks. J. Artif. Intell. Res. JAIR-2000 **13**, 155–188 (2000)
15. Kayaalp, M., Cooper, G.: A Bayesian network scoring metric that is based on globally uniform parameter priors, pp. 251–258 (2002)

16. Zweig, G.G.: Speech recognition with dynamic bayesian networks: Ph.D. dissertation. University of California, Berkeley, p. 169 (1998)
17. Kayaalp, M.M., Cooper, G.F.: Learning dynamic bayesian network structures from data: Ph. D. dissertation. University of Pittsburgh, p. 203 (2003)
18. Codetta-Raiteri, D., Bobbio, A., Montani, S., Portinale, L.: A dynamic Bayesian network based framework to evaluate cascading effects in a power grid. In: Engineering Applications of Artificial Intelligence, vol. 25\4, pp. 683–697 (2012)
19. Jone, T.B., Darling, M.C., Groth, K.M., Denman, M.R., Luger, G.F.: A dynamic bayesian network for diagnosing nuclear power plant accidents. In: Proceedings of the Twenty-Ninth International Florida Artificial Intelligence Research Society Conference, pp. 179–184 (2016)
20. Bescos, C., Schmeink, A., Harris, M., Schmidt, R.: Strategies in the use of static and dynamic bayesian networks in home monitoring. In: IEEE Benelux EMBS Symposium, pp. 31–34 (2007)
21. Hulst, I.R.: Modeling physiological processes with dynamic bayesian networks. Thesis Paper. University of Pittsburgh (2006)
22. de Kock, M., Le, H., Tadross, M., Potgeiter, A.: Weather forecasting using dynamic bayesian networks, Technical report, University of Cape Town (2008)
23. Dean, T., Kanazawa, K.: Probabilistic temporal reasoning (1988)
24. Murphy, K.P.: Dynamic bayesian networks: representation, inference and learning. Thesis Paper. University of California, Berkeley (2002)
25. Pearl, J.: Probabilistic Reasoning in Intelligent Systems: Networks of Plausible Inference. Morgan Kaufmann Publishers Inc., Burlington (1988)
26. van der Gaag, L.C., Coupé, V.M.: Sensitivity analysis for threshold decision making with Bayesian belief networks. In: Lamma, E., Mello, P. (eds.) AI*IA 99: Advances in Artificial Intelligence, Lecture Notes in Artificial Intelligence, pp. 37–48. Springer, Berlin (1999)
27. D. S. Laboratory: GeNIe & SMILE (1998). http://genie.sis.pitt.edu/about.html#genie. Accessed 12 Oct 2017
28. DNV: Det Norske Veritas (2013). http://www.dnv.com/
29. Murphy, K., Russell, S.: Learning the structure of dynamic probabilistic networks. In: Proceedings of the Fourteenth Conference on Uncertainty in Artificial Intelligence (1998)
30. Dempster, A., Laird, N., Rubin, D.: Maximum likelihood from incomplete data via the EM algorithm. J. Roy. Stat. Soc. 1–38 (1977)
31. Friedman, N.: The Bayesian structural EM algorithm. In: Fourteenth conference on Uncertainty in Artificial Intelligence (UAI 1998), Madison, Wisconsin, USA, SF, pp. 129–138. Morgan Kaufmann (1998)
32. Zhang, Z., Kwok, J., Yeung, D.: Surrogate maximization (minimization) algorithms for AdaBoost and the logistic regression model. In: Proceedings of the Twenty-First International Conference on Machine Learning (ICML 2004), p. 117 (2004)

Satellite Dual-Polarization Radar Imagery Superresolution Under Physical Constraints

Sergey Stankevich(✉), Iryna Piestova, Sergey Shklyar, and Artur Lysenko

Scientific Centre for Aerospace Research of the Earth, National Academy of Sciences of Ukraine, Kiev, Ukraine
{st, ipestova, shklyar}@casre.kiev.ua, mercennarius94@gmail.com

Abstract. A novel physically justified method for spatial resolution enhancement of satellite dual-polarization synthetic aperture radar data is proposed. The method starts from the conversion of the specific land surface radar backscattering into the land surface dielectric permittivity in each polarization band separately. Said conversion is founded on a well-known integral equation model of synthetic aperture radar (SAR) backscattering. Transition from raw radar data to dielectric permittivity forms a common unified image field in each polarization band. Due to the SAR platform's own movement, these fields are affected by some subpixel shift to each other. So, the opportunity to apply the superresolution technique over all permittivity fields at once is enabled with considering ones' subpixel shift. Dual-image iterative superresolution based on Gaussian regularization was used. A standalone software module for statistical estimation of inter-images subpixel shift was developed earlier and applied in current research. A noticeable spatial resolution enhancement of the land surface dielectric permittivity field was achieved. This was demonstrated and quantified for actual dual-polarization radar images from the Sentinel-1 European SAR satellite system over two test sites in Ukraine.

Keywords: SAR imagery · Satellite dual-polarization radar · Superresolution · Subpixel shift · Physical constraints · Land surface dielectric permittivity · Land cover types

1 Introduction

Synthetic aperture radar (SAR) data is widely used for various remote sensing applications, like sea surface natural oil seeps detection [1], structural controls on mineralization in desert [2], geological mapping and mineral exploration [3], reservoirs monitoring [4], soil moisture estimation [5], vegetation assessment [6], etc. It has many advantages over optical imagery, in particular cloud-independent operating, high detectability of metalized surfaces, sensitivity to moisture content, and so on [7]. Operational commissioning of the novel European remote sensing satellite system Sentinel-1 equipped with *C*-band dual-polarization SAR has made radar data even more popular and accessible [8]. One of the current Sentinel-1 SAR imagery drawbacks is the

N. Shakhovska and M. O. Medykovskyy (Eds.): CCSIT 2019, AISC 1080, pp. 439–452, 2020.
https://doi.org/10.1007/978-3-030-33695-0_30

lack of spatial resolution of raw radar data. Due to this fact, the radar data spatial resolution enhancement is very important. There are various possibilities to improve the visual resolution of radar imagery, but upon that radiometric values of radar images are often distorted. This is unacceptable if thematic analysis of SAR data is expected thereafter. Another problem is the fundamental inability to improve actual resolution without additional sources of information. Such sources can be both additional images of the study area and the physical properties of the land surface. So, obtaining additional radar images, it becomes possible to apply a well-known superresolution approach to SAR data.

2 State of the Art

A lot of methods for satellite imagery spatial resolution enhancement are known: subpixel-shifted thermal infrared data enhancement [9], which used for thermal fields' micromapping [10], spatial resolution enhancement using land cover's topological features [11], and multiband radiometric conversion [12], Sen2Res software module for multispectral imagery spatial resolution enhancement [13], spatio-temporal adaptive superresolution [14] and many more.

The quality and level of detail of satellite imagery in remote sensing applications is ensured by the spatial resolution of imaging instrument sensor array [15].

The spatial resolution enhancement requires increasing the number of detector elements in sensor array. It leads to design complication and cost rise of imaging system. In addition, the maximum number of detectors in sensor array is rigidly limited by the level of microminiaturization of semiconductor manufacturing.

The subpixel superresolution allows to soften up these limitations. Subpixel superresolution is based on sequential retrieval of multiple images of the same scene, which are shifted relative to each other on the fractional part of the pixel size. The special processing of the input image sequence calculates the image values inside discrete parts of pixels (subpixels).

Superresolution can be carried out by sequential masking of certain parts of photodetectors in sensor array [16], by sequential linear subpixel shifting of image within the focal plane along the rows and columns of a sensor array [17], or by sequential angular displacement of imaging system's optical axes along the rows and columns of a sensor array [18].

Superresolution significantly reduces the technical requirements for imaging system as a result of the registration time elongation and a certain complication of the receiving unit.

Superresolution is the only possible way to enhance the spatial resolution of remote imaging in the case when the both the number of elements in a sensor array and the field of view of imaging system are constrained [19].

The superresolution general methodology sets up the transition from the initial pixel grid of low resolution images, shifted to a fraction of pixel relative to each other, into the joint subpixel grid of higher resolution output image within a common field of view. Next, the enhanced resolution image values inside subpixels are calculated using one of the known reconstruction algorithms [20].

Spatial resolution enhancement consists of the following stages [21]:

- the sub-pixel displacement estimating of low resolution input images;
- a low resolution images combining into a joint interlaced image;
- the enhanced resolution output image restoring by sharpening the interlaced image.

At the first stage, a series of the low resolution images of joint scene is acquired, shifted relative to each other at a certain fraction of pixel. The next step is the combining the subpixel-shifted images into joint interlaced one. Since the point spread function (PSF) of this image remains the same as for a low resolution image, the interlaced image will have a reduced sharpness. Therefore, the image sharpening is necessary. To do this, the subpixel displacement value should already be known.

There are various types of subpixel displacement, including shift, rotation, affine and projective transformations, continuous and sectionally continuous extensions, and more. For linear shift and global affine transformations, the calculation can be performed in frequency domain thorough the discrete Fourier transform (DFT) [22].

In aerospace imaging the sequence of subpixel-shifted images can be obtained with respect to the platform own movement (Fig. 1).

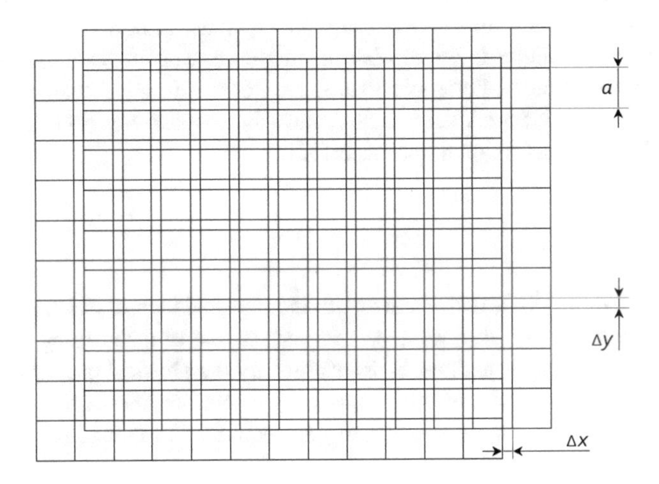

Fig. 1. Subpixel image formation due to the platform movement.

The choice of the image restoration method with superresolution principally depends on the subpixel registration model as well as the mutual arrangement of pixel grids in the low resolution input images. The affine, projective, continuous and sectionally continuous transformations, linear shifts, and rotating shifts are considered separately [23].

The correlation method for estimation of subpixel displacement may be realized in frequency domain. The phase correlation between frequency spectra of input images is estimated, and mutual displacement between them is determined by the maximum likelihood (ML) method [24].

A method for narrow-band interpolation of restored image, known as error-energy reduction (EER), is described in [25, 26]. The spatial restrictions determined by the image values at the points of interpolation, and frequency restrictions are applied one-by-one in this method.

The method of projection on convex sets (POCS) solves a set of linear equations with convex (linear and quadratic) constraints [27]. In the classical POCS method, the sensor PSF included into system of equations.

Ayers and Dainty have been introduced the iterative blind deconvolution (IBD) [28]. This is the most common method for inverse filtering [29] based on DFT. It provides a sufficient resistance to noise and low computational costs. The main drawback of IBD is the non-guaranteed convergence of iteration process.

The method for enhanced resolution image restoring based on wavelet transform is also developed [30]. Its advantage is a native consistency with the interlaced image regularity and relatively small computational costs.

However, there are no analogous methods for different bands joint processing, which are appropriate for radar data. Specifically, the Sentinel-1 satellite SAR imagery consists of two separate bands: both vertical and horizontal polarization data. Because of the intrinsic issues of the land surface radar imaging, the backscattering values in vertical polarization, as a rule, significantly exceed ones in horizontal polarization.

A particular technique is appropriate in this case. Spatial resolution is enhanced through the conversion of heterogeneous source data into an overall single physical surface feature. For infrared data such feature is temperature [31], but microwave remote sensing requires its own approach [32].

3 The Method

For multi-polarization SAR data, such an unified physical feature of land surface can be the dielectric permittivity. It plays a very important role in quantifying the parameters of natural and artificial formations, is determined by the physical properties of the land surface [33] and it, as a rule, is anisotropic [34]. Therefore, the land surface dielectric permittivity is ideal for superresolution of dual-polarization radar imagery.

3.1 Input Data Transform

First, the raw SAR data is converted into backscatter coefficients through radiometric calibration [35].

In the framework of the small perturbation method [36], backscattering coefficients are

$$\sigma^0 = 8k^4 s^2 \cos^4\theta \left|\alpha_p\right|^2 W(2k\sin\theta, 0). \tag{1}$$

where $k = 2\pi/\lambda$ is the wave number, λ is a SAR operating wavelength (5.547 cm for Sentinel-1 C-band SAR), θ is the wave incidence angle, s is the standard deviation of rough land surface irregularities, $W(\cdot)$ is the isotropic irregularities' spectrum for backscattering.

In Eq. (1) index p of the α parameter characterizes polarization type: $p = h$ in case of horizontal polarization, $p = v$ in case of vertical polarization. Backscattering parameters α_h and α_v are calculated as follow [37]:

$$\alpha_h = \frac{\varepsilon - 1}{\left(\cos\theta + \sqrt{\varepsilon - \sin^2\theta}\right)^2}, \tag{2}$$

$$\alpha_v = (\varepsilon - 1)\frac{(\varepsilon - 1)\ \sin^2\theta + \varepsilon}{\left(\varepsilon\cos\theta + \sqrt{\varepsilon - \sin^2\theta}\right)^2} \tag{3}$$

where ε is the land surface dielectric permittivity.

Isotropic irregularities' spectrum for backscattering can be estimated by equation:

$$W(2k\sin\theta, 0) \cong \frac{1}{2} l \exp\left[-(kl\sin\theta)^2\right]. \tag{4}$$

where l is a correlation radius [37].

From Eqs. (1)–(4) it follows that backscattering coefficients are estimated for different polarizations in different ways. The relations (2) and (3) allow to obtain the land surface dielectric permittivity value by a couple of two independent measurements.

Thus, in order to evaluate the surface permittivity, the first equation reduces to linear, and second to quadratic one. The resulting system of equations can be easily solved taking into account the physical constraints (surface permittivity is not greater than the threshold value) using simple mathematical transformations.

3.2 Subpixel Shift Estimation

When land surface dielectric permittivity is calculated by both vertical and horizontal polarization backscatterings separately, then subpixel shift between ones' distributions can be estimated statistically [38]. To simplify later computations, it is practicable to pre-covert images into the frequency domain using DFT.

Subpixel processing is performed for images witch have subpixel shift relatively that is less than pixel dimension. The reconstructed resulting image of higher spatial resolution is the output of this processing. Since stochastic disturbances affect the process of image registration, it is necessary to calculate the subpixel shift between all images that are involved in the reconstruction of the resulting image.

Before starting the restoration, it is necessary to bind the input images to each other. One should be chosen as a base one, and used for transformation the others to achieve pixel alignment accuracy. Since the image is built according to the laws of the central projection, it is possible to achieve such a combination by affine transformation according to the system of reference points of the images [39].

When all the data transformed into the same physical indicator, there are two images with sub-pixel shift. Since its' sub-pixel offset is evaluated, an algorithm based on Gaussian regularization [40] can be applied in a sliding window.

3.3 Data Processing

Thus, in compliance with the technique, described above, the Sentinel-1 SAR imagery processing performed as shown in Fig. 2 dataflow diagram.

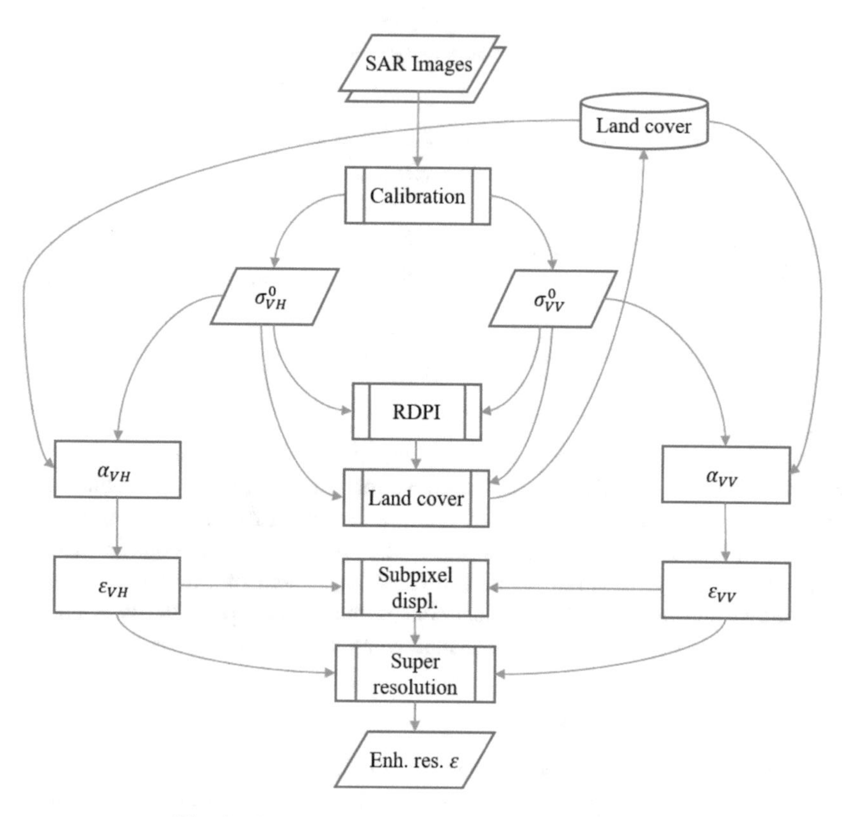

Fig. 2. SAR imagery processing dataflow diagram.

The original dual-polarization SAR image after radiometric calibration is converted to specific radar backscattering, both σ^0_{VH} and σ^0_{VV}. Because the land surface roughness parameters are different for different types of land cover, then their preliminary classification is required. Such classification is based on the calibrated dual-polarization data, but a relative difference polarization index (RDPI) also may be involved to improve classification reliability [6]. The land cover classification ensures the each pixel relation with the land surface roughness parameters from the land cover database. After it, the land surface dielectric permittivity can be calculated from (2) and (3) equations by radar backscattering data using a specially developed software module.

For this purpose, the radar beam incidence angle data and the land cover types' database are used. After the superresolution's process, the land surface dielectric permittivity distribution of enhanced spatial resolution is obtained.

3.4 Satellite Imagery Superresolution Algorithm

The superresolution implementation must perform the coinciding of input images, their centering, the autocovariance matrices calculation, the window-processing operator's definition and modification, and iterative regularization [41].

The algorithm for superresolution image restoring used the two low resolution input images consists of the following steps:

Step 1. Sub-pixel displacement values between input images estimating.
Step 2. The noise image forming.
Step 3. The average values and the autocovariation functions of images and noise calculating.
Step 4. The operator for a regularizing inverse transform inside a sliding window determining.
Step 5. The initial approximation calculating.
Step 6. Iterative image regularization inside a sliding window with the residuals estimating and corrections applying.
Step 7. The iterative procedure completion if the residuals has stop reduced or the maximum number of iterations has been reached.
Step 8. The output image with superresolution obtaining from the final iteration result.

The algorithm for satellite imagery superresolution with iterative regularization inside a sliding window is implemented as a standalone software module with a graphical user interface (GUI), which is shown in Fig. 3.

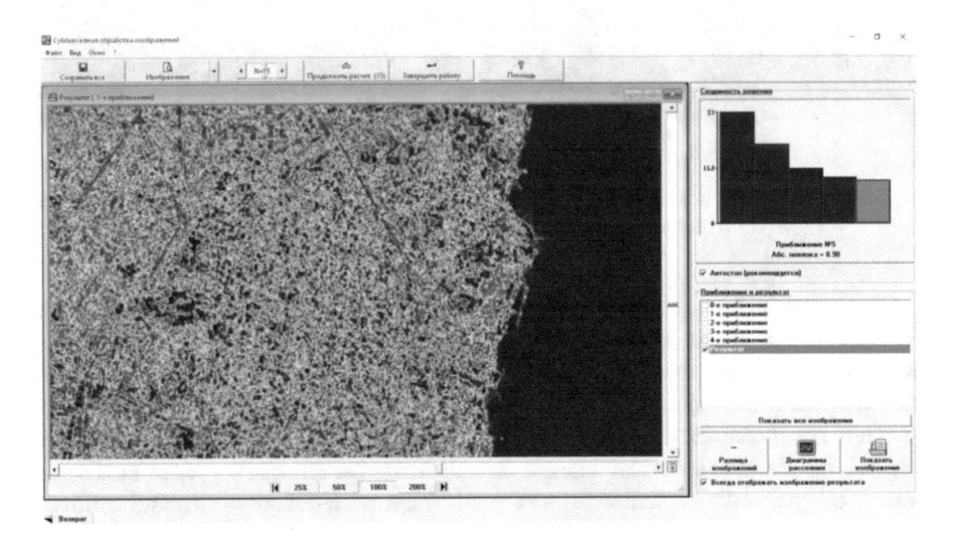

Fig. 3. GUI of software module for satellite imagery superresolution.

Developed software module allows to control the iteration processing, as well as to display the partial results and corresponding statistics if necessary.

4 Results and Discussion

The above-stated method was used to enhance spatial resolution of dielectric permittivity fields over two test areas: Odessa city and Kiev city (Ukraine). Each test site was represented by a 2000 × 2000 pixels (20 × 20 km) size radar image fragment.

Raw Sentinel-1A and Sentinel-1B 10 m spatial resolution Interferometric Wide (IW) swath mode SAR Ground Range Detected (GRD) data product (Fig. 4) provided by Copernicus Scientific Data Hub (https://cophub.copernicus.eu/dhus/) was processed according to the above presented superresolution technique. Radiometric calibration was performed first using open-source Sentinel Application Platform (SNAP) software (https://step.esa.int/main/toolboxes/snap/). The spatial distributions of specific radar backscattering values for vertical and horizontal polarization were evaluated.

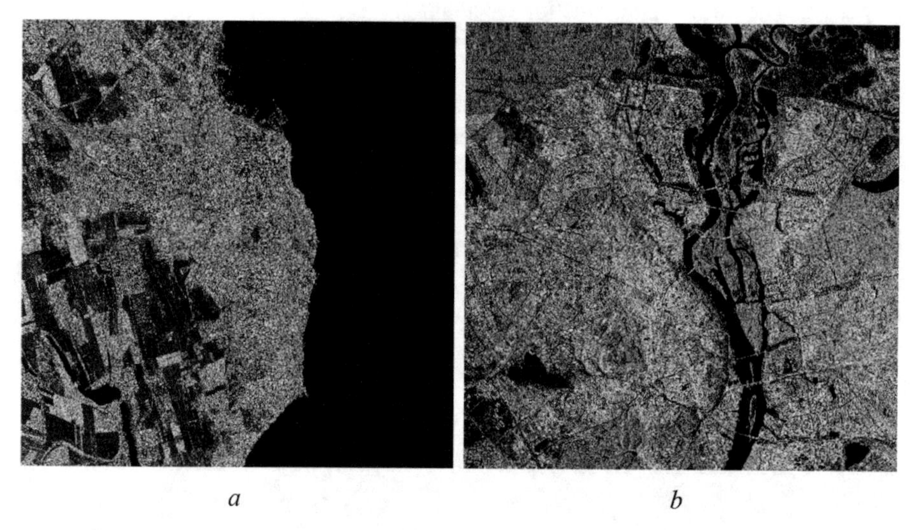

Fig. 4. Input SAR data (horizontal polarization image): *a* – Odessa city (Ukraine), June 15, 2019; *b* – Kiev city (Ukraine), June 17, 2019.

The recalculation the dual-polarization data of the radar backscattering into the physical values of the land surface dielectric permittivity is carried out taking into account the distribution of land cover types (Fig. 5). The last one was obtained by unsupervised land cover classification based on three data layers: σ^0_{VH}, σ^0_{VV} and RDPI.

Land cover classification was conducted according to the nearest neighbor method, which is the simplest and most relevant for radar datasets. The main types of land cover were distinguished: vegetation – trees and grass, water, bare soils, and artificial surfaces.

Once the land cover classification has been completed, all the necessary parameters for further radar data processing have been determined. As a result, the spatial distributions of the land surface dielectric permittivity within the study area (Fig. 6) were obtained.

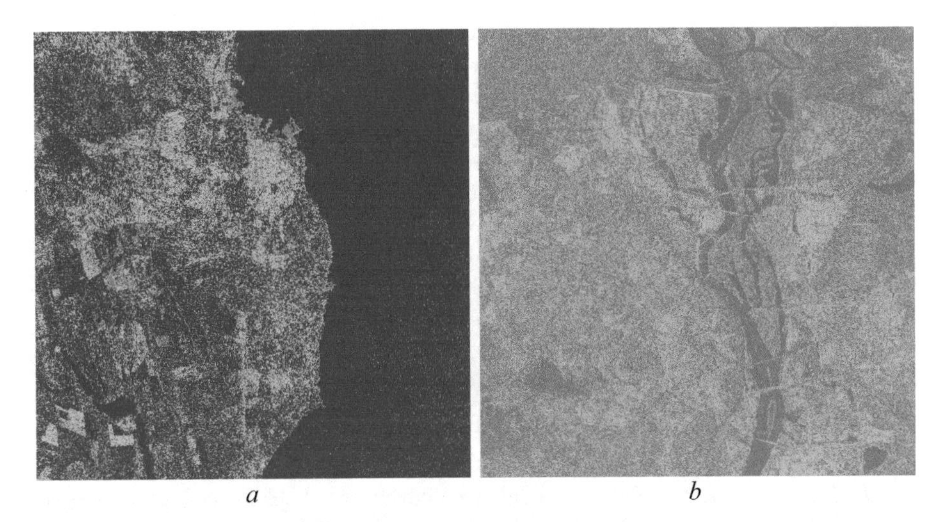

a *b*

Fig. 5. Study area land cover classification: *a* – Odessa city (Ukraine), June 15, 2019; *b* – Kiev city (Ukraine), June 17, 2019.

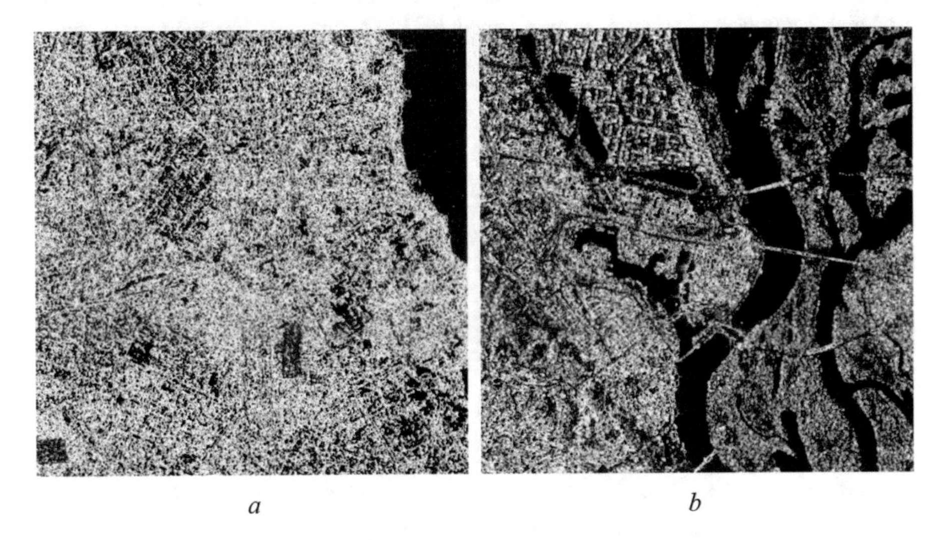

a *b*

Fig. 6. Dielectric permittivity spatial distribution based on vertical polarization data: *a* – Odessa city (Ukraine), June 15, 2019; *b* – Kiev city (Ukraine), June 17, 2019.

The last steps were the calculation of the subpixel displacement between two images based on different polarization permittivity data and restoration the dielectric permittivity spatial distribution with an enhanced spatial resolution (Fig. 7). Since the input images were 2000 $\times$ 2000 pixels size, then after resolution enhancement the output images became 4000 $\times$ 4000 pixels size, i.e. nominally doubled resolution.

Visual comparison of Figs. 6 and 7 image fragments demonstrates some improvement in terrain's detail after the superresolution. To determine the quantitative

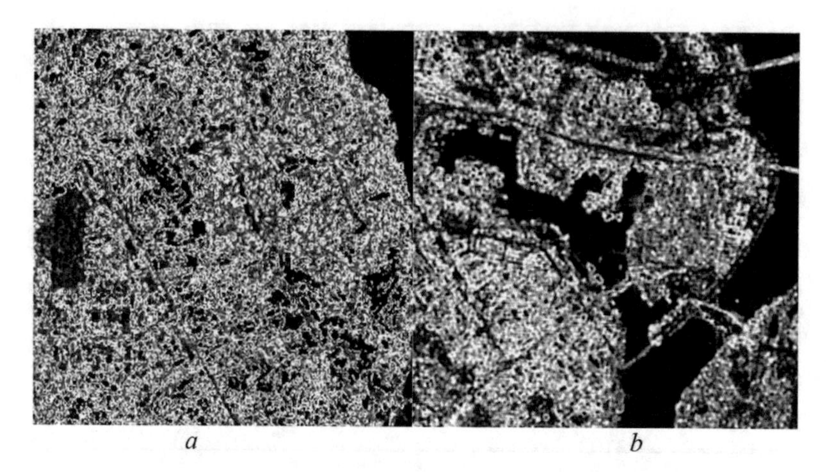

Fig. 7. Enhanced spatial resolution dielectric permittivity spatial distribution: *a* – Odessa city (Ukraine), June 15, 2019; *b* – Kiev city (Ukraine), June 17, 2019.

improvements in spatial resolution of data derived from radar imagery superresolution, a spatial-frequency analysis was performed [42]. The modulation transfer functions (MTF) of input and output images were calculated automatically using special-purpose software (Figs. 8 and 9).

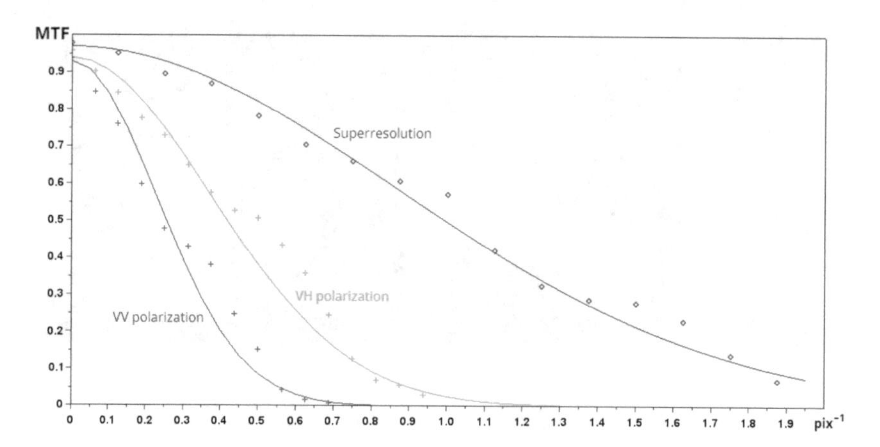

Fig. 8. Experimental MTFs of dielectric permittivity spatial distributions (Odessa city test area, June 15, 2019).

The evaluation was performed with experimental MTFs, which are marked in Figs. 8 and 9 as separate points. Solid lines designate an approximating MTFs. Experimental MTFs demonstrate the 35–65% improvement in resolution between the input dielectric permittivity distributions and superresolution outputs.

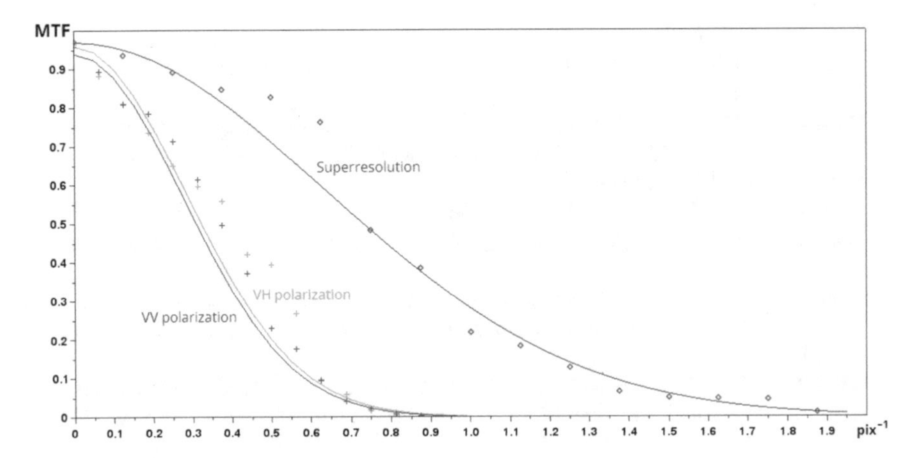

Fig. 9. Experimental MTFs of dielectric permittivity spatial distributions (Kiev city test area, June 17, 2019).

5 Conclusions

The physically grounded method for dual-polarization radar satellite imagery superresolution is offered. This method is used the physical values unifying of the raw radar imagery. In such way two or several separate spatial distributions of the same physiological value, namely – land surface dielectric permittivity, are formed from different polarization images of radar scattering. Since the raw radar images are definitely subpixel-shifted one relative other, this circumstance can be used for the superresolution of the joint dielectric permittivity [43].

The described superresolution technique is well-suited for processing SAR data from the modern Sentinel-1 European satellite system. The resolution improvement in spatial distribution of dielectric permittivity by 35–65% after superresolution was demonstrated experimentally. The proposed approach will be quite useful in many applications of radar remote sensing – in geology and mineral prospecting, in natural resource conservation and environmental protection, in agriculture, in bioproductivity, in carbon cycle analysis, and in others.

In our opinion, future research can be aimed at elaboration of radar backscattering theory to increase the accuracy of the land surface dielectric permittivity calculation, as well as at research and development a comprehensive database on radiometric specifications of various land cover types. Algorithms and software for superresolution need to be refined and highly-developed too.

References

1. Zatyagalova, V.V., Ivanov, A.Y., Golubov, B.N.: Application of ENVISAT SAR imagery for mapping and estimation of natural oil seeps in the south Caspian sea. In: Proceedings of the 'ENVISAT Symposium 2007', Montreux, pp. 1–6. ESA (2007)

2. Kusky, T.M., Ramadan, T.M.: Structural controls on Neoproterozoic mineralization in the South Eastern Desert, Egypt: an integrated field, Landsat TM, and SIR-C/X SAR approach. J. Afr. Earth Sci. **35**, 107–121 (2002)
3. Ramadan, T.M.: Use of ERS-2 SAR and Landsat TM images for geological mapping and mineral exploration of Sol Hamid Area, South Eastern Desert, Egypt. Egypt. J. Remote Sens. Space Sci. **VI**, 13–24 (2003)
4. Amitrano, D., Di Martino, G., Iodice, A., Mitidieri, F., Papa, M.N., Riccio, D., Ruello, G.: Sentinel-1 for monitoring reservoirs: a performance analysis. Remote Sens. **6**(11), 10676–10693 (2014)
5. Gao, Q., Zribi, M., Escorihuela, M.J., Baghdadi, N.: Synergetic use of Sentinel-1 and Sentinel-2 data for soil moisture mapping at 100 m resolution. Sensors **17**(9), 21 (2017). a.1966
6. Stankevich, S.A., Kozlova, A.A., Piestova, I.O., Lubskyi, M.S.: Leaf area index estimation of forest using Sentinel-1 C-band SAR data. In: Proceedings of 5th Microwaves, Radar and Remote Sensing Symposium (MRRS 2017), Kiev, pp. 253–257. IEEE (2017)
7. Ulaby, F., Long, D.G.: Microwave Radar and Radiometric Remote Sensing. University of Michigan Press, Ann Arbor (2013)
8. Geudtner, D., Torres, R., Snoeij, P., Ostergaard, A., Navas-Traver, I., Rommen, B., Brown, M.: Sentinel-1 system overview and performance. In: Proceedings of 'ESA Living Planet Symposium 2013', Edinburgh, 4 p. ESA (2013)
9. Stankevich, S.A., Shklyar, S.V., Podorvan, V.N., Lubskyi, N.S.: Thermal infrared imagery informativity enhancement using sub-pixel co-registration. In: International Conference on Information and Digital Technologies (IDT), Rzeszów, pp. 245–248. IEEE (2016)
10. Piestova, I., Lubskyi, M., Svideniuk, M., Golubov, S., Sedlacek, P.: Satellite imagery resolution enhancement for urban area thermal micromapping. Cent. Eur. Res. J. **4**(1), 35–39 (2018)
11. Zaitseva, E., Piestova, I., Rabcan, J., Rusnak, P.: Multiple-valued and fuzzy logics application to remote sensing data analysis. In: 26th Telecommunications Forum (TELFOR), Belgrade, pp. 1–4. IEEE (2018)
12. Stankevich, S.A., Lubskyi, M.S., Mosov, S.P.: Natural color aerial imagery superresolution with bands radiometric conversion. In: Proceedings of 17th International Conference on Mathematical Methods in Electromagnetic Theory (MMET 2018), Kiev, pp. 99–102. IEEE (2018)
13. Brodu, N.: Super-resolving multiresolution images with band-independent geometry of multispectral pixels. IEEE Trans. Geosci. Remote Sens. **55**(8), 4610–4617 (2017)
14. Zhu, H., Tang, X., Xie, J., Song, W., Mo, F., Gao, X.: Spatio-temporal super-resolution reconstruction of remote-sensing images based on adaptive multi-scale detail enhancement. Sensors **18**(2), 20 (2018). a. 498
15. Zhu, L., Suomalainen, J., Liu, J., Hyyppä, J., Kaartinen, H., Haggren, H.: A review: remote sensing sensors. In: Rustamov, R.B., Hasanova, S., Zeynalova, M.H. (eds.) Multi-purposeful Application of Geospatial Data, pp. 19–42. IntechOpen, London (2018)
16. Karybali, I.G., Psarakis, E.Z., Berberidis, K., Evangelidis, G.D.: An efficient spatial domain technique for subpixel image registration. Signal Process. Image Commun. **23**(9), 711–724 (2008)
17. Ben-Ezra, M., Zomet, A., Nayar, S.K.: Video super-resolution using controlled subpixel detector shifts. IEEE Trans. Pattern Anal. Mach. Intell. **27**(6), 977–987 (2005)

18. Belenok, V.Yu., Burachek, V.G., Zatserkovny, V.I., Popov, M.A., Stankevich, S.A.: Subpixel image acquisition for detailed aerospace imaging. In: Proceedings of the 8th International Conference on Digital Technologies (DT 2011), pp. 190–193. University of Žilina, Žilina (2011)
19. Stankevich, S.A., Shklyar, S.V., Tiagur, V.M.: Subpixel resolution satellite imaging technique. In: Proceedings of the 9th International Conference on Digital Technologies (DT 2013), pp. 81–84. University of Žilina, Žilina (2013)
20. Park, S.C., Park, M.K., Kang, M.G.: Super-resolution image reconstruction: a technical overview. IEEE Signal Process. Mag. **20**(3), 21–36 (2003)
21. Young, S.S., Driggers, R.G., Jacobs, E.L.: Signal Processing and Performance Analysis for Imaging Systems. Artech House, Norwood (2008)
22. Stone, H.S., Orchard, M.T., Chang, E.-C., Martucci, S.A.: A fast direct Fourier-based algorithm for subpixel registration of images. IEEE Trans. Geosci. Remote Sens. **39**(10), 2235–2243 (2001)
23. Vandewalle, P., Baboulaz, L., Dragotti, P.L., Vetterli, M.: Subspace-based methods for image registration and super-resolution. In: Proceedings of the 15th International Conference on Image Processing (ICIP 2008), San Diego, pp. 645–648. IEEE (2008)
24. Shekarforoush, H., Berthod, M., Zerubia, J.: Subpixel image registration by estimating the polyphase decomposition of cross-power spectrum. In: Proceedings of the Computer Society Conference on Computer Vision and Pattern Recognition (CVPR 1996), San Francisco, pp. 532–537. IEEE (1996)
25. Gerchberg, R.: Super-resolution through error energy reduction. Acta Opt. **21**(9), 709–720 (1974)
26. Papoulis, A.: A new algorithm in spectral analysis and band-limited extrapolation. IEEE Trans. Circ. Syst. **22**(9), 735–742 (1979)
27. Stark, H., Oskoui, P.: High-resolution image recovery from image-plane arrays using convex projections. J. Opt. Soc. Am. **7**(11), 1715–1726 (1989)
28. Ayers, G.R., Dainty, J.C.: Iterative blind deconvolution method and its applications. Opt. Lett. **13**(7), 547–549 (1988)
29. Popov, M.A., Stankevich, S.A.: About restoration of the scanning images received onboard a Sich-1M space vehicle by inverse filtering method. In: Proceedings of the 31st International Symposium on Remote Sensing of Environment, Saint Petersburg, pp. 488–490. ISPRS (2005)
30. Nguyen, N., Milanfar, P.: A wavelet-based interpolation-restoration method for superresolution (wavelet superresolution). Circ. Syst. Signal Process. **18**(4), 321–338 (2000)
31. Lyalko, V.I., Popov, M.A., Stankevich, S.A., Shklyar, S.V., Podorvan, V.N., Likholit, N.I., Tyagur, V.M., Dobrovolska, C.V.: Prototype of satellite infrared spectroradiometer with superresolution. J. Inf. Control Manag. Syst. **12**(2), 153–164 (2014)
32. Ulaby, F.T., Moore, R.K., Fung, A.K.: Microwave Remote Sensing: Active and Passive, vol. II. Artech House, Dedham (1982)
33. Ku, C.-S., Chen, K.-S., Chang, P.-C., Chang, Y.-L.: Imaging simulation for synthetic aperture radar: a full-wave approach. Remote Sens. **10**(9), 16 (2018). a. 1404
34. Wang, H.: Soil moisture retrieval from microwave remote sensing observations. In: Civeira, G. (ed.) Soil Moisture, pp. 29–54. IntechOpen, London (2019)
35. Freeman, A.: Radiometric calibration of SAR image data. ISPRS Arch. **XXIX**(B1), 212–222 (1992)

36. Kang, W., Yu, S., Ko, S., Paik, J.: Multisensor superresolution using directionally adaptive regularization for UAV images. In: Toro, F.G., Tsourdos, A. (eds.) UAV Sensors for Environmental Monitoring, pp. 272–294. MDPI, Basel (2018)
37. Soares, J.V., Rennó, C.D.: Soil moisture retrieval from active microwave remote sensing. In: Proceedings of the First Latino-American Seminar on Radar Remote Sensing – Image Processing Techniques, Buenos Aires, pp. 192–203. ESA (1997)
38. Stankevich, S.A., Shklyar, S.V., Lysenko, A.R.: Software module for estimating subpixel shift of images acquired from quadcopter. Ukr. J. Remote. Sens. **17**, 10–13 (2018)
39. Kennedy, R.E., Cohen, W.B.: Automated designation of tie-points for image-to-image coregistration. Int. J. Remote Sens. **24**(17), 3467–3490 (2003)
40. Popov, M.O., Stankevich, S.A., Shklyar, S.V.: An algorithm for resolution enhancement of subpixel displaced images. Math. Mach. Syst. **1**, 29–36 (2015)
41. Liu, H.Y., Zhang, Y.S., Ji, S.: Study on the methods of super-resolution image reconstruction. Int. Arch. Photogramm. Remote Sens. Spat. Inf. Sci. **XXXVII**(B2), 461–466 (2008)
42. Chen, C.H.: Signal and Image Processing for Remote Sensing. CRC Press, Boca Raton (2012)
43. Stankevich, S., Piestova, I., Shklyar, S., Lysenko, A.: Satellite dual-polarization radar imagery superresolution under physical constraints. In: Proceedings of the International Scientific Conference "Computer Sciences and Information Technologies" (CSIT 2019), vol. 3, pp. 228–231 (2019)

Situation Diagnosis Based on the Spatially-Distributed Dynamic Disaster Risk Assessment

Maryna Zharikova(✉) and Volodymyr Sherstjuk

Kherson National Technical University, Kherson, Ukraine
marina.jarikova@gmail.com, vgsherstyuk@gmail.com

Abstract. A dynamic spatially-distributed model of integral risk assessment is represented in the paper. A multi-risk for a valuable object is formed as a combination of four components such as danger, threat potential, threat level, and potential losses. In order to provide comparing the risks from different disasters and assess their joint influence on the valuable object in the form of multi-risk a quantitative value of each risk component is proposed to represent in the form of qualitative value using the appropriate scales. A diagnostic method for disaster response operations based on the spatially-distributed model of integral risk assessment is developed. A hybrid algorithm of identification of the situation in disaster conditions using the case-based and rule-based reasoning is described. The experiment examining the validity and efficiency of the proposed hybrid diagnosis method is described. It's concluded that the proposed method provides sufficient performance for the cell size 5 m and above, so it is acceptable for solving the practical forest fire fighting problems in GIS-based DSS.

Keywords: Forest fire fighting · Disaster · Risk · Intelligent diagnosis method · Symptoms · Situation

1 Introduction

The first decade of the XXI century is characterized by activation of natural disasters that disturb human and natural environments, causing injuries, casualties, property damages, as well as business interruptions. Expanding the scope of disasters occurs due to such factors as rising population, growing urbanization, deepening industrialization, environmental degradation, global warming [1], etc. Thus, at present, the problem of timely and grounded decision making for disaster response, which allows mitigating disaster damage, is in the focus of attention of researchers.

Giving a significant disaster dynamic, a big volume of heterogeneous input data characterized by incompleteness and uncertainty, as well as a significant amount of computations, a decision-maker faces a problem of time shortage that leads to psychological tension and complicates making the adequate decisions in high responsibility conditions. For this reason, taking into account insufficient psychophysiological and heuristic capabilities of the decision-maker, to eliminate the problem of time shortage in disaster conditions the decision support systems can be used.

In order to effectively respond to natural disasters, such systems should be based on the natural risk assessments making at spatial and temporal scales. Such assessments

N. Shakhovska and M. O. Medykovskyy (Eds.): CCSIT 2019, AISC 1080, pp. 453–472, 2020.
https://doi.org/10.1007/978-3-030-33695-0_31

are related to the certain valuable objects situated within an area of interest (AOI) and represent the potentials for the valuable objects to take losses as a result of a disaster.

So, there is a growing interest in developing spatial risk-oriented decision support systems for disaster response. The primary task for decision-making is to diagnose the situation based on the assessments of dangers, threats, or risks to valuable objects. Diagnosis is considered as a process of assigning a given data to some category and then respond appropriately based on such classification [1].

The aim of this paper is to propose a new spatially-distributed dynamic model of disaster risk analysis and a method of the situation diagnosis based on this model.

2 Literature Review

There is a vast literature on risk assessment for disaster response. The majority of the authors assess risk quantitatively, using probabilistic methods of simulation modeling not suitable for the assessment of disaster risk in response time, where probability is out of the question [2–6]. Such methods of risk assessment lead to high computational complexity and reduce system performance significantly, which is not allowed in response time systems.

Literature review [5–7] has shown that the concept of disaster risk should contain the following interrelated groups of components:

1. the components describing the potential of the danger source, i.e. disaster;
2. the components describing the valuable objects being under a destructive influence of disaster.

Let us consider the main approaches to disaster risk assessment that are now in use [8]:

1. the probabilistic approach,
2. the approach based on the event trees,
3. the approach based on the risk matrix,
4. the approach based on the indicators.

The first two approaches are quantitative, and the last two approaches are qualitative.

The probabilistic approach to the risk assessment is based on statistical data processing and requires knowledge of the probabilities of disaster occurrence and spread [8–11]. As a result of the difficulty in obtaining a representative statistical sample, it is taken within the big areas of terrain (often in a national scale) in a big period of time (10–100 years) compensating unrepresentativeness. The methods of simulation modeling linked to a high computational complexity often compensate for an insufficient number of statistical data. Risk assessment obtained using this approach has a static value. The probabilistic approach can make good use for disaster prevention but can't be of value for response time decision making when the decisions are made in time limit conditions.

One of the common methods of risk assessment is using the event trees that allow modeling a sequence of events forming the structures of any complexity level. Event trees can be used for assessing multi-risk in the conditions of several disasters coming one after another ("domino effect"). The limitation of the event trees is the fact that the events are not linked to the terrain [12]. Nevertheless, the method is quite flexible and has a big potential for development. Events georeferencing could allow modeling spatially-distributed disaster risk. Such models can be completed with different ways of estimating the uncertainty (probability theory, fuzzy, rough sets, etc.)

The approach based on the risk matrix enables assessing the risk using certain classes instead of precise values that allows overcoming the problem of quantitative approach in the conditions of an unrepresentative sample. Risk matrixes are built based on the expert knowledge that allows classifying the areas of terrain by the levels of risk on the basis of the information on disaster frequency and expected damages [13]. It's noteworthy that the assessment of such frequency is not always absolute, as different combinations of disaster frequency and expected damages can correspond to the same area of the terrain. The effectiveness of the implementation of this approach depends on the competence of the experts who form the disaster scenarios and rank them in accordance with the frequency and expected damages.

The approach based on the indicators is used in situations when quantitative and semi-quantitative methods can't be employed due to a lack of input data. This approach allows measuring overall risk assessment, taking into account additional components such as social and economic components, as well as the vulnerability of the objects under disaster risk. The drawback of the approach is that the final assessment is relative and doesn't show the information about the expected damages [14].

Based on the review of risk assessment methods provided above we can conclude that existing approaches allow obtaining static risk assessment and can't be implemented in response time systems [15]. Most common risk assessment methods are quantitatively based on statistical data analysis. Giving the fact that it's difficult to collect an abundance of data describing the real disasters, a sample is taken within a big area of terrain and in a big period of time. The methods of simulation modeling connected with conducting a big number of computational experiments compensate for the unrepresentativeness of the sample. In particular, the Monte Carlo method is often used for calculation of disaster occurrence probability. In any case, the statistical methods have essential drawback such as huge computational complexity.

Much has been also written about the problem of diagnosis in the decision support systems for disaster response. In [16] the authors define a diagnosis as a process allowing to identify a problem using the decision criteria (DC) (the data of observation obtained by monitoring or assessments).

There are several basic approaches to designing the diagnosis systems such as model-based approach, an approach based on the heuristic knowledge, expert approach, case-based approach, etc.

The model-based approach is based on the models and is used providing that a model of a system can be built [17, 18]. In most practical tasks it's impossible. The most commonly used approach for such cases is the approach based on heuristic knowledge that describes a problem area that reproduces expert reasoning [19, 20]. Knowledge-based systems have some limitations such as the inability to reproduce creative thoughts of experts in unusual situations and to adapt to environmental changes. The more field-specific approach to the development of diagnosis systems is the expert approach that accumulates specific subject knowledge [21, 22]. The development of the systems based on the expert approach is a labor-intensive and costly knowledge engineering process [23, 24]. Knowledge engineering is connected with some difficulties. On the one hand, the experts not always can explain their logic so that it can be programmed. On the other hand, different experts often have a different understanding of the same problem.

The case-based approach focused on using a repository of the existing solutions (cases) allows eliminating the problem of expert knowledge representation. When a diagnosis task arises the case-based system identifies a scenario with existing cases. Compared to the expert system where knowledge is represented explicitly by experts in the form of heuristic rules, in a case-based system, the knowledge is represented implicitly in the form of cases [25].

Another way to eliminate the problem related to knowledge engineering is the development of the learning system that obtains knowledge automatically [26]. There are several technologies of learning used in such systems [27] such as neural networks [28], genetic algorithms [29], symbolic artificial intelligence [30], statistical pattern recognition, evolutionary computation [31], etc.

Obviously, disasters are processes with unpredictable behavior. Taking into account the uncertainty and unavailability of the input data describing such processes, as well as a lack of scientific understanding of their behavior, the modeling and forecasting of disasters is a challenging task. This fact eliminates a separate using of such approaches as knowledge engineering, as well as the approaches based on the models and rules. The most adequate approach is the case-based approach but its realization depends on the competence of the base of accumulated cases because the oversized base leads to a big computational complexity, and the insufficient volume of the base doesn't cover the whole set of possible situations during classification. The learning systems are nonlinear, they can classify the situations even in an uncertain environment, but they don't provide necessary credibility for a decision-maker.

Thus, giving the fact that none of the above-mentioned approaches to the diagnosis can be used separately, automatic diagnosis of the situation based on the observations obtained during monitoring requires the development of hybrid method differing from classical ones. The authors propose a diagnosis method based on a combination of case-based and rule-based approaches that will provide a reduction of case base volume.

The paper is dedicated to solving the applied research task of response time diagnosis of the situation in the condition of deciduous disasters based on a spatially-distributed dynamic risk model. The authors consider the response time comprehensive risk analysis taking into account the full range of disaster events spreading in space and time.

3 Object Domain Description

We consider risk from disasters, which vary in their origin (e.g. biological, geophysical, hydrological, climatological, and extraterrestrial), intensity, scale, rate of spread, impact on vulnerable objects, as well as controllability.

According to a scale, disasters can cover local areas of terrain (concentrated disasters) such as floods, landslides, or the whole regions and even countries such as drought or the processes connected with the climate change.

According to a spread rate, disasters can be divided into fast-spreading and slow-spreading. The fast-spreading disasters spread over a short period of time (from several seconds to several days), duration of the slow-spreading disasters reaches from a month to several centuries.

According to a level of efficiency of decision-maker intervention (controllability), disasters can be divided into controlled and uncontrolled.

The paper dwells on the response time systems, therefore the authors are limited to a consideration of the fast-spreading, local disasters with the duration being from several hours to several days. On top of it, giving the fact that the risk analysis should be a base of situation diagnosis, the considered disasters should be controlled.

The disasters are connected with vulnerable objects, which fall under their adverse effect and require attention from a decision-maker. As a result of time-limit in the response systems all the objects cannot be considered. Therefore, it's necessary to extinguish a group of target objects being of value for people, which require foremost defense. Target objects can include the residential buildings and structures, the industrial buildings and structures, the transport infrastructure (roads and bridges), the energy infrastructure objects (pipelines, power lines), etc.

Each target object has its value, as well as characterized by vulnerability and resilience. When a disaster affects a target object, its value changes dynamically. The characteristics of vulnerability and resilience determine dynamics of value in reaction to a certain disaster influence. This dynamic can be represented as a function. Each class of target objects is characterized by a set of value dynamics functions for each class of disasters.

Risk in the context of this research is connected to each target object. Risk is a potential of realization of undesired, adverse effects for target objects in disaster conditions. Such effects result in decline in object's value.

Risk can be potential and active. A potential risk for an object is characterized by a potentiality for occurrence of disaster, which can cause damage to this object. Potential risk is described by a danger that is a possibility of occurrence of disaster of a certain intensity within a certain area.

A danger is materialized when disaster starts spreading and during this process can cause damage to a certain target object. In this case, instead of danger, it makes sense to say about a threat that determines active risk for this object. A threat is a dynamic risk component, which describes spatially-temporal relation between disaster contour and vulnerable object. A threat arises at the moment of danger materialization and loses its meaning at the moment when a disaster covers the object.

After a disaster covers the object its value starts decreasing according to a value dynamics function.

Risk is an integral assessment characterizing a potentiality for an object to be damaged as a result of a disaster of a certain type and intensity. Treats and risks are considered for the target objects. Disaster risk for a certain target object is a function of danger, threat, disaster intensity and potential damage expressed as a change in object's value. In the case of several disasters effecting a target object we can say about multi-threat and multi-risk.

In the paper, the authors consider the diagnosis of the situation for disaster response. Such situation is based on one or several disasters spreading within some area of interest. The area of interest contains a set of target objects, for each of which it's necessary to assess multi-risk. As a result, we obtain a dynamic spatially-distributed model of disaster risk analysis.

4 Spatially-Distributed Dynamic Risk Analysis

We consider the disaster risk analysis within a certain AOI that is represented as an open connected subspace X of three-dimensional Euclidean space endowed with the topological properties [7, 32]. Space X is discretized by a grid C of isometric cubic cells $c \in C$. Assume that each cell $c \in C$ is a homogeneous spatial object of minimal size.

Disaster risk analysis is focused on consideration of low-probability random events that give rise to undesirable adverse consequences [33–35]. In the context of this paper, an adverse event is a disaster occurrence within a certain origin cell $c_l \in C$. Then the risk can be represented as a chance of certain losses for a valuable object in the case of disaster occurrence.

Thus, the level of risk for a valuable object will characterize the situation (disaster) with an uncertain outcome and adverse consequences. The source of risk will be a disaster, the occurrence of which is connected with a certain origin cell $c_l \in C$ in the AOI. A set of cells being most distant from the origin cell in each possible directions creates a disaster contour at any time t. In fact, disaster spreading can be described by moving its contour.

Risk has the following properties:

- *uncertainty*: risk exists if and only if there are a lot of possible sequences of events and, consequently, a lot of possible outcomes;
- *the existence of losses*: risk exists if and only if the outcome of the event can lead to the losses (damage) or other negative consequence;
- *the possibility of analysis*: risk exists if a subjective opinion of a decision-maker is formed and quantitative estimation of a future negative event (possible losses) is given;
- *importance*: risk exists if a would-be event has certain practical importance and affects the interests of at least one subject (decision-maker or an object's owner).

Therefore, the risk doesn't exist without belonging, and risk analysis connects a certain (valuable) object with a certain interested subject (decision maker) through the estimation of chance (possibility) of costs for response operations.

The necessity of assessing the possible losses of the object's value under the adverse scenario of events due to disaster occurrence is a crucial feature of disaster risk estimation. It's noteworthy that a consideration of a chance (possibility) of adverse event occurrence for a certain object provides the estimation of danger for this object. A consideration of the chances of negative consequences of the adverse event for the object provides the threat assessment for a given object.

Thus, a danger, threat, and risk assessments are the certain (different) aspects of the same practical problem, namely the problem of assessing the chances and negative outcomes of adverse events, which have a stochastic nature. In this context, all pointed assessments have a certain value for a decision-maker and can be applied together for spatially-distributed dynamic risk assessment.

Usually, the quantitative risk assessment is represented as a multiplication of event probability and losses. At that, the risk probability is calculated based on the frequency of event occurrences [36]. In the context of this paper, the subject of risk analysis is the estimation of the chances of losses resulting from the involvement of the valuable object in the disaster that has been occurred by the time of consideration and is spreading in time and space. In this case, the frequency of event occurrences is not a reliable base for the estimation of the chance. This leads to the conclusion that the estimation of a conditional probability based on events frequency is not appropriate for the response time systems. Obviously, the assessments of possibility using fuzzy or rough methods based on visual observations made by remote sensing technologies is much more reliable. In this case, the chance estimation can be represented as a conditional possibility that a target object is on the disaster.

Another problem is the estimation of losses for the valuable objects, which can not be estimated in the monetary terms, such as the losses of historical, architectural, cultural values. The authors propose a solution to this problem using quantification of quantitative estimations with the help of the ordered scales and switch to the qualitative assessment. The qualitative assessments of danger and threat can be obtained in the same way [7].

The risk R_{o_iF} for a valuable object oi from disaster F with the source u_F can be assessed as a combination of the following components (Fig. 1):

1. the assessment of danger μ_F created by the source u_F;
2. the assessment of the threat potential S_F for the object o_i from the disaster F;
3. the assessment of the threat level ς_{o_iF} for the object o_i from the disaster F;
4. the assessment of potential losses of value ΔV_{o_iF} for the object o_i from the disaster F.

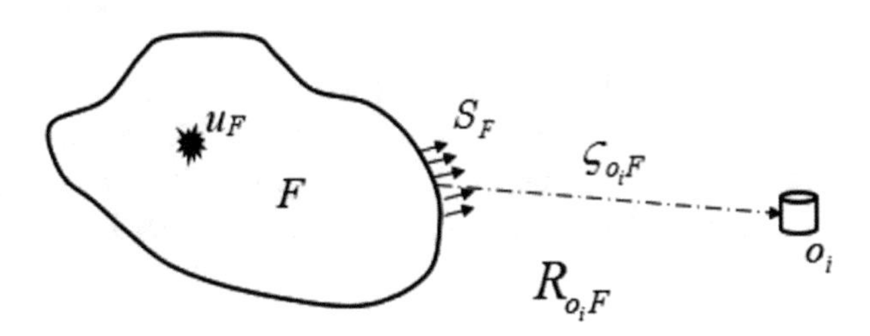

Fig. 1. Risk components

In order to provide comparing the risks from different disasters and assess their joint influence on the valuable object in the form of multi-risk a quantitative value of each risk component P^i such as danger, threat potential, threat level, and potential losses, is proposed to represent P^i first as normalized value from 0 to 1, and then in the form of qualitative value using the appropriate scales.

For each risk component, P^i we introduce a set being a carrier of values $\mathfrak{M}^i = \{\upsilon_1,\ldots\upsilon_m\}$, create a lattice $\mathfrak{L}^i = \langle \mathfrak{M}^i, \preccurlyeq \rangle$ where $\preccurlyeq$ is a partial order relation and define a function $Range\left(dom\left(P^i\right) \xrightarrow{\mathfrak{M}^i} \mathfrak{L}^i\right)$ that divides the range of values of P^i into m intervals.

Mapping the values of each risk component P^i onto a lattice $\mathfrak{L}^i$ allows us switching from a set of values defined on $\mathbb{R}$ to a compact m-limited set of values on $\mathfrak{M}^i$.

Qualitative assessment of a risk component P^i is made using the implementation of the border norms to a scale of influence levels such as minimal, low, acceptable, critical, etc. A number of levels are chosen by experts. Such implementation is performed by the quantification of the quantitative values of risk component P^i on a lattice $\mathfrak{L}^i$ using partial order relation $\preccurlyeq_{P^i}$ such as $\varsigma_{P_0^i} \preccurlyeq_{P^i} \varsigma_{P_1^i} \preccurlyeq_{P^i} \cdots \preccurlyeq_{P^i} \varsigma_{P_Z^i}$ into an appropriate order scale $\c{C}_{P^i} = \left\{\varsigma_{P_0^i}, \varsigma_{P_1^i}, \ldots \varsigma_{P_Z^i}\right\}$ (Table 1).

Table 1. The ordinal scale for risk component assessment

$\varsigma_{P_0^i}$	Minimal value
$\varsigma_{P_1^i}$	Low value
$\varsigma_{P_2^i}$	Acceptable value
$\varsigma_{P_3^i}$	Critical value

Suppose Q_{P^i} is a function that uniquely takes each value of risk component P^i to a certain level of the scale $\c{C}_{P^i}$, such that $Q_{P^i} : P^i \rightarrow \c{C}_{P^i}$. We can quantify the values of each risk component.

To obtain a qualitative risk assessment for a disaster F with a source u_F for an object o_i, we should first obtain a standardized quantitative risk assessment as a product of the normalized component values: $\breve{\mu}_F, \breve{S}_F, \breve{\varsigma}_{o_i F}, \Delta\breve{V}_{o_i F}$. It's difficult to present a threat assessment in the form of one normalized value. It can be presented as an interval value $\hat{\breve{\varsigma}}_{o_i F} = \left[\underline{\hat{\breve{\varsigma}}_{o_i F}}, \overline{\hat{\breve{\varsigma}}_{o_i F}}\right]$.

Taking into account the fact that the threat assessment is an interval value, the risk assessment will also be an interval value [7, 36]:

$$\begin{aligned} \hat{\breve{R}}_{o_i F} &= \breve{\mu}_F \times \breve{S}_F \times \hat{\breve{\varsigma}}_{o_i F} \times \Delta\breve{V}_{o_i F} = \\ &= \left[\breve{\mu}_F \times \breve{S}_F \times \underline{\hat{\breve{\varsigma}}_{o_i F}} \times \Delta\breve{V}_{o_i F}, \breve{\mu}_F \times \breve{S}_F \times \overline{\hat{\breve{\varsigma}}_{o_i F}} \times \Delta\breve{V}_{o_i F}\right] \end{aligned} \quad (1)$$

In the case of several sources of destructive processes that simultaneously affect the valuable object o_i, a multi-risk is created for this valuable object:

$$\hat{\tilde{R}}_{o_i} = \bigoplus_{j=1}^{n_i} \hat{\tilde{R}}_{o_i F_j}, \tag{2}$$

where n_i is a number of risk sources for the valuable object o_i (Fig. 2).

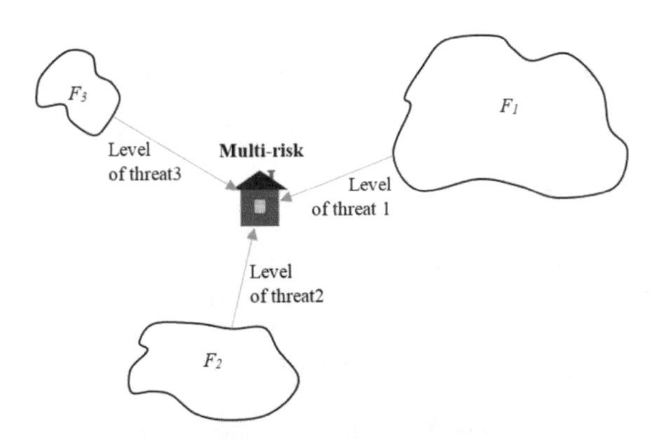

Fig. 2. Multi-risk

The normalized interval quantitative values of the multi-risk can be converted into qualitative values by defining the corresponding ordinal scale $\varsigma_R = \{\varsigma_{R0}, \varsigma_{R1}, \ldots \varsigma_{Rz}\}$ (Table 1) with the partial-order relation $\preccurlyeq_R$ induced on it, such that $\varsigma_{R0} \preccurlyeq_R \varsigma_{R1} \preccurlyeq_R \ldots \preccurlyeq_R \varsigma_{Rz}$ [7] (Table 2).

Table 2. The ordinal scale for risk assessment

ς_{R0}	Absence of risk
ς_{R1}	Low level of risk
ς_{R2}	High level of risk
ς_{R3}	Critical level of risk

For the direct transformation of the normalized interval quantitative estimation of multi-risk into a qualitative assessment, we define a function $\Re$ that implements the mapping $\Re : \hat{\tilde{\mathrm{R}}} \rightarrow \varsigma_R$. Using this mapping, the quantitative interval value of the multi-risk $\hat{\tilde{R}}_{o_i}$ for the valuable object o_i uniquely matches the qualitative assessment R'_{o_i} with respect to the scale ς_R.

A similar approach can be used to assess the multi-risk for any cell c_m of the universe C including the cells that do not belong to any of the valuable objects O^*

(in this case, the value of a cell can be determined by the elements of its state vector, for example, a cell with coniferous vegetation can have a value being equal to the cost of wood, based on the product of the existing volume of wood within the cell at its price). Multi-risk can be expressed quantitatively by the interval $\hat{\tilde{R}}_m$ or qualitatively as R'_m.

A multi-risk trajectory for a certain cell c_m during a time interval $[t_j, \ldots t_k] \in T$ is a continuous sequence $\left[\hat{\tilde{R}}_m(t_j), \ldots, \hat{\tilde{R}}_m(t_k)\right]$.

Accordingly, the dynamic spatially-distributed model of multi-risk for the AOI can be represented as $R = \left[\bigcup_{m=1}^{M} \hat{\tilde{R}}_m(t_j), \ldots, \bigcup_{m=1}^{M} \hat{\tilde{R}}_m(t_k)\right]$, where M is a number of cells in the AOI.

5 Diagnosis of the Situation

The AOI is characterized by an integrated dynamic spatially-distributed estimation of the multi-risk in the conditions of the destructive process at the time t:

$$R^t_\Omega = \{R_i(t) | \forall o_i \in O^*(t)\}. \tag{3}$$

Distributed risk assessment reflects the set of decision criteria (DC) that need to be attributed to a particular class of situations. To do this, we need to define the set of classes of possible situations $\check{S} = \{\check{S}_0, \check{S}_1, \ldots \check{S}_n\}$, such that $\check{S}_1 \cup \check{S}_2 \cup \ldots \cup \check{S}_n = \check{S}$. Suppose p_i, $i = 1, \ldots, n$ is a set of possibilities for their occurrence [15, 36–39], and $X = \{x_1, x_2, \ldots x_m\}$ defines the set of DC associated with the classes of situations $\check{S}$.

Integral risk assessment $R_\Omega \in X$ for each class of situations is included in the set of DC.

Let s^* be a current situation and X^* be its DC vector. Due to restricted visibility, some of the DCs X^* can be uncertain or inaccurate.

Let each DC $x_i \in X$, $i = 1, \ldots, \mathrm{m}$ has a domain of possible values $E_i \cup e_*$, $i = 1, \ldots, m$, and e_* is an uncertain value. DCs from the vector X can be described as the intervals using the rough sets [40], or intervals with the membership functions using the interval fuzzy set, and some of them can be empty.

The diagnostic problem is the task of identifying a possible class of situation $\check{S}^* \in \check{S}$ that can explain the set of uncertain DCs X^* for the current situation s^*. Such a task is the problem of pattern recognition [7].

Each situation corresponds to a certain point or the neighborhood of the point in the Cartesian space of DCs $E_1 \times E_2 \times \ldots \times E_m$.

Obviously, each unidentified situation s^* that has DCs X^* should be compared with a set of classes of possible situations $\check{S} = \{\check{S}_0, \check{S}_1, \ldots \check{S}_n\}$. As a result of the uncertainty of estimates for some DCs, it is not always possible to find the exact match.

The situation in the system should be determined on the basis of the location of the valuable objects that are at maximum risk, as well as the location of the forces and means assigned to eliminate the disaster. Let us denote the set of areas, which contain such forces and means, by $Z = \{z_1, z_2, \ldots, z_n\}$. That is, in order to diagnose the situation in the system at the time t, it is necessary to define [7]:

1. the set of valuable objects, which are in critical risk conditions at the time t: $O*(t) = \{o_1, o_2, \ldots, o_k\}$;
2. the set of disasters: $F(\mathrm{t}) = \{F_1(t), F_2(t), \ldots, F_l(t)\}$;
3. the set of areas of the assigned forces and means location: $Z = \{z_1, z_2, \ldots, z_n\}$.

To diagnose a situation at the time t for each valuable object $o_i \in O*(t)$, we must evaluate the least time for which the contour of some disaster from the set $F(\mathrm{t})$ will reach this object, as well as the least time necessary to move the forces and means from their closest location.

Each valuable object $o_i \in O*(t)$ corresponds to the set of time intervals $T_{o_iF} = \{t_{o_iF_1}, t_{o_iF_2}, \ldots, t_{o_iF_l}\}$, for which the contours of disasters will reach this object, as well as the set of time intervals $\mathrm{T}_{o_iZ} = \{\tau_{o_iZ_1}, \tau_{o_iZ_2}, \ldots, \tau_{o_iZ_n}\}$ required to deliver forces and means from their locations.

The first set is dynamic. After defining these two sets for each valuable object $o_i \in O*(t)$, we need to find the smallest value from each set. After that, each valuable object $o_i \in O*(t)$ will match the couple: $t_{o_iF} = \min(T_{o_iF})$, $\tau_{o_iZ} = \min(\mathrm{T}_{o_iZ})$.

The situation in the system at the time t can be defined as a set of pairs:

$$\check{s}^t = \{(T_{o_iF}(t), \mathrm{T}_{o_iZ}(t)) | \forall o_i \in O^*(t)\}. \quad (4)$$

We distinguish the following classes of the situations in the system:

- $\check{S}_1$: $(\forall o_i \in O^*(t))(t_{o_iF} > \tau_{o_iZ})$ – a class of non-critical situations, where the decision-maker has enough time for the deployment of forces and means for the elimination of the disaster;
- $\check{S}_2$: $(\exists o_i \in O^*(t))(t_{o_iF} \leq \tau_{o_iZ})$ – a class of critical situations, where the task of rescuing objects is too difficult to solve, so, first of all, it is necessary to reassign forces and means to the valuable objects o_i for which $t_{o_iF} \leq \tau_{o_iZ}$;
- $\check{S}_3$: $(\forall o_i \in O^*(t))(t_{o_iF} \leq \tau_{o_iZ})$ – a class of particularly critical situations where the task of rescuing objects may be unattainable.

In order to visualize the information for a certain time t, we draw the risk surface that reflects the normalized assessment of the level of risk for each cell. This surface visually reflects the convex areas of the terrain with a maximal risk. However, in some situations, areas with the highest risk do not always need the attention of the decision-maker.

We must select the areas with the maximum value among them. Therefore, we draw also the value surface in the same coordinate system, but the increase in value will be displayed in reverse order: from 0 to −1.

The areas with maximum value will be displayed with concave areas on the surface (Fig. 3).

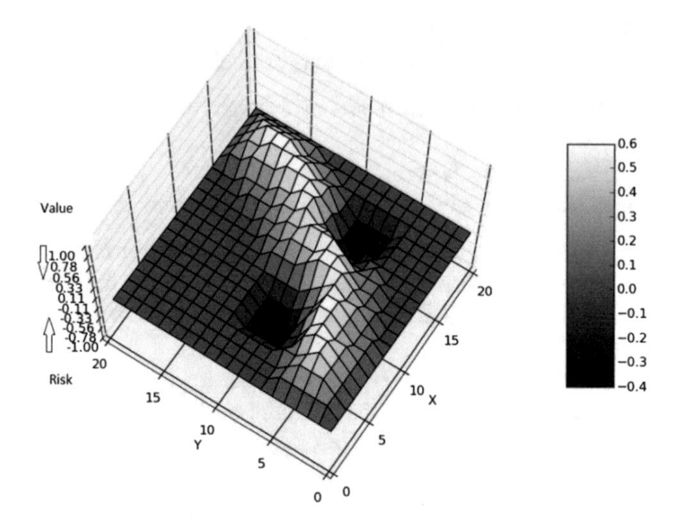

Fig. 3. The surfaces of risk and value

In order to select areas requiring the attention of the decision-maker, it is necessary to draw two cut surfaces that are perpendicular to the *Z*-axis and establish the critical values of value and risk.

Figure 4 shows the cut surface for the surface of values. All areas with the value below this surface, have critical value. The same cut surface is drawn for the surface of risk. All areas with the risk assessment being above this surface are in critical risk conditions.

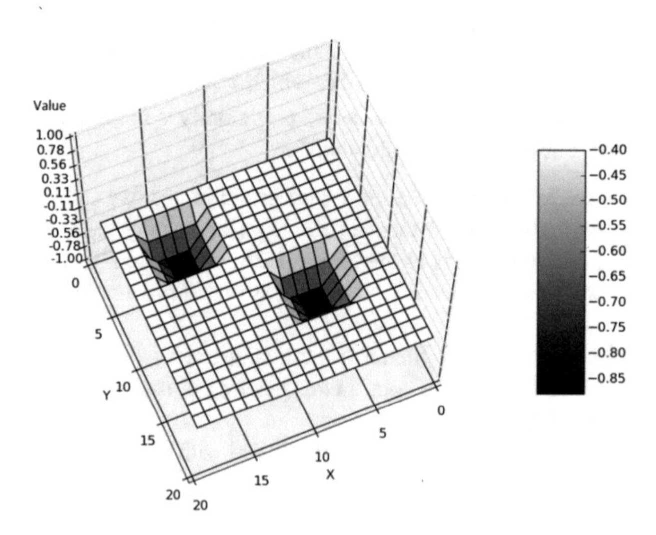

Fig. 4. The cut surface for allocation of valuable objects

The intersection of a set of areas having a critical value and ones being in critical risk requires the attention of the decision-maker (Fig. 5).

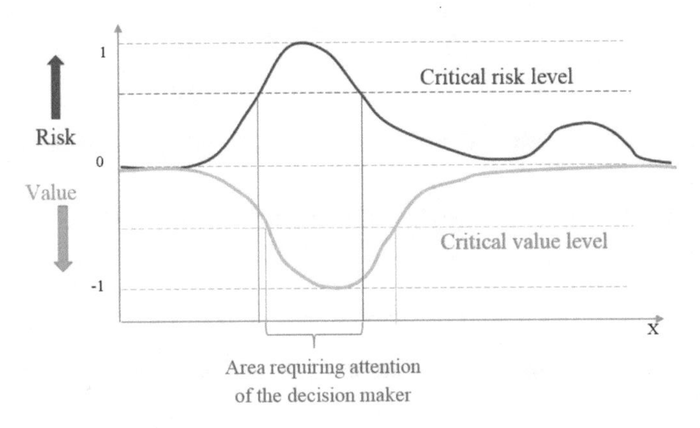

Fig. 5. The critical levels of risk and value

By varying the arrangement of cutting surfaces, we can narrow or expand the set of valuable objects $O^*(t)$ that need protection.

6 Diagnosis of the Situation in Decision Support System

At the moment of risk occurrence, a decision-maker faces a problem situation that must be diagnosed in order to make adequate decisions on risk minimization. To diagnose a situation the areas with the valuable objects being at maximum risk are allocated. In general, a diagnosis system performs the task of assigning a given input described as a set of observable DC or symptoms to some output described as a category (class) of the given situation. Thus, a diagnosis system should be provided with a set of DCs such as objects' and disaster front locations, observed disaster parameters such as spread rate, and estimations obtained as a result of risk analysis such as danger degree, levels of threat and risk.

Diagnosis is focused on finding a class of the current situation s^* using the nearest neighbor method [40, 41]. Each class of the set of possible situations $\check{S} = \{\check{S}_0, \check{S}_1, \ldots \check{S}_n\}$ is considered as a case that should be described in the case base.

To define a class of a situation we enter an identification function $\iota_{\check{S}} : X^* \to \check{S}^*$ which allows us to recognize the class of the situation s^*. The algorithm of defining the class of the situation includes the following steps:

1. Calculation of distances $\{L_1, \ldots L_2, \ldots L_n\}$ between the current situation s^* and the classes of well-defined situations Š = {Š₀, Š₁, … Šₙ}:

$$L_i(s^*, \check{S}_i) = \sqrt{\sum_{j=1}^{m} l_j^2 \left(x_j^*, x_j\right)}, \tag{5}$$

where $l_j(x_j^*, x_j)$ is a proximity measure for a DC j of the current situation s^* and the case $\check{S}_i$.

2. Searching for a minimal distance α between the current situation s^* and the classes $\check{S} = \{\check{S}_0, \check{S}_1, \ldots \check{S}_n\}$:

$$\alpha = \min_{i=(1,n)} \left(L_i\left(s^*, \check{S}_i\right)\right). \tag{6}$$

If $\alpha = 0$ (6) we obtain the actual match of the current situation to one of several classes:

$$\check{S}^* = \check{S}_i \middle| L_i\left(s^*, \check{S}_i\right) = 0. \tag{7}$$

3. If $\alpha \neq 0$ (6), there is incomplete information about the DCs of the current situation. Then, the class of the current situation is described by a subset of classes $\check{S}^{\prime *} \subset \check{S}$ that correspond to an incomplete description of the current situation s^*:

$$\check{S}^{\prime *} = \left\{\check{S}_i \middle| L_i\left(s^*, \check{S}_i\right) = \alpha, i = 1, \ldots, n\right\}. \tag{8}$$

Thus, we range all the situations belonging to $\check{S}_i \in \check{S}^{\prime *}$, $i = 1, \ldots, n$, according to their integrated risk estimations $R_{\Omega i}$ and choose subsets of the most critical classes of the situations $\check{S}^{\prime\prime *}$ from the set $\check{S}^{\prime *}$:

$$\check{S}^{\prime\prime *} \in \check{S}^{\prime *} | \forall \left(\check{S}_i \in \check{S}^{\prime\prime *}\right) \left(R_{\Omega i} = \max_{j=\left(1, \left|\check{S}^{\prime *}\right|\right)} \left(R_{\Omega j}\right)\right). \tag{9}$$

4. Ranking the classes of the situations from the set $\check{S}^{\prime\prime *}$ obtained at the previous step using the possibilities of their occurrence p_i we have the most possible situation:

$$\check{S}_i^* \in \check{S}^{\prime\prime *} \middle| p_i = \max_{j=\left(1, \left|\check{S}^{\prime\prime *}\right|\right)} \left(p_j\right). \tag{10}$$

The proposed algorithm allows matching the current situation to the class of the most possible of the most dangerous situations with respect to the observed DCs X^*.

The case base implies diagnostic knowledge containing the classes of the possible situations (Fig. 6).

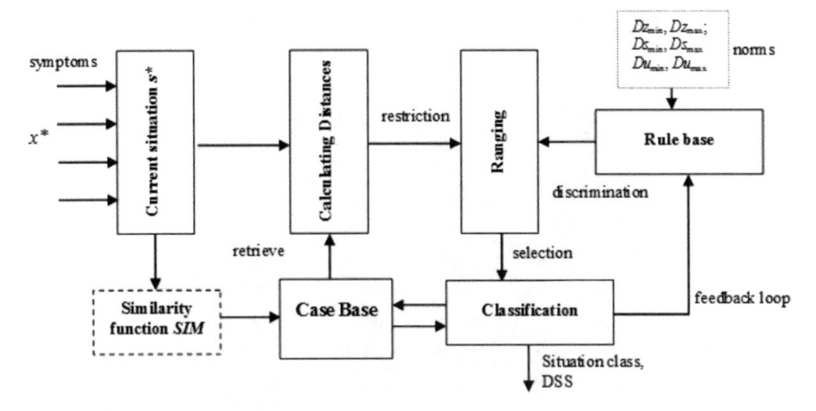

Fig. 6. Realization of the hybrid diagnosis method

A rule base represents knowledge about the problem area and the timing metrics ξ_T on time scale T such that $\forall t_i, t_j \in TD_Z = \{D_{Z1}, \ldots D_{Zj}, \ldots D_{Zm}\}$, where $D_Z = \{D_{Z1}, \ldots D_{Zj}, \ldots D_{Zm}\}$ is a set of time limits. Thus, the rule base contains a set of time limits D_Z necessary for extracting similar situations. Using the time norms, the current situation can be classified regarding its temporal characteristics as normal, critical, dangerous or catastrophic. When the case base becomes too big it can be presented in the form of rules that decreases its capacity.

According to the proposed hybrid diagnosis method, the cases are the points in the Cartesian space of symptoms and their neighborhoods are the sets of constraints composing the rules.

7 The Results of the Research

To examine the validity and the efficiency of the proposed hybrid diagnosis method, we have conducted an experiment based on the collected retrospective information on the series of large-scale forest fires, which had been taken place in Tsyurupinsk and Golopristan forestries, Ukraine, on July 20–31, 2007. The proposed method has been implemented using Visual C++ and tested on computer based on the Pentium i5-7400 3-3,5 GHz processor and 16 GB RAM. The obtained software has been integrated into the GIS-based response time decision support system (DSS) Forest Project, which should evaluate a number of DCs, e.g. danger degrees, threats, and risks, for a given set of target valuable objects, as well as provide the geospatial analysis of risks in response time disaster situations.

Figure 7 depicts the map of Tsurupinsk forestry (Kherson, Ukraine) represented in the DSS Forest Project using the spatial model based on the grid with the variable cell size.

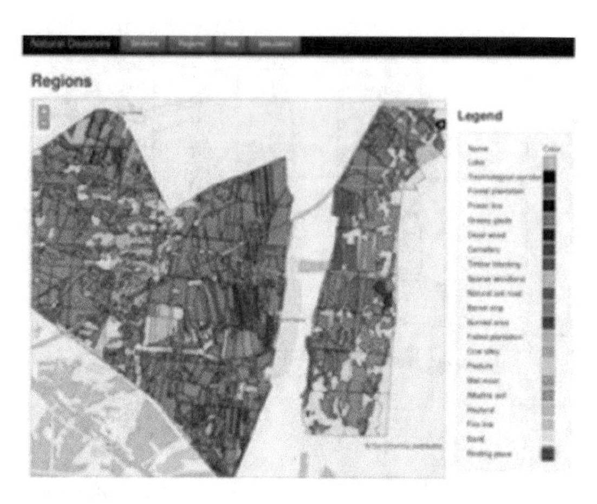

Fig. 7. Representation of the Tsurupinsk forestry in GIS-based DSS Forest Project

We have modeled the ongoing processes of the situation diagnosis during the forest fire propagation performed by the decision-maker manually, as well as performed by own tools of the GIS-based DSS, and using the intelligent diagnosis software integrated into the GIS-based DSS. Within the spatial model the target objects such as roads and recreation camps were highlighted. For each target object we assessed quantitative values of risk components such as fire intensity, danger, threat, and object's value.

During the experiment, we have evaluated the total time of decision-making needed for identifying the problem and classifying the situation. In the case of intelligent diagnosis performed by the proposed method, we evaluate the time needed for building the risk surface over the value surface and obtaining their cut surface to choose spatial areas that require the attention of the decision-maker. In other cases, we evaluate the time necessary to obtain a final assessment of the spatial distribution of risk to respond to the situation appropriately.

Obviously, the experiment results depend on the cell size within the GIS spatial model. The accuracy of the situation classification also depends on the cell size. Therefore, we have varied the cell size to investigate its impact on the situation diagnosis time and compared the results obtained using DSS with the results obtained using manual decision-making. The results of evaluating the time are depicted in Fig. 8, as well as results of evaluating the diagnosis accuracy, are depicted in Fig. 9. Diagnosis accuracy was calculated as a ratio of a number of correctly diagnosed situations to a number of experiment series.

Clearly, the proposed intelligent diagnosis method provides higher accuracy of the situation diagnosis while winning in time. With this new method, the situation assessment can be obtained by the decision-maker in only 150 ms when the cell size is 5 m, which makes the proposed method of practical use. The use of the proposed model makes it possible to accelerate the decision-making process within the forest fire fighting process by about 50% and above.

Thus, it can be concluded that the proposed method provides sufficient performance for the cell size 5 m and above, so it is acceptable for solving the practical forest fire fighting problems in the response time GIS-based DSS.

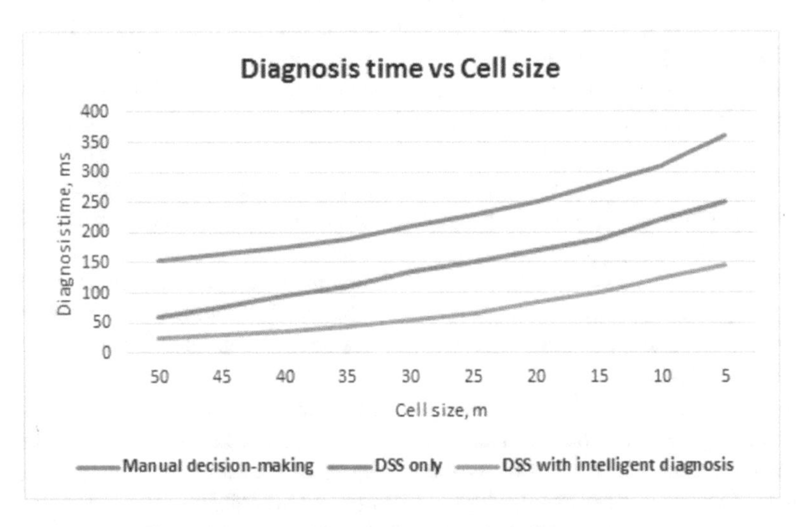

Fig. 8. Influence of the cell size on the diagnosis time

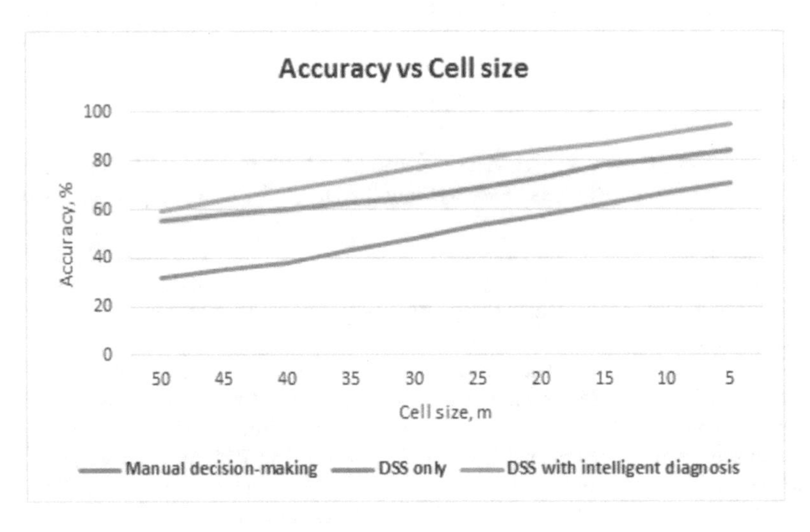

Fig. 9. Influence of the cell size on the diagnosis accuracy

8 Conclusion

The dynamic model of integral risk assessment proposed in the paper allows the disaster risk to be represented as a spatially-distributed process. A multi-risk for a valuable object is formed as a combination of four components such as danger, threat potential, threat level, and potential losses. In contrast to existing approaches to risk assessment, this model gives us a more comprehensive concept of disaster risk nature and stimulates making more grounded decisions [42].

In order to provide comparing the risks from different disasters and assess their joint influence on the valuable object in the form of multi-risk a quantitative value of each risk component is proposed to represent in the form of qualitative value using the appropriate scales.

The hybrid diagnosis method presented in the paper provides planning tactical disaster response operations. The method is capable of identifying the problem, classifying the situation and makes it possible to respond appropriately based on the result of such classification.

Both the dynamic model of integral risk assessment and the hybrid diagnosis method were integrated into the GIS-based response time DSS Forest Project aimed at the diagnosis of forest fires, as well as the planning of decision making for forest fire response. The experiment examining the validity and efficiency of the proposed hybrid diagnosis method has shown that the proposed method provides sufficient performance for the cell size 5 m and above, so it is acceptable for solving the practical forest fire fighting problems in the response time GIS-based DSS.

References

1. Shen, G., Zhou, L., Wu, Y., Cai, Z.: A global expected risk analysis of fatalities, injuries, and damages by natural disasters. Sustainability **10**(7), 2573 (2018). https://doi.org/10.3390/su10072573
2. Thompson, M.P., Haas, J.R., Gilbertson-Day, J.V., Scott, J.H., Langowski, P., Bpwne, E., Calkin, D.: Development and application of a geospatial wildfire exposure and risk calculation tool. Environ. Model Softw. **63**, 61–72 (2015). https://doi.org/10.1016/j.envsoft.2014.09.018
3. Thompson, M.P., Calkin, D.E., Finney, M.A., Ager, A.A., Gilbertson-Day, J.V.: Integrated national-scale assessment of wildfire risk to human and ecological values. Stoch. Environ. Res. Risk Assess. **25**(6), 761–780 (2011). https://doi.org/10.1007/s00477-011-0461-0
4. Thompson, M.P., Zimmerman, T., Mindar, D., Taber, M.: Risk terminology primer: basic principles and glossary for the wildland fire management community. Gen. Tech. Rep. RMRS-GTR-349. Fort Collins, CO: U.S. Department of Agriculture, Forest Service, Rocky Mountain Research Station (2016)
5. Gallina, V., Torresan, S., Critto, A., Sperotto, A., Glade, T., Marcomini, A.: A review of multi-risk methodologies for natural hazards: consequences and challenges for a climate change impact assessment. J. Environ. Manag. **168**, 123–132 (2016). https://doi.org/10.1016/j.jenvman.2015.11.011
6. World Meteorological Organization (WMO): Comprehensive risk assessment for natural hazards. Geneva: WMO/TD No. 955, Switzerland (1999)

7. Zharikova, M.: Methodological basis of geoinformation technology of decision support in combined natural and man-made systems in destructive processes conditions. Doctoral Thesis, Ukrainian Academy of Printing, Lviv, Ukraine (2018)
8. Van Westen, C.J., Shroder, J., Bishop, M.P.: Remote sensing and GIS for natural hazards assessment and disaster risk management. Treatise Geomorphol. **3**, 259–298 (2013). https://doi.org/10.1016/B978-0-12-374739-6.00051-8
9. Calkin, D.E., Thompson, M.P., Finney, M.A., Hyde, K.D.: A real-time assessment tool supporting wildland fire decision making. J. For. **109**(5), 274–280 (2011)
10. Finney, M.: Modeling the spread and behavior of prescribed natural fires. In: Proceedings of the 12th Conference on Fire and Forest Meteorology, Jekyll Island, Georgia, pp. 138–143 (1993)
11. Finney, M.: The challenge of quantitative risk analysis for wildland fire. For. Ecol. Manag. **211**(1–2), 97–108 (2005). https://doi.org/10.1016/j.foreco.2005.02.010
12. Rausand, M., Hoyland, A.: System Reliability Theory: Models, Statistical Methods, and Applications. Wiley-Interscience, Hoboken (2004)
13. Kloprogge, P., Van der Sluijs, J., Petersen, A.: A Method for the Analysis of Assumptions in Assessments. Netherlands Environmental Assessment Agency, Bilthoven (2005)
14. Krishnamoorthi, N.: Role of remote sensing and GIS in natural-disaster management cycle. Imp. J. Interdiscip. Res. **2**(3), 144–154 (2016)
15. Zharikova, M., Sherstjuk, V.: Threat assessment method for intelligent disaster decision support. Advances in Intelligent Systems and Computing, vol. 512, pp. 81–100 (2017). https://doi.org/10.1007/978-3-319-45991-2_6
16. Balakrishnan, K., Honavar, V.: Intelligent diagnosis systems. J. Intell. Syst. **8**(3), 237–290 (1998). https://doi.org/10.1515/JISYS.1998.8.304.239
17. Cheng, T., Kocka, T., Zhang, N.L.: Effective dimensions of partially observed polytrees. Int. J. Approx. Reason. **38**(3), 311–332 (2005). https://doi.org/10.1016/j.ijar.2004.05.008
18. Lucas, P.J.F.: Bayesian model-based diagnosis. Int. J. Approx. Reason. **27**(2), 99–119 (2001). https://doi.org/10.1016/S0888-613X(01)00036-6
19. Bhagwat, A.: Knowledge-based service diagnosis system. Int. J. Comput. Sci. Technol. **3**(5), 182–184 (2015)
20. Ward, M.O., Grinstein, G.G., Keim, D.A.: Interactive Data Visualization - Foundations, Techniques, and Applications. A.K. Peters, Ltd., Natick (2010). https://doi.org/10.1201/9780429108433
21. Amarosicz, M., Psiuk, K., Rogala, T., Rzydzik, S.: Diagnostic shell expert systems. Diagnostica **17**(1), 33–40 (2016)
22. Tan, C.F., Wahidin, L.S., Khalil, S.N., Tamaldin, N., Hu, J., Rauterberg, G.W.M.: The application of expert system: a review of research and applications. ARPN J. Eng. Appl. Sci. **11**(4), 2448–2453 (2016)
23. Martinez, J., Vega-Garcia, C., Chuvieco, E.: Human-caused wildfire risk rating for prevention planning in Spain. J. Environ. Manag. **90**(2), 1241–1252 (2009). https://doi.org/10.1016/j.jenvman.2008.07.005
24. Martinez, M.V.: Knowledge engineering for intelligent decision support. In: Proceeding of the Twenty-Sixth International Joint Conference on Artificial Intelligence (IJCAI 2017), Montreal, pp. 5131–5135 (2017). https://doi.org/10.24963/ijcai.2017/736
25. Rieck, C.Z., Wu, X.W., Jiang, J.C., Zhu, A.X.: Case-based knowledge formalization and reasoning method for digital terrain analysis – application to extracting drainage networks. Hydrol. Earth Syst. Sci. **20**, 3379–3392 (2016). https://doi.org/10.5194/hess-2015-539
26. Rieck, K., Trinius, P., Willems, C., Holz, T.: Automatic analysis of malware behavior using machine learning. J. Comput. Secur. **19**(4), 639–668 (2011)

27. Symeonidisa, A.L., Chatzidimitriouc, K.C., Athanasiadisd, I.N., Mitkas, P.A.: Data mining for agent reasoning: a synergy for training intelligent agents. Eng. Appl. Artif. Intell. **20**(8), 1097–1111 (2007). https://doi.org/10.1016/j.engappai.2007.02.009
28. Goldberg, Y.: A primer on neural network models for natural language processing. J. Artif. Intell. Res. **57**(1), 345–420 (2016)
29. Gong, Y., Li, J., Zhou, Y., Li, Y., Chung, H.S., Shi, Y., Zhang, J.: Genetic learning particle swarm optimization. IEEE Trans. Cybern. **46**(10), 2277–2290 (2016). https://doi.org/10.1109/TCYB.2015.2475174
30. Hoffman, R.: Origins of situation awareness: cautionary tales from the history of concepts of attention. J. Cogn. Eng. Decis. Mak. **9**(1), 73–83 (2015). https://doi.org/10.1177/1555343414568116
31. Fogel, D.B., Fogel, L.J., Porto, V.W.: Evolving neural networks. Biol. Cybern. **63**(6), 487–493 (1990)
32. Allam, A.A., Bakeir, M.Y., Abo-Tabl, E.A.: Some methods for generating topologies by relations. Bull. Malays. Math. Sci. Soc. **31**(1), 35–45 (2008)
33. Scott, J.H., Thompson, M.P., Calkin, D.E.: A wildfire risk assessment framework for land and resource management. Gen. Tech. Rep. RMRS-GTR-315. U.S. Department of Agriculture, Forest Service, Rocky Mountain Research Station (2013)
34. Apostolakis, G.E.: How useful is quantitative risk assessment. Risk Anal. **24**(3), 515–520 (2004). https://doi.org/10.1111/j.0272-4332.2004.00455.x
35. Aven, T., Zio, E.: Model output uncertainty in risk assessment. Int. J. Perform. Eng. **9**(5), 475–486 (2013)
36. Andreu, A., Hermansen-Baez, L.A.: Fire in the South 2: the southern wildfire risk assessment. A report by Southern Group of State Forester, 32 p. (2008)
37. Dubois, D., Prade, H.: Possibility theory, probability theory, and multiple-valued logics: a clarification. Ann. Math. Artif. Intell. **32**, 35–66 (2001). https://doi.org/10.1023/A:1016740830286
38. Dubois, D., Prade, H.: What are fuzzy rules and how to use them. Fuzzy Sets Syst. **84**(2), 169–185 (1996). https://doi.org/10.1016/0165-0114(96)00066-8
39. Dubois, D., Prade, H.: Possibilistic logic: a retrospective and prospective view. Fuzzy Sets Syst. **144**(1), 3–23 (2004). https://doi.org/10.1016/j.fss.2003.10.011
40. Pawlak, Z., Jerzy, W., Slowinski, R., Ziarko, W.: Rough sets. Commun. ACM **38**(11), 88–95 (1995). https://doi.org/10.1145/219717.219791
41. Zharikova, M., Sherstjuk, V.: The hybrid intelligent diagnosis method for the MultiUAV-Based forest fire-fighting response system. In: Proceedings of 13th International Scientific and Technical Conference on Computer Sciences and Information Technologies (CSIT), Lviv, pp. 339–342 (2018). https://doi.org/10.1109/STC-CSIT.2018.8526609
42. Zharikova, M., Sherstjuk, V.: Situation diagnosis based on the spatially-distributed dynamic disaster risk assessment In: Proceedings of the International Scientific Conference "Computer sciences and information technologies" (CSIT 2019), vol. 3, pp. 205–209. IEEE (2019)

Information Technology of Satellite Image Processing for Monitoring of Floods and Drought

Dmitry Mozgovoy[1] and Volodymyr Hnatushenko[2](✉)

[1] Department of Physics, Electronics and Computer Systems, Oles Gonchar Dnipro National University, Gagarin av., 72, Dnipro 49010, Ukraine
m-d-k@i.ua

[2] Department of Information Systems and Technologies, Dnipro University of Technology, av. Dmytra Yavornytskoho, 19, Dnipro 49005, Ukraine
vvgnatush@gmail.com

Abstract. The information technology of automated processing of optical and radar data from Sentinel 1A/B and Sentinel 2A/B satellites in order to ensure all-weather monitoring of floods and droughts was developed and tested. Comparison of the processing results confirmed a rather high efficiency of water object recognition both in the two-polarization radar data of the C-band and in the multispectral data of the visible and near-IR ranges. Both methods showed similar results in quality of recognition of water objects with a small number of unrecognized or falsely recognized objects (on average less than 10% of the area of recognized objects). The information technology for automated processing of optical and radar data from Sentinel satellites for monitoring floods and droughts is independent from weather conditions over the monitored territory (imaging is possible even with 100% cloudiness). Owing to the high degree of automation of data processing in the developed information technology, it can be used to promptly inform about the course and consequences of floods and droughts not only representatives of government services and commercial structures, but also the general public.

Keywords: Satellite imagery · Information technology · Monitoring · Floods · Drought · Radar · Polarization · Image processing

1 Introduction

Among the most large-scale and dangerous natural disasters that occur annually in different countries, apart from fires and earthquakes, floods and droughts are in the first ranks in terms of regularity, affected areas and general material damage [1–4]. A drought can lead to losses to agriculture, affect inland navigation and hydropower plants, and cause a lack of drinking water and famine. A flood is a significant rise of water level in a stream, lake, reservoir or coastal region. Flooding in the cities paralyzes normal operation of almost all municipal services, provokes accidents on the roads, leads to eventual destruction of buildings and structures, and, in particularly severe cases, results in casualties among the population. In rural areas, floods lead to destruction of crops and other vegetation, death of animals and destabilization of natural ecosystems. Droughts,

N. Shakhovska and M. O. Medykovskyy (Eds.): CCSIT 2019, AISC 1080, pp. 473–487, 2020.
https://doi.org/10.1007/978-3-030-33695-0_32

as well as floods, lead to serious environmental and economic consequences. Their main difference lies in the long-term processes initiated by a natural disaster and the complexity or even impossibility of abatement of its consequences.

2 Current State

In recent years, to determine the scale, dynamics and consequences of anthropogenic and natural emergencies, Earth remote sensing data, in particular, satellite imagery, are increasingly being used, which, compared to terrestrial measurement methods, have numerous undeniable advantages [3–8]:

- maximum objectivity and reliability (satellite images help to completely eliminate human errors, as well as deliberate distortion or concealment of important information);
- wide scope and informativeness (you can observe any, even hard-to-reach territory, on the Earth with a coverage of thousands of kilometers);
- multidisciplinarity (the possibility of using the same images to solve a wide range of applied problems);
- almost immediate availability of results (imaging and processing takes less than 1 day);
- synchronization of data acquisition (simultaneous observation of a large number of objects located at a considerable distance from each other);
- complete safety (no risks to human health and life);
- high economic efficiency (significantly lower costs compared to ground methods and aerial surveys);
- no need to obtain legal permits for surveying objects, which allows you to shoot any desired objects;
- maximum confidentiality (ease of obtaining data and minimizing the risks of information leakage).

After commissioning of the second Sentinel-2B satellite in 2017, the Copernicus system has become the most informative and affordable source of remote sensing data of medium spatial resolution. This can be judged by the large number of web services providing a wide range of information products, mainly based on optical images from satellites Sentinel-2A and Sentinel-2B. This can be explained by a successful combination of high-quality characteristics of the images themselves (rather high spatial and radiometric resolution, a variety of spectral channels) and frequent revisiting of same places (5 days for two satellites).

3 Formulation of the Problem

Unfortunately, the most common and informative multispectral satellite imagery in the visible and infrared ranges has two significant limitations during the survey:

- the need for adequate illumination of the surveyed territory (as a rule, high-quality imaging is possible at the Sun angles of more than 30°);

– strong dependence on the presence of clouds over the surveyed territory (the maximum permissible percentage of cloud cover is usually not more than 2 to 5%).

Therefore, the actual frequency of imaging in optical channels (taking into account the clouds) for Sentinel 2A/B satellites, even for small areas, is much lower (sometimes several times). The use of data from other remote sensing satellites, Landsat-7, Landsat-8 and Terra (ASTER multispectral radiometer), as additional free sources is not always possible due to the poorer characteristics of these images, and also due to the presence of clouds.

The main objective of the research is to develop and test the information technology for automated processing of optical and radar data from the Sentinel 1A/B and Sentinel 2A/B satellites to ensure all-weather monitoring of floods and droughts.

4 Possible Ways to Solve the Problem

Therefore, to ensure all-weather satellite monitoring of the dynamics and consequences of floods and droughts, in addition to satellite imagery in the visible and infrared bands, it is necessary to use bipolarization radar imaging data in the C or X bands obtained from various high- and medium-resolution remote sensing satellites equipped with radar with synthesized aperture [9].

In this case, it is possible to partially improve the frequency of obtaining information from the desired territory by radar imaging, one of the main advantages of which is independence from clouds. Obviously, radar data cannot provide a full RGB image in natural colors, and spectral indices (NDVI, SAVI, NBR, NDWI) cannot be calculated. However, preliminary studies confirmed the high effectiveness of using radar images from Sentinel-1A and Sentinel-1B satellites together with optical images from satellites Sentinel-2A and Sentinel-2B (Table 1) for monitoring floods and droughts as an alternative or additional source of data for areas covered with dense clouds [10].

Table 1. Main characteristics of Sentinel 1A/B and Sentinel 2A/B satellites.

Satellite	Date of launch	Spacecraft mass, kg	Orbit altitude, km	Orbit inclination, deg	Revisit time, days (for one)	Prime contractor
Sentinel-1A	2014-04-03	2280	693	98,2	12	Thales Alenia Space (Italy)
Sentinel-1B	2016-04-26	2280	693	98,2	12	Thales Alenia Space (Italy)
Sentinel-2A	2015-06-23	1200	786	98,5	10	Airbus Defence & Space (France)
Sentinel-2B	2017-03-07	1200	786	98,5	10	Airbus Defence & Space (France)

5 Information Technology of Satellite Image Processing

5.1 Choice of the Territory of Observations

The reasons for choosing the region of research and the period of observation are based on the following. The State of California is one of the least abundant with fresh water in the United States, while having high water consumption – the drought in the state has reached catastrophic size. January 2015 was the driest month in California for the whole period of observations since 1895. Two thirds of the state's population depend on the centralized water supply – about 25 million people and more than 400 thousand hectares of agricultural land. Due to pumping groundwater out for agricultural land irrigation the level of ground waters and the snow cover have become extremely low. The water level in the reservoirs of California is close to the critical notch. State authorities are forced to tighten measures to save water, their supply sometimes being insufficient even to satisfy the prime demand of population.

In recent years, the situation in California has become significantly more complicated due to the annual large-scale forest fires [3]. Therefore, areas of freshwater the Anderson Lake near San Hose, California were selected as subjects for research (Fig. 1).

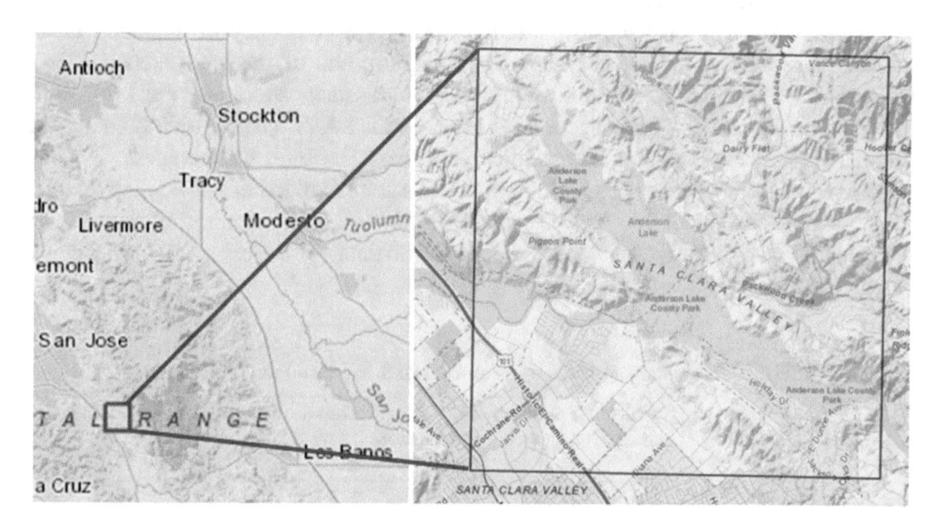

Fig. 1. Fragments of different-time radar images from the Sentinel-1A/B satellites for a given section of Anderson Lake (two-polarization RGB composites)

5.2 The Input Data

For testing proposed technology of automated processing of optical and radar data to ensure all-weather monitoring of floods and droughts used bipolarization images from Sentinel-1A and Sentinel-1B satellites and multispectral images from satellites Sentinel-2A and Sentinel-2B (Figs. 2, 3 and Table 2).

Table 2. Main survey characteristics of Sentinel 1A/B and Sentinel 2A/B satellites.

Satellite	Sensor type	Operating range	Swath, km	Spatial resolution, m	Radiometric resolution, bit	Data storage	Error rate
Sentinel-1A/B	SAR	C-band (5.405 GHz)	80 240	5 × 5 5 × 20	10	1.4 Tbit (EOL)	$<10^{-9}$ (BER)
Sentinel-2A/B	MSI	VNIR SWIR	290	10 20	12	2 Tbit (EOL)	$<10^{-8}$ (FER)

Fig. 2. Fragments of different-time radar images from the Sentinel-1A/B satellites for a given section of Anderson Lake (two-polarization RGB composites)

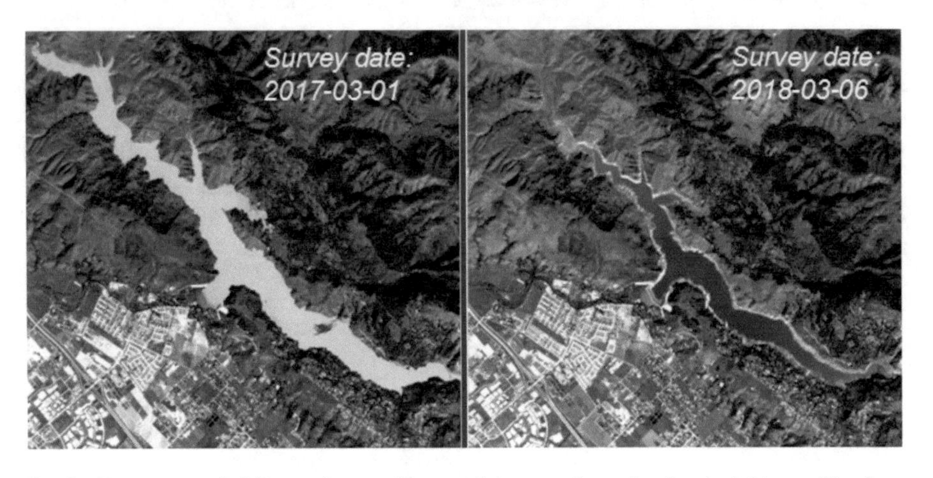

Fig. 3. Fragments of different-time multispectral images from the Sentinel-2A satellite for a given section of Anderson Lake (RGB composites of visible bands)

5.3 Description of the Developed Information Technology

The proposed information technology is implemented in the form of a software package consisting of two software modules:

- a module for processing bipolarized C-band radar data obtained from Sentinel-1A/B satellites [11];
- a module for processing multispectral data of visible and near-IR bands obtained from Sentinel-2A/B satellites.

Procedures for processing and analyzing radar images from Sentinel-1A/B satellites in order to simplify the user's work were divided into the following stages:

- Preliminary operations (Fig. 4) performed in automatic mode (unpacking images of HV and VV polarizations and accompanying metadata, georeferencing by orbital data, radiometric and geometric correction, transformation into a required cartographic projection);
- Thematic processing operations (Fig. 5) performed in a semi-automatic mode (calculation of the ratio of HV and VV polarization channels, formation of the RGB composite and visualization, setting of binarization thresholds, water class selection, morphological filtration and vectorization, formation of a thematic map and export of results into standard pixel and vector formats).

Dual-polarization 24-bit RGB composite image for visualization was calculated by HV and VV bands sigma values (32 bit floating point format) with follow formulas:

$$\begin{aligned} B_{red} &= F_{byte}(F_{norm\_VH}(F_{linear\_db}(VH_{sigma}))), \\ B_{green} &= F_{byte}(F_{norm\_VV}(F_{linear\_db}(VV_{sigma}))), \\ B_{blue} &= F_{byte}(F_{norm\_VH-VV}(F_{linear\_db}\left(VH_{sigma}\right) - F_{linear\_db}(VV_{sigma}))), \end{aligned} \tag{1}$$

were $F_{byte}\ (x)$ – linear transform from float 32 type to byte (8 bit) type; $F_{norm\_VH}\ (x)$ – additive-multiplicative transform VH sigma value (in dB) for better visualization as R-band in the 24-bit RGB composite; $F_{linear\_db}\ (x)$ – non-linear transform VV and VH sigma value from linear to dB (float 32 type); $F_{norm\_VV}\ (x)$ – additive-multiplicative VV sigma value (in dB) for better visualization as G-band in the 24-bit RGB composite; $F_{norm\_VH\text{-}VV}(x)$ – additive-multiplicative VH-VV sigma value (in dB) for better visualization as B-band in the 24-bit RGB composite.

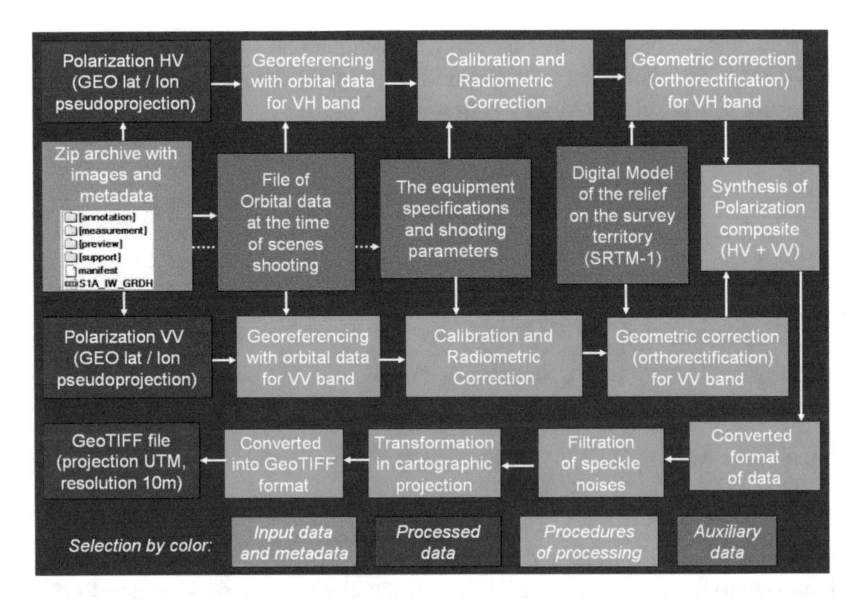

Fig. 4. Block diagram of the preliminary processing of two-polarization SAR imagery from Sentinel-1A/B satellites (GRDH product)

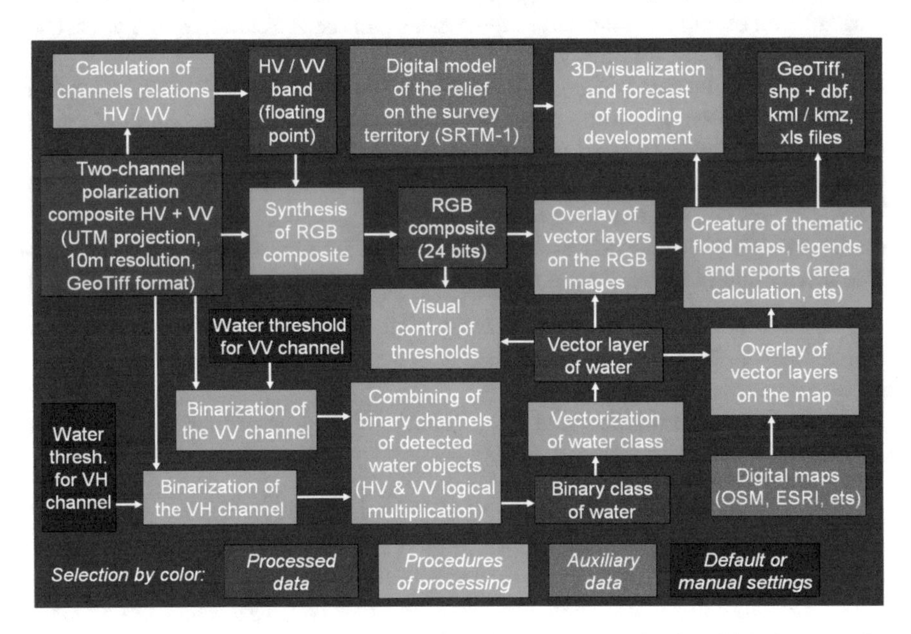

Fig. 5. Block diagram of the thematic processing for two-polarization SAR imagery from Sentinel-1A/B satellites (GRDH product)

Linear additive-multiplicative transform for sigma values for better visualization it in the 24-bit RGB composite was founded empirically with multi-expert assessment. This transform was calculated with follow formulas:

$$VH_{sigma\_norm} = \begin{cases} 0, \text{ if } VH_{sigma\_db} < -25.6; \\ 10 \times \left(VH_{sigma\_db} + 25.6\right), \text{ if } -25.6 \leq VH_{sigma\_db} < 0; \\ 255, \text{ if } VH_{sigma\_db} \geq 0, \end{cases}$$

$$VV_{sigma\_norm} = \begin{cases} 0, \text{ if } VH_{sigma\_db} < -20.6; \\ 10 \times \left(VH_{sigma\_db} + 20.6\right), \text{ if } -20.6 \leq VH_{sigma\_db} < 0; \\ 255, \text{ if } VH_{sigma\_db} \geq 0, \end{cases}$$

$$VH/VV_{sigma\_norm} = \begin{cases} 0, \text{ if } VH_{sigma\_db} < -15.6; \\ 10 \times \left(VH_{sigma\_db} + 15.6\right), \text{ if } -15.6 \leq VH_{sigma\_db} < 0. \\ 255, \text{ if } VH_{sigma\_db} \geq 0. \end{cases}$$

Procedures for processing and analyzing multispectral images from Sentinel-2A/B satellites due to the need to take into account the clouds were divided into five stages:

- spatial selection for all spectral channels (cutting out the desired section manually or using the vector layer) and bringing the result to the same spatial resolution;
- calculation of refined cloud and shadow masks for the selected area of interest using preset or manually set thresholds;
- calculation of the standard normalized differential water index (NDWI) or modified normalized differential water index (MNDWI) and its visualization in the form of a palette image with a legend (standard or user-defined);
- threshold binarization and morphological filtration of recognized water objects using preset or manually defined settings;
- vectorization of recognized water objects, formation of a thematic map and export of results into standard pixel and vector formats.

Block diagram of the automated recognition of water objects from visible and infrared images from Sentinel-2A/B satellites on the EOS web services shown on Fig. 6.

Standard normalized differential water index (NDWI) was calculated by visible and near infrared bands with follow formula:

$$NDWI = \frac{(\text{Green} - \text{NIR})}{(\text{Green} + \text{NIR})} = \frac{(\text{Band4} - \text{Band8})}{(\text{Band4} + \text{Band8})}. \quad (2)$$

Modified normalized differential water index (MNDWI) was calculated by visible and short-wave infrared bands with follow formula:

$$MNDWI = \frac{(\text{Green} - \text{SWIR})}{(\text{Green} + \text{SWIR})} = \frac{(\text{Band4} - \text{Band12})}{(\text{Band4} + \text{Band12})}. \quad (3)$$

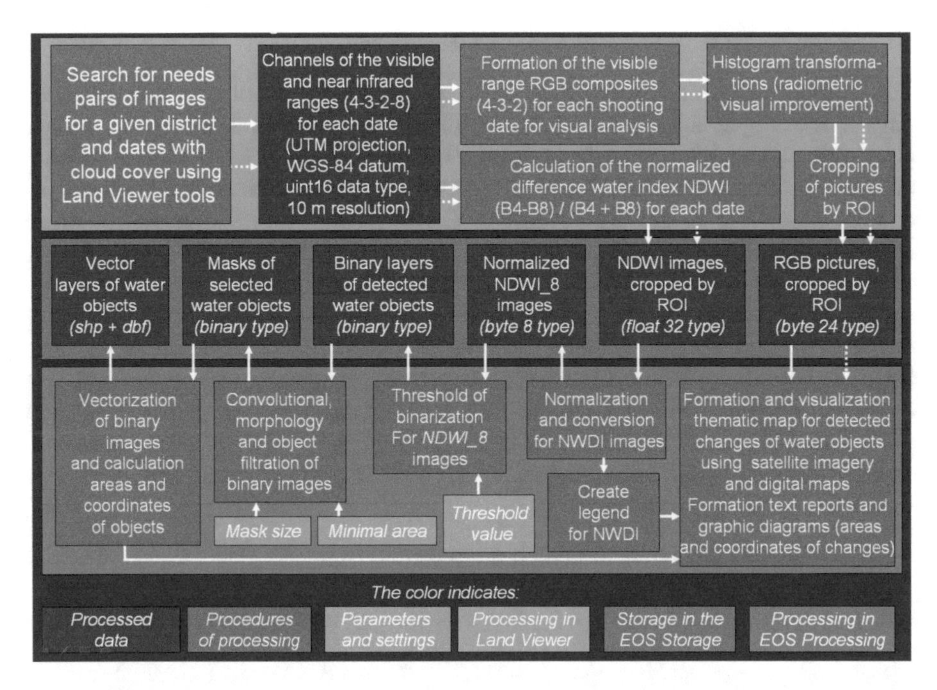

Fig. 6. Block diagram of the automated recognition of water objects from visible and infrared images from Sentinel-2A/B satellites on the EOS web services

For assessment and visualization of the water objects changes between two dates, normalized volatility index (NVI) was calculated with follow formulas:

$$\begin{aligned} NVI_{NDWI} &= (NDWI_{new} - NDWI_{old})/(NDWI_{new} + NDWI_{old}), \\ NVI_{MNDWI} &= (MNDWI_{new} - MNDWI_{old})/(MNDWI_{new} + MNDWI_{old}), \end{aligned} \quad (4)$$

were $NDWI_{new}$ – NDWI image for a newer date; $NDWI_{old}$ – NDWI image for an older date; $MNDWI_{new}$ – MNDWI image for a newer date; $MNDWI_{old}$ – MNDWI image for an older date.

5.4 Applied Metrics for Accuracy Evaluation

In this paper the following well-known indicators of accuracy classification have been chosen: confusion matrix and Kappa coefficient [13, 14].

A confusion matrix for several classes is an instrument that applies a cross-tabulation for the evaluation of correlation among the values of matching classes obtained from different sources.

The kappa (κ) coefficient measures the agreement between classification and ground truth pixels. A kappa value of 1 represents perfect agreement while a value of 0 represents no agreement.

5.5 Requirements to the Software and Hardware

The sizes of remote sensing data files received from Sentinel satellites are usually quite large. For example, a scene of radar data, taken from satellites Sentinel-1A/B in two polarizations, can take several gigabytes. Therefore, for quick processing of such images in real time, it is desirable to use modern computers with multicore processors of Intel class I7 or higher and at least 64 GB of RAM. The software can be either commercial (ERDAS, ENVI, ArcGIS, etc.) or free (SNAP, SAGA, GRAAS, QGIS, etc.), working in both MS Windows and in Linux environments.

To ensure a higher degree of automation of the processing procedures, it is possible to use relevant tools (for example, Imagine Model Maker in the ERDAS package, Graph Builder in the SNAP package) or programming languages (for example, IDL in the ENVI package, Python + GDAL in the QGIS system). However, currently the most efficient way to process and store satellite images is to use specialized web services, for example EOS DA, which, in comparison with traditional software and hardware, have significant advantages, such as:

- high economic efficiency (no purchase of powerful graphic stations and expensive software is required);
- minimum requirements for the level of user training (there is no need to spend time studying large and complicated software packages).

6 Results of the Research

As a result of automated processing of optical and radar data from Sentinel satellites, raster masks (Fig. 7) and vector layers (Fig. 8) of recognized water objects were obtained in the created software package. The accuracy of the classification of water bodies was estimated by comparison with etalon classes obtained by means of a manual vectorization of the origin satellite images [12]. Quantitative assessment of the classification accuracy perform with use confusion matrices and Kappa coefficient [13, 14].

The classification overall accuracy in this case, they were slightly worse and amounted to from 85% to 91% with in kappa coefficient from 0.71 to 0.83.

Comparison of the results of the processing confirmed a rather high efficiency of water object recognition both in the two-polarization radar data of the C-band and in the multispectral data of the visible and near-IR ranges:

- both methods showed similar results in quality of recognition of water objects with a small number of unrecognized or falsely recognized objects (on average less than 10% of the area of recognized objects);
- the accuracy of the delineation of water bodies on the test areas for both methods (Fig. 9) was also quite high (for images of medium spatial resolution);
- both methods showed good repeatability of the results of recognition of water objects on different test sites (with the same settings);
- both methods showed high consistency (stability) of the results of recognition of water objects on images obtained on different dates (with the same settings);

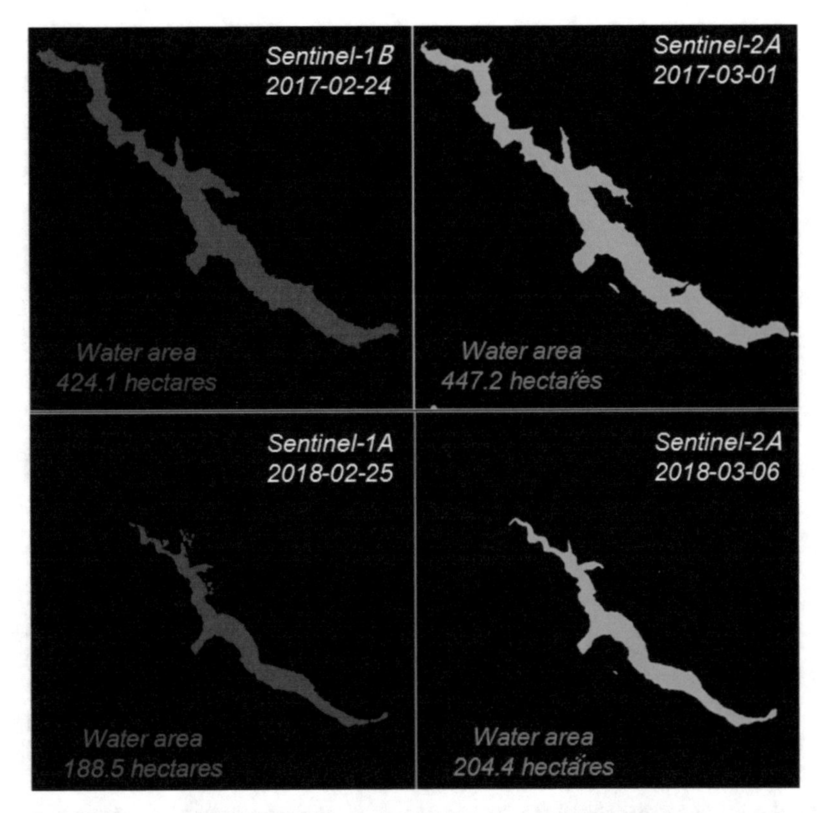

Fig. 7. Comparison of the raster masks of recognized water objects obtained by radar images from the Sentinel-1A/B satellites and by visible and infrared images from Sentinel-2A satellite

– both methods showed a fairly good correspondence of the results of the recognition of water bodies in images from Sentinel satellites with the results of processing images of ultra-high spatial resolution.

The effect of reducing the bit depth of the source data and the use of lossy compression on the recognition accuracy of water bodies are given in Table 3.

As can be seen from the table, reducing the bit depth of data and the use of lossy compression slightly reduces the recognition accuracy, while significantly reducing the amount of stored data, as well as increasing the speed of their processing.

The scope of practical application of the proposed information technology is quite extensive, as it enables automated all-weather satellite monitoring of floods and droughts both in the interests of public services (rescuers, environmentalists, municipal services) and in the interests of commercial structures (insurance companies, travel agencies, farmers, construction firms, etc.).

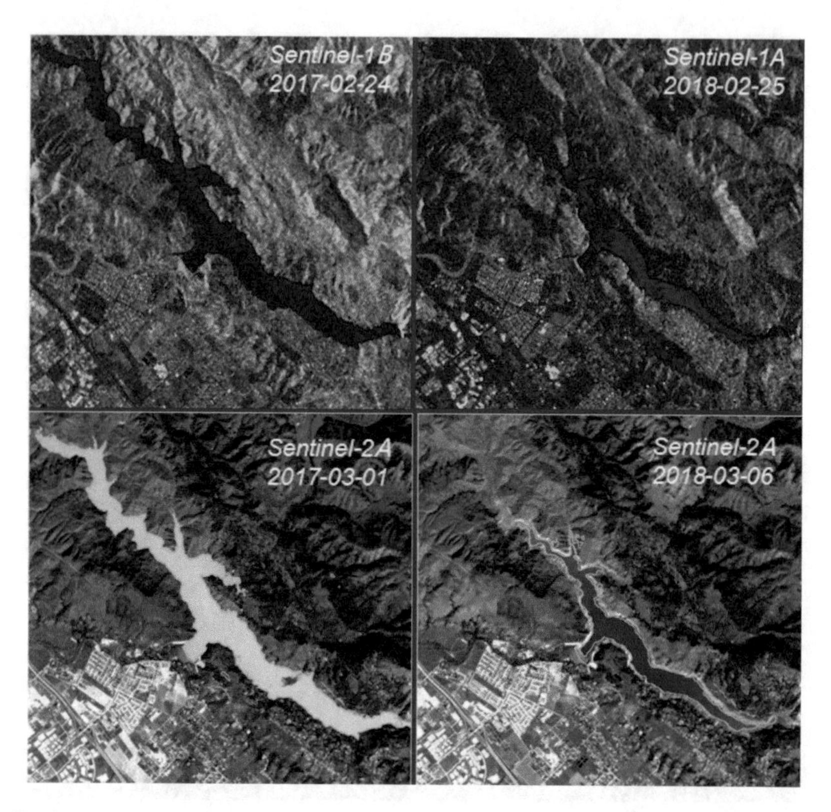

Fig. 8. Accuracy comparison of the vector layers of recognized water objects obtained by radar images from the Sentinel-1A/B satellites and by visible and infrared images from Sentinel-2A satellite

Table 3. Recognition accuracy vs. bit depth of the source data and lossy compression.

Satellite	Float32	Uint16	Uint16 + JPEG2000	Byte	Byte + JPEG2000	Byte + JPEG
Sentinel-1A/B	85…89%	83…88%	82…87%	82…86%	81…85%	81…83%
Sentinel-2A/B	–	89…91%	89…90%	87…88%	86…88%	85…87%

The information technology for automated processing of optical and radar data from Sentinel satellites for monitoring floods and droughts has the following advantages:

- no requirements for sufficient illumination of the surveyed territory (can be imaged both during the day and at night);
- independence from weather conditions over the monitored territory (imaging is possible even with 100% cloudiness – see example on top Fig. 9);

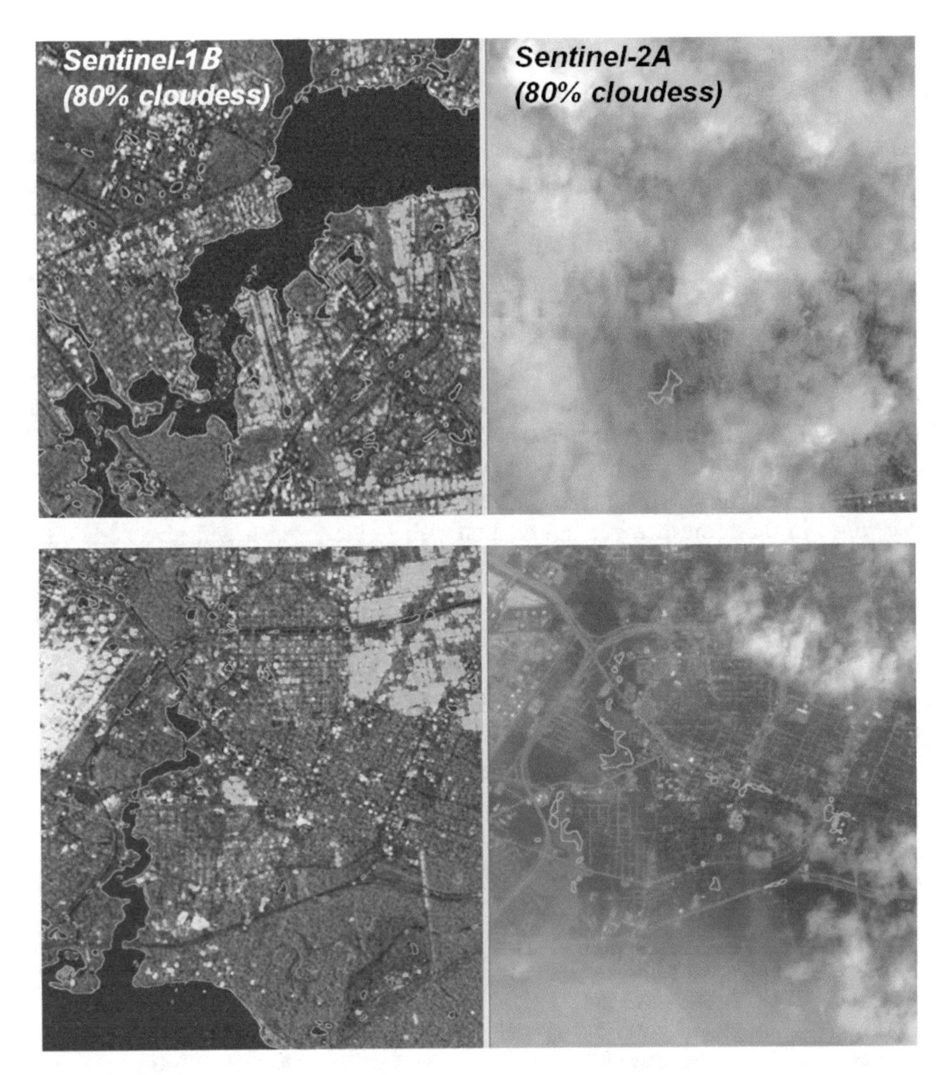

Fig. 9. Comparison of recognized water objects for same territory and imaging date obtained by radar images from the Sentinel-1B satellite and by visible and infrared images from Sentinel-2A satellite

- simplicity and the possibility of rapid mastering by non-specialists in the field of satellite image processing (orientation to a mass user);
- high speed of processing and ease of visualization of results, which is important in solving problems of operational monitoring of emergencies;
- a high degree of automation of processing of radar and optical remote sensing data, which makes it possible to implement this methodology in the form of software for geoinformation web services.

7 Summary and Conclusion

Droughts and floods are among the most devastating natural hazards in the world, claiming more lives and causing extensive damage to economies. Climate change is expected to increase the frequency and magnitude of droughts and floods to which the agriculture sector is particularly exposed. Space technology has made substantial contribution in all-weather monitoring of floods and droughts. As a result of the conducted research, the information technology of automated processing of optical and radar data from Sentinel 1A/B and Sentinel 2A/B satellites was developed and tested in order to ensure all-weather monitoring of floods and droughts. Owing to the high degree of processing automation of radar and optical remote sensing data, the developed technology can be used to promptly inform about the course and consequences of floods and droughts not only representatives of government services and commercial structures, but also the general public, the most numerous user.

Currently, the software implementation of this information technology is being developed in the form of a geo-information web service, which will give additional significant advantages over the existing version, such as:

- work directly in the browser, which does not require additional software installed by the client;
- software and hardware independence, which enables using this web service on mobile devices;
- the results of image processing are stored on the server, which allows all customers to use the web service regardless of their location.

References

1. Ban, H.-J., Kwon, Y.-J., Shin, H., Ryu, H.-S., Hong, S.: Flood monitoring using satellite-based RGB composite imagery and refractive index retrieval in visible and near-infrared bands. Remote Sens. **9**, 313 (2017)
2. Chen, X., Jiang, J., Li, H.: Drought and flood monitoring of the Liao River Basin in Northeast China using extended GRACE data. Remote Sens. **10**, 1168 (2018)
3. Hnatushenko, V.V., Hnatushenko, Vik.V., Mozgovoy, D.K., Vasiliev, V.V:. Satellite technology of the forest fires effects monitoring. Scientific Bulletin of National Mining University, Issue 1 (151), pp. 70–76 (2016)
4. Psomiadis, E., Soulis, K.X., Zoka, M., Dercas, N.: Synergistic approach of remote sensing and GIS techniques for flash-flood monitoring and damage assessment in Thessaly plain area, Greece. Water **11**, 448 (2019)
5. Washaya, P., Balz, T., Mohamadi, B.: Coherence change-detection with Sentinel-1 for natural and anthropogenic disaster monitoring in urban areas. Remote Sens. **10**, 1026 (2018)
6. Hnatushenko, V.V., Mozgovoy, D.K., Vasyliev, V.V.: Satellite monitoring of deforestation as a result of mining. Scientific bulletin of National Mining University - State Higher Educational Institution "National Mining University". Dnipropetrovsk № 5 (161), pp. 94–99 (2017)

7. Jeyaseelan, A.T.: Drought & flood assessment and monitoring using remote sensing and GIS. In: Dun, D. (ed.) Satellite Remote Sensing and GIS Applications in Agricultural Meteorology, p. 291. World Meteorological Organization: Hyderabad, India; Geneva, Switzerland (2003)
8. Hnatushenko, V.V., Mozgovoy, D.K., Vasyliev, V.V., Kavats, O.O.: Satellite Monitoring of Consequences of Illegal Extraction of Amber in Ukraine. Scientific bulletin of National Mining University. - State Higher Educational Institution "National Mining University". Dnipropetrovsk № 2 (158), pp. 99–105 (2017)
9. Mason, D.C., Davenport, I.J., Neal, J.C., Schumann, G.J.-P., Bates, P.D.: Near real-time flood detection in urban and rural areas using high-resolution Synthetic Aperture Radar images. IEEE Trans. Geosci. Remote Sens. **50**, 3041–3052 (2012)
10. Garkusha, I.N., Hnatushenko, V.V., Vasyliev, V.V.: Using Sentinel-1 data for monitoring of soil moisture. In: 2017 IEEE International Geoscience and Remote Sensing Symposium (IGARSS), Fort Worth, TX, USA (2017). https://doi.org/10.1109/IGARSS.2017.8127291
11. Garkusha, I.N., Hnatushenko, V.V., Vasyliev, V.V.: Research of influence of atmosphere and humidity on the data of radar imaging by Sentinel-1. In: IEEE 37th International Conference on Electronics and Nanotechnology (ELNANO) (2017). https://doi.org/10.1109/ELNANO.2017.7939787
12. Mozgovoy, D., Hnatushenko, V., Vasyliev, V.: Accuracy evaluation of automated object recognition using multispectral aerial images and neural network. In: Proceedings of the SPIE Tenth International Conference on Digital Image Processing (ICDIP 2018), 108060H (2018). https://doi.org/10.1117/12.2502905
13. Hnatushenko, V.V., Mozgovoy, D.K., Serikov, I.Ju., Vasyliev, V.V.: Automatic vegetation classification using multispectral aerial images and neural network. System technologies. Dnipro № 6 (107), pp. 66–72 (2016)
14. Mozgovoy, D.K., Hnatushenko, V.V., Vasyliev, V.V.: Automated recognition of vegetation and water bodies on the territory of megacities in satellite images of visible and IR bands. ISPRS Ann. Photogramm. Remote Sens. Spatial Inf. Sci. **IV**(3), 167–172 (2018). https://doi.org/10.5194/isprs-annals-IV-3-167-2018

Inductive Modeling

Modified Method of Structural Identification of Interval Discrete Models of Atmospheric Pollution by Harmful Emissions from Motor Vehicles

Mykola Dyvak, Natalia Porplytsya, and Yurii Maslyiak(✉)

Department of Computer Science, Ternopil National Economic University, Ternopil, Ukraine
mdy@tneu.edu.ua, ocheretnyuk.n@gmail.com, yuramasua@gmail.com

Abstract. The paper deals with a problem of modeling of dynamics of harmful emissions from motor vehicles using a mathematical model in the form of a difference equation. To build such models, a method of structural identification based on the bee colony behavioral models is widely used. It is shown that in order to reduce the time complexity of this method and simultaneously ensure the possibility of finding of a unified model that would be applicable for different points in the city, it is important to ensure the completeness of a set of structural elements. It is shown that in order to increase its efficiency, it is expedient to pre-process the input data obtained in an interval form. It is proposed to use the subtractive clustering method for this purpose. The example of building of model of atmospheric pollution by harmful emissions from motor vehicles using cluster analysis of experimental data to form the initial set of structural elements is considered.

Keywords: Method of structural identification of models · Difference equation · Interval data · Structural elements · Subtractive clustering · Automated formation of structural elements · Atmospheric pollution · Nitrogen dioxide

1 Introduction

Differential equations are often used to model processes of atmosphere pollution by harmful emissions of non-stationary pollution sources, such as motor vehicles [1–4]. However, the building of such models requires a deep analysis of the physics of pollution process, determination of diffusion coefficients, etc. Therefore, in contrast to this approach, discrete dynamic models (DDMs) in the form of difference equations are used for this purpose [5, 6].

The identification of DDMs is executed based on the measurement results of dynamics of harmful emission concentrations which usually are inaccurate due to measuring devices [6, 7]. As a result, structural and parametric identification tasks [8, 9] must be solved based on interval data in order to build a DDM.

A method based on the bee colony behavioral model (BCBM) [8, 10] is used for structural identification of DDM. Let's note that the method also implements principles

N. Shakhovska and M. O. Medykovskyy (Eds.): CCSIT 2019, AISC 1080, pp. 491–507, 2020.
https://doi.org/10.1007/978-3-030-33695-0_33

of inductive modeling for automated building of a model structure in the form of a difference equation (DE) [11–13].

For example, the application of this method for building of the model of daily cycle dynamics of Nitrogen dioxide (chemical formula is NO_2) concentrations at the crossroads of Ruska-Zamkova-Shashkevycha streets in Ternopil city is considered in the paper [7].

However, the results of experiments showed that the model of dynamics of harmful emission concentrations [7] is not unified. This is due to neglect of the influence of various factors, including random ones, which are characteristic for a specific point for which the modeling is executed [14–16].

The results of analysis of this structural identification method showed that the degree of "unification" of the built model, that is, the possibility of its application for the different points in the city, directly depends on the "quality" and completeness of the formed set of structural elements. Let's note that the formation of the initial set of structural elements of the model, in the existing method of structural identification, is executed empirically [10, 17].

At the same time, in the paper [18], it is proposed to execute the procedure of formation of the set of structural elements based on the pre-analysis of the sample of input interval data. Such an analysis is based on subtractive clustering method [19–22]. On the one hand, the application of this procedure ensures the possibility of forming of "qualitative" and complete set of structural elements. On the other hand, it ensures significant reduction of the computational complexity of the method of structural identification.

Therefore, the purpose of this work is conducting of experimental research to verify the effectiveness of the procedure of automated formation of the set of structural elements and building of unified discrete dynamic model of atmosphere pollution dynamics.

The construction of the unified model gives a possibility to reduce the amount of required measurements, as well as the amount of equipment required. In addition, it will provide an opportunity to predict the amount of harmful emissions at any point of Ternopil city.

2 Statement of the Problem

The difference analogs of differential equations are built based on real measurements of harmful emission concentrations as well as parameters of the environment of their dynamics. It should be noted that the accuracy of such data may be low due to errors of gas sensors. Taking this fact into account, it is enough to build a mathematical model with an accuracy that corresponds to an accuracy of a measurement experiment [6]. In this case, the experimental data are represented in the interval form [6, 23–26]. Then, the dynamics of harmful emission concentrations at a specific point for the environmental monitoring task is described by interval discrete dynamic model (IDDM) in the following general form:

$$[\hat{v}_k(\lambda_s)] = [\hat{v}_k^-; \hat{v}_k^+] = \vec{f}^T([\hat{v}_{k-d}]\ldots[\hat{v}_{k-1}], \vec{u}_{p,0}\ldots\vec{u}_{p,k}) \cdot \vec{g} \\ k = d, \ldots, K, \tag{1}$$

where $\vec{f}^T(\bullet)$ is the vector of basis functions, nonlinear in a general case. These basis functions perform conversion of the modeled characteristic values and the values of input variables at discrete moments (periods) of time $k = d, \ldots, K$; $[\hat{v}_k(\lambda_s)]$ is the interval value of modeled characteristic at the moment (period) of time $k = d, \ldots, K$; $\lambda_s = \{f_1(\bullet) \cdot g_1; \ldots; f_{m_s}(\bullet) \cdot g_{m_s}\}$ is the set of elements of a current IDDM structure, $\{f_1(\bullet); \ldots; f_{m_s}(\bullet)\} \subset F$ is the set of structural elements that defines the current s-th IDDM structure, F is the set of all structural elements, m_s is the number of elements in the current structure λ_s; $\vec{u}_{p,0}, \ldots, \vec{u}_{p,k}$ is the known vector of input variables (controls) with number $p = 1, \ldots, P$ at time $k = d, \ldots, K$; d is an order of the difference equation (DE); $\vec{g}$ is unknown vector of parameters of the DE.

As a result of structural identification procedure, $\vec{f}^T(\bullet)$; $\vec{u}_{p,0}, \ldots, \vec{u}_{p,k}$; d are defined [10]. To implement the difference scheme, it is also necessary to set the values of parameters vector $\vec{g}$ components and set the initial conditions. That is, the value of each element from the set $[\hat{v}_{k-d}] \ldots [\hat{v}_{k-1}], \vec{u}_{p,0} \ldots \vec{u}_{p,k}$ for certain discretes, usually, the initial ones. For the parametric identification task, the structure of the difference equation is considered as known.

The conditions of consistency between experimental data represented in the interval form with the data obtained based on the mathematical model (1), are formulated as follows [10, 17]:

$$[\hat{v}_k^-; \hat{v}_k^+] \in [z_k^-; z_k^+], \forall k = 0, \ldots, K. \tag{2}$$

where, z_k^-, z_k^+ are the lower and upper bounds of the interval of possible values of measured concentration of harmful substance.

Substituting, instead of interval estimates $[\hat{v}_k^-; \hat{v}_k^+]$, the expression for their calculation based on IDDM (1) into expressions (2) taking into account given initial numeric intervals of each element from the set $[\hat{v}_0^-; \hat{v}_0^+] \in [z_0^-; z_0^+], \ldots, [\hat{v}_{d-1}^-; \hat{v}_{d-1}^+] \in [z_{d-1}^-; z_{d-1}^+]$ and given vectors of input variables $\vec{u}_{p,0}, \ldots, \vec{u}_{p,k}$, an interval system of nonlinear algebraic equations (ISNAE) is obtained [9].

$$\begin{cases} [\hat{v}_0^-; \hat{v}_0^+] \in [z_0^-; z_0^+], \ldots, [\hat{v}_{d-1}^-; \hat{v}_{d-1}^+] \in [z_{d-1}^-; z_{d-1}^+]; \\ [\hat{v}_k] = [\hat{v}_k^-; \hat{v}_k^+] = \vec{f}^T([\hat{v}_{k-d}] \ldots [\hat{v}_{k-1}], \vec{u}) \cdot \vec{g}, k = d, \ldots, K; \\ \hat{v}_k^- \le \vec{f}^T([\hat{v}_{k-d}] \ldots [\hat{v}_{k-1}], \vec{u}) \cdot \vec{g} \le \hat{v}_k^+, k = d, \ldots, K. \end{cases} \tag{3}$$

So, the task of IDDM (1) parameters identification under conditions (2) is the task of solving of the obtained ISNAE (3). Such a task is NP-complete task. Therefore, to solve it, the method of random search of at least one ISNAE is used [9]. The solution of structural identification task consists in obtaining of such a set of elements λ_0, so for them, such IDDM structure could be formed for which the conditions (2) would be met. To estimate the quality of the current IDDM structure, the value of indicator $\delta(\lambda_s)$ is used. It quantitatively expresses the proximity of the current structure to satisfactory one in the terms of ensuring of conditions (2). The indicator $\delta(\lambda_s)$ is called objective function of optimization task of the IDDM structural identification.

So, the task of IDDM structural identification is formally represented in the form:

$$\delta(\lambda_s) \rightarrow^{\widehat{\vec{g}_s, \vec{f}_s(\bullet)}} \min, \; \vec{f}_s(\bullet) \in F. \tag{4}$$

It is important to note that the task of IDDM structural identification consists in procedure of multiple formation of a structural elements set and solving of parametric identification task, that is, solving of the above mentioned ISNAE (3).

3 Method of Structural Identification for Interval Discrete Difference Model

It is proved that one of the most effective methods for solving of task (3) is the method of IDDM structural identification based on the BCBM [8, 10, 27, 28]. It is based on the principles of inductive modeling [11, 12] and implements the principles of collective search and food provision by a bee colony in the wildlife [29–32]. This method of IDDM structural identification is described in detail in [10].

The essence of this method consists in formation of the initial set of IDDM structures and its subsequent modifications using a set of operators. Let's note that the mentioned operators implement elements of the foraging behavior of honeybees colony and models the processes of nectar sources exhaustion. This, in the context of IDDM structural identification task, means the procedures of formation, modification and selection of the current structures set. Such a modification of the current structures set λ_s is implemented in such a way as to ensure the reduction of the objective function value $\delta(\lambda_s)$ on each iteration of the method. The procedure of structural identification is continued until such an IDDM structure is obtained for which $\delta(\lambda_0) = 0$. Fulfillment of this condition means that the built based on the structure λ_0 IDMM ensures the fulfillment of conditions (2) for the set of all discretes. The results of analysis of computational scheme of the IDDM structural identification method showed that the "quality" of forming of initial set of structural elements significantly affects the computational complexity of its implementation. It is worth to note that it is formed empirically in the existing method. Therefore, further research is aimed at development of method of automated formation of a structural elements set using procedures of pre-analysis of interval data sample based on clustering and interval data analysis.

4 Method of Automated Forming of Structural Elements Set

As it was noted earlier, the computational complexity of the structural identification method significantly depends on the quality and completeness of the formed initial set of structural elements. Therefore, it is important to ensure the formation of a complete set of structural elements, on the one hand. Also, it is worth to ensure the minimal number of structural elements, on the other hand.

As it is known, the uncertainty and inaccuracy of data leads to the need of their interval representation. For example, the interval representation of the observation results in the task of building of IDDM of harmful emission concentrations is related to two groups of factors: measurement errors and errors caused by impossibility to take into account the influence of various factors including random ones. Such important factors for this tasks class are different characteristics of the point for which the dynamic model is built. For example: natural ventilation, presence of vertical and horizontal air streams, etc. Let's assume that the affect of these factors is additive. Then, the dependence is true:

$$[\widehat{v}_k(\lambda_s)] = \sum_{i=1}^{m} f_i^s(\bullet) \cdot g_i^s + G(\lambda_s) \tag{5}$$

where $f_i^s(\bullet)$ are the components of the vector of basis functions, $G(\lambda_s)$ is a function in the structure of the DE, related to not taken into account factors.

Let's note that the components of basic functions vector $f_i^s(\bullet)$ may be dependent on input (taken into account) factors $u_{p,0}, \ldots, u_{p,k}$. In addition, the relationship between the output characteristic $[\widehat{v}_k(\lambda_s)]$ and input (taken into account) factors $u_{p,0}, \ldots, u_{p,k}$ is ambiguous due to the interval representation of the components of the vector $\vec{f}_s^T([\widehat{v}_{k-d}], \ldots, [\widehat{v}_{k-1}], u_{p,0}, \ldots, u_{p,k})$. Such an ambiguity (it may be caused by the measuring errors) is represented by the intervals of possible values of the output characteristic for the given values of the input variables. At this, the presence of an unknown function $G(\lambda_s)$ in the structure of the difference equation (5) significantly complicates the task of structural identification, since it significantly increases the ambiguity of the relationship between the output characteristic $[\widehat{v}_k(\lambda_s)]$ and input (taken into account) factors. One way to account for this uncertainty may be to extend the intervals in the sample interval data, which will necessarily result in loss of accuracy of the model, that is, to extend the range of estimates using the output characteristic $[\widehat{v}_k(\lambda_s)]$. Therefore, such an approach is not always acceptable. Another approach is to eliminate the uncertainty related to the additive component $G(\lambda_s)$ during preprocess of the interval data. For example, such an approach can be realized using the following transformation:

$$\frac{\partial v_k(\lambda_s)}{\partial u_{p,k}} = \frac{\partial(\sum_{i=1}^{m} f_i^s(\bullet) \cdot g_i^s)}{\partial u_{p,k}} + \frac{\partial G(\lambda_s)}{\partial u_{p,k}} \tag{6}$$

Assuming that there is no dependence between $G(\lambda_s)$ and the components of the vector of input (taken into account) factors $u_{p,0}, \ldots, u_{p,k}$, we obtain zero as the second addend (6).

Based on the made assumptions and conducted analysis of uncertainty and ambiguity in the case of the IDDM structural identification, we can formulate requirements to preliminary analysis of the input data sample in order to ensure the "quality" and completeness of the formed structural elements set. Namely, it is advisable to check the existence of a unambiguous relationship between the output characteristics of the object and the factors that affect these characteristics based on a preliminary analysis of the data sample. If there is no such an unambiguous dependence, it is necessary, based on expression (7), during the interval data preprocessing, to ensure the removal of the additive component $G(\lambda_s)$ which defines a part of the structure of the difference equation (5).

In the case of use of discrete data sample, the procedure (6) can be approximately executed in such a way:

$$\frac{\partial v_k(\lambda_s)}{\partial u_{p,k}} \approx \frac{z_{k-1} - z_{k-2}}{u_{p,k-1} - u_{p,k-2}}, k = 1, \ldots, K. \quad (7)$$

Then, during the pre-analysis of the input data sample in order to ensure the high "quality" and completeness of the formed set of structural elements, it is expedient to verify the unambiguous relationship between the output characteristics of the process and the factors that affect these characteristics. To do this, the modified cluster analysis method is used.

Let a set of input data for structural identification is given:$[z_k^-; z_k^+]$ is the interval of possible values of the measured characteristic at the discrete time intervals $k = 0, \ldots, K$, $u_{p,k}$ are the values of input variables at the discrete time intervals $k = 0, \ldots, K$; $p = 1, \ldots, P$ is a number of the input variable. Subtractive clustering method the basis for the clustering of input data. This method does not require a large sample of experimental data and a predetermined number of clusters which greatly reduces the time for its implementation. It should be noted that the number of clusters based on this method is determined by the only parameter which is the cluster radius [22]. In the further consideration, the index p will be ignored when denoting input variables from a set $p = 1, \ldots, P$.

In accordance with the clustering method, in the beginning, the potential cluster centers set is formed from the data matrix strings for the clustering of input variables and the potentials of the cluster centers are calculated using the expression [19]:

$$P_h(c_h) = \sum_{k=0}^{K} \exp(-\alpha \cdot \|\vec{c}_h - \vec{x}_k\|), \quad (8)$$

where $\vec{c}_h = (c_{1,h}, c_{2,h})$ is a potential center of h-th cluster; $\vec{x}_k = (z_k, u_k)$ input data: z_k is a middle of the interval of measured concentration of the harmful emissions at the time period k, u_k is corresponding quantity of motor vehicles that crossed the road at a given point for a given time period k; $\|\vec{c}_h - \vec{x}_k\|$ is a distance between potential center of h-th cluster and input data $\vec{x}_k$, $k = 0, \ldots, K$, $h = 1, \ldots, H$.

The "mountain peaks" coordinates are chosen to be the cluster centers. Scilicet, the cluster center is a point which is characterized by the concentration of harmful emissions in the time interval k and the corresponding number of vehicles, with the highest value of the potential [19]:

$$(\vec{c}_h) = \arg \max_{h=1,\ldots,H} P_h(z_k, u_k). \quad (9)$$

To avoid forming of similar clusters, it is necessary to recalculate values of potentials for other possible cluster centers [20]:

$$P_{h+1}(\vec{c}_{h+1}) = P_{h+1}(\vec{c}_h) - P_h(\vec{c}_h) \cdot \exp(-\beta \cdot \|\vec{c}_{h+1} - \vec{c}_h\|), \\ h = 1, \ldots, H \tag{10}$$

where $P_h(\vec{c}_h)$ is the potential of a possible center of h-th cluster at the h-th iteration; $P_{h+1}(\vec{c}_h)$ is the potential потенціал потенційного цент of a possible center of h-th cluster at the h + 1 iteration; β is the positive constant; $\|\vec{c}_{h+1} - \vec{c}_h\|$ is the distance between possible center of h + 1 cluster and center of formed h-th cluster.

The process of calculation of cluster centers is executed until all the lines of the input variable matrix X, which is represented by the set $z_k, u_k, k = 1, \ldots, K$, are excluded. It is worth to note that unlike the existing subtractive clustering method, the cluster radius is fixed and determined by the uncertainty of the data sample in the proposed one.

This is due to the purpose of cluster procedure application which is ensuring of unambiguous relationship between the output characteristic and the influence factors in the way of transforming of input data sample into a sample of their differential characteristics. Thus, the radius of a cluster will be determined by the expression:

$$r = \frac{\max\limits_{k=1,\ldots,K}(z_k^+ - z_k^-)}{2}. \tag{11}$$

As a result of clustering procedure application, H clusters are obtained.

The unambiguous relationship between the output characteristic of the process and the factors that affect this characteristic is considered as the case when the cluster projections on the axis of input variable do not intersect each other. It means the possibility of finding of unambiguous dependence within the initially given interval width for the output characteristic, that is, the cluster radius, for the clusters of number h and number j. Mathematically, the condition has the following form:

$$\exists\, [u_{proj\,h}^-; u_{proj\,h}^+] \cap [u_{proj\,j}^-; u_{proj\,j}^+] = \varnothing, \\ h = 1, \ldots, H, j = 1, \ldots, J, h \neq j, \tag{12}$$

where $[u_{proj\,h}^-; u_{proj\,h}^+]$ is the interval obtained as a result of projection of the h-th cluster on the axis of input variable; $[u_{proj\,j}^-; u_{proj\,j}^+]$ is the interval obtained as a result of projection of the j-th cluster on the axis of input variable.

Otherwise, the transformation (9) is applied for the data sample and the clustering procedure for the transformed data sample is repeated. If the above procedure is executed at one iteration and the condition (12) is met, then, the following element must be added to the set of structural elements for structural identification task:

$$f_i^s(\bullet) = \frac{z_{k-1} - z_{k-2}}{u_{p,k-1} - u_{p,k-2}}. \tag{13}$$

If condition (12) is not met for the transformed sample, it is obvious that the hypothesis about the additivity of the components in formula (5) is not proved.

For example, as a result of the second iteration execution, such an element of the difference equation must be added o the set of structural elements for the task of structural identification:

$$f_i^s(\bullet) = \frac{z_{k-1} - 2 \cdot z_{k-2} + z_{k-3}}{u_{p,k-1} - 2 \cdot u_{p,k-2} + u_{p,k-3}}. \tag{14}$$

The iterations of the data sample transformation are repeated until an unambiguous relationship is obtained. That is, the fulfillment of the condition (12). At the same time, by the result of each iteration, the elements (13) and (14) (and, finite differences of higher orders, if necessary) must be added to the initial set of structural elements.

5 Example of Application of the Method for Forming of Structural Elements Set

Let's consider the main transformations of the input data sample with the purpose of defining of needed structural elements for the task of IDMM structural identification in the example of modeling of daily cycle dynamics of Nitrogen dioxide (chemical formula is NO_2) concentrations at the crossroads of Ruska-Zamkova-Shashkevycha streets and the crossroads of Chekhova-Za Rudkoyu streets of Ternopil city [7]. As shown in [7], the attempts to build a unified model for these crossroads were unsuccessful.

Based on the theoretical substantiation and the proposed algorithm of cluster analysis, let's find out the essence of the problem and determine the needed structural elements for successful solving of the structural identification task for the mentioned mathematical model.

In these examples, the measurements were executed every second. In order to compensate the random measurement errors, the measured instantaneous values were averaged in the 20 min window. At the same time, the traffic intensity has been fixed. The relative error of the gas sensor is 15% (Y-axis). The errors of measuring of the air temperature and humidity are such low that can be ignored. Intensity of motor vehicles traffic has been fixed without errors (X-axis). The common measurement results normalized to the interval of [0, 1] for the both mentioned crossroads are represented in the Fig. 1.

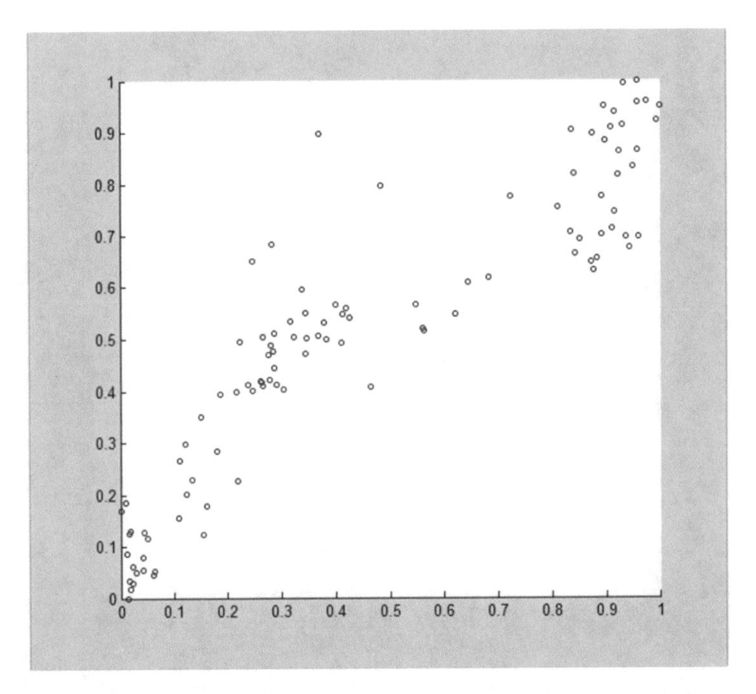

Fig. 1. The illustration of the combined data sample of normalized measured values at the crossroads of Ruska-Zamkova—Shashkevycha and Chehova-Za Rudkoyu streets.

The results of data sample clustering in accordance with the above described clustering method using "subclust" function of the MATLAB software [22] are shown in the Fig. 1.

Let's note that the radius of all clusters is the same. It is calculated based on the analysis of normalized data sample using the expression:

$$r = \frac{\max\limits_{k=1,\ldots,K}\left(z_k^+ - z_k^-\right)}{2} = 0,135. \quad (15)$$

As we can see in the Fig. 2, the obtained clusters intersect. Moreover, it is shown that the projections of the clusters on the input variable (traffic intensity, which are intervals) X-axis also intersect, in the Fig. 3. It means that condition (12) is not satisfied. Therefore, it is necessary to apply the transformation (13) for each value of the data sample and repeat the clustering procedure for the transformed sample.

Obviously, the condition (10) for the input data sample is not fulfilled. Figure 4 demonstrates projections of the clusters on the axis of input variable (motor vehicles traffic intensity) based on the results of the second iteration of the applied method of formation of the initial structural elements set. As we can see from this figure, the condition (10) is met for all intervals.

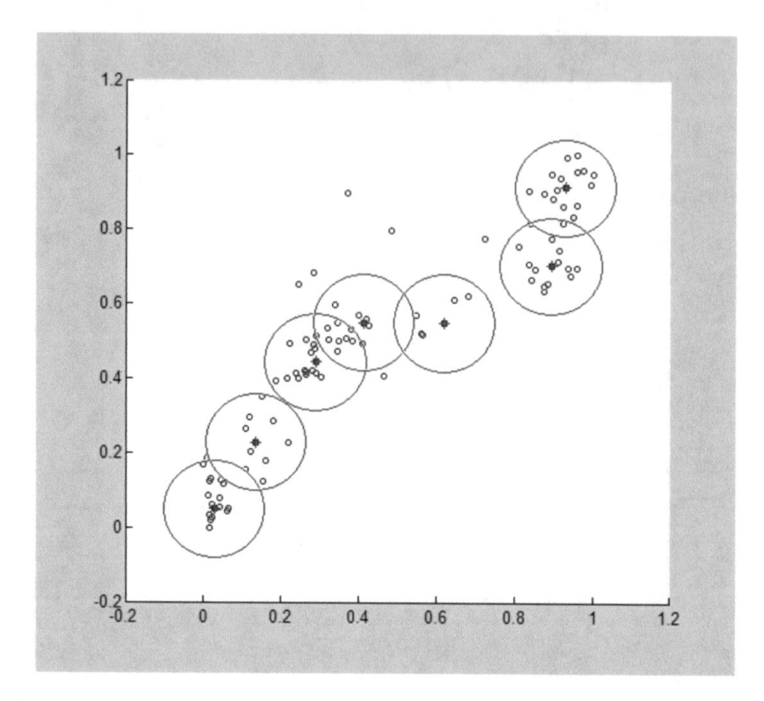

Fig. 2. The result of clustering of the normalized data sample of measured values at the crossroads of Ruska-Zamkova-Shashkevycha and Chehova-Za Rudkoyu streets.

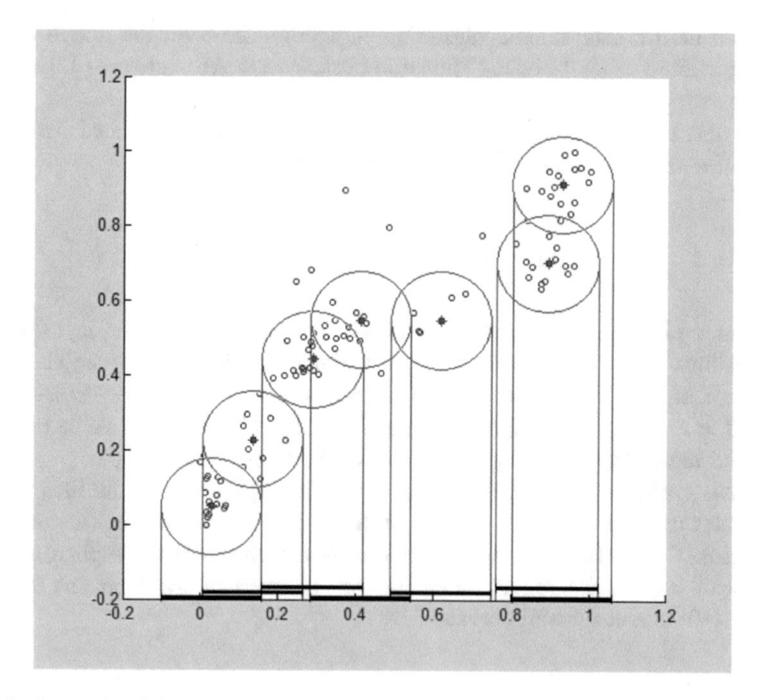

Fig. 3. Intervals of the cluster projections on the axes of input variable (traffic intensity).

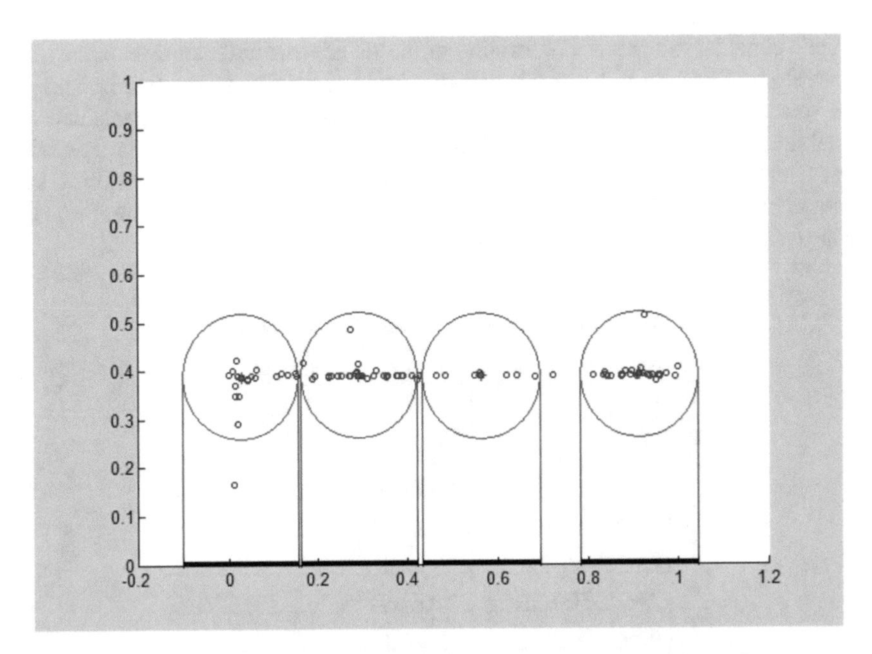

Fig. 4. Intervals of the cluster projections on the axis of input variable for the second iteration.

Thus, as a result of the conducted analysis of the data sample in the considered example, it became possible to find out the reason of impossibility of building of a unified model for NO_2 concentration dynamics at the mentioned city crossroads. The reason is ambiguity of relationship between the concentration of NO_2 and the motor vehicles traffic intensity in various points of the city. The results of the conducted analysis showed that solution of this problem consists in adding of such elements to the initial set of structural elements F in the structural identification task: $\frac{[\hat{v}_{k-1}]-[\hat{v}_{k-2}]}{u_{p,k-1}-u_{p,k-2}}$ and $\frac{[\hat{v}_{k-1}]-2[\hat{v}_{k-2}]+[\hat{v}_{k-3}]}{u_{p,k-1}-2u_{p,k-2}+u_{p,k-3}}$.

6 Comparison Analysis

Let's conduct a comparative analysis of effectiveness of the known method of structural identification of IDDM and modified one with the use of preliminary cluster analysis of the data sample as well as the analysis of the models obtained in both cases in the form (1).

To do this, let's first consider the example of building of IDDM for modeling of a daily cycle dynamics of NO_2 concentrations at the crossroads of Ruska-Zamkova-Shashkevycha streets in Ternopil city using known method of structural identification based on the BCBM. Details of the building of this model and the experimental data sample are represented in [7].

Let's consider the process of forming of the set of structural elements in solving of structural identification task for this example in more details. At the first iteration, the set F was formed. Formed set F is represented in the Table 1 [7]. It contained polynomial functions of not higher than second degree for the DE of not higher than second order. Thus, the cardinality of the set F was $L = 65$. However, the resulting set of structural elements did not ensure the possibility of finding of IDDM in the form (1) for which the conditions (2) would be fulfilled.

Table 1. The fragment of initial set of structural elements F.

№	Structural element	№	Structural element
1	$[\hat{v}_{k-1}]$	25	$[\hat{v}_{k-1}]/[\hat{v}_{k-1}] \cdot u_{k-1}$
2	$[\hat{v}_{k-2}]$	26	$[\hat{v}_{k-2}]/[\hat{v}_{k-2}] \cdot u_{k-1}$
3	$[\hat{v}_{k-1}] \cdot [\hat{v}_{k-1}]$	27	$[\hat{v}_{k-1}]/[\hat{v}_{k-2}] \cdot u_{k-1}$
4	$[\hat{v}_{k-2}] \cdot [\hat{v}_{k-2}]$	…	…
5	$[\hat{v}_{k-1}] \cdot [\hat{v}_{k-2}]$	32	$[\hat{v}_{k-1}]/u_{k-1}$
6	$[\hat{v}_{k-1}] \cdot u_{k-1}$	33	$[\hat{v}_{k-2}]/u_{k-1}$
7	$[\hat{v}_{k-2}] \cdot u_{k-1}$	…	…
8	$[\hat{v}_{k-1}] \cdot u_{k-2}$	45	$[\hat{v}_{k-1}] \cdot u_{k-1}/u_{k-2}$
9	$[\hat{v}_{k-2}] \cdot u_2$	46	$[\hat{v}_{k-2}] \cdot u_{k-1}/u_{k-2}$
10	$[\hat{v}_{k-1}] \cdot [\hat{v}_{k-1}] \cdot u_{k-1}$	…	…
11	$[\hat{v}_{k-2}] \cdot [\hat{v}_{k-2}] \cdot u_{k-1}$	63	$[\hat{v}_{k-1}] \cdot [\hat{v}_{k-1}] \cdot (u_{k-2}/u_{k-1})$
12	$[\hat{v}_{k-1}] \cdot [\hat{v}_{k-2}] \cdot u_{k-1}$	64	$[\hat{v}_{k-2}] \cdot [\hat{v}_{k-2}] \cdot (u_{k-2}/u_{k-1})$
…	…	65	$[\hat{v}_{k-1}] \cdot [\hat{v}_{k-2}] \cdot (u_{k-2}/u_{k-1})$

Therefore, the decision to expand the set F has been made. At the second iteration of the method, the elements of the third degree for the DE of not higher than the third order were added to the current set of structural elements, for example: $[\hat{v}_{k-1}] \cdot [\hat{v}_{k-1}] \cdot [\hat{v}_{k-1}], [\hat{v}_{k-1}] \cdot [\hat{v}_{k-1}] \cdot [\hat{v}_{k-2}]$. Thus, the set F_1 of cardinality $L = 135$ has been formed. However, the set also did not ensure the possibility of building of adequate model in terms of fulfillment of conditions (2). At the third iteration of the method, the elements containing fractions of the measured values of motor vehicles traffic intensity were added to the set F_1, for example: $[\hat{v}_{k-1}] \cdot u_k/u_{k-1}, [\hat{v}_{k-1}] \cdot u_k/u_{k+1}$. Thus, the set F_2 of cardinality $L = 155$ has been formed. As a result, the IDDM has been obtained:

$$\begin{aligned}[\hat{v}_k] = 0,0365 + 0,3541 \cdot [\hat{v}_{k-1}] + 0,118 \cdot [\hat{v}_{k-1}] \cdot [\hat{v}_{k-3}] \\ + 0,5059 \cdot [\hat{v}_{k-1}] \cdot u_k/u_{k-1} - 0,0154 \cdot [\hat{v}_{k-2}] \cdot u_{k-1}/u_{k+1} \\ k = 22, \ldots, 71.\end{aligned} \quad (16)$$

The predicted daily dynamics of NO_2 concentrations based on the IDDM (16) is shown in the Fig. 5. Interval estimates of the measured values of NO_2 concentrations are illustrated by the dashed lines. Their predicted values based on IDDM (16) are illustrated by solid lines.

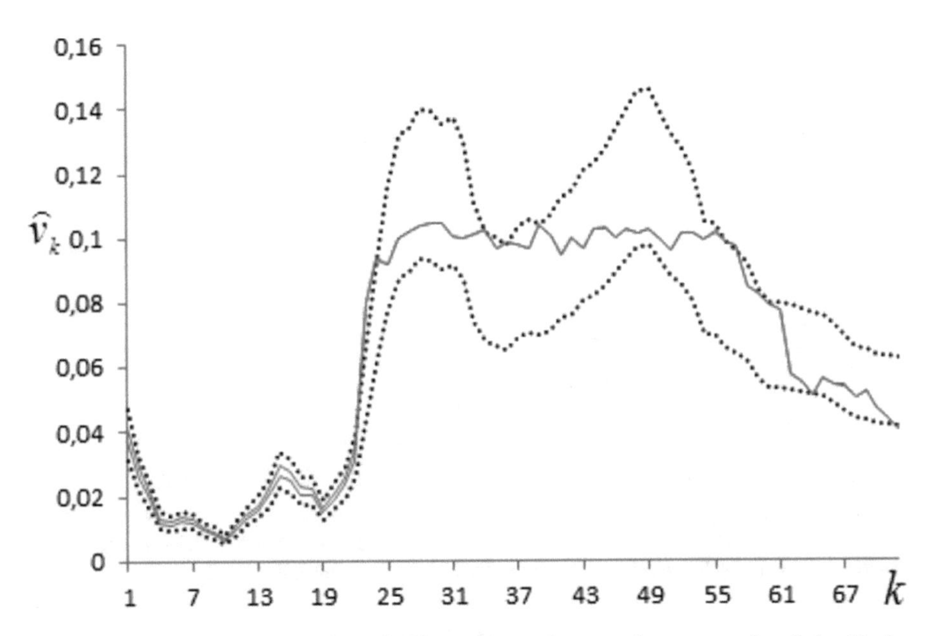

Fig. 5. The daily cycle dynamics of NO_2 concentrations at the crossroads of the Ruska-Zamkova-Shashkevycha streets in Ternopil city obtained by the IDDM (16) [7].

As we can see in the Fig. 5, the predicted intervals of NO_2 concentrations slightly deviate from the experimental data but such a deviation does not exceed 5%. The verification of IDDM (16) for modeling of dynamics of harmful emission concentrations has been executed based on the measured data at another point in the city of Ternopil, in particular, at the crossroads of Chekhova-Za Rudkoyu streets. However, modeling results showed that the IDDM in the form (16) is not suitable for use at any other point in the city.

Then, let's consider the example of building of IDDM for the same point in the city but based on the improved method of structural identification which involves a preliminary cluster analysis of the data sample of the measured concentrations of harmful emissions from vehicles. Let's note that the detailed description of the process of preliminary analysis of experimental data sample using subtractive clustering method is represented in the previous section.

As a result of analysis of the data sample for the considered example the reason of impossibility of building of unified model for prediction of dynamics of NO_2 concentrations at the indicated crossroads has been established. It occurred due to the ambiguity of the relationship between the NO_2 concentration the motor vehicles traffic intensity for different points in the city. The results of the analysis showed that in order to solve this problem, it is necessary to add the following structural elements to the set F in the task of structural identification:

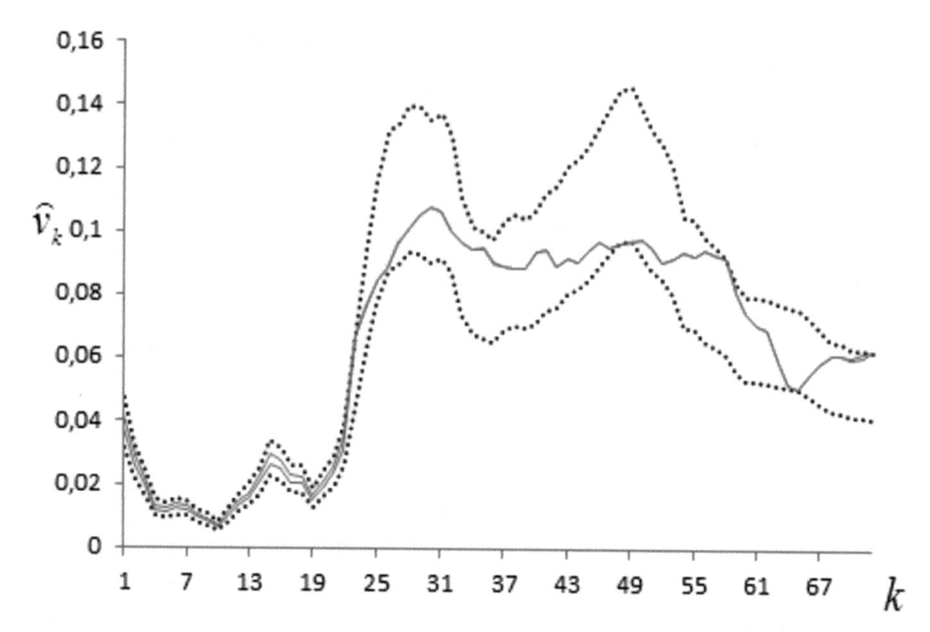

Fig. 6. The daily cycle dynamics of NO_2 concentrations at the crossroads of the Ruska-Zamkova-Shashkevycha streets in Ternopil city obtained by the IDDM (18).

$$\frac{[\hat{v}_{k-1}] - [\hat{v}_{k-2}]}{u_{k-1} - u_{k-2}} \text{ and } \frac{[\hat{v}_{k-1}] - 2 \cdot [\hat{v}_{k-2}] + [\hat{v}_{k-3}]}{u_{k-1} - 2 \cdot u_{k-2} + u_{k-3}}. \tag{17}$$

Therefore, at the first iteration of the method, the set F was created containing the polynomial functions of not higher than the second degree, L = 87. Also, the structural elements identified at the input data preliminary analysis stage (17), were added to the set F. As a result, the IDDM has been obtained:

$$\begin{aligned} [\hat{v}_k] = 0,0674 + 0,0752 \cdot ([\hat{v}_{k-1}] - [\hat{v}_{k-2}])/(u_{k-1} - u_{k-2}) \\ + 0,1722 \cdot [\hat{v}_{k-5}] \cdot [\hat{v}_{k-7}] - 0,1269 \cdot [\hat{v}_{k-1}] + 0,0006 \cdot [\hat{v}_{k-1}] \cdot u_{k-1} \\ - 0,0778 \cdot [\hat{v}_{k-5}] \cdot u_{k-8}/u_{k-2}, k = 22, \ldots, 71. \end{aligned} \tag{18}$$

In the Fig. 6, the predicted daily dynamics of NO_2 concentrations based of the IDDM (18) is demonstrated.

As we can see in the Fig. 6, the resultant IDDM (18) fulfills the conditions (2) on the set of all discretes.

The model verification has been also executed to model the dynamics of NO_2 concentrations based on measured data at another point, in particular, at the crossroads of Chekhova-Za Rudkoyu streets in Ternopil city. The estimated predicted values of the NO_2 concentrations based on IDDM (18) are represented in the Fig. 7.

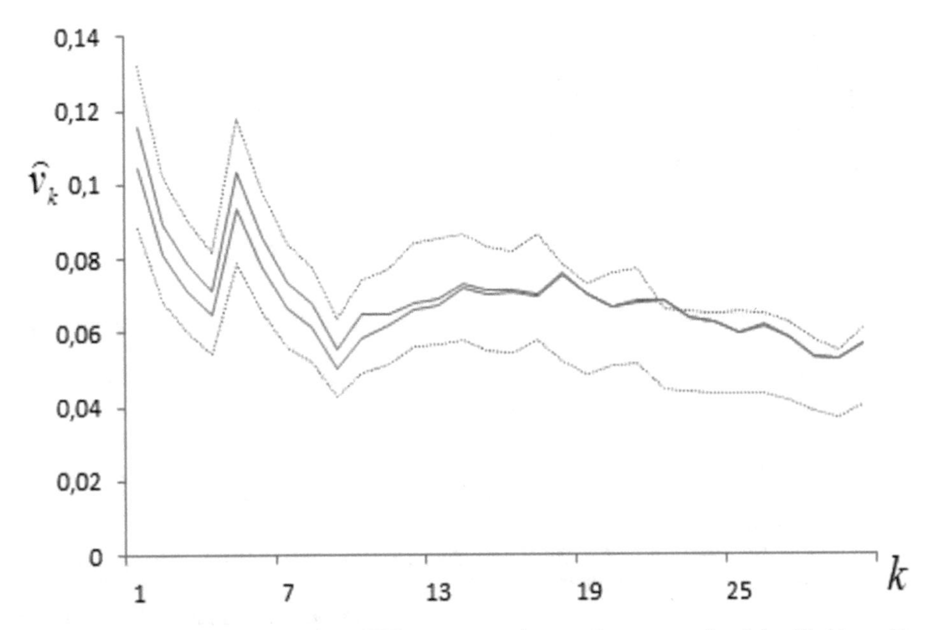

Fig. 7. The daily cycle dynamics of NO_2 concentrations at the crossroads of the Chekhova-Za Rudkoyu streets in Ternopil city obtained by the IDDM (18).

As we can see from the Fig. 7, the predicted values of the NO_2 concentrations are within the range of all discrete values of measured data sample. Consequently, the obtained IDDM (18) is unified one, that is, it can be applied for modeling of dynamics of harmful emission concentrations in different points of the city.

7 Conclusion

It has been established that the "quality" and completeness of the initial set of structural elements is the crucial factor that affects the computational complexity of the IDDM structural identification method based on the BCBM. In the existing method, the above mentioned initial set of structural elements has been formed empirically. It has been proposed to use the procedure of preliminary analysis of the interval data sample for the automated formation of this set which is based on the application of subtractive clustering method.

The method of IDDM structural identification based on BCBM has been improved by adding procedures of automated formation of initial set of structural elements. The procedures are based on the clustering of data sample which ensures improvement of quality of the formed set. It allowed to reduce the computational complexity of the method implementation.

Experimental research has been conducted to analyze the effectiveness of both the known method of the IDDM structural identification and modified one with the use of a preliminary cluster analysis of the input data sample. The results of the research have shown that the substantiated procedure of formation of the set of structural elements in the task of structural identification allowed to significantly reduce the computational complexity – by more than 10 times. In addition, the analysis of the obtained model showed that it is unified and can be used to model the concentrations of harmful emissions in different parts of the city.

References

1. Gromova, O.V.: Analysis of models of propagation of substances in the atmosphere from stationary sources. Sci. Work. UkrRHMI **253**, 173–181 (2004). (in Ukrainian)
2. Ivakhnenko, A.G.: The inductive method of self-organizing models of complex systems. Naukova dumka, Kyiv, Ukraine (1982). (in Russian)
3. Csanady, G.T.: Turbulent Diffusion in the Environment. Springer, Dordrecht (2012)
4. Pant, P., Harrison, R.M.: Estimation of the contribution of road traffic emissions to particulate matter concentrations from field measurements: a review. Atmos. Environ. **77**, 78–97 (2013)
5. Kelley, W.G., Peterson, A.C.: Difference Equations: An Introduction with Applications. Academic Press, New York (2001)
6. Ocheretnyuk, N., Voytyuk, I., Dyvak, M., Martsenyuk, Ye.: Features of structure identification the macromodels for nonstationary fields of air pollutions from vehicles. In: Proceedings of the 11th International Conference on the Modern Problems of Radio Engineering Telecommunications and Computer Science (TCSET 2012), p. 444 (2012)
7. Dyvak M., Voytyuk I., Porplytsya N., Pukas A.: Modeling the process of air pollution by harmful emissions from vehicles. In: Proceedings of the 14th International Conference on Advanced Trends in Radioelecrtronics, Telecommunications and Computer Engineering (TCSET 2018), pp. 1272–1276 (2018)
8. Dyvak, M., Porplytsya, N., Maslyiak, Y., Kasatkina, N.: Modified artificial bee colony algorithm for structure identification of models of objects with distributed parameters and control. In: Proceedings of the 14th International Conference on Experience of Designing and Application of CAD Systems in Microelectronics (CADSM 2017), pp. 50–54 (2017)
9. Dyvak, M., Porplytsya, N., Borivets, I., Shynkaryk, M.: Improving the computational implementation of the parametric identification method for interval discrete dynamic models. In: Proceedings of the 12th International Conference on International Scientific and Technical Conference on Computer Sciences and Information Technologies (CSIT 2017), pp. 533–536 (2017)
10. Dyvak, M., Porplytsya, N.: Formation and identification of a model for recurrent laryngeal nerve localization during the surgery on neck organs. In: Advances in Intelligent Systems and Computing III: Selected Papers from the International Conference on Computer Science and Information Technologies, CSIT 2018, pp. 391–404 (2019)
11. Stepashko, V.: Developments and prospects of GMDH-based inductive modeling. In: Advances in Intelligent Systems and Computing II: Selected Papers from the International Conference on Computer Science and Information Technologies, CSIT 2017, pp. 474–491 (2018)
12. Stepashko V.: From inductive to intelligent modeling. In: Proceedings of the 13th International Scientific and Technical Conference on Computer Sciences and Information Technologies (CSIT 2018), pp. 32–35 (2018)

13. Stepashko, V., Moroz, O.: Hybrid searching GMDH-GA algorithm for solving inductive modeling tasks. In: Proceedings of the First International Conference on Data Stream Mining and Processing (DSMP 2016), pp. 350–355 (2016)
14. Alonso, G., Benito, A., Lonza, L., Kousoulidou, M.: Investigations on the distribution of air transport traffic and CO2 emissions within the European Union. J. Air Transp. Manag. **36**, 85–93 (2014)
15. Carslaw, D.C.: Evidence of an increasing NO_2/NO_x emissions ratio from road traffic emissions. Atmos. Environ. **39**(26), 4793–4802 (2005)
16. Nejadkoorki, F., Nicholson, K., Lake, I., Davies, T.: An approach for modelling CO_2 emissions from road traffic in urban areas. Sci. Total Environ. **406**(1–2), 269–278 (2008)
17. Porplytsya, N., Dyvak, M.: Interval difference operator for the task of identification recurrent laryngeal nerve. In: Proceedings of the 16th International Conference on Computational Problems of Electrical Engineering (CPEE 2015), pp. 156–158 (2015)
18. Dyvak, M., Maslyiak, Y., Voytyuk, I., Maslyiak, B.: Modified method of subtractive clustering for modeling of distribution of harmful vehicles emission concentrations. In: CEUR Workshop Proceedings of the International Conference on Advanced Computer Information Technologies (ACIT 2018), pp. 58–62 (2018)
19. Pal, N.R., Chakraborty, D.: Mountain and subtractive clustering method: improvements and generalizations. Int. J. Intell. Syst. **15**(4), 329–341 (2000)
20. Lee, J.W., Son, S.H., Kwon S.H.: Advanced mountain clustering method. In: Proceedings Joint 9th IFSA World Congress and 20th NAFIPS International Conference, vol. 1, pp. 275–280 (2001)
21. Velthuizen, R.P., Hall, L.O., Clarke, L.P., Silbiger, M.L.: An investigation of mountain method clustering for large data sets. Pattern Recogn. **30**(7), 1121–1135 (1997)
22. Shtovba, S.: Introduction to the theory of fuzzy sets and fuzzy logic. http://matlab.exponenta.ru/fuzzylogic/book1/index.php. Accessed 22 Apr 2019. (in Russian)
23. Moore, R.E.: Reliability in Computing: The Role of Interval Methods in Scientific Computing. Elsevier, Amsterdam (2014)
24. Alefeld, G., Herzberger, J.: Introduction to Interval Computation. Academic Press, New York (2012)
25. Gentleman, R., Geyer, C.J.: Maximum likelihood for interval censored data: consistency and computation. Biometrika **81**(3), 618–623 (1994)
26. Dyvak, M., Pukas, A., Oliynyk, I., Melnyk, A.: Selection the "Saturated" block from interval system of linear algebraic equations for recurrent laryngeal nerve identification. In: Proceedings of the Second International Conference on Data Stream Mining and Processing (DSMP 2018), pp. 444–448 (2018)
27. Akay, B., Karaboga, D.: Artificial bee colony algorithm variants on constrained optimization. An Int. J. Optim. Control. Theor. Appl. **7**(1), 98–111 (2017)
28. Karaboga, D., Basturk, B.: On the performance of artificial bee colony (ABC) algorithm. Appl. Soft Comput. **8**(1), 687–697 (2008)
29. Seeley, T.D., Camazine, S., Sneyd, J.: Collective decision-making in honey bees: how colonies choose among nectar sources. Behav. Ecol. Sociobiol. **28**(4), 277–290 (1991)
30. Price, L.W.: Global optimization by controlled random search. J. Optim. Theory Appl. **40** (3), 333–348 (1983)
31. Couzin, I.D., Krause, J., Franks, N.R., Levin, S.A.: Effective leadership and decision-making in animal groups on the move. Nature **433**, 513 (2005)
32. Dyvak, M., Porplytsya, N., Maslyiak, Yu.: A method of formation of structural elements in the task of structural identification of interval discrete models of the atmosphere pollution processes by harmful emissions of motor vehicles In: Proceedings of International Scientific Conference on Computer Sciences and Information Technologies, vol. 1, pp. 195–194. IEEE (2019)

Homogeneous Space as Media for the Inductive Selection of Separating Features for the Construction of Classification Rules

Tatjana Lange(✉)

University of Applied Sciences, 06217 Merseburg, Germany
tanja.28.lange@gmail.com

Abstract. This paper deals with problems which require the reconstruction of structures of multi-dimensional dependencies from data. From a mathematical point of view these problems belong to the most complicated problems of artificial intelligence, such as reconstruction of the structures of multi-dimensional regressions or difference equations. In classification we meet similar problems when we have to select the space of classification features. Here we consider a special problem of supervisor-based classification that can be solved by the classification method "Alpha-procedure". This problem consists in the following: Normally, the construction of the separating rule is performed during a training phase, where a supervisor defines the belonging of objects to classes by using a training set of data that can be small. But obviously the rich (partly subconscious) experience of the supervisor which is not described quantitatively somehow influences his decision. This may concern the importance, the uselessness, or even the harmfulness of the features. By this reason, the construction of the separation rule directly in the Euclidian data space leads to instability of that rule in certain cases. The paper explains why the Alpha-procedure, that performs an inductive construction of the separating rule in the homogeneous Lorentz space, allows a stable classification of new objects in the application phase without supervisor. It also shows, from the point of view of group transformations and their invariants, the difference between the mathematical apparatus for the search of the decision rule in a fixed feature space and in a space that is constructed by selecting features.

Keywords: Classification · Pattern recognition · Homogeneous Lorentz space · Transformation groups · Invariant · Alpha-procedure · Inductive selection

1 Introduction

The problem of comparing mathematical objects which are described in different reference frames becomes a complicated mathematical task if no common and absolute reference frame is available.

The mathematical objects can be different models of physical objects which depend on variables that are different for each of the models. With respect to pattern recognition, the mathematical object can be the separating rule depending on the different sets of separating features.

N. Shakhovska and M. O. Medykovskyy (Eds.): CCSIT 2019, AISC 1080, pp. 508–522, 2020.
https://doi.org/10.1007/978-3-030-33695-0_34

The problem is how to compare and select mathematical objects of different structure? Which criteria can we use? Which is the kind of geometry to use? These questions define the basis of the inductive search of solutions [1–4].

This paper deals with pattern recognition methods from the point of view of the theory of group transformations and their geometries with the aim to show the peculiarity of the Alpha-procedure.

The Alpha-procedure is a stepwise working inductive classification method proposed by Vassilev [5–8].

Its first distinguishing mark is that it works in the supervisor-based training phase only with the most useful[1] object features that are defined by a special preselection procedure that is part of the first stage of the Alpha-procedure.

The second distinguishing mark is that it builds step-by-step two-dimensional feature planes starting with the most powerful[2] pair of object features. Thereby, it constructs each time a new artificial feature in a way that a line is turned in the plane around the origin by the angle Alpha up to the point where the projections of the object onto this line give the best separation. This line is a new artificial feature axis. It generates, together with the axis of the next initial object feature, a new plane. This procedure is repeated until the objects are completely (or at least in the best way) separated. Figure 1 tries to explain this stepwise approach.

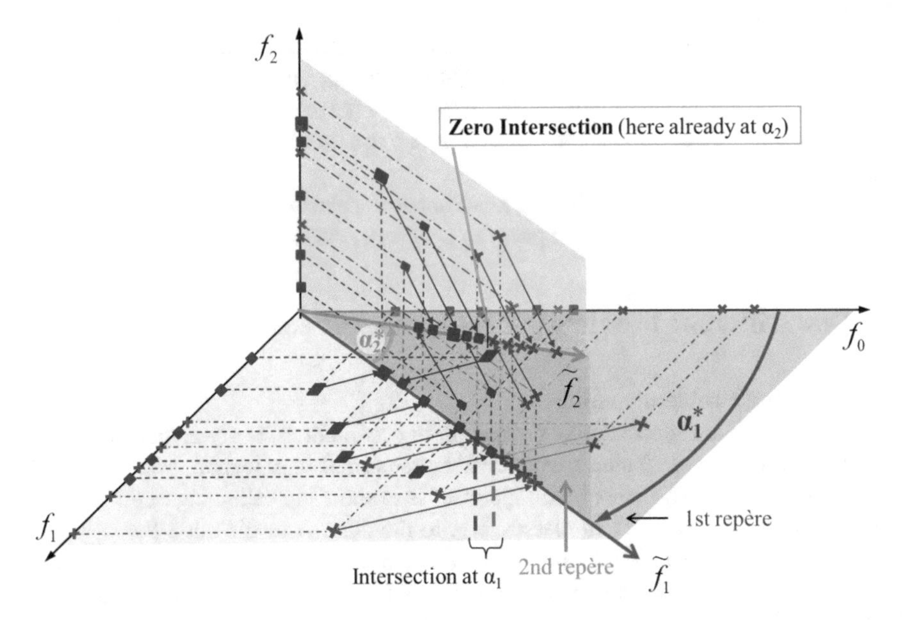

Fig. 1. Principle scheme of the stepwise construction of the separating rule by Alpha-procedure

[1] In the sense of the minimum of the intersection of objects belonging to different classes

[2] see footnote 1

All the successful steps together define the separation rule that will be afterwards used for separating new objects, now without any involvement of a supervisor.

A detailed description of the Alpha-procedure is given, for example, in [9].

The main problem of the multi-step procedure of the inductive selection of the separating decision rule consists in the legitimacy of comparing partial decisions on each step, which must be ensured. That means, the comparison criterion should maintain the relations between the features of the objects of different classes on each step, or with other words, it must not violate the relations of the influence or the non-influence between objects. Yet the criterion should not compare the objects but rather their features. This is what the Alpha-procedure does and it ensures this legitimacy by performing the comparison in the so-called homogeneous space.

Working in the homogeneous space is the most important peculiarity of the Alpha procedure.

It is the aim of the given paper to demonstrate this point and to explain, why just this mathematical apparatus is suited for supervisor-based classification in cases, when the success of the classification lies in the selection of the features but not in the optimization of the separation surface in a fixed coordinate system of features.

This is also valid in such cases, when the given and a priori fixed coordinate system of features does not correspond to the concrete application field in such a range that even a correction of the optimization landscape, i.e. the regularization (e.g. by Tikhonov) does not really help.

The term "homogeneous space" appeared in mathematics as a generalization of the event space that is used in the special relativity theory. It is based on the geometric group of Lorentz transformations [9], which is characterized by the fact that any transformation out of this group maintains the structural relations of points of this space.

In case of pattern recognition **the structural relation is the belonging of the objects** (i.e. of the points in the space) **to a certain class.**

2 Two Different Problems of Supervisor-Based Classification

The Common Problem Position

Normally, a training set (that may be random or non-random) is given that consists of k objects with n different features which can be measured (e.g. weight, seize, ...). There is also a supervisor. The role of the supervisor is to show the belonging of the objects of the training set to the different classes, e.g. to the class A or the class B.

The measured data from the training set are usually represented in form of k "marked" points or vectors in an n-dimensional Euclidean space, which we will call "data space".

The coordinates of the points (or vectors) are the measured values of the object features which we will call "measurable features".

The aim of the training process consists in finding a rule for separating the classes based on the training set. Afterwards this rule will be used for the automatic division of new objects into classes, now without any collaboration of a supervisor.

The mathematical aspects of the problem solution depend on a large scale on the question, which restrictions are imposed on the common problem.

Correspondingly, we will distinguish two types of problems even if we have to deal with an identical common problem. We will call these two types of problems "type I problems" and "type II problems" and describe in the following where we see the differences.

Type I Problems:
It is known (or at least assumed), that

- the axes of the data space truly correspond to the object features that are important for the division of the objects into the given classes,
- there are no features in the data space that are indifferent for the classification,
- there are no harmful features in the data space that possibly undermine the classification,
- there do not exist unknown but important object features that cannot be represented in form of the measured data and which we will call "non-measurable features". Such "non-measurable features" can be connected, for example, with a situation where the supervisor has some of the true information only in an intuitive form in his mind. Sometimes such features could be measured but we do not even suspect that they exist.

Type II Problems:
We look at the same common problem position as before but **one or some** of the restrictions listed above are violated. Even more, the success of the classification completely depends on a comparative analysis of the original measurable features. We will call this comparative analysis of the measurable features "inductive feature selection".

Now, let us consider some mathematical aspects of both types of problems.

3 Inherent Law, Belonging to a Class and the Theory of Transformation Groups

The data points in Euclidean space themselves, which describe the objects, are not so interesting for us. Their direct division into classes in this space can be compared with the work of a sighted robot, which arranges technological parts into the baskets *A* and *B* depending on the colour with which these parts were painted.

But if we understand the term "class" as a certain inherent law that unites the objects of this class into a specific system, then we move into a very advanced chapter of mathematics –into the group theory.

We can say that the theory of transformation groups is a theory about mathematical descriptions of inherent laws maintaining the principal features of the mathematical objects of a concrete group, which can be changed by a transformation of the given group **without changing the common inherent laws of the group**. These preserved features of transformations of a concrete group are called invariant (or invariants) of the given group.

The ways of representing the numerical relations which are inherent in the laws may be different.

Different transformations are used for the description of the inherent laws.

As known, transformations are such correspondences where any element x out of a set X is mapped to a certain element y out of another set Y. But we should not forget that this mapping **can be bijective or not**.

For the description of the inherent laws of dynamic systems, i.e. systems that consist of certain connections of inertial elements (as we have, for example, in two-port electrical networks), we can use differential transformations in form of differential equations. The solution of the differential equation will be the response $y(t)$ to the input $x(t)$ on the initial condition $y(t_0) = C$.

For the description of the same dynamic system we can also use another transformation – the integral transformation of convolution. In this connection there are three different meanings depending on the abstractions which are put into the functions "participating in the convolution".

Even more, for dynamic systems we can also use the Kernel transformation while maintaining the invariant Lebesgue measure. We mean the well-known for a long time fortunate discovery of the mankind – the Fourier transformation (and Laplace transformation), which connects mathematically the descriptions of the inherent law with the help of the convolution on the one hand and with the help of the differential equation on the other hand.

In addition, in optimal control we use the Lie transformation groups – the phase space together with the Poincaré integral invariant and Pontryagin's theory of continuous groups.

Today, with the advanced computer power, we can also use the old Galois group for investigating the stability of dynamic systems.

Why are there so many different mathematical descriptions of one and the same law that is inherent in a dynamic system?

We see two reasons:

First, we expect from any mathematical description its extrapolability beyond the original observation range. For example, when we have the differential equation of a dynamic system then the system response to any stimulus can be calculated. For this it is enough to know the system state in the initial switch-on moment.

Second, the extrapolability can often only be guaranteed when using at the same time more then only one mathematical idealization. These idealizations may be contradictory during the measurements and/or during the extrapolation. Examples are the density of a material point, the value of a continuous function in a point (we can measure only the mean value in an interval around the point) and the so-called "white noise". But the formal introduction of the generalized Dirac delta function $\delta(t)$ as an identity element of the Abelian group in the transformation of the convolution

$$\delta(t) * f(t) = \int_{-\infty}^{+\infty} \delta(\tau) \cdot f(t - \tau) d\tau = f(t) \tag{1}$$

together with the Fourier-Transformation $\mathcal{F}\{\delta(t) * f(t)\} = 1 \cdot F(j\omega)$ revolutionary expanded the possibility of the extrapolation in physics and technology, especially with the todays computer power.

All this is the basis of the present theory of automated control. The use of the term "group" gives us the answer to the following questions:

- How is the precise understanding of what a dynamic system is? Is it allowed, for example, to consider time series as filtered output of a system when the input is also a time series (e.g. a discrete sampled signal)?
- What should we measure and which measurement method should we use in connection with the next question?
- How can we "foresee", i.e. calculate the system response to a continuous input action (now normally in a digital form) when we only have the measured data of the "black box" and we do not have the possibility to compose the differential equation because we do not know the internal physical construction of the system?
- How can we construct a dynamic system with the required quality of precision and speed of system response?

Please, consider the statements above, especially about dynamic systems and their inherent law, as examples for the following:

Let us look, how we can use the experience which we have from the description of an inherent law in form of transformation groups for the construction of a stable decision rule for the recognition of classes of inherent laws, assuming that we have only small data sets available consisting of the measured object features together with the marks of the supervisor with regard to the belonging of the objects to the classes.

4 Geometry and Group Theory

The most important term of the theory of group transformations – the invariant of the group – can be easily demonstrated in a visual manner with the help of classical geometric transformations of planes.

For example, the homothetic transformation as a partial case of the similarity transformation maintains the angles between the curves, and that is the invariant of group of transformations.

In case of affine transformations, which are very important for the pattern recognition problem, the centroid is the main invariant of this group.

The equi-affine transformations in addition maintain the area of the geometric shape.

Thanks to I. Newton and R. Hooke, two opponents in discussions, we have now two complementary descriptions of inherent laws in their original forms: the differential equation and the integral kernel equation (later called Fredholm integral equation). The first was used for the description of the long-distance force effects, the second for the description of the force effects in a solid medium. Moreover, Newton discovered that bodies under the influence of a central force move along a trajectory that is a conic section (ellipse, parabola, hyperbola). But these sections were already described by Euclid.

Two former students of Gauss – A. F. Möbius and B. Riemann – developed these two kinds of describing inherent laws for spaces of dimensions $n \geq 3$.

Möbius investigated the affine geometry and the projective transformations of a cone **to itself**. He represented the double-hyperbola in form of the Möbius strip. This became the theoretical basis for the Lorentz transformation and for Einstein's theory of relativity.

Riemann followed the direction of continuous (topological) transformations. He introduced the term of an "abstract mathematical space" much earlier than the following definitions were worked out: the phase space, the event space, the Hilbert space, the Banach space and others.

Now, we have an advanced definition of a space available, proposed by A.D. Alexandrov. It can be described as follows:

A space is a set of any objects which we will call points. They can be geometric shapes or functions or states of physical systems etc. Considering such a set of objects as a space we distract our attention from all possible kinds of their features and we take into account only those features of the population which are defined by their **relations**. These relations between points and some other shapes, i.e. the relations between sets of points, define the "geometry" of the space. In case of an axiomatic space construction the main features of these relations are expressed by the corresponding axioms.

Now, let us consider the features of some abstract spaces under the perspective of their application to the classification problems of type I and type II.

Let us look at three main aspects of the above two types of classification problems.

The classification problem differs slightly from the problem of describing inherent laws. In case of classification we need to find a stable separating rule for the division of the objects into classes and each of the classes owns its "personal" inherent law.

For the **type I classification problem** there is no need to investigate the "essence of coordinates" of the data space. For example, there is no need for detecting and filtering "harmful" features. For this type of problems we can suggest **two concepts** which underlie the solution algorithm:

The **first concept** proposes to search the separating rule directly in the original Euclidian space of measured data, i.e. in the data space.

The basis of this concept is given with the theorem of A. Novikoff about the convergence of the perceptron and leads us to the search of a saddle point in the data space with the help of minimax algorithms. This is related to groups of affine transformations by the following: In the Euclidian vector space (data space) we have to find the vector that points from a given point towards the "centroid" of the generalized (in the sense of symmetry groups) convex hull of the points of both classes (when we consider only two classes) and we have to maximize the "positive" length of this vector. The minimax vector ρ_0 (see Fig. 2) will be the normal to the separating plane. The centroid is the main invariant of an affine transformation. The initial points of the vector can be changed. Consequently, the mutual directions from the point to the centroid will also change (Vapnik's Support Vector Machine [10, 11])

The **second concept for type I problems** transfers the view of classes into the Hilbert space (e.g. "Method of potential functions" [12]).

The main apparatus of this method is based:

- On the basic idea of the Fourier series, that any periodic function can be characterized by its specific relation between the mean of the infinite series and the mean of its individual terms, which are represented by the coefficients of the series. Today, in the group theory, this is called the nature of the transformation; this uses the exponential function, what is a kind of group invariant.
- On Hilbert's idea that all this can be represented in form of the Hilbert space, and on his theory of the existence of a finite basis.
- On the ergodicity, i.e. equality of time average and ensemble average, whereby the Lebesgue measure is maintained.
- On Rozonoer's idea [12] of simulating each feature of the objects of the data set with the help of a Gaussian bell-shaped function (a positive one for type I problems and a negative one for type II problems). Their ensemble average over all objects produces a separating point for each feature and is a normal to the separating plane in the finite space of features.

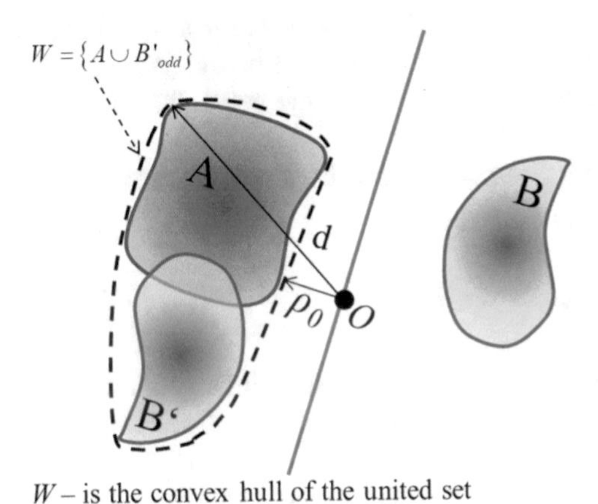

Fig. 2. Geometric interpretation of Novikoff's theorem

Now, let us go to the **type II problems**.

The problem consists in the violation of at least one of the four assumptions of type I problems described in the first chapter of this paper.

The most important problem is the violation of the stability of the decision rule. Here, our understanding of stability follows Tikhonov [13]:

A task of defining the solution $z = R(u)$ out of the space F using the original (measured) data will be a **stable task** in the spaces (F, U), if for any number $\varepsilon > 0$ one can find such a number $\delta(\varepsilon) > 0$ in a way, that from the inequality $\rho_U(u_1, u_2) \leq \delta(\varepsilon)$ follows $\rho_F(z_1, z_2) \leq \varepsilon$,

where $z_1 = R(u_1), \quad z_2 = R(u_2), \quad u_1, u_2 \in U, \quad z_1, z_2 \in F$

ρ_U is the distance in the space U and ρ_F is the distance in the space F.

Instable solutions take us to an incorrect formulated problem, i.e. to a violation of Hadamard's definition of a correct problem:

The task of defining the solution z out of the space F by "source data" u out of the space U is a correctly formulated task on the pair of metric spaces U and F when the following requirements (conditions) are fulfilled:

- a solution z out of the space F exists for each element $u \in U$;
- the solution is unique;
- the task is stable on the spaces (F, U).

Sometimes a solution can be stabilized, e.g. by introducing an additional stabilizing term into the optimization criterion [13] or with the help of validation.

As we can see from the statements above both the definitions of stability and of correctness and also the regularization methods are not suitable for the comparison of coordinate systems and their axes because all they work with metric spaces.

We can reword Hadamard's conditions [14] using a group-based approach:

- There must exist a continuous transformation that transfers the information about the belonging of objects to classes from the metric space U (our data space) into an abstract homogeneous space (implied by Möbius) in which the solution Z exists.
- The definition of the transformation must be unique.
- The transformation owns as invariant the belonging of the object to classes.

5 Homogeneous Space, Lorentz Transformation and Alpha-Procedure

5.1 Möbius Homogeneous Space

Historical Development of the Idea of the Homogeneous Space

The idea of using geometry for the modelling of searching a solution for the "live intellect" **without measurements and calculation** as a mapping of relations by assignment we can find already in the works of Pappus of Alexandria (4th century) and Descartes (17th century).

Möbius constructs his Möbius net on the basis of Pappus' theory in the 19th century. This net demonstrates the geometric convergence of projective mappings, i.e. the bijection of projective transformations, and shows the conditions of solution stability.

Years later we see how the introduction of the Möbius strip leads to the term "homogeneous space".

After that Poincaré formulates the group of Lorentz transformations.

Einstein and Minkowski use the homogeneous space for the comparison of coordinate systems in their special theory of relativity.

Gödel also used the idea of a homogeneous space as an "abstract relation structure" for the proof of his incompleteness theorems. He says analogously: "Classes and terms can be interpreted as .... real objects ... which exist independently of our constructions and definitions. I think that the assumption of such objects is just as legitimate as the assumption of material bodies and that there are just as many reasons to belief in their existence."

Bijection of Projective Transformations and Homogeneous Space

Möbius proved that the projective transformation of a cone into itself is **bijective**.

The idea of that proof is based on the transfer of **one plane in an n-dimensional space into another plan in the same space**.

In analytic geometry this corresponds to a transfer from Cartesian coordinates into homogeneous coordinates.

Homogeneous coordinates own the special feature that the characteristics of the interactions between objects are maintained.

For a bijection it is necessary that any plane of a projective bundle[3] S can by transferred into any other plane.

The bundle contains also parallel planes that intersect at infinity. By this reason the projective plane is complemented by points and lines at infinity. This is expressed in the duality principle and in the incidence of line and point.

Möbius created a model of a projective plane for a hyperbolic intersection of a cone, as shown in Fig. 3. When we "stick together" the points A and A' on the one hand and the points B and B' on the other hand at infinity then the hatched internal areas of the hyperbola will turn into the homomorphic inside of an ordinary circle, but the non-hatched part will turn into the one-side surface of the Möbius strip

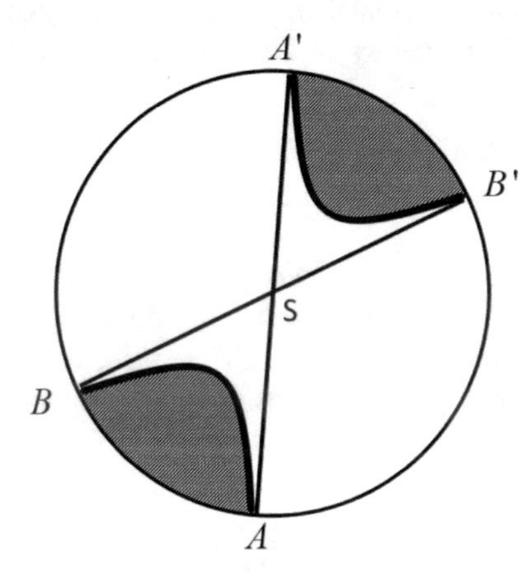

Fig. 3. Model of a projective plane by Möbius

5.2 Lorentz Transformation

The Lorentz transformation is based on projective geometry.

The bijective mapping and the introduction of coordinates by means of projective (non-metric) methods is constructed on the following fundamental ideas:

- principle of duality,
- projective bundle,
- projective plane.

[3] The totality of all lines and (two-dimensional) planes of a space which intersect a given point S of the space is called projective bundle of the lines and planes with the centre S.

We will shortly explain two of these points because they are important for the understanding of the essence of the Lorentz transformation and for the characteristics of the Alpha-procedure:

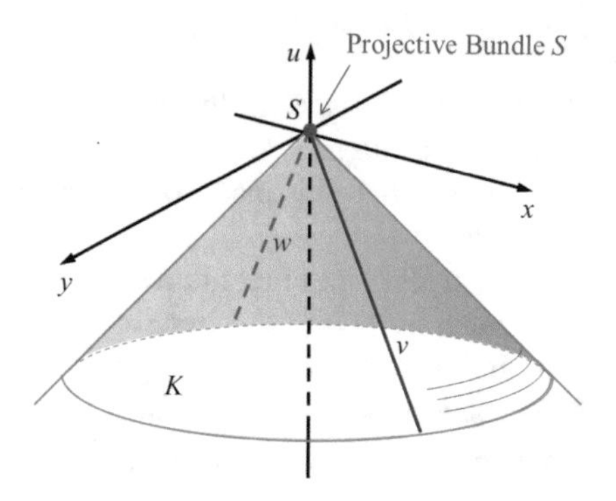

Fig. 4. $\omega-$ transformation: Rotation of the cone as solid body around the axis of the cone

Principle of Duality: The incidence of the geometric objects "straight line" and "point" means the interchangeability of the objects (straight lines and points) in any expression. For example:

- Two points are incident with one (and only one) line.
- Two lines are incident with one (and only one) point.

Projective Bundle: The totality of all straight lines and planes intersecting a given point in the space is called projective bundle with the center *S*.

The **Lorentz transformation** is a projective transformation that transforms a cone into itself (see Figs. 4, 5 and 6). It main invariant are the volume and the bundle *S*.

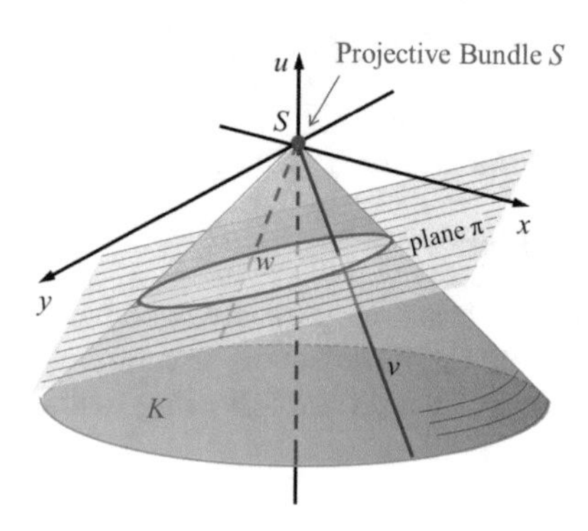

Fig. 5. $\pi-$ transformation: Reflection of the space in any plane π intersecting the axis of the cone

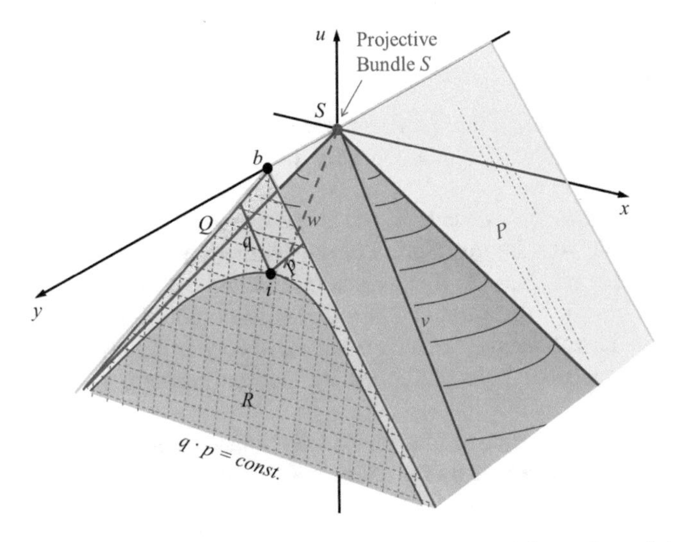

Fig. 6. $L-$ transformation: Description of the projection of the surface of the cone

With respect to Fig. 6 the following explanations are necessary:

The whole surface of the cone consists of hyperbolas as shown in the figure. The lines w and v are two opposite generators of the cone. The "roof" over the cone consists of the planes (or "roof slopes") P and Q. Compressing or stretching of the space along the direction towards or from the planes P and Q does not change the volume of the space of coordinates. The generators w and v are tangent lines where the cone touches the planes P and Q. The intersection line of the planes P and Q (the Ridge line of the "roof") is the coordinate axis Sy.

By a general consideration of the perspective projections as a mapping of two projective planes onto each other, it is obvious that the ellipses, the hyperbolas and the parabolas are nothing else but the results of the projective transformation of a circle. The appearance of the type of the curve depends on the mutual position of the projective plane and the circle.

The projective transformations which transform a circle into a parabola **do not** belong to the Lorentz transformations.

The proof of the existence of the identity Lorentz transformation and its uniqueness can be, as shown by A.D. Alexandrov, based on the transformation of the repère of a projective plane:

$$\Lambda = L_1 \cdot \omega \cdot \pi \cdot L_2^{-1}, \tag{2}$$

where ω, π, L are Lorentz transformations.

5.3 The Difference Between Lorentz Transformation and Ordinary Equi-Affine Transformations

The equi-affine transformations maintain the area or the volume as an additional invariant to the affine transformation, but they do not transfer the planes by pairs into a homogeneous space.

The condition for such a transfer would be the existence of a connecting coordinate, as proven by A.D. Alexandrov. For example, **homogeneous** coordinates on a plane may be the three numbers X, Y, Z which are connected together by the relation $X : Y : Z = x : y := 1$, where x and y are Cartesian coordinates. In this example the coordinate z is the connecting coordinate. In the theory of relativity the fundamental constant of the speed of light multiplied by the continuous time is the connecting coordinate.

As known, the transformation of the Euclidian coordinates $x, y, \ldots, z$ into $x', y', \ldots, z'$ can be expressed by the following system of equations:

$$\begin{aligned} x' &= a_1x + b_1y + \ldots + c_1z \\ y' &= a_2x + b_2y + \ldots + c_2z \\ &\vdots \\ z' &= a_nx + b_ny + \ldots + c_nz \end{aligned} \tag{3}$$

If the determinant of the system is $D = 1$, then we have an ordinary equi-affine transformation. But if one of the coefficients, e.g. $c = c_1 = c_2 = \ldots = c_n$ has an identical value in all equations[4] then we get the homogeneous coordinates $x', y', \ldots, z'$ which have some famous characteristics.

In case of Alpha-procedure the last line and the last column are realized in an algorithmic way by marking the objects of the training set in correspondence to their belonging to different classes. The statement of the supervisor with respect to the belonging of the objects of the training set to certain classes is constant for each coordinate.

We can see the following from the above explanation:

The Alpha-procedure does not search the separating surface (or plane) with the help of a minimax optimization in a fixed space. In the opposite, it performs an inductive search (in a system of bundle *S*) of a space, that is suitable for the selection of classes. The search is performed by a subsequent enrolment of plane in a homogeneous space.

The successful decision way is stored by computer. In the post-training application phase we have to deal with new points (objects) whose belonging to a certain class is a priori unknown. Their classification will be performed on the basis of their position on the axis of the last artificial feature.

The classification way can be represented as a subsequent enrolment of such planes as shown in form of repères in Fig. 7. These repères maintain the main invariant of the Lorentz transformation. – the bundle *S*. The point *S* matches the origin of coordinates in the initial Euclidian data space.

[4] The coefficient $c = c_1 = c_2 = \ldots = c_n$ may be the speed of the light (taken with "minus") in the special theory of relativity.

The search of the sequence can also be performed with the help of parallel programming [15, 16].

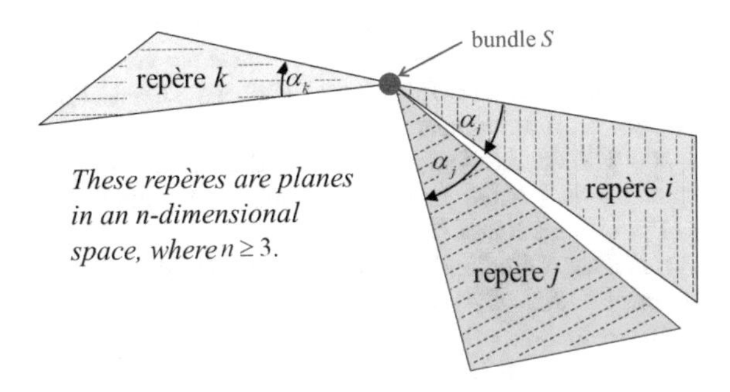

Fig. 7. Representation of planes in the n-dimensional space by repères

6 Summary

The goal of the given paper was to provide a theoretical basis for the Alpha procedure which has been working successfully for a long time.

It explained the connection between the type of the group invariant that is behind a specific classification method and its practical application area, on the basis of group-theoretical considerations.

In case of practical applications where the data set of features is fixed the algorithms of minimax optimization work fast and in a stable way in the initial Euclidian data space. They are based on invariants of affine transformations.

But in cases where a selection of the features their selves is necessary, i.e. where the selection of the system of coordinates is required, we have to use algorithms which are based on special projective equi-affine transformations (Lorentz transformations). These transformations differ from the ordinary equi-affine transformations in the circumstances that they have an additional connecting and stabilising variable with a constant coefficient available in their system of equations.

When we use the Alpha-procedure in pattern recognition the role of the additional variable is played by the process of searching the best class-separating projection of the mutual arrangement of the objects of the training set (which are marked corresponding to their belonging to classes) in the homogeneous space. The homogeneity of the space will be ensured by any of the three Lorentz transformations (ω, π, L - see [9]). The Alpha-procedure uses two of them. This way, the indirect or “hidden” information used intuitively by the supervisor during the training phase will be fixed quantitatively.

In case of the Alpha-procedure the feature space is created “hand-in-hand”, i.e. simultaneously, with the construction of the separating rule. This is the main difference compared to other methods where the separating rule is found in a priori defined and consequently fixed feature space.

References

1. Akaike, H.: Experiences on development of time series models. In: Bozdogan, H. (ed.) Proceedings of the First US/Japan Conference on the Frontiers of Statistical Modeling: An Information Approach, vol. 1, pp. 33–42. Kluwer Academic Publishers, Dordrecht (1994)
2. Lange, T.: New structure criteria in GMDH. In: Bozdogan, H. (ed.) Proceedings of the First US/Japan Conference on the Frontiers of Statistical Modeling: An Information Approach, vol. 3, pp. 249–266. Kluwer Academic Publishers, Dordrecht (1994)
3. Madala, H.R., Ivakhnenko, A.G.: Inductive Learning Algorithms for Complex Systems Modeling. CRC Press, Boca Raton (1994)
4. Stepashko, V.S.: Method of critical variances as analytical tool of theory of inductive modeling. J. Autom. Inf. Sci. **40**(3), 4–22 (2008)
5. Vassilev, V.I.: The reduction principle in pattern recognition learning (PRL) problem. Pattern Recogn. Image Anal. **1**(1), 23–32 (1991)
6. Vassilev, V.I., Lange, T.: The principle of duality within the training problem during pattern recognition. Cybern. Comput. Eng. **121**, 7–16 (1998). (in Russian)
7. Vassilev, V.I., Lange, T., Baranoff, A.E.: Interpretation of diffuse terms. In: Proceedings of the VIII. International Conference KDS 1999, pp. 183–187 (1999). (in Russian) Kaziveli (Krimea, Ukraine)
8. Vassilev, V.I., Lange,T.: Reduction theory for identification tasks. In: Proceedings of International Conference on Control, Automatics-2000, 11–15 September 2000, Section 2, Lviv, pp. 49–53 (2000). (in Russian)
9. Lange, T.: The alpha-procedure as an inductive approach to pattern recognition and its connection with lorentz transformation. In: Shakhovska, N., Stepashko, V. (eds.) Advances in Intelligent Systems and Computing II. CSIT 2017. Advances in Intelligent Systems and Computing, vol. 689. pp. 280–299. Springer, Cham (2018)
10. Vapnik, V., Chervonenkis, A.Ya.: The Theory of Pattern Recognition. Nauka, Moscow (1974)
11. Cortes, C., Vapnik, V.: Support-vector networks. Mach. Learn. **20**, 273–297 (1995)
12. Aizerman, M.A., Braverman, E.M., Rozonoer, L.I.: The Method of Potential Functions in the Theory of Machine Learning. Nauka, Moscow (1970). (in Russian)
13. Tikhonov, A.N., Arsenin, V.Ya.: Methods of Solving Incorrect Tasks. Nauka, Moscow (1974). (in Russian)
14. Hadamard, J.: Sur les problèmes aux dérivées partielles et leur signification physique. Bull. Univ. Princeton **13**, 49–52 (1902)
15. Mukhin, V., Volokyta, A., Heriatovych, Y., Rehida, P.: Method for efficiency increasing of distributed classification of the images based on the proactive parallel computing approach. Adv. Electr. Comput. Eng. **18**, 117–122 (2018)
16. Lange, T.: Transformation of the Euclidian data space into a homogeneous event space for the inductive construction of classification rules. In: Proceedings of International Scientific Conference "Computer Sciences and Information Technologies" (CSOI-2019), vol. 1, pp. 173–178. IEEE (2019)

Improvement of a Sorting-Out GMDH Algorithm Using Recurrent Estimation of Model Parameters

Serhiy Yefimenko(✉) and Volodymyr Stepashko

Department for Information Technologies of Inductive Modelling, International Research and Training Center for Information Technologies and Systems, Ave Glushkov, 40, Kyiv 03680, Ukraine
syefim@ukr.net, stepashko@irtc.org.ua

Abstract. The paper presents and discusses a GMDH algorithm MULTI-R being an improved and revised version of the known multistage algorithm MULTI based on the successive search of model of the globally optimal structure. This means that the MULTI algorithm is intended for discovering the result of exhaustive search by the combinatorial algorithm COMBI GMDH with radically less computations. But this algorithm has some substantial drawbacks, for example, it tends to choose underfitted models in the searching process and is not optimized with respect to the parameter estimation procedures. The new revised version MULTI-R differs from the original algorithm MULTI by using a recurrent procedure of parameters estimation and additional optimizing the model structure to enhance both the computation speed and accuracy of discovering the globally optimal model. The comparative numerical characteristics of the processing speed and structural accuracy of this modified algorithm and the original one are given for several test tasks.

Keywords: Inductive modelling · GMDH · Combinatorial algorithm · Sorting-out · Successive search · Multistage algorithm · Recurrent computations

1 Introduction

GMDH algorithms [1–4] are effective means for constructing models of objects and processes from data observed. They are used for solving such modelling problems as forecasting, extrapolation, classification etc. All the variety of GMDH algorithms can be classified into searching and iterative ones [5] with respect to the type of the generation process of model structures.

One of the most practically applied GMDH algorithms is the combinatorial COMBI [6–8] with exhaustive search of model structures and finding the best one by the value of a given external criterion. The algorithm has exponential dependence of computational complexity on the arguments number therefore the possibilities of the combinatorial algorithm to solve complex applied tasks are limited. COMBI requires high expenses of computational resources since the modelling time is doubled when adding one argument. Even with the use of special means like parallel computations, it does not allow solving tasks with more than 40 inputs (arguments) [9].

N. Shakhovska and M. O. Medykovskyy (Eds.): CCSIT 2019, AISC 1080, pp. 523–534, 2020.
https://doi.org/10.1007/978-3-030-33695-0_35

The aim of successive search GMDH algorithms is to find the optimal solution obtained via exhaustive search algorithms with much less volume of computing expenses. The most typical example of such kind algorithms is the multistage combinatorial-selective algorithm MULTI GMDH [10] based on the principle of non-final decisions and successive search of the best model with a stepwise complication of partial model structures in one argument. This algorithm, examined in [11], searches the best model by highly reduced computations and makes possible to solve modeling tasks with more than 100 arguments. But this algorithm generates a multiextremal function of the criterion change and such drawback produces obstacles when choosing the best model: as a rule, it stops prior to discovering the globally optimal model (produces an underfitted model).

The purpose of this paper is to enhance the possibilities of the successive search algorithm MULTI by applying:

- recurrent estimation of model parameters;
- additional optimization of model structures;
- correction of the algorithm stopping rule

in order to solve more complex problems.

The structure of the paper is as follows. Section 2 describes the problem definition of the best model construction. Section 3 informs on the peculiarities of the known algorithm MULTI GMDH. Section 4 explains the recurrent modification of the classical Gauss algorithm for solving linear algebraic equations in the process of model parameters estimation. Section 5 presents the improved algorithm MULTI-R with recurrent parameters estimation and model structure optimization. The results of testing experiments of the revised algorithm are presented and discussed in the Sect. 6. The last Sect. 7 contains finale conclusions.

2 Problem Definition

Generally, the model generation problem can be stated as finding the optimal model $\hat{y}_f^* = f^*(X, \hat{\theta}_f)$ in a certain set of models Φ by minimization of a given model quality criterion $CR(\cdot)$ as the solution of the discrete optimization task [5]:

$$f^* = \arg\min_{f \in \Phi} CR(y, f(X, \hat{\theta}_f)), \tag{1}$$

where X, $\dim X = n \times m$, is a matrix comprising results of n observations over m inputs (arguments, regressors); y, $\dim y = n \times 1$, is an observation vector of the object output; and the parameters estimation vector $\hat{\theta}_f$, $\dim \hat{\theta}_f = s_f \times 1$, for each function $f(X, \theta_f) \in \Phi$ is the solution of the continuous optimization task:

$$\hat{\theta}_f = \arg\min_{\theta_f \in \Re^{s_f}} QR(y, f(X, \theta_f)), \tag{2}$$

where $QR(\cdot)$ is a given criterion that characterizes the solution quality for the parameter estimation problem.

In a special case of liner models (or more generally models linear in parameters), which is used in the algorithm MULTI, the model set Φ consists of $p_m = 2^m - 1$ all possible models of different complexity containing 1, 2, …, m arguments:

$$\hat{y}_l = X_l \hat{\theta}_l, \quad l = 1, \ldots, 2^m - 1. \tag{3}$$

Than the optimization problem (1) reduces to the following expression:

$$y^* = \arg \min_{l=1,\ldots,2^m-1} CR(y, \hat{y}_l), \tag{4}$$

which can be represented in another equivalent form:

$$y^* = \arg \min_{s=\overline{1,m}} \min_{k=\overline{1,C_m^s}} CR(y, \hat{y}_{sk}), \tag{5}$$

where s is the number of nonzero members of a model, or *model complexity*, and k is the index of a model in a set Φ_s of C_m^s models of the complexity s. In this case, it is evident that the method of least squares is optimal for solving the task (2) when for any s the parameter vector is equal to $\hat{\theta}_s = (X_s^T X_s)^{-1} X_s^T y$ and the number of all possible models can be presented as

$$p_m = \sum_{s=1}^{m} C_m^s = 2^m - 1. \tag{6}$$

Evidently that the value p_m characterises the exponential computational complexity of any algorithm (including COMBI) based on the exhaustive search of models of various structures, and this fact puts substantial obstacles for using them in tasks with more than 30 arguments.

Expressions (4) and (5) both describes the same problem of exhaustive search of all possible models of different structures. Such kind of problems are solved by the COMBI GMDH algorithm of various modifications based on these two representations.

At the same time, the variant (5) helps to explain the core of any successive search algorithm: it should contain some procedure of forming a subset $\Phi_s^* < \Phi_s$ of partial models of the complexity s being the most prospective for directed search of the global minimum. Typical procedure of such type which can be written as follows [5]:

$$\hat{y}_s^l = (X_{s-1}^i | x_s^j) \hat{\theta}_s, \quad s = \overline{1,m}, \quad i, l = \overline{1, F_{s-1}}, \tag{7}$$

where F_{s-1} is a number of best models chosen on the stage $s - 1$ (freedom of choice), j is an index of a vector argument absent in the matrix X_{s-1}^i. This procedure is the base of the MULTI algorithm which has the polynomial complexity instead of the exponential one in COMBI.

3 Successive Search Algorithm MULTI GMDH

The original algorithm MULTI [10] was constructed for reducing the models search. This is a combinatorial and selective algorithm having a finite number of stages (not more than m). The main feature of the algorithm is the use of the principle of non-final decisions. This principle states that not the best one but several models (solutions) are selected and passed to the next stage of the algorithm. It allows increasing the probability of finding the global minimum of the given selection criterion CR.

General procedure of the algorithm is as follows: in the first stage, all models with one argument are built, and all of them are selected; in the second stage, only arguments missing in the model are added by one at a time, and best models of the complexity 2 improving the criterion value are chosen for the next stage, etc. This procedure is performed until the criterion value is decreasing.

The main properties of the multistage algorithm MULTI are the followings [10]:

- it is of combinatorial and selective type, which means that at any stage (a) it is used the complete searching of all arguments being included one-by-one into partial models, and (b) among all generated models, the certain number F of the best ones are selected for the next stage;
- it has a limited number of stages, and the main parameters such as breakpoint and freedom of choice at each stage are determined automatically;
- the arguments number (complexity) of partial models increases by only one argument at each stage;
- it is stopped after finding the true model (in case of noiseless data) or the model of optimal complexity (for noisy data);
- it is of the polynomial complexity m^3 instead of the exponential one 2^m for the exhaustive search.

The comparative dependences of the number of compared models on the number of variables m for the two algorithms, COMBI (curve NC) and MULTI (curve NM), are presented graphically in Fig. 1 [11].

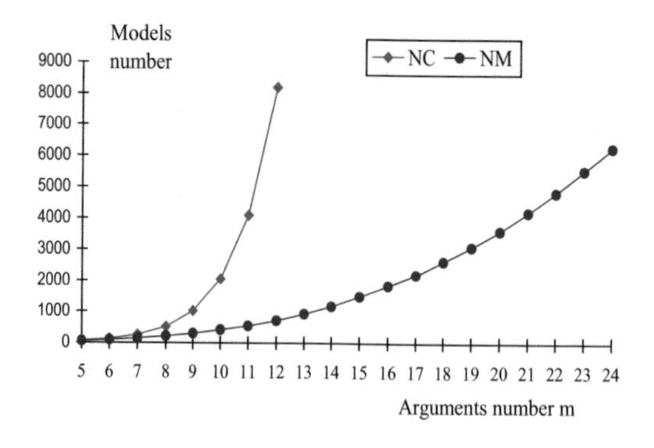

Fig. 1. The dependences of the compared models number on the variables number

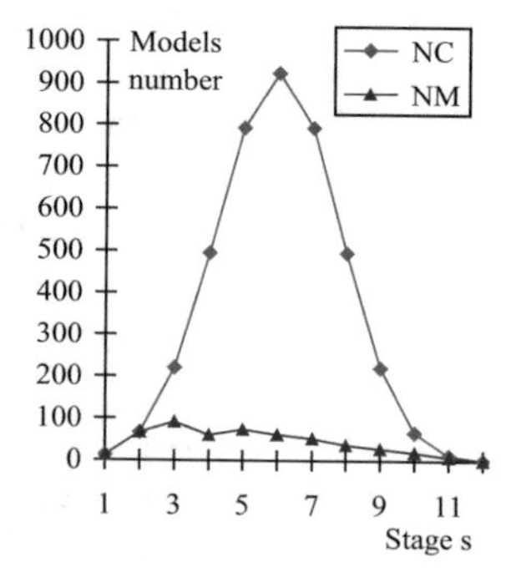

Fig. 2. Number of built models on every stage

Figure 2 compares changes in the number of built models with successive stages of the MULTI and COMBI algorithms [11]. The total number of models built by COMBI (for $m = 11$) is 4095 and by MULTI only 480.

The sequential generating procedure of partial model structures with adding one argument allows using recurrent procedures for parameters estimation. For example, let us have a task with $m = 5$ arguments for second stage of MULTI algorithm. The sequence of all model structure combinations (for the best model $y_1 = a_1x_1$ of the first stage) will be as follows:

$$\begin{aligned} y_{21} &= a_1x_1 + a_2x_2 \\ y_{22} &= a_1x_1 + a_3x_3 \\ y_{23} &= a_1x_1 + a_4x_4 \\ y_{24} &= a_1x_1 + a_5x_5 \end{aligned}$$

It is advisable to estimate the parameters of the indicated models of the second stage by a recurrent procedure (based on the parameters estimates, obtained at the first stage of the algorithm). A similar procedure is applied at the next 3rd stage to any of the best models selected at this 2nd stage, etc.

4 Recurrent Computations When Estimating Parameters

To estimate parameters of a linear model containing s arguments using the least squares method, one need to solve a corresponding system of s linear equations with s unknown values: $X_s^T X_s \theta_s = X_s^T y$. To do that in the case of well-conditioned matrix $H_s = X_s^T X_s$, it is sufficient to use the conventional Gauss method. But the peculiarity of the process of sequential model complications is that any two consecutive models differs only in one argument which makes the application of this method directly to every of these two models very redundant as for the volume of calculations.

One of the most effective ways to enhance possibilities of searching algorithms (in sense of the processing speed up) is the use of recurrent computing [12–14]. Computational complexity of parameters estimation in the process of successive adding a new argument to a model structure is proportional to the square of the model complexity (number of the estimated parameters) for any recurrent algorithm and to the cube for a non-recurrent one.

Short-form description of the recurrent modification [14, 15] of classical Gauss algorithm is done below.

The matrix $H_s = X_s^T X_s$ of the size $s \times s$ is reduced to the superdiagonal form by computing only elements $h_{i\,s}^s$, $i = \overline{2, s-1}$, $h_{s\,i}^s$, $i = \overline{2,\ s}$, and $g_s = X_s^T y$ at every step s, $s = \overline{1,\ m}$ during the direct motion. The elements of the nested matrix H_{s-1} of size $(s-1)\times(s-1)$ (reduced to superdiagonal form on the previous step) remain unchanged. Hence only "bordering elements" (bold fonts) are computed on a step s:

$$\left[\begin{array}{cccccccc|c} h_{11} & h_{12} & h_{13} & \ldots & h_{1s\text{-}1} & \mathbf{h_{1s}} & \ldots & h_{1m} & g_1 \\ h_{21} & h_{22} & h_{23} & \ldots & h_{2s\text{-}1} & \mathbf{h_{2s}} & \ldots & h_{2m} & g_2 \\ h_{31} & h_{32} & h_{33} & \ldots & h_{3s\text{-}1} & \mathbf{h_{3s}} & \ldots & h_{3m} & g_3 \\ \ldots & \ldots & \ldots & \ldots & \ldots & \ldots & \ldots & \ldots & \ldots \\ h_{s\text{-}1,1} & h_{s\text{-}1,2} & h_{s\text{-}1,3} & \ldots & h_{s\text{-}1,s\text{-}1} & \mathbf{h_{s\text{-}1,s}} & \ldots & h_{s\text{-}1,m} & g_{s\text{-}1} \\ \mathbf{h_{s1}} & \mathbf{h_{s2}} & \mathbf{h_{s3}} & \ldots & \mathbf{h_{s,s\text{-}1}} & \mathbf{h_{ss}} & \ldots & h_{sm} & \mathbf{g_s} \\ \ldots & \ldots & \ldots & \ldots & \ldots & \ldots & \ldots & \ldots & \ldots \\ h_{m1} & h_{m2} & h_{m3} & \ldots & h_{m,s\text{-}1} & h_{ms} & \ldots & h_{mm} & g_m \end{array}\right]$$

It should be pointed out that the method does not require the use any additional computer RAM because the same matrix H of the normal system is used to store the information necessary to include an additional argument.

5 Improved Algorithm MULTI-R with Recurrent Parameters Estimation and Model Structure Optimization

The original algorithm MULTI is constructed for sequential complication of models structures and does not exclude irrelevant arguments. However, in the case of a large number of true arguments and their complex multiple dependencies, it is possible to include redundant (false) arguments at the initial stages of the algorithm. So the algorithm demands special optimization of the best selected models structures. To solve this problem, an improved algorithm MULTI-R has been constructed. Figure 3 presents a flow-chart of the MULTI-R algorithm.

The following operations are performed at stage s, $s=\overline{1,m}$, of the algorithm:

1. Partial models are generated by adding one of the missing arguments to each i-th model of the previous stage (that correspond to the zero element of the structural vector d_i of this model). To ensure the non-repeatability of structures, a special check procedure is used which analyzes the structural vectors of the selected best models of the previous stage. This procedure is performed from the second stage because every generated model structures of the first stage are not repeatable.
2. Parameters of every non-repeatable model $\hat{y}_s$ are estimated by the least-squares method (LSM) using the modified Gauss method explained above.
3. The best F models (by the criterion value) are determined among all generated on the stage s to be transferred to the next stage of selection. To get the result of the exhaustive search, F is enough to be equal to arguments number m [11].
4. As the criterion CR, the traditional for GMDH *regularity criterion* is used:

$$AR(s)=\left\|y_B-X_{Bs}\hat{\theta}_{As}\right\|^2,$$

which is based on the data set division into two subsets: $X=(X_AX_B)^T$, $y=(y_Ay_B)^T$.

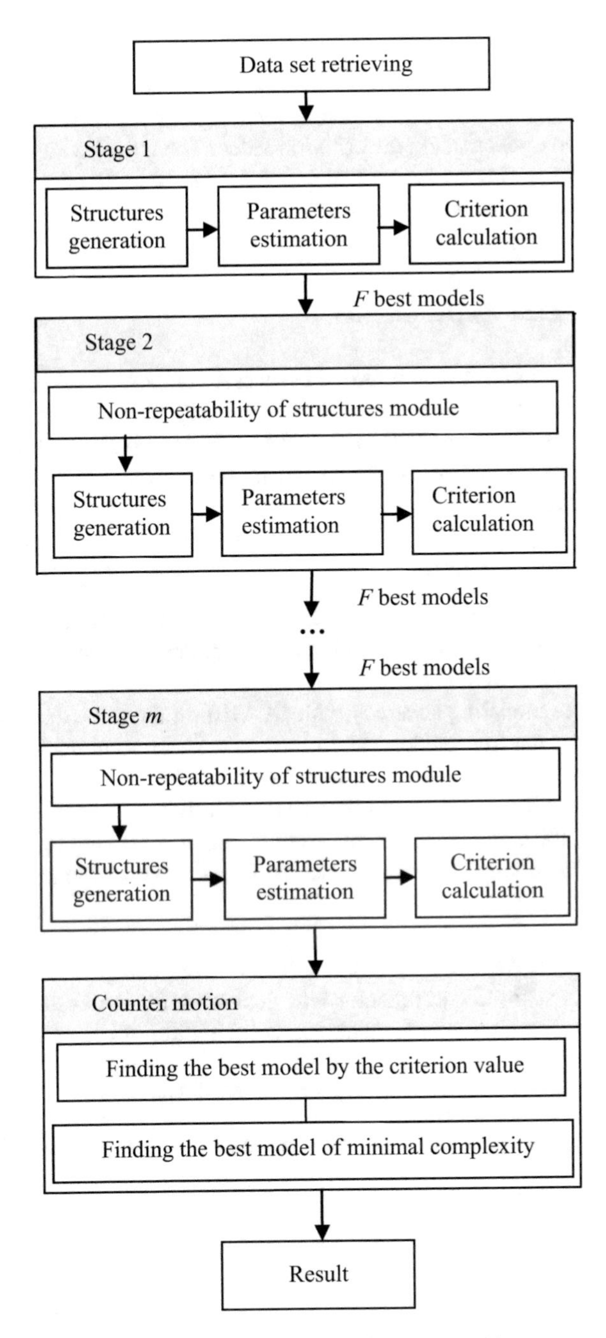

Fig. 3. Flowchart of the MULTI-R algorithm.

5. After finishing all m steps of forward motion (including redundant arguments), the backward motion is performing. From the best model (let it has complexity l, $l \le m$) at the step i, $i = \overline{m, 1}$, one by one arguments are consistently excluded under condition of decreasing the criterion value. The model with minimal selection criterion is transferring to the next step $i - 1$. The best model of the algorithm is selected just during the backward motion.

6 Results of Test Experiments

The study of the properties of the improved successive search algorithm MULTI-R was carried out experimentally on various test tasks in order to determine the effectiveness of the introduced modifications of the original searching algorithm. The modelling time was measured with respect to the condition of finding the result of the exhaustive search using COMBI.

Four experiments for testing the algorithm effectiveness were executed on several artificial tasks of various complexity:

(1) checking the structural accuracy on a simple test: 25 arguments, only 5 of them are informative, i.e. included into the true model; the rest 20 arguments are redundant (uninformative);
(2) comparing the modeling accuracy with COMBI on the same data sample with 25 arguments but the true model was formed as a linear combination of first twenty arguments with adding uniformly distributed noise.
(3) testing the accuracy and computing time on a big generated sample with 300 records and 200 arguments; the true model contains 100 arguments;
(4) evaluating the solving time dependence on the number of arguments: 50, 100, 200 and 400 ones.

Experiment 1. A test task was formed as follows: the design matrix X of the size 50×25 (50 records for 25 arguments) was generated for the system of conditional equations $X\theta = y$. Vector y was formed as a combination of only five arguments:

$$y = x_{11} + x_{12} + x_{13} + x_{14} + x_{15}.$$

The best model by the minimum of the regularity criterion AR was the following one:

$$y_1 = 7 \cdot 10^{-10} x_2 + x_{11} + x_{12} + x_{13} + x_{14} + x_{15} - 3 \cdot 10^{-10} x_{20} + 10^{-11} x_{21}$$

with criterion value $AR_1 = 1.1 \cdot 10^{-17}$.

The best model of minimal complexity (as a result of backward motion):

$$y_2 = x_{11} + x_{12} + x_{13} + x_{14} + x_{15}, \; AR_2 = 1.2 \cdot 10^{-17}.$$

Figure 4 shows changes in the minimum and maximum values of the regularity criterion for all tested models during the sequential execution of stages.

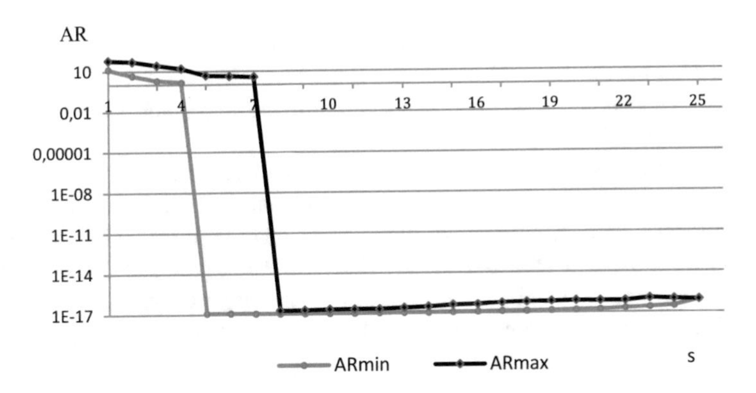

Fig. 4. Maximum and minimum values of the regularity criterion (logarithmic scale).

Experiment 2. Vector y was formed as a linear combination of first twenty arguments from the matrix X (the same as in Experiment 1) with adding uniformly distributed noise. Modelling result was compared with COMBI.

The best model by COMBI algorithm contains all true arguments and additional x_{23}, its criterion value: AR_3 = 1.167. The modeling time was 199 s.

The best model by MULTI algorithm (forward motion) contained all arguments without true one $x_{22,}$ its criterion value: AR_4 = 1.225. The modeling time was 0.03 s.

The use of backward motion has made it possible to get the result of the exhaustive search (by the COMBI algorithm).

The Fig. 5 shows the changes in the minimum and maximum values of the regularity criterion for all tested models during the sequential execution of stages.

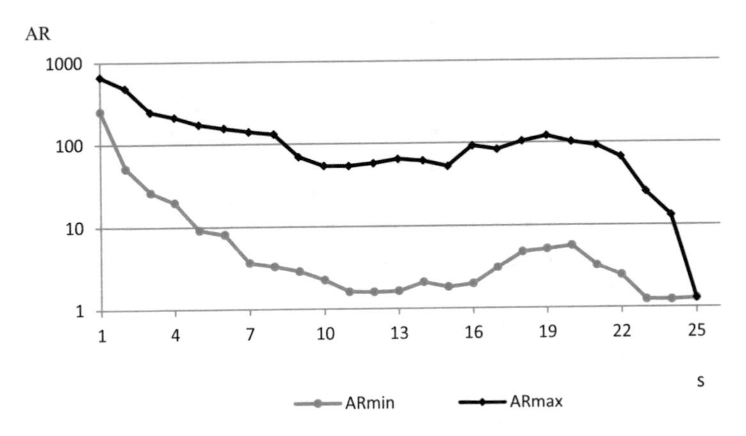

Fig. 5. Maximum and minimum values of the regularity criterion (logarithmic scale).

Experiment 3. The design matrix X of the size 300 × 200 (300 records for 200 arguments) was generated. Vector y was formed as a linear combination of one hundred consecutive arguments (from x_{25} to x_{124}). The best selected model in forward motion contained 125 arguments (all true and 25 false), its criterion value: $AR_5 = 1.9 \cdot 10^{-15}$. The use of the backward motion has made it possible to get the model containing all 100 true arguments (without false). Its criterion value: $AR_6 = 2.5 \cdot 10^{-15}$.

The Fig. 6 shows the changes in the minimum and maximum values of the regularity criterion for this experiment.

The modification of the algorithm presented in this paper differs from the original MULTI algorithm as well as also by the absence of a stopping rule. It has an advantage when the curve of the minimum values of the selection criterion is multiextremal as it is the cases in Figs. 5 and 6. Due to that, the MULTI in the Experiment 2 stops at stage 11 and selects the model containing only 11 out of 20 true arguments. And it stops at stage 43 in the Experiment 3, whereas true model contains 100 arguments.

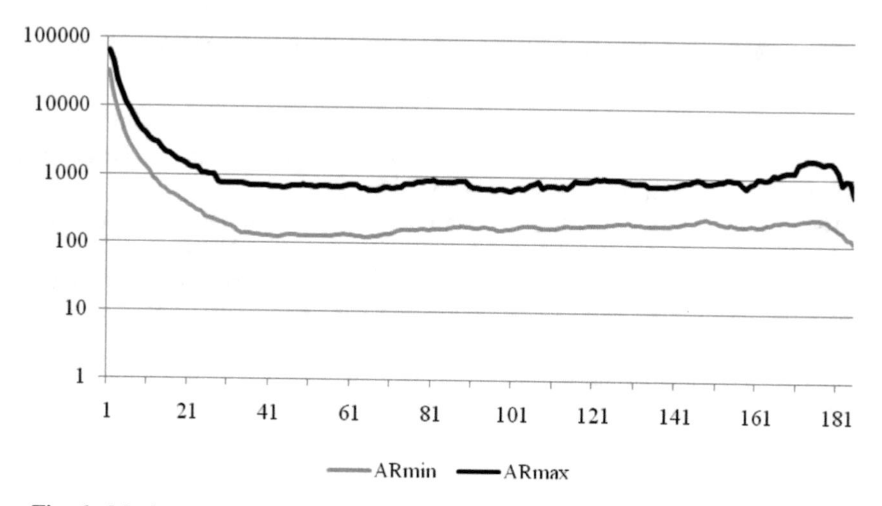

Fig. 6. Maximum and minimum values of the regularity criterion (logarithmic scale).

Experiment 4. Table 1 shows the results of the experiments for evaluation of the dependence of the MULTI-R run-time on arguments number. In these experiments true model contained half of the total number of arguments. Time 1 refers to forward motion and actually this is the run-time of original MULTI algorithm but with the added recurrent parameters estimation. Time 2 refers only to the backward motion and time 3 indicates general run-time of the MULTI-R algorithm. The last column of the table shows a slow increase of run-time with respect to the only forward process typical for the original MULTI algorithm.

Figure 7 presents characteristics of the time growth for the forward and backward processes of the algorithm.

Table 1. Modelling time.

Arguments number	Time, sec.			Increase of time, %
	1	2	3	
50	0.41	0.07	0.48	17
100	10.28	2.25	12.53	22
200	272.49	67.56	340.05	25
400	9480	2893	12373	31

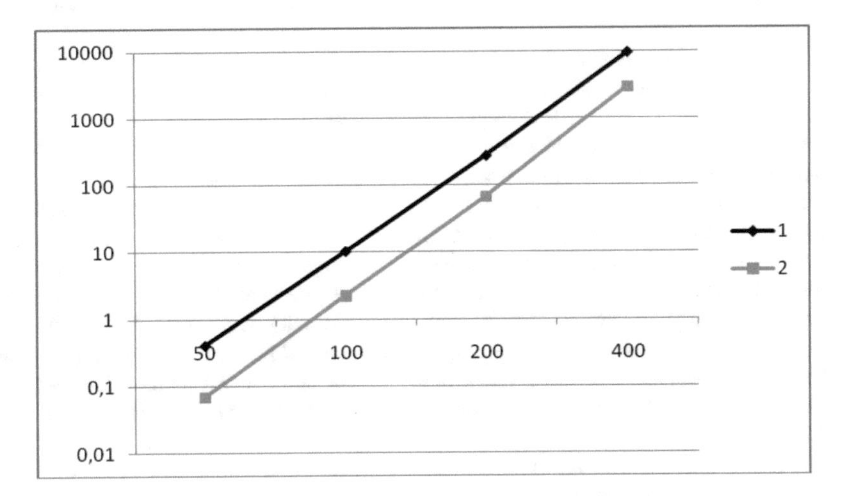

Fig. 7. Modeling time of forward (1) and backward (2) motion, logarithmic scale.

7 Conclusion

Successive search algorithm MULTI GMDH was improved. The main features of the new revised algorithm MULTI-R are the use of recurrent procedures for parameter estimation and extra optimization of the model structures.

The dependence of computational complexity (number of elementary arithmetic operations) of regression coefficients calculation on arguments amount has square character for the recurrent algorithm. This is significantly less than the complexity of non-recurrent scheme.

The results of the test experiments demonstrate that optimization procedure for models structure in the MULTI-R algorithm enables to obtain the result of the exhaustive search provided by the COMBI algorithm. A disadvantage of the algorithm is the slowly increased modelling time which is compensated by the use of recurrent computing.

To further develop the suggested algorithm, we plan to introduce a kind of the recurrent algorithm for the backward successive process and to apply parallel operations for both forward and backward motions. These improvements are very promising for substantial enhancement of the MULTI-R efficiency.

References

1. Ivakhnenko, A.G.: Heuristic self-organization in problems of automatic control. Automatica (IFAC) **6**, 207–219 (1970)
2. Farlow, S.J. (ed.): Self-Organizing Methods in Modeling: GMDH Type Algorithms. Marcel Decker Inc., New York (1984)
3. Madala, H.R., Ivakhnenko, A.G.: Inductive Learning Algorithms for Complex Systems Modeling. CRC Press, New York (1994)
4. Snorek, M., Kordik, P.: Inductive modelling world wide the state of the art. In: Proceedings of the 2nd International Workshop on Inductive Modelling, Prague, pp. 302–304. CTU (2007)
5. Stepashko, V.: Developments and prospects of GMDH-based inductive modeling. In: Shakhovska, N., Stepashko, V. (eds.) Advances in Intelligent Systems and Computing II. AISC, vol. 689, pp. 474–491. Springer, Cham (2018)
6. Stepashko, V.S.: A combinatorial algorithm of the group method of data handling with optimal model scanning scheme. Sov. Autom. Control **14**(3), 24–28 (1981)
7. Ivakhnenko, A.G., Müller, J.-A.: Recent developments of self-organising modeling in prediction and analysis of stock market. Microelectron. Reliab. **37**, 1053–1072 (1997)
8. Anastasakis, L., Mort, N.: The development of self-organization techniques in modelling: a review of the group method of data handling (GMDH). Research Report 813, Department of Automatic Control & Systems Engineering, The University of Sheffield, UK (2001)
9. Yefimenko, S., Stepashko, V.: Intelligent recurrent-and-parallel computing for solving inductive modeling problems. In: Proceedings of 16th International Conference on Computational Problems of Electrical Engineering (CPEE 2015), Lviv, Ukraine, pp. 236–238 (2015)
10. Stepashko, V.S.: A finite selection procedure for pruning an exhaustive search of models. Sov. Autom. Control **16**(4), 84–88 (1983)
11. Stepashko, V.S., Kostenko, Yu.V.: Combinatorial-selective algorithm for consecutive search for a model of optimal complexity. In: Proceedings of the 1st International Conference on Inductive Modeling, ICIM 2002. SRDIII, Lviv, vol. 1(1), pp. 72–76 (2002). (in Russian)
12. Seber, G.A.F.: Linear Regression Analysis. Wiley, New York (1977)
13. Gergely, J.: Matrix inversion and the solution of systems of linear and non-linear equations by the method of bordering. USSR Comput. Math. Math. Phys. **19**(4), 1–10 (1979)
14. Stepashko, V.S., Efimenko, S.N.: Sequential estimation of the parameters of regression model. Cybern. Syst. Anal. **41**(4), 631–634 (2005)
15. Yefimenko, S., Stepashko, V.: Revised successive search GMDH algorithm with recurrent estimating model parameters. In: Proceedings of the International Scientific Conference "Computer Sciences and Information Technologies" (CSIT 2019), vol. 1, pp. 191–194. IEEE (2019)

Solving the Individual Control Strategy Tasks Using the Optimal Complexity Models Built on the Class of Similar Objects

Ie. Nastenko, V. Pavlov(✉), O. Nosovets, K. Zelensky, Ol. Davidko, and Ol. Pavlov

National Technical University of Ukraine “Igor Sikorsky Kyiv Polytechnic Institute”, 37, Prospekt Peremohy, Kyiv 03056, Ukraine
nastenko.e@gmail.com, pavlov.vladimir264@gmail.com, o.nosovets@gmail.com, zelensky126@ukr.net, alexander.davydko@gmail.com, cheshirelk@gmail.com

Abstract. The conventional approach for calculating individual optimal strategies assumes that the best control actions are determined for the same object that has been studied by monitoring or conducting active trials. However, the class of objects for which is impossible to organize repeated tests is widespread. An example is patients with a particular disease, for each of which it is impossible to organize separate trials to study possible strategies for its cure. This paper proposes an approach to formulate the individual strategies optimization task that uses observational data obtained during the monitoring or active experiment on a sample of similar objects. It is proposed to obtain the state models of the optimal complexity object that are nonlinear in the parameters and initial conditions of the object and linear in control actions, to construct an effective calculation technology. As a modeling tool, algorithms of Group Method of Data Handling (GMDH) are used. The optimization task of individual strategies is formed after substituting the individual values of object parameters in the model of functional and models of constraints. The final calculation procedure takes the form of a linear programming problem. Limitations of the approach and an example of calculating the individual strategy are considered.

Keywords: Decision making · Group method of data handling · Optimal strategies · Linear programming · Retest · Clinical trials

1 Introduction

The calculation problem of individual strategies arose for the authors in the applications of the medical field. At present, the problem of developing computational techniques to determine treatment strategies for specific patients in accordance with their case history, parameters, and the current state characteristics has become extremely urgent [1]. The solution, that allow us to move from a general description of the object state management to their particular implementations, and consequently to obtain a more fine-tuning of the model of a particular object, is an urgent approach for many

N. Shakhovska and M. O. Medykovskyy (Eds.): CCSIT 2019, AISC 1080, pp. 535–546, 2020.
https://doi.org/10.1007/978-3-030-33695-0_36

applications including medicine. Nevertheless, the ISPOR report that is devoted to the analysis of optimization methods in medicine notes an almost complete absence of works that are aimed to obtain the best individual solutions for the patient's treatment process. One of the few individual strategies calculation examples for a particular dynamic programming model is given in [2]. However, the paper does not propose a formalization of the general approach to individual strategies definition for the patient's treatment. The analysis of the applied approaches to solution of similar problems in other areas (economics, ecology, social science) showed that the principle of determining the parameters of interest for a particular object based on a model built on a certain set of similar objects, is usually used in the problems of approximation, prediction, classification, but not in the problems of calculating individual solutions. This makes it reasonable to consider and develop technologies for solving the class of individual strategies optimization problems.

2 Problem Statement

There is a Z object for which it is impossible to carry out repeated trials. To describe the Z transition from the initial state $\mathbf{x}^{in}$ in a finite $\mathbf{x}^{out}$, the existence of models of the form is assumed

$$q = \mathrm{G}(\mathbf{x}, \mathbf{u}, \mathbf{x}^{in}), \quad \mathbf{x}^{out} = \mathbf{F}(\mathbf{x}, \mathbf{u}, \mathbf{x}^{in}), \tag{1}$$

where q is the object state criterion, $\mathbf{x}$ is the object state vector.

There is a region of interest ROI$\mathbf{x}^{in}$ of initial object states. A transition from ROI$\mathbf{x}^{in}$ to the region of permissible values $H(\mathbf{x}^{out}) \leq 0$ (or in the particular case $\mathbf{x}^{out}_{min} \leq \mathbf{x}^{out} \leq \mathbf{x}^{out}_{max}$) should occur under the influence of the control actions vector $\mathbf{u}$: $\mathbf{u}_{min} \leq \mathbf{u} \leq \mathbf{u}_{max}$

It is necessary to find the value of $\mathbf{u}$ (in a more general case, the sequence $\mathbf{u}_{i,\ i=1,\ldots,n}$) that transfers object Z from $\mathbf{x}^{in} \in \mathrm{ROI}_{\mathbf{x}^{in}}$ to $\mathbf{x}^{out}$: $H(\mathbf{x}^{out}) \leq 0$ optimally in the sense of q.

Examples of objects in this class of tasks:

- Z is a biological object (human), ROI$\mathbf{x}^{in}$ is a region of initial states for a particular disease, $\mathbf{u}^{opt}$ is the best treatment strategy.
- Z is a specific economic object, ROI$\mathbf{x}^{in}$ is a region of initial states for some crisis period, $\mathbf{u}^{opt}$ is the optimal strategy for overcoming the crisis.

The main problem here that the optimization task of the transition Z from a given $\mathbf{x}^{in} \in \mathrm{ROI}_{\mathbf{x}^{in}}$ to the domain $H_{\mathbf{x}^{out}}$ requires the knowledge of a certain set of object reactions under different control strategies when transferring an object from $\mathrm{ROI}_{\mathbf{x}^{in}}$ to $H_{\mathbf{x}^{out}}$ that are in a matrix $\mathrm{M} = |\mathrm{X}|\mathrm{U}|$ which is used to obtain models (1). However, as mentioned above, object Z cannot be put on repeated tests, that is, a forcible return of Z to the area ROI$\mathbf{x}^{in}$ and the use of $\mathbf{u}$ for transferring Z to $H_{\mathbf{x}^{out}}$. The necessary passive observations can only be obtained randomly.

3 Ways and Difficulties in Solving the Problem

First, we briefly describe the generally accepted approach [3], which is given here with precision to the notation. Let's a block observations matrix **M** = |**X**|**U**| was obtained as a result of object monitoring or an active experiment. Assume we obtained the transition operator for object **F** from the initial state to the final state in the form $\mathbf{x}^{out} = \mathbf{F}(\mathbf{x}, \mathbf{u}, \mathbf{x}^{in})$, according to the observation data. Here, $\mathbf{x}$ is the state vector, $\mathbf{u}$ is the control vector, $\mathbf{x}^{in}$ is the initial state of the object. We know the quality function for the final state of the object $q = G(\mathbf{x}, \mathbf{u}, \mathbf{x}^{in})$ and restrictions on the final state for each coordinate $\mathbf{x}^{out}_{min} \le \mathbf{x}^{out} \le \mathbf{x}^{out}_{max}$(in general $\mathbf{H}(\mathbf{x}^{out}) \le 0$) and restrictions on the control vector $\mathbf{u}_{min} \le \mathbf{u} \le \mathbf{u}_{max}$. Then, it is possible to formulate an optimization problem for this object in the form

$$\begin{cases} \underset{\mathbf{u}}{\text{extr}}\, q = \underset{\mathbf{u}}{\text{extr}}\; G(\mathbf{x}, \mathbf{u}, \mathbf{x}^{in}) \\ \mathbf{x}^{out}_{min} \le \mathbf{F}(\mathbf{x}^{in}, \mathbf{x}, \mathbf{u}) \le \mathbf{x}^{out}_{max}, \\ \mathbf{H}(\mathbf{x}^{out}) \le 0, \;\; \mathbf{u}_{min} \le \mathbf{u} \le \mathbf{u}_{max}. \end{cases} \tag{2}$$

Then, problem (2) can be presented as a nonlinear programming problem with varying complexity depending on the type of operator **F**, the constraints complexity, and the type of functional q. In a number of cases (small dimension of the problem, the convexity of the region of feasibility, etc.) the problem can be effectively solved.

Note that such a conventional approach is implemented only when it is possible to perform repeated tests or observations upon the object during its transition from the initial states of interest $\mathbf{x}^{in}$ to some other output states $\mathbf{x}^{out}$, using a certain value of the control vector **u.** That is, when there are no fundamental restrictions on implementation of repeated tests. Then, it is possible to obtain the necessary matrix $M = |X|U|$, form the required models $G(\mathbf{x}, \mathbf{u}, \mathbf{x}^{in}), \mathbf{F}(\mathbf{x}, \mathbf{u}, \mathbf{x}^{in}), \mathbf{H}(\mathbf{x}^{out})$, and construct a one-stage optimization problem (2) to make a control decision. However, there is a wide class of objects that cannot be re-trialed, or it is impossible to observe repeatedly similar states for them. A typical example is biological objects (in particular people) where it is impossible to carry out repeated clinical trials with deadly disease infection, to study different modes of treatment therapy, and, then, to develop optimal treatment strategy individually.

At the same time, if it is possible to carry out single tests for sufficiently homogeneous groups of objects while fixing the individual object's parameters essential for obtaining the quality models, we can now form a block matrix of observations **M** using the sample of objects. Then, it is possible to construct the models that are common to the group of objects under consideration while forming the optimization problem. By substituting the object's individual parameters and the initial conditions, we obtain and solve a particular variant of optimization problem configured specifically for this object, thereby obtaining the individual strategy.

By implementing the proposed sequence of actions let us consider the special case of the problem (2) - the problem of one-stage optimization when we need to transfer some object with parameters **p** from a given initial state $\mathbf{x}^{in}$ to a final state $\mathbf{x}^{out}$ using the optimal value of the control vector **u**. The vector of the final state is not specified, the range of permissible values is defined for it; the best value of the criterion variable q must be achieved. In the considered formulation, the difference between the term "criterion variable" and "criterion" is that the criterion variable can have a certain value even before making control decisions - the value of q^{in}. The criterion variable value after the control action has been applied is q^{out}.

So, let the matrix of object-properties **M** is obtained as a result of monitoring or active experiment:

$$\mathbf{M} = |\mathbf{P}|\mathbf{X}^{in}|\mathbf{q}^{in}|\mathbf{X}^{out}|\mathbf{q}^{out}|\mathbf{U}|, \tag{3}$$

where $\mathbf{P} = [p_{ij}]$, $i = 1,\ldots,n$, $j = 1,\ldots,s$ is the object parameter matrix where each row contains s corresponding object parameter values, $\mathbf{X}^{in} = [x_{ij}^{in}]$, $\mathbf{X}^{out} = [x_{ij}^{out}]$, $i = 1,\ldots,n$, $j = 1,\ldots,d$, d is the number of object state variables, $\mathbf{q}^{in} = (q_1^{in}, q_2^{in}, \ldots, q_n^{in})^T$, $\mathbf{q}^{out} = (q_1^{out}, q_2^{out}, \ldots, q_n^{out})^T$ is the values of the criterion variable vector before and after the application the control action, $\mathbf{U} = [u_{ij}]$, $i = 1,\ldots,n$, $j = 1,\ldots,h$, h is the number of control variables.

Let the available data allow us to construct adequate models for the objects output states

$$x_i^{out} = F(\mathbf{p}, q^{in}, \mathbf{x}^{in}, \mathbf{u}),\ i = 1,\ldots,d \tag{4}$$

for the criterion variable

$$q^{out} = G(\mathbf{p}, q^{in}, \mathbf{x}^{in}, \mathbf{u}) \tag{5}$$

and for the domain of solution existence

$$\mathbf{H}(\mathbf{p}, q^{in}, \mathbf{x}^{in}, \mathbf{u}) \le 0 \tag{6}$$

Then, we can formally write the following optimization problem:

$$\begin{cases} \underset{\mathbf{u}}{\text{extr}}\, q^{out} = \underset{\mathbf{u}}{\text{extr}}\, \underset{U}{G(\mathbf{p}, q^{in}, \mathbf{x}^{in}, \mathbf{u})} \\ x_{1\,\min}^{out} \le F_1(\mathbf{p}, q^{in}, \mathbf{x}^{in}, \mathbf{u}) \le x_{1\,\max}^{out} \\ \ldots\ldots \\ x_{d\,\min}^{out} \le F_d(\mathbf{p}, q^{in}, \mathbf{x}^{in}, \mathbf{u}) \le x_{d\,\max}^{out} \\ \mathbf{H}(\mathbf{p}, q^{in}, \mathbf{x}^{in}, \mathbf{u}) \le 0 \\ \mathbf{u}_{\min} \le \mathbf{u} \le \mathbf{u}_{\max} \end{cases} \tag{7}$$

By substituting the object data **p**, q^{in}, $\mathbf{x}^{in}$ in (7) we obtain an optimization problem (8) that it is focused on the optimal effect calculation for the selected object:

$$\begin{cases} \min\limits_{\mathbf{u}} q^{out} = \min\limits_{\mathbf{u}} g_u(\mathbf{u}) \\ x_{1\,\min}^{out} \le f_1(\mathbf{u}) + a'_{01} \le x_{1\,\max}^{out} \\ \ldots\ldots \\ x_{d\,\min}^{out} \le f_d(\mathbf{u}) + a'_{0d} \le x_{d\,\max}^{out} \\ \mathbf{f}(\mathbf{u}) \le 0 \\ \mathbf{u}_{\min} \le \mathbf{u} \le \mathbf{u}_{\max} \end{cases} \tag{8}$$

Taking into account well-known difficulties with the solution of nonlinear programming problems, we will try to find ways to obtain (8) in a linear form. Then, to solve these problems it would be possible to use an effective computational tool - linear programming methods.

However, let us point out another problem. If the objects under the consideration are patients, then obtaining adequate models is faced, as a rule, with insufficient level of control variables variability in the experimental data. The problems arise because of the methodology of clinic trails is purposed to use the patients' sample to proof the therapeutic effect importance [4], but not to obtain experimental data, providing the best adequacy of the models. As a result, researchers are forced to deal with either the indicated clinical trial data or simply the monitoring data of the treatment process when solving the optimization problem for a specific patient.

One of the possible mechanisms for increasing the models' adequacy under these conditions is using the optimization models that are tuned to the specific conditions of drugs using. The tuning is performed by parameters and initial conditions of the patient. Note that the parameters and the object initial states after substitution in the model are converted to the fixed values of model parameters. So, to improve the models' adequacy, it is possible to search for their structures as optimal complexity structures [5], that are nonlinear in the mentioned variables, while not going beyond the linear class of optimization problems in control variables.

Let us briefly summarize the justification of GMDH algorithms application [5] for construction the models (4)–(6):

- The data structure describing the state of patients, as a rule, contains more than dozens or even hundreds of attributes, which makes it relevant to use GMDH algorithms to obtain the models of optimal complexity.
- Typically, the variation level in the initial conditions and parameters of patients is significantly higher than the variation level of control actions that the clinical studies data contains. This feature is related to the research purpose aimed to prove the significance of the therapeutic effect. Consequently, it is needed to use the modeling algorithms with promoting of the control variables.
- As we will see further, it is desirable to obtain models in a form, that is linear in controls and nonlinear in other task parameters (patient parameters and his initial states), if the level of a particular task complexity allows that. Then, having nonlinear models as a whole, there is an opportunity to obtain later an optimization problem linear in control. It is possible to satisfy the specified requirements by filtering only the "allowed" combinations of the initial variables in the search tree of generalized variables when modeling.

To build the models of optimal structure based on the mentioned features, it is natural to apply algorithms of GMDH [5].

Further, the development of the approach for calculating individual strategies involves a transition from tasks of the form (7), (8) to optimization problems with more efficient computational procedures. Below we show the main stages of the proposed approach and consider an example for calculating the optimal individual strategy.

4 An Approach to Calculate the Individual Strategy

So, we have a matrix of initial data in the form (3). Let us form the one-stage optimization problem to obtain the final object state in a given area with the best value of the criterion variable. In order to obtain the object state models and the criterion variable with the best adequacy that is linear in the control variables, we will form the generalized variables for the modeling of x_j^{out}, $j = 1,\ldots,d$ and q^{out} as a set of variables that are nonlinear in p, q^{in}, x^{in} and linear in u. Then, using the modeling algorithms, you can get a transition models from the objects state $\mathbf{x}^{in}$, q^{in} to the objects state $\mathbf{x}^{out}$, q^{out} in the following form

$$q^{out} = g(\mathbf{p}, q^{in}, \mathbf{x^{in}}) \cdot \mathbf{u} + g_0(\mathbf{p}, q^{in}, \mathbf{x^{in}}) + c_0, \tag{9}$$

$$x_i^{out} = \mathbf{f}_i(\mathbf{p}, q^{in}, \mathbf{x^{in}}) \cdot \mathbf{u} + f_{0i}(\mathbf{p}, q^{in}, \mathbf{x^{in}}) + a_{0i},\ i = 1, \ldots, d, x + y = z \tag{10}$$

$$\mathbf{H}_u(q^{in}, \mathbf{x}^{in}) \cdot \mathbf{u} + H_0(q^{in}, \mathbf{x}^{in}) + \mathbf{b}_0 \leq 0. x + y = z \tag{11}$$

The criterion optimization task $g(\mathbf{p}, q^{in}, \mathbf{x^{in}}) \cdot \mathbf{u} + g_0(\mathbf{p}, q^{in}, \mathbf{x^{in}})$ with constraints on x_j^{out}, $j = 1, \ldots,\ d,\ u_{ij}, j = 1, \ldots, h$ and the optimization region (11) now has the form:

$$\begin{cases} \min\limits_u g\left(\mathbf{p}, q^{in}, \mathbf{x^{in}}\right) \cdot \mathbf{u} + g_0\left(\mathbf{p}, q^{in}, \mathbf{x^{in}}\right) \\ x_{1\,\min}^{out} \leq \mathbf{f}_1\left(\mathbf{p}, q^{in}, \mathbf{x}^{in}\right) \cdot \mathbf{u} + f_{01}\left(\mathbf{p}, q^{in}, \mathbf{x}^{in^b}\right) + a_{01} \leq x_{1\,\max}^{out} \\ \ldots\ldots \\ x_{d\,\min}^{out} \leq \mathbf{f}_d\left(\mathbf{p}, q^{in}, \mathbf{x}^{in}\right) \cdot \mathbf{u} + f_{0d}\left(\mathbf{p}, q^{in}, \mathbf{x}^{in}\right) + a_{0m} \leq x_{m\,\max}^{out} \\ \mathbf{H}_u\left(\mathbf{p}, q^{in}, \mathbf{x}^{in}\right) \cdot \mathbf{u} + \mathbf{h}_0\left(\mathbf{p}, q^{in}, \mathbf{x}^{in}\right) + \mathbf{b}_0 \leq 0, \\ \mathbf{u}_{\min} \leq \mathbf{u} \leq \mathbf{u}_{\max} \end{cases} \tag{12}$$

To calculate the specific controls of an object "Z", we substitute the values of its parameters and case history p_{Zj}, x_{Zj}^{in}, q_Z^{in}, $j = 1, \ldots, d$ in the obtained models of form (9), (10), (11), adjusting the system of restrictions and criteria to the specified object. Then the problem (9) takes a linear form by the control variables. In the simplified version – when the permissible area is given by the constraints on each coordinate separately – we obtain a problem in the form (13):

$$\begin{cases} \min\limits_{\mathbf{u}} q^{out} = \min\limits_{\mathbf{u}} \ \mathbf{c}_u \cdot \mathbf{u} \\ x_{1\,\min}^{out} \leq \mathbf{a}_1 \cdot \mathbf{u} + a'_{01} \ \leq x_{1\,\max}^{out} \\ \ldots\ldots \\ x_{d\,\min}^{out} \leq \mathbf{a}_d \cdot \mathbf{u} + a'_{0d} \leq x_{d\,\max}^{out} \\ \mathbf{u}_{\min} \leq \mathbf{u} \leq \mathbf{u}_{\max} \end{cases} \quad (13)$$

Thus, it is formed partially nonlinear models descriptions, increasing the forecast accuracy and at the same time, the optimization problem of a separate object state does not go out of the linear optimization class.

In order to be sure in the calculation adequacy of the values $x_{i\,\min}^{out}$, $x_{i\,\max}^{out}$, $j = 1, \ldots, d$, $u_{i\,\min}, u_{i\,\max}$, $i = 1, \ldots, h$, the problem solution search area should be limited by the area where the adequacy of the objects state models are confirmed. Quantitative and partially qualitative adequacy of the object state models can be reached if the models are calculated by GMDH [5] with the proper external criterion and a relevant set of generalized variables.

5 Example of Individual Strategy Calculation

Consider the calculation of treatment strategies for specific patients in the period after coronary artery bypass surgery. Data for modeling were obtained from the results of monitoring the treatment process of 129 patients. The following variables were used for therapeutic effects: u_1 - duration of medication, u_2, u_3, u_4 - doses of therapeutic drugs that have been applied. As a criterion for the task, the patient's life expectancy was chosen. Below we give a description for the patient's parameters p, therapeutic effects u, input x^{in} and output x^{out} of the patient states:

p_1 is the time that patient spent under the doctor's supervision (days);
p_2 is the patient age;
p_3 is the functional class of the patient's heart failure;
p_4 is the final systolic volume upon the patient admission to the hospital;
p_5 is the final systolic cavity size upon the patient admission;
p_6 is the stenotic arteries number;
p_7 is the percentage of viable myocardial tissue;
p_8 is a part of the lactic acid salts in the patient's blood upon admission to hospital;
p_9 is the pulmonary vascular resistance index upon admission;
p_{10} is the number of coronary artery shunts implanted during surgery;
x_1^{in} is the systolic pressure during hospitalization;
x_2^{in} is the saturation of hemoglobin mixed venous blood with oxygen upon the patient admission;
x_3^{in} is the oxygen utilization factor during hospitalization;
u_1 is the medication duration after surgery (days);

u_2 is the dose of the drug A; u_3 is the dose of the drug B; u_4 is the dose of the drug C;
q^{out} is the duration of life after surgery and subsequent treatment (months);
x_1^{out} is the systolic pressure after treatment;
x_2^{out} is the saturation of hemoglobin mixed venous blood with oxygen after treatment;
x_3^{out} is the oxygen utilization rate after treatment.

The task solution must satisfy the following conditions: it is needed to determine the duration and optimal combination of the doses of the therapeutic drug for a particular patient maximizing the life expectancy after treatment in the permissible area of the patient's state.

One of the latest algorithm versions [6] was used for the modeling. Let us denote Δ_w as the normalized relative mean square error (NRMSE) on the Learning sample of w objects. Below we show the obtained models for the functional of the task and the state of patients at the end of the treatment period:

$$\begin{aligned} q^{out} = & -22{,}457 - 155{,}95\frac{p_3}{p_1} - 12.454 \cdot p_2 + 5.708 p_3 + 0.98 \cdot x_2^{in} + (23.26\frac{x_2^{in}}{x_3^{in}} \\ & - 0.0045 \cdot \frac{p_3}{x_2^{in}}) \cdot u_1 + (0.58 \cdot x_2^{in} \cdot x_2^{in} - 0.0005 \cdot p_7) \cdot u_2 - 0.796\frac{x_2^{in}}{p_1} \cdot u_3 + 0.345\frac{p_3}{x_3^{in}} \cdot u_4 \end{aligned} \tag{14}$$

$$\begin{aligned} x_1^{out} = & 68{,}874 - 2{,}75 \cdot p_2 + (15.32\frac{1}{x_1^{in}} - 0.346 \cdot x_2^{in} + 0.102 \cdot p_{10}) \cdot p_3 + (9.315 \cdot \frac{1}{x_1^{in}} - 4.122 \cdot \\ & \frac{1}{p_4}) \cdot x_3^{in} + 0.85 \cdot p_{10} + 12.615\frac{x_2^{in}}{p_1} \cdot u_1 - 0.056\frac{p_4}{p_1} \cdot u_2 - 0.02\frac{p_1}{x_2^{in}} \cdot u_3 - 0.003\frac{p_3}{p_1} \cdot u_4 \end{aligned} \tag{15}$$

$$\begin{aligned} x_2^{out} = & 0{,}705 + 0.065\frac{p_3}{x_1^{in}} - 0.001 * x_1^{in} + 0.006 \cdot p_{10} + 0.0005\frac{x_3^{in}}{p_1} \cdot u_1 \\ & - 0.0001\frac{p_4}{x_3^{in}} \cdot u_2 - 0.0004\frac{x_2^{in}}{p_4} \cdot u_3 - 0.0001\frac{p_3}{p_1} \cdot u_4 \end{aligned} \tag{16}$$

$$\begin{aligned} x_3^{out} = & 12{,}98 + 5.668\frac{p_1}{x_1^{in}} + 3.179 \cdot p_2 + (0.118 \cdot x_2^{in} - 12.3\frac{1}{x_1^{in}}) \cdot p_3 + 0.148 \cdot x_1^{in} \\ & - 0.782 \cdot p_{10} - 0.035\frac{p_1}{x_1^{in}} \cdot u_1 + 0.032\frac{p_4}{x_3^{in}} \cdot u_2 + 0.034\frac{p_1}{p_4} \cdot u_3 - 0.002 \cdot u_4 \end{aligned} \tag{17}$$

NRMSE for (14) is $\Delta_w = 0.155$, for (15) is $\Delta_w = 0.341$, for (16) is $\Delta_w = 0.379$, for (17) is $\Delta_w = 0.355$. Taking into account the found models, we write down the optimization problem (18) corresponding to (12):

$$\begin{cases} \max\limits_u\ q^{out} = \max\limits_u\ -22{,}457 - 155{,}95\frac{p_3}{p_1} - 12.454 \cdot p_2 + 5.708 \cdot p_3 + 0.98 \cdot x_2^{\text{in}} + (23.26\frac{x_2^{in}}{x_3^{\text{in}}} \\ -0.0045 \cdot \frac{p_3}{x_2^{\text{in}}}) \cdot u_1 + (0.58 \cdot x_2^{\text{in}} \cdot x_2^{\text{in}} - 0.0005 \cdot p_7) \cdot u_2 - 0.796\frac{x_2^{\text{in}}}{p_1} \cdot u_3 + 0.345\frac{p_3}{x_3^{\text{in}}} \cdot u_4 \\ x_{1\ \min}^{\text{out}} \le 68{,}874 - 2{,}75 \cdot p_2 + (15.32\frac{1}{x_1^{\text{in}}} - 0.346 \cdot x_2^{\text{in}} + 0.102 \cdot p_{10}) \cdot p_3 + (9.315 \cdot \frac{1}{x_1^{\text{in}}} - 4.122 \cdot \\ \frac{1}{p_4}) \cdot x_3^{\text{in}} + 0.85 \cdot p_{10} + 12.615\frac{x_2^{\text{in}}}{p_1} \cdot u_1 - 0.056\frac{p_4}{p_1} \cdot u_2 - 0.02\frac{p_1}{x_2^{\text{in}}} \cdot u_3 - 0.003\frac{p_3}{p_1} \cdot u_4 \le x_{1\ \max}^{\text{out}} \\ x_{2\ \min}^{\text{out}} \le 0{,}705 + 0.065\frac{p_3}{x_1^{\text{in}}} - 0.001 * x_1^{\text{in}} + 0.006 \cdot p_{10} + 0.0005\frac{x_3^{\text{in}}}{p_1} \cdot u_1 - 0.0001\frac{p_4}{x_3^{\text{in}}} \cdot u_2 \\ -0.0004\frac{x_2^{\text{in}}}{p_4} \cdot u_3 - 0.0001\frac{p_3}{p_1} \cdot u_4 \le x_{2\ \max}^{\text{out}} \\ x_{3\ \min}^{\text{out}} \le 12{,}98 + 5.668\frac{p_1}{x_1^{\text{in}}} + 3.179 \cdot p_2 + (0.118 \cdot x_2^{\text{in}} - 12.3\frac{1}{x_1^{\text{in}}}) \cdot p_3 + 0.148 \cdot x_1^{\text{in}} \\ -0.782 \cdot p_{10} - 0.035\frac{p_1}{x_1^{\text{in}}} \cdot u_1 + 0.032\frac{p_4}{x_3^{\text{in}}} \cdot u_2 + 0.034\frac{p_1}{p_4} \cdot u_3 - 0.002 \cdot u_4 \le x_{3\ \max}^{\text{out}} \\ u_{1\ \min} \le u_1 \le u_{1\ \max},\quad u_{2\ \min} \le u_2 \le u_{2\ \max},\quad u_{3\ \min} \le u_3 \le u_{3\ \max},\quad u_{4\ \min} \le u_4 \le u_{4\ \max} \end{cases} \tag{18}$$

Next, we set limits on the patient's state variables at the end of the treatment period based on the desired results and the assessment of the area of adequate representation of the object by the model. Taking into account the real state of the patient, the solution may not exists to the problem with the desired results. It will be necessary to adjust the restrictions and repeatedly recalculate the task in order to obtain the optimal solution. Restrictions on the control actions should also take into account the existing treatment protocols:

$$64 \le x_1^{\text{out}} \le 120,\ 0.5 \le x_2^{\text{out}} \le 1,\ 21 \le x_3^{\text{out}} \le 50,\ 21 \le x_3^{\text{out}} \le 50,\ 0 \le u_1 \le 135,\ 0 \le u_2 \le 300,$$
$$0 \le u_3 \le 200,\ 0 \le u_4 \le 250.$$

Let us convert (18) to its specific form (19), that is tuned to the patient. We substitute the parameters and initial conditions of the selected patient $p_1 = 32$, $p_2 = 36$, $p_3 = 4$, $p_4 = 183.3$, $p_{10} = 3$, $x_1^{in} = 72$, $x_2^{in} = 0.64$, $x_3^{in} = 36$ in (18). The task takes the form of a linear programming problem:

$$\begin{cases} \max\limits_u\ -11.155 + 0{,}128 \cdot u_1 + 0{,}1133 \cdot u_2 - 0.584 \cdot u_3 + 0.2299 \cdot u_4 \\ 60 \le\ 72.078 + 0{,}1338 \cdot u_1 - 0{,}1493 \cdot u_2 - 0.5096 \cdot u_3 - 0{,}00146 \cdot u_4 \le\ 120 \\ 0.5 \le\ 0{,}672 + 0{,}000367 \cdot u_1 - 0.000363 \cdot u_2 - 0.00000159 \cdot u_3 - 0{,}000048 \cdot u_4 \le 1 \\ 21 \le\ 35.243 - 0.02381 \cdot u_1 + 0.1161 \cdot u_2 + 0.01274 \cdot u_3 - 0{,}002 \cdot u_4 \le\ 50 \\ 0 \le u_1 \le 135,\quad 0 \le u_2 \le 300,\quad 0 \le u_3 \le 200,\quad 0 \le u_4 \le 250 \end{cases} \tag{19}$$

By solving (19), we obtain the drug dose A of 120 units, the drug dose C of 150 units, and medication duration of 135 days. The systolic pressure must be 72, the saturation as 0.67, the oxygen utilization rate as 45.6 at the end of treatment period. The life duration is 65.4 months. The obtained calculation results satisfy the given restrictions. The found values of therapeutic effects can only be considered as a recommendation of the decision support system for the attending doctor. Only the doctor that is responsible for the final result, can decide on application the recommendations in medical practice. At the same time, this tool allows changing the constraints limits of the optimization task and play out the possible variants of the treatment strategy. This significantly expands the possibilities of predictive analysis in the evaluation of different variants of treatment strategies.

6 The Result Discussion and Future Research Directions

Next, we discuss some of the approach problems.

The decision-maker needs to assess how stable the optimal solution is with respect to the bias of the models' parameter estimations due to the statistical nature of the models. In doing so, it is useful to compare the stability margin with the actual variance of the model parameter estimates. It is desirable to provide these analysis functions in any decision support system where statistical models are used.

It is the most correct to obtain the optimal individual strategy by solving the problem (8) or (13) using a simple search on a grid that covers the problem feasibility region if the dimension of the optimization problem allows to do that. If the methods of mathematical programming is used to solve (8) or (13), we allow the transformation of statistical relations by analytical or numerical methods. And the larger is the error of the obtained models, the more incorrect are such transformations. Let us analyze described above from the point of obtaining the results by application of mechanisms of simplex method. The vertex following mechanism of the simplex algorithm in the constraint domain that is aimed to the optimum is an iterative procedure for resolving the corresponding constraint system with respect to a new set of basic variables. At the same time, this procedure is correct only when the corresponding relations are exact expressions. In our case, the initial system as well as the task simplex is formed by statistical expressions that provide the most probable values to the dependent variables. Each relation is transformed relative to another basis variable will differ from the most probable. This will shift the simplex towards the larger errors of the resulting solutions. This "skew" can theoretically be compensated if the initial models are not traditional regressions, but the statistical relationships that provide the best equal NRMSE for each variable when used as a dependent variable. Then, the obtained models can be considered as rough analogs of exact relations, and for them the analytical procedures of moving along the vertices of the simplex, can be considered as valid. Another possible approach is to replace analytical procedures that is used for solving systems of equations with procedures for constructing linear regressions with respect to a new composition of basic variables. However, it seems that such an iterative mechanism is vulnerable from the point of view of the algorithm convergence correctness, because, in fact, we get a new task simplex at each iteration.

The problems of the proposed approach are minimized in accordance with how high is the level of the models' adequacy on which the optimization problem is formalized. Improving the quantitative and qualitative adequacy of the models requires dividing the sample into object subgroups with the necessary degree of the subsamples' homogeneity to ensure the state model adequacy within a single structure with an acceptable degree of the object parameters variance. The required data structuring can be formed by a hierarchical clustering mechanism. It is formed the data relating to the various qualitatively adequate model structures in the clusters of the top level. Data clusters that correspond to the maximum permissible variance of the objects' parameters belonging to the group are formed for each structure at the lowest level. The dispersion value is limited by the permissible error of the modeling.

Frequent insufficient variability of control variables in the data available for the modeling is a separate problem in this approach. As a result, we obtain qualitatively inadequate model structures since real cause-and-effect regularities are not included in the model. In such cases, we have to use algorithms with promotion of control variables and additional mechanisms to control the qualitative adequacy of the models, to ensure that control variables are included in the model. Here, we stress that algorithmic control over the models' qualitative adequacy is an extremely extensive problem and deserves a special attention. Some tools of such control can be implemented as special structures of mathematical programming problems. Many useful ideas in this direction can be found in [7].

In conclusion, let us point out the problem of correspondence of the initial data to the problem of solution individualization. This is a problem of availability of initial data having the necessary volume and variability. The problem is particularly evident when you calculate individual treatment strategies according to the clinical trials data. Obtaining adequate models here is faced with significant difficulties because the variations of parameters and initial state of the patients in the main group is much higher than the variation of therapeutic effects [8]. These data properties arise because the clinical trial objectives are focused solely on proving the significance of the therapeutic effect in the investigated group of patients comparing to the placebo group. This situation can be improved if the principles of multi-level experiment planning will be implemented in the methodology of clinical trials. Then, the results of tests can be effectively used for calculation of the individual treatment strategies.

7 Conclusion

In this work, we have formalized the one-step problem for calculating individual strategies using the models built on the class of similar objects. The calculation method offers the construction of state models and functional as models of optimal complexity linear in control actions and nonlinear in parameters and initial conditions of the object state. An individual strategy is sought as a solution of the standard optimization problem having the form of a linear programming problem after substitution the individual parameters in the constraint models and functional of the object. It is possible to apply principles of dynamic programming to generalize the considered one-step problem for controlling the object at n stages if the functional has the corresponding properties. It is easy to see, since the considered one-step problem has Markov properties: the transition to the $\mathbf{x}^{out}$ state only depends on the object parameters, its state at the previous stage (for the current stage it is $\mathbf{x}^{in}$), and the control actions vector $\mathbf{u}$ that is applied. The conditions of a correct application of proposed optimization models assume obtaining models with a high quantitative and qualitative adequacy. This requires to make significant changes to the existing methodology of clinical trials in medicine in particular. As an example, a calculation of the individual treatment strategy is shown. The proposed approach can be extended to other areas of decision-making problems taking into account existing remarks.

References

1. Crown, W., et al.: Constrained optimization methods in health services research—an introduction: report 1 of the ISPOR optimization methods emerging good practices task force. Value Health **20**(3), 310–319 (2017). https://doi.org/10.1016/j.jval.2017.01.013
2. Denton, B., Kurt, M., Shah, N., Bryant, S., Smith, S.: Optimizing the start time of statin therapy for patients with diabetes. Med. Decis. Making **29**(3), 351–367 (2009). https://doi.org/10.1177/0272989x08329462
3. Tabak, D., Kuo, B.: Optimal control by mathematical programming. Prentice-Hall, Englewood Cliffs (1971)
4. Brody, T.: Clinical Trials, 2nd edn. Academic Press, San Diego (2016)
5. Ivakhnenko, A., Stepashko, V.: Noise–Immunity Modeling, 1st edn. Naukova dumka, Kyiv (1985). (in Russian)
6. Vanin, V., Pavlov, A.: Development and application of self-organization algorithms for modeling of complex processes and objects which are represented by the point former. Proc. Tavria State Agrotechnical Acad. **24**(4), 51–56 (2004). Melitopol (in Ukranian)
7. Zgurovsky, M.Z., Pavlov, A.A.: The four-level model of planning and decision making. In: Combinatorial Optimization Problems in Planning and Decision Making: Theory and Applications, 1st edn. Studies in Systems, Decision and Control, vol. 173, pp. 347–406. Springer, Cham (2019). https://doi.org/10.1007/978-3-319-98977-8
8. Nastenko, Ie., Pavlov, V., Nosovets, O., Zelensky, K., Davidko, Ol., Pavlov, Ol.: Optimal complexity models in individual control strategy task for objects that cannot be re-trialed. In: Proceedings of International Scientific Conference “Computer Sciences and Information Technologies” (CSIT-2019), vol. 1, pp. 207–210 (2019)

Applied linguistics

Lexicographical Database of Frequency Dictionaries of Morphemes Developed on the Basis of the Corpus of Ukrainian Language

Oksana Zuban(✉)

Taras Shevchenko National University of Kyiv, 14 T. Shevchenko Blvd, Kyiv 01601, Ukraine
oxana.mell.zuban@gmail.com

Abstract. The paper presents the pattern of lexicographical computer modelling and structure of the database of electronic frequency dictionaries of morphemes, introduced in linguistic research system called The Corpus of Ukrainian Language, collectively created in Computational Linguistics Laboratory, Institute of Philology, Taras Shevchenko National University of Kyiv. Electronic frequency dictionaries of morphemes within the Corpus of Ukrainian language are the only modern interactive text-oriented research system of linguistic morphemic field of knowledge in the contemporary Ukrainian computer morphemic lexicography that processes linguistic and statistical data about the units of morphemic level of Ukrainian text structure. The paper represents the model of logically structured data of the relational database of frequency dictionaries of morphemes that prefers methodological approach in calculating statistical features of morphemic units in a text according to automatic morphemic segmentation of initial forms, lemmatized by text tokens without morphemic annotation of the latter. The pattern used in lexicographical computer modelling does not reduce efficiency and speed of search and classification options of the created text-oriented system of morphemic text analysis. This pattern also does not minimize the importance of linguistic research results based on the material provided; on the contrary, it increases an explanatory value of linguistic research through systematization and different classification aspects of morphemic information.

Keywords: Database (DB) · Electronic frequency dictionary of morphemes (EFDM) · Corpus of Ukrainian language (The Corpus)

1 Introduction

The practice and theory of creating databases are being successfully developed in the field of morphemic and derivational lexicography in Ukrainian computational linguistics. Nowadays there exist two databases:

(1) Morphemic-Derivational Database of Ukrainian Language [1], created in the Department of Structural and Mathematical Linguistics at Potebnia Institute of

N. Shakhovska and M. O. Medykovskyy (Eds.): CCSIT 2019, AISC 1080, pp. 549–566, 2020.
https://doi.org/10.1007/978-3-030-33695-0_37

Linguistics, Academy of Sciences of Ukraine (nowadays it is a Department of Lexicology, Lexicography, and Structural and Mathematical Linguistics at Potebnia Institute of Linguistics, Academy of Sciences of Ukraine);

(2) Automated System of Morphemic and Derivational Analysis (ASMDA) [2, 3], created by the staff of Computational Linguistics Laboratory, Institute of Philology, Taras Shevchenko National University of Kyiv.

These two databases were created using the experience of morphemic computer lexicography, initiated during compilation of famous morphemic dictionaries: Frequency dictionary "Morphemes of Russian language" by Oliverius [4], Root dictionary "Derivational dictionary of Russian" by Worth [5]; Morphemic Dictionary of Czech edited by Slavícková [6]. Nevertheless, the morphemic-derivational DBs of Ukrainian are based on various principles of computer modelling of morphemic word structure, and they differ in lexicographical structure and the tasks to be solved.

Morphemic-Derivational Database of Ukrainian language is a knowledge base designed to function as a kind of guide for a researcher in a field of linguistics, and it is undoubtedly extremely important for organizing a full-scale study of language, but it is static, it cannot be used in a mode of automated text analysis. There are well-known printed morphemic and derivational dictionaries of Ukrainian based on the Morphemic-Derivational Database of Ukrainian language, including "The Dictionary of Ukrainian affixal morphemes" [7], "Ukrainian Dictionary of Roots and Clusters" [8]. They show statistical data (absolute and relative frequency) of some functional morphemic types in a language system and in Ukrainian belles-lettres texts.

Automated System of Morphemic and Derivational Analysis (ASMDA) is a specialised intelligent system which functions as a search engine on morphemics and word formation in Ukrainian. Initially it was designed to perform automatic text analysis, and nowadays it is used as an automatic morphemic segmentator of the sample text lexicon in the Corpus of Ukrainian language.

Corpus of Ukrainian language [9] is a research tool directed to solve a wide class of linguistic problems, particularly in the field of morphemics and word formation, although most of the Slavic languages corpora do not contain texts parametrized at morphemic level. According to our information, apart from the Corpus of Ukrainian language, the following corpora provide search by morphemes: Russian National Corpus [10], The Computer Corpus of Texts from Russian Newspapers of the End of the 20th century [11] that was later included into the Polistylistic Corpus of Modern Russian Texts [12]. A number of questions arise: is there a need in morphemic (derivational) text annotation in a corpus? What research opportunities does automatic corpus-oriented morphemic and derivational analysis give to a linguist?

The answers to these questions are to be found in academic writings of famous Ukrainian and foreign linguists: Karpilovska [13]; Klymenko [7]; Darchuk [3]; Polikarpov, Kukushkina, Toktonov [14]; Liashevska [15] et al. The first layer of corpus-oriented morphemic-derivational analysis is a study of neological processes in modern lexicon; this develops a new area of linguistic research – neology, and a new area of lexicography – neography. Lexicographical modelling of language innovations in Ukrainian linguistics is not only limited to registration of new words and their interpretation; neographical and neological studies aim to investigate derivational, grammatical, stylistic and other processes projected on development of modern Ukrainian lexicon. When examining modern Ukrainian derivational nomination, L.P. Kysljuk lists a number of ongoing issues resolved in corpus-oriented academic writings only: “The topicality of studying derivational nomination as a main way of updating modern Ukrainian lexicon is caused by search of answers to the following questions: (1) which derivational resources sustain the typological features of Ukrainian nomination in new social and political conditions for the use of Ukrainian language and how do they do that? how do they provide it with new functions in the status of the national language? (2) to what extent do the mechanisms of identity protection of Ukrainian nomination exist in word formation in modern conditions of European integration processes and globalization? (3) what social, cognitive and communicative factors influence the choice and realisation of derivational resources of Ukrainian language system in modern collective, shared (usage) and individual (idiolect) language practice? (4) how does modern language practice – shared and individual – influence the language system (word formation) and modern Ukrainian derivational standards?” [16: 10–11].

The study of derivational nomination on the basis of text corpus broadens direction of research of neological and dynamic processes in the language system with statistical data which are an essential condition of systemic description of language usage in synchrony and diachrony. Taking into account the experience of computer modelling of morphemic word structure during compilation of Chronological Morphemic and Word-Formational Dictionary of Russian [17], laboratory staff of the Laboratory for General and Computational Lexicology and Lexicography at the Philological Faculty of Moscow State University developed a pattern of automatic morphemic annotation in the Polistylistic Corpus of Russian [14], which was also used by the compilers in the Integrated Text-Analytical System “Style-Analyzer2” [18]. Profound linguistic research of morphemic and derivational structure of Russian lexicon was carried out based on data from frequency dictionaries of morphemes; this research is characterized by comprehensive analysis of interplay of factors of new lexicon formation and systemic usage of statistical data [14]. The data of electronic frequency dictionaries of morphemes in the Corpus of Ukrainian language also enabled several statistical studies on idiostyles of Ukrainian poets at morphemic level [19, 20]. These studies prove the relevance of statistical data about morphemic units (roots, prefixes, suffixes) and morphemic structures (morphemic length of a word, morphemic model of a word) in stylometry model of an author’s style.

The compilers of Automatic Morphemic-Derivational Analysis System in Russian National Corpus [10] presented the prospectives of corpus-oriented approach to the examination of derivational processes. “So far the productivity of word formation in a text and in a language was studied mainly in stylistic aspect. However, it seems to be interesting to analyse the correlation of derivational patterns realisation in text with realisation of other structures; how specific derivational patterns combine with others; what is the difference between specific features of verbal and noun roots; how do cognates participate in co-reference investigation; how frequent is one or other pattern in the whole corpus or in this or that genre. It is also interesting to trace microchanges in derivational processes (e.g., what is the speed of inclusion of new words into word formation and so on). All these opportunities are available in derivational mark-up in the corpus; this mark-up was carried out with the help of electronic Morphemic-Derivational Dictionary and provided with search engine" [15: 211]. Linguistic statistical research in the field of morphemics and word formation was carried out on the basis of Russian National Corpus, e.g., academic writings of Tatevosov [21], Pazelskaya [22], which demonstrate efficient selection of research material and possibilities of studying derivational and morphemic processes in correlation with syntactic valence and semantics of described lexemes.

Apart from corpus-oriented linguistic studies, automatic morphemic text analysis opens up new prospects for creating the systems of automatic text contents recognition. Team of scientists from the Russian National Machine Intelligence Research Institute under the leadership of Packin [23, 24] created an electronic hyperdictionary "Ariadna" which structures morphemic, derivational and inflectional information based on lexical vocabulary of the dictionary by Zaliznyak [25] and lists of morphemes from "Dictionary of morphemes" by Kuznetsova and Yefremova [26]. This dictionary is connected with the System of Knowledge Representation and Reasoning "Abrial-2", and is intended to design a semantic network of profound content analysis of Russian text. Creation of resident system dictionary on the basis of morphemes caused considerable reduction of its volume and an increase in system efficiency. "Taking into consideration derivational relations and morphemic structure of Russian words can give significant benefits during formation of semantic networks for all the language. The volume of necessary semantic descriptions and, correspondingly, the amount of work of computer lexicographers in conditions of priority description of morphemes instead of lexemes may result in considerable reduction of work, taking into account average productivity of a Russian root (approximately 13 by A.M. Tikhonov). In view of the fact that the most frequent roots simultaneously are the most productive ones (the power of "нести" is 540), it is difficult to overestimate the efficiency of priority semantic description of morphemes for compact description of semantics used in Russian natural language analysis" [24].

Active usage of corpus morphemic and derivational data in linguistic research in different fields proves the necessity of automatic morphemic text parametrization in research corpora and defines the topicality of subject of our study.

The aim of the paper is to describe the procedure of lexicographical computer modelling and the structure of databases (DB) of electronic frequency dictionaries of morphemes (FDM) compiled on the basis of the sample texts of the Corpus of Ukrainian Language (the Corpus).

2 Automatic Morphemic Analysis in the Corpus of Ukrainian Language

In analyser-module, the texts of the Corpus are parameterized by 4 levels of linguistic analysis:

(1) morphological analysis is a basic stage for all the following levels: it implies identification of words' morphological characteristics (part of speech and

grammatical meanings of each text token), as well as token lemmas (in automatic mode);

(2) morphemic analysis is an identification of lemmas' morphemic organization (initial forms) of lexical register (in automatic mode);
(3) syntactical analysis implies determination of a word combination, its type and kind of syntactical relations (in automatic mode), as well as sentence structure trees (in automatic/automated mode);
(4) semantic analysis is an attribution of a semantic field code of taxonomic classification to each token/lemma (in automatic/automated mode).

All the types of automatic linguistic analysis are carried out in relation to big linguistic databases which have been compiled based on the developed pattern of computer modelling of units at different language levels (computer grammar of Ukrainian). "We created computer grammar for automatic analysis of Ukrainian text; this grammar is a hierarchical set of computer models: morphemic-derivational, morphological, syntactical model based on formal, exact and unambiguous rules. These models can be considered research ones, because algorithmic rules embedded in grammar lead to identification of this or that linguistic phenomenon (morphs, word-forms with their part of speech and categorical characteristics, word combinations, sentence dependency trees etc.). Activity of a linguist is simulated with the help of algorithms: it provides the switch from collection of texts to the system in their basis. We also defined the elementary units and classes of elementary units. The developed models are inductive, non-semantic and deterministic (structural) models of analysis" [3: 28].

Linguistic mark-up of a text – an annotation – is a result of automatic linguistic analysis. Linguistic annotation in the Corpus is carried out in two ways:

(1) total morphological annotation of all tokens on the basis of uploaded texts (automatic attribution of grammatical code to each text token and automatic lemmatization take place during an upload of a text to the corpus);
(2) sample syntactical and semantic annotation of text tokens on the basis of limited sample texts and compilation of autonomous databases.

There is no morphemic annotation at morphemic level of text in the Corpus. Automatic morphemic analysis is performed for initial forms in lemmas register which is compiled as a result of token lemmatization of the sample texts. As a result of automatic morphemic analysis, there appear autonomous DBs compiled on the basis of individual sample texts, which contain systematized information about morphemic structure of words and statistical characteristics of morphemic units (roots, affixes and morphemic word structures). The compilers of the Corpus deliberately renounced morphemic text annotation with the increase in corpus data because a mark-up of a 1-million corpus of text tokens at all levels of analysis requires a very powerful technical support, otherwise the use of corpus makes only slow progress. For example, only derivational annotation of the word *переподготовка* in Russian National Corpus contains 4 lines of derivational tags [15: 217].

Automatic morphemic analysis of initial word forms in the Corpus is performed on the basis of morphemic structures DB (containing ≈ 200 000 words) of the Automated

System of Morphemic and Derivational Analysis (ASMDA) described in several papers of project compilers [2, 27] and in the monograph "Computer Annotation of Ukrainian Text: Results and Prospectives" by Darchuk [3]. Methodology of automated compilation of this DB has become the subject of study in university courses on computer lexicography and automatic morphemic analysis; it was described in university textbooks [28, 29].

A morphemic word structures DB appears to be an automatic dictionary:
чотирибальний,A,RFIGRKSLFN;
шестибальний,A,REIFRJSKFM;
where each word has an attributed linguistic information about: (1) grammatical code of a word (A – adjective); (2) a pattern of morphemic word structure (RFIGRKSLFN; REIFRJSKFM).

Each morph is modelled by two-letter code:

(*e.g.*, *чотир*RF/*и*IG/*баль*RK/*н*SL/*ий*FN)

the first Latin letter denotes a type of a morph: P – prefix, R – root, S – suffix, F – flexion, I – interfix, X – postfix, and the second one denotes morph boundaries with a sequence number (from the beginning of a word) of a terminal letter of each morph. Literal-number boundaries of morphs are converted into Latin script according to the sequence number in simplified alphabetical system:

R(5)I(6)R(10)S(11)F(13) = RF(5)IG(6)RK(10)SL(11)FN(13).

Morphemic pattern defines literal boundaries and functional types of morphs, and at the same time it is a formula for automatic morphemic segmentation of initial forms in lexical register compiled on the basis of sample texts. Electronic frequency dictionaries of morphemes are compiled according to the results of automatic morphemic analysis; these dictionaries function as autonomous lexicographical systems.

3 Lexicographical System of Frequency Dictionaries of Morphemes in the Corpus of Ukrainian Language

There are automatically generated alphabetic-frequency dictionaries of word-forms, lexemes, morphemes, morphemic word structures, and word combinations based on separate sample texts in the section of "Frequency dictionaries" [30] of the Corpus of Ukrainian language:

1. Frequency dictionaries based on stylistic subcorpora:

   - FDMs of poetic texts (as of 2018);
   - FDMs of scientific texts (as of 2018);
   - FDMs of legislative texts (as of 2018);
   - FDMs of artistic prose (as of 2018);
   - FDMs of artistic prose (as of 2012);
   - FDMs of publicistic texts (as of 2018);
   - FDMs of publicistic texts (as of 2012);
   - FDMs of medical texts (endocrinology);
   - FDMs of folklore texts (as of 2014).

2. Frequency dictionaries based on the sample texts of individual authors:
   - Lesya Ukrainka's book of poems ("On the Wings of Songs");
   - the book "Selected Works" by Lina Kostenko;
   - the book of poems "Palimpsests" by Vasyl Stus;
   - the book of poems "Circulation" by Vasyl Stus;
   - the book of poems "The Merry Graveyard" by Vasyl Stus;
   - poetic texts of Taras Shevchenko;
   - the novel "Raven" by Vasyl Shkliar;
   - the novels "Sweet Darusia" and "Apocalypse" by Maria Matios;
   - prose texts by Serhiy Zhadan;
   - poetic texts by Serhiy Zhadan.

Every frequency dictionary represents two types of FDMs: Frequency Dictionary of Root and Affixal Morphemes; Frequency Dictionary of Morphemic Word Structures. Electronic frequency dictionaries of morphemes within the Corpus are the only text-oriented research system in linguistic and morphemic field of knowledge in the contemporary Ukrainian computer morphemic lexicography that systematizes the data of automatic morphemic and statistical analysis on the basis of massive amounts of text. The structure and interactive search options of the Corpus FDMs are described in the series of articles by the project authors [2, 27, 31]. Morphemic dictionaries are compiled by a user according to the selected classification options in interactive mode. Every dictionary has three interrelated functional zones:

(1) morphemic: the register of morphemic units (roots, affixes, morphemic word structures) with statistical features of absolute and average frequencies;
(2) lexical: the realisation of a morphemic unit selected by a user in the words of a sample text;
(3) textual: a concordance of a word selected by a user with a definite morphemic unit and a source from which the contextual usage (sentence) was taken

In FDMDB development, certain sequence of tasks was defined in accordance with the zones of FDM that represent the units of three language levels:

(1) automatic systematization of textual and lexical data of the sample text;
(2) automatic morphemic analysis of the initial forms (lemmas) of the samples and compilation of morphemic units registers;
(3) automatic calculation of frequency features of morphemic units in the sample text.

4 Systematization of Lexical and Textual Data of the Sample Text

In the process of building up an autonomous lexicographical system of electronic frequency dictionaries based on sample texts of the Corpus we set the target of developing a logically structured reference model of the output data of the sample text. This model is to provide automatic compilation not only of lexical frequency dictionaries, but also frequency dictionaries of units of different language levels: lexemes, word-forms, morphemes, morphemic structures, word combinations and semantic taxa. That's why a unified systematization of textual and lexical data is performed for all the types of frequency dictionaries in the Corpus. During the process of the morphological annotation of texts every text token gets an attributed two-letter grammatical code (the 1st letter stands for a part of speech, the 2nd one – for a grammatical meaning of a word-form), lemma and the number of a sentence in the analysed text. E.g., after the procedure of morphological annotation the text segment:

Реве та стогне Дніпр Широкий

will have the following data structure:

Реве,ГЯ,ревти,1/та,СС,та,1/стогне,ГЯ,стогнати,1/Дніпр,йИ,Дніпр,1/широкий, АИ,широкий,1.

Two tables are automatically filled-in after processing the morphologically annotated text:

(1) The table of words from the text "temp_freq_wf" (see Table 1);

Table 1. Data fragment of the "temp_freq_wf" table.

Code	wrd	cls	vib	tid	sentnum
1	реве	ГЯ	1	10295	1
2	та	СС	1	10295	1
3	стогне	ГЯ	1	10295	1
4	Дніпр	йИ	1	10295	1
5	широкий	АИ	1	10295	1

(2) The table of the initial forms "temp_freq" (see Table 2).

Table 2. Data fragment of the "temp_freq" table.

Code	wrd	cls	vib	tid	sentnum
1	ревти	Г	1	10295	1
2	та	С	1	10295	1
3	стогнати	Г	1	10295	1
4	Дніпр	й	1	10295	1
5	широкий	А	1	10295	1

"Temp_freq" is the source table for FDM development. (The structure of the table will be described in part 5).

5 The Database Structure of the Frequency Dictionaries of Morphemes

The interactive feature of the two types of electronic FDMs (morphemes and morphemic word structures) is provided by 6 automatically filled-in tables of the lexicographical DB (the structure of objects and relations between them are shown in Fig. 1). To illustrate the structure of FDMDB, the example is a DB based on sample texts of Shevchenko's poetic [32] writings containing ~ 68295 tokens. According to data pattern, FDMDB is relational. DBMS (Database Management System) is developed using MS Access with additional software (C#).

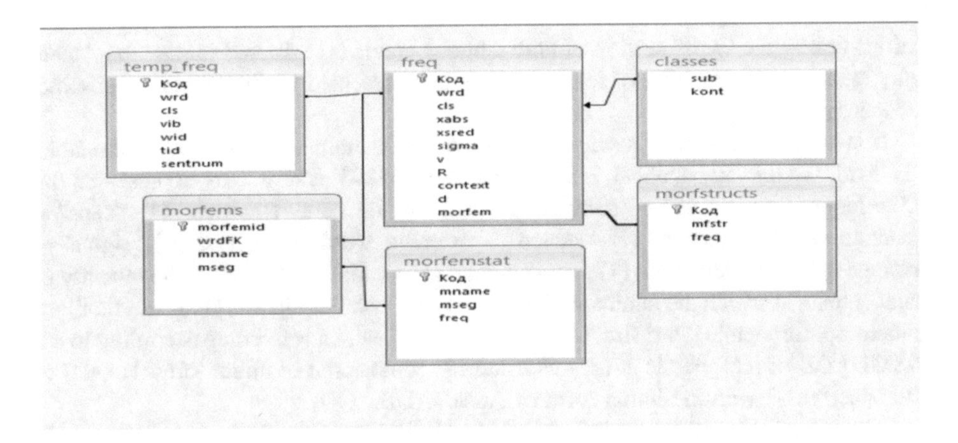

Fig. 1. Relational data pattern diagram of the frequency dictionaries of morphemes.

5.1 The "temp_freq" Table

The "temp_freq" table organizes textual and lexical data in 6 sections (see Table 2): (1) "code" – word number in the table; (2) "wrd" – lemma (*ревти*); (3) "cls" – part of speech (Г – verb); (4) "vib" – the sub-sample text number (the whole sample text of T. Shevchenko's writings was automatically divided on the basis of the first table section for statistical calculation into 68 sub-samples containing 1004 tokens each); (5) "tid" – 10295 (the number of the work in the Corpus); (6) "sentnum" – the number of the sentence in the text. Using the word filter in "wrd" section one can automatically compile a card (separate table) for statistical calculations for every lemma (see Table 3). Lemma is registered in a table as many times as many tokens of this lemma occur in the sample text.

5.2 The "freq" Table

The "freq" table (see Table 4) is compiled automatically on the basis of data from a previous table and structures the alphabetical lemmas register and statistical lemma characteristics that have been automatically calculated. Correlation of the two tables is

Table 3. Data fragment of the "temp_freq" table for the lemma (*ревти*).

Code	wrd	cls	vib	tid	sentnum
1	ревти	Г	1	10295	1
4079	ревти	Г	5	10306	21
7209	ревти	Г	8	10311	49

Table 4. Data fragment of the "freq" table for the lemma (*ревти*).

Code	wrd	cls	xabs	xsred	sigma	v	R	d	morfem
5501	ревти	Г	24	0,338028169014085	0,484228547540031	1,43250945313926	16	8,287824	RF

carried out by the "wrd" section of both tables. Lemma is registered in the "freq" table only once, and the number of lemma tokens is represented by the feature of absolute frequency ("xabs").

The table consists of 10 sections: (1) "Code" is the number of the word in the table; (2) "wrd" – the alphabetical lemmas register (68295 text tokens represented by 7526 lemmas); (3) "cls" – grammatical code of the part of speech; (4) "xabs" – absolute word frequency; (5) "xsered" – average word frequency; (6) "sigma" – average quadratic deviation; (7) "v" – coefficient of variation; (8) "R" – the quantity of sample texts in which the lemma was used; (9) "d" – stability index; (10) "morfem" – a pattern of morphemic word structure of *ревти* – RF, which is build up according to the ASMDADB of morphemic structures using the comparison of initial form (lemma) of the table "freq" with the initial form of ASMDADB.

The "freq" table is the central object in FDMDB (see Fig. 1), and its functions in the lexicographical system of FDM are the following:

(1) it structures the lexical register of sample texts for morphemic segmentation (every unit in the register is an initial form of one lexeme, and is divided into morphemes only once);
(2) it structures statistical characteristics of lexemes that are used for automatic calculation of statistical data about morphemes and morphemic word structures, because absolute frequency of a morpheme/morphemic word structure equals the sum of absolute frequencies of lexemes which contain the defined morpheme/morphemic word structure;
(3) it organizes the data about patterns of morphemic lexeme structures of sample texts for compilation of the "morfstructs" table.

5.3 The "morfstructs" Table

The "morfstructs" table arranges the data about the absolute frequency ("freq" section) of morphemic word structures of the words registered in the "mfstr" section (see Table 5). This table is compiled automatically on the basis of the "morfem" section from the "freq" table (see Table 4). In this section the chosen filter of the pattern of morphemic word structure compiles a sample of all the words in which the selected pattern can be identified. E.g., the PPRF pattern compiles the sample of 12 words (see Table 6).

The "xabs" section of this sample represents data about the absolute frequency of every word of the PPRF pattern. These data are summarized, and the sum is registered in the "freq" section of the "morfstructs" table. E.g., the absolute frequency of the PPRF pattern is 26 (see Table 5).

Table 5. Data fragment of the "morfstructs" table.

Код	mfstr	freq
152	PPRFX	9
153	PPRSF	204
154	RSRF	3
155	**PPRF**	**26**
156	RIF	3

Table 6. The sample fragment of the "freq" table according to PPRF pattern.

Code	wrd	cls	xabs	xsred	sigma	v	R	d	morfem
860	вспомин	Й	2	3,27868852459016E-02	0,253966121062781	7,74596669241483	1	1,110223E-15	PPRF
1329	донедавний	А	1	1,63934426229508E-02	0,126983060531391	7,74596669241483	1	1,110223E-15	PPRF
1884	заповідь	К	1	1,63934426229508E-02	0,126983060531391	7,74596669241483	1	1,110223E-15	PPRF
1886	запопасти	Г	3	4,91803278688525E-02	0,216244359971687	4,39696865275764	3	4,323538	PPRF
2039	здобути	Г	1	1,63934426229508E-02	0,126983060531391	7,74596669241483	1	1,110223E-15	PPRF
2127	знайти	Г	5	8,19672131147541E-02	0,328687502553499	4,00998753115268	4	4,823128	PPRF
3424	недосвіт	Й	1	1,63934426229508E-02	0,126983060531391	7,74596669241483	1	1,110223E-15	PPRF
5188	роздобути	Г	2	3,27868852459016E-02	0,1780783687082	5,43139024560011	2	2,988105	PPRF
5250	розповити	Г	2	3,27868852459016E-02	0,1780783687082	5,43139024560011	2	2,988105	PPRF
5766	сповити	Г	5	8,19672131147541E-02	0,274314762798058	3,3466401061363	5	5,679506	PPRF
5770	сповідь	К	2	3,27868852459016E-02	0,1780783687082	5,43139024560011	2	2,988105	PPRF
6204	увійти	Г	1	1,63934426229508E-02	0,126983060531391	7,74596669241483	1	1,110223E-15	PPRF

The Frequency Dictionary of Morphemic Structures is compiled according to the data of the "morfstructs" table. A pattern of morphemic word structure is a unit in the dictionary register. Information about absolute frequency of the PPRF pattern and its lexical realisation, for instance, contains all the words with morphemic PPRF structure according to the data of the "freq" table.

5.4 The "morfems" Table

The "morfems" table is filled-in automatically on the basis of data in "wrd" section of the "freq" table. Correlation of the tables is carried out by the "wrdFK" section of the "morfems" table and the "code" section of the "freq" table. The "morfems" table (see Table 7) consists of 4 sections and organizes data about morphemic segmentation of the initial word forms. E.g., morphemic segmentation of the word "*ревнути*" is shown in 3 lines in the table:

Table 7. Data fragment of the "morfems" table for the lemma (*ревнути*).

morfemid	wrdFK	mname	mseg
12341	5078	R	рев
12342	5078	S	ну
12343	5078	F	ти

The three lines (12341-12343) represent three morphemes *-рев-*, *-ну-*, *-ти-* (Sect. 4) and their functional type in the 3rd section: R (root); S (suffix); F (flexion). The 2nd section "wrdFK" has the same number 5078 written three times corresponding to the morphemes *-рев-*, *-ну-*, *-ти-*. This means that the defined morphs constitute one word *ревнути*.

Automatic morphemic segmentation of initial word forms of the "wrd" section of the "freq" table is carried out according to ASMDADB of morphemic structures, and is registered in the columns of the "morfems" table using special software (see Fig. 2).

The initial forms register of the "freq" table provides a word which is compared with the same word in the ASMDADB: 5078 ревнути ↔ ревнути, V, RDSFFH. The segmentation of a literal word record is carried out according to the segmentation pattern of dividing words into morphs (RDSFFH) taken from the ASMDADB. RDSFFH model is a combination of two-letter codes, in which the first symbol is the functional type of a morpheme, and the second one is a right-side quantitative-literal boundary of a morph in a word, converted into the Latin script according to the number of a letter in a letter sequence in a simplified alphabetical system: RD – R (root), D (the third grapheme of *рев-нути*); SF – S (suffix), F (the fifth grapheme of *ревну-ти*); FH – F (flexion), H (the seventh grapheme of *ревнути*). The first symbol of a two-letter code is registered in the separate line of an "mname" section. The second symbol of a two-letter code is used by the software to divide the word into parts according to the number of the grapheme in a letter sequence after preliminary calculations of graphemes in a word. The chain of graphemes is delimited after the grapheme defined by the number in a letter sequence. Every part of a divided word is a morpheme which is saved in a separate line of the "mseg" section.

```
bool bitenwrd(string wrd, string morfem, string wrdid, OleDbConnection con)
{
bool w = false;
int st = 0, nd = 0;

OleDbCommand cmd = new OleDbCommand("", con);
for (int i = 0; i < morfem.Length; i += 2)
{
if (morfem[i] == '/' || morfem[i] == '|')
{
break;
}
nd = (int)morfem[i + 1] - (int)'A';
if (nd <= wrd.Length)
{
string seg = wrd.Substring(st, nd - st);

if (seg == "") seg = "0";
cmd.CommandText = "INSERT INTO morfems (wrdFK,mname,mseg) SELECT " + wrdid + ",'" +
morfem[i] + "','" + seg.Replace("'", "''") + "'";
cmd.ExecuteNonQuery();
}
st = nd;
}
```

Fig. 2. Fragment of a Program Code of Automatic Morphemic Word Segmentation

5.5 The "morfemstat" Table

The "morfemstat" table structures the register of morphemes identified in the words of the "morfems" table, and organizes the absolute frequency of these morphemes in Taras Shevchenko's poetry (see Table 8).

Table 8. The sample fragment of the root morphemes of the "morfemstat" table.

Code	mname	mseg	freq
1005	R	зараз	12
1006	R	раз	19
1007	R	ранн	2
1008	**R**	**рев**	**41**
1009	R	регот	12

The table was filled-in automatically according to the data from the "morfems" and "freq" tables. The number in "wrdFK" section of the "morfems" table correlates with the same number in "Code" section of the "freq" table, in which every number (i.e. a lemma) in "xab" section stands for absolute frequency rate of the word. Their sum is automatically saved in the "freq" section of the "morfemstat" table. E.g., the calculation for the root morph -рев- sample is shown in Fig. 3. The total number of words containing -рев- root is 41, so absolute frequency of the root -рев- is 41 (see Table 8).

morfemid	wrdFK	mname	mseg
4516	1902	R	рев
4520	1903	R	рев
12333	5075	R	рев
12341	5078	R	рев
12344	5079	R	рев
12346	5081	R	рев

"morfems" table

Code	wrd	xabs
1902	заревіти	1
1903	заревти	10
5075	ревіти	3
5078	ревнути	3
5079	ревти	22
5081	ревучий	2
		$\sum$ 41

"freq" table

Fig. 3. The calculation sheet for absolute frequency of the root -рев-

The Frequency Dictionary of Morphemes for the chosen functional type of a morpheme (prefix, root, suffix, interfix, postfix) is compiled according to the data of the "morfemstat" table. Morphemes of one functional type are the units in the dictionary register – e.g., roots (see Table 8). Each root is provided with information about absolute frequency and lexical realisation of each root morpheme in the words according to the table "freq" data.

5.6 The "classes" Table

The "classes" table arranges data about grammatical codes meaning: 1st "sub" section contains a list of grammatical codes; 2nd "kont" section – meaningof a grammatical code. E.g. a grammatical code "A" in the 1st section has a value of "adjective" in the 2nd section. The "classes" table is connected with the section "cls" of the "freq" table with the help of the grammatical code symbol (see Fig. 1). The "classes" table is very small (22 lines, 2 sections), and is built up manually. Its function is to create a drop-down list in the frequency dictionary interface, so the user could be able to search for morphemes and morphemic word structures in the lexical samples of specific parts of speech.

6 Quantitative Evaluation of the Quality of the Automatic Morphemic Analysis

The quality of the automatic morphemic analysis in the Corpuscan be estimated once the quantity of the initial forms (lemmas) of a sample text with a defined morphemic structure has been determined. This information is arranged in the "morfem" section of the "freq" table (Sect 5.2): in this section a pattern of morphemic word structure (e.g. for the lemma *ревти*RF (see Table 4) is emerging from comparison of the initial form (lemma) of the "freq" table with the initial form in the DB of morphemic structures according to the data of ASMDADB of morphemic structures (Sect. 2). The patternof morphemic word structure in the "morfem" section is not defined for all the words of the register, e.g. such words of T. Shevchenko's sample text as *ревти-завивати, регот, реєстер, решотка, Ржавиця, рибалонька, ридати-молитися* etc. do not have a pattern of morphemic structure. Lack of a pattern of morphemic word structure is caused by the fact that the initial word forms (lemmas) register of each sample text is automatically generated based on text tokens, and automatic morphemic analysis is performed for only those words that are included into ASMDADB of morphemic structures (compiled on the basis of standard dictionaries of Ukrainian language). There are words in the lexical register of sample texts that have not been documented in

Table 9. Quantitative descriptions of processed/unprocessed lexicon of the four sample texts.

Sample text	Number of tokens in a sample text	Number of units in the initial forms (lemmas) register of a sample text	Lexicon with defined morphemic word structure		Lexicon with undefined morphemic word structure	
			Number of lemmas	%	Number of lemmas	%
T. Shevchenko	68295	7526	6191	82,26	1335	17,74
L. Kostenko	9977	6347	5956	93,84	391	6,16
Lesya Ukrainka	10916	5146	4908	95,38	238	4,62
V. Stus	15897	9733	8029	82,49	1704	17,51

standard dictionaries and have not been included into ASMDADB of morphemic structures that is why a part of register lexicon remains unprocessed. According to the Table 9, the quantity of processed/unprocessed lexicon in the four sample texts may vary, and it defines a different system efficiency of the automatic morphemic analysis in the Corpus. The processed lexicon rate in the sample texts of L. Kostenko and Lesya Ukrainkais $\approx$ 94–95%, and in the sample texts of T. Shevchenko and V. Stus it is $\approx$ 82%.

The chosen filter of the "morphem" section of the "freq" table automatically compiles a register of unprocessed lexicon which contains the words that have not been documented in the dictionaries of Ukrainian language, and are either system errors or stylistically marked vocabulary (archaisms, dialect words, neologisms, nonce words, potential words, regular forms (diminutives, augmentatives, degrees of comparison, participles, transgressives etc.), proper nouns, abbreviations), or not included into the register of the DB of morphemic structures for other reasons.

Usually unprocessed words with absolute frequency(f) $\geq$ 10 are programming errors that require correction and updating of ASMDADB, but most of this sample constitute low-frequency stylistically marked words (for example, in the unprocessed lexicon of T. Shevchenko's sample text: f 1–1024 words; f 2 – 182 words; f 3–79 words). A register of such words is a very important material for editing ASMDADB and studying the peculiarities of style or idiostyle, and it needs a compulsory and deep linguistic research. Thus, the methodology of FDM compilation on the basis of the Corpus of Ukrainian language allows to automatically generate lexical registers which constitute a hypothetical sample of stylistically marked words.

7 Conclusions

The model of logically structured DB of frequency dictionaries of morphemes was tested on the basis of 20 sample texts. It can be used as a reference model in the tasks of computer statistical lexicography of different languages. This model's priority is a methodological approach to calculating the statistical characteristics of morphemic units based on morphemic structure of initial forms of text tokens. The section of "Frequency Dictionaries" of the Corpus [30] represents electronic FDMs with the human and machine interface created in the form of a ASP.Net web-application with SQL queries to the tables of lexicographical DB with the user's forecast requests taken into account. This type of electronic FDM can be compiled for all the authors and texts within the Corpus.

Computer lexicographical system of frequency dictionaries of morphemes in the Corpus of Ukrainian language provides new opportunities in research of morphemic organization of Ukrainian word in a dictionary and in a text. Automatic structurization of different morphemic units and morph-structures according to interactive classification options of frequency dictionaries allows to analyse the inclusion of morphemes and morphemic word patterns into word formation. It also allows to research morphemic length and intensity of words in Ukrainian texts of different functional styles, as well as morphotactics of different morpheme types, and to study statistical parameters of morphemic level organization of Ukrainian texts belonging to different functional styles.

The experience of carrying out the lexicographical computer modelling of Ukrainian morphemic system demonstrates that it is not obligatory to create morphemic or derivational annotation of texts for the purpose of retrieval of relational and functional characteristics of morphemic units from the text. The methodology for automatic morphemic analysis of text lexemes (their initial forms) used in lexicographical computer modelling does not reduce efficiency and immediacy of classification and search options of the created text-oriented system for morphemic text analysis. This methodology also does not minimize the importance of linguistic research results based on the material provided; on the contrary, it increases an explanatory value of linguistic research through structurization and different classification aspects of morphemic information. The approach adopted is based on ontological principle of morphemic word structure of inflected languages. The morphemic structure of the word stem representing the explicit word semantics remains relatively stable during inflection. The quantity of morphemes does not change, only inflectional allomorphy may appear, but it is taken into account during lemmatization and, if necessary, may be automatically ascribed to morphemes as a potential feature.

References

1. Клименко, Н., Карпіловська, Є., Комарова, Л.: Морфемно-словотвірний фонд української мови як дослідницька та інформаційно-довідкова система. In: Карпіловська, Є., Пономарів, О., Савенко, А. (eds.). Клименко Н.Ф. Вибрані праці, pp. 545–558. Видавничий дім Дмитра Бураго, Київ (2014)
2. Zuban, O.: Automatic morphemic analysis in the corpus of the ukrainian language: results and prospects. Jazykovedný časopis/J. Linguist. **68**(2), 415–426 (2017)
3. Дарчук, Н.: Комп’ютерне анотування українського тексту: результати і перспективи. Освіта України, Київ (2013)
4. Оливериус, З.: Морфемы русского языка: частотный словарь. Univerzita Karlova, Praga (1976)
5. Уорт, Д.: Русский словообразовательный словарь. Введение. Новое в зарубежной лингвистике, 14, pp. 227–260. Прогресс, Москва (1983)
6. Slavíčková, E.: Retrográdní morfematický slovník češtiny. EAV, Praha (1975)
7. Клименко, Н., Карпіловська, Є., Карпіловський, В., Недозим, Т.: Словник афіксальних морфем української мови. Київ (1998)
8. Карпіловська, Є.: Кореневий гніздовий словник української мови. Наукова думка, Київ (2002)
9. Корпус української мови. http://www.mova.info/corpus.aspx. Accessed 10 July 2019
10. Национальный корпус русского языка. http://www.ruscorpora.ru. Accessed 10 July 2019
11. Компьютерный корпус текстов русских газет конца XX-ого века. http://www.philol.msu.ru/~lex/corpus/corp_descr.html. Accessed 05 Feb 2019
12. Кукушкина, О., Поликарпов, А., Пирятинская, Е.: Полистилевой корпус текстов современного русского языка. Humlang (2006). http://www.philol.msu.ru/~humlang/articles/polystylcorp.html. Accessed 05 Feb 2019
13. Карпіловська, Є.: Вступ до прикладної лінгвістики: комп’ютерна лінгвістика. ТОВ «Юго-Восток, Лтд» , Донецьк (2006)

14. Кукушкина, О., Поликарпов, А., Токтонов, А.: Анализ системных характеристик словообразовательного процесса (На основе анализа новых лексических единиц газетного материала «Полистилевого корпуса современного русского языка»). Humlang (2006). http://www.philol.msu.ru/~humlang/articles/polystylcorp.html. Accessed 05 Feb 2019
15. Ляшевская, О.: Корпусные инструменты в грамматических исследованиях русского языка. Издательский Дом ЯСК: Рукописные памятники Древней Руси, Москва (2016)
16. Кислюк, Л.: Сучасна українська словотвірна номінація: ресурси та тенденції розвитку. Видавничий Дім Дмитра Бураго, Київ (2017)
17. Поликарпов, А., Богданов, В., Крюкова, О.: Хронологический морфемно-словообразовательный словарь русского языка: создание базы данных и ее системно-квантитативный анализ. Вопросы общего, сравнительно-исторического, сопоставительного языкознания, 32, pp. 172–184 (1998)
18. Поликарпов, А., Поддубный, В., Кукушкина, О., Кубарев, А., Варламов, А., Суровцева, Е., Пирятинская, Е.: Комплексная тексто-аналитическая система «СтилеАнализатор-2», основанная на Web-технологиях: разработка, наполнение данными и тестирование на прикладных задачах. http://www.philol.msu.ru/~lex/pdfs/polikarpov_poddubny_kukushkina_etc_stileanalizator_rffi2011-2013.pdf. Accessed 05 Feb 2019
19. Зубань, О.: Особливості морфемної будови слів у поетичних текстах Т. Шевченка (на матеріалі Корпусу української мови). Українське мовознавство, 44(1), pp. 123–133 (2014)
20. Зубань, О.: Стилеметричні ознаки морфемних структур слів у поетичному мовленні Т. Шевченка (на матеріалі Корпусу української мови). Мовні і концептуальні картини світу, 48, pp. 165–179 (2014)
21. Татевосов, С.: Множественная префиксация и анатомия русского глагола. In: Киселева, К., Плунгян, В., Рахилина, Е., Татевосов, С. (eds.) Корпусные исследования по русской грамматике, pp. 92–156. Пробел-2000, Москва (2009)
22. Пазельская, А.: Модели деривации отглагольных существительных: взгляд из корпуса. In: Киселева, К., Плунгян, В., Рахилина, Е., Татевосов, С. (eds.) Корпусные исследования по русской грамматике, pp. 65–91. Пробел-2000, Москва (2009)
23. Пацкин, А.: Гиперсловари на базе системы "Абриаль". Компьютерная лингвистика и интелектуальные технологии: метриалы ежегод. Международной конференции "Диалог – 2002". http://www.dialog-21.ru/digest/2002/. Accessed 05 Feb 2019
24. Пацкин, А.: Опыт построения полной морфемно-ориентированной семантической сети для русского языка. Компьютерная лингвистика и интелектуальные технологии: метриалы ежегод. Международной конференции "Диалог – 2004". http://www.dialog-21.ru/digest/2004/. Accessed 05 Feb 2019
25. Зализняк, А.: Грамматический словарь русского языка. Русский язык, Москва (1977)
26. Кузнецова, А., Ефремова, Т.: Словарь морфем русского языка. Русский язык, Москва (1986)
27. Зубань, О.: Частотні морфемні словники в Корпусі української мови - джерело стилеметричних досліджень. Acta Universitatis Palackianae Olomucensis Philologica 104 - 2016: UCRAINICA VII: Současná ukrajinistika Problémy jazyka, literatury a kultury, pp. 224–231 (2016)
28. Дарчук, Н.: Комп'ютерна лінгвістика (автоматичне опрацювання тексту): підручник. ВПЦ "Київський університет", Київ (2008)
29. Перебийніс, В., Сорокін, В.: Традиційна та комп'ютерна лексикографія: навч. посібник. Вид. центр КНЛУ, Київ (2009)
30. Частотні словники Корпусу. http://www.mova.info/article.aspx?l1=210&DID=5215. Accessed 05 Feb 2019

31. Зубань, О.: Електронні частотні морфемні словники в Корпусі української мови. Науковий вісник Східноєвропейського національного університету імені Лесі Українки, Серія: Філологічні науки, 3(304), pp. 315–320 (2015)
32. Частотний словник мови Т. Шевченка – "Твори в п'яти томах". http://www.mova.info/cfqsh_2.aspx. Accessed 05 Feb 2019

Improving Keyphrase Extraction Using LL-Ranking

Svetlana Popova[1,2](✉), Vera Danilova[3], Mikhail Alexandrov[3,4], and John Cardiff[1]

[1] Technological University Dublin, Dublin, Ireland
svp@list.ru, john.cardiff@it-tallaght.ie
[2] St. Petersburg State University, St. Petersburg, Russia
[3] RANEPA, Moscow, Russia
maolve@gmail.com, malexandrov@mail.ru
[4] Autonomous University of Barcelona, Barcelona, Spain

Abstract. Keyphrases provide a concise representation of the main content of a document and can be effectively used within information retrieval systems. In the paper, we deal with the keyphrase extraction problem when a given number of keyphrases for a text should be extracted. The research is focused on the keyphrase candidates ranking stage. In the domain, the question remains open of whether the keyphrase extraction quality can be improved by putting limits on the number of phrases of different lengths extracted during candidate ranking. We assume that the quality of resulting keyphrases can be enhanced if we introduce Limitations on the number of phrases of specific Lengths in the resulting set (LL-ranking strategy). The experiments are performed on the well-known INSPEC dataset of scientific abstracts. The obtained results show that the proposed limitations help to significantly increase the quality of extracted keyphrases in terms of *Precision* and $F1$.

Keywords: Keyphrase extraction · Keyphrase candidates ranking · Length feature in keyphrase extraction problem

1 Introduction

The astronomic growth in Internet data traffic pushes the demand for more and more efficient massive data mining and sorting algorithms. Data structuring and enrichment with metadata (e.g. keywords, keyphrases, domain terminology, etc.) allow to simplify the information retrieval process. We focus on the keyphrase extraction problem in our research. Following [1] we define a keyphrase list as "a short list of phrases (typically five to fifteen noun phrases) that capture the main topics discussed in a given document".

Automatic keyphrase extraction is a subtask of automatic keyphrase generation [1]. The generated phrases may not be present in the body of a given document, while keyphrase extraction implies that the extracted phrases appear verbatim in this document. We will distinguish between keyword and keyphrase in the following way: keywords are single words and keyphrases can include from one to three or more words.

N. Shakhovska and M. O. Medykovskyy (Eds.): CCSIT 2019, AISC 1080, pp. 567–578, 2020.
https://doi.org/10.1007/978-3-030-33695-0_38

Several of the most closely related fields are automatic index generation and information extraction [1]. As observed in [1], keyphrase extraction is a more demanding task than automatic index generation. For example, a keyphrase list should be relatively short and contain only the most important and topic-specific phrases for a given document. An index can involve phrases that are less important and less topic-related, and usually is relatively long [1]. Keyphrases can be used in indexing, while having many other applications. Keyphrases must be human-readable, and the index does not have this requirement.

Information extraction is also distinct from keyphrase extraction [1], because it focuses on the extraction of specific types of task-dependent information (e.g. named entities, events, time, location, etc.). The keyphrase extraction goal is to produce topic-specific phrases that may not represent a specific type.

We will establish a distinction between keyphrase extraction, collocation extraction and domain terminology extraction. Collocations can be broadly defined as sequences of lexical items in text that occur more often than it would be expected by chance. Keyphrases can occur with any frequency in a corpus. Domain terms refer to terms describing concepts that are widely used in the domain corpus. Keyphrase extraction produces topic-specific phrases that may not be widely used in the domain corpus. The extracted keyphrases may contain both collocations and domain terminology, but are not limited to them.

Thus, keyphrase extraction problem can be defined as an independent area that is similar to some extent but not the same as the mentioned research domains. Within keyphrase extraction, we concentrate our attention on short scientific texts. These texts are considered due to their abundance in electronic libraries (e.g. abstracts for scientific papers or projects).

Keyphrases can be used to improve features in different applied tasks, in particular, those of information search [2–5], document clustering/categorization [6–8], opinion mining [9] and text summarization [10,11].

The goal of the work is to improve the quality of keyphrase extraction from short texts. We study the possibility of enhancing the results by setting limits on the number of phrases of different lengths produced by keyphrase extraction algorithms. To understand what limits should be set, a training collection is needed. The limits will be different for different datasets. In this research, we use the INSPEC [12] collection.

The rest of this paper is organized as follows. In Sect. 2, the overview of the related works is done. As keyphrase length in words is one of the main features in our research, we focus on the works that involve this feature. In Sect. 3, we describe the methodology of our research: the evaluation strategy, the dataset and proposed algorithm to keyphrase extraction. In Sect. 4, the experiments are described and the results are presented and discussed. Section 5 contains conclusions and ideas for future work.

2 Related Work

2.1 Existing Methods

The existing approaches to keyphrase extraction can be divided into two types:

- KE-1: identification of terms that potentially form a phrase (word ranking/classification) and the subsequent phrase building;
- KE-2: preliminary extraction of candidate keyphrases, candidate ranking/classification and final selection of keyphrases.

KE-1 approach in the first stage mainly employs graph models for word weighting and ranking (e.g. [13,14]) or neural networks and conditional random fields, which produces a decision for each word (e.g., either this word is a part of a keyphrase or is located at the beginning, at the end or outside of it) as in [15–17]. The next step is keyphrase construction on the basis of the results of the previous stage. The length feature is frequently related to the general limitation on the maximum length of phrases (e.g., often the maximum length is set to 4–5 words).

In the KE-2 approach, the following techniques have been proposed for candidate phrase selection: extraction of noun phrases or sequences according to predefined patterns [12,18]; n-gram extraction from text or its parts, as in [12,19,20]; subsequences with the highest frequency of occurrence in the collection [3,4], and their combinations. There are important features that are taken into account, among them the phrase length (the other features include the position of the first occurrence of a given phrase with respect to the document start point, the *tf* (term frequency) and *idf* (inverse document frequency) of the phrase, the sequences of word prefixes and suffixes in the phrase, word embeddings, the nearest context enriched with part-of-speech tags, etc.).

2.2 Length Feature

One of the popular tasks is to extract a given number of keyphrases for a text. In this case, the keyphrases identified applying the KE-1 approach (or the KE-2 approach using classification) undergo additional ranking and a required number of phrases is selected from the obtained set. All candidate phrases are usually ranked in the same way. Next, a predefined number of keyphrases are picked that received the best score during ranking. The final results can represent a mixture of keyphrases of various lengths in arbitrary proportions.

We assumed that the quality of extracted keyphrases can be increased if we split the candidates into groups according to their lengths and select a fixed number of phrases from each of the groups. Commonly, the length is used as a feature at the ranking/classification stage. It means that the length parameter affects the results of weighing and the final decision of the classifier (whether a given word sequence is a keyphrase or not). Despite the fact that the phrase length feature is quite popular in the field, to our knowledge, there are no works mentioning the results of its application directly to the ranking procedure to define the number of phrases in each length group that should form the resulting set.

The authors of [19] emphasize the use of "the length of the term candidate, i.e. its number of words" as one of the features in the evaluation of the candidates. To select the best keyphrases from the candidate set, the experiments with different algorithms are conducted: Decision tree (C4.5), Multi-Layer Perceptron (MLP), Support Vector Machine (SVM) and a combination of these models with boosting and bagging techniques. The resulting candidates are n-grams of up to 5 words with a set of constraints.

The authors of [21] employ an inferred document logical structure to identify the candidates. The extraction is performed using regular expressions. A Naive Bayes model is used for candidate phrase selection. Among the enumerated features is the phrase length in words. The authors experiment with various combinations of features to find the best ones. The resulting best combinations include the feature "keyphrase length".

The authors of [20] point out and explain why the use of *tf-idf* in keyphrase extraction falls short. It is observed that phrases occur much more seldom than single terms within the same document, which gives a bias towards single words. Taking this into account, the authors introduce a boosting factor for compound terms into the *tf-idf* calculation to reduce the bias. Moreover, document frequency depends heavily on keyphrase length and it varies a lot for one-word and compound phrases. For this reason, the authors of [20] propose either to use a large corpus for frequency calculation or to consider each compound phrase as occurring in only one document. For one-word phrases, *idf* is calculated in a regular way. Hence, in [20], the *tf-idf* based ranking (+ position associated factor) takes into account the length of phrases through the use of the boosting factor and making the difference between the ways of *idf* estimation for one-word and multiword phrases.

Keyphrase length is also considered as a parameter in [22]. It is observed that longer n-grams tend to be keyphrases. Consequently, the weight of the term t is boosted by its term length.

In [23], a supervised approach is presented and an emphasis is put on the quality of the training sets annotation. For keyphrase generation, the authors employ gradient boosted decision trees, as well as a rich feature set covering statistical, structural and semantic properties. Keyphrase length is also part of this set.

In [24], the Maui algorithm is presented that is based on the Kea system [2] (one of the pioneer systems in keyphrase extraction). Maui is a tool for automatic tagging on the web similar to keyphrase extraction systems. A machine-learning model is trained to perform candidate selection. The phrase length in words is part of the feature set.

In [15], a combination of part-of-speech pattern matching and heuristic rules is used for candidate generation where the length of keyphrases is limited to up to 5 words. A similar restriction is imposed in [25] and a number of other papers. This use of keyphrase length feature is the most widely spread in the field.

Hence, researchers recognize the importance of the length parameter. In spite of this, it has not been investigated whether the use of constraints on the maximum number of phrases of each length in the resulting keyphrase set impacts the quality. Our present study is an attempt to fill this gap. The use of constraints can potentially contribute to the improvement of the existing keyphrase extraction algorithms. In the paper, we prove experimentally that the achieved results are promising.

3 Method

3.1 Evaluation

As in most works, we use micro-average $F1$ to evaluate the quality of extracted keyphrases. *Precision* is used both as an independent quality evaluation metric and for the $F1$ calculation because we specify the value of k: the number of keyphrases that should be extracted. In this case for the same k, the *Recall* either coincides or shows the same behaviour as *Precision*. For this reason, the *Precision* metric is paid very close attention.

To calculate the *Recall* we take the number of keyphrases in the gold standard as it is, irrespective of k. It allows the comparison of the obtained $F1$ values with the results of other approaches including those where the ranking and selection of a fixed number of phrases are not required.

Notice that although the keyphrase extraction domain exists for a long time, the performance of the existing methods is still fairly poor on the commonly-used evaluation datasets [26]. The authors of the [26] analyzed the errors and observed that most errors belong to four major types, one of which is evaluation errors (the others are: overgeneration errors, infrequency errors and redundancy errors, more details can be found in [26]). An evaluation error happens when a candidate generated by the algorithm, although being semantically equal to the corresponding gold standard keyphrase, fails to be recognized by the scoring program as the semantic equivalent [26].

Thus, the quality evaluation in the keyphrase extraction domain remains an open issue. There were attempts to use such measures from the machine translation and automatic summarization domains as BLEU, METEOR, NIST and ROUGE [27]. Several alternative evaluation metrics were proposed in [27,28]. But these evaluation techniques were not further investigated and $F1$, *Precision* and *Recall* became commonly used in the domain, which allows to use them for the comparison of different keyphrase extraction approaches.

3.2 DataSet

We employ the following subsets of the INSPEC dataset of scientific abstracts [12–14,25,28,29]: INSPEC-trial (1000 documents) and INSPEC-test (500 documents). The dataset includes abstracts from journal papers from the following domains: Computers and Control, Information Technology. Each abstract has two sets of keyphrases: (1) a controlled set of terms assigned by a professional indexer (terms restricted to the INSPEC thesaurus); (2) a set of uncontrolled terms that can be any suitable terms [12]. Phrases from both types of sets may or may not appear in the abstracts. Following other works in the domain, we use the keyphrases from the uncontrolled set as the gold standard. These phrases are largely presented in the abstracts and phrases from the controlled set are not (76.2% as opposed to 18.1%) [12].

Table 1 contains several examples of the abstracts and keyphrases from the uncontrolled gold standard set. Table 2 provides a statistical information for the INSPEC-test.

Table 1. Examples of texts and keyphrases of the "gold standard"

Title and Abstract	Phrases of the "gold standard" (manually assigned phrases)
E-government. The author provides an introduction to the main issues surrounding E-government modernisation and electronic delivery of all public services by 2005. The author makes it clear that E-government is about transformation, not computers and hints at the special legal issues which may arise	E-government, modernisation, electronic delivery, public services, legal issues
Accelerated simulation of the steady-state availability of non-Markovian systems. A general accelerated simulation method for evaluation of the steady-state availability of non-Markovian systems is proposed. It is applied to the investigation of a class of systems with repair. Numerical examples are given	General accelerated simulation method, numerical examples, non-markovian systems, accelerated simulation, steady-state availability

Table 2. Statistical information for the INSPEC-test

	Avg.	Min.	Max.
Document length in words	122	23	338
The number of keyphrases in the uncontrolled set	10	2	31

Notice that one part of the researchers use the full version of the INSPEC gold standard (e.g. [13]), while the other part prefers the short one (e.g. [12]). In the latter, the phrases that do not occur in the abstracts are excluded. Actually, the choice of the gold standard version affects *Recall* and does not influence *Precision*. In our work, we focus on *Precision*. For this reason in our experiments, any version of the gold standard applies, and we use the short one.

In the working version of the dataset, we placed periods between the titles and abstracts.

3.3 Motivation and the Algorithm

Our hypothesis is that by setting explicit constraints on the number of phrases of each length that can be allowed in the resulting keyphrase set we can improve the quality of the obtained phrases. Let:

- the alg_{extr} algorithm that performs the extraction of keyphrase candidates;
- the alg_{rank} algorithm that ranks the candidates to select a prespecified number of phrases;
- the output of alg_{extr} on the training collection T_{train}: for each text, the information on the candidates, their lengths and information whether these candidates are keyphrases or not;
- a text t from the T_{test} collection for which alg_{extr} extracted m candidates.

The task for alg_{rank} is: for text t, select k ($k < (m+1)$) phrases that will form the resulting keyphrase set.

We propose a solution as follows. Introduce the further described constraints into the ranking algorithm alg_{rank}. For text t, pick up to k_1 phrases of length 1 out of the candidate set, up to k_2 phrases of length 2,..., up to k_g phrases of length g, so that $k = k_1 + k_2 + ... + k_g$. To determine the values of k_1, k_2, $k_3, ..., k_g$, the following approach is proposed. It processes the T_{train} collection where, for each text, there is a gold standard annotation - keyphrases selected manually by the experts. Let us consider a keyphrase as a true positive if it was automatically extracted from a given text by the alg_{extr} algorithm and it forms part of the gold standard for the same text.

Hence, the alg_{extr} algorithm calculates the sum of keyphrases of all lengths that were extracted as true positives. For each length i, it counts the total number of times a phrase of length i was extracted as true positive. Next, the ratio of the phrases of each length i to the total number of the extracted true positives is estimated. The resulting percentages define the values of k_1, k_2, $k_3, ..., k_g$ depending on the k total number of keyphrases that we need to retrieve.

Our assumption is that the training set processing gives an idea of the lengths of the keyphrases extracted using alg_{extr} and the accuracy for each length. We estimate the percentages of phrases of each length among all the phrases extracted as true positives. We assume that the proportions will be the same for texts from the T_{test} collection. They are also supposed to remain the same if we need to select k phrases for each text in the test collection. Consequently, we claim that it makes sense to take these proportions into account in the alg_{rank} ranking algorithms to increase the quality of keyphrase extraction. We implemented this idea in LenghLimit-ranking strategy: LL-ranking (the approach to determine the values of k_i).

Algorithm 1 The approach to determine the values of k_i

t - a text from the T_{train} collection
$Mcor_{ki,t}$ - number of true-positive candidates of length i in t extracted by alg_{extr}
k - number of phrases that should be extracted
k_1, k_2, k_3,...,k_g=0 - number of phrases of length 1,2,...,g that should be found
$Mcor_{all}$=0
c_1, c_2, ..., c_g=0
p_1, p_2,..., p_g=0

Start:
for $(0 < i < (g + 1))$ **do:** $c_i = \sum_{t \in T_{train}} (Mcor_{ki,t})$
end for
$Mcor_{all} = \sum_{i,0<i<(g+1)} (c_i)$
for $(0 < i < (g + 1))$ **do:** $p_i = (c_i * 100)/Mcor_{all}$; $ki = (k/100) * p_i$
end for
return k_1, k_2, k_3,...,k_g
end.

4 Experiments

4.1 Description of the Experiments

We set up the following experiment:

1. Training step. INSPEC-trial is used as the training T_{train} collection to find the values of $p_1, p_2, ..., p_g$ (in Algorithm 1) that are needed to calculate k_1, $k_2, ..., k_g$.
2. *Main experiment.* INSPEC-test is used as the T_{test} collection. The candidates are extracted from its texts using alg_{extr}. Next, alg_{rank} selects k keyphrases ($k = 5$, $k = 10$, $k = 15$) where there are no more than k_1 phrases of length 1, no more than k_2 phrases of length 2, etc., k_1, k_2, ..., k_g are calculated for each fixed value of k by using $p_1, p_2, ..., p_g$.
3. *Comparison experiment.* The alg_{extr} algorithm extracts the candidates for the INSPEC-test set. Nextly, alg_{rank} picks k keyphrases ($k = 5$, $k = 10$, $k = 15$). In this case, any number of phrases of any length can be selected.
4. *Baseline experiment.* For the INSPEC-test, the candidates to keyphrases are found using alg_{extr}. After this step, k keyphrases ($k = 5$, $k = 10$, $k = 15$) are selected randomly.

The details of the algorithms that are used for this case study as alg_{extr} and alg_{rank} are as follows.

Keyphrase Extraction Algorithm (alg_{extr}). Alg_{extr} extracts the candidates to keyphrases from a given text as continuous sequences of words that belong to the allowed parts of speech. The punctuation (except hyphen), stopwords and the words that have other than allowed part-of-speech tags function as separators between phrases. The allowed parts of speech are NN (noun, singular or mass), NNS (noun, plural), JJ (adjective).

Ranking (alg_{rank}). The candidate phrases are ranked in accordance with their weight: $v(ph_j) = \sum weight(w)$, where ph_j - phrase, w - words that compose the phrase ph_j, $weight$ - word weight: $weight(w) = tf(w)/df(w)$, where $tf(w)$ - the absolute frequency of occurrence of word w in the given collection, $df(w)$ - document frequency of word w in the same dataset.

In case the rounding of the values $k_1, k_2, ..., k_g$ results in $k < (k_1+k_2+...+k_g)$ or, for a certain text, the number of phrases of length i is less than k_i, the missing quantity is filled from the set of two-word candidates and, if it is still not sufficient, the missing keyphrases are taken from the three-word candidate set, etc.

4.2 Results of the Experiments

Table 3 shows the information for keyphrases that were extracted from INSPEC-trial using alg_{extr}. The following notation is adopted: *length* - the length of phrases in a given column; p - the value of p parameter; *correct* - the number of correctly extracted phrases of a given length. Hence, on the basis of INSPEC-trial we obtained the values of p that will be used in the subsequent step during the INSPEC-test processing.

Table 3. Information for keyphrases that were extracted from INSPEC-Trial

Length	1	2	3	4	5	6	7
Correct	359	1559	610	110	20	3	2
p	13%	59%	23%	4%	1%	0%	0%

Table 4 presents the results obtained on the INSPEC-test collection. The following notation is used: k - the number of keyphrases selected during the ranking process, *Main experiment*, *Comparison experiment*, *Baseline experiment* - as described in Sect. 4.1 (Description of the experiments) above.

The results provided in Table 4 show the following picture. The ranking algorithm used for this case study compares favourably to the random selection of the required number of phrases. Moreover, the fewer the number of phrases to

Table 4. The results obtained on the INSPEC-Test collection

Experiment:	Main experiment		Comparison experiment		Baseline experiment	
k	Precision	F1	Precision	F1	Precision	F1
5	**0.40**	**0.32**	0.28	0.21	0.21	0.17
10	**0.32**	**0.36**	0.26	0.29	0.21	0.23
15	**0.28**	**0.37**	0.24	0.32	0.20	0.26

select, the more the algorithm gains in performance. The introduction of constraints on the number of phrases of each length ensures an additional significant increase in performance quality (up to 0.12 in $Precision$ and up to 0.11 in $F1$). This improvement is considerable and it grows as the value of k decreases. The results show that the assumption of the work can contribute to the state of the art in the domain and further studies in this direction are justified.

4.3 Evaluation of the Results

It is of notice that the purpose of the work is to show experimentally that limiting the number of phrases of different lengths during ranking can significantly enhance the results at the ranking stage. In this work, we employ a simple ranking algorithm to demonstrate the efficiency of the proposed method. The results achieved herein outperform those of the base algorithms (e.g., in [12], the maximum $Precision$ is 29.7 and $F1$ is 0.34 for the short gold standard; in [13], $Precision$ is equal to 31.2 and $F1$ 0.36 for the full gold standard). Our algorithm is inferior to the algorithms that use more complex ranking mechanisms (e.g. in [25], $F1$ is 0.47), however, the proposed constraints can be easily adapted and used in other ranking algorithms.

5 Conclusions

The overview of previous works in keyphrase extraction gives us the following vision of the state of the art. In the domain of keyphrase extraction, there is a question that remains open: in what way do the constraints on the number of phrases of a certain length selected during candidate ranking influence the performance quality? The goal of this work is to conduct a pilot study to answer the above question and decide on whether to impose these constraints. The experiments show that the use of this new feature in the ranking step can significantly increase the performance quality and allows achieving very competitive results using simple keyphrase extraction algorithms.

In this work, we apply a basic method for keyphrase weighing at the ranking step as an example that shows the efficiency of setting limits on the number of selected phrases with the account of the length parameter. The findings reaffirm the importance of further research in this field. Since the results are promising, it makes sense to combine the proposed length constraints with more complex ranking keyphrase extraction algorithms. We consider this as a strategy for our future work.

Acknowledgments. The reported study was partially funded by RFBR (Russian Fund of Basic Research) according to the research projects No. 16-37-00430 mol_a (for Svetlana Popova) and No. 18-07-01441 a (for Mikhail Alexandrov).

References

1. Turney, P.: Learning algorithms for keyphrase extraction. Inf. Retr. **2**, 303–336 (2000)
2. Gutwin, C., Paynter, G., Witten, I., Nevill-Manning, C., Frank, E.: Improving browsing in digital libraries with keyphrase indexes. J. Decis. Support Syst. **27**, 81–104 (1999)
3. Bernardini, A., Carpineto, C.: Full-subtopic retrieval with keyphrase-based search results clustering. In: Web Intelligence and Intelligent Agent Technologies, WI-IAT 2009 (2009)
4. Zeng, H.-J., He, Q.-C.,Chen, Z., Ma, W.-Y., Ma, J.: Learning to cluster web search results. In: Proceeding SIGIR 2004, pp. 210–217 (2004)
5. Popova, S., et al.: Sci-search: academic search and analysis system based on keyphrases. In: The 4th Conference on Knowledge Engineering and Semantic Web, Russia. Communications in Computer and Information Science Series, vol. 394, pp. 281-288. Springer (2013)
6. Popova, S., Danilova, V.: Using extended stopwords lists to improve the quality of academic abstracts clustering. In: OTM 2016 Workshops. LNCS, vol. 10034. Springer, Cham (2017)
7. Hammouda, K., Matute, D., Kamel, M.: CorePhrase: keyphrase extraction for document clustering. In: MLDM 2005. LNCS, vol. 3587, pp. 265–274. Springer, Heidelberg (2005)
8. Hulth, A., Megyesi, B.: A study on automatically extracted keywords in text categorization. In: Proceedings of the Coling/ACL 2006, pp. 537–544 (2006)
9. Berend, G.: Opinion expression mining by exploiting keyphrase extraction. In: Proceedings of the 5th International Joint Conference on Natural Language Processing, pp. 1162–1170 (2011)
10. Jones, S., Lundy, S., Paynter, G.W.: Interactive document summarization using automatically extracted keyphrases. In: HICSS 2002, vol. 4 (2002)
11. DÁvanzo, E., Magnini, B.: A keyphrase-based approach to summarization: the LAKE system at DUC-2005. In: Proceedings of DUC 2005 (2005)
12. Hulth, A.: Improved automatic keyword extraction given more linguistic knowledge. In: Proceedings of the EMNLP 2003, pp. 216–223 (2003)
13. Mihalcea, R.: TextRank: bringing order into texts. In: Proceedings of EMNLP 2004, Barcelona, Spain, pp. 404–411 (2004)
14. Wan, X., Xiao, J.: Single document keyphrase extraction using neighborhood knowledge. In: Proceedings of the 23rd National Conference on Artificial Intelligence, pp. 855–860 (2008)
15. Wang, L., Li, S: PKU ICL at SemEval-2017 task 10: keyphrase extraction with model ensemble and external knowledge. In: Proceedings of SemEval-2017, pp. 934–937 (2017)
16. Marsi, E., Skidar, U., Marco, C., Barik, B., Saetre, R.: NTNU-1@ScienceIE at SemEval-2017 task 10: identifying and labelling keyphrases with conditional random fields. In: Proceedings of SemEval-2017, pp. 938–941 (2017)
17. Prasad, A., Kan, M.-Y.: WING-NUS at SemEval-2017 task 10: keyphrase identification and classification as joint sequence labeling. In: Proceedings of SemEval-2017, pp. 973–977 (2017)
18. Dung, T.N., Kan, M.-Y.: Keyphrase extraction in scientific publications. In: ICADL 2007 Proceedings of the 10th International Conference on Asian Digital Libraries: Looking Back 10 Years and Forging New Frontiers, pp. 317-326 (2007)

19. Lopez, P., Romary, L.: HUMB: automatic key term extraction from scientific articles in GROBID. In: SemEval-2010, ACL 2010, pp. 248–251 (2010)
20. El-Beltagy, S.R., Rafea, A.: KP-Miner: participation in SemEval-2. In: Proceedings SemEval-2010, ACL 2010, pp. 190–193 (2010)
21. Dung, T.N., Kan, M.-Y.: WINGNUS: keyphrase extraction utilizing document logical structure. In: Proceedings SemEval-2010, ACL 2010, pp. 166–169 (2010)
22. Danesh, S., Sumner, T., Martin, J.H.: SGRank: combining statistical and graphical methods to improve the state of the art in unsupervised keyphrase extraction. In: Proceedings of the Fourth Joint Conference on Lexical and Computational Semantics (*SEM 2015), pp. 117–126 (2015)
23. Sterckx, L., Caragea, C., Demeester, T., Develder, C.: Supervised keyphrase extraction as positive unlabeled learning. In: Proceedings EMNLP 2016, pp. 1924–1929 (2016)
24. Medelyan, O., Frank, E., Witten, I.H.: Human-competitive tagging using automatic keyphrase extraction. In: Proceeding EMNLP 2009, vol. 3, pp. 1318-1327 (2009)
25. Tsatsaronis, G., Varlamis, I., Norvag, K.: SemanticRank: ranking keywords and sentences using semantic graphs. In: Proceeding of the Coling 2010, pp. 1074–1082 (2010)
26. Hasan, K., Ng, V.: Automatic keyphrase extraction: a survey of the state of the art. In: 52nd Annual Meeting of the Association for Computational Linguistics, ACL 2014 - Proceedings of the Conference, pp. 1262–1273 (2014)
27. Kim, S.N., Baldwin, T., Kan. M.-Y.: Evaluating N-gram based evaluation metrics for automatic keyphrase extraction. In: Proceedings of the 23rd International Conference on Computational Linguistics (Coling 2010), Beijing, pp. 572–580 (2010)
28. Zesch, T., Gurevych, I.: Approximate matching for evaluating keyphrase extraction. In: International Conference RANLP 2009, Borovets, Bulgaria, pp. 484–489 (2009)
29. Liu, Z., Li, P., Zheng, Y., Sun, M.: Clustering to find exemplar terms for keyphrase extraction. In: Proceedings EMNLP 2009, pp. 257–266 (2009)

Statistical Models for Authorship Attribution

Iryna Khomytska and Vasyl Teslyuk(✉)

Lviv Polytechnic National University, Lviv 79013, Ukraine
Iryna.khomytska@ukr.net, vasyl.m.teslyuk@lpnu.ua

Abstract. A combination of three hypothesis methods has been proposed for authorship attribution. These are: the Student's t-test, the Kolmogorov-Smirnov's test and the chi-square test. This combination of methods can perfect adequacy of authorship attribution. With the help of these methods, the quantity of phoneme groups is minimized. The texts can be differentiated by one or three phoneme groups. The minimization of phoneme groups facilitates authorship attribution. On the basis of the results obtained, statistical models have been built. The models show author-differentiating capability of phoneme groups in authorship attribution. This way the highest author-differentiating capability of phoneme groups is determined. New software realizes the models applied. The program makes transforming a text into transcription faster. The transcription variants of words available in the program are used.

Keywords: Statistical methods · Minimization of phoneme groups · Author-differentiating capability · Authorship attribution

1 Introduction

The authorial style has its definitive distinctions by which authorship attribution is done. The problem has been researched since B.C. time. In every research certain language level and differentiating parameters were chosen to facilitate authorship attribution and make it more efficient. Thus on the phonological level, the phonemes with the highest author-differentiating capability were determined. On the lexical level, a list of the most frequently used words was compiled from the author's works [1–6]. On the syntactical level, the most frequently used syntactical structures were considered author-differentiating [7, 8]. The difficulties the researchers always faced were connected with interpenetration of language units from one authorial style into another one [9–12]. As each authorial style on the lexical and syntactical levels can be considered an open system, in which the number of elements is not stable, the authorship attribution cannot be done with utmost fidelity. In order to avoid instability of elements, it is reasonable to choose the phonological level, on which the number of elements is stable and unchangeable [13, 14]. In our research texts by different authors are differentiated by phonostatistical structures – distribution of phoneme group (PG) frequencies in each text.

The goal of our investigation is to minimize the number of PG by which authorship attribution is done. To solve this problem, the methods, models and software must be developed. As latest publications [15–17] show, a combination of the most efficient statistical methods in authorship attribution must be proposed.

N. Shakhovska and M. O. Medykovskyy (Eds.): CCSIT 2019, AISC 1080, pp. 579–592, 2020.
https://doi.org/10.1007/978-3-030-33695-0_39

2 Mathematical Support of Software System

2.1 The Method Developed

The developed complex method (CM) is an offered combination of the hypothesis methods: the Kolmogorov-Smirnov's test (TKS) [18–20], the chi-square test (TC) and the Student's t-test (TS). This combination of statistical methods has proved to be efficient in differentiating texts by different authors [21–23]. The algorithm of the method CM is as follows:

1. two texts of equal size are chosen;
2. the texts get transcribed;
3. two samples with 51000 consonant phonemes in each are formed;
4. each sample is divided into 51 portions;
5. calculation of phonemes in a sample is done;
6. consonant phoneme mean frequencies $\bar{X}$ are calculated;
7. consonant group mean frequency $\bar{x}_r^\alpha$ is chosen as a differentiation criterion;
8. the Pearson's test is used to prove normal distribution of phoneme group frequencies:

$$f(X) = \frac{1}{\sigma\sqrt{2\pi}} \cdot e^{-\frac{(X-\mu)^2}{2\sigma^2}}, \tag{1}$$

   where $f(x)$ is normal distribution density, μ is a general population mean frequency and σ^2 is a general population variance [24–27];
9. the TS is used for texts differentiation: the difference $\bar{x}_1^\alpha - \bar{x}_2^\alpha$ is estimated for the level of significance 0.05, and the difference $\bar{x}_1^\alpha - \bar{x}_2^\alpha$ is essential if 2Q < 0.05 (2Q is a double-sided significance criterion);
10. the TKS is used for texts differentiation [28, 29]:

$$D_{n,m} = \sup_{-\infty<z<\infty} |F_n(z) - F_m(z)|, \tag{2}$$

   where F_n(z) and F_m(z) are empirical functions of distribution for the two samples (n and m). The texts are differentiated if $\lambda_{n,m} \geq \lambda_\alpha$:

$$\lambda_{n,m} = \sqrt{\frac{nm}{n+m}} D_{n,m} = \sqrt{\frac{nm}{n+m}} \sup_{-\infty<z<\infty} |F_n(z) - F_m(z)|. \tag{3}$$

11. the TC is used for texts differentiation [28–30]:

$$\hat{\chi}_n^2 = \sum_{i=1}^{s} \sum_{j=1}^{k} \frac{\left(\upsilon_{i,j} - \frac{n_j \upsilon_j}{n}\right)^2}{\frac{n_j \upsilon_j}{n}}, \quad \upsilon_j = \sum_{j=1}^{k} \upsilon_{ij}, \tag{4}$$

where $\upsilon_{i,j}$ – a number of *i*-results in *j*-series, s – a number of phoneme groups, k – a number of texts, n_j – a portion quantity in a sample, n – a portions number in two samples. The texts are differentiated if $\hat{\chi}_n^2 \geq \chi_{1-\alpha,(s-1)(k-1)}^2$;

12. the results obtained by the three tests are compared;
13. the author-differentiating capability of phoneme groups is determined.

2.2 The Models Developed

The developed model determines the author-differentiating capability of phoneme groups (APC). The model realizes the statistical methods: the TKS, the TC and the TS. The author-differentiating capability of phoneme groups is got by the expression:

$$APC = \frac{t + \lambda_{n,m} + \chi_n^2}{3} \quad (5)$$

Another developed model determines the highest author-differentiating capability: $n \times sc = const$, where n – a number of essential differences established for the texts compared; sc – author-differentiating capability of certain phoneme groups. If maximum number of essential differences is 10, it takes number 1. In this case the following expression is true: 10 × 1 = const.

2.3 The Developed Software

The developed software is based on the developed methods and models [31–37]. The structure of the differentiation program consists of the following modules:

(1) module of data input/output [34];
(2) module of transforming a text into transcription;
(3) module of calculating the phoneme quantity in a sample;
(4) module of determining the PG average;
(5) module of determining the Pearson's test;
(6) module of determining the Student's t-test;
(7) module of determining the Kolmogorov-Smirnov's test;
(8) module of determining the chi-square test.

The transcription of a word is obtained from the transcription site and is preserved in the data structure HashMap. The transcribed text is saved in the data-base [34, 35]. The data base H2 is integrated into the framework Spring Boot. It is an open x-plat embedded data base on Java [36].

Figure 1 shows the class structure in which a sample is formed.

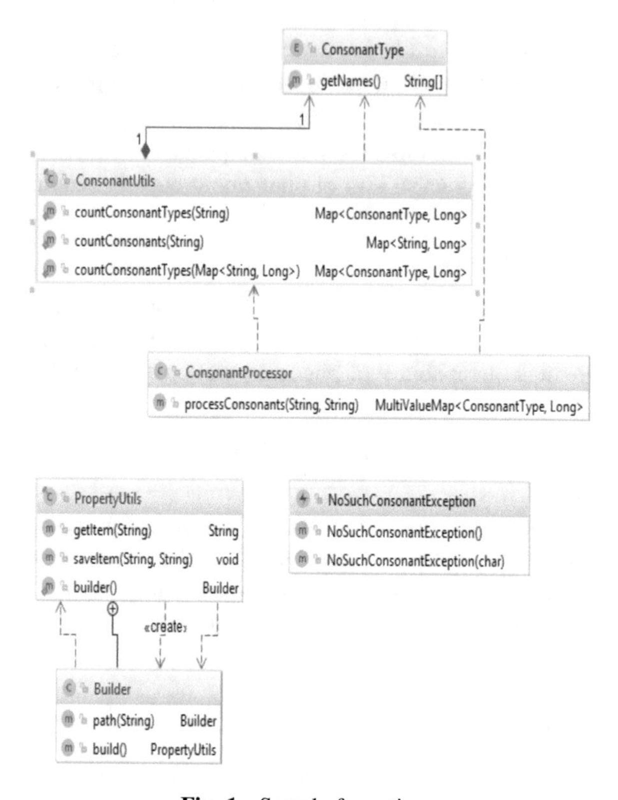

Fig. 1. Sample formation.

The program algorithm is shown in Fig. 2. The algorithm differs from the existing algorithms for authorship attribution [1, 9, 10] because the texts are differentiated on the phonological level. On this language level the researched texts must be transformed into their transcription symbols. Because of this, it is more complicated to write a program on the phonological level. But, as its structure is more strictly arranged, it is possible to enhance validity of authorship attribution. Besides, the applied methods are efficient on the phonological language level and can be less efficient on the other language levels. Consequently, the offered combination of statistical methods is not used in the existing software [11, 12].

The developed algorithm makes it possible to determine the author-differentiating capability of PG and differentiate the researched texts by the PG with the highest author-differentiating capability. The reduced number of PG makes authorship attribution more automated.

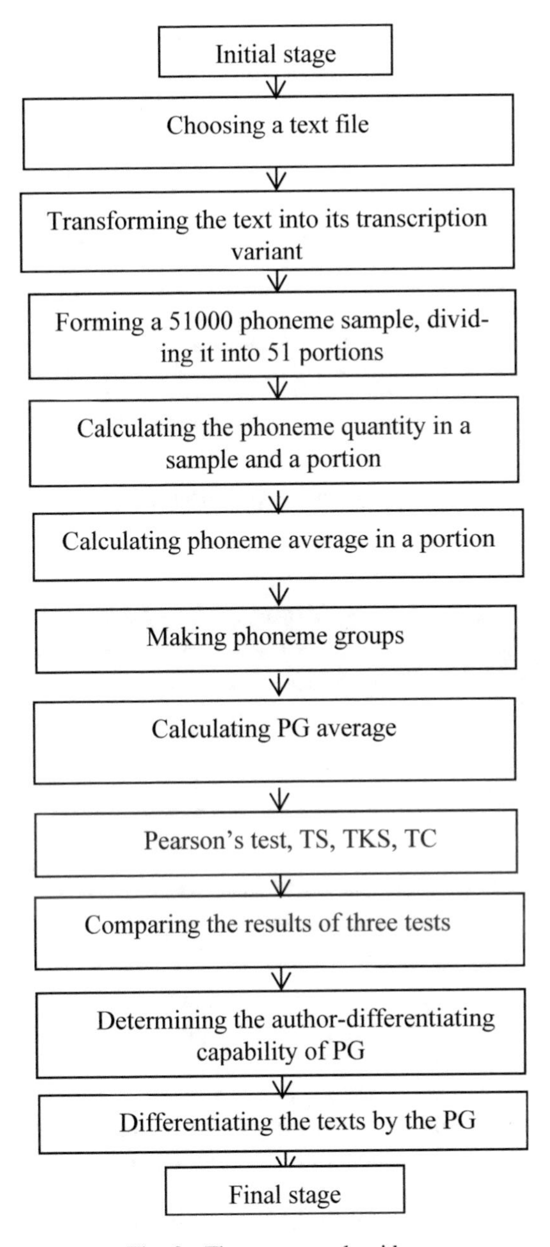

Fig. 2. The program algorithm.

3 Results of the Study

In our research four samples from the publicist style have been chosen for authorship attribution. These are public speeches by two US Presidents: Barack Obama and Donald Trump, and two newspaper articles by Susan Logan and David Webster.

The topic of the samples is similar. The four samples are compared in pairs to characterize the authorial style of each author. In the first comparison, the speeches by D. Trump and the newspaper article by D. Webster are analyzed (Figs. 3, 4, 5, 6 and 7).


```
"file1": "papers by David Webster.doc",
"file2": "PRESIDENT TRUMP.doc",
"sample1": {
    "STOP": "489,504,462,494,495,475,493,498,476,500,477,486,465,453,483,469,493,489,490,508,494,488,497,501,495,505,469,477,495,475,496,499,492,479,474,508,458,473,500,461,505,477,486,441,483,475,469,503,484,495,495",
    "FRICATIVE": "224,222,236,211,223,249,248,222,212,218,220,201,225,238,222,229,221,223,256,230,210,229,218,239,235,228,226,221,220,240,252,223,208,225,225,191,229,247,205,229,219,209,202,243,225,227,229,218,240,249,219",
    "LABIAL": "177,195,208,221,209,224,224,186,161,174,167,172,172,179,154,197,221,205,219,194,171,170,173,190,186,191,209,219,210,213,224,197,164,185,158,173,165,195,179,184,153,176,167,178,168,178,197,216,206,221,179",
    "DORSAL": "17,1,7,4,10,2,0,4,0,7,16,13,14,11,10,9,5,9,2,4,0,5,6,1,5,1,4,6,11,2,0,4,0,5,17,12,12,11,1,4,12,16,10,14,10,10,5,12,3,2,2",
    "SONOROUS": "199,207,218,214,209,207,212,211,201,219,214,218,226,218,212,228,216,215,202,210,204,209,203,163,186,201,220,216,219,208,207,211,202,215,224,203,222,212,209,212,203,212,224,223,220,216,224,220,204,207,213",
    "VELAR": "36,32,30,26,27,34,44,46,32,52,24,21,31,35,37,30,30,26,35,43,30,50,44,25,40,31,31,26,30,33,41,48,30,49,32,21,25,42,35,38,45,22,28,26,40,36,27,32,25,40,38",
    "NASAL": "205,173,170,187,182,158,171,152,154,180,182,166,153,155,150,171,184,179,161,170,160,170,176,190,207,177,169,182,181,162,167,160,159,166,174,186,148,158,163,153,185,176,157,144,164,152,174,184,170,179,154",
    "CORONAL": "629,656,626,615,639,603,621,634,639,660,658,638,621,624,651,643,619,645,623,648,652,644,652,627,628,661,625,615,638,609,627,624,656,625,673,642,630,615,643,621,674,631,644,602,657,632,646,642,628,627,650"
},
```


Fig. 3. The quantity of phonemes in sample 1.


```
},
"sample2": {
    "STOP": "468,491,471,498,504,508,499,463,465,487,471,472,459,491,485,479,514,477,517,500,456,470,491,460,492,451,455,480,499,471,494,509,499,505,473,447,485,436,453,440,487,417,465,474,454,488,493,474,510,488,506",
    "FRICATIVE": "252,211,254,223,205,207,220,265,255,244,262,251,240,235,231,237,218,211,205,219,277,242,248,266,207,250,241,250,218,253,222,208,202,214,253,262,239,241,231,253,238,240,216,254,244,244,221,244,218,212,209",
    "LABIAL": "166,177,159,162,143,183,165,176,193,210,211,168,186,175,167,141,161,159,182,162,180,192,213,211,171,155,179,181,160,162,161,158,176,164,178,190,198,163,186,191,188,196,171,165,182,172,166,152,156,160,183",
    "DORSAL": "6,6,2,2,4,4,3,0,1,1,4,3,4,3,3,1,3,8,3,3,0,1,1,4,2,1,5,2,2,2,2,7,5,2,2,1,1,1,2,4,1,3,1,2,4,2,4,0,4,6,5",
    "SONOROUS": "211,234,232,219,236,223,222,226,234,229,206,214,240,226,235,229,215,236,218,220,225,242,222,207,224,215,242,225,235,219,230,219,232,220,222,244,234,231,249,247,227,268,224,215,241,220,237,227,219,230,220",
    "VELAR": "41,53,41,39,45,43,57,52,37,47,37,43,39,51,50,34,50,35,44,60,47,40,48,36,43,46,39,49,55,35,43,43,44,57,48,40,44,49,35,42,47,36,40,47,41,50,54,33,49,36,44",
    "NASAL": "141,174,148,174,175,193,187,140,159,161,160,155,157,174,154,169,165,172,194,191,129,164,165,152,194,160,150,164,165,155,161,172,189,199,148,150,173,157,144,146,169,132,174,158,154,170,159,161,166,170,196",
    "CORONAL": "639,631,677,675,698,651,651,633,645,633,612,641,635,639,660,695,678,670,651,651,630,649,631,601,640,634,643,637,660,671,684,668,656,650,646,627,647,625,640,616,632,620,627,651,633,639,660,685,684,668,649"
},
```


Fig. 4. The quantity of phonemes in sample 2.

Figure 6 shows that the fricative, velar and nasal PG follow normal probability distribution in sample 1; the stop, fricative, sonorous and velar consonant PG are distributed in the same way in sample 2; the samples have important discrepancies in fricative and velar consonant PG.

The important discrepancies are got in the fricative and velar PG. The important discrepancies are got in eight phoneme groups. The important discrepancies are got in all PG except the coronal phoneme group. The results got by the three tests are shown in Table 1.

In this comparison, the velar and fricative phoneme groups have the highest author-differentiating capability. In the second comparison, the speeches by D. Trump and the newspaper article by S. Lagon are analyzed (Figs. 8, 9 and 10).

The important differences are revealed in the stop and velar PG.

The important differences are established in eight phoneme groups.

Table 1. The author-differentiating capability of phoneme groups.

Phoneme group:	TS	TKS	TC
Labial		+	+
Velar	+	+	+
Fricative	+	+	+
Nasal		+	+
Sonorous		+	+
Coronal		+	
Dorsal		+	+
Stop		+	+


```
23      },
24      "Statistic Criterion": "STUDENT",
25      "Results of Pearson test of sample1": {
26          "STOP": false,
27          "FRICATIVE": true,
28          "LABIAL": false,
29          "DORSAL": false,
30          "SONOROUS": false,
31          "VELAR": true,
32          "NASAL": true,
33          "CORONAL": false
34      },
35      "Results of Pearson test of sample2": {
36          "STOP": true,
37          "FRICATIVE": true,
38          "LABIAL": false,
39          "DORSAL": false,
40          "SONOROUS": true,
41          "VELAR": true,
42          "NASAL": false,
43          "CORONAL": false
44      },
45      "Statistic Results": {
46          "STOP": null,
47          "FRICATIVE": false,
48          "LABIAL": null,
49          "DORSAL": null,
50          "SONOROUS": null,
51          "VELAR": false,
52          "NASAL": null,
53          "CORONAL": null
54      },
55      "Error message": []
56  }
```


Fig. 5. The results obtained by the Pearson's and the Student's t-tests.

The important differences are revealed in all PG except the stop and nasal PG.

The results by the three tests are shown in Table 2.

In this comparison, the velar phoneme group has the highest author-differentiating capability.

In the third comparison, the speeches by B. Obama and the newspaper article by D. Webster are analyzed (Figs. 11 and 12). The essential differences have been obtained by the Kolmogorov-Smirnov's and the chi-square tests.

```
23      },
24      "Statistic Criterion": "KOLMOGOROV_SMIRNOV",
25 ▾    "Statistic Results": {
26          "STOP": false,
27          "FRICATIVE": false,
28          "LABIAL": false,
29          "DORSAL": false,
30          "SONOROUS": false,
31          "VELAR": false,
32          "NASAL": false,
33          "CORONAL": false
34      },
35      "Error message": []
36  }
```


Fig. 6. The results obtained by the Kolmogorov-Smirnov's test.


```
23      },
24      "Statistic Criterion": "CHI_SQUARE",
25 ▾    "Statistic Results": {
26          "STOP": false,
27          "FRICATIVE": false,
28          "LABIAL": false,
29          "DORSAL": false,
30          "SONOROUS": false,
31          "VELAR": false,
32          "NASAL": false,
33          "CORONAL": true
34      },
35      "Error message": []
36  }
```


Fig. 7. Theresults obtained by the chi-square test.


```
23      },
24      "Statistic Criterion": "STUDENT",
25 ▾    "Results of Pearson test of sample1": {
26          "STOP": true,
27          "FRICATIVE": false,
28          "LABIAL": false,
29          "DORSAL": true,
30          "SONOROUS": false,
31          "VELAR": true,
32          "NASAL": true,
33          "CORONAL": false
34      },
35 ▾    "Results of Pearson test of sample2": {
36          "STOP": true,
37          "FRICATIVE": true,
38          "LABIAL": false,
39          "DORSAL": false,
40          "SONOROUS": true,
41          "VELAR": true,
42          "NASAL": false,
43          "CORONAL": false
44      },
45 ▾    "Statistic Results": {
46          "STOP": false,
47          "FRICATIVE": null,
48          "LABIAL": null,
49          "DORSAL": null,
50          "SONOROUS": null,
51          "VELAR": false,
52          "NASAL": null,
53          "CORONAL": null
54      },
55      "Error message": []
56  }
```


Fig. 8. The results obtained by the Pearson's and the Student's t-tests.

The important discrepancies are got in eight PG.

The important discrepancies are got in all PG except the labial and velar phoneme groups.

The results got by the three tests are shown in Table 3.


```
24        "Statistic Criterion": "KOLMOGOROV_SMIRNOV",
25 ▾      "Statistic Results": {
26            "STOP": false,
27            "FRICATIVE": false,
28            "LABIAL": false,
29            "DORSAL": false,
30            "SONOROUS": false,
31            "VELAR": false,
32            "NASAL": false,
33            "CORONAL": false
34        },
35        "Error message": []
36    }
```


Fig. 9. The results obtained by the Kolmogorov-Smirnov's test.


```
23        },
24        "Statistic Criterion": "CHI_SQUARE",
25 ▾      "Statistic Results": {
26            "STOP": true,
27            "FRICATIVE": false,
28            "LABIAL": false,
29            "DORSAL": false,
30            "SONOROUS": false,
31            "VELAR": false,
32            "NASAL": true,
33            "CORONAL": false
34        },
35        "Error message": []
36    }
```


Fig. 10. The results obtained by the chi-square test.

Table 2. The author-differentiating capability of phoneme groups.

Phoneme group:	TS	TKS	TC
Labial		+	+
Velar	+	+	+
Fricative		+	+
Nasal		+	
Sonorous		+	+
Coronal		+	+
Dorsal		+	+
Stop	+	+	

```
23      },
24      "Statistic Criterion": "KOLMOGOROV_SMIRNOV",
25 ▾    "Statistic Results": {
26          "STOP": false,
27          "FRICATIVE": false,
28          "LABIAL": false,
29          "DORSAL": false,
30          "SONOROUS": false,
31          "VELAR": false,
32          "NASAL": false,
33          "CORONAL": false
34      },
35      "Error message": []
```

Fig. 11. The results obtained by the Kolmogorov-Smirnov's test.

```
23      },
24      "Statistic Criterion": "CHI_SQUARE",
25 ▾    "Statistic Results": {
26          "STOP": false,
27          "FRICATIVE": false,
28          "LABIAL": true,
29          "DORSAL": false,
30          "SONOROUS": false,
31          "VELAR": true,
32          "NASAL": false,
33          "CORONAL": false
34      },
35      "Error message": []
```

Fig. 12. The results obtained by the chi-square test.

Table 3. The author-differentiating capability of phoneme groups.

Phoneme group:	Student's t-test	Kolmogorov-Smirnov's test	Chi-square test
Labial		+	+
Velar		+	
Fricative		+	+
Nasal		+	+
Sonorous		+	+
Coronal		+	+
Dorsal		+	+
Stop		+	+

The statistical methods applied made it possible to determine the author–differentiating capability of consonant phoneme groups and build the statistical models (Figs. 13 and 14). The developed method and models made it possible to reduce the number of consonant PG in which the important discrepancies are got and this way improve the efficiency of authorship attribution.

The statistical models are a realization of the developed CM. The method is an offered combination of the hypothesis methods: the TKS, the TC and the TS. The combination of methods has proved to be efficient in a text comparison from the publicist style. By the method applied the author-differentiating capability of PG has been revealed. The greatest number of established important discrepancies in the velar and fricative PG means that these PG have the highest author–differentiating capability and they can be chosen for the authorship attribution. The reduced number of phoneme groups facilitates the attribution process.

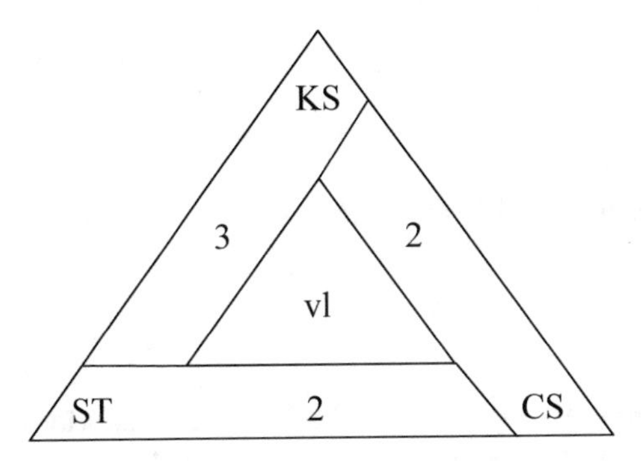

Fig. 13. Scheme of presenting the results of realization of the statistical model of determining the author–differentiating capability of the velar phoneme groups. KS – the Kolmogorov-Smirnov's test, CS – the chi-square test and ST – the Student's t-test; vl – the velar phoneme group; 2, 2, 3 – a number of comparisons, in which the phoneme group has the author-differentiating capability.

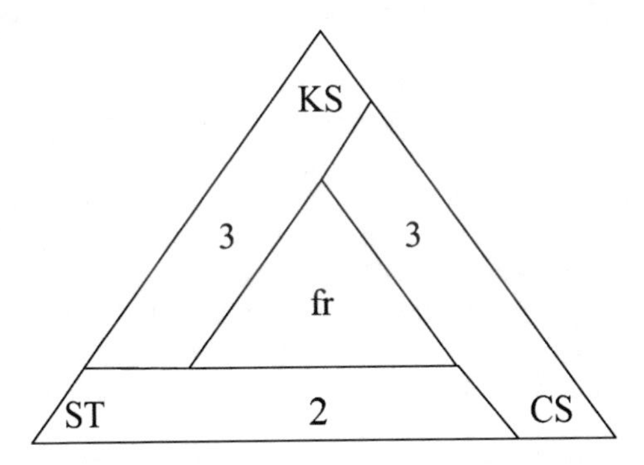

Fig. 14. Scheme of presenting the results of realization of the statistical model of determining the author–differentiating capability of fricative phoneme groups. fr – the fricative phoneme group.

4 Conclusions

The developed statistical models of authorship attribution make it possible to differentiate authorial styles more precisely. To prove the efficacy of the models, more than 500 experiments have been made. The results of some of them are presented in this paper. The Barack Obama's and Donald Trump's speeches, two newspaper articles by Susan Logan and David Webster have been differentiated by consonant phoneme groups. On the basis of the results obtained, the author–differentiating capability of consonant phoneme groups has been determined. For the fricative phoneme group, the important discrepancies are got by the three tests (the TKS, the TC and the TS). For the sonorous and dorsal phoneme groups, the important discrepancies are got by the two tests (the Kolmogorov-Smirnov's test and the chi-square test). The number of consonant phonemes has been reduced to three; the fricative, sonorous and dorsal. These phoneme groups have the highest differentiating capability and by them authorship attribution can be done more efficiently. The developed software on the Java programming language makes transcription process faster and improves efficiency of identifying characteristic features of a particular authorial style. In our future research another statistical method will be tested to improve efficiency of authorship attribution.

References

1. Argamon, S., Koppel, M., Pennebaker, J., Schler, J.: Automatically profiling the author of an anonymous text. Commun. ACM **52**(2), 119–123 (2009)
2. Lytvyn, V.: Development of a method for the recognition of author's style in the Ukrainian language texts based on linguometry, stylemetry and glottochronology. Eastern Eur. J. Enterp. Technol. **4/2**(88), 10–18 (2017)
3. Mubin, S.T., Rajesh, S.P.: Authorship identification with multi sequence word selection method. In: Thermal Stresses—Advanced Theory and Applications, pp. 653–661 (2019). https://doi.org/10.1007/978-3-030-16657-1_61
4. Jamak, A., Alen, S., Can, M.: Principal component analysis for authorship attribution. Bus. Syst. Res. **3**(2), 49–56 (2012)
5. Zhao, Y., Zobel, J.: Searching with style: authorship attribution in classic literature. In: Proceedings of the Thirtieth Australasian Conference on Computer Science, Australian Computer Society, vol. 62, pp. 59–68 (2007)
6. Vysotska, V., Lytvyn, V., Hrendus, M., Kubinska, S., Brodyak, O.: Method of textual information authorship analysis based on stylometry. In: Proceedings of the XIII-th Scientific and Technical Conference, CSIT, Lviv, pp. 9–16 (2018)
7. Bisikalo, O.: Sentence syntactic analysis application to keywords identification Ukrainian texts. Radioelectronics Comput. Sci. Control **3**(38), 54–65 (2016)
8. Altmann, G., Levickij, V., Perebyinis, V.: Problems of Quantitative Linguistics. Ruta, Chernivtsy (2005)
9. Ontika, N.N., Kabir, Md.F., Ashraful, I., Ahmed, E., Huda, M.N.: A computational approach to author identification from Bengali song lyrics. In: Proceedings of International Joint Conference on Computational Intelligence, pp. 359–369 (2019). https://doi.org/10.1007/978-981-13-7564-4_31
10. Burrows, J.: Delta: a measure of stylistic difference and a guide to likely authorship. Literary Linguist. Comput. **17**(3), 267–287 (2002)
11. Stamatatos, E., Fakotakis, N., Kokkinakis, G.: Computer-based authorship attribution without lexical measures. Comput. Humanit. **35/2**, 193–214 (2001)
12. Koppel, M.: Computational methods in authorship attribution. J. Assoc. Inf. Sci. Technol. **60** (1), 9–26 (2009)
13. Shulzinger, E., Bormashenko, E.: On the universal quantitative pattern of the distribution of initial characters in general dictionaries: the exponential distribution is valid for various languages. J. Quant. Linguist. **24**, 273–288 (2017)

14. Hausner, M.: Elementary Probability Theory. Springer, Boston (1995)
15. Martindale, C.: On the utility of content analysis in author attribution: the federalist. Comput. Humanit. **29**(4), 259–270 (1995)
16. Waheed, A., Imran, S.B., Shabana, R.: Design and implementation of a machine learning-based authorship identification model. Sci. Program. 1–14 (2019). https://doi.org/10.1155/2019/9431073
17. Ivanov, L., Aebig, A., Meerman, S.: Lexical stress-based authorship attribution with accurate pronunciation patterns selection. In: Sojka, P., Horák, A., Kopeček, I., Pala, K. (eds.) Text, Speech, and Dialogue, pp. 67–75. Springer, Cham (2018). https://doi.org/10.1007/978-3-030-00794-2_7
18. Kolmogorov, A.N.: Foundations of the Theory of Probability. Chelsea Publishing (1950)
19. Kolmogorov, A.N.: Mathematics and its Historical Development. Nauka, Moscow (1991). Edited by V. A. Uspensky
20. Gnedenko, B.V., Kolmogorov, A.N.: Limit Distributions for Sums of Independent Variables. Addison-Wesley, Boston (1968)
21. Khomytska, I., Teslyuk, V.: Authorship and style attribution by statistical methods of style differentiation on the phonological level. In: Shakhovska, N. (ed.) Advances in Intelligent Systems and Computing III, Lviv, vol. 871, pp. 105–118 (2018)
22. Gomez, P.C.: Statistical Methods in Language and Linguistic Research. University of Murcia, Spain (2013)
23. Watanabe, S.: Probability Theory and Mathematical Statistics. Springer, Heidelberg (1988)
24. Gries, Th.S.: Statistics for Linguistics with R: A Practical Introduction (Trends in Linguistics: Studies & Monographs), p. 348 (2009)
25. Kornai, A.: A Mathematical Linguistics. Springer, London (2008). https://doi.org/10.1007/978-1-84628-986-6
26. Khomytska, I., Teslyuk, V., Holovatyy, A., Morushko, O.: Development of methods, models and means for the author attribution of a text. Eastern Eur. J. Enterp. Technol. **3/2**(93), 41–46 (2018)
27. Rozanov, I.A., Silverman, R.A.: Probability Theory: A Concise Course. Dover Publications Inc., New York (2007)
28. Jorgensen, P.E.T.: Analysis and Probability. Springer, New York (2006)
29. Bhattacharya, R., Waymire, E.C.: A Basic Course in Probability Theory, 2nd edn. Springer, Cham (2017)
30. Khomytska, I., Teslyuk, V., Labinska, L.: Program system of authorship attribution of texts on the phonological level. In: Proceedings of the XIV-th Scientific-Practical Conference "Problems and Perspectives of Development of Economics, Enterprise and Computer Technologies in Ukraine", pp. 15–16. LPNU, Lviv (2018)
31. Bobalo, Yu., Seniv, M., Yakovyna, V., Symets, I.: Method of reliability block diagram visualization and automated construction of technical system operability condition. In: Shakhovska, N. (ed.) Advances in Intelligent Systems and Computing III, vol. 871, pp. 599–610 (2019)
32. Batyuk, A., Voityshyn, V., Verhun, V.: Software architecture design of the real-time processes monitoring platform. In: IEEE Second International Conference on Data Stream Mining & Processing (DSMP 2018), Lviv, Ukraine, pp. 98–101 (2018)

33. Shakhovska, N., Vysotska, V., Chyrun, L.: Intelligent systems design of distance learning realization for modern youth promotion and involvement in independent scientific researches. In: Shakhovska, N. (ed.) Advances in Intelligent Systems and Computing, vol. 512, pp. 175–198 (2016)
34. Denysyuk, P.: Usage of XML for fluidic MEMS database design. In: International Conference on Perspective Technologies and Methods in MEMS Design (MEMSTECH 2006), Lviv-Polyana, p. 148 (2007)
35. Teorey, T.J., Lightstone, S.S., et al.: Database Design: Know it all, 1st edn. Morgan Kaufmann Publishers, Burlington (2009)
36. Horstmann, C.: Java SE 9. Core Java SE 9 for the Impatient. "Вильямс" (2018)
37. Khomytska, I., Teslyuk, V.: Mathematical methods applied for authorship attribution on the phonological level. In: Proceedings of International Scientific Conference Computer Sciences and Information Technologies (CSIT-2019), vol. 3, pp. 7–11 (2019)

Individual Sign Translator Component of Tourist Information System

Olga Lozynska(✉), Valeriia Savchuk, and Volodymyr Pasichnyk

Information Systems and Networks Department,
Lviv Polytechnic National University, Lviv, Ukraine
{Olha.V.Lozynska, Valeriia.V.Savchuk, Volodymyr.V.Pasichnyk}@lpnu.ua

Abstract. This paper is devoted to the problem of communication of people with hearing disabilities during the tourist trip. The Ukrainian Sign Language is taken into account. The analysis of sing language translation systems is given in the paper. The authors are working on the "Tourist sign translation system" as a component of "Mobile information assistant of the tourist", that provides wide range of services on all stages of the trip. The functions and the structure of the system is provided with the help of UML diagrams. The translation system consists of two main components: "Offline Phrasebook" and "Individual Translator", so it can be used as online and offline. The users of the system are tourists, users with hearing disabilities and administrator. This component used the rule-based method of the translation into Ukrainian Sign Language using concept dictionary. As a result, a project of the Ukrainian sign translator is being presented.

Keywords: Ukrainian Sign Language · Translation algorithm · Concept dictionary · Mobile application · Tourist information technologies

1 Introduction

There always will be a problem of providing the high level of life for people with disabilities. Even an average service can be a problem.

In the following research hearing disabilities are taken into account. Regardless, of the economic and social development level, scientists of every country increase the number of information products that take into account sign language. The difficulty is that every language has its own signs and words structure, so it is hard to develop an universal system.

Ukrainian Sign Language (USL) is a natural way of communication of people with hearing disabilities, that is why there is a must to research the problem and improve the processes of rendering and gathering information with the use of this sign language.

Up-to-date sociological data in Ukraine:

- Nearly 400 thousand people are having hearing disabilities;
- 59 special boarding schools are working in Ukraine;
- 20 universities have abilities to teach students with hearing disabilities.

N. Shakhovska and M. O. Medykovskyy (Eds.): CCSIT 2019, AISC 1080, pp. 593–601, 2020.
https://doi.org/10.1007/978-3-030-33695-0_40

Right now, a lot of work is being done to make the processing of the Ukrainian sign language easier and much faster. This article is devoted to the problem of communication of Ukrainian tourists with hearing disabilities during the trip, as communication is an obligatory part of good quality tourist trip.

The scientists are developing a translating component of "Mobile information assistant of the tourist" (MIAT) that takes into account Ukrainian Sign Language. With this need the semantic analysis of the language was previously provided.

As for the MIAT, it is a tourist information system that provides individual support on all stages of the trip (planning, traveling, analysis) taking into account peculiarities of every user. The scientist of the Information systems and networks department of Lviv Polytechnic National University are working on its project.

2 Related Work

Every day information technologies for people with special needs are growing and improving. The most popular of them are mobile applications, mobile dictionaries, Sign Language simulators and machine translation systems of Sign Language that help deaf people to enjoy the communication with other persons (see Fig. 1).

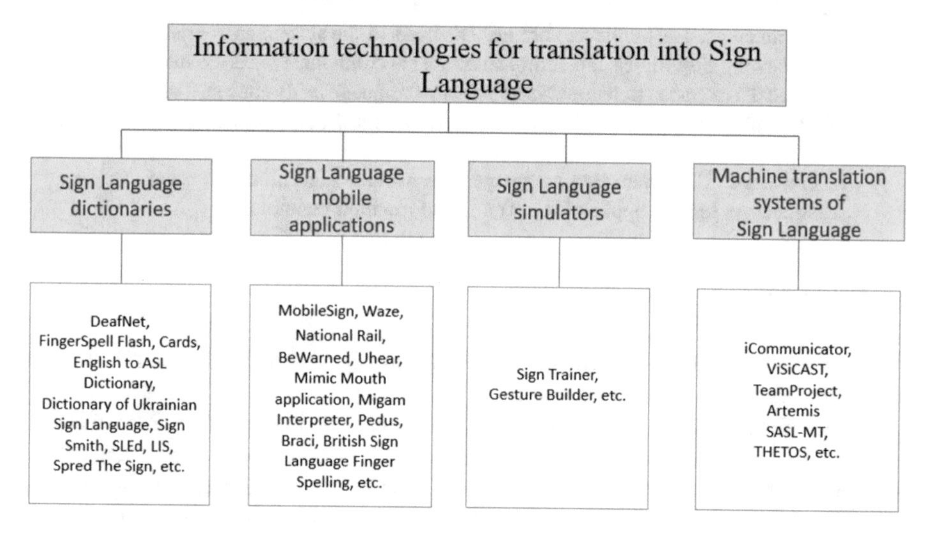

Fig. 1. Information technologies for translation into Sign Language.

MobileSign [1] – designed for learning British Sign Language on the go and consist of more than 4000 signs. User can quickly search the needed sign. This app is free.

Waze [2] & National Rail [3] are the travel apps. They are simple to use, and provide all the information that need for user. Waze is the world's largest community-based navigation app. National Rail is useful to make the train journeys easier.

BeWarned [4] is the free application that include four technical assistants for the deaf: Sound Monitor, Connect, Emergency Call and Dance. Sound Monitor detects sounds of potential danger in real time, for example scream, car honk, dog bark, siren. In case of danger the application warns the user of the vibration and flashing on the smartphone. Connect helps people with hearing disabilities communicate with others - it converts text into speech and vice versa. Dance convert musical tracks into vibration, light signals and a visual equalizer pulsation. Emergency Call helps the user to call their friends when he is in a dangerous situation.

British Sign Language Finger Spelling [5] – the free app designed for the study and the use of sign language on practice.

UHear [6] is the free app for testing people their hearing. This is a hearing loss screening test which allows users to test their hearing to determine if it is within normal range or if you have a potential hearing loss.

Engineers from California University in San Diego [7] developed a glove that recognizes the sign language and translates it into digital text. A glove can connect to a smartphone or computer via Bluetooth and send text to them in real time. The total cost of the components of the device does not exceed 100 dollars.

Mimic Mouth application [8] helps learning to read-from-lips and learn sign language. This app proposes for users a virtual three-dimensional model of the face which will accurately show the movements of the facial muscles and lips correspond to the vowel, word and sentences. It can be used by deaf and hearing impaired people.

Another useful application for deaf people is a sound-recognition platform Braci [9] that can be detecting a wide range of sounds which can be displayed on a smartphone, smart watch or other portable smart devices. Braci converts sounds into notifications that user can see and feel.

Pedus [10] is a free application that helps deaf and hard-of-hearing people to communicate. For this goal, speech-recognition and synthesis technologies to make phone calls are used. Pedius' voice recognition software translates what the person on the other end says and transfers it into a text format.

The Ukrainian scientists Shpinkovsky [11] proposed software system for converting text into sign language. The user can choose the sign language to convert text information (English, Russian or British two-handed sign languages). However, the developers of this system did not specify how many signs has this system and where it can be downloaded.

Migam Interpreter [12] – the free app that unite deaf and hearing people through education and communication. This is a service which enables instant video access to a Sign Language interpreter, via a Web browser, mobile application, or any device equipped with a camera and Internet access. Migam Interpreter translates sign language into Polish, English, Italian and Spanish. The application works on smartphone, tablet or through a browser.

El-Gayyar with other researchers [13] proposed translation from Arabic speech to Arabic Sign Language based on cloud computing. They developed a mobile-based frame work to serve the community of Arabic deaf people. The scientists designed a description of a three-layered architecture (Data Layer, Service Layer and Interface Layer) to help the deaf community in Egypt. To evaluate the user satisfaction, it was selected three groups of users: deaf-specialized teachers, deaf persons and normal people. The deaf users don't like the avatar representation and prefer real human videos.

That is the reason why Deaf people have such difficulties when communicating, with hearing people via writing, with hearing people and the other way around, hearing people also encounter problems with understanding the specific word order, in which Deaf usually write. Pen and paper are not enough to communicate effectively.

3 Main Part

3.1 The Architecture of the Tourist Sign Translation System

To eliminate barriers in communication between deaf and hearing people and to improve the living conditions of deaf people the mobile application that will be translated text to sign language is proposed. The main goal of the paper is to develop the tourist application for mobile devices that could be used by deaf people and people without hearing disabilities taking into account the individual characteristics of these people.

Main tasks of the project:

1. To record and process a video of the most commonly used signs.
2. To conduct sign video processing in a format that is easy to use on mobile devices.
3. Design the architecture of the software application for mobile devices.
4. To develop, implement and test the developed application on mobile devices.

The scientist of the Information systems and networks department of Lviv Polytechnic National University are working on the information system MIAT and "Tourist Sign Translation System" (TGTS) is one of its components [14, 15].

The main purpose of the system is to provide the user speech to signs translations in tourist purposes. The main functions of the system are speech to sign translation, text to sign translation, multilanguage, offline mode and tourist sign phrasebook (see Fig. 2).

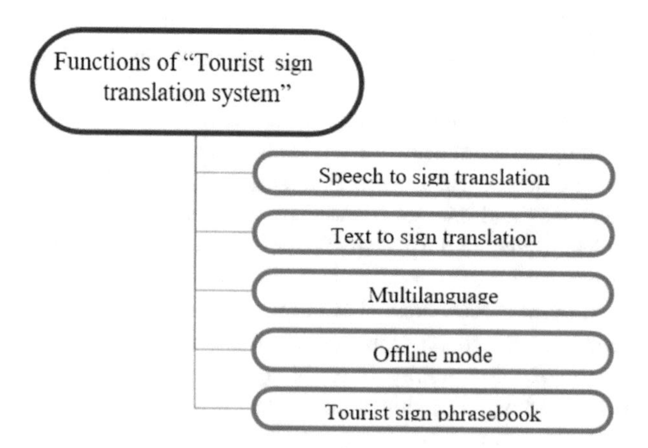

Fig. 2. "Tourist sign translation system" functions

The main user of the system is the tourist or group of tourists both hearing or impaired (for example, for hearing impaired foreign tourists travelling to Ukraine).

The system also has an administrator that provides data into the database and checks the correctness of system work.

The system has two main components (Fig. 3):

- "Offline Phrasebook" – is a component that allows to download, store and search through a text to sign phrasebook with most commonly used phrases during the tourist trip, according to the language that is needed.
- "Individual Translator" – is component, that fulfils the function of translation any sentence to Ukrainian sign language.

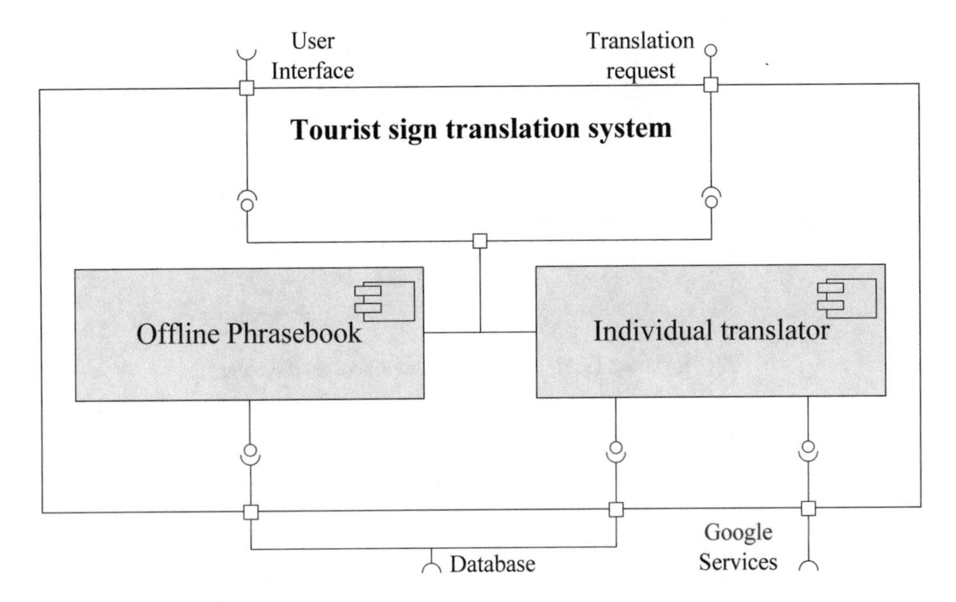

Fig. 3. "Tourist sign translation system" architecture

"Individual Translator" is a complex component. Its structure is presented in the Fig. 4.

The main components of it are:

- "Translator to Ukrainian" – the input data is the sentence the user wants to translate; output is this sentence in Ukrainian. The speech to text is converted via Google services.
- "Sign Translator" – is a component that uses the method of the translation Ukrainian Sign Language based on concept dictionary and converts text to signs.
- "Sign Visualizer" – is a component that converts every sign in its visual description and displays it on the screen of the device.

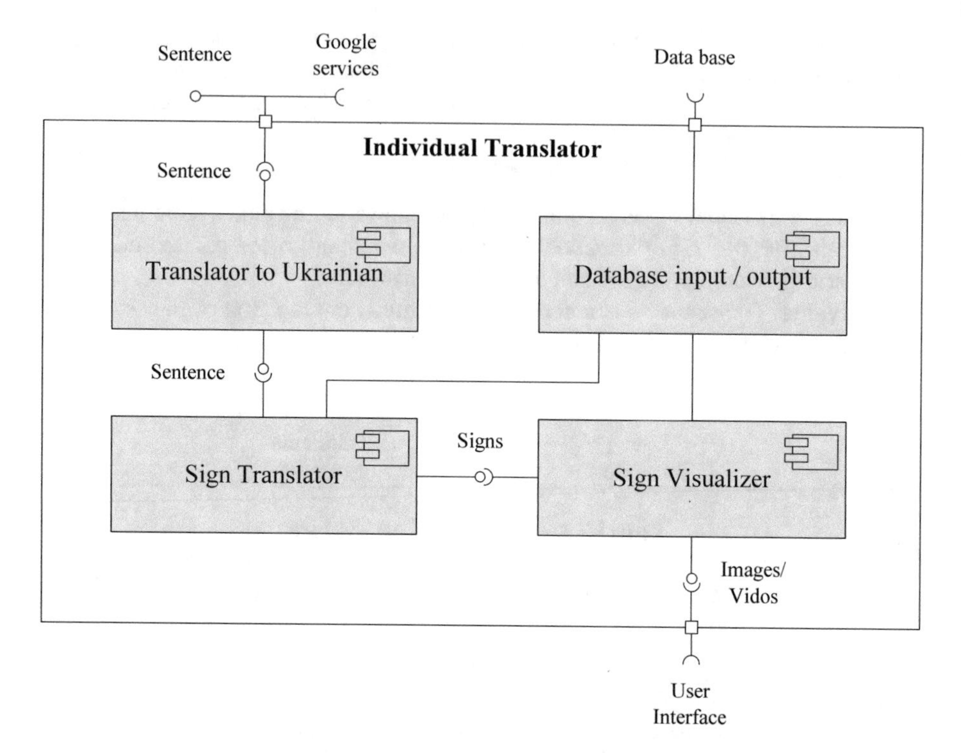

Fig. 4. "Individual Translator" component architecture

3.2 The Algorithm of The Translation into Ukrainian Sign Language Using Method Based on Concept Dictionary

The algorithm of the translation that used in Sign Translator Information System for Tourist is shown on Fig. 5.

At the first step the user must to register for definition of his type (hearing or deaf person). The next step is to input data by typing each word with a screen keyboard or by means of speech.

At the next stage of algorithm, we used the "Sign Translator" component. This component uses the rule-based method of the translation into Ukrainian Sign Language [14] and concept dictionary. And finally, the last step of the algorithm is video representation of translated data with sings. Video representation of most commonly used signs recorded with the use of sign language speakers.

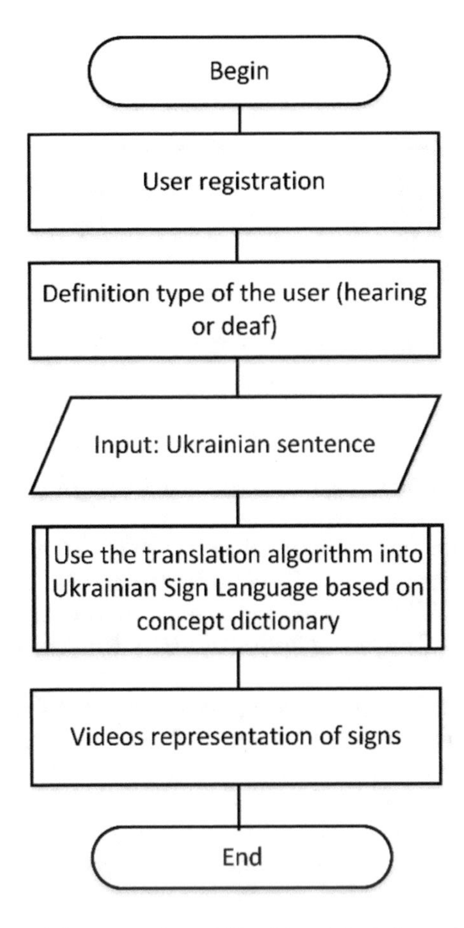

Fig. 5. The algorithm of the translation of Sign Translator Information System for Tourist

4 Results

As a result of the research on this stage the project of the Sign Translator Information System for Tourist was developed. First of all, the functions of the system and its users were defined. Secondly, its structure was developed, taking into account offline and online modes of the system. Thirdly, the general algorithm of the system work was defined and described in the paper.

On the basis of the proposed algorithm for the translation into Ukrainian Sign Language using concept dictionary [16, 17], a component “Sign Translator” was tested. The translation algorithm showed the following results (see Fig. 6):

- 55% of sentences without concepts were translated correctly;
- 32% of the sentences containing the concepts were translated correctly;
- 13% of the sentences were not translated due to lack of word – sign corresponding.

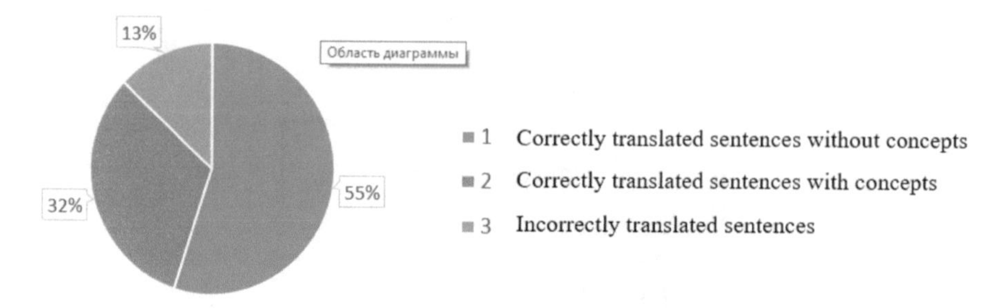

Fig. 6. The diagram of translation results

5 Conclusion

This paper is devoted to the problem of communication of people with hearing disabilities during the tourist trip.

The authors assume, that the sign languages should be taken into account when developing a tourist support system. This is an up-to-date issue according to the growth of the tourist industry. With the reason, the authors are working on a project of Tourist sign translation system. The main function of the system are speech and text to sign translating.

The article is devoted to Ukrainian sign language translating, as nearly 400 thousand citizens of Ukraine are having hearing disabilities and need proper support.

The method of the rule-based translation into Ukrainian Sign Language based on concept dictionary is developed. Using the concept dictionary and the rule-based method of translation into sign language, it was possible to increase the percentage of correctly translated sentences from 63% to 87% (in comparison with the rule-based method).

The functions, user roles and architecture of the Tourist sign translation system is defined and developed with the help of UML diagrams.

The future tasks are devoted to the development of the system database and prototype.

References

1. MobileSign. http://www.mobilesign.org/
2. Waze. https://www.waze.com/en/directions/australia/grafton/pacific-national-rail/100206047.1002191544.3961704.html
3. National Rail. http://www.nationalrail.co.uk
4. BeWarned. https://dou.ua/lenta/articles/dou-projector-bewarned
5. British Sign Language Finger Spelling. https://apps.apple.com/us/app/british-sign-language-finger-spelling/id389417770
6. UHear. https://apps.apple.com/us/app/uhear/id309811822
7. Project "Smart Glove". https://www.kpbs.org/news/2017/jul/12/smart-glove-can-translate-sign-language
8. Mimic Mouth application. https://kolomnie.pl/lodz/restauracja/litera-cafe/63271

9. Platform Braci. http://www.braci.co
10. Pedus. https://www.pedius.org/us
11. Migam Interpreter. http://migam.org/en/content/how-does-migam-interpreter-work
12. Shpinkovsky, O.A., Shpinkovska, M.I., Filipchuk, Y.D.: System for converting text information into sign language. Autom. Technol. Bus.-Process. **3**, 55–59 (2014)
13. El-Gayyar, M.M., Ibrahim, A.S., Wahed, M.E.: Translation from Arabic speech to Arabic Sign Language based on cloud computing. Egypt. Inf. J. **17**, 295–303 (2016)
14. Lozynska, O.V., Davydov, M.V., Pasichnyk, V.V.: Rule-based machine translation into Ukrainian sign language. Information technology and computer engineering. Sci. J. VNTU **1** (29), 11–17 (2014)
15. Vyklyuk, Y., Savchuk, V., Pasichnyk, V., Kunanets, N.: Information Technologies of Personalized Tourist Accompaniment. LAP LAMBERT Academic Publishing, Saarbrücken (2018)
16. Lozynska, O.V., Davydov, M.V., Pasichnyk, V.V.: Entity-relationship model of Ukrainian Sign Language concepts. In: Information Systems and Networks, vol. 805, pp. 279–289. Bulletin of Lviv Polytechnic National University (2014)
17. Lozynska, O., Savchuk, V., Pasichnyk, V.: The sign translator information system for tourist. In: Proceedings of International Scientific Conference on Computer Sciences and Information Technologies (CSIT-2019), vol. 3, pp. 162–165 (2019)

Woman Lingual Cultural Type Analysis Using Cognitive Modeling and Graph Theory

Olena Flys(✉)

Applied Linguistics Department, Lviv Polytechnic National University, Lviv, Ukraine
olenkabondaruk90@gmail.com

Abstract. This paper observes an analysis of lingual cultural type of woman by using graphical modeling approach. The study has been carried out in terms of applied linguistics and illustrates the integration of natural language processing technologies. The main contribution is an attempt to implement Text Mining tools to analyze poetic discourse of the late 16th – early 17th centuries and provide cognitive model of lingual cultural type. The topicality of cognitive modeling application in linguistic studies is the ability to structure and systematize the existing information, identify the scenarios for information system, and predict the relationship dynamics between the components of this system. The paper also examines the possibility of using graph theory to study lexical variables that reflect the objects of human life. Overall, the research findings indicate the most frequent connections between the lingual cultural type of woman and the lexical variables used to depict physical, social or spiritual objects of the reality in English culture.

Keywords: Cognitive modeling · Graphs · Lingual cultural type · Natural language processing · Text mining

1 Introduction

Today, the automatic analysis of data extraction from the text units became one of the most topical areas in artificial intelligence researches. Scientists are working on the development and improvement of automatic analysis systems for solving a number of linguistic issues. These achievements are particularly significant for the study of poetic texts because of their non-structured system of semantic and syntactic relationships, as well as blurred boundaries between the text blocks. The implication of graph theory method is the most frequent technique in the automatic analysis of a text since they are able to produce its compact and formalized presentation.

This paper attempts at the cognitive model development of lingual cultural type "Woman" on the basis of its graphical modeling in a text using Text Mining technologies. The main objective of cognitive modelling is to create and analyze the cognitive map of a situation by using graphs.

In this research, the functioning of lingual cultural type is analyzed on the material of poetic texts, since literature is considered to be a core point of lingual, cultural, and social human activity. It is a synthesis of author's view of the world in accordance to

N. Shakhovska and M. O. Medykovskyy (Eds.): CCSIT 2019, AISC 1080, pp. 602–619, 2020.
https://doi.org/10.1007/978-3-030-33695-0_41

the national experience embedded in culture. Most of the lingual cultural types are based on the prototype images of real people or depict the fictional characters that represent the behavioral character of the society.

The main tasks of Text Mining technologies include the construction of clusters or associations along with the analysis of text features in order to extract knowledge from text arrays.

2 The Researches of Lingual Cultural Type

According to the recent studies, the lingual cultural type has become one of the significant notions in the modern linguistic term base. This term correlates with the postulation systems of cognitive linguistics, lingvopersonology, and lingual culture studies. A lot of attention is devoted to the lingual cultural type [1] as a special type of concept, the most important characteristics of which belong to a certain person that is significant for a particular lingual culture [2]. It is defined as "the cultural significance of a type-bound personality for understanding specific culture and studying it as the linguistic object (i.e. taking into account the designation of the corresponding concept that is embedded in language)" [3:22].

Considering lingual cultural type as a product of cultural and cognitive research of language personality simplifies the human conception as a representative of a certain ethnic group rendering the socially predetermined and recognizable personal characteristics of his/her behavior according to the nationally-bound characteristics [4:3]. This notion is a complex of lingual cultural and communicative values, attitudes and behavioral reactions [5].

Due to culture studies, the lingual cultural types are interpreted as "the recognizable images of particular culture representatives that form the cultures of their community" [6:179]. The representatives of this approach pay particular attention to the investigation of the type typical features, describing its figurative and valuable characteristics.

As the object of cognitive linguistics, the lingual cultural type is considered a mental entity (a concept) that provides a type with the additional conceptual, evaluative and figurative components [7]. According to the cognitive linguistics perspective, the lingual cultural type might be depicted via the following features: (a) general recognition and associativity; (b) recurrence; (c) symbolism; (d) brightness; (e) typicalness; (f) precedence [8].

The linguistic procedure of the lingual cultural type description is based on the extraction of its conceptual, imaginative and value key components. The conceptual component describes the semantic sense of a linguistic unit or a phrase used to refer to the type. It is grounded on the linguistic unit interpretations, its fixed vocabulary definitions and descriptions. The imaginative component is responsible for visual presentation of a lexical unit in linguistic consciousness; it indicates the social, physical, and spiritual characteristics of the type (e.g. age, gender, appearance, social origin, etiquette, religious position, behavior, activities). The value component contains national and group value orientations, it point to a negatively or positively colored assessment of a type in the society, mostly it is expressed with phraseological and metaphorical units.

3 Adjustment of Text Mining Methods to the Analysis of Lingual Data Extraction

The outlined research is initiated with the help of Text Mining technologies aimed at extracting the language units describing a lingual cultural type "Woman". The most relevant experiments based on Text Mining tools include Text Mining Clustering applications [9], investigation of Text Mining visual programming platform [10], computerized systems for natural language processing used to solve linguistic problems [11].

In this paper we render an idea that Text Mining is a natural language processing technique of identifying unknown and useful interpretations of the knowledge in the raw data sets that are necessary for decision making in various fields [12]. The most relevant Text Mining capability in terms of linguistic researches is the segmentation that is aimed at conveying the intellectual data analysis.

Text Mining methods might be adjusted to the linguistic researches due to their capability to solve not only the technical tasks such as the consolidation of similar documents and referencing, but it can also help to split a text into elements. Text Mining processes applied in linguistic purposes facilitates the process of studying large texts arrays, making it possible to focus on the specific segments of a text. Among all the Text Mining tools capable of providing the text analysis procedures we can suggest RapidMiner Studio, SAS Text Miner, GATE, Oracle Text, Knowledge Server, KNAME, Sisense. The advantage of these platforms is the integral ability to process natural language with a graphical output of the results that have to be previously described by a certain set of functions.

For rendering the qualified analysis of the text sets, presumably containing the lingual cultural type "Woman", and further compilation of a list containing the lexical variables used to denote the woman, we have chosen RapidMiner Studio 9.2.001 platform (2019) [13]. It is considered to be the most versatile tool for constructing algorithms to solve the problems of text mining and data stream mining [14]. The main advantages of this environment are the ability to generate applications using a universal tool like the XML code [15] as well as the user-friendly visual graphical interface of the analytical processes implementation.

As RapidMiner Studio is written on Java programming language, it ensures the appropriate environment for providing the machine learning, text analysis, and predictive analytics tasks. This platform is able to accomplish the variety of both experimental and applied tasks regarding intellectual data analysis such as text mining, multimedia mining, and data stream mining. Therefore, it might be used in different spheres of human working activities (business, industry, education, economy) as well as in the developing of computer applications. RapidMiner Studio provides all the stages of data extraction, including their visualization, verification and optimization. RapidMiner provides the analysis of machine learning procedures, including data upload and conversion (transformation of the categorical data into numerical that is extremely crucial when the source data is lexical), data preprocessing and visualization, predicted analysis and statistical simulation, and data evaluation [13].

Inour study, the Process Documents operator consists of 4 subprocesses (see Fig. 1) that are sequentially linked: • tokenization based on the lexical analysis (Tokenize Linguistic (Tokenize)); • Filter Stopwords; • Selection of the word stem (Stem); • bringing the text to a single register (Transform Cases). The results of text data processing facilitates the categorization and extraction of central linguistic units that belong to lingual cultural type "Woman".

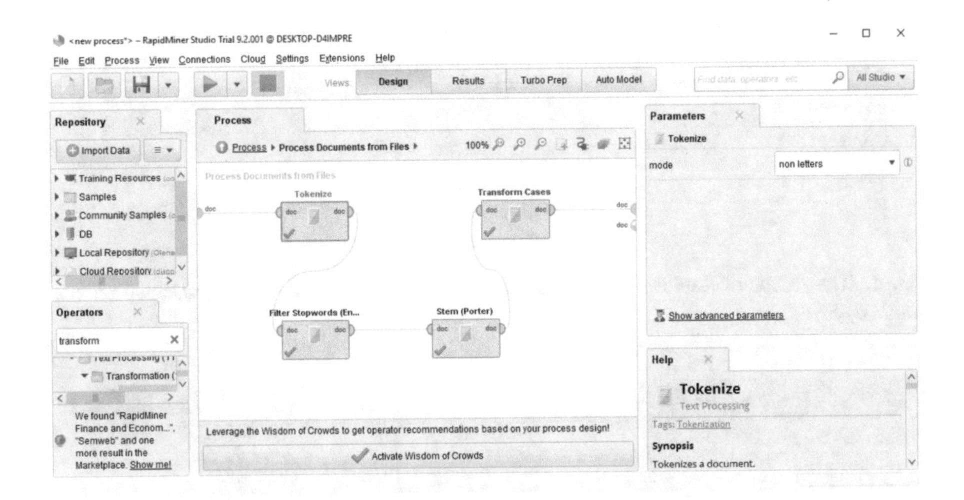

Fig. 1. RapidMiner Studio text document processing of poetic texts

In order to get unbiased results, the lingual corpus of this research consists of 1692 poetic texts that have been verified to be written only during late 16 – early 17 century in England. As the unproven data, authorship or release year of text analyzed are all significant variables that can dramatically affect the output. So, it was crucial for this analysis to collect only the texts from the arranged historical period, as we are interested in revealing lingual and cultural peculiarities of language exactly in Elizabethan era.

Proceeding from the above text investigation, we got the results illustrating the computer analysis of the text documents by means of RapidMiner platform (see Figs. 2 and 3). It has revealed the exact lexical units used by poets to denote the lingual cultural type "Woman". With the help of RapidMiner we have formed the sampling of nouns that nominated women in the uploaded resources. Furthermore, the outcomes of this platform can not only show the total number of word occurrences within the analyzed corpus, but also depict the detailed amount of word occurrences in a separate author's set of a corpus (see Table 1).

On this stage of the research the findings have revealed that the woman type occurs in 1146 verses and is represented by 44 units (*dame, daughter, lady, queen, mistress, princess, countesse, empress, madam, gentlewoman, captainess, conqueress, patronesse, girl, maid, maiden, bonilasse, damsel, lass, mother, granddam, sister, niece, stepdam, kinswoman, fair, goddess(e), belle, bride, wife, widow, silk-wife, housewife,*

Word	Total Occurences	Document Occurences	Word	Total Occurences	Document Occurences
boni	1	1	bricklay	1	1
bonibel	1	1	bridal	16	1
bonilass	2	1	bride	13	4
bonnet	3	3	bridegro...	6	6
bonnev	1	1	bridewel	2	2
bonni	13	6	bridg	5	2
book	74	28	bridl	5	5
bookebind	2	1	brief	6	6
boon	3	2	briefli	1	1
boord	3	3	brier	4	4
boot	16	14	bright	150	49

Fig. 2. The results of RapidMiner Studio text document computer analysis (presented units are *bonilass, bride*)

Word	Total Occurences	Document Occurences	Word	Total Occurences	Document Occurences
countervail	1	1	damag	1	1
countess	5	1	damask	9	7
countest	1	1	dame	67	26
countless	1	1	dammta	1	1
countrei	3	2	damn	12	6
countri	57	30	damnat	2	2
coupl	5	4	damnd	1	1
cour	1	1	damo	1	1
courag	27	15	damon	5	2
couragc	1	1	damp	4	3
couragi	4	4	damsel	13	4

Fig. 3. The results of RapidMiner Studio text document computer analysis (presented units are *countess, dame, damsel*)

matron, dowager, witch, wench, whore, bawd, dowdy, shrow, trull, murderess, tygresse). The total frequency of lingual cultural type “Woman” is 68% that prove the high role of women depiction in literature and the great interest of her activities in the ordinary life.

Further studies on lingual data evaluation and segmentation are aimed at applying the other RapidMiner functions that can illustrate the most often used attributes of lingual cultural type “Woman”.

Table 1. Examples of lexical units frequency based on RapidMiner Studio output

Lexical unit	Total occurence	Poets occurence	Text occurence
Lady	137	35	103
Woman	107	35	78
Queen	94	35	79
Mistress	75	33	72
Mother	72	23	55
Maid	52	26	52
Wife	47	22	39
Dame	67	26	32
Maiden	24	15	23
Sister	23	10	14
Lass	20	15	19
Virgin	20	10	19
Wench	16	6	11
Damsel	13	4	11
Bride	13	4	8
Girl	10	8	10
Whore	12	1	7
Witch	9	6	6
Widow	8	1	2
Womankind	7	5	6
Princess	6	3	5
Countesse	5	2	5
Empress	5	5	5
Trull	4	4	4
Madam	2	2	2
Bonilasse	2	1	1
Matron	2	2	2
Housewife	2	2	2
Bawd	1	1	1
Dowdy	1	1	1
Silk-wife	1	1	1
Dowager	1	1	1
Gentlewoman	1	1	1
Captainess	1	1	1
Conqueress	1	1	1
Patronesse	1	1	1
Shrow	1	1	1
Murderess	1	1	1
Tygresse	1	1	1
Feminine	1	1	1

4 The Application of Graph Theory

Text might be considered as a unit consisted of separate segments (vertices) connected with the help of special bonds (edges) [16]. Accordingly, the vertices of a graph correspond to the key lexical variables, while the edges indicate the peculiarities of semantic ties formed within the text between the variables.

Each graph $G = (V, E)$ is represented as a square adjacency matrix A with the links between vertices and edges [17:83]. In every single matrix the value of element aij is equal to the number of edges from the graph vertex i to the vertex j, where $aij \in \{0, 1\}$. We will assume that any vertex is adjacent to itself. Suppose there is an unoriented graph $G(V, E)$ (see Fig. 4), where:

$$
\begin{aligned}
\mathrm{V} &= \{0, 1, 2, 3, 4, 5\} \\
\mathrm{E} &= \{(0,1), (0,2), (0,5), (1,2), (1,3), (2,3), (3,4), (4,5)\} \\
\mathrm{n} &= 6
\end{aligned}
$$

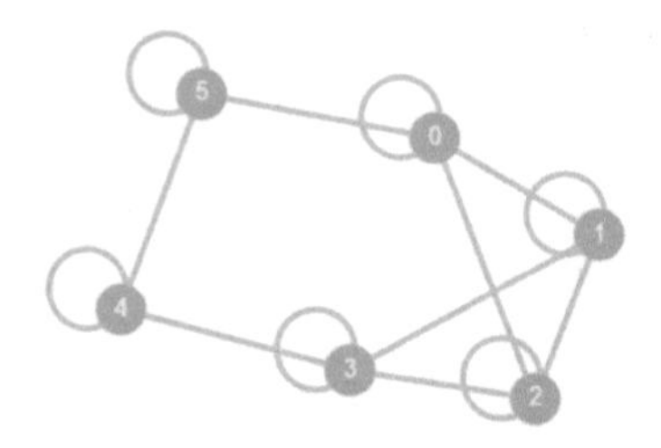

Fig. 4. Unorientated graph *G (V, E)*

The graph adjacency matrix (see Fig. 5) is mostly used to conduct an analysis of information flow patterns in the complex systems. Consequently, it becomes possible to present a text by a corresponding graph and to formalize the process of the text analysis. The conversion of linear text into multidimensional object, i.e. into a graph, is beneficial for the text analysis as it reflects the complexity of human language, perception and people consciousness [18].

In terms of applied linguistic researches, the directed graphs of adjacency matrix are substituted for the application of fuzzy graphs. They are distinguished by a membership function of vertices and edges set values ($0 \leq \mu_{\tilde{A}}(x) \leq 1$) that determine a membership degree of an item to the matrix. If the membership function is equal to zero, the item

$$
\mathrm{A} := \begin{matrix}
1, 1, 1, 0, 0, 1, 0, 0, \\
1, 1, 1, 1, 0, 0, 0, 0, \\
1, 1, 1, 1, 0, 0, 0, 0, \\
0, 1, 1, 1, 1, 0, 0, 0, \\
0, 0, 0, 1, 1, 1, 0, 0, \\
1, 0, 0, 0, 1, 1, 0, 0,
\end{matrix}
$$

Fig. 5. Adjacency matrix A for graph *G (V, E)*

does not strictly belong to a set and is not included in a list. If the value is equal to one, it strictly belongs to a set: E $= \{\langle 3|0,5\rangle, \langle 4|0,8\rangle, \langle 5|1\rangle, \langle 6|1\rangle, \langle 7|0,8\rangle, \langle 8|0,5\rangle\}$.

We decided to apply the graph modeling theory in order to improve the research process of lingual cultural type "Woman". It is conducted through the notion of a lexical variable that differentiates from the numerical one by the variable lingual values (words or phrases) instead of the numbers [19]. According to Zade [20], fuzzy set Ã is a set of pairs $\{(\mathrm{x}, \mu_{\tilde{A}}(x)) : x \in \mathrm{X}\}$, where $\mu_{\tilde{A}}(x) : \mathrm{X} \to [0,1]$ is a membership function of Ã. Whereas the lexical variable is reflected as the {X, T(X), U, V, S} quintuple, where:

- *X* is the variable name that is regarded as $\mathrm{T(X)} = \{x_i, \mathrm{i} = 1, \mathrm{m}\}$ to highlight the set of variable *X* and illustrate the fact that predominantly the linguistic values of variable *X* are characterized by a large set of names;
- *V* is a syntactic rule generating the names of values of the lexical variable *X*;
- *S* is a semantic rule that corresponds each fuzzy variable with a name from *T(X)* to the fuzzy subset of the universal set *U*.

Graph modeling of lingual cultural type "Woman" by using fuzzy sets is carried out through the defined sequence of stages. First, the texts are split into the segments due to the unstructured nature of poetic texts. It will provide the conventional structure of the analyzed fragment and define a set of lexical units. Second, each word is given a membership function $\mu_{\tilde{A}}(x)$, which will take either "0" or "1" value depending if a word occurs in a segment. In case the word is present in more than one segment that is caused by the blurred boundaries within the poetic text blocks, the value will acquire the value between "0" and "1". For instance, if the word "*fair*" is present only in block A, its membership function is $fair = \{\langle a|1\rangle, \langle b|0\rangle, \langle c|0\rangle\}$. If the same word is on the line of block A and block B, the function is $fair = \{\langle a|0,5\rangle, \langle b|0,5\rangle, \langle c|0\rangle\}$. A fuzzy graph of the lexical variable "mother" will have the following formula (the text is split into 4 segments): $mother = \{\langle 1|0\rangle, \langle 2|0,5\rangle, \langle 3|0\rangle, \langle 4|0,5\rangle, \langle 5|0,5\rangle, \langle 6|1\rangle, \langle 7|1\rangle, \langle 8|0,5\rangle\}$.

> *"Come, little babe; come, silly soul, || Thy father's shame, thy mother's grief, || Thou little think'st and less dost know || The cause of this thy mother's moan,|| Thou want'st the wit to wail her woe, || come to mother, babe, and play, || If any ask thy mother's name, || Tell how by love she purchased blame.* (Nicholas Breton)

The graph representation model of textual data is an integral part of applied linguistic researches as it can efficiently depict the relationship and arrange the information. Most of the literary texts are packed with a range of collocations, attributes and metaphors, for that reason there was presented an idea of the collocation relationship between the words indicating the "occurrence of two or more words within a well-defined unit of information" [21]. In order to select the most meaningful and significant collocations within the poetic texts we should consider a, b as the number of sentences containing A and B, k as the number of sentences containing both A and B, and n as the total number of sentences [22]. The significance of these measurements will indicate the probability of collocation relationship items occurrence in terms of a segment.

In order to provide a graphical analysis of lingual cultural type "Woman" we apply Grafoanalizator 1.3.3 platform (see Fig. 6), which can construct and process a graph, as well as present it in an adjustable visual form. This program was developed in 2010 by a group of discrete mathematics scientists.

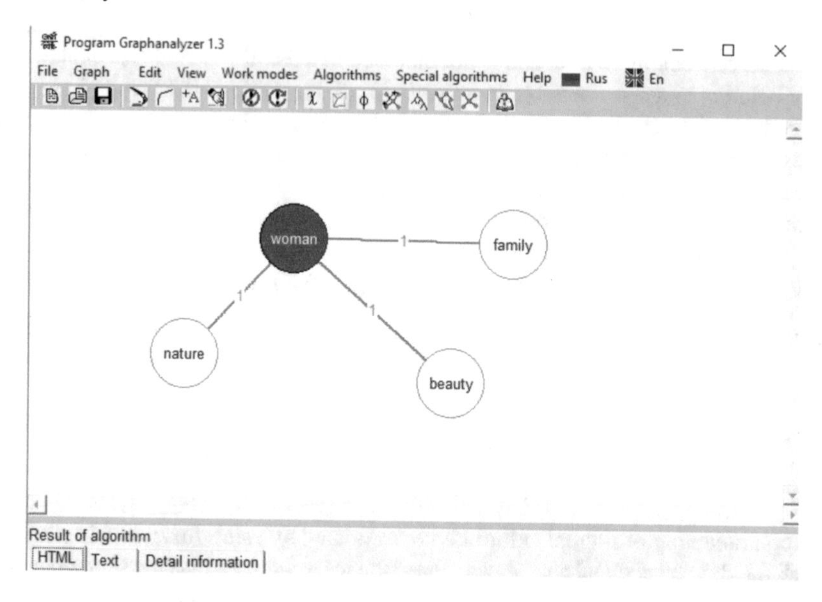

Fig. 6. Program Grafoanalizator 1.3.3 run screen

The advantages of this program are the abilities to work with different graph types (simple, oriented, non-orientated), the implementation of specified algorithms for graphs processing (path search, graph labeling, etc.), the matrix construction, the accurate data visualization [23]. The disadvantages of this platform include the absence of automatic allocation of vertice values from the data set and the inability to work with a big amount of data.

5 Cognitive Modelling of Lingual Cultural Type "Woman"

The main purpose of cognitive modeling [24] is the visualization of analyzed object functioning, which is considered as a structured system of internal and external elements that interact with each other on the basis of cause-and-effect relationships [25].

The cognitive approach activates the human intellectual processes and helps to capture the vision of a situation, phenomenon or subject in the form of a formal model, i.e. "the cognitive map" [26]. This map contains the well-known behavioral patterns of a particular object of the reality in a form of a graph; its vertices depend on this object's actualization scope and may contain some factors, attributes or characteristics, while the edges point at the cause-and-effect relationships between them [27]. The graph itself is a base for constructing a cognitive model that makes it possible to explore an object of the reality through the spectrum of its interactions with the other objects of the same reality.

The cognitive model can be represented as $G = \langle V, E \rangle$ [28], where:

- G is an unoriented graph (i.e. a cognitive map) of the analyzed object;

- V is a set of vertices (i.e. lexical variables) $Vi \in V, i = 1, 2, \ldots, k$ that are the elements of the investigated system (factors of the situation are also included in this set);
- E is a set of edges $e_{ij} \in E, i, j = 1, 2, \ldots, N$ that represent the direct interactions between the variables, considering the cause-and-effect relationships responsible for portraying the influence factors have on each other, in other words they show the relationship between the vertices Vi and Vj.

For the accomplishment of lingual cultural type cognitive modeling we should initially compile a list of lexical variables used to designate a woman in the elaborated poetic texts that has been previously obtained with the help of RapidMiner platform. The graphic model construction of the analyzed type is based on presenting the cause-and-effect relationships between the extracted lexical variables and those variables that represent the physical, social and spiritual objects of the reality of English culture with the help of Grafoanalizator platform. The values of vertices in the constructed graphs will contain the lexical variables indicating the object "woman" and other objects of the reality it interacts with, while the edges will represent the quantitative relationships between these vertices.

The process of data representation in the intelligent systems is hugely influenced by the frame organizations. In order to conduct the cognitive modeling of lingual cultural type we should also take into account the frame theory that is responsible for clarifying the thematic modeling of lexical variables used to nominate objects of human world. According to the computer science terminology, the frame is interpreted as a structure for a declarative, substantive presentation of knowledge about the thematically unified situation [29:16]. We comprehend a frame as "the special organization of knowledge, which is a prerequisite for human ability to understand the words closely related to each other" [30:223]. It is reasonable to depict a frame as a grid, which consists of nodes, united by ties and responsible for certain qualities of a given situation. Thus, frame theory initiates a process of the enhanced division of lexical variables in accordance with their most often used contexts. It is realized because of the frame in-built knowledge about a situation or an object that exist in a certain historical period of a particular society [31:122–144].

An integral part of exemplifying the graph analysis is a clear division of the lexical variables into the groups corresponding people life and activities. Therefore, there is a necessity to accomplish the categorization of the objects that are present in human world with the help of native speaker associative and verbal network. It is influenced of external factors, contains systematized, accumulated knowledge and affects human consciousness. The associative field enables the study of human verbal knowledge, the system of semantic and grammatical relations, as well as the motives and evaluations. The reproduction of associative meanings used by a native speaker allows us to describe the verbal set of units applied to denote the realities of the world and define the peculiarities of words usage. The complexity of associative and verbal network description is connected with its emotional component, as besides the lexical meaning it also contains the emotions of an individual that can determine the way of interpreting the lexical units [32].

The graphic modeling of lingual cultural type “Woman” based on the lexical variable “maiden” (see Fig. 7) illustrates the strong bond between lexical variables used to denote the woman and the society as well her beauty: *When turtles tread, and rooks, and daws, And maidens bleach their summer smocks* (Shakespeare William), *The butler was quick and the ale he did tap, The maidens did make the chamber full gay* (Anonym), *New brooms, green brooms, will you buy any? Come, maidens, come quickly, let me take a penny* (Robert Wilson), *If so you have, maidens, I pray you bring hither, That you and I friendly, May bargain together* (Robert Wilson), *And in a kirtle of green saye, The green is for maidens meet* (Edmund Spenser).

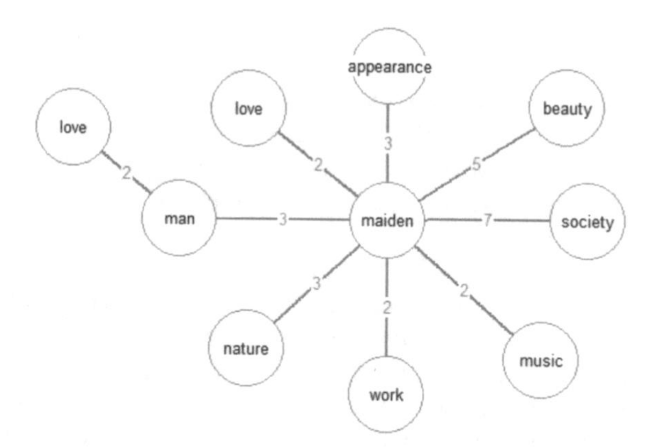

Fig. 7. Graphic modeling of lingual cultural type “Woman” based on the lexical variable “maiden”

The internal connections between the components of the described graph model can be represented as the adjacency matrix (see Fig. 8), where the level of influence between one item on the other one is indicated by at the row and the column section.

	A	mai	love	app	mar	wor	mus	soc	bea	nat	A	love
A		0	0	0	0	0	0	0	0	0	0	0
mai	0		2	3	3	2	2	7	5	3	0	0
love	0	2		0	0	0	0	0	0	0	0	0
app	0	3	0		0	0	0	0	0	0	0	0
mar	0	3	0	0		0	0	0	0	0	0	2
wor	0	2	0	0	0		0	0	0	0	0	0
mus	0	2	0	0	0	0		0	0	0	0	0
soc	0	7	0	0	0	0	0		0	0	0	0
bea	0	5	0	0	0	0	0	0		0	0	0
nat	0	3	0	0	0	0	0	0	0		0	0
A	0	0	0	0	0	0	0	0	0	0		0
love	0	0	0	0	2	0	0	0	0	0	0	

Fig. 8. Adjacency matrix of the linguistic variable “maiden”

The graphic modeling based on the lexical variable "lass" (see Fig. 9) illustrates its strong bond with the lexical variables used to denote the emotion of love and beauty of a woman: *Yet should thilk lass not from my thought, So you may buy gold too dear* (Edmund Spenser); *But now, ye Shepheard lasses! who shall lead, Your wandring troupes, or sing your virelayes?* (Edmund Spenser)*; When in a danceHe falls in a trance, To see his black-brow lass not buss him, And then whines out for death t'untruss him* (John Lyly)*; My bonny lass, thine eye, So sly, Hath made me sorrow so; Thy crimsoncheeks, my dear, So clear, Have so much wrought my woe* (Thomas Lodge); *It was a lover and his lass, It was a lover and his lass, That o'er the green cornfield did pass* (William Shakespeare)

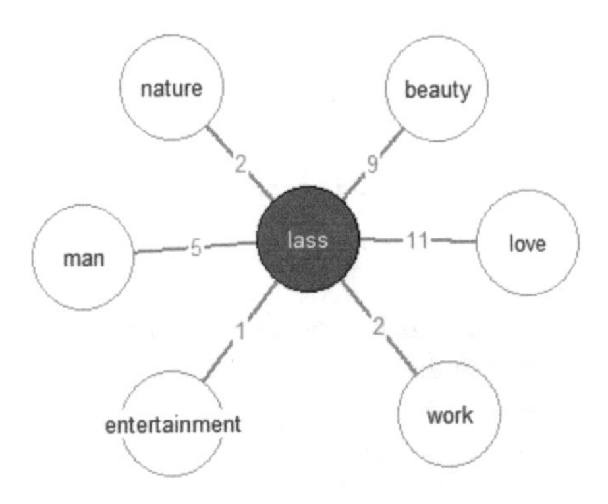

Fig. 9. Graphic modeling of lingual cultural type "Woman" based on the lexical variable "lass"

The next example illustrates more complex graphic modeling of lingual cultural type "Woman" that is based on a lexical variable "mother" (see Fig. 10). First, we might see that the strongest bond between the lexical variables denoting the woman is with the variables used to express the relation with a man. Secondly, there is the simultaneously indissoluble connection between three vertices – love and baby: *Amongst all creatures bearing life, A woman is the worthyest thing: Shee is to man a faythfull wife: Shee mother was to Christ our king* (Humfrey Gifford)*; In my birth my Mother died, Young and faire in heavie plight* (Batrholomey Young)*; Love still a boy, and oft a wanton is, School'd only by his mother's tender eye* (Sir Philip Sidney)*; The winged boy upon his mothers knee, Wantonlie playing neere to Paphos shrine*(Fulke Greville); *He stakes his quiver bow, and arrows, His mother's doves and team of sparrows* (John Lyly); *As child that in the cradle quails, Or else within the mother's wombHath his beginning and his tomb.* (Walter Raleigh); *Mother's wag, pretty boy, Father's sorrow, father's joy. The wanton smiled, father wept; Mother cried, baby lept; He must go, he must kiss Child and mother, baby bliss;* (Robert Green); *'Tis neither love the son, nor love the mother, Which lovers praise and pray to; but that love is, Which she in eye and I in heart do smother.* (Thomas Lodge); *Worthy Mother, High-thoughted, like to her, with bounty laden* (Bartholomew Griffin); *Anger the one, and envy moved the other, To see my love more fair than Love's fair mother* (Richard Lynch); *Whilst the earth, our common mother, Hath her bosom decked with flowers* (Samuel Daniel).

Fig. 10. Graphic modeling of lingual cultural type "Woman" based on the lexical variable "mother"

The next graphic modeling is based on the lexical variable "dame" (see Fig. 11) and illustrates its strong bond with the lexical variables used to denote the emotion of love and the admiration of woman's beauty: *Fine Dame, since that you be so coy* (Queen Elizabeth I)*; Thy birth, thy beauty, nor thy brave attire, Disdainful Dame, which doest me double wrong, Thy high estate, which sets thy heart on fire* (George Gascoigne)*; I know some pepper-nosed dame, Will term me fool, and saucy jack, That dare their credit so defame, And lay such slanders on their back* (Humfrey Gifford); *You modest Dames, inricht with Chastitie, Maske your bright eyes with Vestaes sable vaile* (Richard Barnfield); *But oh coy Dame intollerable smart* (Samuel Daniel)

Fig. 11. Graphic modeling of lingual cultural type "Woman" based on the lexical variable "dame"

The following example demonstrates one more complex graphic modeling of the analyzed lingual cultural type that is based on a lexical variable "daughter" (see Fig. 12). We can notice new edge that correlates to this unit and has the strongest bondwith it – music, which points at the interests and activities of woman in the analyzed historical period: *It was a lording's daughter, the fairest one of three, That liked of her master as well as well might be, Till looking on an Englishman, the fair'st that eye could see, Her fancy fell a-turning* (William Shakespeare)*; She can start our franklin's daughters, In their sleep with shrieks and laughters* (Ben Johnson)*; I pray thee [King Edel], nay I conjure thee, to nourish as thine own, Thy niece, my daughter Argentile, till she to age be grown; And then, as thou receivest it, resign to her my throne* (William Warner)*; Because thou wast the daughter of a king, Whose beauty did all nature's works exceed, And wisdom, wonder to the world did breed* (Henry Constable).

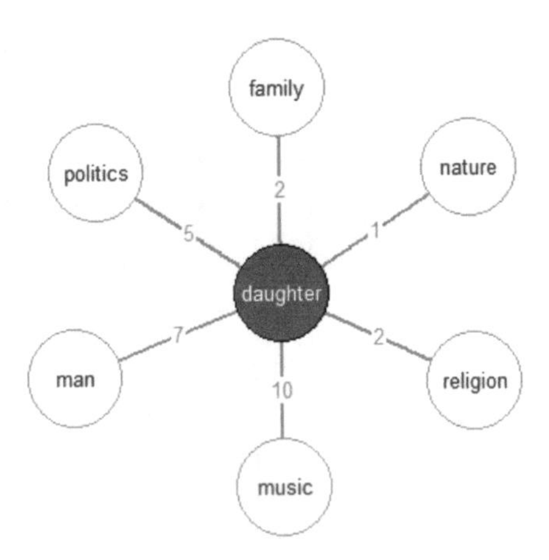

Fig. 12. Graphic modeling of lingual cultural type "Woman" based on the lexical variable "daughter"

The illustrated graphic models contributes to the formation of lingual cultural type "Woman" cognitive modeling. It indicates that the main interrelations of lexical variables used to denote a woman are recognized in the connection with: V1 – society, V2 – beauty, V3 – love, V4 – music, V5 – work, V6 – appearance, V7 – man, V8 – nature, V9 – religion, V10 – entertainment, V11 – mythology, V12 – family, V13 – poetry, V14 – hope, V15 – reign (see Table 2). Each of these groups consists of lexical units combined by leading features that denote different key spheres of human life during the analyzed historical period. Taking into account a big number and frequency of lexical units that designate a woman, we assume that she kept an important role in the society of English country despite the common subjective thoughts on her social position.

The established cognitive model reflects the most often used patterns of depicting the life and behavioral peculiarities of a woman in English language. Each culture cluster has its peculiar criteria of other person estimation that is reflected in language. According to the obtained data, we can observe the interrelationship map indicating at

Table 2. Interrelation output of lingual cultural type "Woman" graphic modeling

	V1	V2	V3	V4	V5	V6	V7	V8	V9	V10	V11	V12	V13	V14	V15
Bawd	•														
Bonilasse		•													
Bride		•	•			•	•								
Captainess	•														•
Conqueress															•
Countesse							•		•				•		•
Dame		•	•	•			•	•			•	•	•		•
Damsel		•	•	•	•										
Dowager					•	•									
Dowdy	•				•					•					
Empress			•				•								•
Feminine						•									
Gentlewoman	•					•	•								•
Girl	•	•	•				•			•					
Housewife	•				•		•					•			
Lady	•	•	•	•	•	•	•	•	•						
Lass	•	•	•			•	•	•		•					
Madam			•												
Maid	•	•	•	•	•	•	•	•	•		•				
Maiden	•	•	•	•	•	•	•	•							
Matron												•			
Mistress	•	•	•	•	•	•	•	•	•		•	•	•		•
Mother		•	•		•		•	•	•		•	•			•
Murderess	•														
Patronesse			•												
Princess		•	•												•
Queen							•								
Shrow	•					•									
Silk-wife					•										
Sister				•			•	•	•		•	•	•	•	
Trull	•					•									
Tygresse			•												
Virgin		•	•				•								
Wench	•	•					•								
Whore	•													•	
Widow	•	•					•							•	•
Wife	•		•				•			•					
Witch						•									•
Woman		•	•				•						•		
Womankind	•				•										

the strong bond between woman and lexical units denoting the society, beauty, and love. So that we can state about the positive appraisal of a woman as well as the admiration of her appearance that is often illustrated in the poetry by describing the parts of her body, clothes, and face.

We can also see that there is a tendency between units' bonds according to the contextual type of the lexical variables denoting a woman (i.e. the words used to denote a person who has more power that others or is a leader – *queen, conqueress, captainess* are connected with the group *power* by the relevant attributes; units that indicate a young, unmarried woman – *girl, lass, maiden* are often linked to the groups *beauty* and *appearance*). Additionally, the data acquired raised the necessity to group the lexical units of woman by a common feature (family, work, etc.) and provide a deeper analysis not only of the separate units but also of the lexical groups.

6 Conclusions

The research findings show that the cognitive model of lingual cultural type "Woman" allows to analyze the actual cultural and verbal relationship between the woman and the objects that are inalienable elements of the human world. The graphic model of the analyzed lingual cultural type helps to determine the patterns of these bonds depending on the exact lexical variable (*maiden, lass, mother, dame, daughter*), as well as their frequency. The importance of investigating the lingual cultural type is connected with the capability to disclose the lingual and cultural pictures of the world, as well as the mentality of a particular nation in a certain historical period.

To summarize the obtained infographics, it should be mentioned that the constructed cognitive models do not reflect all the possible connections between the items due to the possible error in the text data segmentation. At this research stage, the designed models are significant for the identification of common groups of the lexical variables that in the further analysis will be merged according to the key characteristic.

In terms of linguistic researches, the application of graphs might also be valuable for analyzing the problems of genre differentiation and texts attribution. In addition, it might be used to examine the epic and lyrical spiritual poetry with the aim to conduct the comparative analysis of the syntactic structures.

References

1. Maslova, V.A.: Introduction to the Cognitive Linguistics. Flinta, Moscow (2007)
2. Sukalenko, T.: Metaphoricalimage of a woman in Ukrainianlingual consciousness. NaukovyjVisnukVolynskogoNatsionalnogoUniversytetuim. L. Ukrainku, No 2, Lutsk, pp. 54–58 (2008)
3. Karasik, V.I.: Language Circle: Personality, Concept, Discourse, pp. 21–27. Paradigm, Volgograd (2002)
4. Elena, Y.: Lingual-cultural types of "English crank" (dissertation brochure), Volgograd, p. 22 (2005)
5. Olha, D.: Lingual-cultural types of Russia and France of the 19th century (dissertation brochure), Volgograd, p. 24 (2007)

6. Karasik, V.I.: Language Keys, pp. 175–181. Gnosis, Moscow (2009)
7. Popova, Z.D., Sternin, I.A.: Cognitive Linguistics, Moscow (2007)
8. Olha, D.: Lingual-cultural types of Russia and France of the 19th century. Ph.D. thesis, Volgograd (2007)
9. Abakumov, O.: Clustering applications in the Text Mining. Novyie informatsionnyie tehnologii v avtomatizirovannyih sistemah, No. 13, Moscow, pp. 128–129 (2010)
10. Perovšek, M., Kranjc, J., Erjavec, T., Cestnik, B., Lavrač, N.: TextFlows: a visual programming platform for text mining and natural language processing. Sci. Comput. Program. **121**, 128–152 (2016)
11. Kleinberg, J.M.: Authoritative sources in a hyper-linked environment. J. ACM **46**(5), 604–632 (1999)
12. Bird, S., Klein, E., Loper, E.: Natural Language Processing with Python. O'Reilly Media, Sebastopol (2015)
13. RapidMiner Studio 9.2.001 platform. https://rapidminer.com/products/studio/ Accessed 19 Feb 2019
14. Hofmann, M., Klinkenberg, R.: RapidMiner: data mining use cases and business analytics applications. In: Data Mining and Knowledge Discovery Series, p. 525. Chapman & Hall/CRC (2013)
15. Jianchao, H., Rodriguez, J.C., Beheshti, M.: Discovering a decision-based diabetes prediction model. In: Advances in Software Engineering, pp. 99–109. Springer (2009)
16. Tierney, P.J.: Qualitative analysis framework using natural language processing and graph theory. Int. Rev. Res. Open Distance Learn. **13**(5), 173–189 (2012)
17. Manber, U.: Introduction to Algorithms: A Creative Approach, pp. 79–93. Addison-Wesley, Reading (1989)
18. Solé, R.V., Coromin-Murth, B., Valverde, S., Steels, L.: Language networks: their structure, function and evolution. Complexity **15**(6), 20–26 (2010)
19. Olga, P., Evgeniy, K.: Expert Fuzzy Information Processing. Springer-Verlag, Berlin Heidelberg (2011)
20. Zade, L.A.: The Notion of Lingual Variable and Its Application to the Approximate Decision. Mir, Moscow (1976)
21. Gelbukh, A.: Computational linguistics and intelligent text processing. In: 5th International Conference on CIC Ling 2004, Seoul, Korea, vol. 2945. LNCS, 15–21 February 2004
22. Bordag, S., Heyer, G., Quasthoff, U.: Small worlds of concepts and other principles of semantic search. In: LNCS, vol. 2877, pp. 10–19. IICS (2003)
23. Grafoanalizator 1.3.3 platform. http://grafoanalizator.unick-soft.ru/. Accessed 21 Feb 2019
24. Huff, A.: Mapping strategic thinking, pp. 11–49. Chichester (1990)
25. Axelrod, R.: Structure of Decision: The Cognitive Maps of Political Elites. Princeton Legacy Library (1976)
26. Kosko, B.: Fuzzy Thinking: The New Science of Fuzzy Logic. Hyperion (1993)
27. Roberts, F.S.: Discrete Mathematician Models Applied to the Social, Biological, and Economical Tasks, Moscow (1986)
28. Bereza, O.A.: Cognitive model construction in the touristic sphere as the social and economical system. Izvestia YUFY, Tehnicheskie Nauki, No. 3 (92), pp. 7–12 (2009). https://cyberleninka.ru/article/v/postroenie-kognitivnoy-modeli-turistsko-rekreatsionnoy-otrasli-kak-sotsialno-ekonomicheskoy-sistemy-regiona. Accessed 23 Feb 2019
29. Baranov, A.N.: Introduction to the Applied Linguistics, pp. 11–23, Moscow (2001)
30. Charles, J.F.: Frames and the semantics of understanding. Quaderni di semántica **VI**(2), 222–254 (1985)

31. Kulakov, F.M.: Supplement to the Russian edition. Frames for the knowledge presentation, pp. 122–144, Moscow (1979)
32. Flys, O.: Cognitive Modeling of Lingual Cultural Type "Woman". In: Proceedings of International scientific conference on "Computer Sciences and Information Technologies" (CSIT-2019), vol. 3, pp. 24–30. IEEE (2019)

ICT in High Education and Social Networks

Virtual Educational Laboratory for Databases Discipline

Yalova Kateryna(✉) and Yashyna Kseniia(✉)

Dniprovsk State Technical University, Dniprobydivska 2,
Kamyanske 51918, Ukraine
katerynayalova@gmail.com, yashinaksenia85@gmail.com

Abstract. This paper deals with the development of virtual laboratory for Databases discipline, such laboratory can be used by Information Technology students as an auxiliary tool for self-education and acquiring professional skills in Ukrainian universities. The functional and software requirements for the laboratory practicum are given briefly in the paper. The functional and object models for the process of utilizing educational tools are presented in graphical and mathematical forms. The functional model represents the logic of the business process, data domain participants, and data exchange scheme. Business Process Modeling Notation has been used for representing the functional model in graphical form. The object model shows structural features of the data domain describing the entities, their attributes, and their interconnections. Entity Relation Diagram has been used for representing the object model in graphical form. The paper gives a description of the structure of relational database, as well as the features of graphical web-interface which implement the required functionality. The web-interface has been developed accounting for understandability, simplicity, and lowered sensitivity to user's mistakes. At the database and application level, the access rights are divided between user roles. In contrast to the existing ones, the virtual practicum presented herein gives a student the ability to master writing various SQL-queries. The given models of the data domain are universal and can be used for development of e-learning systems for various courses. The distinctive features of the proposed virtual laboratory are given here, as well as improvements that can be achieved when used in educational process.

Keywords: Virtual learning tools · Laboratory practicum · E-learning · Blending learning

1 Introduction

Negative transformations in the field of higher education such as: obsolescence and physical deterioration of laboratory equipment, devices, mechanisms and computer equipment, difficulty in obtaining funding to maintain a decent level of material and technical support of the process of training students of higher education, continuous reduction of classroom hours lead to the need to find effective mechanisms and means to support the appropriate level of acquisition of general and professional competencies of future specialists. In modern conditions, modern computer tools and technologies

N. Shakhovska and M. O. Medykovskyy (Eds.): CCSIT 2019, AISC 1080, pp. 623–636, 2020.
https://doi.org/10.1007/978-3-030-33695-0_42

can be effectively used in the educational process to visualize the processes being studied and provide a detailed illustration of the object of study [1]. The process of acquiring new knowledge and competencies within the discipline is divided into learning of the theoretical lecture material, organization of the process of mastering practical skills and abilities, management of the process of independent work of students, performing control of knowledge acquisition. At each of these stages, information technologies (IT) and computer equipment can act as an effective means of maintaining a high quality of presenting the learning material and be used to motivate students to cognitive activity.

Development of virtual laboratories can be carried out within commercial and noncommercial programming and modeling environment. Scientists involved in the creation and implementation of virtual laboratories in the educational process, such as: Yu. Zharkykh, R. Kolodiy, K. Bobryvnyk, V. Lapynskyi, A. Perekrest, Dongfeng Liu, D. Islek, W. Waldrop etc., note the relevance and prospects of this scientific and practical task. Authors of [2–4] consider the creation of the virtual educational laboratories for students of technical and technological specialties in the form of the web-oriented application using LabView programming environment. The results of the computer laboratory workshop development for the discipline «Digital equipment» on the basis of the NI Multisim modelling environment are presented in the [5]. Authors of the paper [6] present results of the virtual laboratory practice on «Circuit theory» course implementation. This software was developed using Delphi programming language. It allows to perform laboratory works at the virtual test bench which is equipped with the same functionality as real one. Results of the virtual laboratory workshop development for remote sensing and researching of digital transmission systems with visualization by the means of the MOODLE are described in the paper [7]. For distance learning of chemical objects and processes such software as Chermlab and Virtual Chemistry Laboratory are widely used. Despite of this in the work [8] authors demonstrate the electronic educational laboratory creation for studying of chemistry using the JavaScript programming language and the hypertext markup language (HTML). In the work [9] authors discuss positive and negative aspects of the virtual laboratories development for disciplines of the electro mechanics specialty. There are a lot of virtual tools for studying of programming: CodeAcademy, DataCamp, DataQuest, freeCodeCamp, Udacity etc. Comparative characteristic and description of them are presented in the work [10]. In contrast to enumerated educational on-line tools the offered virtual laboratory is designed for obtaining specific professional competencies on the base of university curriculum. The most famous academic virtual educational platform is cloud-based Integrated Development Environment CS50 powered by Cloud 9, developed in Harvard University (USA) [11].

Despite the large number of research papers devoted to the development of virtual laboratory practicums automation of the process of acquiring professional competences remains an actual scientific and practical problem especially important for technical specialties and specialties of practical orientation [12].

The purpose of Databases course is that the students achieve the following competencies:

- ability to conduct a systematic analysis of the data domain (DD);

- ability to make up the infologic model for the DD accounting for data normalizing rules;
- ability to select optimal software technologies for implementing the database physical model;
- ability to create high-quality program code implementing data processing mechanisms: database management operations, CRUD data operations, SQL queries, etc.

The main purpose of the article is to present the results of designing the architecture and web-interface of a virtual laboratory practicum (VLP) for the Databases discipline, which can be used as an additional learning toolkit for students of higher education of the «Information technology» field of knowledge. The scientific research significance consists in the development of models, algorithms and the method for VLP development as a mean of information and communication pedagogical technology. The research results broaden the existing theoretical knowledge about implementation of distance learning tools. They open up prospects for further applied researches and elaborations in pedagogical science and information technologies.

The process of the VLP development is consists of the following tasks:

- formation of functional and program requirements for a virtual learning toolkit;
- development of the AS-IS and AS-TO-BE functional models of the DD describing user roles, actions, interaction scheme functions and data flows movement;
- designing a relational database based on the object model of the DD;
- development of the architecture of a virtual laboratory practicum taking into account the distribution of data access rights by different users;
- designing the web-interface of a virtual laboratory practicum.

In the work authors present the results of the VLP life cycle stages, namely: data domain analysis, requirements formation and design.

2 Functional and Software Requirements for the Virtual Laboratory Practicum

One of the methods of improving the results of students' self-study regarding the discipline chosen is utilization of additional IT tools available remotely for both on-campus and partial-time students [13]. For implementing distance learning, it is reasonable to use e-learning systems, academic MOOC (Massive open on-line courses) platforms, commercial and non-commercial Learning Management Systems. On-line education systems provide access to theoretical materials available in various formats on the Internet. The virtual laboratories can be developed as an integral part of e-learning systems used for improving practical skills and abilities of the students.

VLP should be a set of software and hardware means designed for practical training and improving the level of professional competencies of future specialists. The easiest way to implement a virtual laboratory practicum is to develop a web-based interactive application with distributed data access rights, which involves all the participants of the learning process [14].

The developed practicum meets the characteristics of the system of individual training of a simulator type. And the main purpose of its application is not to obtain theoretical knowledge, but to acquire practical skills in accordance with the requirements of the «Databases» discipline.

Just as for any other piece of software, the process of virtual laboratory development is comprised of five phases: requirements analysis and data domain analysis, design, implementation, testing, and deployment. Data domain analysis gives us the following results: defined requirements for informational, organizational, and software tools, functional and object-based models of the DD. System architecture, database logical structure, as well as user interface prototype are developed at the design phase. Implementation phase implies programming the database and its data processing mechanisms. The purpose of testing phase is to reveal and fix program flaws. The final phase includes deployment and utilization of the virtual laboratory in the educational process. Adoption of spiral lifecycle model in development of the virtual laboratory allows having simultaneously an operable version, updates log, and development plan for the next version.

The VLP is implemented as a combination of training and mentoring programs, with the following basic functional requirements:

- minimally set level of theoretical training of students;
- development of skills in the question – answer mode;
- availability of a meaningful dialogue with the user. Each action of the trainee is accompanied by comments, prompts, algorithms for solving typical tasks;
- possibility of constructing an adaptive individual training scheme based on existing knowledge of the trainee;
- implementation of various ways and forms of presenting tasks of practical orientation;
- motivation and support of cognitive activity of trainees in the course of achieving educational goals;
- availability of pedagogically justified feedback from the system, providing assistance in solving the tasks set and detailed information on the errors made;
- availability of various mechanisms of assessment and self-assessment of acquired knowledge and skills;
- pedagogically justified determination of educational activity, which on the one hand allows managing the learning process in accordance with the working curriculum of the discipline, and on the other hand does not violate the field of student independence.

The architecture of the virtual laboratory shall consist of the following functional modules:

- user data access module providing graphical interface available via a browser;
- common services module providing user identification data storage;
- educational services module intended for creating, editing, deleting, and formatting educational e-content and its structural-and-logical scheme;
- database providing storing all the system data;
- client-server data processing module.

As a language for software implementation of the web-based application, it is advisable to use an open source programming language, for example – PHP. The MySql Server database management system software can be used to create a relational database of the virtual laboratory practicum. All data processing should be carried out using server data processing, the client is provided only with the functions of displaying the user interface. The use of a «thin client» allows for the implementation of the basic principle of distance learning – «to learn anywhere and anytime» [15]. To use the developed practicum, the student should have access to the Internet and technical means of interaction with the system: computer, smartphone, tablet, etc. And most importantly, there is no need to install special software or software platform to support the VLP operation.

The developed software must meet the requirements of high-quality program code, the main of which are: independence from hardware and software platforms; distribution of processes and their autonomy; scalability and flexibility; ability to integrate and expand.

3 Modeling of the Virtual Laboratory Data Domain

Data domain modelling gives a set of models reflecting its structural peculiarities and functional principles. The functional and object models of the DD should be described using minimum number of objects, their properties, functions and relations required and sufficient for solving the problem given. The requirements set for the DD models are as follows:

- formalization providing a comprehensive description of the DD structure;
- understandability for customers and developers that can be achieved by using graphical methods for model presentation;
- feasibility which means there are tools available for physical implementation of the model obtained;
- adequacy and efficiency of the models.

To formalize the results of the domain analysis, it is advisable to present them in the form of mathematical models that display the functional and structural features of the system.

3.1 Functional Data Domain Model

A functional model is a domain description based on the analysis of semantics of objects and phenomena, performed without orientation on the future use of programming or technical computer tools, in which the functional aspect of domain modeling is emphasized.

Within the VLP framework a combined scenario of automated training is used, consisting of software-controlling and logically-controlled components. The software-controlling component allows setting the output sequence and the structure of training materials in the practicum itself. The contents and logical-structural scheme of sessions and tasks are determined by the teacher at the stage of development of training materials. The logically-controlled component of the training scenario provides the student with the opportunity to choose the laboratory session or practical task of interest to him, as well as to make a decision on the need to obtain theoretical information or prompts of the system during its execution.

The functional modelling of the DD is based on analyzing the semantics of the objects and phenomena, as well as on decomposition of actors' functions without orientation towards specific software tools. The functional model of the virtual laboratory allows us to define the structure, amount, and schemes of information data flows from the sources to the consumers. The main data is the educational materials created by teacher and open for the students to use via the virtual laboratory. Definition of the functions and precedents (allowable actions) for all subjects of the DD allows to separate users and define their roles.

There are three user roles with different functions and data access rights, provided in the VLP: administrator, teacher and student. The main actions of users, depending on the role, are as follows:

- administrator: system configuration, work capacity control and debugging of the system operation as a whole, testing of new modules, database administration, access rights management, publication of training materials in the system;
- teacher: development of training materials in the prescribed format, definition of the list of laboratory sessions, practical tasks, prompts and controlling actions for the learning process, definition of grades for each task, setting the level of complexity, overall control of educational activity;
- student: learning of training materials, history of the training progress, creation of the individual learning trajectory within the framework of the proposed content.

Decomposition is the main notion of functional modelling. It means dividing a complex process into comprising actions. Decomposition allows to structure the system model as a hierarchical structure made of individual diagrams describing a specific business-process. Usually, an AS-IS model should be created first that describes system's organization. Nowadays, the Databases laboratory classes are carried out only in a classroom under the supervision of a teacher. The VLP will allow solving practical tasks on-line providing logging of results and selection of operation mode: training without assessment or test. This article presents the results of creating AS-TO-BE models that describe the business processes after reengineering and automation.

The following main functional actions of the administrator regarding the training materials should be outlined: adding, deleting, and editing the existing electronic content. The assignment of functions to edit training materials to the administrator gives the opportunity to ensure the unification and standardization of their presentation, and also increases the tolerance of the system to data input errors of users.

The teacher will be set a task to form electronic training materials: a logical-structural scheme of laboratory sessions, practical tasks, answers, prompts and points for the assessment of tasks.

After the authorization, the student is given the opportunity to choose a needed laboratory work and a practical task to perform. If necessary, the student can refer to a brief theoretical basis, or proceed with the performance of a practical task. After receiving the task, the student has the opportunity to enter the performance result, there is no time limit while doing so. If the student's result corresponds to the reference one, a notification on this is shown, the results are recorded in the database, and the student is able to move on to the next practical task. In the laboratory practicum, the functions of automatic control of both syntax and semantic errors of the database development process and creation of queries to it are implemented. Syntax errors are incorrect writing of the program code. Syntax control of the response is to establish the correspondence of the data received from the student to the complex of syntax rules of the SQL language. Semantic errors are incorrect understanding of the task, errors in the normalization or in the logic of a database design. The peculiarity of the developed practicum is that the verification of completed tasks is aimed at comparing the array of results of the student's sample with the array of data of the correct query. This approach allows avoiding the situation, when the correct program code written by the student, which differs from the reference result only in syntactic construction, is interpreted by the system as incorrect. If the student's error is semantic (the query structure is correct, but the query does not perform the set task), then the student is provided with the structure of the correct sample as a prompt. If the student's error is a syntax one (the query code is incorrect), then the student receives an error message with the text of the comments. Also, the student has the opportunity to work out different versions of the logic of creating SQL-queries. Within the VLP framework, the quality of laboratory works performance is assessed based on the number of points earned and errors made.

Graphical and mathematical methods are available for creating DD models. Graphical methods allow to represent the analysis using graphical notations and rules for displaying them. Their main advantage is simplicity and understandability. Mathematical modelling is less resource-consuming and more formalized method of knowledge presentation. In this paper, Business Process Modelling Notation (BPMN) is used which allows to represent the system as a set of interconnected business-process diagrams. The BPMN methodology contains an alphabet of intuitively understandable elements that allows describing complex semantic structures and interconnections. Moreover, it defines how the diagrams describing the business processes can be transformed into executable models in Business Process Execution Language, XML-based language used for formalized description of business processes and interaction protocols. For example, in Fig. 1 a functional model diagram is given for the process of acquiring practical skills by students.

Similar diagrams have been made for describing the actions and functioning algorithms for each user role. For describing the interactions and data exchange sequences between the users a diagram has been developed which represents swimlanes for main user roles pools, events and message flows.

In the mathematical form, data processes can be represented as follows:

$$C = \langle E, G, S \rangle \tag{1}$$

where C is a DD limited by the requirements of the task to be automated; $E = \{E_1, \ldots, E_N\}$ is a multitude of processes of the DD, which are described by the input and output actions, take into account the rules and algorithms of the system operation and display the resources used in the course of performing an action. G is a multitude of relations between processes, where the result of the i-th process is presented as the

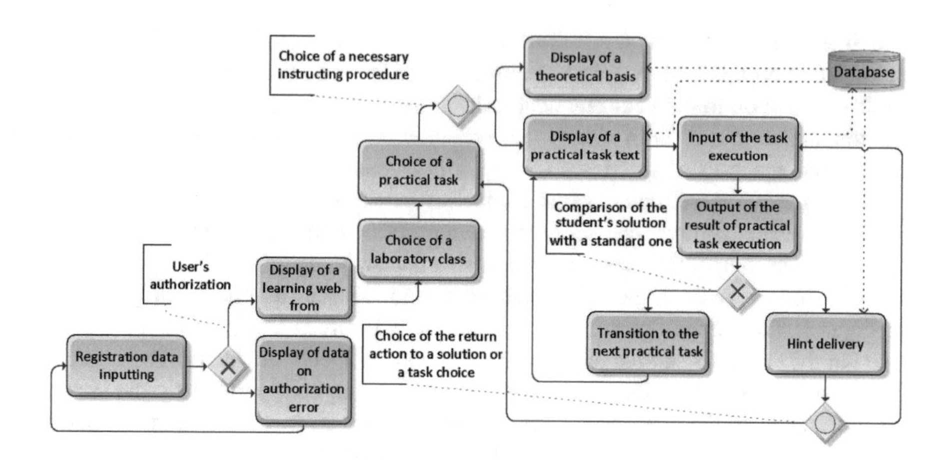

Fig. 1. Diagram for the process of student's getting practical skills

input action on the *j*-th process; *S* is a dictionary of glossaries of each level of the model that forms a unified, semantically significant description of all definitions of the DD and is used in the formation of the natural hierarchical structure of objects and functions of the DD. The sets of input and output dataflows, control actions and mechanisms for each *i*-th process can be described as information, material, labor, or financial resources of the DD. The following constraints have been accounted for while creating the functional model:

- a business process cannot have no results, since its purpose is transforming the input dataflow into the output one;
- a business process cannot execute without resources and control action defining its implementation algorithm.

Input action set *In* for each *i*-th input dataflow of business process E_i is defined as $In_i = \{in_1, \ldots, in_N\}$. Each output dataflow $O_i = \{o_1, \ldots, o_N\}$ of business process E_i describes output data obtained in the course of interactions between the participants of educational process within virtual laboratory. Control action $L_i = \{l_1, \ldots, l_N\}$ is a set of control flows that govern rules, algorithms, and constraints for E_i business process. The resource-flow set $R_i = \{r_1, \ldots, r_N\}$ describes resources either used in virtual laboratory or for data processing. Thus, each business process of the DD can be described as follows:

$$E_i = \langle In_i, O_i, L_i, R_i \rangle \quad (2)$$

«Output-Input» type of relations *G* have been determined within the DD, i.e. the output dataflow set of process E_i are transformed into input dataflows of process E_j, when the latter become available the process E_j can be started, here k_1 is the output (out) of *i*-th process which serves as k_2 input (in) of j-th process:

$$out_i^{k1} = G_{OI}(E_i);\ in_j^{k2} = out_i^{k1} \tag{3}$$

The functional model development gives us the ability to represent the relations between the objects, data exchange schemes and algorithms, define data access rights at user level, design the architecture of the virtual laboratory.

3.2 Object Data Domain Model

Accounting for the functional features of the DD determined earlier we have to describe it in terms of interconnected entities. The object model is created for this purpose – an infologic level model, in which the structural aspect of the system under modelling is emphasized. The object model of the DD allows taking into account the structural features of interrelated and interacting objects. The structural components of object model are entities, their attributes, and interconnections. Data domain entity is a unit that can be static or dynamic, simple or complex, created by summation and aggregation mechanism. Entity features are its description, attributes or properties representing a feature- or a quantity-based values. The connections between the objects have a property of multiplicity, i.e. the number of child entity instances per one parent entity instance. The main data structure used for describing the object model is a hierarchical tree that can be easily implemented in the form of a database.

The object model can be created using both graphical and mathematical methods. For simplicity and the purpose of visualizing the DD analysis, in this paper the Entity Relation Diagram (ERD) is used, which can generate entity specifications and relate them with each other using «one-to-one» , «one-to-many» , and «many-to-many» connections.

The fragment of the created data domain object model which describes the structure of the aggregated object «Discipline» is shown in the Fig. 2. There are only main attributes of such DD objects as: disciplines, teachers, laboratory classes and practical tasks in the presented ERD.

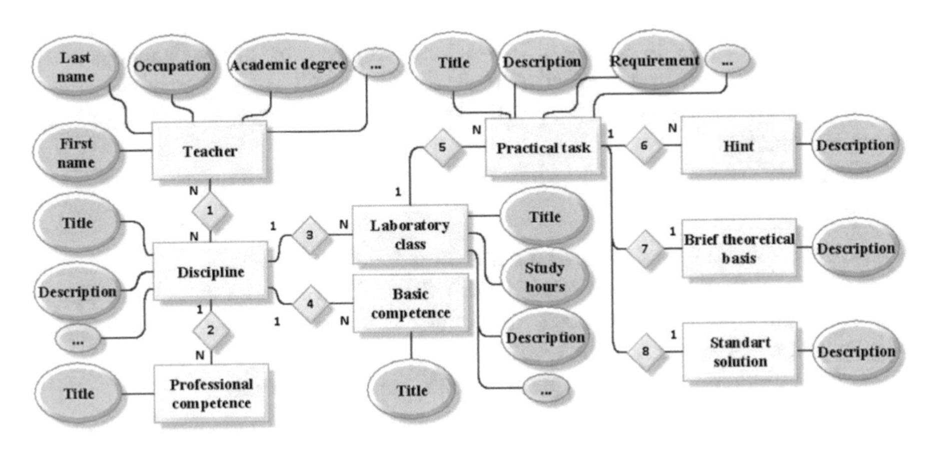

Fig. 2. Fragment of the data domain object model

We can describe an object model mathematically as follows:

$$OM = \langle O, P, R \rangle \tag{2}$$

where OM is an object model of the DD, $O = \{O_1, \ldots, O_N\}$ is a multitude of entities, which consists of a set of all domain objects significant for the problem solving; $P = \{P_1, \ldots, P_N\}$ is a multitude of characteristics of entities of the domain, $R = \{R_1, \ldots R_N\}$ is a multitude of relationships between the objects with a given multiplicity of relationship. In the course of determining a multitude of relationships between entities, the data generalization and aggregation mechanisms were used.

A special feature of the designed object model is its universality, which allows to represent the data and knowledge about the system using standardized, general form that would be understood not only by the developers but also the participants of the DD. The following items can be named among the special features of the model:

- generalized description of the DD using structural relations without applying to any specific university, major, or discipline. This allows for using it for developing various tools of e-learning systems;
- utilization of natural hierarchical object classification, which adequately represents complex data structures as a «one-to-many» connection;
- scalability. The model represents the real objects and their features but is able to be extended in case input conditions of the DD or its environment are changed.

The universal description of the structural features of the DD allows easy and rapid transition to creating database logical model and its software implementation.

4 Database and Web-Interface

The input data for designing a database is the object model of the DD, which displays the quality and quantity attributes of entities, their structure and schemes of interrelation between them, as well as the rules of data normalization. The existing natural structural dependences of objects of the DD are displayed by the mechanisms of the relational database, for example, «discipline» is represented as an aggregated entity of such objects as: laboratory works, practical tasks, prompts, reference solutions, theoretical information.

The general sequence for transition from DD object model to database model is as follows:

- one data domain entity should be saved in one database table;
- one entity feature should be represented with a table field of a given data type;
- connection multiplicity between the entities is shown in the database relational connections based on foreign keys;
- database modelling is carried out using data normalization rules.

The database is developed by modelling the data on logical and physical levels. The logical model of the database is presented as Fuller Attributer Model, which allows to represent the problem domain data as tables showing primary and foreign keys, indices, as well as to show domains and table attribute data type.

The database contains all laboratory sessions provided for the discipline, a brief theoretical basis and a set of practical tasks. The number of laboratory works and practical tasks in each of them is set by the teacher. For each task, depending on its difficulty, the teacher provides several prompts, examples or recommendations, using different formats of their presentation: text, video, animation. Prompts give answers to common errors that may occur during the task performance and ways to correct them. Each practical task stores in the database an array of data obtained when the SQL-query is correctly created, which is used as a reference one in the course of evaluating the correctness of the task performance by the student. Detailed description of the task progress: the number of attempts of performance, the correctness of performance, the correspondence to the reference solution, the number of requests for prompts and assistance, the points received are recorded in the electronic journal of laboratory works performance, which is automatically populated during the interaction of the student with the VLP. Information from the electronic journal is provided to the teacher, on the basis of which the teacher can monitor the educational activity of the student.

User interaction with the virtual laboratory practicum is carried out through the developed and software-implemented web-interface, which was developed taking into account the requirements of clarity, ease of mastering, reducing the sensitivity to user errors. For each user role, the interface of web-forms is different, but they are developed in the same style and with due regard to the requirements of unification of the dialogue with the user. When developing the scheme of the student's dialogue with the training module, the criteria for the interface effectiveness and the psychological principles of the student's interaction with the computer were taken into account, namely:

- involvement of the student in the dialogue is carried out only at the time of reasonable need, so as not to disrupt the course of mental activity;
- the appearance of system messages does not violate the internal dialogue of the student and his field of independence;
- the student has the opportunity to individually refer to the prompts at a time when he realizes his error and agrees to accept help. The system prompts were created taking into account the fact that they should be non-intrusive, but sufficient to support educational activity.

The main elements of the user interface are: the navigation menu (a set of hyperlinks and transitions to choose laboratory works and practical tasks), the elements of displaying tasks, the elements of entering the student's answer, the elements of displaying prompts, the elements of displaying the results of a practical task performance, the elements of displaying reported data, the elements of entering the parameters of search for training materials. The web form interface prototype for realization of practical tasks is shown in the Fig. 3.

There are six main functional fields on the web form (Fig. 3):

- task number and the title of the topic to which it belongs – field 1;
- navigation panel – field 2;

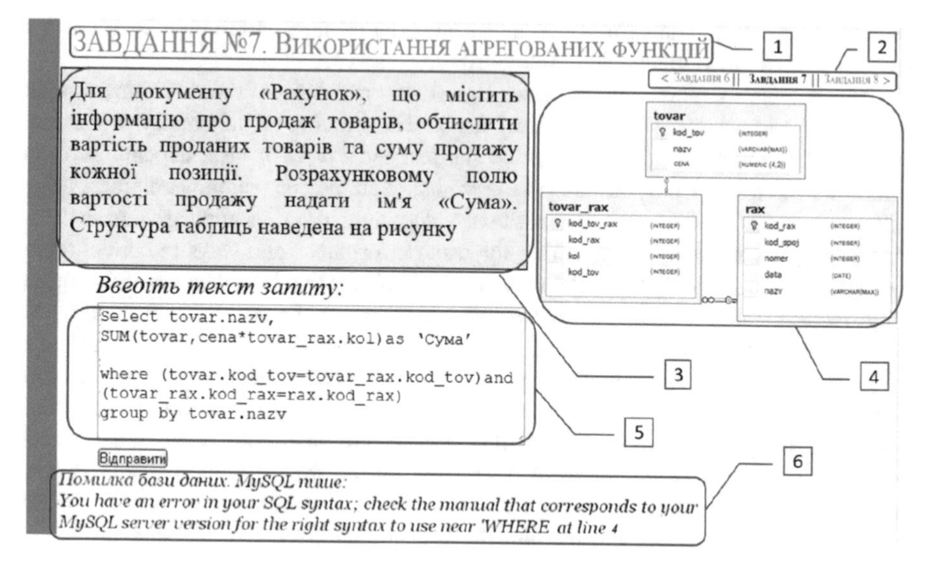

Fig. 3. Web-interface prototype

- description of the task in a verbal form – field 3;
- logical structure of the database tables for which it is necessary to create queries – field 4;
- field for query inputting – field 5;
- field of mistakes representation – field 6.

The electronic content of the practicum is structured on the basis of the logical-structural scheme of the discipline set by the teacher.

5 Conclusions

Reducing hours of the classroom load puts teachers to the task of supporting a high level of professional training of students and their motivation for self-education in accordance with the curriculum. An effective tool to achieve these goals can be the use of training IT, distance and e-learning systems, various training modules and virtual laboratories. For specialties in the field of information technologies, performing laboratory works and practical tasks is the most important pedagogical tool in mastering new skills and acquiring general educational and professional competencies. Development, implementation and successful introduction of virtual toolkits to support independent work of students are complex and relevant scientific and practical tasks.

The authors propose to use the developed virtual laboratory practicum as an additional training tool to acquire practical skills and abilities in developing, debugging and testing of SQL-queries in the «Databases» discipline. The article briefly describes the functional and software requirements for a virtual laboratory practicum. The database structure and the features of the graphical web-interface are presented, which allow implementing the predetermined functionality and take into account the functional and object models of the DD. The developed models of the DD reflect the natural dependences of subjects of the learning process, meet the requirements of versatility and adequacy. The peculiarities of administration, functions and roles of users taking into account the differentiation of access rights to training materials are considered. The draft of the virtual practicum is aimed at a specific discipline and implements its logical-structural scheme of presenting training materials, but the developed models can be used in the implementation of virtual training toolkits of practical orientation for other disciplines and specialties. The presented virtual practicum, in contrast to the existing ones, allows the student to hone the skill of writing various SQL-queries. Practicing in writing SQL queries is available in two modes: training mode without grading and exam mode, the results of the latter can be used by teachers for grading. Verification of correctness of the software implementation of tasks is aimed not only at verifying the syntax of using SQL language keywords, but also at matching the resulting data arrays obtained during the execution of an SQL-query.

The distinctive features of the proposed virtual laboratory are as follows:

- cost-free programming tools for its implementation, low requirements for the hardware and software platforms, no need to use software under commercial licensing;
- implementation of automated laboratory classes for students of various education forms. Client platform requirements: a PC or mobile device with Internet access and any web-browser installed;
- low operational threshold, minimum IT-competencies required from all education process participants;
- utilizing the virtual laboratory does not violate the discipline's logical structure and acts as an auxiliary self-study tool for the students.

The importance of the proposed solutions is that the arrangement of efficient self-study for the students is a highly important task for the teacher, since self-study amounts to 70…90% of study hours in Ukrainian universities.

At the moment, the virtual laboratory development is at the implementation phase. After data inputting and software testing, the authors plan to publish the laboratory practicum on the server of the Dniprovsk State Technical University (Ukraine) for the implementation of blending learning for IT-students. The goal of further researches is to represent the results of numerical assessment of the effectiveness of the developed virtual laboratory practicum use by carrying out its approbation for students of full-time, part-time and extramural forms of education.

References

1. Yefymenko, Yu.: Computer laboratory practicum on digital technology. Digests Berdyansk State Pedagogical Univ. **6**, 5–11 (2013)
2. Bobrivnyk, K., Gladka, N., Kiktev, M.: Designing virtual learning laboratory for students of technical and technological professions. Energy Autom. **3**, 18–23 (2014)

3. Lapinsky, V., Voevodin, S.: Laboratory workshop environment on the base of NI LabView. Comput. Sch. Fam. **6**, 29–33 (2015)
4. Shamshin, A.: Remote lab in physical practice. New Comput. Technol. **15**, 185–188 (2017)
5. Malykhina, T.: Virtual laboratory workshop on the basis of GEANT4 in the Linux environment to explore the processes of radiation interaction with matter. In: Second International Conference on Foss Lviv, Lviv, Ukraine, pp. 81–84 (2012)
6. Zianchuryna, I., Liuliuchenko, T.: Virtual laboratory practice on «Circuit theory» course. Opened Inf. Comput. Integr. Technol. **49**, 228–233 (2011)
7. Kolodiy, R.: Development of a virtual laboratory workshop for remote sensing of digital transmission system in the MOODLE. Bull. Lviv Polytech. Natl Univ. **775**, 63–68 (2013)
8. Ovcharenko, V.: Organization laboratory practical work on medicinal chemistry on modern means of IT. Young Sci. **10**(37), 269–272 (2016)
9. Chornyi, O., Rodkin, D., Yevstifieiev, V., Perekrest, A., Velichko, T.: Virtual laboratory complexes for an educational process and scientific researches. Trans. Kremenchuk Mykhailo Ostrohradskyi Natl. Univ. **3**(50), 28–42 (2008)
10. Theory and practice. https://theoryandpractice.ru/posts/17365-8-onlayn-platform-s-kursami-po-programmirovaniyu. Accessed 20 Aug 2019
11. CS50 IDE. https://cs50.readthedocs.io/ide/online/. Accessed 25 Aug 2019
12. Zharkykh, Yu., Lysochenko, S., Sus, B., Shkavro, A.: Problems of organization of laboratory practicum in the process of e-learning. Sci. Notes Pedagogy **1**, 72–78 (2011)
13. Traxler, J.: Distance learning – predictions and possibilities. Educ. Sci. **8**(35), 1–13 (2018)
14. Porumb, C., Orza, B., Mihon, D., Rad, A.: Virtual laboratory and classware concepts in internship programmes. In: 17th International Conference on IT Based Higher Education and Training, Olhao, Portugal, pp. 211–220 (2018)
15. Yalova, K., Zavgorodnii, V.: Challenges and prospects in development of e-learning system for IT students. Eng. Educ. Life-Long Learn. **1**(26), 25–43 (2016)

Information Modeling of Dual Education in the Field of IT

Roman Holoshchuk[1], Volodymyr Pasichnyk[2], Nataliia Kunanets[2], and Nataliia Veretennikova[2(✉)]

[1] Department of Social Communication and Information Activities, Lviv Polytechnic National University, Lviv, Ukraine
holoshchuk@vlp.com.ua

[2] Information Systems and Networks Department, Lviv Polytechnic National University, Lviv, Ukraine
vpasichnyk@gmail.com, nek.lviv@gmail.com, nataver19@gmail.com

Abstract. Modern approaches to the organization of the educational process generate a need for the introduction of dual education. Such approach is particularly relevant for IT specialties. The main principles of dual education are presented and it is suggested to use the information model of a dual education applicant as a set of knowledge about a person used for organization of effective learning processes, as well as the thematic subject model of a dual education applicant, the functional object model of a dual education applicant, the semantic subject model of a dual education applicant, the procedural subject model of a dual education applicant. It is provided the model of dual education that consists of the following models such as a model of educational material, test question models, models of reference knowledge of an expert (a teacher), a model of knowledge and a model of the practical component of the educational process.

Keywords: Dual education · Model of listener · Dual education applicant

1 Introduction

The advantages of transitioning the organization processes of various forms of education to new information technologies are unquestionable. Transformation processes in the education industry are generated by the need to integrate into the European educational space, as well as to increase the competitiveness of knowledge gained by future specialists. Obtaining the significant achievements in establishing effective educational processes in the IT industry is impossible without the creation of a competitive national training system. It is important to create conditions for combining work with education not only in the form of gaining education by part-time form of training, but also because of introduction of practical training forms for applicants of higher full-time education in cooperation with firms - employers. The Law of Ukraine on Education says that a person has a right to receive education in different forms, without excluding a possibility of combining them and providing the chance to choose the dual one.

N. Shakhovska and M. O. Medykovskyy (Eds.): CCSIT 2019, AISC 1080, pp. 637–646, 2020.
https://doi.org/10.1007/978-3-030-33695-0_43

A dual form of education is a form of education that involves a combination of training people in educational institutions with workplace training in enterprises or organizations for obtaining a certain qualification, usually on the basis of a learning agreement on the dual form of education [1].

1.1 Analysis of Recent Researches and Publications

In publications devoted to the study of educational problems, the main attention is paid to the development of electronic manuals, platform selection, didactics of new educational environments, teaching methodology, etc. However, the issues related to the mathematical modeling of educational processes remain unconsidered, as well as the issues related to the mathematical and software learning processes remain insufficiently described [2–4]. The use of informational social and communication technologies promotes the creation of intelligent systems based on effective mathematical models and can not only provide information and analytical functions, but also create conditions for the operational management of the learning process, form an effective organization environment with access to distributed sources of learning information using the new network-centric technologies [5–7].

2 The Concept of Dual Education

The principle of dual education at the level of vocational education in European countries is known for the ability to provide the labor market with a highly skilled workforce that can adapt to new and changing conditions and introduction of new technologies. At the same time, the introduction of dual education contributes to reducing unemployment, as a skilled workforce, educated and trained in accordance with this concept, has a very high mobility in the labor market.

The term dual system was introduced into pedagogical terminology in the mid-1960s in the Federal Republic of Germany as a new, more flexible form of vocational training. Duality as a methodological characteristic of vocational education involves the coordinated interaction of the educational and manufacturing spheres in the training of skilled personnel of a certain profile within the framework of new organizational forms of training [8, 9].

The German model formed the basis of the concept of dual education, which is being implemented in Ukraine based on vocational education institutions. The project began to be implemented several years ago on the basis of three institutions of vocational education.

The main task of introducing elements of the dual form of education is to eliminate the main disadvantages of traditional forms and methods of training future specialists, to bridge the gap between theory and practice, education and production, and to improve the quality of training qualified personnel, taking into account the requirements of employers in the framework of new organizational forms of education.

Creation of effective distance learning systems is based on effective mathematical models that can not only provide information and analytical functions but also create conditions for the operative management of distributed distance learning processes

[11], serve as an effective learning environment for organizing and managing the process, and putting in place an effective student-oriented curriculum and a convenient for obtaining a dual education schedule.

The design of an integrated student-centered learning environment for dual learning, which includes a distance learning system, is based on the development of a set of mathematical models that combines the structural model of updated training courses, the contents of which are filled in accordance with the requirements of the IT market, the student's knowledge model of dual education and the process model learning based on duality [12].

Consequently, the problem of designing an integrated educational information and communication environment for modern learning systems, based on the modern achievements in the field of didactics, ergonomics and mathematical modeling, is becoming increasingly relevant. The complex of mathematical models proposed in this paper is based on the model of the dual education applicant and their knowledge.

Modeling and designing any training system begin with the definition of its content. The training content is not a given system of knowledge, but it is systematically organized educational activities and a set of knowledge that ensure its effectiveness.

The content of education is determined by modeling the dual education applicant and the system of dual education [13].

3 Model of Dual Education Applicant

The model of dual education applicant is a knowledge about a person used to organize effective learning processes. This is a set of precisely presented (formalized) facts about a listener, which describe the various aspects of their condition, namely knowledge, personal and professional qualities, etc.

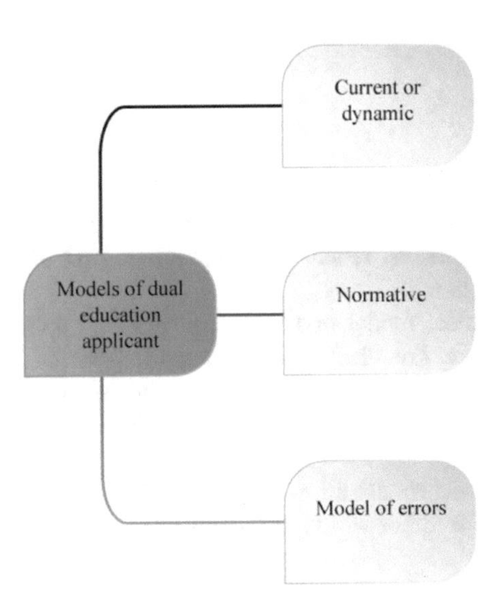

Fig. 1. Model of dual education applicant

The approaches to modeling the dual education applicant or our knowledge about it have not been formed yet. Based on the classical training models, it is proposed to consider the model of dual education (Fig. 1).

The model of a dual education student can vary within the existing three types of education applicant models:

- current, or dynamic (listener in real conditions),
- normative (reference prototype),
- errors (long-term prototype).

The normative model of education applicant is conditioned by the standard of a specialty. Normative model is a collection of:

- subject models (normative models of educational subjects),
- requirements for the professional activity of a specialist,
- professional requirements for a specialist,
- other requirements.

The subject model of an education applicant is the standard of gaining knowledge on a subject. The subject model of education applicant distinguishes the educational areas from all the numerous subject areas; thus, it is a model of the educational subject area or a model of the educational subject.

Introduction of the concept of the subject model of an education applicant allows the modeling of the dual education applicant to be integral, because in this case, the modeling will be for all aspects, including the person who learns.

Modeling of educational subject areas is significantly different from the modeling of other subject areas, because the aims of modeling are different too. Modeling of other subject areas should ensure that socially meaningful results are obtained, while modeling of the subject area is the process of solving educational problems.

The subject model of the dual education applicant consists of five components:

- thematic,
- functional,
- semantic,
- procedural,
- operational.

The thematic subject model of the dual education applicant determines the list of subjects to be studied, that is, reflects thematic structuring of subject knowledge. In fact, this is the curriculum.

The functional subject model of the dual education applicant determines which functions perform subject knowledge and reflects the functional structuring under the headings. For example, in physics it can be distinguished concepts, laws, properties, equations, models, algorithms, etc. This model involves the decomposition of knowledge.

The functional subject model allows you to detail what a student should know at the reproductive level, in other words to remember.

The semantic subject model of the dual education applicant determines declarative or factual knowledge. This is a consistent set of statements, each of which is the single complete thought.

The semantic subject model of the dual education applicant is the basis for the formation of an oriented basis of actions.

Procedural subject model of the dual education applicant contains procedural subject knowledge, that is, rules for the transformation of objects in the subject area. They include:

- instructions,
- algorithms
- methodology,
- decision-making strategies.

Procedural knowledge plays the role of schemes of a targeted basis of general actions. They are implemented with the help of abilities.

Skills relate to behavioral knowledge. The list of subject skills is the operational subject model of the dual education applicant.

Ability to study is a system with a hierarchical structure consisting of several levels. They form skills such as:

- basic,
- methodological,
- general,
- interdisciplinary,
- subject.

Subject skills consist of general, specific and experimental ones.

Examples of general subject knowledge in physics are to reproduce independently phenomena and processes; to use systems of physical units, etc.

Specific subject skills are to find, define, build, count, evaluate, consider, decompose, compile, apply, etc.

Specific subject skills have a horizontal structure, because they are composed.

In the general case, specific subject skills consist of abilities of all levels such as basic, methodological, general and interdisciplinary.

On the basis of the semantic model, diagnostics of knowledge is organized, while on the basis of the operational model it is diagnostics of abilities.

The operating model creates the basis for implementation of the spectral approach in the development of tasks or sequence of tasks.

For the system of dual education, the means and methods of knowledge control are very important. In order to increase the objectivity of control, it is essential to unify the measurement scale.

A convenient tool for pedagogical diagnostics of knowledge, especially for a dual system of education, are computer tests.

We will consider one of the methods of developing and using diagnostic computer tests. The method is based on the principles of "white" and "black box" (Fig. 2). The "white box" uses the semantic model of knowledge, and the "black box" is the investigated system (of the dual education applicant). The input data that interacts with the “black box” is a specially selected test tasks. Pedagogical diagnostics is carried out by comparative analysis of the test results (output data) with reference, obtained from similar effect on the “white box”.

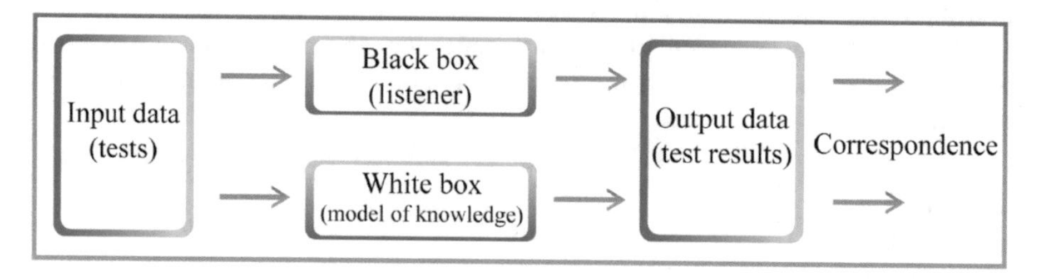

Fig. 2. The process of sounding mathematical formulas in Ukrainian

In order to study the content of the “black box” in detail, it is necessary to create special sets of input data (tests). Diagnosis of students’ knowledge through computer tests can be divided into three stages:

1. Initial diagnosis. Detection of the general didactic state of the dual education applicant and an approximate assessment of their knowledge. For this purpose, tests are used, which are usually composed of a small number of tasks (35–40) of the same complexity (often they have informative and operational character of the base level).
2. General diagnostics of knowledge. It is the determination of the level of learning. It is necessary to have complete tests of the subject area, which include tasks of various complexity with each class of equivalence. Such tests contain a large number of tasks.
3. “Delicate” diagnostics of knowledge. It means the detailed knowledge control on a specific topic or module section. It is advisable to use tests consisting of interdependent set of tasks having different types (informative, operational, algorithmic, creative) and complexity levels (minimal, basic, programmed, deepened).

Designing a model of knowledge plays an important role in the processes of dual education. As a result, the functioning of the learning environment depends on this, and most importantly, the control of learning. In addition, the semantic model of knowledge facilitates the task of compiling a set of test tasks. Using the semantic model of knowledge, experts make test tasks that fill the test space. With the help of expert analysis and test experiment, the test space is filtered, and high-quality test tasks are entered into the database of test tasks.

4 Model of Dual Education System

The main problem that arises from the design of dual education systems is to increase the flexibility of the system by increasing complexity.

The model of dual education consists of the following models:

- a model of educational material;
- a test question model;
- a model of reference knowledge of an expert (teacher)
- a knowledge model
- a model of the practical component of the learning process.

Let's consider some of the above models. The model of educational material will be presented as a network (graph) (Fig. 3).

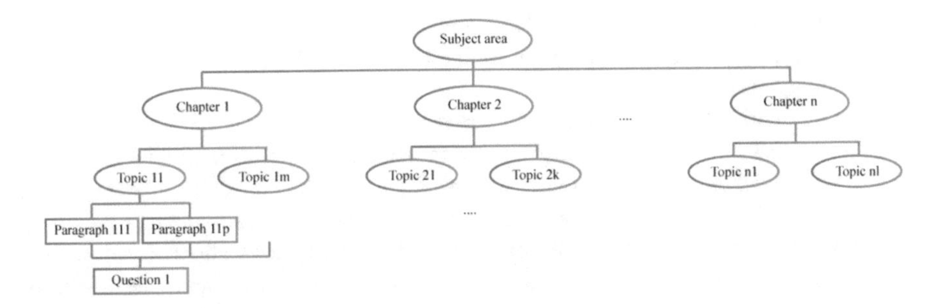

Fig. 3. The graph of a model of educational material

The vertices of the graph are the finite units of the training material (paragraphs, sections, etc.), as well as test questions, which are formed based on a finite set of other vertices of the graph.

The vertices T characterize the following characteristics:

- a set of keywords (V).
- fuzzy characteristic L, which indicates the importance of studying a certain unit of educational material. In the beginning, these values are determined on the basis of the matrix of paired comparisons of expert assessments, and in the process of learning approaches to the formation of dual education system vary depending on the level of its knowledge (it means that the graph of the educational material is adapted to a particular user, and provides the most effective learning process).
- ancestor vertices.
- time for studying material (cost) is defined as the sum of values by descendants (C) for ancestor vertices.

The arcs between the vertices are of two types:

- Arc AND combines the vertices, which are studied simultaneously;
- Arc OR combines the vertices that a student can study on his or her own choice (either one or both).

Therefore, the model of learning material is defined as

$$G = (T, R),$$

where T is the set of vertices, R is the method of bypassing the graph (the way of presenting educational or test material).

Adjusting the graph of the educational material according to the knowledge of the dual education applicants is a function of the type:

$$G' = f(G, P),$$

where G is the initial graph of the educational material;

P is a set of previous knowledge of dual education applicants (obtained as a result of preliminary testing);

G' is an adapted graph of educational material.

In the process of preliminary testing of dual education applicants, fuzzy characteristics of their level of knowledge about certain subject areas are received. The higher the level of knowledge of dual education is, the lower the value of the assessment is. Applying the operation of multivalued OR to the obtained estimates and fuzzy characteristics of the graph's vertices of the educational material with the same names, we obtain a new value of the characteristic. The use of multivalued logic as a mathematical device for processing fuzzy characteristics is expedient to use for a number of reasons [12]:

1. The operation OR chooses larger operand, so the characteristic of "important" themes will not be lowered, and the themes that the student knows badly will have a high fuzzy characteristic that will stimulate the learning process for dual education applicants.
2. Expert assessments used to determine the fuzzy characteristics of the units are used only once during the training of the system, which saves time costs for gathering and processing expert information.
3. Since a relational database is chosen as a model for data storage, the use of multivalued logic will not lead to the expansion of relational algebra.

Linguistic variables are suggested to use for fuzzy characteristics of the graph's vertices of the study material and the results of the test evaluation of dual education applicants, since they allow us to operate terms that are understandable to a teacher and a student.

Test question model is an advanced graph of study material, which, in addition to the names of units of study material, a set of answers to test questions are included.

The construction of such a structure allows organizing some prompts during the testing of dual education students. Prompt is a one-time reference to a test question model (searching a path to the vertex associated with the current one). The depth of the path from a certain question to the answers to it is equal to the number of points that is allocated to this question. One reference to the graph is considered to be the removal of one score from the number of points held in response to the question. In the case of a complete path from the question to the answer it is considered as receiving 0 points.

The model of the teacher's reference knowledge is the display of a set of test questions in the set of correct answers.

$$RA = \gamma(T),$$

where RA is the set of correct answers;
T is the set of test questions.

The model of knowledge of the dual education applicant is the four

$$St = \, <P, G', \eta, R>,$$

where P is the set of previous knowledge of dual education applicant;
η is a function of the transition in the graph of test tasks;

R is the set of estimates obtained during the test.

The function of the transition in the graph of test tasks is to determine the subset of answers for a subset of questions. For dual education applicants, a graph of test questions is a "black" box, the inputs is a set of questions, and outputs is a set of answers. A student can turn to n-times with help, where n is the number of points allocated to this question. As a result, reference shows a part of the path from the question to the answer to a student, where the depth of the path is equal to the number of requests.

A number of previous knowledge of dual education applicants is obtained as a result of previous testing of dual education applicants in the subject area to identify the general level of development and topics that students know the best (the worst).

5 Conclusion

In this paper the mathematical models are proposed and investigated, that are characteristic for processes and subjects of dual education, and the creation of an integrated educational informational and communication environment is based on them.

The approach to the knowledge diagnosis of the dual education applicant is offered, its advantages are the adjustment of the learning process and testing on the potential of each dual education applicant.

The test question model is considered as an expanded graph of the learning material, and it is suggested to use the linguistic variables for the presentation of fuzzy characteristics of the vertex graph of the learning material and the results of the test assessment of the dual education applicant as they allow the use of terms understood by the teacher and the student.

This approach provides an opportunity for effective testing of knowledge of the dual education applicant.

References

1. On approval of the Concept of training specialists for the dual form of education: the order of the Cabinet of Ministers of Ukraine in September 19, 2018, No. 660 (2018). https://zakon.rada.gov.ua/laws/show/660-2018-%D1%80 (in Ukrainian)
2. Koudahl, P.D.: Vocational education and training: dual education and economic crises. Soc. Behav. Sci. **9**, 1900–1905 (2010)
3. Center of the Chinese vocational education, Dual educational system of Germany. Suzhou University Press (1993)
4. Xu, M.: Development of dual educational system of Germany in China. In: Zhang, W. (ed.) Advanced Technology in Teaching. Advances in Intelligent and Soft Computing, vol. 163, pp. 289–293. Springer, Berlin (2012)
5. Barakabitze, A.A., William-Andey Lazaro, A., Ainea, N., Mkwizu, M.H., Maziku, H., Matofali, A.X., Sanga, C.: Transforming African Education Systems in Science, Technology, Engineering, and Mathematics (STEM) Using ICTs: challenges and Opportunities. Education Research International, vol. 2019, 29 p.
6. Ruohonen, M.J.: Network-centric work—implications to professional IT education. In: Juliff, P., Kado, T., Barta, B.Z. (eds.) Educating Professionals for Network-Centric Organisations, vol. 17, pp. 9–17. Springer, Boston (1999)
7. Callagha, V., Chernett, P.: A network-centric approach to computer science education and research based on robotics. Comput. Sci. Educ. **8**(2), 100–117 (1998)
8. Yang, M., Tu, W., Qu, Q., Zhao, Z., Chen, X., Zhu, J.: Personalized response generation by dual-learning based domain adaptation. Neural Netw. **103**, 72–82 (2018)
9. Koudahl, P.D.: Vocational education and training: dual education and economic crises. Procedia Soc. Behav. Sci. **9**, 1900–1905 (2010)
10. Wiesner, S., Horizonte, B.: The development of technicians as a key factor for a sustainable development of renewable energies using an adapted education method based on the successful German dual education (Duale Ausbildung). Energy Procedia **57**, 1034–1036 (2014)
11. Bomba, A., Nazaruk, M., Pasichnyk, V., Veretennikova, N., Kunanets, N.: Information technologies of modeling processes for preparation of professionals in smart cities. In: Hu, Z., Petoukhov, S., Dychka, I., He, M. (eds.) Advances in Intelligent Systems and Computing book series, vol. 754, pp. 702–712. Springer, Cham (2018)
12. Ifenthaler, D., Pirnay-Dummer, P.: Computer-Based Diagnostics and Systematic Analysis of Knowledge. Springer, New York (2010)
13. Holoshchuk, R., Pasichnuk, V., Kunanents, N., Veretennikova, N., Pytlenko, T.: Modeling the distance network-oriented education systems. In: Proceedings of International Scientific Conference Computer Sciences and Information Technologies (CSIT-2019), vol. 3, pp. 191–196. IEEE (2019)

Coordination of Marketing Activity in Online Communities

Oksana Peleshchyshyn[1(✉)] and Tetiana Klynina[2]

[1] Lviv Polytechnic National University, Lviv, Ukraine
oksana.p.peleshchyshyn@lpnu.ua
[2] National Aviation University, Kiev, Ukraine
tklynina@gmail.com

Abstract. Planning marketing activities in online communities depends on the chosen strategy for using virtual communities.

The active involvement of business representatives in communities is achieved by agreeing on the stylistics of communications and the frequency of discussions with the rules and traditions of online communities. The indicator of the communicative effectiveness of the marketer depends on the reaction of society to marketing activities in the community and allows you to assess the effectiveness of the use of important online communities.

In the case of a large number of important online community and human resource constraints, the challenge is to find the optimal distribution of enterprise representatives among communities. The target function of the task should take into account the importance of the online community and the communicative effectiveness of the performers. The data for determining these indicators and the limitations on optimization tasks are derived from the analysis of online community statistics and the evaluation of marketing commentary discussions.

To coordinate the marketing goals of company management and personal goals of representatives in online communities, it is expedient to use the apparatus of the theory of coordination. Then, the iterative process of distributing online communities involves solving local optimization problems by choosing community implementers to participate in them, detecting the inconsistency of the resulting distributions and removing them by changing the task parameters and implementing coordination impacts.

Keywords: Online marketing · Online community · Theory of coordination

1 Introduction

Communication opportunities of today's popular virtual communities are actively used by marketing services of enterprises to disseminate their advertising and explanatory information in order to promote the company and its products on the market, supporting actual and attracting new consumers.

Monitoring the content of the online communities and discussions provides companies with additional information in the form of judgments, feedback, criticisms of community members about the quality of products and services. Analysis of these data allows to take into account the reaction of society in planning the activities of the enterprise and marketing activities in virtual environments on the Internet.

N. Shakhovska and M. O. Medykovskyy (Eds.): CCSIT 2019, AISC 1080, pp. 647–660, 2020.
https://doi.org/10.1007/978-3-030-33695-0_44

In researches of problems of marketing activity of the enterprise in the Internet considerable attention is paid to questions of organization of marketing in online communities [1, 3, 6, 12, 14, 17, 19, 21, 23, 28, 29], analysis of information content of thematic discussions in the communities [7, 8, 10, 11, 13, 18, 20, 25], development of methods for building and managing web communities [2, 5, 15, 16], management of information activities of the company in a virtual environment [4, 9, 20, 22, 26, 27].

It assesses the importance of online communities for marketing, depending on the chosen strategy for using the communication capabilities of virtual communities [8, 24]. However, the question of the optimal allocation of enterprise resources for marketing in online communities has not been explored enough.

2 Optimization and Evaluation of the Effectiveness of Marketing Activities in Online Communities

2.1 Factors Affecting Marketer Activity in Online Communities

The strategy of using online communities is crucial when planning human resources to implement marketing activities in virtual environments. In the case of the passive use of communities for collecting information contained in discussions and relating to marketing topics, the specialist should have extensive knowledge of skills search in the World Wide Web and effectively use the available tools for analyzing and storing data.

Active engagement with communities and participation in discussions bring following additional requirements to marketing professionals:

- adherence to network etiquette during communication;
- taking into account the traditions and rules of the online community during placing marketing messages;
- the timeliness of processing new messages related to marketing topics.

Among the basic strategies for using online communities in marketing [24], the presence of representatives of the company in the life of communities involves representative, active and hyperactive strategies. The realization of each of these strategies implies various marketer's activity in selected online communities by creating topics for discussion and posting, and different requirements for the style of their design.

For different strategies, the definition of the importance of marketing online communities is different. After selecting a set of online communities, which is important to interact with, it is necessary to evaluate the possibility of active participation of representatives of an enterprise. Such a question arises, in particular, in the analysis of highly specialized communities with restrictions on the topic of discussion. For example, the camera forum was chosen as a matter of importance, which, however, applies only to Canon products, and our company sells Pentax products. Information from such a community is very interesting for an enterprise in an analytical way, such as information about competitors, expectations, and consumer requests. However, official participation in discussions to distribute its own information on this forum is not possible.

The activity of forming messages depends on the scope of the enterprise and the strategy of using the online community environment. If an enterprise operates in a fast-growing industry and has many competitors, it is necessary to generate information flows more actively and respond more quickly to customer appeals in order to gain additional benefits, to effectively disseminate marketing information and to stay in touch with important marketing online communities. Less dynamic and less competitive industries may require less activity.

The strategy of using online communities also affects the activity of the marketer and the scope of work. When choosing a representative strategy, the number of new messages depends directly on the planned marketing efforts and may be small. Comments and responses to community members' requests are mainly provided in the case of a community-based consumer support service.

In the case of implementation of an active strategy, the number of messages significantly increases, the requirements for the specialty of marketers grow. The number of online communities that the company actively interacts with grows, but is limited only to the importance of solving marketing problems.

The hyperactive strategy generates the maximum number of posts from representatives of the company, as it takes into account not only important online communities but also the majority of relevant ones. As the main communication medium, online communities in this case require the specialists of the enterprise to react quickly and professionally to consumer appeals.

In general, the number of appeals from members of online communities to the representatives of the company indicates the demand for information and the interest of society to the company and its products, the relevance of marketers to the requirements of virtual environments.

The stylistics of the messages that marketers generate depends on the marketing strategy and online discussion rules. Representative strategy involves announcing mainly advertising and explanatory nature, the purpose of which is to interest the audience, support the position of the enterprise and update information about its products and activities. Sometimes even provocative messages are permissible to stimulate the audience and increase its response to an important issue, of course, if community rules allow such actions.

The active strategy is aimed at long-term engagement with important online communities, so the message should have a well-considered well-informed character. Reports of open-ended advertising can be perceived as spam and reduce the effectiveness of community marketing activities. The image of an expert in its field will be achieved by the marketer through consumer-friendly messages that contain plausible substantiated information and traditions of the online community.

The high activity of a marketer during the implementation of a hyperactive strategy requires the formation of posts and the brief and understandable information that is given to a wide audience. The emergence of conflicts and confrontation between the participants of the discussion is not approved and very detrimental to the image of the enterprise.

For each strategy, the requirements for posting can be determined. One of the options for documenting such requirements is the definition of certain rules, for example, in the form of templates of typical posts, which should be stored in the database as a service information and used during the formation of the structure and content of messages.

It is advisable for enterprise managers to check periodically whether the marketers' messages correspond to the company's chosen style of marketing information to Internet users. If you use database to record marketing communications in online communities, the stylistic message of the representatives of the company can be analyzed, being pre-filtered by the authors of the discussions that are recorded in database. If the activity of marketers in posting in communities is very high, then the number of copyright messages may be large. In practice, you can use a certain sample of posts from this set to simplify the audit.

2.2 Indicators of Marketing Effectiveness in Online Communities

A criterion for the quality of the presentation of marketing information will be considered a reaction of society in various forms on the postings of a marketer. The comments of the online communities members, which give an assessment of the company, its products, that is, an assessment of the object of marketing, are used to analyze the position of the company and its development. Messages that assess the qualifications and acceptability of community members in the style of post marketing, help resolve the issue of a marketing specialist, adjust its actions to get a positive feedback from the audience, the desired effect of the dissemination of information.

To characterize the communicative effectiveness of a marketer's actions in online communities, we define the following estimates of his posts: positive, neutral, negative.

The positive assessment is to indicate the usefulness of the information provided by the marketer to consumers, in the encouraging citation of messages by other community members.

Negative assessment of marketer's messages is manifested in criticizing the form and content of information provided by users, community administration.

An unbiased assessment of the post shows that the stylistics of the presentation of marketer information did not cause a pronounced reaction of society, comments from community members to the message are missing or related only to the object of marketing promotion.

The results of evaluating the response to marketing messages in the online community should be counted together with the data of the message itself for further analysis of the effectiveness of the communication process.

Usually one marketer message can cause a few comments with a different response from contributors. The resulting evaluation of this message can be obtained in several ways:

- definition of an estimate that prevails among commentary ratings for the post;
- analysis of the reaction taking into account the weight of the participants in the discussion (the moderator's response is more important than the comment of the casual guest).

A more complex version of the analysis is to determine the overall assessment of the message by scoring the estimates of different contributors, where the convolution coefficients are determined depending on the weight of the opinions of the commentators.

The weight of the appraiser is determined by the contributors within the discussion or within the community in general.

One negatively evaluated message can give the value the same effect as a few positively evaluated, of course, the opposite one. Therefore, it is advisable to introduce weight factors to mark the influence of the direction of evaluation (positive, negative, neutral) on the total reaction of society to the message of marketers in the community. During the implementation of active strategies, negatively evaluated posts are much more important than positively evaluated ones, the weight of neutral postings is close to zero.

In the absence of positively and negatively evaluated postings of a marketer, the value of effectiveness depends on how the neutral response of the community is measured on the messages in it. If it is assumed that the weighting factor of a neutrally evaluated post is zero, then the actions of the marketer can be considered ineffective, regardless of the number of reports that was generated, and accordingly ineffective. However, if we take into account that even invaluable thematic reports increase the mention of the object of marketing promotion in online environments, it would be useful to establish a positive meaning for the neutral assessment, albeit small, in comparison with the positive weighting factor.

For neutrally rated posts, the value of weight can be both positive and negative. If, for some reason (for example, an organizational one), a marketer's message should always receive a community evaluation, then the lack of response to the stylistic messages of the marketer should be assessed with a negative weight.

Depending on the strategy of using the communities in the marketing activities of the company, the ratio and weight of the scales change for various evaluations by the users of the actions of the marketer. Thus, the weight of the negatively assessed post increases significantly in the case of implementation of an active strategy, compared with the representative one, and in the case of a hyperactive one, it becomes even higher, as the risks of the company compromising accordingly increase due to unskilled discussions.

When distributing the scales, it is expedient to take into account the ratio of active two-way and one-way authoring streams of information placement in the course of implementation of the respective strategies. For example, for a representative strategy, the presence of feedback is not significant, so the weights of positively and neutralized posts are the same.

Taking into account the foregoing in Table 1, an example of the weighting of different ratings for active online community strategies is shown.

Let us consider a set VC of important online communities the company actively interacts with. This set is described below:

$$VC = \{VC_i\}_{i=1}^{N^{VC}} \tag{1}$$

Table 1. An example of estimating scores by marketers posting community members

Strategy	Estimate		
	Positive	Neutral	Negative
Representative	1	1	–2
Active	1	0,1	–6
Hyperactive	2	0,2	–10

Let us define the context-weighted total attraction of the society $TA(VC_i)$ in the community VC_i so:

$$TA(VC_i) = \begin{cases} \sqrt{(Rp_i \cdot Cp_i + Rz_i \cdot Cz_i)^2 + (Rn_i \cdot Cn_i)^2}, \text{if } Rz_i \geq 0; \\ \sqrt{(Rp_i \cdot Cp_i)^2 + (Rn_i \cdot Cn_i + Rz_i \cdot Cz_i)^2}, \text{ if } Rz_i < 0 \end{cases}, \quad (2)$$

where Rp_i, Rz_i, Rn_i – weight of the post estimated positively, neutrally and negatively in the community VC_i; Cp_i, Cz_i, Cn_i – the number of positive, neutral and negative posts in the i-community.

As can be seen from the formula (2), a neutral mark by the market participants of the community signals the value of a positive or negative effect depending on the sign of the weight determined for it.

The value of $TA(VC_i)$ characterizes the degree of perturbation of the online community through the postings of a marketer, regardless of the tone of the response of community members.

The resulting effect of attraction $EA(VC_i)$ from the actions of a marketer in the community VC_i is determined as the result of the convolution of the quantities of differently oriented estimates of the reaction:

$$EA(VC_i) = Rp_i \cdot Cp_i + Rz_i \cdot Cz_i + Rn_i \cdot Cn_i. \quad (3)$$

The communicative effectiveness of a marketer's activity in a separate community $ECM(VC_i)$ characterizes the orientation and completeness of the social attraction of community users and it is defined as the ratio of the resulting impact assessment (3) to the total attraction of the marketing community (2):

$$ECM(VC_i) = \frac{EA(VC_i)}{TA(VC_i)}. \quad (4)$$

The value of communicative efficiency is in the range [–1, 1]. This performance is unmatched in the case of the most useful actions of a marketer in the online community (marketer's posts receive approval regarding their professionalism and community style and provide an additional positive effect). Negative communicative effectiveness is in the case of a predominance of a negative reaction to the style of marketing messages, therefore, the activity of a marketer in the online community is harmful to the enterprise due to the compromise of the professionalism of its representatives.

The definition of the communicative effectiveness of the marketer's actions in the online community with the help of formula (4) makes it possible to assess the effect of marketing activities in the community and, thus, the effect of using the online community in the marketing activities of the enterprise.

The effectiveness of the use of the online community in marketing will be considered as the degree to which the marketing value of the community is used and will be determined in such a dependence:

$$EVC(VC_i) = ECM(VC_i) \cdot IMP(VC_i) \cdot CPM(VC_i), \tag{5}$$

where $ECM(VC_i)$ – the communicative effectiveness of the marketer in the community VC_i; $IMP(VC_i)$ – the marketing importance of the i-community; $CPM(VC_i)$ – the number of postings of a marketer in the i-community.

All evaluated comments of the marketer, regardless of the direction of their assessment are shown in the formula (5):

$$CPM(VC_i) = Cp_i + Cz_i + Cn_i. \tag{6}$$

The indicator of the communication effectiveness of the marketer characterizes the response of an important online community to the messages posted on it by the representatives of the company. With the same communicative efficiency and productivity (number of posts), the greater effect of the dissemination of marketing information comes from participating in the more important community.

At a certain level of communicative efficiency of a marketer, an increase in the effectiveness of participation in the life of an important online community is achieved through the growth of the number of posts. However, this improvement is possible to a certain extent. For each online community, one can determine the frequency of new posts that shows the activity of all members of the community over a period of time. A specialist in an enterprise, who is a new contributor, should consider this factor in planning the posting of messages in order to introduce them organically into the community. In practice, too intense distribution of materials (exceeding the average level of activity) can cause community resistance, perceived as spam, and as a result, negatively affect the image of the writer. Communicative efficiency in this case can be significantly reduced because of the appearance of critical comments or even ignoring the community of marketing expert messages (prevalence of neutralized posts). Accordingly, the effectiveness of the use of the online community will be reduced, despite the formal increase in the number of posts.

In order to avoid this situation, it is advisable to determine the number of new messages for a certain period, based on community statistics, when planning an online community activity. It is important that the intensity of the formation of new posts has a constant and evenly distributed character in time. Once (within one to two days), the high level of marketer activity on the community site, and therefore a long period of silence, usually indicates a formal approach to disseminating information rather than a desire to establish a long-term and effective interaction. The period for analyzing community activity and planning marketing events will be chosen over a week, because weekly planning of marketing work is convenient considering the distribution of activity of other members of the community (during work and weekends).

The number of messages that a marketer can post in the online community for a certain period without threatening their image (to be suspected of spam or malware), we call the **capacity of the online community for the period**. The capacity by expert analysis of the statistics of the online community will be identified here. Of course, this value should not exceed half the average number of new messages among active subscribers for the selected period.

Capacity values need to be determined at the beginning of marketing activity in the online community, and after a certain period of time (and periodically thereafter), it is advisable to review it. The need for such an examination is due to the fact that the community may undergo significant changes in the direction of increasing (or decreasing) the average number of posts per period, consequently, it is necessary to correct the marketing manager's work plan in order to optimize the use of specialists and information resources of the community.

If an enterprise uses several professionals to interact with social environments, then there is a question of sharing the online communities among them. It is clear that it is advisable to use more effective marketers in order to work in more important online communities. The effectiveness of the actions of various marketers in the individual online community may vary due to different levels of training, experience and compliance with the requirements that are placed into the implementation of a particular strategy. Importance is the relevance of the style of writing messages to various marketers in the style of discussion in the online community. In addition, the personal qualities of the staff, in particular, the ability to act on a regular basis and to distribute their working hours among the communities in a balanced manner, also influence the effectiveness.

2.3 Model of the Task of Finding the Optimal Distribution of Human Resources for Marketing in Online Communities

In order to maximize the effect of the participation of business representatives in the lives of selected online marketing communities, it is necessary to allocate optimally human resources - marketers among selected communities.

Set of marketers HR (Human Resource), who work in important online communities, will be given below:

$$HR = \{HR_j\}_{j=1}^{N^{HR}}. \tag{7}$$

The effect of the activity of th HR_j marketer in the VC_i community is determined by the function $F(VC_i, HR_j)$. The variables X_{ij} will be defined as: $X_{ij} = 1$, if j-th marketer works in the i-th online-community; $X_{ij} = 0$, if j-th marketer does not work with i-th community. The mathematical model of the optimization problem is as follows:

$$\sum\nolimits_{i=1}^{N^{VC}} \sum \sum\nolimits_{j=1}^{N^{HR}} F(VC_i, HR_j) \cdot X_{ij} \to max, \tag{8}$$

$$X = \{X_{ij} | X_{ij} \in \{0, 1\}, i = 1, \ldots, N^{VC}, j = 1, \ldots, N^{HR}\}. \tag{9}$$

The effect of the work of a marketer in the online community can be determined by using a formula similar to a formula (5) for the effectiveness of using the online community for a certain period:

$$F\left(VC_i, HR_j\right) = ECM\left(VC_i, HR_j\right) \cdot IMP(VC_i) \cdot CPM\left(VC_i, HR_j\right), \tag{10}$$

where $ECM\left(VC_i, HR_j\right)$ – the communicative effectiveness of the *j*-th marketer in the *i*-th community; $IMP(VC_i)$ – importance of the *i*-th community; $CPM\left(VC_i, HR_j\right)$ – the number of posts of the *j*-th marketer in the *i*-th community for the period.

Then expression (8) will be given in such a way below:

$$\sum\nolimits_{i=1}^{N^{VC}} \sum \sum\nolimits_{j=1}^{N^{HR}} ECM\left(VC_i, HR_j\right) \cdot IMP(VC_i) \cdot CPM\left(VC_i, HR_j\right) \cdot X_{ij} \rightarrow max. \tag{11}$$

For the full use of dedicated human resources, a condition is required for each marketer to participate in a certain number of communities:

$$MinVCtoHR_j \leq \sum\nolimits_{i=1}^{N^{VC}} X_{ij} \leq MaxVCtoHR_j, j = 1, \ldots, N^{HR}, \tag{12}$$

where $MinVCtoHR_j$ i $MaxVCtoHR_j$ – the smallest and largest number of online communities involved in the *j*-th marketer. The values of the $MinVCtoHR_j$ and $MaxVCtoHR_j$ boundaries are determined by the number of marketers and online communities in such a way as to cover the most important ones.

Similarly, in order to capture fully the attention of marketers across the whole range of important online communities, optimization challenges are subject to additional restrictions on the number of marketers that work with each community:

$$MinHRinVC_i \leq \sum\nolimits_{j=1}^{N^{HR}} X_{ij} \leq MaxHRinVC_i, i = 1, \ldots, N^{VC}, \tag{13}$$

$$MinHRinVC_i \geq 1, i = 1, \ldots, N^{VC}, \tag{14}$$

where $MinHRinVC_i$ – the smallest number of marketers to participate in the *i*-th community; $MaxHRinVC_i$ – the largest number of company representatives in the *i*-th community.

The value of $MaxHRinVC_i$ is determined by the expert method, based on the available human resources, the importance of the online community and the activity of its participants. For example, a single representative in a community with a high incidence of new posts may have problems with the timely processing of messages and requests. For very important communities, it may have negative consequences in the form of critical comments from marketers and missing important information.

In addition, the interaction of several marketers on the site of a single community (explicit or implicit) may provide additional benefits in critical and lengthy discussions, since it enables interchange of specialists to perform routine actions, as well as team work on increasing the number of multi-dimensional argument comments, and the revitalization of the discussion. Numerical representation is expedient in the online community of the company itself and in important narrow-professional online communities and communities, on the basis of which support is provided to the consumers of products of the enterprise.

Thus, the model of the task of finding the optimal distribution (9) of marketers in online communities consists of the objective function (11) and the constraints (12)–(13).

The formula (10) for determining the effect does not take into account costly profits of the work performed by specialists of the company in the online community. Realization of the constructed mathematical model allows redistributing resources in situations when:

- cost estimation of posting marketing message in the online community is the same for each marketer and every online community;
- the main issue is the redistribution of workers between the sites (incomplete or overloaded staff), and the change in the cost-effectiveness evaluation does not matter.

In practice, it is not possible to determine the value of the equivalent of the benefit of placing a marketer on the online community site. For approximate ratings, you can use estimates of the cost of advertising on a community site. However, professional author's announcements usually have a much greater effect compared to units of ad. Therefore, for evaluation, it is necessary to take the cost of advertising with a refinement factor, which is approximately five to ten units.

2.4 Coordination of Marketing Actions in Online Communities

An enterprise can use the communication capabilities of virtual environments to distribute marketing information and organize customer feedback. In this case, the main goal is to increase the effect of the participation of representatives of the company in important online communities.

The organizational structure of human resources involved in the process is usually hierarchical. The top level is the management of the company, which sets the task of marketers and implements the overall control of the process. At the bottom level - representatives of the company that work with online communities. It can be both employees of an enterprise and an employee or an organization.

In the case of using third-party services, an enterprise typically does not distribute the specific human resources of the partner organization between online communities, but is limited to setting goals, defining metrics to measure the extent to which the goal is achieved, and generally verifying the activities of marketers in the communities. The task of optimization in this case is transferred to the management and employees of partners.

To control the achievement of the goal, management determines the target function that is used in the task of optimizing the use of human resources. In particular, to evaluate the reaction of society and to solve the problem of distribution of marketers between communities, it is expedient to use the performance indicators mentioned above.

The complexity of the process of organizing the activities of marketers in online communities and analyzing their effectiveness depend on the quantity and information capacity of online communities, available human resources and the activity of other community members. Searching for an optimal solution to the problem of distributing marketers through online communities is significantly complicated if the number of important online marketing communities and performers is high.

The own goals of marketing executives might be another complicating factor. For example, getting a higher salary for doing a larger number of tasks and their own view of communities that may be more efficient due to correspondence of their features and peculiarities to personal preferences and skills of professionals.

Consequently, a two-level organizational system is obtained, its components have their own target functions and interact to achieve a common goal.

In this case, in order to achieve the best distribution of representatives of the company between the sites of online communities and to manage the process of interaction, it is expedient to use the apparatus of the theory of coordination.

In the task of coordination as the target functions of the whole system and the component of the upper level, the coordinator uses the function from the optimization task. In this case, the restrictions imposed on the solutions of the optimization problem are the control parameters of the coordination task.

The solution of the coordination problem is described as a set $X = \{X_{ij} | X_{ij} \in \{0,1\}, i = 1, \ldots, N^{VC}, j = 1, \ldots, N^{HR}.\}$, elements of which represent the involving of *j*-th marketer to work in the *i*-th online community and are defined as a model of the optimization problem (8)–(9).

Solutions of local problems are described by analogous sets $\tilde{X}_j = \{\tilde{X}_{ij} | \tilde{X}_{ij} \in \{0,1\}, i = 1, \ldots, N^{VC}.\}$, elements of which represent the choice of the j-th marketer of the set of online-communities for activities in them.

The process of finding a solution to a coordination task is iterative. First, the system coordinator defines global constraints on solving the problem of finding the best distribution and predicts the value of coordination parameters for local tasks. The coordination parameters for the component j^* are the values of the variables $\{X_{ij}, i = 1, \ldots, N^{VC}, j = 1, \ldots, N^{HR}, j \neq j^*\}$.

Next on each subsequent iteration:

- marketers are solving local problems - choosing online communities to take part in them;
- the coordinator finds the inconsistency of the resulting local solutions between the co-operation and the solution of the global problem;
- the coordinator eliminates the failure of local solutions by changing the parameters of local problems and using control effects.

The purpose of the process and the criterion for its completion is to reconcile global and local solutions and, as a result, to achieve such a distribution of marketers on online communities, which in real time meets best the goals of all process participants.

Global constraints determine the range of permissible values for solutions of local tasks. Such constraints in the task of finding the distribution of marketers on online communities are the conditions (12) and (13).

In addition to the mechanism of global constraints, in order to manage the process of negotiating solutions in certain cases, it is advisable to use direct guidance on the involving a marketer j^* to work in the community i^*:

$$X_{i^*j^*} = 1. \tag{15}$$

It is advisable to use the controlling influence of type (15) coordinator when choosing executors to work with the most important or problematic online communities, if it is impossible to reconcile the solutions in a different way.

Depending on the type of selected target function, it is effective to use additional restrictions on the parameters of the function itself. In particular, if the target function uses the formula for the effectiveness of using the online community for a certain period (10), then the restrictions are imposed on the number of marketing posts in the online community:

$$CPMMin_{ij} \leq CPM\left(VC_i, HR_j\right) \leq CPMMax_{ij}, \forall i = 1, \ldots, N^{VC}, \forall j = 1, \ldots, N^{HR}, \tag{16}$$

where $CPMMin_{ij}$ i $CPMMin_{ij}$ – respectively, the minimum and maximum number of posts j-th marketer in the i-th online community for the period. The boundaries in the condition (16) are determined by the desired activity of the representatives of the enterprise in the community. To avoid risks, it is advisable to set the upper limit not higher than the capacity of the online community.

For certain types of activities, determining the coordinator of such restrictions requires an analysis of the previous activity of participants in online communities. In particular, it concerns the implementation of an online customer service community. The number of tasks in this case depends on the intensity of the Internet users, and the restrictions are determined on the basis of statistical data.

3 Conclusions

The strategy of using online communities in marketing activities of the enterprise is crucial in the planning and evaluation of the effectiveness of marketing activities in the virtual space. An analysis of the reaction of community members to the participation of experts in discussions and the definition of indicators of the communicative effectiveness of marketers makes it possible to assess the effectiveness of the dissemination of marketing information through the online community.

In the planning of communications, we solve the problem of finding optimal distribution of representatives among online communities.

The target function of the proposed mathematical model of the optimization problem takes into account the importance of the online community, the activity and the communicative effectiveness of the marketer's actions. The comprehensiveness of the attention of important communities is achieved through additional restrictions on the solutions.

Searching for optimal representation in virtual communities is complicated when you need to take into account the professionalism and personal goals of the performers. In this case, it is advisable to use the theory of coordination apparatus to search for the distribution of communities among specialists. The iterative process of solving the coordination task involves choosing community implementers to participate in them and coordinating with the coordinator of solutions for global and local optimization tasks. The resulting distribution of marketers through online communities takes into account the goals of all process participants and provides optimal use of online communities in marketing.

References

1. Arthur, D., Motwani, R., Sharma, A., Xu, Y.: Pricing strategies for viral marketing on social networks. In: Leonardi, S. (ed.) Internet and Network Economics (WINE 2009), pp. 101–112. Springer, Berlin (2009)
2. Bacon, J.: The Art of Community: Building the New Age of Participation (Theory in Practice). O'Reilly Media, Sebastopol (2009)
3. Bandias S., Gilding A.: Social media: the new tool in business education. Public Interest and Private Rights in Social Media, Chandos Publishing Social Media Series, pp. 115–128 (2012)
4. Boon, E., Pitt, L., Salehi-Sangaria, E.: Managing information sharing in online communities and marketplaces. Bus. Horiz. **58**(3), 347–353 (2015)
5. Buss, A., Strauss, N.: Online Communities Handbook: Building Your Business and Brand On the Web. New Riders Press, Berkeley (2009)
6. Cheng, F., Wu, C., Chen, Y.: Creating customer loyalty in online brand communities. Comput. Hum. Behav. (2018)
7. Christensen, K., Liland, K., Kvaal, K., Risvik, E., Biancolillo, A., Scholderere, J., Nørskovh, S., Næs, T.: Mining online community data: the nature of ideas in online communities. Food Qual. Prefer. **62**, 246–256 (2017)
8. Fedushko, S., Peleschyshyn, O., Peleschyshyn, A., Syerov, Y.: The verification of virtual community member's socio-demographic profile. Adv. Comput. Int. J. (ACIJ) **4**(3), 29–38 (2013)
9. Hussain, S., Guangju, W., Jafar, R., Ilyas, Z., Mustafa, G., Jianzhou, Y.: Consumers' online information adoption behavior: motives and antecedents of electronic word of mouth communications. Comput. Hum. Behav. **80**, 22–32 (2018)
10. Korzh, R., Peleschyshyn, A., Holub, Z.: Analysis of integrity and coverage completeness of the informational image of a higher education institution. In: Proceedings of the 13th International Conference on Modern Problems of Radio Engineering, Telecommunications and Computer Science, TCSET 2016, p. 825 (2016)
11. Kim, J., Hastak, M.: Social network analysis: characteristics of online social networks after a disaster. Int. J. Inf. Manage. **38**(1), 86–96 (2018)
12. Krush, M., Pennington, J., Fowler III, A., Mittelstaedt, J.: Positive marketing: a new theoretical prototype of sharing in an online community. J. Bus. Res. **68**(12), 2503–2512 (2015)
13. Liu, B.: Web Data Mining: Exploring Hyperlinks, Contents, and Usage Data. Springer, Heidelberg (2011)

14. Messik, R.: Social media: Blessing or Curse? – A business perspective. Public Interest and Private Rights in Social Media, Chandos Publishing Social Media Series, pp. 145–152 (2012)
15. Peleschyshyn, A., Holub, Z., Holub, I.: Methods of real-time detecting manipulation in online communities. In: Proceedings of the 11th International Scientific and Technical Conference on Computer Sciences and Information Technologies, CSIT 2016, p. 15 (2016)
16. Peleshchyshyn, A., Korzh, R.: Basic features and a model of university units: university as a subject of information activity. Eastern Eur. J. Enterp. Technol. **2**(2), 27–34 (2015)
17. Plume, C., Dwivedi, Y., Slade, E.: Online Brand Communities, Social Media in the Marketing Context, A State of the Art Analysis and Future Directions, pp. 41–78 (2017)
18. Rains, S., Brunner, S.: What can we learn about social network sites by studying Facebook? A call and recommendations for research on social network sites. New Media Soc. **17**(1), 114–131 (2014)
19. Rath, M.: Application and Impact of Social Network in Modern Society. In: Hidden Link Prediction in Stochastic Social Networks, pp. 30–49 (2018)
20. Russell, M.: Mining the Social Web: Data Mining Facebook, Twitter, LinkedIn, Google+, GitHub, and More. O'Reilly Media, Inc., Sebastopol (2013)
21. Scott, D.: The New Rules of Marketing and PR: How to Use Social Media, Blogs, News Releases, Online Video, and Viral Marketing to Reach Buyers Directly. Wiley, Hoboken (2010)
22. See-Toa, E., Ho, K.: Value co-creation and purchase intention in social network sites: the role of electronic word-of-mouth and trust – a theoretical analysis. Comput. Hum. Behav. **31**, 182–189 (2014)
23. Seraj, M.: We create, we connect, we respect, therefore we are: intellectual, social, and cultural value in online communities. J. Interact. Mark. **26**(4), 209–222 (2012)
24. Sloboda, K., Peleshchyshyn, P.: Peculiarities of positioning and online marketing of Lviv Polytechnic National University in social networks. Eur. Appl. Sci. **2**(5), 24–32 (2013)
25. Srivastava, A., Sahami, M.: Text Mining: Classification, Clustering, and Applications. Taylor and Francis Group, London (2009)
26. Taiminen, H.: How do online communities matter? Comparison between active and non-active participants in an online behavioral weight loss program. Comput. Hum. Behav. **63**, 787–795 (2016)
27. Teo, H., Johri, A., Lohani, V.: Analytics and patterns of knowledge creation: experts at work in an online engineering community. Comput. Educ. **112**, 18–36 (2017)
28. Weinberg, T.: The New Community Rules: Marketing on the Social Web. O'Reilly Media, Inc., Sebastopol (2009)
29. Zeng, M.: Foresight by online communities – the case of renewable energies. Technol. Forecast. Soc. Chang. **129**, 27–42 (2018)

Increasing the Efficiency of the Processes of Formation of the Informational Image of the HEI

Roman Korzh[1](✉), Andriy Peleshchyshyn[1](✉), Olha Trach[1](✉), and Mikola Tsiutsiura[2](✉)

[1] Lviv Polytechnic National University, Lviv, Ukraine
korzh@lp.edu.ua, apele@ridne.net, olya@trach.com.ua
[2] Kyiv National University of Construction and Architecture, Kyiv, Ukraine
tsiutsiura.mi@knuba.edu.ua

Abstract. The article presents an analysis of the formation of the informational image of the HEI, namely: natural unmanaged process of formation of informational image of HEI, informational image of HEI as an element of the system of rating of higher education, analysis of sources of formation of information image of HEI, management of the process of formation of information image. The stages of the informational image formation process are analyzed, they are shown how they are realized in two models: passive and active. The scenarios of further actions in the community are proposed in cases of low productivity (subjective, aggressive, stationary cases). The typical scenarios of the HEI information activity for the stationary case of low productivity are presented. The analysis of the integrity and completeness of coverage of the information image of the HEI is carried out. The analysis of thematic gaps in the description of the HEI and the informational case of HEI is carried out. The expert analysis of the description of the topics of the HEI, the analysis of the activity of the competitive higher educational institutions, and the analysis of the semantic structure of the WWW information space are considered.

Keywords: Information activity · Information image · Higher education institution · Thematic gaps

1 Introduction

The effectiveness of information activity as one of the forms of interaction between the organization and society is an important element in the organization goals achievement. Compared to other forms of the organization activity, information activity places the highest demands on the need to take into account technical progress and social trends. That is why one of the key factors of successful information activity today is the effective use of the newest environments of information transmission and accumulation, primarily of the Internet network and services based on it.

For the last decade, the active development of socially-oriented Internet services and the formation of the phenomenon of global archives of collective public

N. Shakhovska and M. O. Medykovskyy (Eds.): CCSIT 2019, AISC 1080, pp. 661–679, 2020.
https://doi.org/10.1007/978-3-030-33695-0_45

communications processes (web forums, social networks, specialized services Web 2.0, etc.), which is reflected in important international documents [1], is typical to the Internet.

Universities traditionally play, besides the direct educational role, the role of a center of social life and communication [2]. Universities have always been centers of communication, exchange of ideas, the accumulation of information and knowledge as well as the formation of different directions teams. A similar situation was observed in Ukraine. The importance of the leading universities as centers of social life in the post-Soviet space in general and in Ukraine, in particular, sharply increased in the early 90's of the XX century [3].

However, the rapid development of information and communication technologies, especially the Internet, has become one of the factors of the universities role reduction as centers of social communication [4, 5]. Due to the passivity typical for the HEIs of Ukraine (even for the largest technical universities) in the implementation of information technology and in the issues of communication platforms organization of a new generation, technically equipped young people no longer consider universities as an important center of communication, for the long time prefer internet communication, in which Ukrainian HEIs do not yet play a significant communicative role.

The situation with the removal of HEIs from the processes of knowledge and information sharing was further aggravated with the active transition of Internet users from traditional communication on forum-type sites to communication in modern social networks. The vast majority of Internet users find necessary platforms for communication in the social networks [6].

The HEIs of Ukraine do not practically take part in the formation of such platforms of active systematic, administratively planned participation. Universities have already lost the public role of the platform for the exchange of practical knowledge with the transition from the main communication processes from traditional communication to online communication.

Thus, there is a negative tendency to reduce the HEIs role as centers of public communication against the background of an increase in the factor of Internet users grouping around educational issues and autonomous structuring of communities around HEIs. It means that HEIs have become an object around which social network users are bundled. The technology of the search for group members, friends is largely focused on an educational feature that increases the value of HEIs (as well as schools) for the organization of virtual communities.

Thus, HEIs in Ukraine and in the global dimension can be defined as such social institutions that:

- are the objects around which social network users are united and virtual communities are formed;
- do not systematically affect the communication processes occurring within the virtual communities.

It is important that this trend is characteristic not only for Ukraine but also for many other countries and global regions [7].

As a consequence of such characteristics, the main communicative processes carried out around the HEIs, are not useful to society, and often are harmful to HEIs.

Virtual communities that are formed around the HEIs are mostly used by participants for the exchange of abstracts, course papers, dissemination of information (often inaccurate) on the possibility of the corruption schemes implementation. Information on the educational process organization, located in such environments, is often irrelevant, incomplete, contains inaccuracies. Posts in communities are sometimes saturated with abusive vocabulary and offensive language to the address of teachers.

Formation of communities with a biased negative attitude to HEIs is a logical consequence of the educational establishments passivity. After all, the interested user (student, entrant, young scientist, etc.), having no direct contact with the competent HEIs employees, can not meet their information needs, therefore, he or she does not take active part in it. At the same time, the lack of the HEIs representation, allows unobstructed expression of negative thoughts and this attracts negatively motivated users (for example, students who received low marks at the session).

It is obvious that there is a need for a fundamental change in the situation, when the HEIs from a static observer of network information processes in the global network becomes an active participant and coordinator of the process of useful information and knowledge accumulation in the Internet social media (on which there is a tangible public request in modern conditions [8]).

The essence of the active position of HEIs in information processes in the global environment should not be limited to simple advertising of an educational establishment (such an approach, in particular, is rather risky for the reputation of HEIs), but consists in the formation of socially useful content, of scientific, educational, educational and informational-individual character [9]. It should be noted that despite a significant public request for such content, the Internet already has very large arrays of such information. In fact, the user needs not just information, but qualitative, relevant knowledge presented in a convenient and accessible form in the right context. Only in this case, the content will have genuine social value. Obviously, HEIs generally have the potential to create such content (in fact, universities are the generators of such knowledge for society), but the creation and its placement on the Internet (especially in social environment) is a labor-intensive, complex, costly task, which at the same time often contradicts with individual operational tasks (distribution of commercial printed matter, protection of intellectual property, etc.). The obvious incentive for HEIs in this case is only the formation of their positive information image in society. The possibility to achieve the strategic marketing objective of long-term positioning of the HEIs on the market of scientific and educational services by providing the information needs of the society is the bridge that allows us to combine ordinary Internet users, higher education establishments and society as a whole into a fairly stable macrosystem [10].

A key aspect of this macrosystem is the presence of mutual interest and bidirectionality of communication between the process participants. As a result, the most demanded is the content-based information activity in the form of HEIs participation in the social environments of the Internet (hereinafter the SEI), which inevitably pushes the traditional schemes of information exchange on the Internet in the foreground. Issues of the importance of different social environments and their dominance on the Internet have been investigated and analyzed for quite a long time, therefore, there is no need to analyze it in this paper in detail. In the next subsection, the systemic features of information activities in the SEI and their differences from traditional types of IA are investigated. It should be noted that due to many factors, informational activity in the SEI is fundamentally more complex than usual. Among such factors there are the following: dynamism, reactivity and proactivity of the environment, the probability of direct aggression, direct inverse connection, large volumes of information flows, and others.

Unlike the usual advertising and information activity, activity on creation of socially useful content is complex in terms of organizational and technical point of view. As it will be shown below, it should cover all HEIs subdivisions, impose additional requirements for the IA process management [11], require the integrated use of diverse hardware and software. However, from the point of view of social significance and usefulness for the consumer, the formation of educational content in the dynamic environment of online communities is extremely important, as in this way a socially active self-motivated process of obtaining knowledge is realized as the most effective form of learning [12].

However, the benefits from such activities to HEIs are fundamentally larger than from advertising. After all, the result is socially useful content, which for a long time will serve to users, "live its own life" (will be distributed, updated, quoted, discussed). At the same time, this content will be clearly associated with HEIs and will serve to increase the network and public credibility and attractiveness of HEI among potential users of its services (this effect is described in particular in the following work [13, 14]). Therefore, the transition from advertising to socially-useful content creation is a qualitatively new step in the development of HEIs both in terms of the costs of organizational efforts and resources, and in terms of the benefits derived from it.

2 Processes of HEIs Informational Image Formation

The absence of a single specialized model of the formation process of the information image of the HEIs in the social environments of the Internet limits the possibilities of forecasting the development of this direction and planning of complex activities on the organization of information activities of a higher educational establishment on the Internet, generates certain threats to the HEIs image, in particular due to the omission or incorrect formation of certain types of information activity. In addition, the distribution of efforts at different stages of the information image formation is sometimes irrational, important leadership steps at the early stages are missed.

The process of the HEIs informational image formation, will be analyzed in details, namely, its stages and resources, which are being activated on them.

In the conditions of the WWW social component birth (and other global systems, for example, FidoNet) due to a number of circumstances (first of all – "caste" of authors of information content) a certain ethics was formed regarding the placement of advertising and information materials in the original social-oriented environments. There was no uniform system of rules, but the most significant were restrictions on the subjective information placement ("about him/herself"), the prohibition of advertising (including the "hidden", which was treated fairly broadly), other forms of business interests realization.

In the process of the Internet expansion as a basic social and communicative infrastructure, indicated rules have lost their relevance and have undergone significant transformations. Organizations (including HEIs) have the opportunity to be active participants in the process, with official representation in the virtual communities. The stages of the informational image formation process are analyzed, it is shown how they are implemented in two models [24]:

- passive – HEI is not a direct active participant in information and communication processes;
- active – HEI directly participates in the formation of its own information image at different stages.

2.1 Natural Unmanaged Process of HEIs Informational Image Formation

For a further analysis of the processes of the HEIs information image formation in social environments as well as methods and mechanisms for their management, it is necessary to analyze a certain basic version of the II creation. Today, such an option of the formation is ideal, and in practice, it is not practically implemented for large HEIs, for the following reasons. But this does not diminish the need for its formalization and analysis as the basis for other processes, a specific benchmark for the effectiveness and implications assessment of the HEIs II formation processes. It should be noted that this approach corresponds to the approach proposed in the work [15] concerning the analysis of the territorial social systems state.

The process is considered to be **natural uncontrollable** if a number of conditions are fulfilled:

- HEI do not have a policy and purposeful concrete actions regarding active influence on the information image;
- the actions of Internet users for the formation of an information image are not controlled and not stimulated by the structures of the HEIs, or other interested entities (affiliated organizations, competitors, opponents);
- the motivation of users concerning an information image creation is not prejudiced with respect to the HEIs, but is limited to general-cultural and general-social motivations (experience sharing, mutual assistance, development of the educational and national segments of the Internet);
- absent active functioning associations of users, purpose of which is to coordinate and take actions in order to damage the information image of the HEIs (including targeted information placement in social environments).

In the above-mentioned conditions, it is not argued that targeted information is either a false advertisement (in the case of the HEIs itself or its partners participation in the process) or defamatory, compromising or disinformation materials (in the case of information being placed by competitors or groups directed against HEIs). And in one case and in the second case, the materials can be both reliable and disinformative, and in many cases, the activity of HEIs and groups is useful for society (in fact, the task of this scientific work is to increase the effectiveness of the HEIs activity on the reliable information placement – useful for society activity). Therefore, the purpose of the conditions is to identify artificially created images of HEIs, without taking into account public value or harm.

An information image created in a natural, unmanaged process is a high-quality source of knowledge and experience both for the administration of the HEIs and for ordinary Internet users who interact or plan to interact with the HEIs. The HEIs administration, in particular, may use an image to identify bottlenecks and shortcomings in their work.

2.2 Information Image of the HEIs as an Element of the Higher Education Rating System

As it was stated above, the HEIs II, if created in a natural uncontrolled process, is the source of objective information for consumers of its services. Obviously, the presence of such assessments inevitably leads to the intentions to use them in the processes of the HEIs objective assessments formation. Such assessments are the main tool for the formation of the various currently popular university ratings [16]. A number of ratings of the new generation as the main indicator of the University's activity is used by certain methods used the calculated magnitude of its presence on the Internet. Among other such ratings, the most popular are Webometrics and UniRank. Both of them are focused on typical to classic sites specific techniques for the evaluation of the site complex position, which include, in particular: the total amount of pages on the site, their qualitative characteristics, popularity and citation of the site (it should be noted that to a large extent this approach resonates with the proposed in the paper [182] formal model of the Web site position). Directly, social component is not taken into account (according to official explanations that are sufficiently superficial and do not include the details of the algorithms for the rank values calculation), however, at least three key factors of the rating are largely determined by the characteristics of the HEIs II:

- the attendance of the site is also formed by means of the audience of the Internet social environment;
- the citation of the site is largely carried out in the SEI, the citation of scientific works is often determined by the social activity of a scientist;
- a significant portion of the site content can be generated by affiliated with site communities.

As a result, the information image of the HEIs becomes one of the main factors of high rating in the mentioned ratings.

Ratings will appear in the future which are based exclusively on judgments of social networks users, which will already use direct data. Obviously, in this case, formal indicators (likes, distribution, followers, etc.) as well as semantic analysis of texts will be used. The absence of this type of rating today can be attributed to the considerable technological complexity and the formation cost.

Another, interesting network phenomenon are informal ratings, which are formed on the basis of subjective assessments of students. A typical example of such ratings is the *Studzone*, a specialized site where you can leave your assessment and judgment about the teacher. Such services provide highly distorted and incomplete information about HEIs, but they also have a specific public request. Taking into account the socially oriented nature of such informal ratings, it can be safely asserted that they are directly dependent on or even are a part of the HEIs II.

The peculiarity of all the above mentioned ratings is their base on the conditionally "objective" information image, that is, which was created in a natural unmanaged process. Such a restriction is fundamental and calls into question the expediency of such ratings in the future, as each HEI has already started or is planning to influence its own information image. The essence of this influence will be discussed in the following sections.

Taking into account the certain public interest, and sometimes also the resonance of such ratings, the activities related to the improvement of representation on the Internet are being actualized for the HEIs. In this sense, this scientific work is the basis for such instruments, although the author does not set himself the task of approaches development to artificial improvement of the HEIs rating. Instead, an approach is proposed to raise the level of public awareness on HEIs and effective feedback, which positively reflects on all well-formulated rating systems. In this sense, it is advisable to draw an analogy with the problem of methods classification of site positioning ("white", "gray" and "black" optimization) and the corresponding choice.

2.3 Analysis of the Sources of the HEIs Information Image Formation

However, the processes of the large HEIs information image formation today have already lost their independence from external influences and can not be considered unmanageable, due to the following reasons:

- strong competition between HEIs and the emergence of their own policies and marketing departments;
- high reactivity of students, other affiliated users and availability of groups that are intended to compromise HEIs.

It is expedient to use the technologies of hidden user motivations identification, as well as filters based on the socio-demographic characteristics of users [17–19], performing the task of HEIs modernization and the shortcomings elimination in its work to clear the information image from information created as a result of targeted actions.

Let us consider the more detailed scheme of uncontrolled natural process:

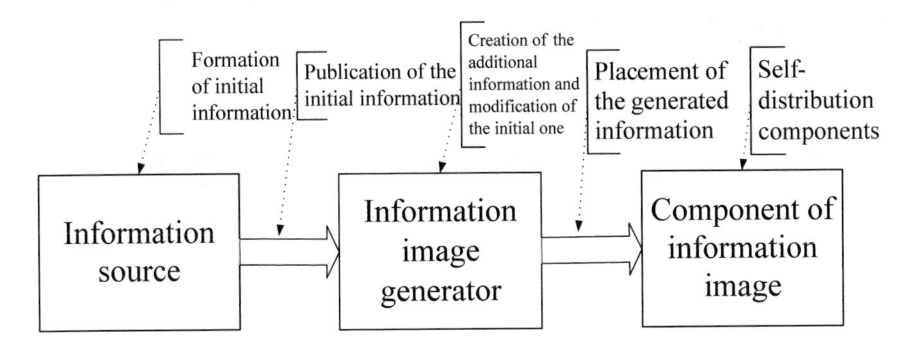

Fig. 1. Three stage process of information image formation

According to the scheme, the components of the information image are formed as a result of individual users activity in interaction with other members of the communities.

The interaction site and the interacting community will be called an information image generator. Typical examples of generators of information image are:

- users groups of social networks;
- small forums with the community of their participants;
- sections or topics of major forums with a community of forum members that show activity in a section or topic;
- thematic interactive sites or blogs and the commentators communities;
- thematic sections of the online media and the commentators community.

Generators of the information image arise and function forward as a reaction of society to certain information drives or initiatives. The sources of such causes or initiatives will be referred to as information sources. Information sources are essentially starting points for the process of information image formation. Often information sources are out of the Internet, directing to it the results of their activities. Examples of such off-line sources: journalistic investigations, activities and preparation of an information report of the HEIs subdivision, etc.

In addition to specified off-line sources, there may also be Internet sources in which an event, which is the source of the starting information for the generator, occurs directly within the Internet. The extreme and, at the same time, a rather common occurrence is the coincidence of the place of the event with the place of generation. That is, an informational source arises from the initiative of one of the subsequent members of the generator community. In fact, in this case, the topic that was invented for the sake of consideration is discussed.

The emergence of such information sources indicates a high public interest in HEIs, and if the origin of the sources is planned and coordinated, it indicates the unnatural nature of the process of HEIs II formation (either for the purpose of compromise or for the purpose of obsessive advertising).

The nature of the information source as information derived from the HEIs resources is investigated further.

In its operation, the information image generator can use several information sources.

We formalize the above concepts. We describe the information image of the HEI as a set of separate components:

$$Img = \{Img_i\}_{i=1}^{N^{(Img)}} \quad (1)$$

where Img_i – *i-th* component of information image; $N^{(Img)}$ – number of components.

Similarly, we describe the plurality of generators of *Gen*'s information image and information sources *Source:*

$$Gen = \{Gen_i\}_{i=1}^{N^{(Gen)}} \quad (2)$$

where Gen_i – *i*-th generator; $N^{(Gen)}$ – quantity of generators.

$$Source = \{Source_i\}_{i=1}^{N^{(Source)}} \quad (3)$$

where $Source_i$ – i-th information source; $N^{(Source)}$ – quantity of sources.

Between these sets there are relations that describe the structure of the system of information image formation.

Ratio $S\_G \subset Source \times Gen$ describes the relationship between information sources and generators, shows what information sources are used in one or another generator. That is:

$$\langle Source_i, Gen_j \rangle \in S\_G, \text{ if } Source_i \text{ is used in } Gen_j \quad (4)$$

Ratio S_G is the ratio "many to many". One generator can use multiple sources. Conversely, one source can be used in several generators.

Ratio $G\_Im \subset Gen \times Img$ describes the relationship between generators and components of the information image, shows what generators form one or another component of the II. That is:

$$\langle Gen_i, Img_j \rangle \in G\_Im, \text{ if } Gen_j \text{ is used in } Img_j \quad (5)$$

Ratio G_Im is the ratio "many to many". One generator can participate in the formation of several components of the information image, and vice versa, one component can be formed from several generators.

2.4 Management of the Information Image Formation Process

In practice, not only formalization of the uncontrolled process of the HEIs informational image formation is important, but also the analysis of the methods of this process management as well as the design of new scientifically justified methods of management, in particular methods for detection and counteracting malicious and asocial actions in the information space.

According to the scheme given in Fig. 1, management can take place in one of three stages of the process:

- Activities of the information source.
- Activities of the information generator.
- Activity on the image formation from the generator's materials.

The management of an information source involves management of the process of information preparation for its further disclosure on the generator. Management can cover all aspects of the information processing before the publication: collecting, presenting, editing, adapting, modifying content and censoring. Additionally, adjustments in the intensity and scope of the information preparation process may take place. However, the management of the process of the information image formation at this stage should take into account the further processing of information from the source on the generator. That is, management at this stage has a postponed character in terms of the result.

Generator management involves a community activity process management that processes information coming from a source. As a rule, the presence of the community means that direct control can not be implemented, otherwise the community inevitably collapses and the generator becomes ineffective. Partial management of the generator is developed within the framework of research of users communities management, including web forums [20]. Generator management is limited to:

- actions towards community – critics of the participants, their support, moderation (in the presence of authority), interaction with the administration and participants;
- accumulation of information – comments, participation in discussions, placement of additional materials.

Management at the stage of the image creation from the generator materials involves management in two directions:

- formation of content – management of the referencing process, quotation, annotation of the created materials, their literary elaboration;
- distribution of the image component – advertising, anti-advertising, copying, replication, etc.

Different control options and differences between the managed process from the unmanaged one are illustrated in Fig. 2.

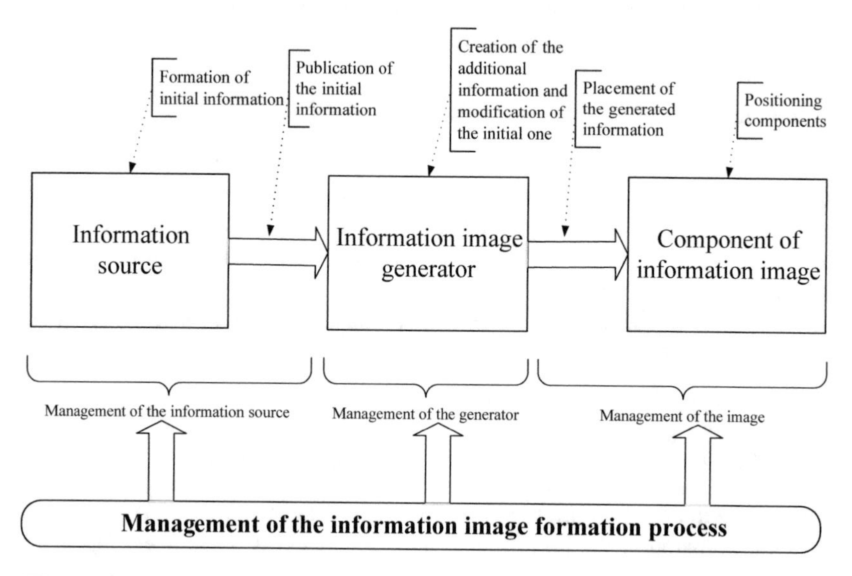

Fig. 2. Step-by-step management of the process of the information image formation

Since the information source and component of the information image, in contrast to the generator, are rather passive objects, their management focuses primarily on changing of the ratio *S_G* and *G_Im*.

The results of management at each stage affect the next stage, but these influences are "underserved" outcome in terms of the character. Thus, changes to the information source control stage without management at the stage of an information image generation can cause certain threats concerning deployment of an information image in an undesirable for the HEI direction. For example, information placement on the disciplines content in a particular specialty without further responding to requests and discussions of users may lead to misleading interpretations that, in turn, will generate criticism or lead to misinformation. As a result, the efforts of the HEIs to improve the image will only make it worse for it; in the eyes of the society, the HEI will act as a backward one, which does not provide modern training or misleads.

Similarly, if all HEIs efforts are focused on the image management (only on reacting on the placed materials), its position in the present conditions will only worsen, since only participation in discussions where the position of the educational institution is confirmed by high-quality content, and not only the official, but also of educational and scientific character.

The worst in terms of image formation can be considered a variant of attempts to control the generator (without appropriate measures in the previous and next stages). In practice, such attempts at this kind of management are reduced to the placement of few press releases useful to the public (often perceived as spam), the reaction on which is then not monitored, and in a communicative sense, "attracts negativity," as well as attempts to pressure users who are members of communities-generators. Such scenarios usually form negative, often sarcastic content about the university, in the worst cases, they may lead to the emergence of groups motivated to information aggression in the direction of the HEIs users.

Therefore, the maximum effect from the actions of the process control is possible only if the continuity of the control process is maintained. Attempts to compensate the lack of effect by additional impacts at the final stage are resource-intensive, have significant risks to HEIs and are limited in result. Similarly, insufficient attention to the final stage is able to completely alleviate the efforts concerning the improvement of the HEIs informational image, which were provided in the first stage.

3 Prediction of the Prospects of Work on Modifications of the Existing Information Image

The presence of factual data on the change of the HEIs information image as a result of the targeted activity of its subdivisions allows to make certain forecasts regarding the future prospects of activity in those or other generators.

The main factor of promising (as already noted above) is the high performance of the IA of the HEIs in a particular generator. Low leads to loss for HEIs in the resource of the performers (for the effect obtaining significant efforts and time are consumed) and in the information resource (to obtain the effect existing texts are used, which of the number of network restrictions is undesirable to replicate between different communities).

Productivity may be low in the following cases:

- **“subjective”** – the low communication quality or quality of the information resource located in the generator;
- **“aggressive”** – the existing aggressive actions of the generator community or negative social processes;
- **“stationary”** – there are high community indicators for HEIs evaluation at the beginning of the period.

Accordingly, for each of the above cases, it is advisable to use a separate scenario for further actions in the community. The identification of each case is carried out by the expert, taking into account the dynamics of the formal indicators change and evaluation of the activity of the subdivision employees.

In the “subjective” case, the basis of the strategy for the situation improvement is either the qualification upgrade and administrative measures concerning the subdivision or redistribution of the responsibility areas.

In the “aggressive” case, it is necessary to involve PR and social communication specialists and to implement special procedures for HEIs II protection from harmful actions.

A “stationary” case is generated in situations when a completely positive view of HEI has already been formed in the generator and further information activities can not substantially increase it. In this case, the following three scenarios are possible (see Fig. 3).

“Conservation of II” – the activity in the generator is reduced to the minimum necessary to maintain communication (often periodic short messages and reaction to information events). It is advisable to use this option for general-purpose communities. More detailed questions of the correct preservation of the information image in the generator are investigated.

“Saturation of II” – despite the impossibility of a significant improvement of the opinion about HEIs II, the generator is used as a platform for the placement of information materials on HEI, in particular, in order to increase the quoting of HEI materials and its mentioning. This option is appropriate for reputable professional communities.

“Participation in development” – participation in the generator transformation. In particular, increasing its popularity thanks to HEIs materials, interaction with active participants, establishment of in-depth partnerships with the administration of the community, other forms of interaction that change the community itself. This option can be used for professional communities that are young enough and show good dynamics of development, but have not yet acquired high credibility. In this scenario, not only the results of the informational nature, but also the administrative ones are possible – for example, the formation of the personnel reserve of the HEI or NGO affiliated with the HEI.

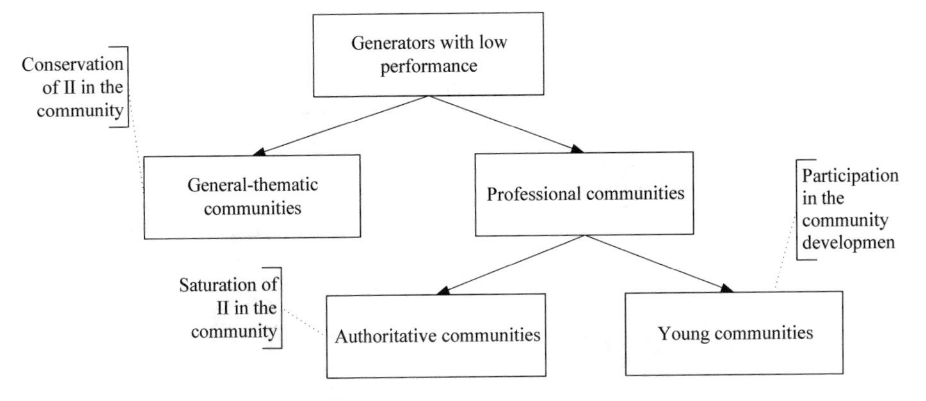

Fig. 3. Typical scenario of HEIs IA for stationary case of low-performance

Determination of the desired development scenario requires an in-depth understanding of the generator development prospects, therefore, it should be carried out by the subdivision responsible for it.

4 Analysis of the Integrity and Completeness of the HEIs Informational Image Coverage

An indicator was introduced *KCover* (*University*), which determines the proportion of communities which are really engaged in active information activities and which form the HEIs II. The indicative values of the indicator are provided, which allow to state whether enough activity is carried out, that is, whether enough "covered" is the information environment in which there is the HEIs II.

However, the indicator *KCover* (*University*) does not solve all the tasks for analyzing the integrity and completeness of the information image coverage in view of the multidimensional nature of the HEIs IA. We distinguish these tasks in the following way:

- the analysis of thematic gaps in the HEIs II is aimed at omissions detection in the coverage of relevant communities, important generators ignoring;
- the analysis of the topics coverage of the HEIs IA is aimed at deficiencies identification in the existing thematic description of the HEIs, important thematic areas ignoring.

The indicated gaps in the HEIs and HEIs II areas are illustrated in Fig. 4.

Let us consider a more detailed tasks of identification and thematic gaps elimination in the HEIs II and the thematic description of the HEIs.

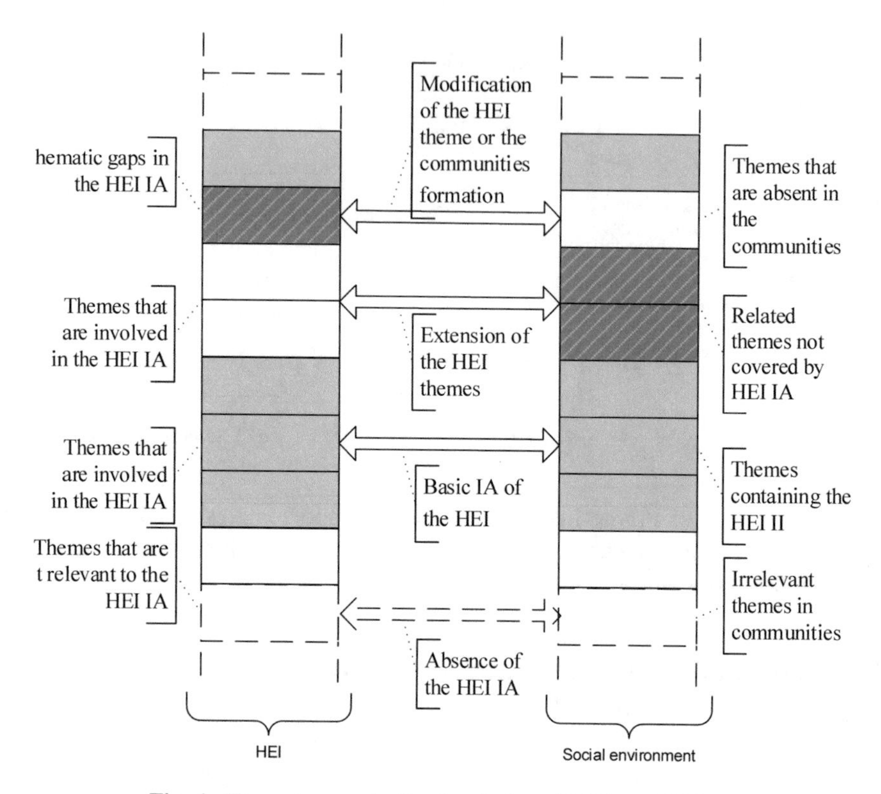

Fig. 4. Thematic gaps in the description of HEIs and HEIs II

4.1 Analysis of Thematic Gaps in the HEIs Information Image

Considerable part of the modern HEIs has a wide range of thematic areas of activity in both the educational and scientific spheres. Accordingly, the expected result of a complex HEIs IA is the presence of components in the information image that reflect or are relevant to each of these areas.

In order to describe the thematic areas of the HEI separate subdivision in [21], the characteristic "General thematic section" $DTTh(Dep_i)$ is introduced. Then we define the **general topic of HEI** as:

$$DTTh(University) = \{DTTh(Dep_i)\}_{i=1}^{N^{(DTTh)}} \tag{6}$$

We determine the criterion that the information image on a particular subject is sufficiently available in social environments. **Let us suppose that there is no thematic gap, if all the leading communities have certain elements of the HEIs II of the defined topic.**

To determine the lead communities, it is expedient to use the indicator of the generator importance [22]. This indicator allows you to form a generators rating by their value for the HEIs IA, in which the first positions are occupied by generators of high importance.

Based on the importance indicator and generators ranking, we will specify the above provided criterion. Taking into account that the importance of generators (as well as other indicators of objects rating in the WWW) have the ability to rapidly fall (possibly under the Zipf law) from the first few positions to the next, we determine that it is sufficient to have the HEIs II in a small number of major generators (practically in 3–7 communities).

Taking into account the nature of the importance decrease, to calculate the extent of the topic coverage of the HEIs II we use the formula:

$$CoverTh(Th_i) = \log_{(1+..Ctop)}\left(\begin{array}{c} Ctop * Eimg(Gen_1) + (Ctop-1)* \\ * Eimg(Gen_2) + .. + Eimg\left(Gen_{Ctop}\right) \end{array}\right) \tag{7}$$

where $Ctop$ – constant, number of important generators; $Eimg\left(Gen_j\right)$ –the function of the availability of the HEIs II in j-th generator (1 – exists, 0 – absent).

The logarithmic method of coverage degree calculation by using the “inverted” ranking position as a multiplier in the sublogarithmic expression is traditional for computing the information completeness for various information objects ranked according to social importance or significance (in particular, it is used in library and search information systems).

If for Th_i subject index is equal or close to zero, then we assume that there is a thematic gap in the HEIs II.

A special case of a thematic gap is a topic that can not be covered because of objective reasons. The main reason for this is the lack of communities that are relevant to the subject matter. In this case, it is expedient to remove or redefine of such a thematic area of the HEIs IA, its association with another one. It is optimal to turn the topic into a subject that is in demanded in social environments.

A particular interesting case is the latest topics that appeared in a short period prior to the analysis (for example, the latest scientific research or technological breakthroughs, new actual social issues). For such topics, the lack of communities is a temporary phenomenon; therefore, it is inappropriate for the HEIs to leave such topics, and vice versa – to initiate the creation and promote the development of new communities that provide image leadership in the subjects.

In addition to the assessment of the coverage level of the HEIs II topics, the rating allows you to identify critically important for the community topics, activities in which are necessary for a fullfledged IA of the HEIs.

4.2 Analysis of the Coverage Completeness According to Topics of HEIs IA

If the preliminary type of the analysis was focused on the identification of the information space thematic areas, which do not mention the HEIs, although HEIs is active in these topics, the analysis of the topics coverage of the HEIs IA is aimed at the subjects identification in social environments that should have been presented in $DTTh(University)$, but are absent indeed.

Let us analyze the task in detail. Obviously, no HEI can cover the entire scientific and educational field in its activities. So, in any case, there are topics that can not be included in the IA of the HEIs in principle. At the same time, there may be themes that are really present in HEIs, but information activity for them is not followed up, or the wording of the topic is so unsuccessful that it does not correspond to the themes of the existing communities. Such gaps occur at the appropriate preparatory stage of the HEIs IA [23] and need to be removed as such which reduce the effectiveness of the HEIs IA and cause additional risks.

There are following ways to solve this problem (Fig. 5):

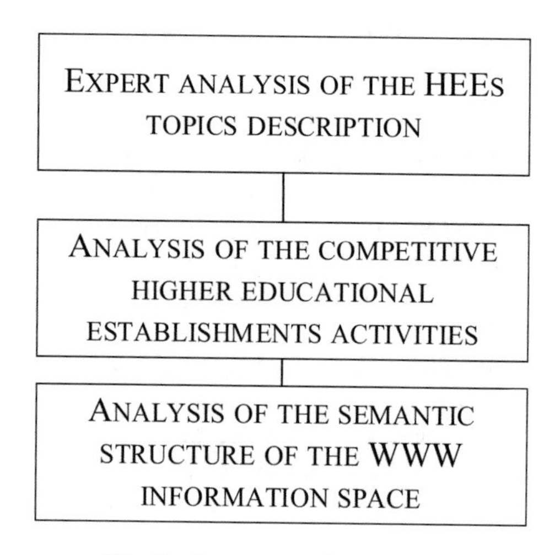

Fig. 5. 3 ways to solve problem.

It is advisable to complement each of the following paths to others to the extent that is available to a specific HEI. Let us consider them in more detail.

An expert analysis of the HEI topic description is to involve network communications specialists and marketers who analyze the relevance of the thematic description of the HEI to the interests of target groups. At the final stage it is advisable to engage applied linguists-terminologists to the task for the formulation of the subject. The main disadvantage of this approach is the lack of a sufficient understanding of high-tech and science-intensive subject areas by experts, which makes it impossible for them to perform their tasks effectively.

Analysis of the competitive higher educational establishments activities eliminates the above-mentioned lack of an expert approach and consists in the study of the thematic structure of the competitive HEIs II. Communities are found in which the information image of the competitor is available and their subject matter is determined. Given the analysis of several competitors, a rather complete picture of the thematic structure of the subject area is obtained. This approach is rather laborious, however, it should be noted that the analysis of competitors activities is an important component of a comprehensive information activity for any powerful organization (not only HEIs). The disadvantage of this approach is its ineffectiveness from the point of view of leading HEIs (“flagships” in its scientific and educational areas), since competitors activities are likely to be weaker and incomplete compared with leading HEIs, so a qualitative picture of the subject area can not be provided.

Analysis of the semantic structure of the WWW information space is based on the use of WWW ontologies that can be correlated with the subject domain. Among these ontologies today we will mark Wikipedia. Catalogs of websites are of particular importance. Another valuable source of information on topics are information accumulation systems focused on social tags (terms that are put directly to resources by users) – bookmark services, collective blogs, etc. This approach provides the most accurate and relevant reflection on the subject of social environments. The disadvantage of the approach is the need for an in-depth stylistic elaboration of the received thematic sections (none of the resources mentioned does not provide a sufficiently structured list of topics).

An important feature of the considered problem of analyzing the completeness of coverage by topics of HEIs IA is its “concealment”. By virtue of their nature, omissions in the fullness of the subject do not generate operational problems for a certain, prolonged period of the HEIs IA. However, gaps eventually lead to the formation of a significant layer of negative or harmful elements of the HEIs II, which may suddenly increase their significance or become a source of information threats. Therefore, the indicated analysis should be carried out regardless of how complete the thematic description of the HEIs is issued on the subjective level.

5 Conclusion

The article analyzes the process of formation of the information image of the HEI, namely, its stages and resources, which are applied to them. The three-step process of forming an information image is presented, where the components of the information image are formed as a result of the activity of individual users in interaction with other members of the communities. The step-by-step management of the information image formation process is presented. The typical scenarios of the HEI information activity for stationary case of low productivity are presented. The analysis of the integrity and completeness of coverage of the information image of the HEI is carried out. Thematic gaps in the description of HEI and informational image of HEI are presented.

References

1. Document WSIS-03/GENEVA/DOC/4-E. Declaration of Principles Building the Information Society: a global challenge in the new Millennium. World Summit on the Information Society, Geneva (2003). Tunis (2005). http://www.itu.int/wsis/docs/geneva/official/dop.html
2. Bottery, M.: The end of citizenship? The nation state, threats to its legitimacy, and citizenship education in the twenty-first. Camb. J. Educ. **33**(1), 101–122 (2003)
3. Kwiek, M.: Transformationen und Anpassungen im polnischen Hochschulwesen. Die Hochschule **1**, 187–199 (2003)
4. Lambert, R., Butler, N.: The Future of European Universities: Renaissance or Decay?, vol. 65. Centre for European Reform, London (2006)

5. Magnusson, J.: Examining higher education and citizenship in a global context of neoliberal restructuring. Can. Ethn. Stud. J. **32**(1), 72–99 (2000)
6. Shih, C.: The Facebook Era: Tapping Online Social Networks to Build Better Products, Reach New Audience, and Sell More Stuff. Mann, Ivanov and Ferber (2010). 304 pages
7. Delanty, G.: Ideologies of the knowledge society and the cultural contradictions of higher education. Policy Futures Educ. **1**(1), 71–82 (2003)
8. Forkun, Yu.: Coordination model for the development of information and documentation support in social and communication systems. Information technologies and systems in the document-research sphere. DonNU, Donetsk (2013). 143 pages
9. Stone, D.: Using knowledge: the dilemmas of 'bridging research and policy'. Compare **32**(3), 285–296 (2002)
10. Metz, A.: The Social Customer: How Brands Can Use Social CRM to Acquire, Monetize, and Retain Fans, Friends, and Followers. McGraw-Hill (2011). 304 pages
11. Tomashevsky, V., Novikov, Y.: Model of motivation of personality training. Scientific reports of the National Technical University of Ukraine "Kyiv Polytechnic Institute", vol. 2, pp. 73–86 (2012)
12. Lande, D., Furashev, V.: Monitoring systems, extracting facts, building links based on the analysis of unstructured texts. Legal Informatics. Research Center for Legal Informatics, vol. 2, no. 26, pp. 3–9 (2010)
13. Demirhan, K., Çakır-Demirhan, D.: Political Scandal, Corruption, and Legitimacy in the Age of Social Media. IGI Global, Hershey (2016). 295 pages
14. Deželan, T., Vobič, I.: (R)evolutionizing Political Communication through Social Media. IGI Global, Hershey (2016). 333 pages
15. Zgurovsky, M., Gavrish, O., Voitko, S.: Development of methodology for determining the level of threats to the sustainable economic development of Ukraine. Econ. Bull. NTUU KPI Collect. Sci. Works **8**, 26–33 (2011)
16. Buschev, S., Biloshchitsky, A., Gogunsky, V.: Know-metric bases: characteristics, possibilities and tasks. Manage. Dev. Complex Syst. **18**, 145–152 (2014)
17. Korzh, R., Peleshchyshyn, A., Syerov, Yu., Fedushko S.: Principles of university's information image protection from aggression. In: Proceedings of the XIth International Scientific and Technical Conference (CSIT 2016), pp. 77–79. Lviv Polytechnic Publishing House, Lviv (2016)
18. Trach, O., Peleshchyshyn, A.: Development of models and methods of virtual community life cycle organization. In: Hu, Z., Petoukhov, S., He, M. (eds.) Advances in Artificial Systems for Medicine and Education II, AIMEE 2018. Advances in Intelligent Systems and Computing, vol. 902, pp. 473–483. Springer, Cham (2019)
19. Korzh, R., Fedushko, S., Trach, O., Shved, L., Bandrovskyi, H.: Detection of department with low information activity. In: Proceedings of the XIth International Scientific and Technical Conference Computer Sciences and Information Technologies, pp. 224–227 (2017)
20. Fedushko, S., Ustyianovych, T.: Predicting pupil's successfulness factors using machine learning algorithms and mathematical modelling methods. In: Hu, Z., Petoukhov, S., Dychka, I., He, M. (eds.) Advances in Computer Science for Engineering and Education II, ICCSEEA 2019. Advances in Intelligent Systems and Computing, vol. 938, pp. 625–636. Springer, Cham (2019)
21. Korzh, R., Fedushko, S., Trach, O., Shved, L., Bandrovskyi, H.: Detection of department with low information activity In: Proceedings of the XIth International Scientific and Technical Conference Computer Sciences and Information Technologies (CSIT-2017), pp. 224–227 (2017)

22. Korzh, R., Peleshchyshyn, A., Syerov, Yu., Fedushko, S.: University's information image as a result of university web communities' activities. In: Shakhovska N. (eds) Advances in Intelligent Systems and Computing: Selected Papers from the International Conference on Computer Science and Information Technologies, CSIT 2016, vol. 512, pp. 115–127. Springer, Cham (2017)
23. Fedushko, S., Syerov, Yu., Korzh, R.: Validation of the user accounts personal data of online academic community. In: Proceedings of the XIIIth International Conference Modern Problems of Radio Engineering, Telecommunications and Computer Science (TCSET 2016), pp. 863–866 (2016)
24. Korzh R., Peleshchyshyn A., Trach O., Tsiutsiura M.: Analysis of the integrity and completeness of the higher education institution informational image coverage In: Proceedings of International Scientific Conference Computer Sciences and Information Technologies (CSIT-2019), vol. 3, pp. 48–50. IEEE (2019)

Determination of Measures of Counteraction to the Social-Oriented Risks of Virtual Community Life Cycle Organization

Olha Trach and Solomia Fedushko(✉)

Lviv Polytechnic National University, Lviv, Ukraine
olya@trach.com.ua, felomia@gmail.com

Abstract. The article identifies socially-oriented risks in the life cycle organizing of the virtual community. Namely: the appearance of a negative audience; reducing the quality of information content; anti-legal materials and community activities; loss of community control. The indicators of the community's entry into the area of socially-oriented risks for effective and successful management of the virtual community are determined. The algorithm of determination of level of intensity of measures of counteraction to socially-oriented risks is developed, based on the analysis of socially-oriented risks of the virtual community, which, unlike existing ones, are inherent to virtual communities only. That made it possible to increase the efficiency of creating a virtual community and to improve the functioning of the operation throughout its existence, ensuring achievement of goals and development of the virtual community. The result of the implementation of the algorithm is the classification of the risk indicator for: high, medium and low levels. Measures to counteract socially-oriented risks will help predict and structurally create and manage the community and improve the overall process of creating a virtual community.

Keywords: Virtual community · Life cycle · Risks · Project · Socially-oriented risks

1 Introduction

Taking into account the current trends and popularity of the Internet more often when creating a new brand, event or product, its preliminary presentation takes place on the Internet. The same thing happens with the finished product. The best platform for this is virtual communities. In addition, it is advertising and marketing strategy, which covers the broad category of various age groups.

Virtual communities form a large part of the web space, which provides opportunities to meet the information needs and interaction of the participants, and every day their quantity becomes larger, the existing ones are rapidly developing.

In the basis of any process of creating and managing information systems, certain basic standards and processes are laid. The virtual communities are no exception. This set of standards and processes is reflected in the virtual community life cycle. The virtual community life cycle as an object is important in the following areas of research:

N. Shakhovska and M. O. Medykovskyy (Eds.): CCSIT 2019, AISC 1080, pp. 680–695, 2020.
https://doi.org/10.1007/978-3-030-33695-0_46

- security and information wars in virtual communities [1];
- creating and managing virtual communities [2, 3];
- manage and create the content of the virtual community [4];
- attract users of virtual communities [5];
- interaction of users in social Internet services [6];
- marketing and advertising in the virtual community [7, 8];
- educational and scientific processes in virtual communities [9–12].

Actually developing methods and tools of virtual community life cycle organization is an important and actual task, as it is the basis for increasing the efficiency of creating a virtual community and improving its functioning throughout its existence, ensuring achievement of goals and development of the virtual community. Therefore, studying the risks and methods of their counteraction during of virtual community life cycle organization is an actual and necessary task.

2 Risks in Project Management

When performing any steps of virtual community life cycle organization situations may appear in which individual indicators will go beyond the predicted values, worsening the overall status of the project. These situations are in fact the risks that arise during the project implementation. Risk investigations when creating and managing a virtual community are incomplete. In project management on topical issues is the management of project risks. The researchers pay considerable attention to the risks in project management, studies are conducted in such areas as: methods and tools for risk analysis in projects [12–15], risk management methodology [16, 17], measures to counteract risks [18–20]. Thus, it is necessary to determine the socially-oriented risks of virtual community life cycle organization, identification of risk response measures and developing an algorithm for determining the level of intensity of measures to counteract socially-oriented risks during of virtual community life cycle organization.

The number of risks that arise is significant. For further risk analysis, it is advisable to consider the typical, most commonly occurring risks according to the experience of community management specialists. Typical project management risks, which occur during of the virtual community life cycle organization is:

- occurrence of design error;
- risk of loss of workforce (performers of stages and directions);
- time risks;
- administrator and moderator errors during community management, etc.

However, one of the inherent virtual communities is socially-oriented risks.

3 Socially-Oriented Risks of Virtual Community Life Cycle Organization

Socially-oriented risks – this is the emergence of the situation, in which individual parameters of the indicators will go beyond the predicted values, worsening the overall status of the project.

During of the virtual community life cycle organization, we highlight such socially-oriented risks [21, 22]:

- the appearance of a negative audience;
- reducing the quality of information content;
- anti-legal materials and community activities;
- loss of community control.

The system of socially-oriented risks for virtual community life cycle organization is described by the tuple (1):

$$SOR(Com_i) = \left\langle \begin{matrix} NegativeAud^{(User)}(Com_i), InfContent^{(Inf)}(Com_i), \\ Antilegal^{(User,Inf)}(Com_i), Control^{(User)}(Com_i) \end{matrix} \right\rangle \quad (1)$$

where $NegativeAud^{(User)}$ – the risk of appearance of a negative audience;
$InfContent^{(Inf)}$ – the risk of reducing the quality of information content;
$Antilegal^{(User,\ Inf)}$ – the risk of anti-legal materials and community activities;
$Control^{(User)}$ – the risk of losing control of the community.

During the implementation of design work for the formation of the virtual community, it is necessary to provide protection against the occurrence of a risk, which involves the following countermeasures (Fig. 1).

3.1 The Risk of Appearance of a Negative Audience

The risk of appearance of a negative audience is that in the virtual community a subset of users appears, who are negatively tuned to the activities of the community or other members of the community. To protection, measures against the risk of a negative audience belong to:

- avoiding participants of provocateurs;
- removing flames;
- prevention of cyberbullying.

Avoiding Participants of Provocateurs

The measure of counteraction is to identify the participants, which create a provocative situation, a negative situation with respect to other participants or the activities of the virtual community.

Provocateur – is a member of the virtual community, which by its actions artificially creates a negative environment in the virtual community.

To provocateurs can be attributed:

Troll – member of the virtual community, who joins the virtual community to commit provocative actions against other members or community activities. The community participant carries out the role of a troll by manual control and with the help of specialized software tools - robots, or using these two methods.

A user with an unstable mental-emotional state – members of the virtual community, which by virtue of their mental disorders do actions aimed at provoking other participants.

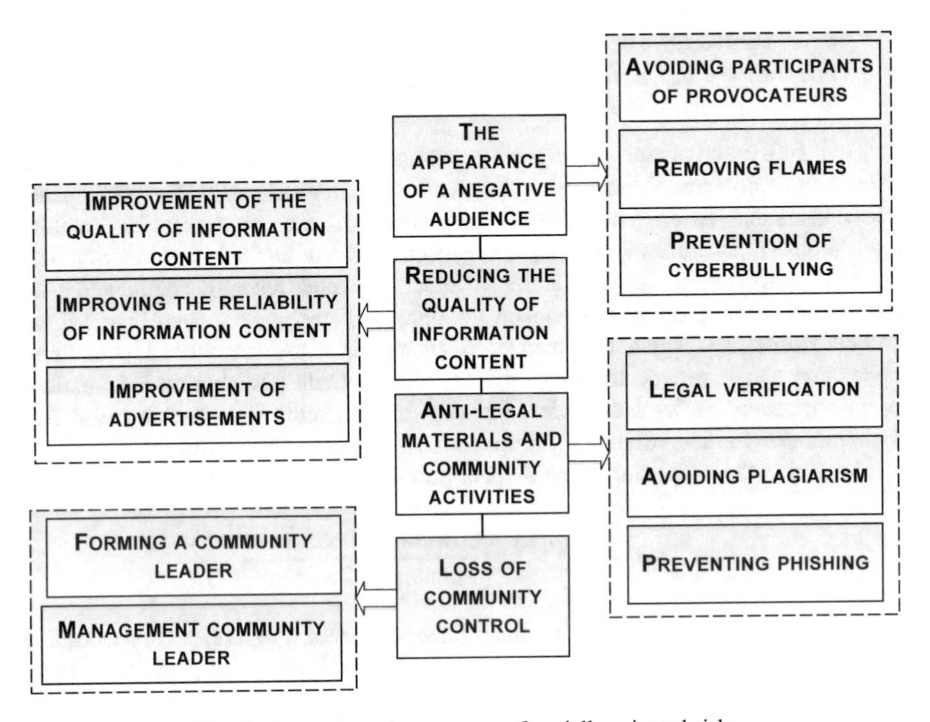

Fig. 1. Protection of emergence of socially-oriented risks

Participants-provocateurs (in particular trolls) in the virtual community you can identify by:

1. Rating:
   - Troll-member with high rating. Member of the virtual community has a high rating, but his posts, avatar and nickname show that he is a troll.
   - Troll-member with low rating. Another sign troll-member – this is a negative rating, lowered for violations.

2. Style of communication:
   The participant-provocateur is different from the other members of the virtual community in style of communication, which violates the communicative atmosphere. Most often a provocateur writes provocative posts, gives sharp answers to the remarks, does not support a polite conversation, avoids an honest answer.
3. Time spent in the community:
   - regular participant-provocateur – member of the virtual community, which often acts in provocative manner in various sections of the virtual community.
   - "Thematic" participant-provocateur – member of the virtual community, which creates provocative actions only in a certain section. This member can leave only one message in the virtual community, causing them a negative reaction from other members of the community.

To identify and avoid participants-provocateurs and there are several ways to prevent this problem. «Manually» – the moderator himself tracks the participants-provocateurs and performs actions to remove them from the community. By installing applications with the discovery of a participant-provocateur on the keywords, which are negative. Applying to the moderator methodology and recommendations for the detection of trolls, programs developed by a group of researchers from Stanford and Cornell University, which is intended as an auxiliary tool for virtual community moderators. Often, provocateurs create fake accounts (with false contact information), to prevent such participants it is necessary to use tools for checking a person (example, «Hoverme», «Identify», «Pipl.com», «Spokeo», «WebMii»).

Determine the proportion of participant-provocateurs in the virtual community:

$$PartProvocateur^{(User)} = \frac{Quantity(Provocateur^{(User)})}{Quantity(Common^{(User)})} \quad (2)$$

where $Quantity(Provocateur^{(User)})$ – the number of detected participants-provocateurs in the virtual community; $Quantity(Common^{(User)})$ – total number of participants in the virtual community.

Removing Flames

Remedies for the removal of flames are detected and blocked by the participants-flames. Flame – a discussion in which community members move from the usual and the original topic to personal image, controversy and quarrels. Participants who distribute the flame called flames. The appearance of the graphic designers in the virtual community leads to conflicts between the participants. This often happens based on differences of political, religious, social and national views. Which leads to the appearance of a negative audience virtual community.

In the virtual community, it is worthwhile to highlight two types of participants-flames:

- the participant who himself declares the flame;
- other members or community moderator who noticed a flame.

To protect the virtual community from participants-flames, the administrator and moderator of the virtual community need to identify the flame participants with the help available studies of linguistic peculiarities of texts and linguistic and communicative indicators. Participants-flames, similarly to the participants-provocateurs create fake accounts (with false contact information). In order to prevent such participants-flames, it is necessary to apply tools for person verification.

Determine the proportion of participant-flames in the virtual community:

$$PartFlame^{(User)} = \frac{Quantity(Flame^{(User)})}{Quantity(Common^{(User)})} \quad (3)$$

where $Quantity(Flame^{(User)})$ – the number of detected participants-flames in the virtual community; $Quantity(Common^{(User)})$ – total number of participants in the virtual community.

Prevention of Cyberbullying

This method of counteraction is to detect cyberbullying. Cyberbullying – these actions are aimed at persecuting people using the Internet and electronic equipment. Often, cyberbullying is used to humiliate, harassment or persecution of other members of the virtual community. The appearance of such participant negatively tweaks other users of the virtual community and has negative consequences.

Cybersecurity measures in the virtual community should be used as a refinement in the community rules and identify participants through existing recommendations for identifying pest-members. Pest-members, similar to provocateurs and flames, create fake accounts (with false contact information), to prevent such participants it is necessary to use tools for checking a person.

Determine the proportion of participant, who engaged in cyberbullying in the virtual community:

$$PartKiberbulinh^{(User)} = \frac{Quantity(Kiberbulinh^{(User)})}{Quantity(Common^{(User)})} \quad (4)$$

where $Quantity(Cyberbullying^{(User)})$ – the number of detected participant, who engaged in cyberbullying in the virtual community; $Quantity(Common^{(User)})$ – total number of participants in the virtual community.

Indicator of entry of the virtual community into the risk area of a negative audience of the virtual community, the basis for determining this indicator is the proportion of harmful participants:

$$NegativeAud^{(User)} = (1 - w_{Pr} * \frac{Quantity(Provocateur^{(User)})}{Quantity(Common^{(User)})} + \\ + w_F * \frac{Quantity(Flame^{(User)})}{Quantity(Common^{(User)})} + w_K * \frac{Quantity(Cyberbullying^{(User)})}{Quantity(Common^{(User)})}) \quad (5)$$

where $\{w_{\mathrm{Pr}}, w_F, w_K\} \in w$ – weight coefficients of each indicator respectively, are determined by the manager of virtual community life cycle organization, $0 \geq w_i \leq 1$, $w_i \in w$, $\sum_{w_i \in w} w_i = 1$; *Quantity(Provocateur*$^{(User)}$*)* – the number of detected participants-provocateurs in the virtual community; *Quantity(Flame*$^{(User)}$*)* – the number of detected participants-flames in the virtual community; *Quantity(Cyberbullying*$^{(User)}$*)* – the number of detected participant, who engaged in cyberbullying in the virtual community; *Quantity(Common*$^{(User)}$*)* – total number of participants in the virtual community.

3.2 Reducing the Quality of Information Content

The essence of the risk of lowering the quality of content is that in the virtual community the level of reliability decreases, literacy and quality of information content. The content attribute of virtual communities is posts and comments.

Security measures to reduce the risk of content belong:

- improvement of the quality of information content;
- improving the reliability of information content;
- improvement of advertisements.

Improvement of the Quality of Information Content

Security measure to check the quality of information content is to refine the rules in terms of requirements for information content. Introducing the rules for posting messages and comments in the virtual community.

Also an important part of the quality of information content is the prevention of flood, offtopik, wipe, overquoting (Internet etiquette of communication). Flood – is a message, post or comment in the content of which there is no useful or relevant information. Offtopik – is a message, post or comment, which is not related to the topics of the virtual community. Wipe – creating new meaningless discussions, for moving up-to-date information down the page. Overquoting – too large a quote that is uncomfortable for to understanding. Another component of quality information content is competent and easy to perceive written post.

Security measures are ongoing monitoring of the administrator and moderator of the virtual community to identify the above components of bad etiquette on the Internet. Application of information content management system (cms-system). Also, one of the options for improving the quality of information content is the involvement of external copywriters. Determine the proportion of low-quality information content in the virtual community:

$$Content^{(Inf)} = \frac{Quantity(Content^{(Inf)})}{Quantity(Common^{(Inf)})} \tag{6}$$

where *Quantity(Content*$^{(Inf)}$*)* – the quantity of low-quality informational content, executed not according to the rules of the virtual community; *Quantity(Common*$^{(Inf)}$*)* – total number of information content for the virtual community.

Improving the Reliability of Information Content

The essence of the measures to protect the authenticity of information content is verification of the authenticity of information content, which is published in the virtual community. These measures are necessary to prevent fake (false) information, which can cause panic among community members, manipulation, incitement of hostility between members of the community, taming someone's reputation, etc.

To protect the virtual community from fake information there are many services and applications to verify the authenticity of the content, as text data and multimedia. Image Verification Services: Findexif.com, Foto Forensics, Google Search by Image, Jeffrey's Exif Viewer, JPEGSnoop, TinEye. Services for checking test information: Snopes.com, PeopleBrowsr, HuriSearch, Geofeedia, Verify.org, ua, Lazy Truth, Trooclick, etc.

Determine the proportion of inaccurate information content in the virtual community:

$$Certainty^{(Inf)} = \frac{Quantity(Certaintity^{(Inf)})}{Quantity(Common^{(Inf)})} \tag{7}$$

where $Quantity(Certaintity^{(Inf)})$ – the quantity of inaccurate information content of the virtual community; $Quantity(Common^{(Inf)})$ – total number of information content for the virtual community.

Improvement of Advertisements

The essence of the measures to protect the verification of advertisements is to tracking to a community did not get advertising, which will upset the participants or does not relate to the content and activities of the virtual community. Also, the excessive amount of advertisements overwhelms the virtual community by violating the communicative atmosphere.

The measures to protect the community from excessive advertising include the use of the mechanism of Internet advertising - targeting, regulation of advertising messages relative to participants. A ban on advertisements created by a member of the virtual community.

Determine the proportion of negative advertising in the virtual community:

$$Blurb^{(Inf)} = \frac{Quantity(Blurb^{(Inf)})}{Quantity(Common^{(Inf)})} \tag{8}$$

where $Quantity(Blurd^{(Inf)})$ – the quantity of negative advertising in the virtual community; $Quantity(Common^{(Inf)})$ – total number of information content for the virtual community.

Indicator of entry of the virtual community into the risk area of reducing the quality of information content of the virtual community:

$$InfContent^{(Inf)} = w_{Con} * \frac{Quantity(BadContent^{(Inf)})}{Quantity(Common^{(Inf)})} + \\ + w_{Cer} * \frac{Quantity(BadCertaintity^{(Inf)})}{Quantity(Common^{(Inf)})} + w_{Bl} * \frac{Quantity(Blurb^{(Inf)})}{Quantity(Common^{(Inf)})} \quad (9)$$

where $\{w_{Con}, w_{Cer}, w_{Bl}\} \in w$ – weight coefficients of each indicator respectively, are determined by the manager of virtual community life cycle organization, $0 \geq w_i \leq 1$, $w_i \in w$, $\sum_{w_i \in w} w_i = 1$; *Quantity(Content*$^{(Inf)}$*)* – the quantity of low-quality informational content, executed not according to the rules of the virtual community; *Quantity(Certaintity*$^{(Inf)}$*)* – the quantity of inaccurate information content of the virtual community; *Quantity(Blurd*$^{(Inf)}$*)* – the quantity of negative advertising in the virtual community; *Quantity(Common*$^{(Inf)}$*)* – total number of information content for the virtual community.

3.3 Anti-legal Materials and Community Activities

The essence of the risk lies in the fact that in the virtual community there is an informational content that does not comply with the current legislation, which entails criminal liability.

Anti-litigation materials and activities of the community belong to the measures of protection against risk:

- legal verification;
- avoiding plagiarism;
- preventing phishing.

Legal Verification

The essence of the counteraction is to clarify the rules of the virtual community regarding the spirit and content of the community in accordance with the current legislation. Failure to comply with the rules of the virtual community brings with it the prosecution. It is necessary to avoid the information content of the virtual community, containing sensitive information (confidential, service, personal), materials prohibited by law, viruses, etc.

Determine the proportion of violation of the rules in the legal field in the virtual community:

$$Crime^{(Inf)} = \frac{Quantity(Crime^{(Inf)})}{Quantity(Common^{(Inf)})} \quad (10)$$

where *Quantity(Crime*$^{(Inf)}$*)* – the quantity of informational content of the virtual community that does not comply with current legislation; *Quantity(Common*$^{(Inf)}$*)* – total number of information content for the virtual community.

Avoiding Plagiarism

The essence of the response is to track and ban the publication of information content that contains signs of plagiarism. Failure to comply with copyright and related rights provides for criminal liability that is a threat to the reputation and activities of the virtual community. To prevent the publication of a content that contains signs of plagiarism, it's up to the moderator to use anti-plagiarism programs before posting to the virtual community (example, StrikePlagiarism.com, Etxt-Антиплагиат (AntiPlagiarasm.net), «AdvegoPlagiatus 1.3.1.7» etc.).

Determine the proportion of informative content containing plagiarism in the virtual community:

$$Plagiarism^{(Inf)} = \frac{Quantity(Plagiarism^{(Inf)})}{Quantity(Common^{(Inf)})} \tag{11}$$

where *Quantity(Plagiarism*$^{(Inf)}$*)* – the quantity of informational content of the virtual community with content of plagiarism; *Quantity(Common*$^{(Inf)}$*)* – total number of information content for the virtual community.

Preventing Phishing

This method is to protect personal information and confidential data of users of the virtual community and preventing phishing. Phishing – it is a misrepresentation (illegal reception) of personal data of participants of the virtual community.

The proportion of participants who commit phishing action calculated by the formula:

$$Phishing^{(User)} = \frac{Quantity(Phishing^{(User)})}{Quantity(Common^{(User)})} \tag{12}$$

where *Quantity(Phishing*$^{(User)}$*)* – the quantity of virtual community users who commit phishing; *Quantity(Common*$^{(User)}$*)* – total number of participants in the virtual community.

The rate of entry of the virtual community into the area of the risk of anti-legal information content and the activities of the virtual community:

$$AntiLegal^{(User,Inf)} = 1 - \left(w_{Cr} * \frac{Quantity(Crime^{(Inf)})}{Quantity(Common^{(Inf)})} + \right. \\ \left. + w_{Pl} * \frac{Quantity(Plagiarism^{(Inf)})}{Quantity(Common^{(Inf)})} + w_{Ph} * \frac{Quantity(Phishing^{(User)})}{Quantity(Common^{(User)})}\right) \tag{13}$$

where $\{w_{Cr}, w_{Pl}, w_{Ph}\} \in w$ – weight coefficients of each indicator respectively, are determined by the manager of virtual community life cycle organization, $0 \geq w_i \leq 1$, $w_i \in w$, $\sum_{w_i \in w} w_i = 1$; *Quantity(Crime*$^{(Inf)}$*)* – the quantity of informational content of the virtual community that does not comply with current legislation; *Quantity(Plagiarism*$^{(Inf)}$*)* – the quantity of informational content of the virtual community with content of plagiarism; *Quantity(Common*$^{(Inf)}$*)* total number of information content for the virtual community; *Quantity(Phishing*$^{(User)}$*)* – the quantity of virtual community users who commit phishing; *Quantity(Common*$^{(User)}$*)* – total number of participants in the virtual community.

3.4 Loss of Community Control

The essence of the risk is that the moderator may lose control over the management of the virtual community.

The measures to protect against the risk of loss of control over the community belong:

- forming a community leader;
- management community leader.

Forming a Community Leader
The essence of community protection lies in the formation and management of a participant-leader in his favor.

$$NewLeader^{(User)} = \frac{Quantity(NewLeader^{(User)})}{Quantity(Common^{(User)})} \tag{14}$$

where *Quantity(NewLeader*$^{(User)}$*)* is quantity of leaders created by community administrators; *Quantity(Common*$^{(User)}$*)* is total number of virtual community participants.

Management Community Leader
The essence of risk lies in identifying leaders in the virtual community and collaborating with them.

$$Leader^{(User)} = \frac{Quantity(Leader^{(User)})}{Quantity(Common^{(User)})} \tag{15}$$

where *Quantity(NewLeader*$^{(User)}$*)* is the quantity of leaders of the virtual community; *Quantity(Common*$^{(User)}$*)* is total number of participants in the virtual community.

The rate of entry of the virtual community into the zone of risk of loss of control over the virtual community:

$$\begin{aligned} Control^{(User)} &= w_{NL} * \left(1 - \frac{Quantity(NewLeader^{(User)})}{Quantity(Common^{(User)})}\right) + \\ &+ w_L * \frac{Quantity(Leader^{(User)})}{Quantity(Common^{(User)})} \end{aligned} \tag{16}$$

where $\{w_{NL}, w_L\} \in w$ – weight coefficients of each indicator respectively, are determined by the manager of virtual community life cycle organization, $0 \geq w_i \leq 1$, $w_i \in w$, $\sum_{w_i \in w} w_i = 1$; Quantity(NewLeader(User)) – the quantity of leaders created by community administrators; Quantity(Leader(User)) – the quantity of leaders of the virtual community; Quantity(Common(User)) – total number of participants in the virtual community.

4 Algorithm for Determining the Level of Intensity of Measures to Counteract Socially-Oriented Risks

The algorithm is designed to implement the proposed measures to protect the virtual community from socially-oriented risks, which provide continuous and high-quality management of the virtual community, support for the topic and reputation of the virtual community.

The input data for the algorithm, namely content and user data are (Fig. 2):

$$SOR_i = \sum SOR(Com_i) \tag{17}$$

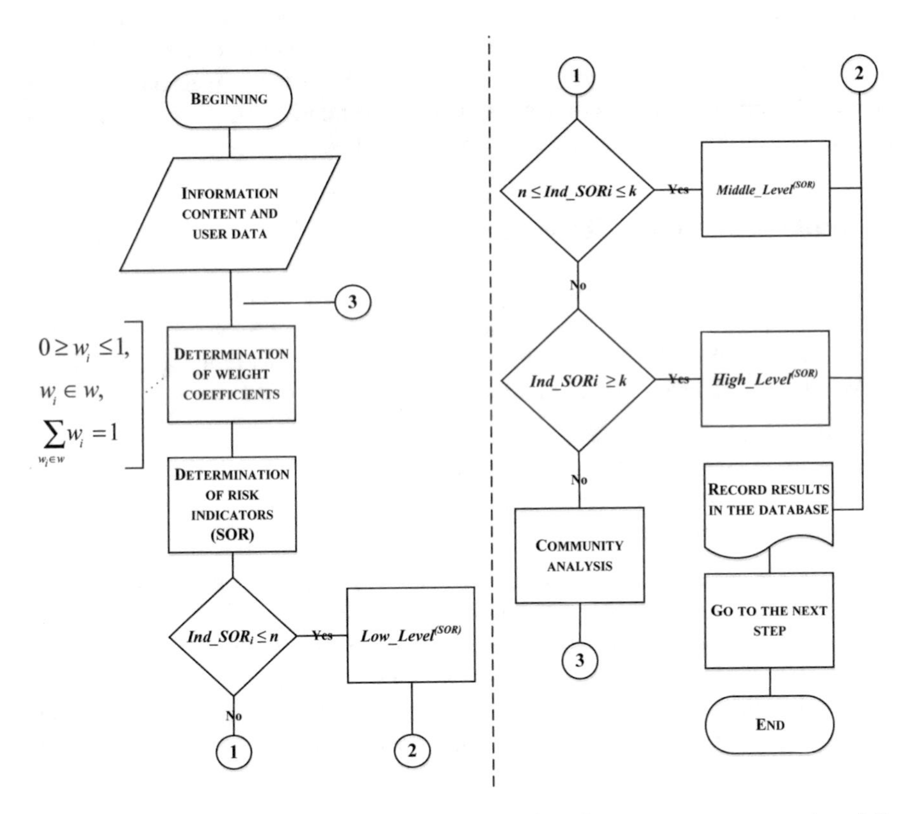

Fig. 2. Algorithm for determining the level of intensity of measures to counteract socially-oriented risks

The result of the implementation of the algorithm is the classification of the risk index:

- **High level** (Ind $\leq$ n, n determined by the manager of the life cycle of the virtual community).
  For a high level of occurrence of the indicator in the risk zone, it is necessary to analyze the community (task for analytics) and identify gaps in the organization's lifecycle management of the virtual community.
- **Middle level** (n $\leq$ Ind $\leq$ k, n and k are determined by the manager of the life cycle of the virtual community).
  The average level of occurrence of the indicator in the risk zone requires the adoption of measures to combat socially-oriented risks.
- **High level** (Ind $\geq$ k, k determined by the manager of the life cycle of the virtual community).
  Due to the low level of entry into the risk zone, the community can function fully.

The community analysis contains actions aimed at a thorough analysis of the virtual community. Recording of the results in the database of the virtual community life cycle organization (data of entry into the risk zone, mark on the implementation of this algorithm). Formation of data (documents) about the transition to the next stage of the organization of the life cycle of the virtual community.

5 Implementation of Research Results

Implementation of the proposed measures to protect the virtual community from socially-oriented risks was held at the official virtual community of the Lviv Polytechnic National University. The implementation took place during 2017–2019.

Measures were taken to protect all risks, namely:

- appearance of a negative audience;
- reducing the quality of information content;
- anti-legal materials and community activities;
- loss of community control.

Figure 3 is presented the timetable for community of the Lviv Polytechnic National University entry into the area of socially-oriented risks.

Figure 4 presents the statistics of community entry into the area of socially-oriented risks in the virtual community of Lviv Polytechnic in January–May 2019.

Consequently, community statistics show that the most dangerous socially-oriented risks are appearance of a negative audience and reducing the quality of information content.

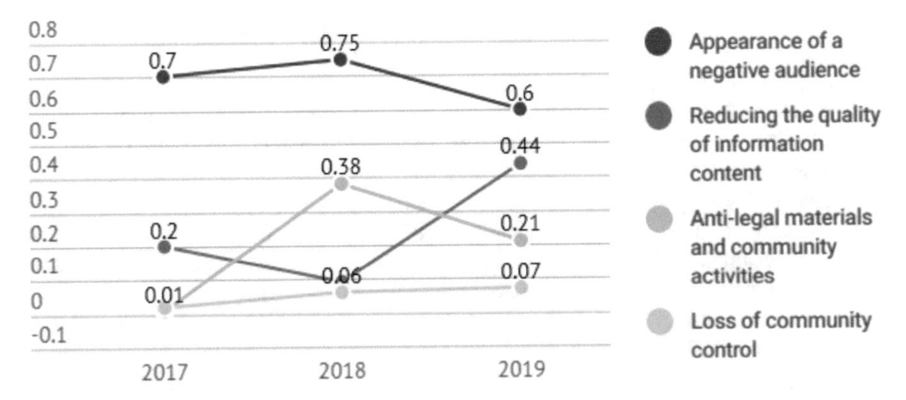

Fig. 3. Statistics from 2017 to 2019 years

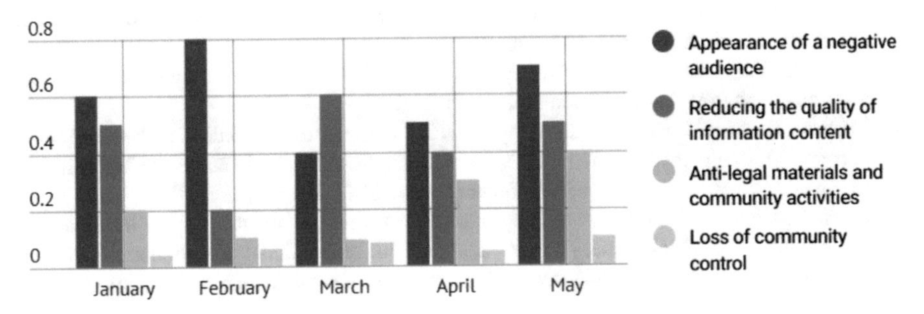

Fig. 4. Statistics for 2019 year

6 Conclusion

The article defines socially-oriented risks during the organization of the life cycle of the virtual community and indicators of the community's entry into the area of socially-oriented risks. The indicators presented are important for effective and successful management of the virtual community. The algorithm of determination of level of intensity of measures of counteraction to socially-oriented risks is developed. The algorithm provides an opportunity to increase the effectiveness of creating a virtual community and improve the functioning throughout its existence, ensuring the achievement of goals and the development of the virtual community.

References

1. Tamjidyamcholo, A., Bin Baba, M.: Evaluation model for knowledge sharing in information security professional virtual community. Comput. Secur. **43**, 19–34 (2014)
2. Porter, C., Devaraj, S., Sun, D.: A test of two models of value creation in virtual communities. J. Manag. Inf. Syst. **30**(1), 261–292 (2013)

3. Ostrand, V.: Creating virtual communities that work: best practices for users and developers of E-collaboration software. Int. J. E-collaboration **12**(4), 41–60 (2016)
4. Zhezhnych, P., Markiv, O.: Linguistic comparison quality evaluation of web-site content with tourism documentation objects. In: Advances in Intelligent Systems and Computing II. CSIT 2017. Advances in Intelligent Systems and Computing, vol. 689, pp. 656–667. Springer, Cham (2018)
5. Fedushko, S., Ustyianovych, T.: Predicting pupil's successfulness factors using machine learning algorithms and mathematical modelling methods. In: Advances in Computer Science for Engineering and Education II. ICCSEEA 2019. Advances in Intelligent Systems and Computing, vol. 938, pp. 625–636. Springer, Cham (2019)
6. Tsai, H., Pai, P.: Explaining members' proactive participation in virtual communities. Int. J. Hum Comput Stud. **71**(4), 475–491 (2013)
7. Shen, K., Khalifa, M.: Effects of technical and social design on virtual community identification. Behav. Inf. Technol. **32**(10), 986–997 (2013)
8. Anisimova, O., Vasylenko, V., Fedushko, S.: Social networks as a tool for a higher education institution image creation. In: 1st International Workshop on Control, Optimisation and Analytical Processing of Social Networks, COAPSN 2019. CEUR Workshop Proceedings, vol. 2392, pp. 54–65 (2019)
9. Wang, J., Yang, J., Chen, Q., Tsai, S.: Creating the sustainable conditions for knowledge information sharing in virtual community. Springerplus **5**, 9 (2016)
10. Berezko, O., Zhezhnych, P.: Rethinking the NGO website from the knowledge management perspective. In: Proceedings of the 12th International Scientific and Technical Conference on Computer Sciences and Information Technologies, CSIT 2017, pp. 389–392 (2017)
11. Peleshchyshyn, A., Vovk, N.: The problem's formation of the entire informational environment at the long existed universities. In: 12th International Scientific and Technical Conference on Computer Sciences and Information Technologies, vol. 1, pp. 139–142 (2017)
12. Artem, K., Holoshchuk, R., Kunanets, N., Shestakevysh, T., Rzheuskyi, A.: Information support of scientific researches of virtual communities on the platform of cloud services. In: Advances in Intelligent Systems and Computing III. CSIT 2018. Advances in Intelligent Systems and Computing, vol. 871, pp. 301–311. Springer, Cham (2019)
13. Ansah, R., Sorooshian, S., Bin Mustafa, S., Oludapo, O.: Constructions project management risks' framework. Qual. Access Success **18**(158), 90–95 (2017)
14. Doskocil, R., Lacko, B.: Risk management and knowledge management as critical success factors of sustainability projects. Sustainability **10**(5), 13 (2018)
15. Heravi, G., Gholami, A.: The influence of project risk management maturity and organizational learning on the success of power plant construction projects. Project Manag. J. **49**(5), 202–237 (2018). https://doi.org/10.1177/8756972818786661
16. Keshk, A., Maarouf, I., Annany, Y.: Special studies in management of construction project risks, risk concept, plan building, risk quantitative and qualitative analysis, risk response strategies. Alexandria Eng. J. **57**(4), 3179–3187 (2018). https://doi.org/10.1016/j.aej.2017.12.003
17. Moeini, M., Rivard, S.: Sublating tensions in the IT project risk management literature: a model of the relative performance of intuition and deliberate analysis for risk assessment. J. Assoc. Inf. Syst. **20**(3), 243–284 (2019). https://doi.org/10.17705/1jais.00535
18. Muriana, C., Vizzini, G.: Project risk management: a deterministic quantitative technique for assessment and mitigation. Int. J. Project Manag. **35**(3), 320–340 (2017). https://doi.org/10.1016/j.ijproman.2017.01.010
19. Pimchangthong, D., Boonjing, V.: Effects of risk management practices on it project success. Manag. Prod. Eng. Rev. **8**(1), 30–37 (2017)

20. Tavares, B., Silva, C.: Practices to improve risk management in agile projects. Int. J. Software Eng. Knowl. Eng. **29**(3), 381–399 (2019)
21. Peleshchyshyn, A., Trach, O.: Determining the elements of socially-oriented risks when organizing the life cycle of the virtual community. Inf. Secur. **23**(2), 130–135 (2017)
22. Trach, O., Peleshchyshyn, A.: Development of directions tasks indicators of virtual community life cycle organization. In: XIIth International Scientific and Technical Conference "Computer Sciences and Information Technologies" (CSIT-2017), Lviv, pp. 127–130 (2017)

The Intellectual System Development of Distant Competencies Analyzing for IT Recruitment

Antonii Rzheuskyi[1(✉)], Orest Kutyuk[1], Orysia Voloshyn[1], Agnieszka Kowalska-Styczen[2], Viktor Voloshyn[3], Lyubomyr Chyrun[3], Sofiia Chyrun[3], Dmytro Peleshko[3], and Taras Rak[3]

[1] Lviv Polytechnic National University, Lviv, Ukraine
antonii.v.rzheuskyi@lpnu.ua, bliZZman96@gmail.com
[2] Silesian University of Technology, Gliwice, Poland
[3] IT STEP University, Lviv, Ukraine
box@itstep.org

Abstract. AHP study methods of solving problems to use AHP comparing alternatives (TOPSIS) and supervised learning algorithm using linear regression are proposed. To meet the challenges these methods were chosen language PHP, its own established framework CollEntRes, additional tools that helped during the programming process to detect errors and correct them (Debug), popular modern framework to work with the client part Angular, relational control systems, databases MariaDB HeidiSQL and its client, the technology to work with the client part of the site HTML and CSS and additional tools such as Bootstrap and SCSS. As a result, within practical implementation of software tool has been developed whose main goal is an analysis of matrix distant for recruitment in IT sector. The software tool covers general information and requirements for software, described in detail basic functionality of the software analysis and test cases for all kinds of users is presented in the article.

Keywords: AHP · TOPSIS · Machine learning · Distant · Distance learning · Competencies · IT recruitment · Intellectual system

1 Introduction

Socio-economic change and fast transition in the planning of industrial information society that takes place in modern Ukraine require fundamental changes in various spheres of government. In the foreground is a reform of the education system [1–7]. The national program "Education. Ukraine XXI century" provides for the development of the educational sector based on new advanced concepts, introduction to the educational process new technologies to work with students and scientific and methodological advances, the introduction of information support for the educational sector, the entry of Ukraine into the transcontinental system computer cybernetic information [2]. The development of education in our country should lead to:

N. Shakhovska and M. O. Medykovskyy (Eds.): CCSIT 2019, AISC 1080, pp. 696–720, 2020.
https://doi.org/10.1007/978-3-030-33695-0_47

- Appearance features that update learning content and teaching methods of disciplines dissemination of knowledge [3];
- Reduce the need for high demands for access to all levels of education, implementation capabilities and provide a large number of young people, taking into account those who cannot study at universities/institutes for traditional forms of education due to lack of physical or economic opportunities, professional employment, distance of cities, prestigious and well-known schools, etc. [4];
- The appearance of continuous education "through life", including all types of education directions [5];
- The appearance of individual approaches to education for large numbers of people.

To achieve these goals, it is necessary the rapid development of distance education, implementation of which is regulated by the National Program of Informatization of Ukraine. Distance learning is a form of learning that is becoming more and more popular in Ukraine [6]. Taking the example of the United States and developed countries in Europe, where this form of learning is already long familiar to the public and educational organizations in our country the best trying to adapt it is into a modern community [7]. This form of training is available in almost all segments of the population. Its methods are widely used in all levels of education and training in the systems for the management [8]. In contrast, distance learning, leading control study of materials that occur in time (e.g. during exam period), distance education provides knowledge control after each reading course and has a flexible schedule to write the final test. In addition, one of the advantages of distance education is unparalleled support faculty, each student is attached to the teacher who monitors his progress, consults and helps in any educational issues [9]. It is also important to define the factor of discipline, which is interested student, sequence and pace of learning. A student who is studying in distance form of education called distant [10]. He is available opportunity to choose the sequence of study courses and define for them. Low price is a big plus of distance education. "Educate yourself" in order to have the knowledge of particular communication activities, expanding and adding to your knowledge base - the main purpose of distance education. The goal of the work is development of intelligent system matrix analysis distant for recruitment in the IT sector. To achieve the goal must perform the following tasks: to conduct a study of the literature and identify which ones are best suited to the subject of work; to conduct a study of existing software and decide which ones are relevant to the theme of work; perform system analysis about 'the object of the study; develop optimal solutions for tasks; design and develop a software interface to work with the client system [11]; develop an algorithm to calculate customer data systems; Design and develop a software interface for visualization client system. About object of research is analysis distant knowledge. Subject of study is intelligent analysis matrix distant for recruitment in the IT sector. The paper improved analysis algorithms AHR and TOPSIS, which can significantly improve the accuracy of the assessment. Output can be used to simplify the process of finding and hiring candidates of IT recruiters and get them a full analysis of the relevant knowledge distant positions.

There are systems that implement distance learning for students of any study areas [1–3]. These systems are public and open to all comers [4–9]. However, none of the systems makes it possible to analyze data of their students and to operate them in external purposes [10–15]. In this system, there are two problem to be addressed the most optimal algorithm, calculation competence of students in their technical knowledge and analysis of their personal qualities [16–21]. Without addressing these problems, recruiters are unable to automatically choose the students for further work [22–25]. Implementation AHR of student's knowledge will allow recruiters IT companies selecting IT profession, students determine the best course of promising individuals and fill its own database for possible projects [26–31]. For the students this assessment will enable the analysis of their abilities and asks vector of development and specific field [31–37]. Implementation of personal analysis as student recruiters will determine whether a person is ready to work in a team or student all necessary options to work with a particular client [38–41]. Such information may be sent to the department of their company with people who determine suitability for specific qualities of certain objects, or they need to improve a particular property for future employment in the company [42–46]. For students such information is extremely important because the system will issue him feedback from all levels of training (teachers, classmates, most rating systems) and receiving data such student alone can determine which traits should be improved and on what terms should concentrate [47–49]. Along with information about the success of the object, the recruiter will get a complete picture for further work on employment distant [50–53].

2 A Systematic Analysis of the Research Object and Subject Area

System analysis is scientific method of cognition, which is a sequence of algorithmic installation of structural relationships between the elements or variables of the system investigated, based on a set of experimental mathematics, general, statistical and natural methods [54]. System analysis is closely related operations research and analysis requirements. In alternative definition, analysis interpreted as explicit formal inquiry held to assist in determining the best and most optimal course of action and the adoption of the most appropriate solution [55]. Development of intelligent system that uses a computer includes a step of system analysis. This improves design data model, visionary creation or expansion of the database. System analysis is a set of mechanisms. In practice, research systems analysis applied by using the following techniques [56]:

- Procedure theory of operations research that allows obtaining quantitative evaluation of research;
- Analysis of research facilities in the face of uncertainty;
- Systems engineering, which includes the design and analysis of complex systems in the study of their functioning (planning and evaluation of the economic efficiency of technological processes and so on.).

Using the methods of decision-making procedures should. Adhere to its basic stages:

1. Formalization of problematic situations;
2. Setting goals;
3. Methods for obtaining the objectives of criteria;
4. Designing models to determine decisions;
5. Defining the acceptable (optimal) variant solution;
6. Approval decisions;
7. Preparation of solutions before implementation;
8. The promotion decisions;
9. Monitoring the implementation of decisions;
10. Describe of efficiency solutions.

In the process of designing intelligent system, we need to reveal its internal structure and analyze the purpose of the system [57–68]. One of many ways of solving this problem is to build a tree of objectives. The tree needs is model that shows orderliness and hierarchy of goals, their relationship and provides visualization of their distribution and gradation (sub) and tasks. Wood goals are a system for the purposes of any object that has its own structure. The basic algorithm for the design to the construction is wood goals decomposition. Decomposition - the process of breaking skaldic system or the system into parts that are easier to understand, program, maintain and fix their problematic side. Separation system occurs under certain common features. In the design objectives tree, the decomposition process is used to mix the main objective of the ways to achieve it, which are formulated as individual task performers. The main requirement for design objectives tree is breaking a complex system, process or phenomenon into simpler components. To implement such a rule is necessary implementation system approach [69–75]:

- higher target in a tree hierarchy subject to decomposition to achieve lower on the hierarchy of objectives;
- Lower in the hierarchy goals - a way to achieve higher goals that need no introduction as a subsystem of the main goals.

General purpose of the intelligent system is to determine the level of knowledge of student distant and dynamics of its internal qualities IT fields. During the design objectives tree, the main goal is recursively divided into sub-goals:

1. Incoming criteria for analyzing knowledge distant
   (a) determining the criteria for appropriate analysis distant
   (b) determining IT profession as an object of analysis of student
   (c) sending a request to the server for processing
2. Asked to take account of the internal dynamics of a student
   (d) receiving a request from a client system
   (e) Inquiry and making uniform criteria for student
   (f) generation output data

3. Calculation of students' knowledge of the use of AHR of the input parameters
   (g) receiving the request with all parameters
   (h) processing data according to criteria distant
   (i) data entry of students in AHR algorithm to calculate the level of knowledge in the relevant distant IT profession generation of output data
4. Display data of the analysis on the website
   (j) output overall success distant
   (k) output data distant knowledge relative to the respective IT profession
   (l) output data internal qualities distant

Each intelligent system is able to accept and give that pass through certain processes, estimated by some algorithms may be stored in databases or data warehouse and generated a resulting data according to these parameters request. For the best picture of all processes on objects of intellectual system to use UML diagrams. Diagram of use cases or use case diagram - a diagram for depicting the relationship between the actors in the precedents and intelligent systems. Use case diagram is a graph consisting of a set of actor's precedents (use cases) that limited border system (rectangle), associations between actors and precedents, relations between precedent and generalized relationship between actors (Fig. 1).

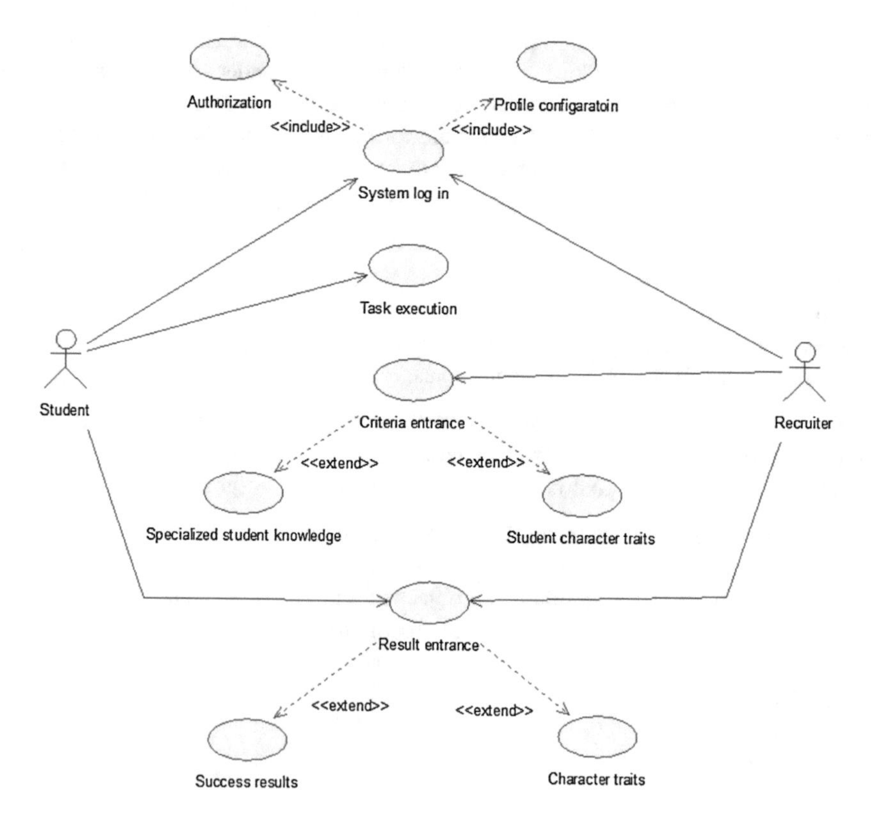

Fig. 1. Use case diagram

The purpose of this chart - designed system is portrayed as actors or entities that interact with the system using use cases. Usage needed to describe the services that the system provides the actor. That is, each use case defines a set of actions performed by the system in the dialogue with the actor. Figure 6 shows a diagram for the use of the IP system. Class diagram is a representation structure in the form of a static model of the system. It reflects the declarative elements such as data types, classes, and their related content. This chart is able to contain designation packages of some elements of behavior, but the dynamics in this diagram is not disclosed to this, there are other kinds of charts. Class diagram is depicting the current state static structure model of classes in the aspects of object-oriented programming. It shows display interfaces, classes, cooperation and objects and their relationships. Figure 2 shows a diagram for this class intellectual system.

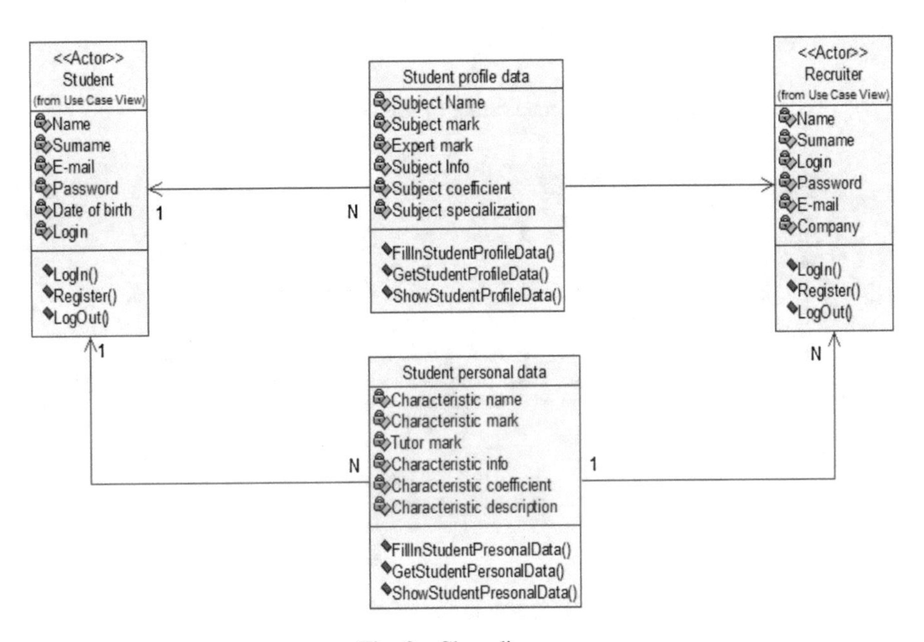

Fig. 2. Class diagram

Chart communication or cooperation diagram - a diagram that shows the interaction between the components of the composite structure and roles of cooperation. In contrast diagram sequence diagram for cooperation clearly indicate the relationship between objects. In UML, diagrams are four types of interaction diagrams of synchronization, communication, sequence, interaction. Figures 3 and 4 show the object of cooperation "Student" and "Recruiter" for this intelligent system, respectively. State diagram is a chart that shows the state changes of objects in the time interval. The object is presented as makers of automata theory with standardized symbols. Figure 5 shows the sequence for this intelligent system.

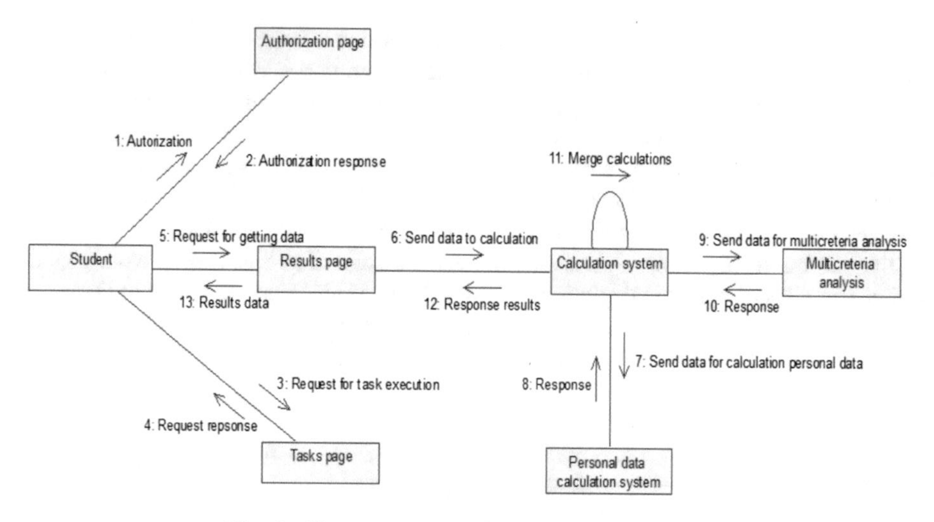

Fig. 3. Chart cooperation for the object "Student"

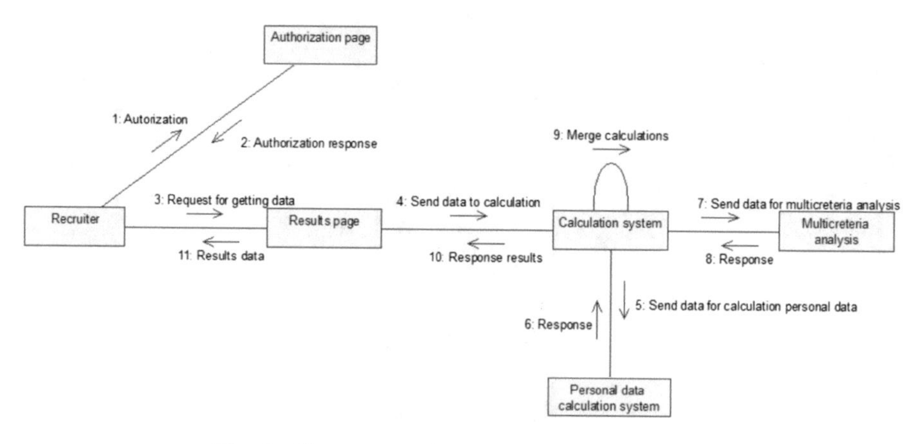

Fig. 4. Chart cooperation for the object "Recruiter"

Activity Diagram is diagram representing the count activities. Count of activities - a finite automaton graphs whose vertices are defined by actions, and transitions between them occurring after the action. Figure 6 chart shows the activity for this intelligent system. Component Diagram is a chart that displays the components, relationships and dependencies between them. The diagram shows the component interdependencies between the components of the program, given the source code components, and executable binary components. Software modules presented as a component. Some components exist at compile-time, part - in the layout, and the rest - in the course of the program. Figure 7 shows a diagram of the intellectual component system.

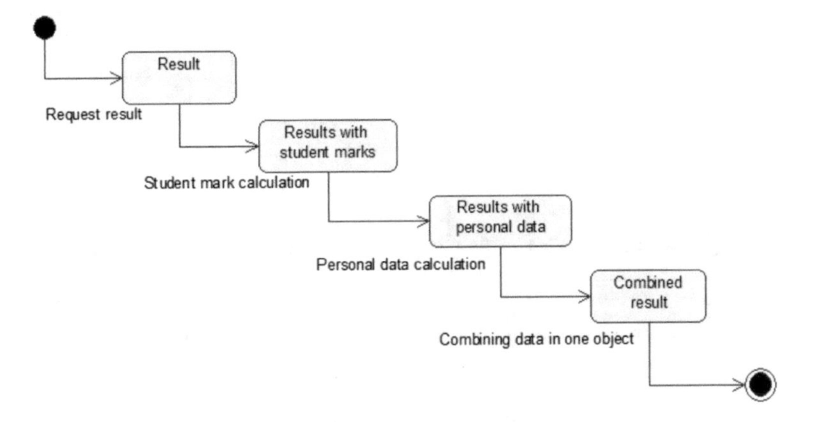

Fig. 5. State diagram

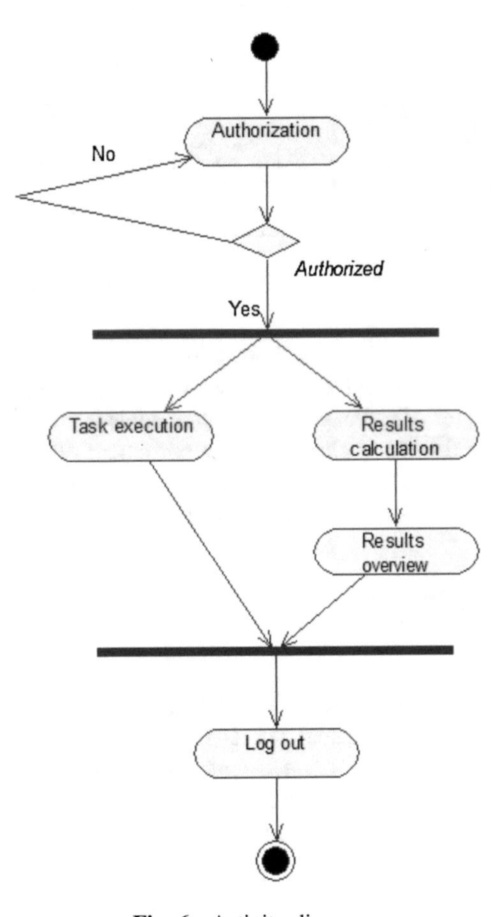

Fig. 6. Activity diagram

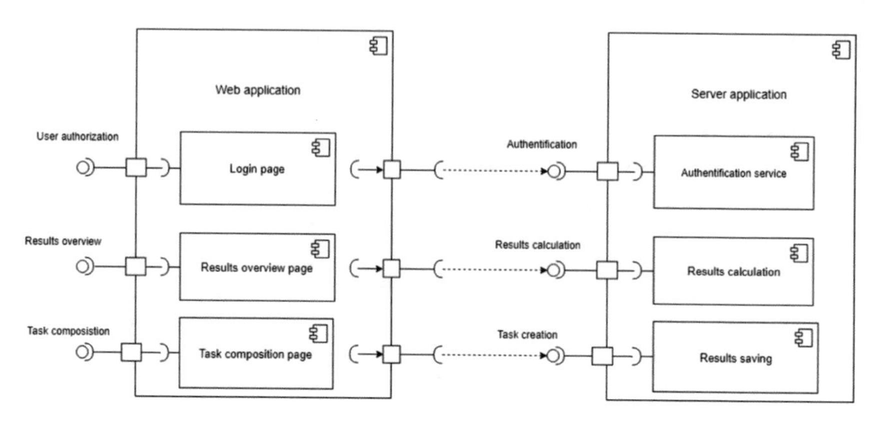

Fig. 7. Component diagram

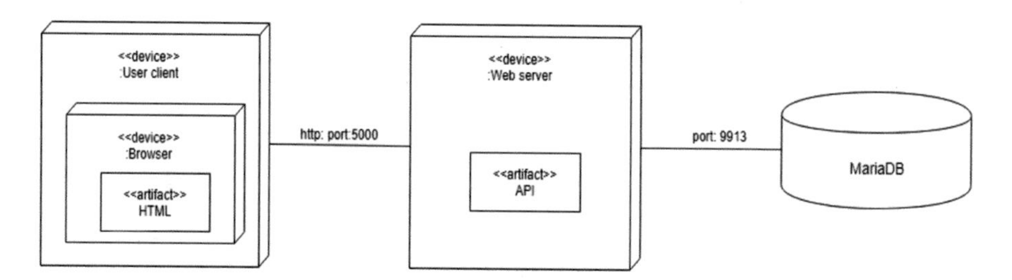

Fig. 8. Deployment diagram

Deployment diagram is a chart that illustrates the calculation process in the course of the program components and features that run in these processes. Components presented working copies of pieces of code. Components that are not represented in the work program for the following charts is not represented; instead, they can be represented in diagrams component. Deployment diagram illustrates the working copies of the component. Figure 8 shows a deployment diagram for this intelligent system.

3 Choice and Justification of the Methods to Solve the Problem

For distant competencies analysis needs to implement two modules: the calculation of student learning outcomes and calculates its ranking of personal characteristics. To implement the first module was selected analysis comparing alternatives (TOPSIS) which was partially described in the first chapter. Categories offered tuning program, sorted by importance in descending order [1–5]:

- motivate people skills and achieve common goals;
- the ability to provide and support quality evaluation of the work done;
- adaptation to new situations;

- a reference to the preservation of environment;
- skills in handling information and communication technologies;
- autonomy at work;
- understanding and awareness of multiculturalism and differences;
- persistence and determination;
- be socially responsible and to have public awareness;
- the ability to manage and develop projects;
- ability to work in concert;
- have an entrepreneurial spirit, be proactive;
- have abstract thinking, synthesize and analyze;
- ability to communicate with experts from other fields;
- act, relying on ethical foundations;
- ability to work with foreigners;
- knowledge of the subject area;
- teamwork skills;
- skills in conducting research;
- make informed decisions;
- application of knowledge in practice;
- identification, formalization and solving problems;
- focus on safety;
- ability to find, process and analyze data from different sources;
- the ability to produce new ideas (creativity);
- be aware of equal opportunity and gender issues;
- the ability to accept criticism;
- the ability to operate over time;
- fluency in their native language;
- desire to learn;
- Foreign language.

Because research professional expert group relative relationship between the components of the educational program and software Tuning competences of the program in this paper are not available, they assume that their weight equivalent [11].

After analyzing the weight, the expert group assesses weights courses or semesters. The set M is learning outcomes by topic. $M = \{r_1, r_2, \ldots, r_N\}$. G is result study competence category $G = \{g_1, g_2, \ldots, g_N\}$ [11–13]. Figure 9 depicts the relationship between tree analysis disciplines, themes, courses and learning outcomes. To determine a more accurate assessment of student knowledge TOPSIS method should be used trapezoidal fuzzy numbers. In the process of finding correspondence between the used scale ECTS [14–19] and numerical estimates, which are presented in Table 1.

In this paper, sorted by priority list of 31 competencies ($K = \{k_1, k_2, k_3\}$) under the program tuning. Priority is calculated by the method of analytic hierarchy process (AHP) [1–7, 18]. Each competence belongs to one of $\{A, B, C\}$ [2]. Criteria competence K_i sorted by categories. Each category has its ratio relative importance of the criteria of competence (as average priority categories) [19]. The method TOPSIS [2–7, 20, 21] evaluated the relative importance of factors hidden competency criteria and weights w_{ij} partial criteria of competence $w_j = w_i * w_{ij}$ [2–7, 22]. While studying shaped profile student distant S is taking into account each of the levels of competence KAS, KBS, KCS in the range of [0, 1]. A student studying for some specialty $P = \{p_1, p_2, p_N\}$. For statistics students have the minimum number required to complete the courses $C = \{c_1, c_2, \ldots, c_N\}$ or $\{SP_i = SP_{i1}, SP_{i2}, SP_{i3}\}$. The course consists of a plurality of subjects $D = \{d_1, d_2, \ldots, d_N\}$. Each discipline is a block topics $L = \{l_1, l_2, \ldots, l_N\}$ or $\{SD_i = SD_{i1}, SD_{i2}, SD_{i3}\}$. From this it was concluded that $K_i \Rightarrow (D_{i1}, D_{i2}, \ldots, D_{iN})$, is expert group remains the choice of how best to design an educational program for each of the specialties. Each discipline has a certain weight of each of the competencies performed by a group of experts. For each of the courses conducted construction table relationships between software components of competence and professional educational programs with a definite specialty (column is the list of disciplines lines is useful list). Example, 121 for specialty "computer science" [1–7, 23–25] Expert Group identified a set of 46 subjects and 24 competencies described relationship between software components of competence and professional educational programs of certain specialties. Each cell of this table is the presence or absence of a particular competence in a particular discipline.

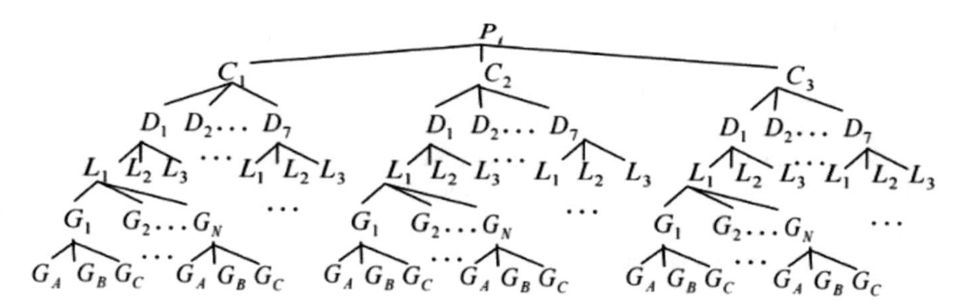

Fig. 9. Hierarchy analysis of learning outcomes

Table 1. Linguistic value and corresponding trapezoidal fuzzy numbers

N	Linguistic value	For ECST	Fuzzy	Rating	Fuzzy numbers	Symbol
1	badly	F	0.15	0–25	(0,0,1,2)	1
2	unsatisfactorily	FX	0.35	26–49	(1,2,2,3)	2
3	enough	E	0.55	50–60	(2,3,4,5)	4
4	satisfactorily	D	0.65	61–70	(4,5,5,6)	6
5	okay	WITH	0.75	71–79	(5,6,7,8)	8
6	very well	B	0.85	80–87	(7,8,8,9)	9
7	perfectly	A	1	88–100	(8,9,10,10)	10

Table 2 shows a list of top positions in the IT cluster in Ukraine. List sorted and formed with the help of key personnel Lviv IT Cluster. Unfortunately, to date there is no explicit requirements for each position. Each company determines requirements and correlates priorities relative positions. For example, in some of them and to the position of Tester enough to know the basics of testing and theoretical basis, and the position of Software Engineer should know the basics of the theory of algorithms.

Table 2. The coefficients of the relative importance of the recommended core IT occupations

N	Position	Important factor	The boundaries of the assessment
1	Student	0	50
2	Assistant	0.007092	50.3546
3	Tester	0.008599	50.78455
4	Recruiter	0.011141	51.3416
5	Associate Software Engineer	0.014037	52.04345
6	Administrator	0.018602	52.97355
7	Programmer	0.023866	54.16685
8	Software Engineer	0.031904	55.76205
9	Senior Software Engineer	0.041297	57.8269
10	Team Leader	0.055204	60.5871
11	Associate Manager	0.071152	64.1447
12	Manager	0.092353	68.76235
13	Senior Manager	0.115327	74.5287
14	Senior Executive	0.142542	81.6558
15	Systems architect	0.170746	90.1931
16	Postgraduate	0.196136	99.9999

To implement the second module was selected supervised learning, which uses TOPSIS algorithm. Simple TOPSIS is a regression model with one independent variable. This means that it is considered in two-dimensional space of the sample, which is formed with one independent variable and one dependent variable (usually x and y are coordinates in a Cartesian coordinate system). This model is designed to find a linear function that can predict the most exact values of the dependent variable as a function of the independent variable. The most fully developed theory and implemented in software, TOPSIS analysis when the least squares method is applied to the equation in the form of a linear dependence of the effective rate of independent variables and the estimated parameters. If the linearity is missing, it AHR be obtained by replacing the output factor new variables and output parameters of the transition from b to new factors in the transformation equations whole. Sometimes are the need and replace, and transformation. Models that are reduced to standard linear form, quite a lot, it can be used TOPSIS analysis in solving applications. The most common examples of variable substitution is replacing variables x_k in polynomic equations and replace the value of 1/x in the hyperbolic functions. In the first case, polynomial regression function of the pair is a linear function regression. Converting the exponential function to a standard type carried out by the logarithm of the left and right sides of the equation. The same is true for the exponential and exponential functions. For logarithmic parabola required as well as the replacement of variable x_2 to a new variable. For feature requires replacement variable Inf on new variable. A somewhat more complicated situation reverse model. It is adjusted to the standard view when accessing and left and right sides of the equation. The result is $1/y = b_0 + b_1 * x$. Then the value of 1/y replaced with new: $z = 1/y$.

4 Description of Implementation Tasks

For the program success is a necessary condition for the continued support of the server that executes the program and manages the database. The server part of mandatory application must be installed apache main components is a Web server that allows you to host web sites on the Internet. In addition, Apache provides support protocol http, of which the lion's share of requests architecture program.

Business logic programs written using language PHP. Querying the database is also written in PHP language with technology PDO. The program interface is written using a markup language and cascading style sheets. For adaptive layout library site uses Twitter Bootstrap. The program is designed to assist in making decisions regarding the candidate's student recruiter Lviv IT companies. The system enables to monitor future IT industry employees using multi algorithms and visualization tools.

In this program there are no restrictions on users who use it need Internet access and any device that has built-in functionality to work with websites? Optimize RAM costs achieved by using its own framework and PHP libraries that decrease the performance of queries to the database, so even on devices with little RAM website work without heavy loads. Designed intelligent system developed based on the popular algorithm TOPSIS. This algorithm was developed for calculating structured distant success. Algorithm development was due to the need to replace point scoring.

The basic idea TOPSIS method is the best alternative that must have not only the proximity to the ideal solution, but also to be away from each alternative to inappropriate decisions. Here, the optimal solution is a vector containing the maximum values for each of the criteria for all alternatives and unacceptable (worse) solution is the vector containing the minimum values for each criterion. As the nature of TOPSIS method, using the latter can be quite effectively solve the problem of fuzzy multi-objective optimization that make up the mathematical basis for decision support in the tasks of human resources management. In a multi-objective optimization decision, theory refers to choosing the best solution among the possible alternatives. This information system is represented as a website, so no need to install additional files. Necessary condition for successful operation is access to the Internet and set a modern browser. Recommended browsers are the latest versions of Google Chrome, Mozilla Firefox and Safari. When using older browsers it AHR not display accurate baseline data on the client side. The link to the site home page opens, which is description of the user on the website. The data in the system are divided into several categories:

- inputs that make users of the system;
- output data generating system;
- integrated data, access to which have only web resource administrator/moderator.

Input that make users is a position chosen (recruiter IT company) selected student or group distant (representatives of the university) and selected students to see the success or possible employment in the IT sector (registered students). Background is a response to customer requests search site. The answers can be as successful and not successful (if incorrectly given input). Integrated data is data that directly affect the performance of the algorithm: student assessment, evaluation experts, and evaluation criteria. Access to editing and adding these data should only moderator and administrator web resource.

5 Analysis of the Results

The IS of analysis of competencies is a Web resource that requires Internet access to its users. The system can be used with not only home computers and laptops but also mobile… The main object of this program is students who need to be registered in the system and provide the results of training, employees of universities and the main audience of the site, recruiters IT companies that are able to conduct search students for future employment in IT companies. To get started, you need to go to the website and go through the authorization process. This is the only condition for the fulfillment of which the user will be able to watch any necessary information. The user will get its rights in the system that allow or prohibit performing certain actions according to its status. Selecting some options, the client system receives a response to its request. For distant table it can be relatively competencies selected IT occupations, which own data quality or the results of his university studies. For university staff it can be a table of learning outcomes distant. For recruiters - it is possible to monitor distant that best suited to their chosen IT profession. In case of incorrect data entry of any incorrect data, users get an error message. According to the text message, the user can understand the cause of the error and correct input. The program entitled "Analysis of competencies." In their own language framework using PHP, JavaScript and a large number of aids (frameworks, libraries, and plug-ins) are implemented. The main purpose of the website is the realization of the objectives hiring of selected professions in the IT sector. The program also effectively used in monitoring the market of IT specialists. As described above, the system is a prerequisite authorization process. This process is identical for all types of users. A prerequisite for registration is e-mail username and password to verify. For register user must click on Register at the top right of the website (Fig. 10).

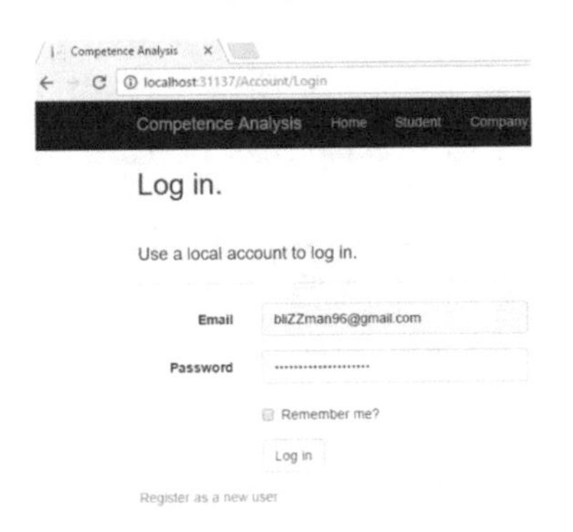

Fig. 10. Authorized users

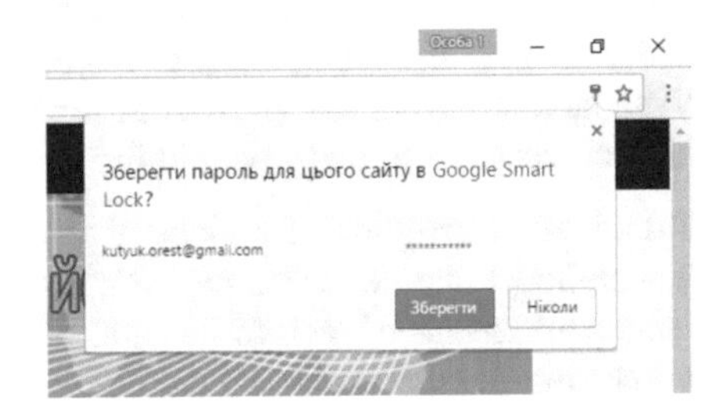

Fig. 11. Save password website

To improve the security of the system used by the system complicated passwords because user need take account of the fact that the password must contain uppercase and lowercase letters, numbers, symbols, such as, for example, @. If users do not comply, it wills error, which contain additional supporting information should be used characters (Fig. 11). After the authorization, select website of the homepage, which contains information about the Web resource (which is a site for and what it is) (Fig. 12). From the top, the so-called Heder site contains navigation menus. The menu consists of six buttons:

- Home - Home on which the description and purpose of the program;
- Student - page that marked for Students;
- Company - page that is designed for recruiters system;
- University - the page that is intended for workers University web-site;
- About - page on which the description of algorithms for solving means used by intelligent;
- Contact - the page that contains the contacts and employees Developer website that you can ask a question, e-mail using.

Fig. 12. Page «Home» Web resource

The site can run three types of users: recruiter, distant and university. Consider the work of everyone. For work distant in the system is page Student. In distant can choose two options:

- Success of training for university disciplines. To do this, the student must select yourself from the dropdown list and click Submit. After completion of the steps under the button Submit, a list of required data.
- View table the necessary knowledge for the proposed IT specialists. For this option, distant necessary to carry out similar actions will result in a table of competence.

Additional description of options of student is on the same page as (Fig. 13). For work recruiter in the system is page Company. Recruiter can be acquainted with the relevant requirements of each IT position. The only options that can benefit recruiter, a selection from the dropdown list the desired position and clicking Submit, right table appears students' knowledge and successful results, which are relevant to post. Sorted in descending order is. Additional description recruiter action is also on the same page (Figs. 14 and 15).

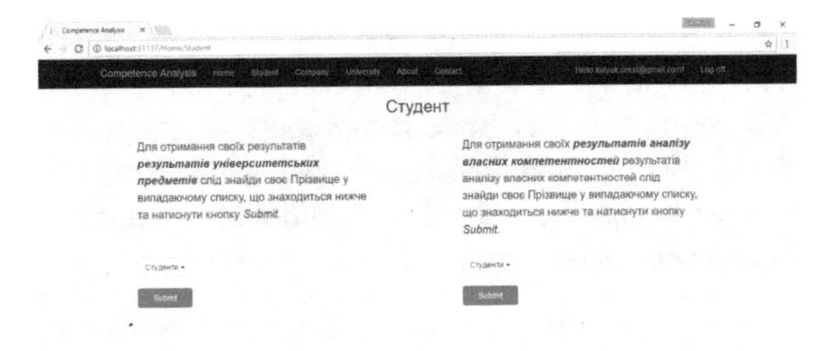

Fig. 13. Page of distant

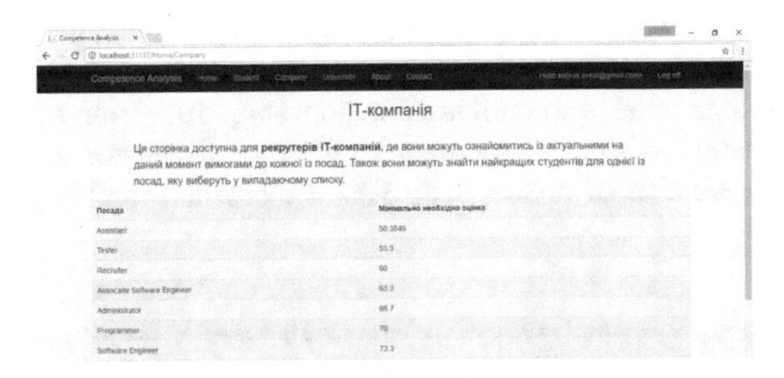

Fig. 14. Table rating required for IT majors

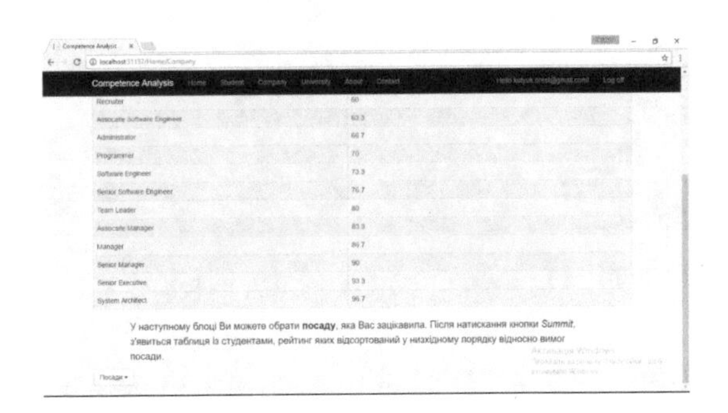

Fig. 15. Option for recruiters

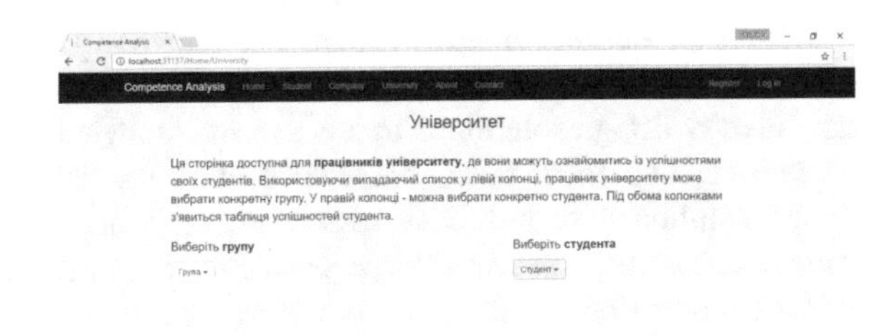

Fig. 16. Pages of University employee

For work University employee in the system is page University. In university employee have the ability to view, analyze and compare the success of students of the faculty/institute. To begin, the user must determine the group in which the student learns desired and click Submit, and then distant data, which it aims to review and click Submit. Students sorted in alphabetical order. Additional description recruiter action is also on the same page (Fig. 16).

6 Analysis of Test Case

For a demonstration of the software, give parkland use for all types of users, such as university staff, student and recruiter. To work, the student should go to the Student tab and option selects their success at university or option (Figs. 17 and 18), who will own matrix of knowledge relative to current IT majors (Fig. 19). To use recruiters need to go on the Company tab and select option knowledge distant for one of the jobs are listed (Fig. 20). Because of action, will distant table that qualify for this position (Fig. 21).

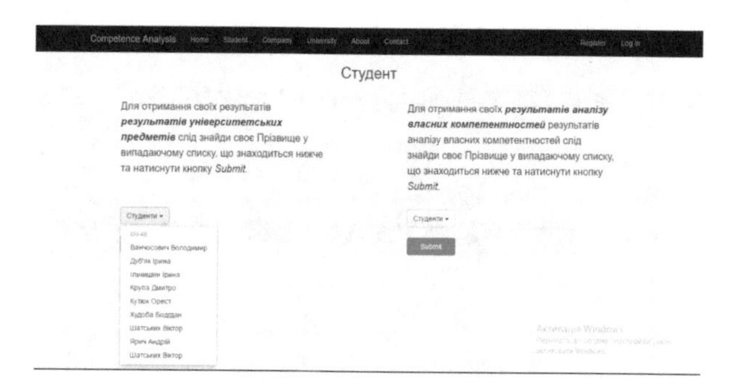

Fig. 17. Choice own knowledge matrix view

For workers should go to the tab and select option University matrix knowledge distant selecting the first group, where he trained, and most distant (Fig. 22). Following the action table appears distant knowledge matrix (Fig. 23).

The purpose of the master's qualification work is to develop intelligent system matrix analysis distant for recruitment in the IT sector. The designed system is designed for ease of recruiters IT companies is analyzing new staff in the face of distance education students. Novelty design solution is to analyze the knowledge and character of students distant and calculation of their competence in IT specialties. Organizing effect of using design solution is to introduce a more accurate analysis of students' knowledge and their cooperation with recruiting agencies for future employment. The economic effect of the introduction of design solution is to reduce the time that is spent searching for candidates, ordering and formalization of the data collected on certain criteria. In this system, there are two problem to be addressed the most optimal algorithm, calculation competence of students in their technical knowledge and

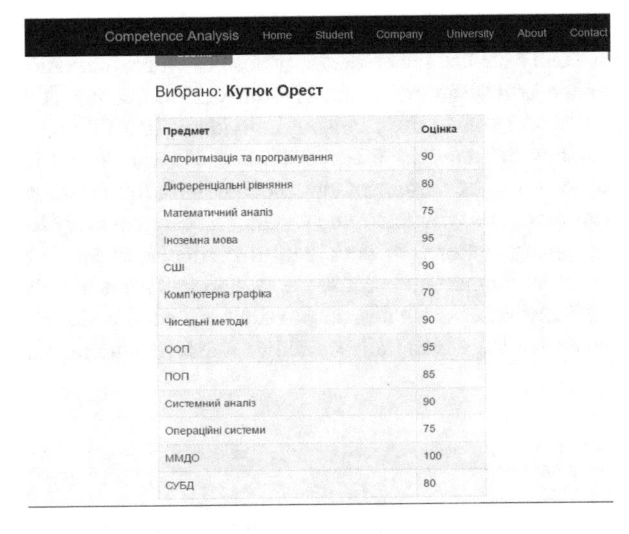

Предмет	Оцінка
Алгоритмізація та програмування	90
Диференціальні рівняння	80
Математичний аналіз	75
Іноземна мова	95
СШІ	90
Комп'ютерна графіка	70
Чисельні методи	90
ООП	95
ПОП	85
Системний аналіз	90
Операційні системи	75
ММДО	100
СУБД	80

Fig. 18. Result request the student to review their knowledge matrix

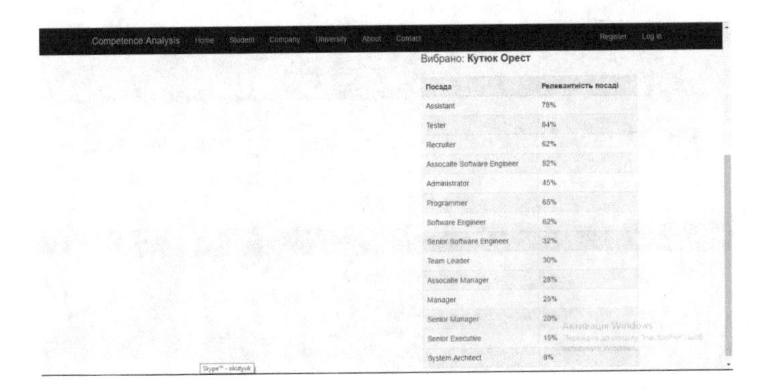

Посада	Релевантність посаді
Assistant	78%
Tester	84%
Recruiter	62%
Assocaite Software Engineer	82%
Administrator	45%
Programmer	65%
Software Engineer	62%
Senior Software Engineer	32%
Team Leader	30%
Assocaite Manager	28%
Manager	25%
Senior Manager	20%
Senior Executive	15%
System Architect	8%

Fig. 19. The result of the request to review student knowledge matrix

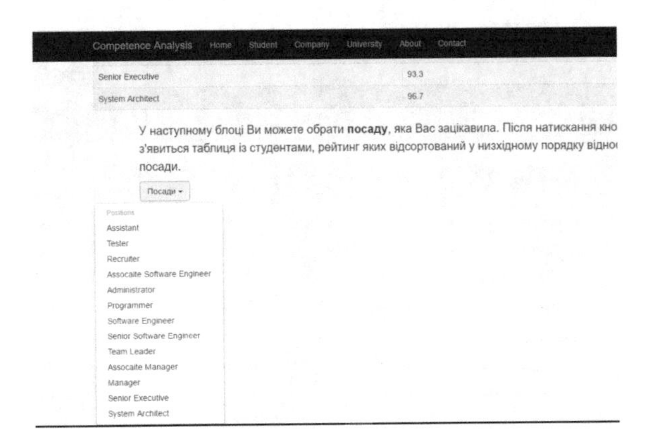

Fig. 20. List IT jobs

analysis of their personal qualities. Without addressing these problems, recruiters are unable to automatically choose the students for further work. Implementation AHR of student's knowledge will allow recruiters IT companies selecting IT profession, students determine the best course of promising individuals and fill its own database for possible projects. For the students this assessment will enable the analysis of their abilities and asks vector of development and specific field. Implementation of personal analysis as student recruiters will determine whether a person is ready to work in a team or student all necessary options to work with a particular client. For students such information is extremely important because the system will issue him feedback from all levels of training (teachers, classmates, most rating systems) and receiving data such student alone can determine which traits should be improved and on what terms should concentrate.

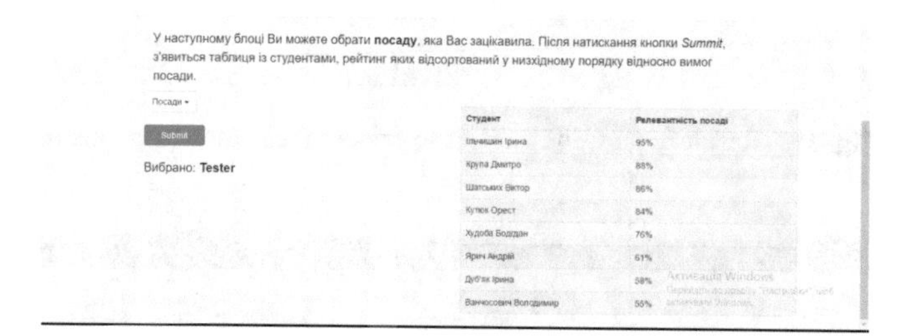

Fig. 21. Matrix Knowledge distant relatively profession Tester

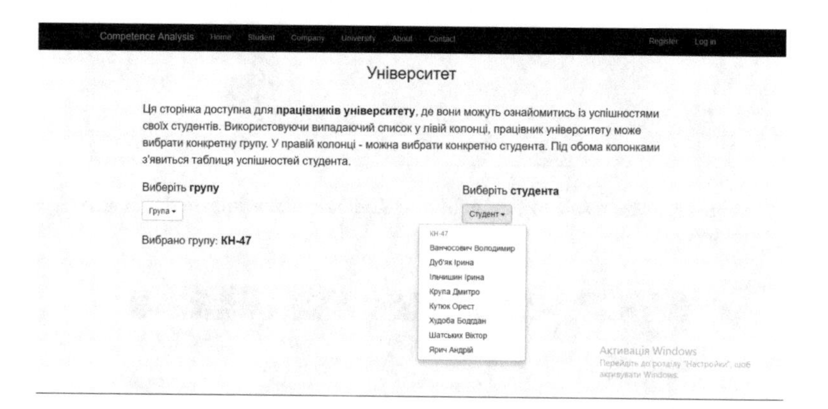

Fig. 22. Available distant group KNSSH-21

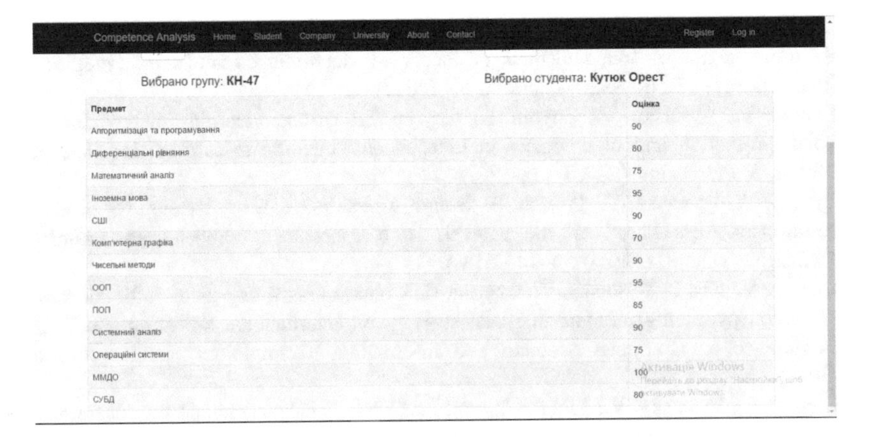

Fig. 23. Results of studies student group KNSSH-21 Kutyuka Orest

7 Conclusions

It is presented detailed algorithmic intellectual component system, namely, an improved algorithm TOPSIS. Intelligent system was designed for multiple users and groups, described in detail for each of the groups. A reference example of the program for all possible types of user of information system is shown in a work. It provides an economic rationale for the system. A software tool that is able to analyze knowledge distant as a web resource is received. This web site is adapted to the different modern devices, providing ease of use. As established software product can improve the administrative side of the web resource automating some actions that are currently forced to perform administrator and moderator. For a more accurate analysis of the competence necessary to increase is number of courses and number of experts.

References

1. System Administrator Leveling Matrix. https://docs.google.com/spreadsheets/d/1FBr20VIOePQH2aAH2a_6irvdB1NOTHZaD8U5e2MOMiw/pub?output=html
2. Haugeland, J.: Programmer competency matrix. perspectives on software, technology and business. http://sijinjoseph.com/programmer-competency-matrix/
3. Abramson, D., Krishnamoorthy, M., Dang, H.: Simulated annealing cooling schedules for the school timetabling problem (1997). http://www.rdt.monash.edu.au/~davida/papers/cool.ps.Z
4. Shakhovska, N., Vysotska, V., Chyrun, L.: Intelligent systems design of distance learning realization for modern youth promotion and involvement in independent scientific researches. In: Advances in Intelligent Systems and Computing, vol. 512, pp. 175–198 (2017)
5. Shakhovska, N., Vysotska, V., Chyrun, L.: Features of E-learning realization using virtual research laboratory. In: Computer Science and Information Technologies, CSIT, pp. 143–148 (2016)

6. Bobalo, Y., Stakhiv, P., Mandziy, B., Shakhovska, N., Holoschuk, R.: The concept of electronic textbook "Fundamentals of theory of electronic circuits". In: Przegląd Elektrotechniczny, 88 NR 3a/2012, pp. 16–18 (2012)
7. Lytvyn, V., Vysotska, V., Chyrun, L., Chyrun, L.: Distance learning method for modern youth promotion and involvement in independent scientific researches. In: Data Stream Mining & Processing (DSMP), pp. 269–274 (2016)
8. Lytvyn, V., Vysotska, V., Pukach, P., Bobyk, I., Pakholok, B.: A method for constructing recruitment rules based on the analysis of a specialist's competences. Eastern Eur. J. Enterp. Technol. **6**(2(84)), 4–14 (2016)
9. Chyrun, L., Kis, I., Vysotska, V., Chyrun, L.: Content monitoring method for cut formation of person psychological state in social scoring. In: International Scientific and Technical Conference on Computer Sciences and Information Technologies, CSIT, pp. 106–112 (2018)
10. Chyrun, L., Vysotska, V., Kis, I., Chyrun, L.: Content analysis method for cut formation of human psychological state. In: International Conference on Data Stream Mining and Processing, pp. 139–144 (2018)
11. Kanishcheva, O., Vysotska, V., Chyrun, L., Gozhyj, A.: Method of integration and content management of the information resources network. In: Advances in Intelligent Systems and Computing, vol. 689, pp. 204–216. Springer (2017)
12. Kowalik, D., Rusyn, B.: Innovative vocation didactics aimed at the preparation of staff according to Industry 4.0 and Europe 2020. In: International Conference on Education Reform and Modern Management, pp. 12–17 (2017)
13. Vysotska, V., Burov, Y., Lytvyn, V., Oleshek, O.: Automated monitoring of changes in web resources. In: Lecture Notes in Computational Intelligence and Decision Making, vol. 1020, pp. 348–363 (2019)
14. Demchuk, A., Lytvyn, V., Vysotska, V., Dilai, M.: Methods and means of web content personalization for commercial information products distribution. In: Lecture Notes in Computational Intelligence and Decision Making, vol. 1020, pp. 332–347 (2019)
15. Lytvyn, V., Vysotska, V., Mykhailyshyn, V., Rzheuskyi, A., Semianchuk, S.: System development for video stream data analyzing. In: Lecture Notes in Computational Intelligence and Decision Making, vol. 1020, pp. 315–331 (2019)
16. Su, J., Sachenko, A., Lytvyn, V., Vysotska, V., Dosyn, D.: Model of touristic information resources integration according to user needs. In: Computer Sciences and Information Technologies, pp. 113–116 (2018)
17. Mukalov, P., Zelinskyi, O., Levkovych, R., Tarnavskyi, P., Pylyp, A., Shakhovska, N.: Development of system for auto-tagging articles, based on neural network. In: CEUR Workshop Proceedings, vol. 2362, pp. 116–125 (2019)
18. Shakhovska, N.B., Noha, R.Y.: Methods and tools for text analysis of publications to study the functioning of scientific schools. J. Autom. Inf. Sci. **47**(12), 29–43 (2015)
19. Arzubov, M., Shakhovska, N., Lipinski, P.: Analyzing ways of building user profile based on web surf history. In: CSIT, vol. 1, pp. 377–380 (2017)
20. Shakhovska, N., Shvorob, I.: The method for detecting plagiarism in a collection of documents. In: Computer Sciences and Information Technologies (CSIT), pp. 142–145 (2015)
21. Rusyn, B., Vysotska, V., Pohreliuk, L.: Model and architecture for virtual library information system. In: Computer Sciences and Information Technologies, pp. 37–41 (2018)
22. Rusyn, B., Lytvyn, V., Vysotska, V., Emmerich, M., Pohreliuk, L.: The virtual library system design and development. In: Advances in Intelligent Systems and Computing, vol. 871, pp. 328–349 (2019)

23. Lytvyn, V., Vysotska, V., Dosyn, D., Burov, Y.: Method for ontology content and structure optimization, provided by a weighted conceptual graph. Webology **15**(2), 66–85 (2018)
24. Lytvyn, V., Peleshchak, I., Vysotska, V., Peleshchak, R.: Satellite spectral information recognition based on the synthesis of modified dynamic neural networks and holographic data processing techniques. In: International Scientific and Technical Conference on Computer Sciences and Information Technologies, CSIT, pp. 330–334 (2018)
25. Gozhyj, A., Kalinina, I., Vysotska, V., Gozhyj, V.: The method of web-resources management under conditions of uncertainty based on fuzzy logic. In: International Scientific and Technical Conference on Computer Sciences and Information Technologies, pp. 343–346 (2018)
26. Gozhyj, A., Vysotska, V., Yevseyeva, I., Kalinina, I., Gozhyj, V.: Web resources management method based on intelligent technologies. In: Advances in Intelligent Systems and Computing, vol. 871, pp. 206–221 (2019)
27. Lytvyn, V., Vysotska, V., Dosyn, D., Lozynska, O., Oborska, O.: Methods of building intelligent decision support systems based on adaptive ontology. In: International Conference on Data Stream Mining and Processing, DSMP, pp. 145–150 (2018)
28. Burov, Y., Vysotska, V., Kravets, P.: Ontological approach to plot analysis and modeling. In: CEUR Workshop Proceedings, vol. 2362, pp. 22–31 (2019)
29. Lytvyn, V., Vysotska, V., Peleshchak, I., Rishnyak, I., Peleshchak, R.: Time dependence of the output signal morphology for nonlinear oscillator neuron based on Van der Pol Model. Int. J. Intell. Syst. Appl. **10**, 8–17 (2018)
30. Lytvyn, V., Vysotska, V., Veres, O., Rishnyak, I., Rishnyak, H.: The risk management modelling in multi project environment. In: International Scientific and Technical Conference on Computer Sciences and Information Technologies, CSIT, pp. 32–35 (2017)
31. Lytvyn, V., Vysotska, V., Pukach, P., Vovk, M., Ugryn, D.: Method of functioning of intelligent agents, designed to solve action planning problems based on ontological approach. Eastern Eur. J. Enterp. Technol. **3**(2(87)), 11–17 (2017)
32. Vysotska, V., Lytvyn, V., Burov, Y., Gozhyj, A., Makara, S.: The consolidated information web-resource about pharmacy networks in city. In: CEUR Workshop Proceedings, vol. 2255, pp. 239–255 (2018)
33. Lytvyn, V., Kuchkovskiy, V., Vysotska, V., Markiv, O., Pabyrivskyy, V.: Architecture of system for content integration and formation based on cryptographic consumer needs. In: Conference on Computer Sciences and Information Technologies, CSIT, pp. 391–395 (2018)
34. Lytvyn, V., Vysotska, V., Kuchkovskiy, V., Bobyk, I., Malanchuk, O., Ryshkovets, Y., Pelekh, I., Brodyak, O., Bobrivetc, V., Panasyuk, V.: Development of the system to integrate and generate content considering the cryptocurrent needs of users. Eastern Eur. J. Enterp. Technol. **1**(2–97), 18–39 (2019)
35. Lytvyn, V., Vysotska, V., Demchuk, A., Demkiv, I., Ukhanska, O., Hladun, V., Kovalchuk, R., Petruchenko, O., Dzyubyk, L., Sokulska, N.: Design of the architecture of an intelligent system for distributing commercial content in the internet space based on SEO-technologies, neural networks, and Machine Learning. Eastern Eur. J. Enterp. Technol. **2**(2–98), 15–34 (2019)
36. Mukalov, P., Zelinskyi, O., Levkovych, R., Tarnavskyi, P., Pylyp,A., Shakhovska, N.: Development of system for auto-tagging articles, based on neural network. In: CEUR Workshop Proceedings, vol. 2362, pp. 106–115 (2019)
37. Shakhovska, N., Basystiuk, O., Shakhovska, K.: Development of the speech-to-text chatbot interface based on Google API. In: CEUR Workshop Proceedings, pp. 212–221 (2019)

38. Korobchinsky, M., Vysotska, V., Chyrun, L., Chyrun, L.: Peculiarities of content forming and analysis in internet newspaper covering music news. In: International Scientific and Technical Conference on Computer Sciences and Information Technologies, pp. 52–57 (2017)
39. Vysotska, V., Chyrun, L.: Analysis features of information resources processing. In: Proceedings of the International Conference on Computer Science and Information Technologies, CSIT, pp. 124–128 (2015)
40. Vysotska, V., Chyrun, L., Chyrun, L.: The Commercial content digest formation and distributional process. In: Proceedings of the XI-th International Conference on Computer Science and Information Technologies, CSIT 2016, pp. 186–189 (2016)
41. Su, J., Vysotska, V., Sachenko, A., Lytvyn, V., Burov, Y.: Information resources processing using linguistic analysis of textual content. In: Intelligent Data Acquisition and Advanced Computing Systems Technology and Applications, Romania, pp. 573–578 (2017)
42. Vysotska, V., Rishnyak, I., Chyrun L.: Analysis and evaluation of risks in electronic commerce. In: 9th International Conference on CAD Systems in Microelectronics, pp. 332–333 (2007)
43. Vysotska, V., Chyrun, L., Chyrun, L.: Information technology of processing information resources in electronic content commerce systems. In: Computer Science and Information Technologies, CSIT 2016, pp. 212–222 (2016)
44. Rzheuskyi, A., Gozhyj, A., Stefanchuk, A., Oborska, O., Chyrun, L., Lozynska, O., Mykich, K., Basyuk, T.: Development of mobile application for choreographic productions creation and visualization. In: CEUR Workshop Proceedings, vol. 2386, pp. 340–358 (2019)
45. Vasyl, L., Vysotska, V., Dosyn, D., Roman, H., Rybchak, Z.: Application of sentence parsing for determining keywords in Ukrainian texts. In: Computer Science and Information Technologies, CSIT, pp. 326–331 (2017)
46. Vysotska, V., Hasko, R., Kuchkovskiy, V.: Process analysis in electronic content commerce system. In: International Conference on Computer Sciences and Information Technologies, CSIT, pp. 120–123 (2015)
47. Vysotska, V., Fernandes, V.B., Emmerich, M.: Web content support method in electronic business systems. In: CEUR Workshop Proceedings, vol. 2136, pp. 20–41 (2018)
48. Lytvyn, V., Sharonova, N., Hamon, T., Vysotska, V., Grabar, N., Kowalska-Styczen, A.: Computational linguistics and intelligent systems. In: CEUR Workshop Proceedings, vol. 2136 (2018)
49. Lytvyn, V., Vysotska, V., Burov, Y., Demchuk, A.: Architectural ontology designed for intellectual analysis of E-tourism resources. In: International Scientific and Technical Conference on Computer Sciences and Information Technologies, CSIT, pp. 335–338 (2018)
50. Lytvyn, V., Vysotska, V., Rzheuskyi, A.: Technology for the psychological portraits formation of social networks users for the IT specialists recruitment based on big five, NLP and Big Data analysis. In: CEUR Workshop Proceedings, vol. 2392, pp. 147–171 (2019)
51. Lytvyn, V., Vysotska, V., Rusyn, B., Pohreliuk, L., Berezin, P., Naum, O.: Textual content categorizing technology development based on ontology. In: CEUR Workshop Proceedings, vol. 2386, pp. 234–254 (2019)
52. Vysotska, V., Lytvyn, V., Burov, Y., Berezin, P., Emmerich, M., Basto Fernandes, V.: Development of information system for textual content categorizing based on ontology. In: CEUR Workshop Proceedings, vol. 2362, pp. 53–70 (2019)
53. Zdebskyi, P., Vysotska, V., Peleshchak, R., Peleshchak, I., Demchuk, A., Krylyshyn, M.: An application development for recognizing of view in order to control the mouse pointer. In: CEUR Workshop Proceedings, vol. 2386, pp. 55–74 (2019)

54. Kaminskyi, R., Kunanets, N., Pasichnyk, V., Rzheuskyi, A., Khudyi, A.: Recovery gaps in experimental data. In: CEUR Workshop Proceedings, vol. 2136, pp. 170–179 (2018)
55. Kazarian, A., Holoshchuk, R., Kunanets, N., Shestakevysh, T., Rzheuskyi, A.: Information support of scientific researches of virtual communities on the platform of cloud services. In: Advances in Intelligent Systems and Computing III, vol. 871, pp. 301–311 (2018)
56. Rzheuskyi, A., Kunanets, N., Stakhiv, M.: Recommendation system virtual reference. In: Computer Sciences and Information Technologies (CSIT), pp. 203–206 (2018)
57. Kaminskyi, R., Kunanets, N., Rzheuskyi, A., Khudyi, A.: Methods of statistical research for information managers. In: International Scientific and Technical Conference on Computer Sciences and Information Technologies, pp. 127–131 (2018)
58. Tomashevskyi, V., Yatsyshyn, A., Pasichnyk, V., Kunanets, N., Rzheuskyi, A.: Data warehouses of hybrid type: features of construction. In: Advances. In Intelligent Systems and Computing, pp. 325–334 (2019)
59. Rzheuskyi, A., Matsuik, H., Veretennikova, N., Vaskiv, R.: Selective dissemination of information – technology of information support of scientific research. In: Advances in Intelligent Systems and Computing III, vol. 871, pp. 235–245 (2019)
60. Naum, O., Chyrun, L., Kanishcheva, O., Vysotska, V.: Intellectual system design for content formation. In: Proceedings of the International Conference on Computer Science and Information Technologies, CSIT, pp. 131–138 (2017)
61. Lytvyn, V., Vysotska, V., Burov, Y., Veres, O., Rishnyak, I.: The contextual search method based on domain thesaurus. In: Advances in Intelligent Systems and Computing, vol. 689, pp. 310–319 (2018)
62. Lytvyn, V., Pukach, P., Bobyk, I., Vysotska, V.: The method of formation of the status of personality understanding based on the content analysis. Eastern-European Journal of Enterprise Technologies **5**(2(83)), 4–12 (2016)
63. Lytvyn, V., Vysotska, V., Veres, O., Rishnyak, I., Rishnyak, H.: Classification methods of text documents using ontology based approach. In: Advances in Intelligent Systems and Computing, vol. 512, pp. 229–240 (2017)
64. Lytvyn, V., Vysotska, V, Veres, O., Rishnyak, I., Rishnyak, H.: Content linguistic analysis methods for textual documents classification. In: Proceedings of the XI-th International Conference on Computer Science and Information Technologies, CSIT 2016, pp. 190–192 (2016)
65. Kochan, R., Lee, K., Kochan, V., Sachenko, A.: Development of a dynamically reprogrammable NCAP. In: Proceedings of the IEEE Instrumentation and Measurement Technology Conference, pp. 1188–1193 (2004)
66. Hiromoto, R.E., Sachenko, A., Kochan, V., Koval, V., Turchenko, V., Roshchupkin, O., Yatskiv, V., Kovalok, K.: Mobile ad hoc wireless network for pre- and post-emergency situations in nuclear power plant. In: Proceedings of the 2nd IEEE International Symposium on Wireless Systems within the Conferences on Intelligent Data Acquisition and Advanced Computing Systems, pp. 92–96 (2014)
67. Lytvynenko, V., Wojcik, W., Fefelov, A., Lurie, I., Savina, N., Voronenko, M., et al.: Hybrid methods of GMDH-neural networks synthesis and training for solving problems of time series forecasting. In: Lecture Notes in Computational Intelligence and Decision Making, vol. 1020, pp. 513–531 (2019)
68. Babichev, S., Durnyak, B., Pikh, I., Senkivskyy, V.: An evaluation of the objective clustering inductive technology effectiveness implemented using density-based and agglomerative hierarchical clustering algorithms. In: Lecture Notes in Computational Intelligence and Decision Making, vol. 1020, pp. 532–553 (2019)

69. Bidyuk, P., Gozhyj, A., Kalinina, I.: Probabilistic inference based on LS-method modifications in decision making problems. In: Lecture Notes in Computational Intelligence and Decision Making, vol. 1020, pp. 422–433 (2019)
70. Vysotska, V.: Linguistic analysis of textual commercial content for information resources processing. In: Modern Problems of Radio Engineering, Telecommunications and Computer Science, TCSET 2016, pp. 709–713 (2016)
71. Lytvyn, V., Vysotska, V.: Designing architecture of electronic content commerce system. In: Computer Science and Information Technologies, CSIT 2015, pp. 115–119 (2015)
72. Veres, O., Rusyn, B., Sachenko, A., Rishnyak, I.: Choosing the method of finding similar images in the reverse search system. In: CEUR Workshop Proceedings, pp. 99–107 (2018)
73. Mukalov, P., Zelinskyi, O., Levkovych, R., Tarnavskyi, P., Pylyp, A., Shakhovska, N.: Development of system for auto-tagging articles, based on neural network. In: CEUR Workshop Proceedings, vol. 2362, pp. 106–115 (2019)
74. Basyuk, T.: The main reasons of attendance falling of internet resource. In: Proceedings of the X-th International Conference on Computer Science and Information Technologies, CSIT 2015, pp. 91–93 (2015)
75. Rzheuskyi, A., Kutyuk, O., Vysotska, V., Burov, Ye., Lytvyn, V., Chyrun, L.: The architecture of distant competencies analyzing system for IT recruitment In: Proceedings of International Scientific Conference “Computer sciences and information technologies” (CSIT-2019), IEEE v. 3, pp. 254–261 (2019)

A Complex System for Teaching Students with Autism: The Concept of Analysis. Formation of IT Teaching Complex

Vasyl Andrunyk[1], Volodymyr Pasichnyk[1], Natalya Antonyuk[2], and Tetiana Shestakevych[1](✉)

[1] Lviv Polytechnic National University, Lviv, Ukraine
{vasyl.a.andrunyk, tetiana.v.shestakevych}@lpnu.ua, vpasichnyk@gmail.com

[2] Department of Regional Studies and International Tourism, Ivan Franko National University of Lviv, Lviv, Ukraine
nantonyk@yahoo.com

Abstract. The modeling of a complex system for teaching students with autism demands a systematic approach. After the consideration of this system as a set of components, it is necessary to analyze various factors and formal connections between them. The structural analysis of the system reveals its composition and internal structure. As a tool of the functional structure presentation, it is advisable to use a unified modeling language. The system of formation of IT teaching complex is one of the subsystems of teaching students with autism. A recommender system seems to be a helpful tool to synthesize an information and technology complexes for teaching students with autism. It assumed, that such recommender system generates a proposal for the composition of the relevant information and technology complex, the components of which are the components of the IT platform, selected and combined in accordance with the requirements and needs of the paraprofessionals for the effective resolution of the formulated educational task.

Keywords: Student with autism · Complex system · Recommender system · Information and technology complex

1 Introduction

With the increasing importance of inclusion in the life of society, there is a natural need to explore, design, predict and manage such a complex system. The task of modeling complex systems is now perceived differently. Researchers of complex systems deviate from understanding the model of a complex system as a set of characteristics important for the researcher, the model is now a mean of research, it became a tool for the experimental verification of design decisions. It is relevant to thoroughly apply the systematic approach to the analysis of a complex system of inclusion, taking into account its peculiarities as a complicated system.

N. Shakhovska and M. O. Medykovskyy (Eds.): CCSIT 2019, AISC 1080, pp. 721–733, 2020.
https://doi.org/10.1007/978-3-030-33695-0_48

2 Developing a Complex Educational System for Students with Autism: A System Approach

In [1], the authors agreed that teaching of people with special needs is a complex, multifactorial, multi-stage cyclical process, which, among others, is a part of a complex system of social and educational inclusion [2, 3]. Such a complex system is open, it is evolving, it has a large number of components, as well as lack of central management, it also has a feedback, such system is hierarchical, variable, self-organizing, varied, dynamical, non-linear, etc. [3].

To apply the methodology of a systematic approach to creating a complex system for teaching students with autism, we use the following approaches [2, 4].

1. At the first stage, the system for teaching autistic students will be considered as a set of components, various factors and formal connections between them, their changes under the influence of external conditions will be analyzed, as well. To do this, it is necessary to involve the relevant specialists and use knowledge in various sciences and spheres of life, i.e. medicine, psychology, pedagogy, etc.
2. In the study of such a system, it is necessary to take into account the conditions of uncertainty, i.e. that different characteristics of the system can be varied. The general aim of the system functioning may change, for example, such aim may be the develop of the cheapest approach to providing educational services, or shortening the term of teaching if the educational level is reached, or the optimal combination of educational material to ensure correction of the psychophysical features of the student with autism, etc. The parameters of the external environment of the educational system for students with autism may also change, for example, legal aspects of education, changes in educational policy in the state, etc. Also, the behavior of the participants of the system may be different, for example, the unequal level of pedagogical skill of paraspecialists, the availability of various types of information technologies (virtual reality, complemented reality).
3. Due to the complexity of the system for teaching students with autism, it will be necessary to decompose the system into subsystems in different directions, which will enable to divide a complex task into a series of smaller, simpler related tasks. For example, the task of forming an individual learning trajectory should be divided into several subsystems, such as the establishment of a corrective and pedagogical component of the curriculum for a student with autism, taking into account his psychophysical features. It is also possible to apply aggregation methods when it is expedient to combine the results of psychophysical diagnosis of a person, the information technology of the formation of a personal teaching trajectory, the methods and means of educational support, the information technology for evaluating the results of teaching and the proposal of the next educational stage in the integrated IT support system of teaching [1].
4. To describe a complex teaching system for students with autism, it is advisable to use the standard language of system simulation, for example, a unified modeling language UML for a graphical description of a system. Structure, Behavior, and Interaction diagrams will allow us to conveniently describe the characteristics of the system for teaching students with autism. A unified modeling language will enable

professionals from different fields to collaborate on research and modeling of such a complex system.

The main feature of the system approach in the analysis and synthesis of complex systems is the need for iterations with new data [4].

Based on the approaches to creating a complex system, the general steps of research and modeling of a complex system for teaching autistic students are as follows.

1. Analysis of the functioning of the teaching system for students with autism.
2. Structural analysis of the system, allocation of its main subsystems.
3. Formation of the functional structure of the system.
4. Development of the mathematical model of the system.
5. Formation of ways to intensify processes in the system.
6. Definition of system management criteria.
7. Development of the model of functioning of the system.
8. Evaluation of the developed system to meet the requirements.
9. Estimation of optimization level, determination of the optimal level of automation.

The system created in this way will become the basis for the development of an information system of supporting the teaching of students with autism. The IT support system will be improved along with the improvement of the model of the complex system.

Within this work, we plan to analyze the functioning of a complex system for teaching students with autism, conduct a structural analysis of such a system, form its functional structure, and also develop a mathematical model of one of the subsystems. For a graphical description of the system, we use the unified modeling language (UML).

2.1 Analysis of the Functioning of the Educational System for Students with Autism

Features of Teaching Children with ASD. *Autism* or *autism spectrum disorders* is a common term used to group some brain disorders, including Asperger's syndrome, autistic disorders, disintegration disorder, intellectual, and other common disorders. Children with autism experience difficulties with communication and social interaction, have emotional issues, limited imagination, and so on. At the same time, learning is considered one of the best ways of socialization of a person with special needs. But traditional teaching techniques are based on communication and social interaction, which complicates the teaching of children with autism. Such students experience significant stress during communication, socialization, during sensory or visual contact, which may lead to inadequate behavior [5]. An optimal way to teach such students is to dip them into a personalized educational environment, created according to their personal needs and capabilities.

The process of education of children with autism implicates the autistic student, his/her parents, paraprofessionals, such as inclusive teachers, school administration, etc. The process depends on external legislative, medical, educational regulations, financial factors, as well as internal characteristics of a process (for example, the pedagogical talent of a teacher). It is also proved to be useful to use the information technologies to support the teaching of such children [6, 7]. The use of such technologies is even more preferable because of the peculiarities of the autistic disorder and the way the students interact with gadgets. Using such technologies enables the creation of the simple, predictable, repetitive action, in a habitual, safe environment, with minimal changes, with the possibility of verbal and nonverbal communication [5, 6].

The basic conceptual scheme of the functioning of the educational system for students with autism is presented in Fig. 1.

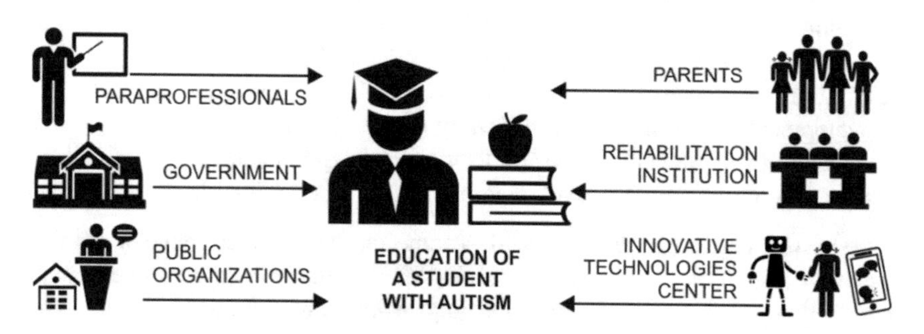

Fig. 1. Conceptual scheme of education of students with autism

2.2 Structural Analysis of the System

The development of the organizational structure of the complex system for teaching for students with autism allows to describe the composition of the subsystems and the connections between them, it establishes the functions of the subsystems, their internal structure at different levels of detail, describe the material, data, and information flows, and construct the general information structure.

The technical structure of the complex system for teaching students with autism reflects the basic technical means for obtaining information and its processing, as well as appropriate devices and technologies for providing network connections.

The functional structure of the complex system for teaching students with autism makes it possible to distinguish functions in the studied system, to select functions that are appropriate to automate, to develop a hierarchy of management tasks and corresponding models [4] (Fig. 2).

2.3 Formation of the Functional Structure of the System

Based on the above concept of a complex system for teaching autistic students, we present the functional structure of such a complex system, and for this purpose use the UML use case diagram. This diagram is convenient for visualizing of the participants of the process of teaching students with autism, as well as for the allocation of basic functions of the studied system.

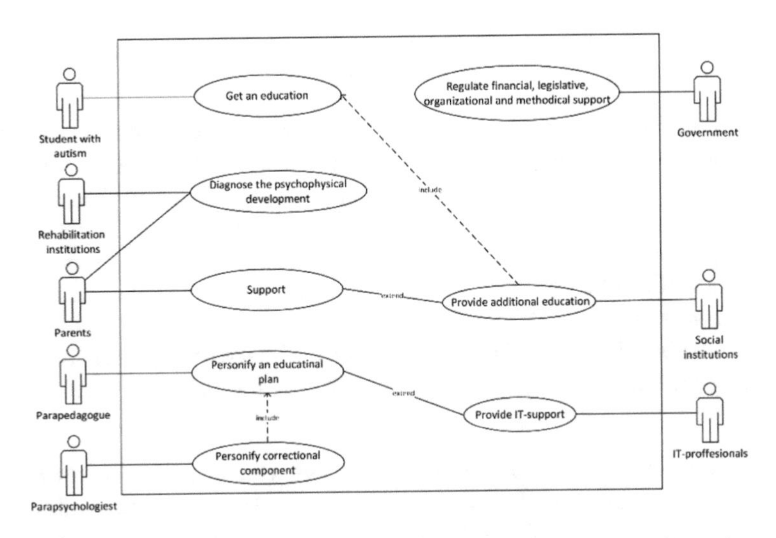

Fig. 2. The use case diagram of the process of teaching students with autism

Undoubtedly, this diagram provides the most general understanding of the basic functions of the studied system. Further detail is expedient, depending on the aspect of the analysis of such a complex system. As an example, we shall consider the subsystem of the formation of the IT-complex for teaching students with autism. In Fig. 3, there is a functional model of such a subsystem. We used the UML use case diagram to present it.

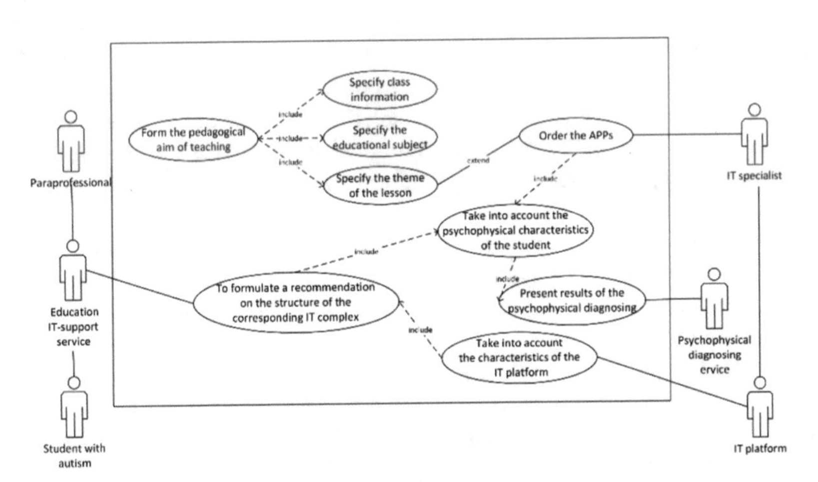

Fig. 3. Use case diagram of the process of formation of the IT-complex for teaching of students with autism

Into the system of formation of the information and technology complex of teaching students with autism, we aggregated paraprofessionals and the autistic students, as well as services for the study of the results of psychophysical research and IT-teaching. In fact, the system of formation of the IT complex for teaching of students with autism is a technological object of management with the given properties.

2.4 Developing the Mathematical Model of the System of Formation of the IT Complex for Teaching of Students with Autism

To develop such a model, it is necessary to select a mathematical mean that will enable the most complete description of the features of processes in the system and would be a convenient tool for research of such a system. During the study of the process of teaching students with autism, its characteristic features were identified, which include a strict sequence of stages of the studied process, as well as the need for parallel implementation of certain educational tasks. For the formal presentation of such claims, we use Petri nets. The advantages of using this mathematical abstraction are the ability to map the parallel phenomena in complex systems with the help of causative links, and a visual presentation of processes in complex systems, to which the system for teaching students with autism belongs to.

Petri net for the modeling of parallelism in the system of formation of the IT complex for the education of students with autism. The Petri net provides a visual, formalized representation of the behavior of parallel systems with asynchronous interactions [8]. The Petri net provides a compact representation of the structure of relations between elements of the system and the dynamics of changes in its states under given initial conditions.

Based on the formal description of the processes of inclusive education, we model the process of teaching students with autism using Petri nets. We assign such a network both graphically and analytically, i.e. as a finite sets of positions P, transitions T, the input functions of I and output functions O. Transitions in such a Petri net are interpreted as events, and positions are interpreted as the conditions of such events occurrence. Events in the process of teaching of students with autism are the fulfillment of certain educational tasks, the consequences of the implementation of such tasks are conditions for the onset of subsequent events. The sequence of the implementation of educational tasks is reflected by the operation of the Petri net transitions. Matching rules for triggering transitions is a way of expressing cause-and-effect relationships between conditions and events in the system.

The Petri net $C = (P, T, I, O)$, that models the system of formation of the IT complex for the education of students with autism, is given at Fig. 4. Here the set of positions, $P = \{p_0, p_2, ..., p_{22}\}$, set of transitions $T = \{t_1, t_2, ..., t_{13}\}$; initial marking μ_0, *and* one token in position p_0.

$I(t_1) = \{p_0\}$; $I(t_2) = \{p_1, p_2, p_3\}$; $I(t_3) = \{p_4\}$; $I(t_4) = \{p_5\}$; $I(t_5) = \{p_5, p_6\}$; $I(t_6) = \{p_7\}$; $I(t_7) = \{p_8\}$; $I(t_8) = \{p_9\}$; $I(t_9) = \{p_{10}\}$; $I(t_{10}) = \{p_{11}\}$;

$O(t_1) = \{p_1, p_2, p_3\}$; $O(t_2) = \{p_4\}$; $O(t_3) = \{p_5, p_6\}$; $O(t_4) = \{p_5\}$; $O(t_5) = \{p_7\}$; $O(t_6) = \{p_7\}$; $O(t_7) = \{p_8\}$; $O(t_9) = \{p_{12}\}$; $O(t_8) = \{p_{12}\}$; $O(t_{10}) = \{p_{12}\}$.

We shall determine the positions of the Petri net and their content in terms of teaching students with autism. We introduce the corresponding formal designations (Table 1).

Before the first triggering of transition t_1, there is one token in p_0 position. Consistently performing permissible transitions, the chip moves through the network, forming on each iteration the educational strategy of a student with autism.

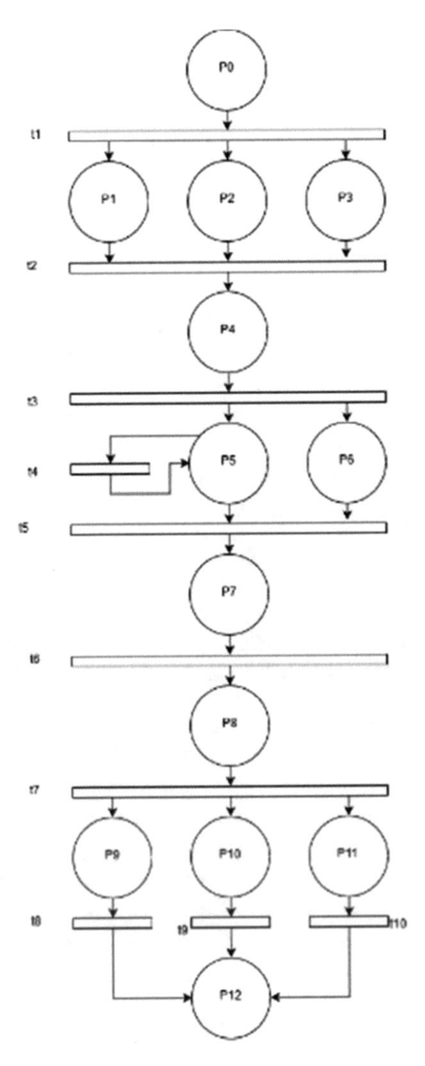

Fig. 4. Petri net as a model of parallelism in the system of formation of the IT complex for teaching of students with autism

The proposed model of the system for the formation of the IT complex for teaching of students with autism is convenient to implement as an appropriate recommender system. The basic concept of such a system is shown in Fig. 5.

Figure 6 provides an algorithm for the functioning of such a recommender system.

Table 1. Positions of Petri net as a model of parallelism in the system of formation of the IT complex for teaching of students with autism, and their semantical interpretation

Position	Position interpretation
p_0	The need for forming an IT teaching complex for students with autism
p_1	Class
p_2	Discipline
p_3	Theme
p_4	Educational aim
p_5	Database of IT-platform components
p_6	Data warehouse of the psychophysical diagnosing results
p_7	Characteristics of components of the teaching IT complex
p_8	Project of the teaching IT complex for students with autism
p_9	Results (feedback) of the teaching IT complex implementation
p_{10}	Psychophysical characteristics of the student being educated
p_{11}	Academic educational results
p_{12}	Readiness for the next exploitation of recommender system

Table 2. Transitions of Petri net as a model of parallelism in the system of formation of the IT complex for teaching of students with autism, and their semantical interpretation

Transition	Transition interpretation
t_1	Initiate exploitation of the recommender system
t_2	Form the education aim of teaching a student with autism
t_3	Initiate work with data warehouses and databases of autistic students educational and psychophysical diagnosing results
t_4	Development of new software that meets the requirements of paraprofessionals, autistic students, and software standards for people with special needs
t_5	Set characteristics of components of the IT complex
t_6	Consolidate the characteristics and formulate a recommendation of the structure of the IT teaching complex
t_7	Transfer the project of the IT teaching complex to the system for teaching children with autism
t_8	Analyze and update the data warehouse of a comprehensive assessment of the student's psychophysical development
t_9	Analyze and update the database of the student's academic results
t_{10}	Analyze and update the data warehouse of the IT teaching complex

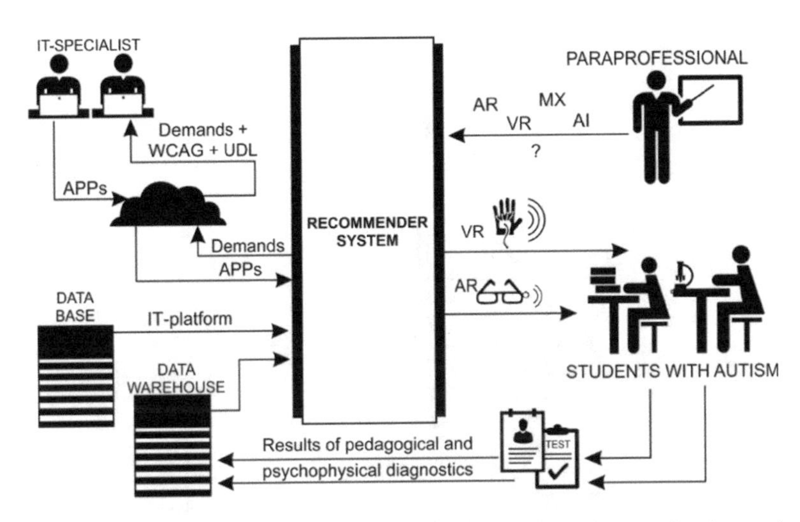

Fig. 5. The basic scheme of the functioning of the recommender system for the synthesis of information and technology complexes for the students with autism education

A formal model of the system for the synthesis of information and technology complexes for teaching students with autism. The above-mentioned recommender system synthesizes a set of information and technology complexes for teaching students with ASD. Formally this process can be described as a function

$$\varphi : U \times T \times Q \longrightarrow M,$$

where U is a set of educational aims, $U = \left\{u_{l,z}^{k} \,|\, l = \overline{1, L},\ z = \overline{1, Z},\ k = \overline{1, K}\right\}$, and $u_{l,z}^{k}$ is a k-th theme of a l-th subject for form z;

T is a set of personalized demands to the software and hardware, which are obtained as a result of the analysis of psychophysiological and pedagogical diagnosis X, $T = \left\{t_{w}^{i,r} \,|\, i = \overline{1, I},\ r = \overline{1, R},\ w = \overline{1, W}\right\}$, where $t_{w}^{i,r}$ is a demand to r-th component of i-th technology type for student w, W is the number of students at the *form z*. The set X is obtained as a result of analysis of psychophysical and pedagogical diagnosis, which is presented as a function $\phi : X \longrightarrow T$, where $X = \left\{x_{s}^{h,g} \,|\, h = \overline{1, H},\ g = \overline{1, G},\ s = \overline{1, S}\right\}$, and $x_{h,g}^{w}$ is a h-th result *of* a g-th test (psychophysiological or educational) for student w; and S is a number of records in the data warehouses of the psychophysiological and educational results, and $S \geq W$;

M is a set of recommended IT-complexes, $M = \left\{m_{w} | w = \overline{1, W}\right\}$, where m_{w} is an IT-complex, proposed for student w;

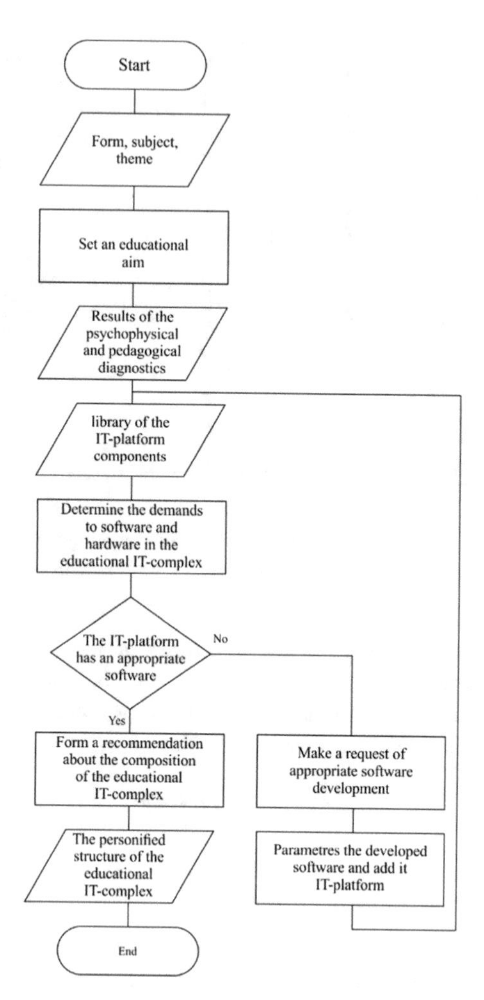

Fig. 6. An algorithm for the functioning of the recommender system for the synthesis of information and technology complexes for teaching of the students with autism

Q is a set of informational and technological means, that are available for the paraprofessional, it is more convenient to set as a set of three subsets, $Q = Q_1 \cup Q_2 \cup Q_3$, $Q = \{q_{i.j} \,|i = \overline{1,3}, j = \overline{1,J}\}$, and $q_{i.j}$ is a j-th IT mean of an i-th technology type,

$$i = \begin{cases} 1, for\ the\ Augmented\ reality \\ 2, for\ the\ Virtual\ reality \\ 3, for\ the\ Mixed\ reality \end{cases}.$$

Here $Q_1 = \{q_{1.1}, q_{1.2}, q_{1.3}, q_{1.4}, q_{1.5}\}$ is a set of augmented reality means, $Q_2 = \{q_{2.1}, q_{2.2}, q_{2.3}, q_{2.4}, q_{2.5}, q_{2.6}\}$ is a set of virtual reality means, $Q_3 = \{q_{3.1}, q_{3.2}\}$ is a set of mixed reality means. The detailed description of each of the sets is given in the Table 3.

Table 3. Means of information technologies for teaching children with autism

Technology type	Item	Example
Augmented reality	$q_{1.1}$	Smartphone
	$q_{1.2}$	Tablet
	$q_{1.3}$	Video camera
	$q_{1.4}$	VR glasses
	$q_{1.5}$	Software
Virtual reality	$q_{2.1}$	Stand-alone VR glasses or VR helmet
	$q_{2.2}$	Glasses or helmet as an addition to a PC
	$q_{2.3}$	Glasses or helmet as an addition to a smartphone
	$q_{2.4}$	Sensor-manipulator
	$q_{2.5}$	Full-functioning VR gloves
	$q_{2.6}$	Software
Mixed reality	$q_{3.1}$	VR glasses
	$q_{3.2}$	Software

The set of technologies for teaching children with ASD refers to those types of technologies that have proved to be most effective in working with such children [9–17].

3 Conclusions and Future Works Prospects

The provision of informational and technological support for inclusion and, in particular, teaching of students with autism is an urgent task that the government and individual organizations and centers (social, medical, rehabilitation) are interested in. Specialists from different spheres of life – physicians, psychologists, educators, IT specialists and others – are involved into this task. We can achieve a better result if we have the opportunity to coordinate processes in such a complex system of social and educational inclusion. Using a systematic approach to the analysis and modeling of a complex system for teaching students with autism will enable the use of such a model to improve the system itself.

In this paper, authors analyzed the functioning of a complex system for teaching students with autism, conducted a structural analysis of such a system, and formed its functional structure. The authors also develop a mathematical model of one of the subsystems, i.e. system for teaching students with autism. For modeling, the UML was used, as well as the mathematical apparatus of the Petri nets. The Petri net allowed to display in the model not only its functionality but also structural properties, which greatly simplifies the process of modeling of parallelisms in educational processes. In the following papers, the authors plan to investigate the ways to intensify processes in the system, and lay the foundation of such complex system management.

References

1. Shestakevych, T.: The method of education format ascertaining in program system of inclusive education support. In: IEEE 2017 12th International Scientific and Technical Conference on Computer Sciences and Information Technologies (CSIT 2017), Lviv, pp. 279–283 (2017)
2. Shestakevych, T., Pasichnyk, V., Kunanets, N., Medykovskyy, M., Antonyuk, N.: The content web-accessibility of information and technology support in a complex system of educational and social inclusion. In: IEEE 2018 XIIIth International Scientific and Technical Conference on Computer Sciences and Information Technologies (CSIT 2018), Lviv, pp. 27–31 (2018)
3. Shestakevych, T., Pasichnyk, V., Nazaruk, M., Medykovskiy, M., Antonyuk, N.: Web-products, actual for inclusive school graduates: evaluating the accessibility. In: Shakhovska, N. (eds.) Advances in Intelligent Systems and Computing, Advances in Intelligent Systems and Computing, vol. 871. Springer, Cham (2019)
4. Ladanyuk, A., Smityukh, Ya., Vlasenko, L., et. al.: System analysis of complex control systems, NUKhT, Kyiv, p. 274 (2013)
5. Tsiopela, D., Jimoyiannis, A.: Pre-vocational skills laboratory: development and investigation of a web-based environment for students with autism. Procedia Comput. Sci. **27**, 207–217 (2014)
6. Lucas, M., Gonçalves, D., Guerreiro, T., Plácido da Silva, H.: A web-based application to address individual interests of children with autism spectrum disorders. Procedia Comput. Sci. **14** (2012)
7. Ploog, B.O., Scharf, A., Nelson, D., Brooks, P.J.: Use of computer-assisted technologies (CAT) to enhance social, communicative, and language development in children with autism spectrum disorders. J. Autism Dev. Disord. **43**(2), 301–322 (2013)
8. Rozenburg, G., Engelfriet, J.: Elementary net systems. In: Reisig, W., Rozenberg, G. (eds.) Lectures on Petri Nets I: Basic Models – Advances in Petri Nets. Lecture Notes in Computer Science, vol. 1491, pp. 12–121. Springer (1998)
9. Andrunyk, V., Shestakevych, T., Pasichnyk, V.: The technology of augmented and virtual reality in teaching children with ASD. Econtechmod **7**(4), 59–64 (2018)
10. Aresti-Bartolome, N., Garcia-Zapirain, B.: Technologies as support tools for persons with autistic spectrum disorder. A systematic review. Int. J. Environ. Res. Public Health http://www.mdpi.com/journal/ijerph
11. Syahputra, M.F., Arisandi, D., Lumbanbatu, A.F., Kemit, L.F., Nababan, E.B., Sheta, O.: Augmented reality social story for autism spectrum disorder. In: 2nd International Conference on Computing and Applied Informatics, 2017 IOP Publishing IOP Conference Series, Journal of Physics: Conference Series, vol. 978, p. 120 (2018)
12. Fletcher-Watson, S.: A targeted review of computer-assisted learning for people with autism spectrum disorder. Towards a consistent methodology. https://link.springer.com/article/10.1007/s40489-013-0003-4
13. Ganz, J.B., ReaHong, E., Goodwyn, F.D.: Effectiveness of the PECS Phase III app and choice between the app and traditional PECS among preschoolers with ASD. Res. Autism Spectrum Disord. **7**(8), 973–983 (2013)

14. Mesa-Gresa, P., Gil-Gómez, H., Lozano-Quilis, J.-A., Gil-Gómez, J.-A.: Effectiveness of virtual reality for children and adolescents with autism spectrum disorder: an evidence-based systematic review. Sensors (Basel) **18**(8), 2486 (2018)
15. Ke, F., Im, T.: Virtual-reality-based social interaction teaching for children with high-functioning autism. J. Educ. Res. **106**, 441–461 (2013)
16. Bai, Z., Blackwell, A.F., Coulouris, G., Coulouris, G.: Using augmented reality to elicit pretend play for children with autism. Trans. Vis. Comput. Graph. **21**(5), 598–610 (2015)
17. Andrunyk, V., Pasichnyk, V., Shestakevych, T., Antonyuk, N.: Modeling the recommender system for the synthesis of information and technology complexes for the education of students with autism. In: IEEE 2019 14th International Scientific and Technical Conference on Computer Sciences and Information Technologies (CSIT 2019), Lviv, pp. 183–186 (2019)

The Mobile Application Development Based on Online Music Library for Socializing in the World of Bard Songs and Scouts' Bonfires

Bohdan Rusyn[1], Liubomyr Pohreliuk[1], Antonii Rzheuskyi[2](✉), Roman Kubik[2], Yuriy Ryshkovets[2,3], Lyubomyr Chyrun[4], Sofiia Chyrun[4], Anatolii Vysotskyi[5], and Vitor Basto Fernandes[6]

[1] Karpenko Physico-Mechanical Institute of the NAS Ukraine, Lviv, Ukraine
rusyn@ipm.lviv.ua, liubomyr@inoxoft.com
[2] Lviv Polytechnic National University, Lviv, Ukraine
{antonii.v.rzheuskyi,yuriy.v.ryshkovets}@lpnu.ua, roman89kubik@gmail.com
[3] SoftServe, Lviv, Ukraine
[4] IT Step University, Lviv, Ukraine
[5] Anat Company, Lviv, Ukraine
[6] University Institute of Lisbon, Lisbon, Portugal

Abstract. The aim of the current work is to create a mobile application that helps bard singers and guitar players sing the same songs and play the same chords. The development of this software product was conducted under the operating system Android using Java programming languages and Kotlin. SQLite database management system was selected for storing data. XML is the main markup language for the development of the graphical interface. Pure architecture as the basic architecture in conjunction with the MVP pattern has been chosen. The developed mobile application is simple and easy to use and provides the following basic functions: display a list of songs with chords to them, search for songs by name, detailed song display, ability to show - hide chords, add our own songs to the songwriter, automatically detect and mark chords in song lyrics, dynamically show chords when we click on it in a song, create a list of our favorite songs, change the tone of the song, show chords for the musical instrument chosen by user.

Keywords: Mobile application · Java · Kotlin · SQLite · Database management system · Android · Software

1 Introduction

The aim of this work was to create a software product that should help users to socialize, as well as unify knowledge of how to play songs for musicians [1–6]. Many people coming for the first time on bard singing do not feel comfortable, because there

N. Shakhovska and M. O. Medykovskyy (Eds.): CCSIT 2019, AISC 1080, pp. 734–756, 2020.
https://doi.org/10.1007/978-3-030-33695-0_49

are a lot of songs they have never heard [7–9]. This is due to the fact that such types of events gather people from a certain circle of acquaintances, with their preferences and a certain song reserve [10–13]. Because there are a lot of songs, it is logical that it is difficult for a person to remember them all. The next problem is that people who play musical instruments often can not remember how to play a particular song. Also, if there are several musicians on bard singers, then there is a problem that everyone knows the different interpretation of a song. In this case, only one musician will play, and others will adjust to his game [14–19]. The ideal way to solve this problem is to give each participant a bard singing guide with songs and chords to them. However, the printing of singers is expensive and resource-intensive. So a good solution to these problems might be his electronic version. In addition, this product will have the ability to upgrade without re-publishing a new paper singer [20]. The user will only need to update the app and new songs, and the bug fix will be applied to the mobile app. One more significant advantage of an electronic singer is that he is always at hand because people are now keeping their mobile phone on their own. One of the biggest advantages of an electronic version of the songwriter is that the cost of distributing a singer is minimal. That is, in order to distribute an electronic songwriter only access to the Internet and mobile phone is necessary. Although there are several analogue competitors in the market with a similar functionality, each of them has many drawbacks [21]. Therefore, the task is to fix all the shortcomings that are in the analogues, as well as add a unique function. The software will be easy to use and solve problems, if the products of the analogues do not solve, and therefore should become successful.

2 Features of Bards Singing Evenings and Using of Software to Be Carried Out

Bard evening is a social event aimed at recreating its participants from everyday worries and singing songs in the circle of friends and acquaintances. Also no less important in a bard evening is not only singing songs, but also playing them on guitar or other musical instruments. Most often, bard parties have lovers without special skills in singing and without skills to play any musical instrument [1, 22–27]. The singer is a collection of songs, sometimes chords or notes are added to the songs. An analogue to the textbook in music. Most often singers are paper-based, in the form of a book. The singers are usually selected thematically - for example: pop songs, folk songs, carols and so on [28–32]. Depending on the theme of the bard evening, the organizers will usually issue paper singers who will collect songs that will be sung during the evening. Of course, frequent participants of the bard's singing evenings themselves buy singers who have a more general set of songs that does not correspond to the theme of a particular bardevening. Also, the owners of the singers often adjust the songs in the singers or add their own songs or songs that have not been added to the songwriter.

In order to successfully hold a bard evening, we need to plan the whole program of the event correctly. One of the most important components of the Bard's evening, without which this event can not be called a bard evening, is their own songs and songs. In order for all participants to be able to sing songs, even the ones they hear for the first time, organizers of bard's evenings often create thematic singers, which are distributed to the participants of the event [33–41]. Because of this, to take part in bard singing, we often need to insert a tab that covers the cost of composing and printing a collection of songs. However, such a scheme generates many problems that need to be addressed. The first problem is the large number of singers who will have a frequent participant in the bard's evenings. After all, every time, coming on a bard evening, the organizers give the participant a singing artist. However, the participant is not always required because the frequent members already have several copies of the singers, as well as frequent participants of such events know most of the songs that will sing. This problem also raises another problem. A frequent participant of bard evenings must buy a singer every time he comes to such an event. This problem can be solved not by compulsory purchase of a singer. That is, do not give the singer to each participant, and make a stand, after which the singers will sell all who want to [42–47].

However, the solution to the previous problem may cause the following problem. It is commonplace for people to forget, therefore there will be such cases, when going on a bard evening, people will forget to take a singer on it. They will have the opportunity to buy it on the spot, but few will do it, because buying a singer is a certain expense, and what to buy is what lies at our home. And so, people who forgot to bring the singers will not be able to sing songs together with all other participants of the bard evening. The next problem is the creation of a paper singer. Namely its styling and formatting. Often the compilers of a paper singer do not think about its convenience. They simply paste from him the songs copied from the Internet. The compilers of the singers forget to format the inserted text. If the singer, besides the songs also contains chords or notes to the song, the compiler, being incompetent in this, can insert chords that do not match this song. This together creates many problems for the user of the songwriter. Another problem with using a paper singer is the lack of the ability to edit a song. Add her contributions to her. There is also no possibility of automatic transposition of chords. Transposition or transposition is the transfer of a musical composition from one tonality to another tone of the same tilt without any change. Transposition is used both in tonal music (so-called "tonal transposition") and in the atonal. For example, in dodecaphony, transposition is defined as the transfer of a series of sounds to one of the 12 degrees of the tempered volume. In online sources, we can find songs at once with chords, but they can often contain misspellings or write in a tone that is unusual for we. A simple software tool will be able to perform this task instantaneously and without any human effort [48–53].

The last, but equally important issue of paper singers is the consumables for their creation. After all, it requires paper to be created, and paper is made of wood. Cutting down wood humanity is harmful to ecology. To solve this problem, we need to refuse paper consumption wherever possible. That is why, as an alternative to a paper singer, software that has a similar functional with a paper singer can be used [54–62].

The main kind of software in this subject area is in the form of mobile applications and web services. This is quite logical reasons because we always carry the mobile phone with us, and it is also convenient to keep important information in one place. Web services are great for watching songs from home as well as creating our own collection of songs. However, it often happens that places where bard evenings are held are in a zone where there is no mobile coverage. Then comes the help of mobile apps that have saved songs that the user can access, even without being connected to the

Internet. At the moment there are not so many mobile singers. After reviewing foreign and domestic counterparts, I have chosen the best and most convenient applications and web services [63–77].

(1) Ukrainian Songs website [4]. The largest Ukrainian - language web service with Ukrainian songs. The purpose of the site "Ukrainian Songs" is a collection of the best Ukrainian songs from ancient times to the present, as well as information about their authors and performers. Most of the songs include chords for accompaniment on the guitar, and it is also planned to place notes and music tracks in midi and mp3 formats. Advantages:

- A huge database of songs and chords to them
- Search for songs
- Chord index
- Forum
- A large union of people constantly updating content of the site and completing it with new songs.

Disadvantages: only the web site version is available.

(2) Mobile application "At the Watra - Singer" [5]. The only Ukrainian application is a singing artist. There is a paid and free version of the application. Contains more than 300 Ukrainian songs. Advantages:

- Offline work
- Chord Show
- Change the font size of the text
- Day and night mode

Disadvantages:

- Outdated design
- Chord index for guitar only
- Lack of ability to edit songs
- Lack of ability to add your own songs
- Lack of search for a song
- Lack of the ability to create a list of your favorite songs

(3) Mobile application "Songs for guitar Rus" [6]. A mobile application, a Russian developer, that uses web service queries to download songs and view them offline. There is an option to edit a song. Advantages:

- Downloading songs from the internet
- Offline mode
- List of chords
- List of favorite songs
- Creating playlists

Disadvantages:

- Many ads
- Inexpensive search
- There is a paid version with additional functionality
- There is no automatic chord definition
- A small number of Slavic songs.

The creation and filling of the singer requires a lot of knowledge in the music field. Often the authors of the singers do not take into account all these factors, but make them faster. Also, composers often forget about the musicians who will use their singers and do not take into account all the needs. Despite the large number of electronic analogue singers on the market, they all have both their own pros and cons. Therefore, the lyceum vocalist's mobile application is a rather promising project and will be widely used among Slavic bard evening enthusiasts or ordinary people.

3 Statement of the Problem of Development of the Mobile Application "Bard's Signer"

The task will be to create mobile software that will show songs and chords to them. This product will have the function of auto-detect chords in the song, and the ability to click on the chord to see its tab. Another important function will be the transposition of chords in the song. Also, the mobile app should work offline so that access to songs could be even in a place without a mobile coverage. This software product should be simple and easy to use. To do this, we need to develop a mobile application, design a database and fill it with data.

To design and create a mobile application interface, the Sketch software tool, the graphical user interface layout constructor [7], was used. The selected software is simple and flexible to use, since it allows us to drag items to the work surface by simply dragging. Also a software tool allows us to create and apply styles to a group of graphic elements right away. To work with prototypes, directly during the implementation of the software, all prototypes were transferred to the Zeplin software, which is more convenient from the point of view of the developer [8].

The development of a mobile app was decided to run under the operating system Android OS. As the programming language for developing a mobile application, the object-oriented Java language was selected, as well as the statically-typed programming language Kotlin. The combination of these two programming languages has allowed us to develop a flexible yet stable architecture. I decided to develop an application under the Android operating system for several reasons:

(1) Android is the operating system and platform for mobile phones and tablets, created by Google based on the Linux kernel.
(2) Android SDK - has open source, it can be downloaded for free, and used.
(3) The presence of a large number of open libraries.
(4) Android is the most popular operating system in the world, with a monthly number of active users close to 2 billion.

(5) Android OS has built-in support for a SQLite database on a kernel, ensuring high performance of the database application.

As a development environment, I selected the Android Studio developed by Jet-Brains, since this environment is completely free, and is also the official development environment for the Android operating system.

The main purpose of creating a mobile application "Bard's singer" is to create an application that can be used to view songs and chords to them. It should also provide an opportunity to show the chord for the selected musical instrument, as well as provide a convenient interface to add their own song to the singer. Another important function is automatic chord definition in the song: as well as their marking. Comparison of the main analogues products is presented in the Table 1.

Table 1. Comparison of analogues

Features	Slavic songs	At bonfire - the singer	Guitar songs (Rus)
Ability to add your own song	+	+	–
Ability to add your own song	–	–	+
Chord Show	–	–	+
List of favorite songs	+	–	+
Work offline	–	+	+
Define chords from text	–	–	–

Since the product is created taking into account all the needs of musicians and participants of the bard evenings, as well as being able to do offline and the ability to add their own songs, this product has a great chance to become popular as well as useful.

The software product will perform the following main functions:

- Display a list of songs with chords to them
- Search for songs by name
- Detailed song display
- Ability to show - hide chords
- Add your own songs to the songwriter
- Automatically detect and mark chords in song lyrics
- Dynamically show chords when you click on it in a song
- Create a list of your favorite songs
- Change the tone of the song
- Show chords for the user-selected musical instrument

From the client side, you need a mobile phone based on the operating system Android version 4.1 (API 15) or higher.

4 The Characteristics of Systems

Req 1. Display the list of songs with chords to them
Description: This function displays all the songs in the mobile singer.
Priority: High
User action: The user opens the application and goes to the section with all the songs
System: Opens a screen with a list of songs
Message: List of songs
Functional requirements: View all songs from the songwriter
Req 2. Search for songs by name
Description: This function searches for songs by their name
Priority: High
User action: Enter a song title or part name
System: Searches all songs
Message: List of songs with matching in the title, or empty list
Functional requirements:
1. Find a song among all songs by the songwriter by its name
2. Ability to choose a song found
Functional requirements of the program
Req A.1 Detailed song display
Description: This feature displays the song in full and displays the chords to her
Priority: High
User action: Selects a song from the list of songs
System: Opens a song in a new window, analyzes the chords in it
Message: The song and chords that are in the song
Functional requirements:
1. Open a song in a new window
2. Definition and formatting of chords
Req A.2. Ability to show - hide chords
Description: This function allows you to change the presentation of the song in the application
Priority: Low
User action: Disables the chord in the settings
System: Analyzes the song and hides chords from there
Message: A song without chords
Functional requirements:
1. Analysis of the chords in the song
2. Removing chords from the song
Req A.3 Add your own songs to the songwriter
Description: This feature allows you to add your own songs to the songwriter
Priority: High
User action: Inserts the title and the lyrics
System: Creates a new record in the database
Message: Successfully creating a new song
Functional requirements: Create a new song

Req A.4 Automatically detect and mark chords in the lyrics
Description: This function allows you to automatically detect and mark the chords in the lyrics
Priority: Medium
User Action: Creates a new song
System: Creates a database entry
Message: Successfully creating a new song
User action: Causes the auto-determination of chords
System: Analyzes the lyrics for chords and tags them
Message: Successfully marking chords
Functional requirements:
1. Create a new song record
3. Analyze the text and mark the chords in it
Req A.5. Dynamically show chords when you click on it in a song
Description: This function displays a tabular chord when pressed on it
Priority: High
User action: Click on the chord
System: Retrieves information about the chord
Message: Chord Tab Tab Dialog
Functional requirements: Show the corresponding chord
Req A.6 Create a list of your favorite songs
Description: This feature allows the user to create a list of favorite songs
Priority: Medium
User Action: Moves to song details
System: Opens a window with details of a song
Message: Details of the song
User action: Adds to your favorites list
System: Creates a record of your favorite song in the database
Message: The song is labeled as your favorite
Functional requirements: Create a record of your favorite song in the database
Req A.7 Change the tone of the song
Description: This function changes the chords in the song
Priority: Low
User Action: Moves to song details
System: Opens a window with details of a song
Message: Details of the song
User action: In the settings, the tone changes to the song
System: Analyzes all chords and accordingly changes their tone, stores them change database
Message: A song with a changed tonality chord
Functional requirements:
1. Analysis of chords and change their tonality
2. Saving changes in the database
Req A.8 Show chords for user-selected music instrument
Description: This function allows you to choose which tool will be shown tabs chords
Priority: Medium

User action: In the settings, select a musical instrument
System: Saves user selection
Message: Displays chords for the selected tool
Functional requirements:
1. Saving user selection
2. Display the chords for the selected tool

The main element of navigation is the elements of the program as well as the hardware button "Back". When you open a program, the user can select the group of songs he wants to watch, or all the songs. Next, the user can view the details of the song and the chords to it. The entire design is based on the concepts of Google Material Design.

5 Development of Mobile Application "Bard's Signer"

For writing applications, the Android operating system uses various architectural patterns. All of them have their own advantages and disadvantages. After analyzing most of them, I decided to use a clean architecture (Clean Architecture [9]) and an MVP pattern (Model - View - Presenter [10]) for the presentation level. The main benefits of combining these two architectures are the light extensibility and component independence of each other. Pure architecture was first proposed by Robert Martin in the article "The Clean Architecture" [9] in 2012, which describes the main approaches to this architecture. The most pure architecture is a simple set of recommendations and rules for building an application. The main idea behind this architecture is to split the application into three layers: presentation layer, data layer, and layer of usage cases (domain layer). Presentation layer - At this level all graphical components of the application are located. The graphic level is responsible for displaying data for the user, as well as for engaging with the user. Data layer is this level is responsible for obtaining data. It knows how to get data and how to deal with them. Implements user-level interfaces. Domain layer is this level describes the use cases of the program. That is, at this level realized the business logic of the project. This level knows nothing about where the data comes from. It only has interfaces that show that the data should be received. How exactly these data will be received is unknown. At this level, this information is not important. The only important thing is how these data are processed and transmitted to the presentation level. This layer should not depend on the frameworks used. Another advantage of pure architecture is the easy testing of the code, since all components are independent of each other. In addition to pure architecture, the MVP template for the presentation layer was implemented in the project.

MVP is a pattern consisting of three components:

1. Model is the data that needs to be displayed.
2. View (view window) is where the viewer views the user.

Presenter is a component that responds to View events and decides how to process these events and display them on View. All communication between the presenter and the presentation takes place at the level of interfaces. The advantage of this approach is to break the code into logical units. Since there are no authorizations in the thesis project, only one type of user is available in it. Each user will have a full set of software functionalities. In the future, it's possible to develop an application and add paid features to it: which will be allowed only for a certain group of users.

The use case diagram (Fig. 1). Depicts the actor (user) as well as the functionality that he can use. As can be seen from the precedent diagram, some precedents may include additional functionality, such as searching for songs by name, including the listing of songs, and some expanding with additional actions such as adding a song, expanding it further, and automatically determining chords in it.

To design the behavior of the program, an UML sequence diagram was created (Fig. 2). This diagram reflects the interaction of objects arranged in time. The diagram shows a user, system (own application), and a database.

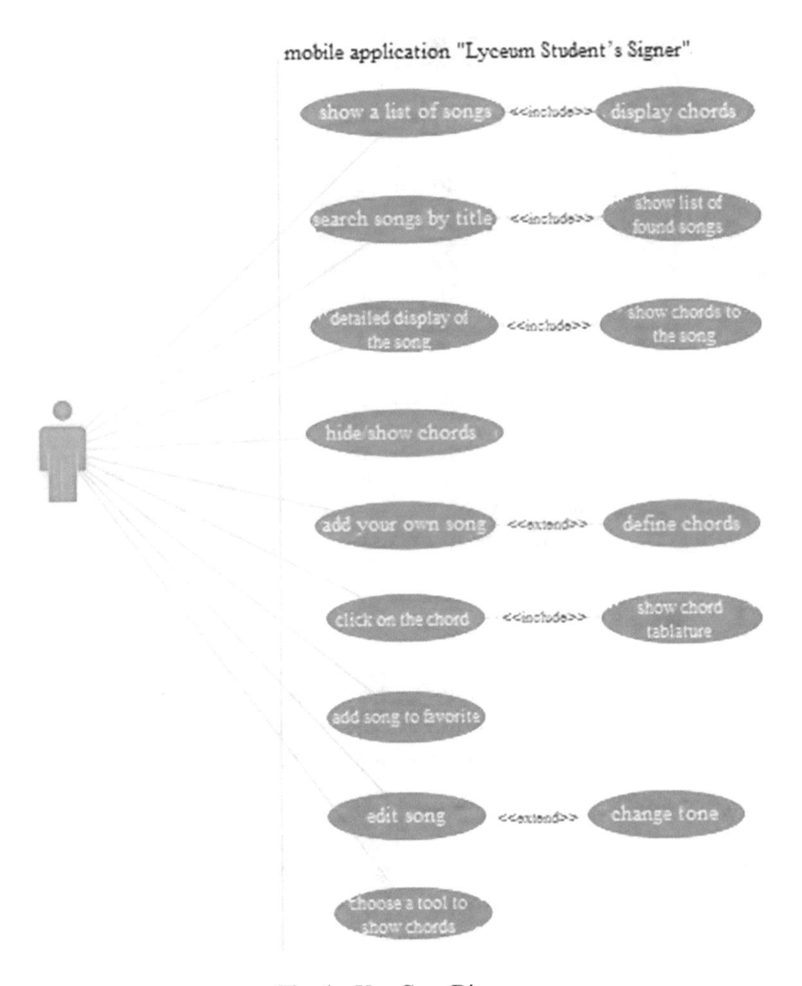

Fig. 1. Use Case Diagram

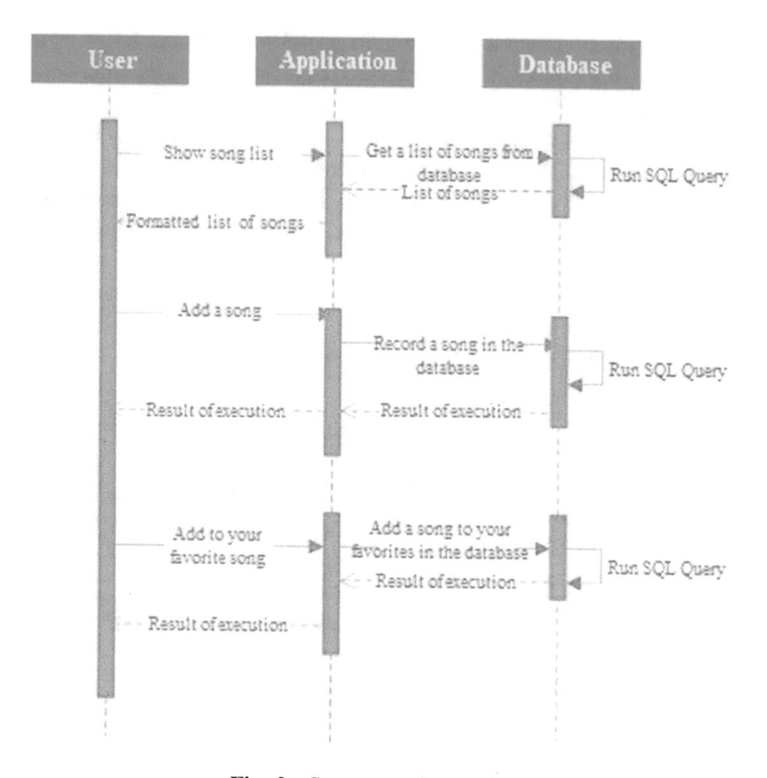

Fig. 2. Sequence of execution

To work, the user interacts with the system, which in turn generates a query to the database. After receiving the query results, the database sends a response to the system that formats the data and displays it to the user.

6 Designing the Database

To create a database logical model (Fig. 2), the CA ERwin Data Modeler environment was used - a simple and convenient tool for designing a database. The database of a developed mobile application will not be complicated, and will contain the following entities:

- Song. Contains information about a song, namely: unique identifier, song title, lyrics with song chords, and link to category number.
- Category. Each song will have its own category. In the meantime, only three categories will be available, namely: Patriotic songs, At Vatry, Interesting foreign. An entity category will have such fields as a unique identifier and a category name.
- Favorite. In this essence, your favorite songs will be saved. Essences will have the following fields: a unique identifier and a pointer to a favorite song.

- History. This entity is intended to display the last songs that have been viewed by the user. This entity will contain the following attributes: a unique identifier and a link to the song.

Since no logic "many to many" has been created during logical designing, the logical scheme and the physical circuit are identical. The schema of the database is presented in Fig. 3.

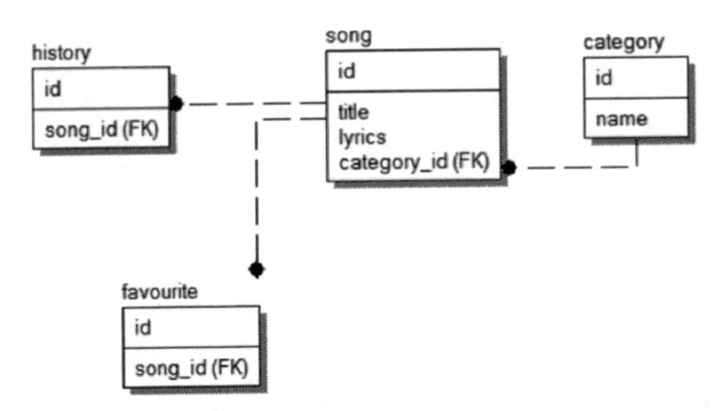

Fig. 3. Database schema

After defining the main functionality of the software being developed, I created the initial map of the mobile application, which will demonstrate the structure of the future system. The main parts are:

- The main window of the program
- Settings window
- A box with a list of songs and a search option
- Details of the song
- Window to add songs

Next, the first prototypes of the mobile application "The Bard's singer" were created. The main window of the program was developed (Fig. 4). The user opens the program to the main window of the program. From here, he can fall into any of the program's sections: open a list of all songs, or a song by category. Similarly, in the side menu there is an opportunity to go to the application setup page and the page to add a new song. When a user selects a particular category of songs, or all songs, he falls into the window with a list of songs. This window shows the names of the songs and the first 4 lines. This search box allows you to search for songs by name. In order to get to the detail of the song with chords, the user only needs to click on the song of his choice. After the user selects a song, the application redirects it to the page with the song of his choice (Fig. 5).

On this page, the user sees the full text of the lyrics of the song, as well as the marked chords to it. The user has the ability to click on the chord and see his tab. A free version of the Sketch software was used to design the user interface. Prototypes of all other main windows of the program are added in Figs. 6 and 7.

Fig. 4. Prototype of the main program window

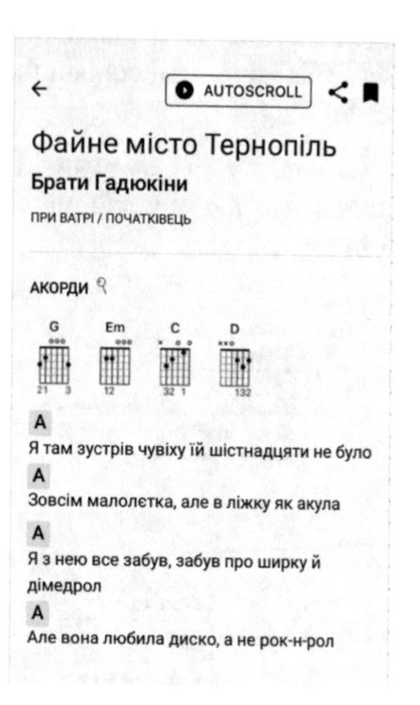

Fig. 5. Prototype of a typical song viewer

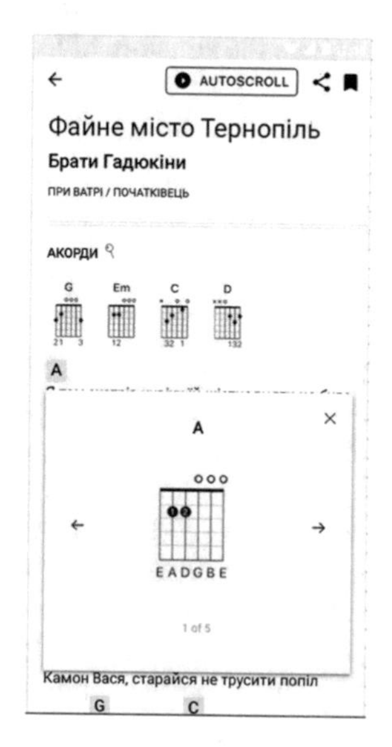

Fig. 6. Tabs of chords

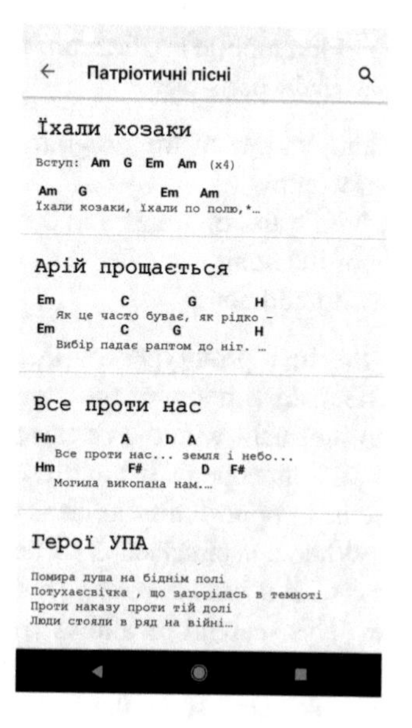

Fig. 7. List of songs

7 Realization and Testing of the Mobile Application "Bard's Signer"

To create SQLite database models, a convenient and free version of the software, under the MacOS operating system, was used because the development was carried out on Mac OS, SQLiteStudio. With this program, a database was created and filled with its data. To migrate the SQLite database file and integrate it into the program, the basic mechanisms used to work with files on the Android operating system were used. To provide access to a database from a mobile application, an object-relational projection, from Google, called Room, was used. The main benefits of Room Library are:

- Small amount of writing a monotonous code, to perform simple queries.
- Verification of SQL queries during compilation, which means there is no risk of error during the execution of the program, which can lead to the termination of the application.

After creating models it was necessary to organize the process of data exchange between the application and the database. Interactive and repositories were created to organize this process. The repository contains a specific implementation of access to the database application. For each table, a separate repository was created: songs, categories, and favorites. Each required request for correct work is ordered separately using SQLite queries. A separate DAO class is created for each table. DAO - (data access object) is intended for access to the database. The Room library is a set of methods in the interface to access the database.

```
public interface CategoryRepository {
Single<List<Category>> getAll();
Maybe<Category> getById(int id)}
```

Interactors prescribe the main business logic of the product, as well as the process of data exchange between models. The advantage of pure interactors is that it is easy to test. That is, to write and run a unit test, just the java virtual machine, without the creation of the Android environment. This repository uses DAO to access data from the database. An example of the DAO interface for accessing a table with the categories is below:

```
@Dao
public interface CategoryDao {
@Query("SELECT * FROM category")
Single<List<CategoryEntity>> getAll();
@Query("SELECT * FROM category WHERE category.id = :id")
Maybe<CategoryEntity> getById(int id);}
```

After creating a component for working with the database, and after describing all cases of use (i.e., interactors) and business logic, appropriate graphical representations were created. Graphic views in the Android operating system are Activity and Fragment classes. For the development of the mobile application "Lyceum Singer" it was chosen to use Activity. Each activity has its own content, which can be set by program, or described in a.xml file. As planned, the user interface was designed to be simple and user-friendly. The main colors are gray, black and olive. It is the three colors that form the main color palette according to the Material Design Guidelines. In the main window of the program (Fig. 8) there is a quick access to the functions of the program, such as: a list of all songs, songs by category, a list of your favorite songs, the song is selected at random. Also from this window there is an opportunity to go to the settings menu, where the user will be able to hide the chord's show to the song. Another option in the settings menu is the choice of a musical instrument for which the chord tab is displayed.

Fig. 8. The main window

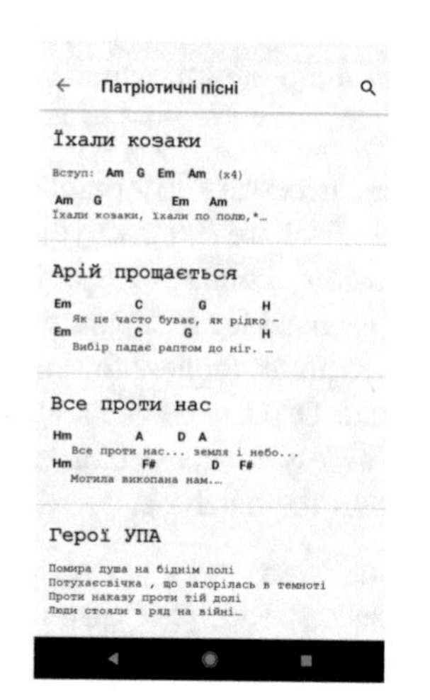

Fig. 9. Example of viewing songs

Fig. 10. Example of the chords in the song

When you select a specific category, the user goes to the program window, which shows the list of songs, according to the category chosen by the user. The list of songs is depicted in the form of the title of the song and its first four lines (Fig. 9). After a user presses a song, it will be redirected to the song window, where the title of the song, the category to which it belongs, her words and chords are displayed. The Courier New font was selected to display the song, since it is easy to read, and all characters have the same width in it. To emphasize the chords in the song, they were highlighted in bold, and a gray background was added (Fig. 10). This design was designed in such a way as to enable it to be used horizontally in a mobile phone. The algorithm for determining and marking chords in a song is at the level of use cases, as this is the usual business logic of the project, which should not depend on the chosen implementation technologies, as well as on the graphical representation. Since there is no specific algorithm for such a task, the recommendations of the teacher of the class of guitar, from music school № 5 were used to create it. The basis of this algorithm was the use of regular expressions and their search in the text. Testing the developed tool. This section provides information about the testing, test results, and a quality assessment. In the process of testing, there were two steps: testing the correctness of algorithms for computing and working with the database, testing the user interface and its ease of use. To create test cases, it was decided to divide them into two parts: checking the correct execution of the code (checking algorithms) and validating the presentation of the data to the user (checking the interface). In order to test individual functions, and separate, independent parts of the code, unit tests were created. To do this, a testing framework for the Java programming language - JUnit4 was connected. Table 2 presents the modalities for using modular testing.

Table 2. Module testing options

Usage options	Test cases	Test data
Chord Recognition	4	27
Removing chords from text	4	27
Search for a song by title	2	2
Add a new song	2	2
Edition of a songs	2	2
Get a list of songs	1	6
In general	15	66

The main purpose of testing the graphical user interface is to check the correct display of the lyrics with chords, as well as to check the correct navigation on the mobile application. I have formed a set of test cases, a list of which is given in Table 3. During the execution of the unit tests, there was no malicious error, which means that all algorithms work correctly. At this stage, there were no discrepancies with the above-mentioned options for using the graphical user interface testing. Because there are no functional safety requirements for this software, security testing is unnecessary. Since all the test conditions were successful: the test cases for the algorithms were tested, the user interface testing was performed (Table 4).

Table 3. Usage options for testing the interface

Usage option	Test data
Checking the main window of the program	5
Checking the window with a list of all the songs	10
Checking the window with a list of all songs by category	10
Check the window with details of the song	2
Checking the window with the selected chord display	5
Check the box for adding your own song	2
Checking the editing window of the selected song	2
In general	36

Table 4. Test cases for the algorithms

Test № 1: View all songs list	
Description: Checking the list of all songs in the list	
Steps:	Expected results
(1) Select a category for all songs (2) See all songs	(1) All songs from the songwriter
Test № 2: Choosing a random song	
Description: Check the random song	
Steps:	Expected results
(1) Choose the category "Astonish Me" (2) View the song (3) Go back (4) Select the category "Astonish Me" (5) Watch the song	(1) Opening a random song (2) Discovery of another random song
Test № 3: Searching for songs	
Description: Checking the search for songs by name	
Condition: test № 1 passed	
Steps:	Expected results
(1) Go to the song list (2) Click on the search for songs (3) Enter the name or part of the title of song	(1) List of songs in which the title has the desired value
Test № 4: View the song with chords	
Description: Check the correctness of the song display with the chords	
Steps:	Expected results
(1) Go to the window with details of the song (2) Select chord and click on it	(1) Song with labeled chords (2) Details of the chord with tablature
Test № 5: Hiding chords	
Description: Checking the correctness of hiding chords in songs	
Condition: test № 1 passed	
Steps:	Expected results
(1) Go to the settings window (2) Disable chord show (3) Return to the main window programs (4) Go to the song list (5) Choose a song	(1) Display the list of songs without chords (2) Display the detail of the song without chords

Test Case Resolution – Passed. All errors that occur during the program will automatically be sent to the developer in the form of a bug report. In case of questions, the user may contact the developer by sending him an email. Errors will be resolved in subsequent versions of this software product.

8 Conclusions

Within the research, the subject area was analyzed. An analysis of existing analogue singers in the market was conducted. Despite their small number on the market, they all have both their own pros and cons. Next, the task was set and a specification of the program requirements was formulated, which had to be solved during the execution of the work. It was determined that the application would be developed under Android operating system. After the task was formed, the design of the mobile application began. A multi-layered clean architecture with a MVP template at the presentation level has been chosen as the main architecture. A database and basic prototypes of the user interface were also designed. At this stage, the main stack of technologies to be used for the implementation of this project was also selected. To implement the project, the mobile application "Bard Signer" was chosen to combine the two Java programming languages because of its high stability and reliability, and Kotlin because of its novelty and ease. When the mobile application was written, several test steps were conducted, namely modular, for testing algorithms and testing the graphical user interface. During the modular testing, there were no defects in the operation of the algorithms and the system as a whole. Also, testing the graphical user interface has shown that the interface is convenient for end users. An economic analysis of the work was carried out, and it was determined that the project is pay-back, so it is expedient to develop it. We consider the bard's singer mobile application is a rather promising project and will be widely used among Slavic bard's enthusiasts or ordinary people who love singing songs at the company. In the future the program will be designed so that it will be possible to add songs of other countries.

References

1. Rusyn, B., Vysotska, V., Pohreliuk, L.: Model and architecture for virtual library information system. In: Computer Sciences and Information Technologies, pp. 37–41 (2018)
2. Rusyn, B., Lytvyn, V., Vysotska, V., Emmerich, M., Pohreliuk, L.: The virtual library system design and development. In: Advances in Intelligent Systems and Computing, vol. 871, pp. 328–349 (2019)
3. Shakhovska, N., Vysotska, V., Chyrun, L.: Features of E-learning realization using virtual research laboratory. In: Proceedings of the XI-th International Conference on Computer Science and Information Technologies, CSIT 2016, pp. 143–148 (2016)
4. Shakhovska, N., Vysotska V., Chyrun, L.: Intelligent systems design of distance learning realization for modern youth promotion and involvement in independent scientific researches. In: Advances in Intelligent Systems and Computing, vol. 512, pp. 175–198. Springer, Cham (2017)
5. Naum, O., Chyrun, L., Kanishcheva, O., Vysotska, V.: Intellectual system design for content formation. In: Proceedings of the International Conference on Computer Science and Information Technologies, CSIT, pp. 131–138 (2017)
6. Korobchinsky, M., Vysotska, V., Chyrun, L., Chyrun, L.: Peculiarities of content forming and analysis in internet newspaper covering music news. In: Proceedings of the International Conference on Computer Science and Information Technologies, CSIT, pp. 52–57 (2017)

7. Kanishcheva, O., Vysotska, V., Chyrun, L., Gozhyj, A.: Method of integration and content management of the information resources network. In: Advances in Intelligent Systems and Computing, vol. 689, pp. 204–216. Springer (2018)
8. Lytvyn, V., Vysotska, V.: Designing architecture of electronic content commerce system. In: Proceedings of the X-th International Conference on Computer Science and Information Technologies, CSIT 2015, pp. 115–119 (2015)
9. Vysotska, V.: Linguistic analysis of textual commercial content for information resources processing. In: Modern Problems of Radio Engineering, Telecommunications and Computer Science, TCSET 2016, pp. 709–713 (2016)
10. Mukalov, P., Zelinskyi, O., Levkovych, R., Tarnavskyi, P., Pylyp, A., Shakhovska, N.: Development of system for auto-tagging articles, based on neural network. In: CEUR Workshop Proceedings, vol. 2362, pp. 106–115 (2019)
11. Shakhovska, N., Basystiuk, O., Shakhovska, K.: Development of the speech-to-text chatbot interface based on Google API. In: CEUR Workshop Proceedings, vol. 2386, pp. 212–221 (2019)
12. Su, J., Vysotska, V., Sachenko, A., Lytvyn, V., Burov, Y.: Information resources processing using linguistic analysis of textual content. In: Intelligent Data Acquisition and Advanced Computing Systems Technology and Applications, Romania, pp. 573–578 (2017)
13. Lytvynenko, V., Savina, N., Krejci, J., Voronenko, M., Yakobchuk, M., Kryvoruchko, O.: Bayesian networks' development based on noisy-MAX nodes for modeling investment processes in transport. In: CEUR Workshop Proceedings, vol. 2386, pp. 1–10 (2019)
14. Lytvynenko, V., Lurie, I., Krejci, J., Voronenko, M., Savina, N., Taif., M. A.: Two step density-based object-inductive clustering algorithm. In: CEUR Workshop Proceedings, vol. 2386, pp. 117–135 (2019)
15. Rzheuskyi, A., Gozhyj, A., Stefanchuk, A., Oborska, O., Chyrun, L., Lozynska, O., Mykich, K., Basyuk, T.: development of mobile application for choreographic productions creation and visualization. In: CEUR Workshop Proceedings, vol. 2386, pp. 340–358 (2019)
16. Lytvyn, V., Vysotska, V., Rzheuskyi, A.: Technology for the psychological portraits formation of social networks users for the IT specialists recruitment based on big five, NLP and Big Data analysis. In: CEUR Workshop Proceedings, vol. 2392, pp. 147–171 (2019)
17. Vysotska, V., Chyrun, L., Chyrun, L.: Information technology of processing information resources in electronic content commerce systems. In: Computer Science and Information Technologies, CSIT 2016, pp. 212–222 (2016)
18. Lytvyn, V., Vysotska, V., Chyrun, L., Chyrun, L.: Distance learning method for modern youth promotion and involvement in independent scientific researches. In: Proceedings of the IEEE First International Conference on Data Stream Mining & Processing (DSMP), pp. 269–274 (2016)
19. Vysotska, V., Rishnyak, I., Chyrun, L.: Analysis and evaluation of risks in electronic commerce. In: 9th International Conference CAD Systems in Microelectronics, pp. 332–333 (2007)
20. Vysotska, V., Chyrun, L.: Analysis features of information resources processing. In: Proceedings of the International Conference on Computer Science and Information Technologies, CSIT, pp. 124–128 (2015)
21. Vysotska, V., Chyrun, L., Chyrun, L.: The commercial content digest formation and distributional process. In: Proceedings of the XI-th International Conference on Computer Science and Information Technologies, CSIT 2016, pp. 186–189 (2016)
22. Bobalo, Y., Stakhiv, P., Mandziy, B., Shakhovska, N., Holoschuk, R.: The concept of electronic textbook "Fundamentals of theory of electronic circuits". In: Przegląd Elektrotechniczny, 88 NR 3a/2012, pp. 16–18 (2012)

23. Lytvyn, V., Vysotska, V., Pukach, P., Bobyk, I., Pakholok, B.: A method for constructing recruitment rules based on the analysis of a specialist's competences. Eastern Eur. J. Enterp. Technol. **6**(2(84)), 4–14 (2016)
24. Chyrun, L., Kis, I., Vysotska, V., Chyrun, L.: Content monitoring method for cut formation of person psychological state in social scoring. In: International Scientific and Technical Conference on Computer Sciences and Information Technologies, CSIT, pp. 106–112 (2018)
25. Chyrun, L., Vysotska, V., Kis, I., Chyrun, L.: Content analysis method for cut formation of human psychological state. In: International Conference on Data Stream Mining and Processing, pp. 139–144 (2018)
26. Mukalov, P., Zelinskyi, O., Levkovych, R., Tarnavskyi, P., Pylyp, A., Shakhovska, N.: Development of system for auto-tagging articles, based on neural network. In: CEUR Workshop Proceedings, vol. 2362, pp. 116–125 (2019)
27. Shakhovska, N.B., Noha, R.Y.: Methods and tools for text analysis of publications to study the functioning of scientific schools. J. Autom. Inf. Sci. **47**(12), 29–43 (2015)
28. Shakhovska, N., Shvorob, I.: The method for detecting plagiarism in a collection of documents. In: Computer Sciences and Information Technologies (CSIT), pp. 142–145 (2015)
29. Arzubov, M., Shakhovska, N., Lipinski, P.: Analyzing ways of building user profile based on web surf history. In: Computer Sciences and Information Technologies (CSIT), vol. 1, pp. 377–380 (2017)
30. Lytvyn, V., Vysotska, V., Burov, Y., Veres, O., Rishnyak, I.: The contextual search method based on domain thesaurus. In: Advances in Intelligent Systems and Computing, vol. 689, pp. 310–319 (2018)
31. Lytvyn, V., Vysotska, V., Veres, O., Rishnyak, I., Rishnyak, H.: Classification methods of text documents using ontology based approach. In: Advances in Intelligent Systems and Computing, vol. 512, pp. 229–240 (2017)
32. Su, J., Sachenko, A., Lytvyn, V., Vysotska, V., Dosyn, D.: Model of touristic information resources integration according to user needs. In: Computer Sciences and Information Technologies, pp. 113–116 (2018)
33. Lytvyn, V., Vysotska, V., Dosyn, D., Burov, Y.: Method for ontology content and structure optimization, provided by a weighted conceptual graph. Webology **15**(2), 66–85 (2018)
34. Lytvyn, V., Peleshchak, I., Vysotska, V., Peleshchak, R.: Satellite spectral information recognition based on the synthesis of modified dynamic neural networks and holographic data processing techniques. In: International Scientific and Technical Conference on Computer Sciences and Information Technologies, CSIT, pp. 330–334 (2018)
35. Gozhyj, A., Kalinina, I., Vysotska, V., Gozhyj, V.: The method of web-resources management under conditions of uncertainty based on fuzzy logic. In: International Scientific and Technical Conference on Computer Sciences and Information Technologies, pp. 343–346 (2018)
36. Gozhyj, A., Vysotska, V., Yevseyeva, I., Kalinina, I., Gozhyj, V.: Web resources management method based on intelligent technologies. In: Advances in Intelligent Systems and Computing, vol. 871, pp. 206–221 (2019)
37. Lytvyn, V., Vysotska, V., Dosyn, D., Lozynska, O., Oborska, O.: Methods of building intelligent decision support systems based on adaptive ontology. In: International Conference on Data Stream Mining and Processing, DSMP, pp. 145–150 (2018)
38. Burov, Y., Vysotska, V., Kravets, P.: Ontological approach to plot analysis and modeling. In: CEUR Workshop Proceedings, vol. 2362, pp. 22–31 (2019)

39. Lytvyn, V., Vysotska, V., Peleshchak, I., Rishnyak, I., Peleshchak, R.: Time dependence of the output signal morphology for nonlinear oscillator neuron based on Van der Pol Model. Int. J. Intell. Syst. Appl. **10**, 8–17 (2018)
40. Lytvyn, V., Vysotska, V., Veres, O., Rishnyak, I., Rishnyak, H.: The risk management modelling in multi project environment. In: International Scientific and Technical Conference on Computer Sciences and Information Technologies, CSIT, pp. 32–35 (2017)
41. Lytvyn, V., Vysotska, V., Pukach, P., Vovk, M., Ugryn, D.: Method of functioning of intelligent agents, designed to solve action planning problems based on ontological approach". Eastern Eur. J. Enterp. Technol. **3**(2(87)), 11–17 (2017)
42. Vysotska, V., Lytvyn, V., Burov, Y., Gozhyj, A., Makara, S.: The consolidated information web-resource about pharmacy networks in city. In: CEUR Workshop Proceedings, vol. 2255, pp. 239–255 (2018)
43. Lytvyn, V., Kuchkovskiy, V., Vysotska, V., Markiv, O., Pabyrivskyy, V.: Architecture of system for content integration and formation based on cryptographic consumer needs. In: Conference on Computer Sciences and Information Technologies, CSIT, pp. 391–395 (2018)
44. Lytvyn, V., Vysotska, V., Kuchkovskiy, V., Bobyk, I., Malanchuk, O., Ryshkovets, Y., Pelekh, I., Brodyak, O., Bobrivetc, V., Panasyuk, V.: Development of the system to integrate and generate content considering the cryptocurrent needs of users. Eastern Eur. J. Enterp. Technol. **1**(2–97), 18–39 (2019)
45. Lytvyn, V., Vysotska, V., Demchuk, A., Demkiv, I., Ukhanska, O., Hladun, V., Kovalchuk, R., Petruchenko, O., Dzyubyk, L., Sokulska, N.: Design of the architecture of an intelligent system for distributing commercial content in the internet space based on SEO-technologies, neural networks, and Machine Learning. Eastern Eur. J. Enterp. Technol. **2**(2–98), 15–34 (2019)
46. Vasyl, L., Vysotska, V., Dmytro, D., Roman, H., Rybchak, Z.: Application of sentence parsing for determining keywords in Ukrainian texts. In: Computer Science and Information Technologies, CSIT, pp. 326–331 (2017)
47. Vysotska, V., Hasko, R., Kuchkovskiy, V.: Process analysis in electronic content commerce system. In: International Conference on Computer Sciences and Information Technologies, CSIT, pp. 120–123 (2015)
48. Vysotska, V., Fernandes, V.B., Emmerich, M.: Web content support method in electronic business systems. In: CEUR Workshop Proceedings, vol. 2136, pp. 20–41 (2018)
49. Lytvyn, V., Sharonova, N., Hamon, T., Vysotska, V., Grabar, N., Kowalska-Styczen, A.: . In: Computational Linguistics and Intelligent Systems. CEUR Workshop Proceedings, vol. 2136 (2018)
50. Lytvyn, V., Vysotska, V., Burov, Y., Demchuk, A.: Architectural ontology designed for intellectual analysis of E-tourism resources. In: International Scientific and Technical Conference on Computer Sciences and Information Technologies, CSIT, pp. 335–338 (2018)
51. Lytvyn, V., Vysotska, V., Rusyn, B., Pohreliuk, L., Berezin, P., Naum, O.: Textual content categorizing technology development based on ontology. In: CEUR Workshop Proceedings, vol. 2386, pp. 234–254 (2019)
52. Vysotska, V., Lytvyn, V., Burov, Y., Berezin, P., Emmerich, M., Basto Fernandes, V.: Development of information system for textual content categorizing based on ontology. In: CEUR Workshop Proceedings, vol. 2362, pp. 53–70 (2019)
53. Zdebskyi, P., Vysotska, V., Peleshchak, R., Peleshchak, I., Demchuk, A., Krylyshyn, M.: An application development for recognizing of view in order to control the mouse pointer. In: CEUR Workshop Proceedings, vol. 2386, pp. 55–74 (2019)

54. Sachenko, A., Kochan, V., Turchenko, V.: Intelligent distributed sensor network. In: Instrumentation and Measurement Technology Conference IMTC/98, pp. 60–66 (1998)
55. Kochan, R., Lee, K., Kochan, V., Sachenko, A.: Development of a dynamically reprogrammable NCAP. In: Proceedings of the IEEE Instrumentation and Measurement Technology Conference, pp. 1188–1193 (2004)
56. Hiromoto, R.E., Sachenko, A., Kochan, V., Koval, V., Turchenko, V., Roshchupkin, O., Yatskiv, V., Kovalok, K.: Mobile ad hoc wireless network for pre- and post-emergency situations in nuclear power plant. In: Proceedings of the 2nd IEEE International Symposium on Wireless Systems Within the Conferences on Intelligent Data Acquisition and Advanced Computing Systems, pp. 92–96 (2014)
57. Leoshchenko, S., Oliinyk, A., Skrupsky, S., Subbotin, S., Zaiko, T.: Parallel method of neural network synthesis based on a modified genetic algorithm application. In: CEUR Workshop Proceedings, vol. 2386, pp. 11–23 (2019)
58. Romanenkov, Y. Pasichnyk, V., Veretennikova, N., Nazaruk, M., Leheza, A.: Information and technological support for the processes of prognostic modeling of regional labor markets. In: CEUR Workshop Proceedings, vol. 2386, pp. 24–34 (2019)
59. Berko, A., Alieksieiev, V., Lytvyn, V.: Knowledge-based Big Data cleanup method. In: CEUR Workshop Proceedings, vol. 2386, pp. 96–106 (2019)
60. Veretennikova, N., Lozytskyi, O., Vaskiv, R., Kunanets, O., Leheza, A., Lozynska, O., Kunanets, N.: Information and technology support for the training of visually impaired people. In: CEUR Workshop Proceedings, vol. 2386, pp. 307–320 (2019)
61. Baran, I., Kunanets, N., Matsiuk, H., Mytnyk, M., Shunevich, K., Skorenkyy, Y., Yaskilka, V.: Open online training courses for engineering purpose. In: CEUR Workshop Proceedings, vol. 2386, pp. 331–339 (2019)
62. Kunanets, N., Matsiuk, H.: Use of the smart city ontology for relevant information retrieval. In: CEUR Workshop Proceedings, vol. 2362, pp. 322–333 (2019)
63. Levchenko, O., Romanyshyn, N., Dosyn, D.: Method of automated identification of metaphoric meaning in Adjective+Noun word combinations (Based on the Ukrainian language). In: CEUR Workshop Proceedings, vol. 2386, pp. 370–380 (2019)
64. Bisikalo, O., Ivanov, Y., Sholota, V.: Modeling the phenomenological concepts for figurative processing of natural-language constructions. In: CEUR Workshop Proceedings, vol. 2362, pp. 1–11 (2019)
65. Shepelev, G., Khairova, N., Kochueva, Z.: Method “Mean – Risk” for comparing poly-interval objects in intelligent systems. In: CEUR Workshop Proceedings, vol. 2362, pp. 12–21 (2019)
66. Khairova, N., Kolesnyk, A., Mamyrbayev, O., Mukhsina, K.: The aligned Kazakh-Russian parallel corpus focused on the criminal theme. In: CEUR Workshop Proceedings, vol. 2362, pp. 116–125 (2019)
67. Yurynets, R., Yurynets, Z., Dosyn, D., Kis, Y.: Risk assessment technology of crediting with the use of logistic regression model. In: CEUR Workshop Proceedings, vol. 2362, pp. 153–162 (2019)
68. Vysotska, V., Burov, Y., Lytvyn, V., Oleshek, O.: Automated monitoring of changes in web resources. In: Lecture Notes in Computational Intelligence and Decision Making, vol. 1020, pp. 348–363 (2019)
69. Demchuk, A., Lytvyn, V., Vysotska, V., Dilai, M.: Methods and means of web content personalization for commercial information products distribution. In: Lecture Notes in Computational Intelligence and Decision Making, vol. 1020, pp. 332–347 (2019)
70. Lytvyn, V., Vysotska, V., Mykhailyshyn, V., Rzheuskyi, A., Semianchuk, S.: System development for video stream data analyzing. In: Lecture Notes in Computational Intelligence and Decision Making, vol. 1020, pp. 315–331 (2019)

71. Lytvynenko, V., Wojcik, W., Fefelov, A., Lurie, I., Savina, N., Voronenko, M., et al.: Hybrid methods of GMDH-neural networks synthesis and training for solving problems of time series forecasting. In: Lecture Notes in Computational Intelligence and Decision Making, vol. 1020, pp. 513–531 (2019)
72. Babichev, S., Durnyak, B., Pikh, I., Senkivskyy, V.: An evaluation of the objective clustering inductive technology effectiveness implemented using density-based and agglomerative hierarchical clustering algorithms. In: Lecture Notes in Computational Intelligence and Decision Making, vol. 1020, pp. 532–553 (2019)
73. Bidyuk, P., Gozhyj, A., Kalinina, I.: Probabilistic inference based on LS-method modifications in decision making problems. In: Lecture Notes in Computational Intelligence and Decision Making, vol. 1020, pp. 422–433 (2019)
74. Kulchytskyi, I.: Statistical analysis of the short stories by Roman Ivanychuk. In: CEUR Workshop Proceedings, vol. 2362, pp. 312–321 (2019)
75. Basyuk, T.: The popularization problem of websites and analysis of competitors. In: Advances in Intelligent Systems and Computing, vol. 689, pp. 54–65. Springer, Cham (2018)
76. Basyuk, T.: Innerlinking website pages and weight of links. In: International Scientific and Technical Conference on Computer Science and Information Technologies, pp. 12–15 (2017)
77. Basyuk, T.: Popularization of website and without anchor promotion. In: Computer Science and Information Technologies (CSIT-2016), pp. 193–195 (2016)

Software Engineering

Analysis of the Architecture of Distributed Systems for the Reduction of Loading High-Load Networks

Yurii Kryvenchuk[1(✉)], Pavlo Mykalov[1], Yurii Novytskyi[1], Maryana Zakharchuk[1], Yuriy Malynovskyy[1], and Michal Řepka[2]

[1] Lviv Polytechnic National University, Lviv 79013, Ukraine
yurkokryvenchuk@gmail.com, pmykalov@gmail.com, yuranov_lpnu@ukr.net, maryana.zk@gmail.com, inem.news@gmail.com

[2] Institute of Technology and Businesses in České Budějovice, Ceske Budejovice, Czech Republic

Abstract. In high-capacity networks, there is always a problem of delaying the receipt of packets between a client and a server. The load distribution should be made automatically based on the analysis of the distributed system state, since in the processing of Large Data it is necessary to analyze flows in a distributed, open dynamic system with a variable structure in real time. A distributed system for the task of reducing the load in high-capacity networks has been developed. An architectural scheme of "entering the remainder" is applied by introducing the new essence of the "last message". This allows us to write the following message in the field of correspondence in the field. Therefore, we will be able to receive the latest message of any correspondence, but now, after each message arrives, it will be necessary to record it in two places. The cascade time synchronization scheme is proposed. The accuracy of time is important in distributed systems and allows you to synchronize the process. To do this, the Marzullo algorithm was used. This made it possible to establish a logarithmic relationship between the efficiency indicator and the number of machines. In this regard, it is important not to use too many computers with an algorithm that cannot provide efficient computer management. Improved messaging scheme. This allows you to define the entities used in this approach and to find references to each other. Query distribution managers send requests not only to each machine in sequence, but in real time recognize the one that is least downloaded and select it to handle the most demanding queries. This allows you to polynomically reduce the computation time.

Keywords: High-load systems · Distributed system · Architecture · Thread queue · Multithreading · Marzullo algorithm

N. Shakhovska and M. O. Medykovskyy (Eds.): CCSIT 2019, AISC 1080, pp. 759–770, 2020.
https://doi.org/10.1007/978-3-030-33695-0_50

1 Introduction

Industry 4.0, like medicine, goes a long way in the big steps, both industries have a large amount of data that needs to be processed. Measuring equipment receives measurement information from sensors, puts it and sends it to the server. Servers exchange information received. Due to the huge amount of real-time measurements, monitoring of the controlled parameters of each production process in real time, there is a problem with the overload of networks, servers.

The urgency of work in this direction is conditioned by the fact that due to the overload of networks and servers, solutions that are accepted in real time, may be based on incomplete information. Information delay has a great influence on the quality of high-tech operations. Particularly relevant is the processing of information using the technology of the Great Data. In this case, it is necessary to analyze the flows in a distributed open dynamic system with a variable structure in real time.

2 State of Arts

First of all, it's worth considering what a high-capacity system is and what problems arise in such systems.

So, let's look at the very term of the web program. To begin with, it's worth explaining the structure of the client-server web projects in order to further clarify and clear all the explanations. The essence of this scheme is to send requests from the client to the server and send back server responses to the client. In the works [1–3], the types of client-server interactions were analyzed (Fig. 1). [2] provides an example of using the client-server architecture in applications of a smart home, but does not analyze network traffic. In [3], we consider ways of processing multidimensional data in a distributed environment, but the rate of exchange of messages is not analyzed.

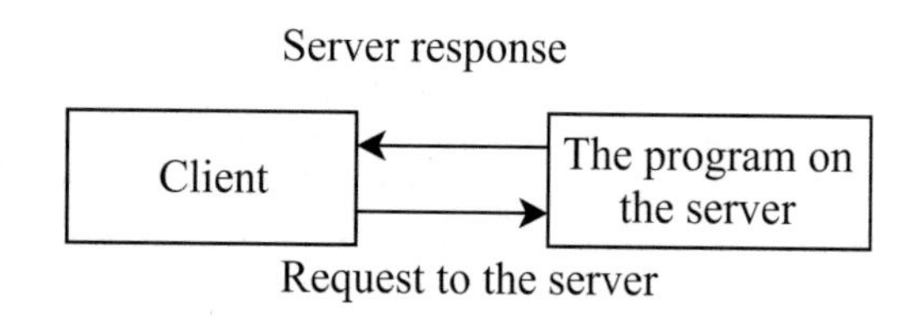

Fig. 1. Scheme of interaction between the client and the server

This scheme is the easiest example, since there may be a lot of customers, these clients send requests independently of each other. Consequently, the server must be able to handle several queries simultaneously, an example of such tasks is given in [4]. In this case, their processing should occur asynchronously. For this problem there is the concept of an asynchronous web server, analyzed in [5]. For each request, the asynchronous web server creates a new thread, the threads work asynchronously. This means working out their own requests independently of each other.

It is necessary to explain the term of the stream. Flow is a separate set of commands that are executed by kernels or logic processors. If the number of threads is larger than the number of processors, the commands of different threads will be executed one processor one by one (Fig. 2). This slowly slows down the process of processing data and makes it impossible to use it in on-line mode, and as shown in [6], this makes the processing of information in a highly loaded mode inappropriate.

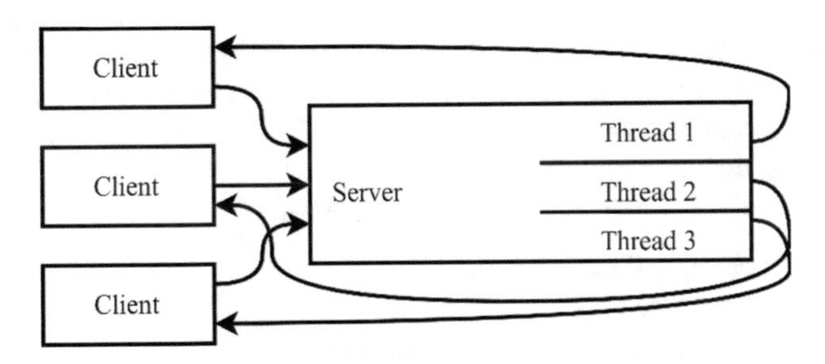

Fig. 2. Structure queue threads.

There is a better tool than multithreading - it's asynchronous execution. The main advantage of asynchronous servers is the efficient use of all computing resources. This approach is presented in [7, 8] for processing data in social networks and open distributed systems in general. When the thread executes the CRUD (creation, reading, update, deletion) of the response waiting for the external process, the processor where the thread is running is not used and simply waits for the end of the operation. However, in asynchronous execution, the processor will start serving other threads, so time will not be spent.

The problem with many clients is solved in [9], however, practice shows that a number of requests per unit time can reach such a speed that even the most powerful server cannot cope with this load (methods for determining the load are given in [10]). One example of such a problem is the Slashdot effect - a powerful influx of website attendance, basically not great after the information on this resource appeared in the newsletter of a well-known blog.

On the other hand, there are web services that can handle the entire data stream, for example, Stack Overflow processes 61 million requests per day, not including regular customers. For comparison, an ordinary static site may fall if about 150 visitors visit it at the same time. However, Stack Overflow supports 700 visitors per second, with a database and a lot of information logic on the server side. This result can be achieved by assigning tasks to several machines and their synchronous work with the database system, which is common to all servers or the successful choice of the architecture of the system and optimization of its work, proposed in the works [10–13]. Problems of processing the queue of flows are given in [14], but methods of optimization over time are not investigated.

Consequently, the problem of choosing the optimal architecture of distributed systems for reducing the load in high-capacity networks remained unsolved. The deployed load must be carried out automatically on the basis of the analysis of the state of the distributed system, since in the processing of Large data it is necessary to analyze flows in a distributed open dynamic system with a variable structure.

3 The Aims and Objectives of the Study

The purpose of the research is to solve the problems of high load of the system through the processing of distributed queries. In order to solve such problems, smart "request distribution managers" are offered. Managers send requests not only to each machine in sequence, but in real time they recognize the one that is least loaded and select it to handle the most demanding queries.

To achieve the goal, the following tasks were set:

- to develop a cascade time synchronization scheme;
- Improve messaging scheme
- Improve the requests of distribution managers.

4 Distributed Multiplayer Games with Time Synchronization

Since the advent of computers, the task of synchronizing time has become very important not only in the local network, it is necessary that time is the same for all systems on the planet. In particular, the accuracy of time is important in distributed systems, in which the importance of the order of calculation of tasks and data processing.

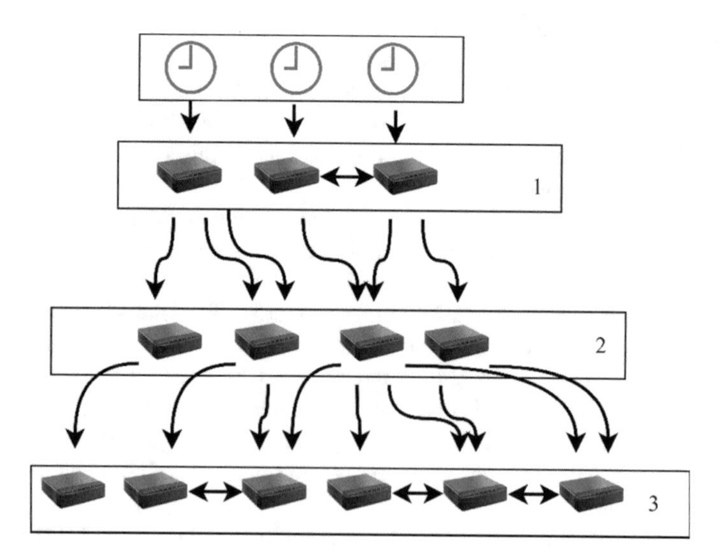

Fig. 3. Synchronization of cascading time at levels of the hierarchy 1–3

For example, when executing transactions, to ensure synchronization of the update and the integrity of the data, or all updates must be completed or none of them. This stores the transaction log. This log records information about all the requested changes (Fig. 3).

The upper part of the circuit has extremely accurate atomic clocks, after which the first layer of nodes appears, the second layer and so on. For synchronization with the general time, each node asks several other nodes in its parent layer and in its own layer (memorizes the excess element) and takes several interval pairs [t1, t2], one from each node of its neighbor. Thus, the Mansoullo algorithm [9] gives one way to select an interval - it simply selects an interval that is a subset of the largest number of return intervals (Fig. 4).

Now, in order to find the desired interval that appears in the largest number of returned intervals, it is necessary to consider each possible time interval or even the subinterval and calculate which is active for the maximum number of return intervals. Each interval has the form [start, end]. First, we represent each interval in the form [start, +1], [end, −1].

Now we have a list of tuples, the first element of which is the time. Sort these items in time, regardless of what is the second element. Thus we obtain [t (i), val (i)], [t (i + 1), val (i + 1)], … where val (i) = ±1. Initialize two variables count = 0 and max = 0. Now skip the sorted tuples and set count = count + val (i). Compare if count > max then, set max = count and set response = [t (i), t (i + 1)].

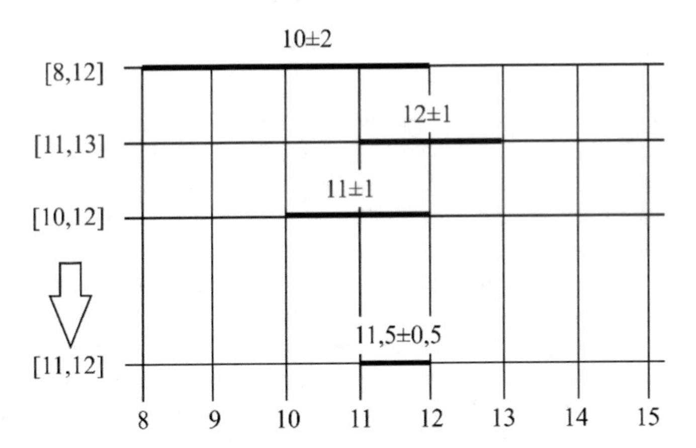

Fig. 4. Consideration of the time interval.

In general, synchronization of time is very important when creating web games designed to allow them to handle a large number of players. For example, we can take any massive multiplayer online role-playing game (MMO RPG), which can be played by thousands of people at a time. Obviously, the game differs from web pages through the mass of clients, since all graphical computations occur precisely on it at the expense of a graphics processor (GPU). Despite the fact that we use computer players for complex tasks to interact with the virtual world filled with other players, it is necessary to somehow reconcile the information about all objects of the game world among all the players of the server.

Processing of this data takes place on remote gaming servers. Because in these games, many players and distribute all requests for different machines, as shown in the figure, for such a difficult task is incredibly difficult, so the developers have decided to distribute players to gaming servers. Therefore, now in the whole gaming system there will be N virtual worlds, each of which will be supported by its own server. Each of these servers is able to cope with its load, especially if the number of users is limited (Fig. 5).

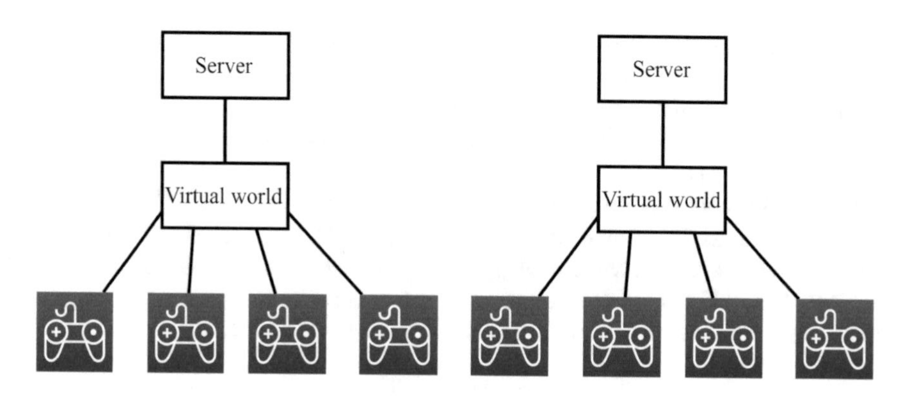

Fig. 5. MMORPG environment layout

Such systems are used in most modern multiplayer games. A striking example is well known to all, this is a Counter Strike in this game, players share a maximum of 10 people per server. For such games, the speed of data transmission is extremely important, as the share of the game is affected even by a fraction of a second. Obviously, the time the request is sent to the server and the response is affected by the distance between the client and the hosting machine. This time is called ping, a term used not only in games but also in multiplayer shooters, is a good example for demonstrating its principles. Ping depends not only on distance, but also on the load of your Internet channel, and sometimes on the speed of data exchange on the Internet (if it is very slow). In order to make this distance as small as possible, servers are located at different hosts around the world. For example, while playing Counter Strike in Ukraine, you can see the difference in delay depending on the location of the server. At the American server, the delay is quite large and makes from 200 to 300 mls, and the delay in the game on the servers of the CIS countries - no more than 30 ml.

Consequently, the problems with the delay are solved, but all servers have a common database containing information about the players, so each server must with it and enter the information in it. Since servers are located in different parts of the world, login times will only be relevant for the server that hosts it directly. Here you need an NTP protocol that allows you to log information, such as starting a game search by a certain player in the database, which can be anywhere with the right time.

The conclusion can be made one-time synchronization is incredibly important in creating a distributed system in which data transfer can include transactions and the need for logging.

5 Messaging Scheme

Highly loaded systems are an extremely complex section in the field of software development, often even experienced programmers are not sure about this area. Therefore, it is easy to guess that the development of a highly loaded system is a very difficult task.

First of all, you need:

- clearly understand the task that the system will perform;
- what data do we actually work with;
- how much data should be processed;
- how much resources need to be processed;
- whether the system is distributed;
- how should the load change due to the optimal use of machine resources.

For example, let's take a rather trivial task - to make a simple social network with minimal functionality. However, it soon becomes clear that solving this problem is not completely trivial.

First, you need to understand the information that is being processed. Obviously, the task of any social network is to provide communication between users. For example, you need to create a chat between two users. In addition, messages must be stored and displayed as "previous messages" during a new conversation (Fig. 6).

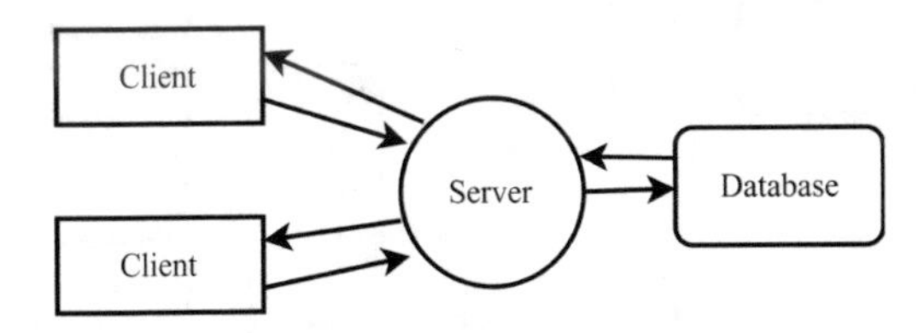

Fig. 6. Chats exchange scheme.

From the scheme, one can see that the message will be sent from one client to another as a server response. These messages will be displayed in the chat window. In addition, all messages will be written to a database in a special model, which will represent data throughout the system.

In order to understand the preservation of data, it is necessary to determine the essence used in this approach and how these entities depend on each other. The following entities are used: first user, second user, message and correspondence.

In Fig. 7 shows that each message is linked to the user as one to many, and compliance, in fact, is only a specific set of messages provided in strict order. The analysis of the interaction between correspondence and users showed: if the matching is the same for each pair of users, then we conclude that there is no additional functionality that does not depend on the main correspondence. Then this means that it is possible to get the matching name from the names of the two users.

Now you have to select containers in which the states of entities will be stored. The easiest message is a small text that can be written in bai data as text. In addition, the message has a receipt time that can be stored in variables of type short int and in the database as a date. As for correspondence, two user identifiers identify the message table, so the database will have the following form.

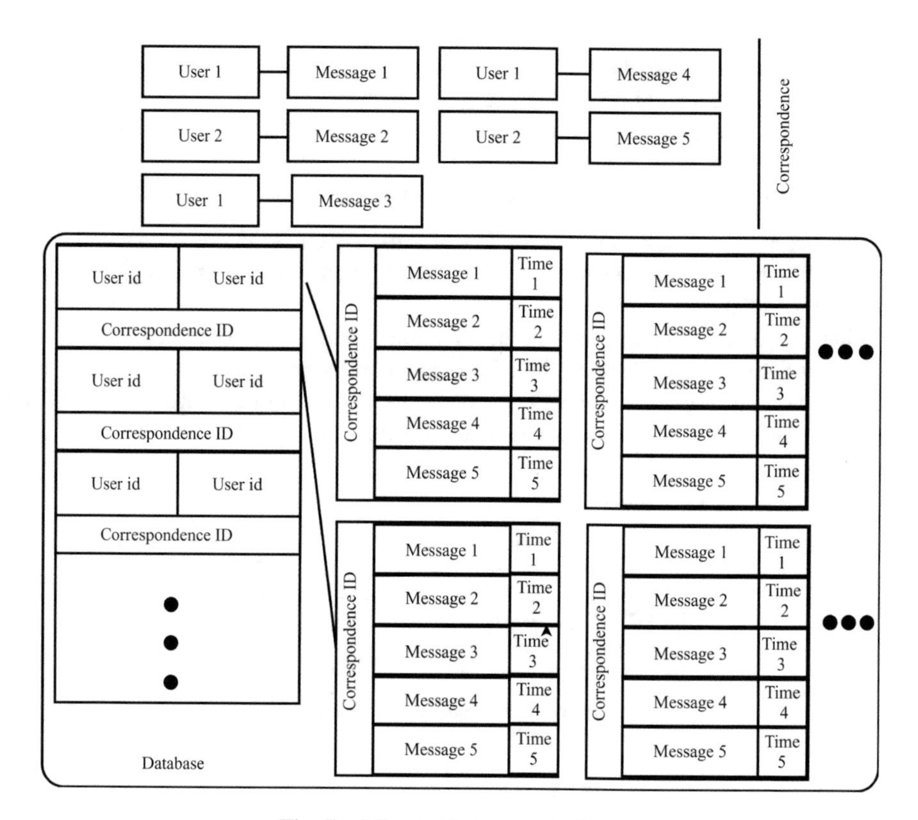

Fig. 7. Scheme of message exchange

The match ID, as already mentioned, can be obtained from a pair of user IDs. Accordingly, there is another table in which two user IDs and a match identifier are stored, through which access to spreadsheets with messages is obtained. It would seem that such a solution should work, but still not effective in accessing the last message. To this end, it is suggested to use an architectural sample of the "input remainder" by entering a new "last message", which is written in the field next to the field of correspondence. This way, it allows you to quickly receive the last message of any correspondence, but now, after each message arrives, it will be necessary to record it in two places.

6 Message Distribution Manager

High loads can be distributed on different computers, which makes the system able to handle much more information per unit time. The question now is how to distribute the load on machines. One of the most common methods is Round Robin, which is used by most cloud platforms, such as Microsoft Azure. This method is quite simple. In turns, a request is generated for each working machine, although the machines do not interact with each other. However, one machine runs more than one process. Because of this, various processes will be loaded, in addition, various requests for complexity are formed. For example, there are such requests that one machine will process in a few seconds, but there are those which will take a few minutes (in complex calculations).

To solve such problems there are intelligent "request distribution managers". These managers send requests not only to each machine in sequence but in real time recognize the one that is least loaded and select it to handle the most demanding queries (Fig. 8).

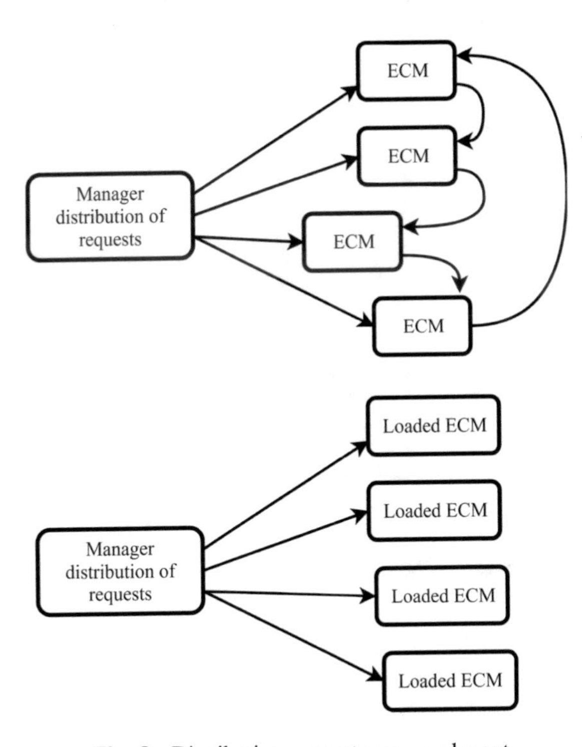

Fig. 8. Distribution request manager layout

Stack queues are used to determine the system load. If we take the ratio of the number of queries that are now in queue to the requests that the server processes for a certain period of time, we get a ratio that we can characterize the power of this server at the moment. This way of leveling the load much better than the previous, but this method is also not always effective, especially in mass computing and interaction with the database. The fact is that if the requests are small, but they will come in large numbers, and managers will accept all the resources of one machine for one-computational calculations, both methods lose their relevance. One such query loads one computer at a time, since others will not do anything that is incredibly inefficient use of computing resources. To do this, there are algorithms that allow you to handle the same query on different machines before dividing the query into subtasks, each of which executes separately. Where exactly the requests will be executed, identifies an additional service that actually allocates the load.

7 Results of Load Research in Distributed Systems

From the experiment (Fig. 9), we see that after the distribution, the system's performance increases and requests are processed better than a system operating on a single system. But after adding more machines, the system starts to spend more time on data synchronization and load sharing.

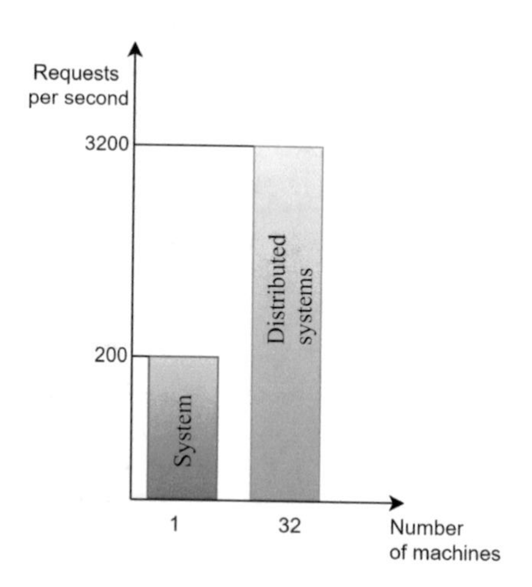

Fig. 9. Dependence of the efficiency indicator on the number of machines.

Dependence of the indicator on the number of machines resembles logarithmic dependence. Therefore, it is important not to use too many computers with an algorithm that cannot provide efficient computer management.

8 Conclusions

The scheme of time synchronization is proposed. The accuracy of time is important in distributed systems and allows you to synchronize the process. An algorithm was used for this purpose. This made it possible to establish a logarithmic relationship between the index of effectiveness and the number of machines.

Improved messaging scheme. This allows you to define the entities used in this approach and to find references to each other. This way, it allows you to quickly receive the last message of any correspondence, but now, after each message arrives, it will be necessary to record it in two places.

Query distribution managers improved. Managers send requests not only to each machine in sequence, but in real time they recognize the one that is least loaded and select it to handle the most demanding queries. This allows you to polynomically reduce the computation time.

References

1. Brin, S., Page, L.: The anatomy of a large-scale hypertextual web search engine. Comput. Netw. ISDN Syst. 107–117 (1998). https://doi.org/10.1016/s0169-7552(98)00110-x
2. Kryvenchuk, Y., Shakhovska, N., Shvorob, I., Montenegro, S., Nechepurenko, M.: The smart house based system for the collection and analysis of medical data. In: CEUR, vol. 2255, pp. 215–228 (2018)
3. Melnykova, N., Marikutsa, U., Kryvenchuk U.: The new approaches of heterogeneous data consolidation. In: XIIIth International Conference on Scientific and Technical Conference Computer Sciences and Information Technologies (CSIT), pp. 408–411 (2018). https://doi.org/10.1109/stc-csit.2018.8526677
4. Boyko, N.: A look trough methods of intellectual data analysis and their applying in informational systems. In: XIth International Scientific and Technical Conference on Computer Sciences and Information Technologies (CSIT), pp. 183–185 (2016). https://doi.org/10.1109/stc-csit.2016.7589901
5. Intanagonwiwat, C., Govindan, R., Estrin, D., Heidemann, J., Silva, F.: Directed diffusion for wireless sensor networking. Trans. Netw. 2–16 (2003). https://doi.org/10.1109/tnet.2002.808417
6. Levis, P., Lee, N., Welsh, M., Culler, D.: TOSSIM: accurate and scalable simulation of entire TinyOS applications. In: Proceedings of the 1st International Conference on Embedded Networked Sensor Systems, pp. 126–137 (2003). https://doi.org/10.1145/958491.958506
7. Goh, K.I., Oh, E., Kahng, B., Kim, D.: Betweenness centrality correlation in social networks. Phys. Rev. E **67**(1), 017101 (2003)
8. Vito, L., Massimo, M.: A measure of centrality based on the network efficiency. New J. Phys. **9**, 1–29 (2007). https://doi.org/10.1088/1367-2630/9/6/188
9. Jianwei, W., Tianzhu, G.: A new measure of node importance in complex networks with tunable parameters. In: WiCOM, Beijing (2008). https://doi.org/10.1109/wicom.2008.1170
10. Zheng, C., Dong, J.: Sliding window calculating method of time synchronization based on information fusion. In: Tan, H. (ed.) Knowledge Discovery and Data Mining, vol. 135, pp. 687–691. Springer, Heidelberg (2012). https://doi.org/10.1007/978-3-642-27708-5_95

11. Olexa, R.: Implementing 802.11, 802.16, and 802.20 Wireless Networks: Planning, Troubleshooting, and Operations. Elsevier (2004)
12. Kryvenchuk, Y., Shakhovska, N., Melnykova, N., Holoshchuk, R.: Smart integrated robotics system for SMEs controlled by Internet of Things based on dynamic manufacturing processes. In: Conference on Computer Science and Information Technologies, pp. 535–549 (2018). https://doi.org/10.1007/978-3-030-01069-0_38
13. Peleshko, D., Ivanov, Y., Sharov, B., Izonin, I., Borzov, Y.: Design and implementation of visitors queue density analysis and registration method for retail videosurveillance purposes. In: First International Conference on Data Stream Mining and Processing, pp. 159–162 (2016). https://doi.org/10.1109/dsmp.2016.7583531
14. Melnykova, N., Melnykov, V., Vasilevskis, E.: The personalized approach to the processing and analysis of patients' medical data. In: IDDM, pp. 103–112 (2018)
15. Khavalko, V., Khudyy, A.: Application of neural network technologies for information protection in real time. In: First International Conference on System Analysis and Intelligent Computing, pp. 173–177 (2018)
16. Khavalko, V., Tsmots, I.: Image classification and recognition on the base of autoassociative neural network usage. In: 2nd Ukraine Conference on Electrical and Computer Engineering, pp. 1118–1121 (2019)

An Approach to Multiple Security System Development Using Database Schemas

Pavlo Zhezhnych[1(✉)] and Teodor Burak[2(✉)]

[1] Department of Social Communication and Information Activities, Lviv Polytechnic National University, Lviv, Ukraine
pavlo.i.zhezhnych@lpnu.ua

[2] Information Center, Lviv Polytechnic National University, Lviv, Ukraine
teodor.o.burak@lpnu.ua

Abstract. Information security is a key issue in an Enterprise Information System (EIS) development. It is important characteristic of the entire EIS and all EIS's information subsystems. Information security effectiveness affects adequacy of enterprise decision making at all management levels and especially depends on database security. So, it is a good practice to develop a unified relational database for several subsystems of EIS. This paper discusses an approach to multiple security system development for several subsystems using one or several schemas of the unified database. The key peculiarity of the approach is an ability to evaluate "similarity" of database security systems. The "similar" database security systems should be united into the common security system, otherwise they must be separated. The "similarity" is calculated as weighted correlation between sets of user roles permissions defined as functional on sets of database tables, data operations and user roles. The proposed approach was tested on a production database of University Management Information System that allowed optimizing of its data access control through several database schemas. Also, the approach allows automation of determining the feasibility of creating new database schemas in the further development of the EIS.

Keywords: Database · Database schema · Information security · Access control · SQL · Information system

1 Introduction

Enterprise Information System (EIS) is an information system (IS) that supports the automation of business processes at an enterprise (or at a corporation) by integrating several information subsystems. Regularly, EIS provides processing of big volume of data to support a large and/or complex enterprise or organization. That is, EIS should be used in all parts and at all levels of the enterprise [12].

Obviously, effective functioning of EIS is based on a usage of one database (DB) or more. Almost certainly the DB is a relational database and is managed by a DB management system (DBMS) like Oracle Server, MS SQLServer, IBM DB/2 etc. Each of the DBMSs provides rich means and several approaches to security system development. Applying of the approaches depends on peculiarities of IS and information

N. Shakhovska and M. O. Medykovskyy (Eds.): CCSIT 2019, AISC 1080, pp. 771–780, 2020.
https://doi.org/10.1007/978-3-030-33695-0_51

security still remains a key issue in EIS development. The EIS security system must be flexible and should provide high performance in using and maintenance [15].

EIS supports the enterprise decision making which adequacy depends on data confidentiality, integrity and availability [18] as well as on data authenticity, reliability and timeliness consolidated from different information sources [14]. Respectively, development of a unified DB for EIS is a good practice to avoid different problems of EIS complexity including data integration. And subsystems of EIS should rather process data within the unified DB than within separate DBs (Fig. 1).

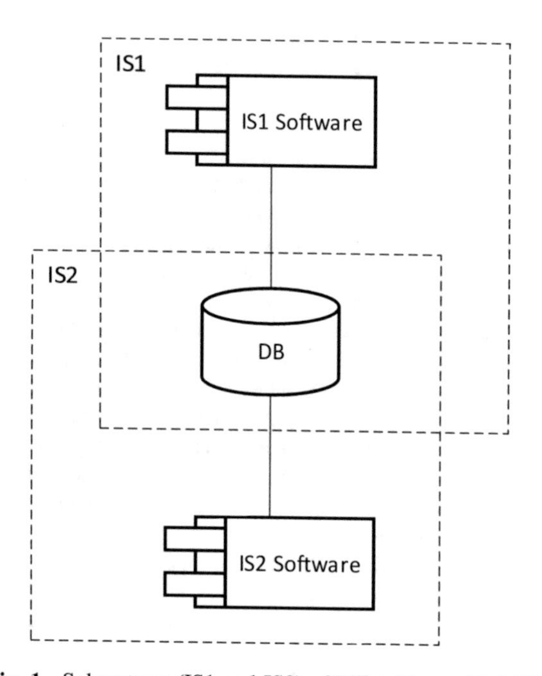

Fig. 1. Subsystems (IS1 and IS2) of EIS with a unified DB.

Therefore, the EIS data security system should be developed jointly for different subsystems that operate the unified DB.

One of the approaches to implementing the DB security system is to use a concept of "schema". In relational DB the schema defines tables (with attributes, relationships, indexes, integrity constraints, etc.), views (as representation of tables), program units (packages, procedures, functions, triggers), and other elements [1]. In this case, elements of one schema can refer to elements of another one.

Consequently, the mechanism of DB schemas allows development of the DB security system for several EIS subsystems providing access to the unified DB for all subsystems at one time. In this case, it is advisable to highlight two most appropriate ways of schemas using:

- one DB schema is used for several subsystems of EIS;
- each subsystem of EIS uses its own DB schema.

Unfortunately, the expediency of using of one or more schemas for EIS subsystems is currently determined at the level of the art of DB designing.

2 Literature Review

The main purpose of a DB security system is to control data access by DB users. Even though pure Role Based Access Control (RBAC) models can lead to "role-explosion" in the form of huge number of separate roles, modifications of RBAC models are considered as models with the best ratio of "flexibility/complexity" [13]. Most RBAC modification are directed to fine-grained access control with minimal quantity of engaged roles [3]. This is achieved by assigning a set of role's permissions to certain data values. In [9] the RBAC model is extended to support attribute-based access control (ABAC). [5] introduced support of user's organizational attributes as well as [4] introduced support of hierarchical management structures attributes. In [17] the RBAC model is modified for temporal access control with time attributes. For more accurate and complex access control based on peculiarities of SQL-queries several approaches are proposed like Disclosure Control System [2], Fine-Grained Access Control model (called Truman) and Non-Truman models [7]. So, the studies listed above demonstrate broad possibilities for adapting RBAC models to EIS security system development. However, they do not touch upon an implementation of the EIS security system with DB schemas.

The DB schema is a powerful mean for developing a DB access control system. In [10] it is a mean of tree-structured data representation. In [11] schemas are arranged as XML documents that control access to data based on the requesting user's identification. Also, DB schema is considered as layer of DB logical representation. In [6] data protection tools are implemented in DB schema with universal data model. [17] proposes to control data access through a set of views implemented at the DB logical level. [16] describes ontology-based approach for establishing a security management framework with schemas to manipulate classes modelling objects represented therein. [8] proposes to configure relational DB schema for the object-oriented management service to permit manipulation of security classes that provide security features pertaining to use of relational DB system. Consequently, these studies consider DB schemas as a means of implementation of access control systems. Unfortunately, they do not offer approaches to development a single access control system for several ISs that operate unified DB.

3 An Approach to Determining the Dependencies Between Security Systems of EIS

This paper proposes a formalized approach to solving the problem of multiple security system development. The proposed approach is based on the ability of relational DBMSs to create multiple schemas for the unified DB.

Therefore, the problem of development of EIS security system with the use of DB schemas can be reduced to adoption of an "optimal" decision on the number of schemas of the unified DB necessary for functioning of several EIS subsystems.

In this paper the case with the only two EIS subsystems is considered. However, the proposed approach can be applied to any number of EIS subsystems.

3.1 "Similarity" of DB Security Systems

To determine the "optimal" decision on the number of schemas of the unified DB we introduce the concept of "similarity" of security systems. That is, two security systems of EIS subsystems are considered as "similar" if these subsystems use similar permissions of users' roles.

On Figs. 2 and 3 two variants of DB access control containing "similar" and "dissimilar" security systems are represented.

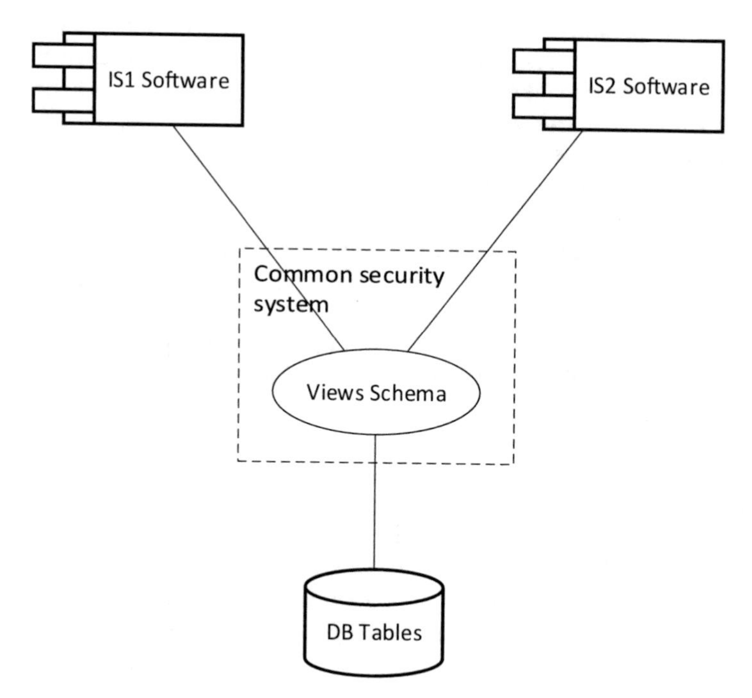

Fig. 2. Subsystems (IS1 and IS2) of EIS with "similar" security DB systems united into the common security system.

So, if security systems of two EIS subsystems are "similar" then they should be united into the common security system (Fig. 2). Otherwise, the DB security systems of the EIS subsystems must be separated (Fig. 3).

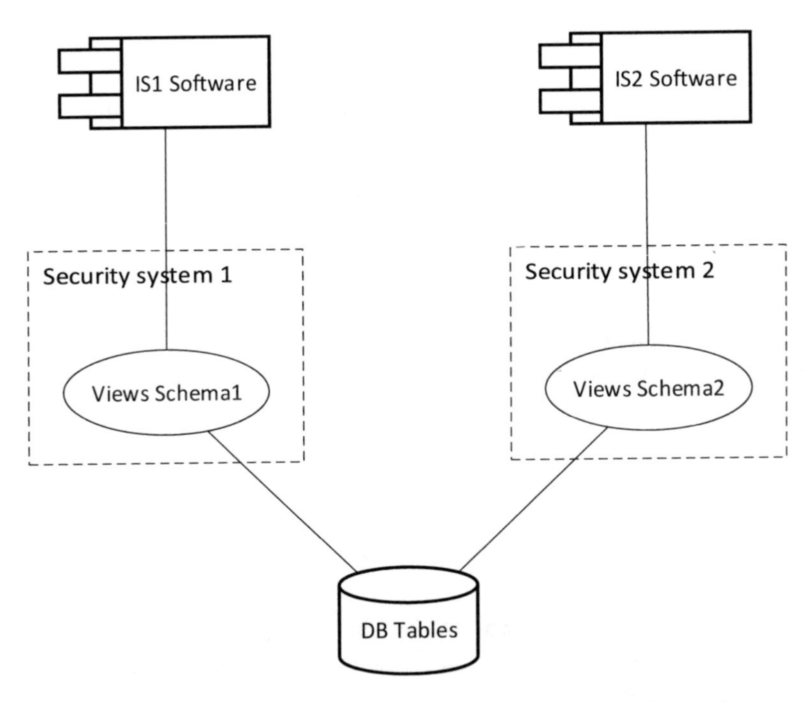

Fig. 3. Subsystems (IS1 and IS2) of EIS with "dissimilar" separate DB security systems.

3.2 Method of Determination of DB Security Systems "Similarity"

The developed method of determination of DB security systems "similarity" is applicable to data access control implementation with a set of views [17] that should be located in one or two DB schemas.

The basis of the method is the calculation of the weighted correlation between sets of user roles permissions of two EIS subsystems. These permissions relate to abilities to perform CRUD operations which can be granted at the DBMS level and can be set at the logical level of DB [17].

Let's introduce the following notation:

- $Table = \{Table_i\}_{i=1}^{N^{(Table)}}$ a set of DB tables that are simultaneously used by the two EIS subsystems, $N^{(Table)}$ - number of the tables;
- $Operation = \{Operation_j\}_{j=1}^{N^{(Operation)}}$ - a set of data operations, $N^{(Operation)}$ - number of the operations;
- $Role^{(1)} = \left\{Role_k^{(1)}\right\}_{k=1}^{N^{(Role1)}}$, $Role^{(2)} = \left\{Role_l^{(2)}\right\}_{l=1}^{N^{(Role2)}}$ - sets of user roles of the first and second EIS subsystems, $N^{(Role1)}$, $N^{(Role2)}$ - numbers of roles of the subsystems.

Each operation $Operation_j$ belongs to one of CRUD-operations or its restricted set [19]. The following is true:

- the CUD operations are typical for relational DBs, that is, they correspond to CREATE, UPDATE, DELETE queries on DB tables;
- the R-operation corresponds to a SELECT query that takes into account user's role permissions on the according DB table.

To identify user's role permissions for applying operations on DB tables, we introduce the following functional:

$$P : \langle Table, Operation, Role \rangle \rightarrow \{0; 1\} \tag{1}$$

The permission functional returns the value 1 if the operation on the table for the corresponding role is allowed. Accordingly, this functional returns the value 0 if the operation is not allowed. These values are denoted as follows:

$$P_{ijk}^{(1)} = P\left(Table_i, Operation_j, Role_k^{(1)}\right) \tag{2}$$

$$P_{ijl}^{(2)} = P\left(Table_i, Operation_j, Role_l^{(2)}\right) \tag{3}$$

For each pair of table-operation $\langle Table_i, Operation_j \rangle$, let's introduce a weight w_{ij}, which means importance of applying the operation $Operation_j$ on the table $Table_i$. And the following condition is fulfilled:

$$\sum_{i=1}^{N^{(Table)}} \sum_{j=1}^{N^{(Operation)}} w_{ij} = 1 \tag{4}$$

For each role of the first and the second EIS subsystem, let's define the mean values of the permission functionals:

$$\bar{P}_k^{(1)} = \sum_{i=1}^{N^{(Table)}} \sum_{j=1}^{N^{(Operation)}} w_{ij} P_{ijk}^{(1)} \tag{5}$$

$$\bar{P}_l^{(2)} = \sum_{i=1}^{N^{(Table)}} \sum_{j=1}^{N^{(Operation)}} w_{ij} P_{ijl}^{(2)} \tag{6}$$

Then weighted correlation of values of the permission functionals (2) and (3) with respect to the k-role of the first EIS subsystem and the l-role of the second subsystem is determined as follows:

$$r_{kl} = \frac{\sum_{i=1}^{N^{(Table)}} \sum_{j=1}^{N^{(Operation)}} w_{ij} \left(P_{ijk}^{(1)} - \bar{P}_k^{(1)}\right)\left(P_{ijl}^{(2)} - \bar{P}_l^{(2)}\right)}{\sqrt{\sum_{i=1}^{N^{(Table)}} \sum_{j=1}^{N^{(Operation)}} w_{ij} \left(P_{ijk}^{(1)} - \bar{P}_k^{(1)}\right)^2 \sum_{i=1}^{N^{(Table)}} \sum_{j=1}^{N^{(Operation)}} w_{ij} \left(P_{ijl}^{(2)} - \bar{P}_l^{(2)}\right)^2}} \tag{7}$$

If the weighted correlation value (7) is sufficiently close to 1 (that is $r_{kl} \geq \alpha$, where $\alpha \in (0; 1)$ is an admissibility indicator of using the common security system) then the corresponding roles of each EIS subsystem are intended to perform similar functions, that is, there is a dependency between the roles.

4 Practical Applying of the Approach

The practical applying of the proposed approach depends on accurate definition of the permission functional (1) and the corresponding sets of tables and operations. For a better distribution of the correlation values (7) it is expedient to remove those items from sets T and O for which the permission functional value is constant for all roles of R.

The proposed approach was tested on a production DB of University Management Information System (UMIS) that contains 253 core tables and covers most business processes of the university. The carried-out calculations concerned of a staff management IS of UMIS and its subsystems.

To estimate DB security systems "similarity", 39 DB tables that are directly used to process staff data were analyzed. There are 4 operations on the tables belonging to the limited set [19] (no DELETE operation is applicable): Create, Read all (SELECT all records), Read own (SELECT records visible to a user), restricted Update (full UPDATE of new records and UPDATE of processing stage attributes for submitted records).

Table 1 shows values of the equally weighed correlation (7) for 5 user roles of 4 subsystems of staff management IS. The role names are given in the format "Rx.x" where the first digit represents the subsystem number and the second one denotes the role number of the subsystem.

Table 1 Equally weighed correlation between sets of permission functionals for UMIS subsystems' roles

Correlation	R1.1	R2.1	R3.1	R3.2	R.4.1
R1.1	1,000	0,356	0,435	0,408	0,151
R2.1		1,000	0,818	0,873	0,425
R3.1			1,000	0,937	0,348
R3.2				1,000	0,371
R.4.1					1,000

From the correlation values in Table 1, it follows that the security systems of the subsystems #2 (the role R2.1) and #3 (the roles R3.1 and R3.2) are sufficiently "similar" (these values are highlighted in gray). Therefore, for these subsystems, it is expedient to use one DB schema to control data access. It is also obvious that subsystems #1 and #4 should have separate DB schemas with separate security systems.

Figure 4 shows how to build security systems for UMIS subsystems #1-4 according to the weighted correlation calculations between sets of permission functionals in Table 1.

Note that the approach to implementing the security systems, presented in Fig. 4, is not static, and may change over time. For example, a common security system for subsystems #2 and #3 can be divided into two security systems, if their roles become insufficiently "similar". Also, the security system of subsystems #1 or #4 can be combined with the security system for subsystems #2, 3, if the corresponding roles become sufficiently "similar".

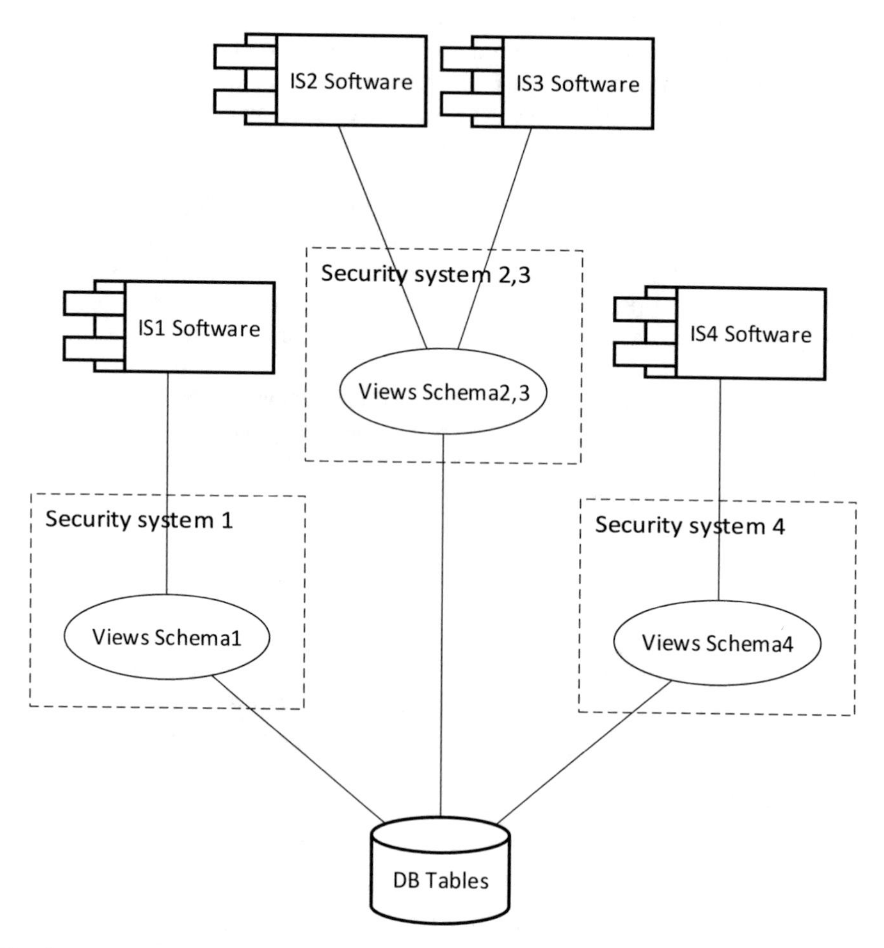

Fig. 4. DB security systems for UMIS subsystems #1–4 according to the weighted correlation calculations between sets of permission functionals.

5 Conclusions

This paper describes the approach to multiple security system development for several EIS subsystems using one or several schemas of the unified DB. The key peculiarity of the approach is an ability to evaluate "similarity" of DB security systems. If security systems of two EIS subsystems are "similar" then they should be united into the common security system. Otherwise, the DB security systems of the EIS subsystems must be separated.

Determination of DB security systems "similarity" is performed by calculation of weighted correlation between sets of user roles permissions of two EIS subsystems. The user roles permissions are considered as the permission functional on sets of DB tables, data operations and user roles. The permission functional denotes an ability to perform some CRUD operation or its restricted set on the DB table by the user's role granted at the DBMS level. If the weighted correlation value between sets of user roles permissions of two EIS subsystems is sufficiently close to 1 then the corresponding roles of each EIS subsystem are intended to perform similar functions. And their security systems should be combined into the common security system.

The proposed approach was tested on a production EIS – UMIS, its subsystems and corresponding schemas of unified DB. The carried-out calculations concerned of subsystems of the staff management IS of UMIS. The calculated values of equally weighed correlation between sets of permission functionals for 5 user roles of 4 subsystems confirmed necessity to create 3 DB schemas that provide separate data access control (the first DB schema for the subsystem #1, the second DB schema for the subsystems #2 and #3, and the third DB schema for the subsystem #4).

Also, this approach allows automation of determining the feasibility of creating new DB schemas as well as combining or dividing existing DB schemas in the further evolution and development of EIS subsystems.

References

1. Ben-Natan, R.: Implementing Database Security and Auditing: A Guide for DBA's, Information Security Administrators and Auditors. Elsevier Digital Press, eBook, Burlington (2009)
2. Bender, G., Kot, L., Gehrke, J.: Explainable security for relational databases. In: Proceedings of the 2014 ACM SIGMOD International Conference on Management of Data, Snowbird, Utah, USA (2014). https://doi.org/10.1145/2588555.2593663
3. Bertino, E., Ghinita, G., Kamra, A.: Access control for databases: concepts and systems. Found. Trends® Databases **3**(1–2), 1–148 (2011)
4. Blazhko, A.A., Antoshchuk, S.G., Saoud, E.: Automated design method of hierarchical access control in database. In: Proceedings of 5th IEEE International Workshop on Intelligent Data Acquisition and Advanced Computing Systems: Technology and Applications, 21–23 September, Rende (Cosenza), Italy, pp. 361–363 (2009)
5. Brodersen, K., Rothwein, T.M., Malden, M.S., Chen, M.J., Annadata, A.: Database access method and system for user role defined access. United States Patent, No.US6732100B1, 4 May 2004. https://patents.google.com/patent/US6732100B1/en

6. Grachev, V.M., Esin, V.I., Polukhina, N.G., Rassomakhin, S.G.: Data security mechanisms implemented in the database with universal model. Bull. Lebedev. Phys. Inst. **41**(5), 123–126 (2014)
7. Guarnieri, M., Basin, D.: Optimal security-aware query processing. Proc. VLDB Endow. **7** (12), 1307–1318 (2014). https://doi.org/10.14778/2732977.2733003
8. Kagalwala, R.A., Thompson, J.P.: Database schema for structured query language (SQL) server, United States Patent, No.USOO7653652B2, 26 January 2010. https://patents.google.com/patent/US7653652B2/en
9. Kuhn, D.R., Coyne, E., Timothy, R.W.: Adding attributes to role-based access control. IEEE Comput. **43**(6), 79–81 (2010)
10. Kuji, K.: Database access system and database access method. United States Patent, No. US007778955B2, 17 August 2010. https://patents.google.com/patent/US7778955B2/en
11. Lucovsky, M.H., Pierce, S.D., White, S.D., Movva, R., Kalki, J., Auerbach, D.B., Ford, P.S., Jacobs, J.C., Steckler, P.A., Hsueh, W.C., Keil, K.D., Gopal, B., Kannan, S., Yi-Wen Guu, George, S.J., Hoffman, W.R., Smoot, P.M., Fang, L., Taylor, M.B., Wu, W.C., Leach, P.J., Ward, R.B., Yuan, Y.-Q.: Schema-based services for identity-based data access. United States Patent, No.USOO7302634B2, 27 November 2007. [https://patents.google.com/patent/US7302634B2/en
12. Olson, D.L., Subodh, K.: Enterprise information systems: contemporary trends and issues. World Scientific, 579 (2009)
13. Sandhu, R., Bertino, E.: Database security-concepts, approaches, and challenges. IEEE Trans. Dependable Secure Comput. **2**, 2–19 (2005)
14. Shakhovska, N.: Consolidated processing for differential information products. In: Proceedings of the VIIth International Conference “Perspective Technologies and Methods in MEMS Design”, Polyana, Ukraine, pp. 176–177 (2011)
15. Shastri, A.A., Chatur, P.N.: Efficient and effective security model for database specially designed to avoid internal threats. In: Proceedings of the International Conference Smart Technologies and Management for Computing, Communication, Controls, Energy and Materials (ICSTM). IEEE (2015)
16. Tsoumas, B., Dritsas, S., Gritzalis, D.: An ontology-based approach to information systems security management. In: Computer Network Security (MMM-ACNS 2005), LNCS, vol. 3685, pp. 151–164. Springer, Berlin (2005)
17. Zhezhnych, P., Burak, T., Chyrka, O.: On the temporal access control implementation at the logical level of relational databases. In: Proceedings of the 11th International Scientific and Technical Conference on Computer Sciences and Information Technologies (CSIT), pp. 84–87 (2016)
18. Zhezhnych, P., Tarasov, D.: Methods of data processing restriction in ERP systems. In: Proceedings of the IEEE 13th International Scientific and Technical Conference on Computer Sciences and Information Technologies (CSIT), Lviv, Ukraine, vol. 1, pp. 274–277 (2018)
19. Zhezhnych, P., Tarasov, D.: On restricted set of DML operations in an ERP System’s database. In: AISC Systems and Computing III, vol. 871, pp. 256–266. Springer, Cham (2018)

The Special Ways for Processing Personalized Data During Voting in Elections

Nataliia Melnykova[1(✉)], Mykola Buchyn[1], Solomia Albota[1], Solomia Fedushko[1], and Swietlana Kashuba[2]

[1] Lviv Polytechnic National University, Lviv, Ukraine
melnykovanatalia@gmail.com, solomie4ka@gmail.com, felomia@gmail.com, buchyn@ukr.net
[2] University of Economy, Bydgoszcz, Poland
swietlana.kashuba@byd.pl

Abstract. In the article, the authors analyze the existing information technologies of electronic voting in elections and determine their advantages and disadvantages. They offer ways of processing personal information about the voter. Voter personal data is formalized, which allowed estimating his position and predicting the results of voting at the expense of Bayesian optimization.

Keywords: Electronic voting · Elections · Information technologies · The formalization of personalized data · Processing data

1 The Problems Personalized Data Processing in Spheres of Public Life

In the modern era of the information society, we observe the active influence and use of information technologies in all spheres of public life. The sphere of politics is not an exception, in which the use of information technologies led to the emergence of such phenomena as electronic democracy, electronic government, electronic elections, electronic voting, etc. This necessitates the intensification of efforts and the mobilization of political will in this direction. Electronic voting as a component of electronic democracy is not an exception. Its use is actively observed in various public projects, however, at the level of the official electoral process, the use of electronic voting is at the initial stage. The intentions for its implementation exist only in the form of proposals and initiatives of individual politicians or scholars.

At the same time, taking into consideration the lack of adequate democratic governance experience, as well as the events observed during the recent electoral practices in the world, it becomes apparent: the problem of electronic voting introducing in Ukraine requires integrated and balanced consideration, taking into account all advantages, disadvantages and risks. The foreign experience of using electronic voting is characterized by a variety of forms of its implementation, the availability of various electronic systems (models) that are used during the election process.

Individual approach for decision-making and processing personal data touches the problem of personalization that is one of the vectors of the development of many

N. Shakhovska and M. O. Medykovskyy (Eds.): CCSIT 2019, AISC 1080, pp. 781–791, 2020.
https://doi.org/10.1007/978-3-030-33695-0_52

problem areas of human activity. This is observed in business, education, medicine, consulting and marketing research and in the process of voting that [1–3]. Personalization is updated due to the importance of decision-making by the participants in the process during the elections, taking into account its individual peculiarities.

To evaluate the results and predict the decision of the respondent it is expedient to apply methods of artificial intelligence, namely: methods of cluster analysis search of associative rules, Bayesian approach [4].

2 Types and Main Systems of Electronic Voting

Electronic voting implies the concept that «…combines several different types of voting, covering both the process of electronic voting and the process of automatic votes counting using electronic devices and special software» [1].

Among the types of electronic voting, in our opinion, it is necessary to allocate two groups: remote and direct (stationary). Remote e-voting does not require a voter to attend a polling station during the exercise of the will. First of all, it includes online voting and voting using a mobile phone (SMS voting). Also, remote voting can be conducted with territorial restrictions (the voter must be present on a certain specified territory during elections) and without being bound to the place of stay (in this case, voters can vote from any point of the planet) [5].

Direct (stationary) electronic voting requires the voter's direct presence on the polling station during the act of will expression. Such voting can be either completely electronic (using electronic machines), or in a certain way, the combination of electronic and traditional expression of will (voting in ordinary paper ballots with subsequent reading and counting by electronic means).

Technologically, electronic voting is carried out using a certain type of information system, depending on the type of voting. The foreign scholar, A. Sopuyev identifies two types of electronic systems that are used to implement direct electronic voting [6]:

- Information systems of ballots optical scanning. Such systems are the result of a combination of traditional and electronic voting. They presuppose the availability of electronic complexes at the polling stations, which consist of a scanning device and a ballot boxes located below it [6].
- Information systems of direct voice recording. Such systems presuppose to have electronic complexes at polling stations that enable voters to vote with touch screens or mechanical buttons. This kind of information systems is characterized by the paper ballots absence, the results of voting are accumulated on special media carriers (usually removable). Frequently, information direct voice recording systems are connected to the electronic network and can immediately transfer data to the electronic center at the higher-level election commission for verification and counting [5, 7].

Through a qualitative assessment of the varieties and models of electronic voting, we would like to note that each of them has both its strengths and weaknesses in comparison with each other and in comparison with the traditional voting at the polling station using paper ballots.

3 Qualitative Assessment of Remote Electronic Voting

The following key benefits of remote e-voting compared to traditional ones can be highlighted [7]:

- Quickness of votes counting and determining the results of voting.
- Convenience for voters, the possibility to vote without leaving home.
- Accessibility for voters with disabilities, as well as for citizens who are staying or living abroad on Election Day.
- Reduction of financial expenses, spent time volumes and involved labor resources for the preparation and holding elections.
- Removing the «human factor» and, as a consequence, reducing inaccuracies in the votes counting that may be the result of errors or intentional facts.
- Reducing of the opportunities for pressure on voters during the expression of will, including reducing the administrative resource's influence.
- Potential increase in the level of citizens' involvement in the elections (first of all – young people).
- The increase in the level of citizens' awareness of the electoral process due to the concentration of information in one information system and the voters' ability to get complete information about the course of the election process.
- Environmental friendliness as a result of the lack of use of wood for the paper production and the printing of ballots.
- Possibility of holding elections in conditions of political and social instability, which is especially important for Ukraine.
- Elimination of such negative events of election practice as multiple voting, damages of voting papers, manipulation during the transfer of votes from polling stations to higher levels, etc.
- Possibility to use information systems and software for other public spheres and pro-vision of various services [6].

Giving the evaluation to the effectiveness and feasibility of the e-voting introduction, it becomes apparent that there are certain disadvantages and threats that are present or may potentially appear during the expression of the wills by means of information technologies. Among the most significant «weak» sides of the electronic voting, in our opinion, we can note the following:

Subjective Disadvantages and Threats. This concerns, in particular, the negative attitude of a large number of voters to the electronic voting system. This may be due both to the low level of computer literacy of citizens, and to the lack of confidence in e-voting, in its ability to ensure the democratic process of expression of will. It is intensified by the complexity of e-voting compared to traditional voting for voters and for members of election commissions, and necessitates a large-scale information and awareness campaign [6, 7].

Objective Disadvantages and Threats. To this category of factors that may adversely affect the prospects for the implementation of e-voting the poor level of information technology development in a particular country, the lack of proper provision of citizens with computers and access to the Internet should be included. Also, objective threats include potential hacker attacks, the presence of hostile countries that aim at interfere into the election process, etc. [7].

Disadvantages and Threats Related to the Operation of the Electronic System through which electronic voting is carried out. This includes the potential vulnerability of the electronic system to external influences and interference. Also, this category of risks includes the problem of continuity and autonomy of the electronic system functioning [6].

Legal Disadvantages and Threats. This category of threats includes, in particular, the lack of a proper regulatory framework that does not allow to regulate the mechanisms of electronic voting effectively. In addition, one of the most important components of any vote is to ensure democratic election principles. In this context, it should be noted that the potential electronic voting introduction may violate most of the important democratic electoral principles:

- free elections (the opportunities for putting pressure on voters are potentially increasing in the absence of control);
- general and equal elections (due to different access to information technologies and different levels of computer literacy, not all citizens will be able to exercise their voters rights, voters will be in an unequal position to each other);
- one-time voting (there is a probability of multiple expressions of one person through various channels for voting); personal voting (there is no guarantee that the act of will expression is carried out personally by the voter, and not on his behalf by another person);
- secret voting (there is no guarantee that the act of will expression will not be linked to a voter who may subsequently lead to persecution of citizens for their political convictions);
- a publicity of elections (the problem of the election process and specific election procedures control becomes extremely complex and can lead to numerous abuses), etc. [7].

4 The Formalization of the Voter Personalized Data

In order to take into account the peculiarities of electronic voting, it is necessary to have a personalized approach to the processing of his data. This will help to determine the peculiarities of voters and the electorate's tendencies.

We propose the formalization of personalized voter data, where PD is a set of data whose elements are the subset of time-independent data (A_{in}) and time-dependent data (A_t) of the voter, which characterize its general status.

$$PD = \{\{A_{in}\}, \{A_t\}\}. \tag{1}$$

The elements of a subset of time-dependent data A_t, for example: personal data that defines the place and conditions of the voter's voting (the region of residence, polling station), personal data that determine his belonging to a certain category of electorate (age, profession, occupation, political level awareness, etc.

$$A_t = \{a_{t1}, a_{t2}, \ldots, a_{tn}\}, \quad (2)$$

The elements of the set time-independent data A_{in}, include initials, category of birth region, the name of the birth place, date of birth, unique record in the Demographic Unified State Register and so on.

$$A_{in} = \{a_{in1}, a_{in2}, \ldots, a_{inm}\}, \quad (3)$$

where

$$A \cup S \rightarrow PD. \quad (4)$$

Given the voter's personal data, there is a need for formalizing the voter's state (VS) based on time-dependent (A_t) and time-independent data (A_{in}).

$$VS(t) : A_{in} \rightarrow A_t, \quad (5)$$

where

$$A_t = A_t(t). \quad (6)$$

For the process of personalizing decisions on choosing the result of voter, the estimated function F(PD), which is formed on the basis of the Bayesian theorem, is determining. The weight of the appearance of the next event corresponds to the largest value of the a posteriori probability of the appearance of the next state, taking into account the time-dependent input parameters, that is,

$$F(A_t) = P(VS|A_t). \quad (7)$$

Then, according to the Bayesian formula, we obtain the aposterior probability of occurrence of the event VS at the appearance of the event A_t:

$$P(VS|A_t) = (P(A_t|VS)P(VS))/(P(A_t)) \quad (8)$$

where:

- $P(A_t|VS)$ is the probability that the voter belongs to the some age group and lives in the some region of Ukraine, which is established on the basis of input parameters;
- $P(VS|A_t)$ - the probability that a resident of the some region of the state and belongs to the some age group, which is determined by the input parameters of the voter;
- $P(A_t)$ - the probability that the voter belongs to the some age group.

After receiving the input parameter A_t and calculating the probability of the Bayesian formula, we can write it to the place P(VS). The appearance of a new input parameter results in a new update (increase or decrease) of this probability. Each time the current value of this probability will be considered apriori value when applying the Bayesian formula.

There is a problem in management the logical conclusion, namely, we have the set of hypotheses Z in the knowledge base that is the results of decision-making:

$$Z_1, Z_2, \ldots\ldots\ldots, Z_i \tag{9}$$

and a finite set of indicators (input time-dependent parameters):

$$A_{t1}, A_{t2}, \ldots\ldots\ldots, A_{tn}, \tag{10}$$

where i is the total number of hypotheses, and n is the total number of time-dependent parameters.

The each hypothesis Z corresponds to its subset of the associated indexes (input parameters). According to the strategy of algorithms of ordered search in the space of states of the system of decision making, one cans consistently processing the entire list of possible parameters. In the final result, having calculated all the hypotheses that take into account the appearance of all possible input parameters, we get the final result, as the maximum probability is assigned to the value of the estimated function F(At), which is the determined factor for determining the next state of the search.

$$F(A_t) = max(P(VS|A_t). \tag{11}$$

The weight of the appearance of the next event corresponds to the largest value of the a posteriori probability of the appearance of the next state, taking into account the time-dependent input parameters, for the process of personalized search, the intermediate results of which are obtained as a result of determining the estimated function F (PD) using Bayesian theorem.

Figure 1 shows a fragment of the search solution tree using the evaluation function.

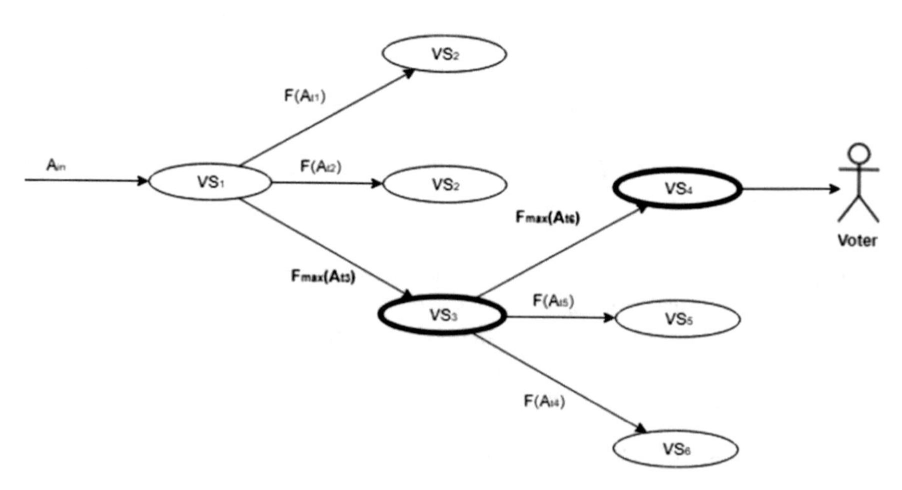

Fig. 1. The finding the target value of the voter decision support system output.

5 The Strategy of Electronic Voting Introduction in Ukraine

Based on the mentioned above, we are inclined to constructively treat the prospects for the introduction of electronic voting in Ukraine provided by the balanced and coherent approach, the development of a comprehensive strategy and a phased implementation of it. In our opinion, the electronic voting method in Ukraine should be based on a model of electronic expression of will, and the strategy for the introduction of electronic voting should include the following phased tasks [8, 9]:

- development of a comprehensive legal framework, on the basis of which the introduction of electronic voting will take place;
- development of an effective and secure electronic system through which electronic voting will take place; protection voters' data and their decisions during all process of electronic voting; distribution of information technologies throughout the territory of Ukraine;
- introduction a large-scale information and awareness campaign aimed at forming a high level of computer literacy confidence in the mechanism of electronic voting; introduction of electronic voting as one of the additional procedures to the traditional voting for the expression of citizens' will [10–12].

Given the trends in the implementation of electronic voting strategy should provide development features of a decision support system for forecasting trends voters will of citizens [13, 14]. For this we offer the architecture of the system, which will include a block of communication of the system users, a block of logical data processing using artificial intelligence methods and warehouse data block, Fig. 2.

6 Experiment's Results

In order to study the peculiarities of electoral voting, we conducted an experiment on the election of the student self-governance head of the Institute of Computer Science and Information Technologies. In the context of this event, students had to become voters at the time of the election campaign. To do this, the following steps were taken regarding the quality of voting.

The authentication of the participants was aimed at providing an individual login for them. This approach allows each voter to be distinguished in order to prevent fraudulent manipulations, to ensure the mass participation, and the quality of data processing.

The vote was held at a certain period of time, which allowed to outline the time limits for this event. As a result, the categories of voting participants were formed. During the time allowed, students checked for the presence of the registration for this event, confirmed their awareness of participation. The next step was to get the voting personal keys. Accordingly, each participant could go to his office and use the key to log into the system for voting and choose to confirm his choice.

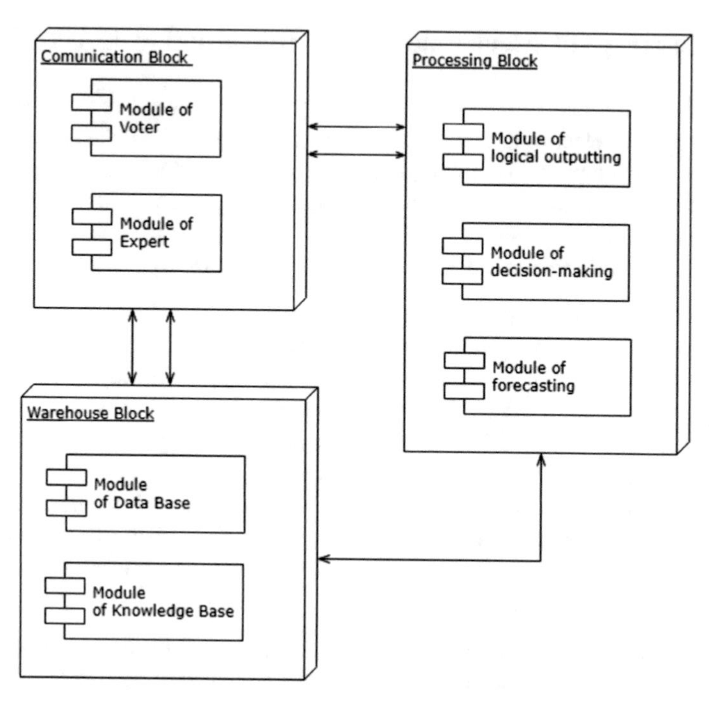

Fig. 2. The architecture of decision support system voters on forecasting trends expression of the will

The next step was to check the activity of the participants, confirmation of the fact of voting on the results of voting. Conduct of voting results and statistical analysis of voting results.

Consumer of experiment's results, we were offered the main stages of the process and identified dependencies between them.

The results of the voting processing of students, that numbering more than 1000. The voters had the opportunity to choose one of the four applicants and get acquainted with their program, each of which included strategic steps in the development of student self-government. By the results of the voting, we evaluated the needs of students accordingly.

The results of the voting processing of students, that numbering more than 1000. The voters had the opportunity to choose one of the four applicants and get acquainted with their program, each of which included strategic steps in the development of student self-government. According to the results of the voting, we assessed the needs of students by the applicants' programs (Figs. 3 and 4).

Taking into account that students who voted: 1 course - 400 people, 2 courses - 350 people, 3 courses - 231 people, 4 courses - 164 people.

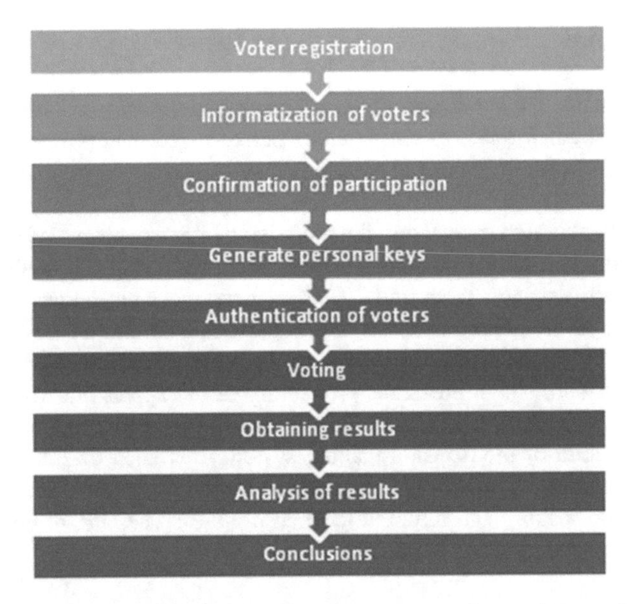

Fig. 3. The architecture of decision support system voters on forecasting trends expression of the will

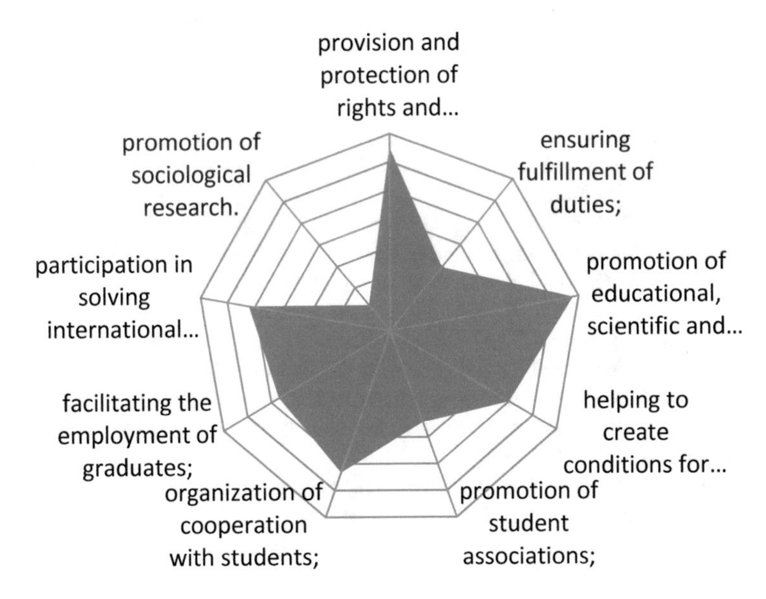

Fig. 4. The diagram of the will of students is presented, taking into account the results of individual voting.

7 Conclusions

Given the presence of a number of weaknesses and threats and current trends of organization of human relations and features personalized data processing worldwide introduction of electronic voting is evolutionarily necessary in view of an integrated and coherent strategy.

Based on the research carried out, the processing of personalized information on the prediction of the will of the citizens has been proposed using the method of orderly search, based on decision trees and Bayesian network. This allows us to find the summary value of the solutions based on the personalized data of respondents of the voting process by searching for the maximum evaluation function, which corresponds to the largest value of the a posteriori probability of the appearance of the next state, taking into account the time-dependent input parameters of respondents.

The architecture of the decision-making system for forecasting tendencies during the voting was proposed, which is characterized by the presence of logical outputting module, decision making and forecasting to ensure the processing of personal data of voters and the peculiarities of expression of will.

References

1. Melnykova, N.: The basic approaches to automation of management by enterprise finances. In: 2017 12th International Scientific and Technical Conference on Computer Sciences and Information Technologies (CSIT), Lviv, pp. 288–291 (2017). https://doi.org/10.1109/STC-CSIT.2017.8098788
2. Boyko, N., Sviridova, T., Shakhovska, N.: Use of machine learning in the forecast of clinical consequences of cancer diseases. In: 2018 7th Mediterranean Conference on Embedded Computing (MECO), Budva, pp. 1–6 (2018). https://doi.org/10.1109/MECO.2018.8405985
3. Shakhovska, N.: The method of big data processing. In: 2017 12th International Scientific and Technical Conference on Computer Sciences and Information Technologies (CSIT), Lviv, pp. 122–126 (2017). https://doi.org/10.1109/STC-CSIT.2017.8098751
4. Shakhovska, N., Kaminskyy, R., Zasoba, E., Tsiutsiura, M.: Association rules mining in big data. Int. J. Comput. **17**(1), 25–32 (2018). EID: 2-s2.0-85045215369
5. Yaryhin, G.O.: Internet voting and prospects for its implementation for Russian citizens abroad. In: ANI: Economics and Management, vol. 5, no. 3(16), pp. 261–268 (2016)
6. Sopuyev, A.A.: Automation of electoral processes in the Kyrgyz Republic. Science and new technologies, no. 2, pp. 197–205 (2014)
7. Peskova, O.Y., Fateeva, S.V.: Risks and e-voting issues , in Informative counteraction to the threats of terrorism, no. 23, pp. 152–164 (2014)
8. Melnykova, N., Marikutsa, U., Kryvenchuk, U.: The new approaches of heterogeneous data consolidation. In: 2018 IEEE 13th International Scientific and Technical Conference on Computer Sciences and Information Technologies (CSIT), Lviv, pp. 408–411 (2018). https://doi.org/10.1109/STC-CSIT.2018.8526677
9. Mulesa, P., Perova, I.: Fuzzy spacial extrapolation method using manhattan metrics for tasks of medical data mining. In: Computer Science and Information Technologies CSIT 2015, Lviv, Ukraine, pp. 104–106 (2015). https://doi.org/10.1109/STC-CSIT.2015.7325443

10. Perova, I., Pliss, I., Churyumov, G., Eze, F.M., Mahmoud, S.M.K.: Neo-fuzzy approach for medical diagnostics tasks in online-mode. In: 2016 IEEE First International Conference on Data Stream Mining & Processing (DSMP), Lviv, Ukraine, pp. 34–38 (2016). https://doi.org/10.1109/DSMP.2016.7583502
11. Kryvenchuk, Y., Shakhovska, N., Shvorob, I., Montenegro, S., Nechepurenko, M.: The smart house based system for the collection and analysis of medical data. In: CEUR, vol. 2255, pp. 215–228 (2018)
12. Kryvenchuk, Y., Shakhovska, N., Melnykova, N., Holoshchuk, R., Smart integrated robotics system for SMEs controlled by internet of things based on dynamic manufacturing processes. In: Conference on Computer Science and Information Technologies, pp. 535–549 (2018). https://doi.org/10.1007/978-3-030-01069-0_38
13. Bodyanskiy, Ye., Perova, I., Vynokurova, O., Izonin, I.: Adaptive wavelet diagnostic neuro-fuzzy system for biomedical tasks. In: Proceedings of 14th International Conference on Advanced Trends in Radioelectronics, Telecommunications and Computer Engineering (TCSET), Lviv-Slavske, Ukraine, 20–24 February, pp. 299–303 (2018)
14. Perova, I., Litovchenko, O., Bodyanskiy, Ye., Brazhnykova, Ye., Zavgorodnii, I., Mulesa, P.: Medical data-stream mining in the area of electromagnetic radiation and low temperature influence on biological objects. In: Proceedings of 2018 IEEE Second International Conference on Data Stream Mining & Processing (DSMP), 21–25 August, 2018, Lviv, Ukraine, pp. 3–6 (2018)

Geoinformation Technology for the Determination of Carbon Stocks in the Dead Organic Matter of Forest Ecosystems in the Ukrainian Carpathians

Olha Tokar[1], Mykola Gusti[1](✉), Olena Vovk[2], Serhii Havryliuk[3], Mykola Korol[3], and Halyna Tobilevych[4]

[1] International Information Department, Lviv Polytechnic National University, Lviv, Ukraine
tokarolya@gmail.com, kgusti@yahoo.com
[2] Artificial Intelligence Systems Department, Lviv Polytechnic National University, Lviv, Ukraine
olenavovk@gmail.com
[3] Forest Inventory and Forest Management Department, Ukrainian National Forestry University, Lviv, Ukraine
serhiy_havrylyuk@nltu.edu.ua, nikkorol@ukr.net
[4] Department of Fine Arts, Design and Methods of Their Teaching, Ternopil Volodymyr Hnatiuk National Pedagogical University, Ternopil, Ukraine
levych@ukr.net

Abstract. An improved model for determining carbon stocks in dead organic matter (DOM) of forest ecosystems has been proposed, and respective geoinformation technology has been developed. The geoinformation technology integrates input data collection in the field, processing the data, a mathematical model for estimation of the carbon stocks in DOM and tools for forming thematic maps and graphs for presenting the results. The proposed geoinformation technology enhances the accuracy of estimation of the carbon stock in the lying dead wood and litter in forest ecosystems and obtaining a spatially explicit estimate of DOM carbon stocks of forest ecosystems.

Keywords: Forest ecosystem · Dead organic matter · Litter substrate · Lying wood · Carbon · Carbon stock · Geoinformation technology

1 Introduction

Forest ecosystems play a special role in stabilizing the concentration of carbon dioxide, one of the most common greenhouse gases in the atmosphere. Therefore, studies of the dynamics of carbon stocks in various reservoirs of forest ecosystems, in particular in Ukraine, are extremely relevant today [2, 3].

The carbon content in forest ecosystem reservoirs and emissions can be calculated according to three levels of complexity proposed by the International Panel on Climate

N. Shakhovska and M. O. Medykovskyy (Eds.): CCSIT 2019, AISC 1080, pp. 792–803, 2020.
https://doi.org/10.1007/978-3-030-33695-0_53

Change (IPCC): Tier 1, Tier 2 and Tier 3, and three approaches can be applied for land use representation: Approach 1, Approach 2 and Approach 3 [9]. Since IPCC methodology is universal and does not take into account the specifics of individual regions, this leads to an increase of uncertainties in estimates of the carbon stocks in forest ecosystems, in particular in the dead organic matter (DOM). Although, preliminary results of studies on the carbon stocks in soils and litter for some forests in Ukraine have already been obtained [4, 12–16], currently there is not enough empirical information for performing calculations of the carbon stocks in forest ecosystems such as dead wood, soil, and litter, based on national parameters.

Thus, the development of new approaches to the assessment of carbon stocks in various reservoirs of forest ecosystems, that are harmonized with the IPCC methodology and take into account specifics of each region are relevant.

IPCC encourages countries to develop and apply geographically explicit approach for greenhouse gas (GHG) emission inventories in the Land Use, Land Use Change and Forestry Sector that corresponds to the Approach 3 of the land use representation [9]. Geographically explicit GHG inventory systems for the Energy, Industry and Agriculture Sectors in Ukraine as well as a first prototype for the Land Use, Land Use Change and Forestry Sector have been developed [5, 6, 8]. Tokar et al. [17] continue development of the geoinformation technology for studying carbon budget of forest ecosystems. However, the component integrating the input data collected in sample plots, processing and storing of the data and mathematical models for calculating the carbon stocks in deadwood and litter of forest ecosystems has not been developed yet.

The aim of the study is to improve the mathematical model for determining the carbon stocks in DOM of forest ecosystems, creating a software complex for the spatial assessment of carbon stocks in DOM of forest ecosystems based on this model and develop a geoinformation technology for integrating the field measurements data, processing and storing the data, the mathematical model for calculating the emissions and outputting the results in a form of thematical maps. The developed geoinformation technology is applied in the experimental plots of the Spas Forest District.

2 Method

Dead organic matter includes lying wood and litter. The national forest accounting system in Ukraine does not provide reliable information on stocks and the dynamics of the amount of dead wood in the national forests. The study presents methods for calculating the stock of dead lying wood for forest areas. Such estimate of carbon dynamics in DOM reservoirs improves the accuracy of determining carbon emissions and removals in forest ecosystems.

2.1 Geoinformation Technology Overview

The processes of carbon sequestration in dead organic matter in forest ecosystems depend on many factors. Therefore, to assess the dynamics of carbon stocks in DOM, one needs to have detailed information about the dead wood and litter in the forest ecosystem, parameters for calculating the mass of carbon in dead wood and litter, characteristics of dead wood and litter, and mathematical models. Then, using a geoinformation approach to estimating carbon stocks in forest ecosystems, one can conduct a spatial inventory of carbon stocks in DOM of forest ecosystems and present the results on a map. All these requirements are realized in the developed geoinformation technology for estimating carbon stocks in the forest ecosystem of DOM shown in Fig. 1.

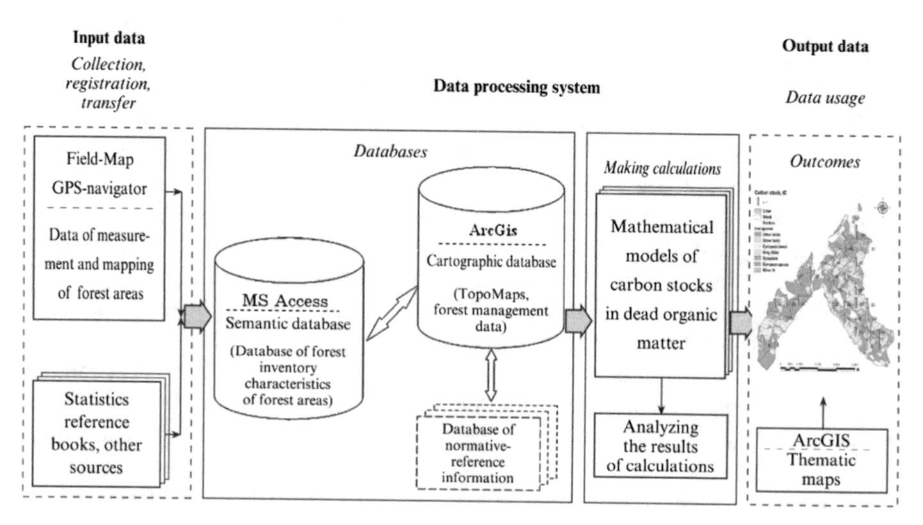

Fig. 1. The structural scheme of geoinformation technology for assessing the carbon stocks of dead organic matter in forest ecosystems (schematic representation of the relationships between the main components of geoinformation technology).

The developed information technology consists of the following components: input information (data collection, data registration), data processing system (organization of data storage and database processing) and output information (database reports, thematic maps in ArcGIS). Field-Map is an information technology for field data collection. It consists of a set of tools for dendromentric measurements, laptop and a geoinformation software for processing the measurements (https://www.fieldmap.cz/). The forest measurements, forest related statistics and other data required for calculations are entered into the Database of forest inventory characteristics. The input data are entered into the database through the forms, that reflect the layouts with which forest experts work; data of normative-reference information is entered directly in the database tables. The corresponding queries are formed on the basis of the developed and improved mathematical models, and then the results are presented in the form of reports and the obtained results are visualized on digital maps. The user has an option to view the results in the form of layers of a digital map or records of a geospatial database, where each record contains information about the parameters and indicators of carbon stocks for some elementary area (experimental plot). There is also the possibility of forming SQL-queries to a geospatial database based on the numerical characteristics of the objects and their geographical coordinates. It is important that the resulting database and the digital maps after respective converting can be used in other applications.

2.2 Experimental Plots

Landsat satellite images of medium resolution (30 m) were used for the classification of the forest cover of the Spas Forest District and development of the "forest mask". QuickBird satellite images of high resolution (0.61 m) are used for clarification of the classification results. The images of the two Landsat satellite images of medium resolution (30 m) were used for the classification of the forest cover of the Spas Forest District and development of the "forest mask". QuickBird satellite images of high resolution (0.61 m) are used for clarification of the classification results. The images of the two satellites are applied in the study because the resolution of Landsat images is 28.5 × 28.5 m/pixel, that is, an area of about 812 m^2 corresponds to one pixel on the screen. Such spatial resolution is not enough, given that the area of experimental plots established in the research are about 500 m^2.

A raster grid with a step of 1 × 1 km was used to mark the locations of the experimental plots (Fig. 2).

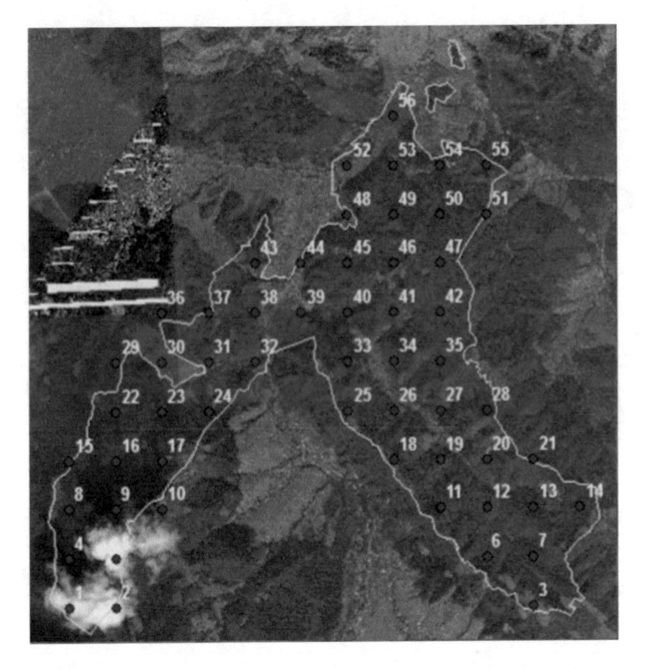

Fig. 2. Numbered experimental plots on the territory of the Spas Forest District.

As a result, 56 experimental plots were obtained. Circular plots of a constant radius (r = 12.62 m) with an area of 500 m^2 (0.05 ha) were laid at the intersection of the raster, which is approximately equal to the size of the elementary pixel of the satellite image. The use of this type of placement of the experimental plots ensures the randomness of coverage of the territory, which is important in statistical studies. If the center of the experimental plot (PFZ) is on land not covered with forest vegetation, the coordinates of the center of the experimental plot and the category of land cover were recorded. Statistical inventories were performed at the experimental plots including measurements of diameter and decomposition stage of standing and lying dead trees following the methodology described in [7].

For estimating the volume of dead lying wood in forest stands we use the method given in [1, 11]. Lying dead trees are counted on three linear transects of 15 m length (horizontal position). Transects start at a distance of 1 m from the center of the experimental plot and are laid with azimuths 35°, 170° and 300° (Fig. 3).

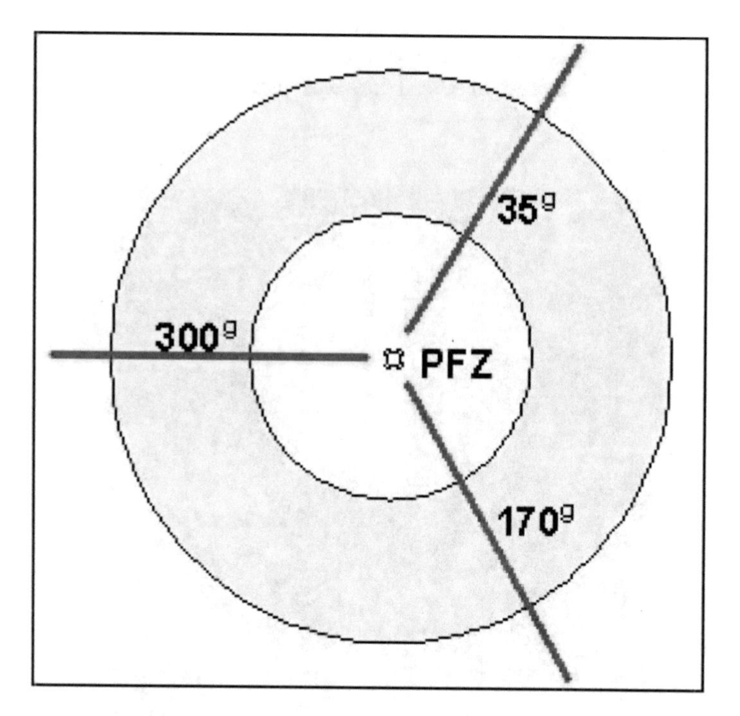

Fig. 3. Transects on an experimental plot for measurements of the laying deadwood (PFZ – center of the experimental plot).

Measurements on the transects are taken in order, one after another. Lying dead wood is taken into account when its average diameter at the intersection with the transect is at least 7 cm (diameter is measured on the surface perpendicular to the central axis of the lying trunk).

All parts of dead wood that cross the transect line and meet the criteria are taken into account. To calculate the volume of dead wood, it is necessary to measure the average diameter of the log at the intersection with the transect and the angle of the intersection. The length of the log is not measured.

For each log crossing the transects, the following attributes are recorded: log number, log diameter, angle of a part of dead wood, degree of decomposition of dead wood. The degree of decomposition of dead wood is determined at the intersection with the transect line using a knife [7]:

(1) fresh (the cambium may still be living);
(2) firm deadwood;
(3) beginning decay (the knife enters easily >1 cm deep when jabbed in parallel to the wood fibers);
(4) advanced decay, wood is soft (the knife enters easily >1 cm deep, even at right angle to the wood fibers);
(5) rotten (all wood is very soft and falling apart).

2.3 Mathematical Models for Determining the Mass of Carbon in Dead Wood and Litter

For estimating the dynamics of carbon of dead organic matter, one can apply the Tier 2 or Tier 3 methods. Then the estimate of annual carbon stock changes in DOM reservoirs will be presented as the sum of changes in the reservoirs of dead wood and litter. In general, the calculation process is as follows:

$$\Delta C_{DOM} = \Delta C_{DW} + \Delta C_{LT},$$

where C_{DOM} – annual changes of carbon stocks in dead organic matter (including dead wood and litter), tC/year;

ΔC_{DW} – annual changes of carbon stocks in dead wood, tC/year;
ΔC_{LT} – annual changes of carbon stocks in litter, tC/year.

Due to their versatility, the IPCC methods are not adapted to fully take into account regional peculiarities of the country, and this in turn leads to uncertainties in the estimates of carbon stocks in reservoirs, in particular in the dead organic matter reservoir and an increase in uncertainty in calculating carbon stocks in forest ecosystems in general. Therefore, in this study we use the national approach, which is based on the IPCC methodology and takes into account the characteristics of the forests of Ukraine [3, 4, 14–16].

Litter. For calculating the carbon stock in litter of forest stands we use an approach described in [3]:

$$C_{LT} = k_1\left(k_2A^2 + k_3A + k_4\right)P^{1.2}S,$$

where k_1, k_2, k_3, k_4 – coefficients, values of which for different tree species are presented in Table 1 [3];

A – forest stand age;
P – relative stocking of the forest stand;
S – area of the forest stand.

Table 1. Coefficients for calculating carbon stock in litter [3]

Tree species group	k_1	k_2	k_3	k_4
Coniferous	0.48	−0.003	0.51	0.63
Oak, beech, hornbeam, ash, walnut	0.42	−0.0012	0.24	0.25
Other deciduous	0.37	−0.0016	0.24	0.1

Dead Wood. The volume of dead wood lying on the experimental plots are calculated according to the formula (based on approach presented in [1]):

$$V_{DW} = \frac{1}{h_j}\sum_{k=1}^{h}\frac{\pi^2}{8L_k}\sum_{i=1}^{S_k}\left(\frac{d_{1i}+d_{2i}}{2}\right)^2\frac{1}{cos\alpha_i}$$

where V_{DW} – volume of dead lying wood on the experimental plot, m^3/ha;
h_j – number of transects on the experimental plot;
L_k – total length of kth transect lines on the experimental plot, m;
d_{1i}, d_{2i} – measured diameters of dead lying timber in two dimensions, cm;
a_i – angle between the measured dead log and the transect, deg.;
S_k – number of dead lying logs on kth transect.

Then the mass of carbon in dead lying wood on the experimental plot is estimated by multiplying the volume, V_{DW}, by the wood density and carbon content, e.g. taken from [9]. The default values for the parameters are 0.47 t d.m./m^3 for wood density and 0.5 tC/t d.m. for carbon content, however, tree species specific values, forest type specific values [9] and decomposition stage specific values can be applied as well [13]. Depending on the decomposition stage the wood density varies from 0.45 t d.m./m3 for wood of alive trees, increases to 0.48 t d.m./m^3at the first decomposition stage, then steadily decreases to 0.19 t d.m./m^3at the fourth decomposition stage [13].

3 Application of the Developed Geoinformation Technology

In the study the data from the experimental plots of the Spas Forest District (circular experimental plots of 0.05 ha) [10] and materials of the forest inventory of the Broshniv State Forestry Enterprise of the Ivano-Frankivsk Region are used for the estimation of carbon stocks in dead wood and litter of forests. The data have been collected in 2011–2014.

Measurements on the experimental plots and mapping of the horizontal placement of trees were performed using Field-Map technology. The geo-location of the trees was performed using a GPS receiver. Field data were processed in Microsoft Excel and promptly entered into the Microsoft Access database (see Fig. 1).

Using the methodology described above, we calculated the carbon stocks in litter and lying and standing dead wood in the experimental plots of the Spas Forest District (Tables 2 and 3) and presented the results on a digital map (Fig. 4).

Table 2. Carbon stock in litter and lying dead wood of forest stands in the Spas Forest District (based on the field data collected in 2011–2014)

№ exp. plot	Dead organic matter				Carbon stock on experimental plot
	Litter		Lying dead wood		
	Carbon stock	Carbon stock per unit area	Carbon stock	Carbon stock per unit area	
	t	t/ha	t	t/ha	t
4	0.310	6.200	0.018	0.350	0.328
5	0.150	3.010	0.032	0.640	0.182
6	0.230	4.590	0.016	0.310	0.246
7	0.300	6.030	0.035	0.690	0.335
11	0.200	4.070	0.015	0.290	0.215
12	0.190	3.850	0.001	0.020	0.191
13	0.150	3.070	1.240	24.800	1.390
17	0.120	2.410	0.008	0.160	0.128
18	0.180	3.570	0.330	6.600	0.510
19	0.080	1.680	0.000	0.000	0.080
20	0.340	6.860	0.012	0.230	0.352
22	0.190	3.720	0.440	8.800	0.630
23	0.300	5.900	0.056	1.110	0.356
24	0.200	3.920	0.012	0.240	0.212
25	0.120	2.420	0.025	0.490	0.145
26	0.170	3.310	0.005	0.090	0.175
27	0.250	5.020	0.014	0.270	0.264
28	0.230	4.580	0.010	0.190	0.240
33	0.300	6.040	0.007	0.130	0.307
35	0.160	3.130	0.008	0.150	0.168
37	0.090	1.890	0.025	0.490	0.115
38	0.300	6.010	0.010	0.190	0.310
40	0.110	2.210	0.011	0.210	0.121
41	0.180	3.690	0.029	0.580	0.209
44	0.080	1.630	0.020	0.400	0.100
45	0.200	3.970	0.032	0.640	0.232
47	0.200	3.990	0.010	0.200	0.210
49	0.290	5.790	0.000	0.000	0.290
51	0.250	4.920	0.012	0.230	0.262
54	0.320	6.480	0.000	0.000	0.320

Table 3. Carbon stock in standing dead trees of forest stands in the Spas Forest District (based on the field data collected in 2011–2014)

№ exp. plot	Tree species	Number of trees	Carbon stock, t	Carbon stock, t/ha
4	European spruce	17	0.35	7
5	Silver birch	1	0.64	13
	European spruce	1		
6	European beech	2	0.31	6
	European spruce	1		
7	European spruce	15	0.69	14
11	European beech	1	0.29	6
	Silver fir	1		
12	European spruce	1	0.02	0.4
13	European beech	1	1.24	25
	European spruce	7		
17	European beech	1	0.16	3
18	European spruce	1	0.33	7
20	European spruce	5	2.23	5
	Silver fir	1		

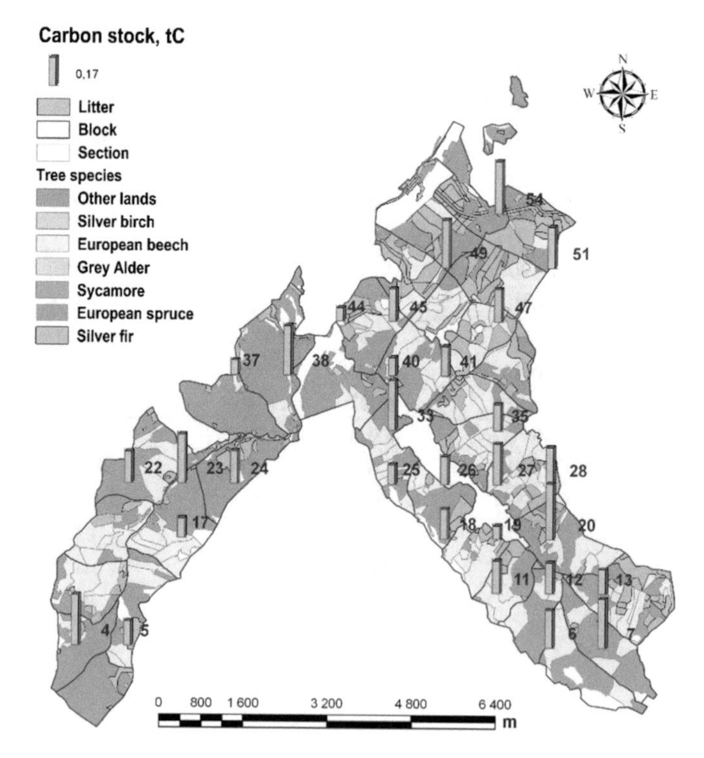

Fig. 4. Carbon stocks in litter of forest on experimental plots of Spas Forest District, tC (based on the field data collected in 2011–2014).

Thus, carbon stocks in litter of coniferous forests are the largest – 6 tC/ha, lower carbon stocks are in litter of deciduous forests – 1.17 tC/ha, and on average, 4.2 tC/ha accumulates in the litter of the Spas Forest District.

For comparison, Fig. 5 shows the carbon stocks per unit area in the litter of a number of forestry areas in the mountainous part of the LvivRegion [15], current calculations for the Spas Forest District in the Ivano-Frankivsk Region and the average value for Ukraine [4]. The carbon stocks per unit area in the litter of Spas Forest District, as estimated in current study, is within the range of the estimates for other mountainous forestry areas.

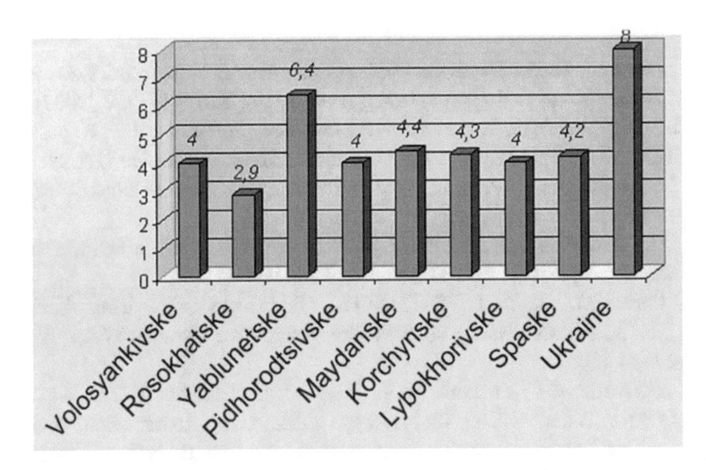

Fig. 5. Carbon stocks in litter of forestry areas in the mountainous part of the Lviv Region [14], current calculations for the Spas Forest District (based on the field data collected in 2011–2014) and the average value for Ukraine [4], tC/ha.

Carbon stocks in standing dead trees at the experimental plots of the Spas Forest District is presented in Table 3. The carbon stock varies with the stand tree species composition considerably.

4 Conclusion

The paper presents an improved mathematical model for determining the carbon stocks in DOM of forest ecosystems; respective geoinformation technology for integrating the field measurements, processing and storing the data, the mathematical model and outputting the results in a form of thematical maps; and results of application of the developed geoinformation technology in the experimental plots of the Spas Forest District.

The proposed improved multi-level model for determining carbon stocks in lying dead wood and litter of forest ecosystems increases the overall accuracy of carbon stock estimates, while the developed geoinformation technology is used to carry out spatial assessment of carbon stocks in dead organic matter of forest ecosystems. The geoinformation technology applies the principle 'from one object to many', thus utilizing geo-referenced input data in contrast to the usual approach utilizing average values. The information technology corresponds to the "Approach 3" for land representation and Tier 2/3 for complexity mentioned in the IPCC Guidelines [9] as the high-level spatially explicit approach for achieving the accurate greenhouse gas inventories. Besides that, such information technology can be an effective toolkit to support decision making in forestry and nature conservation.

References

1. Boehl, J., Braendli, U.-B.: Deadwood volume assessment in the third swiss national forest inventory: methods and first results. Eur. J. Forest Res. **126**, 449–457 (2007)
2. Buksha, I.F., Butrim, O.V., Pasternak, V.P.: Inventory and monitoring of greenhouse gases in sector LULUCF. National Agrarian University, Kharkiv, Ukraine (2008)
3. Buksha, I.F., Pasternak, V.P.: Inventory and monitoring of greenhouse gases in forestry. National Agrarian University, Kharkiv, Ukraine (2005)
4. Buksha, I.F., Raspopina, S.P., Pasternak, V.P.: Stocks of organic carbon in soils and litter at forest monitoring sites. For. For. Melioration **120**, 106–112 (2012)
5. Bun, R., Gusti, M., Kujii, L., Tokar, O.B., Tsybrivskyy, Y., Bun, A.: Spatial GHG inventory: analysis of uncertainty sources. a case study for Ukraine. Water, Air, Soil Pollut. Focus **7**, 63–74 (2007)
6. Bun, R.A., Gusti, M.I., Dachuk, V.S., Kujii, L.I., Oleksiv, B.Ya., Stryamets, H.V., Stryamets, S.P., Tokar, O.Ye., Tsybrivskyy Ya.B. (eds.): Information Technologies for Greenhouse Gas Inventories and Prognosis of the Carbon Budget of Ukraine. Ukrainian Academy of Printing, Lviv, Ukraine (2004)
7. Commarmot, B., Tinner, R., Brang, P., Brändli, U.: Stichprobeninventur im Buchen-Urwald Uholka-Schyrokyj Luh – Anleitung für die Inventur 2010. Eidg. Forschungsanstalt für Wald, Schnee und Landschaft WSL, Birmensdorf, Switzerland (2010). https://www.dora.lib4ri.ch/wsl/islandora/object/wsl:10266
8. Danylo, O., Bun, R., See, L., Charkovska, N.: High-resolution spatial distribution of greenhouse gas emissions in the residential sector. Mitigation and Adaptation Strategies for Global Change (2019, In Press). https://doi.org/10.1007/s11027-019-9846-z
9. Eggleston, H.S., Buendia, L., Miwa, K., Ngara T., Tanabe, T. (eds.): 2006 IPCC Guidelines for National Greenhouse Gas Inventories, vol. 4: Agriculture, Forestry and Other Land Use. IGES, Japan (2006)
10. Havryliuk, S.A., Khomiuk, A.H.: Realization of statistical inventory of the virgin beech forest in Uholsko-Shyrokoluzhanskyi massif. For. For. Paper Woodworking Ind. **37**(2), 38–41 (2011)
11. Keller, M.: Schweizerisches Landesforstinventar. Feldaufnahme-Anleitung 2013. Forschungsanstalt fur Wald, Schnee und Landschaft WSL, Birmensdorf, Switzerland (2013)
12. Pasternak, V.P., Yarotskiy, VYu.: Carbon stock and dynamic assessment in the forests of North-East of Ukraine. Sci. Bull. UNFU **23**(6), 57–62 (2013)
13. Rozhak, V.P.: Pecularities of formation of dead wood stocks of forest ecosystems of Stryy-Sian Verkhovyna (Ukrainian Carpathians). Scientific Issues Ternopil Volodymyr Hnatiuk National Pedagogical University Series: Biology **2**(59), 18–24 (2014)

14. Rozhak, V., Kozlovskyi, V.: Stock and composition of fresh litter and litter in forest ecosystems in the Stryy-Sian Verkhovyna (Ukrainian Carpathians). Visnyk of Lviv University. Biological series, vol. 62, 160–169 (2013)
15. Shpakivska, I.M., Maryskevych, A.H.: Assessment of carbon stocks in forest ecosystems of the Eastern Beskydy. For. For. Melioration **115**, 176–180 (2009)
16. Shpakivska, I.M., Rozhak, V.: The dynamics of wood litter in the forest ecosystems of the Stryy-Sian Verkhovyna (Ukrainian Carpathians). Journal of scientific publications of graduate and doctoral students. Biological sciences series, vol. 1, 175–179 (2014)
17. Tokar, O., Lesiv, M., Korol, M.: Information technology for studying carbon sink in stemwood of forest ecosystems. Econtechmod **1**(1), 113–120 (2014)

A Smart Home System Development

Vasyl Lytvyn[1,2(✉)], Victoria Vysotska[1], Nataliya Shakhovska[1], Vladyslav Mykhailyshyn[1], Mykola Medykovskyy[1], Ivan Peleshchak[1], Vitor Basto Fernandes[3], Roman Peleshchak[4], and Serhii Shcherbak[5]

[1] Lviv Polytechnic National University, Lviv, Ukraine
{Vasyl.V.Lytvyn, Victoria.A.Vysotska, Nataliya.b.shakhovska}@lpnu.ua, vladyslavmykhailyshyn@gmail.com, peleshchakivan@gmail.com
[2] Silesian University of Technology, Gliwice, Poland
[3] University Institute of Lisbon, Lisbon, Portugal
[4] Ivan Franko Drohobych State Pedagogical University, Drohobych, Ukraine
rpeleshchak@ukr.net
[5] EPAM, Lviv, Ukraine

Abstract. The intelligent system of a smart house, which is designed to create from any house, office, or building a smart room, was created in the overall process. Recent trends in technology development demonstrate that more and more things are being automated around us. Even ordinary things become «smarter» and open a new functional use. This has led to an increase in demand for solutions that can provide a convenient and secure way to manage such devices. Having analyzed the literary and Internet sources, it was discovered that there is a large number of analogy systems, which nevertheless have some differences. Taking into account the analysis, it was decided to distinguish the market share through the introduction of intellectual algorithms, namely, algorithmic face recognition, and decision support system for shortcut service. Such solutions are not offered by any of the representatives of analogues, which will allow to receive the market share without strict competition with other manufacturers. Analysis of the analogues made it possible to define their weaknesses and strengths. As a result the experience of analogues was used while the designing and development of the system, and a lot of problems were avoided. In addition, a complex analysis of the software product was conducted, which allowed to see design of its structure, modules and their interconnection in detail. A hierarchy of tasks was also built according to the level of processes importance. An optimal alternative was identified for allocating resources between the main processes in the operating system with the use of analytical hierarchy method. The optimal tools were chosen for the development of the system, which allowed to create a fast, reliable, optimized and user-friendly system with a comfortable mobile and web-based user interfaces.

Keywords: Smart house · User interface · Intelligent device · Control panel · Intelligent system · Internet of Things · Mobile communication networks ·

N. Shakhovska and M. O. Medykovskyy (Eds.): CCSIT 2019, AISC 1080, pp. 804–830, 2020.
https://doi.org/10.1007/978-3-030-33695-0_54

Smart house system · System analysis · Convolutional neural network · Decision support system · Intellectual building · Apple Home Kit · Analytical hierarchy method

1 Introduction

Any building (administrative, industrial, or residential) consists of a set of subsystems, which are responsible for the realization of some functions, which solve different problems in the process of functioning of this building. As these subsystems become more complex and the quantity of tasks they are responsible for is increasing, it gets harder to manage them. In addition, costs for keeping service stuff as well as subsystems repair and maintenance grow fast [1]. For the first time these problems appeared during exploitation of big administrative and industrial complexes [2].

A modern building of such type is like a miniature city [3]. Actually all the services, which previously were the indispensable attributes of cities, function here [4]. Usually there is an administrative service or an administrator, who use and sustain these systems almost round the clock, in such buildings [5]. However, there are many automation means, which cope with such tasks as heating, ventilation, microclimate support, lighting, fire alarm, entry/exit control etc., but these systems service and managing requires the presence of administrative staff [6]. Its duty is controlling these subsystems and taking action in case of breakdown [7]. However, in some situations even qualified staff actions can be inefficient [8]. These are the cases of emergence of global building (and people inside) threats such as fire, earthquake and other natural disasters [9]. Then immediate extraordinary actions are required. People's actions and correctness of their actions can be insufficient in a critical situation. In the past traditional systems for sustaining different aspects of life were projected [10] as automated. Such systems were created separately for each function and were united for any part of the building. Systems, which were placed into the buildings, possessed only that functionality and complexity which were required in the moment of building constructing [11]. Further system expansion and modernization were complex and expensive tasks due to a set of various factors. Exploitation costs of such systems are enormous due to a necessity to support qualified administrative staff for supervising and observation of all the elements. Thus, a question about creating an independent intellectual system, which would control all the components independently, arose a long time ago. In addition, such system will become a mean for centralization and granting a common interface for all the gadgets, sensors etc. in the case of private houses and flats. Houses and flats with such systems realized are called *smart* [12].

Smart House is a residential house of a modern type, organized for comfortable life by means of up-to-date high-tech devices. The term «smart house» was formulated by the Institute of Intellectual Building in Washington, DC in the 1970s: a building that provides productive and efficient use of the working space. The principle of «Intellectual management system of a building» provides a completely new approach to the organization of life support of the building, which significantly increases the efficiency and reliability of management of all systems and building executive devices due to software and hardware complex. A «smart house» is a system that should be able to recognize some specific situations, which occur in the building, and respond appropriately: one system can control the behaviour of other systems in accordance with previously developed algorithms. The main feature of the intellectual building is the unification of separate subsystems into a single managed complex. An important feature and characteristic of the «smart house», which distinguishes it from other ways of organizing living space, is that it is the most progressive concept of human interaction with living space, when a person sets the desired environment with one command and automatics sets and tracks working modes of all engineering systems and electrical appliances according to external and internal conditions. In this case, the need to use several remote controls when viewing the TV, dozens of switches when controlling lighting, separate units in the management of ventilation and heating systems, video surveillance and signaling systems, gates and others is eliminated [13].

In a house equipped with the «smart house» system, just one click on the wall key (or remote control, touch panel, etc.) is enough to choose one of the scenarios. The house itself will adjust the work of all systems according to wishes, time of day, position in the house, weather, outdoor lighting, etc. to provide comfort inside the house.

Smart house systems become more and more popular, but this is a new industry direction, which is just beginning to develop and therefore needs constant improvement. Such decisions are quite relevant in Ukraine, because the market has enough necessary components for their creation [14]. However, almost all systems are products that support only components developed by the same company as the product itself, which greatly reduces the scope of the use of such systems [15]. Thus, there is a need to create a system that on the one hand, would support as many devices as possible for smart homes that are available on the market (and would not depend on their producers), but on the other hand, would have a convenient interface and extensive options for tuning, automation and management [16].

Research purpose is the creation of an intelligent system of a smart house. The following tasks are needed to be solved to achieve this goal:

- References analysis;
- Subject area system analysis;
- Choice of implementation methods and means;
- Software product implementation.

Research object is the market of system solutions for smart houses. *Research subject* are the ways of device management aggregation for smart houses from different producers, user interface adaptation in accordance with modern needs, automatization of processes, which can run without human intervention.

2 Intellectual Building Concept

Here intelligence means the ability to recognize certain situations and somehow respond to them (of course, the degree of this skill can be different, even very high) [1]. At the same time the intellectual building can be interpreted as «intelligently constructed» [2]. This means that the building should be designed in such a way that all services could integrate with each other with minimal cost (in terms of finance, time and labour), and their service could be organized in the best way [3].

The concept of an intellectual building consists of such statements [1–3, 17–21]:

- Creation of building management integrated system - system with the ability to provide complex work of all building engineering systems: lighting, heating, ventilation, air conditioning, water supply, access control and many others.
- Removal of all maintenance staff of the building, delegating control functions and decision-making to the subsystems of the integrated management system of the building. The «intelligence» of the building is being set in these subsystems - how it will respond to changes in the parameters of the system sensors and other extraordinary situations.
- Implementation of the mechanism of immediate shutdown and transfer in case of need to control a person by any subsystem of the intellectual building. Also a person should be provided with convenient and equal access to the management and depiction of all subsystems and parts of the «Intellectual building».
- Providing the correct functioning of individual subsystems in case of failure of the general control system or other parts of the system.
- Maintenance cost minimization and building systems modernization, which should be provided by the use of common standards in the construction of subsystems, the automatic configuration and detection of new devices and modules when added to the system.
- The presence of a built-in communication environment in the building for connecting devices and system modules. Along with this, the possibility of using an intellectual building as a communication environment in the control system of various types of physical channels: low-voltage lines, power lines, radio channel.

3 Intellectual Building Features

An intelligent building has many advantages. The management system allows owners to create complex and intelligent functioning procedures, because all executive systems can work in a coherent and collaborative way. Hence, it is possible to implement many resource-saving procedures such as [22–27]:

- access control and security assurance,
- instantaneous response to their critical changes,
- remote control and management of the building, because all information and control channels in such a system are digital.

An empty house can be transformed into in a fully functioning one with just one touch: lighting will be turned on, a comfortable microclimate will be installed, curtains will be lowered, etc.

Lighting Management. You can create light scenarios from an unlimited number of light sources with different brightness, you can include them simultaneously or with a delay. Using special light regulators you can not only change the brightness of the lamps, but also the time for which this brightness will be achieved after turning on. The function of illumination constant control was intended mainly for offices. This function makes it possible to maintain a certain illumination of the working surface regardless whether the sun is shining hidden by clouds. The function of constant control of illumination, intended mainly for office premises, makes it possible to maintain a given illumination of the working surface regardless whether the sun or sky is hidden by clouds. The automatic inclusion of outdoor lighting according to the time of day and the presence of people not only provides additional comfort, but also frightens unwanted guests [28–32].

Microclimate Management. The system continuously monitors the temperature in each room individually and maintains it at a predetermined level, controlling the radiator valves or air conditioning valves directly and automatically activates or deactivates ventilation if necessary. The system helps to save money every day due to different modes of the system: comfortable mode, night mode, mode «no one is in the house». The modes change according to the schedule or on command. It is enough to set the temperature for each mode only once. The heating and air conditioning systems can be switched off automatically to save energy if the windows of the room are open for ventilation [33–37].

Jalousie. In the summer, their angled slats automatically return at a certain angle and prevent the room from getting into excess sunlight without reducing the light flow. Thus, they prevent redundant heating of the room and help save energy that gets consumed by the air conditioner.

Security System. A smart house logs all the events that took place during the absence of the owners: who came and when, how much time they were in the house, what suspicious persons spun around the house for a long time [38–41]. Their faces and actions are recorded in the memory. When someone enters the house the system notifies you by telephone and calls security. In case of emergency (for example, the flow of water), the system not only informs the appropriate service, but also takes the necessary measures to localize the accident (stop the flow of water). During your absence, the house can simulate the habitual lifestyle of the owners, including evening light and music, thereby creating the effect of presence [42–47].

4 Analogues Review

Smart house systems are fairly widespread in the market, most major manufacturers of electronic devices have such systems, but all systems have certain disadvantages that prevent them from becoming a universal solution [48–66].

One of the smart home systems is the CLAP product, which offers a complex solution for home/office devices and a management program (Fig. 1).

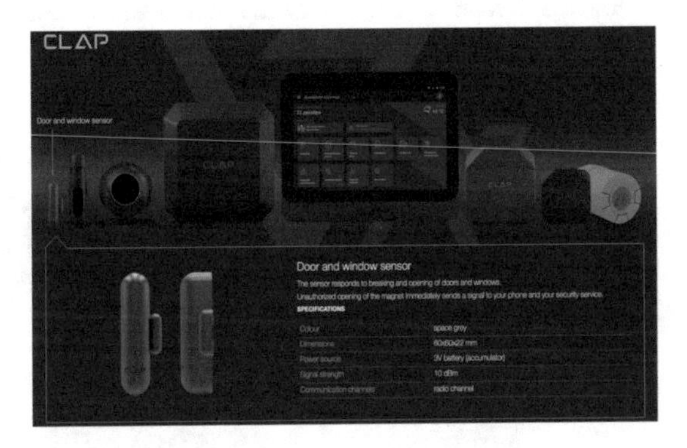

Fig. 1. CLAP intelligent system interface

The system disadvantages are the following: some devices should be mounted in walls or ceilings and all devices are manufactured by the same company and as a result the user can not use other devices for system expansion of in case of shutdown. In addition, the system does not have a built-in recognition system for people.

Apple HomeKit is a platform built on the iOS platform that manages your smart home. The idea is simple: instead of having tons of various clever home-based programs that do not communicate with each other, HomeKit brings them together, offers a control center on your devices and uses Siri on your iPhone, iPad, Apple. Watch or HomePod for managing. The Apple Home system allows the user to manage a smart house through various types of interfaces (Fig. 2), as well as with help of the intelligent voice assistant Siri. This product is a quite widespread solution, as it is developed by one of the leading IT companies in the world. Despite this, The Apple Home has some disadvantages.

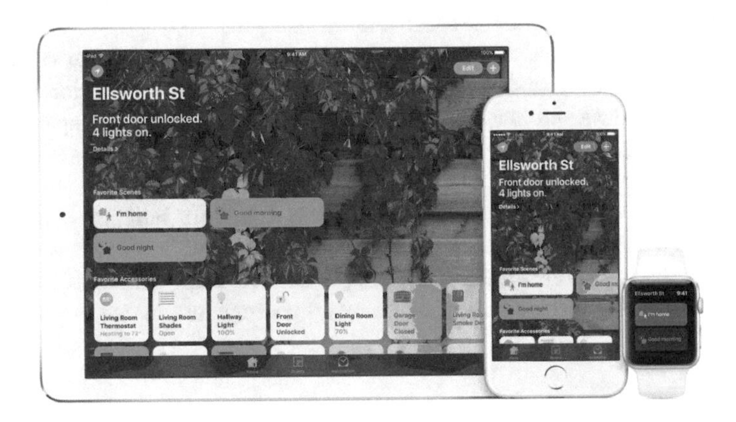

Fig. 2. Apple Home Kit interface

One of the problems of the Apple Home Kit is that this system only supports devices that are specifically designed to be compatible with it (Fig. 3), that's why all devices must have the same API for management, which significantly reduces the number of possible devices, because every manufacturer wants to use their own control system.

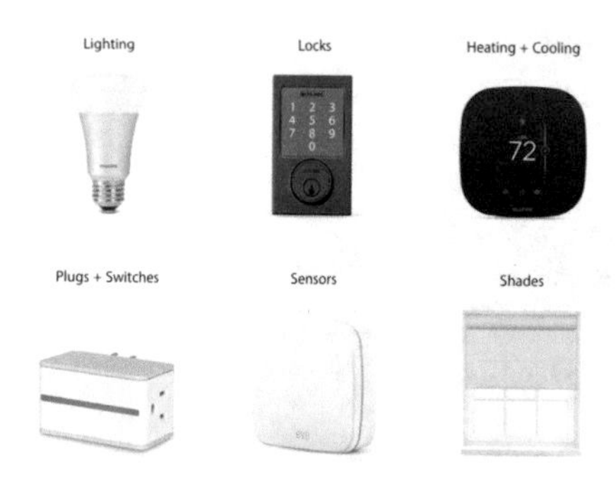

Fig. 3. Apple Home Kit devices

Fig. 4. Xiaomi devices for a smart house

Therefore, Apple Home Kit system is a popular solution with good support, which has an idea similar to the system being developed, but there are things that are different. The benefits of the Apple Home Kit are popularity, powerful support, constant updates and adding more and more devices as well as the presence of the own voice assistant. However, this product also has drawbacks in comparison with the system being developed, for example, the system does not have a built-in face recognition algorithm, as well as an automatic rule generation system, based on previous user actions. Thus, although the systems are fairly similar, they will still gain popularity on somewhat different markets, as they have a different approach to increasing the number of supported devices, as well as different software mechanisms.

Mi Home is a universal application that allows you quick and convenient management of the smart house devices. Designed by Xiaomi, the application is designed to create a smart house and manage it with a smartphone. Since this company is one of the largest manufacturers of devices for a smart home (Fig. 4), it was obvious that it will also create its own version of the system for working with them.

All devices have minimalistic design, which is a plus for building a smart house, because they are almost invisible. Many other smart house systems use Xiaomi devices. The company has developed the MiHome application to work with devices, as well as to manage and monitor the status of all data and parameters (Fig. 5).

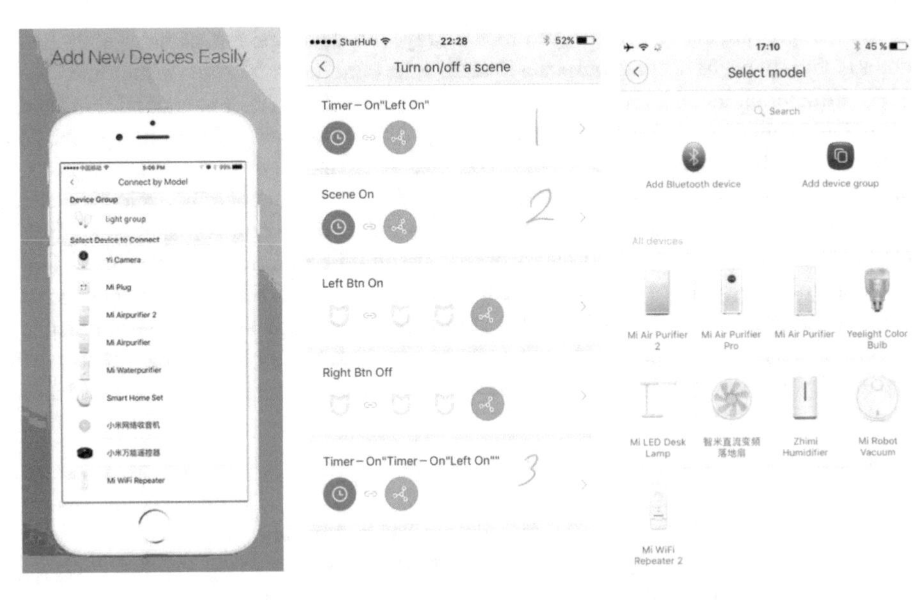

Fig. 5. MiHome application **Fig. 6.** Scene setup **Fig. 7.** Creation of groups

This application allows the user to add devices, manage them, set up scenes (Fig. 6) (certain rules for executing commands), and also combine devices into certain groups (Fig. 7). Added devices can be divided into 2 groups - those that work independently, and those who need the Xiaomi Gateway gateway - these devices work with the special ZigBee protocol. MiHome application is a very popular market solution, its developer, Xiaomi, is also one of the largest device makers for this kind of solution. The functionality of the application is quite similar to the functional of the system being developed. For example, MiHome also supports certain sets of rules (scenes), but these rules should always be set manually, without the help of the system. Another disadvantage of this product is that it does not have an interface version, so it can be inconvenient to use for users with a small screen size of the smartphone, or to perform some actions (for example, watching videos from cameras setup in the house).

5 System Analysis of the Research Object

System analysis is a methodology for studying properties and relationships in objects that are difficult to observe and difficult to understand. The feature of system analysis is the appliance of analysis and synthesis methods. System analysis provides tools for identifying and eliminating uncertainty when solving complex problems based on finding the best solution from existing alternatives [19]. The necessity of studying the methodology, mastering the theory and practice of system analysis is determined by the ever-increasing need for solving complex interdisciplinary tasks for various purposes. These needs determine not only the rapid development of world globalization, the high pace of improvement of science and technology, the development of innovative and other technologies of various purposes, but also the conditions for the constant increase of environmental threats, man-caused, natural and other disasters. These conditions and factors determine the operational urgency and practical necessity of training specialists who possess the apparatus of solving complex systems of timely prediction tasks, objective forecasting and system analysis of available socio-economic, scientific and technical and other problems, problems and situations, possible man-made, ecological, natural and other accidents and catastrophes. It should be emphasized that the effectiveness and reliability of timely prediction, objective forecasting, system analysis of various alternatives to possible complex solutions and strategies for action in practice largely depend on the ability of the system researcher to master and efficiently use the capabilities of the methodology of system analysis in a timely manner. The basic principles of system analysis are [20]:

- principle of unity: consideration of the system as a whole and as a set of components;
- principle of the global goal: the global goal has an absolute priority;
- principle of connectivity: each part of the system is considered together with its connections with the environment;
- modularity principle: decomposition of the system on the module and consideration of it as a set of constituents;
- principle of functionality: the structure of the system and its functions should be considered together with the priority of the functions over the structure;
- hierarchy principle: the implementation of the hierarchical construction of the components of the system by importance;
- principle of development: it is necessary to take into account the system's ability to develop and expand;
- uncertainty principle: uncertainties and coincidences should be taken into account when determining the system's strategy.

Relevance tree is a graphical schema of subordination and interrelation of goals, which demonstrates the division of the overall goal into tasks and subtasks. The main idea of the «relevance tree» is decomposition, this method shows the ways to achieve the goal. The main goal of the developed system is the implementation of a smart house system. To achieve the main goal, it is necessary to specify it (Fig. 8). This method allows us to obtain a hierarchical structure containing purposes, directions and problems in a convenient form. Microsoft Visio environment is used for building the relevance tree, Microsoft Visio provides all the necessary elements in accordance with the standard.

To achieve the goal, you have to complete a number of subtasks, namely:

- Creating an interface for management. To complete this task, you need to complete the following steps: develop a web application and create a mobile application;
- Creating a service for device management. This subtask includes is implementation of access interfaces for all possible devices;

- Creating a video management service. Contains the following sub-targets is developing a neural network for recognizing people and using a system shortcut.
- Creating an authorization service.

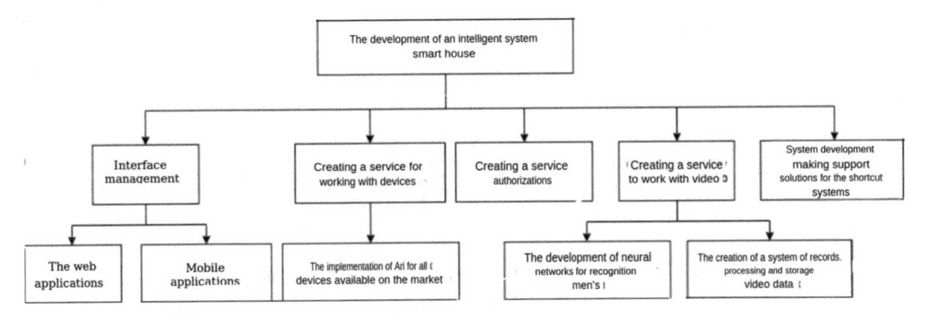

Fig. 8. Relevance tree

For a better understanding of the processes that ensure the interaction of the system with the environment and occur directly within the system itself, it is necessary to construct a UML diagram, which allows to show graphically various processes and approaches of work with the system and, accordingly, identify possible shortcomings.

Unified Modeling Language is a standard visual modelling language designed for:

- modelling business and similar processes;
- analysis, design and implementation of software systems

UML is a common language for business analysis and software architecture that is used to describe, instruct, design, and document existing or new business processes, the structure and behaviour of software systems. UML can be applied to a variety of areas (for example, banking, finance, the Internet, aerospace, healthcare, etc.). It can be used with all major methods of object developing and component software and for different implementation platforms (for example, J2EE, .NET). UML is a standard modelling language, not a software development process. Its functionality is following:

- giving instructions in the order of the team activities;
- indicating which components should be developed;
- directing the tasks of individual developers and teams in general
- proposing criteria for monitoring and measurement of products and project activities.

UML was designed in a deliberately independent process and can be applied in the context of various tasks. However, it is best suited for using random, iterative, and additional development processes. It is necessary to build a class diagram (Fig. 9) of the service in order to understand the architecture of the service, the relationships between classes and interfaces, as well as the processes of their interaction.

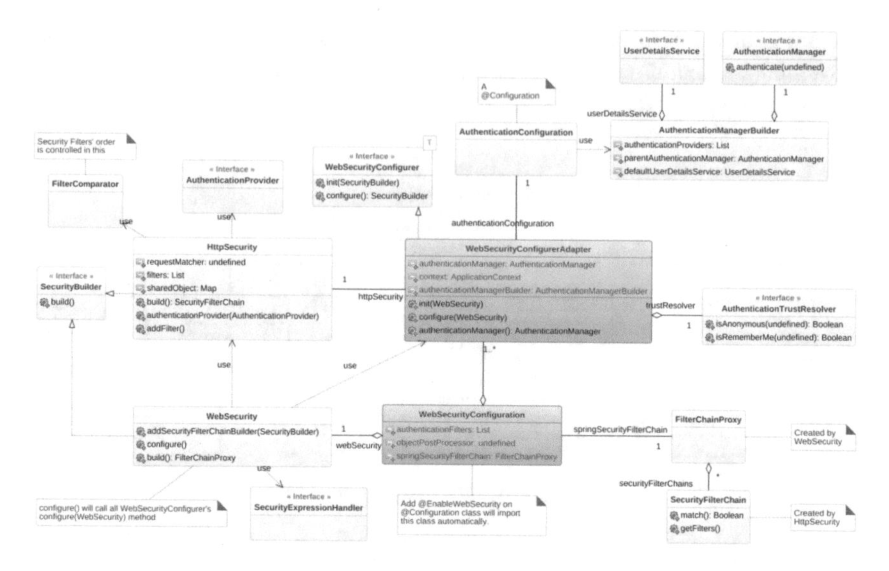

Fig. 9. Class diagram

As you can see from the diagram, many classes are responsible for the security of the system, because this is one of the most important tasks, since the loss of control over services that manage the events in the whole house can be extremely dangerous.

The next step is to consider the possible interaction of users with the system. This functionality is in the use case diagram (Fig. 10), which describes the possible use cases of the system (as a black box) from the user's point of view. As you can see from the diagram, a lot of different functional options for working with the system are available for the user. Such options can be either Registration or Authorization into the system, Giving orders, Adding new devices, or Viewing videos that we saved by the video service. For a better understanding and debugging of the system, it is also necessary to construct a state diagram (Fig. 11), which shows all possible states in which the system can be at a certain point in time; it can help to test the system.

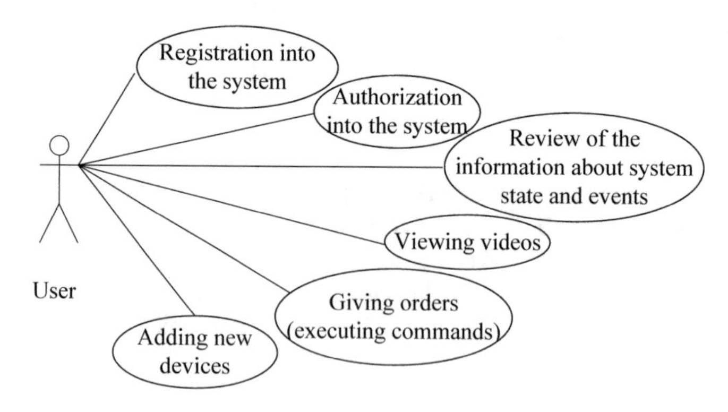

Fig. 10. Use case diagram

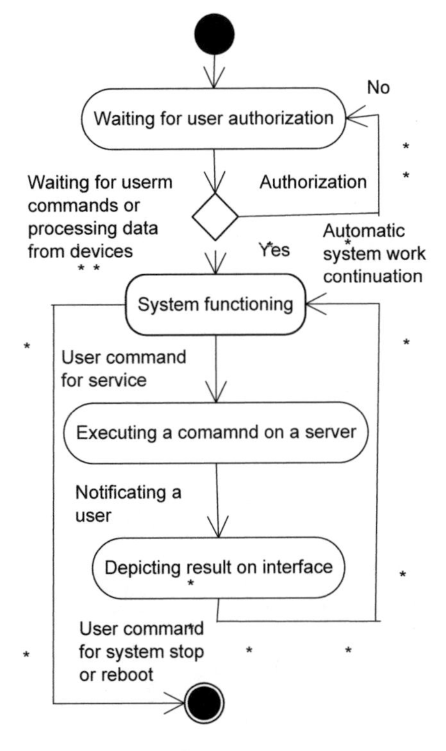

Fig. 11. System state diagram

The diagram shows that once launched, the system will work automatically, simultaneously receiving tasks from the user and executing them, as well as monitoring and displaying information about changes in the system. A stop or restart of the system is possible only after a compulsory user command. Sequence diagram (Fig. 12) will allow you to see the logical sequence of all inquiries and processes inside the system, this diagram will be especially relevant for the authorization service, which requires extremely high clarity and stability of operations.

As you can see from the diagram, there are several different scenarios of event development, depending on whether the user was able to authorize into the system or not. This information will help in further testing, and system users' support.

After developing the system, it needs to be deployed and run. In order to better understand the necessary infrastructure, and in the case of microservice it is especially important, it is necessary to construct a deployment diagram (Fig. 13). As you can see from the diagram, the system itself can be deployed on two servers - a web server for services, and one more server for databases. Graphical interfaces for user access will be created as a website or mobile application, depending on the management device.

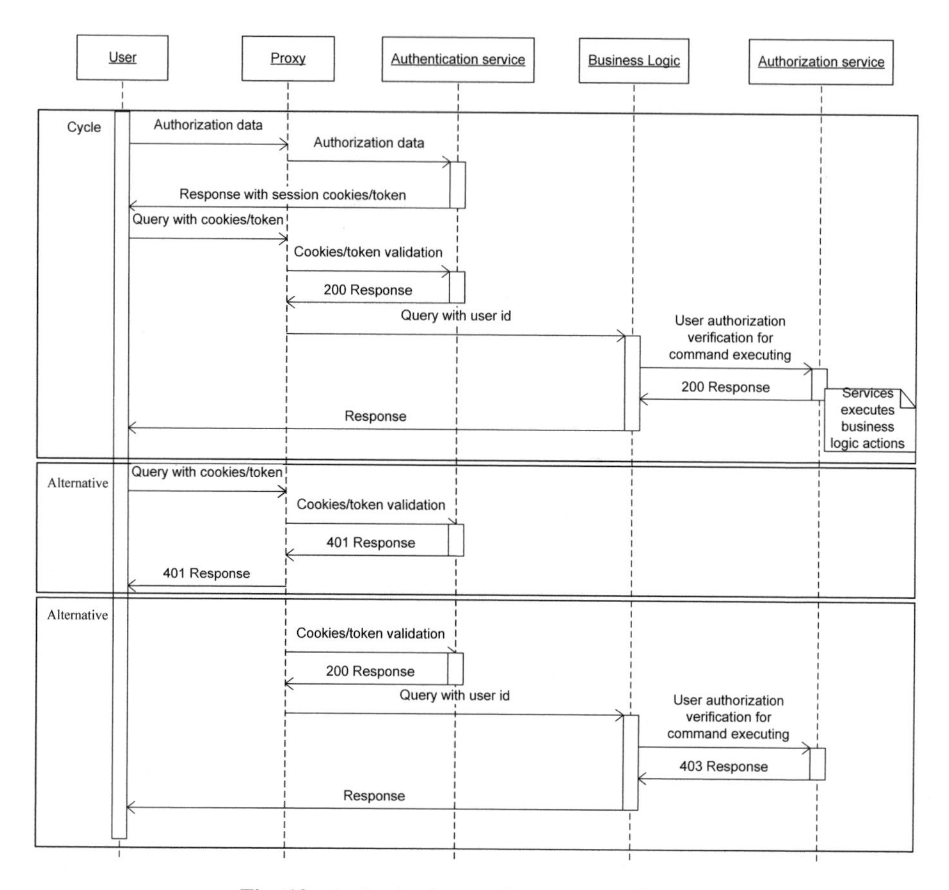

Fig. 12. Authorization service sequence diagram

So, the UML diagrams help to see the system from different sides, better understand the principles of its work, and how to use it. Data Flow Diagram shows how the system communicates with the environment [4], and how the data is processed in the system from the point of view of inputs and outputs. From its name it is clear that the focus is on the flow of information, what data enters the system, where it goes and how it will eventually be saved. When constructing diagrams we use the notion of Jordan. First, we build a context diagram, which depicts the external flows of data that enter the system, and which return information from it (Fig. 14).

As you can see, the system accepts user commands (for example, turning various devices on or off), in addition, the system automatically collects user data for further processing using cameras. As a result of the system's operation, the user receives information about the events, the state of the system and its operation through the graphical interface and the direct operation of devices. Now, for more detailed

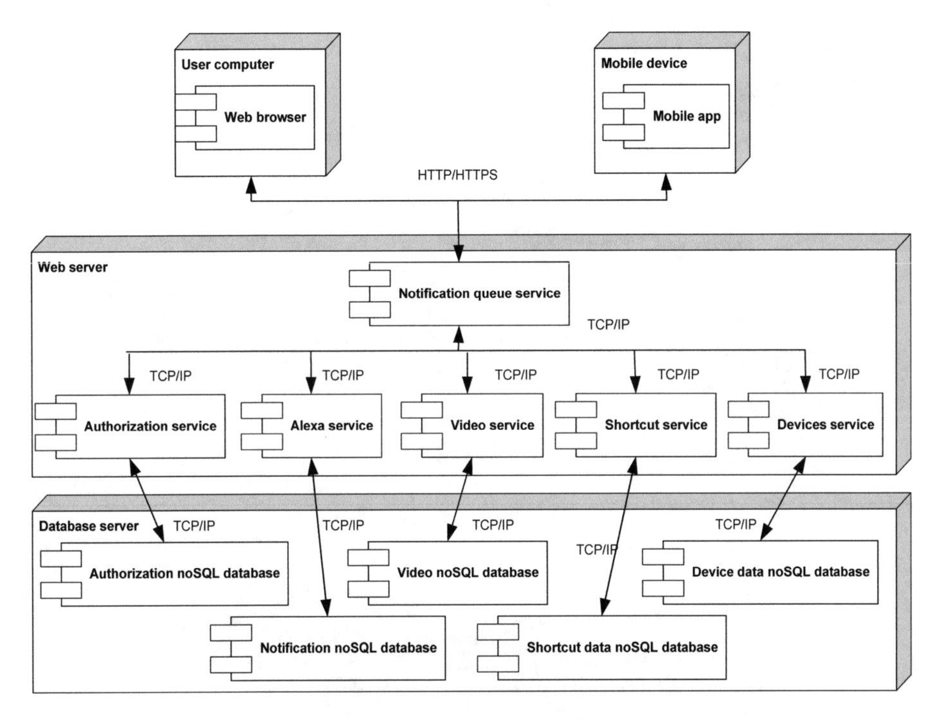

Fig. 13. System deployment diagram

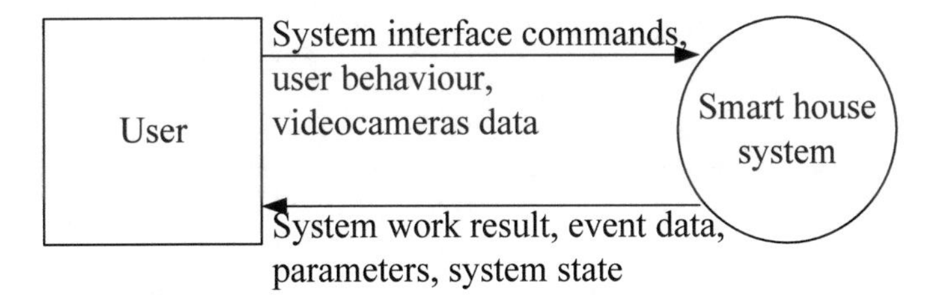

Fig. 14. Smart house system context diagram

information on the processes that occur within the system itself, it is necessary to construct a first-level decomposition diagram (Fig. 15). At this stage, you can see data flows that occur between the main components of the system. Let's perform a decomposition for them (Figs. 15, 16, 17 and 18). From the internal video data flows decomposition diagram it is clear that the service is connected to the neural network for searching and recognizing faces in the video. Also, the service stores video data and allows the user to view them in the future.

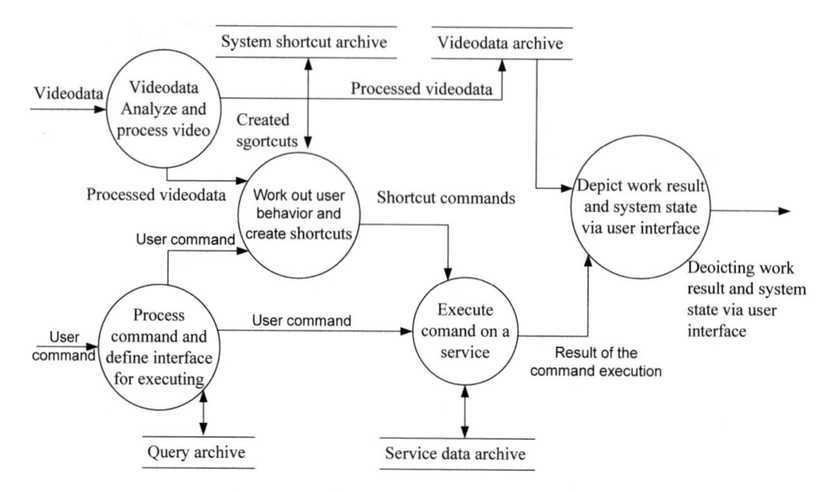

Fig. 15. The 1st level decomposition diagram

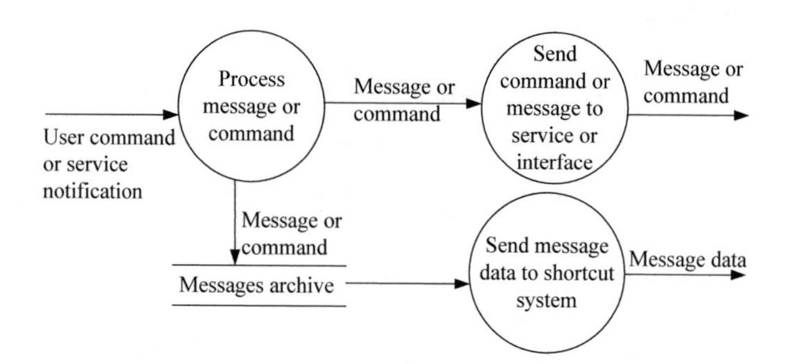

Fig. 16. Working process of messages queue decomposition

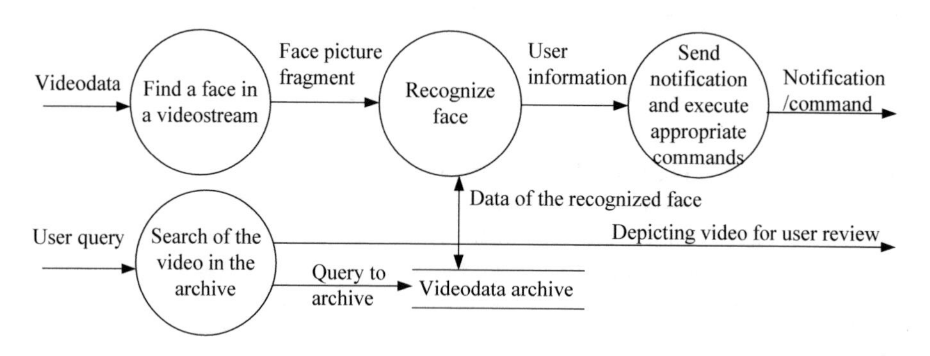

Fig. 17. Video analysis decomposition

This service is intended for analysis of all processes and events occurring in the system, and the creation on the basis of the data obtained certain patterns of behaviour of the system without user intervention (shortcuts), or offering possible options to the user depending on the previous usage. Other services are relatively simple and do not require additional analysis of data flows passing through them.

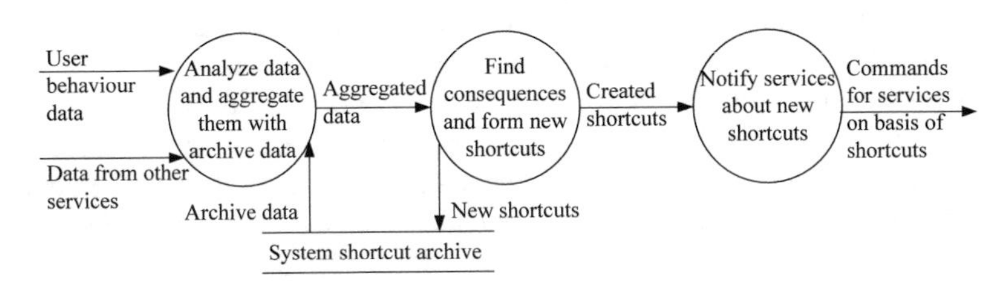

Fig. 18. Shortcut system process decomposition

Thus, after decomposing of each of the main system process, it became possible to examine and analyze their subprocesses more precisely, as well as data flows that ensure the interaction of all processes and subprocesses, both inside and outside the system. Also, the construction of such charts allowed to understand the behavior of certain services, which will be extremely useful in testing the system.

6 Formulation and Reasoning of the Problem

The main tasks while designing and developing the system will be choosing the right system architecture, determining the correct system characteristics, and the model of its functioning.

The main intellectual solutions of the system being developed will be:

- face recognition service;
- decision support system for the system of shortcuts.

It is necessary to specify the requirements for each service, as well as to the infrastructure and business logic of the system being developed for a successful choice of methods for solving the problem. First of all, it is necessary to determine the requirements for the architecture of the system, since after the start of development, changes in the architecture will be an extremely difficult task.

The main requirement for architecture will be the independence of individual tasks, because apart from the heavy intelligent solutions (in terms of software resources), services will still work for other components of the system. In this case, the dependence of all the parts between them can lead to significant product hangs and the long delays while waiting for the user to respond. The best solution in this case will be the micro-server architecture - that is, all the services of the system (certain functional units of the system, separated by the purpose of business logic) will function independently, and communicate with each other only if necessary. Now we need to describe the requirements for intelligent system services. There will be two of them - video service with the built-in face recognition system, and service shortcut (service for building certain rules for the system, based on previous user actions).

Firstly, we need to describe the requirements for face recognition system. An important criterion for choosing a particular method is that cameras are usually located near the ceiling in smart houses, that is, the user's face will fall into the camera's lens at a certain vertical angle. The chosen system should be able to work with images from this angle, and be highly precise, as this system will also be part of the home security system and will provide access to many product control functions.

Since smart house systems do not provide a large number of users (typically 5–7 people), the primary parameter will be not the speed but the accuracy of the recognition (the valid recognition time is 1–2 s). The next step is to define the requirements for a decision support system. The system is intended to help the user using the system, and will be a smart promoter that will create certain rules of behaviour based on past user actions. So the system should focus on historical data (commands given by the user, events occurring in the system at certain moments of time, etc.). In addition, it should be able to allocate certain patterns from these data, combine them on a certain attribute, which will act as a trigger for this rule (for example, the response time, or the distance of the user to a reasonable house).

Consequently, the main requirements and problems that need to be addressed when designing and developing a smart home system were highlighted. The next step is to choose methods to solve these problems.

7 Selection and Reasoning of Methods for Solving the Problem

Regarding the used methods of development, it is especially important to pay attention to the intellectual components of the system - the neural network for face recognition and the decision support system for shortsacks.

Deep convolutional neural network was used to recognize faces.

The Convolutional Neural Network (CNN) is a special architecture of artificial neural networks, proposed by Jan Lekun in 1988 for effective image recognition, part of the Deep learning technology. It uses some features of the visual cortex with so-called simple cells that respond to straight lines at different angles and complex cells, their response is associated with the activation of a certain set of simple cells. Thus, the idea of convolutional neural networks is to alternate the Convolution layers and sub-sampling layers. The network structure is unidirectional (without feedback).

Convolutional neural networks have shown the best results in the field of recognition of individuals, they are the logical development of the ideas of such architectures as a cognitron and neocognitron. The success is caused by the possibility of recording a two-dimensional image topology, in contrast to the multilayer perceptron. The convolutional neural networks provide partial resistance to scale changes, landslides, turns, angles, and other distortions. Convolutional Neural Networks combine three architectural ideas to ensure invariance to scale, rotation, and spatial distortion:

- local receptor fields (provide a local two-dimensional connection of the neurons);
- general weight coefficients of synapses (provide detection of some features in any place of the image and reduce the total number of weight coefficients);
- hierarchical organization with dimensional subsamples.

At the moment, convolutional neural networks and their modifications are considered to be the best concerning accuracy and speed algorithms for finding objects on stage. A reverse error propagation method was used for training. The network automatically learns to identify 128 points on a person's face, which is unique for everyone, and on the basis of this data it is determined whether this person is a user or not. The Data Driven model of the system is based on historical data (on the data obtained during operation of the system). Also, for the analysis of these data, the k-medium algorithm was used. The k-medium method creates k-groups from a set of objects in such a way that the members of the group are most homogeneous. This is a popular cluster analysis technique for studying a data set. Cluster analysis is a family of algorithms designed to form groups in such a way that group members are most similar to each other and are not like elements that are not in the group. The cluster and group are synonyms in the world of cluster analysis. With this method, you can select the frequency of repeating certain commands, or operations, and then allows you to create a certain regular rule.

8 Software Product Implementation Description

The developed system is called «Uncle Tom's Cabin».

The following software is required to operate the system:

- any operating system (Linux, Windows, Mac OS);
- web browser (Chrome, Firefox, IE version not lower than 6.0), for working with graphic interface.

The product is designed in IntelliJ IDEA, PyCharm, VS Code environments using the following programming languages:

- Java - message queue service;
- Python - video service, shortcut service, device service;
- C#, .Net Core - authorization service;
- HTML and CSS - visualization of the web interface;
- Typescript, Angular6 - frontend part of the system.

The system of a smart house is designed to solve the following tasks:

- management of «smart» devices within a smart house;
- security of a smart house;
- face recognition and the corresponding reaction to authorized users or other people;
- intelligent user support (creating new rules based on user behavior);
- providing a convenient system management interface.

Functional constraints of this system are:

- absence of open-loop API in the first devices created for smart house systems;
- necessity of round-the-clock functioning server (at local deployment of the system).

Since the system was developed as cross-platform, it can be launched on any operating system. The hardware requirements of the system are the following: 2.4 GHz processor frequency, 2 GB of RAM, 60 GB of storage space, Internet connection speeds of at least 10 MB/s. Despite the quantity of intelligent services developed while designing the system, its structure is quite easy and understandable. Thus, the NoSql database used to store data services consists of only one document, for each database, and has no additional links and dependencies. This allows you to get and operate on data as quickly as possible, which greatly increases the system performance as a whole. The architecture of the whole system is constructed in such a way that communication between services takes place using the message queue - this is done in order not to lose a given command or a sent notification (for example, with because of a problem on some service). So, the system architecture (Fig. 19) and databases are selected in such a way as to use the system resources as efficiently as possible, in addition, the processes occurring in the system are made as independently as possible, which avoids blockages due to dependencies overlapping between different services.

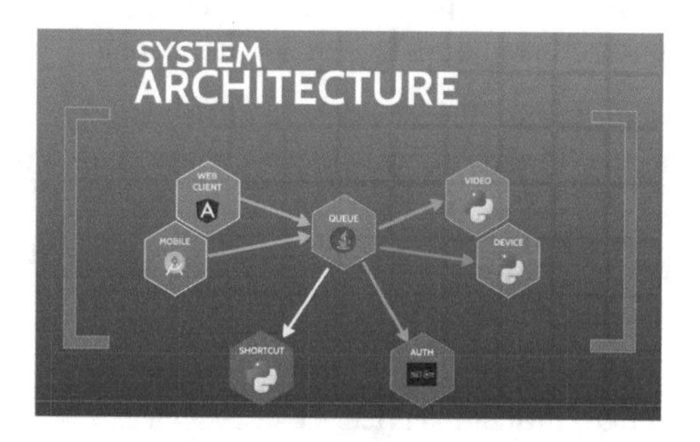

Fig. 19. System architecture

It is necessary to analyze the system and ensure that it meets the requirements and criteria after completion of the system development design. The user needs to register in the system to start working with it, for this you need to open the web page of the system interface, and then go to the registration page (Fig. 20). If the user has already registered - he can authorize using his email and password (Fig. 21).

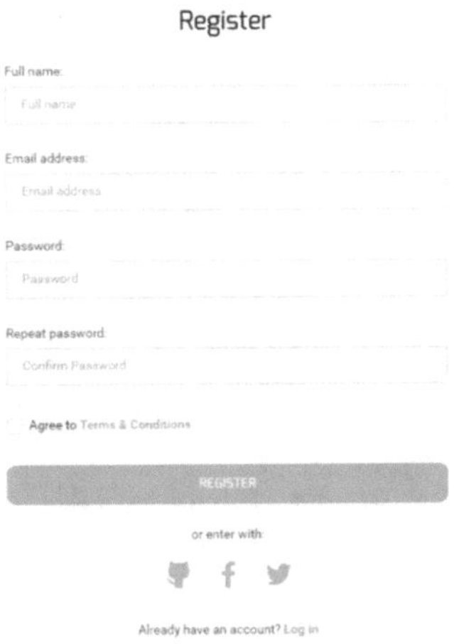

Fig. 20. System registration page

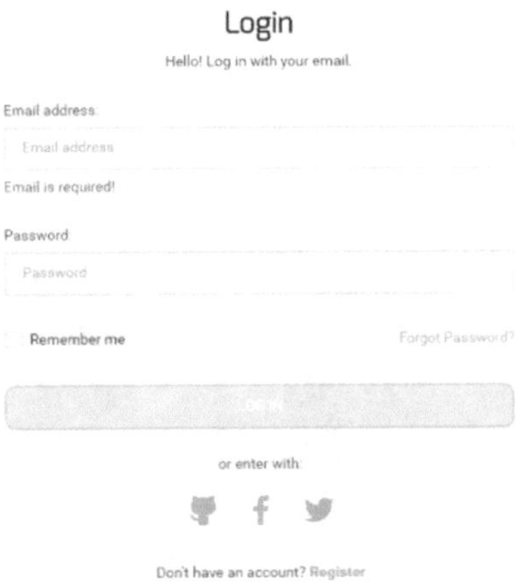

Fig. 21. Authorization page

After authorization, the user will be able to access the main page (Fig. 22) with all the necessary controls - device management, data from sensors and counters, video from video service, etc. A device control panel is on the main page of the user interface (Fig. 23). Using this panel the user can turn on/off devices, view their status, change it, and access all connected devices. Parameters of different sensors (Fig. 24), for example, air temperature or humidity within a reasonable house are located separately. The interface allows you to view metrics and set the desired parameters using the same slide bar. When the user changes current value, the air conditioner, the heater or the humidifier, which will set the temperature, or the humidity level at the user-specified value will be automatically switched on.

Fig. 22. System main page

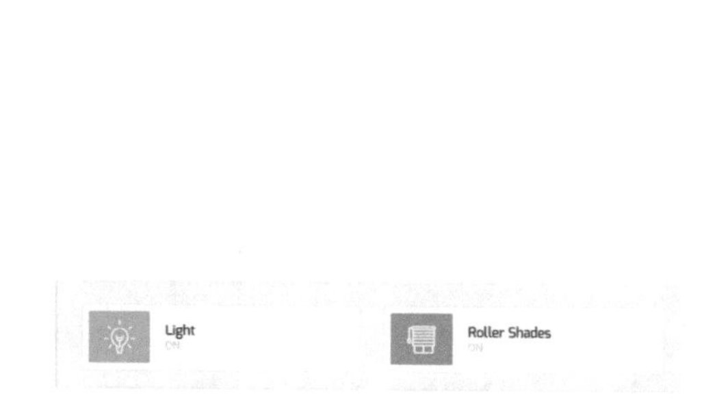

Fig. 23. Device control panel

Fig. 24. Temperature and humidity control panel

Also, the house plan is available in the system (Fig. 25), by means of which it is possible to separate the devices in the house into rooms, and manage them not separately, but within the room. For example, turn on/off the lighting. The system also allows you to connect to smart meters of many manufacturers and get data from them. For user convenience, these data are displayed in the form of a chart (Fig. 26), which specifies the period and volume of consumption of a particular type of service.

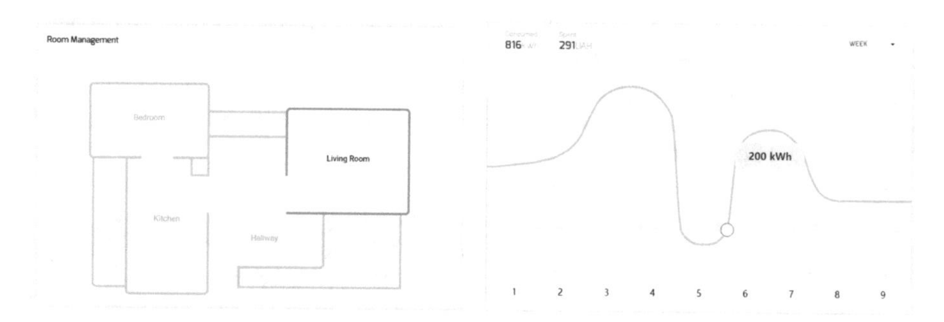

Fig. 25. House plan on the control interface

Fig. 26. Services consumption chart

In addition, the interface also provides the ability to view videos online from any camera connected to the system (Fig. 27). Also, the user can view event logs that were captured by cameras, or view records made earlier.

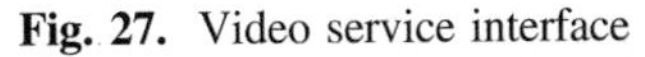
Fig. 27. Video service interface

Fig. 28. Mobile application

The system also supports the mobile management interface in a form of a mobile application (Fig. 28). The application duplicates the basic management features that are present in the web interface, and adds the ability to create certain shortcuts based on the geolocation of the mobile device on which it is installed. Obviously, the functioning of such a system, in particular its intellectual components, can not be fully disclosed through images, since the main processes are related to work in real time.

9 Conclusion

The intelligent system of a smart house, which is designed to create from any house, office, or building a smart room, was created in the overall process. Recent trends in technology development demonstrate that more and more things are being automated around us. Even ordinary things become «smarter» and open a new functional use. This has led to an increase in demand for solutions that can provide a convenient and secure way to manage such devices. So, when designing a smart house system, you need to consider the ability to support as many devices as possible, as this is one of the key benefits of systems of this kind. You also need to create different system management interfaces (website and mobile application). The key difference of system being developed from existing analogues is the use of intelligent mechanisms to facilitate the operation of the system, as well as the provision of additional functionality. Face recognition system will turn into such mechanisms that can be used both in the security system and for tracking actions within the house, as well as a system for automatically generating certain rules and commands based on previous user actions. Such solutions will allow you to adapt to the methods of using the system more easily, as well as significantly simplify the interaction mechanism of system users. So, several different technologies were chosen for the creation of the information system - in particular C#,

Java and Python, each of which was used to perform the tasks which they are best for. The Angular6 framework was used to create the front part of the system as the system was developed as a web application. We used HTML markup language and CSS stylesheets to visualize web pages. Specific methods were used to create intelligent services. A k-medium algorithm was used to create a decision support system, to analyze events occurring in the system and executed commands and attempt to create certain rules for automating individual processes. Face Detection uses a deep convolutional neural network to analyze and recognize people who get into a smart house video camera. The use of these tools made it possible to design the system correctly, make it fast and reliable, and provide a proper visual interface for the user. The optimal tools were chosen for the development of the system, which allowed to create a fast, reliable, optimized and user-friendly system with a comfortable mobile and web-based user interfaces. After the development and testing of the system, the optimal hardware and software requirements for the complete and uninterrupted operation of the product were identified. In addition, for the convenience of users, all the technical requirements and system parameters were described. After completion of the system, a control sample which showed that the system performs all the functions provided and works properly. Despite the fact that the system has all the functions, it is necessary to improve it, because solutions implemented in it will soon be implemented by competitors. So you need to think about possible options for improving and expanding the system. For example, the use of various voice assistants, as another interface management system, adding support for new devices, adding the ability to deploy the system in cloud services, etc. To sum up, a software product was designed and created (namely the intelligent system of a smart house) during the execution of the work. Its architecture, infrastructure and interconnections between services were designed. All the services from which the system was built and their communication system were created and tested. While developing and testing of the system, the hardware and software required for functioning were determined, their basic requirements were determined. After completion of the development, a control case analysis was carried out, which showed that system is fully functional and performs all set task. Finally, we can conclude that system functions are designed and implemented properly, system performs all set tasks steadily and quickly.

References

1. Kok, K., et al.: Smart houses for a smart grid. In: International Conference and Exhibition on Electricity Distribution-Part 1, pp. 1–4 (2009)
2. Shakeri, M., et al.: An intelligent system architecture in home energy management systems (HEMS) for efficient demand response in smart grid. Energy Build. **138**, 154–164 (2017)
3. Sun, Q., et al.: A multi-agent-based intelligent sensor and actuator network design for smart house and home automation. J. Sens. Actuator Netw. **2**, 557–588 (2013)
4. Nascimento, G., Ribeiro, M., Cerf, L., Cesário, N., Kaytoue, M., Raïssi, C., Meira, W.: Modeling and analyzing the video game live-streaming community. In: Latin American Web Congress, pp. 1–9 (2014)

5. Lypak, H., Rzheuskyi, A., Kunanets, N., Pasichnyk, V: Formation of a consolidated information resource by means of cloud technologies. In: International Scientific-Practical Conference on Problems of Infocommunications Science and Technology (2018)
6. Rzheuskyi, A., Kunanets, N., Stakhiv, M.: Recommendation system: virtual reference. In: 13th International Scientific and Technical Conference on Computer Sciences and Information Technologies (CSIT), pp. 203–206 (2018)
7. Kaminskyi, R., Kunanets, N., Rzheuskyi, A.: Mathematical support for statistical research based on informational technologies. In: CEUR Workshop Proceedings, vol. 2105, pp. 449–452 (2018)
8. Obermaier, J., Hutle, M.: Analyzing the security and privacy of cloud-based video surveillance systems. In: Proceedings of the 2nd ACM International Workshop on IoT Privacy, Trust, and Security, pp. 22–28 (2016)
9. Xu, D., Wang, R., Shi, Y.Q.: Data hiding in encrypted H.264/AVC video streams by codeword substitution. IEEE Trans. Inf. Forensics Secur. **9**(4), 596–606 (2014)
10. Saxena, M., Sharan, U., Fahmy, S.: Analyzing video services in web 2.0: a global perspective. In: Proceedings of the 18th International Workshop on Network and Operating Systems Support for Digital Audio and Video, pp. 39–44 (2008)
11. Brône, G., Oben, B., Goedemé, T.: Towards a more effective method for analyzing mobile eye-tracking data: integrating gaze data with object recognition algorithms. In: International Workshop on Pervasive Eye Tracking & Mobile Eye-Based Interaction, pp. 53–56 (2011)
12. Reibman, A.R., Sen, S., Van der Merwe, J.: Analyzing the spatial quality of internet streaming video. In: Proceedings of International Workshop on Video Processing and Quality Metrics for Consumer Electronics (2005)
13. Perniss, P.: Collecting and analyzing sign language data: video requirements and use of annotation software. In: Research Methods in Sign Language Studies, pp. 56–73 (2015)
14. Tran, B.Q.: U.S. Patent No. 8,849,659. U.S. Patent and Trademark Office, Washington, DC (2014)
15. Badawy, W., Gomaa, H.: U.S. Patent No. 9,014,429. U.S. Patent and Trademark Office, Washington, DC (2015)
16. Badawy, W., Gomaa, H.: U.S. Patent No. 8,630,497. U.S. Patent and Trademark Office, Washington, DC (2014)
17. Golan, O., Dudovich, B., Daliyot, S., Horovitz, I., Kiro, S.: U.S. Patent No. 8,885,047. U.S. Patent and Trademark Office, Washington, DC (2014)
18. Chambers, C.A., Gagvani, N., Robertson, P., Shepro, H.E.: U.S. Patent No. 8,204,273. U.S. Patent and Trademark Office, Washington, DC (2012)
19. Maes, S.H.: U.S. Patent No. 7,917,612. U.S. Patent and Trademark Office, Washington, DC (2011)
20. Peleshko, D., Ivanov, Y., Sharov, B., Izonin, I., Borzov, Y.: Design and implementation of visitors queue density analysis and registration method for retail video surveillance purposes. In: Data Stream Mining and Processing (DSMP), pp. 159–162 (2016)
21. Maksymiv, O., Rak, T., Peleshko, D.: Video-based flame detection using LBP-based descriptor: influences of classifiers variety on detection efficiency. Int. J. Intell. Syst. Appl. **9** (2), 42–48 (2017)
22. Rusyn, B., Lutsyk, O., Lysak, O., Lukeniuk, A., Pohreliuk, L.: Lossless image compression in the remote sensing applications. In: DSMP, pp. 195–198 (2016)
23. Kravets, P.: The control agent with fuzzy logic. In: Perspective Technologies and Methods in MEMS Design, MEMSTECH 2010, pp. 40–41 (2010)
24. Babichev, S., Gozhyj, A., Kornelyuk, A., Litvinenko, V.: Objective clustering inductive technology of gene expression profiles based on SOTA clustering algorithm. Biopolym. Cell **33**(5), 379–392 (2017)

25. Nazarkevych, M., Klyujnyk, I., Nazarkevych, H.: Investigation the Ateb-Gabor filter in biometric security systems. In: Data Stream Mining and Processing, pp. 580–583 (2018)
26. Emmerich, M., Lytvyn, V., Yevseyeva, I., Fernandes, V.B., Dosyn, D., Vysotska, V.: Preface: modern Machine Learning Technologies and Data Science (MoMLeT&DS-2019). In: CEUR Workshop Proceedings, vol. 2386 (2019)
27. Vysotska, V., Burov, Y., Lytvyn, V., Demchuk, A.: Defining author's style for plagiarism detection in academic environment. In: Proceedings of the 2018 IEEE 2nd International Conference on Data Stream Mining and Processing, DSMP 2018, pp. 128–133 (2018)
28. Lytvyn, V., Peleshchak, I., Vysotska, V., Peleshchak, R.: Satellite spectral information recognition based on the synthesis of modified dynamic neural networks and holographic data processing techniques. In: International Scientific and Technical Conference on Computer Sciences and Information Technologies (CSIT), pp. 330–334 (2018)
29. Su, J., Sachenko, A., Lytvyn, V., Vysotska, V., Dosyn, D.: Model of touristic information resources integration according to user needs. In: International Scientific and Technical Conference on Computer Sciences and Information Technologies, pp. 113–116 (2018)
30. Rusyn, B., Vysotska, V., Pohreliuk, L.: Model and architecture for virtual library information system. In: Computer Sciences and Information Technologies, CSIT, pp. 37–41 (2018)
31. Lytvyn, V., Sharonova, N., Hamon, T., Cherednichenko, O., Grabar, N., Kowalska-Styczen, A., Vysotska, V.: Preface: computational linguistics and intelligent systems (COLINS-2019). In: CEUR Workshop Proceedings, vol. 2362 (2019)
32. Burov, Y., Vysotska, V., Kravets, P.: Ontological approach to plot analysis and modeling. In: CEUR Workshop Proceedings, vol. 2362, pp. 22–31 (2019)
33. Vysotska, V., Lytvyn, V., Burov, Y., Berezin, P., Emmerich, M., Basto Fernandes V.: Development of information system for textual content categorizing based on ontology. In: CEUR Workshop Proceedings, vol. 2362, pp. 53–70 (2019)
34. Lytvyn, V., Vysotska, V., Kuchkovskiy, V., Bobyk, I., Malanchuk, O., Ryshkovets, Y., Pelekh, I., Brodyak, O., Bobrivetc, V., Panasyuk, V.: Development of the system to integrate and generate content considering the cryptocurrent needs of users. Eastern Eur. J. Enterp. Technol. **1**(2–97), 18–39 (2019)
35. Lytvyn, V., Kuchkovskiy, V., Vysotska, V., Markiv, O., Pabyrivskyy, V.: Architecture of system for content integration and formation based on cryptographic consumer needs. In: Computer Sciences and Information Technologies, CSIT, pp. 391–395 (2018)
36. Lytvyn, V., Vysotska, V., Demchuk, A., Demkiv, I., Ukhanska, O., Hladun, V., Kovalchuk, R., Petruchenko, O., Dzyubyk, L., Sokulska, N.: Design of the architecture of an intelligent system for distributing commercial content in the internet space based on SEO-technologies, neural networks, and machine learning. Eastern Eur. J. Enterp. Technol. **2**(2–98), 15–34 (2019)
37. Chyrun, L., Gozhyj, A., Yevseyeva, I., Dosyn, D., Tyhonov, V., Zakharchuk, M.: Web content monitoring system development. In: CEUR Workshop Proceedings, vol. 2362, pp. 126–142 (2019)
38. Bisikalo, O., Ivanov, Y., Sholota, V.: Modeling the phenomenological concepts for figurative processing of natural-language constructions. In: CEUR Workshop Proceedings, vol. 2362, pp. 1–11 (2019)
39. Babichev, S., Taif, M.A., Lytvynenko, V., Osypenko, V.: Criterial analysis of gene expression sequences to create the objective clustering inductive technology. In: IEEE 37th International Conference on Electronics and Nanotechnology, pp. 244–248 (2017)
40. Kazarian, A., Kunanets, N., Pasichnyk, V., Veretennikova, N., Rzheuskyi, A., Leheza, A., Kunanets, O.: Complex information e-science system architecture based on cloud computing model. In: CEUR Workshop Proceedings, vol. 2362, pp. 366–377 (2019)

41. Veres, O., Rishnyak, I., Rishniak, H.: Application of methods of machine learning for the recognition of mathematical expressions. In: CEUR Workshop Proceedings, vol. 2362, pp. 378–389 (2019)
42. Zdebskyi, P., Vysotska, V., Peleshchak, R., Peleshchak, I., Demchuk, A., Krylyshyn, M.: An application development for recognizing of view in order to control the mouse pointer. In: CEUR Workshop Proceedings, vol. 2386, pp. 55–74 (2019)
43. Lytvyn, V., Vysotska, V., Dosyn, D., Lozynska, O., Oborska, O.: Methods of building intelligent decision support systems based on adaptive ontology. In: Proceedings of the 2018 IEEE 2nd International Conference on Data Stream Mining and Processing, DSMP 2018, pp. 145–150 (2018)
44. Vysotska, V., Lytvyn, V., Burov, Y., Gozhyj, A., Makara, S.: The consolidated information web-resource about pharmacy networks in city. In: CEUR Workshop Proceedings, pp. 239–255 (2018)
45. Kravets, P.: The control agent with fuzzy logic, perspective technologies and methods. In: MEMS Design, MEMSTECH 2010, pp. 40–41 (2010)
46. Lytvyn, V., Vysotska, V., Rusyn, B., Pohreliuk, L., Berezin, P., Naum O.: Textual content categorizing technology development based on ontology. In: CEUR Workshop Proceedings, vol. 2386, pp. 234–254 (2019)
47. Lytvyn, V., Vysotska, V., Rzheuskyi, A.: Technology for the psychological portraits formation of social networks users for the IT specialists recruitment based on big five, NLP and big data analysis. In: CEUR Workshop Proceedings, vol. 2392, pp. 147–171 (2019)
48. Vysotska, V., Burov, Y., Lytvyn, V., Oleshek, O.: Automated monitoring of changes in web resources. In: Lecture Notes in Computational Intelligence and Decision Making, vol. 1020, pp. 348–363 (2020)
49. Demchuk, A., Lytvyn, V., Vysotska, V., Dilai, M.: Methods and means of web content personalization for commercial information products distribution. In: Lecture Notes in Computational Intelligence and Decision Making, vol. 1020, pp. 332–347 (2020)
50. Vysotska, V., Mykhailyshyn, V., Rzheuskyi, A., Semianchuk, S.: System development for video stream data analyzing. In: Lecture Notes in Computational Intelligence and Decision Making, vol. 1020, pp. 135–331 (2020)
51. Lytvynenko, V., Wojcik, W., Fefelov, A., Lurie, I., Savina, N., Voronenko, M., et al.: Hybrid methods of GMDH-neural networks synthesis and training for solving problems of time series forecasting. In: Lecture Notes in Computational Intelligence and Decision Making, vol. 1020, pp. 513–531 (2020)
52. Babichev, S., Durnyak, B., Pikh, I., Senkivskyy, V.: An evaluation of the objective clustering inductive technology effectiveness implemented using density-based and agglomerative hierarchical clustering algorithms. In: Lecture Notes in Computational Intelligence and Decision Making, vol. 1020, pp. 532–553 (2020)
53. Bidyuk, P., Gozhyj, A., Kalinina, I.: Probabilistic inference based on LS-method modifications in decision making problems. In: Lecture Notes in Computational Intelligence and Decision Making, vol. 1020, pp. 422–433 (2020)
54. Chyrun, L., Chyrun, L., Kis, Y., Rybak, L.: Automated information system for connection to the access point with encryption WPA2 enterprise. In: Lecture Notes in Computational Intelligence and Decision Making, vol. 1020, pp. 389–404 (2020)
55. Kis, Y., Chyrun, L., Tsymbaliak, T., Chyrun, L.: Development of system for managers relationship management with customers. In: Lecture Notes in Computational Intelligence and Decision Making, vol. 1020, pp. 405–421 (2020)
56. Chyrun, L., Kowalska-Styczen, A., Burov, Y., Berko, A., Vasevych, A., Pelekh, I., Ryshkovets, Y.: Heterogeneous data with agreed content aggregation system development. In: CEUR Workshop Proceedings, vol. 2386, pp. 35–54 (2019)

57. Chyrun, L., Burov, Y., Rusyn, B., Pohreliuk, L., Oleshek, O., Gozhyj, A., Bobyk, I.: Web resource changes monitoring system development. In: CEUR Workshop Proceedings, vol. 2386, pp. 255–273 (2019)
58. Gozhyj, A., Chyrun, L., Kowalska-Styczen, A., Lozynska, O.: Uniform method of operative content management in web systems. In: CEUR Workshop Proceedings, vol. 2136, pp. 62–77 (2018)
59. Veres, O., Rusyn, B., Sachenko, A., Rishnyak, I.: Choosing the method of finding similar images in the reverse search system. In: CEUR Workshop Proceedings, vol. 2136, pp. 99–107 (2018)
60. Mukalov, P., Zelinskyi, O., Levkovych, R., Tarnavskyi, P., Pylyp, A., Shakhovska, N.: Development of system for auto-tagging articles, based on neural network. In: CEUR Workshop Proceedings, vol. 2362, pp. 106–115 (2019)
61. Basyuk, T.: The main reasons of attendance falling of internet resource. In: Proceedings of the X-th International Conference on Computer Science and Information Technologies, CSIT 2015, pp. 91–93 (2015)
62. Rzheuskyi, A., Gozhyj, A., Stefanchuk, A., Oborska, O., Chyrun, L., Lozynska, O., Mykich, K., Basyuk, T.: Development of mobile application for choreographic productions creation and visualization. In: CEUR Workshop Proceedings, vol. 2386, pp. 340–358 (2019)
63. Sachenko, S., Pushkar, M., Rippa, S.: Intellectualization of accounting system. In: IEEE International Workshop on Intelligent Data Acquisition and Advanced Computing Systems: Technology and Applications, Dortmund, Germany, pp. 536–538, 6–8 September 2007
64. Sachenko, S., Rippa, S., Krupka, Y.: Pre-conditions of ontological approaches application for knowledge management in accounting. In: IEEE International Workshop on Intelligent Data Acquisition and Advanced Computing Systems: Technology and Applications, pp. 605–608 (2009)
65. Sachenkom, S., Lendyuk, T., Rippa, S.: Simulation of computer adaptive learning and improved algorithm of pyramidal testing. In: International Conference on Intelligent Data Acquisition and Advanced Computing Systems (IDAACS), vol. 2, pp. 764-770 (2013)
66. Sachenko, S., Lendyuk, T., Rippa, S., Sapojnyk, G.: Fuzzy rules for tests complexity changing for individual learning path construction. In: Intelligent Data Acquisition and Advanced Computing Systems: Technology and Applications, pp. 945–948 (2015)

Online Tourism System Development for Searching and Planning Trips with User's Requirements

Nataliya Antonyuk[1,2], Mykola Medykovskyy[3], Liliya Chyrun[3], Mykola Dverii[3], Oksana Oborska[3](✉), Maksym Krylyshyn[3], Artem Vysotsky[4], Nadiia Tsiura[5], and Oleh Naum[6]

[1] Ivan Franko National University of Lviv, Lviv, Ukraine
nantonyk@yahoo.com
[2] University of Opole, Opole, Poland
[3] Lviv Polytechnic National University, Lviv, Ukraine
nick.dveriy95@gmail.com, oksana949@gmail.com,
maksum99@gmail.com
[4] Variancy Lviv, LLC, Lviv, Ukraine
[5] IT Step University, Lviv, Ukraine
[6] Ivan Franko Drohobych State Pedagogical University, Drohobych, Ukraine
oleh.naum@gmail.com

Abstract. The aim of the work is to simplify the process of finding and planning travels for users through the use of the intellectual system of integration and content creation. Information sources that comprehensively describe the concept of tourist information resources are considered. An overview and comparison of certain indicators of existing tourism information and service provider services that can be considered conceptually identical to the system being developed are carried out. The existing data integration technologies were identified and analyzed, their advantages and disadvantages were examined, and the feasibility of using each of the technologies in the developed system is considered. The purpose of the system is analyzed, which is to simplify and speed up the travel planning process for the average user by providing him with all the necessary information. The boundaries of the system and the external environment in which it will function are identified; the components of the system, the main business processes, and data streams have been analyzed. A conceptual model of the system was created with the help of object notation. The list of requirements for the system, its input and output data, as well as the form of their display and dynamic characteristics are described. This work describes and analyzes the methods of data integration and content creation. The means of solving the problem were analyzed and the optimal means for the realization of the intellectual system of the formation and integration of the content is chosen taking into account the needs of the user. To implement the database, the MongoDB DBMS is selected. This article describes the structure of the system, the logic of the functions created software and processes and the results of their implementation. It is confirmed the correctness of the functioning of this system and the compliance of the results of the task. To do this, the

N. Shakhovska and M. O. Medykovskyy (Eds.): CCSIT 2019, AISC 1080, pp. 831–863, 2020.
https://doi.org/10.1007/978-3-030-33695-0_55

program has been tested and shown in the drawings of all stages of its work. With manual testing, the program proved to be convenient and easy to use.

Keywords: Consolidated information Web resource · Online tourism · Web content · Information system · Web recourses · Data integration · Content search · Content analysis

1 Introduction

Modern civilization in the present conditions was at the junction of epochs when information technology came to replace post-industrial society and obsolete technologies [1]. Nowadays doctrines and theories exhaust themselves. The achievements of the scientific and technological revolution are applied in all spheres of human life, from material production to the sphere of services [2]. Most scientists define as the main factor shaping the socio-economic foundations of the modern social system - information technology [3]. While initially they were the driving force for traditional areas of production, later, due to the development and increase in the number of information flows, they were increasingly used in the services market, including in the study of tourism and the formation of tourism business.

Today, even small travel companies are ready to apply the latest technology that allows you to instantly provide the desired information, such as transport, possible options for booking a home and automatically take management decisions related to increased interest in the tourism of ordinary users [4].

After uniting the planet into a holistic communication computer network in the '90 s of the XX century, the latter became the foundation of the world's communications and information exchange system. Information in modern conditions is an important and valuable resource for the evolution of society, increasing the competitiveness of tourist firms and the introduction of innovations. Due to this, communication between customers and sellers of tourist information can be implemented virtually, where providers of products for tourists are transported carriers, hotels, tour agencies, etc. [9].

Tourism and information are always inseparable [10]. On the basis of certain information, a decision is made on the trip [11]. Choosing a tour to buy is also information. Successful activity of a travel company requires the use of a constant stream of true and relevant information for the adoption of important management decisions, the purpose of which is to obtain the desired end result – profit [12]. Due to the fact that tourism industry participants exchange information throughout the day, there is a need for the ability to collect and process it. From this, it follows that the development of information technology (IT) in the field of tourism should be a priority [13].

The information and technology revolution has changed the principles and means of doing business [14]. Therefore, for the prompt and absolute control, a perfect analysis of the current situation, speed, and quality of customer service expected and the inevitable introduction of intelligent information systems [15].

One of these can be an intelligent system of integration and the formation of content, taking into account the needs of the user, which includes methods and tools for the search for information and further formation of it in the form of a response to a user's request. Taking into account the constant growth of demand for travel services and travel in Ukraine and abroad, as well as the importance of tourism for the economy of the country and the world in general, it can be concluded that the availability of intellectual systems in this field is a necessary component of its existence and development. Despite a large number of similar information systems and services on the Internet, the relevance of the developed intellectual system remains high, as this system will provide new features and meet the ever-increasing demands of users [16].

The aim of the work is to simplify the process of finding and planning travels for users through the use of the intellectual system of integration and content creation. One of the main tasks is to search and return their results to the user for further choosing the best options for air carriers, hotels, car rental services, etc. Aggregation of all user data will significantly reduce the time it takes to find the information you need when planning a trip by the user [17]. The results obtained during the search give you the opportunity to analyze and compare all available options and choose the best one.

Thus, the object of research is the process of searching, integrating and creating content that will be useful for tourists. The subject of research is the intellectual system of integration and formation of content, taking into account the needs of the user.

The scientific novelty of results is to apply the latest technologies for the implementation of the intellectual system, as well as to use and compare methods of data integration from various Internet resources to ensure the rapid and reliable operation of the system. The work has explored and used methods of content integration to generate results in response to a user's request, in particular, to provide quick access to all the necessary information that may be needed by the user in the process of planning a tourist trip. The system that is being developed is in the testing stage, which detects the maximum number of errors before the product is introduced into the market. In the future, this system can be used as the basis for a global search system and the formation of tourist content that will be used all over the world [18].

2 Analytical Review of Literary and Other Sources

A significant feature of tourism is that at the time of purchase of a certain service, it exists only on the site of reservation, where the client only virtually uses this tourist product [1–3, 19]. The specificity of this approach is that the desired product is impossible to try, say, taste or touch before deciding whether or not to buy the product data, and access it only at the location [20]. Because of this, the term "information tourism resources" becomes more difficult by adding it to this particular component of "virtuality", which transforms it into "information-virtual tourist resources" [21]. Such resources are repositories of information about tourism objects (possibilities of placement of clients, cultural and architectural monuments, monuments, customs of peoples, habitation, etc.); assortment and prices of tourist services; available infrastructure; sources available electronically - web pages, interactive maps and maps; payment and booking methods used to meet all possible needs of a modern travel services user [22]. Listed resources, as well as the whole set of technical means (from electronic equipment and communication networks to information processing programs), form a global tourism information environment in which the latest information technologies provide access to the following information-virtual resources that stimulate the development of tourism [23]:

- cognitive resources (web pages and tours in virtual reality, during which the tourist can pre-go to the place of travel, get acquainted with places of interest, service, and prices);
- auxiliary resources (electronic maps, electronic road maps with access to GPS devices, terrain maps, etc.);
- organizational resources (electronic booking methods in hotel rooms, transport, electronic payment methods).

The first group of informational and virtual resources - cognitive resources, includes means and methods by which a travel services user can obtain all the necessary information about the desired tourism object and can subsequently use it when planning the journey itself [24]. This set of resources allows you to choose a place to travel, to identify interesting and well-known tourist infrastructure, and usually includes firms that provide travel services, as well as regions that are interested in expanding their own tourism market [25]. The auxiliary information and virtual resources that make up the second group may be useful to users while travelling. Examples of these resources are firms with a narrow specialization, for example, firms involved in cartography and programming [26]. These include electronic maps and atlases: local, providing information directly on specific tourist infrastructure sites, and global (for example, Google Earth) - road and topographic maps, topical photos of the streets and cities of an entire planet in good quality and enormous detail [27]. The components of the third group are companies and firms that provide services related to local "tourism" objects (hotels, carriers, etc.). Ability to book a room online through the Internet, airline tickets, saves time spent on the organizational processes of travel services users and save money [1–3, 28]. Such a diversity of information resources, as well as the division of them according to certain specifics, often leads to a situation where the user is forced to view a huge amount of sites and resources in order to get complete information about the forthcoming trip [29]. Again, since the sample of each of the following groups of tourist information resources is enormous, for example, let's look at resources from the last group, namely organizational resources [30].

2.1 An Overview of Information Resources to Find Available Airfares

Airplanes are the fastest modern means of travel, so all journeys between countries or even between cities begin with the search for available airline tickets and flights. There are more than 150 different airlines, both international and local within the country [31]. Each major airline has its own site, where you can search and book your desired ticket. However, this approach takes a lot of time, because each site needs to specify certain search parameters and wait for the result. That is why in the modern Internet network there is a large number of so-called sites-aggregators, which act as intermediaries between the user and the airline's websites. Using such a site, it is enough just once to specify the necessary information for the search, and the system itself will extract and analyze data from dozens of sources and return the result [32].

Consider the first example of this site https://www.skyscanner.com.ua/. Skyscanner is the leading global search site where people plan and order from millions of travel options at the best prices. Skyscanner is always free for travelers. Skyscanner is collaborating with airlines and partners in online travel in several ways: some partners agree to provide Skyscanner with a commission for each reservation made through them, or each time visitors visit a partner website. Other partners place ads on the Skyscanner website and applications (Fig. 1).

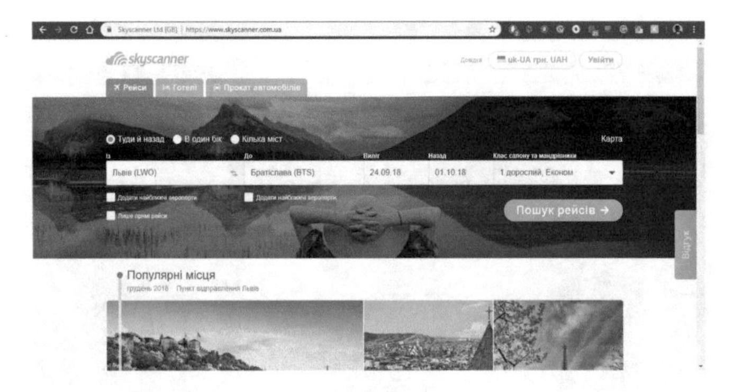

Fig. 1. The main window of Skyscanner

Next, we will consider service-aggregator https://avia.tickets.ua/.

Tickets.ua is a unique service that includes redundancy, payment systems, and business logic. To provide high-quality service to users, this platform introduces the latest technological developments. Tickets.ua meets the international PCI DSS protection standard. This guarantees the complete security of the payment card data used in the process of buying a ticket on Tickets.ua. Price analysis deals with airlines, ongoing searches for new partners and compromises with all market participants involved in pricing, to minimize the cost of airfare for the end user (Fig. 2).

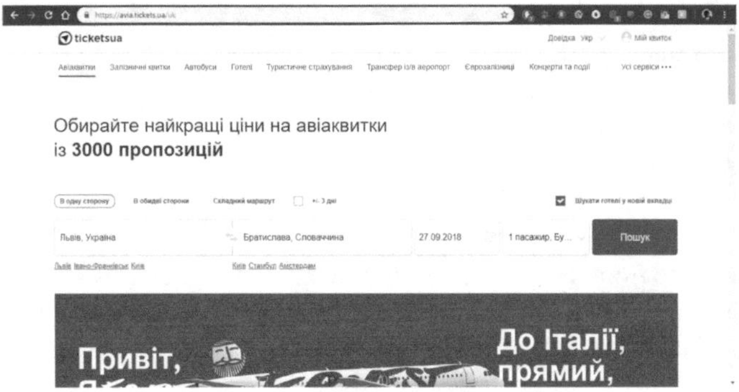

Fig. 2. The main window of Tickets.ua

The latest example will be a service that provides more functions for finding the right information than the previous ones - https://www.momondo.ua/. momondo is a comprehensive search engine for travel services that finds and compares cheap airfares, hotels, and car rental locations. The service provides up-to-date information on available travel options without additional fees. The site does not sell airline tickets, does not reserve hotels and does not rent a car - only shows affordable prices, suggesting to choose the best option. The service itself simply redirects to the chosen site and where it is possible to make a booking procedure. momondo is awarded numerous honors for the fastest meta-search system, the best use of social networks and the most convenient site for finding airline tickets (Fig. 3).

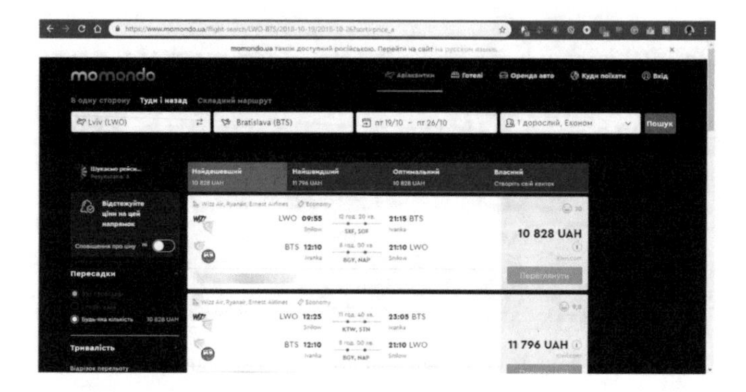

Fig. 3. The main window of momondo

We will conduct a comparative description of these three services on the following indicators: completeness of information, speed of search, user-friendliness, availability of search results for the same query (Table 1).

Table 1. Comparison of indicators of similar information resources

Function	Product		
	Skyscanner	Tickets.ua	Momondo
Information completeness	+	+	–
Searching speed	+	–	+
Convenience of use	–	–	+
The presence of search results	+	+	+
Completeness of the functional	–	–	–

As you can see from the table, each service has both advantages and disadvantages. Basically, there are disadvantages in terms of the user interface and the functionality that provides this or that service.

2.2 Analysis of Modern Data Integration Technologies

Data Integration. The process of data integration is typically linked to data warehousing and involves the consistent execution of operations such as extraction, transformation, loading, acceleration of ETL (Extraction, Transformation, Loading) of information from various sources into a single database for processing and research of these data. A prerequisite for such an integration is to undertake an in-depth analysis, firstly, of the constituent systems and data to determine the relevant information, which will continue to undergo extraction and transformation procedures with subsequent necessary "purifications" of such data, and secondly, the finite structures, in which the received data will be saved. Reporting is performed using analytical tools that allow each time to evaluate newly collected data in a new way, thus helping to obtain the information needed for making a decision [4, 33]. Traditional ETL tools were created for physical data transfer using a single integral transaction, rather than generating a virtual representation of collectible data that is available in real-time. Nevertheless, in our time, more and more products are becoming more modern, providing access to real-time data. ETL applies the physical movement of information from one repository to another while backing up data [34]. Typically, these duplicate data are finite, and, basically, all the component data used for the summary is not available for further processing. It's clear that this type of integration is needed because data integration is primarily used to process and analyze historical data to identify certain patterns that can not be set differently. Another obvious application of this type of integration is the execution of "what if" queries - while defining a set of certain parameters to predict yet unknown possibilities [35].

Integration of Information. Enterprise Information Integration is the collection of data from many systems in one definite, precise and unified form for further study and processing of data. EII technology applies a distributed query to obtain and integrate data from different resources. Usually, such a request is called federated. In this case, queries are divided by sources of information, and the performance results are combined with each other. The purpose of these integration methods is to provide access to information in various information systems in real time. Indexing, caching, and optimizing distributed queries are the main methods that provide real-time access to information. XML, JSON and WEB services are becoming a standard for creating these products. EII is a technology for obtaining information (pull), in which the distributed query searches for the necessary data and represents them in a certain user-friendly form [5, 36].

Integration of Metadata. The computational complexity of integration is greatly reduced if you create a certain structure and perform standardization of data from accessible sources of information [37]. Any resource contains a brief description of the properties and the content of its information. These data are called metadata. Creating, saving and managing metadata makes it possible to process huge volumes of data available to a modern person in any electronic form. There are three ways to use metadata in accordance with the chosen technological way of integrating resources:

- Passive provides clear information about the structure, processes of developing and applying an information resource. Information should be freely accessible to all users;
- Active save specific semantic aspects (for example, conversion rules) as metadata that can be interpreted and applied during execution. In this case, the integration process is driven by metadata. The code (i.e., active metadata) and additional information are organized and systematically stored in one repository, with the relevance of the documentation being increased;
- Semi-active saves static information (for example, defining structures, specification configurations), which will subsequently be read by another software component at runtime. For example, to check for certain parameters when processing requests, metadata is required [6, 38].

An important stage in the integration of metadata is the standardization of lists of categories of subject areas as an important component of the initial search without the use of keywords. Integration of metadata, in addition to supporting the dynamic integration of IS, will serve as a basis [39]:

- increasing the adaptive properties of the common information environment;
- strengthening security mechanisms;
- automation of the administration of through-the-line information processes;
- support analysis based on the reuse of existing analytical applications and accelerating the development of new ones.

As the most responsible functional data integration feature, the requirements for metadata management tools and IS data modelling include [40]:

- automated detection and retrieval of metadata from data sources, programs, and other tools;
- creating and maintaining a data model;
- mutual reflection of the physical and logical data model and their rationalization;
- definition of model-to-model relationship at attribute level through their graphical representation;
- the presence of open metadata repositories with the ability to exchange metadata in both directions with other tools;
- automated synchronization of metadata in multiple instances of metadata management tools;
- the ability to expand the metadata repository with metadata attributes and customer relationships [7, 41].

Choice of Technological Approaches to Data Integration. The choice of technology for data integration depends entirely on the level of development and requirements of the activity, organizational structure, level of autonomy of the organizational structure units and need for analytical data and applications [42]. So, ETL technology is most suitable when there are many information repositories and a large amount of stored historical data, and for the successful work of analytical programs, it is necessary to create a common repository of reliable and verified data for further research and multidimensional queries. ETL technology is also suitable for integrating important reference data, fixing and removing duplicate data, verifying data quality, and other important tasks. The advantage of technology is the ability to transform and move large volumes of data, while making the processes of reconciliation, purification and aggregation in the process of transmission from the source to the repository. ETL is the main method of data integration in multi-level vertical integration systems, which need to integrate large volumes of data and combine integration tools with existing analytical tools and applications. EAI technology is most useful for combining different applications in real time to automate common functions or procedures (for example, viewing data from different programs in a unified form through a common interface). The second case of using EAI is the need for cross-cutting data relevance, that is, to change the data in one application (usually a small set of records), they were reflected in everyone else. This technology works well when using the method of logging changes and then transferring them to corresponding programs or systems.

The EII technology at the general level is used if necessary to create a global system of corporate data based on a set of specific sources. The sources can serve as specialized data warehouses, different in the structure of the database, corporate data warehouses and user files [8, 43]. All sources can be geographically and organizationally distributed, but data, reports, and other information are exchanged inside the storage hierarchy. At the same time, the principle of autonomy of corporation units is realized and a single control is carried out through the allocation of access rights. Such a structure is called a global data warehouse [44].

3 System Analysis and Substantiation of the Problem

The system analysis can better justify the decision that is taken and minimize the complexity associated with the understanding and problem statement. System analysis of software development helps to better understand the working principle and the composition of the system being developed [45]. At the stage of system analysis, the system is reviewed at the conceptual level, without identifying the technical details, which allows you to identify possible problems, as well as find methods for their solution [46]. The level of development of information resources in a particular industry determines the overall degree of automation, informatization, and relevance of this industry [47–51]. Informatization is called a set of scientific and technical and socio-economic processes, which are aimed at creating favorable conditions for satisfying user desires, which in the field of tourism are reduced to a single cognitive need in the tourist service [52–58]. Tourist informatization is closely connected with the use of computer systems, which allowed to change the process of creation and evolution of travel companies. This was made possible by the use of centralized storage and processing systems, as well as control and planning systems. Thus, the introduction of the latest information technologies into the network of hotels has greatly simplified the process of booking rooms, improved the procedure for conducting settlements with customers, improved control over profits, employment, remuneration, improved information on transport, etc. [9, 42, 59–68]. It is in the tourism three classes of computer systems on the functions performed [45]:

- The main information systems that provide processing of customer requests through access to the central reservation system.
- Auxiliary systems that automate the processing of documents within a travel firm.
- Management systems - information systems that were created to summarize information about the work of the firm and make managerial decisions [10].

The World Tourism Organization has repeatedly tried to create a single standardized information graphic system for tourists from all over the world. But because the same objects are different in different countries, it is still not created. The basis for this kind of information system can be the tourist information centres that exist in most cities and even villages with some interactive tourist resources. Such centres exist and continue to be created in Ukraine. As a result, questions may arise, which can be considered informational and informational-virtual tourist resources?

Scientists V. Kvartalnov and I. Zorin refer to informational tourist resources information about certain objects that are on a tourist route and differ with artistic, historical, cognitive or scientific value, and contain information on the origin of cities, villages and farms connected with their myths and legends devoted to the artwork, scientific articles, maps, art albums, photo, audio and video materials, even those who are the owners of information and can convey it in a form that is interesting to customers [11]. The term "information tourism resources", Lviv scientists - authors of the textbook "International Tourism and Services" (K., 2008) define as "a set of forms and types of information about a particular territory or objects, the history of the territory, culture, nature and the population received by tourists directly during a trip in the process of preparation for it or after a trip" [12]. The Internet and other interactive multimedia systems have great importance for tourism development. So, in the current information age, user inquiries have radically changed. People no longer want to find the right tourist information in paper sources, but require an immediate response from the Internet. Creating a system to solve a specific user problem is the primary purpose of system analysis. Such a system contains components that significantly affect the result. The basis of the system analysis consists of the following concepts: system, element, function, purpose, external environment [13].

In general, the concept of "system" can be defined as a plurality of objects, together with the interconnections between these objects and their properties.

Integration and formation of the content are carried out with the help of the user's work (forming request), third-party web resources (providing information), the developed system (the formation of the result). The purpose of search and content development is to facilitate the travel planning process for the user. Content search and formation contains the basic properties of the system: it consists of interacting components, has a certain structure and purpose of the system's existence. From this, we can conclude that integration and the formation of content is a system [14].

An element of the system is a certain part of the system that is considered integral since it is impractical to make its decomposition. Elements of the system of integration and formation of content are data of various types, which are on various Internet resources. The system being developed allows you to do the following: search, integrate, formulate and output the results.

The main purpose of the system being developed is to simplify and speed up the process of finding information when planning a tourist trip or just a trip.

The complexity of the system is divided into simple and complex. Simple systems that have no complicated structures and are constructed of a small number of links and elements are simple. Complex systems are characterized by a large number of components, internal and external bonds that are heterogeneous and perform complex functions [15]. The complexity of the system of integration and the formation of content can be attributed to the complex, as it possesses certain properties of a complex system: integrity, universality, potentiality, urgency.

The Purpose of the Development. The result of the work is ready to use the intelligent system of integration and the formation of content, taking into account needs of the user, which will allow you to search for certain parameters set the necessary information to user and will form the found results in a standardized, fit to process and save the look. The future system should receive data from the external environment, that is, from the user, to store personal data of the user for further use of them in search, if this is required by the user. A system user should be able to access the system from anywhere in the world using the authentication system. All requests must be executed in a stable and correct manner. The number of search and computing operations should be minimal, as well as the time of their execution. Therefore, the purpose of the developed system is to simplify and speed up the process of finding tourist information to reduce the time required to obtain a complete set of data about a particular place of travel.

Description of the System's Purpose. To describe the appointment of an intelligent system of integration and content formation, we apply the object approach, named the UML diagram of usage options. This use-case-diagram shows the relationships between actors and functions in the system. The diagram reflects the typical user-system interaction. Before you build a diagram of use options, you need to define all external entities. An external entity is a role that a user has in the system [16]. In the developed system, you can distinguish the following roles:

- Anonymous user is a user who is not yet authorized in the system.
- Registered user is a user who has undergone the registration process, has verified its data in the system and is authorized by the system

Usage options are functions and actions that can be performed using the system:

- Registration and Authorization;
- Specifying and preserving certain parameters in the system to personalize the search process;
- Direct search process and Saving results.

Having a list of all the functions of the system, we can build a diagram of options for use (Fig. 4) for each of the user roles. Having considered a more detailed usage diagram, you can note, that work with the system is possible from two states - authorized and anonymous. An anonymous user can log in if he already has an account created on the system. In the absence of an account before the authorization, you must complete the registration procedure in the system. When logging in, you need to enter a login and password, if the login and password do not match, the user will receive an error message. The process of information retrieval can be initiated by any user by presetting certain parameters to reduce the search area, such as the dates of the planned trip, the number of people, the possible budget, etc. An authorized user also has the opportunity to personalize the search system for yourself. To do this, you need to fill in certain fields in your personal cabinet, the values of which will subsequently be used in the search process, so that users will not have to set invariable search parameters each time. These can be such data as the budget framework (how much money the user is ready to use for the trip), the number of people (comfortable for family travel), comfort level (economy, business, luxury), etc. The system being developed is very narrowly specialized in tourism, and also tries to combine the ease of use and the completeness of the functions provided. Therefore, the number of roles in the system is not large, since there is no need for some kind of leadership roles (administrator, manager, etc.).

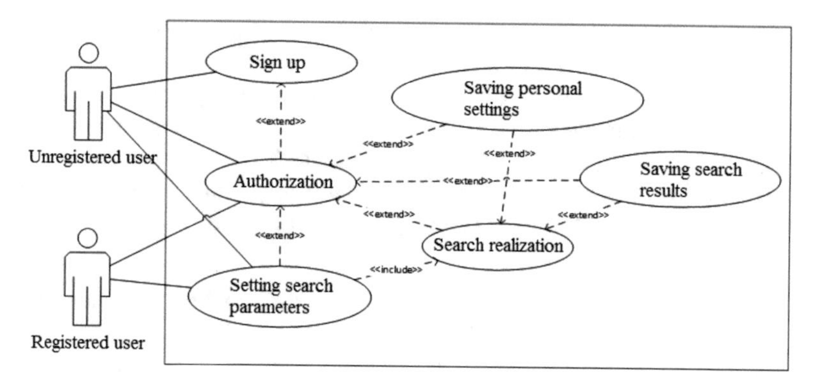

Fig. 4. Diagram of options for use

System Application Place. The intellectual system of integration and content creation is designed for use in the field of tourism, namely, tourism information technologies. Since the knowledge of the world and the environment is one of the oldest features and needs of people, and to meet this need, people are ready to lose money, using the services of travel agencies, we can say that this solution will be interesting for those tourists who already have some experience in drawing up plans, trips, as well as for those who are just beginning to travel, but do not want to pay for travel companies and, at the same time, wants to save time and money. The system being developed is a web application, so you need to have a connection to the Internet to work with. The web application does not have high hardware performance requirements, so it can also work on mobile devices and tablets.

Justification, Development, and Implementation of the System. The planning process is generally complicated because it must take into account many factors that have a certain effect on the final result. The key to successful planning is the availability of as much information as possible and its further analysis. The described approach is completely transposed to the planning process of travel, as there is also a huge number of parameters that need to be taken into account in order to get a good result - prices, quality of service, location, accessibility, and other equally important factors. The complexity of the planning process, route planning, optimal budgeting, quality hotel selection, and car rental is one of the main problems that arise after making a trip decision. The importance of solving this problem is due to the influence of the process on the success of the entire trip. Thus, the actual problem of obtaining the maximum amount of necessary relevant information is solved, which consists of aggregation and the formation of a single access point, which saves the time of the user and his money.

Expected Effects from the Implementation. Since this system is focused primarily on ordinary users, not on travel companies, so there is no need to expect any large-scale effects in one or another field, but certain effects will still take place. After system implementation, the following effects are possible for individual users:

- Economic effect is reducing user's costs by providing information about different travel options.
- Growth in productivity is the main effect of tourism in general and tangential to the work of system in particular is a person who regularly meets his need for rest, works more productively than one who does not rest at all.
- Social effect is again, as one of the main effects of tourism, this effect means the overall improvement in the quality of user's life.

Development of the Conceptual Model of the System. One of the most complex and key tasks in developing systems is to create a conceptual model. This phase helps to identify the main causes and consequences of the type of cause and effect, the accounting of which is required to achieve the desired results, as well as to delineate the scope of the subject area. For conceptual design, it is necessary to describe the subject domain with the terms of the formal language, therefore an object approach to data modeling was chosen for this. This approach focuses on the behavior of objects and data management tools [17]. To construct a conceptual model of the intellectual system of integration and content formation, it is necessary to determine which input (Table 2) and output data (Table 3) are necessary for the functioning of the system and will be the results of its work. The input data is all information that comes from the system environment. Accordingly, the source data is the result of the system's operation: data that are issued to the user after processing by their system [18].

Table 2. Input system data

Name	Description	Limitation
Login and Password	Data for authorizing the user in the system	Number of characters: 8 to 20
Search options	Data to specify for search	Data is required and may not be empty
Personal settings of the user	The data the user specifies in his personal profile and may later use them in the search	These data are optional but contain certain limitations on the format (dates, budget, etc.)

Table 3. Output system data

Name	Description	Limitation
Error message when logging in	The system issues this message if there is no matching pair of values "login-password"	None
Search results	Contains data that matches the search query	Can be empty if the result is absent
Saved search result	A PDF file that a user can save for himself	Can not be empty

Sequence Diagram. In order to reflect the interaction between the components of the system being developed, it is necessary to use a sequence diagram that allows you to visualize the relationships between messages transmitted over a certain period of time, using the conditional time axis. In this case, the interaction between the elements is depicted in the form of a time graph of the set of all interconnected objects [19]. For a sequence diagram, it's enough to display objects that interact without displaying their interactions with any other components. This diagram can be read in two directions:

- First – vertical dashed lines that represent the lifetime of interconnected objects and the time of the existence of objects in the system;
- Second – the moment of management in the form of elongated narrow rectangles, which indicate the period of time during which the object is active, that is, it performs a certain action. In this case, the upper side of the rectangle is the beginning of the active stage, and the bottom – at its end.

Arrows indicate and, according to the transmission time, arrange the messages that objects transmit. There are three types of messages:

- Painted arrow is a message that is intended to perform certain actions and operations, call features or display individual control flows; often goes along with the acquisition of focus control;
- Not painted arrow is message that is used to display a simple asynchronous message that is transmitted at any time;
- Dot-dash arrow is message when returning from the procedure [20].

To construct a sequence diagram (Fig. 5), it is necessary to simulate the lifeline of the actor "User", which is the main one in the system. The diagram shows all possible actions in the user's time.

Component Diagram. The Component Diagram provides a reflection of the transition from a logical representation to a specific implementation of the system using a program code and using specific elements. The graphical constituents of this diagram are components, dependencies, and interfaces [21]. The Component Diagram (Fig. 6) is used to display the physical organization of system components, often they can be files. One of elements of components chart is interface. It is depicted as a circle and can be connected to a component that implements this interface with a solid line without arrows. The connection between components is presented in form of a dotted line with an arrow directed from the dependent element to parent. These dependencies may

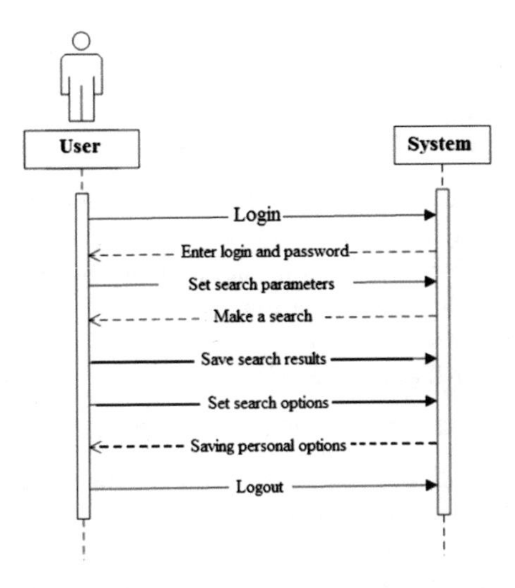

Fig. 5. Sequence diagram of the user's actions in the system

indicate that the independent component contains classes that are used in constructed objects. In the presence of an arrow from the component to the model interface, it is assumed that the component uses the interface at runtime [22].

As a template for the system being developed, the Design Model-Submission-Controller (MVC) design pattern was chosen, which is often used in web programming. The basic idea behind this template is the separation of business logic and the user interface to create opportunities for changes and independent development of individual parts of the system without altering others [23].

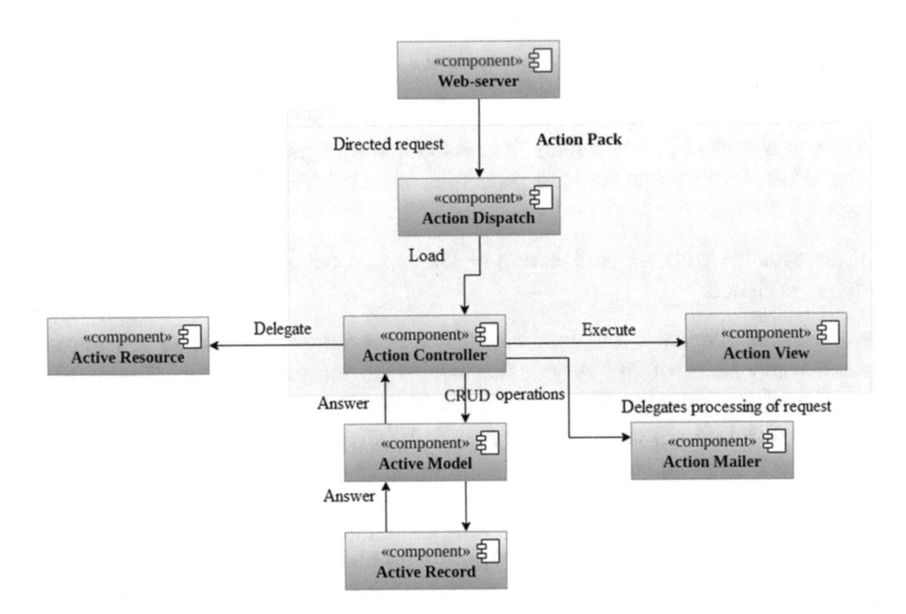

Fig. 6. Component diagram

Class Diagram. The class diagram is a logical model that allows considering the architectural features of the structure of the system and relation between individual entities with their attributes and possible functions.

A class diagram can contain the following structural elements:

- Class is a generalized description of a set of homogeneous entities having the same properties and operations. The class is designated in the form of a rectangle, where the name of the class is indicated from above, in the middle of its properties (which must be inherent to all instances of the class), at the bottom of its function (the possible actions of the class object);
- Relation is the relationship between classes.

The relationship between classes can be of four types:

- Association relation indicates the presence of a certain connection between entities, denoted as a solid line with additional symbols;
- Relation of generalization – indicates the relationship between parent and child objects, is indicated by a solid triangular arrow pointing to the parent class;
- Relation of aggregation is the form of association that reflects the "whole-part" relationship and is denoted by a solid line with an empty rhombus pointing to the whole class;
- Relation of composition is a subspecies of aggregation, which is characterized by the fact that parts of the whole exist only as long as there is a whole. This connection is manifested by the solid line with the rhombus, indicating the class-part [24].

The main classes (Fig. 7) that will be implemented in the business logic of the web system are:

- User is a class that has a description of all the necessary information about the user. It also contains the methods necessary to authorize the user.
- Source is a class that contains a general description of the source of information that will be processed during the search.
- Request is a class that represents direct search instances with all input parameters.
- Settings is a class that describes the personified user preferences.
- The result is a class that contains properties of a search result and how to work with them.

In accordance with the architecture of the system being developed, the following chart was designed:

State Diagram. The state diagram contains a description of possible sequences of transitions and states of the system that determine the generalized behavior of the system throughout its entire life cycle. By describing the reactions of objects to certain events, this diagram represents their dynamic behavior. That is, these diagrams represent the models of all possible changes in the states of specific entities.

A state is a certain characteristic position of an object in its life cycle, in which it performs a certain action, or expects a certain event that will cause a change in its state. Separate types of states are initial and final. The initial state does not have its own internal actions and is a pseudo-state. In the beginning, the objects are in the initial state

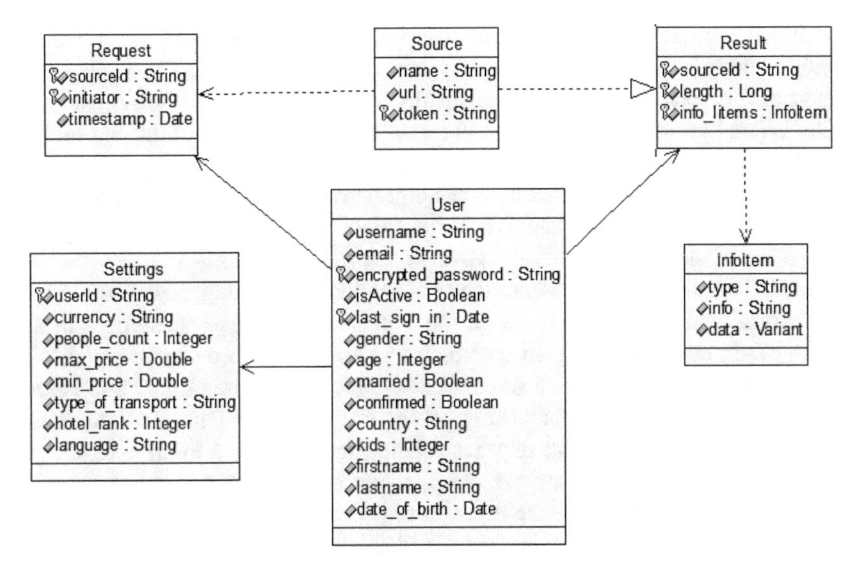

Fig. 7. Web application class diagram

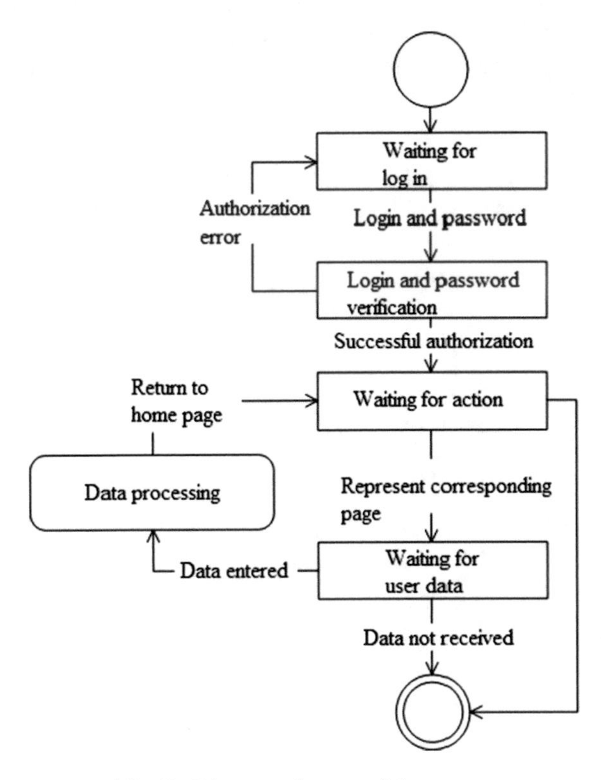

Fig. 8. Diagram of states of the system

at the initial moment of the life cycle. The initial state allows you to see where the change of states begins and is graphically indicated by the circle. The final state does not have any internal actions and is graphically indicated as a thick circle, painted in the middle. When the system finishes, the object enters the final state at the end time. This means that the process of changing states is complete [25].

It is logical to understand that the transition between states is a change in the status of an object from one to another. While the object is in the first state, it can perform certain actions and will be able to move to the next state only if these actions are performed. To move an object from the original state to the target, you need to make the transition work. To do this, one of the events is necessary: receiving a message, receiving a signal, completing an action. It is also possible that there are restrictive conditions that allow a transition only after obtaining this condition of true value.

Diagram of states of the system (Fig. 8) shows that this system starts the execution process in its original state and requires logging in (entering a login and password). After verifying that the login and password are correct, the system either returns to the waiting state of the input (if the incorrect data is entered) or goes to the state of waiting for the choice of action from the user in the event of a successful authorization. Selecting a specific action displays the corresponding page of the system; the system itself waits for further input of data. In the absence of data, the system goes into the final state. In the case of entering the relevant data, they are processed in accordance with the request and the final results of work are output. The next step is to return to the main page, where you can continue working with the system. Otherwise, the system will turn to the final state.

Activity Diagram. The UML activity diagram depicts the behavior of the system in the form of functions that can be performed by the system and the user. The sequence of work can depend on certain conditions. The activity on the diagram is a rectangle with

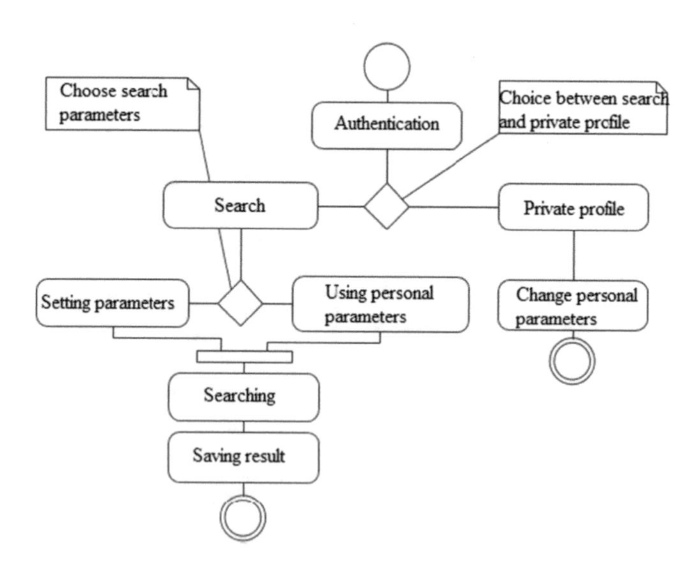

Fig. 9. Activity diagram

rounded corners. The arrows between these rectangles represent the flow of control. The decision process is depicted as two arrows emerging from the corresponding rectangle. Each of them may have the text of a condition that it expresses. Several actions can be performed in parallel and end at a synchronization point [26].

Figure 9 shows a chart of activity for an intelligent system for the formation and integration of content, taking into account the needs of the user. Consider the activities registered in the user system. The first step is to get the system login data - login and password. After receiving this data, the system searches for the corresponding records in the database, if the corresponding record is found, then the system login is executed. In the absence of such compliance, the user receives an error message.

Consider the case when the user logged in, then he can go to the information search or to his private cabinet. When switching to "My Profile", a page opens with information about that user that allows you to modify and store personal information and personalized search settings. When switching to the "Search" activity, a form is opened where you must specify certain required parameters necessary for this process, such as travel dates, number of people, desirable place and place of the start of the trip. When you switch to "View Search Results," the opened page displays the results of the current search. This activity allows you to proceed to save the results.

Deployment Diagram. The UML deployment diagram displays the model of the physical deployment of the system on the nodes. For example, in order to describe an abstract web application, the diagram should show the existing hardware components, the software components working on those nodes, and how those nodes interconnect.

The nodes are represented in the form of rectangular parallelepipeds with internal components, depicted in the form of rectangles. The nodes may have their own internal nodes, which are fed as embedded rectangular parallelepipeds. One node in the deployment diagram can generally express many physical nodes, such as a distributed server cluster. There are two types of nodes: hardware and software. Hardware node – is a physical device with its own memory and means for performing software such as

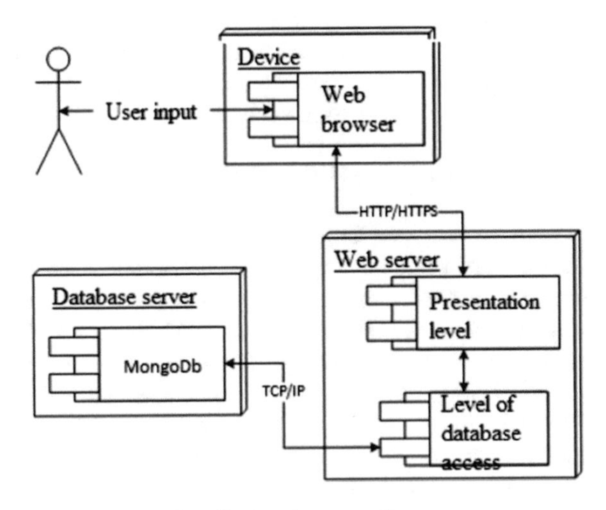

Fig. 10. Deployment diagram

conventional computers, mobile phones. Program node – is a programmed resource of a runtime environment that runs inside an external hardware node and is a service that provides execution of other program elements [27].

Figure 10 shows the deployment diagram for the intellectual system for the formation and integration of content creating a specification of software requirements, taking into account the needs of the user. This diagram consists of three device nodes:

- Device is any device equipped with a web browser that acts as a connecting interface between system and user.
- Web server is a server that actually contains the business logic of the system, accepts and processes user requests and interacts with a database server.
- Database server is a server, that is an interface between a web server and a database, so it responds to queries from the web server side with certain changes in the database.

4 Choice and Justification of Problem-Solving Methods

Let's consider the generalized methods of forming integral data for Internet systems based on the distributed integration tools of the disparate data, their structure, and syntax.

Data Integration Model. The development of a strategy for modelling the stages of data integration is possible by replacing the scheme's time as an object of integration in the standard model for the term of the data set. The section will describe a model that develops and summarizes the formal model of data integration. The generalization is that the integration process takes into account not only the data schema, but also data in particular, as a set of values taken and arranged in a certain way, having certain functions and applications, and which are depicted by means of special means. Each dataset is presented as a schema symbiosis as some formal description of the data structure and a set of values that are formed according to the scheme. Thus, the conditional components of a model are a set of input data sets, an initial set of integrated data, and a certain mapping that performs the matching between the components of both sets. Such a model can be represented as a form

$$< VM^L, Map(VM^L, VM^I), VM^I > \quad (1)$$

where $VM^L = \{<V_i, \Sigma_i> \mid i = 1,\ldots,N\}$ is set of local input samples, Σ_i is the schema of the i-th local data set, created using the language of description of local circuits L^L, V_i is a set of values formed according to the structure of the scheme, $VM^I = <V^I, \Sigma^I>$ is the initial set of integrated data, Σ^I is schema of the original global set of integrated data, created using the language of description of global schemes L^I, V^I is set of values of the initial set of data, $Map(VM^L, VM^I)$ display of input data into the resulting set of integrated data. The main difference between the extended models of integration from the formal model is the notion of a finite set of integrated data as a result of integration. This set can be formed by passing the input values to the shared environment and then displaying them through virtual elements and structures. In general, this model is much more relevant to the real stages of integration than the formal model. Using such a model allows you to create a fairly accurate and adequate description of the stages of data integration. In general, the definition of a certain set of *VM* data can be submitted as the following system

$$VM = <V, G, M, P>, \tag{2}$$

where V is set of values of certain concepts of the chosen subject area, G is a formal syntax of the data, M is a formal description of the data structure, P is standardized image data semantics. In this way, a formed image of a plurality of data in the form of a tuple $<V, \Sigma>$, where V is set of values, Σ is the data schema changes to the look $<V, \Theta>$, where $\Theta = <G, M, P>$, a formal representation of the syntax, structure, and semantics of the data in this set, which is called meta scheme. The schema is a generalization of the scheme's term, which was created by expanding the description of the structure and data constraints by means of describing their semantics and syntax. The meta-schema allows you to create a wider and more detailed description of the data properties at the integration stages, unlike the schema. In general, the process of data integration consists of a sequence of actions that ensure the transformation and creation of new data from the original. The integration process acts in the form of a certain orderly set of actions that must be performed over the data, namely, the coordination, transformation, association and filtration of data. The ultimate goal of this process is to create a global *VM* data set based on a set of initial sets of data that can be formally depicted as

$$VM = I(VM_1, VM_2, \ldots, VM_N), \tag{3}$$

where VM_1, VM_2, …, VM_N are set of input source sets, N is number of sets to be integrated, I is an integration operator denoting the correspondence between the set of incoming sets of data and the output final data set and defines the necessary sequence of changes for obtaining integration results.

In general, such sets of data may include duplicate values, i.e.

$$V_1 \cap V_2 \cap \ldots \cap V_N \neq \varnothing \tag{4}$$

Taking into account the previously described data model, based on the definition of their semantics, syntax, and structure

$$VM = <V, \Theta> = <V, G, M, P>. \tag{5}$$

The integration process can be represented as operations on these components, changing the description of the sets of data to describe all of their constituents

$$<V^I, \Theta^I> = <V^I, G^I, M^I, P^I> = I(<V_i, \Theta_i> \mid i = 1, \ldots, N) \tag{6}$$

$$= I(<V_1, G_1, M_1, P_1>, <V_2, G_2, M_2, P_2>, \ldots, <V_N, G_N, M_N, P_N>),$$

where $<V_i, G_i, M_i, P_i>$, $i = 1, 2, \ldots, N$ is detailed presentation of the i-th data set.

Thus, the data integration process can be divided into separate components of the integration of structure, syntax, values, and semantics of data. The global operator of data integration and in this case, we will depict the form

$$I = <I^V, I^G, I^M, I^P>, \quad (7)$$

where I^V is indicates the relationship between the set of data values in incoming sets and values in the global set; I^G is specifies the match between the syntax components of the local data and the syntax of the global plural; I^M is specifies the correspondence between the description of the structures of the local data sets and the global structure of the finite data; I^P is denotes the correspondence between the components of the semantics description of input kits and the description of the semantics of the source data. The integration phase is broken down into smaller components:

$$\begin{aligned} <V^I, G^I, M^I, P^I> = {} & <I^V(V_1, V_2, \ldots, V_N), I^G(G_1, G_2, \ldots, G_N), \\ & I^M(M_1, M_2, \ldots, M_N), I^P(P_1, P_2, \ldots, P_N)>. \end{aligned} \quad (8)$$

This image of integration forms a certain sequence of processes is the integration of syntax, structure, and semantics of data.

Integration of Syntax. The issue of syntax integration is fundamental compared to the integration of other components of the data. Solving the problem of creating a homogeneous structure and semantics of data is really only under the condition of the existence of a unified standardized system of their designation. Determining the syntax of data in itself is compiled and includes various aspects of their display in databases, repositories, documents, etc. Therefore, the syntax of data can be represented as a combination of three components

$$G = <A, T, L>, \quad (9)$$

where A is alphabet, T is set of data types, L is set of syntactic constraints.

The alphabet contains a set of characters that are used to display data in a defined environment. Typically, the alphabet includes letters, numbers, special characters, and business symbols. However, the composition of the alphabet is influenced, in particular, by factors such as the geographical location of the data processing environment, the type of task for which data is used, the particularities of the methods for storing, sending and processing, the specificity of the interpretation and the use of different data values. In modern systems, along with the usual methods of designating data, we use sound, graphic and other elements for their display and processing. Also, complex and complex types of data, dynamic and static data are used, which adds interference to the creation of a standardized representation of such data. The data type is defined as the result of the data split by the methods of display and processing. In our time, there are many new specific types of data that describe the peculiarities of the processing, use, and content of the data. These are, for example, scalar types such as HyperLink,

Currency, Object, and other complex types - Array, Document, Set, XML and JSON documents, etc., and user-defined data types. Such a large number of data types provides additional possibilities for displaying and processing information, but at the same time, complicates the process of data storage, procedures for their sharing, transformation, and generalization. The limit, as part of the syntax of data, is used to summarize the forms of display of data and values corresponding to the quantities and concepts that they describe. The syntax limit is expressed by means of quantitative indicators, formats, templates, rules for creating values, matching subsets of allowed symbols, etc. The described restrictions can be defined both at the information processing stage and at the level of display of the data to the user.

The process of syntax integration consists of a set of integration processes of the alphabet, types, and limits. The syntax for displaying the values of the integrated set of G^I data is formed by combining three components

$$G^I = <A^I, T^I, L^I>, \quad (10)$$

where $A^I = I^A(A_1, A_2, \ldots, A_N)$ is the alphabet of an integrated set of data created from alphabets of input data sets, $T^I = I^T(T_1, T_2, \ldots, T_N)$ is a set of data types used in an integrated set obtained by adding input types, $L^I = I^L(L_1, L_2, \ldots, L_N)$ is the set of limitations for an integrated set of data, comprised of limitations of input data, I^A, I^T, I^L are operators integration alphabets, types and limits.

Integration of Data Structures. The heterogeneous data model of the information resource of Internet systems is based on the integration of structured information stored in databases and nonrelation data, which includes poorly structured data, data without structure descriptions, dynamic and static data, procedures data, etc. A generalized model of the structure of such a set of heterogeneous data can be depicted as

$$C^I = <D, <ND_1, ND_2, \ldots, ND_k>, <J^D, J^N, J^{DN}>>, \quad (11)$$

where C^I is generalized structure of integrated data, D is the structure of relational data depicted in the form of database tables; ND_1, ND_2, …, ND_k are the structure of non-relational data of different types, J^D is set of relations between relational components, J^N is a set of relations between non-relational components, J^{DN} is a set of relations between relational and non-relational components.

Integration of Data Semantics. The integration of the semantics of data in our time can be accomplished through a variety of approaches, but the ontology-based integration is most promising. Under the term, "ontology" one can consider a holistic standardized description of a particular subject area, which should provide a stable interpretation of the data on this area both from the point of view of the person and from the point of view of the computer. When integrating data, a certain information resource serves as a source of a description made in the form of ontology, so you can identify a specific type of the latter is the ontology of data. The generalized standardized image of an ontology can be considered an expression

$$O = \ <P, R, F> \ , \tag{12}$$

where P is a finite set of concepts of the domain with their attributes, R is a finite set of relationships between terms, F is a finite set of interpretation functions.

Methods for integrating the semantics of data during the development of information resources of Internet systems require the creation of for each incoming set of data $V_i(i = 1, 2, \ldots, n)$ separate ontology $O(V_i)$, which forms a complete, coherent and understandable description of the semantics of the resource and its individual components

$$O(V_i) = \ <P(V_i), R(V_i), F(V_i)> \ , \tag{13}$$

where $P(V_i)$ is a set of approaches that describe the amount of data, their content, characteristics and belonging to a particular category or class, $R(V_i)$ is a set of relationships between the data governing the nature of their interaction and sharing, $F(V_i)$ is a set of semantic limits and interpretation functions that connect them with real terms and elements of the subject area, as well as determine the order of establishing such relations. Such ontology defines the semantic connection of established and specific data objects with the terms of the domain, describing the monolithic data-content structure. Since an instance of the description of ontology in semantic integration is data, then it can be defined as an applied ontology and depicted as a system of metadata of a special type. Thus, the task of semantic integration of data of information resources can lead to the creation of some generalized ontology of information resource through a combination of the set of input ontology's, the detection of dependencies between them, as well as the removal of contradictions in the content between ontology's. To achieve the described result, the semantic integration of data is based on compliance with certain predefined conditions that should provide a content-richly correct combination of input data in the source resource.

One of the significant advantages of semantic integration on the basis of ontology's to other approaches is its ability to create an unambiguous display of data content both at the human level and at the computer level. From this follow the following important characteristics of this integration, namely:

- Creation of a content-complete and orderly description of the semantics of incoming sets of heterogeneous data and the resulting integrated information internet resource;
- The possibility of using standardized tools and functions independent of the application of the system and its subject area;
- Getting a semantically correct result without a person's direct presence.

Accordingly, as a result of semantic integration, a semantically ordered Internet resource is formed that combines the content of the set of unordered input data.

5 Choice and Justification of Solutions to the Problem

In order to accomplish the task, the programming language of the web applications JavaScript is selected. This choice was made due to the fact that this language is suitable for the implementation of both the server and client parts, which greatly simplifies and accelerates the development process. This language is a simple non-strictly typed, yet powerful and large enough to develop web-services with the possibility of scaling. JavaScript is classified as a prototype (a subset of object-oriented), scripting programming language with dynamic typing. In addition to the prototype, JavaScript also partly supports other programming paradigms (imperative and partially functional) and some relevant architectural properties, including dynamic and weak typing, automatic memory management, prototype imitation, functions as first-class objects. It is possible to note the following advantages of this language:

- JavaScript provides plenty of opportunities to handle a variety of tasks. Flexible language allows you to use many programming templates according to specific conditions.
- The popularity of JavaScript opens in front of the program a considerable number of ready-made libraries, which allow you to significantly simplify the writing of the code and level up the imperfections of the syntax.
- Ability to use in many areas. Extensive JavaScript capabilities give programmers a chance to test themselves as the developer of a wide variety of applications, and this, of course, raises interest in professional activities.

When writing web applications, JavaScript programming is used most often. If you briefly list the key features of this language, you should highlight the following:

- Object Orientation. The implementation of the program is the interaction of objects.
- Bringing data types is automatic.
- Functions are objects of the base class. This feature makes JavaScript similar to functional programming languages, such as Lisp and Haskell.
- Automatic memory cleaning. The so-called garbage collector makes JavaScript similar to C# or Java [28].

In order to fully delve into the development process and facilitate the work of programmers in our time, there are a large number of so-called frameworks to develop the client part of the system and environments for the server. In developing this system, one of the most popular tools used to execute JavaScript on the server side, namely Node.js, was used. Node.js is a platform built on the JavaScript engine in Chrome, designed to make fast, scalable network applications easy to create. Node.js uses managed events, a non-blocked I/O model, making it easy, efficient and ideal for systems that handle a large amount of real-time data and work on distributed devices. With Node.js, it's easier to scale the system. When simultaneously connecting to the server thousands of users, the Node works asynchronously, that is, prioritizes and distributes resources more competently. Java, for example, allocates a separate stream for each connection. In this language you can write cross-platform applications - in a bundle of mobile + desktop Node helps to achieve synchronization. For example, when you write a message from the phone, it immediately appears on the laptop and the Internet [29]. JetBrains WebStorm 2017, the free version of which is enough for complete development, has been chosen as the development environment. WebStorm supports the development of web-applications in various scripting languages, and also allows you to develop a client interface since it supports *. HTML. Also, built-in support for version control systems allows you to manage the history and versions of the written code. To implement the database, the NoSQL approach was used. NoSQL (Eng. Not only SQL, not just SQL), in computer science, is a term that denotes the number of approaches aimed at implementing database repositories that have significant differences from the models used in traditional relational DBMSs with access to data by means of the language of SQL. Applies to databases that attempt to solve scalability and accessibility problems due to the atomicity and consistency of data. The main reasons for using this approach are:

- Increase the amount of saved data.
- Interrelated data. Each piece of knowledge is somehow associated with data in other information repositories. Pages on the Internet refer to other pages. Ontology establish interconnections between different terms, and so on.
- Use of poorly structured information. Take a simple example: a description of the car. In this case, the number of parameters can reach up to several dozen. Moreover, a different set of parameters will be used for different cars. In such circumstances, it is extremely difficult to determine in advance the structure of the table, in which the description of the product is stored.
- Architecture. Previously, a typical architecture used one major computer (mainframe) and one database. Nowadays, we actively use web services, each with its backend (roughly speaking, with its own database) and other distributed solutions.

As MongoDB database was selected. MongoDB is a document-based open source database management system that does not require a table schema description. Classified as NoSQL, uses JSON-like documents and a database schema that supports the flexibility of the structure of documents and the simplicity of their modification and transfer [30].

5.1 Description of the Task Realization

This section describes the creation of an intellectual system of formation and integration of content created in the framework of the project taking into account the needs of the user. The program is designed to simplify the travel planning process and find relevant information that may be useful in this journey. To create a program used programming language JavaScript, and server platform Node.js. Because the application is web service for processing data, it was decided to select the MVC software architecture (Fig. 11). The figure shows the mechanism of each of the components of MVC, as well as the interaction between these components. The process of running MVC can be expressed as a change in the status of a Web application: receiving the first request from the user, redirecting a request to a particular handler, creating this query handler, creating a controller, executing processing by the controller, calling the function of business logic and returning the result. After receiving the first request in the routes file. js elements of the route are added to the object routeTable. Next, routing is performed - redirection of the requested route. At this stage, the controller is dispatcher uses the first record of the route from the routeTable collection to create a routeData object, which is then used to initialize the router object. This creates a requestHandler MVC. The router object creates an instance of the request handler class and passes the object routeContext to the explorer. Then controller creation is made - the object router based on the class an application controller creates an instance of the controller. After that, the controller begins to operate, and the router r meter calls the controller's method of execution. As a result, the result is returned; The called method takes the value sent from the user, processes the relevant data, and then generates the result and returns the result data. The basic classes responsible for the business logic of the web application are: User, Source, Request, Settings, InfoItem, Result. Since the designed destination system for operation in the web browser environment, the HTML markup language and cascading CSS stylesheet was used to create the user interface.

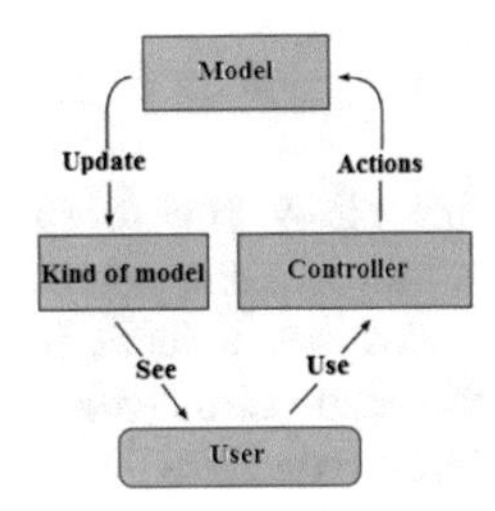

Fig. 11. MVC architecture

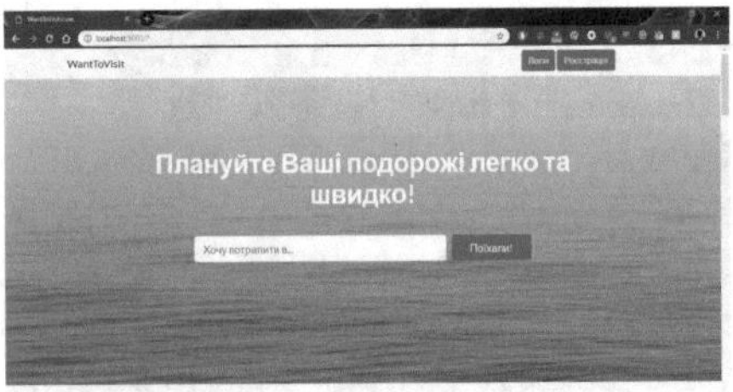

Fig. 12. Start page

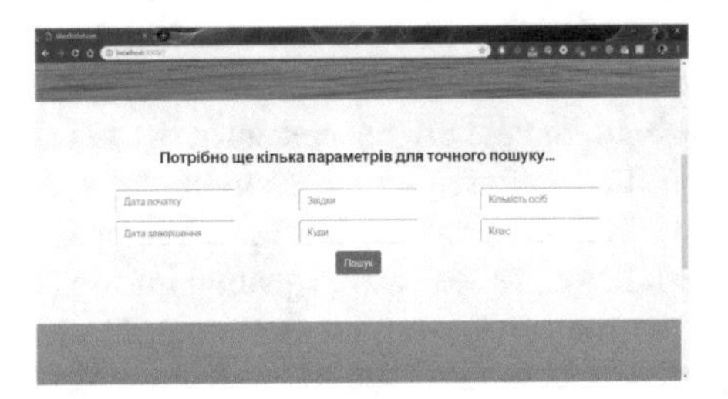

Fig. 13. Form for specifying search parameters

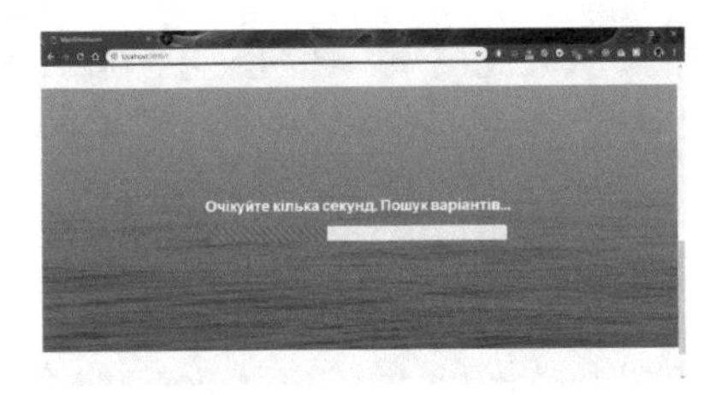

Fig. 14. Search execution status

Fig. 15. Message about no search results

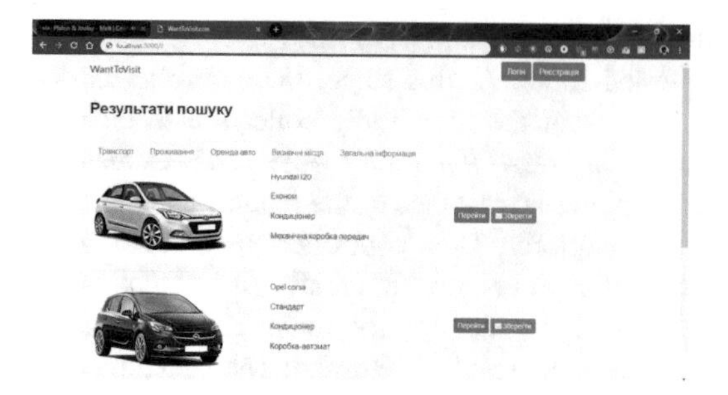

Fig. 16. Auto rental page

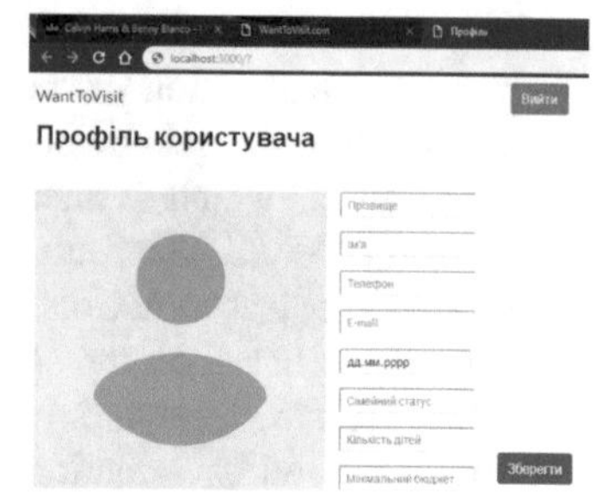

Fig. 17. User's personal cabinet

Also, a free open source library for creating websites that contain the already described CSS components and HTML - Bootstrap has been used to simplify the process of creating the interface, marking and aligning elements, creating inscriptions, shapes, buttons and other components. Index - a prototype of the program's initial window. From here, the user can go to the login page in his personal cabinet, or immediately start the search procedure (Fig. 12).

Here is a form (Fig. 13) that contains elements that allow you to specify the necessary search parameters, or if the user is logged on to the system, you can choose personal settings. By launching a search process that analyzes a large number of sources and cannot be instantaneous, the user will see the status of its execution - the so-called progress line (Fig. 14). After completing the search procedure, in the absence of results, the user will see a corresponding message with the suggestion to refine or change the search parameters to obtain the result (Fig. 15). If the search is successful, the user will get the result. Since the search is carried out in several categories at once, such as transport, accommodation, car rental, points of interest, general information, then the result will be divided into several subpages according to the categories where the information on each of them will be presented as a list whose items will be displayed depending on the category. As an example, let's take a look at the rental category of the car (Fig. 16). Here we see the list of companies providing car rental services at the travel destination, as well as the general information about the company, it shows its rating and the limits of prices for car rental. In order to get more detailed information about a particular company, you just need to select it from the list, then the user will go to a page where there will already Also, the user can go to his personal cabinet (Fig. 17) from any other point of the system using the main menu. It shows all personal settings and user information - name, phone number, email, birthday, family status, availability and number of children, optimal travel budget, whether if a user has a driving license, and more, be a list of specific cars available for rent in the destination.

6 Conclusions

The paper considers the main methodological and means for the integration and formation of content, as well as discussing the importance of using information technology in the field of tourism. It has been found that there are a lot of scattered resources with the information and services needed for travel, but there are very few services available to provide a full range of tourist information support. As a result, a web application written in the JavaScript programming language using the Node.js platform and the HTML + CSS link was created. Architectural template MVC was taken as the basis of architecture. The implementation tool was the WebStorm development environment. The search, viewing and saving functions of the result are implemented. Authentication and authorization functions are also implemented with the appropriate level of available functionality. The search result is presented in the form of lists for each category of information found. Manual testing of Web application features is performed. The result of the program confirmed the correctness of its work. In the future, it is planned to improve the old methods and add new features. It is planned to expand the source base to increase the number of results, and in the future, the system can be developed into a complete platform for ordering, renting and purchasing all necessary resources at each stage of the trip. In the future, the web application can be used for business purposes in the case of agreements with travel agencies and agencies about the use of the system in their work.

References

1. Zhezhnych, P., Markiv, O.: Linguistic comparison quality evaluation of web-site content with tourism documentation objects. In: Shakhovska, N., Stepashko, V. (eds.) Advances in Intelligent Systems and Computing, vol. 689, pp. 656–667. Springer, Cham (2018)
2. Zhezhnych, P., Markiv, O.: A linguistic method of web-site content comparison with tourism documentation objects. In: CSIT 2017, pp. 340–343 (2017)
3. Zhezhnych, P., Markiv, O.: Recognition of tourism documentation fragments from web-page posts. In: International Conference on Advanced Trends in Radioelectronics, Telecommunications and Computer Engineering, pp. 948–951 (2018)
4. Mukalov, P., Zelinskyi, O., Levkovych, R., Tarnavskyi, P., Pylyp, A., Shakhovska, N.: Development of system for auto-tagging articles, based on neural network. In: CEUR Workshop Proceedings, vol. 2362, pp. 116–125 (2019)
5. Shakhovska, N.B., Noha, R.Y.: Methods and tools for text analysis of publications to study the functioning of scientific schools. J. Autom. Inf. Sci. **47**(12) (2015)
6. Shakhovska, N., Shvorob, I.: The method for detecting plagiarism in a collection of documents. In: Computer Sciences and Information Technologies, pp. 142–145 (2015)
7. Arzubov, M., Shakhovska, N., Lipinski, P.: Analyzing ways of building user profile based on web surf history. In: Computer Sciences and Information Technologies, pp. 377–380 (2017)
8. Gozhyj, A., Vysotska, V., Yevseyeva, I., Kalinina, I., Gozhyj, V.: Web resources management method based on intelligent technologies. In: Shakhovska, N., Medykovskyy, M. (eds.) Advances in Intelligent Systems and Computing, vol. 871, pp. 206–221. Springer, Cham (2019)
9. Gozhyj, A., Kalinina, I., Vysotska, V., Gozhyj, V.: The method of web-resources management under conditions of uncertainty based on fuzzy logic. In: International Scientific and Technical Conference on Computer Sciences and Information Technologies, pp. 343–346 (2018)
10. Lytvyn, V., Vysotska, V., Burov, Y., Demchuk, A.: Architectural ontology designed for intellectual analysis of e-tourism resources. In: International Scientific and Technical Conference on Computer Sciences and Information Technologies, pp. 335–338 (2018)

11. Vysotska, V., Fernandes, V.B., Emmerich, M.: Web content support method in electronic business systems. In: CEUR Workshop Proceedings, vol. 2136, pp. 20–41 (2018)
12. Kanishcheva, O., Vysotska, V., Chyrun, L., Gozhyj, A.: Method of integration and content management of the information resources network. In: Shakhovska, N., Stepashko, V. (eds.) Advances in Intelligent Systems and Computing, vol. 689, pp. 204–216. Springer, Cham (2018)
13. Lytvyn, V., Vysotska, V., Pukach, P., Vovk, M., Ugryn, D.: Method of functioning of intelligent agents, designed to solve action planning problems based on ontological approach. Eastern-Eur. J. Enterp. Technol. **32**(87), 11–17 (2017)
14. Chyrun, L., Kis, I., Vysotska, V., Chyrun, L.: Content monitoring method for cut formation of person psychological state in social scoring. In: International Scientific and Technical Conference on Computer Sciences and Information Technologies, CSIT, pp. 106–112 (2018)
15. Chyrun, L., Vysotska, V., Kis, I., Chyrun, L.: Content analysis method for cut formation of human psychological state. In: International Conference on Data Stream Mining and Processing, pp. 139–144 (2018)
16. Rusyn, B., Vysotska, V., Pohreliuk, L.: Model and architecture for virtual library information system. In: Computer Sciences and Information Technologies, pp. 37–41 (2018)
17. Rusyn, B., Lytvyn, V., Vysotska, V., Emmerich, M., Pohreliuk, L.: The virtual library system design and development. In: Shakhovska, N., Medykovskyy, M. (eds.) Advances in Intelligent Systems and Computing, vol. 871, pp. 328–349. Springer, Cham (2019)
18. Zdebskyi, P., Vysotska, V., Peleshchak, R., Peleshchak, I., Demchuk, A., Krylyshyn, M.: An application development for recognizing of view in order to control the mouse pointer. In: CEUR Workshop Proceedings, vol. 2386, pp. 55–74 (2019)
19. Vysotska, V.: Linguistic analysis of textual commercial content for information resources processing. In: Modern Problems of Radio Engineering, Telecommunications and Computer Science, TCSET 2016, pp. 709–713 (2016)
20. Rzheuskyi, A., Kunanets, N., Stakhiv, M.: Recommendation system virtual reference. Comput. Sci. Inf. Technol. **1**, 203–206 (2018)
21. Lypak, H., Rzheuskyi, A., Kunanets, N., Pasichnyk, V.: Formation of a consolidated information resource by means of cloud technologies. In: Problems of Infocommunications Science and Technology, PIC S and T, pp. 157–160 (2018)
22. Kaminskyi, R., Kunanets, N., Rzheuskyi, A.: Mathematical support for statistical research based on informational technologies. In: ICT in Education, Research and Industrial Applications. Integration, Harmonization and Knowledge Transfer, vol. 2105, pp. 449–452 (2018)
23. Rzheuskyi, A., Matsuik, H., Veretennikova, N., Vaskiv, R.: Selective dissemination of information – technology of information support of scientific research. In: Shakhovska, N., Medykovskyy, M. (eds.) Advances in Intelligent Systems and Computing III, vol. 871, pp. 235–245. Springer, Cham (2019)
24. Wiegers, K.: Software Requirements, 2nd edn. Microsoft Press, Redmond (2002)
25. ISO/IEC 25010:2011: Systems and software engineering - Systems and software Quality Requirements and Evaluation (SQuaRE) - System and software quality models 2011 (2011). https://www.iso.org/standard/35733.html
26. Laplante, P.A.: Requirements Engineering for Software and Systems. CRC Press, Boca Raton (2009)
27. Cockburn, A.: Writing Effective Use Cases. Pearson Education, London (2001)
28. Su, J., Sachenko, A., Lytvyn, V., Vysotska, V., Dosyn, D.: Model of touristic information resources integration according to user needs. In: International Scientific and Technical Conference on Computer Sciences and Information Technologies, pp. 113–116 (2018)

29. Lytvyn, V., Vysotska, V., Demchuk, A., Demkiv, I., Ukhanska, O., Hladun, V., Kovalchuk, R., Petruchenko, O., Dzyubyk, L., Sokulska, N.: Design of the architecture of an intelligent system for distributing commercial content in the internet space based on SEO-technologies, neural networks, and machine learning. Eastern-Eur. J. Enterp. Technol. **2**(2–98), 15–34 (2019)
30. Berenbach, B., Paulish, D., Katzmeier, J., Rudorfer, A.: Software & systems requirements engineering: In: Practice. McGraw-Hill Professional, New York (2009)
31. Davis, A.M.: Just Enough Requirements Management: Where Software Development Meets Marketing. Dorset House (2005)
32. Burov, Y., Vysotska, V., Kravets, P.: Ontological approach to plot analysis and modeling. In: CEUR Workshop Proceedings, vol. 2362, pp. 22–31 (2019)
33. Lytvyn, V., Vysotska, V., Dosyn, D., Burov, Y.: Method for ontology content and structure optimization, provided by a weighted conceptual graph. Webology **15**(2), 66–85 (2018)
34. Backlund, A.: The definition of the system. Kybernetes **29**(4), 444–451 (2000)
35. Lytvyn, V., Vysotska, V., Veres, O., Rishnyak, I., Rishnyak, H.: The risk management modelling in multi project environment. In: Proceedings of the International Conference on Computer Science and Information Technologies, CSIT, pp. 32–35 (2017)
36. Gemino, A., Parker, D.: Use case diagrams in support of use case modeling: deriving understanding from the picture. J. Database Manage. **20**(1), 1–24 (2009)
37. Naum, O., Chyrun, L., Kanishcheva, O., Vysotska, V.: Intellectual system design for content formation. In: Proceedings of the International Conference on Computer Science and Information Technologies, CSIT, pp. 131–138 (2017)
38. Vysotska, V., Rishnyak, I., Chyrun L.: Analysis and evaluation of risks in electronic commerce. In: 9th International Conference on CAD Systems in Microelectronics, pp. 332–333 (2007)
39. Vysotska, V., Lytvyn, V., Burov, Y., Berezin, P., Emmerich, M., Basto Fernandes, V.: Development of information system for textual content categorizing based on ontology. In: CEUR Workshop Proceedings, vol. 2362, pp. 53–70 (2019)
40. Lytvyn, V., Peleshchak, I., Vysotska, V., Peleshchak, R.: Satellite spectral information recognition based on the synthesis of modified dynamic neural networks and holographic data processing techniques. In: International Scientific and Technical Conference on Computer Sciences and Information Technologies, CSIT 2018 – Proceedings, vol. 1, pp. 330–334 (2018)
41. Lytvyn, V., Vysotska, V., Uhryn, D., Hrendus, M., Naum, O.: Analysis of statistical methods for stable combinations determination of keywords identification. Eastern-Eur. J. Enterp. Technol. **2/2**(92), 23–37 (2018)
42. Lytvyn, V., Vysotska, V., Rusyn, B., Pohreliuk, L., Berezin, P., Naum, O.: Textual content categorizing technology development based on ontology. In: CEUR Workshop Proceedings, vol. 2386, pp. 234–254 (2019)
43. Vysotska, V., Lytvyn, V., Burov, Y., Gozhyj, A., Makara, S.: The consolidated information web-resource about pharmacy networks in city. In: CEUR Workshop Proceedings, vol. 2255, pp. 239–255 (2018)
44. Korobchinsky, M., Vysotska, V., Chyrun, L., Chyrun, L.: Peculiarities of content forming and analysis in internet newspaper covering music news. In: Proceedings of the International Conference on Computer Science and Information Technologies, CSIT, pp. 52–57 (2017)
45. Chyrun, L., Burov, Y., Rusyn, B., Pohreliuk, L., Oleshek, O., Gozhyj, A., Bobyk, I.: Web resource changes monitoring system development. In: CEUR Workshop Proceedings, vol. 2386, pp. 255–273 (2019)

46. Chyrun, L., Kowalska-Styczen, A., Burov, Y., Berko, A., Vasevych, A., Pelekh, I., Ryshkovets, Y.: Heterogeneous data with agreed content aggregation system development. In: CEUR Workshop Proceedings, vol. 2386, pp. 35–54 (2019)
47. Gozhyj, A., Chyrun, L., Kowalska-Styczen, A., Lozynska, O.: Uniform method of operative content management in web systems. In: CEUR Workshop Proceedings, vol. 2136, pp. 62–77 (2018)
48. Chyrun, L., Gozhyj, A., Yevseyeva, I., Dosyn, D., Tyhonov, V., Zakharchuk, M.: Web content monitoring system development. In: CEUR Workshop Proceedings, vol. 2362, pp. 126–142 (2019)
49. Bobalo, Y., Stakhiv, P., Mandziy, B., Shakhovska, N., Holoschuk, R.: The concept of electronic textbook "Fundamentals of theory of electronic circuits". Przegląd Elektrotechniczny **88**, 16–18 (2012). NR 3a/2012
50. Vysotska, V., Burov, Y., Lytvyn, V., Oleshek, O.: Automated monitoring of changes in web resources. In: Lytvynenko, V., Babichev, S., Wójcik, W., Vynokurova, O., Vyshemyrskaya, S., Radetskaya, S. (eds.) Lecture Notes in Computational Intelligence and Decision Making, vol. 1020, pp. 348–363. Springer, Cham (2020)
51. Demchuk, A., Lytvyn, V., Vysotska, V., Dilai, M.: Methods and means of web content personalization for commercial information products distribution. In: Lytvynenko, V., Babichev, S., Wójcik, W., Vynokurova, O., Vyshemyrskaya, S., Radetskaya, S. (eds.) Lecture Notes in Computational Intelligence and Decision Making, vol. 1020, pp. 332–347. Springer, Cham (2020)
52. Vysotska, V., Mykhailyshyn, V., Rzheuskyi, A., Semianchuk, S.: System development for video stream data analyzing. In: Lytvynenko, V., Babichev, S., Wójcik, W., Vynokurova, O., Vyshemyrskaya, S., Radetskaya, S. (eds.) Lecture Notes in Computational Intelligence and Decision Making, vol. 1020, pp. 315–331. Springer, Cham (2020)
53. Lytvyn, V., Vysotska, V., Rzheuskyi, A.: Technology for the psychological portraits formation of social networks users for the IT specialists recruitment based on big five, NLP and big data analysis. In: CEUR Workshop Proceedings, vol. 2392, pp. 147–171 (2019)
54. Rzheuskyi, A., Gozhyj, A., Stefanchuk, A., Oborska, O., Chyrun, L., Lozynska, O., Mykich, K., Basyuk, T.: Development of mobile application for choreographic productions creation and visualization. In: CEUR Workshop Proceedings, vol. 2386, pp. 340–358 (2019)
55. Lytvyn, V., Pukach, P., Bobyk, I., Vysotska, V.: The method of formation of the status of personality understanding based on the content analysis. Eastern-Eur. J. Enterp. Technol. **5/2** (83), 4–12 (2016)
56. Vysotska, V., Hasko, R., Kuchkovskiy, V.: Process analysis in electronic content commerce system. In: Proceedings of the International Conference on Computer Sciences and Information Technologies, CSIT 2015, pp. 120–123 (2015)
57. Su, J., Vysotska, V., Sachenko, A., Lytvyn, V., Burov, Y.: Information resources processing using linguistic analysis of textual content. In: Intelligent Data Acquisition and Advanced Computing Systems Technology and Applications, Romania, pp. 573–578 (2017)
58. Lytvyn, V., Vysotska, V., Burov, Y., Veres, O., Rishnyak, I.: The contextual search method based on domain thesaurus. In: Shakhovska, N., Stepashko, V. (eds.) Advances in Intelligent Systems and Computing, vol. 689, pp. 310–319. Springer, Cham (2018)
59. Lytvyn, V., Vysotska, V., Chyrun, L., Chyrun, L.: Distance learning method for modern youth promotion and involvement in independent scientific researches. In: Proceedings of the IEEE First International Conference on Data Stream Mining & Processing (DSMP), pp. 269–274 (2016)

60. Mukalov, P., Zelinskyi, O., Levkovych, R., Tarnavskyi, P., Pylyp, A., Shakhovska, N.: Development of system for auto-tagging articles, based on neural network. In: CEUR Workshop Proceedings, vol. 2362, pp. 106–115 (2019)
61. Shakhovska, N., Basystiuk, O., Shakhovska, K.: Development of the speech-to-text chatbot interface based on Google API. In: CEUR Workshop Proceedings, vol. 2386, pp. 212–221 (2019)
62. Shakhovska, N., Vysotska, V., Chyrun, L.: Intelligent systems design of distance learning realization for modern youth promotion and involvement in independent scientific researches. In: Shakhovska, N. (ed.) Advances in Intelligent Systems and Computing, vol. 512, pp. 175–198. Springer, Cham (2017)
63. Shakhovska, N., Vysotska, V., Chyrun, L.: Features of e-learning realization using virtual research laboratory. In: Proceedings of the XI–th International Conference on Computer Science and Information Technologies, CSIT 2016, pp. 143–148 (2016)
64. Lytvyn, V., Vysotska, V.: Designing architecture of electronic content commerce system. In: Proceedings of the X-th International Conference on Computer Science and Information Technologies, CSIT 2015, pp. 115–119 (2015)
65. Vysotska, V., Burov, Y., Lytvyn, V., Demchuk, A.: Defining author's style for plagiarism detection in academic environment. In: Proceedings of the 2018 IEEE 2nd International Conference on Data Stream Mining and Processing, DSMP 2018, pp. 128–133 (2018)
66. Chyrun, L., Chyrun, L., Kis, Y., Rybak, L.: Automated information system for connection to the access point with encryption WPA2 enterprise. In: Lecture Notes in Computational Intelligence and Decision Making, vol. 1020, pp. 389–404 (2020)
67. Kis, Y., Chyrun, L., Tsymbaliak, T., Chyrun, L.: Development of system for managers relationship management with customers. In: Lytvynenko, V., Babichev, S., Wójcik, W., Vynokurova, O., Vyshemyrskaya, S., Radetskaya, S. (eds.) Lecture Notes in Computational Intelligence and Decision Making, vol. 1020, pp. 405–421. Springer, Cham (2020)
68. Vysotska, V., Chyrun, L., Chyrun, L.: Information technology of processing information resources in electronic content commerce systems. In: Computer Science and Information Technologies, CSIT 2016, pp. 212–222 (2016)

Application of the Lisovporiadnyk Software for Management of the Forest Fund of a Forestry Enterprise

Ihor Aleksiiuk[1], Heorhiy Hrynyk[2](✉), and Tetiana Dyak[2]

[1] State Enterprise Biological Resources of Ukraine, Kyiv, Ukraine
aleks.win@i.ua
[2] Ukrainian National Forestry University, Lviv, Ukraine
h.hrynyk@nltu.edu.ua, tanyushkadyak@yahoo.com

Abstract. The paper presents the software for work with attribute and cartographic databases of Ukrderzhlisproekt Forestry Production Association. The main functional capabilities of the Lisovporiadnyk software and application of its basic functions in production are considered. The possibility of introducing changes by forest users to attribute and cartographic information in the database of individual survey units of forestry enterprises' funds is proposed. Some examples of the use of separate components of the software for conducting specific calculations, actualization of the geodetic data concerning the boundaries of separate plots and technological schemes of developing the felling areas are given.

Keywords: Attribute information · Database · Forestry · Forest fund

1 Introduction and Aim of the Research

The main source for obtaining data on the state and dynamics of the forest fund of a forestry enterprise is the implementation of forest management works that presents a set of measures aimed at providing scientific substantiation and effective organization of forest management, including the restoration of the enterprise's boundaries, the inventory with the definition of the species composition and other qualitative and quantitative indicators, the identification of forest stands that require economic measures, the reasonable division of forests based on the functions they perform, the determination of volumes of works on forest regeneration and afforestation, protection and conservation of forests, etc. [5].

The aim of the research are as follows:

- to propose and develop the system of the current changes in the forest fund data for permanent forest users in the existing Lisovporiadnyk software;
- to develop mechanisms for obtaining data on current changes in the forest fund, namely the input of the information on felling, the creation of forest cultures on areas that require reforestation, the transfer of unclosed forest cultures and natural regeneration in forested areas, and also the development of appropriate forms for obtaining information on the survey units which need to be monitored;

N. Shakhovska and M. O. Medykovskyy (Eds.): CCSIT 2019, AISC 1080, pp. 864–878, 2020.
https://doi.org/10.1007/978-3-030-33695-0_56

- to create the mechanisms for visualization of the data on estimated survey units which potentially require inspection and monitoring;
- to systematize the data on intentions of implementation of economic activities and mechanisms for prevention of violations of current laws.

2 Related Works

At the same time, the organization of forest management activities includes a number of measures concerning the functional dividing of forests into categories, defining of exploitable age of the stand and norms of its use, conducting forest management, a national forest inventory and forest assessment, as well as forest certification and monitoring. A significant part of these activities is undertaken by relevant public and private organizations, which usually use information, achievements and experience of scientific institutions and forest users [5, 8, 9].

Modern geoinformation systems (GIS) used, for example, in agriculture, allow monitoring the performance of operations, the movement of machinery during works, the calculation of fertilizer rates, the state of seedlings, etc., which saves time and money. Forestry has its own peculiarities unlike agriculture, since the period of forest cultivation ranges from 60 to 120 years. Existing GIS systems that are used in forestry practice do not allow for long-term planning of forest management measures, since forestry works being performed for a period of 10 years do not enable obtaining reliable data for a certain period of time. Climate changes create favourable conditions for the propagation of forest pests, which results in massive damage to pine forests by bark beetle observed in recent years. Changes that occur in the forest fund require the development of long-term forecast systems for stand growth and reliable information support, which will enable timely responses to negative changes in forest ecosystems [1, 2, 6].

Forest management is a costly and complex process that requires an individual approach to solving certain problems. The development of automated control systems (ACS) of forest management, which ensure obtaining high-quality relevant data on the conditions and dynamics of state forests, has appeared to be a promising direction [1, 2].

Considering the advantages of using ACS, the development and improvement of this system remains an important issue. The main disadvantages of the domestic ACS are the complexity of its use. The data obtained, as a rule, cannot be used promptly in the workplace due to the lack of means for its presentation. Significant distribution of software products (GIS systems) that enable the conditions and dynamics of the forest fund to be shared have insufficient functional support, and the use of several software products leads to difficulty in obtaining the data required [2, 5].

The development of software designed to work with forest management data is quite a promising direction. Such software products should be easy to use so that they can attract a large number of users. In terms of functionality, such programs must fulfill the main tasks that forestry employees face [1].

Lisovporiadnyk software is a set of separate programs written independently. Any existing analogue of software is not borrowed for authentic development.

At the same time, the following software products should be distinguished from existing analogues:

RUMB – Felling Area Plan-Scheme allows creating the plan of the felling area scheme during felling area allocation. This is quite a simple solution that allows printing a logging plan. All you need is an electronic map of a forest tablet or another map [10].

WebULR is a web resource developed by the Department of Algorithmization and Programming of the Information Computing Centre of Ukrderzhlisproekt. It allows creating random sampling of information from electronic forest taxation description with its display on a map. Annual data update is foreseen. The main drawback is the lack of the ability to enter additional data necessary for users [13].

Forests of Ukraine GeoPortal is a web resource developed by the Laboratory of New Information Technologies of G.M. Vysotsky Ukrainian Research Institute of Forestry and Agroforestry. It provides public and authorized access for users to information concerning the Forest Fund of Ukraine. It also allows building logging plans and contains all the valuation features.

The attribute information of some separate survey units plays a decisive role in the Lisovporiadnyk software unlike GIS systems used, where cartographic information is predominant. The principal difference between the Lisovporiadnyk software and other software products is the ability to avoid restrictions related to the use of cartographic data stored in the program along with the attribute information [1, 2].

Working with the attribute information of the relational structure database is usually a rather labour-intensive and complex process and requires users to know the structure of the database (names (codes) of the tables, links between them, names (codes) of fields, etc.).

3 Results

Lisovporiadnyk software is developed in Delphi environment and provides proper use of the forest management database. Due to dynamic requests to the database, the program also eliminates the possibility of indicators that do not have a key value in the tables. In order to facilitate the idea of growing plants, two fields were created on the basis of the indicators, which were contained in the corresponding tables. First of all, these are fields with the composition of the planting, as well as the number of elements of the forest. The program has a very simple interface, which greatly expands the number of users compared to the basic access program [7]. Search results are sent to Microsoft Excel or Microsoft Access, which can be used by other programs.

Lisovporiadnyk software uses Ukrderzhlisproekt databases based on Microsoft SQL Server. The bases connected to the SQL server have code names that are to be used by the program. The base with code name 5600 is responsible for Rivne Region, 0700 is the code of Volyn Region respectively (see Fig. 1). The program has a complete list of domains, as well as the names of temporary databases that can be used under the appropriate names on both the server and the program. The base of forest management is presented by 37 tables, which names can be seen as sections of tree indicators.

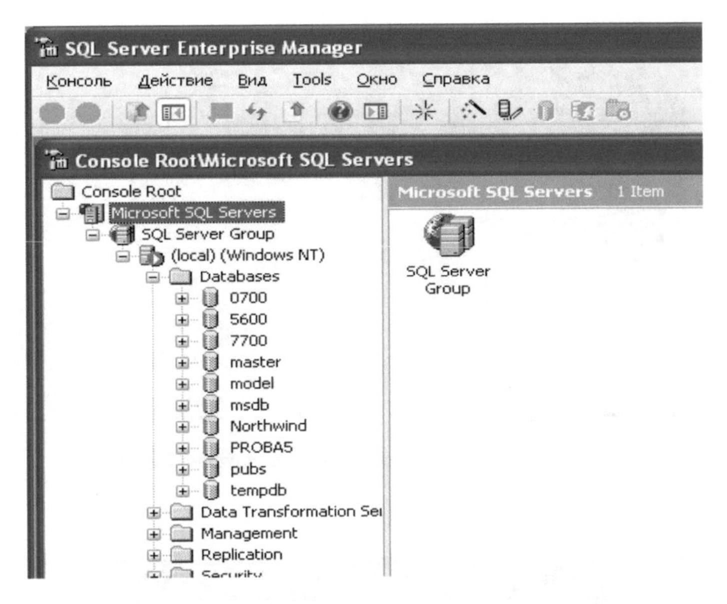

Fig. 1. SQL Server

We should note that the existing foreign analogues of the software for the processing of the database of Ukrderzhlisproekt Forestry Production Association cannot be used without significant alterations. Firstly, forest management itself and, accordingly, other technical and economic indicators are different [3, 12]. Secondly, the data of Ukrderzhlisproekt database, connected to SQL Server, is stored for several 10-year revision periods. Transferring all archives to a new environment can lead to their loss or may cause confusion in the underlying concepts [4]. Firstly, the use of GIS systems requires the application of a high-value licensed software product, which also needs the appropriate level of computer equipment [11]. Secondly, it is also necessary to provide rather long-term training of the staff, which mainly have basic forestry education at the state forest enterprises [11]. Moreover, the development of a software product on the basis of systems requires a unified scheme for the whole state. Therefore, there are certain inconveniences and discrepancies in the use of certain existing programs, in particular for calculation of expected monetary value of felling areas. In any case, the adaptation of existing analogues or the use of software developed on the basis of GIS systems takes a lot of both time and finances.

Figure 2 shows a window for downloading cartographic data in a shapefile format. Besides the loading of maps, there is also a set of other useful functions for displaying cartographic information. Application of Lisovporiadnyk software enables exporting data from databases (see Fig. 3) and performing the division of the object into smaller structures (regional forestry and hunting enterprises - into separate state forestry enterprises – into separate forestries). Lisovporiadnyk software enables transferring of certain divisions and departments to other enterprises replacing their numbering.

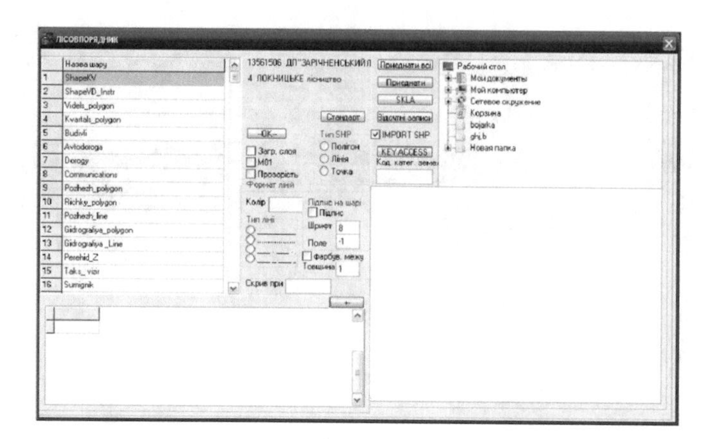

Fig. 2. Downloading cartographic data to attribute information

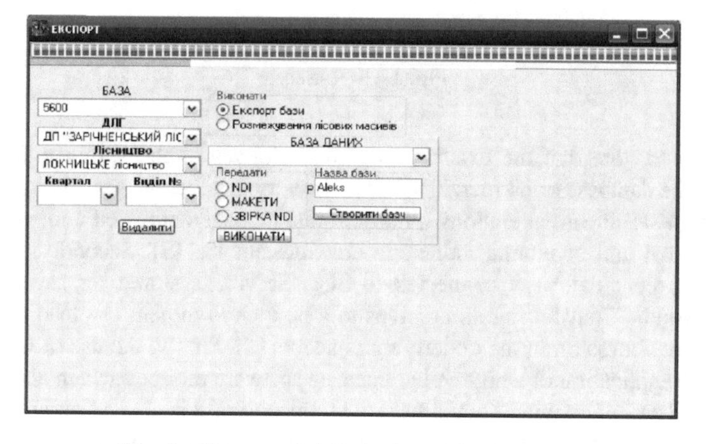

Fig. 3. Export and delimitation of forest massifs

Before loading, the cartography is undergoing a process of verification that is checking whether there is insufficient or excessive information in the appropriate relational database. Data can be stored in two types of databases: SQL Server 2000 or Microsoft Office Access. The software code also provides for the use of attribute information in the absence of a cartographic one.

Working with attribute information of a relational database is usually a laborious and complicated process and requires users to know the database structure (names (codes) of the tables, links between them, names (codes) of fields, etc.). Lisovporiadnyk software allows creating arbitrary queries (see Fig. 4), from finding mature tree stands within the enterprise to obtaining average valuation features according to species in the country as a whole. Users who have a structured query language SQL can independently edit the queries created by the program (see Fig. 5) or use them in other software tools.

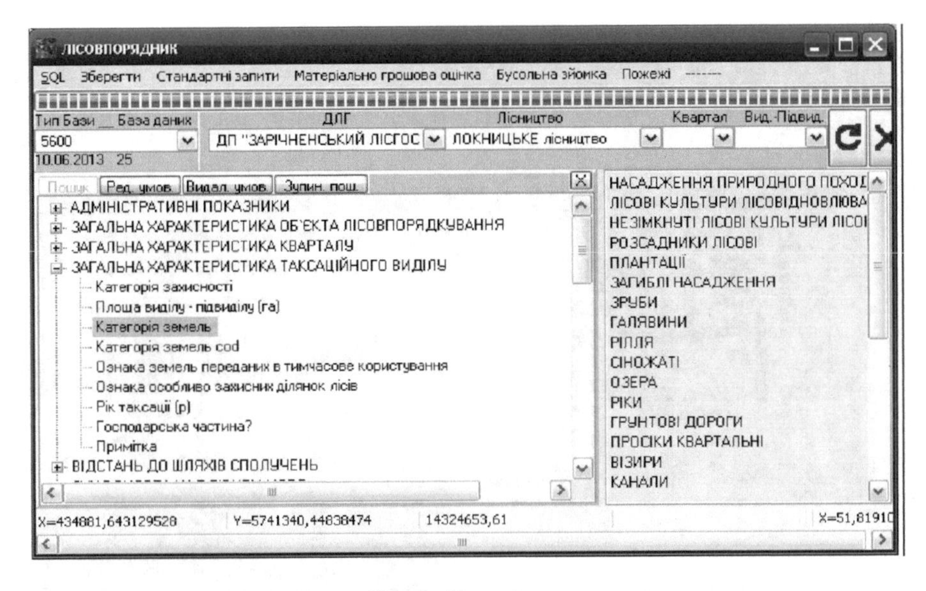

Fig. 4. Query menu

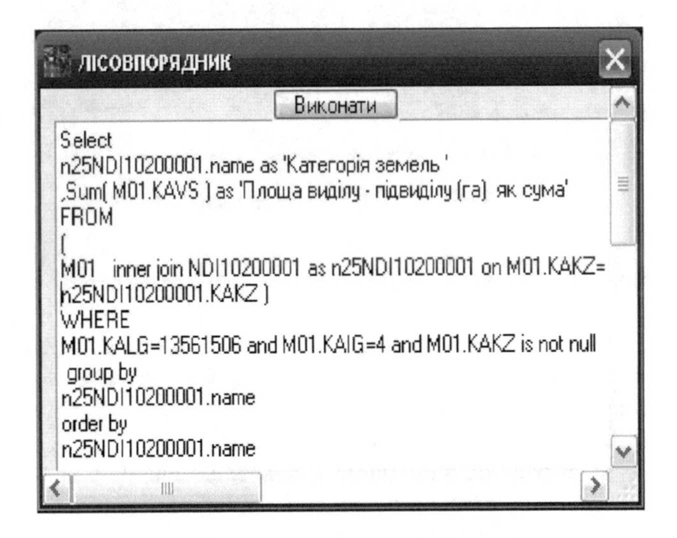

Fig. 5. SQL query editor

The data received from the created request, on the user demand, can be displayed on the map (if it is available), or transmitted to Microsoft Excel for further processing.

In order to work with maps, the program provides a menu that allows as follows: open (close) a map, zoom in, zoom out, move, display information about the taxonomic description of the selected department. The use of colors for the symbols of the selected categories is of a random character, which, if desired, can be changed (see Fig. 6).

The excessive number of categories can often complicate accepting of information on the map. In case when the total number of categories exceeds 200 variants, mapping does not occur.

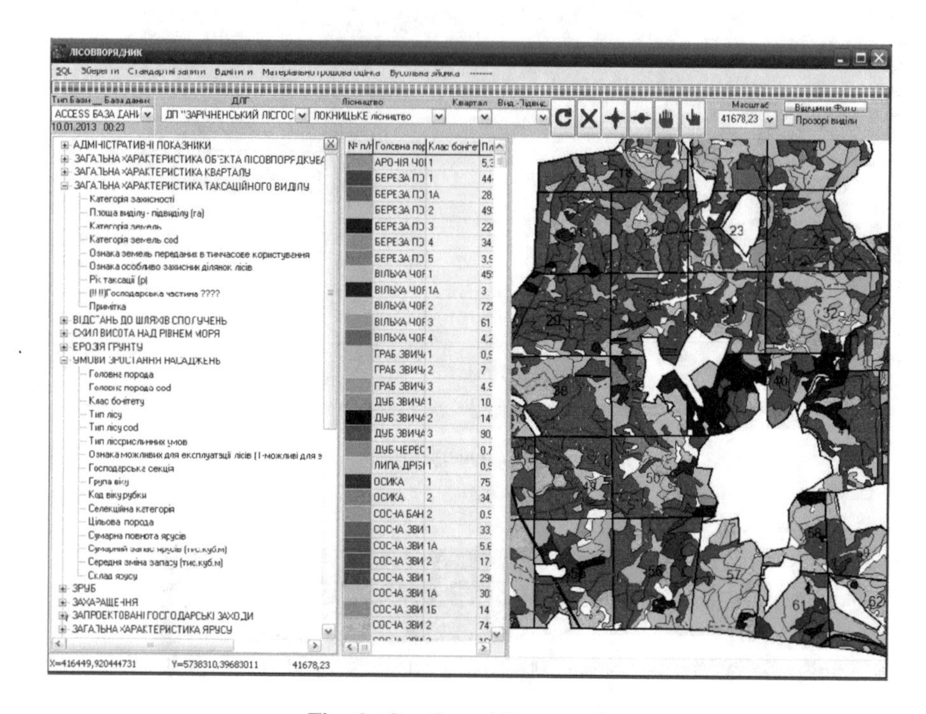

Fig. 6. Creating arbitrary queries

If there is no need to create arbitrary search conditions, it is possible to perform standard queries, including those during cadastral work (see Fig. 7). For standard search terms, there are corresponding queries that are executed automatically. The result of the processing can be obtained by displaying information on a map or in the form of a report that can be transmitted to Microsoft Excel, if desired, or another compatible MS Office application on the user demand.

Being developed in 2010, Lisovporiadnyk software allows accumulating information on the conducted economic activities, and displaying a lot of various data on the maps. This software product was substantially improved on the basis of the Separated Subdivision of National University of Life and Environmental Sciences of Ukraine "Boyarka Forest Research Station", which enabled developing a functional that is of everyday practical value. In general, the software functional allows carrying out the development of the plan for plotting (see Figs. 8 and 9) and calculating the "Material and Financial Assessment of the Felling Area" (see Fig. 10) of the extract of "Felling Tickets" (see Fig. 11), drawing of thematic maps and making arbitrary selection of information along with its displaying on the map, and also providing technological mapping for felling areas [1, 6] (see Fig. 12).

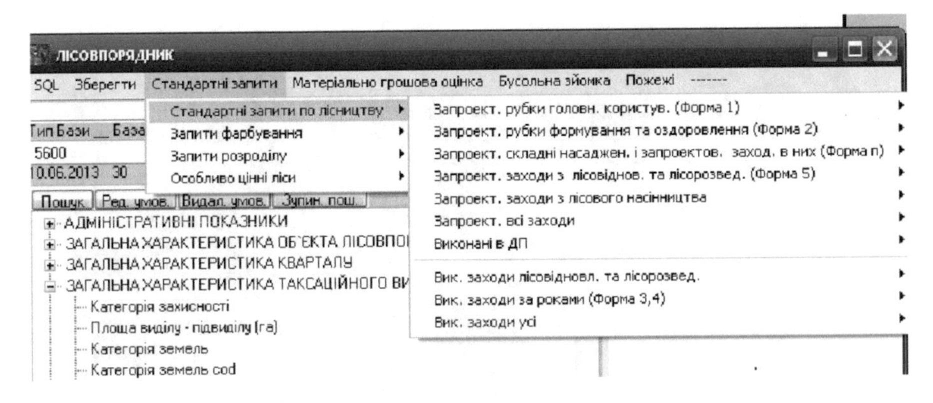

Fig. 7. Performing standard queries

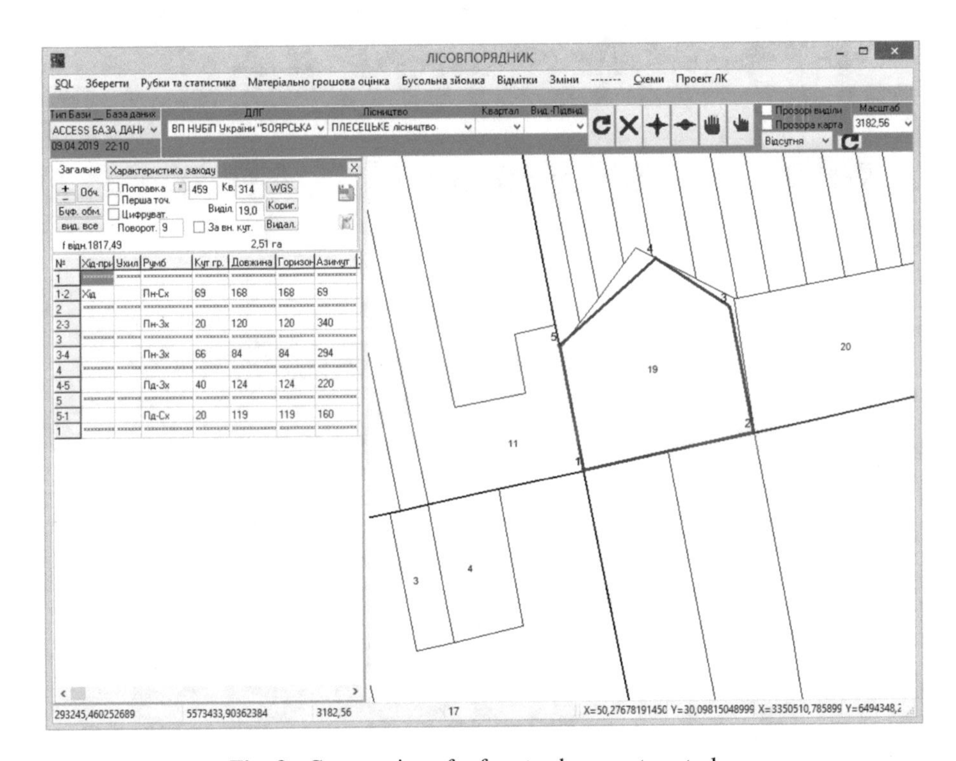

Fig. 8. Construction of a forest subcompartment plan

Withdrawal and assessment of a felling area fund is an important link in the process of harvesting wood during the implementation of different types of felling. The annual size of the felling area fund is approved by the Cabinet of Ministers of Ukraine, based on the size of the established calculating felling area. Due to software processing, more precise calculations of the "Material and financial assessment of a felling area", which

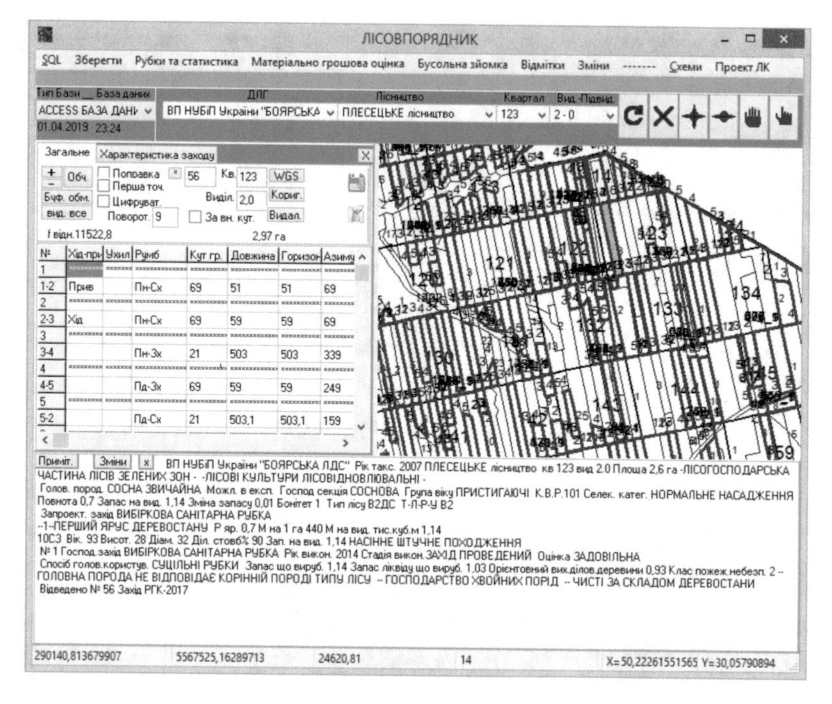

Fig. 9. Archive of plans

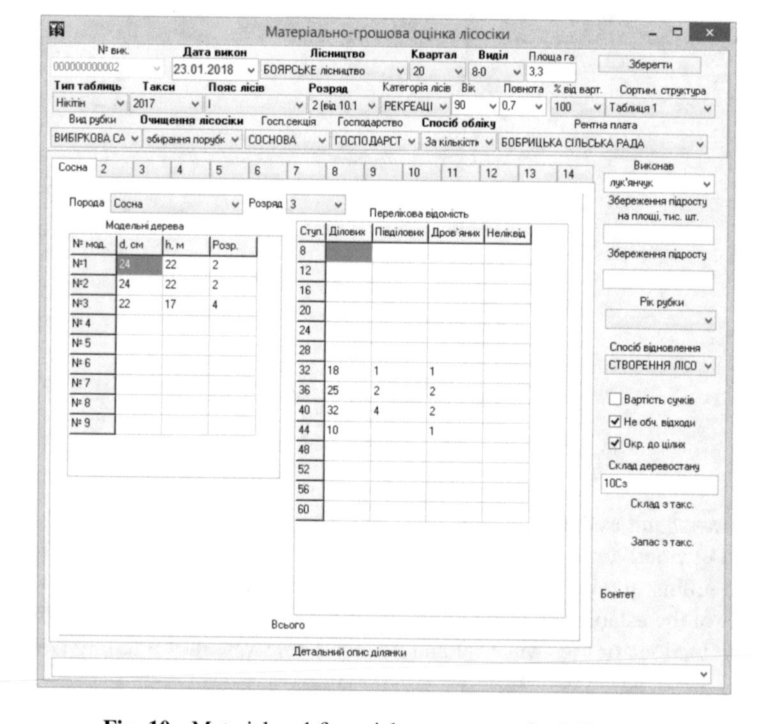

Fig. 10. Material and financial assessment of a felling area

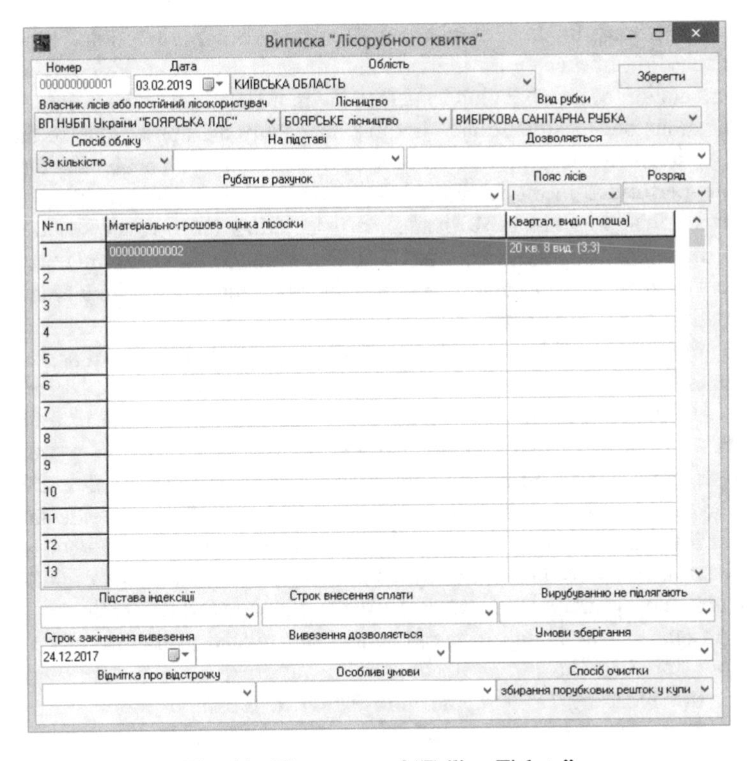

Fig. 11. The extract of “Felling Tickets”

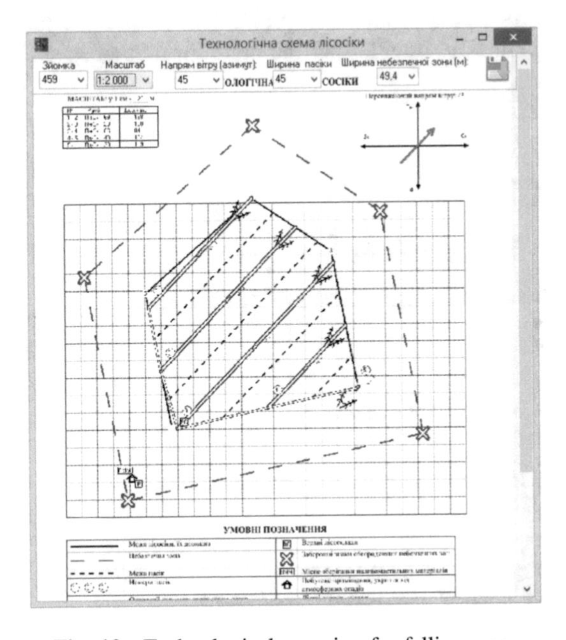

Fig. 12. Technological mapping for felling areas

were time-consuming for their implementation, have become possible. Among them there is a selection of a category individually for each degree of thickness, as well as the implementation of interpolation to determine more accurately the volume of wood. These operations do not require user involvement, only the appropriate settings when choosing options. In the program, Modes of computing, which are similar to those calculations performed in production, took the main place in the program.

A part of this program is dedicated to calculating tax cost of wood, with the possibility of replacing price categories. The forest-tax belt and tax rate are automatically determined as well. "Material and financial assessment of a felling area" does not require significant OS resources.

There are some issues we consider to be still urgent. Among them are as follows: the implementation of current changes in the forest fund concerning the change in the categories of land ranging from areas covered with forest vegetation to felling areas; reforestation of felling areas by creating forest cultures or leaving a separate plot for natural regeneration; transferring of unclosed forest cultures and natural regeneration in the forested areas; planning economic activities and fixing their implementation.

The attribute information on some separate survey units plays a decisive role in Lisovporiadnyk software unlike GIS systems used, where cartographic information is predominant. The principal difference between Lisovporiadnyk software and other software products is the ability to avoid restrictions related to the use of cartographic information stored in the program along with the attribute data. The implemented system for downloading cartographic information to the survey unit database eliminates any inconvenience for users and allows them to quickly obtain the necessary information from the database. Along with loading maps in the SHP format, there are also a lot of other useful functions for displaying cartographic information.

In order to accomplish these tasks, a module for entering current changes has been developed that allows editing a vector map (SHP file) based on the coordinates of the points of the plot plan. The fragment of the code for editing is as follows:

```
SQL.Add('order by AM99.KAIG,AM99.KAWN,AM99.KAWN,A
Active:=true;
Form2.Map1.ClearDrawings;
Form2.Map1.NewDrawing($00000001);
xodd:=false;
a:=-1;
for i:=0 to RecordCount-1 do
begin
 if pruu(Fields.Fields[8].AsString)='Хід ' then
 xodd:=true;
 if xodd then
 begin
  a:=a+1;
  Pochki[a].Poi:=CoPoint.Create;
  Pochki[a].Poi.x:=Fields.Fields[6].AsFloat;
  Pochki[a].Poi.y:=Fields.Fields[7].AsFloat;
  Pochki[a].Num:=Fields.Fields[5].AsString;
 end;
 Next;
end;
if (Pochki[0].Poi.x<>Pochki[a].Poi.x)
 or (Pochki[0].Poi.y<>Pochki[a].Poi.y) then
 begin
```

A significant part of the current changes occurring in the forest fund requires only changes in the attribute information of the survey unit. This concerns the fixation of the conducted economic activities, the transfer of unclosed forest cultures into the forested areas, etc. The fragment of the code is as follows:

```
Active:=false;
SQL.Clear;
SQL.add('Select SFLV,KALG,KAIG,KAWN,KAVN,KARN');
SQL.add(' FROM M88 WHERE M88.KALG='+llg+' and M88.KA
SQL.add(' and  M88.KAWN='+kw1+' and M88.KAVN='+vvd+'
SQL.add(' order by SFLV,KALG,KAIG,KAWN,KAVN,KARN ');
Active:=True;
if RecordCount=0 then
Insert;
begin
 Edit;
 if ZrSpZal1.ItemIndex=1 then
 Fields.Fields[0].asstring:='3211' else
 if ZrSpZal1.ItemIndex=2 then
 Fields.Fields[0].asstring:='3285';

 Fields.Fields[1].asstring:=llg;
 Fields.Fields[2].asstring:=lisn_;
 Fields.Fields[3].asstring:=kw1;
 Fields.Fields[4].asstring:=vvd;
 Fields.Fields[5].asstring:=pvv;
 Post;
end;
```

For the visualization of the system of current changes, a form that is divided into several sections has been developed. “MVCF” allows displaying on the map the most valuable forest sites for conservation and representative areas. A part of the fuctional is designated to reflect the economic activities (the tab “Economic Activities”), which are projected, executed or undertaken (see Fig. 13).

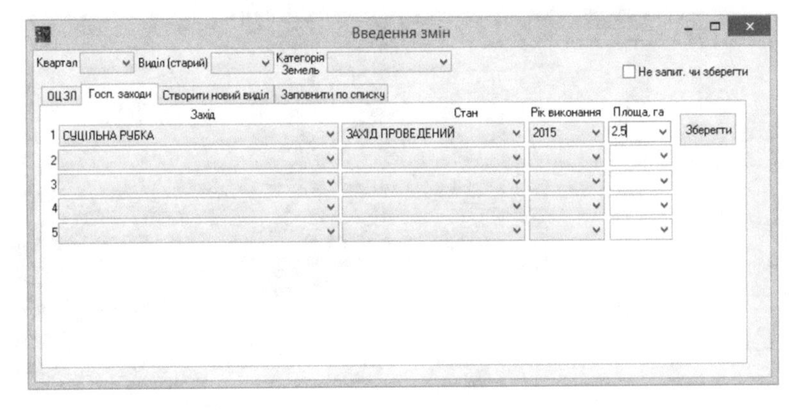

Fig. 13. The form for entering current changes

Concerning the availability of existing data on undertaking of economic activities that is stored in other software products, the fuctional for data loading through the clipboard has been developed, which has significantly reduced the time required for data processing (see Fig. 14).

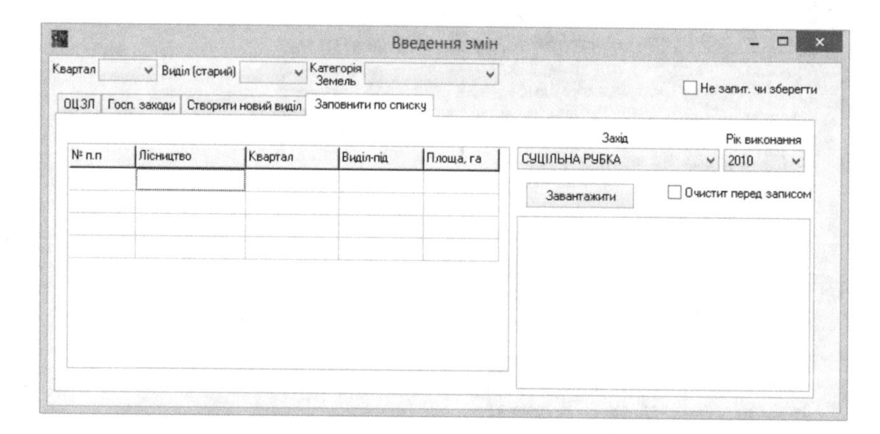

Fig. 14. Loading information from the clipboard

For a convenient and fast search of the necessary information, a functional has been developed in the menu of Lisovporiadnyk software, which allows displaying on the map the most valuable forest sites for conservation, representative areas, designed and undertaken economic activities, felling areas that require artificial reforestation and those left for natural regeneration (see Fig. 15).

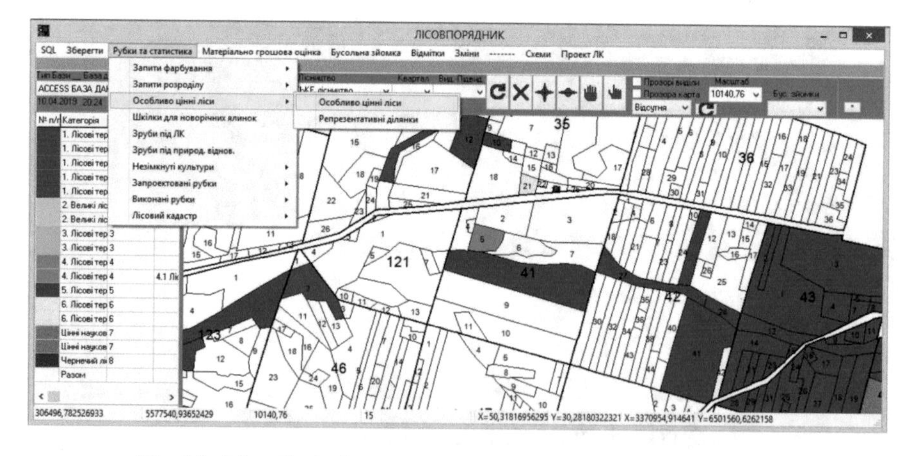

Fig. 15. Map displaying most valuable for conservation forest sites

Unclosed forest cultures require an annual survey and, if necessary, supplement. We have foreseen the possibility to display unclosed forest cultures depending on their age (1, 2, 3… 5 years old) by the years and of all unclosed cultures that need to be examined in general (see Fig. 16).

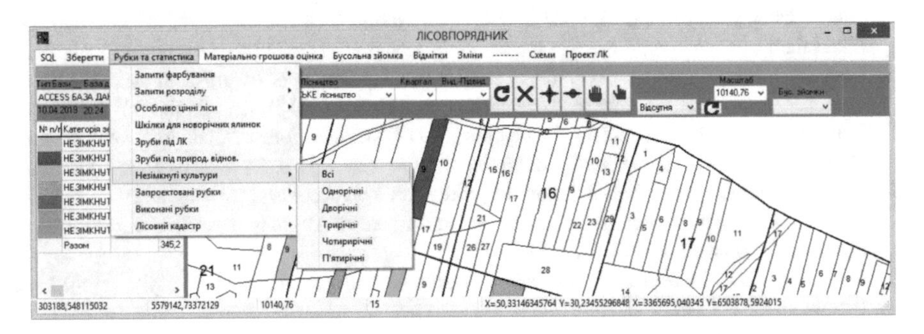

Fig. 16. Unclosed forest cultures

The developed functional allows changing the attribute information of a separate survey unit. After checking for adequacy, the program saves the changes.

4 Conclusions

Lisovporiadnyk software is an alternative to working with Ukrderzhlisproekt databases. The software functions provide significant relief for the daily work of forestry workers. Creating a database for using the software has neither significant time expenditures nor material costs, which greatly reduces the cost of provision and use of information.

There are still some issues related to the association of cartographic materials necessary for a complete and high quality cadastral assessment of forest areas for further research and development.

References

1. Aleksiiuk, I.L., Hrynyk, H.H.: The Lisovporiadnyk software as an interactive tool for working with databases of FPA “Ukrderzhlisproekt”. Sci. Bull. UNFU **21**(14), 345–355 (2011). (in Ukrainian)
2. Aleksiiuk, I.L., Lakyda, P.I., Hrynyk, H.H.: The Lisovporiadnyk software as a system for processing the database of the Forestry Fund of Ukraine. Sci. Bull. UNFU **23**(15), 308–316 (2013). (in Ukrainian)
3. Bank Danych o Lasac. https://www.bdl.lasy.gov.pl/portal. Accessed 21 June 2019
4. Geoportal Forests of Ukraine. http://forestry.org.ua/. Accessed 21 June 2019
5. Girs, O.A., Novak, B.I., Kashpor, S.M.: Forest Management, 384 p. Kyiv (2004). (in Ukrainian)

6. Karpuk, A., Moroziuk, O., Myroniuk V., Aleksiiuk, I.: Methodical recommendations for the allocation of the most valuable forest sites for conservation with using GIS-tech, 20 p. Kyiv (2015). (in Ukrainian)
7. Kimeychuk, I.V.: Information base and the formation of a digital map of scientific objects of heritage of forest cultural. In: 9th Proceedings on International Scientific and Practical Conference "New Technologies in Geodesy, Land Management, Forest Management and Nature Management", pp. 279–285. Uzhgorod National University, Uzhgorod (2018). (in Ukrainian)
8. Lakyda, P.I., Bala, O.H.: Actualization of growth parameters of artificial oak forests of Ukraine's Forest-steppe, 196 p. Korsun-Shevchenkivsky (2012). (in Ukrainian)
9. Lakyda, P.I., Terentjev, A.Yu., Vasylyshyn, R.D.: Artificial forest stands of Scotch pine of Ukrainian Polissya – forecast of growth and productivity, 171 p. Kyiv (2012). (in Ukrainian)
10. RUMB – Felling Area Plan-Scheme. http://fes.com.ua/. Accessed 22 Aug 2019
11. Software for digital cryptography and land management Digitals. http://www.geosystema.net/digitals/?act=index. Accessed 21 June 2019
12. Teslenko, G.S.: Information systems in agrarian management, 232 p. Kyiv (1999). (in Ukrainian)
13. WebULR. http://www.lisproekt.gov.ua/webulr. Accessed 23 Aug 2019

Cyber-Physical Systems

Selection of the Reference Stars for Astrometric Reduction of CCD-Frames

Vadym Savanevych[1], Volodymyr Akhmetov[1,2], Sergii Khlamov[1,2(✉)], Eugene Dikov[1], Alexsander Briukhovetskyi[1], Vladimir Vlasenko[1], Vladislav Khramtsov[1], and Iana Movsesian[3]

[1] Science Department of EOS Data Analytics, Kharkiv, Ukraine
akhmetovvs@gmail.com, sergii.khlamov@gmail.com
[2] V. N. Karazin, Kharkiv National University, Svobody Sq. 4, Kharkiv 61022, Ukraine
[3] Kharkiv National University of Radio Electronics, Nauky Ave. 14, Kharkiv 61000, Ukraine

Abstract. In this paper we presented the new method for selection of the reference stars to carry out the astrometric reduction of sky images. The presented algorithm takes into account the main features of the formation of astronomical measurements in CCD-frames. Also to improve the accuracy and speed up processing the proposed method includes verification of candidates to the reference stars and excluding them if needed based on the specific rule. This computational method was successfully tested in the software for automated search and detection of asteroids, comets and satellites in a series of CCD-frames within the CoLiTec project. The comparative analysis of the accuracy of astronomical measurements of the Solar System small bodies by CoLiTec and Astrometrica software was performed as a result of research. The developed method for selection of the reference stars provides the high accuracy of measurements of celestial objects in digital images.

Keywords: Big data · Catalogues · Reference stars · Position observation · Astrometry · Reduction · Digital frames

1 Introduction

Recently, with the development of technological capabilities and computing technology, we are seeing an exponential growth of the observational data. The tasks of fast and automated processing of such data in order to search for the moving objects (asteroids, comets, satellites) and following cataloging theirs are topical now [1–3]. The coordinates determining of these objects is usually performed by the relative method [4, 5] using stars from the reference catalogue, and the requirements for accuracy of their determination increase every year.

At present, when modern telescopes cover large areas of the sky, there can be from tens to several hundred thousand of reference objects in the image. This allows carrying out the astrometric reduction at a high level of accuracy, excluding systematic errors induced by imperfections of the optical system and radiation detectors.

N. Shakhovska and M. O. Medykovskyy (Eds.): CCSIT 2019, AISC 1080, pp. 881–895, 2020.
https://doi.org/10.1007/978-3-030-33695-0_57

CoLiTec (Collection Light Technology) software (http://neoastrosoft.com) [6–8], Astrometrica (http://astrometrica.at) [9, 10] and others include the astrometric and photometric data reduction using the modern catalogues. Methods for determining of the objects positions in frames are described and enough completed [4]. In the modern literature there is not enough information about the methods for selecting of reference objects on digital frames to perform the astrometric reduction [11]. Earlier, this can be explained by a small number of stars in the reference catalogues which were ranged from a few thousand up to a million stars that distributed at the whole sky. In this case, the 'Generalization of the dependencies' method in photographic astrometry [4] was developed to take into account an irregular distribution of stars.

The modern reference catalogues such as GaiaDR2 [12, 13], PMA [14, 15], HSOY [16] and UCAC5 [17] contain astrometric and photometric information about hundreds of millions of stars. All such astronomical catalogues are included in a big database VizieR (http://vizier.u-strasbg.fr). It is a joint effort of CDS (Centre de Données astronomiques de Strasbourg) and ESA-ESRIN (Information Systems Division). VizieR has been available since 1996, and includes more than 18 thousand catalogues that are available from CDS [18].

However, the increasing of number of the reference objects requires the verification of quality, which can negatively affect the reduction results. Different factors such as stellar blending, saturation, stars on tracks and on diffraction artifacts lead to both random and systematic distortions in measurements of the investigated object positions. The measuring of the moving objects coordinates in a series of frames requires the high accuracy of reduction not only in the center of frames but also at the edges. Therefore, it is necessary to provide the uniform distribution of reference stars in frame and to improve the computational methods for selecting reference stars in order to perform the astrometric reduction with a high accuracy for any part of the frame.

In this paper we present a new algorithm for selection of the reference stars to carry out the astrometric reduction of sky images into the system of modern astronomical catalogues. The proposed algorithm is based on an iterative method for selection of the reference stars in digital frames with a stepped increasing of the reduction polynomial order. This method allows us to take into account all the features of forming process of an astronomical image and to exclude the effects of different facts on the accuracy of astrometric reduction.

The data mining techniques and intelligent management technologies of data analysis are rapidly evolving, but automatic data selection is still one of the main steps of any modern pipeline for data analysis or reduction.

The structure of our paper is as follows. We described a problem statement for the modern astrometric reduction, methods for preliminary or emergency identification in Sect. 2. The task solution and new method for selection of the reference stars in digital frames are described in Sect. 3. The information about excluding of candidates to reference stars to improve accuracy and speed up processing is provided in Sect. 4. The comparative analysis of the accuracy of astronomical measurements of the solar system small bodies by CoLiTec and Astrometrica software as a result of research is provided in Sect. 5. Concluding remarks and discussion are given in Sect. 6.

2 Astrometric Reduction

The measurements of stellar objects are presented in each frame, as well as motionless objects (stars, galaxies, quasars), moving objects (objects with near-zero [19, 20] and non-zero apparent motion, such as asteroids, comets, satellites [21]) and false objects (noises on highly saturated objects, hot pixels, false objects on satellite or meteors tracks, spike-like and halo diffraction patterns). It should not be more than one measurement of celestial object in one frame. The coordinates of measurements of false and moving objects are uniform distributed and mutually independent.

Special attention is paid to the automatic binding of digital images to the star catalogues, namely, the determination of the objects position in space according to the image data.

The absolute and relative methods are used for determining the angular position of objects in space [22]. The absolute method is presented in Fig. 1 and depends on the mechanics of the observational tool (telescope).

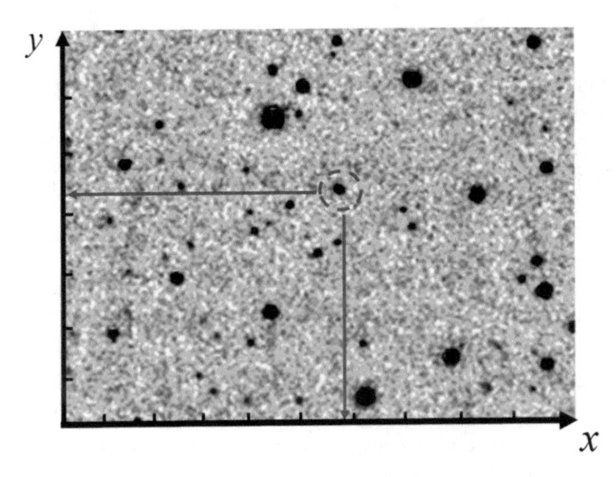

Fig. 1. Absolute method for determining the angular position of objects in space using image data.

The accuracy of astronomical observations is very high so using of the absolute method for determining the objects position in space according to their position in the frame is long gone.

To determine the equatorial coordinates, both asteroids and other celestial objects, the relative method is used as shown in the Fig. 2. It should be based on a quality selection of reference stars.

The accuracy of determining the position of celestial objects in a digital image essentially depends on the accuracy of determining the position of reference stars themselves. This explains the relevance of a separate, independent research of the accuracy of estimates of the angular position of reference stars.

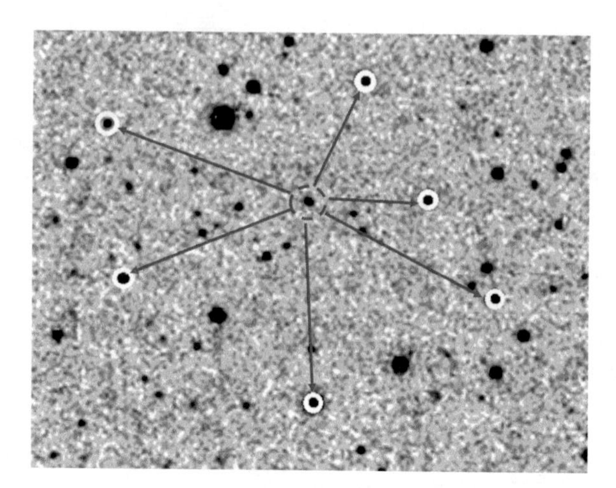

Fig. 2. Relative method for determining the angular position of objects in space using image data.

With the relative method, position of the investigated objects in focal plane of the optical system is measured in relation to the objects whose position in the image is known with high accuracy. Such objects are called reference.

The star catalogue used to identify measurements in frame is called as star reference catalogue. The information about each cataloged celestial object of the set of objects in the star catalogue is organizationally summarized in special forms.

To implement the relative method, it is necessary to identify the objects in frame with the catalogue forms (Fig. 3) and select the reference objects from the identified objects.

For the frame identification it is necessary to find a pairwise correspondence between the set of measurements formed in frame and the set of objects from the star catalogue belonging to the same region of celestial sphere as the formed digital frame [23]. In other words, the identification of the frame measurement and the star catalogue forms can be represented as the establishment of a one-to-one correspondence between the measurement and the star catalogue form that is the establishment of identity of the information contained in measurement and this form. This means that the information of a certain measurement and form corresponds to the same celestial object.

Information about the famous celestial objects is contained in the star reference catalogues [13–17]. The basic information contained in such astrometric catalogues is the accurate coordinates of stars and the speed of their changes (so-called as stellar proper motions).

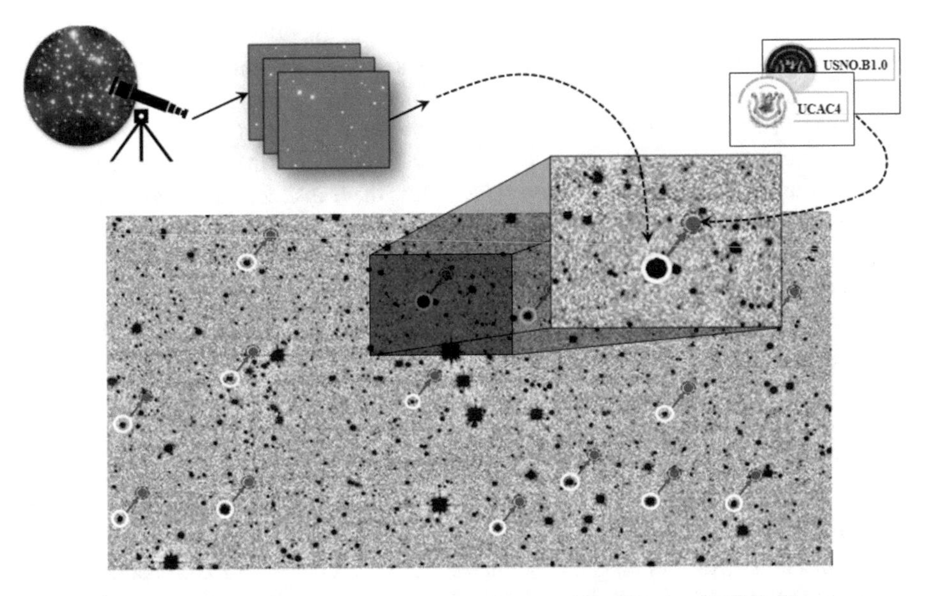

Fig. 3. Graphic representation of the measurements identification in digital image.

The methods for preliminary or emergency identification of such celestial objects allow obtaining the linear plate constants (a_0, a_1, a_2) and (b_0, b_1, b_2) that define the simplest (without taking into account various aberrations) relationship between the tangential (ideal) coordinate system and the digital frame coordinate system:

$$\xi_i = a_0 + a_1 x_i + a_2 y_i; \tag{1}$$

$$\eta_i = b_0 + b_1 x_i + b_2 y_i, \tag{2}$$

where ξ_i and η_i are the tangential (ideal) coordinates of reference stars; x_i, y_i are the measured coordinates of reference stars in the digital frame coordinate system.

The plate constants allow obtaining the estimates of equatorial (α_i, δ_i) coordinates of objects in frame using the following equation [4]:

$$\alpha_i = \alpha_0 + arctg(-\xi_i / \cos\delta_0 - \eta_i sin\delta_0); \tag{3}$$

$$\delta_i = arcsin((\eta_i cos\delta_0 + sin\delta_0) / \sqrt{(1 + \xi_i^2 + \eta_i^2)}, \tag{4}$$

where α_0, δ_0 are the equatorial coordinates of the optical center in CCD-frame.

In most cases, the linear plate constants are not enough for the identification of objects from the reference catalogue and measurements in frame for the wide-field of view. As a rule, their using for only definition stars that corresponds to the central part of frame.

In most cases the cubic model of plate constants is used in the final conversion of the objects coordinates from digital frame to the equatorial coordinate system. For the wide-angle and survey telescopes, such as Thirty Meter Telescope (TMT) [24] and Large Synoptic Survey Telescope (LSST) [25] it is necessary to take into account the five order distortion. This approach provides a reliable finally measuring of the object positions in the whole frame and global cross-match between other catalogues [26, 30].

More data pairs from the reference catalogue and corresponding to them measurements in CCD-frame are needed in order to reliably obtain the cubic and higher degree plate constants. However, the number of pairs required for this cannot be identified using only linear plate constants. Thus, there is a contradiction between the need to use a cubic (and higher) model of the plate constants and a small number of identified pairs 'measurement – reference catalogue' at the first stages or preliminary identification of measurements in frame and reference catalogue.

Therefore, we have developed an iterative method for selection of the reference objects and a step-by-step changing of the degree of astrometric reduction polynomial.

3 Reference Stars Selection

The method for selection of the reference stars in digital frames is based on the multistage complication of parameters of plate constants (increasing of the order of plate constants with refining of estimates of their values and exclusion of non-significant components) with a synchronous increasing of the number of pairs 'measured – reference objects from catalogue' used to calculate parameters of plate constants.

The method for selection of the reference stars, along with the methods for determining the angular position of objects in space using image data, determines the accuracy of the indicated astrometric observations of stars.

The identification of pairs at each stage is performed in the digital frame coordinate system, in which the equatorial coordinates of stars are converted using the plate constants. For conversion of measurements in the equatorial coordinates are used linear, and later cubic (or fifth degree) plate constants.

Linear plate constants are not enough to identify catalogue forms and frame measurements for the wide field of view. With their using, as a rule, only stars are identified that correspond to images in the central part of frame. Most often, in the final recalculation of coordinates of the digital frame coordinate system to the equatorial coordinates, a cubic model of plate constants is used, which ensures reliable identification and position measuring in the entire frame. In turn, for the reliable obtaining of cubic plate constants, more pairs of 'catalogue form – measurement in frame' are needed.

However, the required number of such pairs cannot be identified with using of the linear plate constants. So, in the current version of method for selection of the reference stars three stages are used.

The N_m measurements in CCD-frame and N_{st} stars from the reference catalogue are selected on the first stage. The number of measurements increases at ΔN_m measurements and ΔN_{st} stars at the each next step. The set of reference stars is ordered by brightness for the selection of catalogue stars at the stages of the method for selecting of the reference stars.

It is obvious that the concentration of reference stars in some part of the frame (for example, in center as demonstrates Fig. 4) improves the accuracy of measurements in this part of the frame by decreasing it in the other areas of frame.

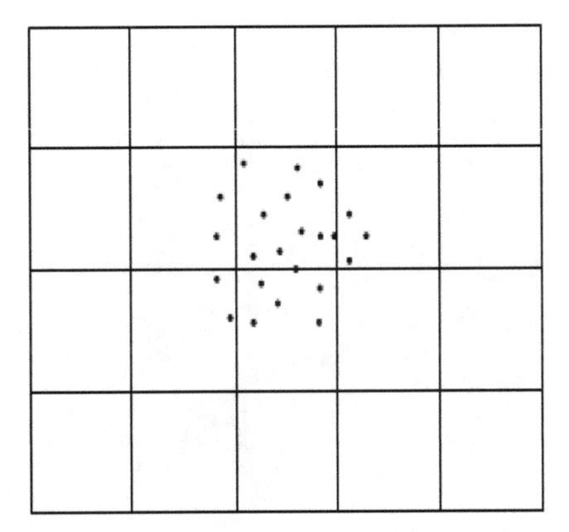

Fig. 4. Positions of the brightest measurements ordered by brightness in CCD-frame.

To provide almost equal accuracy of measurements of the objects coordinates in the whole frame the uniform distribution of reference stars in CCD-frame was performed, as shown in Fig. 5.

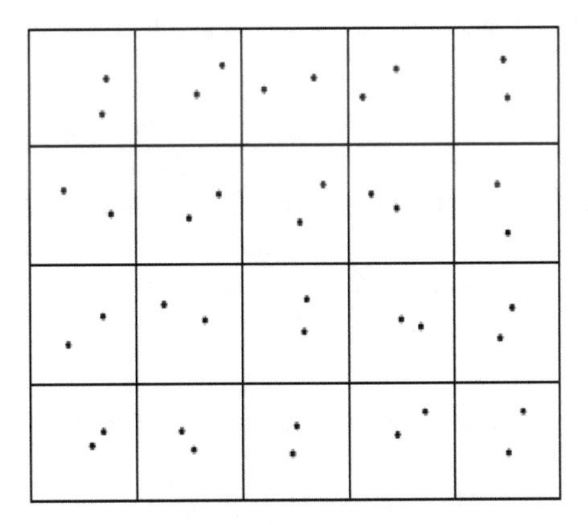

Fig. 5. Uniform distribution of reference stars in CCD-frame.

In other words, the number of reference stars in a particular area of the frame can be considered as its weight in the sum of squares of deviations that are used in the calculation of plate constants. In addition, the uniform distribution of identified pairs in frame helps to avoid cases when the large amount of 'brightest', blending, false measurements/stars is presented in one area of frame. An example of such case is the presence of very bright star or bright satellite track in the frame (Fig. 6).

If these cases are not excluded from consideration, then often enough take place failures during the identification/detection/search of known objects and the identification of whole frames becomes unreliable or significant errors in determining of plate constants are possible.

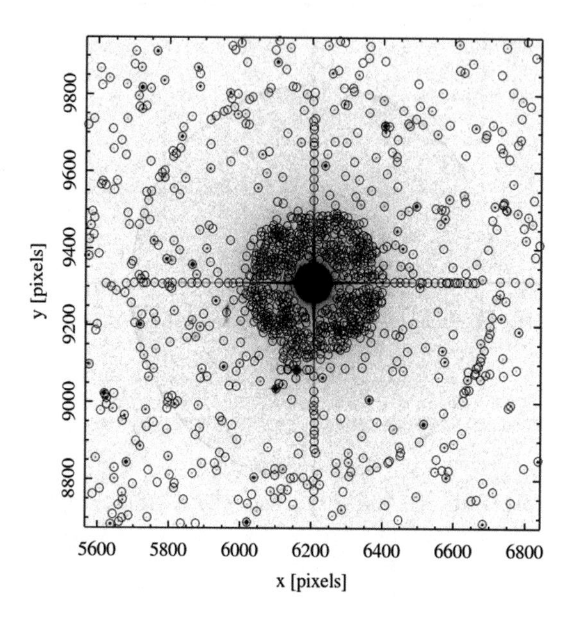

Fig. 6. False objects on tracks, spike-like and halo diffraction patterns in CCD-frame.

In the proposed method the fragmentation of frame is performed by means of uniform distribution of the identified pairs in frame during selection of the candidates to reference stars to calculate the plate constants. The same number of stars is selected in each fragment of frame. In other words, the frame is divided into *MxM* (M^2) parts. The example of such frame division into 81 fragments (*M* = *9*) and the number of reference stars in the each fragment is presented in the Table 1.

Table 1. Frame division into 81 fragments with number of reference stars in each fragment.

9	9	8	13	6	10	10	8	8
7	7	9	11	9	8	8	17	9
7	12	8	12	11	8	10	7	9
10	10	7	10	11	10	11	11	8
11	8	6	6	5	9	9	6	5
10	8	7	11	11	13	3	11	12
7	11	17	9	10	12	10	14	13
7	10	7	9	10	10	10	10	11
13	6	13	13	12	9	5	9	10
0	3	0	0	0	0	0	0	0

The specified number of measurements in CCD-frame N_m and the number of reference catalogue stars N_{st} are distributed to the frame's fragments. From each such fragment the N_m/M^2 and N_{st}/M^2 numbers of the brightness measurements/stars in frame/stars from catalogue are selected.

At the second, third and n-th stages of method, the $\Delta N_m/M^2$ and $\Delta N_{st}/M^2$ numbers of measurements from the frame and reference catalogue accordingly are additionally selected from each fragment of frame.

In Fig. 7 you can see the same CCD-frame as in Fig. 6, but with marked reference stars without false objects on tracks, spike-like and halo diffraction patterns after processing with the proposed method.

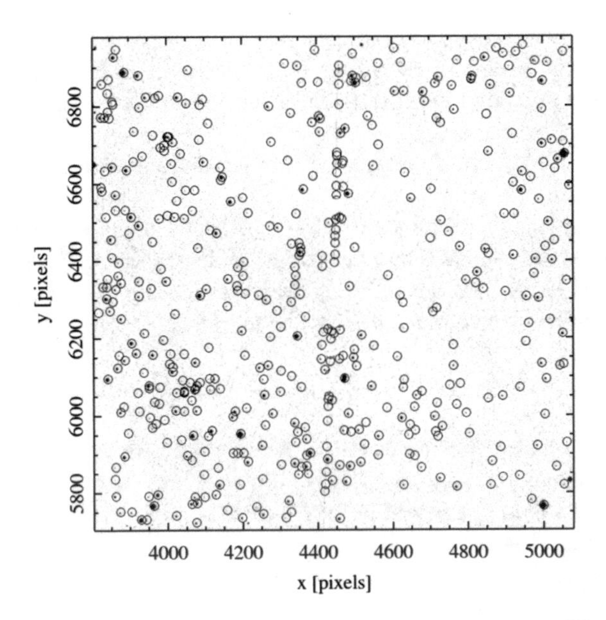

Fig. 7. Reference stars without false objects on tracks, spike-like and halo diffraction patterns in CCD-frame after processing with the proposed method.

Therefore, the larger average number of the reference stars in each fragment, the more accurate the plate constants will be calculated and the more accurate frame identification will be.

4 Excluding of Candidates to Reference Stars

At each stage of the method for selection of the reference stars from a set of candidates, the measurements of nearby objects are excluded. The distance between objects is more than R_{mgroup}. All measurements in frame are excluded from candidates to reference stars according to the condition:

$$\sqrt{\left((x_i - x_m)^2 + (y_i - y_m)^2\right)} < R_{mgroup}. \tag{5}$$

where x_i, y_i are the objects coordinates contained in measurements of the nearby objects in CCD-frame coordinate system. Also the pair is excluded from consideration if it corresponds to a close objects in catalogue with same or more brightness. The stars of used catalogue are excluded according to the data of stellar reference catalogue according to their affiliation to bunches/clusters/compact groups of stars with distance less than the specified $R_{stargroup}$.

Also the stellar detections in images without peaks (saturation objects) are excluded from consideration.

The rejection of significant amount of identified pairs after solving the identification problem improves the estimates accuracy of the plate constants. The final statistics for rejection of pairs is the total deviation σ_k between the estimates of equatorial coordinates in the identified 'measurement – catalogue' pairs:

$$\Delta_{\alpha\delta ijk} = \sqrt{((\alpha_{catj(k)} - \alpha_{meainfr(k)})^2 + (\delta_{catj(k)} - \delta_{meainfr(k)})^2)}. \tag{6}$$

where $\alpha_{catj(k)}$, $\delta_{catj(k)}$ are estimates of the object's right ascension and declination in the j-th form of catalogue; $\alpha_{meainfr(k)}$, $\delta_{meainfr(k)}$ are estimates of the right ascension and declination of the i-th measurement in the N_{fr}-th frame; k is a number of the identified 'measurement – catalogue' pair.

These pairs are rejected if the value of total deviation more than critical value:

$$\Delta_{\alpha\delta ijk} > K_{rej}\Delta_{\alpha\delta}. \tag{7}$$

where K_{rej} is the coefficient of rule for rejection of 'measurement – catalogue' pairs from a set of reference stars; $\Delta_{\alpha\delta}$ is the average absolute value of deviation of the identified pairs in equatorial celestial coordinate system in the set of selected identified pairs.

The average absolute value of deviation of the identified pairs $\Delta_{\alpha\delta}$ can be defined using the following equation:

$$\Delta_{\alpha\delta} = \sqrt{((1/N_{cou})\sum((\alpha_{catj(k)} - \alpha_{meainfr(k)})^2 + (\delta_{catj(k)} - \delta_{meainfr(k)})^2))}. \tag{8}$$

where N_{cou} is the amount of identified 'measurement – catalogue' pairs used to calculate the plate constants;

To calculate the plate constants, only pairs with the deviation of which $K_{rej} = 1$ times less than the mean value of deviation of the identified pair in equatorial coordinates in the selected set of identified pairs are selected. The final calculation of the plate constants is performed using the formed and not rejected 'measurement – reference catalogue' pairs at the end of the third stage of selection of reference stars.

Research on frames that are smaller than 1×1 degree and consists of several million pixels, the following values of the developed computational method parameters were assumed: a number of the brightest frame measurements and catalogue forms that take part in selecting candidates to the reference stars is $N_m = 400$, $N_{st} = 600$, and the maximum allowable distance between neighboring objects (stars) of a group of close objects $R_{stargroup} = 5$ arcsec. The number of fragments into which a frame is divided by each coordinate when selecting reference stars $M = 4$. The number of measurements increases by $\Delta N_m = 300$ measurements and $\Delta N_{st} = 500$ stars at each next step, the maximum allowable distance between neighboring objects (size) of a group of close objects $R_{mgroup} = 20$ pixels; parameter of the rule of pairs rejection of 'measurement – reference' $K_{rej} = 1$.

5 Results

The developed method for selection of the reference stars is implemented to the CoLiTec software for automated search and detection of asteroids and comets in a series of CCD-frames [6] as well as for identifying stars for the light curves creation in CoLiTecVS software [27]. This software uses the online access to the different astronomical catalogues provided by VizieR [18], so it can perform data analysis and selection of the reference stars from these catalogues. The results of the CoLiTec software confirm the efficiency and stability of the developed method.

The International Scientific Optical Network of New Mexico (ISON NM) observatory with CoLiTec software according to the total results for 2011 and 2012 years (based on data from the Minor Planet Center at Harvard University [28]), took seventh place in the world, both in the amount of measurements and in the number of asteroid discoveries.

A comparative analysis of the accuracy of astronomical measurements of the solar system small bodies by CoLiTec and Astrometrica software was performed in research [10]. The same test CCD-frames series were processed using CoLiTec and Astrometrica software. These CCD-frames were obtained at the ISON-NM observatory (MPC code - H15). The observatory is located at Mount Joy (Meyhill), New Mexico, United States, and uses the 40 cm telescope SANTEL400AN and the CCD-matrix FLI ML0900065 (3056 × 3056 PCL, pixel size of 12 μm). The exposure time was 150 s. In total, 19 series of four frames each were used and 2002 positional CCD-measurements of numbered known asteroids were selected [10].

So, the input data for research was as following:

- 2002 positional CCD-measurements from 19 series;
- cubic model for the coordinate reduction;
- UCAC4 catalogue with reference stars;
- HORIZONS service [29] with the investigated objects positions.

All 2002 positional CCD-measurements were completely made and processed by the CoLiTec software, but the Astrometrica software made only 1749 of 2002 CCD-measurements (253 less than CoLiTec software). For these 253 CCD-measurements the Astrometrica software indicated that it could not ensure a reliable CCD-positioning of the object (error: Centroid = 1). This is often associated with an attempt to measure the position of blurred objects, star trails, or faint objects that are in the image next to the bright stars. More than half of them have SNR not greater than 3.5.

General characteristics (root mean square – RMS) of the positional accuracy (right ascension – RA, declination – DE) of CCD-measurements of numbered known asteroids are shown in Table 2. For all apparent brightness ranges the CCD-measurements made by the Astrometrica software have 30–50% greater RMS than CoLiTec software.

Table 2. Deviations of equatorial coordinates for the apparent brightness ranges.

Magnitude, ^{m}		Number of CCD-measurements	Astrometrica		CoLiTec	
Min	Max		RA RMS	DE RMS	RA RMS	DE RMS
11.1	13.0	12	0,38	0,23	0,18	0,12
13.0	14.0	4	0,25	0,53	0,05	0,02
14.0	15.0	4	0,04	0,20	0,03	0,01
15.0	15.5	12	0,41	0,25	0,06	0,06
15.5	16.0	22	0,26	0,07	0,38	0,09
16.0	16.5	25	0,13	0,09	0,14	0,10
16.5	17.0	40	0,18	0,10	0,16	0,19
17.0	17.5	105	0,20	0,09	0,21	0,11
17.5	18.0	204	0,29	0,19	0,24	0,17
18.0	18.5	263	0,37	0,27	0,30	0,23
18.5	19.0	340	0,44	0,39	0,48	0,35
19.0	19.5	414	1,22	1,11	0,54	0,39
19.5	20.0	406	0,87	0,60	0,66	0,54
>20.0		151	0,94	1,03	0,73	0,63
Total		**2002**	**0,77**	**0,67**	**0,50**	**0,39**

Using the Astrometrica software processing, some CCD-frames did not have a reliable identification. The reason for this is that in non-identified frames the reference stars were only in one half of the CCD-frame after processing using Astrometrica software, as shown in Fig. 8, where the reference stars are marked with the white circles.

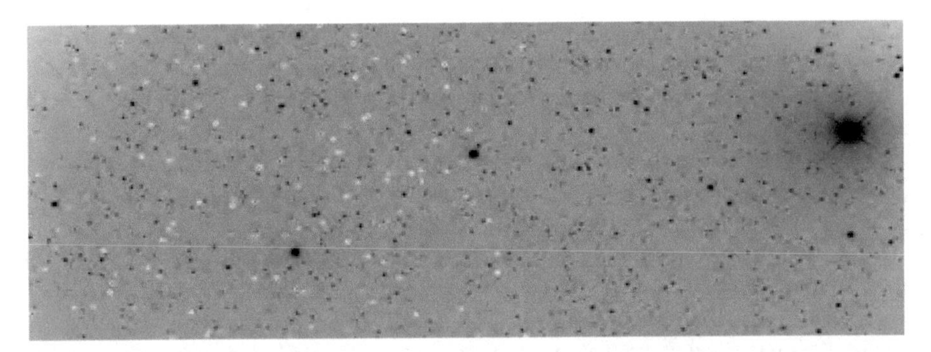

Fig. 8. Reference stars in CCD-frame selected by Astrometrica software (only half of CCD-frame was processed).

The Fig. 9 shows the uniform distribution in CCD-frame of the reference stars that were selected using CoLiTec software with implemented proposed method.

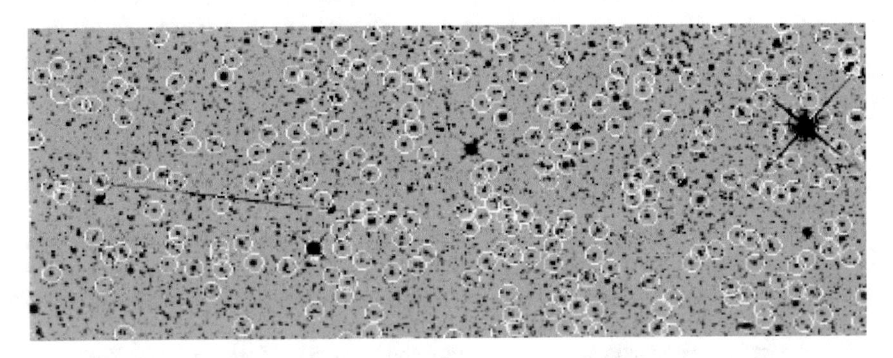

Fig. 9. Reference stars in CCD-frame selected by CoLiTec software with implemented proposed method (whole CCD-frame was processed).

The analysis showed that the frequency of critical errors in CoLiTec software is lower than that in Astrometrica software. It means that the method for selection of the reference stars is well implemented in the CoLiTec software.

6 Conclusions

In this paper, we developed a multi-stage computational method for selection of the reference stars, which takes into account the main features of the formation of astronomical measurements in CCD-frame.

To provide the high accuracy of astrometric reduction the CCD-frame was divided into fragments with an independent selection of candidates to reference stars from each of them. On the first stage the linear plate constants are formed using stars in the center of the frame. On the next identification steps a cubic model (or higher degree) of plate constants was used in order to provide the high astrometry accuracy of objects in the whole frame.

Modern reference catalogues contain a large number of 'measurements – reference' pairs at least half of the previously identified measurements of candidates to reference stars are rejected in order to the increasing of accuracy and reliability of the positional measurements.

The developed computational method for selection of the reference stars was implemented to the CoLiTec software as a part of identification process, which is required for the objects detection in CCD-frame with the high accuracy.

The comparative analysis of accuracy was performed between CoLiTec and Astrometrica software and showed that the frequency of critical errors in CoLiTec is lower than that in Astrometrica.

So, the developed computational method for selection of the reference stars will be very useful for the different software for automated processing of CCD-frames to perform astrometric and photometric reduction.

Acknowledgment. The authors thank CDS (Strasbourg, France) who provided online access to the different astronomical catalogues by VizieR (http://vizier.u-strasbg.fr) [18]. We especially thank all creators of astronomical catalogues and Astrometrica software that described in the paper. We are grateful to the reviewers for their helpful remarks that improved our paper.

References

1. Ericson, J.: Asteroids, Comets, and Meteorites. Cosmic Invaders of the Earth. Fact on File, New York (2003)
2. Kortencamp, S.: Asteroids, Comets, and Meteorids, p. 360. Capstone Press, Mankato (2012)
3. Rivkin, A.: Asteroids, Comets And Dwarf Planets. Greenwood Press, Santa Barbara (2009)
4. Kiselev, A.: Theoretical fundamentals of photographic astrometry. Moscow, Izdatel'stvo Nauka, 264 p. (1989). (In Russian)
5. Kallenberg, O.: Foundations of Modern Probability, p. 535. Springer, New York (1997). ISBN: 0387949577
6. Khlamov, S., Savanevych, V., Briukhovetskyi, O., Pohorelov, A., Vlasenko, V., Dikov, E.: CoLiTec software for the astronomical data sets processing. In: Proceedings of the IEEE 2nd International Conference on Data Stream Mining and Processing, DSMP 2018, pp. 227–230 (2018)
7. Khlamov, S., Savanevych, V., Briukhovetskyi, O., Pohorelov, A.: CoLiTec software – detection of the near-zero apparent motion. In: Proceedings of the International Astronomical Union, vol. 12(S325), pp. 349–352. Cambridge University Press (2017)
8. Kudzej, I., Savanevych, V., Briukhovetskyi, O., Khlamov, S., Pohorelov, A., Vlasenko, V., Dubovský, P., Parimucha, Š.: CoLiTecVS – a new tool for the automated reduction of photometric observations. Astron. Nachr. **340**(1–3), 68–70 (2019)
9. Raab, H.: Astrometrica: Astrometric data reduction of CCD images. Astrophysics Source Code Library, record: 1203.012 (2012)
10. Savanevych, V., Briukhovetskyi, A., Ivashchenko, Yu., Vavilova, I., Bezkrovniy, M., Dikov, E., Vlasenko, V., Sokovikova, N., Movsesian, I., Dikhtyar, N., Elenin, L., Pohorelov, A., Khlamov, S.: Comparative analysis of the positional accuracy of CCD measurements of small bodies in the solar system software CoLiTec and Astrometrica. Kinematics Phys. Celestial Bodies **31**(6), 302–313 (2015)

11. Iatsenko, A., Rybka, S.: Criteria for the selection of reference stars according to their proper motions. Astrometriia i Astrofizika **51**, 75–79 (1984). (In Russian)
12. Gaia Collaboration: Summary of the astrometric, photometric, and survey properties. A & A **595**(A2), 23 (2016)
13. Gaia Collaboration: The Gaia mission. A & A **595**(A1), 36 (2016)
14. Akhmetov, V., Fedorov, P., Velichko, A., Shulga, V.: The PMA Catalogue: 420 million positions and absolute proper motions. MNRAS **469**(1), 763–773 (2017)
15. Fedorov, P., Akhmetov, V., Velichko, A.: Testing stellar proper motions of TGAS stars using data from the HSOY, UCAC5 and PMA catalogues. MNRAS **476**(2), 2743–2750 (2018)
16. Altmann, M., Roeser, S., Demleitner, M., Bastian, U., Schilbach, E.: Hot Stuff for One Year (HSOY). A 583 million star proper motion catalogue derived from Gaia DR1 and PPMXL. Astron. Astrophys. **600**, L4–4 (2017)
17. Zacharias, N., Finch, C., Frouard, J.: UCAC5: new proper motions using gaia DR1. Astron. J. **153**, 166 (2017)
18. Ochsenbein, F., Bauer, P., Marcout, J.: The VizieR database of astronomical catalogues. Astron. Astrophys., Suppl. Ser. **143**(1), 23–32 (2000)
19. Savanevych, V., Khlamov, S., Vavilova, I., Briukhovetskyi, A., Pohorelov, A., Mkrtichian, D., Kudak, V., Pakuliak, L., Dikov, E., Melnik, R., Vlasenko, V., Reichart, D.: A method of immediate detection of objects with a near-zero apparent motion in series of CCD-frames. Astron. Astrophys. **609**(A54), 11 (2018)
20. Khlamov, S., Savanevych, V., Briukhovetskyi, O., Oryshych, S.: Development of computational method for detection of the object's near-zero apparent motion on the series of CCD–frames. Eastern-Eur. J. Enterp. Technol. **2**(9(80)), 41–48 (2016)
21. Savanevych, V., Briukhovetskyi, O., Sokovikova, N., Bezkrovny, M., Vavilova, I., Ivashchenko, Yu., Elenin, L., Khlamov, S., Movsesian, I., Dashkova, A., Pogorelov, A.: A new method based on the subpixel Gaussian model for accurate estimation of asteroid coordinates. MNRAS **451**(3), 3287–3298 (2015)
22. Karttunen, H., Kroger, P., Oja, H., Poutanen, M., Donner, K.: Fundamental Astronomy, 4th edn. Springer, Berlin (2003)
23. Mills, P.: Efficient statistical classification of satellite measurements. Int. J. Remote Sens. **32** (21), 6109–6132 (2011)
24. Skidmore, W., et al.: Thirty meter telescope detailed science case: 2015. Res. Astron. Astrophys. **15**(12), 1945–2140 (2015)
25. Tuell, M., Martin, H., Burge, J., Gressler, W., Zhao, C.: Optical testing of the LSST combined primary/tertiary mirror. In: Proceedings of the SPIE 7739, Modern Technologies in Space- and Ground-based Telescopes and Instrumentation, 77392 V, 23 July 2010
26. Akhmetov, V., Khlamov, S., Dmytrenko, A.: Fast coordinate cross-match tool for large astronomical catalogue. Adv. Intell. Syst. Comput. **871**, 3–16 (2019)
27. Parimucha, Š., et al.: CoLiTecVS–a new tool for an automated reduction of photometric observations. Contrib. Astron. Obs. Skalnaté Pleso **49**, 151–153 (2019)
28. IAU Minor Planet Center. http://www.minorplanetcenter.net. Accessed 21 June 2019
29. HORIZONS System. http://ssd.jpl.nasa.gov/?horizons. Accessed 21 August 2019
30. Savanevych V., Dikov Eu., Bryukhovetsky A., Vlasenko V., Akhmetov V., Khlamov S., Khramtsov V., Movsesian I.: New approach to select reference stars for astrometric reduction of CCD-frames. In Proceedings International Scientific Conference "Computer sciences and information technologies" (CSIT-2019), vol. 2, pp. 110–113 (2019)

Astrometric Reduction of the Wide-Field Images

Volodymyr Akhmetov, Sergii Khlamov(✉), Vladislav Khramtsov, and Artem Dmytrenko

V. N. Karazin Kharkiv National University,
Svobody Sq. 4, Kharkiv 61022, Ukraine
akhmetovvs@gmail.com, sergii.khlamov@gmail.com

Abstract. In this paper we presented the new algorithm for astrometric reduction of the images received from modern large telescopes with very wide field of view. This algorithm is based on the iterative using of the method of ordinary least squares (OLS) and statistical Student t-criterion. The paper contains information about selecting the appropriate reduction model and system of conditional equations for determination of the reduction model parameters with the aim to solve it using the OLS method. The proposed algorithm also provides the automatic selection of the most probabilistic reduction model. At the first iteration the fifth degree polynomial is used, which provides 21 constant plates. The reduction error from the each iteration is used in the next iteration to eliminate reference objects whose residuals more than three sigma. In the developed algorithm the statistical Student t-criterion of reliability is applied after several iterations of getting rid of the noisy objects. This approach allows us to eliminate almost all systematic errors that are caused by imperfections of the optical system of modern large telescopes. The developed software based on the proposed algorithm was applied to perform the astrometric reduction of all measured positions of objects on the digitized photographic plates of SuperCOSMOS data. The research results showed that the new proposed algorithm allows performing the reduction into the system of reference catalogue with the highest accuracy level.

Keywords: Database · Big data · Catalogue · Astrometry · Reduction · Wide field · Data analysis

1 Introduction

In recent years, with the growth of technological capabilities and computer technology, the automation and computerization of the processing has become more widespread in the different areas of the human activity. Astronomy, in particular astrometry, is not an exception too.

Because of the construction of large ground-based telescopes, launching of the space missions, there is a very rapid increasing of the amount of observational data. These data at the first stage are usually represented as images of the stellar sky that should be processed, measured and cataloged.

N. Shakhovska and M. O. Medykovskyy (Eds.): CCSIT 2019, AISC 1080, pp. 896–909, 2020.
https://doi.org/10.1007/978-3-030-33695-0_58

The image data contains useful information about positions and brightness of the sources of some epochs, and thus can be used for:

- creating the catalogues of proper motions in case when two or more catalogues of positions are of different epochs, and the proper motions of common sources can be calculated;
- calculating the orbital properties of moving objects – asteroids, comets, space satellites, etc.;
- investigating of the offsets between optical and radio positions of radio stars and galaxies with active nuclei to analyze their physical properties;
- measuring of the weak lensing of distant sources by massive objects or their groups (by galaxies or clusters of galaxies);
- searching for the candidates in new gravitationally lensed quasars, where measuring and comparing of the offsets of photocentres in two or more particular bands of any gravitationally lensed quasar system are crucial.

For all of these purposes, first of all, it is necessary to perform an astrometric reduction of all objects in images. The last one should be performed as accurately as possible because it can be used in the different software for automated image processing, such as CoLiTec (Collection Light Technology) software (http://neoastrosoft.com) [1, 2], Astrometrica [3] and others. That is why the highest accuracy level of the astrometric reduction is necessary in the modern astronomy.

Also, until recently, when the telescopes were not so advanced, the field of view (FOV) of telescopes was not so wide and the number of the reference stars was measured by the first dozens and hundreds.

Presently, when the modern telescopes cover large areas of the sky and have wide FOV (more than 1 degree square), there are from tens to several hundred thousands of the reference objects that will be appeared in the image. This requires performing the astrometric reduction with the highest accuracy level and eliminating all systematic errors caused by imperfections in the optical system. Therefore, it is necessary to improve computational methods that will make possible to perform the assigned researching tasks with a high level of accuracy during the minimum of computational time.

In the most cases, for performing the astrometric reduction the classical computational methods are used, in particular the method of ordinary least squares (OLS). OLS is one of the linear least squares methods for estimating the unknown parameters within a linear regression model [4]. So, before using OLS method for the modern astrometric reduction, the model and system of conditional equations should be properly selected [5]. OLS chooses the parameters of a linear function of a set of explanatory variables by the principle of least squares: minimizing the sum of squares of differences between the observed dependent values of variable in the given dataset and those predicted by the linear regression model [6].

As we know, for a sufficiently large number of iterations, as in our case of astrometric reduction, the estimates by the OLS method coincide with the estimates by maximum likelihood method, because of the additional assumption that errors are normally distributed [7]. The OLS estimator is consistent when the regressors are exogenous, and optimal in the class of linear unbiased estimators when the errors are homoscedastic and serially uncorrelated. So, the OLS method provides the minimum-variance mean-unbiased estimation when the errors have finite variances [4]. Also in comparison with maximum likelihood method, the OLS method does not have the following disadvantages:

- incomplete evidence for using the maximum likelihood criterion when some parameters are unknown [4, 8];
- necessary to select the value of boundary decisive statistics [8].

Therefore, the authors suggest also using the statistical criterions, such as Student *t*-criterion or Fisher *f*-criterion, instead of using the maximum likelihood criterion, as it was done in the works [9, 10].

In this paper we propose the new algorithm for astrometric reduction of the wide-field images into the system of the modern astronomical catalogues. Proposed algorithm is based on the iterative using of the OLS method and statistical Student *t*-criterion, which allows with the high accuracy to determine all significant coefficients of decomposition that best describe the systematic errors contained in the measured positions of objects.

The structure of our paper is as follows. We described a problem statement for the modern astrometric reduction, a system of conditional equations and solving of it using the OLS method in Sect. 2.

The task solution and new algorithm of coordinate reduction method with its advantages are described in Sect. 3.

The astrometric reduction of big sets of test data and the precision analysis of calculations are provided in Sect. 4.

Concluding remarks and discussion are given in Sect. 5.

2 Astrometric Reduction

The classic astrometric reduction was developed for the classic type astrographs with a relatively small and flat FOV (see, for example, König A. [11]). These telescopes provide the image of the stellar sky, which with a good approximation can be regarded as the central projection of the sphere onto the plane. For such telescopes, the basic concepts of Gaussian optics are fair.

However, for super wide-angle and super high-aperture systems such as the modern forthcoming large telescopes Thirty Meter Telescope (TMT) [12] and the Large Synoptic Survey Telescope (LSST) [13], the concepts of geometrical optics lose their meaning. Nevertheless, in this case the central projection is the closest mathematical model of real astrophotography too. With help of this model, we still can solve many astrometric tasks.

In the each specific case, it is necessary to choose the appropriate reduction model, and to complement it with the necessary terms for minimization of the discrepancies between the model and the real image of the stellar sky.

So, the systems of conditional equations for determination of the parameters of reduction model can be presented in the array form [14]:

$$X_{Nm}A_{m1} = \Xi_{N1}; \quad (1)$$

$$X_{Nm}E_{m1} = \Theta_{N1}, \quad (2)$$

where the X_{Nm} array has N rows and m columns that are predetermined by the coefficients (measured coordinates of N reference stars) of the conditional equations; A_{m1}, E_{m1} – the vectors that are composed of the predetermined parameters a_i and e_i; $i = 1, 2, \ldots, m$; Ξ and Θ – the vectors that are composed of the tangential coordinates ξ_i, η_i of reference stars.

Solving the system of conditional equations using the OLS method provides the most probable unbiased values of unknown parameters:

$$\tilde{A}_{m1} = C_{mm}^{-1} X_{mN} \Xi_{N1}; \quad (3)$$

$$\tilde{E}_{ml} = C_{mm}^{-1} X_{mN} \Theta_{N1}, \quad (4)$$

where $\tilde{A}_{m1}$ and $\tilde{E}_{ml}$ – vectors of the estimates of unknown parameters; a_i and e_i; $i = 1, 2, \ldots, m$; C_{mm}^{-1} – symmetric array that is inverse to the matrix of normal equations:

$$C_{mm} = X_{mN} X_{Nm}. \quad (5)$$

Objects estimates $\tilde{A}_{m1}$ and $\tilde{E}_{ml}$ are determined taking into account the weighting of the tangential coordinates that depend of the accuracy of their spherical coordinates.

The estimates of reduction errors that are based on the internal convergence of the initial data with taking into account the correlations of the parameters a_i and e_i is represented as:

$$\sigma = \sqrt{\sum v_i^2 / N - m}. \quad (6)$$

The correlation arrays of the parameters a_i and e_i:

$$K(a_i a_i); \quad (7)$$

$$K(e_i e_i). \quad (8)$$

The influence of the configuration of reference and determined stars are estimated by the following equation:

$$X_{1m}^0 \, C_{mm}^{-1} \, X_{m1}^0, \quad (9)$$

where $X_{1m}^0 = \left\| 1, x_0, y_0, x_0^2, \ldots x_0^{pq}, y_0^{rs} \right\|$; $X_{m1}^0 = (X_{1m}^0)^T$.

Parameter X_{1m}^0 allows noticing and correcting of the individual contribution of each reference star into the total reduction error. The index ^{0} indicates the measured coordinates of the investigated object.

By using the Student *t*-criterion, we can estimate the significance of the expansion for truncated model:

$$t = (x' - \mu)\sqrt{m}/s, \tag{10}$$

where x' is the selective mean and equals to:

$$x' = 1/m \sum x_i, \tag{11}$$

and s^2 is the selective dispersion and equals to:

$$s^2 = 1/(m-1) \sum (x_i - x')^2. \tag{12}$$

When using the classical OLS method for the performing a reduction, it is necessary to know in advance degree and type of the reduction model. This model should best describe all distortions caused by the optical system of the telescope. In advance, it is not always possible to choose the degree of a polynomial that will describe all the systematic components.

Also, when determining the decomposition coefficients using only iteration, all random errors that hide in the data due to imperfections in the optical system, will not be discarded from used data. This will lead to a distortion of the final result and incorrect interpretation of the obtained data.

3 New Algorithm of Coordinate Reduction

We offer an improved algorithm of the OLS method, one of the features of which is the automatic selection of the degree and shape of the reduction polynomial, which will best describe the experimental data.

There are two equations for each reference object. Because of the large number of the reference objects N, which sometimes reaches hundreds of thousands, the solving a system of equations with high degree polynomials directly through the X_{mN} array and its transposed X_{Nm} array is not justified.

Therefore, we immediately create an array C_{mm} of the normal equations and a vector with multiplying of the tangential coordinates and X_{mN}:

$$\begin{pmatrix} n & \sum_{i=1}^{n} x_{i1} & \cdots & \sum_{i=1}^{n} x_{ip} \\ \sum_{i=1}^{n} x_{i1} & \sum_{i=1}^{n} x_{ip}^2 & \cdots & \sum_{i=1}^{n} (x_{i1} \cdot x_{ip}) \\ \cdots & \cdots & \cdots & \cdots \\ \sum_{i=1}^{n} x_{ip} & \sum_{i=1}^{n} (x_{i1} \cdot x_{ip}) & \cdots & \sum_{i=1}^{n} x_{ip}^2 \end{pmatrix} \begin{pmatrix} \sum_{i=1}^{n} y_i \\ \sum_{i=1}^{n} (y_i \cdot x_{i1}) \\ \cdots \\ \sum_{i=1}^{n} (y_i \cdot x_{ip}) \end{pmatrix}. \tag{13}$$

Such approach makes it possible to quickly obtain estimates of all coefficients of the polynomial reduction by OLS method for a large number of the reference objects. At the first iteration of the solution, we use a polynomial of degree 5, which provides 21 constant plates. Also, we get a reduction error, which we use in the next iteration to eliminate reference objects whose residuals more than 3σ.

After several iterations of getting rid of the noisy objects, in our OLS method we apply the statistical Student *t*-criterion of reliability. This allows excluding some insignificant coefficients of the reduction model, thereby increasing the accuracy of determining the significant one. The elimination of insignificant members of a polynomial is also carried out several times, until the general residual σ will not reach a minimum. As practice shows, in most cases, for this you need to perform 7–12 iterations.

The main advantage of the proposed approach is that we can obtain high-precision values of the desired coefficients of expansion. They will be much more accurate than if they were obtained using a classical approach, which uses only iteration with the reduction model that is not always correct.

The distribution of residuals after performing the astrometric reduction using the standard reduction model (third-degree polynomial) for the photographic plates of Schmidt type telescope is shown in Fig. 1.

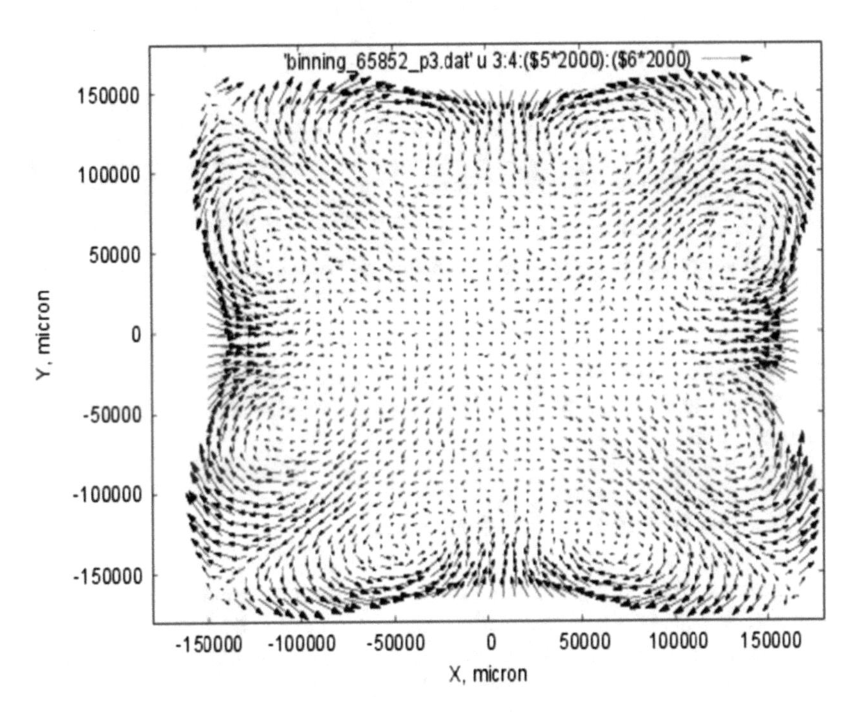

Fig. 1. Mean vector residual map using a standard third-degree-order polynomial plate model.

Unfortunately, we can see a significant systematics residual as a function of stellar position on frame. This indicates to the incorrect selection of the reduction model and, accordingly, the incomplete elimination of systematic errors in the measured positions of the reference objects.

Mean vector residual map in Fig. 2 shows the distributions of residuals after using our method by the fifth-degree-order polynomial plate model. Any significant symmetric errors in measured positions of the reference objects are absent after performing the astrometric reduction using our method by the fifth-degree polynomial.

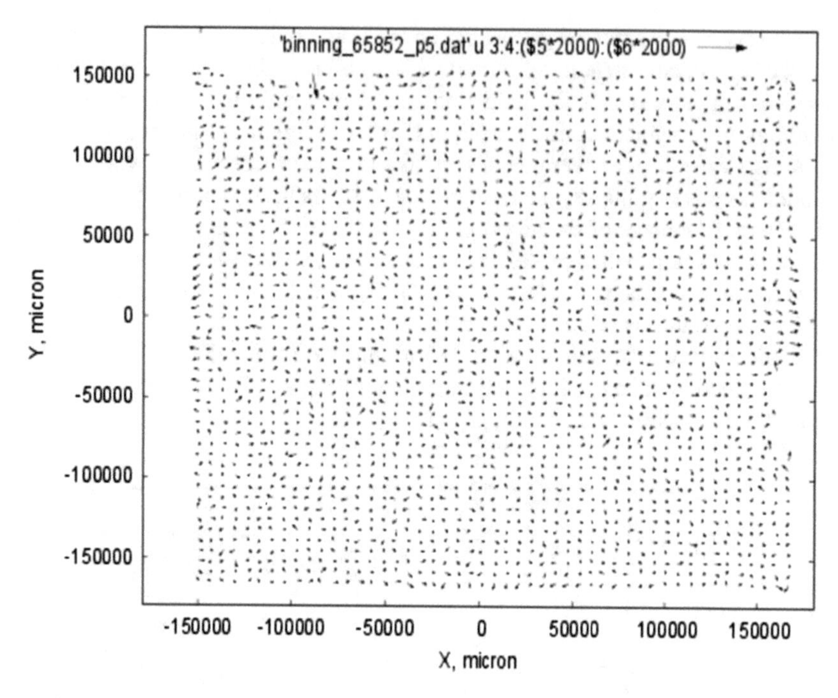

Fig. 2. Mean vector residual map using our method by the fifth-degree-order polynomial plate model.

This approach allows carrying out the astrometric reduction for wide-field images with the highest accuracy level and excludes all coordinate-dependences errors in the observational data.

4 Reduction of SuperCOSMOS Data

We created the software for performing the astrometric reduction of the big sets of data. The software is based on C++ realization. Using the developed software the reduction of all measured positions of objects on the digitized photographic plates of SuperCOSMOS survey [15] into the reference PMA catalogue [16] was performed.

The PMA catalogue contains about 420 million of the reference objects up to 21 optical G magnitude and absolute proper motions of these objects [17]. This catalogue contains objects of the whole celestial sphere that were combined from Gaia DR1 [18] and 2MASS [19] catalogues. The most of systematic errors in positions that are inherent in 2MASS catalogue were eliminated in PMA catalogue. The absolute calibration procedure was carried out using about 1.6 million positions of extragalactic sources and in the magnitude range from 14 to 21; PMA catalogue represents an independent realization of a quasi-inertial reference frame in the optical and near-infrared wavelength range [16].

The SuperCOSMOS is an advanced digitizing machine for the photographic plates. It systematically digitizes sky survey plates taken from UK Schmidt telescope (UKST), European Southern Observatory (ESO) Schmidt telescope, and Palomar Schmidt telescope.

The SuperCOSMOS project aims are to digitize the whole sky in three colors (B, R and I), and one color (R) at two epochs, via automatic scans of sky photographic plates. This dataset is the largest available catalogue of positions and multiband photometric information of the whole sky in optical wavelength range.

It should be noted that the photographic plates of sky of SuperCOSMOS represent a vast archive of measurements for billions of objects for the different information such as object position and brightness.

SuperCOSMOS survey was proposed to be used in the searching for candidates in halo cool white dwarfs, analyzing two-point angular correlation function for galaxies, investigation the objects exhibiting extreme variability, and counting the number-magnitude dependence for galaxies.

All received and digitized data becomes public after SuperCOSMOS processing. It is particularly valuable for astronomers in the world because of the following several reasons [15]:

• 10 micron (0.7 arcsec) pixel size, which is much smaller than the DSS-I (25 micron) and the DSS-II (15 micron);

• images and whole sky object catalogues (to the full plate depth) are available together in a single Flexible Image Transport System (FITS) file [20];

• color indexes, proper motions, and variability parameters are available for the first time in digitized sky surveys.

The images of a piece of sky have 0.7 arcsec pixels and are returned in FITS format with a built-in World Coordinate System (WCS), as the right handed cartesian coordinate system of the celestial sphere [21]. The FITS file automatically contains FITS tables listing the associated object catalogues that can be browsed simultaneously in any FITS viewer. Also the catalogue can be downloaded separately.

The object from the catalogue can be returned either as FITS tables, ASCII lists, or tab-separated lists. Various database query and analysis packages can understand all of these formats. The objects parameters that included into the FITS file are as follows: Right Ascension (RA), Declination (DE), magnitude, color, proper motion, but also some less obvious but often useful parameters such as ellipticity, image area, quality, blend flags, etc. [15].

There are from tens to several hundred thousands of the reference objects on each digitized photographic plate of the SuperCOSMOS survey [22]. Field of view (FOV) of each such plate is about 6 $\times$ 6 degrees, which is commensurable with the working FOV of the LSST telescope [13].

There are more than 5978 such digitized photographic plates of the SuperCOSMOS survey in different photometric filters (B, R, I). All observations from these plates were performed in the different observational epochs for both the northern and southern hemispheres and provided in the Table 1.

Table 1. SuperCOSMOS surveys that were used in this work for deriving the objects positions at an epoch of their observation.

Survey name	Epoch min	Epoch max	FOV, °	DEC min, °	DEC max, °	Fields	Color	Mag limit
SERC-J	1974.463	1994.856	6 × 6	−90	3	894	Blue	23.0
SERC-R	1984.546	2001.289	6 × 6	−90	3	894	Red	22.0
SERC-I	1978.890	2003.000	6 × 6	−90	3	894	Near-IR	19.0
POSS I-R	1949.112	1957.970	6 × 6	2	90	824	Red	20.0
POSS II-B	1985.000	2002.208	6 × 6	2	90	824	Blue	22.5
POSS II-R	1986.000	1999.000	6 × 6	2	90	824	Red	20.8
POSS II-I	1987.000	2002.000	6 × 6	2	90	824	Near-IR	19.5

As it turned out, the use of a large number of high-precision positions of the reference objects it possible to detect the coordinate-photometric dependence of some of the reduction model coefficients. This dependence is called a regular magnitude equation.

All reference stars have been binned in magnitude range with step of one magnitude for obtaining parameters of polynomial model reduction for each bin. The parameters of astrometric reduction model were determined with regard to weighting of the tangential coordinates that depends on accuracy of their spherical coordinates given in the reference catalogues.

Figures 3 and 4 show some values of the different coefficients (constant and fifth degree) of the reduction model as a function of stellar magnitude in range from 10 to 21 magnitudes.

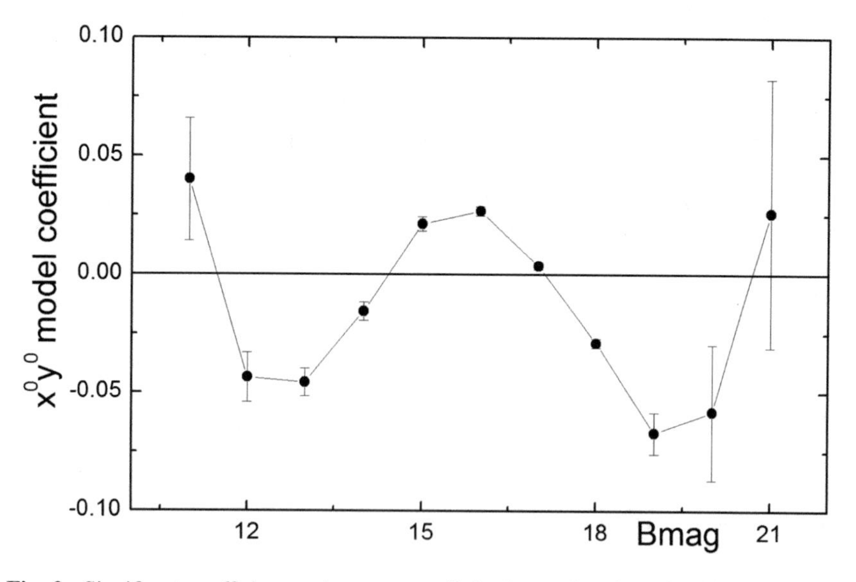

Fig. 3. Significant coefficient a_1 (constant coefficient) as a function of stellar magnitude.

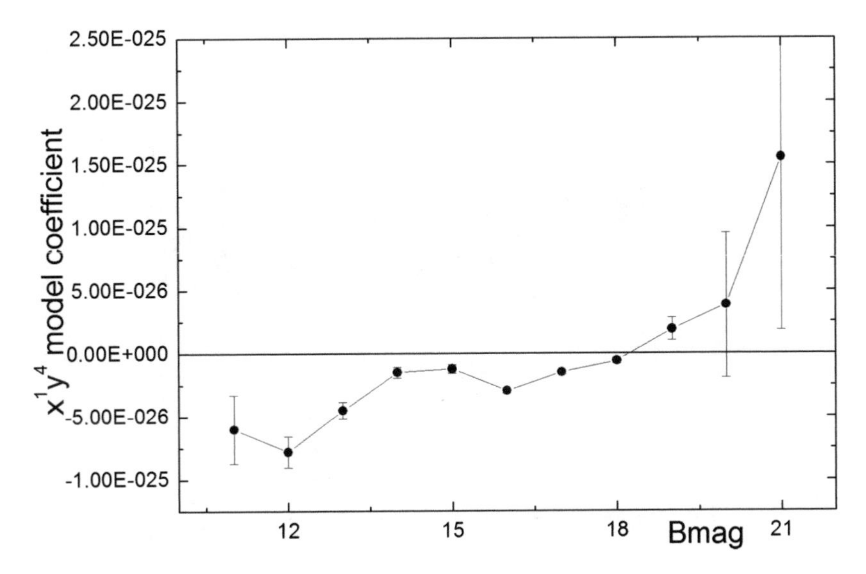

Fig. 4. Significant coefficient a_5 (fifth degree) as a function of stellar magnitude.

The error of the obtained coefficients of the reduction model depends on the stellar magnitude. Only in the brightest and weakest region where the number of reference objects is not enough, we have received large errors in determining the reduction coefficients.

Solving the system of conditional equations using the numerical computations with arbitrary precision for the least squares method provides the most probable unbiased values of unknown quantities. In our case we used the GNU Multiple Precision Arithmetic Library [23] GMP «Arithmetic without limitations» for C/C++ projects.

The differences in declination of reference objects depending on right ascension before and after the astrometric reduction procedure according to the algorithm described above are presented in Fig. 5. As shown in it (top) there are a lot of significant systematic positions distortions before applying the astrometric reduction procedure. But all coordinate-photometric errors were excluded after applying the astrometric reduction procedure (bottom).

Using proposed method for astrometric reduction, the high-accurate positions for all objects of SuperCOSMOS survey were calculated. Total count of these objects for both the northern and southern hemispheres is more than 1 billion. These data and proposed method are very helpful in the process of creating the high-density astrometric and photometric catalogues for the modern big data surveys that includes telescopes with wide field of view.

So, such high-density astrometric and photometric catalogues can be used as a basis and reference catalogues in the different software for automated image processing, astrometric and photometric reduction, such as CoLiTec [24, 25], Astrometrica [26, 27] and others. The regular magnitude equation of some of the reduction model coefficients in photographic plates of Schmidt type telescope using presented method were detected and excluded.

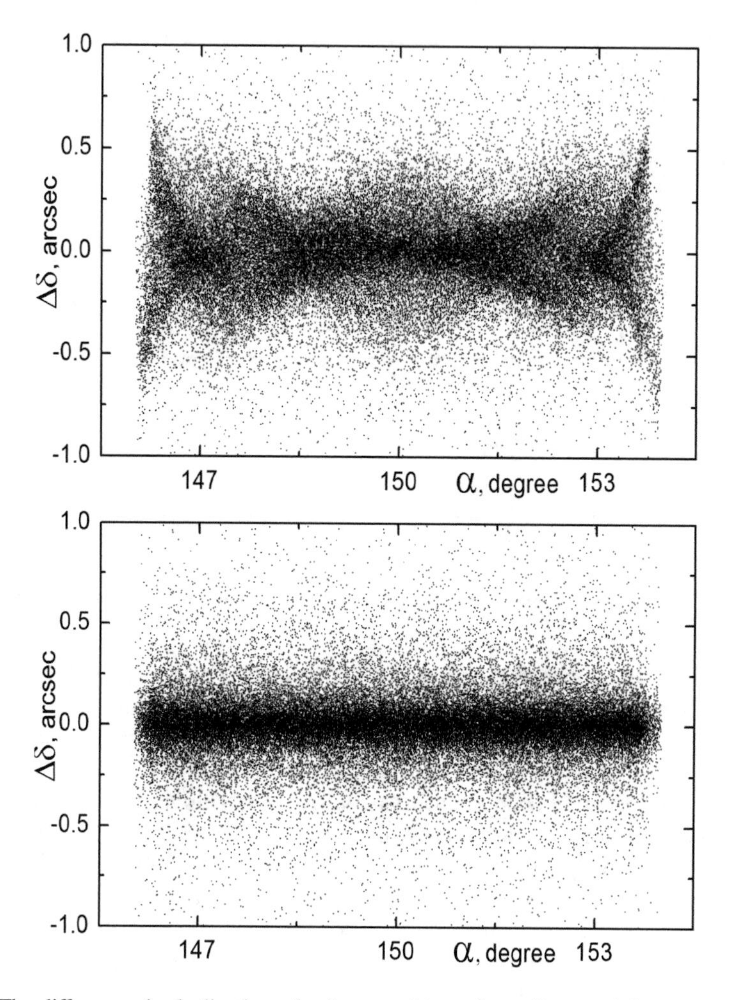

Fig. 5. The differences in declination of reference objects depending on right ascension before (top) and after (bottom) the astrometric reduction.

For the created high-density reference frame the new reduction data of the SuperCOSMOS was used in work [28] as well as an especial tool for the cross-match [29].

5 Conclusions

The increasing of accuracy and number of objects in the modern reference catalogues requires the development of new and modernization of existed methods and approaches for astrometric reduction of wide-field stellar images.

In this paper, we developed the new algorithm for astrometric reduction of the wide-field images using the iterative method of the ordinary least squares (OLS) and statistical Student t-criterion. We performed analysis for how to select the appropriate reduction model and system of conditional equations for determination the parameters of reduction model. The proposed algorithm allows the automatic selection of the most probabilistic reduction model.

Using C++ programming language we developed software that implements the proposed algorithm. It is helpful for the performing astrometric reduction in the different ranges of stellar magnitudes. The developed software was tested using all measured positions of objects on the digitized photographic plates of SuperCOSMOS data. Also, the regular magnitude equation of some of the reduction model coefficients in photographic plates of Schmidt type telescope using the modern reference catalogues in the presented method were detected and excluded.

The developed method allows excluding of all coordinate-photometric errors in the measured positions of objects even at the edges of images received from modern large telescopes with very wide field of view.

Using of the developed method has shown that when performing the reduction of large astronomical catalogues, the new proposed algorithm allows performing the reduction into the system of reference catalogue with the highest accuracy level. Also, the proposed algorithm suppresses all random noise that is presented in the input data and is caused by imperfections in the optical system of modern large telescopes.

Acknowledgment. This research has made using the data obtained from the SuperCOSMOS Science Archive, prepared and hosted by the Wide Field Astronomy Unit, Institute for Astronomy, University of Edinburgh, which is funded by the UK Science and Technology Facilities Council.

This work has made using the data from the European Space Agency (ESA) mission Gaia [30], processed by the Gaia Data Processing and Analysis Consortium (DPAC) [31]. Funding for the DPAC has been provided by national institutions, in particular the institutions participating in the Gaia Multilateral Agreement [32].

References

1. Khlamov, S., Savanevych, V., Briukhovetskyi, O., Pohorelov, A., Vlasenko, V., Dikov, E.: CoLiTec software for the astronomical data sets processing. In: Proceedings of the IEEE 2nd International Conference on Data Stream Mining and Processing, DSMP 2018, pp. 227–230 (2018)
2. Khlamov, S., Savanevych, V., Briukhovetskyi, O., Pohorelov, A.: CoLiTec software – detection of the near-zero apparent motion. In: Proceedings of the International Astronomical Union, vol. 12(S325), pp. 349–352. Cambridge University Press (2017)
3. Raab, H.: Astrometrica: Astrometric Data Reduction of CCD Images. Astrophysics Source Code Library, record: 1203.012 (2012)
4. Masson, M.E.J.: A tutorial on a practical Bayesian alternative to null-hypothesis significance testing. Behav. Res. Methods **43**, 679–690 (2011)
5. Morey, R.D., Wagenmakers, E.-J.: Simple relation between Bayesian order-restricted and point-null hypothesis tests. Stat. Probab. Lett. **92**, 121–124 (2014)
6. Gunawan, S., Panos, Y.P.: Reliability optimization with mixed continuous-discrete random variables and parameters. J. Mech. Des. **129**(2), 158–165 (2006)

7. Savanevych, V., Briukhovetskyi, O., Sokovikova, N., Bezkrovny, M., Vavilova, I., Ivashchenko, Yu., Elenin, L., Khlamov, S., Movsesian, Ia., Dashkova, A., Pogorelov, A.: A new method based on the subpixel Gaussian model for accurate estimation of asteroid coordinates. MNRAS **451**(3), 3287–3298 (2015)
8. Savanevych, V., Khlamov, S., Vavilova, I., Briukhovetskyi, A., Pohorelov, A., Mkrtichian, D., Kudak, V., Pakuliak, L., Dikov, E., Melnik, R., Vlasenko, V., Reichart, D.: A method of immediate detection of objects with a near-zero apparent motion in series of CCD-frames. A & A **609**(A54), 11 (2018)
9. Khlamov, S., Savanevych, V., Briukhovetskyi, O., Oryshych, S.: Development of computational method for detection of the object's near-zero apparent motion on the series of CCD–frames. Eastern-Eur. J. Enterp. Technol. **2**(9(80)), 41–48 (2016)
10. Miura, N., Kazuyuki, I., Naoshi, B.: Likelihood-based method for detecting faint moving objects. Astron. J. **130**, 1278–1285 (2005)
11. König, A.: Astrometry with Astrographs. University Chicago Press, Chicago (1964). Edited by Hiltner, W.A. Chapter 20
12. Skidmore, W., et al.: Thirty Meter Telescope Detailed Science Case: 2015. Res. Astron. Astrophys. **15**(12), 1945–2140 (2015)
13. Tuell, M., Martin, H., Burge, J., Gressler, W., Zhao, C.: Optical testing of the LSST combined primary/tertiary mirror. In: Modern Technologies in Space- and Ground-based Tele-scopes and Instrumentation, p. 77392V (2010)
14. Kiselev, A.: Theoretical fundamentals of photographic astrometry, 264 p. Moscow, Izdatel'stvo Nauka (1989). (in Russian)
15. Hambly, N., et al.: The Super COSMOS Sky Survey – I. Introd. Descr. MNRAS **326**(4), 1279–1294 (2001)
16. Akhmetov, V., Fedorov, P., Velichko, A.: The PMA catalogue as a realization of the extragalactic reference system in optical and near infrared wavelengths. In: Proceedings of the IAU, vol. 12(S330), pp. 81–82. Cambridge University Press (2018)
17. Akhmetov, V., Fedorov, P., Velichko, A., Shulga, V.: The PMA catalogue: 420 million positions and absolute proper motions. MNRAS **469**(1), 763–773 (2017)
18. Collaboration, G.: Summary of the astrometric, photometric, and survey properties. A & A **595**(A2), 23 (2016)
19. Cutri, R., Skrutskie, M., Van, D., Beichman, C., Carpenter, J., Chester, T., Cambresy, L., Evans, T., Fowler, J., Gizis, J., Howard, E., Huchra, J., Jarrett, T., Kopan, E., Kirkpatrick, J., Light, R., Marsh, K., McCallon, H., Schneider, S., Stiening, R., Sykes, M., Weinberg, M., Wheaton, W., Wheelock, S., Zacarias, N.: The 2MASS All-Sky Catalog of Point Sources. CDS/ADC Collection of Electronic Catalogues, p. 2246 (2003)
20. Wells, D., Greisen, E., Harten, R.: FITS: A Flexible Image Transport System. Astron. Astrophy. Suppl. Ser. **44**, 363–370 (1981)
21. Greisen, E., Calabretta, M.: Representations of world coordinates in FITS. A & A **395**(3), 1061–1075 (2002)
22. Hambly, N., Irwin, M., MacGillivray, H.: The SuperCOSMOS Sky Survey – II. Image Detect. Parametrization Classif. Photom. MNRAS **326**(4), 1295–1314 (2001)
23. The GNU Multiple Precision Arithmetic Library GMP «Arithmetic without limitations». https://gmplib.org. Accessed 11 July 2019
24. Kudzej, I., Savanevych, V., Briukhovetskyi, O., Khlamov, S., Pohorelov, A., Vlasenko, V., Dubovský, P., Parimucha, Š.: CoLiTecVS – a new tool for the automated reduction of photometric observations. Astron. Nachr. **340**(1–3), 68–70 (2019)
25. CoLiTec – Collection Light Technology. http://www.neoastrosoft.com. Accessed 11 July 2019

26. Savanevych, V., Briukhovetskyi, A., Ivashchenko, Yu., Vavilova, I., Bezkrovniy, M., Dikov, E., Vlasenko, V., Sokovikova, N., Movsesian, Ia., Dikhtyar, N., Elenin, L., Pohorelov, A., Khlamov, S.: Comparative analysis of the positional accuracy of CCD measurements of small bodies in the solar system software CoLiTec and Astrometrica. Kinematics Phys. Celestial Bodies **31**(6), 302–313 (2015)
27. Astrometrica. http://www.astrometrica.at. Accessed 11 July 2019
28. Fedorov, P., Akhmetov, V., Shulga, V.: The reference frame for the XPM2. MNRAS **440**(1), 624–630 (2014)
29. Akhmetov, V., Khlamov, S., Dmytrenko, A.: Fast coordinate cross-match tool for large astronomical catalogue. Adv. Intell. Syst. Comput. **871**, 3–16 (2019)
30. European Space Agency (ESA) Gaia Science Community. https://www.cosmos.esa.int/web/gaia, Accessed 21 Aug 2019
31. DPAC Consortium. https://www.cosmos.esa.int/web/gaia/dpac/consortium. Accessed 21 Aug 2019
32. Akhmetov, V., Khlamov, S., Khramstov, V., Dmytrenko, A.: New algorithm for astrometric reduction of the widefield images. In: Proceedings of International Scientific Conferece "Computer Sciences and Information Technologies" (CSIT-2019), pp. 106–109. IEEE v. 2 (2019)

Laser Based Technology of Monitoring the Dynamic Displacements of Objects Spatial Structures

Lubomyr Sikora[1], Natalya Lysa[2], Roman Martsyshyn[1], Yulia Miyushkovych[1](✉), and Bohdana Fedyna[2]

[1] Lviv Polytechnic National University, 12, Bandera street, Lviv 79013, Ukraine
lssikora, mrs.nulp@gmail.com, jmiyushk@gmail.com
[2] Ukrainian Academy of Printing, 19, Pid Holoskom street, Lviv 79013, Ukraine
lysa.nataly@gmail.com, fedynabogdana@gmail.com

Abstract. The article describes the method of creating a laser measuring system to control the level of dynamic displacements of spatial structures of objects. In the considered system for vibration control, the creation of a laser vibrometer has been described and the advantages of its use have been substantiated. Using the method of laser sensing allows monitoring of complex technological objects under the influence of various disturbing factors (man-made and natural). Remote sensing in the area of vibration of structures can be carried out in two ways: the method of processing reflected laser beam or the method of direct projection sounding with a laser beam. Described models of a laser vibrometer illustrate the principle of the system. The results of the experiments illustrated the performance of the proposed method. The proposed system of laser control helps to raise the level of reliability of man-made systems (taking into account the development of real dynamic situations).

Keywords: Construction · Vibration · Laser · Signal · Dynamic processes · Active factors · Data · System · Information · Project · Risks · Accident

1 Introduction

At present, quite a lot of methods are known for monitoring and evaluating the parameters of technological systems (the structure of the system, its dynamic parameters, determining deviations from normal operation modes, etc.).

Actuality. When working with spatial structures (bridges, large structures), the problem of quality control indicators is not completely solved. Special problems arise during the creation and operation of large objects with a complex structure (bridges, high-rise buildings, etc.). This is due to the fact that such objects are subject to heavy load, operating over a long period of operation. Destruction of such objects under the combined action of dynamic and static factors leads to destruction, accidents and human losses. If errors were made in the design of the object (for example, the energy aspect of the factors affecting the object was not taken into account), then this may contribute to the destruction of the object (sometimes with human victims). A striking example of such errors can be

N. Shakhovska and M. O. Medykovskyy (Eds.): CCSIT 2019, AISC 1080, pp. 910–919, 2020.
https://doi.org/10.1007/978-3-030-33695-0_59

the destruction of large bridges (for example, the destruction of the bridge in the city of Genoa, Italy in 2015). Accounting methods S. Tymoshenko [1] in the design of complex structures can significantly increase the period of safe operation (with appropriate maintenance). The problem of assessing the stability of complex structures (spatial) to vibration is still not fully resolved. That is why the problem of creating new methods for controlling the vibration of objects is urgent.

2 Problem Task Review

The problem of monitoring and evaluating the stability of spatial structures is relevant for a long time. Taking into account the possibility of destruction of spatial structures under the influence of various factors (vibration, soil and earthquake shrinkage, aging of components and materials) [2] is an integral part of the problem of controlling and evaluating the stability of spatial structures.

The main research in the problem area is carried out in the following areas of statistics and system theory, which are based on [3–6]:

- probable models of random processes and fields for describing object vibration and influence factors;
- procedures for detecting, recognizing, evaluating parameters and filtering signals based on selected dynamics models (which reflect the state of the technical system or spatial object in the current time);
- spatial-temporal signal processing algorithms taking into account the stochastic structure of distribution channels (and perturbation models for the estimation of trajectories and trends in real-time dynamic parameters change);
- procedures of multicriteria optimization of the decision-making process for control in the conditions of data fuzziness (which changes the load mode);
- procedures dynamic assessment of the situation in the energy active objects;
- algorithms for image recognition (spatial-temporal, situational) formed from data streams in different modes of operation of the object;
- procedures of analysis and synthesis of information-measuring systems for assessing the state of spatial constructions of technological objects with varying degrees of controllability;
- selection of indicators of signs for the detection of the boundary and emergency modes of the current dynamic situation (in the designs relative to the target area of admissible parameters).

The classical approach to the structural synthesis of the information control system (ICS) is that the development of the structure is carried out on the basis of a technical task in the framework of existing methods of analysis and synthesis (based on the given model of the measuring system without taking into account the target orientation) [7, 8].

At the same time (not always in full), information on the structure of the research object, the conditions for its functioning with restrictions on resources, observation and reliability are taken into account. In the first place, when implementing the procedure for the synthesis of ICS, it is necessary to keep in mind the whole functioning of the technological object [8, 9]. This allows to build a content model and form quantitative optimization criteria in the form of a system of quality functionals.

With the stochastic nature of the object's operation, the control system often has to deal with a situation of insufficient a priori information. Especially this problem arises when observing the state of technological spatial structures with an unidentified structure and functions, which are unstable in time and blurred by priorities of local goals (they do not have strategic directions) [10]. The problem is exacerbated when decision-making procedures do not have systemic and effective technological support. In these cases, the principle of dual process control is used to make decisions. It provides for the simultaneous use of signals as a means of studying a technological object (behavior trajectories under the influence of perturbing factors).

But there are conditions under which optimal observation and control becomes impossible. This situation occurs when resource constraints or dynamic disturbances significantly exceed the level of the informative useful signal. This leads to disorientation ICS and wrong decisions, and boundary modes – to occurrence of an emergency. In these circumstances, it lost the robustness and performance of ICS (constructed on the basis of the classical filtration theory and the theory of automatic systems with feedback using Mesarovic hierarchical structures).

Laser monitoring systems are used in many areas: for the monitoring of hydroacoustic vibrations [2], for monitoring the parameters of structures [3, 4], for the creation of military images [5], etc.

One of the least investigated areas is the factors of cognitive and knowing flaws and errors that arise during the design and subsequent operation with the participation of operational personnel and designers [6–10]. The development of methods and tools (as components of knowledge) for controlling projects and vibration control systems for complex spatial structures is a continuing problem [11, 12]. This requires an integrated approach using the theory of signals, the theory of data processing, data interpretation and situations, and decision-making.

But despite the available solutions, the development of new laser based methods for remote control of dynamic regimes of large spatial structures (under the influence of active dynamic in time and in the space of factors) is an important problem.

3 Models of Dynamic Factors that Influence the Spatial Structure

Since dynamic factors [1–3] have an energy-active structure [6], the failure to take into account their essence leads to the collapse of the mechanical spatial structure due to oscillations.

Research of their dynamics requires the use and creation of new methods and control systems based on laser remote sensing. The use of such a control system makes it possible to detect the variation of the spatial structure of structures. To solve the above problematic tasks we need to create:

1. Model of n-dimensional spatial variation for a long section of a bridge (100 m).
2. The model of the energy soliton for the opposite flows of transport (as disturbing factors $\left(\xrightarrow{\vec{n}^2} \uparrow \vec{n}_n\right)$ on the vertical and along along the supports).

3. Model of wind load with variable speed (as a factor of transverse oscillation perturbation).
4. Transport streams as excitatory factors with a continuous and discrete structure (homogeneous, heterogeneous, group (one-sided, counter) (1)–(2).

$$TF_1^d = \{(m_i, V_i)\}_{Tn} \quad (1)$$

$$TF_2^d = \left\{ \sum_{i=1}^{K} (m_j, V_j)_{ti} (t_i \in T_n) \right\} \quad (2)$$

In (1)–(2): TF_i – traffic flow, m_i – mass of transport unit of movement, V_i – speed of movement, T_n – group time.

5. The method of laser sensing of the spatial structure (of the research object). The method is the basis for the development of information-measuring systems (Fig. 1).

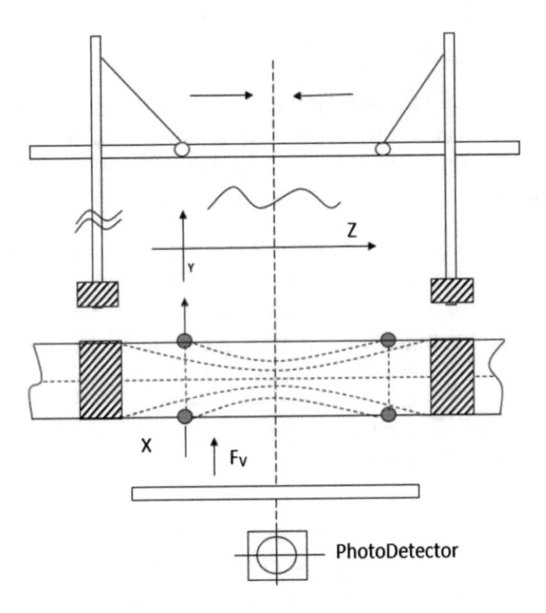

Fig. 1. Structural diagram of the method of laser sensing of the spatial structure

3.1 Active Factors Influencing the Dynamic and Structural Stability of Spatially-Distributed Objects

According to the analysis, it is possible to distinguish systemic active factors of action:

1. Conflicts and incompleteness of knowledge that contribute to the occurrence of errors at the design stage of the spatial structure:
    - incompleteness of the data on the object, structures, materials, dynamics, factors, load, destructive factors, load, destructive forces;
    - gaps in the system of knowledge of designers leads to system and structural errors.
2. Conflicts that arise during the operation (in case of incomplete data and knowledge of the personnel):
    - the structure of dynamic loads and their changes in long and short intervals of time;
    - seasonal, natural factors, cataclysms, which lead to damage to structures.
3. Transport flows, as agents of oscillation of spatial structures of bridges:
    - changes in the level of reliability and aging metal and concrete supports;
    - dynamic destruction of materials, inadequate design of the structure and the object to the requirements and trends of traffic changes and its mass parameters and reliability.

The lack of effective control systems for the structure and reliability of the object's structures, the dynamics of destruction due to deformation displacements under the influence of active factors (automobile and traffic flows and natural dynamic factors) may contribute to the occurrence of accidents and crisis situations.

4 Laser Sounding of a Spatial Structure Oscillations (a Bridge with a Long Span)

To detect and identify spatial oscillations of transport infrastructure objects (and large building structures), a method of projection laser sensing for changing the trajectory of elements at certain critical points of structures was developed. According to [1], the spatial displacements of the coordinates of the support structures can be represented as trajectories (Fig. 2).

To receive a spatial stream of laser signals, the photodetector matrix (PD on Fig. 2) should have a 4-square structure for estimating the dynamics of point displacements along vectors according to the difference equation [11]:

$$trak\ \Delta U^{t}_{\mathrm{var}}(\vec{n}_X \Delta x) = K_M\left(U^{+}_{xt} - U^{-}_{xt}\right) = K_M K_{YS}\left(P^{+}_{Si} - P^{-}_{Si}\right)^{t} \quad (3)$$

In (3): P^{+}_{Si}, P^{-}_{Si} – the power of the received laser beam $\Delta U_{\text{var}}(\vec{n}_X \Delta x)$; K_M, K_{YS} – matrix transformation and beam scattering coefficients, U^{+}_{xt}, U^{-}_{xt} – voltage variation at the output of the channel measuring the control point oscillations along a vector $(\vec{n}_{Xi})$.

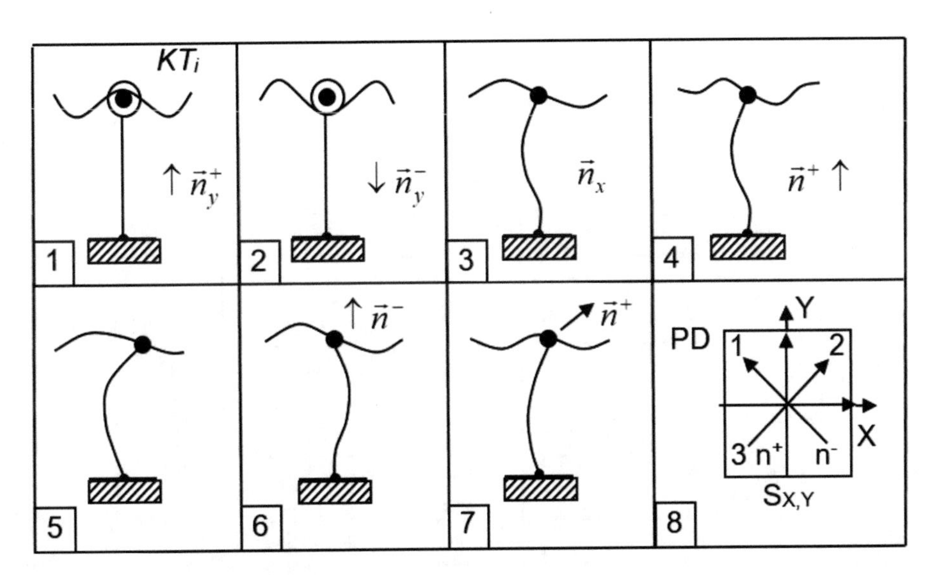

Fig. 2. The spatial orientation of the active action vectors on the $S_{X,Y}$ plane on the bridge supports with a span length from 20 m to hundredths more than meters.

Figure 3 represents a diagram of possible oscillations of the bridge platform and the selection of data by laser sensing of displacement control areas.

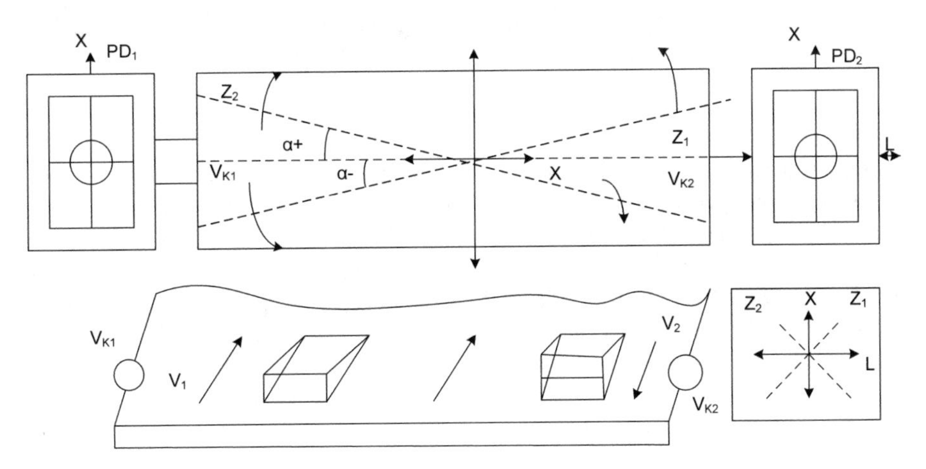

Fig. 3. Oscillations of the blade at the oncoming traffic with speed V(t) and ground displacement of the supports (scheme)

According to Fig. 3 and models of informational transformations, a structural-informational scheme of the method of projection laser sensing of oscillations of a control point of a structure is formed (Fig. 4).

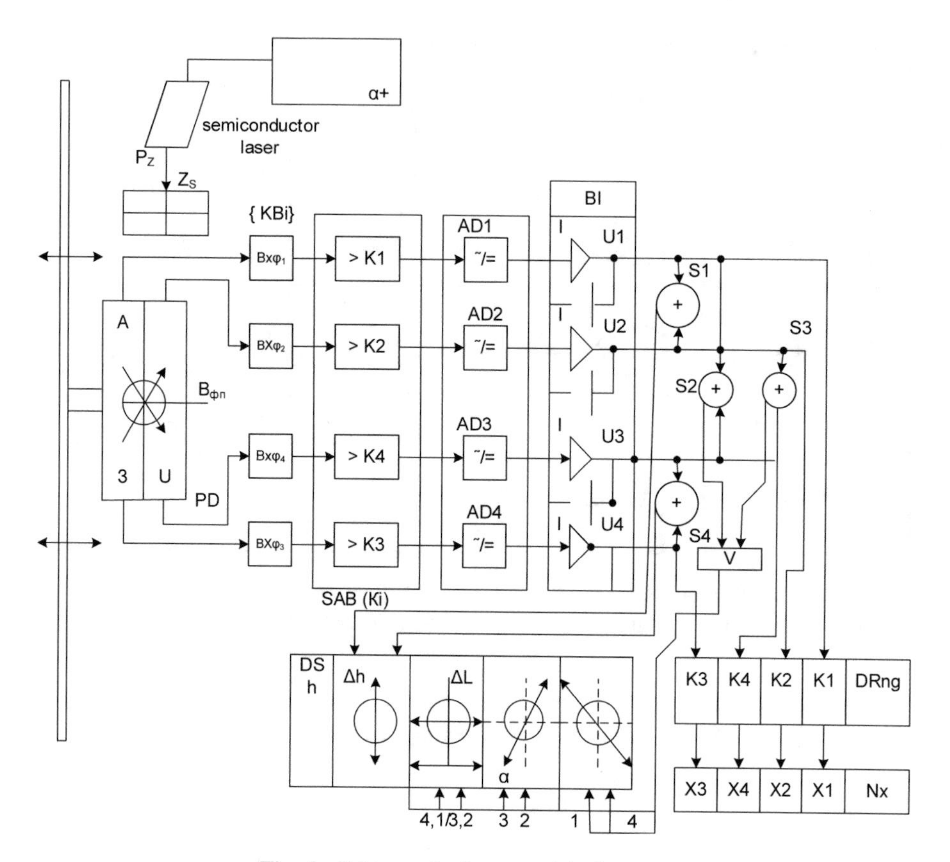

Fig. 4. Scheme of a laser spatial vibrometer

Designation in Fig. 4: PD – 4-square photodetector matrix; P_z – laser sensing power; P_s – received signal; $\{kB_i\}$ – control channels of oscillation dynamics in the basis $\{x, y, n\}$ with input filters of signals $\{B_X \Phi_i(f_M, \Delta f)\}$ with frequency f_M and transmitter Δf; $\{\mathrm{SAB}(K_i)\}$ – signal amplifier block with a coefficient K_i; ADi – analog signal detectors; $BI(U_S)$ – block of signal integrators; $\{S_i\}$ – operational signal adders; $DRng$ – discrete rank load classifier; N_X – digital display; DS_n display integrated indicator of dynamic offsets along the axes (X, Y, Z^+, Z^-).

5 Realization of Laser Measuring System of Vibrations

The model of the construction of the laser remote sensing system of the vibration area described above has been implemented on a laboratory stand (remote sensing with reflected beam).

The designed system was tested on experimental models for experimental confirmation of theoretical substantiation. In Fig. 5 shows the results of laser sensing the level of vibration of the object being studied. The data was rendered on tape by recorders which received data from the laser system.

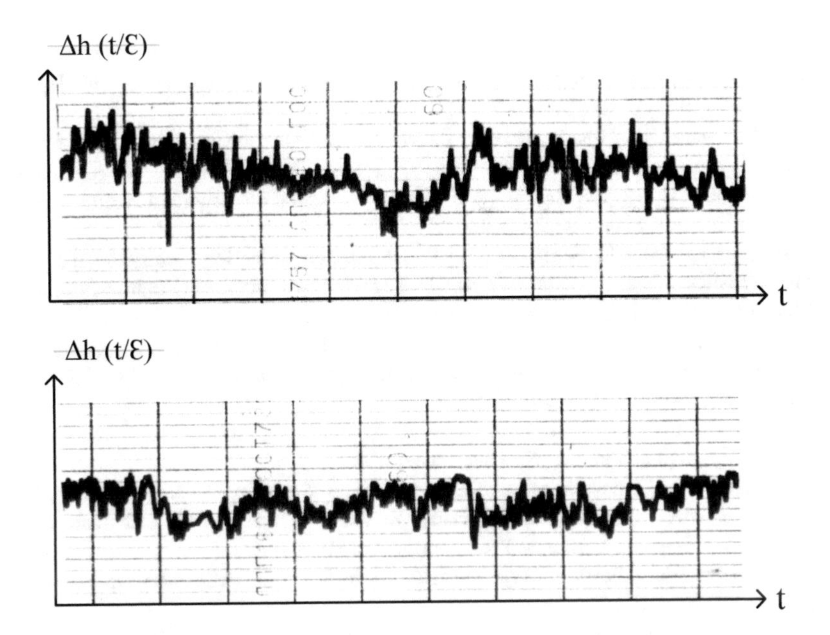

Fig. 5. The result of measuring the level of vibration of an object by a laser monitoring system (recorder tape fragment)

The obtained results (graphs of the trajectory of the displacement of the surface in the sensing zone) clearly reflect changes in the mode of operation of the object.

6 Conclusion

To ensure a high level of reliability of man-made systems, it is necessary (in the design process) to take into account active, informational and cognitive factors of influence on the development of the project and its implementation. The integration of methods (system analysis of information technology of laser measuring systems and surface soliton physics) made it possible to develop a new method for designing tools for the selection of data on the dynamic vibration of complex spatial structures.

Particularly important is the consideration of the development of real dynamic situations, for example, vibrations. The proposed system of laser distance monitoring of dynamic displacements of the sensing area of a spatial structure allows to track such factors of influence. The use of laser remote sensing has become the basis of the newly developed method. Also algorithms for intelligent processing and interpretation of data were built. The use of the developed method has made it possible to increase the effectiveness of the process of controlling the dynamic vibration of complex spatial structures.

References

1. Weaver Jr., W., Timoshenko, S.P., Young, D.H.: Vibration Problems In Engineering. Wiley, New York (1990). ISBN 978-0-471-63228-3
2. Moll, J., Bechtel, K., Hils, B., Krozer. V.: Mechanical vibration sensing for structural health monitoring using a millimeter-wave doppler radar sensor. In: Cam, L., Vincentand Mevel, L., Schoefs, F. (eds.). EWSHM – 7th European Workshop on Structural Health Monitoring, Jul 2014, Nantes, France <hal-01022029>
3. Luzi, G., Crosetto, M., Fernández, E.: Radar Interferometry for monitoring the vibration characteristics of buildings and civil structures: recent case studies in Spain. Sensors (Basel). **17**(4), 669 (2017). https://doi.org/10.3390/s17040669
4. Shimada, Y., Kotyaev, O.: Remote sensing of concrete structure using laser sonic waves. In: Fukuchi, T., Shiina, T. (eds.) Industrial Application of Laser Remote Sensing, Bentham Science, Sharjah, pp. 153–169 (2012)
5. Lutzmann, P., Frank, R., Hebel, M., Ebert, R.: Potential of Remote Laser Vibration Sensing for Military Applications, p. 37, 1 December 2005
6. Schiehlen, W.O.: Dynamics of High-Speed Vehicles. Springer, Wien-New York (1982)
7. Mesarovic, M.D., Macko, D., Takahara, Y. (eds.): Theory of Hierarchical Multilevel Systems, p. 294+xiii. Academic Press, New York (1970)
8. Ishimaru, A.: Wave Propagation and Scattering in Random Media. Academic, New York (1978)
9. Laser Beam Propagation in the Atmosphere. Springer Verlag, York (1978)
10. Sikora, L., Martsyshyn, R., Miyushkovych, Y., Lysa, N., Yakymchuk, B.: Systems approaches of providing the guaranteed functioning of technological structures on the basis of expert coordination of local strategies. In: 2015 Xth International Scientific and Technical Conference on Computer Sciences and Information Technologies (CSIT), Lviv, pp. 166–168 (2015). https://doi.org/10.1109/stc-csit.2015.7325458
11. Sikora, L., Lysa, N., Martsyshyn, R., Miyushkovych, Y.: Models of combining measuring and information systems for evaluation condition parameters of energy-active systems. In: 2016 IEEE First International Conference on Data Stream Mining & Processing (DSMP), Lviv, pp. 290–294 (2016). https://doi.org/10.1109/dsmp.2016.7583561
12. Sikora, L., Martsyshyn, R., Miyushkovych, Y., Lysa, N.: Methods of information and system technologies for diagnosis of vibrating processes. In: 2017 12th International Scientific and Technical Conference on Computer Sciences and Information Technologies (CSIT), Lviv, pp. 192–195 (2017). https://doi.org/10.1109/stc-csit.2017.8098766

13. Drahan, Y.A.P., Sikora, L.S., Yavors'kyy, B.I.: Systemnyy analiz stanu ta obgruntuvannya osnov suchasnoyi teoriyi stokhastychnykh syhnaliv: enerhetychna kontseptsiya; matematychnyy substrat; fizychne tlumachennya. L'viv: NVF «Ukrayins'ki tekhnolohiyi», p. 240 (2014)
14. Guedes Soares, C., Modarres, M., Kaminskiy, M., Krivtsov, V.: Reliability engineering and risk analysis: a practical guide. In: Modarres, M., Kaminskiy, M., Kritsov, V. (eds.) 1999 Reliability Engineering and System Safety, vol 77, pp. 207–208. Marcel Dekker Inc., New York (2002). https://doi.org/10.1016/s0951-8320(02)00008-x
15. Lonngren, K., Scott, A.: Solitons in Action, p. 300. Academic, New York (1978)
16. Oleg, R., Yurii, K., Oleksandr, P., Bohdan, B.: Information technologies of optimization of structures of the systems are on the basis of combinatorics methods. In: 2017 12th International Scientific and Technical Conference on Computer Sciences and Information Technologies (CSIT), Lviv, pp. 232–235 (2017). https://doi.org/10.1109/stc-csit.2017.8098776
17. Sikora, L., Lysa, N., Fedyna, B., Durnyak, B., Martsyshyn, R., Miyushkovych, Y.: Technologies of development laser based system for measuring the concentration of contaminants for ecological monitoring. In: 2018 IEEE 13th International Scientific and Technical Conference on Computer Sciences and Information Technologies (CSIT), Lviv, pp. 93–96 (2018). https://doi.org/10.1109/stc-csit.2018.8526602
18. Sikora, L., Lysa, N., Martsyshyn, R., Miyushkovych, Y., Tkachuk, R., Durnyak, B.: Information technology of laser measurement system creation for automated control dynamics of glue drying in polygraphy. In: 2018 IEEE 13th International Scientific and Technical Conference on Computer Sciences and Information Technologies (CSIT), Lviv, pp. 89–92 (2018). https://doi.org/10.1109/stc-csit.2018.8526683
19. Sikora, L., Lysa, N., Martsyshyn, R., Miyushkovych, Y., Dragan, Y., Fedyna, B.: Technology of monitoring the dynamic displacements of objects spatial structures using a laser surface sensing system. In: Proceedings of International Scientific Conference on Computer Sciences and Information Technologies (CSIT-2019), vol. 2, pp. 9–12. IEEE (2019)

Research of Servers and Protocols as Means of Accumulation, Processing and Operational Transmission of Measured Information

Yurii Kryvenchuk[1(✉)], Olena Vovk[1], Anna Chushak-Holoborodko[1], Viktor Khavalko[1], and Roman Danel[2]

[1] Lviv Polytechnic National University, Lviv 79013, Ukraine
yurkokryvenchuk@gmail.com, olenavovk@gmail.com, annachushak7@gmail.com
[2] Institute of Technology and Businesses in České Budějovice, Ceske Budejovice, Czech Republic

Abstract. The article describes approaches to the system of data accumulation and storage. The analysis of the possibility of accumulation and processing of data on the local server, as well as in the cloud. The study of the dependence of file transfer time on buffer size for cloud computing and processing technology has been carried out.

Keywords: Server · Industry 4.0 · Data transfer · Time dependence

1 Introduction

Industry 4.0 provides for the accumulation and processing of data received in real time. The number of such data for today is quite large, so it's not enough to use primitive workstations as a computer system, or the storage and accumulation of data. Therefore, to solve this problem it is expedient to use a server. The server is understood by a computer that provides its resources to other computers that are called clients. In essence, the server processes and stores the basic information on the computer network. Due to the variety of information used and types of processing, there are different types of servers, the most common of which is a file server. The file server understands the computer that is connected to the network and is used to store the data files to which the workstations access. From a user's perspective, the file server is considered to be central archives, which stores information common to all workstations. In more complex networks, other file servers can be present besides file servers, for example: print server, database server, Web server, mail server and others. Server equipment. According to the hardware, servers do not differ much from the workstations, but the equipment itself has more requirements. This is due to the fact that the file server must process a large number of requests from all workstations quickly enough. Servers are equipped with high-performance processors to provide the necessary performance. It is possible to use systems with several processors simultaneously. In order to increase the productivity of the servers, cache memory is widely used. This super-fast memory is designed to temporarily store the commands and data that is most commonly called. To prevent the loss of information when working with hard disks in the servers used

N. Shakhovska and M. O. Medykovskyy (Eds.): CCSIT 2019, AISC 1080, pp. 920–934, 2020.
https://doi.org/10.1007/978-3-030-33695-0_60

system RAID - redundant arrays of low-cost disks. The RAID system includes a set of hard drives, while different modes of simultaneous recording of the same information on several hard drives are implemented. This allows you to restore data from the backup copy on the other disk in case of a hard disk failure.

In order to ensure the normal operation of the computer network and to prevent loss of information when the power supply is suddenly turned off, the server must be powered from a UPS (Uninterruptible Power Supply) source. The uninterruptible power source uses a rechargeable battery to maintain the computer's (server's) performance over a period of time sufficient to store data, close programs, and shut down normally. There are "reasonable" sources of uninterrupted power supply, which themselves correctly turn off the work of the server, turn it off, and when the power supply volumes automatically restart the system. Based on this principle, local servers are organized, which usually accumulate information from different sensors within the same enterprise and synchronize it with the cloud.

2 State of Arts

Recently, cloud-based technologies have been widely used. The cloud storage and storage system is an abstract concept that is consistent with the system of data accumulation and storage, which allows you to administer yourself through a variety of interfaces. Such an interface abstracts the geographical location of the system, so it becomes irrelevant whether it is remote, local, or hybrid. Cloud storage infrastructures generate new architectures that allow you to work at different levels of service, encompassing a potential large group of users and geographically dispersed storage [1]. An important niche in such services is the ability of the client to manage and control the process of data storage. Many cloud service providers provide the ability to manage services so that the client can manage their costs and control them properly.

The effectiveness of data storage is an essential assessment of the cloud storage infrastructure, especially considering its emphasis on saving. In order to make the system saving effective, you need to save more data. If we talk about reducing physical space, then you need to reduce the amount of output data. To achieve this goal, two methods are used, the first of which is compression, which is the encoding of data using a variety of representations, as well as deduplication, which sums up the elimination of all data twins. Although both the first and second methods may be useful, compression means that the data will be processed by processing (transcribing information into and out of the infrastructure), and deduplication is the signature calculation to find duplicates.

The ability to provide savings can be considered one of the most characteristic features of cloud storage of data. This savings achieved in the acquisition of drives, their power supply, debugging, and, moreover, in managing savings. If we analyze cloud storage from this angle (including SLA and high conservation performance), it can be profitable if we talk about certain application models (Fig. 1). The access repository API is considered one of the most important component of the service. Numerous applications require access to a service storage using an API that is optimized for a particular data storage system, either on its hardware or cloud-based. Thus, the Amazon S3 API cloud storage system provides SDK developers for .Net and Java, as well as libraries for languages and additional platforms. These interfaces typically use REST and Simple Object Access Protocol (SOAP) [2] web services.

Considering working parameters of architecture is necessary when considering it. These parameters are different architectural features such as cost, performance, remote access, etc.

The architecture of cloud storage is, above all, the supply of demand-side storage resources in a multi-intangible and highly scalable environment. The external interface provided by the API for access to drives is, in general, an architecture for cloud storage (Fig. 2). New protocols appear in the cloud, despite the fact that traditionally data storage systems have a SCSI protocol. Among the new protocols, external protocols of Web services, as well as file protocols and even more traditional external interfaces (Internet SCSI, iSCSI, etc.) can be distinguished.

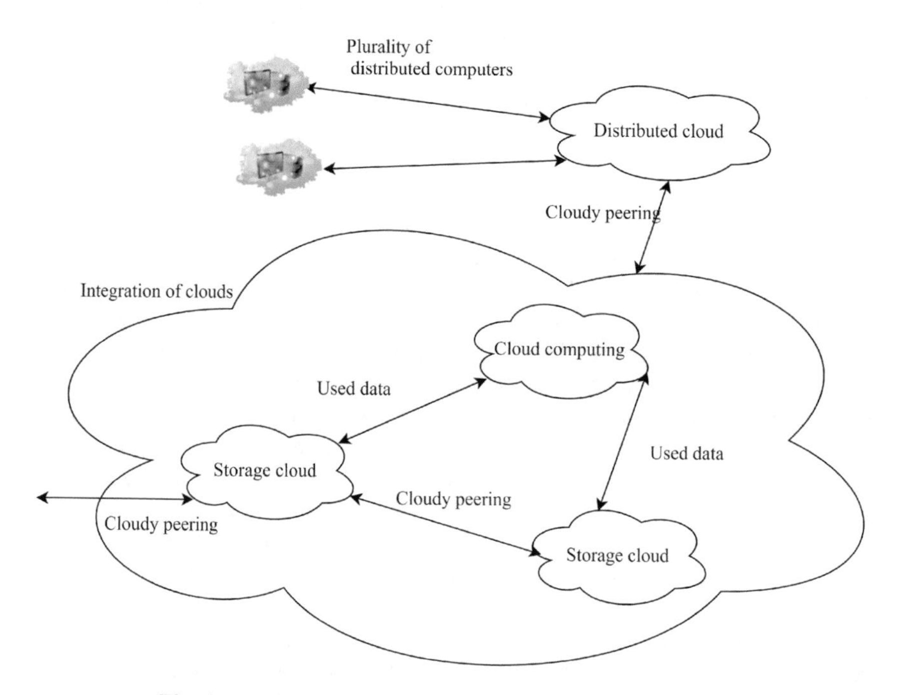

Fig. 1. Models of cloud ecology, built according to data [3].

The logic of data storage is called the level of intermediate software located on the external interface. The logic of data storage implements a number of functions, namely: replication of data and reduction of their volume, according to traditional methods of data placement, taking into account geographic location. Physical storage also organizes the internal interface. This requires a traditional server with physical disks or an internal protocol that implements specific functions. The single structure shown (Fig. 2) gives an opportunity to note certain properties of today's cloud-storage data architecture. These properties do not belong only to a specific degree, but may belong to many (Table 1).

Access to cloud and traditional storage systems is one of the most striking differences (Fig. 3). A large part of suppliers provides a variety of admission methods, but

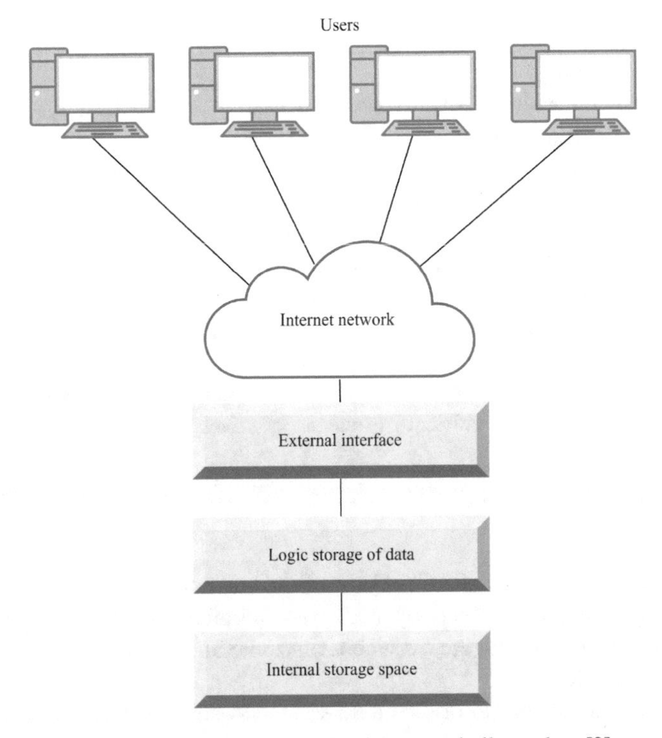

Fig. 2. The architecture of cloud storage, built on data [3].

Web Services APIs are generally recognized. Most of them are executed on the principles of REST, which is based on the object-oriented scheme, created over the HTTP protocol (using TCP as a transport).

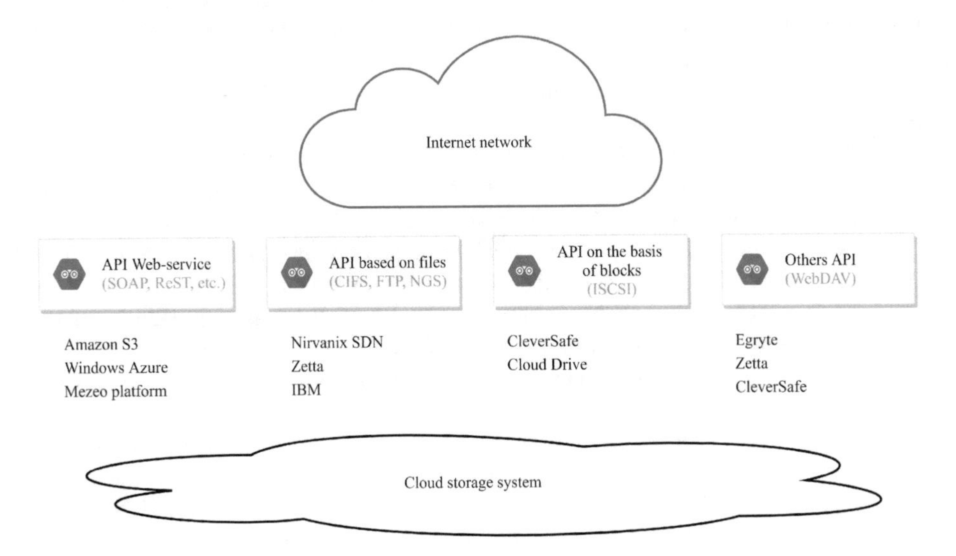

Fig. 3. Methods of access to cloud storage systems, built according to [3].

Table 1. Characteristics of the architecture of cloud storage data, built according to [3].

Characteristics	Description
Controllability	With minimal resources there is the ability to manage the system
Method of access	Protocol for the provision of cloud storage services
Productivity	Calculated by bandwidth and time delay
Multiplayer	Ability to use by many users (tenants)
Scalability	Possibility to use by many users (tenants). Perspective of successive build-up to provide new requirements or handling of increased load
Readiness of data	It is calculated by the system's time-consuming system
Management	Choosing value, performance, or other characteristics - system management
Storage efficiency	The degree of efficiency of the use of drives
Price	Measure the price of data storage (mainly in dollars per gigabyte)

The basics of building a REST-API with no memory status are elementary and effective. They are implemented by numerous suppliers of cloud storage services, namely Amazon Simple Storage Service (Amazon S3), Windows Azure, and Mezeo Cloud Storage Platform.

In order to take advantage of the cloud storage system, Web services APIs require integration with the application, which poses a serious problem. For this reason, the only methods of admission are using cloud storage systems to provide direct integration. For example, file-based protocols such as NFS/Common Internet File System (CIFS) or FTP, or block-based protocols such as iSCSI. Similar admission methods allow Nirva-nix, Zetta, Cleversafe and other cloud storage service providers.

The above protocols are more common, but others are also suitable for cloud storage. Web-based Distributed Authoring and Versioning (WebDAV) is one of the most exciting. WebDAV is furthermore based on HTTP and makes it possible to use the Web as a resource for reading and writing. Among the well-known suppliers that use WebDAV are Zetta, Cleversafe, and others.

You can find solutions that support a number of access protocols. For example, IBM Smart Business Storage Cloud allows you to use protocols in the same file-based storage virtualization infrastructure (NFS and CIFS) and SAN-based protocols.

However, when using different protocols for data exchange among users and cloud storage facilities, there is a large number of obstacles for the rational use of high-quality services. In general, these problems are combined with the inappropriateness of the existing structure of the Internet to provide such services.

You can analyze the difficulties of providing high-quality cloud storage services from different perspectives.

There are many nuances to represent performance, but the main purpose of a cloud storage system is to move data from the user and to a remote cloud service provider. The issue is hidden in the transport protocol TCP, which is considered the main protocol of the Internet. TCP manages the flow of information based on the receipt of packet acknowledgment from the remote host. Loss or deceleration of packets allow you to regulate overload, which further limits the efficiency of evasion from the massive network constraints. TCP fits perfectly for moving small amounts of data through the World Wide Web, but not for the purpose of delivering large-size data - in this case, the time of data exchange (RTT) is increasing.

Amazon with software support for Aspera Software has solved this issue by removing the TCP degree. Fast and Secure Protocol (FASP) - The new protocol was designed to force the transfer of various information in the circumstances of a significant period of response and loss of packets. The basis is UDP, which is considered an additional TCP transport protocol. UDP allows you to control traffic by referring it to the FASP application layer protocol (Fig. 4).

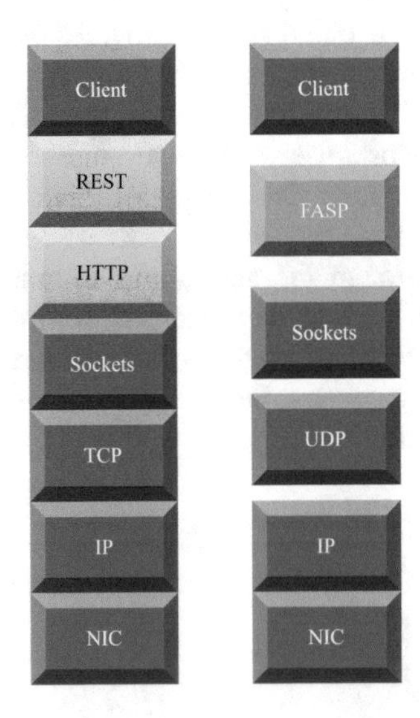

Fig. 4. The architecture of the Fast and Secure Protocol, built according to [3].

FASP effectively uses the application-accessible bandwidth and eliminates the major limited areas of traditional large-scale data circuits thanks to working with conventional network adapters (without acceleration).

3 Multi-use in Cloud Technologies

Multi-user can be called one of the main distinguishing features of the cloud storage architecture (it is used by numerous users). This distinctive feature belongs to different levels of the cloud storage system - from the application level, where users allocate an isolated namespace to a storage level when individual users or categories of users can get independent physical drives. The multi-user environment extends to the network infrastructure that brings together users with drives and ensures good service quality as well as the designated bandwidth for a particular user.

Scalability can be analyzed on some points, but data storage is considered as the allocation of cloud storage resources according to the request.

When dynamically changing storage resources (regardless of whether they increase or decrease) means improved financial performance for the user and high complexity for those who supply cloud services.

Scalability should be ensured not only for the storage system itself (functional scaling), but also for its bandwidth (scaling load). Scalability must be guaranteed not only for the system of conservation (multifunctional scale selection), but also for its bandwidth (scale load selection). Geographic scalability (geographic distribution) is also the main feature of cloud storage, which makes it possible to place data as close as possible to the user, thanks to the group of cloud storage centers (by moving). Replication and distribution are likely if data is "read-only" (Fig. 5).

When storing user data, the cloud service provider must take into account the fact that the user may ask to return the data, and therefore the supplier should have such an opportunity. The complexity of fulfilling this condition in a safe and deterministic way arises through a simple network, user errors and other circumstances.

Dissemination of information is a modern high-availability high-availability scheme that has emerged recently. In the case of loss of data fragments, their restoration is carried out through the algorithm (Information Dispersal Algorithm - IDA), which acts as a "cutting" of data using Reed-Solomon codes and raises the availability of data in the event of physical failures and network failures. Like the RAID, IDA allows you to recover information from a subset of the initial data for certain costs of the error code (Fig. 6).

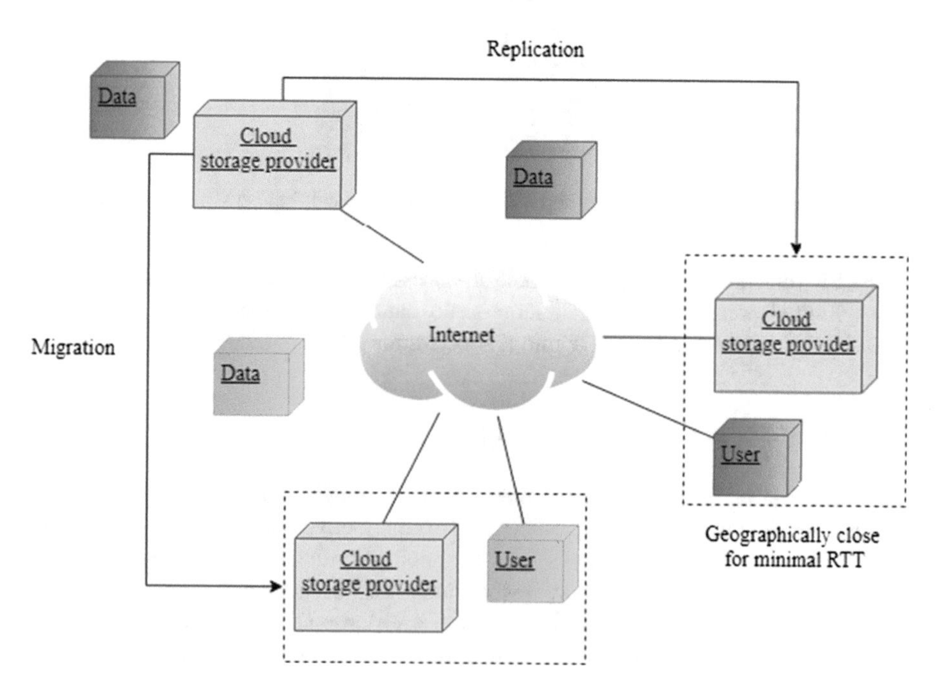

Fig. 5. Scalability of the cloud storage system, built according to [3].

Geographic distribution of drives is possible, since data retrieval using Reed-Solomon correction codes is available. So if p is the number of particles, and m is the permissible number of failures, the resulting overhead costs are $p/(p-m)$. Thus, in the case, shown in Fig. 5, overhead costs for the storage system at $p = 4$ and $m = 1$ will acquire 33%.

Active processing in the absence of hardware acceleration is another IDA reverse side. Also, the required method used by many cloud service providers is replication. This method is simple and effective, despite the fact that the costs are quite significant (100%).

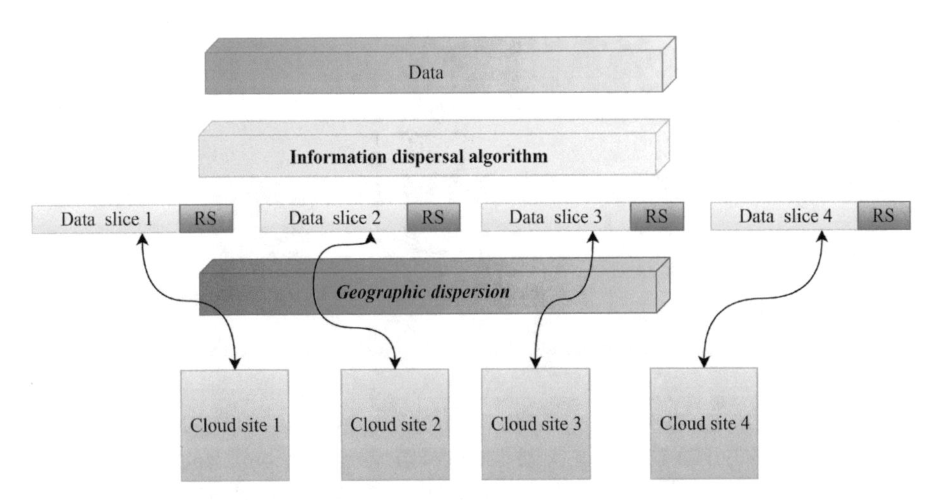

Fig. 6. The approach of Cleversafe to ensure high availability of data, built according to [3].

If until now the suppliers of cloud-based storage of data are mostly dispersed, then it is necessary to consider cloud models, which enable users to control their own data. Storage is deployed in three directions, one of which allows the combination of the other two in order to achieve financial performance and security [4].

Providing infrastructure to suppliers of available cloud storage systems for data storage is through lease (funds for long-term or short-term storage of information and bandwidth of the network). Public-private and private clouds are used by some concepts, but private use them so that the infrastructure of the user's personal network can be reliably built-in. Ultimately, mixed cloud storage systems make it possible to combine these two modifications, setting the principles that set what information should be kept individually, and which can be secured within the public clouds (Fig. 7).

An interesting direction in the development of storage models can be called cloud storage of data. It uncovers the latest perspectives for building, accessing and administering storage systems at enterprises and institutions. Despite the fact that today the cloud storage system is positioned as consumer technology, it is also rapidly improving for enterprises. Mixed modifications of clouds will enable institutions to maintain their own privacy within the local data processing cells, giving less secret information to the cloud in order to save costs and geographic protection.

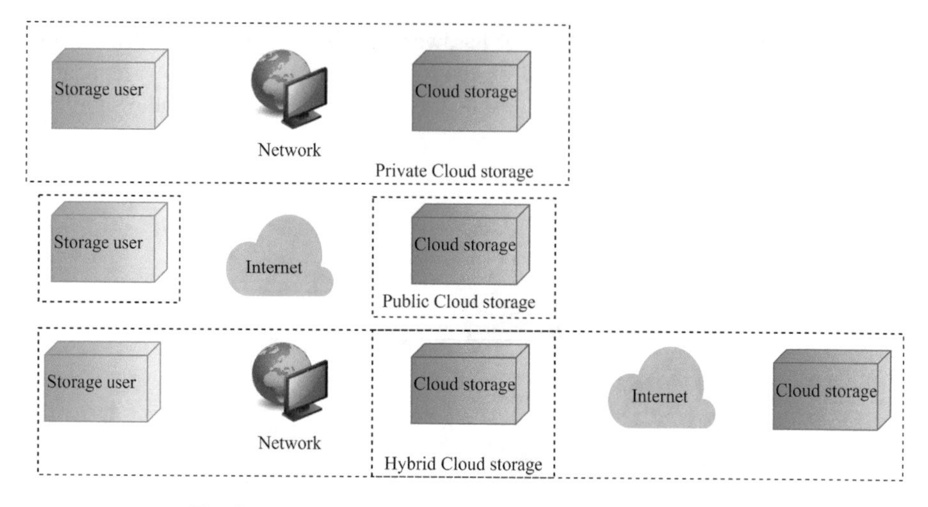

Fig. 7. Temporary storage models, built according to [3].

However, with the introduction of cloud storage in the business, there are problems, since there is a large number of architectural solutions and a significant horizon for the use of such systems. The main negative issues are the lack of generally accepted standards in this area and the impossibility of guaranteeing the quality of service, although the concept of "Cloud Computing" is also of great interest.

According to a survey conducted by IT executives, most of them voted in favor of considering that security is a serious problem for Cloud Computing. Productivity and availability respectively ranked second and third in the list of problems.

In order for cloud computing to be used by customers, security standards must be set up to begin with, the migration of applications between cloud platforms, SLAs (Service-level agreement), and solutions to other problems [5]. The examples provided will enable clients to apply all sorts of necessary combinations of applications, platforms and resources.

The leading areas of Cloud Computing, which require standardization, can be counted:

Cloud middleware or Cloud OS is software designed to allow queries to create instances of cloud-based instances, access resources, and manage their life span. This intermediate level system is the base for service management. Examples of such software are Google App Engine and Amazon EC2/S3. APIs for applications that provide access to resources such as computing power, storage, and virtual server image management systems. These interfaces are required to run applications in network clouds. Currently, most vendors offer their own incompatible APIs that prevent the sharing of resources across networks.

API interfaces for applications that provide access to resources such as computing power, storage resources, and virtual server image management systems. These interfaces are needed to allow applications to work in "network clouds". In our time, most vendors provide personal APIs that are incompatible and can't simultaneously utilize the resources of various networks. At this stage, the market seeks to closely integrate network infrastructures that support API interfaces, but network designers limit their role in this area with basic functions.

Resource management is an important field for creating cloud computing architecture. Typically, network tools are given to the user statically, while in the cloud computing power, the resources of the application's storage should be given dynamically, according to the request. Network tools should be provided separately and do not depend on the wires of applied resources. If you are talking about a regular corporate IT environment, there are a large number of administrative domains that are completely separate and function quite independently. However, in cloud computing, administrative constraints are a major negative phenomenon that is rapidly delaying the allocation of resources, since virtual systems are generated on demand, dynamically, at the very least, and quickly dispersed.

The virtualization method is used in the market for a long time and is rapidly expanding in data processing centers and telecom operators. Despite this, there are a number of inappropriate virtualization technologies among themselves. The lack of common standards makes it difficult to create an affordable Cloud Computing system that will be compatible with other "network clouds" and a large range of educational and informational resources.

In order to ensure interoperability among networks and the collective use of communication and computing resources belonging to diverse owners, "network clouds" require a standard level of management. As an example, you can consider the level of management embedded in the middleware software (cloud middleware) or in the extension of the BGP (Border Gateway Protocol). The use of protocols such as Extensible Messaging and Presence Protocol as the standard interface between networks of different operators will allow the various "network clouds" to use common naming and access control functions as well as user-access rules that will allow you to create compatible services.

And for applications that work in the corporate sphere and those that require a dynamic management of policies at different levels in the "network cloud." When you create a "cloud", you must consider for all resources the rules of the high level, of which this instance is composed. Principles of managing network resources must be consistent with the instructions for applications.

All without exception, the difficulties of cloud technologies arise out of remoteness, distributivism, parallelism, abstraction. In the process itself, it is impossible to call this a problem. As they say, the medal has two sides, that is, good and bad traits can be found.

4 Cloud Data Warehouse Model

Based on the model data warehouse, it is advisable to use the data storage model proposed by Robinson: $S = \langle F, D, G, C, L \rangle$, where

$Z = \{z_1, z_2, \ldots, z_n\}$ – set of data elements,
$z = \{r_1, r_2, \ldots, r_m\}$ – set of data packets
$V = \{v_1, v_2, \ldots, v_k\}$ – set of storage devices,
$Q : Z \to V$ – location of storage devices,
$J : V \to T$ – capacity of storage devices,
$E : V \to T$ – storage capacity of storage devices.

For the needs of modeling cloud storage data, it better to use its scalable version introduced by Petrov:

$$W_m = \langle Z, V(t), Q, J, E \rangle \quad (1)$$

Then the model of the cloud data warehouse is presented as:

$$W_{cloud} = \langle V, V_{free}, W_{ms} \rangle \quad (2)$$

where $V_{free} \subseteq V$ – a subset of free storage devices, $W_{ms} = \{W_{m1}, W_{m2}, \ldots, W_{ml}\}$ – multiple scalable repositories.

Scalable devices in this case are devices from a plurality of generic devices that do not include a subset of free devices $V_i(t) = V \backslash V_{free}$. Scalable devices do not have shared storage devices:

$$\forall t, i, j, i \neq j \Rightarrow V_i(t) \cap V_j(t) = \varnothing \quad (3)$$

The model of the cloud storage as an algebraic system is improved:

$$J_{dw} = \langle W_{cloud\_m}; O; E \rangle, \quad (4)$$

$$W_{cloud\_m} = \langle V, V_{free}, W_{ms}, US \rangle, \quad (5)$$

$$O = \{P_{cc,} P_{mpp}, P_{mpd}\} \quad (6)$$

where, P_{cc} – method of selecting a gateway complexity inquiry
P_{mpp} – method of multiprotocol transmission of stream data,
P_{mpd} – multiplexing method of different data sources for simultaneous transmission,
US – data transfer protocol,
E – predicate workload W_{cloud}.

Data item $z_i \in Wt \cup SemWt \cup UnWt$ can be represented by structured, weakly structured and non-structured data [6].

The predicate of the load capacity of the cloud storage is given as the ratio of the load of the cloud storage at the time points t_1 end t_2. For its determination, the data traffic in the cloud storage is investigated, aggregated data flows are analyzed, the dependence of the fractality level of the total flow is established:

$$E\left(W_{cloud\_m_{t1}}, W_{cloud\_m_{t2}}\right) \rightarrow T \tag{7}$$

Parameters of the cloud storage W_{cloud_m}: inbound/outbound traffic, number of running processes, load and simple processors, average CPU load and cache size.

5 Investigation of the Speed of Data Transmission in the Clouds

Due to the rapid development of the services industry based on the Internet network, there is a need for fast data transfer protocols. An alternative direction to developing such protocols is the superstructure over the UDP protocol to ensure guaranteed data delivery. Using UDP-based protocols compared to TCP is more efficient over long distances, that is, with large delays [7].

The most popular representatives of such add-ons over UDP are UDT [4] and μTP [5]. However, the use of these protocols also raises difficulties in their settings to the real state of the network. One of the most urgent problems in researching the effectiveness of data transfer protocols is the correct configuration of protocols (input data) when transmitting data over large and very long distances. The purpose of the work is to consider two of the most popular protocols, study their effectiveness and analyze the most promising protocol and its optimal parameters.

When transmitting data at small distances with small delay times, basically, only the TCP protocol was used, which proved itself to be insufficiently effective for long distances of data transmission.

To use modified protocols for long-range data transmission, it is necessary to investigate the influence of the parameters of these protocols on the total transmission rate. On the example of a real cloud data warehouse, data protocols were tested at different buffers of these protocols, and data rates were transmitted over time slots, which makes it possible to analyze data transmission rate differences and overall speed. These parameters allow you to determine the effective protocol and the input parameters of this protocol for the most efficient data transfer.

To compare the efficiency of protocols, it is necessary to examine the common configuration parameters of these protocols. For the purity of the research, only common (similar) parameters for two protocols were changed in the work, which is the buffer size when sending and receiving data. To ignore errors in the data transmission time, it was decided to use the parameters for buffers multiplied by 512 bytes (buffer size of the file system read /write buffer on the servers).

The purpose of the study was to determine the effectiveness of data transfer protocols at different buffer sizes in the intercontinental network of cloud storage.

To achieve this goal, the following tasks must be solved:

- investigate the data rate at different buffer values;
- analyze the changes in the speed of data transmission during the transmission of one data block;
- establish the dependence of the data transfer rate on the buffer size;

To conclude on the efficiency of the protocols being investigated. For the study, cloud storage servers located in the United States and the Netherlands were selected. The link between storage servers is 1 GB. The "black box" in this study has a connection, namely its stability and all possible firewalls on this communication channel. Experiments were carried out on files of large sizes, because just when they transmit problems with speed of transmission. A 200 MB file was selected for the experiment. All measurements were performed for each test at least three times and averaged indicators were taken for analysis. Since the experiments were conducted on the same data, traffic for all experiments with a certain probability is similar. Data transmission was carried out both in one way or another, which allowed the introduction of symmetry in the results obtained. As a result of the test tests, the dependencies of the file transfer time on the buffer size were obtained (Table 2).

Table 2. Depending on the time of file transfer from buffer size.

Buffer size	Transmission time µTP (sec)	Transmission time UDT (sec)
16384	2439,09	2151,4
32768	1114,08	1811,28
65536	527,74	1655,22
131072	269,1	649,27
262144	136,58	337,64
524288	77,58	169,57
1048576	67,15	67,54
2097152	66,77	29,51
4194304	67,12	13,05
8388608	86,92	6,5
16777216	78,6	6,0
33554432	97,22	6,0
67108864	97,24	6,51
134217728	67,43	5,03
268435456	67,21	6,38
536870912	67,44	5,5

As can be seen from the graph of the dependence of the transmission time on the size of the buffer (Fig. 8), the speed for both protocols falls under the exponential law, which should be purely theoretical considerations. Since the intercontinental distributed network is a packet switching network, the mathematical model is represented by the Markov process. The work of this process is modeled by the Poisson distribution:

$$F(x) = \frac{\lambda^x e^{-\lambda}}{x!}, \tag{8}$$

and, accordingly, the distribution function for time modeling in the Poisson process is an exponential function:

$$F(x) = \zeta e^{-\lambda x} \tag{9}$$

According to this model, the trend line for the UDT protocol:

$$F(x) = 2777e^{-0.47x}, \tag{10}$$

$$R^2 = 0.89 \; and \; for \; \mu TP \, R^2 = 0.86.$$

Although, with buffer sizes up to 1048576, the transmission time by the protocol μTP is less than the UDT transmission time, after further increasing the buffer size, the UDT protocol continues to tend to reduce the transmission time, and the μTP - time practically does not change within a certain value corridor (Fig. 8) This may indicate that there is saturation of the μTP protocol and it cannot substantially increase the transmission speed. For the UDT protocol, there is also such a point, but at the buffer size level of 8388608. In this case, the transfer of test files at optimal parameters of both protocols is 10 times smaller for the UDT protocol, which already suggests the feasibility of using it in comparison with μTP.

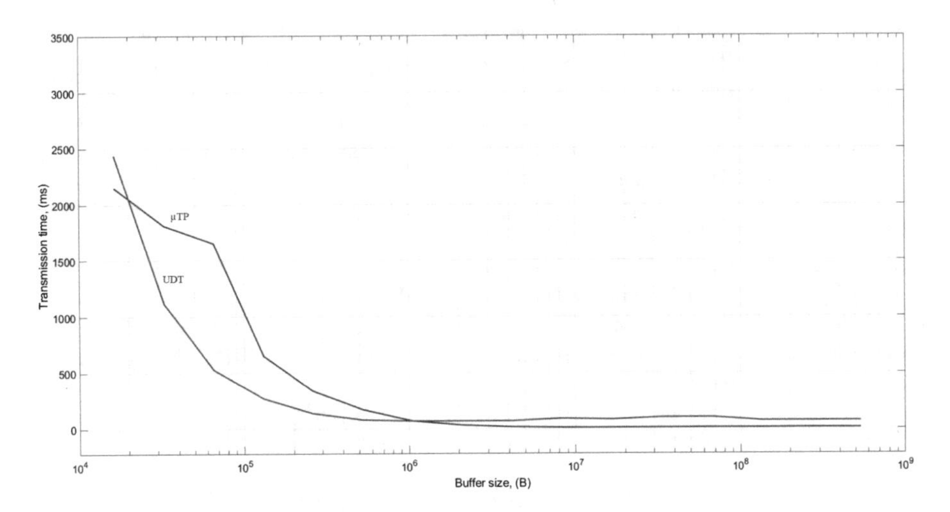

Fig. 8. Dependence of transmission time on buffer size.

6 Conclusions

Since Industry 4.0 technology is a data accumulation from a large number of sensors, there is a need to use large storage facilities for storing them, as well as expeditious access to them at an appropriate time. Also, there is a need to process this information, which entails the use of large computing power. The analysis showed that for these purposes cloud services with their big functionality are best suited. However, network bandwidth plays an important role in transmitting data in global networks that combine sensors, local servers with clouds. As evidence of a study, the most important role is played by data transfer protocols. Having investigated the bandwidth of the most used protocols for this purpose, the following conclusions were obtained: the UDT protocol has an advantage over the μTP protocol since it is more reliable and also has a higher throughput rate of 0.03 units, so it is advisable to use cloud technologies coupled with the UDT protocol to obtain a reliable high speed communication.

References

1. Strabitsky, P., Shakhovska, N.B.: An analysis of approaches to modeling cloud data warehouses. Curr. Probl. Econ. Sci. Econ. J. **11**, 263–269 (2013)
2. Chun, B.G., Dabek, F., Haeberlen, A.: Efficient replica maintenance for distributed storage systems. NSDI **6**, 45–58 (2006)
3. Jones, M.T.: Anatomy of a cloud storage infrastructure. IBM developer works (2010). http://www.ibm.com/developerworks/cloud/library/cl-cloudstorage/cl-cloudstorage-pdf.pdf
4. Huo, Y., Wang, H., Hu, L., Yang, H.: A cloud storage architecture model for data-intensive applications. In: Computer and Management (CAMAN), pp. 1–4 (2011)
5. Wieder, P., Butler, J.M., Theilmann, W., Yahyapour, R.: Service Level Agreements for Cloud Computing, p. 358. Springer (2011)
6. Kryvenchuk, Y., Shakhovska, N., Shvorob, I., Montenegro, S., Nechepurenko, M.: The smart house based system for the collection and analysis of medical data. In: CEUR, vol. 2255, pp. 215–228 (2018)
7. Kryvenchuk, Y., Shakhovska, N., Melnykova, N., Holoshchuk, R.: Smart Integrated Robotics System for SMEs Controlled by internet of things based on dynamic manufacturing processes, pp. 535–549. Springer, Cham (2018)
8. Chang, F., Dean, J.: A distributed storage system for structured data. ACM Trans. Comput. Syst. **26**(2), 1–26 (2008)
9. Dean, J., Ghemawat, S.: Simplified data processing on large clusters. In: Proceedings of the 6th Symposium on Operating System Design and Implementation, pp. 137–150 (2004)
10. White, T.: The Definitive Guide. O'Reilly Media, United States of America (2009)
11. Borthakur, D.: Distributed File System: Architecture and Design [EB/OL]. http://hadoop.apache.org/common/docs/r0.16.0/hdfs_design.html
12. Melnykova, N., Melnykov, V., Vasilevskis, E.: The personalized approach to the processing and analysis of patients' medical data. In: IDDM. pp. 103–112 (2018)
13. Melnykova, N., Marikutsa, U., Kryvenchuk, U.: The new approaches of heterogeneous data consolidation, In: 13th International Scientific and Technical Conference on Computer Sciences and Information Technologies, Lviv, pp. 408–411 (2018)

System for Monitoring the Technical State of Heating Networks Based on UAVs

Artur Zaporozhets(✉), Svitlana Kovtun, and Oleh Dekusha

Institute of Engineering Thermophysics of NAS of Ukraine, Kyiv, Ukraine
a.o.zaporozhets@nas.gov.ua

Abstract. The article presents the causes of defects in pipelines of the centralized heat supply. The possibilities of thermal aerial photography for detecting different types of defects on pipelines in a functioning state are explored. The characteristics and capabilities of the proposed set of devices for monitoring thermal losses in pipelines based on quadrocopters are considered. A method for monitoring the technical condition of pipelines using UAVs is presented. A method for processing thermal images for highlighting anomalous areas is presented. The created hardware-software complex for monitoring the state of trunk pipelines of heat networks based on the UAV is considered. Experiments on the use of UAVs for monitoring heating networks have been conducted. The obtained experimental results, confirming the possibility of differences in the technical condition of pipelines.

Keywords: Heating network · Main pipelines · Thermal image · Aerial photography · Monitoring · Image processing · Quadcopter

1 Introduction

Ensuring comfortable living and working conditions for people in the settlements of Ukraine during the heating season is carried out by heat supply systems, where purified water is used as the heat carrier. Supply of heat carrier from boiler houses to consumers is carried out by heating networks, the total length of which in Ukraine in two-pipe equivalent is 35700 km. Due to the fact that the total deterioration of heating networks is about 70%, the heat loss during transportation by networks reaches more than 15%, and the loss of water is more than 30%, which significantly exceeds the standard level [1]. In this situation, the state and consumers incur large losses, spending significant funds on the additional acquisition of energy carriers and the maintenance of poor-quality and unreliable heating networks [2]. In addition, the old heating systems require frequent repairs and partial replacement of damaged areas, which leads to constant disconnections of consumers from the heat supply [3]. All this entails additional financial costs, which within the state amount to millions of dollars, and their size is constantly growing.

Monitoring the technical condition of the heating network makes it possible to determine the presence of damage to pipelines and their insulation. Basically, the consequences of damage of the heating network during operation is the destruction of metal pipelines as a result of internal and external corrosion, as well as deterioration of the characteristics of thermal insulation and waterproofing [4, 5].

N. Shakhovska and M. O. Medykovskyy (Eds.): CCSIT 2019, AISC 1080, pp. 935–950, 2020.
https://doi.org/10.1007/978-3-030-33695-0_61

The main defects of metal pipelines are: crack, rupture of the metal, thinning of the wall due to mechanical stress, the effects of corrosion or delamination. Defects such as metal rupture also include fistulas occurring in welded joints of pipelines. Defects of thermal insulation is moisture insulation and partial or complete destruction of it. Defects of metal pipelines and thermal insulation can be local or extended.

In the urban landscape, leaks from heating networks are reflected in the thermal field of the earth's surface even in the presence of a solid road surface [6, 7].

After analyzing the features of the functioning of underground heating networks, we can distinguish two types of leaks. The first is the accidental leakage that occurs during the heat conductor ruptures and are accompanied by the outpouring of large amounts of mains water. Water fills the channel, goes out through the cracks, erodes the soil, forming cavities in it, sometimes goes to the surface. Of course, such leaks have a significant impact on the geological environment, being the main cause of flooding of urban areas. In addition, they lead to disruption of heat supply and cause significant damage to the urban economy, and sometimes cause accidents with tragic consequences. There are cases when people and equipment fell into the voids with hot water (Fig. 1).

Fig. 1. Breakthrough of the main pipeline (November 2017, Kyiv)

Therefore, operational services are trying to quickly localize the emergency sections of the network and eliminate leaks [8]. The occurrence of an emergency leak is recorded by a sharp drop in pressure in the supply line, and therefore is determined quite quickly [9, 10].

Leaks of the second type, which can be called permanent, have a different character. Even with the normal operation of heating networks, there are leaks, the magnitude of which does not exceed technically permissible limits. They are associated with errors in the connections of pipelines, in seals, compensators, regulating devices, with the occurrence of small end-to-end violations in the walls – fistulas. Until recently, there was no reliable way to detect such leaks, as they practically do not manifest themselves in control systems (unlike emergency ones).

There are many systems for diagnosing heating equipment. Some of them are given in [11–14]. Further thermal imaging of pipelines will be considered.

The task of identifying permanent leaks is of great practical importance. They are not only long-lasting sources of environmental impact, but also nascent foci of destruction of pipeline walls (accidents) due to the acceleration of the corrosion process. In addition, small leaks lead to a sharp increase in heat loss through damp insulation, losses of heating water, which have to be replenished. According to data from domestic heat power companies, the total actual amount of leakage in large cities is on average 2 l/s per 1 km^2.

2 Research Methods

2.1 Thermal Aerial Photography Features

Thermal infrared (IR) aerial photography is the only remote method that allows, like X-rays, not only to "see" almost all underground heating networks (pipes with a diameter of 50 mm or less), but also to qualitatively evaluate their condition. This is explained as follows. A conductive heat flux spreads from a hot heat conductor, due to which a "thermal trace" forms on the surface of the earth, where the intensity of IR radiation is higher than the background. It is quite obvious that the geometrical parameters and expressiveness of such a trace depend on the diameter of the heat pipe, the temperature of the heat carrier, the method and the depth of the laying.

At the stage of experimental and methodical work, a complex of ground-based thermometric observations synchronous with aerial surveys was performed. Comparison of aerial survey infrared materials and a priori information about design features, laying depth, temperature in networks and condition of heat pipes allowed to get an idea of the nature of the manifestation in the thermal field of tracks of different diameters in different states – from normal to emergency.

The data obtained in the future are used as standards in the interpretation of thermal will and assess the status of thermal networks. Naturally, this is a qualitative assessment, since remotely measuring the absolute thermodynamic temperature is impossible in principle.

The data obtained in the future are used as standards in the interpretation of the thermal field and the assessment of the state of thermal networks. Naturally, this is a qualitative assessment, since remotely measuring the absolute thermodynamic temperature is impossible in principle.

2.2 Processing of Thermal Images

Next will be considered the input thermal image of rank 0, obtained from the thermal imaging device. If its dimensions are 640 × 480 (307200) pixels, then transformation to rank 1 will give an image with dimensions of 320 × 240 (76800) pixels.

The color image is brought to grayscale by any of the known methods:

1. according to the CCIR-601 standard

$$F[y,x] = 0.299R + 0.587G + 0.114B; \tag{1}$$

2. by the arithmetic mean value of the color components of the three channels

$$F[y,x] = (R + G + B)/3; \tag{2}$$

3. fast (using an algorithm with green pixels)

$$F[y,x] = G. \tag{3}$$

where R – red, G – green, B – blue (digital image components).

Further, it is rational to treat an image as a rectangular matrix of $n \times m$ elements whose values lie in the range from 0 to 127. Matrix filters are used to solve image pre-processing tasks that perform the convolution operation, which allows obtaining response values, taking into account the values of the surrounding pixels, within the dimension kernels. For highly noisy thermal images, it is necessary to use masks of relatively high resolution [15]. For the analysis of thermal images, it is proposed to conduct a discrete Laplacian with a filter of 11 $\times$ 11 dimensions

$$D^2_{y,x} = \begin{matrix} -1 & -2 & -3 & -3 & -3 & -3 & -3 & -3 & -3 & -2 & -1 \\ -2 & -4 & -6 & -6 & -6 & -6 & -6 & -6 & -6 & -4 & -2 \\ -3 & -6 & -9 & -9 & -9 & -9 & -9 & -9 & -9 & -6 & -3 \\ -3 & -6 & -9 & 0 & 9 & 18 & 9 & 0 & -9 & -6 & -3 \\ -3 & -6 & -9 & 9 & 27 & 45 & 27 & 9 & -9 & -6 & -3 \\ -3 & -6 & -9 & 18 & 45 & 72 & 45 & 18 & -9 & -6 & -3 \\ -3 & -6 & -9 & 9 & 27 & 45 & 27 & 9 & -9 & -6 & -3 \\ -3 & -6 & -9 & 0 & 9 & 18 & 9 & 0 & -9 & -6 & -3 \\ -3 & -6 & -9 & -9 & -9 & -9 & -9 & -9 & -9 & -6 & -3 \\ -2 & -4 & -6 & -6 & -6 & -6 & -6 & -6 & -6 & -4 & -2 \\ -1 & -2 & -3 & -3 & -3 & -3 & -3 & -3 & -3 & -2 & -1 \end{matrix}$$

Further, the central element of the filter is superimposed on the studied pixel. The remaining elements are also superimposed on neighboring pixels [16]. Next, the sum is calculated, where the terms are the multiplied values of the pixels and the values of the cell of the core that covered the given pixel:

$$G[y,x] = \left(\sum_{dy=-5}^{5} \sum_{dx=-5}^{5} \left(\begin{matrix} F^i[y+dy, x+dx] \times \\ \times D^2_{y,x}[dy, dx] \end{matrix} \right) \right) shr2, \tag{4}$$

where F^i – rank matrix $I \in [0,1]$, $shr2$ – operation of logical shift right by two digits.

A logical shift to the right by two bits (division by 4) of the result of the convolution is made to bring it into the range [−32768, 32767], what twice reduces the amount of necessary memory and allows to get rid of the reaction to noise in the image.

The greater the dimension of the convolution kernel, the more accurate the response can be expected from the current pixel, since the set of neighboring pixels are also involved in the convolution operation, which leads to a large number of calculations.

Image processing by 11×11 filter implies a large number of multiplication operations, and there are two-digit numbers in this matrix, which will lead to an increase in time to calculate the value of one response. As a result, to reduce the computational cost, approximately 3 times, instead of the 11×11 operator, you should use an operator with a kernel size of 7×7 elements and then double the operator with a core of 3×3 elements, which will give an equivalent result when processed by one 11×11 operator.

The method of competitive analysis gives a good result for pattern recognition on non-noisy images [17]. However, in the conditions of noise, poor visibility or the presence of foreign objects it is also possible to apply the contour method using masks of a higher dimension, as well as introducing the definition of "singular points" – extreme response values on such images.

To determine the specific points, the algorithm of the Moravec detector [18] is used as a basis, which compares the extremes at the corners of the image using local detectors. A black-white image arrives at the detector input. At the output, a matrix is formed with elements whose values determine the degree of plausibility of finding the angle in the corresponding pixels of the image. The threshold value allows to "drop" the pixels, the degree of likelihood of which is less than the threshold. The remaining points are special or extremes. The Moravec detector is a simple angle detector, estimating the change in pixel intensity (y, x) by offsetting a square window centered in the current pixel (y, x) by one pixel in each of the 8 directions [19]. This method is implemented as follows:

- for each direction of displacement $F(u, v) \in \{(1,0),(1,1),(0,1),(-1,1),(-1,0),(-1,-1),(0,-1),(1,-1)\}$, the change in intensity is calculated

$$V_{u,v}(y,x) = \sum\nolimits_{\forall a,b} (I(x+u+a, y+v+b) - I(x+a, y+b))^2, \quad (5)$$

  where $I(x + a)$ is the intensity of a pixel with coordinates (y, x) in the source image;
- builds a map of the probability of finding the angles in each pixel of the image by calculating the estimated function. Essentially, the direction is determined, which corresponds to the smallest change in intensity, because the corner must have adjacent edges;
- pixels are cut off, in which the values of the evaluation function are below a certain threshold value;
- recurring corners are removed using the NMS procedure (non-maximal suppression);
- non-zero elements correspond to the angles in the image.

The use of the Moravec detector makes it possible not to calculate the change in intensity, but to immediately perform an analysis on the generated response matrix. The table of directions will determine the maximum and minimum values of the response value located in the center of the window:

$$f = \begin{cases} G[y,x] > 0 \; AND \; G[y,x] > G[y+dy, x+dx] \\ G[y,x] < 0 \; AND \; G[y,x] < G[y+dy, x+dx] \end{cases} \quad (6)$$

for all extremums $dy = dx \in [-1;0) \; AND \; (0,1]$. Extremes can be in maximum proximity to each other through one element of the response, therefore, in order to reduce the number of iterations when an extremum is found, the following, in the direction of the sweep, the value is excluded from the comparison procedure.

Approaching the solution of the segmentation problem, namely the definition of related groups of elements in the image, it is necessary to select a common feature or threshold value that will allow to divide the desired signal into classes. The threshold separation operation is to compare the brightness value of each pixel of the image with the specified threshold value, and can be represented by a filter:

$$B[y,x] = \begin{cases} 1, if \; A[y,x] \geq \Delta T; \\ 0, if \; A[y,x] < \Delta T, \end{cases} \quad (7)$$

where ΔT is a threshold value.

On images with monochromatic objects, the binarization threshold is selected from the histogram of brightness, examples of which are given in Subsect. 2.4. As a rule, light (warmer) areas appear as objects or search areas on thermal images. Depending on the temperature, areas can be a part of the wall, a staircase, and other geometric figures of a certain size, which fall, for example, under a warm stream of air or have heat-generating elements in their physical structure. The difference in brightness can lie in the whole range [0, 255]. That is why there is a need to get rid of noise, reducing the maximum value to 128.

In some cases, the Otsu method can be used to find the threshold, which determines the threshold that minimizes the variation of the pixel brightness in an object. The threshold is chosen between the highest pixel values in the histogram.

Having a generated table of extrema G^E and image matrix F^1, which is formed as a result of applying the Laplacian mask, the binarization threshold can be found by the formula:

$$\Delta T = \frac{\sum_{i=1}^{n} G^E R_i |\sigma|}{n\bar{X}}, \quad (8)$$

where R_i is the i-th element from the extremum list. The coefficient of variation is necessary in order to characterize the one-sidedness of data, values and stability of processes, reflect the degree of variation of values regardless of the scale of measurements. This method of processing thermal images is described in more detail in [20].

One pixel in the image that is not tied to a coherent group of similar pixels of the analyzed segment is estimated by the system as noise and will not be an area due to which it was necessary to reduce the brightness range by one row to the right. Due to this, the Laplacian mask will not react to not so significant single pixels. Theoretically, the input image of 320 × 240 pixels allows to contain 8480 areas (the minimum size is 2 × 2 pixels with a minimum distance between them of one pixel), for a finite number of areas it can be take the value 0xFFFD.

Binarization in the response matrix is performed by cutting the response over the threshold and generating two matrices of domains of the same dimension for responses with positive (G^P) and negative (G^N) values with the conditions:

$$G^P[y,x] = \begin{cases} 0xFFFD, & if\ G[y,x] \geq \Delta T; \\ 0x0000, & if\ G[y,x] < \Delta T, \end{cases}$$
$$G^N[y,x] = \begin{cases} 0xFFFD, & if\ G[y,x] \leq \Delta T; \\ 0x0000, & if\ G[y,x] > \Delta T. \end{cases} \quad (9)$$

2.3 UAV Based Monitoring System

To carry out experimental studies of the monitoring system of the technical condition of heat pipelines, a multi-rotor type unmanned aerial vehicle, model MJX BUGS 3, was used. This quadcopter will also be used to study the effect of air dilution on the concentration of air components [21, 22] in the following works.

The MJX BUGS 3 quadrocopter (Fig. 2) is a world-renowned radio control toy manufacturer Meijiaxin Toys. The company is positioning this drone designed for both aerial photography and dynamic flights.

Fig. 2. MJX BUGS 3

The features of this quadrocopter include:

- support for 3S batteries;
- control of battery charge and flight distance;
- long flight time;
- axle suspension;
- control at 2.4 GHz;
- 360° flip;
- LED-backlight.

The quadcopter case MJX BUGS 3 is made of nylon fiber, has established itself as a reliable and durable material, while the supports are made of ordinary plastic.

The quadcopter MJX BUGS 3 is equipped with brushless motors of type MT1806 with a capacity of 1800 kV. The manufacturer describes them as economical and efficient among the same type of brushless motors. Each motor provides 230 grams of traction.

Also in quadcopter available speed controllers ESC with automatic anti-jamming, eliminating the possibility of burnout engines.

Included with the drone is a axle-free suspension with manual vertical adjustment, adapted for installing a small load. The distance from the ground to the suspension is 80 mm.

The quadcopter is powered by a 2S Li-Po battery with a capacity of 1800 mAh with a discharge current of 25C and an XT30 connector. According to the specification, the battery provides 19 min of continuous flight.

The quadcopter kit also includes hardware operating at 2.4 GHz. Its distinctive feature is the function of intelligent remote control, reports a low battery or a long distance of the drone from the equipment. It is powered by 4 AA batteries. The maximum distance of the drone from the equipment is 300–500 m [23].

During flight tests, the MJX BUGS 3 shows good flight performance even on the type 2S battery included in the kit. The 6-axis gyroscope works smoothly. In practice, the flight time of the quadcopter with a maximum load was 8 min. At a distance of 300 meters, the quadcopter clearly performs the specified flight directions [24].

The main advantages and disadvantages of the quadrocopter MJX BUGS 3 are shown in Table 1.

Table 1. Advantages and disadvantages of the model MJX BUGS 3

Advantages	Disadvantages
• ease of using • power • feedback function with low charge and critical distance • suspension for cargo • compatible with 3S-batteries • LED backlight • price	• lack of dynamism • charged battery as standard

To perform experiments with thermal imaging of heat supply pipelines on the basis of the UAV, a compact thermal imaging camera manufactured by Seek Thermal ™ (USA) (Fig. 3) was installed, which has a wide-angle lens with a total size of 2.5 × 4.4 × 2.5 cm resolution of 320 × 240. The greatest shooting distance is 610 meters, and the closest distance is 15 cm.

Thermal Compact XR was used as a thermal imaging camera. Its technical characteristics are given in Table 2.

Table 2. Thermal camera parameters

Resolution	206 × 156
Working distance	to 550 m
Viewing angle	20°
Pixel size	12 μ
Spectral range	7.5–14 μ
Palettes	9

Fig. 3. XR Compact thermal imaging camera (Seek Thermal)

To implement the method of thermal aerial photography at the Institute of Engineering Thermophysics of the National Academy of Sciences of Ukraine an equipment complex was developed for diagnosing the state of heat equipment (boilers [25, 26] and main heat pipelines). Its appearance is shown in Fig. 4, where 1 – UAV, 2 – a thermal imaging camera, 3 – a communicator (smartphone).

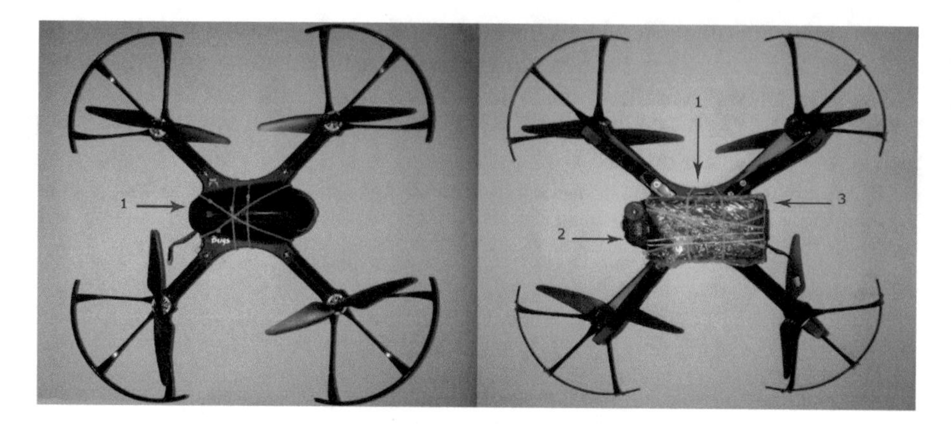

Fig. 4. Hardware complex for monitoring the technical condition of main pipelines (left – top view; right – bottom view): 1 - UAV, 2 - thermal imaging camera; 3 - communication device

2.4 The Results of Experimental Studies

During the initial stage of the research, studies were carried out on the ground parts of the main pipelines of heating networks. Fluke Ti50FT was used as a thermal imaging camera on a UAV. Figures 5 and 6 show a section of a land trunk pipeline of a heating network obtained in the visible and infrared range.

Figure 7 shows the histogram of pixel distribution of Fig. 6 by temperature. Figure 8 depicts a 3D model of Fig. 6 taking into account the temperature characteristics. The parameters of Fig. 6 are shown in Table 3.

Fig. 5. Photo of the main pipeline of the heating network in the visible range

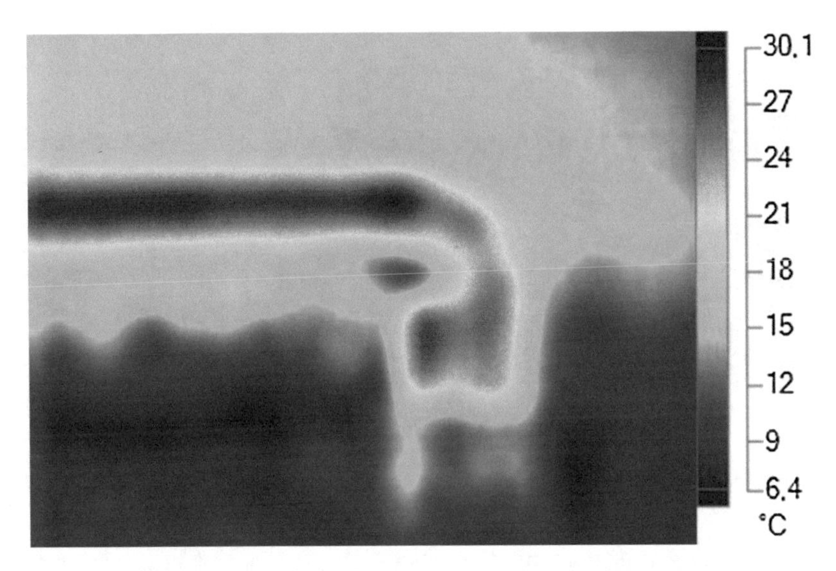

Fig. 6. Photo of the main pipeline of the heating network in the IR range (Fluke Ti50FT)

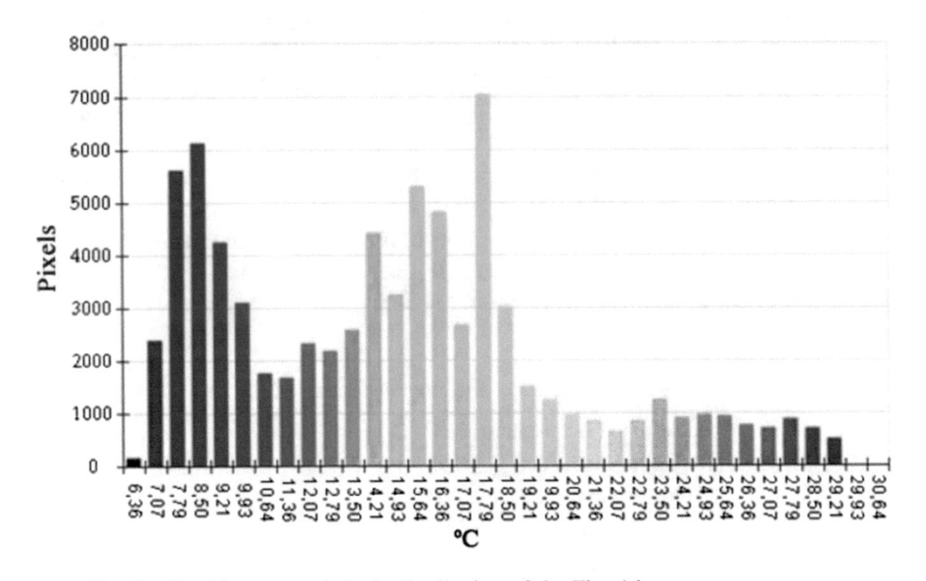

Fig. 7. The histogram of pixels distribution of the Fig. 4 by temperature ranges

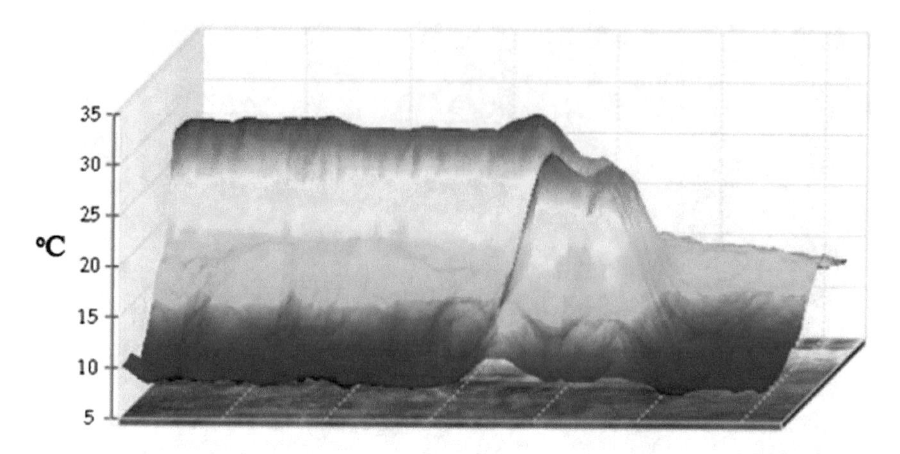

Fig. 8. 3D model of Fig. 4 taking into account the temperature characteristics

For monitoring extended objects (in our case, pipelines of heat networks) it is proposed to fly around the test object using a multi-rotor type UAV [27]. This allows to get high-quality photos and thermal images of the site of the heating system as an object of control for further analysis. The software allows you to use any topographic basis as a map. Binding can be done at two or more points. It is also possible to use as a topological basis of electronic maps [28, 29]. The program provides input, automatic control and editing of the route of the flight. An elevation can be specified for each waypoint.

The results of measuring the thermal state of the heat grids were carried out using a thermal imaging camera, mounted on a UAV.

Figure 9 shows thermal images of sections of the heat network where experimental studies were conducted. The shooting was performed on November 22, 2018 at 18 p.m. on a cloudless sky at an air temperature of −4 °C.

Table 3. Parameters of thermographic image (Fig. 6)

Background temperature	7.00 °C
Radiation coefficient	0.95
Transmission coefficient	1.00
Average temperature	14.87 °C
Image borders	from 6.38 °C to 30.06 °C
Camera model	Ti50FT
IR sensor size	320 × 240
File name	IR20091117_0340.is2
Humidity setting	0 RH % 0 m

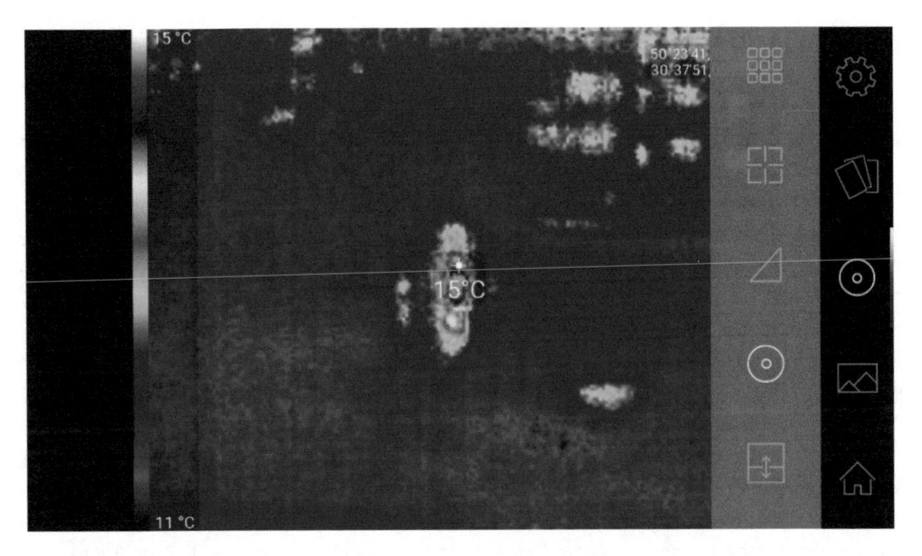

Fig. 9. The working environment of the software complex based on Seek Thermal

Figure 9 clearly shows the possibility of identifying defective areas of main pipelines using low power UAVs.

The use of UAVs allowed us to distinguish 4 technical conditions of the main pipelines of the heating networks shown in Table 4.

Table 4. Assessing the state of the main heating networks with UAVs

Type	% heat losses	Characteristic
Normal	5–10	Dry and integral insulation of pipelines, minimum heat flow from the heat carrier to the earth's surface
Increased	10–15	Wet or broken insulation of pipelines, contributes to the nucleation of corrosive damage; in the thermal field can be displayed by a clear anomaly of the average brightness level and an increased width of the thermal trace
High	15–20	Disturbed and damp isolation of pipelines, the canal is often filled with water from neighboring water pipelines, groundwater or melt water; in the thermal field is displayed as a high-contrast anomaly with a width several times larger than normal
Emergency	>20	Violation of the integrity of the pipeline with the heat carrier spill. Thermal field anomalies have a very high contrast and a broad, diffuse shape, due to the microrelief features

3 Conclusions

Today we can state the rapid development of UAVs, which are mainly used in military operations. The list of subject areas of the use of UAVs for various other studies, operations, conventionally called non-military, is essentially limited. First of all, this limitation is due to the lack of created, developed and manufactured technical tools for conducting diverse studies, primarily measuring tools. It can be predicted that such an imbalance will soon be broken and the process of creating appropriate equipment for the UAV will be adjusted to conduct a wide range of research in various subject areas, among which the defense industry of the states will be priority and prospective.

The creation of mobile information-measuring systems based on UAVs makes it possible to diagnose the state and dynamics of the characteristics in time and space of the studied environment, both in on-line modes and other modes. The on-line mode is especially effective in case of accidents in areas of spatially branched heat networks. With normal operation of the objects under study, the current remote control is the most economical compared with other means of control. This allows to use such measuring tools to create the necessary databases for diagnostics the characteristics of the thermal state of heat networks to predict their dynamics.

References

1. Babak, S., Babak, V., Zaporozhets, A., Sverdlova, A.: Method of statistical spline functions for solving problems of data approximation and prediction of object state. In: CEUR Workshop Proceedings, vol. 2353, pp. 810–821 (2019). http://ceur-ws.org/Vol-2353/paper64.pdf
2. Babak, V.P.: Hardware-Software for Monitoring the Objects of Generation Transportation and Consumption of Thermal Energy. Institute of Engineering Thermophysics of NAS of Ukraine, Kyiv (2016)
3. Prudyus, I., Shkliarskiy, V., Turkinov, G., Storozh, V., Kril, Y.: Analysis and choice of methods of monitoring technical condition of the heating systems. In: International Conference on "Modern Problems of Radio Engineering, Telecommunications and Computer Science (TCSET)", p. 154 (2008)
4. Johnson, B., Barth, R., House, P., Kuntscher, J.: Controlling pipe and equipment operating temperatures with trace heating systems. In: PCIC Europe, pp. 1–10 (2013)
5. Friman, O., Follo, P., Ahlberg, J., Sjökvist, S.: Methods for large-scale monitoring heating systems using airborne thermography. IEEE Trans. Geosci. Remote Sens. **51**(8), 5175–5182 (2014)
6. Shang-bin, J., Le, F., Peng-yue, W., Lei, Q., Yong-ze, J., Qian-kun, S.: Assessment and prediction of leakage degree of expansion joints in underground network heating pipe network based on LS-SVM and ARIMA model. In: 13th IEEE Conference on Industrial Electronics and Applications (ICIEA), pp. 1697–1702 (2018)
7. Jiao, S.-B., Fan, L., Wu, P.-Y., Qiao, L., Wang, Y., Xie, G.: Assessment of leakage degree of underground heating primary pipe network based on chaotic simulated annealing neural network. In: Chinese Automation Congress (CAC), pp. 5895–5900 (2017)
8. Babak, V.P., Zaporozhets, A.A., Kovtun, S.I., Sergienko, R.V.: Diagnosing methods analysis of bulk heating systems technical condition. Sci. Heritage **1**(14), 59–65 (2017)

9. Teng, Q., Wang, W.: The optimization and management research for central heating system. In: IEEE Workshop on Advanced Research and Technology in Industry Applications (WARTIA), pp. 175–177 (2014)
10. Fukuoka, M., Tang, L.: Wireless pipe inspection control system. In: The 5th International Conference on Automation, Robotics and Applications, pp. 497–502 (2011)
11. Babak, V., Zaporozhets, A., Kovtun, S., Serhiienko, R.: Methods and means of heat losses monitoring for heat pipelines. Int. J. "NDT Days" **1**(2), 213–221 (2018)
12. Zaporozhets, A.A., Eremenko, V.S., Serhiienko, R.V., Ivanov, S.A.: Development of an intelligent system for diagnosing the technical condition of the heat power equipment. In: IEEE 13th International Scientific and Technical Conference on Computer Sciences and Information Technologies (CSIT), pp. 48–51 (2018)
13. Zaporozhets, A., Eremenko, V., Serhiienko, R., Ivanov, S.: Methods and hardware for diagnosing thermal power equipment based on smart grid technology. In: Shakhovska, N., Medykovskyy, M. (eds.) Advances in Intelligent Systems and Computing III, vol. 871, pp. 476–489. Springer, Cham (2019)
14. Eremenko, V., Zaporozhets, A., Isaenko, V., Babikova, K.: Application of Wavelet Transform for Determining Diagnostic Signs. In: CEUR Workshop Proceedings, vol. 2387, pp. 202–214 (2019). http://ceur-ws.org/Vol-2387/20190202.pdf
15. Kosar, O., Shakhovska, N. An overview of denoising methods for different types of noises present on graphic images. In: Shakhovska, N., Medykovskyy, M. (eds.) Advances in Intelligent Systems and Computing III, vol. 871, pp. 38–47. Springer, Cham (2019)
16. Gauci, J., Falzon, O., Formosa, C., Gatt, A., Ellul, C., Mizzi, S., Mizzi, A., Delia, C.S., Cassar, K., Chockalingam, N., Camilleri, K.P.: Automated region extraction from thermal images for peripheral vascular disease monitoring. J. Healthc. Eng. 5092064 (2018)
17. Duarte, A., Carrao, L., Espanha, M., Viana, T., Freitas, D., Bartolo, P., Faria, P., Almeida, H. A.: Segmentation algorithms for thermal images. Procedia Technol. **16**, 1560–1569 (2014)
18. Chen, X., Liu, L., Song, J., Li, J., Zhang, Z.: Corner detection and matching for infrared image based on double ring mask and adaptive SUSAN algorithm. Opt. Quant. Electron. **50**, 194 (2018)
19. Wang, X., Song, H., Cui, H.: Pedestrian abnormal event detection based on multi-feature fusion in traffic video. Optik **154**, 22–32 (2018)
20. Loginov, I.D.: Processing and segmentation of thermal images. Young Sci. **13**(147), 62–71 (2017)
21. Babak, V.P., Mokiychuk, V., Zaporozhets, A., Redko, O.: Improving the efficiency of fuel combustion with regard to the uncertainty of measuring oxygen concentration. Eastern-Eur. J. Enterp. Technol. **6**(8(84)), 54–59 (2016)
22. Zaporozhets, A.O., Redko, O.O., Babak, V.P., Eremenko, V.S., Mokiychuk, V.M.: Method of indirect measurement of oxygen concentration in the air. Naukovyi Visnyk Natsionalnoho Hirnychoho Universytetu **5**, 105–114 (2018)
23. Devitt, D., Morozov, R., Medvedev, M., Shapoval, I., Konovalov, G.: Implementation of the hybrid technology for quadcopter motion control in a complex non-deterministic environment. In: 18th International Conference on Control, Automation and Systems (ICCAS), pp. 451–456 (2018)
24. Talha, M., Asghar, F., Rohan, A., Rabah, M., Kim, S.H.: Fuzzy logic-based robust and autonomous safe landing for UAV quadcopter. Arab. J. Sci. Eng. **44**(3), 2627–2639 (2018)
25. Zaporozhets, A.: Development of software for fuel combustion control system based on frequency regulator. In: CEUR Workshop Proceedings, vol. 2387, pp. 223–230 (2019). http://ceur-ws.org/Vol-2387/20190223.pdf
26. Zaporozhets, A.: Analysis of control system of fuel combustion in boilers with oxygen sensor. Periodica Polytechnica Mech. Eng. (2019). https://doi.org/10.3311/ppme.12572

27. Gomez, C., Green, D.R.: Small unmanned airborne systems to support oil and gas pipeline monitoring and mapping. Arab. J. Geosci. **10**, 202 (2017). https://doi.org/10.1007/s12517-017-2989-x
28. Allred, B., Eash, N., Freeland, R., Martinez, L., Wishart, D.: Effective and efficient agricultural drainage pipe mapping with UAS thermal infrared imagery: a case study. Agric. Water Manag. **197**, 132–137 (2018). https://doi.org/10.1016/j.agwat.2017.11.011
29. Zaporozhets A., Kovtun S., Dekusha, O.: Determination of the technical condition of heating networks based on the processing of thermal imaging. In: Proceedings of International Scientific Conference Computer Sciences and Information Technologies (CSIT-2019), vol. 2, pp. 5–8 (2019)

MEMS Gyroscopes' Noise Simulation Algorithm

Dmytro Fedasyuk and Tetyana Marusenkova(✉)

Lviv Polytechnic National University, Lviv, Ukraine
fedasyuk@lp.edu.ua, tetyana.marus@gmail.com

Abstract. MEMS gyroscopes are advantageous devices that promote a wide range of applications, however, they suffer from various stochastic errors some of which accumulate over time (angle random walk, bias random walk). To be able to use such devices, one should apply mathematical models of stochastic processes and hardware-software tools for investigation into noise, figure out the noise characteristics and develop an appropriate method of adaptive signal filtering. The Allan deviation plot is considered the most common tool for studying noise spectral characteristics. However, when distorted by unexpected noise components, the Allan deviation plot becomes difficult to interpret. The aim of this work is to present an algorithm for generating noise typical of real MEMS gyroscopes and its implementation as a part of a complex hardware-software tool for investigation into inertial measurement units, being developed by the authors. With such a tool for simulating noise with specific spectral characteristics, the researcher will be able to understand and explain the behavior of a MEMS gyroscope and thus fit a reasonable filtering method for it.

Keywords: MEMS gyroscope · Inertial measurement unit · Measurement error · Noise synthesis · Random walk

1 Introduction

The work deals with the problem of analyzing noise in MEMS gyroscopes. MEMS (Micro-Electro-Mechanical System) gyroscopes and accelerometers are the basis of integrated IMU (inertial measurement units) sensors [1]. The latter are used for navigation in a wide range of areas including avionics [2], car electronics [3], virtual reality, medicine and rehabilitation [4], etc.

This branch of sensors is developing within two concepts: Intelligent Sensor [5] and Sensor Fusion [6]. Along with gyroscopes and accelerometers, navigation systems may contain magnetometers [7], atmospheric pressure sensors and specialized microprocessors. Such navigation systems are widely used in smartphones [8].

The problem considered in this work is a part of complex research into development of MIS-IDE (Measurement Inertial System Integrated Development Environment) that has been conducted by a team of scholars (including the authors) within a series of commercial projects, which somewhat restrict disclosure of details. MIS-IDE is intended for enhancing the efficacy of synthesizing firmware of navigation systems based on IMU sensors. The scope of the complex work covers developing: models of

N. Shakhovska and M. O. Medykovskyy (Eds.): CCSIT 2019, AISC 1080, pp. 951–967, 2020.
https://doi.org/10.1007/978-3-030-33695-0_62

simulating signals from IMU sensors in accordance with some preset motion scenario, hardware-software tools for testing IMU sensors; algorithms for parametric analysis and synthesis of IMU sensor models, mathematical models for synthesis of noise and measurement errors, algorithms for selection of the most suitable measurement modes, firmware for microprocessors inside embedded systems based on IMU sensors.

2 Literature Overview

Design of devices based on MEMS IMUs is tightly coupled with problems of analyzing their measurement errors and noise terms [9, 10], calibration [11], and compensation for the drift of the transfer function [12]. Solutions to these problems assume that one should use mathematical models of stochastic processes and hardware-software tools for investigation into noise [13]. Upon the investigation results, one figures out the noise characteristics [14] and develops an appropriate method of adaptive signal filtering [15]. The most common tool for noise analysis is the Allan variance [16, 17] and its variations. Adaptive filtering is usually done using variations of the Kalman filter [18].

The random drift of MEMS-based gyroscopes limits their applications. Hence, studies pay attention to developing various models in order to compensate for the random drift and thus improve the performance of the MEMS-based gyroscopes.

The paper [19] presents a self-constructing Wiener-type recurrent neural network with its false nearest-neighbors-based self-constructing strategy and recursive recurrent learning algorithm to model the random drift of the MEMS-based gyroscopes and then compensate them from the calibrated gyroscope measurement.

Two methods for identifying the MEMS gyroscope, the Allan variance method and AR model, are introduced in [20]. Through the two methods, the statistical mathematical properties and dynamic system properties of MEMS gyroscopes are proclaimed.

The paper [21] presents noise analysis of closed-loop MEMS vibratory gyroscopes whose noise characteristics are dominated by the mechanical-thermal noise of the sensor's vibrating structure as well as the electrical noise associated with the pickoff signal conditioning electronics in the sense channel. The mechanical-thermal noise and the electrical noise are represented as uncorrelated additive wideband disturbances dominant at the sensor's input and output, respectively.

Further, [22] presents analysis of the noise spectra of closed-loop mode-matched vibratory gyros. Closed-form expressions for the noise-equivalent angular rate spectrum as well as the integrated angular rate (angle) variance are derived to explore the effects of modal frequency mismatch, closed-loop bandwidth, and the spectra of noise sources appearing at the sensor's input and output. It is shown that noise sources located at the output of the sensor's electromechanical transfer function create angle white noise in the closed-loop sensor.

The model [23] is made in view of the MEMS gyroscope random error, which is applied to error compensation with the Kalman filter. The main noise sources that affect measurement accuracy are determined via the Allan variance method. The principal factors that affect the performance of a MEMS gyroscope are confirmed with the analyses of MEMS gyroscope noise items and the coefficients of various noise sources are compared before and after filtering using Allan variance method. The experiment shows that error model significantly improved the precision of the measurement of a MEMS gyroscope.

In [24] the authors present a heterogeneous environment for modeling and simulations created using Coventor MEMS+ and Matlab/SIMULINK software. The advantage of this solution is a possibility to merge it with Cadence software, which is much expedient for modeling, simulation and design MEMS structure with ROIC (Read Out Integrated Circuit).

A lifting wavelet and a wavelet neural network are proposed in [25]. The lifting wavelet is employed to eliminate MEMS noise MEMS first, then the wavelet neural network is introduced to model and compensate for the random drift. The simulation results show that the proposed methods can effectively eliminate the noise and compensate for the MEMS's random drift.

In view of the large output noise and low precision of the micro-electro-mechanical system gyroscope, the virtual gyroscope technology is used to fuse the data to improve its output precision. The virtual gyroscope technology with high-precision dynamic filtering is proposed in [26]. The high-accuracy model of the real dynamic angular rate based on Taylor formula is established, and the random error model of the MEMS gyroscope array is improved based on the real angular rate modeling results.

Due to the inherent errors of MEMS inertial sensors and their stochastic nature, which is difficult to model, the Kalman filtering with its linearized models has limited capabilities in providing accurate positioning. So, particle filtering has recently been suggested as a nonlinear filtering technique to accommodate arbitrary inertial sensor characteristics, motion dynamics, and noise distributions [27].

However using the above-mentioned algorithms for modeling noise terms of MEMS gyroscopes is rather complicated and problematic when it comes to implementation of MIS-IDE (Measurement Inertial System Integrated Development Environment), due to the difficulties in verification of these models with experimental results obtained from real gyroscopes.

Thus, there exists a problem of developing a MEMS gyroscopes' noise simulation algorithm that would allow us to change the parameters of the mathematical models of MEMS gyroscopes quickly and efficiently [31].

3 Problem Analysis

All the errors in MEMS gyroscopes fall into two large groups: deterministic and stochastic. The former can be partially eliminated due to a proper calibration process. The latter cannot be compensated for so easily, but nevertheless they should be taken into account.

Stochastic errors of a gyroscope occur due to thermo-mechanical noise and flicker noise (both are physical phenomena intrinsic for MEMS structures) and manifest themselves primarily as angle random walk and bias drift.

Even if a gyroscope measures a constant signal, there will be inevitable measurement noise, mainly due to thermo-mechanical processes. The noise is likely to be a white sequence, i.e., it is zero-mean with some finite variance, σ^2. Since any white sequence is uncorrelated, using standard formulae $E(aX + bY) = aE(X) + bE(Y)$ and $Var(aX + bY) = a^2 Var(X) + b^2 Var(Y) + 2ab \cdot Cov(X, Y)$, one can conclude that a white sequence, when integrated over the time space t with a sampling period δt results in another noise sequence that is zero-mean as well but has a variance $\delta t \cdot t \cdot \sigma^2$ [28]. The latter is typical of a random walk. A random walk is defined as a stochastic process consisting of a series of steps, whose direction and size are randomly determined (a popular definition is that each step takes the value of -1 or $+1$ with equal probabilities). That is why angles figured out from integrated readings of a MEMS rate gyroscope are characterized by angle random walk noise, or ARW for short (defined in $^\circ/\sqrt{h}$). It is worth bearing in mind that a random walk is a pure mathematical concept, which just happens to be suitable enough for description of noise processes in MEMS structures. A bias is the value of the gyroscope output signal when there is no input signal. One might expect a zero output but typically it is not the case. It is common to observe some offset when working with any sensor. Usually offsets can be compensated for by simply subtracting them from output signals. However, it is not that simple when working with gyroscopes.

The bias of any gyroscope has several components caused by several reasons. The bias is not constant – on the contrary, it changes over time even under the same temperature and other internal and external conditions. With each power-up, the bias starts with some value, somewhat different from what might have been observed before, i.e., the bias of the average gyroscope experiences some slight changes when the device is powered off and powered on again. This bias component is referred to as turn-on to turn-on bias. Besides, the bias is influenced by temperature. The datasheet of the average gyroscope typically states the bias value at 25 °C. That is why one should use lookup tables that stores the bias values for a discrete set of temperature values. Along with temperature and power-on/power-off cycles, vibration and shock influence the gyroscope bias significantly. Not all gyroscopes are equally susceptible to vibration. On the contrary, fiber-optic ones are intrinsically immune to it. However, MEMS-based gyros depend badly on shock and vibration due to the very design principles of these sensors. That is because MEMS gyroscopes work on Coriolis effect using vibratory excitation. Most MEMS gyroscopes suffer from G dependency also called acceleration effect, i.e., their bias depends on how their mass experiences acceleration along the sensing axis.

The bias of any MEMS gyroscope wanders over time owing to flicker noise in the electronics and in other components that may be influenced by random flickering. Flicker noise has a power spectral density typical of pink noise (1/f) and that is why the terms "flicker noise" and "pink noise" are sometimes erroneously used interchangeably. The former relates to physical phenomena, whereas the latter refers to the mathematical function that describes the relationship between frequencies and amplitudes of noise. As one can judge by the function, 1/f, flicker noise tends to show up at low frequencies. At higher frequencies it is masked ("overshadowed") by white noise.

Bias drift caused by flicker noise is often modeled as a random walk, however, it can be approximated by some another model as well. Thus, they say about bias random walk in gyroscopes (not to be perplexed with angle random walk). Probably, a random walk is not the best choice for description of a process whose variance does not grow over time without limitations. However, for relatively short time periods the random walk model provides rather accurate results. A sample alternative approach is to use a first-order Markov chain [29] $b_i = Scale \cdot b_{i-1} + \varepsilon(i)$ where the current state is determined only by the previous bias value and randomness is represented by a zero-mean white Gaussian noise process whose variance is unknown.

If not properly taken into account, the above-stated errors accumulate quickly and make the gyroscope readings inapplicable.

Two common tools for noise spectral analysis are power spectral density and Allan variance. The Allan variance method was initially developed by David W. Allan for analysis of frequency stability in oscillators (clocks), but now it is widely used for processing other data, including readings of MEMS accelerometers and gyroscopes. The method (or, to be more precise, its overlapping version) is composed by the following steps. Keeping a MEMS gyroscope steady, one acquires an equidistant time series, $\Omega(t)$ of length N, with the sample period τ_0. If the collected time series turns out to be non-equidistant, its analysis will be much more complicated. For some averaging factor m, one takes all possible overlapping sample clusters of period $\tau = m\tau_0$. For example, for $m = 3$, the first sample cluster contains readings #1, #2, #3 and #4, with the total time elapsed between readings #1 and #4, equal to $3\tau_0$. The second sample cluster is comprised by readings #2, #3, #4 and #5, and so on. The last sample cluster for the chosen m contains readings #$(N - 3)$, #$(N - 2)$, #$(N - 1)$ and #N. Each pair of neighboring sample clusters is separated by the sample period, τ_0. Once all the possible $(N - m)$ sample clusters have been formed for the selected value of m (and τ), the Allan variance itself should be calculated as a function of τ. There are several formulae, which can be transformed into each other. One of these formulae is:

$$\sigma^2(\tau) = \frac{1}{2\tau^2(N - 2m)} \sum_{k=1}^{N-2m} (\Theta_{k+2m} - 2\Theta_{k+m} + \Theta_k)^2, \tag{1}$$

where k is an integer number between 1 and N.

Then the dependency of the Allan deviation, which is the square root of the Allan variance, on τ is plotted in log-log format. Judging by the form of the obtained curve, which is called the Allan deviation function, one can figure out the noise characteristics. Each region of the plot is responsible for some noise component. For example, white noise is characterized by the slope −1/2. The slope +1/2 defines random walk noise. The zero slope relates to the bias. A sample plot is shown in Fig. 1.

Figure 2 shows a system of motion sensors placed on a human body and a simplified example of a simulation and measurement result for one IMU sensor. Simulation is aimed at discovering the relationships between errors in measuring the attitude position and preset parameters of IMU sensors, measurement modes and motion scenarios. Upon comparison of experimental and simulation results, one selects an optimal measurement mode and digital filter that would minimize measurement errors. The work solves a problem of developing hardware-software tools for simulation and verification of noise characteristics of MEMS gyroscopes inside integrated IMU sensors.

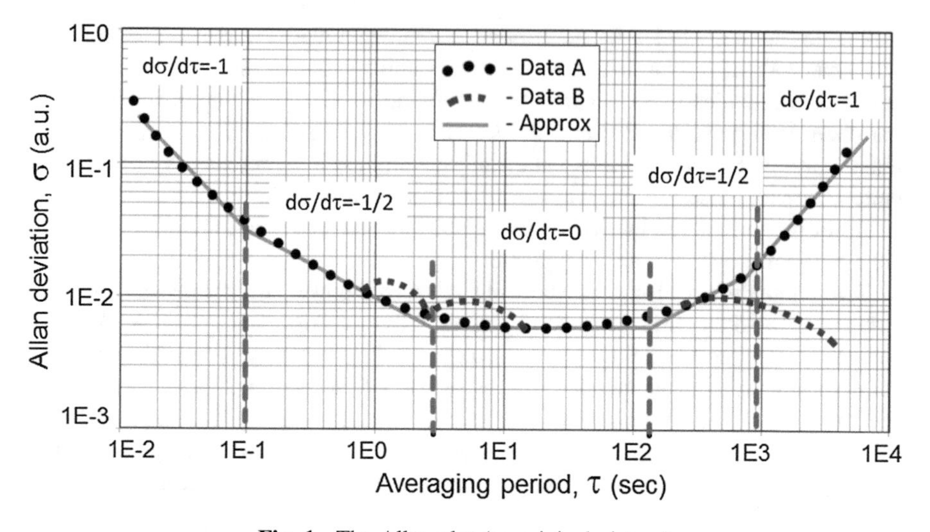

Fig. 1. The Allan plot (an original picture)

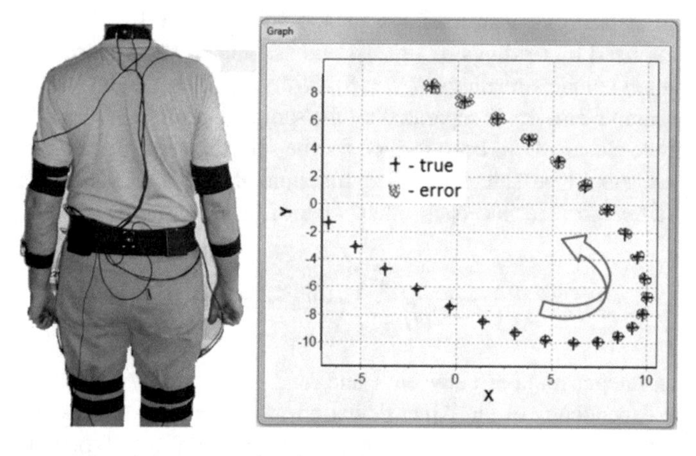

Fig. 2. A system of motion sensors placed on a human body and a sample simulation result for an ideal (true) and real (error) motion scenario

4 Mathematical Model and Algorithm of Synthesizing MEMS Gyroscope Noise

An algorithm for synthesizing noise typical of MEMS gyroscopes is a task of key importance, which is explained by substantial complexity of algorithmic optimization of firmware running in embedded systems based on IMU sensors. As has been shown previously, development of devices based on integrated IMU sensors is tightly coupled with problems of their measurement errors analysis, periodical re-calibration, compensation for the transfer function drift, signal filtering, etc. These problems are solved using mathematical models of stochastic processes and hardware-software tools for investigation into noise.

In order to optimize filtering processes and an algorithm for calculation of the attitude/position upon data arrays from integrated IMU sensors, one uses motion simulation models. Upon these models, the researcher figures out how measurement errors depend on parameters of integrated IMU sensors, noise amplitude-frequency characteristics and measurement modes. Simulation is based mostly on the Monte-Carlo method. The latter, in conjunction with motion simulation models, allows us to compute probable motion trajectories. Thus, in order to perform simulation using the Monte-Carlo method, we need to generate arrays of noise typical of integrated IMU sensors (flicker noise, angle random walk, bias random walk, etc.).

Within our complex work on developing MIS-IDE we proposed several algorithms for synthesizing noise of integrated IMU sensors. This paper presents one of the algorithms, which is based on integrating pseudorandom harmonic signals. The advantage of this algorithm is the possibility to synthesize noise, S_N, with the preset spectral characteristics. The synthesis algorithm is depicted in Figs. 3 and 4. Figure 5 presents the windows of our software where the researcher can choose a noise model, preset noise frequency characteristics (Noise simulation) and visualize a noise sample itself, its histogram and the Allan deviation plot.

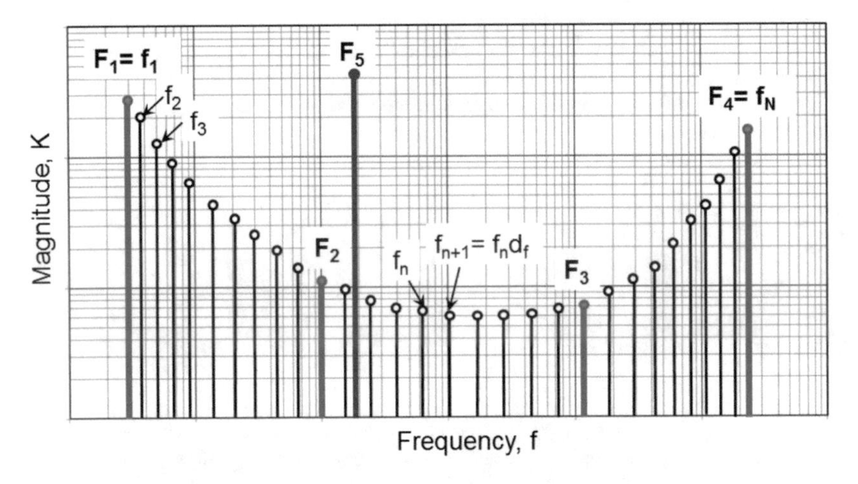

Fig. 3. Distribution of frequencies in the noise synthesis model

In order to form the required amplitude-frequency characteristic we use four basic frequencies, F_1, F_2, F_3, F_4 with the corresponding coefficients K_1, K_2, K_3, K_4. The latter represent the normalized magnitudes of harmonic oscillations at the above-stated frequencies. The value F_1 represents low-frequency noise components, F_4 – high frequency ones whereas F_2 and F_3 stand for mid-range noise components. Noise synthesis is done by summing up harmonic signals of a chosen frequency set f_n whose logarithmic values are distributed evenly in the range F_1 … F_4. The amount of items in frequency set f_n is defined by parameter d_f (denoted by $\%F$ in a window given in

Fig. 5). The frequencies are constructed using the following dependencies: $f_{n+1} = f_n d_f$, $f_1 = F_1, f_4 \approx F_4$ (Fig. 3). Two variants of synthesizing noise S_N are assumed, one with the fixed phase shift φ_n in frequency set f_n (model number NM1) and another with randomly generated phases φ_n (model number NM2).

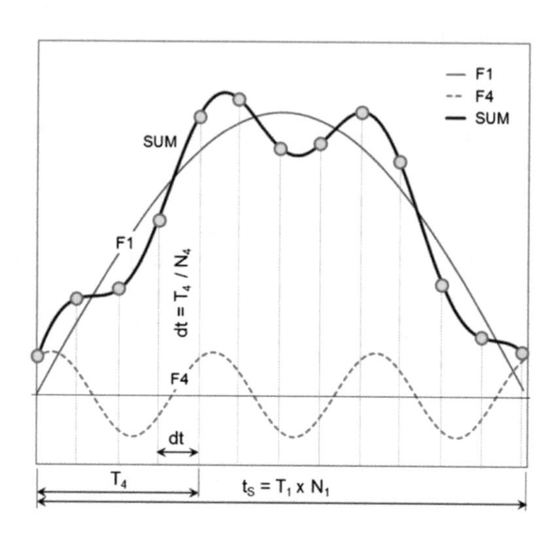

Fig. 4. Forming samples in the noise synthesis model

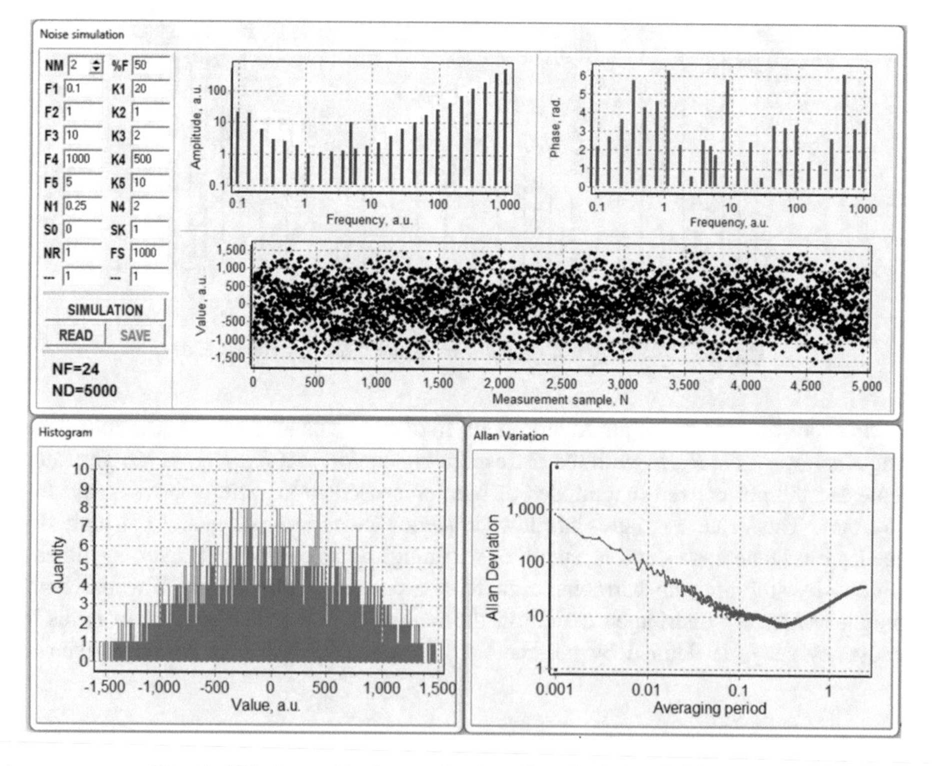

Fig. 5. Windows of noise synthesis and analysis for IMU sensors

Model NM2 is used more frequently as it generates random noise values with each iteration. On the contrary, model NM1 is used in the case when one has to study some model parameters at fixed phases φ_n in frequency set f_n. In order to form the dominating noise harmonic, which can be observed in IMU sensors due to the own oscillation frequency of the MEMS structure or some external parasitic vibration, we use a dedicated frequency, F_5, with the corresponding coefficient K_5.

In accordance with the proposed algorithm for noise synthesis, at the first stage one calculates the approximation function k, defined by four above-stated basic frequencies and their respective coefficients: $k(F_1) = K_1$, $k(F_2) = K_2$, $k(F_3) = K_3$, $k(F_4) = K_4$. At the second stage one figures out frequency set f_n. As has been already said, the frequency distribution over the range $F_1 \ldots F_4$ meets the condition $f_{n+1} = f_n d_f$. At the next stage one computes the normalized array of time intervals t_k, to synthesize noise samples $S_N(t_k)$. The values of the time intervals are ruled by the dependency $t_{k+1} = t_k + dt$, where dt is defined by the amount of samples N_4 per period T_4 at the highest frequency F_4 (Fig. 4), i.e., $dt = T_4/N_4$.

The total amount of samples in the array of synthesized noise $S_N(t_k)$ is defined by dt and total duration t_s. The last one is determined by parameter N_1 with normalization by the lowest frequency F_1 (Fig. 4).

Additionally, the noise model has the following parameters: S_0, the coefficient of the DC (Direct Current) component, which is defined by the averaged offset value of the sensor; S_K, the coefficient of the multiplicative component, defined by the averaged noise magnitude; F_S, the sampling rate whose value transforms time intervals t_k into real time measurement characteristics; N_R, the amount of repetitions in the array of synthesized noise caused by different sampling rates in the integrated IMU sensors.

The effect of repeated array items is shown in Fig. 6. Here the signal from a magnetometer is sampled at a frequency much lower than the signal from a gyroscope.

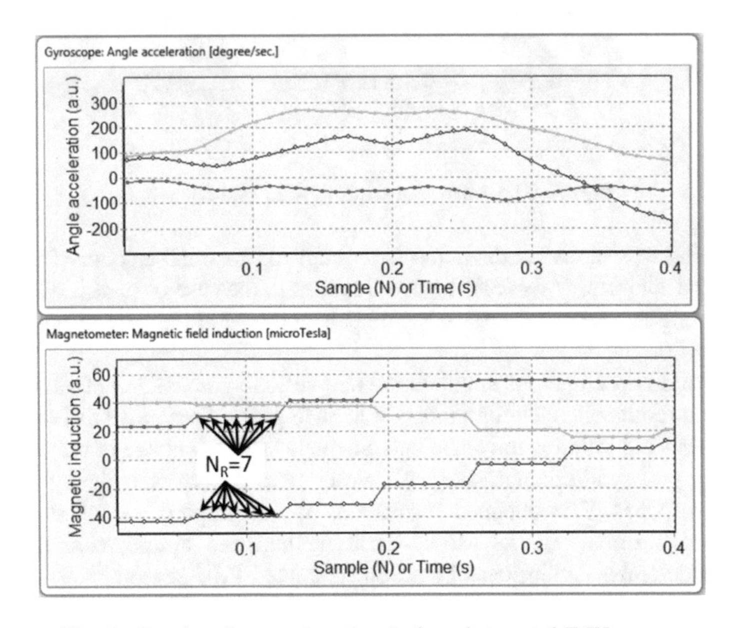

Fig. 6. Results of measuring signals from integrated IMU sensors

Therefore, in accordance with the proposed IMU sensor noise model the array of the synthesized noise values is defined by the normalized sum:

$$S_N(t_k) = \frac{S_K}{N_F}\sum_{n=1}^{N_F}(K_n \sin(2\pi f_n t_k + \varphi_n)) + K_5 \sin(2\pi F_5 t_k + \varphi_5) + S_0 \qquad (2)$$

The terms of (2) were described in detail previously in the text. All the above-stated parameters are specified in window Noise simulation of our integrated development environment MIS-IDE. The values of the synthesized array along with their histogram and Allan deviation plot can be visualized. As can be seen from Fig. 5, the Allan deviation plot provides an adequate characteristic of noise terms typical of gyroscopes in general and conforms to the experimental noise measurements conducted for the gyroscope L3G4200D particularly as will be shown later in the next chapter.

The obtained arrays of synthesized noise are dumped into files, which are then used for simulation of noisy signals from IMU sensors in accordance with the preset motion scenario. Typical Allan deviation plots for synthesized noise arrays at the presence of some artifacts are given in Figs. 7, 8 and 9.

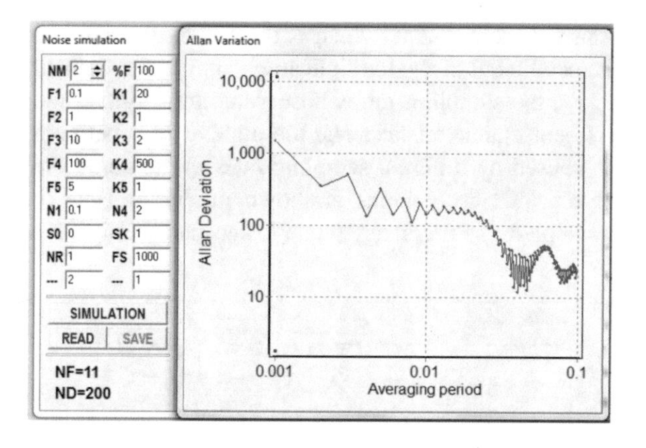

Fig. 7. The Allan plot with $N_F = 11$ and $N_D = 200$

Figure 7 shows the Allan deviation plot obtained at a small amount of frequencies, $N_F = 11$, and samples, $N_D = 200$. Figure 8 relates to the case of parasitic influence of the dominating frequency, $K_5 = 30$. Figure 9 is obtained at the case of repeated array items, $N_R = 20$.

Approbation on numerous simulation examples and comparison of simulation and experimental results obtained for a range of MEMS gyroscopes prove that the proposed noise synthesis algorithm is universal and adequate. The advantage of our algorithm is the possibility of synthesizing noise with preset spectral characteristics typical of real MEMS gyroscopes. The obtained results are of key importance for simulation of measurement errors using the Monte-Carlo method and optimization of firmware running inside embedded systems based on integrated IMU sensors.

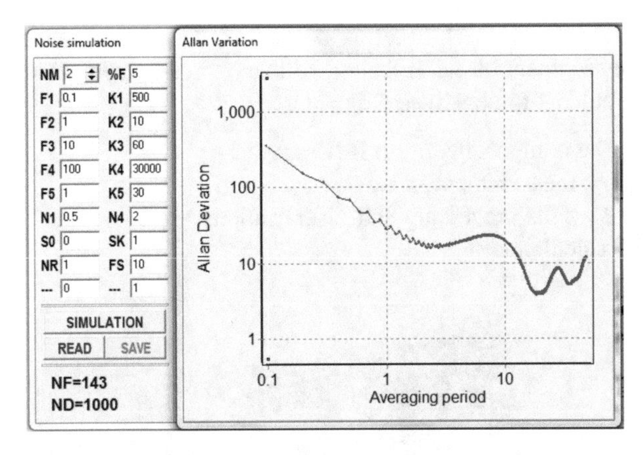

Fig. 8. The Allan plot strongly influenced by the dominating frequency, $K_5 = 30$

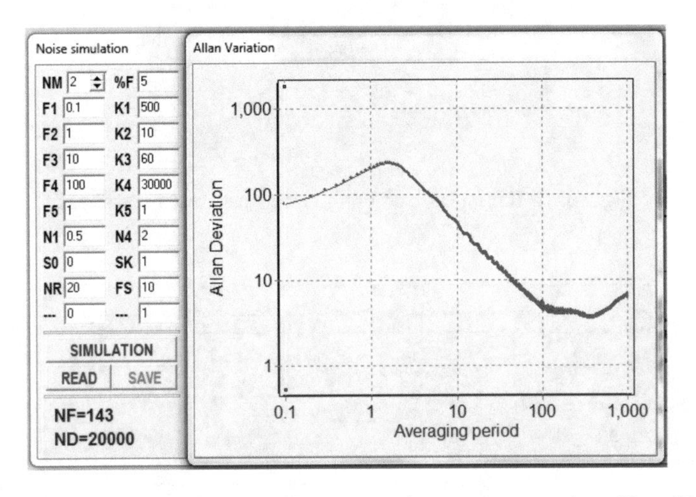

Fig. 9. The Allan plot obtained at repeated measurement points, $N_R = 20$

5 Experimental Study and Model Verification

In order to verify the parameters of the above-mentioned model experimentally we have developed our own hardware-software tool, IMUTester. Its hardware part is based on M5Stack module [30] shown in Fig. 10 ("1" marks the device board, "2" denotes the IMU sensor being studied and "3" marks the accumulator). The M5Stack module is based on the SoC (System on Chip) ESP32 developed by Espressif Systems. The SoC is controlled by a microprocessor Xtensa dual-core 32-bit LX6 performing at up to 600 DMIPS with an ultra-low-power coprocessor. Its wireless connectivity is provided by Wi-Fi 802.11 b/g/n, Bluetooth v4.2 BR/EDR and BLE. The M5Stack corresponds with the modern concept of Internet of Things. The module size is 54 × 54 × 21 mm. Depending on the use requirements, the IMU sensor being studied may be placed inside the package or next to it.

The software of the IMU tester is a part of the above-mentioned integrated development environment MIS-IDE being worked on by our team. One can distinguish five kinds of problems to be solved (Fig. 11):

1. Configuring the components of an IMU sensor.
2. Controlling the measurement modes.
3. Measurement results processing and visualization.
4. Motion path calculation.

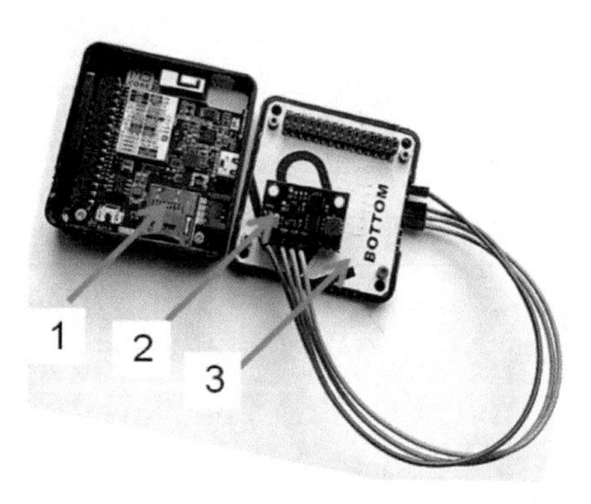

Fig. 10. IMU tester based on the integrated system M5Stack

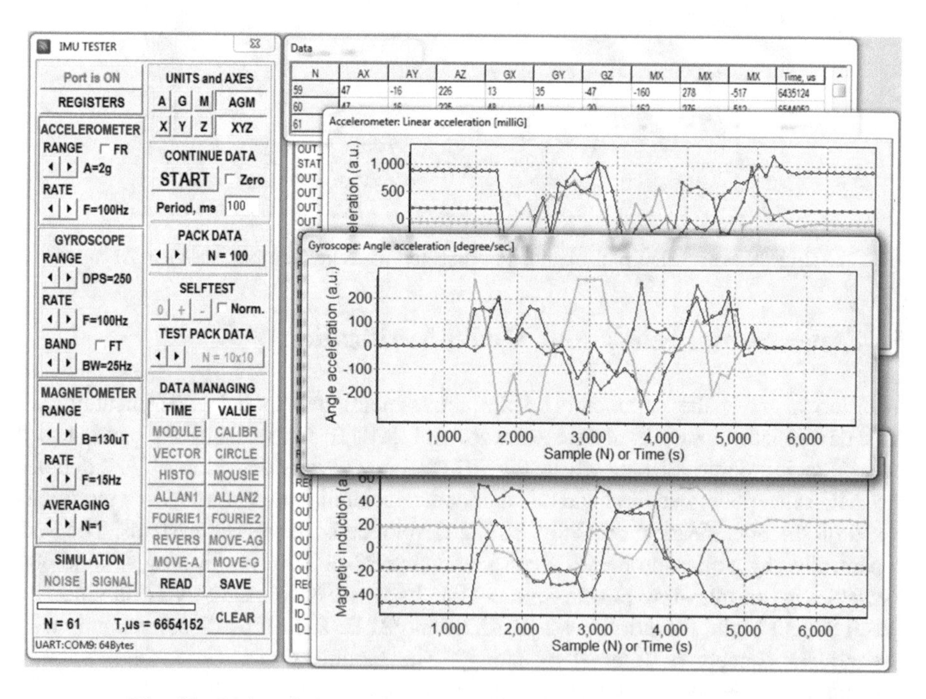

Fig. 11. Main windows of IMU tester and arrays of measured signals

5. Modeling signals and noise.

Figure 12 shows some measured signals (Angle acceleration) and calculated angles (Angle movement) of the MEMS gyroscope being studied at the following settings: measurement range RANGE = 250 DPS (degrees per second), rate of measurement conversion RATE = 100 Hz, frequency band BAND BW = 25 Hz.

In accordance with the goal of the paper, here we consider the results of noise measurement and analysis for a gyroscope being kept steady. Figure 13 shows sample results of investigation into noise of a not calibrated gyroscope L3G4200D.

Activation of the module components and measurement channels on axes X, Y and Z can be done in UNITS and AXES tab (Fig. 11) by clicking the following buttons: A – accelerometer, G – gyroscope, M – magnetometer, AGM – all the components simultaneously, X, Y, Z – selection of a particular axis, XYZ – all the axes are selected at once. The parameters of the accelerometer, gyroscope and magnetometer may be configured in ACCELEROMETER, GYROSCOPE or MAGNITOMETER tab correspondingly. Measurement ranges and sample frequencies can be controlled by pressing buttons < >RANGE and < >RATE correspondingly whereas buttons <BAND> and <> AVERAGING are intended for management of filter parameters (frequency bands and the amount of measurements taken for averaging).

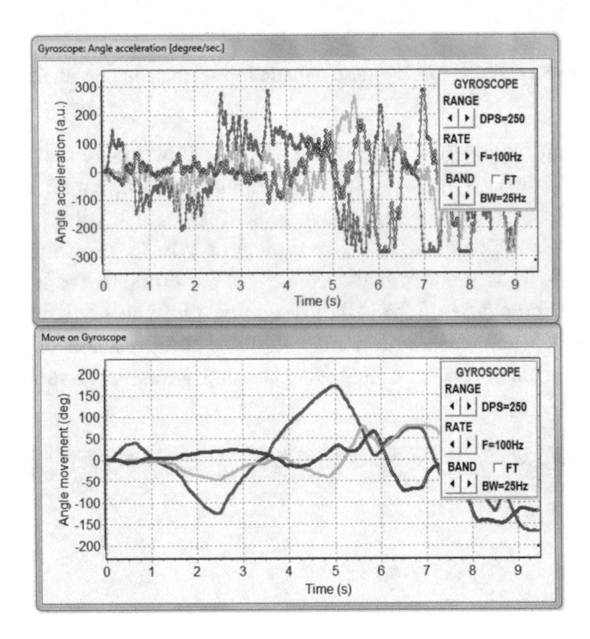

Fig. 12. Gyroscope signals (on top) and calculation of rotation angles (below)

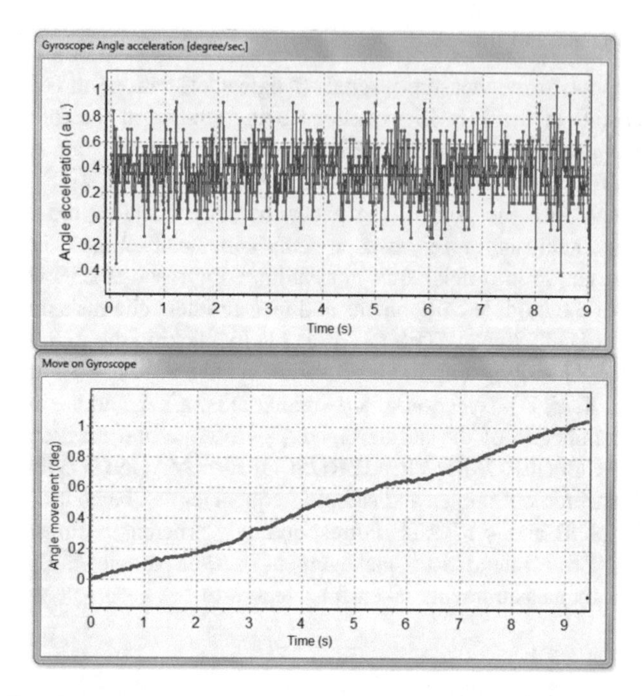

Fig. 13. Gyroscope noise (on top) and calculation of the rotation angle errors (below)

As can be seen, such noise causes the transfer function to drift approximately 1 degree per 10 s. Figure 14 presents the Allan deviation plot, which allows us to judge of noise spectral characteristics and other indicators of instability of a measurement process. One can recognize noise terms so typical of MEMS gyroscopes – flicker noise, angle random walk, bias random walk, etc. Upon experiments we have conducted we can conclude that the shape of the Allan deviation plot can be distorted a lot (e.g. as shown in Fig. 15). Such a distortion may indicate the presence of dominating sinusoidal noise or correlated noise, taking into account parasitic vibrations.

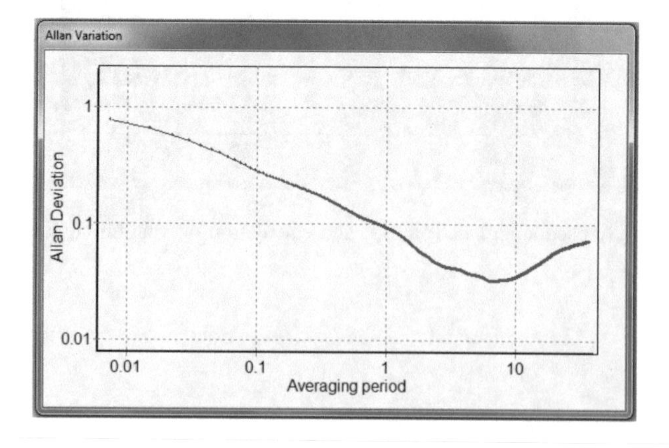

Fig. 14. The Allan deviation plot obtained upon measurement results (variant 1)

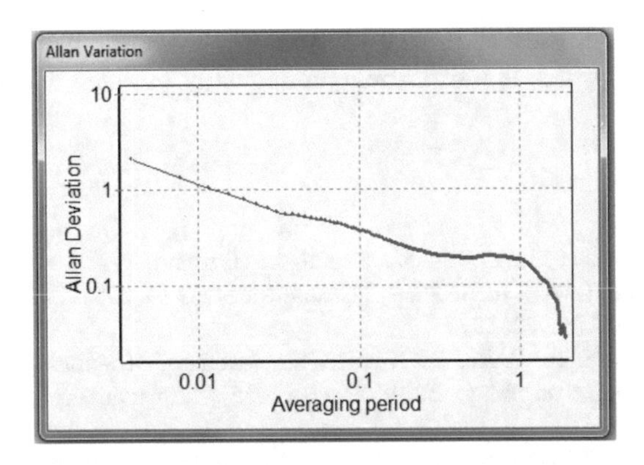

Fig. 15. The Allan deviation plot obtained upon measurement results (variant 2)

Verification of our mathematical model consists in fitting such coefficients that the Allan deviation plot calculated upon experimental data is as close as possible to that one obtained from the model. The coefficients can be quite easily found, hence the proposed MEMS gyroscopes' noise simulation algorithm is expedient to use in MIS-IDE (Measurement Inertial System Integrated Development Environment) that we have been developing.

6 Conclusions

The work presents a solution to the problem of modeling noise characteristics of real MEMS gyroscopes. The latter are susceptible to errors, some of which quickly accumulate over time and, if not properly dealt with, make gyroscope readings inapplicable. Typically researchers use the Allan deviation plot to examine the noise spectral characteristics of a MEMS gyroscope upon its time history collected over time spans long enough for statistical processing. When the Allan deviation plot has its classical shape, there is no much difficulty in extracting main noise terms from it. However, the plot may be severely distorted by components of sinusoidal noise and noise due to parasitic vibrations, which make the Allan plot difficult or impossible to interpret. That is why being able to synthesize noise typical of MEMS gyroscopes is of key importance. Having simulated noise of some preset spectral characteristics, one compares the results with the Allan deviation plot obtained upon real measurements, and if the simulation results conform to the experimental ones, the researcher gets a clue on better understanding and explaining the gyroscope behavior. Therefore, the results are crucial for selection of an appropriate adaptive signal filtering and an algorithm for calculating the attitude/position of an object being tracked with the assistance of MEMS gyroscopes.

We have proposed an algorithm for synthesizing noise with preset spectral characteristics based on integrating pseudorandom harmonic signals. The algorithm was implemented and verified in our own hardware-software tool for testing IMU sensors, IMUTester, based on the module M5Stack with SoC ESP32. Our long-range goal is to develop an IDE for synthesis and optimization of firmware for embedded systems built upon integrated IMU sensors. Thus, the presented part of the work is a necessary step in studying characteristics of inertial sensors. The obtained results conform to the experimental ones, which is proven by reproducibility of Allan deviation plots.

References

1. Höflinger, F., Müller, J., Zhang, R., Reindl, L., Burgard, W.: A wireless micro inertial measurement unit (IMU). IEEE Trans. Instrum. Meas. **62**(9), 2583–2595 (2013). https://doi.org/10.1109/TIM.2013.2255977
2. Blasch, E., Kostek, P., Pačes, P., Kramer, K.: Summary of avionics technologies. IEEE Aerosp. Electron. Syst. Mag. **30**(9), 6–11 (2015). https://doi.org/10.1109/MAES.2015.150012
3. Ahmed, H., Tahir, M.: Accurate attitude estimation of a moving land vehicle using low-cost MEMS IMU sensors. IEEE Trans. Intell. Transp. Syst. **18**(7), 1723–1739 (2017). https://doi.org/10.1109/TITS.2016.2627536
4. Buke, A., Gaoli, F., Yongcai, W., Lei, S., Zhiqi, Y.: Healthcare algorithms by wearable inertial sensors: a survey. China Commun. **12**(4), 1–12 (2015). https://doi.org/10.1109/CC.2015.7114054
5. Nemec, D., Janota, A., Hruboš, M., Šimák, V.: Intelligent real-time MEMS sensor fusion and calibration. IEEE Sens. J. **16**(19), 7150–7160 (2016). https://doi.org/10.1109/JSEN.2016.2597292
6. Lima, P.: A Bayesian approach to sensor fusion in autonomous sensor and robot networks. IEEE Instrum. Meas. Mag. **10**(3), 22–27 (2007). https://doi.org/10.1109/MIM.2007.4284253
7. Holyaka, R., Marusenkova, T.: Split Hall Structures: Parametric Analysis and Data Processing. Lambert Academic Publishing, Norderstedt (2018)
8. Shin, B., Kim, C., Kim, J., Lee, S., Kee, C., Kim, H., Lee, T.: Motion recognition-based 3D pedestrian navigation system using smartphone. IEEE Sens. J. **16**(18), 6977–6989 (2016). https://doi.org/10.1109/JSEN.2016.2585655
9. Zekavat, S., Buehrer, R.M.: Handbook of Position Location – Theory, Practice, and Advances. Wiley, New Jersey (2019). https://doi.org/10.1002/9781119434610.ch2
10. Daroogheha, S., Lasky, T., Ravani, B.: Position measurement under uncertainty using magnetic field sensing. IEEE Trans. Magn. **54**(12), 1–8 (2018). https://doi.org/10.1109/TMAG.2018.2873158
11. Li, Y., Georgy, J., Niu, X., Li, Q., El-Sheimy, N.: Autonomous calibration of MEMS gyros in consumer portable devices. IEEE Sens. J. **15**(7), 4062–4072 (2015). https://doi.org/10.1109/JSEN.2015.2410756
12. Latt, W.T., Tan, U.-X., Riviere, C.N., Ang, W.T.: Transfer function compensation in gyroscope-free inertial measurement units for accurate angular motion sensing. IEEE Sens. J. **12**(5), 1207–1208 (2012). https://doi.org/10.1109/JSEN.2011.2165057
13. Huang, J., Soong, B.: Cost-aware stochastic compressive data gathering for wireless sensor networks. IEEE Trans. Veh. Technol. **68**(2), 1525–1533 (2019). https://doi.org/10.1109/TVT.2018.2887091
14. Shmaliy, Y., Zhao, S., Ahn, C.: Optimal and unbiased filtering with colored process noise using state differencing. IEEE Signal Process. Lett. **26**(4), 548–551 (2019). https://doi.org/10.1109/LSP.2019.2898770
15. Lin, X., Jiao, Y., Zhao, D.: An improved Gaussian filter for dynamic positioning ships with colored noises and random measurements loss. IEEE Access **6**, 6620–6629 (2018). https://doi.org/10.1109/ACCESS.2018.2789336

16. Allan, D., Levine, J.: A historical perspective on the development of the Allan variances and their strengths and weaknesses. IEEE Trans. Ultrason. Ferroelectr. Freq. Control **63**(4), 513–519 (2016). https://doi.org/10.1109/TUFFC.2016.2524687
17. Guerrier, S., Molinari, R., Stebler, Y.: Theoretical limitations of Allan variance-based regression for time series model estimation. IEEE Signal Process. Lett. **23**(5), 597–601 (2016). https://doi.org/10.1109/LSP.2016.2541867
18. Won, S., Melek, W., Golnaraghi, F.: A Kalman/particle filter-based position and orientation estimation method using a position sensor/inertial measurement unit hybrid system. IEEE Trans. Industr. Electron. **57**(5), 1787–1798 (2010). https://doi.org/10.1109/TIE.2009.2032431
19. Hsu, Y., Wang, J.: Random drift modeling and compensation for MEMS-based gyroscopes and its application in handwriting trajectory reconstruction. IEEE Access **7**, 17551–17560 (2019). https://doi.org/10.1109/ACCESS.2019.2895919
20. Wang, Y.: Stochastic and dynamic modeling of MEMS gyroscopes. In: 2012 IEEE International Conference on Mechatronics and Automation, Chengdu, China, 5–8 August 2012. https://doi.org/10.1109/icma.2012.6285749
21. Kim, D., M'Closkey, R.: Noise analysis of closed-Loop vibratory rate gyros. In: 2012 American Control Conference (ACC), Montreal, QC, Canada, 27–29 June 2012. https://doi.org/10.1109/acc.2012.6314985
22. Kim, D., M'Closkey, R.: Spectral analysis of vibratory gyro noise. IEEE Sens. J. **13**, 4361–4374 (2013). https://doi.org/10.1109/JSEN.2013.2269797
23. Cao, H., Lv, H., Sun, Q.: Model design based on MEMS gyroscope random error. In: 2015 IEEE International Conference on Information and Automation, Lijiang, China, 8–10 August 2015. https://doi.org/10.1109/icinfa.2015.7279648
24. Nazdrowicz, J., Napieralski, A.: Modelling, simulations and performance analysis of MEMS vibrating gyroscope in coventor MEMS+ environment. In: 2019 20th International Conference on Thermal, Mechanical and Multi-Physics Simulation and Experiments in Microelectronics and Microsystems (EuroSimE), Hannover, Germany, 24–27 March 2019. https://doi.org/10.1109/eurosime.2019.8724520
25. Liu, Q., Han, B., Xu, J., Wu, M.: Random drift modeling for MEMS gyroscope based on lifting wavelet and wavelet neural network. In: 2011 International Conference on Electric Information and Control Engineering, Wuhan, China, 15–17 April 2011. https://doi.org/10.1109/iceice.2011.5778009
26. Song, J., Shi, Z., Du, B., Wang, H.: The filtering technology of virtual gyroscope based on Taylor model in low dynamic state. IEEE Sens. J. **19**, 5204–5212 (2019). https://doi.org/10.1109/JSEN.2019.2902950
27. Georgy, J., Noureldin, A., Korenberg, M., Bayoumi, M.: Modeling the stochastic drift of a MEMS-based gyroscope in Gyro/Odometer/GPS integrated navigation. IEEE Trans. Intell. Transp. Syst. **11**, 856–872 (2010). https://doi.org/10.1109/TITS.2010.2052805
28. Woodman, O.: An Introduction to Inertial Navigation. University of Cambridge Computer Laboratory, Cambridge (2007)
29. Barrett, J.M.: Analyzing and modeling low-cost MEMS IMUs for use in an inertial navigation system. Master of Science degree thesis, Worchester Polytechnic Institute (2014)
30. M5Stack Documentation. https://buildmedia.readthedocs.org/media/pdf/m5stack/latest/m5stack.pdf
31. Fedasyuk, D., Marusenkova, T.: Analyser and mathematical model for synthesizing noise of MEMS gyroscopes In: Proceedings of International Scientific Conference "Computer Sciences and Information Technologies" (CSIT-2019), vol. 1, pp. 109–112. IEEE (2019)

Author Index

N. Shakhovska and M. O. Medykovskyy (Eds.): CCSIT 2019, AISC 1080, pp. 969–971, 2020.
https://doi.org/10.1007/978-3-030-33695-0